T5-CCM-738

MICROBIOLOGY
CONCEPTS AND APPLICATIONS

MICROBIOLOGY

CONCEPTS AND APPLICATIONS

PAUL A. KETCHUM
Oakland University

JOHN WILEY AND SONS

NEW YORK • CHICHESTER • BRISBANE • TORONTO • SINGAPORE

Cover and Interior Designed by Carolyn Joseph

Library of Congress Cataloging in Publication Data:

Ketchum, Paul A. (Paul Abbott), 1942–
Microbiology : concepts and applications.

Includes bibliographies and index.
1. Microbiology. I. Title. [DNLM: 1. Microbiology.
QW 4 K43mb]
QR41.2.K46 1988 576 87-34607
ISBN 0–471–88897–4

Printed in the United States of America

10 9 8 7 6 5 4 3 2 1

**To Nancy,
Stephen, and Jennifer**

PREFACE

Microbiology: Concepts and Applications presents a balanced coverage of microbiology for students in a wide variety of programs, including biological sciences, health sciences, agricultural sciences, natural resources, food sciences, home economics, and liberal arts. The book was written with the knowledge that some students have had no previous course in biology or chemistry. Therefore the basic principles of biology and chemistry are introduced in the early chapters. Subjects requiring chemical explanations are presented with enough detail to demonstrate the importance of chemistry to microbiology without overpowering the course.

ORGANIZATION

Some basic concepts and the importance of microbiology are introduced through a discussion of its history. This discussion naturally leads to presentations of the basic principles, concepts, and terms needed to describe and understand the biology of microorganisms. Applications of microbiology to everyday life are integrated into the text to enliven the factual material and to facilitate learning. Additional historical developments are described where the text occasions their introduction. They provide students with a perspective on how microbiology has matured into a major scientific discipline. Short essays, both current and historic, and reports from the Centers for Disease Control are highlighted in boxes throughout the text. The topics treated have been specifically selected to stimulate student interest and convey the current impact of microbiology on our modern world.

FEATURES

My goal was to write a book that is both a pleasure to read and an effective learning tool. A number of features help to enliven the material and to assist the learning process.

1. **Outlines** begin each chapter so that the reader can see at a glance the organization of the material.
2. **Focus of This Chapter** captures the flavor of the topics and principles covered.
3. Pronunciations of scientific terms are given where a term first appears in the text. *Bergey's Manual* and *Dorland's Illustrated Medical Dictionary* were the references used for pronunciations.
4. **Key Points** appear throughout the text to focus the reader's attention on important concepts.
5. **Summary Outlines** at the end of each chapter aid review.
6. **Questions and Topics for Study and Discussion,** which follow Summary Outlines, focus attention on important concepts.
7. **Further Readings** are annotated to motivate students to read more about the topics discussed.

Reference material is included in the appendixes of the book. **Units of Measure and Mathematical Expressions** (Appendix 1) contains conversion factors and a description of how logarithms are used in microbiology. The **Classification of Bacteria** according to *Bergey's Manual of Systematic Bacteriology* is outlined in Appendix 2. **Common Word Roots Used in Scientific Terminology** (Appendix 3) will help students learn the meaning of words through their derivations. The appendixes are followed by a **Glossary** containing definitions of all the important terms used in the text.

CHAPTER SEQUENCE

The organization of the book has evolved from my own experience in teaching microbiology for over 15 years and from the suggestions of reviewers. Each chapter is written as an independent unit to accommodate instructors who may wish to present the material in other sequences or to combine chapters. For example, Control of Microorganisms (Chapter 8) can be presented with Antimicrobial Agents and Chemotherapy (Chapter 21). The discussion on bacterial diversity was purposefully divided into two chapters so that instructors can emphasize Bacteria of Ecological, Industrial, and General Significance (Chapter 14) or Bacteria of Medical Importance (Chapter 15). Or some instructors may want to select topics from both chapters. The body-system approach to infectious diseases used in Part 4 readily lends itself to selective coverage and reorganization. Within each chapter on infectious diseases, the responsible bacteria, viruses, fungi, and helminths are presented sequentially.

SUPPLEMENTARY MATERIALS

A *Study Guide* to accompany *Microbiology: Concepts and Applications* has been written by Dr. Josephine Smith to act as a tutorial, review, and study aid. The *Study Guide* is available from the publisher. An *Instructor's Manual* authored by Dr. Josephine Smith and a package of 100 acetate transparencies selected from the book's two-color line drawings are available to instructors upon written request.

ACKNOWLEDGMENTS

Illustrations are extremely valuable for introducing the concepts of microbiology to introductory students, and I am grateful to all the scientists who kindly permitted me to use their illustrations and micrographs. Other scientists have contributed their knowledge, talents, and energy to bring this project to its final form, and I am deeply indebted to them. Selected chapters were read and critiqued by my colleagues at Oakland University, Professors Robert Douglas Hunter, Charles Lindemann, and Satish Walia.

I am also indebted to the following reviewers who contributed both specific suggestions and a broad perspective on the design and execution of the manuscript: Professor W. Murray Bain, University of Maine at Orono; Professor Douglas P. Bingham, West Texas State University; Professor Lester E. Casida, Jr., The Pennsylvania State University; Dr. William Clark, University of California, Los Angeles; Dr. Donald Emmeluth, Fulton-Montgomery Community College; Dr. David Filmer, Purdue University; Professor Frederick Goetz, Mankato State University; Professor Ronald Hinsdill, University of Wisconsin, Madison; Professor Walter Hoeksema, Ferris State College; Professor John Holt, Iowa State University; Professor Thomas Jewell, University of Wisconsin; Ilsa Kaattari, Oregon State University; Dr. Bruno J. Kolodziej, Ohio State University; Professor Allan Konopka, Purdue University; Professor E. R. Leadbetter, University of Connecticut, Storrs; Dr. Frank Mittermeyer, Elmhurst College; Dr. Remo Morelli, San Francisco State University; Dr. Robert Nauman, University of Maryland; Dr. William O'Dell, University of Nebraska, Omaha; Professor Dennis Opheim, Quinnipiac College; Richard Renner, Laredo Junior College; Dr. Karl J. Siebert, Director of Research, the Stroh Brewery Company, Detroit; Professor Jay Sperry, University of Rhode Island; Professor Daniel R. Tershak, The Pennsylvania State University; Dr. Richard Tilton, College of Medicine, University of Connecticut, Farmington; Dr. William Todd, College of Medicine; University of Tennessee, Memphis, Professor Fred Williams, Iowa State University.

Production of a modern science textbook requires the expertise of many talented people. I am indebted to Priscilla Todd for editing the manuscript, Pamela Pelton for supervising production, Carolyn Joseph for the design of the book and its cover, John Balbalis for coordinating the illustration program, and the Wiley Photo Research Department for gathering many of the photographs. I especially want to thank Bernice Heller who was the consulting development editor. Her considerable expertise in publishing and her interest in microbiology had a profound impact on the pedagogy and development of the text.

PAUL A. KETCHUM
Rochester, Michigan

BRIEF CONTENTS

C O N T E N T S

PART I

CHAPTER 25
GASTROINTESTINAL INFECTIONS AND FOOD POISONING 596

CHAPTER 26
INFECTIOUS DISEASES OF BLOOD, LIVER, AND LYMPH 622

CHAPTER 27
INFECTIOUS DISEASES OF THE NERVOUS SYSTEM 646

MICROBIOLOGY

CONCEPTS AND APPLICATIONS

PART 1

Principles of Microbiology

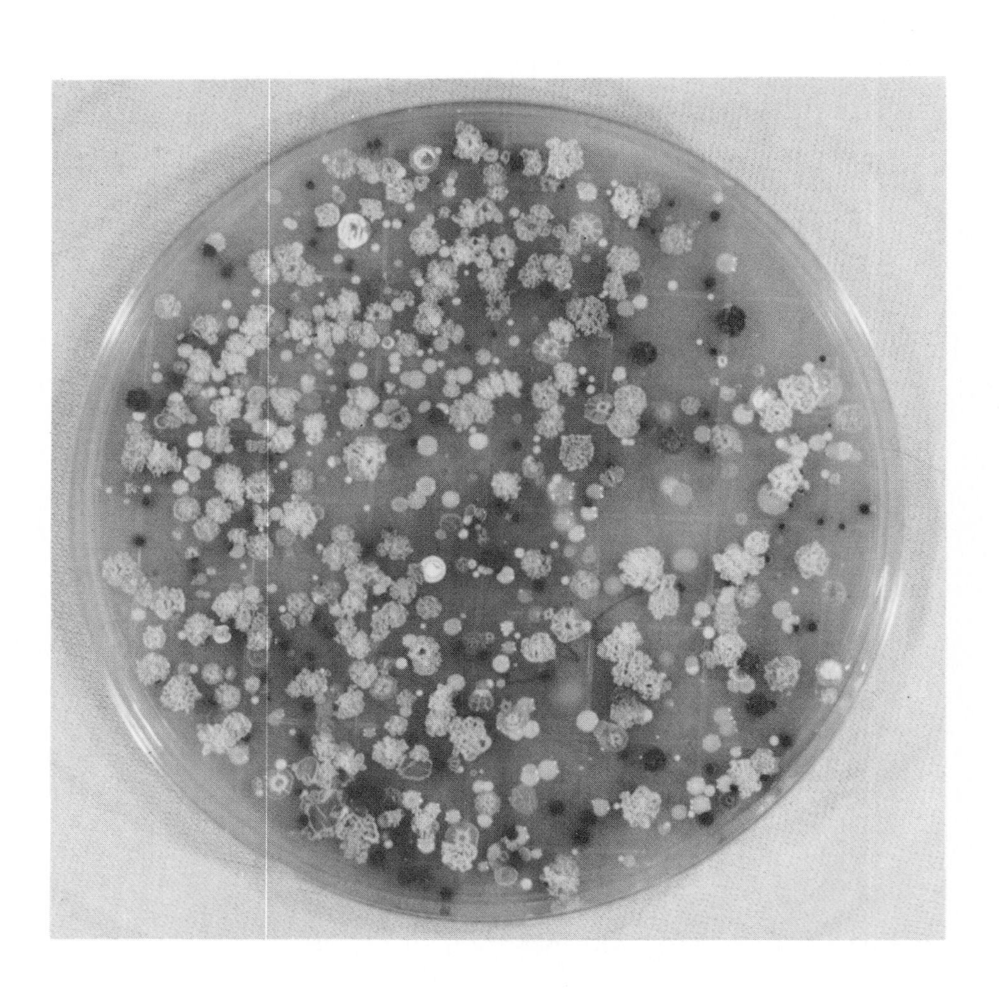

Microorganisms, living creatures too small to be seen with the unaided eye, have a profound influence on daily life—often beneficial, sometimes harmful. Bread is made with the help of yeasts that ferment sugar and carbonate the dough before it is baked; cheeses, yogurt, and sour cream require the presence of microorganisms that sour the milk from which they are made; some foods are treated to prevent microbial spoilage; antibiotics cure such bacterial diseases as "strep" throat, bacterial pneumonias, gonorrhea, and syphilis. Both children and adults are immunized against various diseases, including tetanus, diphtheria, whooping cough, measles, and mumps. Newspapers and broadcast journalists frequently report on AIDS, influenza, or other public health problems caused by microorganisms and viruses. The profound ways in which microorganisms and viruses affect life will become evident as the story of microbiology unfolds, in the following compendium.

CHAPTER 1

The Origins of Microbiology

OUTLINE

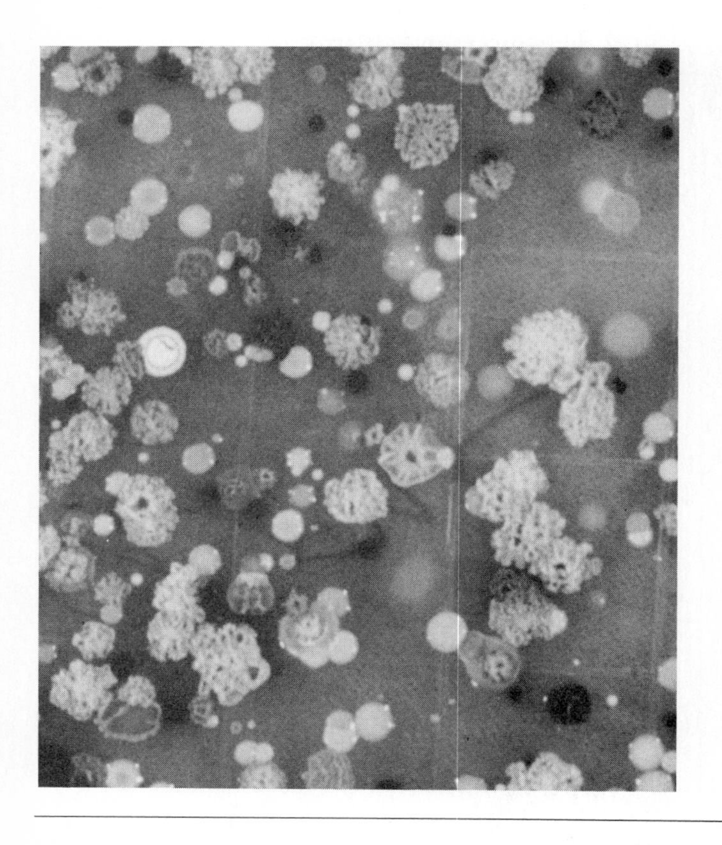

FOCUS OF THIS CHAPTER

Microbiology developed into a science as inquisitive people devised experiments to answer questions about everyday events. Why do milk, meat, and wine become unfit to eat? Why do people who have recovered from cowpox not get smallpox? What are the causes of anthrax, tobacco mosaic disease, foot-and-mouth disease? Where do the microorganisms that contaminate broths come from? Experiments that answered these questions established the basic concepts upon which microbiology developed.

Microbiology did not formally begin until the 1600s. This was long after some other sciences, such as anatomy and physics, had become well established. Microbiology got its late start because light microscopes powerful enough to reveal microscopic forms of life were not invented until the mid-1600s. We owe this momentous development mainly to two scientists, Robert Hooke and Anton van Leeuwenhoek.

DISCOVERY OF MICROORGANISMS

ROBERT HOOKE (1635–1703)

Robert Hooke became the first applied microscopist of note while he was the curator of experiments for the Royal Society of London. He spent many hours using his compound microscope to examine various materials, and he presented his observations at meetings of the Royal Society. Their enthusiastic response encouraged him to publish (1664) his observations in a book he titled *Micrographia*. This book contains detailed drawings of insects, insect larvae, seeds, eggs, feathers, hairs, rocks, and cork. In addition, Hooke described the microscopic forms of fungi commonly called molds (Figure 1.1) as

long cylindrical transparent stalks, not exactly straight, but a little bended with the weight of a round white knob that grew on the top of each of them.

FIGURE 1.1 Robert Hooke observed molds growing on the leather covering of a book. Hooke suggested that the morphological variation in the knobs (A, B, C, D, E) were seed cases similar to those found on the tops of mushrooms. (Robert Hooke, *Micrographia*; reprint ed., Dover, New York, 1961.)

In this first reported observation of molds, Hooke likened the observed knobs to the common spore-bearing cups of the common white mushroom. This insight was remarkable since both the knobs and the caps of mushrooms contain spores, which are asexual reproductive cells of a mold that can grow into new individuals.

ANTON VAN LEEUWENHOEK (1632–1723)

Hooke contributed greatly to knowledge through his observations of tissues and the compartmentalized structure of living things now known as cells; however, he did not observe objects as small as bacteria.

Credit for this discovery is given to Anton van Leeuwenhoek (Lay'wen-hook), a contemporary of Hooke, who was born in and spent almost all of his life in Delft, Holland (Figure 1.2).

At the age of 16, Leeuwenhoek apprenticed to a linen draper in Amsterdam. He became the proprietor of a draper's shop in Delft about the time of his marriage in 1654. At that time, Delft was a major center of commerce and Leeuwenhoek became a successful businessman. Although his main business was managing a drapery shop he was also a qualified surveyor, and in 1660 he was appointed chamberlain to the chief judge, the sheriffs, and the law officers of the city of Delft, a position he held for 39 years.

Leeuwenhoek developed the ability to grind glass lenses and to mount them into brass contraptions (Figure 1.3) that he called microscopes. Samples were placed on the adjustable needle mounted above the lens. He was able to see the specimens on the needle by holding the microscope close to his eye and looking through the lens at an angle to the light source. This created a type of darkfield microscopy, in which very small microbes appear as bright objects against a dark background—similar to the way stars appear to us on a dark night. Microscopy was a fascinating hobby for Leeuwenhoek, and he spent the greater part of his long life making microscopes, observing the microscopic world, and keeping detailed records of his observations in simple, unsophisticated language. When he was finally asked to communicate his observations to the Royal Society in London, he responded in this letter of 1673:

FIGURE 1.2 Anton van Leeuwenhoek (1632–1723) was the first person to observe and describe bacteria. (The Bettmann Archive, Inc.)

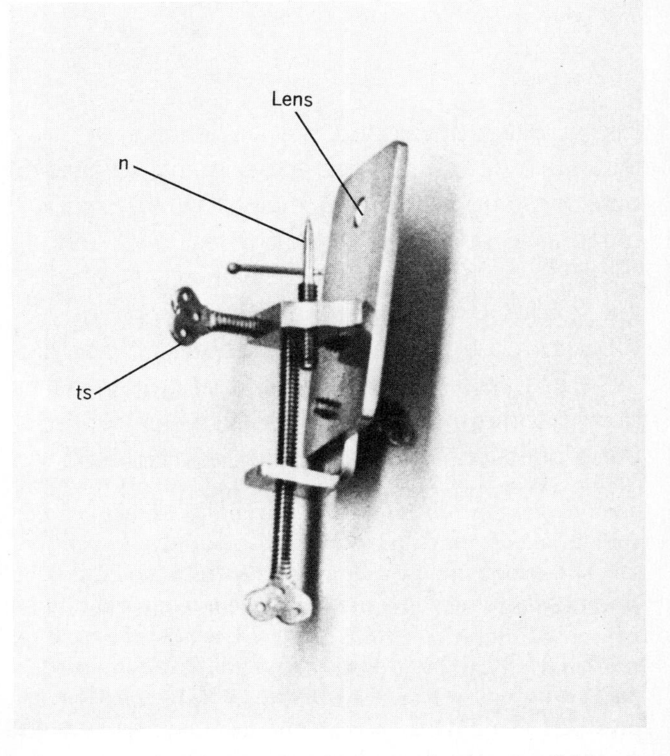

FIGURE 1.3 Anton van Leeuwenhoek ground his own lenses and mounted them in brass microscopes. Specimens are placed on the needle (n) and manipulated by the three thumb screws (ts) of this replica of a Leeuwenhoek microscope.

I have oft-times been besought by divers gentlemen, to set down on paper what I have beheld through my newly invented microscopia: but I have generally declined; first because I have no style, or pen wherewith to express my thoughts properly; secondly, because I have not been brought up to languages or arts, but only to business, and in the third place, because I do not gladly suffer contradiction or censure from others. This resolve of mine however, I have now set aside at the entreaty of Dr. Reg. de Graaf and I gave him a memoir on what I have noticed about mould, the sting and sundry little limbs of the bee and also about the sting of the louse.

Leeuwenhoek attached this humble message to his first communication to the Society. Many letters describing all kinds of microscopic things followed. His eighteenth letter, sent to the Society on October 9, 1676, is the first description of unicellular animals known as protozoa (pro-toh-zo′ah) and contains the first reported observations of the even smaller microorganisms known as bacteria. Leeuwenhoek had looked at water samples from the freshwater river Maas, and from wellwater, rainwater, and seawater. He called the creatures that he observed under his microscope "animalcules" and compared their size to the eye of a louse or a grain of fine sand. His descriptions were so accurate that we can identify the water flea *Daphnia* (daf′nee-ah) and the stalked *Vorticella* (vor-ti-sel′la) among the animalcules he saw. In many of his observations he referred to organisms that were barely discernible, although he was able to describe them in remarkable detail.

Pepper Infusions and the Discovery of Bacteria

Leeuwenhoek wondered why pepper tasted hot. He thought that animalcules like those he observed in other specimens could be responsible, so he mixed pepper grains with three-year-old snowwater and looked for animalcules under his microscope. At first he saw no living creatures; however, on the tenth day they were readily apparent. He commented:

I discovered very many exceedingly small animalcules. Their body seemed, to my eye, as long as broad. Their motion was very slow and oft times roundabout . . .

Nineteen days later he reexamined the same pepper water, and wrote:

I saw still more of the oval animalcules and some of the most exceeding thin little tubes, which I had also seen many a time before this.

Here is a later observation on the pepper water:

The same day, I discovered some very little round animalcules that were about 8 times as big as the smallest animalcules of all. These had so swift a motion before the eye, as they darted among the others that 'tis not to be believed.

Although he didn't discover why pepper tasted hot, his persistent and detailed observations led him to be the first person to observe bacteria, the smallest microorganisms known. For 50 years, Leeuwenhoek made scientific observations and reported them in writings to the Royal Society. One of Leeuwenhoek's most publicized observations was of bacteria from the human mouth (Figure 1.4), which he wrote about in a 1683 letter. He described rod-shaped bacteria, given the name **bacilli** (bah-sil′leye), spherical bacteria known as **cocci** (kok′seye), and bacteria whose spiral bodies resemble corkscrews—called either **spirochetes** (spi′roh-keets), or **spirilla** (spi-ril′ah).

Many of his observations went unconfirmed for years, because no one was able to make microscopes of the quality needed to observe bacteria. Although his microscopes were sent to the Royal Society in London, nobody followed in Leeuwenhoek's footsteps, and the science of microbiology lay dormant for many years after his death.

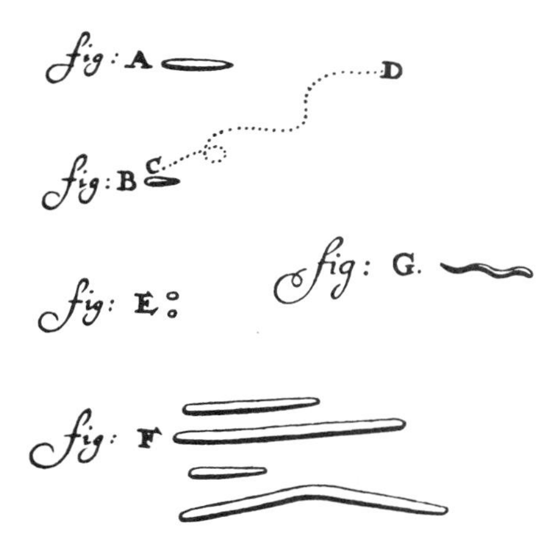

FIGURE 1.4 Bacteria from the mouth as sketched by Anton van Leeuwenhoek. (From Clifford Dobell, ed. & trans., *Antony Van Leeuwenhoek and His "Little Animals,"* 1932. Reissued with a new introduction by Cornelis B. van Niel, Russell & Russell, New York, 1958.)

MICROBIOLOGY IN THE EIGHTEENTH CENTURY

Even though eighteenth-century biology was dominated by the struggle over the larger question of spontaneous generation—life originating from nonliving matter—there was speculation on the relationship between microorganisms and disease. An Italian physician, Marcus A. Plenciz (1705–1786), proposed that all diseases were caused by microorganisms. Since no evidence was presented to support this theory, Plenciz's views were largely ignored. In 1720, Benjamin Martin published a book, *A New Theory of Consumptions: More Especially of Phthisis or Consumption of the Lungs*. Here he proposed that consumption (tuberculosis) was caused by animalcules transferred from infected people to others and then transmitted by blood to the lungs, that these animalcules were especially suited for growth in the lungs, and that other animalcules were responsible for other human illnesses, such as venereal disease. Clearly, Martin did not believe in spontaneous generation; on the contrary, he believed that all living creatures must be produced from eggs. For years, science ignored the speculations of Martin and Plenciz mainly because no evidence was presented to support them. The experimental evidence needed to prove the infectious nature of certain diseases was not to be forthcoming for another hundred years.

SPONTANEOUS GENERATION

Scientists and philosophers were in major disagreement over the controversial concept of spontaneous generation or abiogenesis (ay-bye-o-gen′uh-sis). In the seventeenth and eighteenth centuries, many people believed in the ability of living creatures to arise from nonliving matter such as dead animals, meats, or broths made from meats, hay, or gravy. Abiogenesis is not inconsistent with the biblical rendition of earth's creation, and was supported by many theologians. However, experimentalists believed that living creatures originated only from ova or from other living creatures.

Francesco Redi (1626–1697), an Italian priest, pondered the question of abiogenesis. Redi knew that maggots developed in unprotected meat; however, he surmised that the maggots did not arise spontaneously. He thought that they originated from the flies that landed on the meat. To prove his hypothesis, Redi placed linen cloths over jars containing fresh pieces of meat and left other jars uncovered. Flies were attracted to the meat in both jars, but they could lay their eggs only on the uncovered meat or on the cloth covering the jar. Maggots developed both in the uncovered meat and on the cloth of the covered meat, but they did not develop in the covered meat. These simple experiments showed that maggots grow from fly eggs and are not able to develop spontaneously from meat.

Another Italian priest, Lazzaro Spallanzani (spal-lan-za′ny) (1729–1799), who lived a century after Redi, was a contemporary of the English priest John Needham (1713–1781). These two men took opposite sides in the controversy over spontaneous generation.

John Needham performed experiments with mutton gravy that supported his belief in spontaneous generation. He placed hot mutton gravy in flasks, sealed them with corks to exclude air, and then heated them in "violently hot ashes." The flasks were cooled and then exposed for days to the heat of summer before microscopic observations were made. Needham observed microorganisms in all his flasks and claimed that they arose spontaneously from the mutton gravy.

Spallanzani was deeply troubled by Needham's conclusions because they contradicted Redi's work, so Spallanzani began his own investigation. He set up four groups of flasks: group 1 was made airtight by melting the glass neck of the flask, group 2 was stoppered with cotton, group 3 was stoppered with wooden corks to simulate Needham's experiment, and the last group was left open to the air. All the flasks were filled with seeds and vegetable matter and then heated for one hour at the start of the experiment. Spallanzani set them aside for 25 days before microscopically investigating their contents. The open flasks were teeming with life; the corked flasks contained thousands of swimming animalcules, just as Needham had claimed; but the flasks that were stoppered with cotton and the hermetically (airtight) sealed flasks contained only a few animalcules.

Spallanzani concluded that the number of animalcules in a flask increased with exposure to the air. The animalcules in Needham's corked flasks must have entered from the air and this could be prevented with an airtight seal. This explanation did not satisfy the proponents of spontaneous generation, and some were quick to discredit Spallanzani's experiments. They charged that organisms need air to live and could not originate in Spallanzani's sealed flasks. There was no immediate counterargument to this criticism so the controversy over abiogenesis lasted well into the middle of the nineteenth century.

LOUIS PASTEUR (1822–1895) BECOMES A MICROBIOLOGIST

Science in the nineteenth century was making great advances in chemistry and physics, and like most science students of his time, Louis Pasteur (Figure 1.5) was educated in these disciplines. Pasteur entered school in Paris to study chemistry and made significant contributions to organic chemistry. His knowledge of organic molecules soon led him into discussions with Theodor Schwann, who had demonstrated (1837) that yeasts were responsible for the formation of alcohol in wine and beer fermentations. (A fermentation can be thought of as the decomposition of organic substances by the action of microbes.) Schwann's discovery was even more remarkable because it was contradictory to the chemical theory of fermentation presented by Justin Liebig, who was a respected German chemist. Liebig had proposed that fermentations were the result of the normal chemical decomposition of organic matter.

FIGURE 1.5 Louis Pasteur (1822–1895) was one of the great nineteenth-century microbiologists. (Courtesy of the National Library of Medicine.)

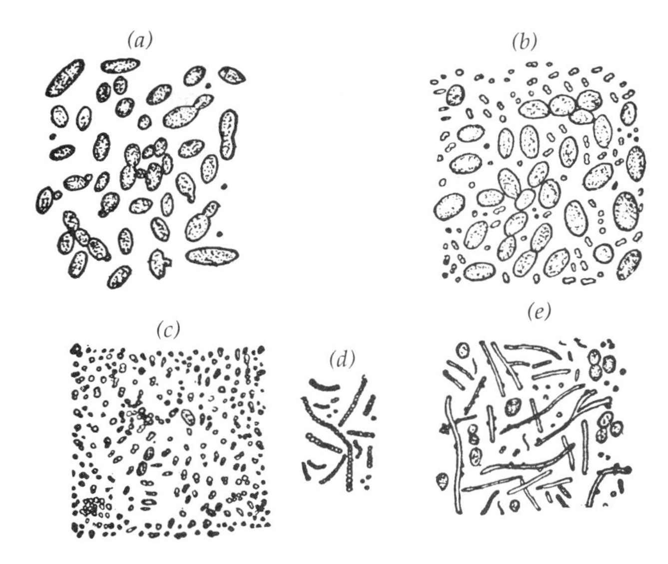

FIGURE 1.6 Diseases of wines. Pasteur observed different organisms in wines as they turned "bad": *(a)* yeast fermentation, *(b)* acid wine in an early stage, *(c)* acid wine in a later stage, *(d)* ropy wine, and *(e)* bitter wine. (From S. J. Holmes, *Louis Pasteur,* Dover, New York 1924; reissued 1961).

Pasteur was attracted to this controversy and investigated many different fermentations. He noticed that yeasts were present in all fermentations that resulted in the production of alcohol. When sour-tasting lactic acid or butyric acid with its pungent aroma was produced, the ferments contained bacteria in addition to yeast cells. Even when wine turned "bad," bacilli were present in addition to the yeast (Figure 1.6). Pasteur concluded that the sugar of the ferment served as a food for the microorganisms. He professed that each ferment is caused by a specific organism that develops and grows only when the special requirements for its well-being are met. Pasteur then made a defined medium with sugar and salts and showed that the sugar was converted by the yeast into alcohol or cell material and that the nitrogen of the cells was derived from the ammonium salts. These remarkable conclusions supported Schwann's experiments (which had been largely ignored) and served as the basis for the study of microbial physiology.

Aerobic and Anaerobic Growth

Aerobes (air′obes) are organisms that require oxygen in order to grow. Since animals require oxygen for growth, scientists in the middle of the nineteenth century assumed that microorganisms also require oxygen. However, Pasteur's observations on the organisms responsible for the butyric acid fermentation convinced him that life also exists and thrives in the absence of oxygen. Microscopic observation of the microbes from the ferments showed that they rapidly moved about; however, this motility stopped when they were exposed to air. Pasteur also observed that the fermentation ceased when the cultures were aerated. He coined the term **anaerobe** (an′air-obe) to describe organisms that grow in the absence of oxygen.

Pasteur Disproves Spontaneous Generation

Pasteur's brilliance, combined with his knowledge of fermentation, led him into the midst of the controversy over spontaneous generation. He reasoned that each ferment is caused by a given microbe. Therefore, if the entrance of microbes into a suitable broth is prevented, no growth should occur, provided the broth was first adequately heated. Pasteur heated flasks of broth and placed guncotton on the tops of the flasks. This, he reasoned, would allow the free passage of air into the flask, a condition necessary to overcome Needham's criticisms of Spallanzani's experiments. Indeed, the flasks remained uncontaminated until the cotton was removed.

Pasteur went a step further and invented the swan-necked flask (Figure 1.7). This flask allows the unencumbered entrance of air, while all dust particles and contaminants settle in the bend of the neck. Heated flasks of this type remained clear and free of ferments until Pasteur broke the neck off. Flasks with broken necks turned cloudy within a few days. Pasteur ex-

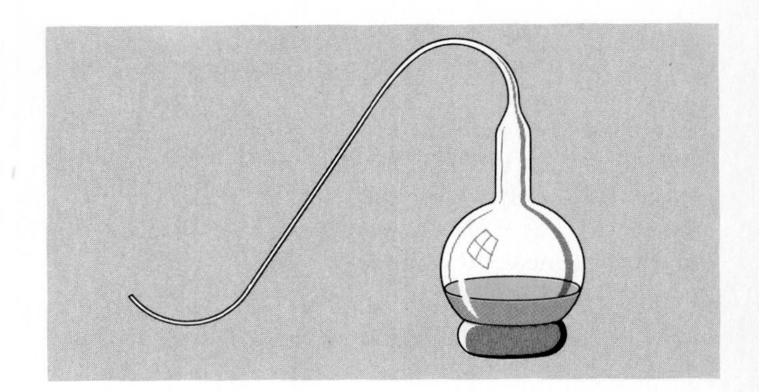

FIGURE 1.7 Pasteur used swan-necked flasks to disprove the theory of spontaneous generation.

plained that organisms capable of causing ferments exist in the air. When precautions were taken to prevent airborne organisms from falling into a broth, ferments did not take place. Pasteur rationalized that different air samples should contain different microorganisms. By opening large numbers of flasks in different places—mountains, lakes, attics, cellars—and immediately resealing the flasks, Pasteur showed that the number of organisms in different air samples varied from place to place.

Problems with Bacterial Spores

Pasteur assumed that the boiling of broth for one hour would kill all living microbes; however, this eventually proved to be incorrect. This fact allowed proponents of spontaneous generation to attack Pasteur's experiments as Spallanzani's experiments had been attacked 100 years earlier. Indeed, some microbes in hay infusion were not killed when the hay was boiled for one hour. For this reason, the beautiful experiments of Pasteur succeeded with yeast extract broths, but not with broths made from hay.

In 1877, Ferdinand Cohn (1828–1898) published a paper on the biology of the rod-shaped bacteria that produce resting forms known as endospores. Bacterial endospores, or simply spores, are extremely resistant to heating and are easily identified under the light microscope. Unlike growing bacterial cells, spores are refractile, appearing as bright objects under the microscope. Cohn's interest in heat-resistant bacteria emanated from the classic experiments of Pasteur and others and from the emerging canning industry, which sought ways to prevent putrefaction of foods. Cohn found that alkaline hay infusions were the most difficult solutions to sterilize. These infusions always became contaminated with the rod-shaped bacterium, *Bacillus subtilis* (sut′ill-is). He further showed that the spores of *B. subtilis* were extremely resistant to heating. When these spores were transferred to a fresh hay infusion, the spores developed into actively growing and reproducing vegetative cells. Obviously, Ferdinand Cohn recognized the source of the problem that had plagued microbiologists for years, but he did not know how to kill the spores.

The solution to this problem was presented by the English physicist John Tyndall (1820–1893) in his paper on sterilization, which he read before the Royal Society in London in 1877. Tyndall recognized, as did Cohn, that hay infusions contained heat-resistant spores. He observed that the spores germinated into heat-sensitive vegetative cells following a brief exposure to heat. Subsequent boiling killed the newly formed vegetative cells. From these observations, he devised an alternating sequence of heat and growth that would ultimately kill all the bacteria present. This technique of sterilizing a liquid broth became known as **tyndallization** (tin-dahl-i-zay′shun).

The experiments of Pasteur and Tyndall demonstrated that microorganisms do not arise spontaneously. No scientist today believes in the ability of humans to produce life by manipulating inert material. However, there are theories concerning the origin of life that include the concept of spontaneous generation. Theoretically, life may have arisen from inert material on the earth when the physical conditions were greatly different from current conditions. The disproof of spontaneous generation set the stage for the discovery of microorganisms as the causative agents of disease.

KEY POINT

Microbiology was unable to develop as a science until the theory of spontaneous generation was disproved. If spontaneous generation were valid, infectious diseases could spontaneously appear in any victim on a random, indiscriminate basis, and experiments with microorganisms would be impossible to repeat. A basic tenet of microbiology, as of all of science, is that experiments can be replicated, or repeated, by other investigators.

ROBERT KOCH (1843–1910) DEMONSTRATES THE BACTERIAL CAUSE OF ANTHRAX

Anthrax, a fatal disease of domestic sheep and cattle, caused economic hardship for German farmers late in the nineteenth century. Robert Koch (Figure 1.8) was a country doctor whose interest in diseases led him to investigate anthrax. This disease appeared sporadically among local flocks of sheep and herds of cattle without an apparent cause or source. Before Koch began his work, there were indications anthrax was caused by a bacterium. Rod-shaped bacteria had been observed in the blood of dead animals, and one microbiologist had transferred anthrax by inoculating animals with dried or fresh blood containing bacteria. Still, nobody had demonstrated that the bacteria actually caused anthrax.

FIGURE 1.8 The German microbiologist Robert Koch (1843–1910) showed that anthrax is caused by a bacterium. (Courtesy of the National Library of Medicine.)

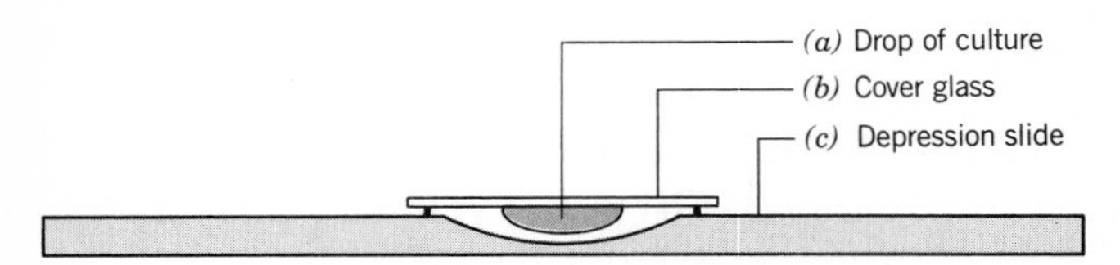

FIGURE 1.9 Robert Koch used the hanging drop technique to observe the growth of anthrax bacteria in fluid taken from the eye of an ox.

Robert Koch quickly ruled out using sheep or cattle as experimental animals in his study of anthrax because these animals were too expensive. He chose to work with mice, which were susceptible to anthrax and could be raised in his laboratory. He took blood from an infected sheep that had died of anthrax and injected it into a series of mice. When one of the mice died, he took its blood and injected it into a healthy mouse. Koch repeated this experiment 20 times and each time the healthy mouse eventually died from anthrax. Examination of the dead mice revealed swollen spleens and the presence of bacilli in both spleen and blood.

Koch tried to grow these bacilli in various broths, without success, until he used the aqueous humor (liquid from an eyeball) from the eye of a cow. He placed a drop of aqueous humor on a thin-glass coverslip and inoculated it with a piece of infected spleen. Koch inverted the coverslip over a depression slide to make a hanging drop preparation (Figure 1.9). He then observed the bacilli grow out from the spleen tissue as vegetative cells before many of them formed refractive spores. He inoculated a mass of these spores into a fresh drop of aqueous humor and observed their germination into vegetative cells. Once he was able to grow the infectious agent of anthrax outside of its animal host, Koch knew how to proceed. He inoculated the spores into healthy mice; the more spores he used, the more quickly the mice died of anthrax. Once again bacilli were found in spleen and blood. These experiments, reported in 1877, conclusively proved that microorganisms can cause infectious disease. Finally, in 1884 Koch summarized his method of proving that a microbe is the causative agent of an infectious disease, a method known as **Koch's postulates:**

1. The microorganism must be demonstrable in all cases of the disease.
2. The microorganism must be isolated from the diseased animal and grown in pure culture.
3. The microorganism from this pure culture must cause the same disease when inoculated into a healthy animal.
4. The experimentally infected animal must contain the microorganism (usually demonstrated by pure culture isolation).

By this time, Robert Koch was recognized as the leader of the German school of microbiology in Berlin. With his assistants, he discovered several species of bacteria that cause human disease and made many contributions to the techniques of microbiology.

PURE CULTURE TECHNIQUES BECOME ESSENTIAL TO MICROBIOLOGY

Technical advances in obtaining pure cultures of bacteria rapidly followed Koch's work with anthrax. Koch noticed that the flat surface of an unrefrigerated boiled potato developed spots of various colors. Microscopic observation of material from these spots revealed the presence of bacteria. One spot contained **rods,** one was composed of the spherical cells known as **cocci,** and a third contained yeast cells. The chance physical sepa-

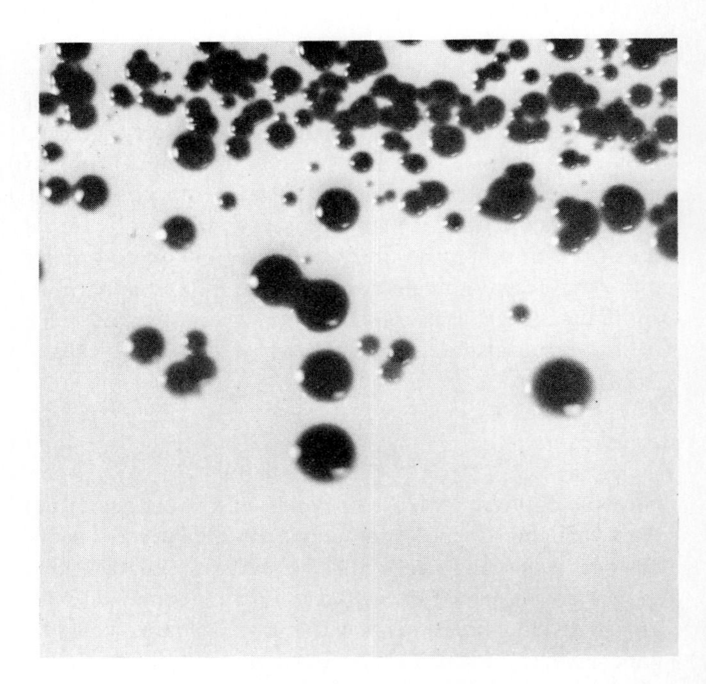

FIGURE 1.10 Bacterial colonies growing on an agar plate. To be visible, a colony must contain at least one million cells.

ration of contaminating microbes on the potato surface permitted the growth of a single cell into a **colony** of many progeny cells (Figure 1.10). From these observations, Koch developed techniques that enabled microbe hunters to grow pure cultures of microorganisms.

In his early experiments Koch used gelatin as a solidifying agent to create a firm surface on which to streak microorganisms. The gelatin was effective at room temperatures, but liquefied at temperatures above 30°C. In 1881, Dr. Walter Hesse and his wife Fanny introduced agar, which Fanny had previously used as a solidifying agent to harden jelly. Agar is extracted from seaweeds, specifically red algae, that grow in warm coastal waters; it is an ideal solidifying agent for microbiology media because after solidifying at about 44°C, it remains solid even above 70°C. Once scientists were armed with the techniques for obtaining pure cultures and the approach dictated by Koch's postulates, the search for the causative agents of disease began in earnest.

KEY POINT

Pure cultures arise from a single organism, which means that all the cells in the culture are identical. In general, experimental microbiology requires the use of pure cultures so that results can be attributed to a single microbe.

INFECTIOUS BACTERIAL DISEASES

Many of the bacterial agents that cause infectious human diseases were discovered between 1877 and 1898 (Table 1.1). The approach that Robert Koch developed to demonstrate the causative agent of an infectious disease and the use of pure culture techniques led to a major breakthrough in this important medical area. (Notice how the development of pure culture techniques, combined with Koch's postulates, stimulated a flurry of scientific discoveries. The use of penicillin as a chemotherapeutic agent in the 1940s and developments in genetic engineering are further examples of how advances in one area of science can stimulate work in other areas.)

Knowing the cause of a disease was not enough to prevent disease. Some bacterial diseases could be prevented by controlling the spread of the infectious agent. For example, cholera was shown by John Snow in 1853 to be spread by water contaminated by human feces, and cholera epidemics could be controlled by prohibiting the consumption of these waters. Another approach to controlling infectious diseases was to immunize people with inactivated toxins (toxoids) or attenuated bacteria unable to cause disease (vaccines). Diphtheria and tetanus were the first toxigenic diseases

TABLE 1.1 Discovery of major bacteria associated with human diseases

DISEASE	ORGANISM	DISCOVERER	YEAR REPORTED
Anthrax	*Bacillus anthracis*	Robert Koch	1877
Gonorrhea	*Neisseria gonorrhoeae*	Albert Neisser	1879
Pyogenic infections	*Staphylococcus aureus*	Alexander Ogston	1881
Tuberculosis	*Mycobacterium tuberculosis*	Robert Koch	1882
Erysipelas	*Streptococcus pyogenes*	Friedrich Fehleisen	1882
Diphtheria	*Corynebacterium diphtheriae*	Theodor Klebs	1883
Tetanus	*Clostridium tetani*	Arthur Nicolaier	1884
Cholera	*Vibrio cholerae*	Robert Koch	1884
Typhoid fever	*Salmonella typhi*	Georg Gaffky	1884
Brucellosis	*Brucella melitensis*	David Bruce	1887
Gastroenteritis	*Salmonella enteritidis*	August Gaertner	1888
Gas gangrene	*Clostridium perfringens*[a]	William Welch	1892
Bubonic plague	*Yersinia pestis*[a]	Alexandre Yersin	1894
Botulism	*Clostridium botulinum*	Emile van Ermengem	1897
Dysentery	*Shigella dysenteriae*	Kiyoshi Shiga	1898

[a]Original names were *Bacterium welchii* and *Bacterium pestis*.

to be controlled by immunizations. Nevertheless, most bacterial diseases remained serious medical problems until the development of antibiotics in the middle of the twentieth century.

DISCOVERY OF VIRAL DISEASES

Some human diseases resisted the Koch approach because they are caused by viruses, which are too small to be seen with the light microscope and do not grow on artificial media. **Viruses** are noncellular submicroscopic entities that grow only inside living cells. Adolf Mayer was the first person to describe the infectious nature of a viral disease. He demonstrated (1886) that tobacco mosaic was an infectious disease of tobacco plants. Another major breakthrough in virology occurred when Dimitri Ivanovski discovered (1892) that tobacco mosaic virus (TMV) could pass through the filters that were used at that time to remove bacteria from solutions. Viruses therefore became defined as filterable infectious agents. Martinus W. Beijerinck (By′jer-inck) (1851–1931) studied TMV and was the first to propose that viruses grow in the living protoplasm of the cell. The filterable nature of the plant disease was soon complemented by Loeffler and Frosch's discovery that the causative agent of foot-and-mouth disease also passed through bacterial filters; it was the first filterable virus shown to cause disease in animals.

By the early 1900s, the important human diseases known to be caused by filterable agents included yellow fever, rabies, measles, and poliomyelitis. About the same time, scientists showed that chicken leukemia and chicken sarcoma (both cancers) were caused by filterable agents; yet, paradoxically, viral causes of human cancers remained undetected for over 50 years. The ability to work effectively with human animal viruses awaited improvements in the techniques for growing viruses in tissue cultures, which became available during the 1950s. Even though virology had a slow beginning, there were some remarkable approaches to preventing viral diseases that predated Koch's discoveries on bacterial diseases (see Jenner's work on smallpox below).

KEY POINT

Koch's postulates can be followed in regard to most infectious diseases; however, there are specific disease situations that violate one or more of the postulates. The bacteria that cause syphilis and leprosy are known, yet neither species has been grown in pure culture outside of animal tissue. Certain bacteria produce toxins that cause disease symptoms, for example botulism; since the bacteria themselves need not be present in an infected person, Koch's fourth postulate does not apply. A corresponding situation is seen with certain cancer-causing viruses that exist only as part of an involved cell's hereditary material in that the virus cannot be isolated from the affected person. Finally, animals are not always available for the experiments required to fulfill Koch's postulates. Nevertheless, the causes of most infectious diseases can be proven by these most important postulates.

PHAGOCYTOSIS AND THE IMMUNE SYSTEM

The start of immunology can be traced to an English country doctor, Edward Jenner (1749–1823), and his work on cowpox. Smallpox (variola) was a dreaded disease that either killed its victims or left them scarred for life. The similar but less severe disease of cowpox was prevalent in Gloucestershire, England, where Jenner lived. Apparently, cowpox begins as an infection of horses' hooves that is transferred to human hands and then to milk cows during milking. Dairy maids contracted cowpox from infected cows that had cowpox lesions on their teats. Jenner's patients who had recovered from cowpox never contracted smallpox, even though other members of the same household were severely ill from smallpox. Jenner became convinced that patients who recovered from cowpox were able to resist smallpox. He documented his observations with numerous patient histories, and finally proved his theory with an experiment that required considerable courage.

On May 14, 1796, Jenner inoculated an 8-year-old boy with cowpox-lesion material taken from the hand of a dairy maid. Seven days later, the boy experienced a mild illness, but soon had completely recovered. On

July 1 and again during September, Jenner infected the boy with material from a smallpox lesion. The boy remained healthy following both exposures to smallpox. Inoculations with material that confers resistance (immunity) to infection became known as a **vaccination** (L. vacca, cow). Jenner's experiments (Figure 1.11) laid the groundwork for protection against infectious agents by artificial immunization.

Louis Pasteur applied those concepts to his study of chicken cholera, a disease that was caused by the bacterium *Pasteurella multocida* (pas-teu-rel′la mul-to-ci′dah). This bacterium, isolated from very sick chickens, had a high virulence, or ability to cause disease. Pasteur grew the bacterium in cultures repeatedly for increasingly long periods and found that its virulence decreased over time. He labeled later cultures **attenuated**—weakened in ability to cause disease. Pasteur perceived that chickens recovered from cholera were resistant to subsequent cholera infection and surmised that inoculation of healthy chickens with attenuated strains of *P. multocida* would prevent chicken cholera. His experiments (published in 1881) proved his hypothesis to be correct and marked the beginning of Pasteur's extensive work on vaccines, including the development of the first vaccine against rabies.

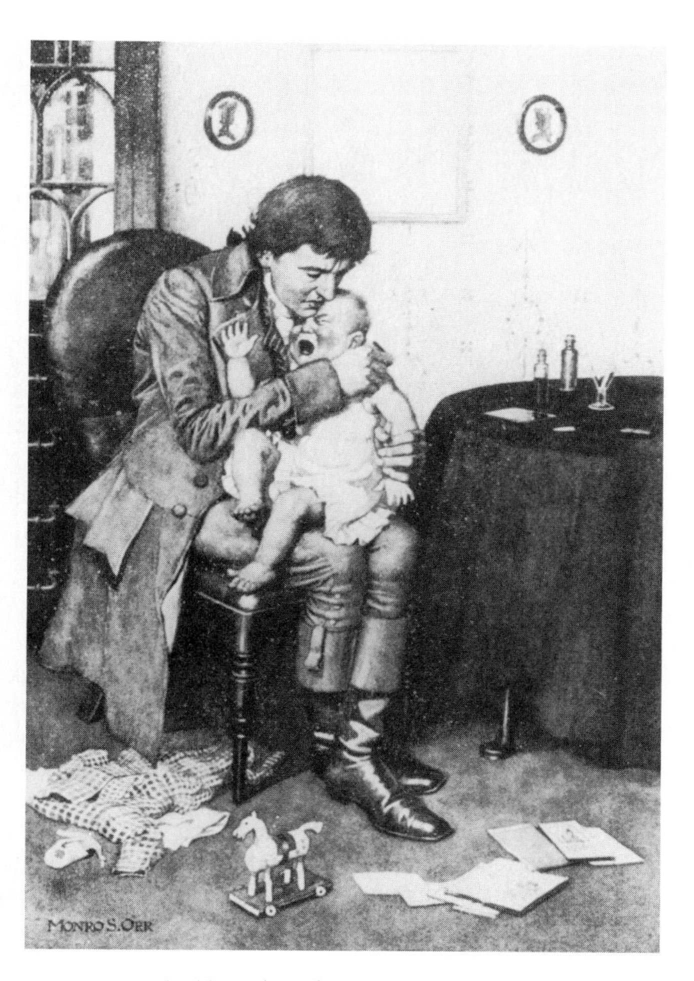

FIGURE 1.11 Edward Jenner (1749–1823) inoculating his 18-month old son with cowpox. (Courtesy of the National Library of Medicine.)

Elie Metchnikoff (1845–1916) was the first scientist to appreciate that certain animal cells had the capacity to engulf and destroy foreign materials. For example, white blood cells remove foreign matter, including bacteria and viruses, from the body and thus contribute to resistance to infection. *Daphnia,* the water flea, was Metchnikoff's experimental vehicle. He infected the flea with fungus spores, and observed the flea's blood cells first bind to the spores, then engulf them, and finally destroy them. He labeled the process **phagocytosis** (pfag′oh-sy-toh′sis) (phagocyte = eating cell). Metchnikoff extended his findings to vertebrates, proposing that their white blood cells are phagocytic cells that protect them against infectious agents.

DISCOVERY OF ANTITOXINS

The specific role of blood in the immune response became better understood through the work of Emil von Behring and Shibasaburo Kitasato. Kitasato had discovered that tetanus is caused by a bacterial toxin produced by *Clostridium tetani* (klaw-stri′dee-um te′tan-eye). He and von Behring demonstrated that rabbits immunized with inactivated tetanus toxin (now called a toxoid) were rendered immune to tetanus. Blood taken from them and mixed with tetanus toxin destroyed the toxin. Moreover, when this immune serum (serum = blood minus cells and clotting factors) was injected into mice, the mice were protected against tetanus. This was the first demonstration of passive immunization; immune serum contains **antitoxin,** a blood protein that specifically inactivates toxin.

Emil von Behring performed similar experiments using diphtheria toxin and was able to produce a tox-

A Centennial Celebration: Pasteur and the Modern Era of Immunization

On July 6, 1885, Louis Pasteur and his colleagues injected the first of 14 daily doses of rabbit spinal cord suspensions containing progressively inactivated rabies virus into 9-year-old Joseph Meister, who had been severely bitten by a rabid dog 2 days before. This was the beginning of the modern era of immunization, which had been presaged by Edward Jenner nearly 100 years earlier.

Pasteur's decision to treat the child followed 4 years of intensive research, culminating in the development of a vaccine capable of protecting experimentally challenged rabbits and dogs. His decision was difficult: "The child's death appeared inevitable. I decided not without acute and harrowing anxiety, as may be imagined, to apply to Joseph Meister the method which I had found consistently successful with dogs." The immunization was successful, and the Pasteur rabies immunization procedure was rapidly adopted throughout the world. By 1890, there were rabies treatment centers in Budapest, Madras, Algiers, Bandung, Florence, Sao Paulo, Warsaw, Shanghai, Tunis, Chicago, New York, and many other places throughout the world.

The "Pasteur Treatment," based on brain tissue vaccine with the addition of formaldehyde, was used in many countries for more than a hundred years. This treatment involved immunizations given daily for 14 to 20 days. As in Pasteur's day, some patients reacted negatively to the brain tissue in the vaccine and suffered serious neurological sequelae. In the United States and other developed countries, more potent, safer, but expensive cell-culture-based rabies vaccines are now used in combination with hyperimmune globulin. The efficacy of such regimens for postexposure treatment has been well proven.

In celebrating the Pasteur centennial, we recognize the preeminent role of vaccines in the control of infectious diseases. As René Dubos states,

Even granted that the antirabies treatment had saved the lives of a few human beings, this would have been only meager return for so much effort. . . . It is on much broader issues that Pasteur's achievements must be judged. He had demonstrated the possibility of investigating by rigorous techniques the infectious diseases caused by invisible, noncultivable viruses; he had shown that their pathogenic potentialities could be modified by various laboratory artifices; he had established beyond doubt that solid immunity could be brought about without endangering the life or health of the vaccinated person. Thanks to the rabies epic . . . immunization [has] become recognized as a general law of nature. Its importance for the welfare of man and animals is today commonplace, but only the future will reveal its full significance in the realm of human economy.

Adapted from *Morbidity and Mortality Weekly Report*, Volume 34, No. 26, July 5, 1985.

oid to immunize humans against diphtheria. He proved that the serum of immunized animals transferred to susceptible animals protected them against the toxin. For this outstanding work von Behring was awarded the first Nobel Prize in Physiology or Medicine.

ENRICHMENT CULTURES OPEN THE DOOR TO GENERAL MICROBIOLOGY

Even before Koch developed the pure culture technique, Ferdinand Cohn (1875) recognized that the different forms of bacteria—rods, cocci, vibrios, and spirilla—are distinct organisms, not merely stages in the life cycle of a single organism. Once microbiologists began to identify different bacterial species and to grow them in pure cultures, they realized that these organisms played different physiological roles in the environment.

Both Martinus Beijerinck and Sergei Winogradsky (Vi-no-grad'sky) (1856–1953) contributed to the development of enrichment culture techniques and to an understanding of general microbiology. In 1888, Beijerinck showed that the bacteria present in the root nodules of leguminous plants (such as peas, clover, and alfalfa) were derived from the soil. Although the physiological role of these bacteria remained a mystery, he realized that the nodules contained pure cultures of a

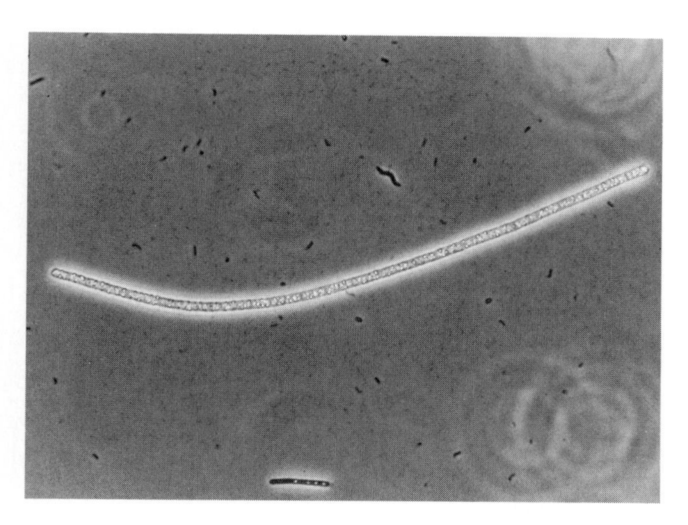

FIGURE 1.12 The long filaments of *Beggiatoa* containing numerous sulfur granules are common bacteria in marine sediments. (Courtesy of Dr. Paul W. Johnson/BPS.)

single bacterium. Equally important were Winogradsky's studies on sulfur metabolism by the bacterium *Beggiatoa* (beg-gi-ah-to′a). Winogradsky showed that the sulfur (S) granules visible in the cytoplasm of *Beggiatoa* (Figure 1.12) were derived from hydrogen sulfide (H_2S) present in the medium. Hydrogen sulfide gas is a natural environmental pollutant. The sulfur granules inside the cells are eventually converted to sulfate (SO_4^{2-}), which is released into the medium. The organisms participate in the natural cycling of sulfur. Winogradsky also made major discoveries concerning the role of bacteria in the nitrogen cycle.

The approach of Beijerinck and Winogradsky to the study of bacteria evolved into the concept of selective enrichments. In 1901, Beijerinck described **enrichment cultures** as those that mimic conditions in nature such that one organism is sufficiently advantaged to be able to outgrow others present. Once the target organism becomes predominant, it is possible to obtain a pure culture by isolating individual colonies. Selective enrichment cultures have been used to isolate thousands of bacteria from natural environments.

KEY POINT

Specimens from natural environments always contain a variety of microorganisms. Microbiologists selectively grow the microbe they are interested in by providing a growth medium and incubation conditions that favor that microbe's growth.

CELL THEORY LINKS THE BIOLOGICAL SCIENCES

Even before Leeuwenhoek's first letter to the Royal Society, Robert Hooke had observed the compartmentalized structure of living systems and had named these structures **cells.** The theory that all living things, except viruses, are composed of cells was first presented in 1838, when Matthias Schleiden proposed that all plant tissue was composed of cells. Theodor Schwann expanded the cell theory by suggesting that animal tissues are also composed of cells. Now we know that the cell is the basic structure of all organisms, except viruses. The form and structure (morphology) of individual cells can vary greatly within a given organism, and morphological variations between single-celled organisms are common. The variations in cell morphology led early investigators to group cells according to size and shape. As many of these early scientists were trained in classical botany, it was natural that they would classify microorganisms on the basis of the hierarchy developed by Carolus Linnaeus.

THE BINOMIAL SYSTEM

Carolus Linnaeus (1707–1778) was the first person to develop a "shorthand" system for naming organisms. He began by describing the plants he collected and then he classified them into species. A **species** comprises those organisms that have the same form and structure and that reproduce their own kind. Linnaeus's system of nomenclature gives each organism a latinized name consisting of two words, a binomial system that remains in use. The first word indicates the taxonomic group, or **genus** (pl. genera), to which a species belongs. The second word is the specific epithet (adjective or modifier) that defines the species of the genus. By convention, the proper name of a species is printed in italics, and the first letter of the genus is capitalized.

Vibrio cholerae is the name of the comma-shaped organism (vibrio) that causes cholera. The generic name

Vibrio refers to this and all other similar vibrioid microorganisms. The epithet cholerae modifies the genus name and designates a specific type of *Vibrio,* namely *Vibrio cholerae.* Note that the epithet (Table 1.1) can also be the name of a disease or the basis of a disease name, such as *Bacillus anthracis,* anthrax; *Clostridium tetani,* tetanus; *Clostridium botulinum,* botulism. The binomial nomenclature enables microbiologists to name organisms and to define them in a systematic manner. *Bergey's Manual of Systematic Bacteriology* is the standard reference book for the classification of bacteria.

The study of *systematics* goes one step further and organizes all organisms into a hierarchy based on their interrelatedness. The earliest attempts to classify bacteria were published in 1786 by Otto F. Müller. From this early beginning, the systematics of microorganisms continues as a dynamic field of microbiology. There is a constant need to revise and update microbial classification as new microorganisms are discovered and new information is reported.

THE MICROBIAL WORLD

Small cellular organisms and viruses comprise the microbial world. Systematists recognize five major groups among microorganisms: viruses, bacteria, protozoa, algae, and fungi. Viruses are the only noncellular living entities and their evolutionary relationship to other organsims is unknown. Bacteria, protozoa, algae, and fungi are cellular microorganisms that exist as single cells or as aggregates of cells having the same morphologies.

Microscopists can readily distinguish between two types of cells: those that contain a nucleus and those that lack a nucleus (Figure 1.13). So far as is known, there are but two major types of cells on earth: cells whose genetic information is enclosed in a membrane-bound nucleus are **eucaryotes** (yoo-kar′ee-ahts), while **procaryotes** (pro-kar′ee-ahts) have no nucleus, and their hereditary information is found in the nuclear region of the cytoplasm. Procaryotes comprise *all* bacteria (including the blue green algae now known as cyanobacteria) and are deemed the earliest cellular forms of life to have arisen on earth. All other organisms are eucaryotes. The specific characteristics of the procaryotic cell and the eucaryotic cell are discussed in Chapters 4 and 5.

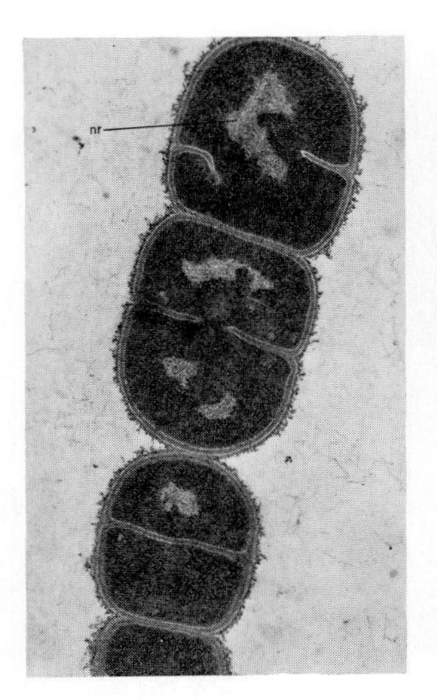

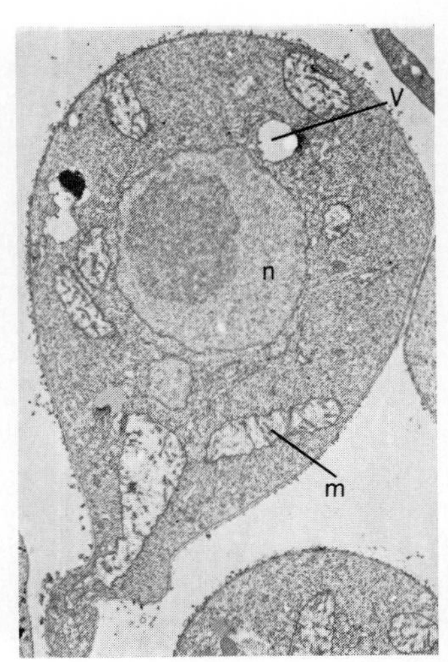

FIGURE 1.13 Procaryotic and eucaryotic cells have different subcellular organizations. Transmission electron micrographs of *(a)* cross section of the bacterium *Caryophanon latum* showing the bacterial nuclear material localized in the nuclear region (nr). (Courtesy of Dr. R. K. Nauman.) *(b)* Cross section of the fungus *Blastocladiella emersonii* showing its membrane-bound nucleus (n), vacuoles (v), and mitochondria (m). (Courtesy of Dr. J. S. Lovett. From J. S. Lovett, *Bacteriol. Revs.*, 39:345–404, 1975, with permission of the American Society for Microbiology.)

THE FIVE KINGDOMS

Most plants and animals are easily recognized as such and are classified in the kingdom **Plantae** or **Animalia.** The discovery of bacteria confused this simple classification because bacteria don't appear to belong to either kingdom. In addition, many eucaryotes share characteristics with both plants and animals and do not easily fit the two-kingdom system.

To deal with this perplexing problem, E. M. Haeckel (1834–1919) proposed that protozoa, algae, fungi, and bacteria be placed in a third kingdom, which he called **Protista** (pro-tis′ta), meaning the first organisms in the sequence of evolution. This remained a viable alternative to the two-kingdom system until electron microscopy clearly revealed the differences between the eucaryotic and the procaryotic cell types. This led to the four-kingdom system, which separates the bacteria from the Protista and classifies them in a fourth kingdom, **Procaryotae.**

The five-kingdom system (Figure 1.14) proposed by Robert H. Whittaker (1924–1980) places the fungi in a

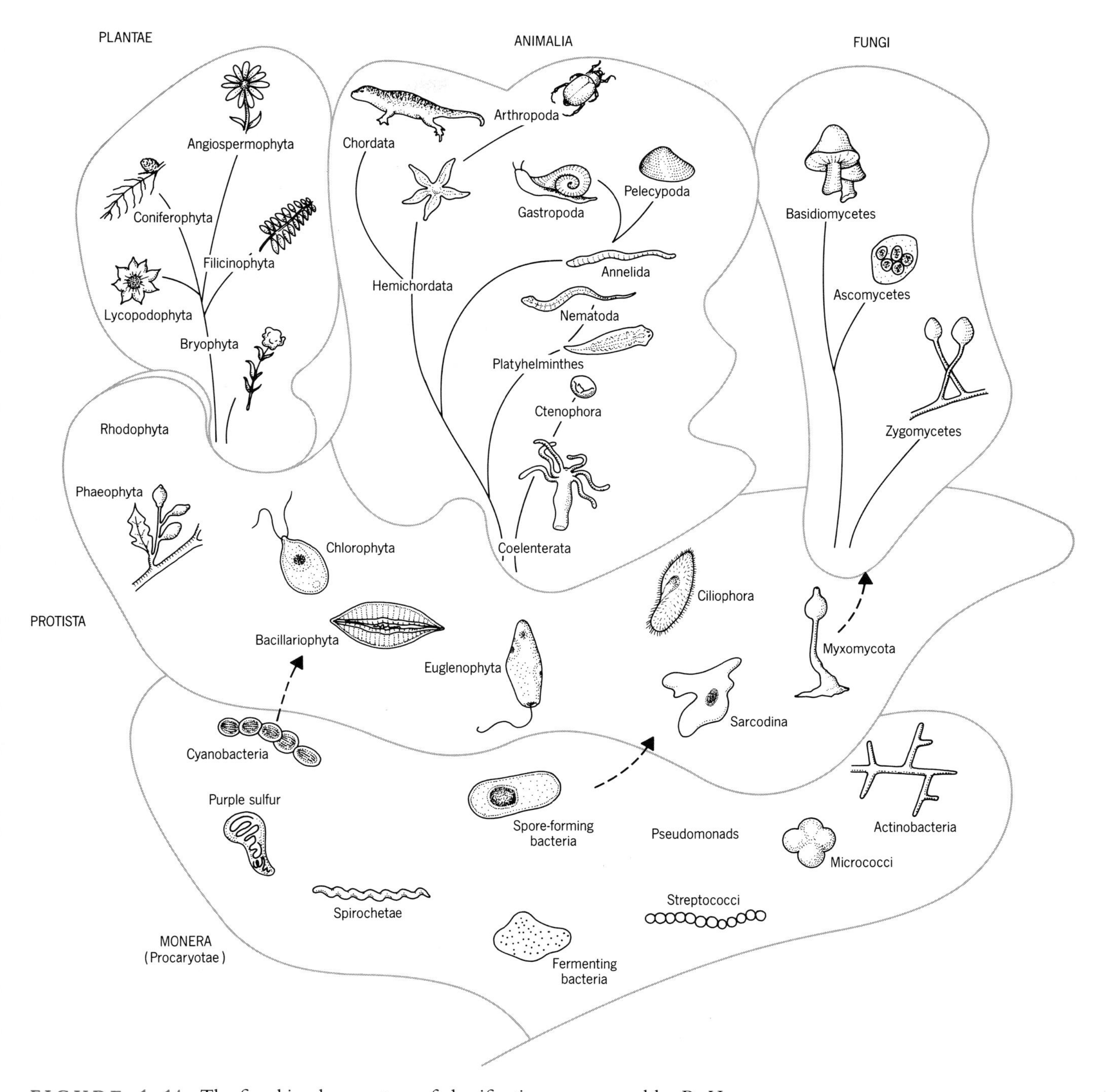

FIGURE 1.14 The five-kingdom system of classification as proposed by R. H. Whittaker. Plants, animals, and fungi are classfied in the separate kingdoms of Plantae, Animalia, and Fungi. Eucaryotic microorganisms are classified in the kingdom Protista and procaryotes are classified in the kingdom Monera, renamed the kingdom Procaryotae.

separate kingdom. In the five-kingdom system, the kingdoms Plantae and Animalia are retained, and bacteria are classified in a separate kingdom, Procaryotae (Whittaker's the **Monera**). The Whittaker scheme emphasizes how organisms acquire their food: plants use photosynthesis, animals ingest food, and fungi are saprophytes that live on dead organic matter. For this reason, Whittaker placed the fungi in a separate kingdom, **Fungi** (fun'jeye). The remaining organisms are classified in the kingdom Protista, a kingdom that is defined more by exclusion than by inclusion. Members of the Protista are all unicellular eucaryotes and include some algae, flagellated water molds, slime molds, and protozoa. As you can see, the task of assigning living creatures to a small number of kingdoms is extremely difficult, and the system is still undergoing revision.

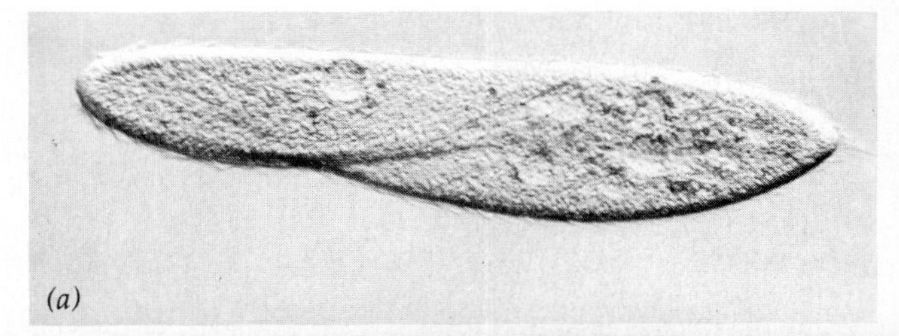

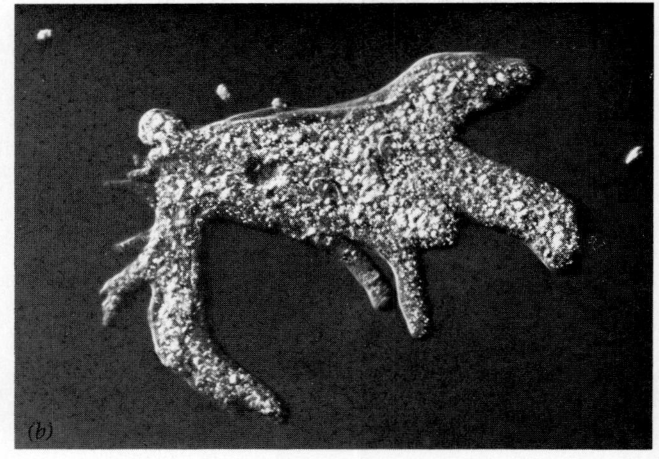

FIGURE 1.15 Micrographs of *(a) Paramecium* and *(b) Amoeba proteus:* two examples of unicellular, eucaryotic microorganisms. (Courtesy of Carolina Biological Supply Company.)

Representatives of the Five Kingdoms

Procaryotae **Bacteria** are the simplest cellular organisms known and are believed to have been the first forms of life on earth. They are greatly diversified in size, shape, and means of gaining energy; yet they all have the procaryotic cell structure. Bacteria are unicellular organisms and only a few species ever form aggregates of many cells. Most bacteria divide by the asexual process of transverse **binary fission** (by'ner-ee fizh'en), which produces daughter cells of equal size. A few bacteria reproduce by budding to produce a daughter cell or bud that is significantly smaller than the parent cell. For many years the cyanobacteria were classified with the eucaryotic algae (blue-green algae) because they, like algae and green plants, produce oxygen during photosynthesis. However, cyanobacteria all have a procaryotic cell structure, so they are now classified with the bacteria.

Protista The groups of microorganisms classified in the kingdom Protista are the protozoa, some algae, flagellated water molds, and slime molds. **Protozoa** are the eucaryotic unicellular animals. They are usually larger than bacteria, and many protozoa actually feed on bacteria. Representative protozoa include the *Paramecium* which moves by cilia (Figure 1.15); and the common *Amoeba* (Figure 1.15), which moves by extending its cytoplasm in the form of a pseudopodium (soo-doh-pod'ium). Most protozoa are harmless, but a few can cause serious human diseases. For example, giardiasis (jee-ahr-deye'ah-sis) is a common intestinal infection caused by *Giardia* found in untreated drinking water, and malaria, one of the most prevalent infectious diseases of the tropics, is caused by a protozoan transmitted by mosquitoes.

Slime molds grow in damp places such as soil and decaying vegetable matter. **Cellular slime molds** grow as single ameboid cells in the soil until nutrients become scarce. Then the amebae aggregate into multicellular slugs that eventually come to rest and differentiate into a stalk and fruiting body (Figure 1.16). The spores thus produced are dispersed and form vegetative amebae upon germination.

Algae are unicellular or multicellular eucaryotic cells that use light energy to grow—the process called photosynthesis. (Figure 1.17). Algae are largely responsible for the primary productivity of aquatic ecosystems. Although most algae are harmless, the red tide dinoflagellate (dye'no-flag'il-late), *Gonyaulax* (gon-yawl'ox) is one that poses a public health problem. These algae produce a potent nerve toxin that is poi-

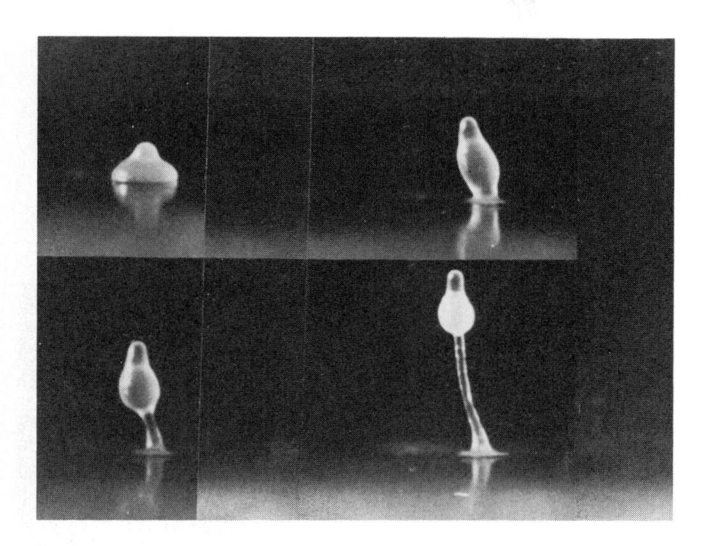

FIGURE 1.16 Formation of the fruiting body in the slime mold, *Dictyostelium discoideum*. Individual amebae aggregate to form a slug. The slug develops a stalk upon which is mounted the spores of the fruiting body.

sonous to fish and is concentrated in shellfish living in the same waters. Humans who eat the shellfish become very sick. Growth of *Gonyaulax* in coastal waters often necessitates the closing of beaches and posting of notices prohibiting shellfishing. A number of algae provide a useful source of commercial products such as agar and carrageenan (kar-ah-gene′ahn), which is used to thicken ice cream, custards, and other foods. Classification of algae is based largely on their photosynthetic pigments, cell wall composition, and type of stored food (see Chapter 16).

Fungi Microorganisms with plant-like characteristics are divided into two groups: the nonphotosynthetic organisms known as the fungi, and the photosynthetic organisms known as algae (Figure 1.17). **Fungi** are a diversified group of nonphotosynthetic eucaryotic cells classified by their mode of sexual and asexual reproduction. Molds are common and are visible to the unaided eye. They grow in damp places,

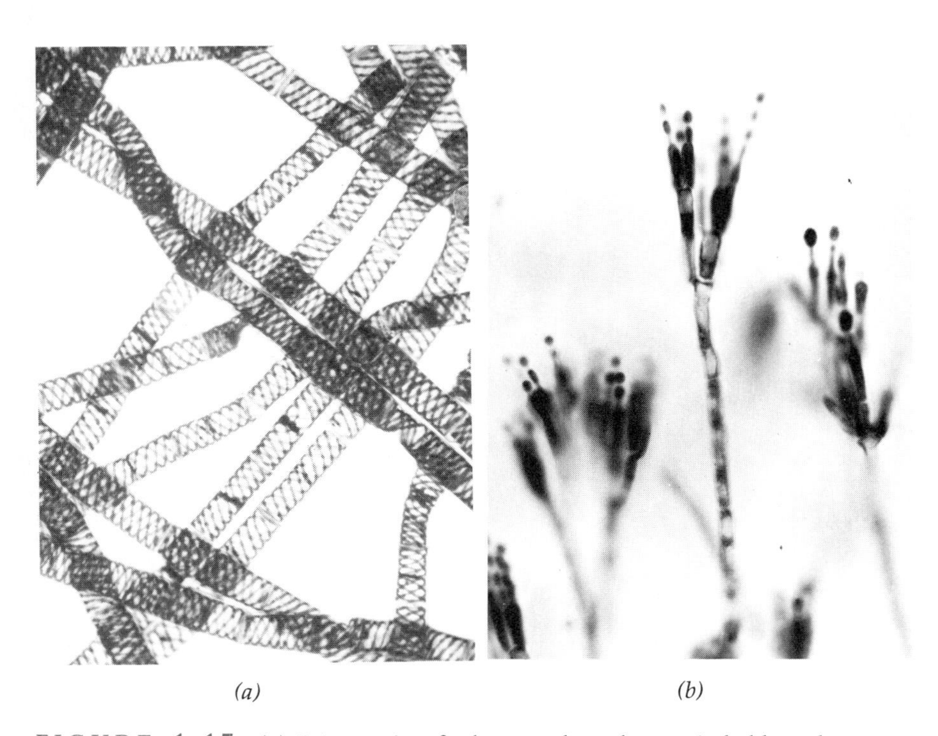

(a) (b)

FIGURE 1.17 *(a) Spirogyra* is a freshwater alga whose spiral chloroplasts are readily visible in individual cells. *Penicillium (b)* is a representative mold that produces spores at the tips of its hyphae. (Courtesy of *(b)* Carolina Biological Supply Company, *(a)* Hugh Spencer.)

such as bathrooms and basements, as well as on food such as bread and cheeses. Yeasts are single-cell molds, some of which play a part in the production of beer, wine, and bread. Edible mushrooms grow wild in the woods and in lawns; some are cultivated, harvested, and sold in grocery stores. The molds (Figure 1.1) seen by Hooke are also fungi. A small group of fungi cause superficial human infections such as athlete's foot and "jock itch." Others can cause serious systemic infections in humans if their spores are inhaled.

Summary Outline

DISCOVERY OF MICROORGANISMS

1. Robert Hooke and Anton van Leeuwenhoek lived in the seventeenth century and used primitive microscopes to observe microorgansims.
2. Hooke was the first to describe simple molds.
3. Leeuwenhoek made remarkable observations of protozoa and was the first to observe bacteria, which he saw in pepper infusions.

MICROBIOLOGY IN THE EIGHTEENTH CENTURY

1. Plenciz and Martin proposed that microorganisms caused diseases such as tuberculosis, but presented no scientific evidence to document their claims, and were ignored.

SPONTANEOUS GENERATION

1. The controversy over spontaneous generation limited the development of microbiology during the eighteenth century.
2. John Needham claimed his experiments with mutton gravy proved spontaneous generation.
3. Lazzaro Spallanzani claimed to disprove spontaneous generation since microbes did not grow in his hermetically sealed (airtight) glass flasks.
4. Louis Pasteur disproved spontaneous generation with his swan-neck flasks, which allowed air to enter the flasks.
5. John Tyndall resolved the problems of bacterial spore contamination by developing a method of killing spores called tyndallization.

DISCOVERY OF INFECTIOUS BACTERIAL DISEASES

1. Robert Koch grew the anthrax bacillus in pure culture and showed that it caused anthrax in animals.
2. Koch's postulates became the scientific approach for demonstrating the infectious agent responsible for a disease.
3. Techniques for obtaining pure cultures of bacteria were developed in Koch's laboratory. Fanny Hesse introduced agar as a solidifying agent for microbiological media.
4. Many of the major human disease-causing bacteria were discovered between 1880 and 1900.

DISCOVERY OF VIRAL DISEASES

1. Adolf Mayer discovered the infectious nature of tobacco mosaic disease and Ivanovski showed that the agent was filterable.
2. Loeffler showed that hoof-and-mouth disease (animal virus) was caused by a filterable agent, and soon many human diseases were known to be caused by filterable viruses.

PHAGOCYTOSIS AND THE IMMUNE SYSTEM

1. Jenner used material from a cowpox lesion to protect patients against smallpox in 1796.
2. Louis Pasteur developed a vaccine against chicken cholera.
3. Elie Metchnikoff observed phagocytosis by cells in the water flea and postulated the same process in humans.
4. Emil von Behring made toxoids of the diphtheria and tetanus toxins and pioneered the use of antitoxins.

ENRICHMENT CULTURES OPEN THE DOOR TO GENERAL MICROBIOLOGY

1. Ferdinand Cohn recognized that morphologically diverse bacteria represent distinct species.
2. Beijerinck discovered the root-nodule bacteria in essentially pure cultures, and Winogradsky devel-

oped his concept of how bacteria grow in inorganic media.

3. Beijerinck and Winogradsky developed the concept of selective enrichment cultures, which eventually led to the isolation of hundreds of microbes.

CELL THEORY LINKS THE BIOLOGICAL SCIENCES

1. All organisms have the cell as the basic unit of life; viruses are noncellular.
2. The binomial system of nomenclature assigns a genus and a species name to each distinct organism.
3. All species belong to one of five kingdoms: Animalia, Plantae, Fungi, Protista, and Procaryotae.
4. All the bacteria, including the cyanobacteria, belong to the Procaryotae.
5. Protozoa, some algae, flagellated water molds, and the slime molds belong to the Protista.
6. Animals, plants, and fungi are classified in the kingdoms Animalia, Plantae, and Fungi.

Questions and Topics for Study and Discussion

QUESTIONS

1. Describe three key events that helped microbiology to develop into a science?
2. How were human diseases first demonstrated to be caused by infectious agents? Explain.
3. What was the rationale behind the attack on Spallanzani's experiments with the hermetically sealed flasks? Who eventually won this argument and what experiments were done to prove the point?
4. What role did tyndallization play in resolving the controversy over spontaneous generation?
5. How did the development of pure culture and enrichment culture techniques advance microbiology?
6. What methods were available for treating and/or preventing bacterial diseases prior to the development of antibiotics?
7. On what basis does the five-kingdom system of classification assign microorganisms to separate kingdoms? List some organisms that do not easily fit into this scheme.
8. All the various forms of bacteria seen in infusions were considered by some early workers to be stages in the life history of one and the same organism. How was this problem resolved so that the different morphological forms became recognized as distinct species?

DISCUSSION TOPICS

1. What impact did the controversy over spontaneous generation have on microbiology?
2. What scientific advances led to the discovery of a large number of bacterial diseases during the last 20 years of the nineteenth century?

Further Readings

BOOKS

Brock, T. D. (ed.), *Milestones in Microbiology,* American Society for Microbiology, Washington, D.C., 1975. Translations of and brief commentaries on selected scientific articles by famous microbiologists including Leeuwenhoek, Pasteur, and Koch.

DeKruif, P., *Microbe Hunters,* Harcourt, Brace, New York, 1953. This popular history of microbiology uses the story format to describe the major scientific contributions of some important microbiologists.

Dubos, R., *Louis Pasteur: Free Lance of Science,* Little, Brown, Boston, 1950. One of the better biographies of this famous microbiologist, written by a respected microbiologist and scholar.

Singleton, P., and D. Sainsbury, *Dictionary of Microbiology,* Wiley, New York, 1978. An important reference source for definitions and explanations of scientific terms and concepts used in microbiology.

CHAPTER

2

Principles of Chemistry

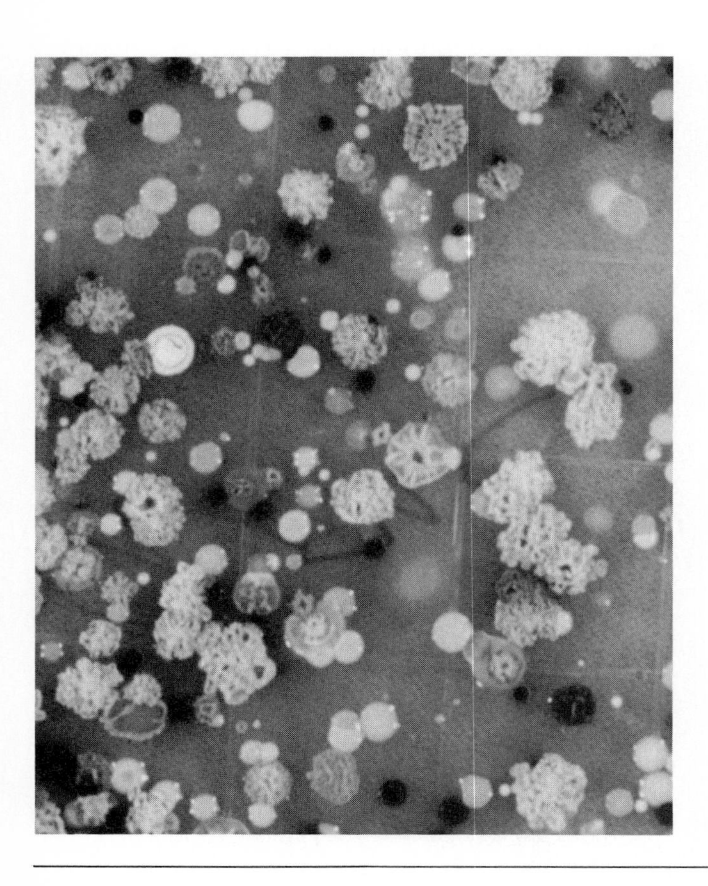

OUTLINE

FOCUS OF THIS CHAPTER

Many biological phenomena can be explained by chemistry, the science of the structure, composition, and properties of substances and their transformations. This chapter reviews the basic chemistry of substances found in living matter and explains the chemical principles that account for the behavior of these substances and how cells convert these substances into molecules.

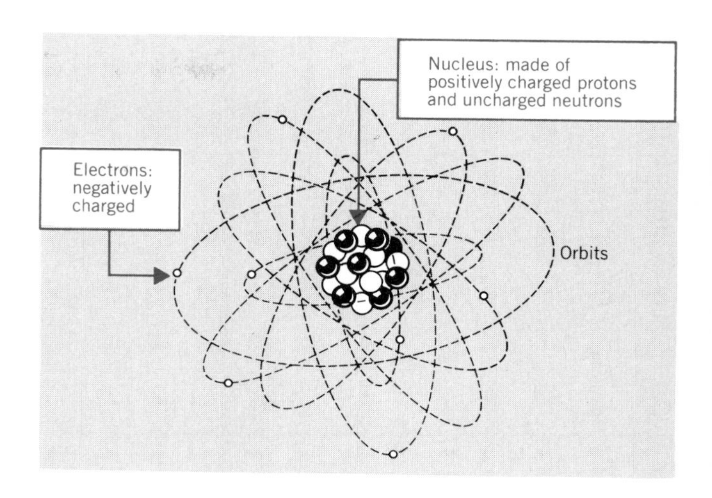

FIGURE 2.1 Neils Bohr proposed the planetary model of the atom in 1913. This schematic representation of the Bohr atomic model has electrons orbiting about a dense nucleus. The British physicist, Lord Rutherford, is given credit for the concept of the atomic nuclei.

For many students, this chapter on the principles of chemistry covers material previously studied; for others the chapter represents a first encounter with chemical principles. Each topic included was selected for its relevance to an understanding of the biochemical phenomena presented in later chapters.

As microorganisms grow and replicate, they produce new cellular materials from the substances present in their immediate environment. Some of these substances are directly incorporated into cell material, while others must be converted by one or more chemical reactions into the biomolecules that constitute living matter. Biomolecules are composed of elements that can be arranged in a limited number of configurations based on the laws of chemistry.

THE BASIC STRUCTURE OF MATTER

All matter is composed of atoms. Although atoms are too small to be seen, they are assumed to be the absolute basic structure of matter since they cannot be broken down by ordinary means. It is only under extreme and unusual conditions that atoms are destroyed, which is fortunate since atomic reactions release tremendous amounts of energy that can cause devastation if not controlled. Controlled atomic reactions occur in atomic reactors, and the energy thus released is used to produce electricity. Just as atoms are almost indestructible, it is impossible to make more atoms than already exist.

Atoms consist of a nucleus surrounded by orbiting electrons (Figure 2.1) that spin around the nucleus, as planets of our solar system orbit the sun. Each nucleus is composed of a specific number of **protons,** which are positively charged particles, and a specific number of **neutrons,** which are uncharged or neutral particles. The particles that spin about the nucleus are negatively charged and are called **electrons.** Each electron spins around the nucleus in a prescribed space or **electron orbital** (Figure 2.1). The number of negatively charged electrons in an atom equals the number of positively charged protons, thus atoms have a neutral charge.

CHEMICAL ELEMENTS

The many chemicals on earth owe their differences to their atomic structures. There are a set number of **chemical elements,** which are made up of atoms having the same number of protons, neutrons, and electrons. Hydrogen is the chemical element with the simplest atomic structure—one proton and one electron.

TABLE 2.1 Electronic configuration of some biologically important elements

ELEMENT (SYMBOL)	NUMBER OF NEUTRONS	NUMBER OF PROTONS	ATOMIC WEIGHT	NUMBER OF ELECTRONS AT EACH LEVEL (*n*)			
				1	2	3	4
Hydrogen (H)	0	1	1.0	1			
Carbon (C)	6	6	12.0	2	4		
Nitrogen (N)	7	7	14.0	2	5		
Oxygen (O)	8	8	16.0	2	6		
Sodium (Na)	12	11	23.0	2	8	1	
Phosphorus (P)	16	15	31.0	2	8	5	
Sulfur (S)	16	16	32.0	2	8	6	
Chlorine (Cl)	18	17	35.0	2	8	7	
Calcium (Ca)	20	20	40.1	2	8	8	2

Carbon is more complex, with six protons, six neutrons, and six electrons (see Table 2.1). The mass or weight of each element varies depending on the number of its protons and neutrons. This is reflected in the element's **atomic weight,** which is essentially equal to the combined number of protons and neutrons in the nucleus. Since electrons weigh 1/1837 the weight of a proton or neutron, they are not included in typical weight calculations.

ELECTRON CONFIGURATION OF ELEMENTS

The reactivity of an element depends on its structure and the arrangement of electrons orbiting the atomic nucleus. Chemists describe electron orbitals in mathematical terms, rather than in physical space, since it is impossible to describe the position of an electron at any given instant. The electron is held in the orbital by the force exerted between the electron and the nucleus. Each electron in a given atom possesses a specific amount of energy that is related to the distance between the electron and the nucleus. Since the electrons in an atom are restricted to individual orbitals, they are said to occupy an **energy level.** Electrons in the energy level closest to the nucleus have the highest energy, that is, it takes more energy to remove electrons from this first energy level than to remove electrons from any of the others.

Each energy level is limited in the number of electrons it can contain. The maximum number of electrons that can be in an energy level (n) is equal to $2 \cdot n^2$. According to this equation, the first level can contain two electrons $(2 \cdot 1^2)$, the second level can contain eight $(2 \cdot 2^2)$, the third level 18 and so on. The electronic configuration of some biologically important elements is shown in Table 2.1. Note that the number of electrons is equal to the number of protons in the atom's nucleus and that the atomic weight equals the number of neutrons and protons in the atom.

ISOTOPES ARE ALTERNATIVE FORMS OF AN ELEMENT

The nuclei of certain elements can acquire or lose neutrons from their nucleus; this changes the atom's weight. These different forms of the same element are called **isotopes.** Carbon has an atomic weight of 12 and its nucleus contains six neutrons and six protons. Carbon 14 (^{14}C) is an isotope of carbon that has six protons and eight neutrons. Some isotopes are stable, whereas others gradually disintegrate by giving off radiation in the form of nuclear particles and/or electromagnetic waves. Unstable isotopes decay at a known rate measured as the half-life of the isotope (Table 2.2); after one half-life only 50 percent of the isotope will remain.

Unstable isotopes are useful as chemical tracers in biological research and in diagnostic medicine, because even low levels can be detected by the radiation they emit. One of the most widely used isotopes in biochemical studies is ^{14}C, an isotope that can be incorporated into many biomolecules. The advantages of using isotopes in biological research are that (1) the chemical behavior of the compound containing the isotope is the same as that of the nonradioactive molecule, and (2) instruments are widely available that can detect

TABLE 2.2 Properties of biologically useful isotopes

ELEMENT	AVERAGE ATOMIC WEIGHT	ISOTOPE	ATOMIC MASS	HALF-LIFE
H	1.01	^{2}H	2.014	Stable
		^{3}H	3.016	12.26 years
C	12.01	^{13}C	13.00	Stable
		^{14}C	14.00	5730 years
N	14.01	^{15}N	15.00	Stable
O	16.00	^{18}O	18.00	Stable
P	30.97	^{32}P	32.00	14.3 days
S	32.06	^{35}S	35.00	88.0 days

extremely low levels of radioactivity, so only minute quantities of a substance need be present. The radioactive isotopes used in biological research usually have a long half-life. Both ^{14}C and ^{3}H (tritium) have half-lives measured in years (Table 2.2). In addition, the isotope of phosphorus (^{32}P) is used to label nucleic acids, and the isotope of sulfur (^{35}S) is used to label proteins. Some isotopes occur naturally, but only in very small quantities. Most of the isotopes used in biological research and medicine are manufactured in nuclear reactors.

IONS ARE CHARGED ATOMS OR MOLECULES

Ions are electrically charged atoms or molecules that often are formed when a compound, such as table salt (NaCl), is dissolved in water (Figure 2.2). The most stable form of an element is one that has filled its outermost energy level (outer shell) with the maximum number of electrons. Elements that contain one or two electrons in their outermost shell (look at sodium and calcium in Table 2.1) tend to lose electron(s) from their outer shell. Atoms with less than a full outer shell (see chlorine in Table 2.1) tend to accept electrons.

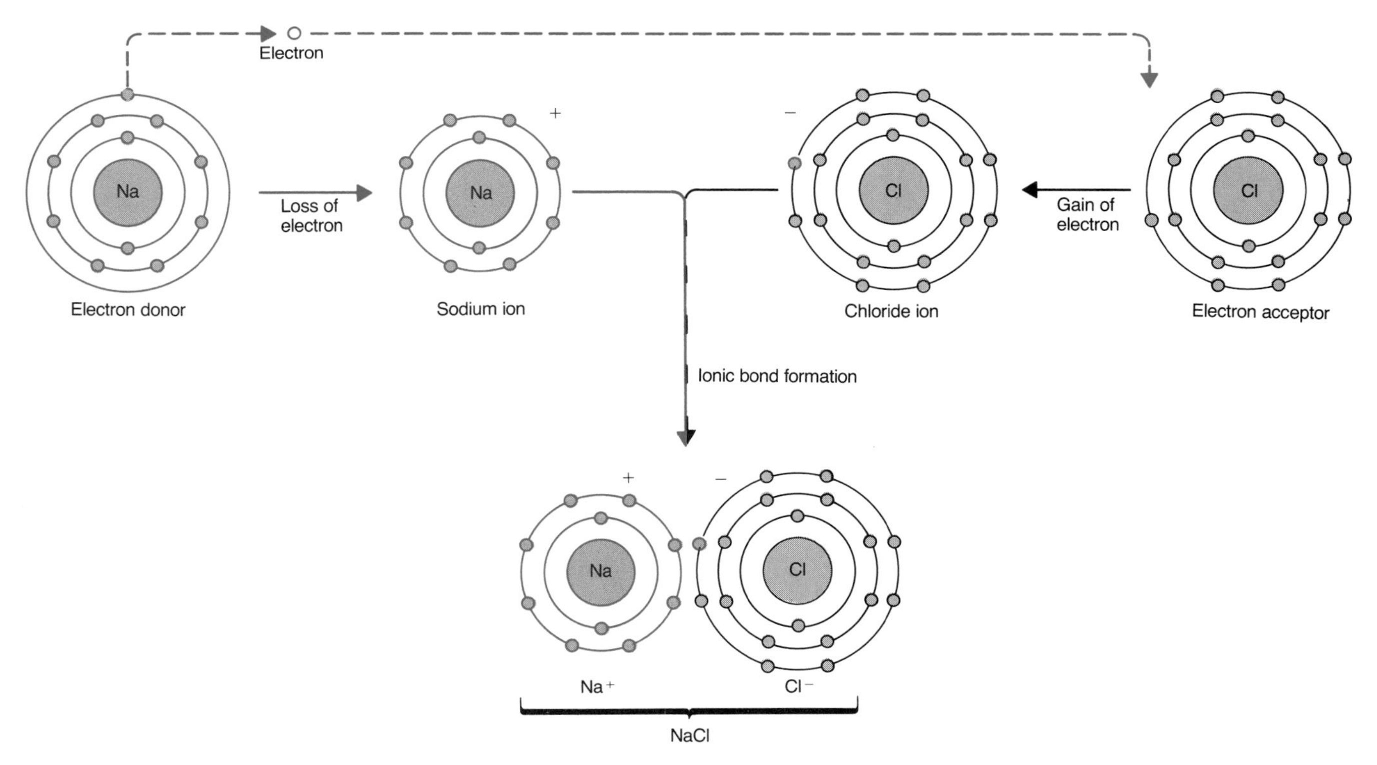

FIGURE 2.2 When sodium chloride is placed in water, the sodium atom gives up an electron to become Na^{+} and the chlorine atom accepts an electron to become $C1^{-}$. Sodium and chloride ions will form crystalline sodium chloride (NaCl), a molecule held together by ionic bonding.

Think about what happens to table salt when it is added to water. Sodium gives up an electron in level 3 to become Na^+, a positively charged cation (Figure 2.2). A **cation** is formed when an element gives up one or more electrons from its outer energy level to become positively charged. Another example is calcium, which can give up two electrons in level 4 to become Ca^{2+}. Note that the atom becomes positively charged when it loses the negatively charged electron. These electrons must go somewhere—and in fact they are accepted by another element trying to fill its outer energy level. In the case of NaCl, chlorine accepts one electron to complete its outermost electron level with the eighth electron. Chlorine is now a negatively charged ion (Cl^-) or an anion. An **anion** is formed when an element accepts one or more electrons into its outer energy level to become a negatively charged ion.

ATOMS COMBINE TO FORM MOLECULES

Matter in all its different forms is constructed from atoms joined together by chemical bonds. The chemical and physical characteristics of the substances formed can be explained on the basis of the elements involved and the strengths of their chemical bonding. The allowable bonding between elements depends on the atomic structure of each element.

ATOMS FORM COVALENT BONDS BY SHARING ELECTRONS

Certain elements complete their outermost electron level by sharing electrons with either identical or dissimilar atoms. Oxygen, for example, will form oxygen

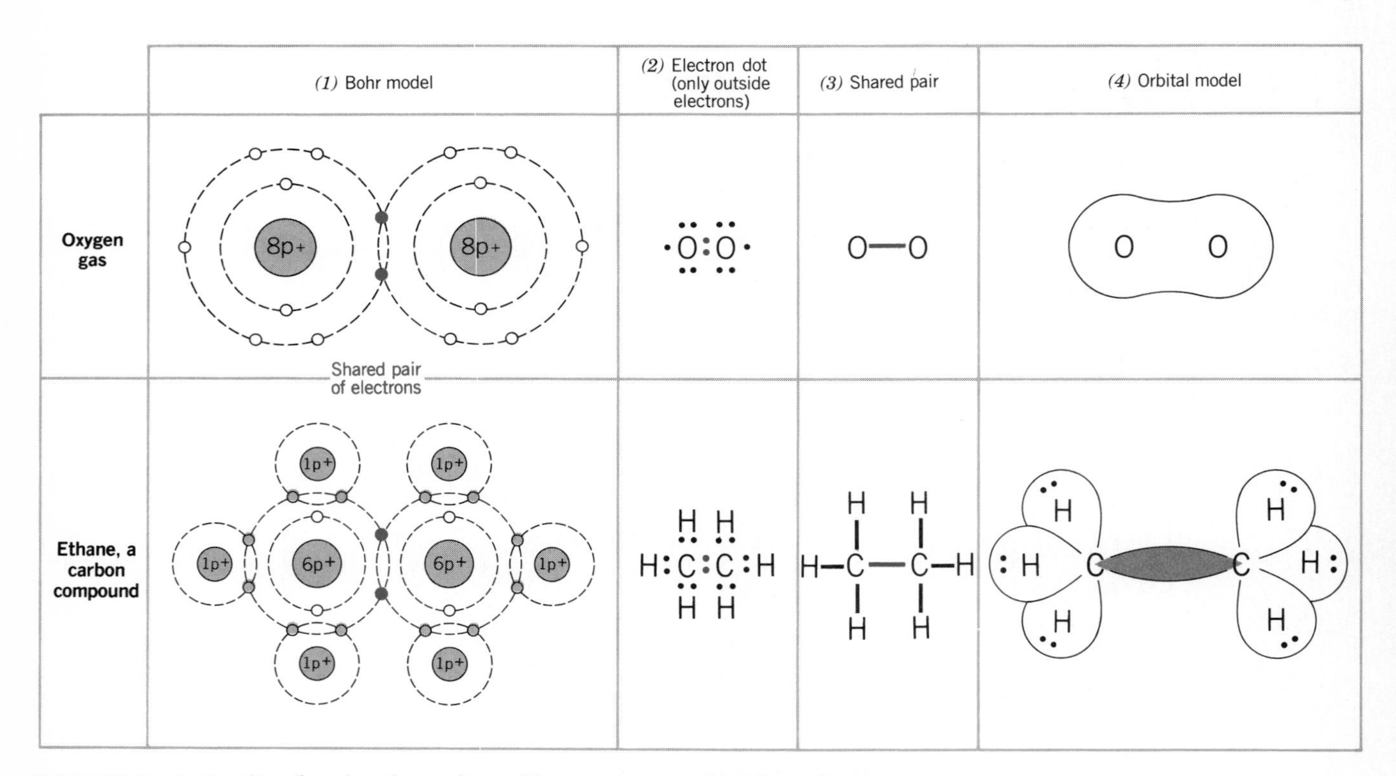

FIGURE 2.3 Covalent bonds are formed between atoms that share electrons. The Bohr model (1) accounts for all the electrons, whereas the electron dot model (2) shows only the outer shell electrons. The shared pair model (3) uses a bar to represent each covalent bond. The orbital model (4) shows the distribution of electrons around each atom.

gas (O_2), in which each atom shares one electron to complete the outside ring (Figure 2.3). Likewise, nitrogen atoms share three electrons to form the stable nitrogen gas (N_2). When electrons are shared by two atomic nuclei, the attractive force between the two atoms is called a **covalent bond.** The amount of energy of a covalent bond depends on the strength of the attractive forces between the atoms. This bond energy is released when the covalent bond is broken.

Organic compounds are those that contain at least one carbon and a hydrogen atom joined by a covalent bond. Carbon has six protons in its nucleus and six electrons; two electrons are in the first energy level and four electrons are in the second energy level (Table 2.1). To complete the second energy level's complement of eight electrons, each carbon atom shares four electrons with other atoms. Stated another way, *carbon atoms in stable compounds always form four covalent bonds.* These bonds can be formed by sharing electrons with hydrogen and adjacent carbon atoms as shown in Figure 2.3. A carbon atom can form covalent bonds with oxygen, nitrogen, phosphorus, sulfur, and with other carbon atoms. The compounds formed with carbon are extremely varied because the number of combinations that result in the formation of stable carbon compounds is tremendous. There are so many carbon-containing compounds that they are studied as a subdiscipline of chemistry called **organic chemistry,** and, as we shall see, organic compounds play a major role in cellular chemistry.

		Kilocalories mole^{-1}
Covalent bonds	N—H (0.102 nm)	93.4
	C—H (0.107 nm)	98.8
	C=O (0.123 nm)	171
	C=N— (0.127 nm)	147
	C=C (0.133 nm)	147
	—C—O— (0.143 nm)	84
	—C—N (0.148 nm)	69.7
	—C—O— (0.154 nm)	83.1
Hydrogen bonds	H—O⋯O	2–10
	H—N⋯O	
	H—N⋯N	

FIGURE 2.4 The distances (in nanometers, 10^{-9} meter) between atomic nuclei joined by covalent bonds vary. The energy needed to break these covalent bonds depends on the atoms involved. Hydrogen bonds are significantly weaker than covalent bonds.

Covalent bonds are relatively strong bonds that are stable at normal biological temperatures. Covalent bonds can be broken at very high temperatures or at body temperatures in the presence of enzymes or other catalysts (see below). There are small differences in the strengths of covalent bonds, varying with the elements involved and the number of electrons shared by the two atomic nuclei.

The distances between the two atomic nuclei in a compound vary depending on the atoms' attractive forces for electrons. The spacing between the nuclei joined by covalent bonds is shown in Figure 2.4. The strength of a covalent bond can be determined by calculating the amount of heat needed to break that bond. The heat is expressed as kilocalories per mole (kcal/mole) of substance. (A **kilocalorie** is the amount of heat required to raise 1 L of water 1°C (centigrade).* A **mole** is the gram molecular weight of the substance in question.)

Unequal Sharing of Electrons

When covalent bonds are formed between different elements, the electrons are often shared unequally. Certain atoms easily give up electrons, whereas other

*In this text the centigrade, or Celsius scale, is used. Conversion from centigrade to Fahrenheit can be approximated by doubling the centigrade number and adding 32. The formula for this conversion is given in Appendix 1.

elements are more reluctant to give them up. This is because each atomic nucleus differentially attracts its own electrons depending on the number of protons in the nucleus and the arrangement of the electrons in their respective energy levels.

Let us look at what happens to the electrons shared by the oxygen and carbon atoms in carbon dioxide. The oxygen is reluctant to give up its electrons, so the shared electrons are closer to the oxygen atoms than the carbon atom. Thus we write: $O \overset{\leftarrow}{=} C \overset{\rightarrow}{=} O$, in which arrows indicate the attractive force of the oxygen atoms. Since carbon dioxide is a linear molecule, the attraction of the charged shared electrons in the molecule is canceled.

Other covalent bonds have an uneven distribution of charge that results in a charged molecule. Nitrogen has a greater attraction for shared electrons than hydrogen does, so the covalent bond between nitrogen and hydrogen ($^{-}N{-}H^{+}$) is polar. We call this a **polar covalent bond** since the nitrogen atom is more negatively charged than the hydrogen atom. Oxygen atoms have a strong attraction for electrons and form polar covalent bonds with carbon ($^{-}O{-}C^{+}$), hydrogen ($^{-}O{-}H^{+}$), and nitrogen ($^{-}O{-}N^{+}$). The polar covalent bonds in biomolecules are extremely important since these charged molecules are usually soluble in water.

Elements that form bonds with themselves are, of course, nonpolar. Examples of **nonpolar covalent** bonds include nitrogen gas (N_2) and the carbon–carbon bonds in hydrocarbons. The absence of charged groups makes nonpolar molecules less soluble in water than polar molecules. For example, long-chain hydrocarbons contain a series of nonpolar covalent bonds

$$\left(\begin{array}{c} \;\;\mathrm{H}\;\;\;\mathrm{H}\;\;\;\mathrm{H} \\ \;\;|\;\;\;\;\;|\;\;\;\;\;| \\ -\mathrm{C}-\mathrm{C}-\mathrm{C}- \\ \;\;|\;\;\;\;\;|\;\;\;\;\;| \\ \;\;\mathrm{H}\;\;\;\mathrm{H}\;\;\;\mathrm{H} \end{array}\right)$$

that contribute greatly to the low solubility of fats and oils in water.

IONIC BONDS

Objects with dissimilar charges attract each other. Na^{+} and Cl^{-} ions in solution will attract each other and, when present in appropriate concentrations, will form crystal structures of sodium chloride (NaCl) held together by ionic bonds (Figure 2.2). The sodium donates an electron to the chloride atom, leaving each atom with eight electrons in its outermost orbital. Sodium chloride is an **ionic compound** because it is electrically neutral and is composed of an orderly array of oppositely charged ions. Other examples of ionic compounds are baking soda (sodium bicarbonate, $NaHCO_3$) and lye (sodium hydroxide, NaOH). The attracting force that holds the negatively and positively charged ions in a crystal structure is called an **ionic bond.** Ionic bonds are also important in the very large molecules present in living matter and in many of the chemical reactions that occur in living cells.

WATER AND THE HYDROGEN BOND

Water is the major compound found in all living organisms. Bacterial cells are about 70 percent water and eucaryotic cells are about 90 percent water. Water is a polar molecule because of its nonlinear molecular geometry and the presence of polar covalent $^{-}O{-}H^{+}$ bonds. The oxygen atom of water attracts the shared electrons, causing the hydrogen atoms to function as positively charged centers and the oxygen atom to function as a negatively charged center. Water molecules readily interact with one another because the positively charged hydrogen protons on one water molecule are attracted to the negatively charged oxygen nucleus of a second water molecule. The attractive force between these two charged centers is an example of a hydrogen bond.

A **hydrogen bond** is formed between a hydrogen atom involved in a polar covalent bond and the negative charge on another polar covalent bond. Hydrogen bonds are significantly weaker than covalent bonds and ionic bonds (Figure 2.4); so weak that they are easily disrupted at the temperature of boiling water. Hydrogen bonds play an important role in the structure of the macromolecules produced by all cells.

Water is an extremely good solvent for charged compounds because it exists in a charged state. Since water is the solvent for living systems, most of the molecules they depend on are charged compounds. These charged compounds include the ions of salts, acids, bases, carbohydrates, and proteins, all of which are soluble in water. Compounds readily soluble in water are termed **hydrophilic,** or water loving. The nonpolar compounds such as fats, oils, and waxes are insoluble in water. These compounds are **hydrophobic,** or water fearing. Hydrophobic compounds serve a special function in the formation of cellular membranes.

IONIZATION OF WATER

Water almost never exists in a pure molecular form because it contains ions that are the reaction products of itself. When two water molecules collide, they can interact to form **hydroxide ions** (OH^-) and **hydronium ions** (H_3O^+).

$$H_2O + H_2O \leftrightarrows H_3O^+ + OH^-$$

Pure water contains the same number of hydronium ions and hydroxide ions, so the positive and negative charges are neutralized. Pure water is said to be neutral.

ACIDS AND BASES

If a compound that ionizes, to produce either a proton (H^+) or a hydroxide ion, is added to water, the water will no longer be neutral. **Acids** are substances that increase the concentration of hydronium ions when added to water. Acids impart a sour or tart taste to the solution by releasing protons (H^+) that are, in fact, hydrogen ions (remember that a hydrogen atom is a proton and an electron). Strong acids such as hydrochloric acid completely ionize in water. Hydrochloric acid ionizes to form protons and chloride ions according to the equation: $HCl \rightarrow H^+ + Cl^-$. The protons react with water to form hydronium ions.

Bases are substances that increase the concentration of hydroxide ions when added to water. The presence of hydroxide ions makes solutions feel slippery and soapy and imparts a bitter taste to foods. Bases are easily detected because they turn common litmus paper blue. Sodium hydroxide is a strong base that completely ionizes to form hydroxide ions according to the equation: $NaOH \rightarrow Na^+ + OH^-$.

Measuring Acidity on the pH Scale

The **pH scale** is a convenient measure of the concentration of hydronium ions in solutions (Table 2.3). The term pH comes from the *p*otential of *H*ydrogen. The scale is defined as the negative logarithm of the hydrogen ion concentration (see Appendix 1 for explanation of logarithms):

$$pH = -\log [H^+]$$

On this scale, neutrality is defined as pH 7 because this

TABLE 2.3 The pH scale

pH	HYDROGEN ION CONCENTRATION $[H^+]$	HYDROXIDE ION CONCENTRATION $[OH^-]$	
14	1×10^{-14}	1	↑
13	1×10^{-13}	1×10^{-1}	
12	1×10^{-12}	1×10^{-2}	
11	1×10^{-11}	1×10^{-3}	Basic
10	1×10^{-10}	1×10^{-4}	
9	1×10^{-9}	1×10^{-5}	
8	1×10^{-8}	1×10^{-6}	
7	1×10^{-7}	1×10^{-7}	Neutral
6	1×10^{-6}	1×10^{-8}	
5	1×10^{-5}	1×10^{-9}	
4	1×10^{-4}	1×10^{-10}	
3	1×10^{-3}	1×10^{-11}	Acidic
2	1×10^{-2}	1×10^{-12}	
1	1×10^{-1}	1×10^{-13}	
0	1	1×10^{-14}	↓

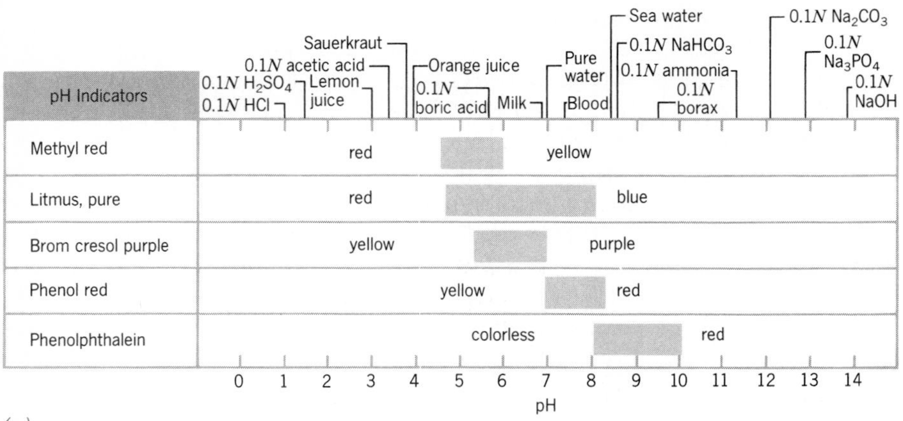

(*a*)

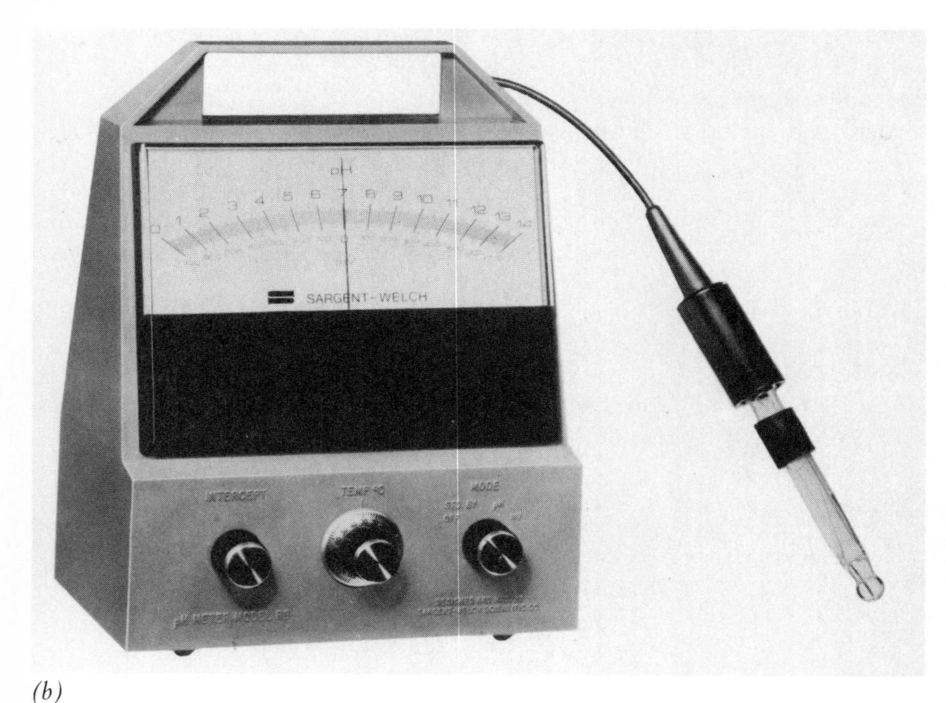

(*b*)

FIGURE 2.5 (Top) The pH scale goes from 0 to 14. Neutrality is pH 7.0, solutions with pH values between 0 and 7 are acidic, and solutions with a pH above 7.0 are alkaline. The pH indicators listed are widely used in microbiology. Each indicator has a pH transition zone in which it undergoes a color change as the pH of the solution changes between alkaline and acidic. (Bottom) Absolute pH values are determined using a pH meter equipped with pH electrodes. (Diagram modified from J. R. Holum, *Fundamentals of General, Organic, and Biological Chemistry*, Wiley, New York, 1978. Photograph courtesy of Sargent-Welch Scientific.)

is the negative log of the hydrogen ion concentration (10^{-7}) of pure water. This means that at neutrality, one liter of water contains 10^{-7} moles of hydrogen ion; or that 10^7 liters of water are needed to contain one mole of hydrogen ion. Values between pH 7 and pH 14 are termed **alkaline** and are obtained by adding a base to the solution. Values between pH 7 and 0 are termed **acidic** and are obtained by adding acids to the solution. Examples of acidic and basic solutions are listed in Figure 2.5. Organic compounds that undergo a color change at specific pHs are used to estimate the pH of solutions. Brom cresol purple, methyl red, litmus, and phenol red (Figure 2.5) are pH indicators widely used in microbiology. When the precise pH of a solution must be known, a pH meter equipped with a suitable electrode can be used.

KEY POINT

The pH of a medium or a natural environment has a major influence on which organisms are able to grow in it. Although most microbes prefer an environment with a nearly neutral pH, some bacteria grow in acid mine waters and others grow in very alkaline environments. During growth, some bacteria produce organic acids that drastically lower the pH. The lactic acid produced by bacteria during the production of yogurt is responsible for the typical tart taste of unflavored yogurt.

BUFFERS RESIST pH CHANGES

Buffers are substances that tend to resist changes in the pH of a solution. There are many chemicals in cells that function as buffers. Inorganic buffers, amino acids, and other organic acids all contribute to the buffering capacity of a system. Inorganic buffers are often used to stabilize the pH of microbiological media during microbial growth. Phosphate buffer composed of $HPO_4^{2-}/H_2PO_4^-$ (monohydrogen phosphate ion/dihydrogen phosphate ion) is one example. The phosphate system works as follows:
Addition of base (OH^-)

$$H_2PO_4^- + OH^- \rightarrow HPO_4^{2-} + H_2O$$

Addition of acid (H^+)

$$HPO_4^{2-} + H^+ \rightarrow H_2PO_4^-$$

Note that the right-hand side of these reactions contains neither hydronium ions nor hydroxyl ions. Cells contain natural buffers, which are necessary to maintain an optimal interior pH and to guard against major environmental pH changes.

CHEMICAL REACTIONS

Molecules are combinations of two or more atoms that form a specific chemical compound. The chemical industry takes the molecules found in raw materials and makes useful products from them. For example, nylon, Dacron, Lucite, and plastic wraps are polymers manufactured from organic molecules. New molecules are formed during chemical reactions by the rearrangement of chemical bonds between atoms or molecules.

The starting material or **reactant(s)** in a chemical reaction are converted to the **product(s)** of the reaction. The products usually have chemical properties that are different from their substrates. Since atoms are neither created nor destroyed during a chemical reaction, the atoms of the reactants can always be found in the products of the reaction.

SYNTHETIC REACTIONS: ANABOLISM

Cells need to make many constituents in order to grow and reproduce. Chemical reactions that result in the formation of compounds needed for cellular growth are called **synthetic reactions.**

$$A + B \longrightarrow C$$

In this reaction the reactants A and B combine to form a new compound, C. Some synthetic reactions involve the transfer of a functional group from one compound to another to make a third compound.

$$AB + C \longrightarrow A + BC$$

In this **transfer reaction,** the B group is transferred from AB to C resulting in the creation of a new compound, BC. Transfer reactions are used to synthesize certain metabolites such as amino acids.

Most synthetic reactions in a cell utilize reactants the cell acquires from its immediate environment or from cellular metabolites. Every cell contains a variety of **metabolites,** the products of chemical reactions within the cell. All of the synthetic reactions occurring in a cell constitute its **anabolism.** These synthetic reactions produce proteins, membranes, and cell walls at the expense of energy.

DECOMPOSITION REACTIONS: CATABOLISM

Many cells gain energy needed for growth by breaking chemical bonds. These chemical reactions are called decomposition reactions and function to produce energy and to recycle cellular materials. Large molecules are broken down into smaller molecules and atoms by chemical reactions:

$$AB \longrightarrow A + B$$

Often the chemical energy released from the reaction can be captured and utilized for synthetic chemical reactions, or for doing useful work, such as movement. **Catabolism** refers to all the chemical reactions of a cell that break down compounds.

THE NATURE OF CHEMICAL REACTIONS IN BIOLOGICAL SYSTEMS

Many chemical reactions occur spontaneously only at unsuitably high temperatures for living cells. These same reactions can occur at lower temperatures, however, if a catalyst is present. **Catalysts** are substances that speed up the rate of a chemical reaction by decreasing the amount of energy required for the reaction to begin.

Living organisms possess unique catalysts known as enzymes. An **enzyme** is a protein that speeds up the rate of a chemical reaction without being used up in the process. Every cell makes many different enzymes, each able to catalyze only one reaction or at most a few reactions. Stated another way, an enzyme is specific for its substrate because the enzyme must physically bind to its substrate in order to act as a catalyst. Enzymes are usually better catalysts at the temperature that is optimum for their cell's growth.

BIOLOGICALLY IMPORTANT INORGANIC COMPOUNDS

All compounds can be divided into two classes: inorganic compounds and organic compounds. **Inorganic compounds** are those molecules that do not contain carbon, such as metals, salts, most gases, and water. Carbon dioxide and carbonate are the only inorganic compounds that contain carbon, and are considered to be inorganic because they lack hydrogen. Organic molecules always contain carbon (C) and hydrogen (H) and have an important function in biological systems, in part because of their great variety.

Examples of inorganic gases that provide essential elements for microbes include oxygen (O_2), nitrogen gas (N_2), and carbon dioxide (CO_2). Examples of inorganic ions that are important to microorganisms include potassium (K^+), calcium (Ca^{2+}), sodium (Na^+), phosphate (PO_4^{3-}), sulfate (SO_4^{2-}), sulfide (S^{2-}), nitrate (NO_3^-), nitrite (NO_2^-), ammonium (NH_4^+), carbonate (CO_3^{2-}), magnesium (Mg^{2+}), and manganese (Mn^{2+}). In addition, most of the trace metals required by growing microbes are present as inorganic ions, such as ferric iron (Fe^{3+}) and molybdate (MoO_4^{2-}).

CLASSES AND FUNCTIONS OF INORGANIC COMPOUNDS

A typical bacterial cell has three classes of inorganic compounds: compounds used to manufacture biomolecules, bulk elements that exist as ions in the cell, and trace elements. All of these inorganic compounds are supplied either by the atmosphere to which the cell is exposed or are present in the cell's immediate environment.

Phosphorus, nitrogen, and sulfur exist in various chemical forms that serve as substrates for chemical reactions within the cell. Phosphate is essential to the structure of the genetic material (nucleic acids) and is the key element in energy transfer within the cell. Nitrogen is the fifth most abundant element in living material. Reduced nitrogen ($R{-}NH_2$) is present in amino acids, purines, pyrimidines, and some vitamins. Ammonia, nitrate, and nitrogen gas (N_2) are forms of nitrogen taken up from the environment and converted by chemical reactions into cellular material. Another essential inorganic element is sulfur, which is found in certain vitamins, amino acids, and peptides. Both sulfate and hydrogen sulfide are natural forms of sulfur that can be utilized by microorganisms. The metabolism of nitrogen and sulfur is discussed in Chapters 14 and 28.

The **bulk elements** include calcium (Ca^{2+}), potassium (K^+), sodium (Na^+), chlorine (Cl^-), and magnesium (Mg^{2+}). These elements behave as ions within cells where they bind to many different biomolecules by ionic and covalent bonds, and/or they exist as free ions in the cytoplasm. All cells require moderate concentrations of the bulk elements. In addition to the normal requirement for bulk elements, marine bacteria require sodium chloride and high concentrations of Ca^{2+} and Mg^{2+} ions in their growth environment.

Trace elements are present in very low concentrations in cells, where they are involved in electron transfer reactions and as components of enzymes. Examples of trace elements include iron, zinc, molybdenum, copper, magnesium, cobalt, selenium, and nickel. Tap water usually contains sufficient concentrations of these elements to supply the needs of a microorganism. Often the problem with experimentally demonstrating a microbe's requirement for a trace element is the interference from trace-metal contamination of the water used to prepare the culture medium.

OXIDATION–REDUCTION REACTIONS

Many inorganic compounds are converted from one form to another by biochemical reactions known as oxidation–reduction reactions. An element is oxidized when its oxidation state increases or becomes more positive. For example, the nitrogen atom of nitrite is oxidized when nitrate is formed in the reaction $\overset{+3}{N}O_2^- + 1/2O_2 \rightarrow \overset{+5}{N}O_3^-$, where the oxidation state of the nitrogen changes from +3 to +5. Conversely, an element is reduced when its oxidation state decreases or becomes more negative: $NO_3^- + H_2 \longrightarrow NO_2^- + H_2O$. Stated another way, an element or a compound that gives up or donates electrons becomes oxidized, and an element or a compound that accepts electrons becomes reduced. In the following reaction iron (III) is reduced to iron (II), while copper (I) is oxidized to copper (II).

$$Fe^{3+} + Cu^+ \longrightarrow Fe^{2+} + Cu^{2+}$$

Remember that electrons are negatively charged so the loss of an electron increases the element's oxidation number. Although oxidation or reduction reactions can be written as separate, independent reactions, in biological systems an oxidation reaction is almost always coupled to a reduction reaction. These coupled reactions are called **oxidation–reduction reactions,** or simply **redox reactions.**

BIOLOGICALLY IMPORTANT ORGANIC MOLECULES

Organic molecules contain carbon and hydrogen and are studied in the branch of chemistry known as organic chemistry. After oxygen, carbon is the most prevalent element in all living things. Carbon's unique chemical properties enable it to exist in a great diversity of molecular structures. The carbon atom forms four covalent bonds with other atoms in order to complete its outer shell. Carbon forms these covalent bonds with another carbon atom or with hydrogen, oxygen, nitrogen, and/or sulfur. Compounds whose formulations include carbon and hydrogen are called **organic molecules.** Living systems contain a great array of organic molecules that serve an equally diverse number of functions.

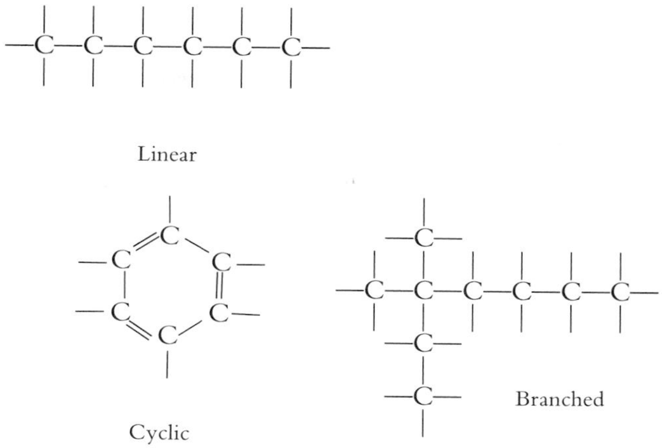

Linear

Cyclic

Branched

FIGURE 2.6 Carbon compounds exist as linear, branched, or cyclical molecules. Each carbon atom in an organic molecule forms four covalent bonds.

Many organic molecules are constructed on linear, branched, or cyclic carbon skeletons like the ones shown in Figure 2.6. By convention, the single lines between carbon atoms represent one pair of shared electrons that comprise one covalent bond. Carbon can form a single covalent bond (C—C), a double covalent bond (C═C), or a triple covalent bond (C≡C) with another carbon atom. These bonds are represented by the number of lines drawn between the carbon atoms.

When atoms other than carbon are attached to carbon skeletons, the number of potential organic molecules becomes astronomical. To organize this vast array of compounds, organic chemists have named the functional groups formed by combinations between carbon and other elements. Some of the important functional groups of biomolecules are shown in Figure 2.7. A functional group can be attached to the organic molecule (represented by R_1 and R_2) to form larger

FUNCTIONAL GROUP	NAME	WHERE FOUND
$R—NH_2$	Amino	Amino acids, amino sugars
R—OH	Hydroxyl	Sugars, alcohols
$R—CH_3$	Methyl	Attached to many classes of organic compounds
R—SH	Sulfhydryl	A few amino acids
$R—C(=O)—OH$	Carboxyl	Organic acids (gives up one proton, $—COO^- + H^+$)
$R—C(=O)—H$	Aldehyde	Sugars and aldehydes
$R—C(=O)—R_2$	Keto or carbonyl	Sugars and some organic acids
$R—C(=O)—O—R_2$	Ester	Lipids

FIGURE 2.7 Functional groups of organic compounds.

biomolecules. The simplest formulation of an organic molecule would be to add a hydrogen to a methyl group ($—CH_3$) to form methane (CH_4) gas. Ethane ($H_3C—CH_3$) is formed when two methyl groups are joined. If a carboxyl group were attached to a methyl group,

$$H_3C—\overset{\overset{\displaystyle O}{\|}}{C}—OH$$

the result would be acetic acid, the major component of vinegar. Acetaldehyde (CH_3CHOH) is formed when the carboxyl group of acetic acid is replaced by an aldehyde group. These are just a few of the ways in which organic molecules can be constructed.

CONFORMATIONS OF ORGANIC MOLECULES

In nature, organic molecules exist as three-dimensional structures, not as the simple two-dimensional structures shown in diagrams. When all the groups attached to a carbon atom are the same, as is the case with methane (CH_4), the molecule is symmetrical and all methane molecules are identical. However, if an organic molecule contains one carbon atom with four different groups attached to it, then there are two different three-dimensional formulations (Figure 2.8) for that molecule. The molecules are known as **isomers,** which are alternative forms of a molecule containing the same atoms. The molecules shown in Figure 2.8 are called **optical isomers** because they differ in their ability to rotate the plane of polarized light, or **stereoisomers** because they are mirror images of one another. Both forms of a stereoisomer pair participate in identical chemical reactions; however, biological systems recognize the different isomers and often only one member of the pair is biologically active.

All amino acids, except for glycine, exist as stereoisomers. Alanine is the simplest example of an amino acid that exists in two isomeric forms. The second car-

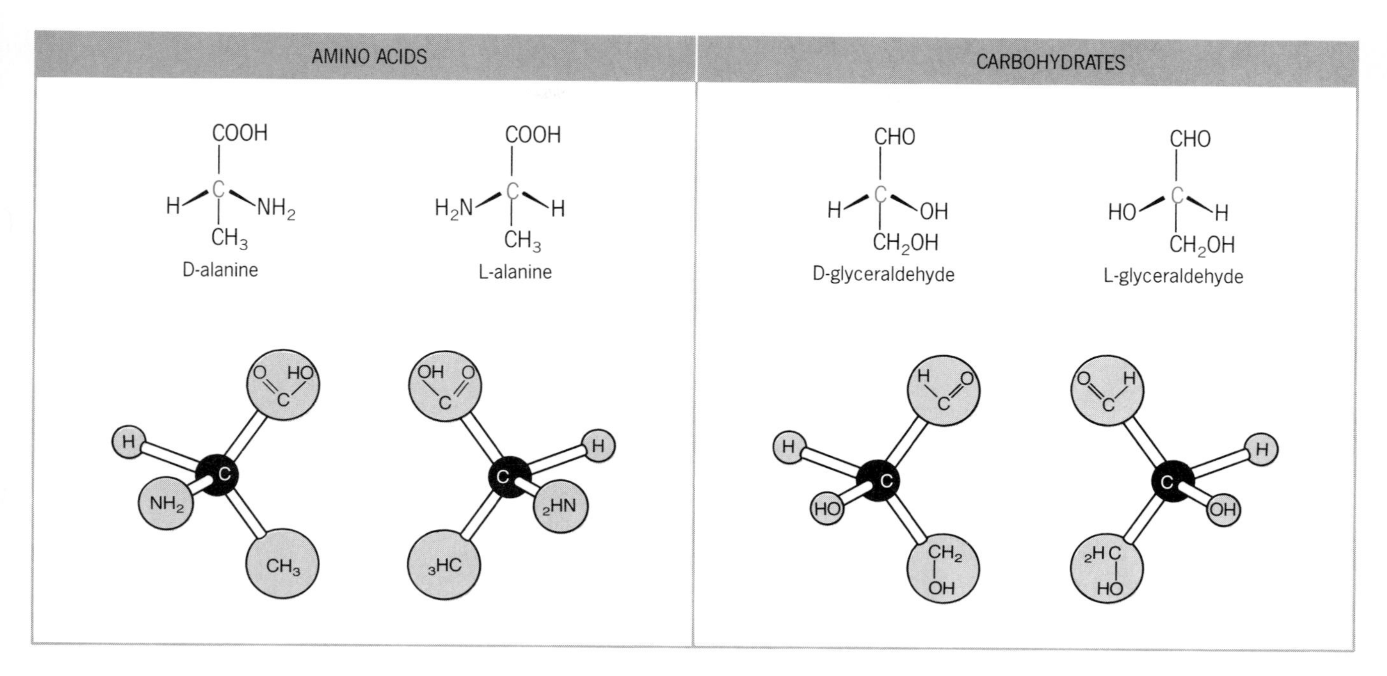

FIGURE 2.8 Stereoisomers are compounds with the same chemical formula that possess different spatial conformations. Each stereoisomer has an asymmetrical carbon atom to which four different groups are attached. The stereoisomers of alanine and glyceraldehyde are shown and are the basis for designating D and L forms of the amino acids.

bon atom of alanine is asymmetric, which means that four different groups (Figure 2.8) are attached to the same carbon atom. When alanine is drawn with the carboxyl group at the top and the methyl group at the bottom, a choice must be made on which side to place the amino group. You would have the same problem if you tried to draw the hands of a friend: the thumb on one hand would have to appear as the mirror image of the thumb on the other hand.

If the amino group is placed on the right, then the hydrogen atom must be placed on the left, giving the D-alanine isomer (D = dextro, right). The mirror image of this structure switches the hydrogen with the amino group: it is called the L-alanine isomer (L = levo, left). All of the amino acids naturally found in proteins are L-amino acids. The difference between the D and L forms of the amino acids is recognized by enzymes involved in protein synthesis as well as by the proteolytic enzymes that break down proteins and peptides. The only natural structures containing D-amino acids are the peptidoglycan of bacterial cell walls, certain bacterial capsules, and some bacterial antibiotics.

MAJOR CLASSES OF BIOORGANIC MOLECULES

Small organic molecules exist in cellular metabolic pools where they are available for synthesizing macromolecules or as substrates of the cell's metabolic processes. The major classes of these molecules are shown in Figure 2.9. **Amino acids** have an amino group ($-NH_2$) and a carboxyl group (—COOH) that is the acidic part of the amino acid. Many cells can use amino acids as their sole source of nitrogen and some can also use them for energy. **Fatty acids** are composed of a long chain of carbon atoms (hydrocarbon) that ends in a carboxyl group. Fatty acids combine with glycerol to form fatty substances known as lipids.

Carbohydrates are organic molecules often referred to as sugars. Sucrose is a common carbohydrate that we use as a sweetening agent. Individual carbohydrate units known as monosaccharides contain between three and seven carbon atoms. The five-carbon monosaccharides called **pentoses** and six-carbon monosaccharides called **hexoses** are important in cellular metabolism and structure.

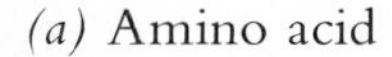

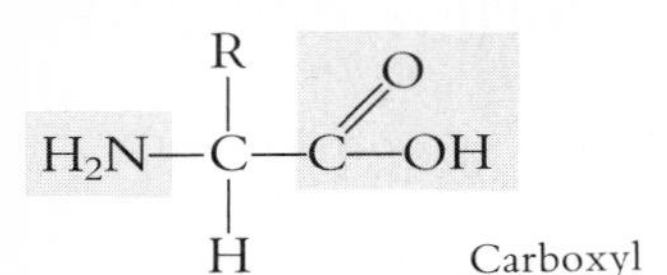

(b) Fatty acid

$R{-}CH_2{-}CH_2{-}CH_2{-}CH_2{-}CH_2{-}CH_2{-}CH_2{-}CH_2{-}CH_2{-}C(=O){-}OH$

Carboxyl

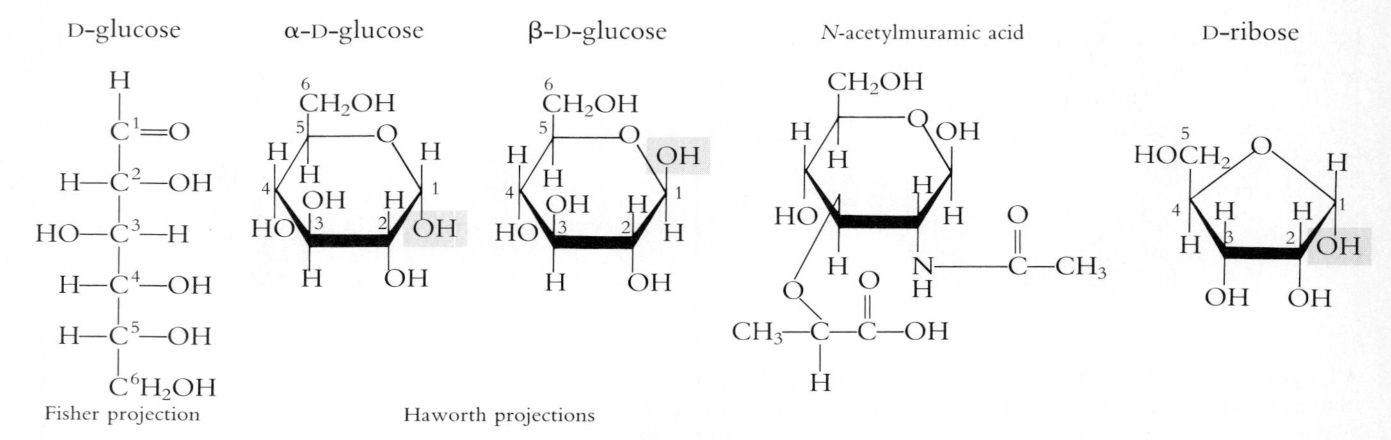

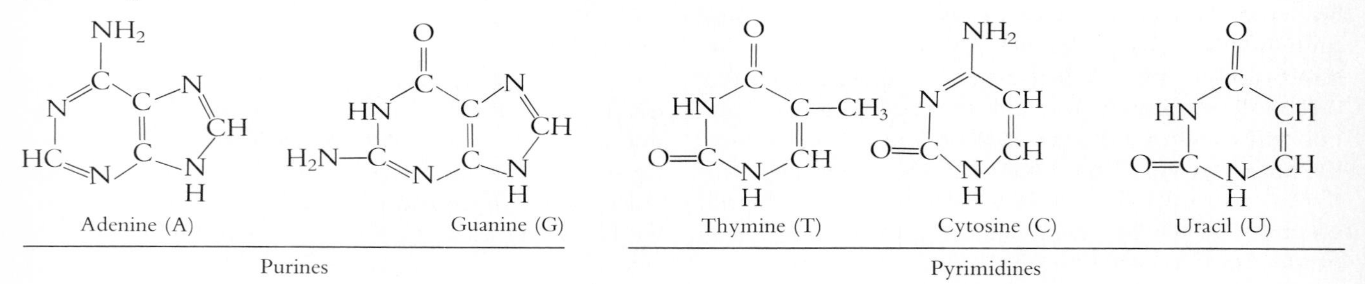

FIGURE 2.9 Representative organic molecules that are important in biology. The letter R indicates the portion of the molecule that can vary in structure. Can you name the biologically important macromolecules that are made from each of the groups shown?

Purines and pyrimidines are heterocyclic compounds that contain both nitrogen and carbon atoms in their ring (Figure 2.9). The purines and pyrimidines are the nitrogenous bases used by cells to form nucleic acids, the informational molecules of heredity.

Amino Acids Make Proteins

Proteins and peptides are polymers of amino acids. **Amino acids** have at least one amino group and one carboxyl group attached to an asymmetric carbon atom. Glycine is the exception since it does not have an asymmetric carbon atom. The 20 naturally occurring amino acids are shown in Figure 2.10. They are organized according to the properties of their R groups, which confer a unique chemistry on each of the amino acids. Alanine, valine, leucine, isoleucine, proline, methionine, phenylalanine, and tryptophan have nonpolar R groups that are hydrophobic. Glycine, serine, threonine, cysteine, asparagine, glutamine, and tyrosine have neutral (uncharged) polar R groups that hydrogen-bond with water. They are more soluble in water than the nonpolar amino acids in the first group. Arginine, lysine, and histidine possess positively charged R groups and are basic amino acids. Aspartic acid and glutamic acid are acidic amino acids because their R groups have a carboxyl function. The chemical properties of the amino acid R groups are very important to their chemical behavior both as metabolites and as components of proteins.

Fatty Acids Are Cellular Lipids

All cells contain fats, which when found in cellular structures are called lipids. **Lipids** are water-insoluble organic molecules that can be extracted from cells by nonpolar solvents, such as the acetone found in nail polish remover. Cell membranes and granules containing fats or oils are the main cellular location of lipids.

Long chain fatty acids build the lipids of cell membranes. Each fatty acid has a chain of between 4 and 24 hydrocarbons (Figure 2.9). The longer hydrocarbon chains are hydrophobic and make the fatty acid shun water, whereas fatty acids with short hydrocarbon chains have limited solubility in water. The major cellular use of fatty acids is to form complex lipids and the major chemical property of lipids is the hydrophobic nature of their fatty acid R group. This property makes lipids ideal building materials for cell membranes.

Carbohydrates: Sugars

Organic compounds that contain carbon and water are known as the hydrates of carbon or simply as carbohydrates. **Carbohydrates** are also referred to as sugars or saccharides. The individual unit of a carbohydrate is known as a **monosaccharide.** Monosaccharides can contain either three carbons (trioses), four carbons (tetroses), five carbons (pentoses), six carbons (hexoses), or seven carbons (heptoses).

Monosaccharides are important metabolites in all cells where they function as substrates for energy-yielding catabolic reactions and as substrates for anabolic reactions. The most abundant monosaccharide in nature is glucose (Figure 2.9), a hexose (six carbons) that is the repeating unit of cellulose, starch, and glycogen. Many microorganisms can utilize glucose as a carbon and energy source.

Carbohydrates with five carbons are pentoses. Ribose and deoxyribose are the pentoses present in ribonucleic acid and deoxyribonucleic acid respectively. In addition, cells contain xylulose and ribulose (Figure 2.9) as metabolites of carbohydrate catabolism.

Monosaccharides can be modified by attaching noncarbohydrate groups to the basic sugar. *N*-acetylglucosamine is made by attaching an acetyl group to glucosamine. *N*-acetylglucosamine and *N*-acetylmuramic acid (Figure 2.9) are modified carbohydrates found in bacterial cell walls.

Nitrogenous Bases Make Nucleotides

Nucleotides, the basic units from which cells build their nucleic acids, contain a nitrogenous base. **Nitrogenous bases** are heterocyclic organic molecules (Figure 2.9) that contain both nitrogen and carbon. The five nitrogenous bases that are found in nucleic acids are adenine (A), guanine (G), thymine (T), cytosine (C), and uracil (U). Thymine, cytosine, and uracil are pyrimidines (Figure 2.9), whereas adenine and guanine are purines. The nitrogenous bases are incorporated into nucleotides (Figure 2.11) when they are combined with phophorylated deoxyribose or ribose. Adenine, thymine, guanine, and cytosine are the bases found in deoxynucleic (dee-ox-y-nu-clay′ic) acid (DNA). Ribonucleic acid (RNA) also contains adenine, guanine, and cytosine, but in RNA thymine is replaced by uracil.

MACROMOLECULES ARE MADE FROM CELLULAR METABOLITES

Amino acids, sugars, purines and pyrimidines, and fatty acids are metabolites that are building blocks for making the biopolymers essential to living cells. Biopolymers are large molecules (macromolecules) made by joining simple organic molecules with covalent bonds. RNA and DNA are polymers of nucleotides

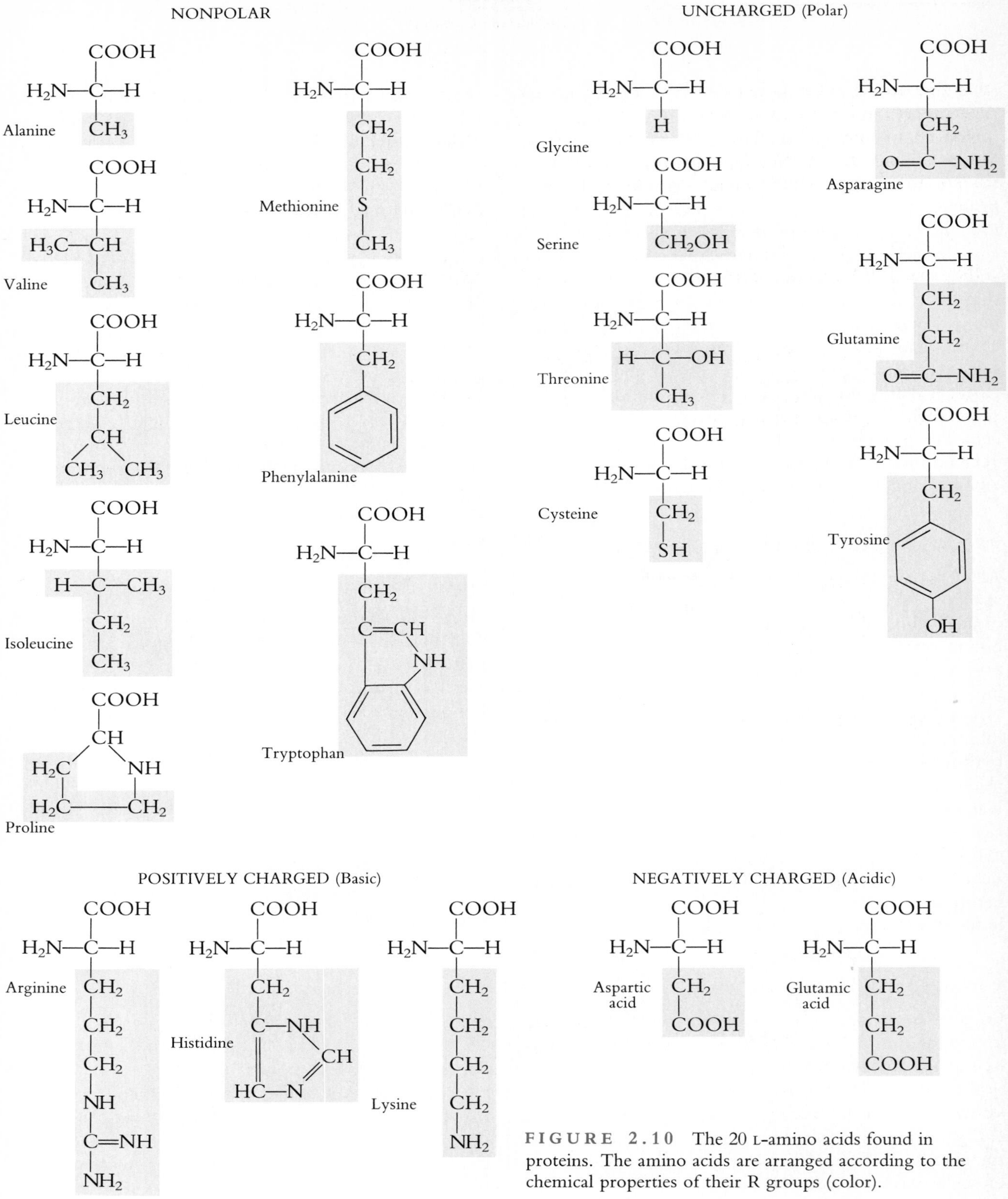

NONPOLAR

COOH
$H_2N—C—H$
CH_3
Alanine

COOH
$H_2N—C—H$
$H_3C—CH$
CH_3
Valine

COOH
$H_2N—C—H$
CH_2
CH
CH_3 CH_3
Leucine

COOH
$H_2N—C—H$
$H—C—CH_3$
CH_2
CH_3
Isoleucine

COOH
CH
H_2C NH
H_2C CH_2
Proline

COOH
$H_2N—C—H$
CH_2
CH_2
S
CH_3
Methionine

COOH
$H_2N—C—H$
CH_2
Phenylalanine

COOH
$H_2N—C—H$
CH_2
C=CH
NH
Tryptophan

UNCHARGED (Polar)

COOH
$H_2N—C—H$
H
Glycine

COOH
$H_2N—C—H$
CH_2OH
Serine

COOH
$H_2N—C—H$
H—C—OH
CH_3
Threonine

COOH
$H_2N—C—H$
CH_2
SH
Cysteine

COOH
$H_2N—C—H$
CH_2
$O=C—NH_2$
Asparagine

COOH
$H_2N—C—H$
CH_2
CH_2
$O=C—NH_2$
Glutamine

COOH
$H_2N—C—H$
CH_2
OH
Tyrosine

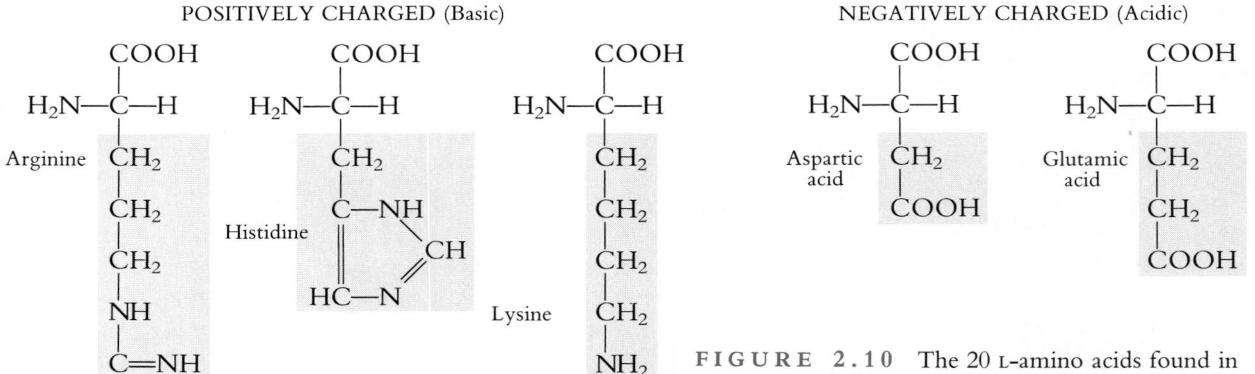

FIGURE 2.10 The 20 L-amino acids found in proteins. The amino acids are arranged according to the chemical properties of their R groups (color).

that function as informational molecules within cells. Proteins are polymers of amino acids and polysaccharides are polymers of monosaccharides. The key lipids are combinations of fatty acids and glycerol. These biopolymers are the chemical essence of living systems.

Nucleotides Make RNA and DNA

DNA and RNA are the information molecules of cells and viruses. Cellular DNA is composed of two very long polymers of deoxyribose connected with phosphates (Figure 2.12). A nitrogenous base is bound to each deoxyribose residue in a specific alignment that encodes biological information. The two strands of DNA are held together by hydrogen bonding between adjacent nitrogenous bases.

Purines and pyrimidines are made into nucleotides before they are used in the synthesis of DNA and/or

FIGURE 2.11 *(a)* The components of the nucleotide AMP are adenine (purine), ribose (sugar), and phosphate. *(b)* Adenosine 3′,5′-cyclic phosphate (cAMP) is a regulatory nucleotide, and adenosine 5′-triphosphate (ATP) is important in cellular energy transfer.

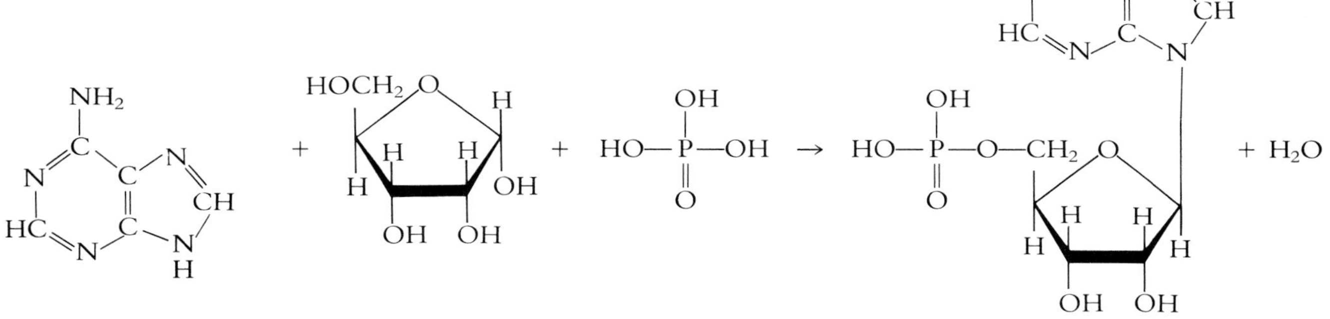

Adenine + α-D-ribose + Phosphoric acid → Adenosine 5′-phosphoric acid (AMP) + H_2O

(a) Components of the nucleotide AMP

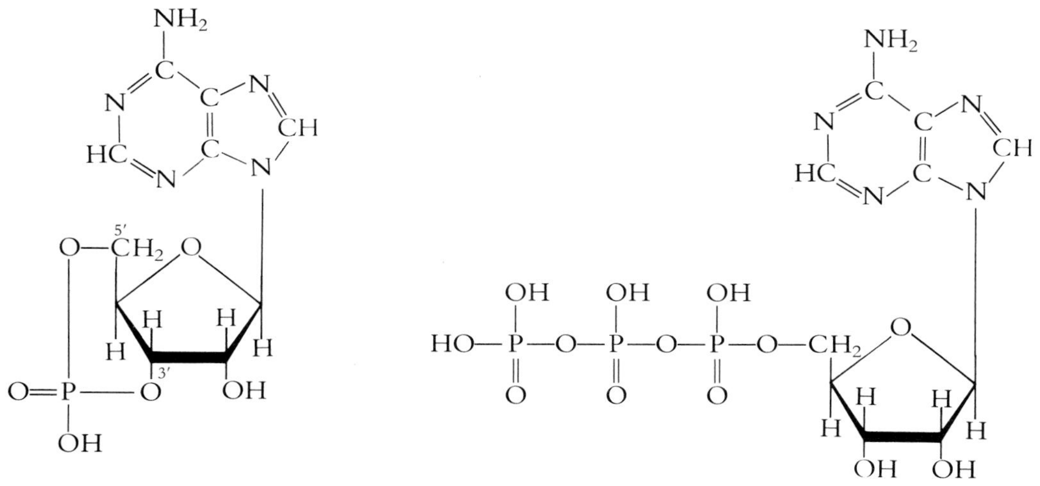

Adenosine 3′,5′-cyclic phosphoric acid (cyclic AMP; cAMP)

Adenosine 5′-triphosphate (ATP)

(b) Important nucleotides

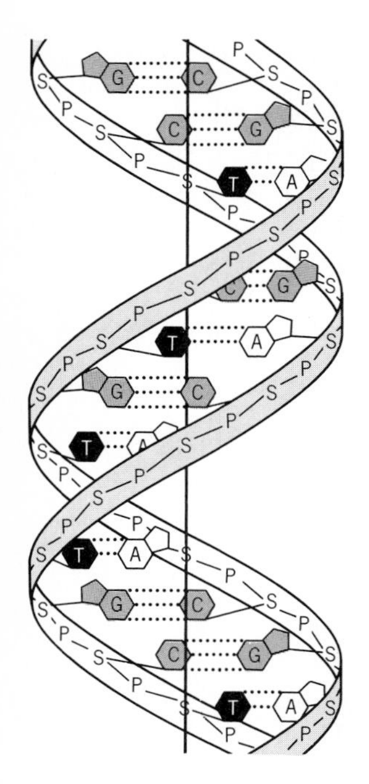

FIGURE 2.12 Cellular DNA is a double-stranded helix composed of two sugar (S)–phosphate (P) backbones joined together through the hydrogen bonding between nitrogenous bases, adenine (A), thymine (T), guanine (G), and cytosine (C).

RNA. Each **nucleotide** is formed from a sugar, a nitrogenous base, and one or more phosphate groups (Figure 2.11). The sugar deoxyribose forms the carbohydrate backbone of DNA, whereas ribose serves this function in RNA. A purine or pyrimidine base is bound to the first carbon of these pentoses (deoxyribose or ribose). One or more phosphate groups bonded to the 5′ carbon of the pentose completes the nucleotide. The details of the complex processes of DNA and RNA synthesis are discussed in Chapter 9.

Nucleotides also participate in metabolic reactions. One of the most important nucleotides in metabolism is adenosine 5′-triphosphate (ATP), which functions in cellular energy transfer reactions (Figure 2.11) in addition to its role in RNA synthesis. Deoxyadenosine triphosphate (dATP), which is identical to ATP except for the replacement of the hydroxyl at the second ribose carbon with a hydrogen atom, is used to synthesize DNA.

Adenosine 5′-monophosphate (AMP) is a precursor of ATP that serves many other functions in cells. AMP is a component of RNA, a metabolite, and a component of a number of coenzymes. **Coenzymes** are small, reusable, organic molecules that play an accessory role in enzymatic reactions. The coenzymes nicotinamide adenine dinucleotide NAD^+, nicotinamide adenine dinucleotide phosphate ($NADP^+$), coenzyme A, and flavine adenine dinucleotide (FAD) all contain AMP as a basic structural component. Such coenzymes are essential to the enzymatic reactions of intermediary metabolism (see Chapter 6).

Lipids Are Hydrophobic Molecules

Lipids are biomolecules that are soluble in nonpolar solvents. These organic molecules are subclassified as waxes, neutral lipids, and phospholipids. Waxes are produced primarily by plants, animals, and insects. Beeswax and paraffin are common waxes that have many practical uses. Cells produce and use the neutral lipids (triglycerides) and the phospholipids as components of cell membranes.

Neutral triglycerides (Figure 2.13) are formed by joining the carboxyl group of the fatty acid with a hydroxyl group of the glycerol in an ester bond. **Esters** are organic compounds formed by the reaction of an alcohol and an acid with the removal of water. When a fatty acid is joined with a glycerol phosphate molecule, a phospholipid is formed. The ester bond is formed when the carboxyl group of a fatty acid and an alcohol group of the glycerol join, with the removal of water.

$$R\text{—}CH_2\text{—}OH + HO\text{—}\overset{\overset{\displaystyle O}{\|}}{C}\text{—}R' \rightarrow$$

$$R\text{—}CH_2\text{—}O\text{—}\overset{\overset{\displaystyle O}{\|}}{C}\text{—}R' + H_2O$$

Glycerol is highly soluble in water; however, after the fatty acids are attached it becomes hydrophobic (insoluble in water). Neutral lipids shun water because the nonpolar hydrocarbon chains of the fatty acids are hydrophobic. Phospholipids are ideally suited to membranes since their charged phosphate group (Figure 2.13) interacts with water while the nonpolar (fatty acyl) end seeks the hydrophobic center of the membrane.

Polysaccharides Are Polymers of Sugars

Combinations of two sugar molecules are called **disac-**

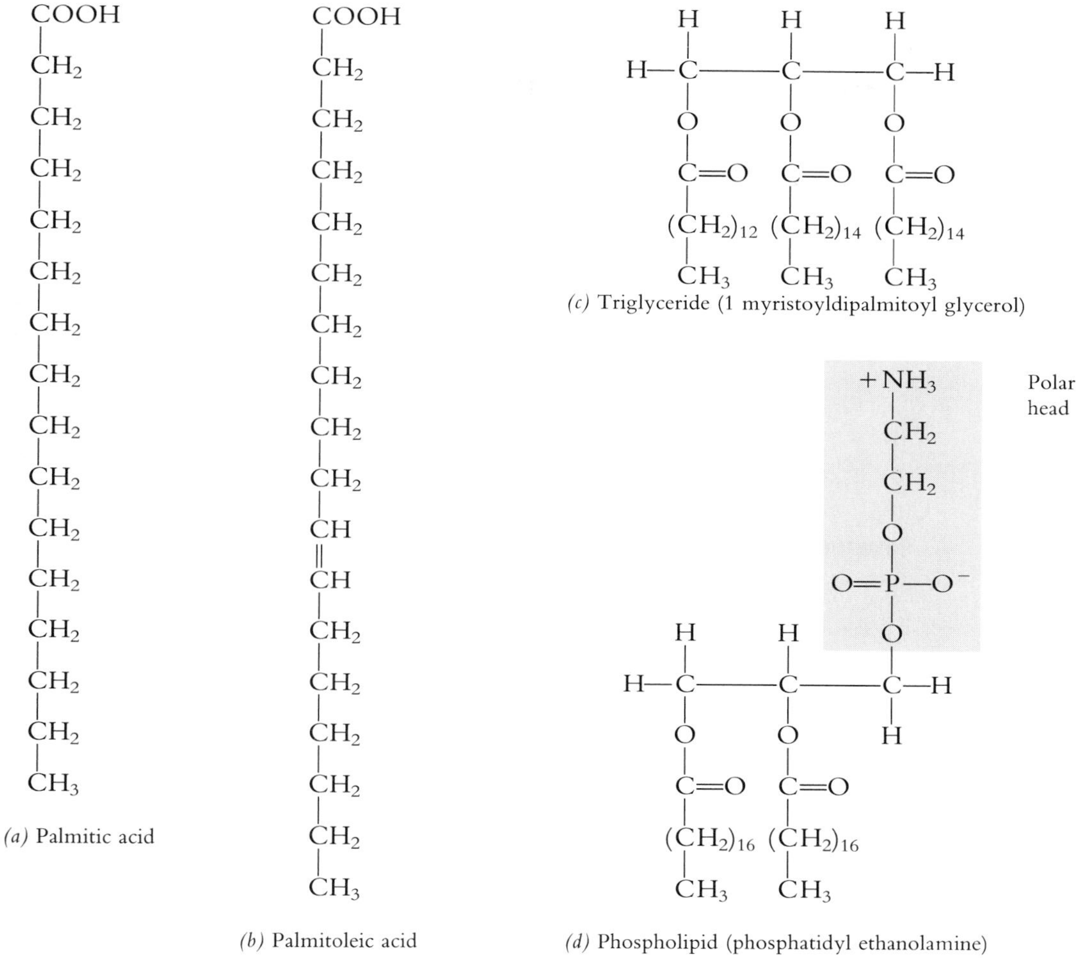

(*a*) Palmitic acid

(*b*) Palmitoleic acid

(*c*) Triglyceride (1 myristoyldipalmitoyl glycerol)

Polar head

(*d*) Phospholipid (phosphatidyl ethanolamine)

FIGURE 2.13 Bacterial lipids. Palmitic acid *(a)* and palmitoleic acid *(b)* are two common bacterial fatty acids. Fatty acids are combined with glycerol to form the neutral triglycerides *(c)*. Phosphatidyl ethanolamine *(d)* is a phospholipid composed of fatty acids, glycerol, phosphate, and ethanolamine.

charides; combinations of three or more sugar molecules are called **polysaccharides.** Disaccharides are formed when two sugars are joined together by a **glycosidic bond.** Sucrose, common table sugar produced from sugar beets and sugar cane, is the disaccharide formed by joining glucose to fructose through a glycosidic bond with a 1—2 α linkage (Figure 2.14). This means that the glycosidic bond is formed between the number one carbon of glucose and the number two carbon of fructose. It is an alpha (α) linkage because the hydroxyl group on the first carbon of glucose is in the down (α) position (see Figure 2.9). *Lactose,* a disaccharide found only in milk, has a beta (β) linkage since the hydroxyl group on the first carbon of galactose is in the up (β) position. Its structure is written galactose β 1—4 glucose. Maltose (glucose β 1—4 glucose) is the disaccharide from barley, which is fermented to make beer. Sucrose, lactose, and maltose (Figure 2.14) are used by bacteria as substrates for growth.

The very large polysaccharides of living organisms are polymers of repeating sugar units joined by glycosidic bonds. Starch, glycogen, and cellulose are all

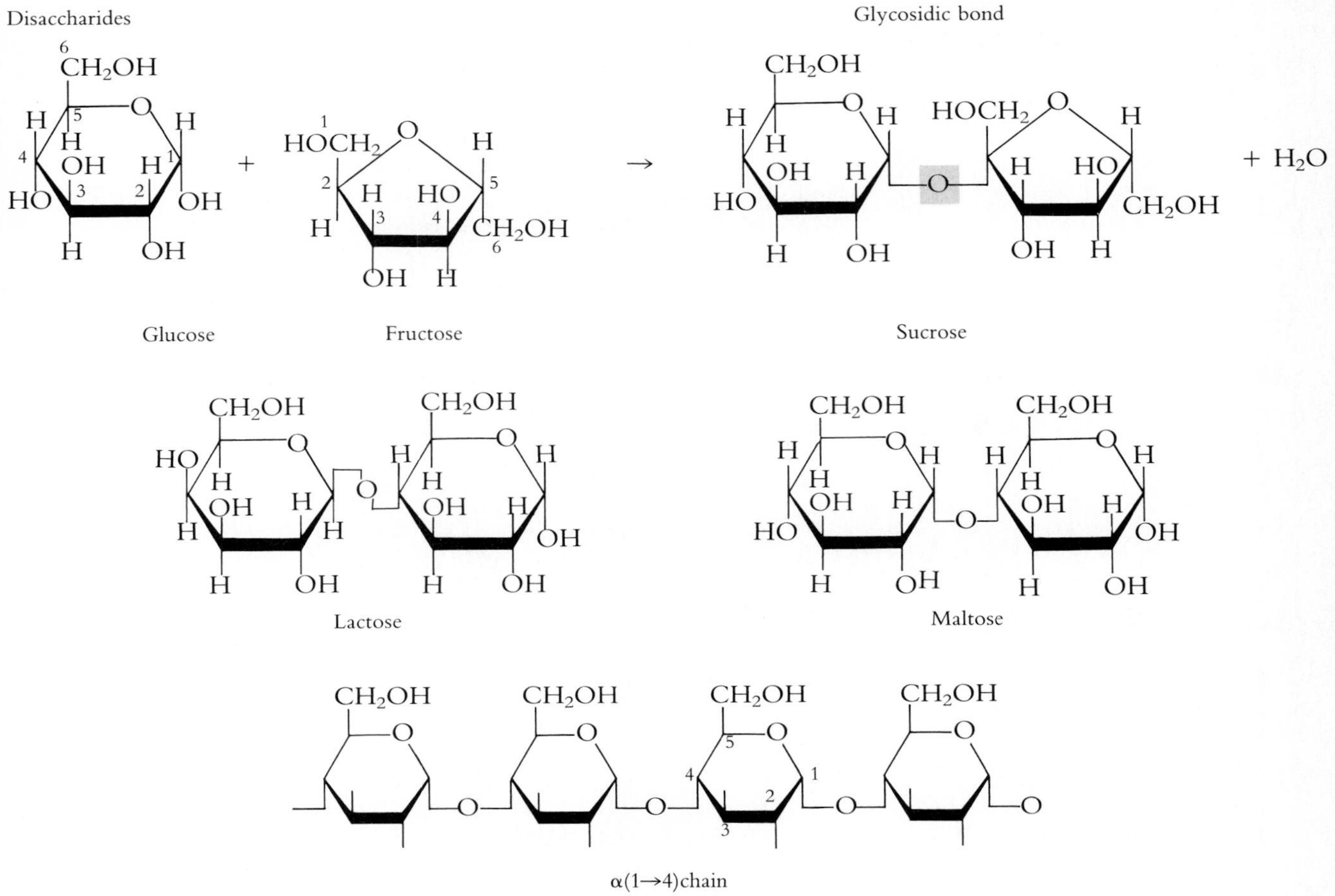

Disaccharides

Glycosidic bond

Glucose

Fructose

Sucrose

Lactose

Maltose

α(1→4)chain

FIGURE 2.14 Disaccharides are formed by a glycosidic bond between two monosaccharides. Sucrose is a soluble disaccharide composed of glucose and fructose. Lactose and maltose are two other important disaccharides. Polysaccharides are composed of repeating monosaccharide units joined by glycosidic bonds. Polysaccharides can be linear (as shown) or branched.

polymers of glucose that vary in the specific carbons used to form the glycosidic bond and in the degree of branching. The linear polysaccharides are represented by **cellulose,** which is a β 1—4 glucose polymer, whereas **glycogen** is a branched polysaccharide of α 1—4 D-glucose with α 1—6 branches. Starch is the polysaccharide stored by plant cells as a food reserve. Potatoes are loaded with starch, and cornstarch is often used in cooking as a thickener. Cellulose is a major component of plant cell walls, exemplified by the cellulose content of wood. Glycogen is the main storage polysaccharide of animal cells.

Proteins Are Polymers of Amino Acids

The 20 naturally occurring amino acids are the building blocks of proteins (Figure 2.10). Polymers of amino acids form when amino acids are combined with a peptide bond. The molecule is called a **peptide** if two or more amino acids are joined together and a **polypep-**

tide when many amino acids are involved. The peptide bond is formed between the carboxyl carbon of one amino acid and the amino nitrogen of the adjacent amino acid upon the removal of water.

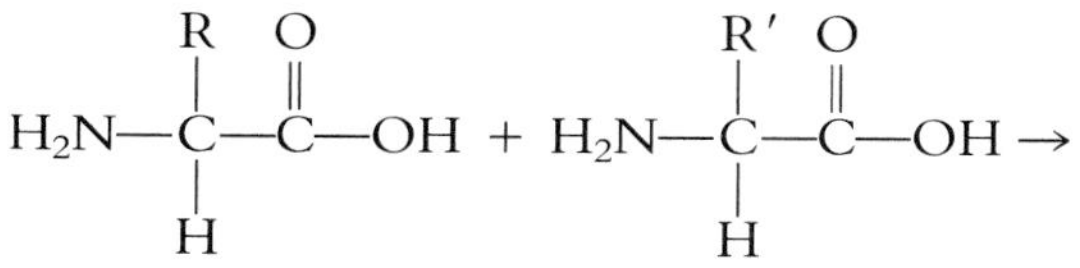

$$\begin{array}{ccccccccccccccc} & & R & & O & & & & & & R' & & O & & \\ & & | & & \| & & & & & & | & & \| & & \\ H_2N & — & C & — & C & — & OH & + & H_2N & — & C & — & C & — & OH \rightarrow \\ & & | & & & & & & & & | & & & & \\ & & H & & & & & & & & H & & & & \end{array}$$

Amino acid R + Amino acid R′⟶

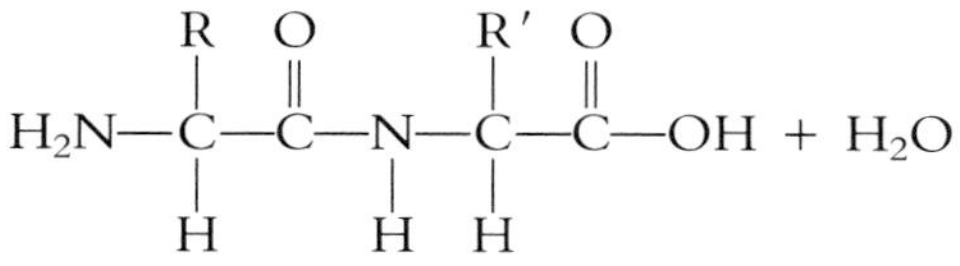

$$\begin{array}{cccccccccccccc} & & R & & O & & & & R' & & O & & & \\ & & | & & \| & & & & | & & \| & & & \\ H_2N & — & C & — & C & — & N & — & C & — & C & — & OH & + H_2O \\ & & | & & & & | & & | & & & & & \\ & & H & & & & H & & H & & & & & \end{array}$$

Peptide + Water

The following are key characteristics of all peptides:

1. The peptide has two ends: an amino-terminal and a carboxyl-terminal end.
2. The R groups can be identical or they can represent any one of the 20 amino acids.
3. The chemical properties of the peptide will depend on the sequence of the amino acids in the peptide.

Proteins are large peptides, or to say it another way, proteins have many amino acid residues. The molecular weights range from very small polymers of 6000 to large proteins over 1,000,000 daltons. A **dalton** is an arbitrary unit of mass essentially equal to the mass of one hydrogen atom. If each amino acid residue has an average molecular weight of 100, then proteins contain between 60 and 10,000 amino acid residues. Cells are remarkable in that they are able to make many copies of very large proteins all with the same amino acid sequence. To understand the complexities of proteins, it is necessary to look at their structure at four levels of organization.

Four Levels of Protein Structure The sequence of amino acid residues in a protein is called its **primary structure** (Figure 2.15). In most proteins this is dictated by the information encoded in the cell's DNA (gene) for that protein (discussed in Chapter 9). A given protein will always have the same amino-terminal end, the same carboxyl-terminal end, and the same sequence of amino acid residues in between. Once the primary sequence is established, the properties of the protein usually result from the chemical and physical interactions of these residues.

Proteins have a natural tendency to form hydrogen bonds between residues of the amino acid polymer. When proteins form hydrogen bonds between closely spaced residues, the peptide spirals into an **alpha helix** structure (Figure 2.15). The hydrogen bonding between amino acid residues of a polypeptide chain determines the **secondary structure** of the protein.

Folding of the polypeptide on itself is called **tertiary structure** and contributes to the compactness of proteins in solution. Certain amino acid residues contain reactive groups that can interact with one another. These interactions cause the polypeptide chain to fold back on itself. Interaction of hydrogen bonds between R groups, hydrophobic interactions between nonpolar R groups, covalent bonds between the sulfurs of cysteine residues, and the ionic bonding between charged R groups stabilize the tertiary structure of protein (Figure 2.15).

The fourth level of protein structure results from the noncovalent interaction of either similar or dissimilar polypeptide chains. This protein organization, called **quaternary structure,** is important to structural proteins as well as to the catalytic function of enzymes, which are proteins. Protein viral coats, bacterial flagella, ribosomes, and many enzymes depend on quaternary interactions for their structure and function.

Proteins are obviously very complex macromolecules, yet scientists have been able to analyze their structure and to determine how they function. Figure 2.16 shows space-filling models of the yeast enzyme hexokinase before and after it has bound its substrate glucose. Like all proteins, this enzyme is composed of a linear sequence of amino acids. The enzyme has a unique active site that binds glucose as it catalyzes the phosphorylation of this carbohydrate. This active site is located in the cleft created when yeast hexokinase assumes its three-dimensional conformation. Note that the conformation of the enzyme changes when it is bound to its substrate. Knowledge of the structure of proteins is an important aspect of understanding their function. The role of proteins in cell structure and metabolism is described in Chapters 4, 5, and 6.

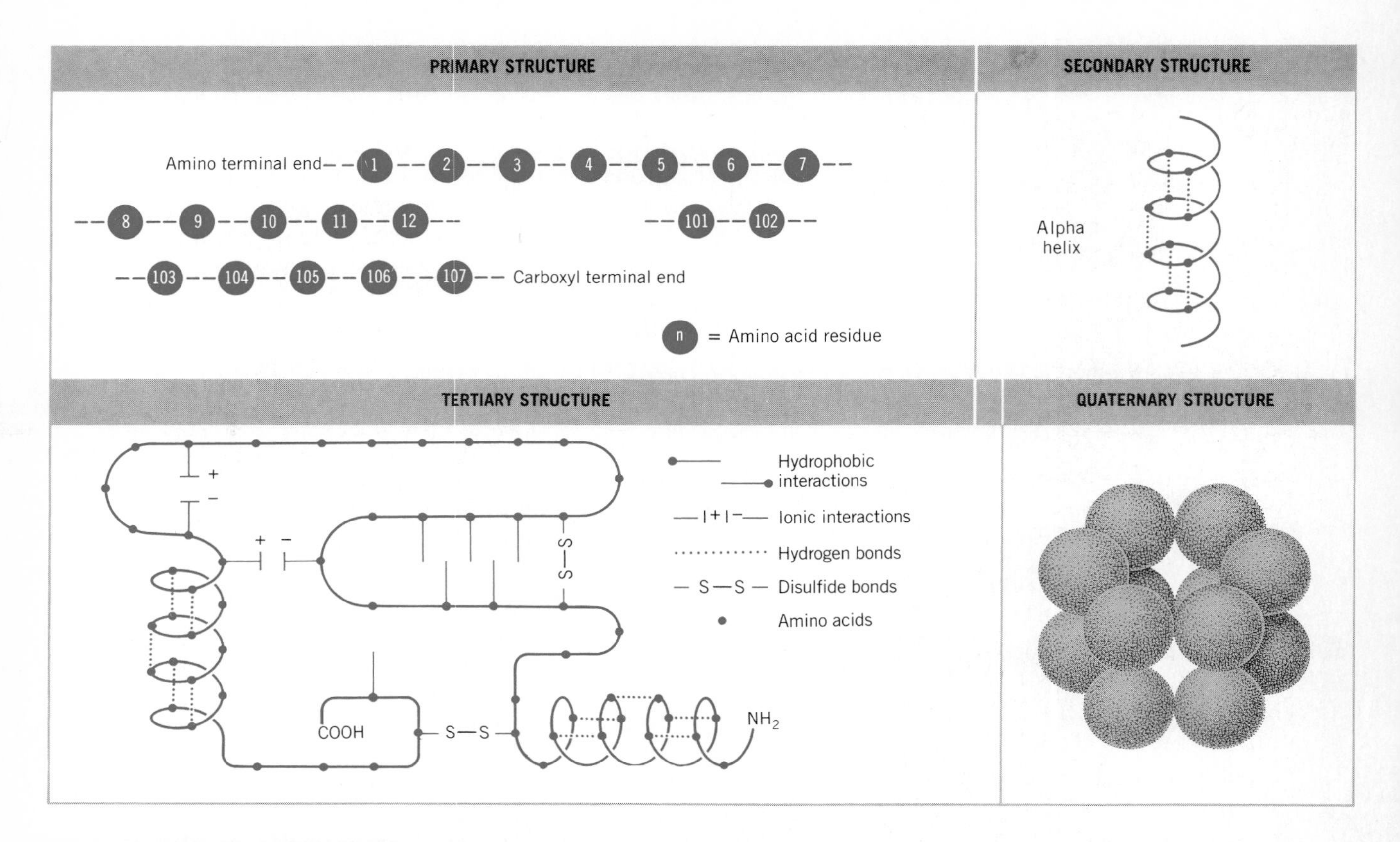

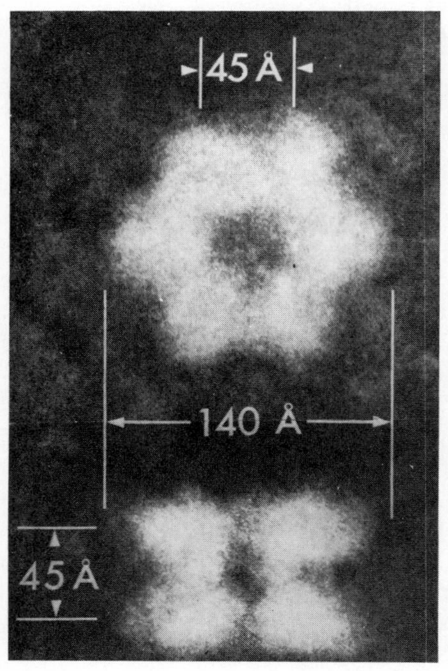

FIGURE 2.15 Proteins are amino acid residues joined by peptide bonds. This string of amino acids can interact within itself to form helices and can fold back on itself to form a tight globular structure. Globular protein subunits can interact to form multimeric proteins. The electron micrograph of glutamine synthetase shows the arrangement of its 12 identical protein subunits. (Courtesy of Dr. Earl Stadtman.)

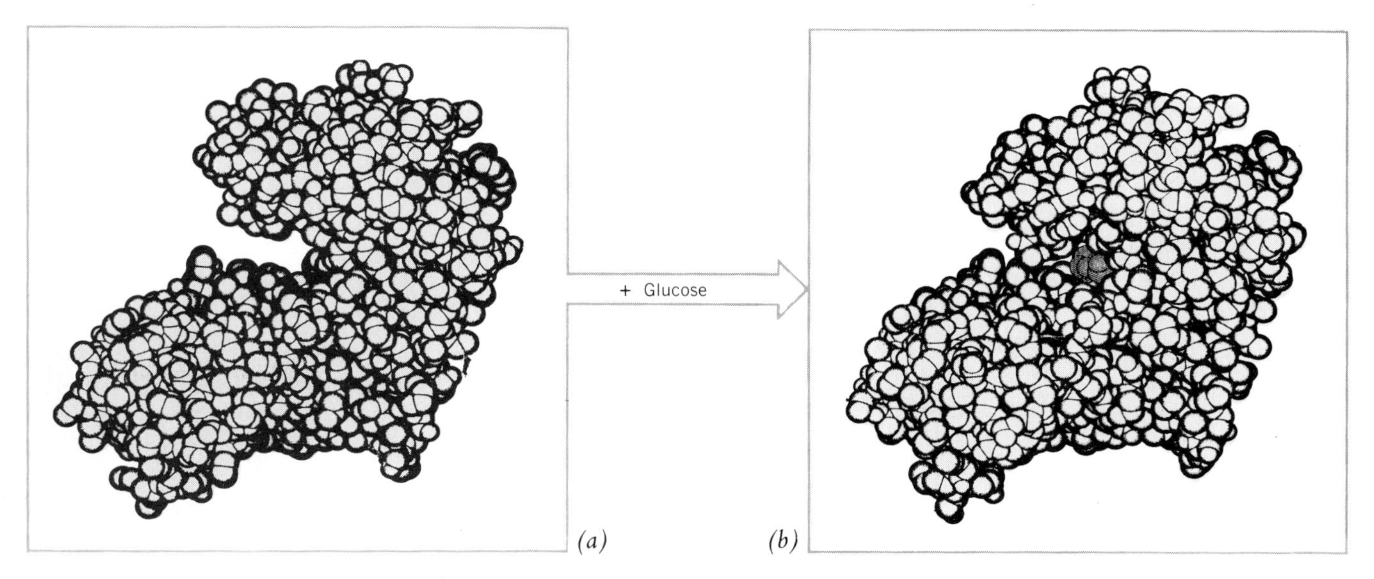

FIGURE 2.16 Space-filling model of yeast hexokinase (*a*) in the native state and (*b*) when bound to its substrate, glucose. (Courtesy of Dr. T. A. Steitz. From C. M. Anderson, F. H. Zucker, and T. A. Steitz, *Science,* 240:375–380, 1979, copyright © 1979 American Association for the Advancement of Science.)

Summary Outline

THE BASIC STRUCTURE OF MATTER

1. Atoms are composed of a nucleus surrounded by orbiting electrons, which are negatively charged particles.
2. The atomic nucleus contains a specific number of positively charged protons and a specific number of uncharged neutrons.

CHEMICAL ELEMENTS

1. Each chemical element has a different atomic structure.
2. An element's atomic weight is essentially equal to the number of protons and neutrons in its nucleus.
3. Electrons orbit the atomic nucleus in electron orbitals that correspond to energy levels. Uncharged atoms have the same number of electrons as protons.
4. Isotopes are created when an atom acquires or loses one or more neutrons. Isotopes decay at a set rate and are used as chemical tracers in biological research and medical diagnosis.
5. Ions are electrically charged atoms or molecules often formed when salts are dissolved in water. Cations are elements that give up one or more electrons from their outer electron level and anions are elements that accept one or more electrons into their outer energy level.

ATOMS COMBINE TO FORM MOLECULES

1. Covalent bonds are formed between atoms that share electrons. Organic compounds contain at least one carbon and a hydrogen atom joined by a covalent bond.
2. Polar covalent bonds are formed by atoms that share the electrons unevenly.
3. Ionic bonds are the attractive forces between negatively and positively charged ions.

WATER AND THE HYDROGEN BOND

1. Water is the major compound found in all living organisms. This polar molecule is an ideal solvent for charged biomolecules.
2. Hydrogen bonds form between hydrogen atoms involved in polar covalent bonds and negatively charged atoms of another polar covalent bond.
3. Water ionizes to form hydroxide (OH^-) and hydronium (H_3O^+) ions.
4. Acids are substances that increase the concentration of hydronium ions when added to water. Bases are substances that incease the concentration of hydroxide ions when added to water.
5. pH is a measure of the acidity/alkalinity of a solution.
6. Buffers are pairs of inorganic or organic compounds that resist changes in pH.

CHEMICAL REACTIONS

1. Molecules are combinations of two or more atoms that form a specific chemical bond.
2. Molecules are formed by chemical reactions that convert reactants to new or altered products.
3. Anabolism is the sum total of the synthetic reactions of a cell; catabolism is the chemical breakdown or degradation of compounds by cells.
4. Chemical reactions catalyzed by enzymes proceed at biological temperatures.

BIOLOGICALLY IMPORTANT INORGANIC COMPOUNDS

1. Inorganic compounds are those molecules that do not contain carbon, such as metals, salts, most gases, and water.
2. Phosphorus, nitrogen, and sulfur are assimilated into organic biomolecules by enzymatic reactions.
3. Bulk elements such as calcium, potassium, sodium, chloride, and magnesium exist in cells as ions.
4. Trace elements are required for the proper functioning of many electron carriers and enzymes.

OXIDATION–REDUCTION REACTIONS

1. Chemical reactions that involve the transfer of electrons are called oxidation–reduction reactions.
2. An element or compound that gives up or donates an electron becomes oxidized, and an element or compound that accepts electrons becomes reduced.

BIOLOGICALLY IMPORTANT ORGANIC MOLECULES

1. Organic molecules are compounds whose formulations include carbon and hydrogen.
2. Molecules with an asymmetric carbon atom can exist in two spacially different forms known as stereoisomers. D- and L-amino acids are examples of biologically important stereoisomers.
3. There are 20 naturally occurring L-amino acids that are found in proteins.
4. Neutral lipids are composed of fatty acids and glycerol. The hydrophobic hydrocarbon chain of the fatty acid makes it an ideal molecule for constructing membranes.
5. Phospholipids have hydrophobic hydrocarbon chains as well as a charged polar end that is hydrophilic.
6. Carbohydrates are used by cells as a source of cellular energy and are structural components of cell walls and nucleic acids.
7. Nucleotides are composed of a nitrogenous base (adenine, guanine, cytosine, thymine, or uracil), either ribose or deoxyribose, and phosphate; they are the building blocks of DNA and RNA.
8. The nucleotide adenosine 5′-phosphoric acid (AMP) is a component of certain coenzymes and is involved in energy transfer reactions.
9. Peptide bonds join the 20 naturally occurring amino acids into polypeptides and proteins. The sequence of amino acids in the peptide constitutes its primary structure. Each polypeptide has an amino terminal and a carboxy terminal end.
10. The hydrogen bonds formed between amino acid residues in the polymer creates the secondary structure, and the tertiary structure results from the folding of the polypeptide on itself. The noncovalent interaction of either similar or dissimilar polypeptide chains creates quaternary protein structure.

Questions and Topics for Study and Discussion

QUESTIONS

1. Explain the differences between atoms, isotopes, ions, and elements.
2. Name the large biomolecules that contain phosphorus or sulfur and explain how a biologist could use isotopes to detect these compounds in cellular material?
3. Describe four types of chemical bonds and give a biological example of each.
4. Diagram the basic structure of an amino acid, a fatty acid, a carbohydrate, a purine, and a pyrimidine, and label each functional group.
5. Explain the esterification reaction used by cells to form a triglyceride.
6. Describe environmental sources for three of the trace metals that contaminate normal tapwater.
7. What color would the following solutions be if they contained *(a)* methyl red or *(b)* phenol red: lemon juice, orange juice, sea water, blood, borax, and O.1*N* $NaCO_3$?
8. Indicate the organic/inorganic nature of the following compounds: hydrogen sulfide (H_2S), methane (CH_4), acetate (CH_3COOH), sodium bicarbonate ($NaHCO_3$), sodium thiosulfate ($Na_2S_2O_3$), ethane (C_2H_6), ethanol (CH_3CH_2OH), sodium nitrate ($NaNO_3$), and glucose ($C_6H_{12}O_6$).
9. Draw the structure for guanosine 5′-triphosphate and highlight the 3′ position of the ribose.
10. Describe the four levels of structural organization in proteins. What level(s) of organization would affect the active site of an enzyme?

DISCUSSION TOPICS

1. The amino acid R groups influence the structure of proteins. Which amino acids would you expect to find on the surface of a protein that is in contact with the interior of a cell membrane?
2. Water is released when carbohydrates form disaccharides and when amino acids are joined to form a polypeptide. What chemical reactions do you envision being involved in the breakdown (hydrolysis) of these molecules?

Further Readings

BOOKS

Holum, J. R., *Fundamentals of General, Organic, and Biological Chemistry,* Wiley, New York, 1978. A basic college chemistry textbook for students entering allied health professions.

Ingraham, J. L., O. Maaløe, and F. C. Neidhardt, *Growth of the Bacterial Cell,* Sinauer Associates, Inc., Sunderland, Mass., 1983. The authors present an excellent description of the chemical composition of bacterial cells that includes a discussion of the relative amounts of the biomolecules present.

Lehninger, A., *Biochemistry,* 3rd ed., Worth Publishers, New York, 1983. A very readable textbook on biochemistry.

Sackheim, G. I., *Introduction to Chemistry for Biology Students,* 2nd ed., Educational Methods, Chicago, 1977. A self-teaching refresher course in chemistry for biology students.

CHAPTER

3

The Microbial World under the Microscope

OUTLINE

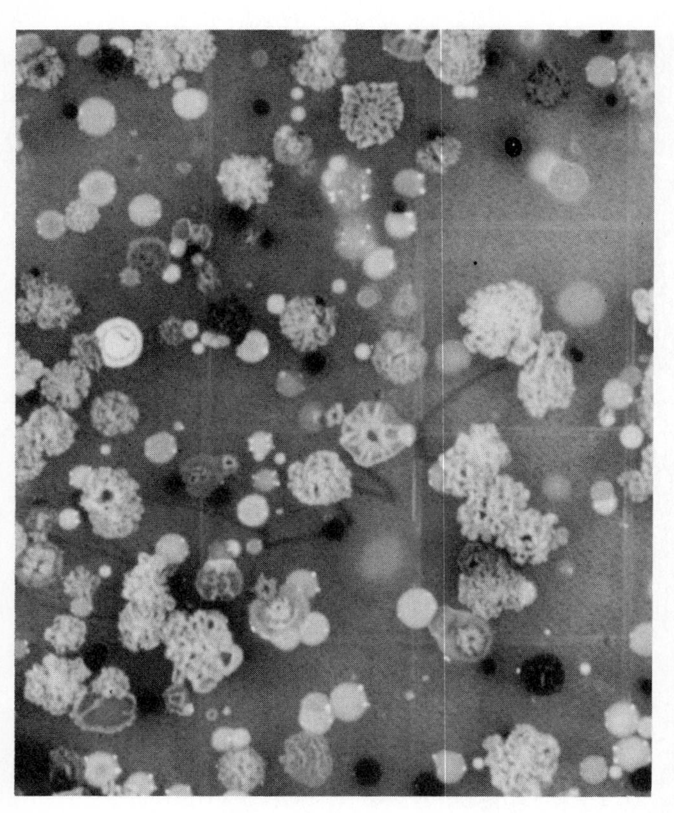

FOCUS OF THIS CHAPTER

Microbiology began with the development of simple microscopes and remains a science whose daily practice depends on microscopy.

Microbiologists use a variety of microscopes, each with specific advantages and limitations, to study microorganisms. This chapter presents the principles of microscopy and applies their use to the study of viruses, microbial cells, and their subcellular structures.

Light microscopy is the cornerstone of microbiology for it is through the microscope that most scientists first become acquainted with microorganisms. Almost every natural environment abounds with a diverse population of microbial life. Samples of pond water, suspensions of soil, and material removed from the digestive tracts of animals contain numerous microbes ranging in size from large protozoa and algae to small bacteria. Most observers are immediately struck by the diversity in size, shape, motility, and behavior of the microorganisms. As one becomes more discerning about these observations, the microscope becomes the tool that can be used to measure the size, determine microbial shapes (Figure 3.1), study the function of subcellular components, and to observe microbial behavior.

METRIC UNITS OF MEASURE

Microbiology is the systematic study of small organisms and viruses, so the very definition of this science implies an ability to measure sizes of organisms. The size of an object visible in a microscope can be measured by comparing it to a grid of known spacings or by knowing the magnification. The dimensions of organisms and their subcellular components are measured by metric units (Table 3.1). The micrometer (μm) is the common unit of measurement of cellular microorganisms such as bacteria: one **micrometer** equals one-millionth of a meter (1 meter $\times 10^{-6}$). To put this in perspective, 1000 bacteria, each 1 μm long and lying end-to-end, would span a space of 1 mm; 10,000 bacteria would span a space of 1 cm. Not all bacteria are the same size; in fact there is great variation in the size of different bacterial species. The smallest bacteria are less than 0.5 μm in width while the largest bacteria have widths of 6 to 12 μm (Figure 3.2). Bacteria are large enough to be seen with the light microscope.

Bacteria contain subcellular components that can be seen only with the aid of an electron microscope. Some of these structures are one thousand times smaller than the cell itself, so their dimensions are given in nano-

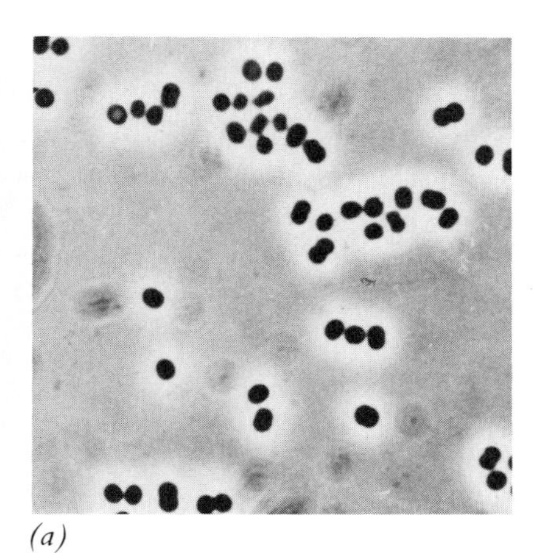

(a)

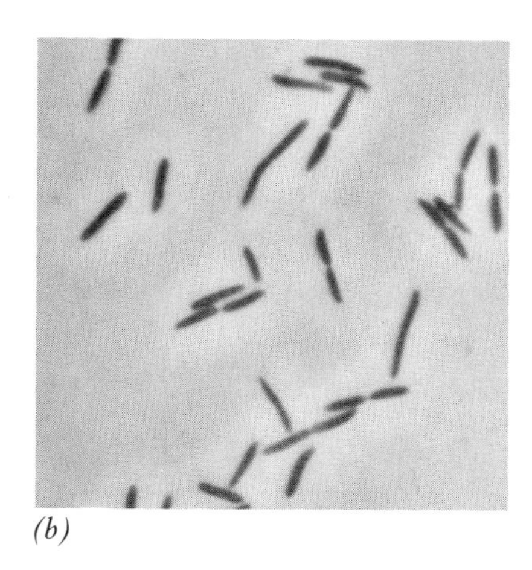

(b)

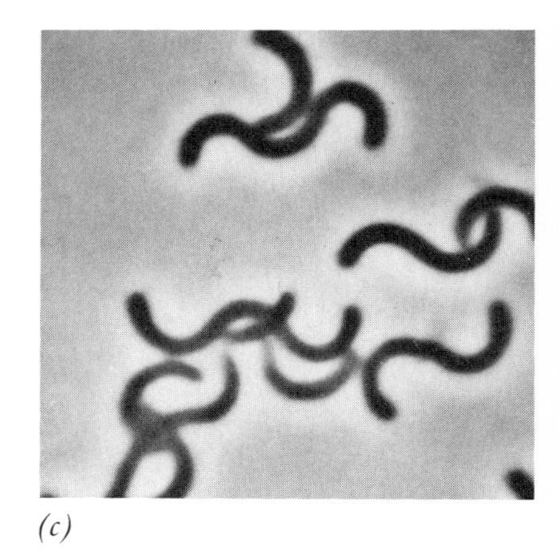

(c)

FIGURE 3.1 The light microscope can be used to distinguish bacterial cell shapes: *(a)* cocci, *(b)* bacilli, and *(c)* spirilla. (From R. Y. Stanier et al., *The Microbial World*, 3rd ed., Prentice-Hall, Englewood Cliffs, N.J., 1970.)

TABLE 3.1 Metric units of measure

LINEAR MEASURE[a]	RELATIVE SIZE
1 meter (m)	39.27 inches = 1 meter
1 centimeter (cm)	1/100 meter or 10^{-2} meter
1 millimeter (mm)	1/1000 meter or 10^{-3} meter
1 micrometer (μm)	1/1,000,000 meter or 10^{-6} meter
1 nanometer (nm)	1/1,000,000,000 meter or 10^{-9} meter
1 angstrom (Å)	1/10,000,000,000 meter or 10^{-10} meter
VOLUME MEASURE[b]	**RELATIVE CAPACITY**
1 liter (1)	0.2642 gallon
1 milliliter (ml)	1/1000 liter
1 microliter (μ1)	1/1,000,000 liter

[a]Most microorganisms are measured in micrometers because their largest dimension is often only one millionth of a meter. One micrometer (sometimes called a micron) is near the limit of resolution by the light microscope. Transmission electron microscopy is capable of resolving structures that are 3 to 8 Å in size. Atoms and the length of bonds between atoms are in the range of 1 to 3 Å. Only the largest atoms have been seen in the electron microscope.
[b]The volume of a typical bacterial cell is estimated to be 1 to 2 × 10^{-12} ml.

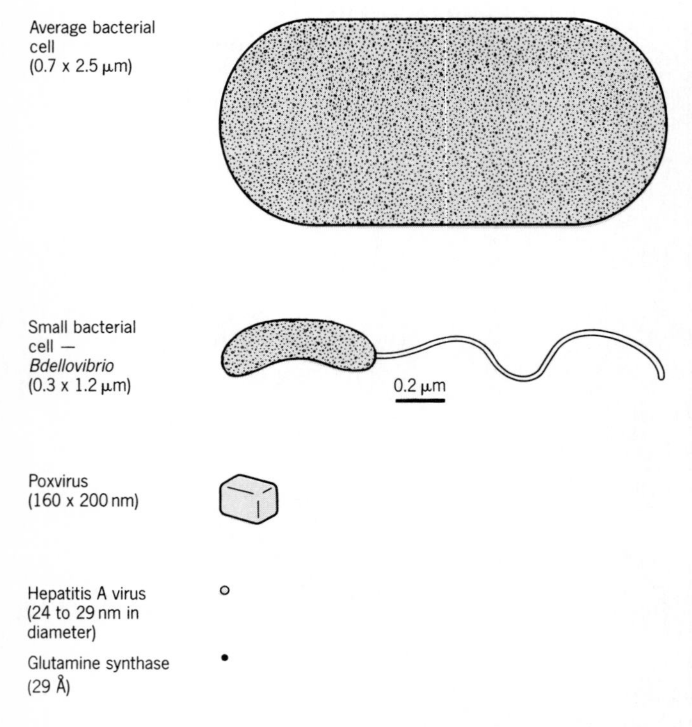

FIGURE 3.2 Relative sizes of cells, viruses, and macromolecules present in the microbial world.

meters (nm). One **nanometer** is equal to one thousandth of a micrometer or one billionth of a meter (10^{-9} meter). Examples of subcellular components include ribosomes and cell membranes. Viruses, because of their small size, are also measured in nanometers (Figure 3.2). The size of animal viruses ranges from the small picornaviruses with diameters of 27 to 30 nm to the large poxviruses, with dimensions of 225 to 300 nm.

Membranes, ribosomes, and viruses are composed of macromolecules. When visible, these macromolecules are measured in angstroms (Å): one **angstrom** is equal to one-tenth of a nanometer or 10^{-10} meter. Macromolecules are at the lower limit of the microscopic world and only the larger macromolecules are visible under the electron microscope.

KEY POINT

Microbiology is the study of small organisms, specifically those forms of life that are too small to be seen with the unaided eye. The human eye can resolve objects 0.1 mm in size, so microorganisms are broadly defined as organisms of less than 0.1 mm in any dimension. This includes bacteria, protozoa, algae, and fungi.

OVERVIEW OF MICROSCOPY

Cytologists are scientists who study the structure of cells for the purpose of understanding the relationships between cell structures and their biological function. Cytologists formulate scientific questions from microscopic observations of living cells. Colonies of microorganisms, helminths, some algae, and fungi can be viewed at low magnifications with a dissecting microscope. Higher magnifications are obtained with a compound microscope, which long ago superseded the simple, single-lens Leeuwenhoek microscopes. The compound light microscope (Figure 3.3) can be used to observe both living and stained preparations of microorganisms. Light microscopes range in quality and cost

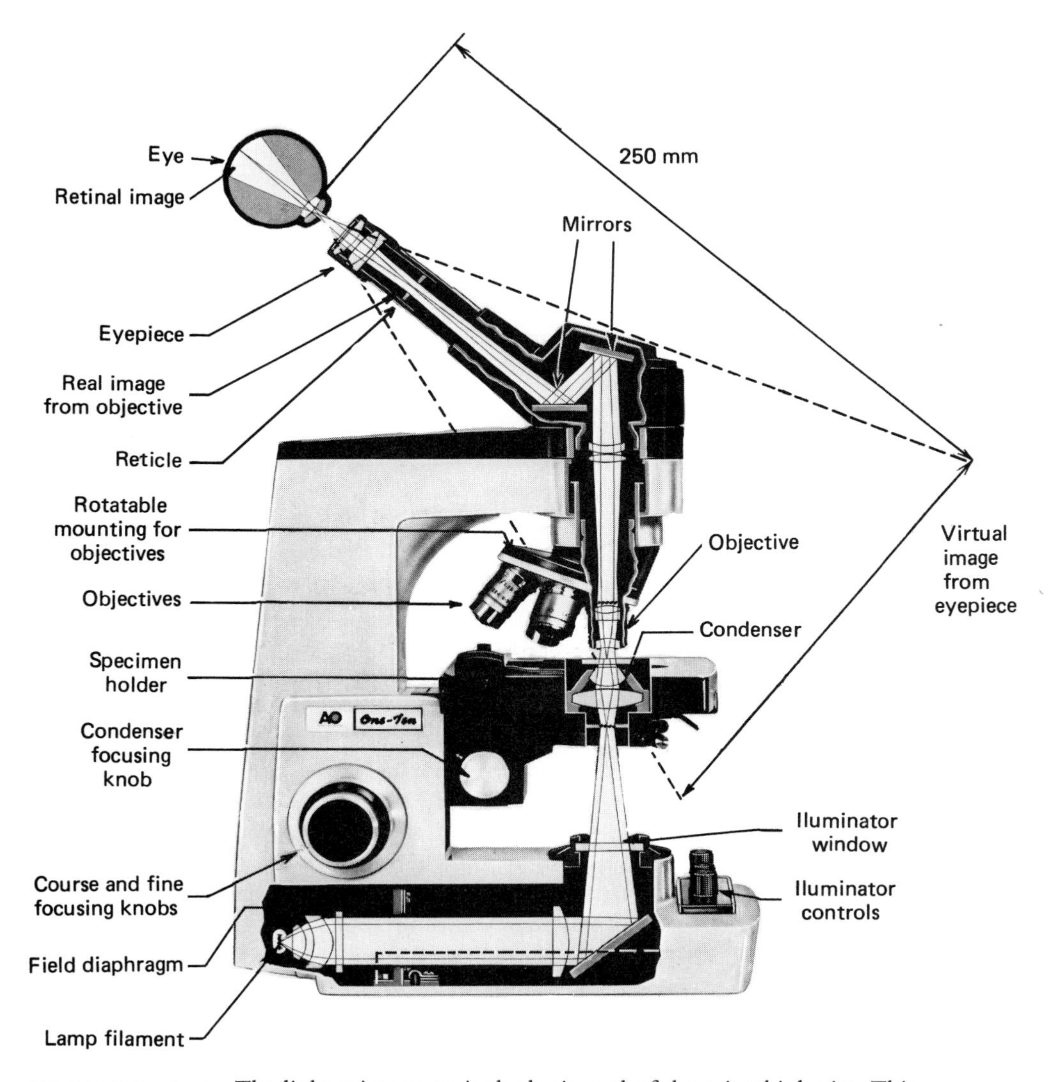

FIGURE 3.3 The light microscope is the basic tool of the microbiologist. This cutaway view traces the light path from the lamp, through the specimen, and then to the human eye. The light microscope can attain magnifications of 1000× and can resolve points separated by 0.2 μm. (Courtesy of American Optical Company.)

from the inexpensive microscopes used in toy biology kits for children to the sophisticated research microscopes that have special optics and built-in cameras.

The light microscope has a limited ability to resolve subcellular structures and cannot resolve viruses. The study of viruses and structures within cells requires the higher magnifications obtained with the transmission electron microscope (TEM). Transmission electron microscopy combined with physiological studies of isolated cellular structures has provided biologists with a detailed understanding of the organization and biological function of the subcellular structures found in procaryotic and eucaryotic cells. More recently, the scanning electron microscope (SEM) has been used to study the structures of cell surfaces. With this array of microscopes, microbial cytologists have provided us with a detailed picture of the internal workings of the major types of microbial cells.

THE LIGHT MICROSCOPE

The typical light microscope (Figure 3.3) is composed of a condenser lens that collects the light from a light source (lamp), and directs it toward the specimen; an objective lens that collects the light coming from the specimen and focuses it to form a primary image of the specimen; and an ocular lens that allows the observer to look at the primary image and to magnify it further. Such an instrument is called a **compound light microscope** (Figure 3.3), since the light from the specimen travels through a series of glass lenses before the image is seen by the observer.

LENSES USED IN LIGHT MICROSCOPES

A compound microscope has two sets of magnifying lenses (Figure 3.3): the ocular lenses (eyepiece) are what the observer looks through, and the objective lenses are positioned close to the specimen. Many light microscopes are equipped with paired ocular lenses (binocular) to reduce eyestrain. Ocular lenses typically have a magnification value of 10×, which means the image of the object being viewed is ten times larger than the object itself. Most microscopes are equipped with at least three objective lenses of different magnifying powers. A typical microscope has objectives with powers of 10×, 40×, and 100× respectively.

The quality of a lens depends on its design, the materials used in its manufacture, and the degree to which optical aberrations are corrected. Inexpensive lenses often create distorted images caused by the shape and nature of the glass from which they are made. More expensive lenses are corrected for at least two defects, spherical and chromatic aberration. **Spherical aberration** is the inability of a lens to be focused on the entire field. The center of the field will be in focus, while the periphery of the field will be out of focus. Another problem is caused by the different colors (wavelengths) present in white light. As light passes through glass it is bent at varying angles, depending on its wavelength. This is how a prism separates white light into the colors of the rainbow. When the same light passes through the microscope's lenses, it is also bent and distorts the image of the specimen to cause **chromatic aberration.** Both spherical and chromatic aberration can be corrected with specific designs and lens-manufacturing techniques.

Resolving Power of the Light Microscope

The lenses used in the light microscope are designed to maximize the magnification of the image without causing distortion. However, there are physical laws that restrict the magnification that is useful to the observer.

The maximum useful magnification is limited by the microscope's **resolving power,** which is the minimal distance between two points that can be distinguished by the observer. The resolving power of a light microscope depends on the wavelength of light used and the **numerical aperture** (NA) of the objective lenses (Figure 3.4). The numerical aperture ($n \sin \theta$) is a measure of the ability of the objective lens to capture light (Figure 3.4). In this equation, n is the refractive index of the material between the specimen and the lens. The numerical aperture of a lens can be increased by (1) increasing the size of the lens opening, and/or (2) increasing the refractive index of the material between the lens and the specimen. The larger the numerical aperture, the better (lower) the resolving power. These parameters are related by the mathematical definition of resolving power *(d)*

$$d = \frac{\lambda}{2 \times \text{numerical aperture}}$$

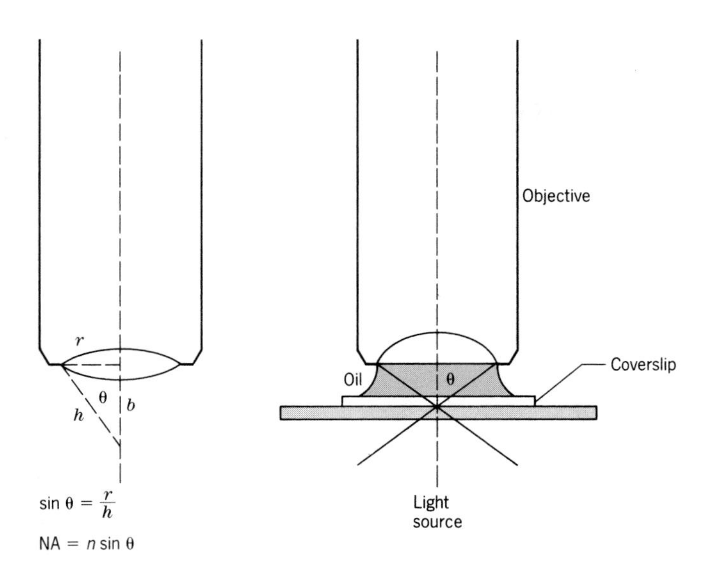

FIGURE 3.4 The resolving power of a light microscope is physically constrained by the numerical aperture (effective opening) of the objective lens. The numerical aperture is directly related to the refractive index of the material between the specimen and the lens and the angle θ. The *refractive index (n)* of a medium is the extent to which the speed of light is slowed by the new medium. Immersion oil (n = 1.52) is used between the specimen and the objective to maximize the refractive index of the system.

TABLE 3.2 Dimensions of selected bacteria[a]

NAME OF ORGANISM	DIMENSION(S), μm
Small Bacteria	
Bdellovibrio bacteriovorus	0.2–0.5 × 0.5–1.4
Bordetella pertussis	0.2–0.5 × 0.5–1.0
Coxiella burnetii	0.2–0.4 × 0.4–1.0
Treponema pallidum	0.1–0.18 × 6–20
Medium Size Bacteria	
Bacillis subtilis	0.7–0.8 × 2–3
Escherichia coli	0.4–0.7 × 1.0–3.0
Staphylococcus aureus	0.5–1.0 diameter
Streptococcus pyogenes	0.5–1.0 diameter
Large Bacteria	
Achromatium oxaliferum	5 × 100
Azotobacter chroococcum	1.5–2 × 5
Beggiatoa gigantea	26–55 × 5–13
Spirochaeta plicatilis	0.75 × 90–250
Thiovulum majus	6 × 12

[a]Data compiled from *Bergey's Manual of Systematic Bacteriology,* Vol. 1 (1984) and Vol. 2 (1986), Williams and Wilkins Co., Baltimore, Md.

where λ is the wavelength of light measured in nm. Although there are physical as well as economic limitations to the size of an objective lens, *the ultimate resolving power of the light microscope is limited by the wavelength of light used.*

Visible light spans the wavelengths from 400 to 700 nm; for reference, green light has a wavelength (λ) of 500 nm. Applying the above formula: the theoretical resolving power *(d)* of a light microscope with an objective lens possessing a numerical aperture of 1.25 when green light is used would be 200 nm (500 nm/2 × 1.25 = 200 nm). Converting 200 nm, to micrometers (μm) we find that the resolving power of the light microscope is 0.2 μm.

Most bacteria have minimum dimensions of about 0.5 μm (Table 3.2) so they can be seen in the light microscope. In contrast, subcellular structures of bacteria and bacterial appendages are not usually visible in the light microscope since their largest dimension is smaller than 0.2 μm.

Objective Lenses

Two basic types of objective lenses are used by microbiologists: the high-dry lens and the oil immersion lens. The **high-dry lens** is used to locate samples on a slide, to observe large microorganisms, and sometimes to study stained bacterial preparations. A high-dry lens doesn't come in contact with the specimen or the coverslip. Since the light passes through an air interface (n = 1.0) between the sample and the lens, the numerical aperture is never greater than 1.0 × sin θ.

The **oil-immersion** lens is designed to be in direct contact with the oil (n = 1.52) placed on the coverslip (Figure 3.4). Immersion oil increases the numerical aperture of the objective lens because of the oil's high refractive index *(n)* (Figure 3.4). The numerical aperture of an oil-immersion objective usually falls between 1.0 and 1.3. The greatest useful magnifications in light microscopy are obtained with oil-immersion lenses because they have the highest numerical aperture. Very expensive oil-immersion objectives can have numerical apertures above 1.3.

PHASE-CONTRAST MICROSCOPY

Most microorganisms appear to be transparent when viewed under the compound microscope because there is almost no contrast between the cell and its liquid surroundings (Figure 3.5*a*). By **contrast** is meant the differences between light and dark actually observed. The ordinary compound microscope depends on light absorption by the cell to create contrast between the specimen and the background. Since microorganisms are composed mainly of water and possess only small amounts of absorbing material capable of interfering with the passage of light, it is difficult to see them in the common brightfield microscope.

Light that passes through a cell is altered by the composition of the cell, even though human eyes cannot distinguish these changes. The phase of the light wave is retarded by the density of the cytoplasmic material. Humans cannot see phase variations, but can see differences in the amplitude of the light wave, which are interpreted as contrast.

In the 1930s, Frederick Zernike devised a method of converting phase changes into differences in light intensity. This invention is called the **phase-contrast microscope** and it enables the user to see otherwise transparent objects. The phase-contrast microscope contains an annular aperture in the condenser that forms a hollow cone of light rays, which in turn is focused by the condenser on the phase plate located in the back of the objective lens. The phase plate is actually a ring of material that retards light focused on it. When properly aligned, all the light not deflected by the specimen passes through the phase plate and is retarded in phase, whereas light passing inside the ring is not affected.

Light rays that pass through the specimen are de-

(a)

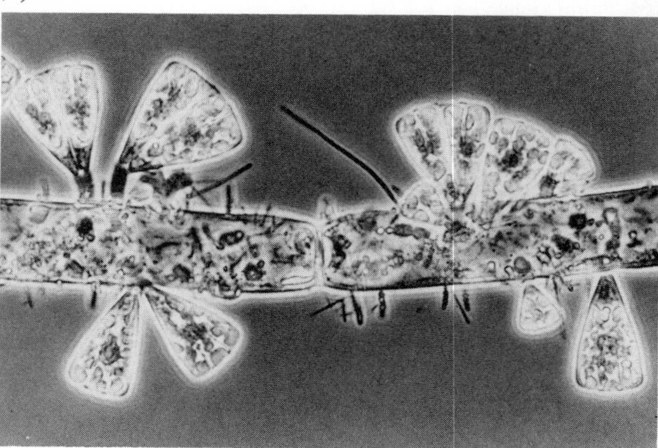
(b)

(c)

FIGURE 3.5 A preparation of epiphytic bacteria and diatoms (*Licmophora* spp.) on a filamentous alga (*Pylaiella littoralis*) was photographed using *(a)* brightfield, *(b)* phase-contrast, and *(c)* darkfield microscopy. (Courtesy of Dr. Paul W. Johnson/BPS.)

flected and pass outside the phase plate. The lens focuses this scattered light to form a primary image of the specimen. Because of the optical effects of the phase plate, the unscattered light will be phase shifted relative to the scattered light coming from the specimen. This shift in phase is just enough to cause the new image of the specimen to be out of phase with the background illumination, causing the light from the image and from the background to "cancel" each other. The resulting darkening of the image increases the contrast seen by the human eye (Figure 3.5*b*). The phase-contrast microscope enables microscopists to study living bacteria that are difficult to see in brightfield microscopy.

DARKFIELD MICROSCOPY

Bacteria less than 0.2 μm in width are invisible in a brightfield microscope (Table 3.2); however, these bacteria can be seen in light microscopes modified to use darkfield illumination (Figure 3.6). In a darkfield microscope, organisms appear as bright spots against a black background, just as the stars appear on a dark night.

Darkfield illumination is accomplished by modifying the condenser so that rays from the light source pass through the specimen only at an oblique angle. The light rays coming directly from the light source are blocked by the iris diaphragm in the objective lens (Figure 3.6). When looking at pure water, the observer sees only a dark field of view. Objects present in the light path scatter (diffract) the light so that some diffracted light rays enter the objective. This light is focused into an image that appears bright against a dark background (Figure 3.5*c*). With darkfield illumination, the observer sees only the light that is scattered by the specimen. Darkfield microscopy is used to observe very thin bacteria, such as *Treponema pallidum* (tre-po-nee'mah pal'li-dum), that are invisible in the brightfield microscope. Recently, microscopists have used darkfield microscopy to observe the blurred motion of bacterial flagella that are only 20 nm in diameter.

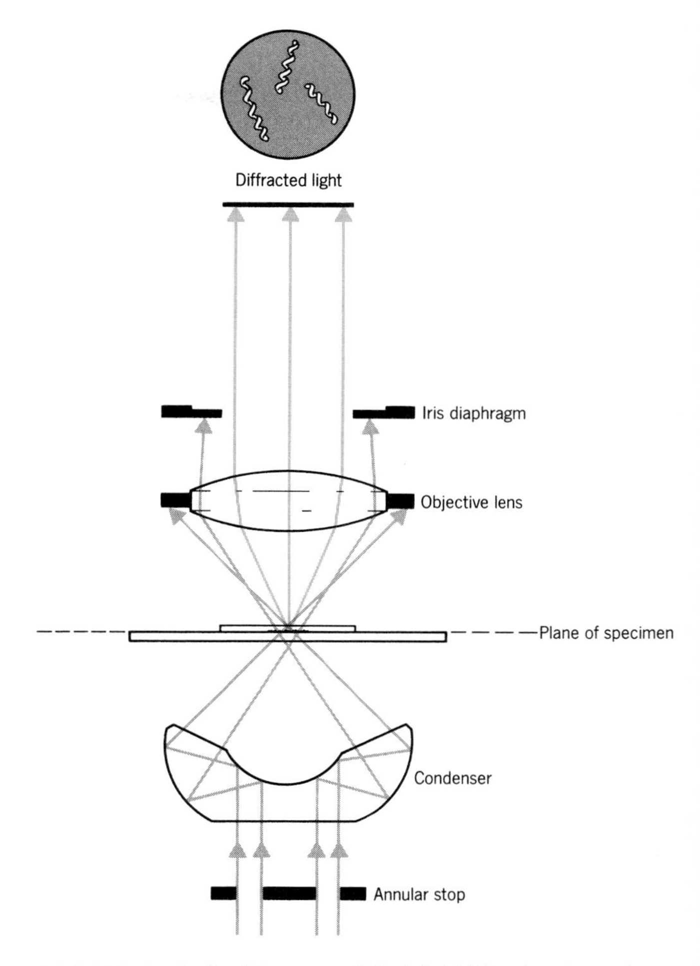

FIGURE 3.6 Diagram of darkfield illumination using the darkfield condenser. Only the light diffracted by the specimen is seen by the observer. Organisms appear as bright objects in'a dark background.

FLUORESCENCE MICROSCOPY

Certain chemical substances are fluorescent (floor-es'ant), which means they absorb one color of light and then, after a delay, emit light of another color. The emitted light is known as **fluorescence.** Microorganisms may possess fluorescent compounds; if they do not, they can be stained with fluorescent dyes that can be seen with the fluorescence microscope. Fluorescence microscopy is commonly used for cytological investigations and in diagnostic microbiology for bacterial identification.

The fluorescence microscope has a special light source that illuminates the specimen with a short wavelength of light (often ultraviolet light) that will be absorbed by fluorescent compounds. The specimen then emits fluorescence that forms the image seen by the observer. The fluorescence is seen against a dark background created by darkfield illumination. Because ultraviolet light causes eye damage, the fluorescence microscope contains a barrier filter in its light path that absorbs the short wavelength excitation light.

ELECTRON MICROSCOPY

The resolving power of the light microscope is limited by the wavelength of visible light. The same physical principles apply to electron microscopy; however, the wavelength of the electron beam is very small (0.04 nm), so structures as small as 5 to 8 Å in size can be resolved using the electron microscope. This enables cytologists to study viruses and large macromolecules such as nucleic acids and proteins. The largest atoms have been seen with the electron microscope, but this rare event is possible only with the most advanced instruments.

TRANSMISSION ELECTRON MICROSCOPE

The transmission electron microscope (TEM) is an expensive research instrument used by biologists to study cell structures (Figure 3.7). Instead of using light, the transmission electron microscope employs an electron beam generated by a tungsten filament, similar to the tungsten filament in an incandescent light bulb (Figure 3.8). This filament must be kept in a vacuum to prevent it from burning out. The vacuum also allows the electrons released from the filament to be accelerated by the powerful electron field created within the microscope.

Before the internal structures of microorganisms can be observed in a TEM, the cells must be fixed, sectioned into thin slices, and then stained. Observation of external appendages such as flagella can be done without making thin sections. All samples are stained with heavy metals such as lead, uranium, and tungstate, to increase electron absorbance. Absorbance of electrons from the electron beam is what produces the contrast in electron microscopy. Because of the sample preparation required, living material cannot be viewed with the transmission electron microscope.

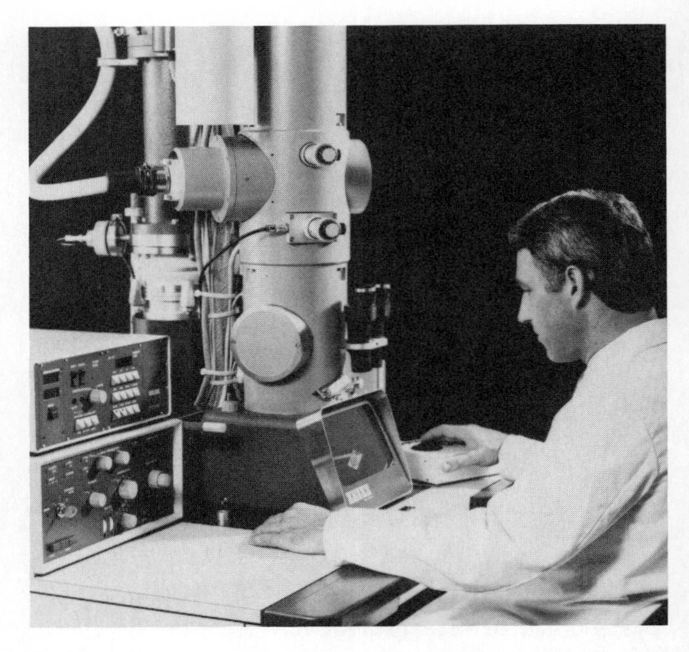

FIGURE 3.7 The transmission electron microscope is an essential tool of the microbial cytologist. A transmission electron microscope can resolve structures as small as 5 to 8Å. (Courtesy of Carl Zeiss, Inc.)

The electron beam is focused with a series of electromagnetic lenses so that it passes through the specimen before it is focused on a viewing screen. Transmission electron micrographs are made by exposing a photographic film to this electron image (Figure 3.9). Black and white prints are then made from the developed film.

Freeze-Etching Reveals Surface Structures

The transmission electron microscope can be used to observe the surface topology of samples using the freeze-etching technique (Figure 3.10). The topography of the surfaces of deep-frozen cells is observed by viewing a carbon replica of the sample in a TEM. The cells are frozen in liquid nitrogen at −196°C and then fractured with a cold mechanical arm. The fracturing removes surface layers, such as the cell wall, from the frozen cell surface. A carbon replica of this fractured surface is then made by coating the sample with carbon and a heavy metal. The cellular material is chemically

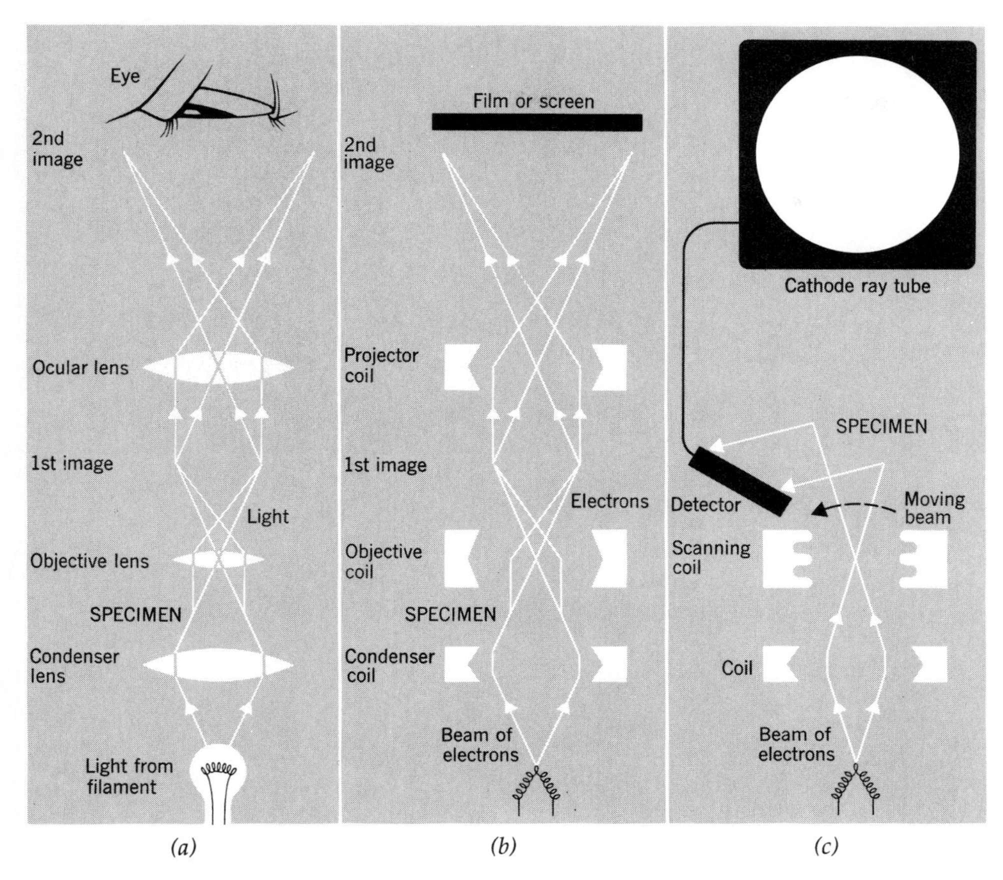

FIGURE 3.8 Diagramatic comparison of *(a)* light microscope, *(b)* transmission electron microscope (TEM), and *(c)* scanning electron microscope (SEM). The light microscope uses glass lenses for focusing the light rays, which pass through the specimen, into the real image seen by the human eye. The TEM *(b)* uses magnets to focus its electron beam. The electrons pass through the specimen before they are focused on an electron-sensitive plate to form an image that can be visualized or photographed. The SEM *(c)* scans the surface of a specimen with electron beams. Deflected electrons are collected by a detector, which send signals to a cathode ray tube (television screen) on which an image is produced. (From G. Stephens and B. North, *Biology,* Wiley, New York, 1974.)

removed from the carbon replica. This replica then becomes the specimen observed in the transmission electron microscope.

The carbon replica photographed in Figure 3.10 shows the different layers of cell wall lying on top of the cell membrane. This technique enables one to see the regular array of macromolecules that create the complex cell wall structure of this bacterium.

SCANNING ELECTRON MICROSCOPE

Remarkable three-dimensional pictures of biological specimens full of crisp detail and sharp relief (Figure 3.11) are obtained with a **scanning electron microscope** (SEM). The scanning electron microscope is a research instrument used by microbiologists in laboratories devoted to cytology. It is called a scanning microscope because the image is created as the specimen

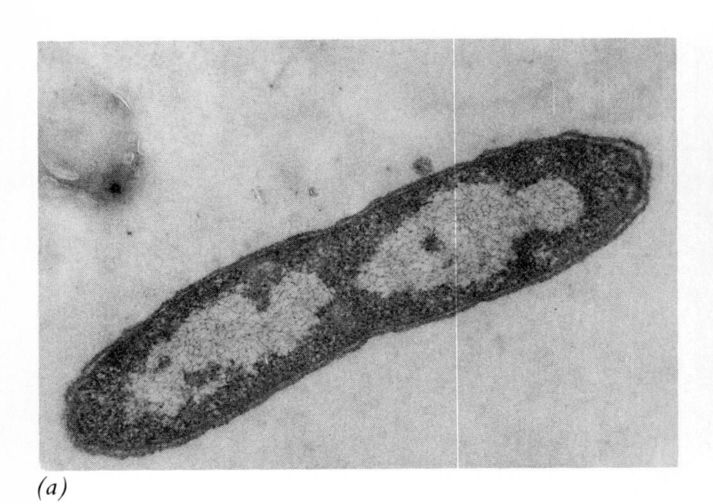

(a)

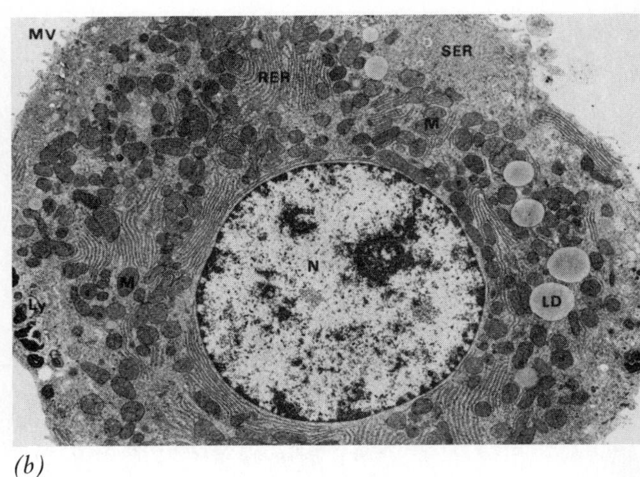

(b)

FIGURE 3.9 Transmission electron micrograph of thin sections through (a) *Pseudomonas* in the process of cell division and (b) a rat liver cell. Note the differences in the nuclear and cytoplasmic organizations of these two distinct cell types. (Courtesy of (a) Dr. J. W. Costerton, (b) Dr. J. D. Cunningham.)

(a)

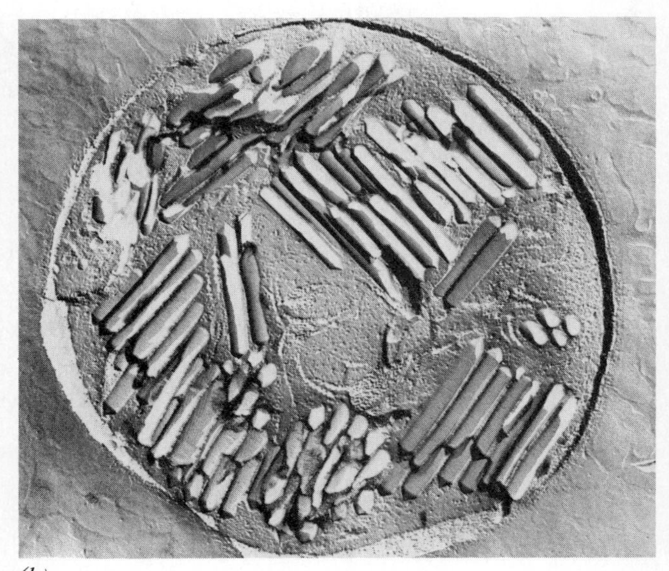

(b)

FIGURE 3.10 The freeze-etching technique is used to study surface structures. Cells are frozen, then etched; a carbon replica is then made of their etched surface structures. (a) A transmission electron micrograph of a freeze-etched preparation of *Nitrosococcus oceanus,* showing its cell-wall (cw) structure and particles on its cytoplasmic membrane (cm). (b) Freeze-etching of *Nostoc muscorum,* showing cytoplasmic gas vacuoles. (Courtesy of (a) Dr. Stanley W. Watson, (b) Dr. J. Robert Waaland/BPS.)

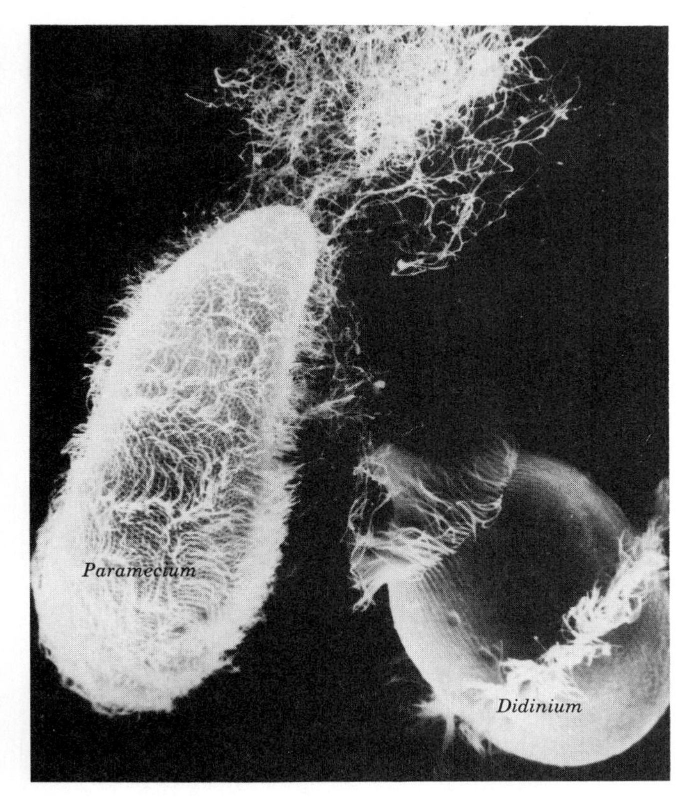

FIGURE 3.11 Scanning electron micrograph of a *Paramecium* about to be engulfed by *Didinium nasutum*. The surface structures of these microbes are clearly shown in this scanning electron micrograph. (Courtesy of Dr. Gregory Antipa.)

is scanned with a beam of electrons. The specimen to be studied is coated with a thin layer of gold or palladium to increase the scattering of the electrons. These scattered electrons are collected by electronic devices and displayed on a cathode ray tube, which is actually a small television screen. Scanning electron microscopy yields an image of the surface structures and topology of a specimen with remarkable detail. One limitation of the SEM is its resolution, which is only ten times greater than that obtained with the light microscope. So scanning electron micrographs show surface structures at relatively low magnification.

Each microscope has its particular advantages. The light microscope can be used to observe living cells. Transmission electron microscopy has the best resolving power and is used to visualize subcellular structures and macromolecules. Scanning electron microscopy is used to observe surface structures. The combined use of these instruments has enabled microbial cytologists to provide a detailed understanding of the structures of microbial cells.

APPLICATIONS OF THE LIGHT MICROSCOPE

The light microscope is a standard instrument in the microbiology laboratory. It is routinely used to observe preparations of living cells and to check for the morphological purity of a given culture. Observations of cell size, shape, motility, spore-forming ability, and reaction to specific stains can all be made with a light microscope. Indeed, the light microscope is the instrument of choice for studying living cells and for identifying many microorganisms.

WET MOUNT AND HANGING-DROP PREPARATIONS

The light microscope is a major tool for studying the behavior of living microorganisms. The samples can be prepared easily as a **wet mount** by placing a drop of the microbial suspension on a microscope slide. The drop is covered with a thin square glass called a coverslip; the coverslip is lowered onto the slide without trapping air between it and the slide. The wet mount can be observed with either the high-dry or the oil-immersion objective of a light microscope, be it a brightfield, phase-contrast, darkfield, or fluorescence microscope.

Specimens can be exposed to air during microscopic observation by means of a **hanging-drop** slide. A drop of the microbial suspension is placed on a coverslip, which is then inverted over the dip in a depression slide. The sample is then viewed with either the high-dry or the oil-immersion objective. Both the wet mount and the hanging-drop preparations are used to observe the behavior of living cells. Motile organisms will dart about or move against the normal flow of the inert particles seen in the preparation. Inert particles or nonmotile organisms will move with the currents of their suspending medium. In addition they may display vibrational motion known as **Brownian movement.**

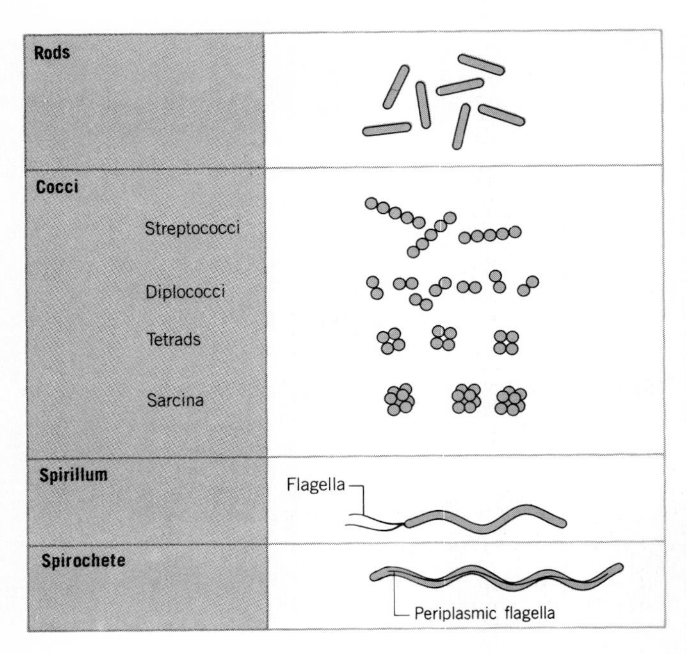

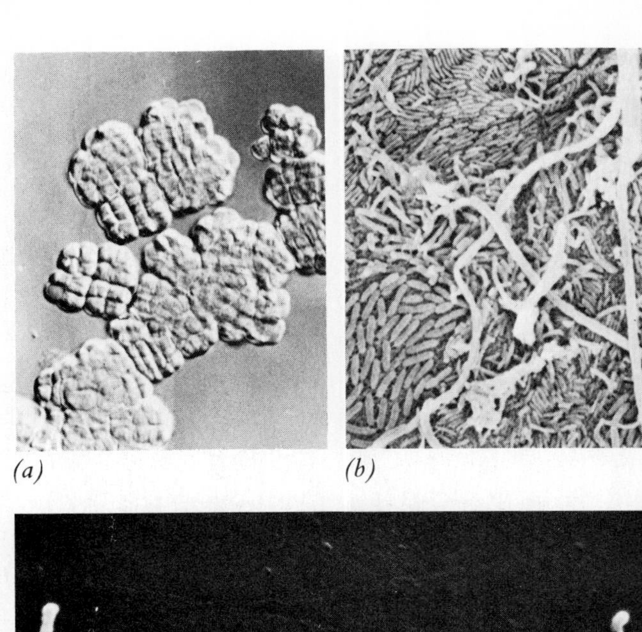

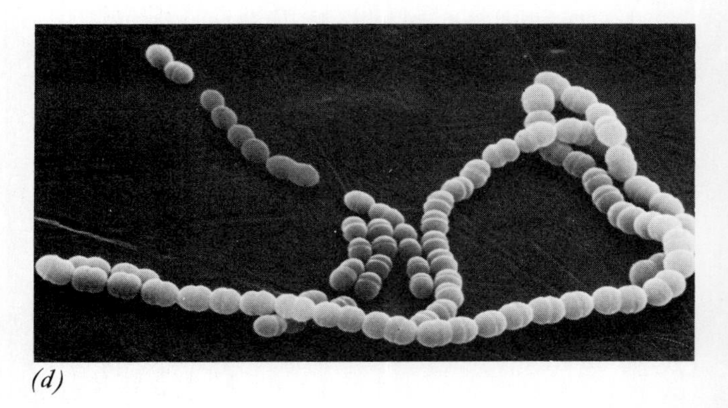

FIGURE 3.12 The four basic shapes of bacteria. Phase contrast micrograph of *(a) Sarcina ventriculi,* which grows in packets of eight cells. Scanning electron micrograph of *(b)* rod-shaped bacteria in the gut of the termite. (Courtesy of Dr. J. A. Breznak. From J. A. Breznak and H. S. Pankratz, *Applied Eviron.Microbiol.,* 33:406–426, 1977). *(c)* A single spirochete. (Courtesy of Dr. N. Charon. From O. Carleton, N. Charon, P. Allender, and S. O'Brien, *J. Bacteriol.,* 137:1413–1416, 1979). *(d)* Dividing streptococci. (Courtesy of Dr. G. Shockman. From D. L. Shungo, J. B. Cornett, and G. D. Shockman, *J. Bacteriol.,* 138:598–608, 1979, *(b–d)* with permission of the American Society for Microbiology.)

This movement is caused by molecular collisions at the submicroscopic level and can be distinguished from bacterial motility by its random, nondirectional character.

OBSERVING BACTERIA WITH THE LIGHT MICROSCOPE

The shape and size of a bacterial cell are the first clues to its identification. A bacterium can have the basic shape of a sphere, a rod, a comma, or a spiral (Figure 3.12). The shape of a bacterium and its cellular arrangement in chains, clusters, or tetrads are useful characteristics in bacterial identification and classification.

Gross Bacterial Morphology

Most bacteria divide by transverse **binary fission** (Figure 3.13), a process of division that results in two equal

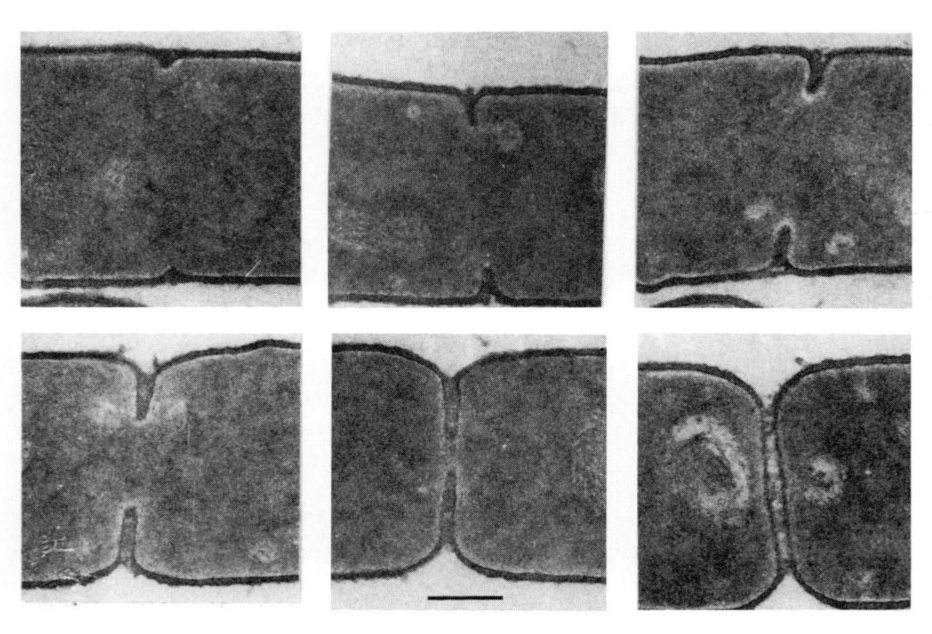

FIGURE 3.13 Binary fission is demonstrated in this series of electron micrographs of cross-wall formation in *Bacillus subtilis*. The sequence begins at the top left. The bar represents 0.2 μm. (Courtesy of Dr. N. Nanniga. From N. Nanniga, L. H. Koppes, and F. C. deVries-Tijssen, *Arch. Microbiol.*, 123:173–181, 1979, with permission of Springer-Verlag, New York.)

daughter cells. Spherical cells, called **cocci** (cock'si), can divide to form pairs, chains, tetrads, or packets of eight cells, depending on whether the cell divides in one, two, or three planes (Figure 3.12). When a coccus divides in one plane it forms a **diplococcus;** division at right angles to the first plane results in **tetrad** formation; and division in the third plane results in packets of eight cells known as **sarcina** (sahr'si-nah).

Rod-shaped bacteria occur in various sizes and shapes ranging from football-shaped to long, slender filaments. Cross walls are often observable in these long filaments indicating the presence of individual cells. Some bacteria are **pleomorphic,** meaning that they can assume different shapes. Pleomorphism is a characteristic of specific bacterial genera and is observed more commonly in physiologically old cultures. Pleomorphism among the bacteria of a culture can mislead the observer to suspect the presence of a contaminant. An example of pleomorphism is demonstrated by **Arthrobacter** (ar-thro-bac'ter), which change from bacilli to cocci as the culture ages. Although most bacteria divide by transverse binary fission, a few rod-shaped bacteria can divide by **budding,** to produce daughter cells of unequal size. A few bacteria have stalks or prostheca for attaching to surfaces. A **prostheca** is a cell-wall-limited, narrow extension of the cell's cytoplasm. Shape, size, prostheca, and budding are morphological characteristics that help microbiologists identify bacteria since these characteristics are genetically passed to succeeding generations.

Bacterial Motility

Bacterial cells with a helical or curved morphology are called spirilla, vibrios, or spirochetes. **Spirilla** are long, curved, or helix-shaped cells whose motility is generated by polar flagella. **Vibrios** are similar to spirilla except that they are much shorter; actually they are comma-shaped cells. Both spirilla and vibrios possess external flagella that extend from the cell surface into the adjacent medium. In contrast, spirochetes are helix-shaped cells that have **periplasmic flagella** (once called **endoflagella** or **axial filaments**) instead of external flagella (Figure 3.12). Periplasmic flagella lie along the protoplasmic cylinder of the spirochete and are responsible for its twisting spiral motility.

Sizes of Bacteria

The dimensions of a "medium-sized" rod-shaped bacterium are 0.6 μm wide by 1 to 3 μm long, whereas the diameter of a "medium-sized" spherical bacterium is about 1.0 μm (Table 3.2). The microbial world also contains bacteria that are too small to be seen clearly with the ordinary light microscope. *Bdellovibrio bacteriovorus* (del-lo-vib'ri-oh bac-ter-i-o-vor'us), a bacterium that is parasitic on other bacteria, rarely grows to be more than 1.0 μm in length. *Treponema pallidum* is too thin a spirochete to be seen with the brightfield microscope, but can be seen with a darkfield microscope. At the opposite end of the size spectrum are *Azotobacter chroococcum* (a-zo'toh-bac'ter chro-o-coc'-cum), a large, oblong soil bacterium, and *Beggiatoa,* which grows as long filaments composed of individual cells. Large bacteria can be 5 to 12 μm in their smallest dimension and can be as long as 100 μm (Table 3.2).

Simple Stains

Bacterial structures are stained with chemical dyes to make them more visible in the brightfield microscope. Procedures in which a single dye is used to stain a cell are called simple stains. Many of the useful bacterial stains are positively charged (basic) chemicals, such as crystal violet, basic fuchsin, and methylene blue. These basic dyes react with negatively charged bacterial components to color the cell.

Simple stains are used to stain whole cells or to stain specific cellular components. Sudan black, for example, is a lipid-soluble compound that stains lipid inclusions, which function as storage granules. Spores can be stained with malachite (mal'ah-kyt) green or carbolfuchsin (kar'bol-fook'sin), but only after significant heat is applied to enable the dye to penetrate the spore. Bacterial flagella can be stained with a basic dye in conjunction with the fixative tannic acid. Sufficient stain

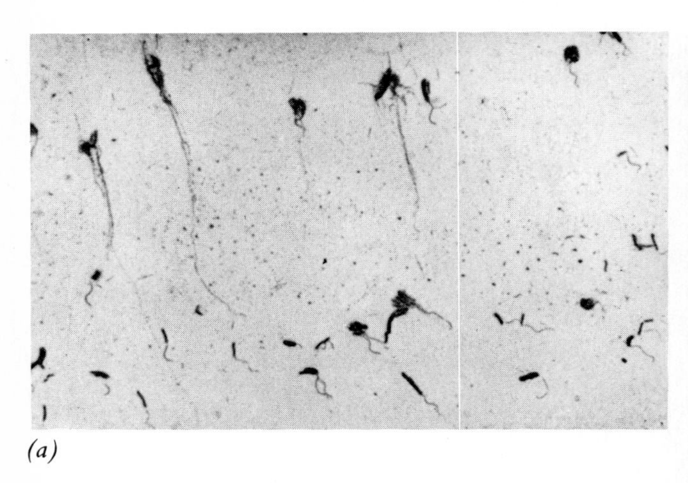

(a)

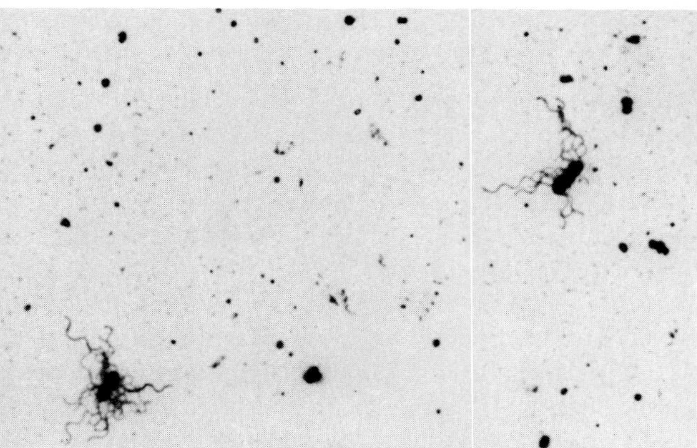

(b)

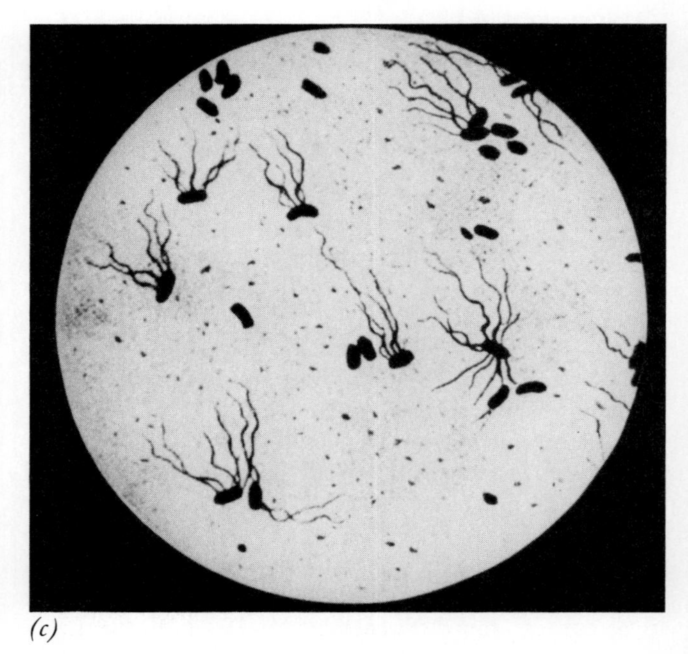

(c)

FIGURE 3.14 Flagella stains of *(a) Vibrio cholerae, (b) Clostridium novyi,* and *(c) Salmonella typhi.* (Courtesy of the Centers for Disease Control.)

accumulates on the flagella to make the filaments visible under the light microscope (Figure 3.14). Simple stains are also used to stain the background of the slide to produce an effect similar to a photographic negative. For example, India ink is used to **negative-stain** preparations of capsulated bacteria. Because the capsules surrounding the bacteria do not react with India ink, they appear as a halo around the cell and stand out against the dark India-ink background.

Differential Stains

Many chemical dyes stain all cells equally, whereas differential staining techniques are designed to distinguish between two or more distinct cell types. These stains usually involve the sequential application of more than one chemical dye and depend on structural differences between groups of bacteria. The Gram stain and the acid-fast stain are examples of differential stains.

Gram Stain Hans Christian Gram developed his differential staining procedure in 1884 as a means of identifying cocci in lung tissue taken from patients who had died of pneumonia. The **Gram stain** is used extensively today because it distinguishes between two cell types, designated gram-negative and gram-positive.

The bacteria to be stained are spread on a microscopic slide and air dried. The slide is then gently heated over a low flame to fix the bacteria to the slide so that they will not wash off during subsequent steps. A drop of crystal violet is applied to the smear for a short time before it is washed off the slide with water. Both cell types are stained blue by the crystal violet (Figure 3.15). Gram's iodine solution is then added to the slide. Gram's iodine acts as a mordant (fixative) by combining with the crystal violet, which becomes fixed inside gram-positive cells. The slide is again washed with water before it is dried and treated with an organic solvent (alcohol or acetone). This is called the decolorization step because the blue crystal violet–iodine complex is washed out of the gram-negative cells. The organic solvent does not decolorize the gram-positive cells, which remain blue. The slide is next counterstained with safranin.

Gram-positive cells remain blue-violet throughout this procedure, since they retain the crystal violet stain. Gram-negative cells initially stain blue, but lose the crystal violet when washed with the organic solvent. The safranin counterstain converts the colorless gram-negative cells to a red color that makes them visible under the light microscope. The Gram-stain reaction of a species is an important criterion in bacterial identification because it differentiates between two distinct types of bacterial cell walls (see Chapter 4). Most bacterial species are either gram-positive or gram-negative; however, some are normally gram-variable.

Steps	Gram-negative Cell	Gram-positive Cell
1. Cells are heated to fix them to the slide. All cells stain blue with crystal violet (1 min).		
2. Iodine is applied (1 min), resulting in no color change.		
3. Cells are washed with alcohol. Gram− cells are decolorized.		
4. Cells are counterstained with safranin. The colorless Gram− cells turn red.		

FIGURE 3.15 Steps in the Gram-stain procedure. In practice, gram-positive cells are blue (shaded) and gram-negative cells are red.

KEY POINT

The Gram stain of an unknown bacterium should be done with known controls on the same slide. This internal control is a check for variations in individual staining techniques. In addition, the physiological state of bacterial cells can alter their reaction to the Gram-stain procedure. Older cultures of gram-positive bacteria that have entered the stationary phase of growth often appear to be gram-negative in this staining reaction. This change occurs because the bacteria are no longer able to maintain intact cell-wall structures as the culture ages. Thus, practice and judgment are necessary for the proper use and interpretation of the Gram stain.

Acid-Fast Stain The **acid-fast stain** is a differential stain used in the identification of *Mycobacterium* species. Acid-fast bacteria contain wax-like material that binds a primary dye, such as carbol fuchsin, even when they are washed with acidified alcohol. In the Ziehl–Neelsen acid-fast stain, the cells are treated with hot carbolfuchsin. Nonacid-fast bacteria are initially stained with the red dye, but are decolorized when they are washed with acidified alcohol. The preparation is then counterstained with methylene blue, which stains the nonacid-fast bacteria blue. When an acid-fast stain is done on a sample of human sputum, the mycobacteria are red, while the tissue cells and nonacid-fast bacteria are blue. This differential stain is used to identify acid-fast bacteria including the Mycobacteria, Actinomycetes, and Nocardiae.

Summary Outline

METRIC UNITS OF MEASURE

1. The dimensions of bacteria are reported in micrometers; one micrometer equals one millionth of a meter.
2. Viruses and subcellular structures are measured in nanometers (10^{-9} meter), while macromolecules are measured in angstroms (10^{-10} meter).

OVERVIEW OF MICROSCOPY

1. Cytologists study the structure of cells by means of various types of light and electron microscopy.

THE LIGHT MICROSCOPE

1. The typical compound light microscope is composed of a condenser lens that collects the light rays and focuses them on the specimen; an objective lens that collects the light coming from the specimen; and an ocular lens that, together with the objective lens, magnifies the image.
2. The ocular lens (10×) and an oil-immersion objective lens (100×) combine to produce a magnification of 1000×.
3. Objective lenses can be corrected for spherical aberrations and chromatic aberrations.
4. The resolving power, which is the minimal distance between two points that can be distinguished by an observer, of the light microscope depends on the wavelength of light and the numerical aperture of the objective lens. A typical light microscope has a resolving power of 0.2 μm.
5. An oil-immersion lens has a higher numerical aperture than the typical high-dry lens and provides higher magnifications.
6. The phase-contrast microscope enables the microscopist to see transparent specimens by altering the phase of the light waves passing through the specimen. These phase shifts are used to increase the contrast between the living cell and its background.
7. Light diffracted by the organism creates the image observed by the microscopist using a darkfield microscope. Organisms appear as bright objects against a dark background. Darkfield illumination is also used in fluorescence microscopy.
8. Fluorescence microscopy uses the light emitted by fluorescent compounds to identify cells or microscopic structures.

ELECTRON MICROSCOPY

1. The transmission electron microscope is used to observe cellular structures and macromolecules as small as 5 to 8 Å. Sample preparations for TEM include fixing, staining with heavy metals, and thin sectioning.
2. The freeze-etching technique enables microscopists to observe surface structures at the macromolecular level. The carbon replica of the etched surface of a frozen specimen is viewed in the TEM.
3. Three-dimensional views of biological specimens can be obtained with the scanning electron microscope. The resolution of the SEM is only ten times greater than that of the light microscope.

APPLICATIONS OF THE LIGHT MICROSCOPE

1. Wet mounts and hanging-drop preparations are used to study living microorganisms.
2. The light microscope is used to observe gross bacterial morphology, including the arrangement of cells, cell shape, size, motility, and means of cellular reproduction.
3. Simple stains are used to increase contrast and to visualize specific cellular components, such as lipid inclusions, spores, and flagella. Negative stains reveal capsules.
4. Differential stains, such as the Gram stain and the acid-fast stain, distinguish between two or more distinct cell types.

Questions and Topics for Study and Discussion

QUESTIONS

1. What part of a meter is a micrometer; an angstrom; a nanometer? What are the size ranges for bacteria and viruses?
2. What factors affect the resolution of a light microscope?
3. Explain the meanings of the terms resolution and magnification.
4. Name a biological application for the brightfield, phase-contrast, darkfield, and fluorescence microscopes.
5. Compare the limitations and uses of the scanning electron microscope to those of the transmission electron microscope.
6. Describe the normal sizes and shapes of bacteria. How does the plane of division in a coccus influence its cellular arrangement?
7. What are the advantages of using an oil-immersion lens to observe bacteria?
8. Name and describe two simple stains and two differential stains. To a cytologist, what structural details about a cell will these stains reveal?

DISCUSSION TOPICS

1. How would you demonstrate the function of a structure you observe in the electron microscope, remembering that the material you are viewing is nonliving?

Further Readings

BOOKS

Ford, Brian J., *Single Lens: The Story of the Simple Microscope,* Harper and Row, New York, 1985. This history of the development of the microscope is brought to life by the author who in 1981 discovered some of Leeuwenhoek's original specimens. Highly recommended for anyone interested in microscopy and the origins of microbiology.

Gabriel, B. L., *Biological Electron Microscopy,* Van Nostrand Reinhold, New York, 1982. This is a step-by-step guide to the use of the electron microscope and the preparation of biological specimens.

Grimstone, A. V., *The Electron Microscope in Biology,* 2nd ed., Edward Arnold, London, 1976. A short paperback that presents the basic techniques for and applications of the electron microscope.

ARTICLES AND REVIEWS

Everhard, T. E., and T. L. Hayes, "The Scanning Electron Microscope," *Sci. Am.,* 226:54–69 (January 1972). An introduction to the technical design and experimental use of the scanning electron microscope.

C H A P T E R

4

Procaryotic Cells: Organization, Structure, and Function

OUTLINE

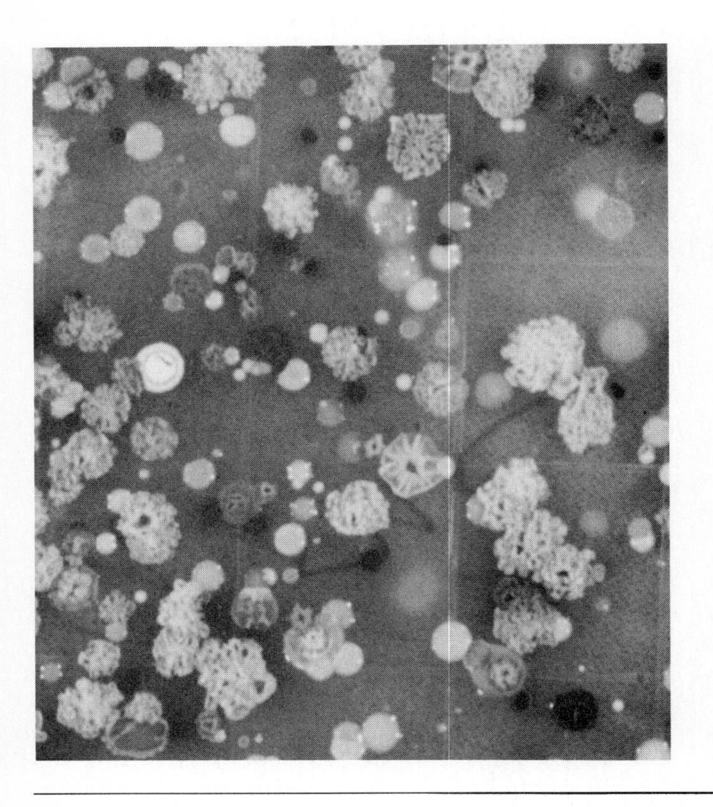

FOCUS OF THIS CHAPTER

Bacteria have a cellular organization and structure that separates them from all other living organisms. This chapter describes the structures present in procaryotic cells and explains how they contribute to the growth, reproduction, survivability, and diversity of bacteria.

Robert Hooke was the first microscopist to recognize the cell as a structural unit of living material. When he viewed a cross section of cork, he saw that it was divided into compartments separated by a thick wall. Long after Hooke's work, Matthias Schleiden (1804–1881) recognized the same compartmentalization of plant material and proposed in 1838 that all plants are composed of cells. The next year Theodor Schwann (1810–1882) expanded the cell theory to cover animals as well as plants. This unified cell theory remained intact until the development of the electron microscope, which demonstrated two distinct cell types—procaryotic cells and eucaryotic cells.

Animals, plants, fungi, and protozoa are known as **eucaryotes** because their cells contain a true nucleus. The nucleus is the sign of cellular complexity that marks the structural sophistication and evolutionary advancement of eucaryotic cells. Eucaryotes may have evolved from the bacteria, whose remains are found in fossils. For over two billion years bacteria appear to have been the only cells on earth. The bacteria retain a simple cell organization: they contain neither a nucleus nor other subcellular organelles and so are known as **procaryotes** (*pro* meaning "first" or "primitive"). Even with their simple cell organization, bacteria are capable of all the essential functions of life.

THE BACTERIAL CYTOPLASM

Every cell is surrounded by a cytoplasmic membrane that separates the interior of the cell from direct access to its external environment Figure 4.1. The cytoplasmic membrane is semipermeable, which means that it selectively serves as a barrier to the flow of nutrients into and out of the cell. The **cytoplasm** of a cell refers to all the structures and components inside this cell-limiting membrane. It is in the cytoplasm of a procaryotic cell that metabolism, protein synthesis, and DNA synthesis occur. The cytoplasmic structures that participate in these functions include the DNA that encodes the cell's genetic information, ribosomes, and the numerous enzymes involved in anabolic and catabolic reactions necessary for cellular growth and reproduction. A few bacteria contain membranes that penetrate the cytoplasm and participate in specific metabolic functions. Unlike eucaryotes, procaryotic cells do not have membrane-bound organelles such as the mitochondria and chloroplasts.

THE NUCLEAR REGION CONTAINS THE CELL'S DNA

During reproduction each cell replicates and transfers its genetic information to its progeny. The hereditary information of a cell is contained in DNA, the macromolecule whose chemical name is **deoxyribonucleic acid.** A bacterial cell's DNA occupies a major portion of its cytoplasm (Figure 4.1). Microscopists first observed the cellular location of bacterial DNA in cells stained with the DNA-specific Feulgen (foil′gen) reagent. The area of the cell that was stained was referred to as the **nuclear region** to distinguish it from the membrane-bound nucleus seen in eucaryotic cells. The nuclear region lacks a limiting membrane (Figure 4.1), and so it is in direct contact with the other components of the cell's cytoplasm.

Experiments have shown that bacteria contain a single, circular, double-stranded DNA molecule that is constantly being replicated in growing cells. Bacterial DNA is released from the cell (Figure 4.2) during cell lysis (cell disruption). The DNA released from *E. coli* has a circumference about 1000 times the length of the cell itself. This DNA molecule is called the **bacterial chromosome** since it contains all the genetic information necessary to perform the cell's life functions.

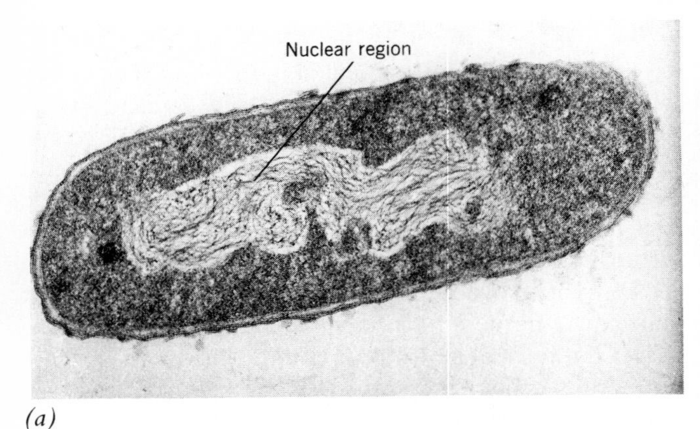

(a)

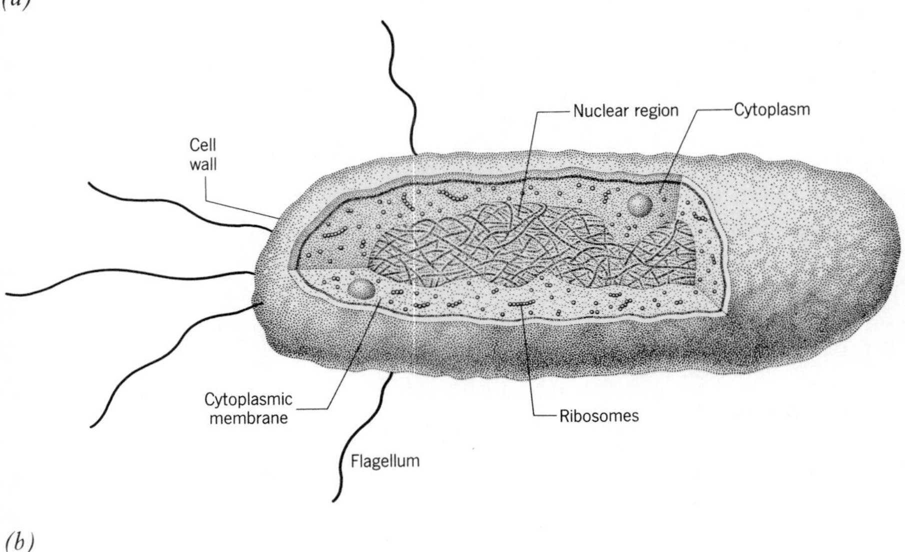

(b)

FIGURE 4.1 *(a)* Electron micrograph of *Escherichia coli* and *(b)* drawing of the key features of this bacterium. (Courtesy of Dr. G. Cohen-Bazire.)

The bacterial chromosome has been isolated as an intact structure from cells by gently disrupting the cell (Figure 4.3). This isolated material is 80 percent DNA, 10 percent RNA, and 10 percent protein. Most of the protein is the enzyme that synthesizes RNA from the DNA template. The chromosome isolated in this fashion is called the **nucleoid** (noo′klee-oid), a term that has replaced other descriptive terminology for the structure of the bacterial nuclear material.

The isolated nucleoid (Figure 4.3) is supercoiled upon itself and contains numerous loops. Experiments suggest that the bacterial chromosome exists in this nucleoid structure inside the cell, that it is organized into at least 43 ± 10 loops of supercoiled DNA, and that the DNA is attached to the cell membrane at strategic locations during replication. This attachment is proposed to separate the two copies of the replicated DNA into the daughter cells. (The complex process of DNA synthesis and the function of the nucleoid are discussed further in Chapter 9.)

In addition to the bacterial chromosome, many bacteria possess one or more small pieces of DNA called **plasmids.** These molecules are physically separate from, and replicate independently from, the cell's chromosome. Plasmids are nonessential to the viability of a cell, since cells can lose their plasmid(s) and continue to grow and reproduce. Plasmids often carry useful genetic information, for example plasmids code for antibiotic resistance, degradative enzymes, or virulence factors (see Chapter 10). Bacterial plasmids influence

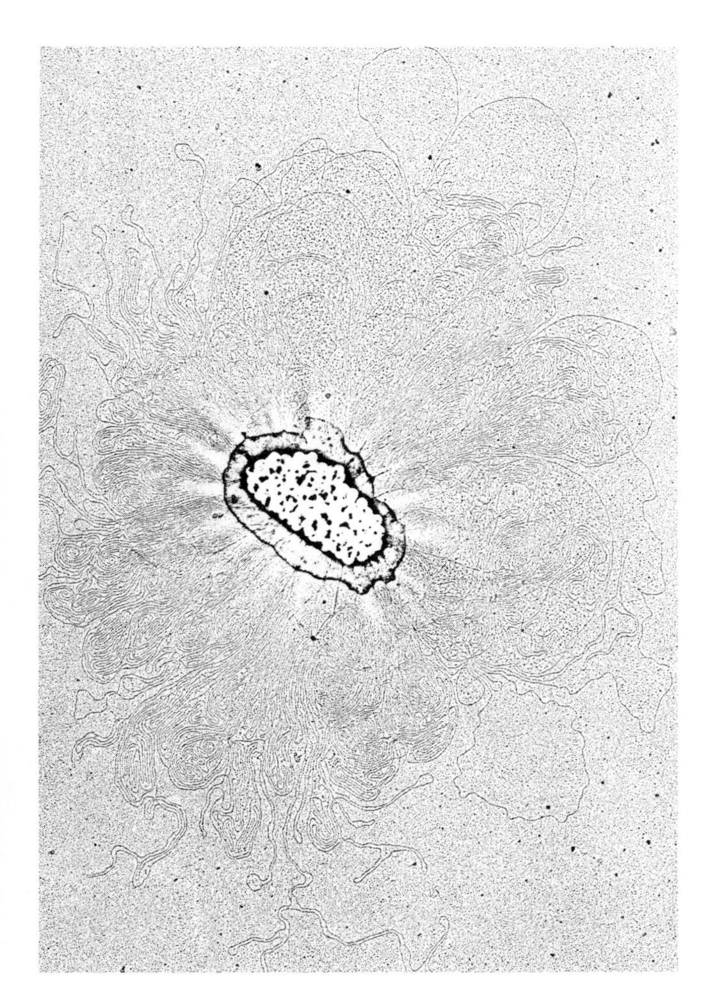

FIGURE 4.2 Electron micrograph of bacterial DNA released from a lysed cell of *Haemophilus* spp. The bacterial chromosome is a closed circular DNA molecule that is approximately 1 mm long. (Courtesy of Dr. L. A. MacHattie. From L. A. MacHattie, K. I. Berns, and C. A. Thomas, Jr., *J. Mol. Biol.*, 11:648–649, 1965, Copyright © Academic Press, London, Ltd.)

many aspects of microbiology, having an impact on the spread of infectious diseases and antibiotic resistance, decontamination of environmental pollutants, and genetic engineering.

KEY POINT

Bacteria contain one copy of their essential genetic information located on a single, circular chromosome.

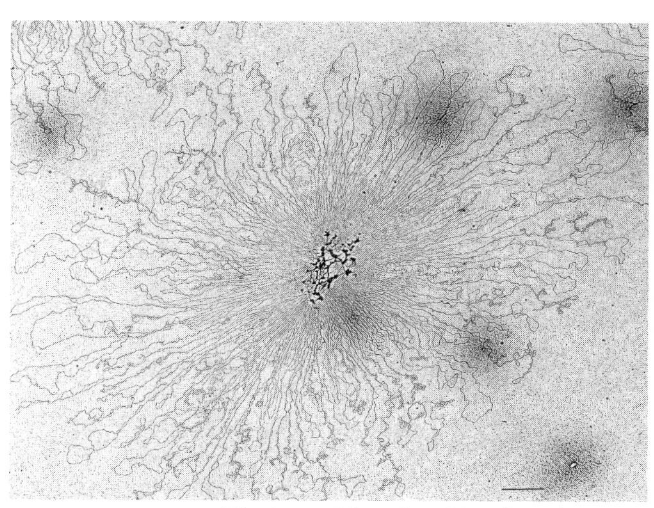

FIGURE 4.3 The bacterial nucleoid is released from procaryotic cells upon gentle rupture of the cell. The nucleoid is 80 percent DNA arranged in numerous loops, 10 percent RNA, and 10 percent protein. Bar equals 1 μm. (Courtesy of Dr. Ruth Kavenoff, Designergenes Ltd./BPS.)

RIBOSOMES ARE THE SITES OF PROTEIN SYNTHESIS

The sequences of amino acids found in proteins are ordered by the directions contained in DNA. Protein synthesis begins when information in the DNA is copied into RNA messages. These messages are continuously produced by the cell in the form of **messenger RNA** molecules. The final step in protein synthesis is the translation of the messenger RNA into protein.

Protein synthesis occurs in the bacterial cytoplasm on small particles called **ribosomes.** Ribosomes are composed of protein (40 percent) and ribosomal RNA (60 percent). Bacterial ribosomes exist free in the cytoplasm, are rarely attached to membranes, and are easily isolated from disrupted cells (Figure 4.4).

The structure of isolated ribosomes has been analyzed by ultracentrifugation. The sedimentation of macromolecules and small particles in high-speed centrifuge rotors is related to their size and shape. As the rotor spins, tremendous centrifugal forces are produced. These forces affect the macromolecules or small particles in a spinning centrifuge tube and accelerate them toward the bottom of the tube. The rate at which particles sediment is measured as a **Svedberg unit** (S), which reflects the particles' size and shape.

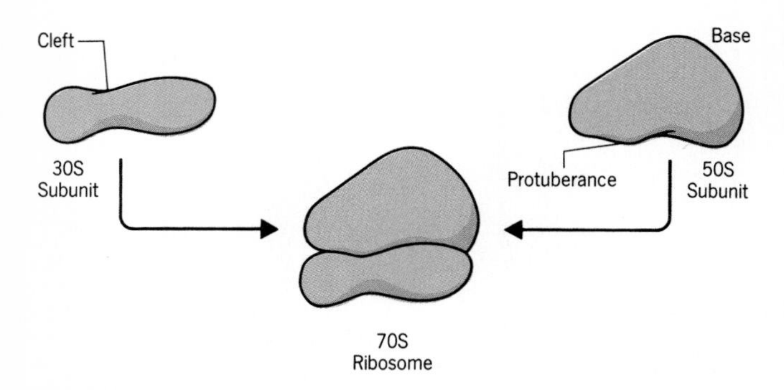

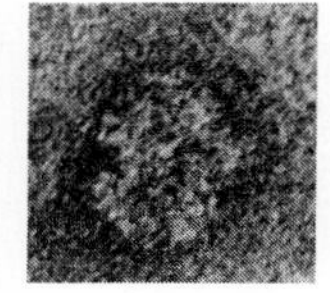

FIGURE 4.4 Structure of the bacterial ribosome. Insert, electron micrograph of an isolated 70S bacterial ribosome. (Courtesy of Dr. Miloslav Boublik.)

Bacterial ribosomes sediment in such a centrifuge as 70S particles. These 70S particles are actually made of two subunits, a 30S subunit and a 50S subunit (Figure 4.4). Since Svedberg units relate to both shape and size, they cannot be added arithmetically (30S + 50S ≠ 70S).

Isolated ribosomes have been chemically analyzed and found to contain protein and RNA. The 30S subunit of the bacterial ribosome is composed of one 16S ribosomal RNA molecule and 21 polypeptides (Figure 4.4). The 50S subunit is composed of 31 polypeptides that combine with a 5S and a 23S ribosomal RNA molecule. When these components are all present, the subunits self-assemble into their respective structures. In the cell cytosol (the fluid portion of a ruptured cell) there is a pool of 30S and 50S subunits that sequentially bind to messenger RNA during the initiation of protein synthesis. Often more than one ribosome attaches to a messenger RNA molecule to create an mRNA-ribosome complex called a **polysome.** Polysomes are found in the bacterial cytoplasm in close proximity to the nucleoid. In fact, the ribosomes can attach to one end of the messenger RNA while the other end is being made on the DNA.

KEY POINT

Bacterial ribosomes are smaller than the ribosomes responsible for protein synthesis in animal cells. The functional consequence of this is that a number of antibiotics inhibit protein synthesis on bacterial ribosomes without killing animal cells. Antibiotics that act by inhibiting protein synthesis in this way include erythromycin and the tetracyclines.

INCLUSION BODIES PRESENT IN THE BACTERIAL CYTOPLASM

Storage products of bacterial metabolism are often sequestered in cytoplasmic granules. The presence of these granules is related to the nutrients available to the cell. Some granules function as storage depots for energy; some contain specific proteins; and gas vesicles provide bouyancy for cells living in aquatic environments. Bacterial inclusion bodies are never enclosed by a bilayer membrane; however, some inclusion bodies are surrounded by a membrane-like protein layer.

Polyhydroxybutyrate (PHB), cyanophycin, and glycogen are substances bacteria store in inclusion bodies (Figure 4.5). **Polyhydroxybutyrate** (pol-ly-hy-droxy-bew-ty'rayt) exists as a polymer of hydroxybutyrate and is deposited in cytoplasmic granules that stain with lipid dyes. Under certain growth conditions, up to 50 percent of a cell's mass can be comprised of PHB. PHB is a reserve material that helps the cell to survive exposure to low-nutrient (starvation) environments. **Glycogen** is a highly branched polymer of glucose that cells use to store carbohydrate. Cyanobacteria produce **cyanophycin,** a polymer of the amino acids arginine and aspartic acid. Cyanophycin is deposited in cytoplasmic granules, and functions as a nitrogen reserve for the cell. These granules are surrounded by a membrane-like layer and are clearly seen with the phase-contrast microscope.

Many bacteria can also accumulate and store phosphate or polyphosphate in inclusion bodies known as **metachromatic granules** (Figure 4.5*b*). These granules derive their name from the fact that they change toluidine blue to a reddish violet color (*metachromatic,* able to change color). Cytologists speculate that the polyphosphate in metachromatic granules functions as a phosphate reserve for cell synthesis. Certain bacteria gain energy by oxidizing hydrogen sulfide (H_2S) to molecular sulfur (S), which they deposit in cytoplasmic granules. After the environmental H_2S is depleted, the cells oxidize the deposited sulfur to sulfate.

Gas vesicles are spindle-shaped hollow structures present in the cytoplasm of numerous bacteria that live in aquatic environments (Figure 4.5*c*). Each vesicle is

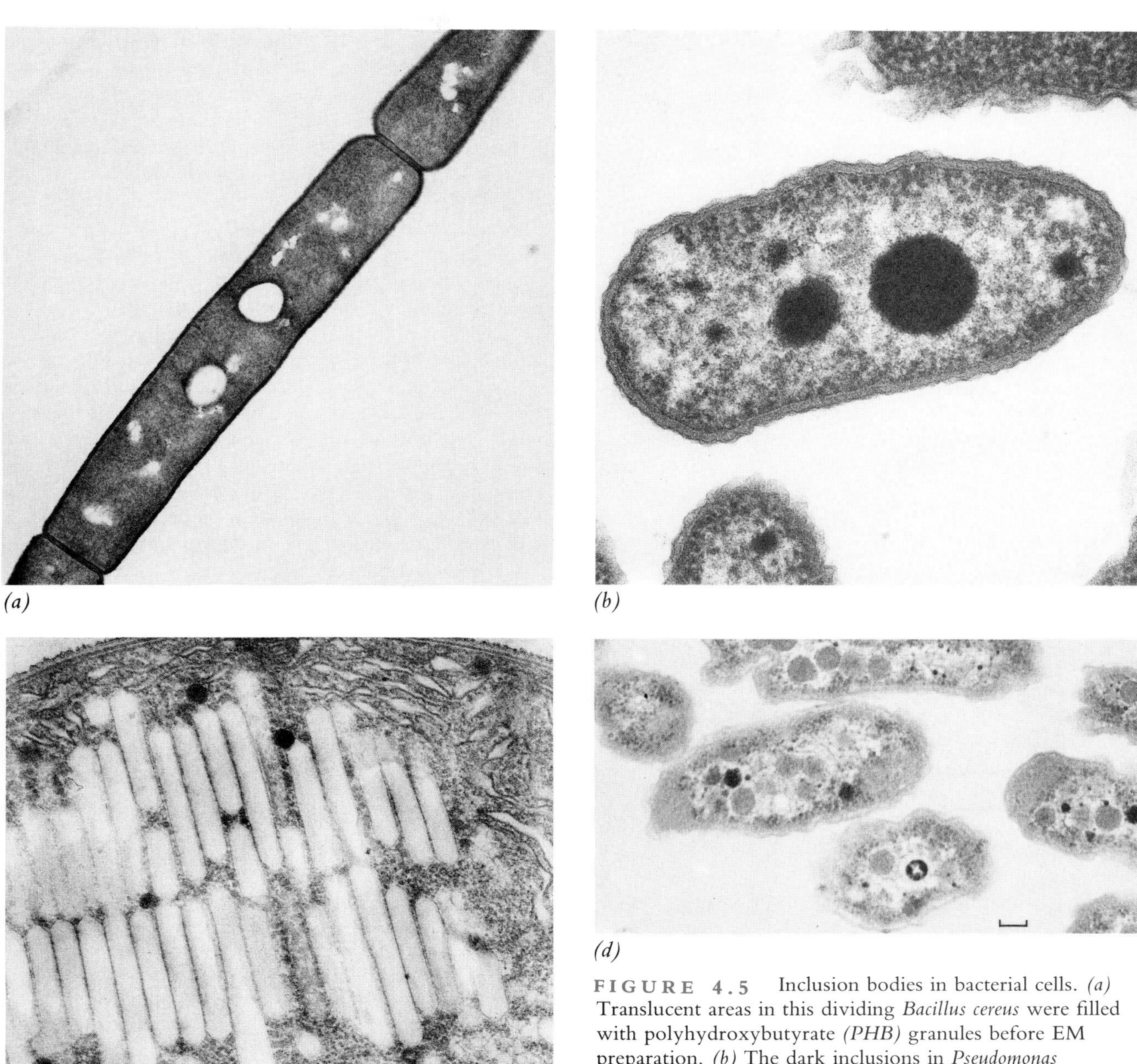

(a) *(b)* *(c)* *(d)*

FIGURE 4.5 Inclusion bodies in bacterial cells. *(a)* Translucent areas in this dividing *Bacillus cereus* were filled with polyhydroxybutyrate *(PHB)* granules before EM preparation. *(b)* The dark inclusions in *Pseudomonas aeruginosa* are polyphosphate bodies. *(c)* Electron micrograph of gas vacuoles inside the cyanobacterium, *Nostoc carneum*. *(d)* Carboxysomes in *Thiobacillus*. (Courtesy of *(a, b)* Dr. T. J. Beveridge/BPS, *(c)* Dr. Thomas E. Jensen, *(d)* Dr. J. M. Shiveley.)

composed of a series of hollow cylinders with a constant diameter of 75 nm and lengths between 200 and 1000 nm. Gas vesicles have conical caps at both ends. The vesicles have an outer protein layer that keeps out water and ions, but permits gases to enter. The vesicles filled with gas give the cell bouyancy, causing it to rise vertically in the water column. Gas vesicles are found in cyanobacteria, photosynthetic bacteria, and in certain chemotrophic bacteria that live in aquatic environments.

Carboxysomes are polyhedral inclusion bodies observed in organisms that depend on CO_2 as a sole carbon source. Carboxysomes (Figure 4.5*d*) are packed with protein granules and are surrounded by a membrane-like layer. The carboxysome is the location of ribulose 1,5-diphosphate carboxylase, an enzyme essential to the conversion of CO_2 to organic compounds.

BACTERIAL MEMBRANES AND CELL WALLS

One can think of the cytoplasm as the contents of a sac that is enclosed by a cell envelope. The bacterial **cell envelope** is composed of the cytoplasmic membrane and cell wall. Bacterial cell walls lie outside the cytoplasmic membrane and contain a rigid layer that protects the fragile cytoplasmic membrane from rupturing and maintains the cell's shape. Passing through the cell wall are surface protrusions such as flagella and pili. The cytoplasmic membrane, cell wall, and the attached protrusions are functionally very important to bacteria because these structures link the metabolic machinery in the cell's cytoplasm to the environment in which it must survive.

THE CYTOPLASMIC MEMBRANE

All cells are surrounded by a semipermeable membrane known as the **cytoplasmic membrane.** This membrane acts as a physical barrier between the cell's cytoplasm and the external environment and selectively controls the movement of substances into and out of the cell. Thus it is said to be semipermeable. The cytoplasmic membrane has a unit-membrane structure, as can be seen in the electron micrographs of *Escherichia coli* (Figure 4.1). The **unit membrane** appears in stained and fixed cells as a double-track structure, 7 to 8 nm wide, that consists of a light central core sandwiched between two dark bands.

The chemical composition of the cytoplasmic membrane is approximately 70 percent protein and 30 percent lipid by mass. In some bacteria, such as *Escherichia coli,* all the lipid in the cytoplasmic membrane is phospholipid. The hydrophobic fatty acyl ends of the phospholipids interact in the center layers of the membrane, whereas the hydrophilic ends (charged phosphate groups) are aligned on the two surfaces of the membrane (Figure 4.6), where they interact with the aqueous cytosol or with the external environment.

The sidedness (inside–outside) of a membrane is determined by the membrane proteins. Proteins that protrude on the inside of the membrane include the catalytic portion of the enzyme ATPase that is involved in energy transformations. Other proteins protrude on the cell-wall side of the membrane, and some are buried inside the bilipid layer to form pores or channels for solutes to flow through. According to the fluid–mosaic membrane model of biological membranes, there is no fixed position for these proteins in the two-dimensional architecture of the membrane. Each protein or protein complex has some lateral movement and some can alter their inside–outside orientation within the lipid bilayer. As a class, membrane proteins perform a variety of tasks. They are involved in energy conversion, the transportation of ions and metabolites into and out of the cell, and in the synthesis of extracellular structures.

The Signal Hypothesis for Membrane Protein Synthesis

What makes a membrane protein different from a cytoplasmic protein? It appears that membrane proteins have a unique *N*-terminal amino acid sequence that "instructs" the cell to insert this protein into the membrane. The synthesis of these proteins occurs on membrane-bound ribosomes with the *N*-terminal signal sequence always being synthesized first (Figure 4.7). There are two basic models that explain what happens next. Model 1, developed from experiments with eucaryotic cells, proposes that the *N*-terminal end of the protein is translocated across the membrane as soon as it is synthesized. The signal sequence is then cleaved off by an enzyme located on the opposite side of the

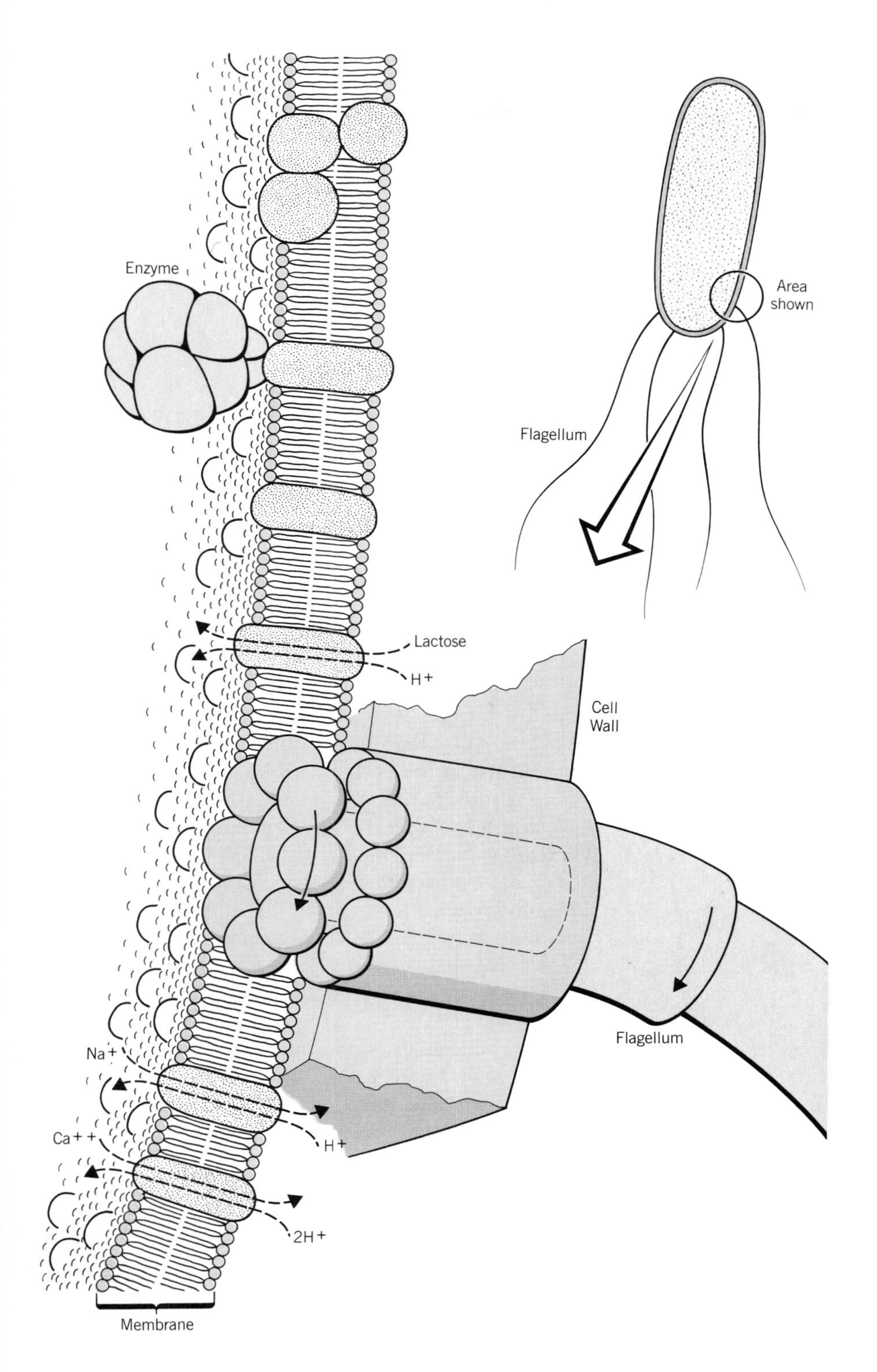

FIGURE 4.6 Diagram of a bacterial membrane showing the flagella atttachment, membrane proteins responsible for transport and enzymatic reactions, and the phospholipid bilayer. (From Peter C. Hinkel and Richard E. McCarty, "How Cells Make ATP," *Sci. Am.*, 232:104–123, 1978, copyright © 1978 Scientific American, Inc. All rights reserved.)

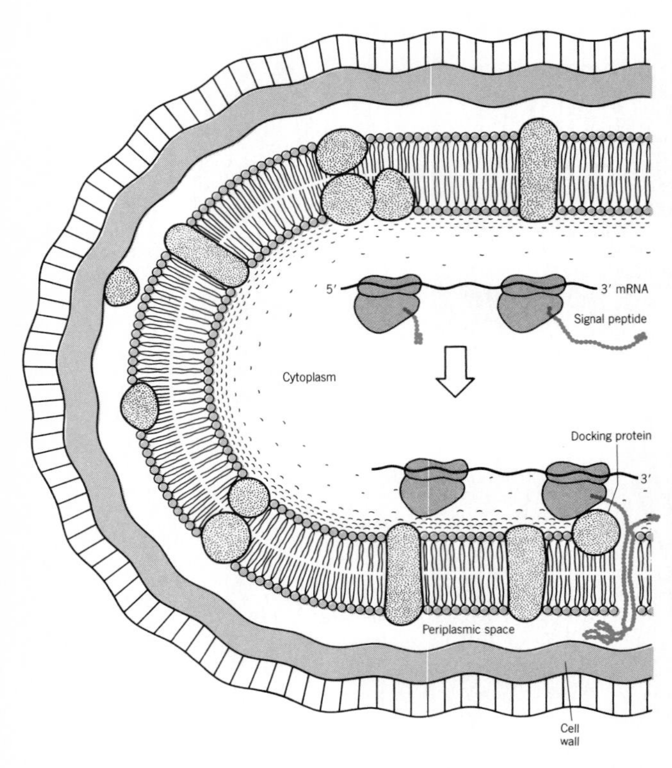

FIGURE 4.7 The signal hypothesis explains how bacterial proteins are transported across the cytoplasmic membrane or integrated into it. Proteins to be translocated have a signal sequence on their amino-terminal end that binds to a docking protein on the inner surface of the cytoplasmic membrane. As the polypeptide is synthesized, it is translocated through the membrane.

membrane (side away from protein synthesis). Model 2, based on experiments with *Escherichia coli,* proposes that the signal sequence binds to the membrane near the site of synthesis and that only a loop of the polypeptide is translocated. In this model, the *N*-terminal sequence is cleaved by an enzyme located on the inside (cytoplasmic side) of the membrane after most of the polypeptide has been translocated. Both models emphasize the importance of the signal sequence, which explains why membrane proteins are different from proteins in the cytoplasm.

Certain membrane proteins are never completely translocated; instead, they protrude from both sides of the membrane. This phenomenon is explained by the presence of amino acid sequences in the protein that trigger a stop to the translocation process. Membrane proteins with a stop signal in the middle of their polypeptide chain have part of the polymer translocated, while the remainder stays on the cytoplasmic side of the membrane. The signal hypothesis provides an explanation of how proteins synthesized in the cell's cytoplasm can function as cell membrane proteins or be exported outside the cell.

Membrane Are Selectively Permeable to Small Molecules

Cytoplasmic membranes are semipermeable or selectively permeable barriers between the inside and the outside of the cell. This barrier prevents cellular metabolites from leaking out of the cell and gives the cell control over which molecules enter its cytoplasm. There are three processes that govern how molecules enter a cell: simple diffusion, facilitated diffusion, and active transport (Figure 4.8).

Simple diffusion is the random movement of molecules from a location of high concentration to one of low concentration. For example, perfume permeates a room by diffusion; a drop of dye in a flask of water will soon color the entire solution. In general, molecules diffuse across cell membranes if they are small and have no charge. Oxygen gas (O_2), nitrogen gas (N_2), ammonia (NH_3), and water diffuse into cells, whereas charged small molecules such as sodium (Na^+) and potassium (K^+) ions do not. The diffusing substances are not concentrated within the cell, instead they tend to establish equal concentrations on both sides of the membrane (Figure 4.8*a*). The rate of diffusion depends largely on the temperature of the system.

Larger molecules and charged molecules (sugars and amino acids) enter cells only by specific transport mechanisms. **Facilitated diffusion** is mediated by a membrane-bound carrier protein (Figure 4.8*b*) that is specific for the transported molecule. Like simple diffusion, the transported molecule is moved from the side of high concentration to the side of lower concentration, but is not concentrated within the cell. The membrane-bound carrier protein (sometimes called a permease) specifically binds with the nutrient or ion to be transported, very similar to the enzyme-substrate

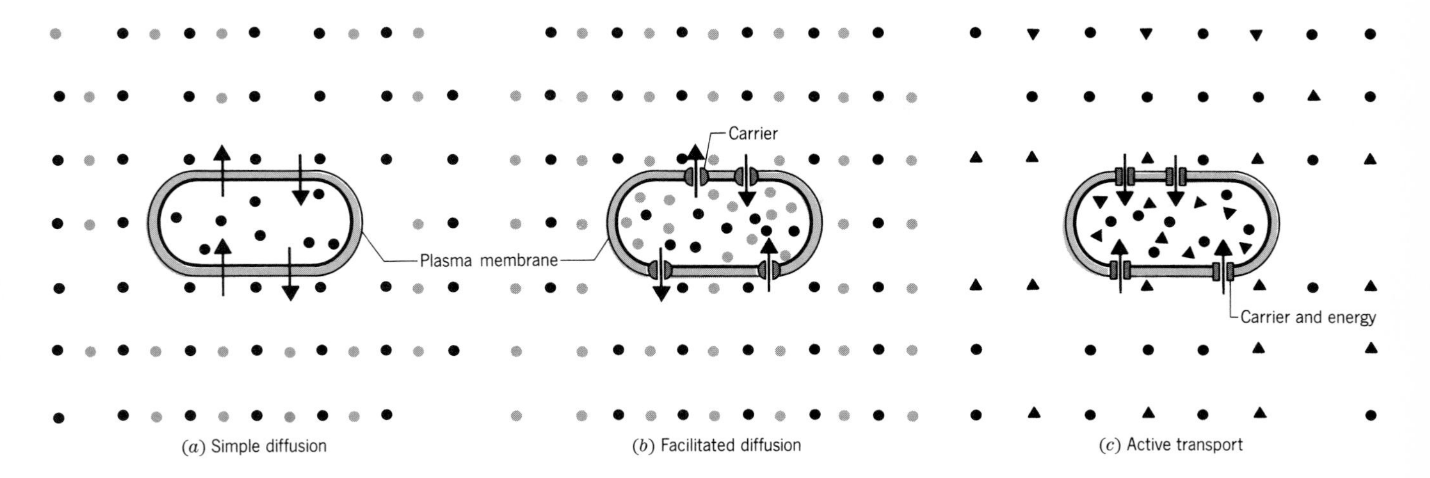

FIGURE 4.8 Compounds are transported across biological membranes by *(a)* simple diffusion, *(b)* facilitated diffusion, and *(c)* active transport. Small uncharged molecules enter cells by simple diffusion. Facilitated diffusion and active transport selectively transport molecules that bind to membrane-bound carrier proteins or permeases. Active-transport systems can concentrate a nutrient inside the cell at the expense of energy.

binding involved in enzyme-catalyzed chemical reactions. Although common in eucaryotes, facilitated diffusion is rarely encountered in procaryotes. Glycerol is the only compound known to be transported by this mechanism in *E. coli*.

A cell uses active transport to build up the cytoplasmic concentration of specific molecules. **Active transport** is mediated by a membrane-bound carrier protein (permease) that specifically binds and transports a molecule against the established concentration gradient, with the expenditure of energy (Figure 4.8*c*). The energy for active transport comes either from the hydrolysis of ATP or another source of phosphate bond energy, or from the proton motive force described in Chapter 6.

Membrane Integrity

Water molecules move across a semipermeable membrane from the region of low solute concentration to the region of high solute concentration by the process of **osmosis** (Figure 4.9). The influx of water creates pressure on the cytoplasmic membrane known as **osmotic pressure.** Many cells live in **hypotonic** environments, where the solute concentration is lower than the solute concentration in the cell, so they must deal with the influx of water and the resulting osmotic pressure. The rigid cell wall of most bacteria resists this pressure and protects the integrity of the plasma membrane.

Bacterial cells that lack a cell wall (protoplasts, L-forms) tend to accumulate water and burst unless they are placed in an isotonic environment. The concentration of solutes in an **isotonic** environment is equal to the concentration of solutes within the cell (Figure 4.9*a*). Under these conditions there is no net movement of water into the cell. Isotonic media are used to preserve the integrity of bacteria that lack cell walls.

Cells placed in solutions with high solute concentrations, a **hypertonic** environment, undergo plasmolysis. As the cell loses water, its cytoplasm shrinks and it becomes unable to grow (Figure 4.9*b*). **Plasmolysis** created by hypertonic solutions has practical applications for food preservation. If salt is added to red meat or fish, or fruits are packed in syrup, their storage life is increased greatly. Because of plasmolysis, most spoilage bacteria are unable to grow in solutions with a high solute concentration.

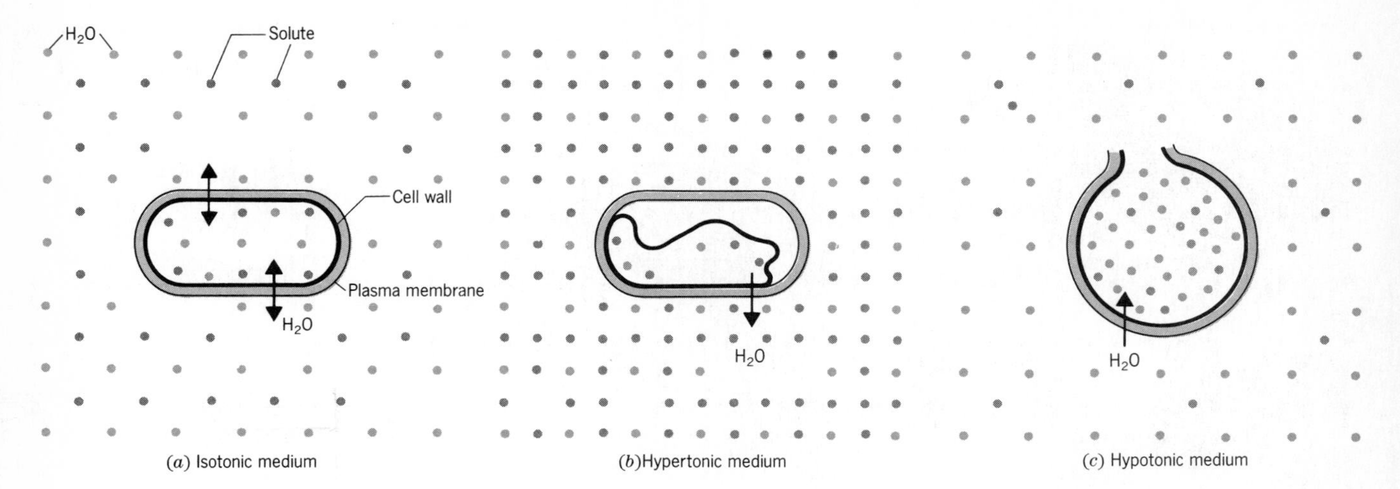

FIGURE 4.9 Water flow across a semipermeable membrane depends on the relative concentration of solutes on the inside and outside of the plasma membrane. In an isotonic medium *(a)*, the solute concentration inside the cell is the same as it is outside the membrane. In a hypertonic medium *(b)*, the outside solute concentration is greater than that of the cell and water leaves the cell. Very low outside solute concentration *(c)* creates a hypotonic medium and water enters the cell creating osmotic pressure, that under certain circumstances can cause the cell to burst.

Special Membrane Functions

Certain chemical reactions occur only on cell membranes. Since the surface area of the cytoplasmic membrane is fixed by the cell's size and shape, bacteria that need additional membrane area must produce membranes that intrude into the cytoplasm. The intracytoplasmic membranes are associated with specific physiological functions and in most bacteria are continuous with the cytoplasmic membrane.

Photosynthetic bacteria have intracytoplasmic membranes that are the site of the light reactions of photosynthesis. They are called bacterial photosynthetic membranes because they contain the bacterial carotenoids and bacteriochlorophylls. The morphology of photosynthetic membranes varies among bacterial species. The green photosynthetic bacteria contain **chlorosomes** (klo′roh-soams), which are independent vesicles in the cytoplasm containing photosynthetic pigments. Other photosynthetic bacteria contain photosynthetic pigments in invaginations (infolding) of their cytoplasmic membrane, which often take the form of lamellae (stacks of membranes), tubes, constricted tubes, or spheres (Figure 4.10). Spherical photosynthetic membranes called **chromatophores** appear as round vesicles in the cells cytoplasm and are released as independent vesicles upon cell lysis. Most cyanobacteria contain an extensive cytomembrane system of flattened sacs, called thylakoids, that are independent of the cytoplasmic membrane (Figure 4.10). Photosynthetic bacteria are discussed in greater detail in Chapter 14.

Chemolithotrophs are aerobic bacteria that oxidize inorganic compounds as their source of energy. The chemolithotrophs display a wide variety of extensive intracytoplasmic membranes (Figure 4.11). Isolated membranes from selected chemolithotrophs are rich in specialized proteins called cytochromes. These membranes probably function in energy-transforming reactions. The architecture of intracytoplasmic membranes in the chemolithotrophs is as varied as the photosynthetic membranes of the photosynthetic bacteria. In chemolithotrophs the intracytoplasmic membranes can grow into the cytoplasm (invaginate) from the cytoplasmic membrane, can exist free within the cytoplasm, or can intersect the cell and divide it into compartments (Figure 4.11).

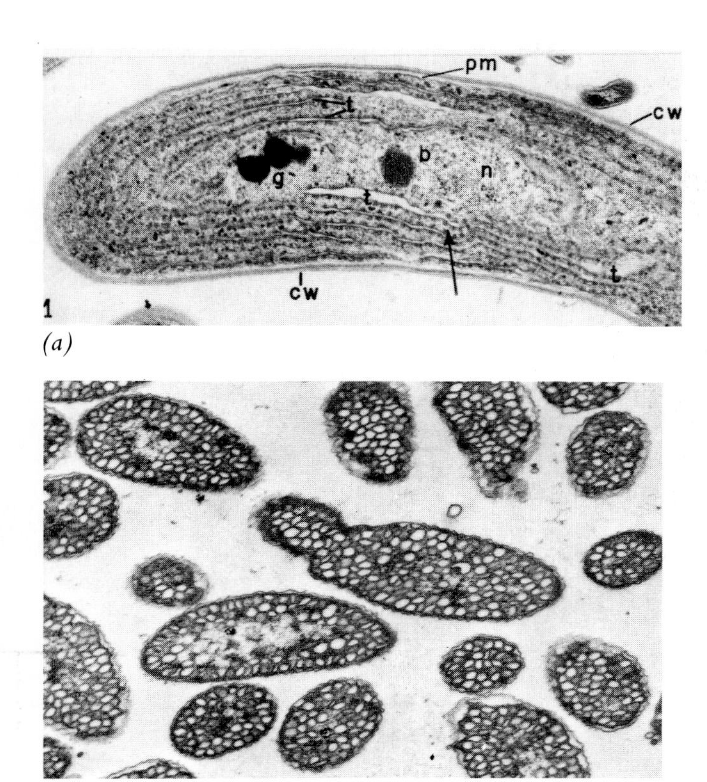

FIGURE 4.10 Photosynthetic bacteria have specialized intracytoplasmic membranes that function in photosynthesis. *(a)* A thin sectin of the cyanobacterium, *Synechococcus* spp. showing its photosynthetic (thylakoid) membranes (t), cell wall (cw), cytoplasmic membrane (pm), and nuclear material (n). (Courtesy of Dr. Stanley C. Holt.) *(b)* The chromatophores of *Rhodospirillum rubrum* contain carotenoids and bacteriochlorophyll pigments. (Courtesy of Dr. S. C. Holt. From S. C. Holt and A. G. Marr, *J. Bacteriol.*, 89:1402–1412, 1965, with permission of the American Society for Microbiology.)

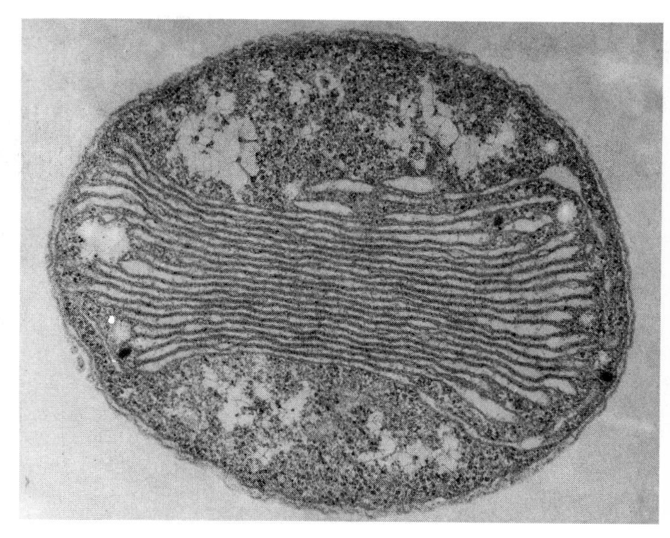

FIGURE 4.11 *Nitrosococcus oceanus* is an example of a chemolithotrophic bacterium that possess extensive intracytoplasmic membranes. The cytochrome-rich membrane system is involved in the oxidation of ammonia to nitrite. (Courtesy of Dr. S. W. Watson. From S. W. Watson and C. C. Remsen, *J. Ultrastruct. Res.*, 33:148–160, 1970, with permission of Academic Press, New York.)

Mesosomes are invaginations of the cytoplasmic membrane (Figure 4.12) that occur in many different bacteria. They are more prevalent in gram-positive bacteria than in gram-negative bacteria, but they are seen in both. Mesosomes are most often associated with cross-wall (septum) formation. The function of mesosomes is unknown; however, there is circumstantial evidence that DNA attaches to the mesosome during cell division. Therefore, mesosomes may be involved in the partitioning of DNA between daughter cells.

A word of caution is in order because bacterial cytologists do not agree on the significance or function of mesosomes. Mesosomes are not seen in all bacterial cells, and they have been observed only in bacterial cells subjected to chemical treatments. This had led some cytologists to postulate that mesosomes are artifacts of the chemical fixation process used to prepare bacteria for electron microscopy.

In all bacteria, the cytoplasmic membrane is the demarcation layer between the inside and the outside of the cell. This membrane can be the site of respiratory enzymes (if present), the generator of intracytoplasmic membranes, and the semipermeable barrier that physiologically limits the cell. Some bacteria contain no structures exterior to the cytoplasmic membrane, but such a condition is the exception. Most bacteria contain a sequence of complex layers external to the cell mem-

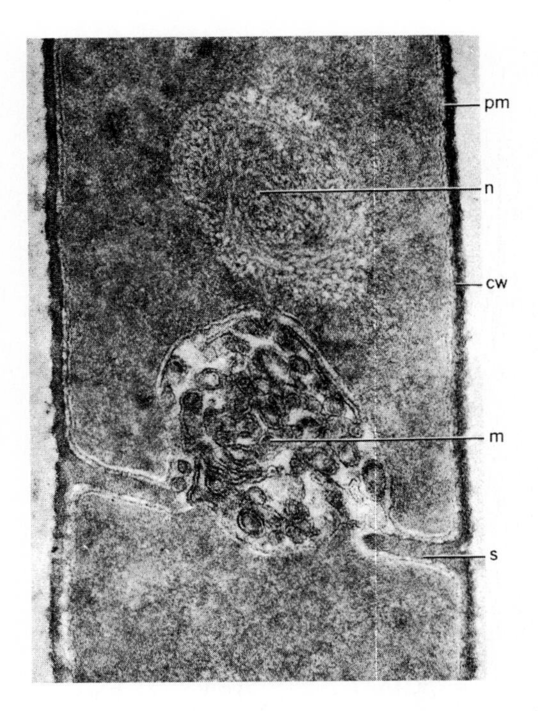

FIGURE 4.12 Electron micrograph of *Bacillus subtilis* showing its mesosome (m), nuclear region (nr), cell wall (cw), septum (s), and plasma membrane (pm). (From P. Sheeler and D. Bianchi, *Cell Biology*, Wiley, New York, 1980.)

brane that protect the cell from external influences and provide rigidity to the cell's shape. These layers comprise the cell wall.

BACTERIAL CELL WALLS

The cell walls of bacteria are chemically complex, protective structures that surround the cell. Although bacterial cell walls serve as a permeability barrier against large molecules, their main function is to constrain the cell so that it doesn't explode from a buildup of osmotic pressure. The rigidity of the cell wall resists osmotic pressure and contributes to the shape of the cell. The cell wall is vulnerable to certain antibiotics that inhibit cell-wall synthesis.

Bacterial cell walls can be divided into three major groups: cell walls of gram-negative bacteria, gram-positive bacteria, and archaebacteria. The basic morphology and chemical structure of the cell walls of gram-positive and gram-negative bacteria are much more constant than are the cell walls of the archaebacteria.

Overview of Bacterial Cell-Wall Structure

Gram-positive bacteria have a thick cell wall that appears to be homogeneous and lies exterior to their cytoplasmic membrane (Figure 4.13). The thick homogeneous layer is composed of a carbohydrate–peptide polymer known as the **mucocomplex** (also referred to as the murein or the mucopeptide layer). The mucocomplex layer in gram-positive cells rests directly on the outer surface of the cytoplasmic membrane. Teichoic (ty-ko'ic) acids penetrate the cell wall and attach to the mucocomplex and to the cytoplasmic membrane.

In contrast, the cell walls of gram-negative bacteria are multilayered. The mucocomplex in gram-negative bacteria is very thin and lies above (not on) the cytoplasmic membrane. Connections between the cytoplasmic membrane and the mucocomplex occur sporadically at sites of adhesion. Beyond the mucocomplex is the **outer membrane** (also called the outer cell wall) that appears to have a unit-membrane structure (Figure 4.13). It is permeable to small molecules, but is impermeable to larger molecules. The area between the cytoplasmic membrane and the outer membrane is functionally defined as the **periplasmic** (per-ee-plas'mic) **space.** The concept of a periplasmic space applies only to gram-negative bacteria since gram-positive bacteria have no outer membrane in their cell-wall.

Peptidoglycan (pep-tid-oh-gly'kan) is the repeating macromolecule that comprises the mucocomplex. **Glycan** refers to the carbohydrate backbone of peptidoglycan and is made of *N*-acetylmuramic acid joined to *N*-acetylglucosamine through a beta (β) 1—4 glycosidic bond (Figure 4.14). Adjacent polymers of the glycan are joined by peptide bridges that attach to the free carboxyl group on *N*-acetylmuramic acid. This peptide contains four or more different amino acids including D-alanine, D-glutamic acid, and one diamino amino acid (either diaminopimelic acid or lysine). The composition of the peptidoglycan is unique to bacteria because *N*-acetylmuramic acid is never found in eucary-

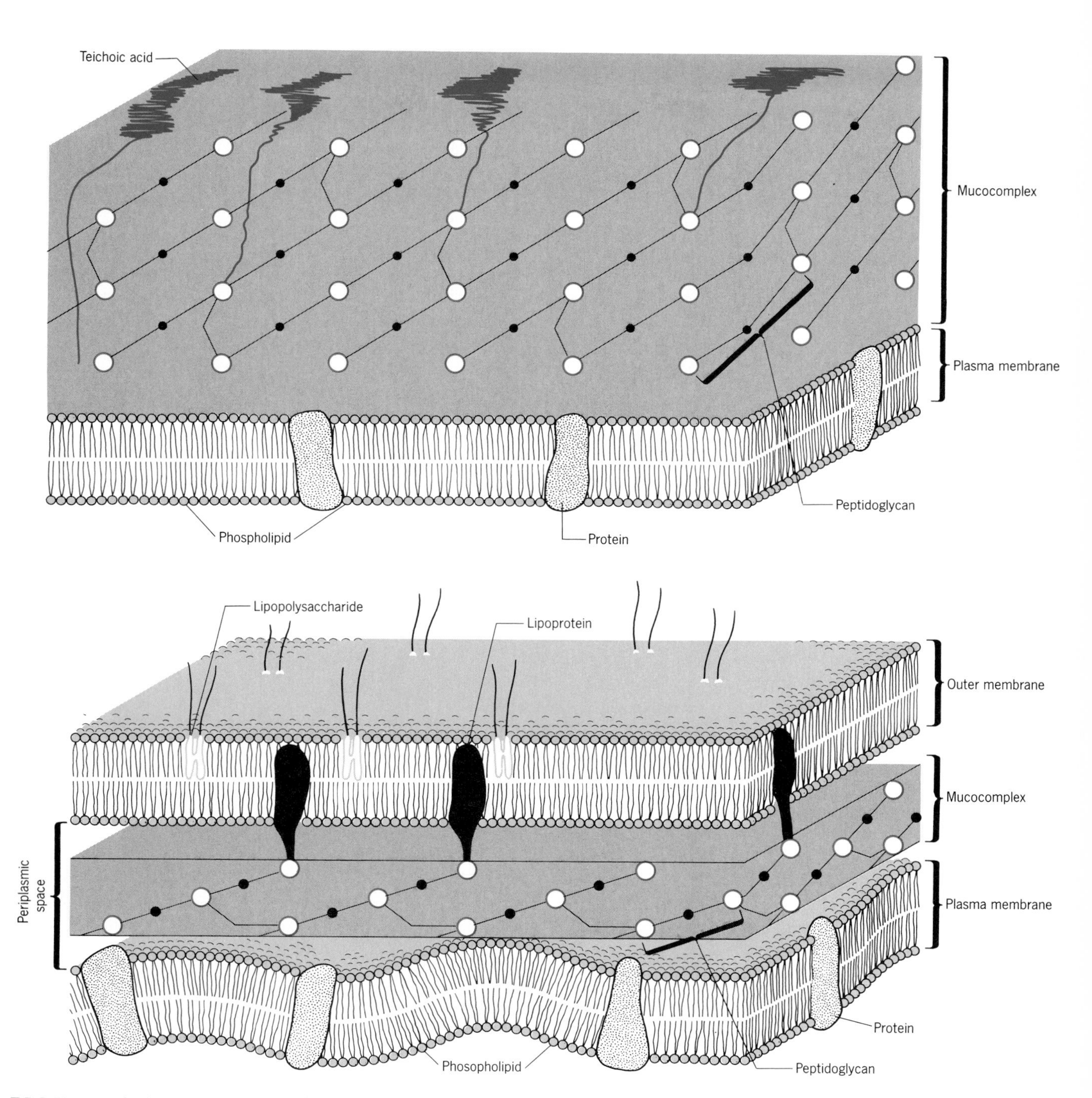

FIGURE 4.13 Comparison of the structures and the composition of the gram-positive (top) and gram-negative (bottom) cell walls.

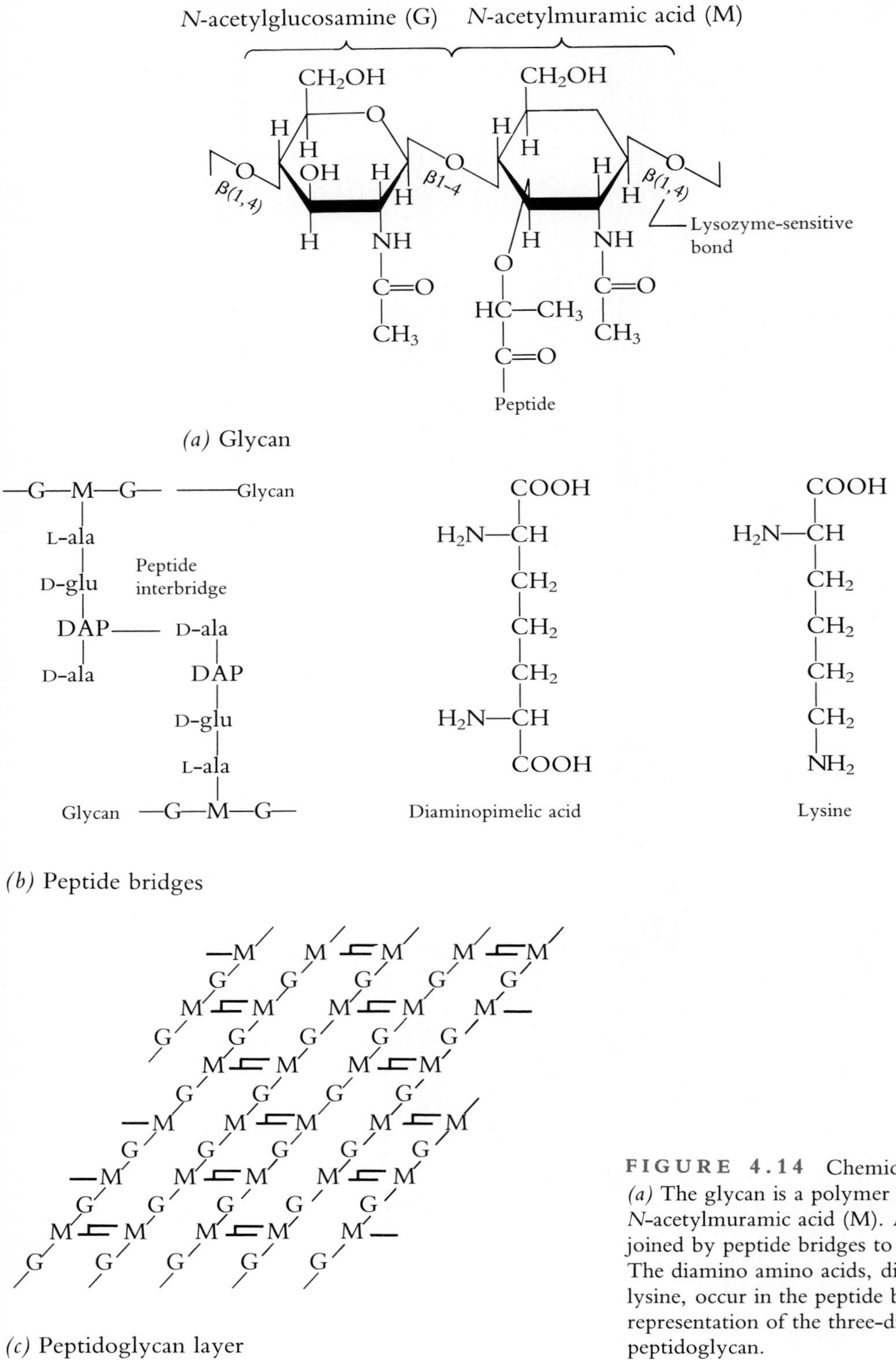

N-acetylglucosamine (G) N-acetylmuramic acid (M)

Lysozyme-sensitive bond

Peptide

(a) Glycan

Glycan

Peptide interbridge

L-ala, D-glu, DAP, D-ala

Diaminopimelic acid

Lysine

(b) Peptide bridges

(c) Peptidoglycan layer

FIGURE 4.14 Chemical structure of peptidoglycan. *(a)* The glycan is a polymer of *N*-acetylglucosamine (G) and *N*-acetylmuramic acid (M). Adjacent glycan molecules are joined by peptide bridges to form the peptidoglycan. *(b)* The diamino amino acids, diaminopimelic acid (DAP) and lysine, occur in the peptide bridge. *(c)* Diagramatic representation of the three-dimensional nature of peptidoglycan.

otic cells. Moreover, **diaminopimelic acid** is found only in the peptidoglycan, and the D-isomers of glutamic acid and alanine are rare in biological structures other than the peptidoglycan. (Some *Bacillus* species produce a capsule of D-glutamic acid.) The peptide extends from adjacent *N*-acetylmuramic acid residues to join in the middle with a peptide bond. The cell walls of gram-positive bacteria are joined by intervening short peptides called **peptide interbridges.** Simple peptide interbridges have also been observed in the peptidoglycans of some gram-negative bacterial.

KEY POINT

Lysozyme is an enzyme found in human nasal secretions, tears, and in the white of chicken eggs. Lysozyme specifically breaks the β 1—4 glycosidic bond between *N*-acetylmuramic acid and *N*-acetylglucosamine in the bacterial cell wall. The bacteria lyse when the disintegrating peptidoglycan layer can no longer resist the osmotic pressure within the cell.

The mucocomplex exists as a three-dimensional structure that forms a sac around the bacterium. This cell wall structure can be composed of many layers of peptidoglycan or it can be only a few layers thick. Depending on its thickness, the mucocomplex acts as a sieve that prevents large molecules from passing through it. The thickness of the mucocomplex and the resulting filtering action differ between gram-negative and gram-positive bacteria.

Cell Walls of Gram-Positive Bacteria

The mucocomplex is the most prominent component of the cell wall in gram-positive bacteria. It lies directly on the cell membrane (Figure 4.13) and is constructed of multiple layers of peptidoglycan interconnected by peptide interbridges. This network of interconnecting polymers forms a rigid sac that constrains the bacterium and dictates its shape. Attached to the peptidoglycan layer and the cell membrane are teichoic acids, which are found only in gram-positive bacteria. **Teichoic acids,** such as polyribitol and polyglycerol (Figure 4.15), are negatively charged acidic polymers of a sugar or an alcohol joined by phosphate ester bonds. The cell wall teichoic acids are covalently attached to the 6-hydroxy position of *N*-acetylmuramic acid in the peptidoglycan and extend to the exterior of the cell wall. In addition, many gram-positive cell walls are coated by a layer of protein, such as the M protein of the streptococci.

Cell Walls of Gram-Negative Bacteria

Gram-negative cells have a thin mucocomplex layer that is externally covered by a complex outer membrane (Figure 4.13). The mucocomplex layer confers shape and rigidity to the cell, even though in gram-negative bacteria the peptidoglycan may be only a single layer thick *(Escherichia coli)*. The **outer membrane** is composed of lipoprotein, lipopolysaccharide (LPS), and proteins. The protein portion of the lipoprotein is covalently bound to the peptidoglycan. This leaves the hydrophobic lipid portion of the lipoprotein to bury itself in the outer membrane.

The **lipopolysaccharides** of the outer membrane are complex molecules composed of three subunits: the O-specific side chain, core, and lipid A. Because the structure of the O-specific carbohydrate side chain varies greatly from species to species, these polymers are important in identifying bacteria, especially species of *Salmonella* (sal-mohn-el′ah). One end of this carbohydrate polymer extends into the surrounding medium while the other end is anchored to the core of the lipopolysaccharide. Cell walls of gram-negative bacteria are toxic to animals. The responsible agent, called endotoxin, is the **lipid A** portion of the lipopolysaccharide. Endotoxin causes fever in animals and serious illness in humans.

In addition to the lipoproteins, outer membranes contain at least 20 proteins. Two of the 20 proteins are called **porins** because they form channels in the outer membrane through which water and metabolites can pass. The outer membrane is selectively permeable to molecules based on their molecular size and charge. Hydrophilic molecules with molecular weights greater than 600 to 900 are prevented from entering, whereas small molecules enter through the pores in the outer membrane. Low quantities of numerous other proteins can be detected in the outer membrane, but their function is unknown.

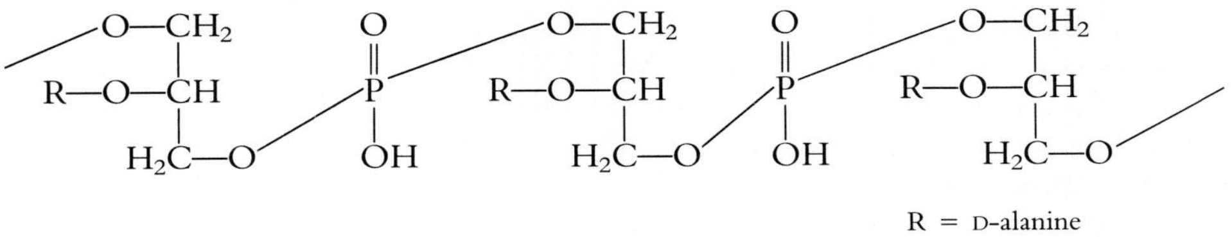

R = D-alanine

(a) Glycerol teichoic acid

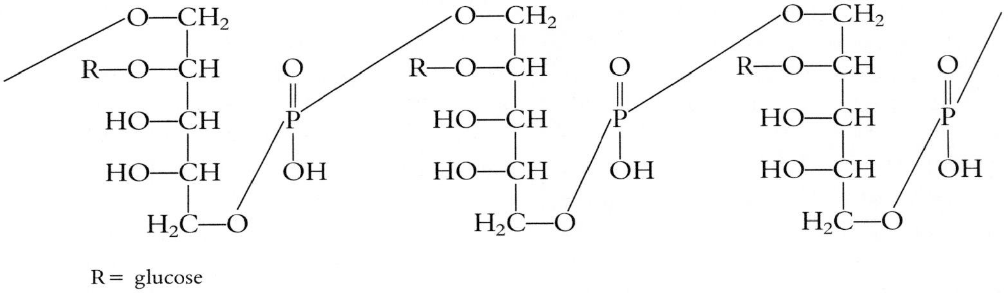

R= glucose

(b) Ribitol teichoic acid

FIGURE 4.15 The teichoic acids found in gram-positive bacteria are polymers of *(a)* polyglycerol phosphate and *(b)* polyribitol phosphate.

Between the cytoplasmic membrane and the outer membrane of gram-negative bacteria is the periplasmic space. This region contains small molecules, binding proteins involved in transport and motility, enzymes involved in cell-wall synthesis, and enzymes that protect the cell against unwanted chemical attack. Many of the proteins in the periplasmic space function in a role analogous to the lysosome of eucaryotic cells (see Chapter 5). They break down macromolecules into metabolites that are either excreted or used as cellular nutrients. The periplasmic space is also used by the cell as a staging area for the synthesis of the peptidoglycan layer.

Upon close examination, the cytoplasmic membrane appears to make direct contact with the outer membrane at numerous places in the cell envelope. These zones of adhesion between the two membranes are known as **Bayer's junctions.** The absence of peptidoglycan in these zones enables the outer membrane to be in direct contact with the cytoplasmic membrane, permitting materials produced in the cytoplasm to have access to the outer membrane. The existence of Bayer's junctions conceptually explains how complex macromolecules produced in the cell's cytoplasm can find their way into the outer membrane.

Antibiotics that Inhibit Cell-Wall Synthesis

Antibiotics are natural substances that kill or inhibit the growth of another organism. The penicillins and cephalosporins (see-fahl-oh-spor′inz) are antibiotics that interfere with the synthesis of the peptidoglycan layer of the bacterial cell wall. Cells unable to produce the rigid cell-wall component are unable to divide and multiply, and eventually lyse. Penicillins and cephalosporins are important therapeutic drugs because of their low toxicity in animals and their effectiveness in treating bacterial infections (see Chapter 21).

Bacteria That Have No Cell Wall

The significance of the cell wall becomes evident when cells lacking a cell wall are studied. Such cells can be created experimentally by removing the cell wall by enzymatic digestion. This is done by treating cells with **lysozyme** (ly'so-zyme), the mammalian enzyme that hydrolyzes the β 1—4 bond between *N*-acetylmuramic acid and *N*-acetylglucosamine in the peptidoglycan. Once the peptidoglycan is hydrolyzed, the cell loses its rigid shape and becomes sensitive to osmotic pressure.

Lysozyme treatment can remove the entire cell wall of a gram-positive bacterium, resulting in an osmotically sensitive cell called a **protoplast.** Protoplasts always take the form of a sphere because the internal osmotic pressure pushes equally in all directions on the internal surface of the malleable cytoplasmic membrane. Because they lack a rigid cell wall, protoplasts are very fragile and must be maintained in isotonic solutions to prevent them from lysing.

Removal of cell walls from gram-negative bacteria is more complicated because the lysozyme cannot readily penetrate the outer membrane. To overcome this problem, gram-negative bacteria are first treated with a chelating agent (key'layt-ing), such as EDTA (ethylene diaminetetraacetic acid), to disrupt their outer membranes. This enables the lysozyme to attack the peptidoglycan layer, which lies beneath the outer membrane. The cells lose their rigid shape and become spherical, as the peptidoglycan layer is digested away. The resulting cells, called **spheroplasts,** are osmotically sensitive even though patches of the wall remain attached to the cell membrane. Unlike protoplasts, spheroplasts retain certain cell-wall markers that are biologically functional.

Certain bacteria never produce cell walls because they are unable to synthesize muramic acid and/or diaminopimelic acid. The mycoplasmas *(Mycoplasma, Ureaplasma,* and *Acholeplasma)* are the major groups of bacteria that have never produced a cell wall. Instead of having a rigid cell structure, the mycoplasmas assume many shapes varying from small cocci to extended tubules or filaments. In past years they were thought to be viruses because their small size and plasticity enabled them to pass through filters that retained other bacteria.

Some mycoplasmas compensate for the absence of a cell wall by incorporating sterols into their cell membranes. The sterols make the cytoplasmic membrane stronger and more able to resist osmotic pressure. **Mycoplasma** acquire sterols from their growth medium or from the animals they infect. Other mycoplasmas grow in natural environments and contain no sterols—a fact used in their classification. Many mycoplasmas live in association with living material and can cause infectious diseases of plants and animals, including humans.

L-forms are bacteria that have lost their ability to form complete cell walls. They were first observed in 1935 at the Lister Institute in Paris, from which they derive their name. L-forms arise from both gram-positive and gram-negative bacteria, often while they are growing in the tissue of their animal host. Some L-forms make partial cell walls, whereas others are completely cell-wall deficient. All L-forms require a very rich growth medium with a high osmolarity to prevent them from lysing. L-forms often have a morphologic resemblance to mycoplasmas, but the two groups of bacteria are unrelated.

Cell Walls of Archaebacteria

Microbiologists have discovered that certain bacteria, the methanogens, halophiles (require high salt), and two thermo-acidophiles, do not contain peptidoglycan in their cell walls. These bacteria collectively are called archaebacteria (ark'ee-bak-ter'-ee-ah). Their cell walls vary greatly in chemical composition, and representatives have been found that contain protein, glycoprotein, or polysaccharides as the major cell-wall constituent. This chemical diversity is reflected in the morphology of their cell walls. For example, the cell-wall morphology of *Methanospirillum hungatii* (methan'oh-spi-ril-lum hun-gat'ee-eye) is distinctly different from the typical gram-positive and gram-negative bacterial cell wall (Figure 4.16). The cell wall of *M. hungatii* has a double-track appearance, with two dark-staining layers and two light-staining regions. In contrast, *Methanobacterium* has a cell-wall morphology that is very similar to the cell wall of a gram-positive bacterium, but it lacks muramic acid. Other bacteria in this group possess no cell-wall polymers that even resemble peptidoglycan, and yet their rigid cell walls give them shape and protect them against osmotic pressure. The archaebacteria are distinctly different

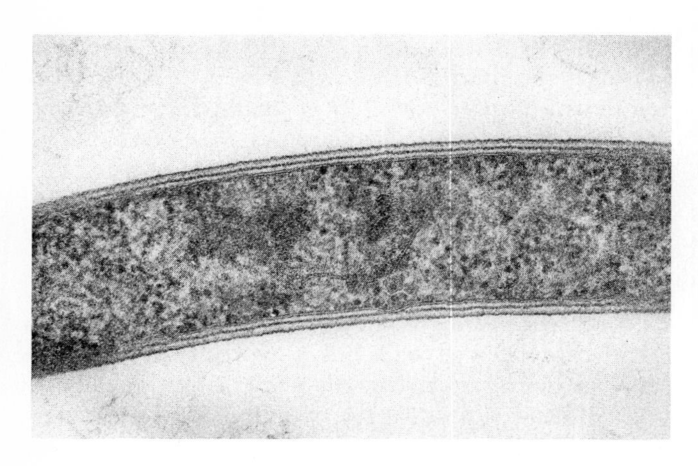

FIGURE 4.16 Electron micrograph of the unique cell wall structure of *Methanospirillum hungatii*. These cell walls do not have peptidoglycan—a characteristic shared by all archaebacteria. (Courtesy of Dr. T. J. Beveridge/BPS.)

from other procaryotes because of the differences in their cell walls, lipids, and genes for the 16S ribosomal RNA. They are now classified as Mendosicutes, the fourth division of the Procaryotae.

THEORY OF THE GRAM-STAIN REACTION

The differential staining characteristics of gram-positive and gram-negative bacteria is related to differences in the structure of their cell walls. Both cell types are stained with the crystal violet-iodine complex. However, when the stained cells are washed with a nonpolar solvent (acetone or alcohol), the blue dye is lost from the gram-negative cells only. This is because the nonpolar solvent removes the outer membrane that is present only in the gram-negative cell wall. While the mucocomplex layer continues to provide structural rigidity to the gram-negative cells, it is too thin to retain the blue crystal violet–iodine complex, which is leached out with the solvent wash. In contrast, gram-positive cell walls have a thick mucocomplex layer containing small pores that prevent the blue dye from being washed out. The result is that gram-positive bacteria are stained blue, whereas gram-negative bacteria are colorless until they turn red on counterstaining with safranin.

Terrance Beveridge and J. A. Davies in 1983 experimentally confirmed this explanation of the Gram-stain mechanism originally proposed by Salton in 1963. They substituted platinum for the iodine in the Gram-stain procedure to make a platinum–crystal violet complex that is visible in transmission electron micrographs. They then showed that the platinum–crystal violet complex remained inside the gram-positive cells, but was removed from the gram-negative cells during ethanol decolorization.

BACTERIAL CAPSULES AND SLIME LAYERS

Many bacteria manufacture and export high molecular weight polymers that adhere to the exterior of the cell wall to form a capsule or slime layer. **Glycocalyx** (gly-koh-kay′licks) is a general term for the polysaccharide capsules or slime layers attached to the outer surface of bacterial cell walls. This term tries to avoid the difficulty in making morphologic distinctions between capsules and slime layers. In general, a bacterial **capsule** has a uniform thickness (Figure 4.17) and can be thicker than the cell itself, whereas material adhering to the cell wall in a diffused arrangement is termed a **slime layer.** Capsules and slime layers are produced by cells only when the proper growth conditions and substrate concentrations prevail. For example, some lactic acid bacteria produce dextran only when excess sucrose is available. Although some **Bacillus** species produce capsules of poly D-glutamate, most bacterial capsules or slime layers are made from a variety of polysaccharides.

Slime layers help bacteria to colonize surfaces. Many bacteria that contribute to dental caries produce slime that helps them adhere to the tooth surface. Bacteria that cause human infections often possess capsules that protect the cell against phagocytosis by white blood cells. These capsules play a significant role in the ability of bacteria to cause infectious diseases (see Chapter 17).

HOW BACTERIA MOVE

Movement of procaryotic cells is mediated by flagella, periplasmic flagella, or the flexible movement of the entire cell (gliding motility). Flagellar motility occurs

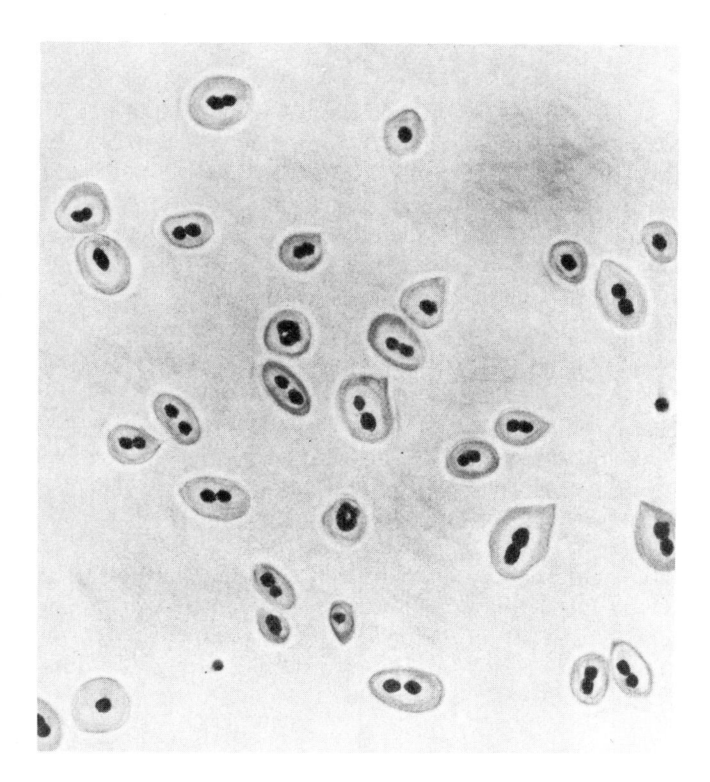

FIGURE 4.17 Capsules surround *Streptococcus pneumoniae*, which grows as a diplococcus. (Courtesy of Centers for Disease Control.)

in liquid media, while gliding motility is observed only on solid surfaces. Bacterial motility is an important determinant in classification. The best-understood form of motility in bacteria is flagellar motility.

BACTERIAL FLAGELLA

Bacterial flagella (sing., flagellum) are external appendages (Figure 4.18) that can be located on one end of an organism **(monotrichous flagella)** (mo-not′ri-kus), on both ends **(amphitrichous flagella)** (am-fit′ri-kus), or dispersed around the cell's exterior **(peritrichous flagella)** (pe-rit′ri-kus). Bacteria that have polar tufts of flagella move by **lophotrichous flagella** (lo-fo′tri-kus). Flagella are thin proteinaceous structures not visible with the ordinary light microscope unless they are first stained.

The flagella of gram-negative bacteria cross the bacterial cell wall and are embedded in the cytoplasmic membrane. The typical flagellum is composed of a **basal body,** a **hook** section, and a **filament** (Figure 4.19). The basal body is composed of paired rings that surround the hollow filament of the flagellum. This structure is inserted into the cell envelope with the bottom or M-ring embedded in the cytoplasmic membrane in a manner that allows the entire structure to rotate. The outer rings of the basal body coincide with the peptidoglycan layer and the outer membrane of the cell wall.

The basal body is composed of between 10 and 13 distinct proteins. One of these proteins aggregates to form the flagella hook, which creates the bend in the basal body at the point where the flagellum enters the cell wall. The filament extends outward from the hook as a long hollow tube (20 nm in diameter) made of the protein flagellin. The flagellin of all bacteria that have been studied is a single protein, except in *Caulobacter*,

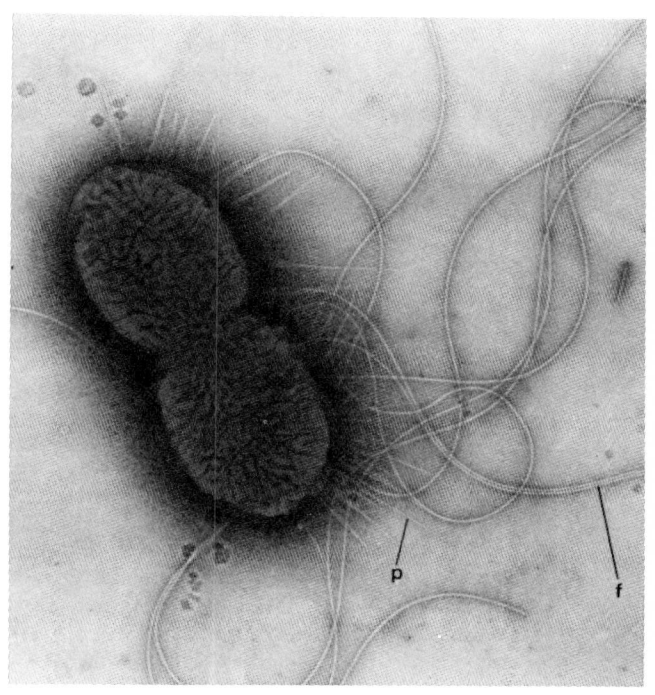

FIGURE 4.18 Electron micrograph of a negatively stained, dividing cell of *Proteus vulgaris* showing flagella (f) and short pili (p). (Courtesy of Dr. R. K. Nauman.)

which has a filament composed of two flagellins. Individual flagellin molecules aggregate (quaternary protein structure) to form the hollow flagellar filament. The filament appears to be formed by molecules that travel from the cytoplasm, through the hollow basal body, and finally to the distal end of the filament where the molecules polymerize to extend the filament.

The structure of flagella in the gram-positive bacterium *Bacillus* is similar, except that the basal body contains only two rings: one lies in the cytoplasmic membrane, and the other is associated with the mucocomplex of the cell wall.

Flagella generate movement when the basal body rotates and turns the flagella filament. This movement can be observed by attaching the filament to a microscopic slide and observing the bacterium. Under these conditions the bacterial cell rotates in place. The force that drives the flagellum appears to be generated by the M-ring, which rests on the cytoplasmic membrane. The **M-ring** rotates either clockwise or counterclockwise, driven by energy of the proton motive force generated at the cytoplasmic membrane.

Chemotaxis and Phototaxis

Bacteria can respond to chemical and physical stimuli and can move toward or away from them. **Chemotaxis** is the movement toward a chemical attractant or away from a chemical repellent (Figure 4.20). **Phototaxis** is movement in response to light. A photosynthetic bacterium's movement toward a light source depends on the wavelength and the intensity of the light. The basic mechanisms of chemotaxis have been elucidated through biochemical and genetic experiments.

Bacteria move by rotating their flagella in a counterclockwise direction. The flagella of the bacterium come together at one end of the cell to form a stable bundle (Figure 4.20) and rotate to propel the bacterium forward in a smooth swimming motion called a **run.** This directional movement is interrupted by clockwise rotation of the flagella. This clockwise rotation unwinds the stable bundle, and the cell tumbles without direction. The tumbling enables the bacterium to change direction randomly, then it is quickly off on another run. Since the runs toward an attractant are longer in duration than movement in other directions, the net move-

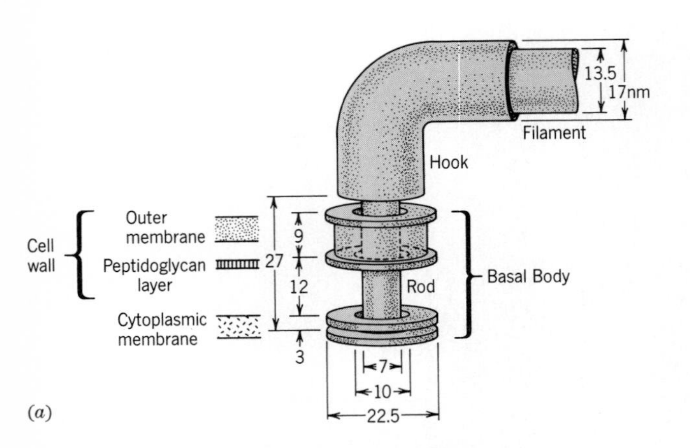

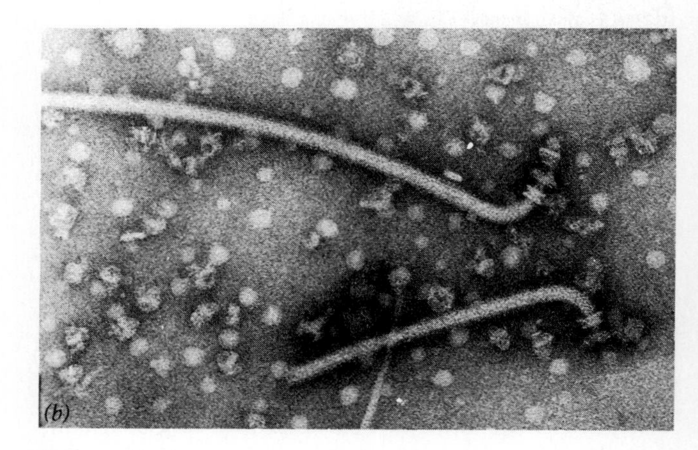

FIGURE 4.19 The bacterial flagellum. *(a)* Diagram of a bacterial flagellum showing the basal body, hook, and filament. (Courtesy of Drs. M. L. DePamphilis and J. Alder. From M. L. DePamphilis and J. Alden *J. Bacteriol.*, 105:384–395, 1971, with permission of the American Society for Microbiology.) *(b)* Electron micrograph of flagella isolated from *Salmonella*. (Courtesy of Dr. K. Kutsukake. From K. Kutsukake, T. Suzuke, S. Yamaguchi, and T. Iino, *J. Bacteriol.*, 140:267–275, 1979, with permission of the American Society for Microbiology.)

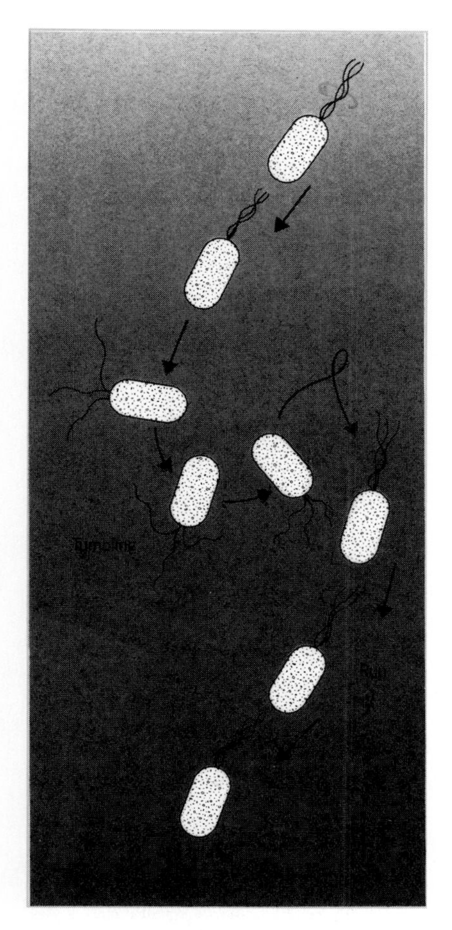

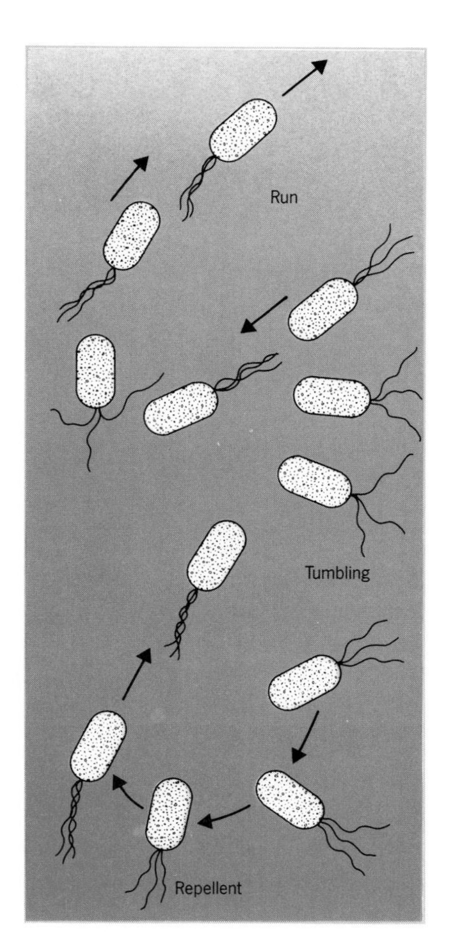

FIGURE 4.20 Bacterial movement is directed by chemotaxis. Bacteria move toward a chemoeffector with a smooth swimming motion or run driven by the counterclockwise rotation of their flagella. The runs are interspersed by tumbling caused by clockwise flagella rotation. It is through tumbling that the bacteria change direction. If the new direction is away from the attractant, the run will be of short duration. Their net movement is a zigzag course toward attractants or away from repellents.

ment of the bacterium is toward the attractant. The same principle explains how a bacterium moves away from a repellent, except the runs away from the repellent are of a longer duration.

Bacteria respond to a number of chemicals called chemoeffectors. Each effector combines with a receptor protein in the periplasmic space or in the cytoplasmic membrane. Through these receptors, bacteria recognize concentration gradients of chemoeffectors and inherently move toward or away from them.

THE PERIPLASMIC FLAGELLA OF SPIROCHETES

Spirochetes are propelled by two or more *periplasmic flagella* that lie on the cell surface beneath the spiro-

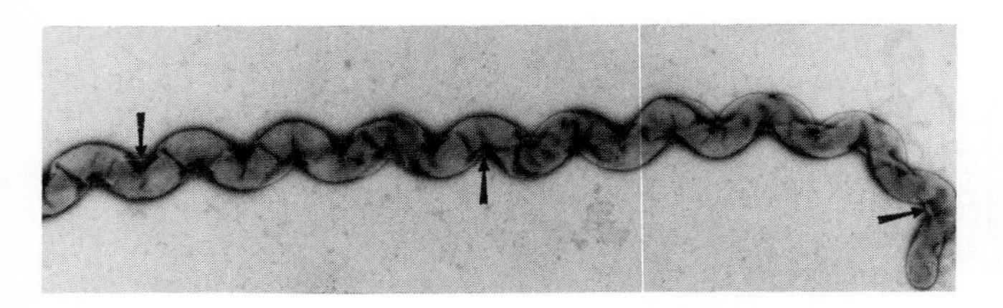

FIGURE 4.21 Electron micrograph of a spirochete, *Leptospira*, showing its periplasmic flagella surrounded by the cell envelope. (Courtesy of Dr. R. K. Nauman. From R. K. Nauman, S. C. Holt, and C. D. Cox, *J. Bacteriol.*, 98:264–280, 1969, with permission of the American Society for Microbiology.)

chete's outer sheath and its gram-negative cell wall. Each flagellum originates from an insertion point located at the end of a nondividing cell and extends toward the midpoint of the protoplasmic cylinder (Figure 4.21). In some spirochetes as many as a hundred flagella are wrapped around the cell's protoplasmic cylinder. In theory, periplasmic flagella rotate around their insertion point, which in turn causes the spirochete to propel itself with a rotary motion.

GLIDING MOTILITY

Gliding bacteria are able to move over solid surfaces without the aid of an obvious locomotor organelle. The cells may bend, but otherwise there is no discernible cellular movement—no wriggling, contraction, or peristalsis. Most gliding bacteria have an axial symmetry and move in a direction parallel to their long axis. They have a typical gram-negative cell wall and all produce an extracellular slime, which may serve as an adhesive. Other than these general characteristics, the gliding bacteria are an extremely diverse group of organisms.

Gliding is manifested by single cells, swarms of cells, or trichomes (try′komes). Single cells of some *Flexibacter* (flex-i-bak′ter) and *Cytophaga* (cy-toph′a-gah) display a flexing movement while gliding on solid surfaces. *Beggiatoa* and many cyanobacteria grow as filaments of cells called **trichomes.** Rotation of the trichome along its long axis is commonly observed during the gliding motility of these filamentous bacteria. The fruiting myxobacteria form swarms of cells that glide to a common location from which they form a fruiting body. The mechanism(s) of gliding remains unknown even though extensive investigations into the possibilities that pili, slime secretion, microfilaments, microtubules, and/or contractile proteins are involved. Further descriptions of the gliding bacteria are presented in Chapter 14.

EXTERNAL APPENDAGES: PILI

Pili (sing., pilus) are hollow, proteinaceous filaments (20 nm in diameter) that protrude from the bacterium's surface (Figure 4.22). Pili have two distinct functions*: (1) they help the bacterium to adhere to surfaces or to each other and (2) they are involved in the transfer of genetic information between cells and during some viral infections.

*Some authors make a distinction between these functions and use the term **fimbriae** to describe the structures with function 1 and pili to describe those with function 2.

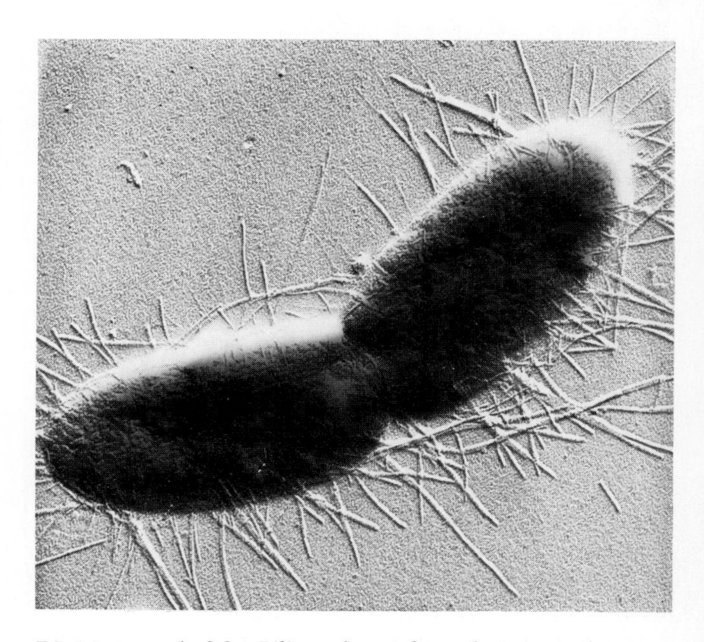

FIGURE 4.22 Pili on the surface of a negatively stained preparation of *Escherichia coli.* (Courtesy of Dr. D. C. Old. From S. Clegg and D. C. Old, *J. Bacteriol.*, 137:1008–1012, 1979, with permission of the American Society for Microbiology.)

Pili can contribute to a bacterium's ability to cause disease by helping it to adhere to cell surfaces, teeth, and tissues. Pili are also involved in the formation of bacterial mats that are seen as pellicles, or thin films on liquid surfaces. Other pili act as receptors for some bacterial viruses.

Genetic transfer between bacteria in physical contact with each other occurs during bacterial conjugation (see Chapter 10). Male cells (donors) possess sex pili coded for by sex plasmids, whereas female cells (recipients) lack pili. The sex pili extending from the surface of the male cell attach to a female cell. The pili then shorten and draw the cells together such that genetic material (DNA) can be transferred from the male cell to the female cell.

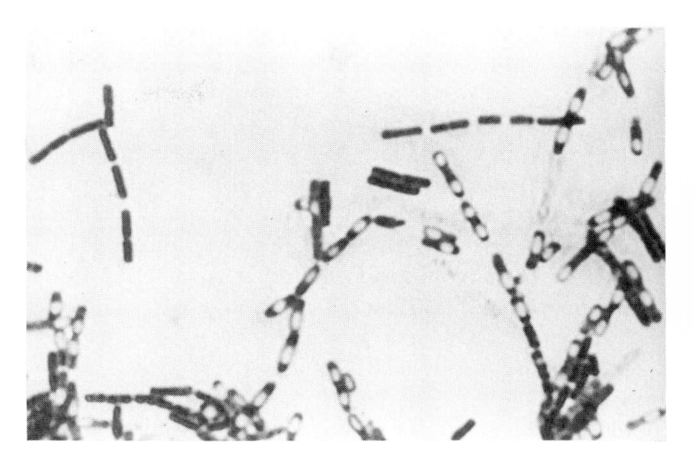

FIGURE 4.23 Phase contrast micrograph of the refractile endospores in *Bacillus* sp.

SPORES AND OTHER RESTING FORMS OF BACTERIA

Bacterial spores are of paramount importance to microbiologists because their ability to survive exposure to high temperatures makes them very difficult to control. Although other microorganisms form spores, bacterial spores are the most heat-resistant forms of a cellular organism known. Bacterial spores are often called **endospores** because they are formed within the vegetative cell. Endospores are ubiquitous in our environment and are often encountered as contaminants in the bacteriology laboratory.

The vegetative cells of endospore-forming bacteria begin to sporulate when exposed to adverse environmental conditions. As the spore develops within the cell it becomes refractile and shines as a bright light in the field of a light microscope (Figure 4.23). Although the vegetative cells of sporeforming bacteria are readily stained with vital dyes, the endospore can only be stained with special spore stains.

Endospores are complex structures formed within a cell during **sporogenesis** (spor-oh-gen′e-sis). A copy of the cell's DNA is sequestered in one pole of the cell as the invaginating plasma membrane forms the spore septum (Figure 4.24). The plasma membrane continues to surround the spore protoplast and eventually the spore is enclosed by a double membrane. The exosporium develops as the outermost covering of the spore as the outer spore membrane disappears. The spore coat composed largely of protein forms beneath the exosporium and surrounds the spore cortex (Figure 4.25), which is composed of a unique peptidoglycan.

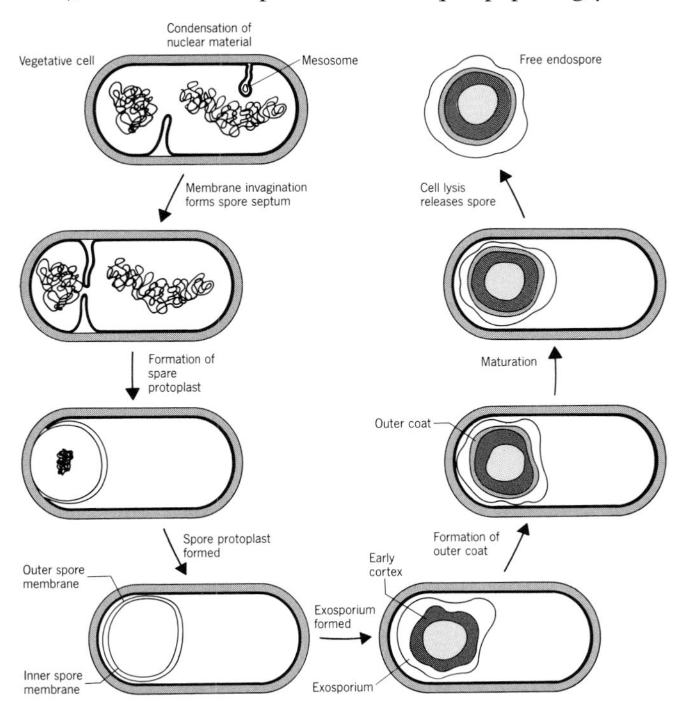

FIGURE 4.24 Stages in the formation of a bacterial endospore.

Meanwhile, the core of the spore condenses as the core wall is formed outside the core membrane. These structures enclose the spore protoplasm with its ribosomes and nucleic acids.

At some point during sporogenesis, the spore becomes refractile, which is indicative of its ability to resist heat. The refractile spore develops its thermostability concurrent with the deposition of dipicolinic (dy-pic-oh-lin′ik) acid and the accumulation of Ca^{2+} in the spore protoplast. During the final stage of sporogenesis, the vegetative cell structure dissolves (autolysis) and the spore is released into the environment. Endospores are highly dehydrated (15 percent water), have no detectable metabolic activity, and are able to survive for years as resting forms of the cell.

Activated spores undergo the process of **germination** when they encounter suitable growth conditions. Spores can be experimentally activated by exposure to heat (65°C), low pH, or strong oxidizing agents. These treatments activate enzymes and initiate outgrowth from the spore state. During the stages of initiation and outgrowth, the spore structures disintegrate and the spore protoplast emerges as a growing vegetative cell.

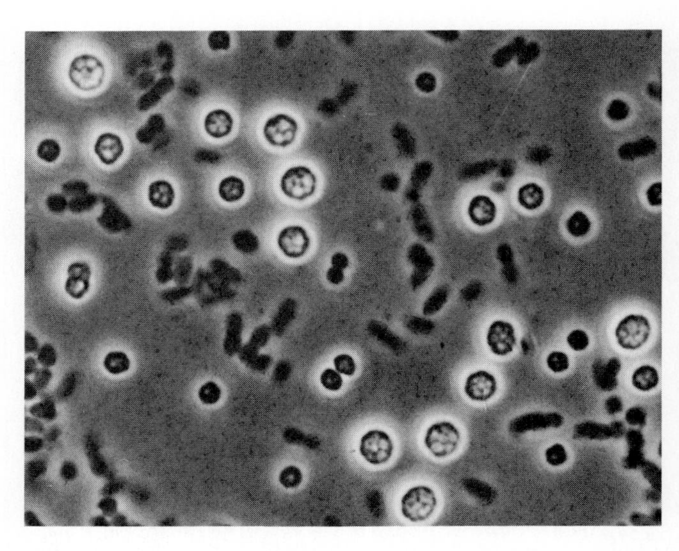

FIGURE 4.26 *Azotobacter vinelandii* is a soil bacterium that resists adverse conditions by forming cysts. The entire cell becomes the refractile cyst (s), which is more resistant to heat, chemicals, and drying than vegetative cells. (Courtesy of J-H. Becking.)

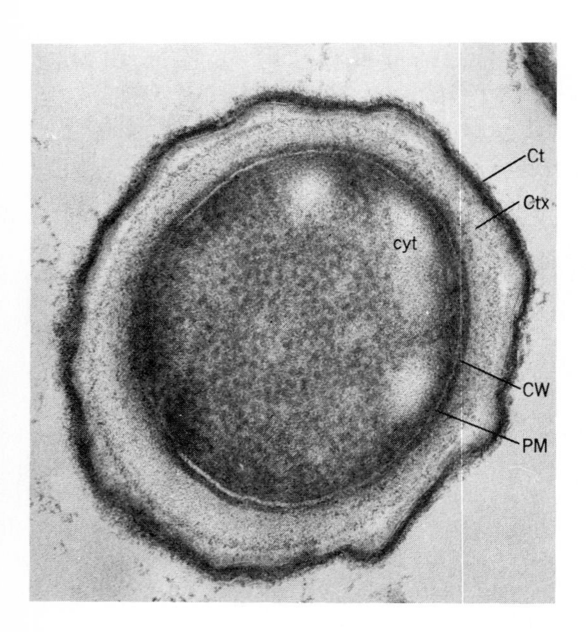

FIGURE 4.25 Electron micrograph of a thin section through an intact spore of *Bacillus megaterium* showing the spore coat (Ct), cortex (Ctx), core wall (CW), plasma membrane (PM), and cytoplasm (Cyt). (Courtesy of Dr. P. C. Fitz-James. From P. C. Fitz-James *J. Bacteriol.*, 105:1119–1136, 1971, with permission of the American Society for Microbiology.)

Endospores are formed by six bacterial genera: *Clostridium, Bacillus, Sporolactobacillus, Sporosarcina, Thermoactinomyces,* and *Desulfotomaculum*. *Sporosarcina* is the only spore-forming coccus known and *Thermoactinomyces* form tubular cells known as a mycelium. Rod-shaped bacteria form distinctively shaped spores in either a central, subterminal, or terminal position in the cell. The morphology of the spore and its cellular location are used to classify bacteria within the spore-forming genera (see Chapter 14).

Bacterial **cysts** are different from endospores in that the entire vegetative cell becomes a cyst (Figure 4.26). Cysts are more resistant to heat, sonic vibration, ultraviolet irradiation, and desiccation than their vegetative counterpart, but they are not as resistant to these factors as the bacterial endospore. Most cyst-forming bacteria are soil organisms such as *Azotobacter* and the myxobacteria. A number of protozoa produce cysts that enable these eucaryotic organisms to survive in adverse environments (see Chapter 16).

Summary Outline

THE BACTERIAL CYTOPLASM

1. All bacteria are procaryotic cells because they possess a nuclear region instead of a nucleus.
2. The bacterial cytoplasm contains soluble enzymes, ribosomes, the cell's DNA, and pools of small molecules required for biosynthetic reactions.
3. The bacterial chromosome is a circular DNA molecule seen as fibrillar material in the nuclear region of the cytoplasm. When isolated, the chromosome is associated with a small amount of RNA and protein and is supercoiled upon itself in the structure known as the nucleoid.
4. Protein synthesis occurs on 70S ribosomes, which are free in the bacterial cytoplasm.
5. The bacterial cytoplasm can contain inclusion bodies of polyhydroxybutyrate, glycogen, cyanophycin, or polyphosphate. Many aquatic bacteria contain gas vesicles, while carboxysomes are present in some bacteria that fix carbon dioxide.

BACTERIAL MEMBRANES AND CELL WALLS

1. The semipermeable cytoplasmic membrane separates the cell's interior from the external environment and makes the cell sensitive to osmotic pressure.
2. Membrane proteins functions as carrier protiens, while others form pores through which solutes flow. A polypeptide signal sequence determines which proteins will be inserted into the membrane.
3. The cytoplasmic membrane is permeable to the small uncharged molecules that enter the cytoplasm by diffusion; most other molecules enter the bacterial cytoplasm by active transport.
4. Aerobic respiration occurs on the bacterial cytoplasmic membrane, whereas intracytoplasmic membranes are the cellular location of the respiratory enzymes of many chemolithotrophs and the site of photosynthesis in photosynthetic bacteria.
5. Bacterial cell walls form a rigid, continuous cell covering that maintains the cells shape even in the presence of high osmotic pressure.
6. Gram-positive bacteria have a thick mucocomplex layer composed of numerous layers of peptidoglycan. Teichoic acids are bonded to the cell membrane and peptidoglycan layer extending to the cell-wall surface.
7. Gram-negative bacteria have a thin peptidoglycan layer that lies beneath the cell's outer membrane, which is composed of lipoprotein and lipopolysaccharide. The periplasmic space is the area between the cytoplasmic membrane and the outer membrane: it contains hydrolytic enzymes, transport enzymes, and chemotactic receptors.
8. Penicillin and other antibiotics kill bacteria or inhibit their growth by interfering with cell-wall synthesis.
9. Mycoplasmas lack a cell wall. Protoplasts and spheroplasts are made by enzymatically digesting away the cell walls of gram-positive and gram-negative bacteria respectively.
10. The cell walls of the archaebacteria are chemically diverse; they do not contain peptidoglycan.

BACTERIAL CAPSULES AND SLIME LAYERS

1. Glycocalyx is the general term for polysaccharide slime layers and capsules found outside the cell wall. Capsules protect cells against phagocytosis and slime layers help bacteria adhere to the surfaces they colonize.

HOW BACTERIA MOVE

1. The bacterial flagellum is composed of a filament, a hook section, and a basal body that is connected to the cell wall and that interacts with the cytoplasmic membrane.
2. Motion is generated when the entire flagellum rotates in response to chemotactic effectors; counterclockwise rotation propels the organism forward, and clockwise rotation results in tumbling.
3. Spirochetes have periplasmic flagella that are structurally identical to flagella except that they lie on top of the protoplasmic cylinder and beneath the outer sheath.

4. Gliding bacteria move over solid surfaces without the aid of visible appendages.

EXTERNAL APPENDAGES: PILI

1. Pili protrude from the bacterial cell wall as extensions that help the bacterium adhere to surfaces or to each other, and/or are involved in the transfer of genetic material.

SPORES AND OTHER RESTING FORMS OF BACTERIA

1. Six genera of bacteria form endospores, which are resting forms of bacteria that are resistant to high heat, sonic vibration, drying, radiation, and chemicals.
2. Some bacteria convert their entire cell to a cyst, which is more resistant to physical and chemical treatments than vegetative cells, but less resistant than endospores.

Questions and Topics for Study and Discussion

QUESTIONS

1. Describe the organization of bacterial genetic information.
2. What is the chemical composition, size, and cellular location of bacterial ribosomes?
3. Describe the chemical structure and function of the cytoplasmic membrane. Which bacteria have specialized intracytoplasmic membranes and what function do these membranes perform?
4. Compare the cell-wall structure(s) of gram-positive bacteria to those of gram-negative bacteria.
5. What is the importance of the mucocomplex to the bacterial wall? How does this structure interact with the outer membrane of the gram-negative bacterial cell wall?
6. What storage products do bacteria produce? What function do gas vesicles play in aquatic bacteria?
7. Define or explain the following terms:

active transport	nucleoid
capsule	osmotic pressure
chlorosomes	periplasmic space
chromatophores	plasmids
endospore	slime layer
facilitated diffusion	teichoic acids
lysozyme	

8. How does the bacterial flagellum impart locomotion to the bacterial cell? How does the bacterium "know" which way to go?
9. Describe the bacteria that lack cell walls. How are protoplasts and spheroplasts made?
10. How do endospores differ from vegetative cells?

DISCUSSION TOPICS

1. What are the major differences between the two forms of the signal hypothesis? What experimental approaches might resolve these differences?
2. If most bacteria contain peptidoglycan in their cell walls, what is the significance of the chemical diversity in the cell-wall composition of the archaebacteria?
3. How can material manufactured in the cytoplasm of the cell become integrated into the outer membrane of gram-negative bacteria?

Further Readings

BOOKS

Ingraham, J. L., O. Maaløe, and F. C. Neidhardt, *Growth of the Bacterial Cell,* Sinauer Associates, Sunderland, Mass., 1983. Detailed chapters on the genetics, synthesis, and assembly of the structures of the bacterial cell.

Rogers, H. J., *Bacterial Cell Structure,* American Society for Microbiology, Washington, D.C., 1983. A textbook supplement that expands the topics discussed in this chapter.

ARTICLES AND REVIEWS

Burchard, R. P., "Gliding Motility of Prokaryotes: Ultrastructure, Physiology, and Genetics," *Ann. Rev. Microbiol., 35*:497–529 (1981). The possible mechanisms for gliding in bacteria are reviewed in the manner that demonstrates the complexities of this problem and the diversity of the gliding bacteria.

Costerton, J. W., G. G. Gusey, and K. J. Cheng, "How Bacteria Stick," *Sci. Am., 238*:86–95 (January 1978). Describes the role of surface polysaccharides (glycocalyx) in infectious processes.

Dworkin, M., "Spores, Cysts, and Stalks," in J. R. Sokatch and L. N. Ornston (eds.), *The Bacteria,* Vol. 7, pp. 2–77, Academic Press, New York, 1979. A well-illustrated review of the structure and formation of bacterial spores, cysts, and stalks.

Greenawalt, J. W., and T. L. Whiteside, "Mesosomes: Membranous Bacterial Organelles," *Bacteriol. Rev., 39*:405–463 (1975). A comprehensive review of the early work on mesosomes.

Hazelbaurer, G. L., and S. Harayama, "Sensory Transduction in Bacterial Chemotaxis," *International Review of Cytology,* 81:33–65 (1983). A general review of bacterial chemotaxis.

Sokatch, J. R., "Roles of Appendages and Surface Layers in Adaptation of Bacteria to Their Environment," in J. R. Sokatch and L. N. Ornston (eds.), *The Bacteria,* Vol. 7, pp. 229–289, Academic Press, New York, 1979. An illustrated review of bacterial flagella, pili, fimbriae, cell walls, and envelopes.

Stanier, R. Y., and C. B. van Neil, "The Concept of a Bacterium," *Archiv. fur Mikrobiologie,* 42:17–35 (1962). A classic paper that presents the arguments for recognizing procaryotic cells as one of the two major types of cells.

Woese, C., "Archaebacteria," *Sci. Am.,* 244:98–122 (June 1981). The archaebacteria are having an important impact on the concepts of bacterial evolution and classification.

C H A P T E R

5

Eucaryotic Cells: Organization, Structure, and Function

OUTLINE

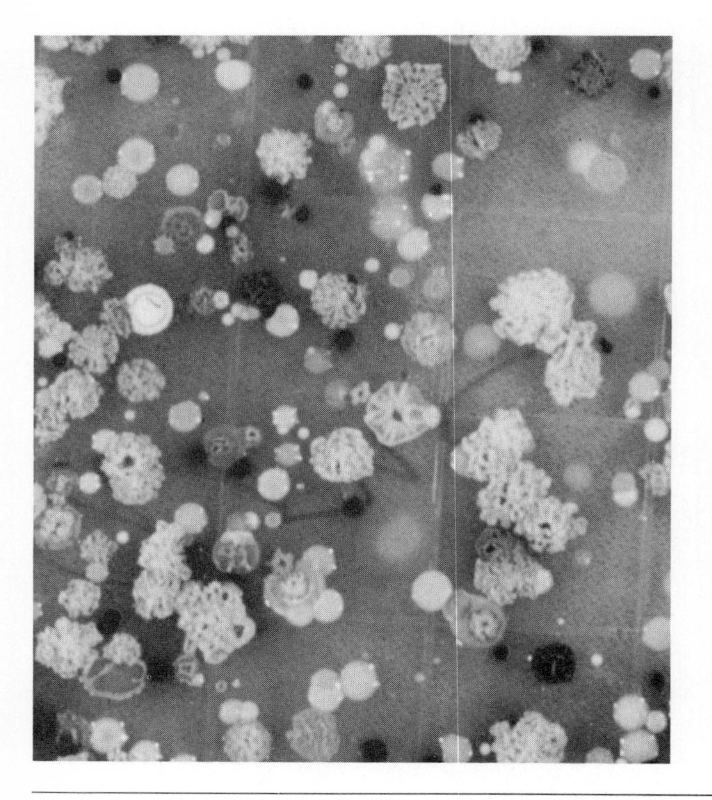

FOCUS OF THIS CHAPTER

Chapter 4 described the procaryote—the simple cellular organization found in bacteria. We come now to the far more complex eucaryotic cell structure found in animals, plants, fungi, and protozoa. The specific functions of eucaryotic cells are performed by subcellular structures called organelles, and the cell's chromosomes are contained in the nucleus, which has a complex mechanism for its replication. This chapter focuses on the specialized structures of eucaryotic cells and their functions.

Plants, animals, and all other higher forms of life have the eucaryotic cell structure. Many of the biological functions of these cells are carried out by specific organelles located in the cell's cytoplasm, including digestion, ATP production, protein synthesis, and heredity. Because of their complexity, eucaryotic cells require much more DNA than do bacteria. The DNA is organized into more than one chromosome, and these are localized in the cell's nucleus. The membrane-bound nucleus is often the most dominant organelle in the eucaryotic cell.

Eucaryotic cells vary greatly in size, shape, and purpose—there is no typical eucaryotic cell. The ova or eggs of birds are examples of the largest eucaryotic cells, whereas amebae and yeasts are examples of unicellular organisms that can be seen only with a microscope. Rather than attempt to describe a variety of eucaryotic types, we present a composite eucaryotic cell and explain its structures and functions.

OVERVIEW OF THE EUCARYOTIC CELL

Eucaryotic microorganisms include protozoa, unicellular algae, yeasts, and many types of multicellular fungi. Not all the algae and fungi are microscopic; some grow to be very large multicellular organisms such as kelp and mushrooms. Amebae are very different from rat hepatocytes (liver cells), and the fungal mycelium is so different from a unicellular alga that it is hard at first glance to recognize the similarities of their cell structure—yet they all are eucaryotes.

The chromosomes of the eucaryotic cell are localized in the nucleus. When a eucaryotic cell divides, it allocates a copy of each chromosome to its daughter cells through nuclear division, or mitosis. The nuclear location of the hereditary material and mitosis are characteristic features of eucaryotic cells.

Many of the cell's physiological functions are organized in discrete organelles (Figure 5.1). Mitochondria are the organelles responsible for aerobic respiration; the nucleus is the site of DNA synthesis; protein synthesis occurs on the rough endoplasmic reticulum; and the packaging of cellular products occurs in the Golgi (gol'jee) apparatus. These organelles are present in the small unicellular eucaryotic microorganisms as well as in the cells that comprise the differentiated tissue of higher plants and animals.

Key features of every eucaryotic cell are the presence of at least one nucleus and a limiting plasma (cytoplasmic) membrane (Figure 5.1). The material inside the plasma membrane, called the **protoplasm** (proh'-toh-plaz-em), is divided into the **karyoplasm** (kar'ee-oh-plaz-em), which is the material inside the nuclear envelope, and the **cytoplasm,** which is the material between the nuclear envelope and the plasma membrane. The cytoplasmic organelles that perform specialized functions are suspended in the fluid portion of the cell, known as the **cytosol** (sy'toh-sol). The function of each organelle in the eucaryotic cytoplasm has been determined through cytological and biochemical studies of isolated organelles.

CELL WALLS AND THE PLASMA MEMBRANE

All eucaryotic cells contain a plasma membrane that surrounds the cytoplasm. Plants, algae, and fungi also

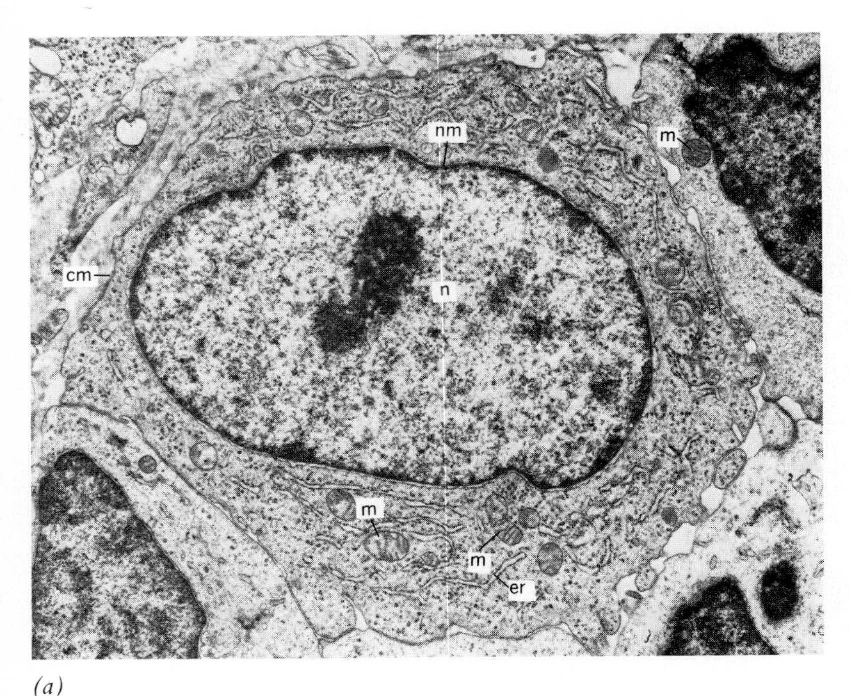

(a)

FIGURE 5.1 Eucaryotic cells have numerous subcellular organelles. *(a)* Electron micrograph demonstrates key features of a typical eucaryotic animal cell; nuclear membrane (nm), mitrochondria (m), endoplasmic reticulum (er), and cytoplasmic membrane (cm). *(b)* Three-dimensional diagram shows the relationships and the relative sizes of the subcellular components in this animal cell. (Diagram from N. Thorpe, *Cell Biology,* Wiley, New York.)

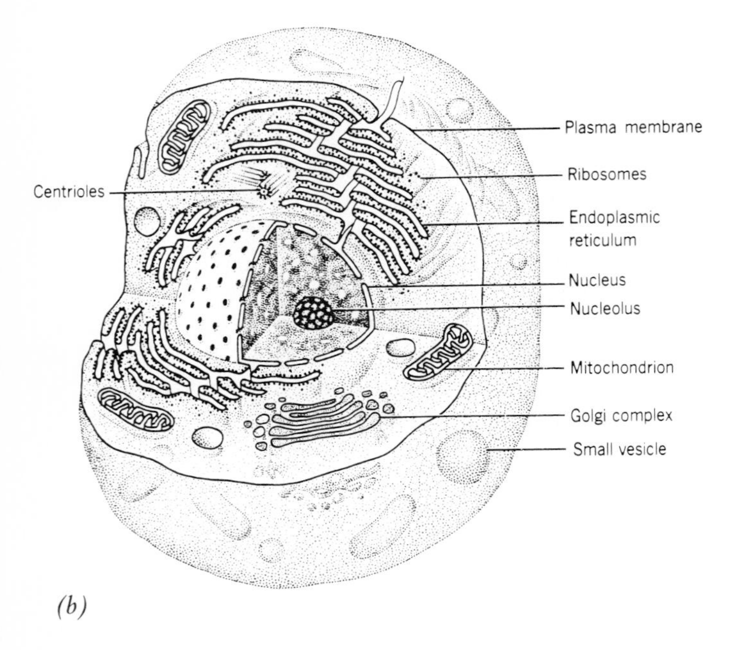

(b)

possess a cell wall outside the plasma membrane, whereas in animal cells the plasma membrane is the cell's outermost structure. The cell wall maintains the cell's shape and prevents it from bursting through osmotic presssure.

CELL WALLS OF PLANTS, FUNGI, AND ALGAE

The cell walls of plants, fungi, and algae differ in composition and physical structure from bacterial cell walls. In green plants, the cell wall is rigid and is composed mainly of polysaccharides, including cellulose (the most abundant polysaccharide), hemicellulose, and pectin. For example, cellulose accounts for some 45 percent of the dry weight in a typical tree trunk. Cellulose and the other polysaccharides protect the cell against lysis or breakdown and provide support to the entire plant.

The filamentous fungi have cell walls composed of

chitin (ky'tin) and beta (β) glucans. Chitin's role in fungal cell walls corresponds to that of cellulose in plants. It is also found in the hard exoskeleton of crustaceans and insects. The unicellular yeasts have rigid cell walls composed of mannan, a polysaccharide that is often covalently linked to proteins. Although chemically complex, these cell walls are distinctly different from the cell walls of most bacteria. These walls give the cell shape and protect the cell against osmotic pressure, invading microbes, and viruses.

KEY POINT

Eucaryotic cells either lack cell walls or possess walls that don't contain the bacterial peptidoglycan. Eucaryotic cell functions are not inhibited by penicillin and other antibiotics that prevent bacterial cell-wall synthesis.

THE PLASMA MEMBRANE

The eucaryotic cytoplasm is surrounded by a **plasma membrane.** This semipermeable membrane is the primary barrier between the cell's interior and exterior environment. It is a lipid bilayer that is morphologically and functionally similar to the cytoplasmic membranes of bacteria. The surface of the membrane is speckled with proteins that may protrude on one side or the other, or traverse the membrane to protrude on both sides.

In addition to lipids and proteins, the plasma membranes of animal cells contain sterols, mainly cholesterol, and glycolipids. The sterols intermesh in the lipid bilayer and provide additional strength to the membrane. Animal cells need this extra strength, since they lack a rigid external cell wall. The glycolipids located on the outside surface of the membrane create a unique surface landscape. These proteins govern the interactions between animal cells and enable the cells to receive messages such as hormones. Other membrane proteins extrude on both the interior and exterior surfaces (transmembrane proteins). Some of these proteins create pores through which nutrients diffuse into the cell; others act as permeases, which transport specific metabolites across the membrane.

Transport of Materials through the Plasma Membrane

Animal cells transport large molecules and particles through the cell membrane in vesicles by a process called endocytosis. **Endocytosis** (Figure 5.2*a*) is the formation of a vesicle around a substance as the substance is being taken into a cell. The transported material can then be digested and assimilated as a source of nutrients. Protozoa form these vesicles around bacteria and other matter that is ingested as a source of food.

Phagocytosis is the endocytotic process for engulfing large particles, such as bacteria (Figure 5.2*b*). Macrophages and other white blood cells remove microorganisms from the body tissue by phagocytosis. In most instances the bacteria are destroyed by the cell; however, some bacteria survive phagocytosis and replicate intracellularly. These bacteria cause serious infectious diseases and are difficult to kill with drugs.

Amebae form phagocytic vesicles around foreign matter that is taken into the cell. Once in the cytoplasm, the phagocytic vesicle joins with a lysosome containing hydrolytic enzymes, and digestion begins. Only specialized animal cells are competent to perform phagocytosis.

KEY POINT

Viruses enter animal cells by endocytotic mechanisms; however, instead of being digested, their protein coats are removed, releasing their nucleic acid so viral reproduction can begin. The reverse of endocytosis, called exocytosis, is a mechanism by which enveloped viruses are released from animal cells.

Many animal cells take up small solutions of large molecules by **pinocytosis** or "cell drinking." During pinocytosis, the plasma membrane folds inward, virtually sucking in liquid and suspended material in the process. Once inside the cytoplasm, the membrane pinches off to form a closed vesicle. The contents of these vesicles are digested by cellular enzymes usually after the vesicle fuses with a lysosome.

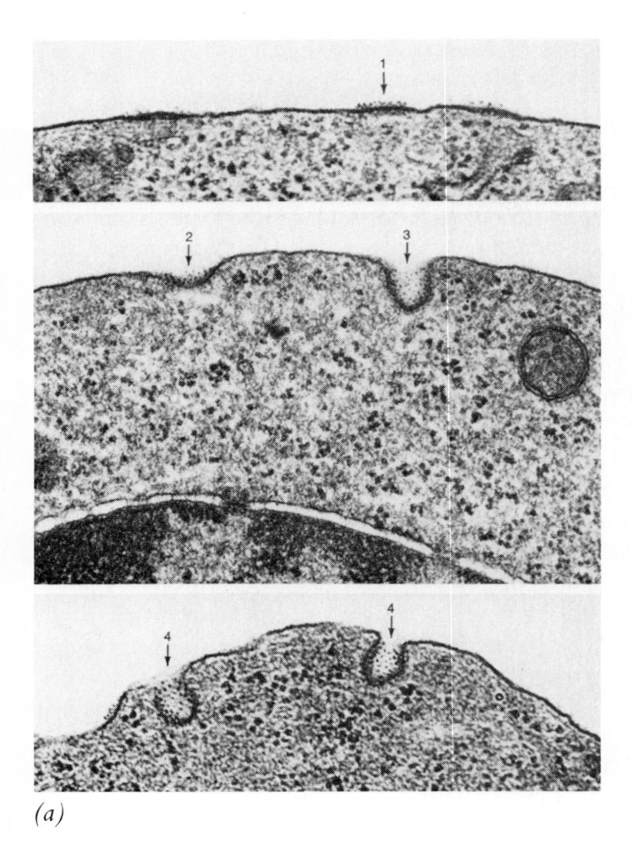

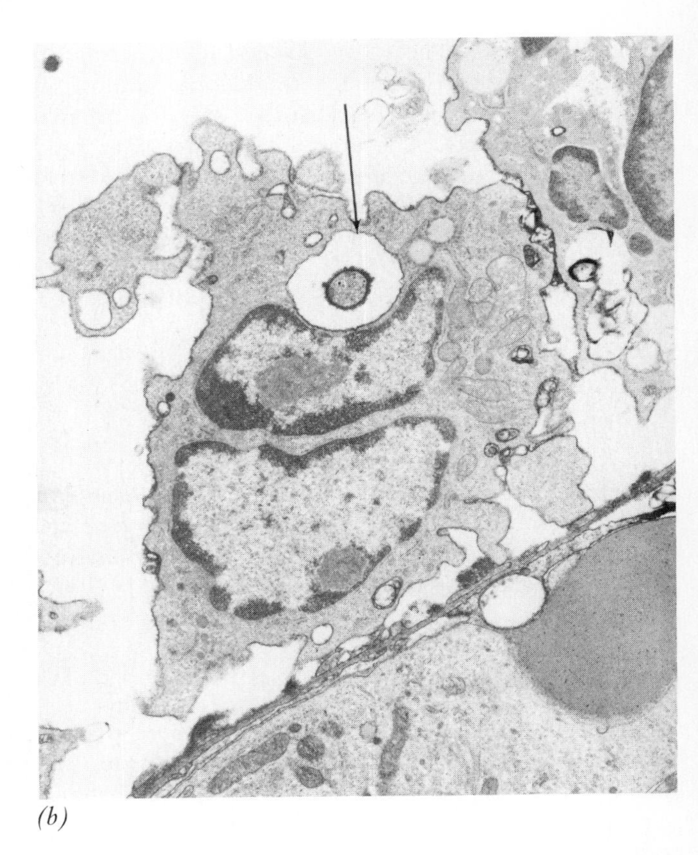

(a) (b)

FIGURE 5.2 Endocytosis is the mechanism by which eucaryotic cells take up material from the environment. *(a)* Pinocytosis occurs when the plasma membrane forms a vesicle around ferritin or other small molecules. The sequence of events in pinocytosis is shown for an erythroblast from guinea pig bone marrow. (Courtesy of Dr. Don W. Fawcett.) *(b)* Large particles or organisms are engulfed during phagocytosis. This mononuclear phagocyte has engulfed the bacterium, *Streptococcus pyogenes* (arrow), which is being digested in a phagocytic vesicle. (Courtesy of Dr. C. L. Sanders, Battelle Pacific Northwest Laboratories/BPS.)

CYTOPLASMIC ORGANELLES

The cytoplasm of the eucaryotic cell contains an extensive interconnecting network of membranes, vesicles, storage granules, and organelles. The cell's role in multicellular organisms is dictated essentially by the organelles present in its cytoplasm. For example, plasma cells have an extensive rough endoplasmic reticulum for synthesizing antibodies; muscle cells are packed with mitochondria to produce the enormous supply of energy needed for muscle contraction; and plant cells contain chloroplasts that convert light energy into chemical energy.

ENDOPLASMIC RETICULUM

The most extensive membrane structure in the eucaryotic cytoplasm is the **endoplasmic reticulum** (en-doh-plaz′mick ri-tik′yu-lem). This membrane network (Figure 5.3) consists of flattened sacs and tubules that are often connected to both the plasma membrane and the nuclear envelope. The endoplasmic reticulum exists in either a rough or a smooth form and is involved in biosynthetic reactions, transport, and packaging.

Rough endoplasmic reticulum is studded with ribosomes (Figure 5.3) that participate in protein synthesis. The proteins formed are either released into the cytosol or transported into the interior channels of the endo-

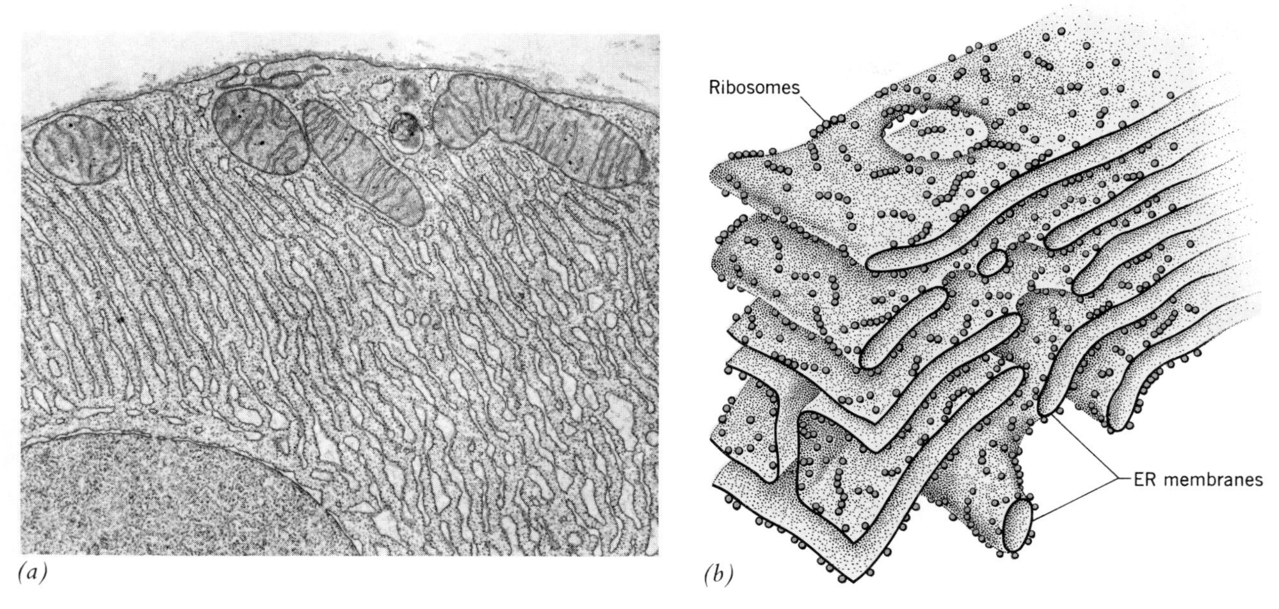

FIGURE 5.3 *(a)* Electron micrograph of the extensive rough endoplasmic reticulum in an animal cell. Numerous ribosomes are attached to this membrane system that is the site of protein synthesis. *(b)* Diagram of the rough endoplasmic reticulum. (From Sheeler and Bianchi, *Cell Biology*, 2nd edition, John Wiley & Sons, 1983.)

plasmic reticulum. Once inside, the proteins can move through the channels of the endoplasmic reticulum to other parts of the cell where they participate in anabolic reactions. The endoplasmic reticulum creates an extensive network of membrane channels throughout the cytoplasm with connections to the nuclear and plasma membranes.

Smooth endoplasmic reticulum is a membrane network involved in glycogen, lipid, and steroid synthesis and in drug detoxification. Although it is often associated with the rough endoplasmic reticulum, it is distinguished by its smooth surface that is devoid of ribosomes. Both the amount and function of smooth endoplasmic reticulum appear to depend on the cell type. Smooth endoplasmic reticulum is more abundant in steroid-secreting cells such as those present in the human adrenal cortex than in cells devoted primarily to protein synthesis. The internal spaces of the smooth endoplasmic reticulum also serve as a transport system for lipids and steroids.

RIBOSOMES FUNCTION IN PROTEIN SYNTHESIS

Eucaryotic cells have 80S cytoplasmic ribosomes that are either attached to the cytoplasmic surface of membranes or are free in the cytoplasm. These ribosomes are significantly larger than ribosomes present in bacteria. Each 80S ribosome is composed of a 40S subunit and a 60S subunit. The 40S subunit contains one 18S ribosomal RNA molecule complexed with 33 polypeptides. The 60S subunit contains 49 polypeptides and three molecules of ribosomal RNA—28S, 5.8S, and 5S ribosomal RNA. All ribosomes perform the same function in protein synthesis; however, significant biochemical differences exist between the protein synthesizing systems of eucaryotic and procaryotic cells.

KEY POINT

There are a number of antibiotics that prevent protein synthesis on 70S bacterial ribosomes without affecting protein

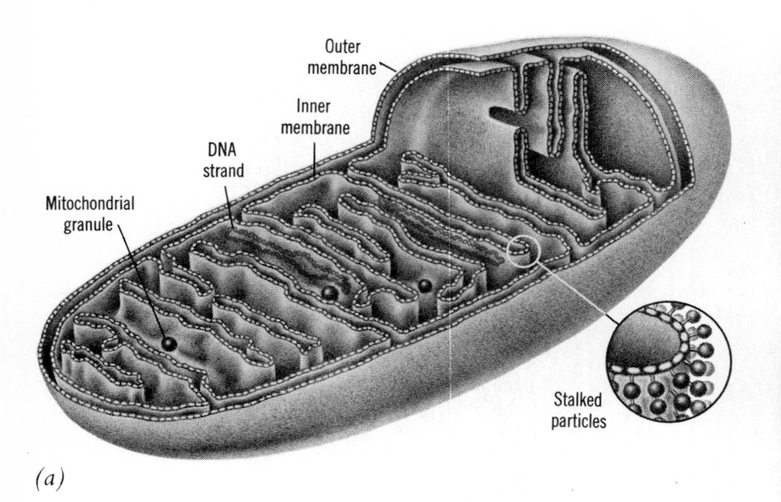

(a)

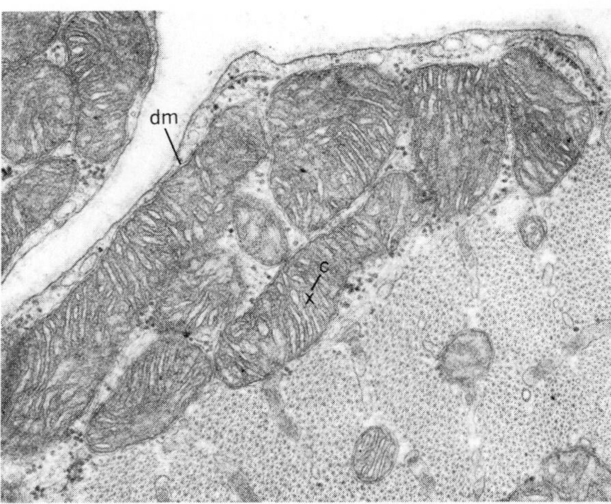

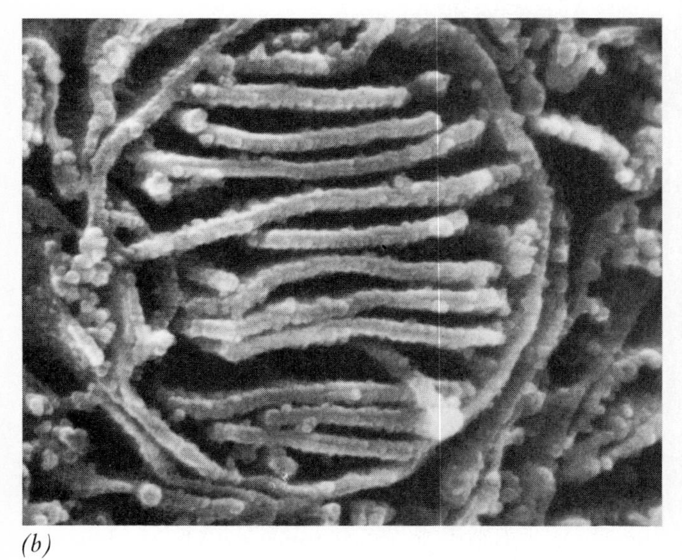

(b)

FIGURE 5.4 *(a)* Mitochondria visible in electron micrograph of skeletal muscle showing the double membrane (dm) and internal cristae. *(b)* Micrograph of a freeze-etched mitochondrion shows its extensive cristae. (Diagram from E. J. Gardner and D. P. Snustad, *Principles of Genetics,* 6th ed., Wiley, New York, 1981. Photographs courtesy of *(a)* Dr. A. Nag, *(b)* Dr. Sheeler.)

synthesis on 80S ribosomes. These antibiotics are excellent drugs for fighting human bacterial infections because they have a low toxicity in humans.

ATP IS PRODUCED IN MITOCHONDRIA

Mitochondria (my-toh-kon′dree-ah) are membrane-bound cytoplasmic organelles whose function is to convert chemical energy into ATP during aerobic respiration. Each mitochondrion is about the size of a bacterium, and in fact some theorists propose that mitochondria have evolved from a symbiotic association with aerobic bacteria.

Mitochondria are the "power houses" of the cell because they transform chemical energy into ATP. This process occurs on the inner membrane of the mitochondria, whose function is similar to that of the plasma membrane of aerobic bacteria. In addition, mitochondria are the site for the oxidation of pyruvic acid, fatty acids, and amino acids, and contain all of the enzymes of the tricarboxylic acid (TCA) cycle.

The typical mitochondrion (Figure 5.4) is the size of an average bacterium—0.5 to 1 μm in diameter and several micrometers long. Mitochondria are often spherical to oblong in shape; however, they can be extensively branched or they may be pleomorphic. Each mitochondrion has an outer membrane and a convoluted inner membrane that is folded into cristae (Figure 5.4*b*). This inner membrane is packed with proteins (80 percent of the membrane is protein) that function in the redox reactions involved in ATP production. By invaginating itself into cristae, the surface area of the inner membrane is greatly increased, thus providing more space for respiratory functions. The cristae of the mitochondrion contain complex proteins that possess ATP synthetase activity; ATP synthetase is the enzyme responsible for ATP formation.

Mitochondria contain DNA and ribosomes, and divide by binary fission, just as bacteria do. The mitochondrial DNA is a circular double-stranded molecule that varies in size from one eucaryotic species to the next. Mitochondrial DNA is usually about 1/200 the size of bacterial DNA. This DNA contains the information necessary to synthesize a limited number of proteins, which are synthesized on mitochondrial ribosomes. These ribosomes in most species are the size of bacterial ribosomes (70S), but the ribosomes of both mammalian and amphibian mitochondria are smaller (60S). Mitochondrial reproduction is dependent on the hereditary information in both the nuclear chromosomes and the mitochondrial DNA. Because of their dependence on information contained in the cell's nucleus, mitochondria are unable to divide outside of their cell's cytoplasm.

CHLOROPLASTS ARE THE SITES OF PHOTOSYNTHESIS

Plant cells have two types of energy-generating organelles: mitochondria and chloroplasts. Chloroplasts are membrane-bound cytoplasmic organelles of plant and algae where the reactions of photosynthesis occur. The chloroplast (Figure 5.5) is a banana-shaped organelle bound by a double membrane and is typically 2 to 3 μm thick and 5 to 10 μm long. The interior of the chloroplast is occupied by the stroma, which contains chromosomal DNA, ribosomes, and the enzymes needed to fix CO_2 to carbohydrate. Permeating the stroma are membranes derived from the inner mem-

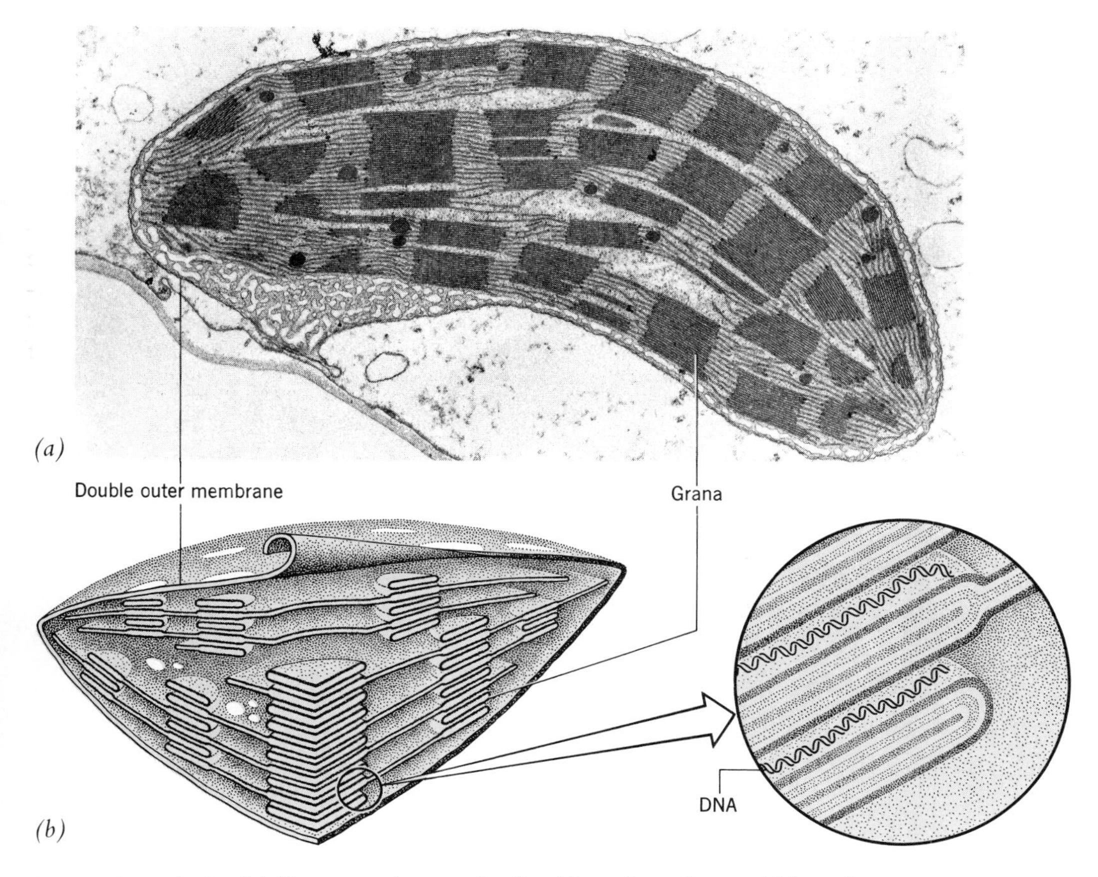

FIGURE 5.5 *(a)* Electron micrograph of a chloroplast of corn. Chloroplasts are the photosynthetic organelles of plant cells. They contain membrane-bound chlorophyll and DNA. *(b)* Enlarged drawing shows the DNA and the stacks of chlorophyll-containing membranes known as grana. (Courtesy of *(a)* L. K. Shumway, *(b)* Dr. E. Weier.)

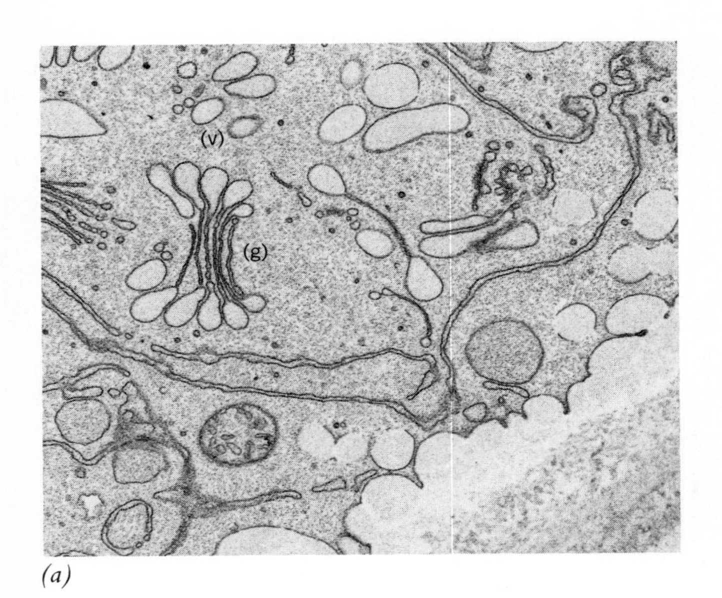

(a)

brane of the chloroplast. The chloroplast membranes are arranged like stacks of flattened sacs, called **grana,** interconnected by the **stroma lamella** (Figure 5.5*b*). Each granum consists of five to 30 flattened membrane sacs called **thylakoids,** which are the location of the pigments (chlorophylls and carotenoids) involved in photosynthesis.

Chloroplasts divide by binary fission in the cell's cytoplasm. Like mitochondria, the circular DNA located in the chloroplast stroma codes for proteins that are synthesized on the 70S chloroplast ribosomes. Each chloroplast contains one or more copies of this DNA, which is used in its replication. Chloroplast replication also requires information present in the cell's nuclear DNA; therefore the chloroplast cannot replicate outside of its host cell's cytoplasm.

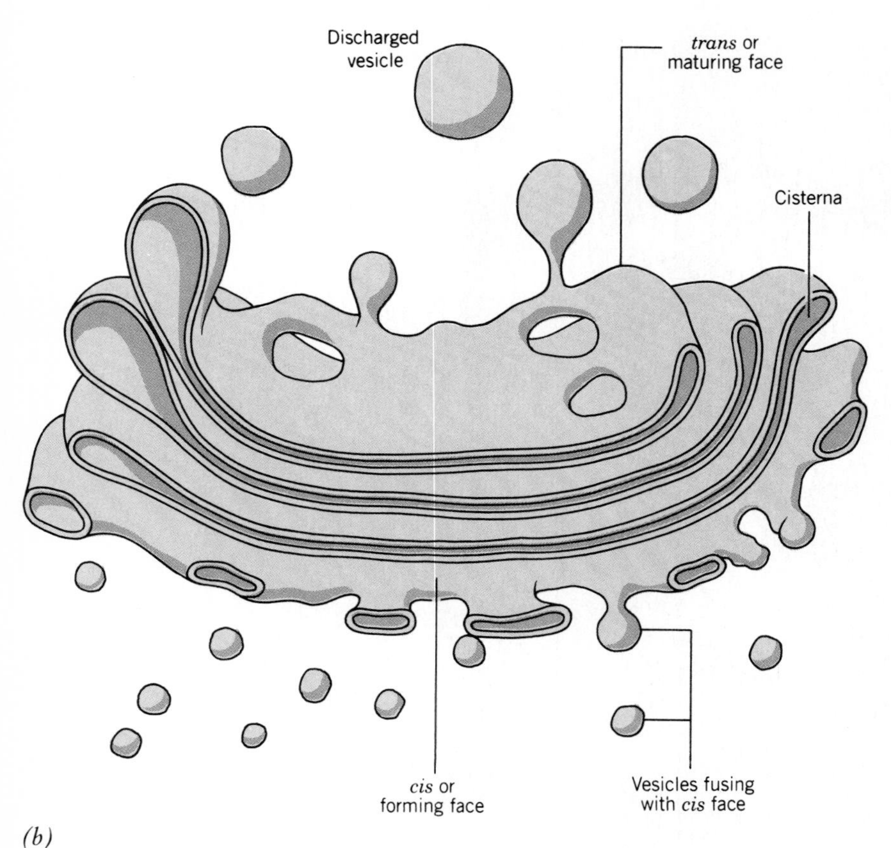

(b)

FIGURE 5.6 *(a)* The Golgi apparatus (g) of this animal cell is in actvely forming vesicles (v). *(b)* Diagram of vesicle formation by the Golgi apparatus. (Photograph courtesy of Dr. Hilton H. Mollenhauer, diagram from Sheeler and Bianchi, *Cell and Molecular Biology,* 3rd edition, John Wiley and Sons, 1987.)

GOLGI APPARATUS: CELL PACKAGING CENTER

The packaging center of virtually all eucaryotic cells is the **Golgi apparatus.** It is comprised of flattened membrane sacs and associated smooth membrane vesicles, first seen in 1898 by Camillo Golgi. Polysaccharides and glycoproteins destined to be secreted by the cell are packaged into vesicles by this membrane complex (Figure 5.6). After the vesicle's contents migrate to the cell's plasma membrane and fuse with it, they are released to the cell's exterior by the process of exocytosis.

Glycosyl (gly′co-sil) transferases are specialized enzymes that act on polysaccharides destined to be excreted. These transferases, localized in the Golgi apparatus, are biochemical markers that identify this organelle. Some glycosyl transferases can also transfer glucose residues to proteins. The proteins are made on the rough endoplasmic reticulum, transported to the Golgi in membrane-protected structures, and then glycosylated in the Golgi where they become glycoproteins. Another function of the Golgi is to package hydrolytic enzymes into lysosomes. Instead of being excreted, the lysosomes remain in the cytoplasm where they participate in cytoplasmic digestion.

LYSOSOMES FUNCTION IN DIGESTION

Eucaryotic cells produce hydrolytic enzymes that recycle or digest proteins, nucleic acids, polysaccharides, and other large molecules. Cells protect themselves from the action of these enzymes by packaging them in membrane-bound organelles called **lysosomes.** Hydrolytic enzymes are produced on the ribosomes of the rough endoplasmic reticulum, sorted out, and transported to the Golgi apparatus where they are packaged into lysosomes (Figure 5.7). The hydrolytic enzymes of the lysosome include proteases, nucleases, glycosi-

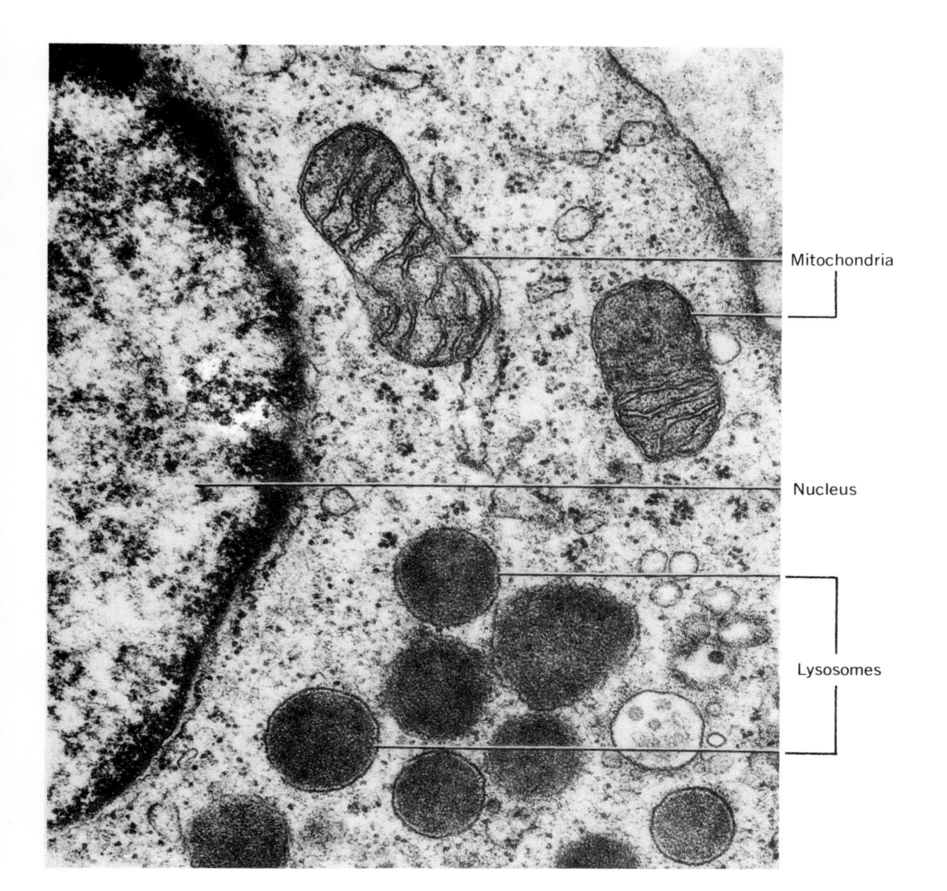

FIGURE 5.7 Electron micrograph of lysosomes in an animal cell. (From Sheeler and Bianchi, *Cell Biology,* 2nd edition, John Wiley and Sons, 1983. Courtesy of R. Chao.)

dases, sulfatases, lipases, and phosphatases. Unlike mitochondria and chloroplasts, lysosomes do not have a distinctive morphology, so they are identified by biochemical and cytochemical techniques. The presence of acid phosphatase is the most widely used marker for their identification.

Another name for lysosomes is "suicide packets," since the enzymes they contain would digest the cell's own protoplasm if they were not sequestered inside the lysosome. Some animals actually utilize their lysosomes to destroy cellular structures during development. An example is the polliwog tail, which is reabsorbed as the polliwog changes from a tadpole to an adult frog.

Lysosomes release their hydrolytic enzymes into the endocytic vesicles with which they fuse (Figure 5.8). Lysosomes specifically recognize phagocytic vesicles and pinocytic vesicles and do not fuse with just any membrane system. The degradative enzymes are released into the newly formed digestive vacuole, where they hydrolyze macromolecules into nutrients that are assimilated into the cytoplasm. This process is the major mechanism by which certain protozoa (amebae) digest their food. Another role of lysosomes is the re-

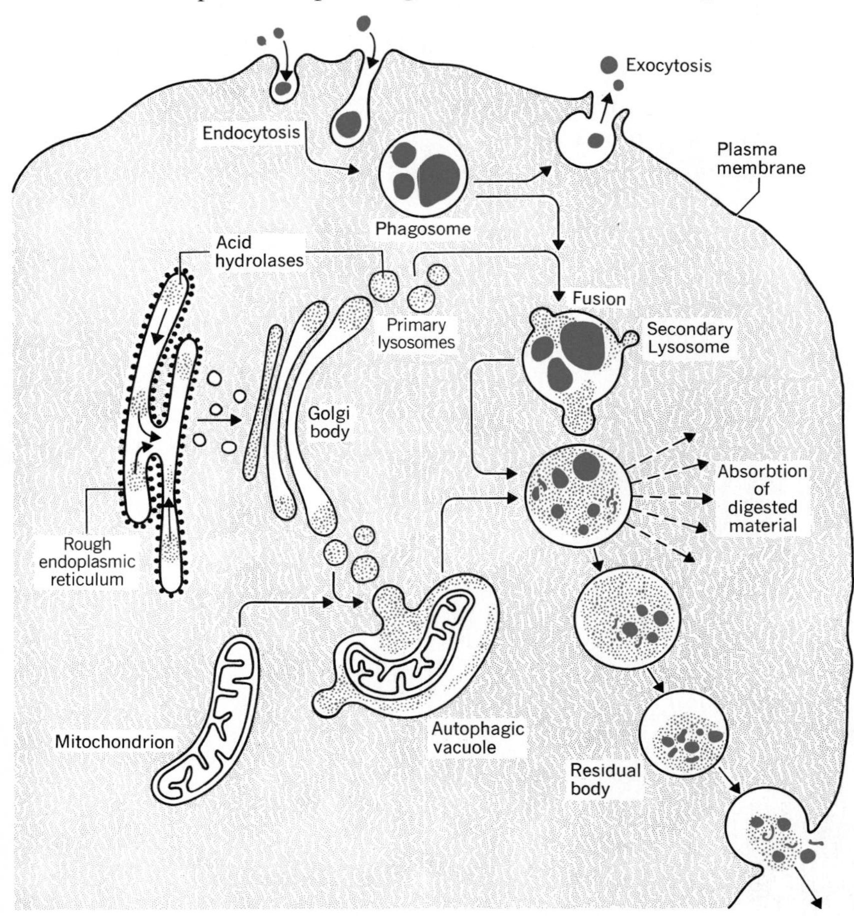

FIGURE 5.8 Lysosomes are involved with the digestion of material within cells. Foreign matter engulfed by endocytosis into a phagosome combines with a lysosome produced by the Golgi body. The hydrolytic enzymes in the lysosome are released into the phagosome upon fusion. Undigested material is released from the cell by exocytosis. (From Sheeler and Bianchi, *Cell Biology*, 2nd edition, John Wiley & Sons, 1983.)

moval of foreign matter from animal tissues. Human macrophages and neutrophils phagocytize microorganisms and viruses, which are destroyed after the phagosome combines with a lysosome.

OTHER CYTOPLASMIC STRUCTURES

Eucaryotic cells are large enough to contain a wide variety of cytoplasmic inclusions. **Peroxisomes** are cytoplasmic organelles composed of a fine granular matrix surrounded by a single membrane (Figure 5.9). They contain at least some peroxide-producing enzymes, such as urate oxidase and D-amino acid oxidase, and contain most of the cell's catalase. Peroxisomes are a major site for aerobic oxidation of metabolites. Such oxidations generate H_2O_2, which is broken down by catalase. Although they are found in virtually all eucaryotic cells, the enzymes they contain vary widely depending on cell type. In addition to peroxisomes, some protozoa and plant cells contain food and water vesicles, which are limited by a single membrane. Animal cells regulate their internal osmotic pressure with water vesicles, which collect excess water in the cytoplasm and then expel it by exocytosis. In mature plants, food material is often stored in the double-membrane-bounded **leucoplast.** A similar plant organelle is the **chromoplast,** which contains the pigments responsible for the color in fruits and vegetables.

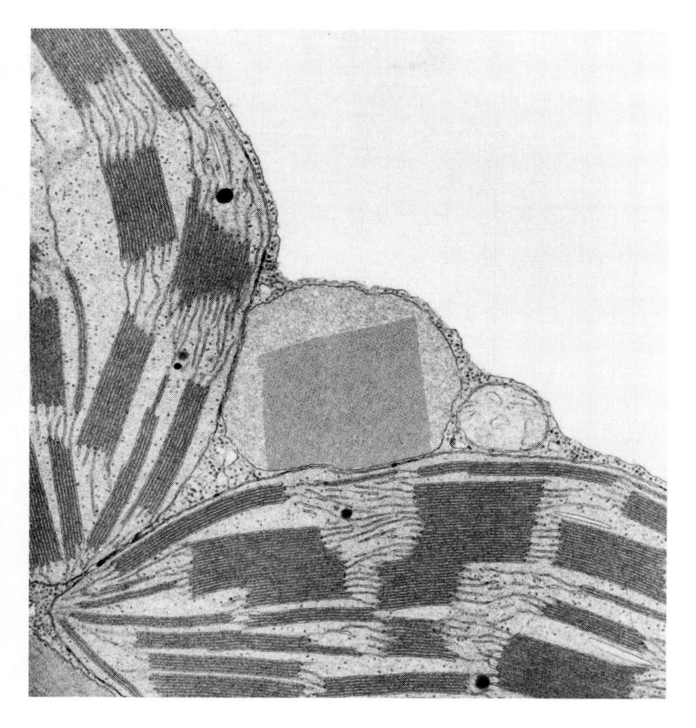

FIGURE 5.9 Leaf peroxisome in contact with two chloroplasts. Peroxisomes are found in both plant and animal cells and contain oxidative enzymes that participate in degradative metabolism and remove toxic products from cells. (From S. E. Frederick and E. H. Newcomb, *J. Cell Biol., 43:*343 1969.)

MOVEMENT IN EUCARYOTIC CELLS

Movement in eucaryotic cells takes many different forms. Cell motility can be driven by external appendages known as cilia and flagella or by internal changes in the protoplasm that produce pseudopodia. Organized internal movement occurs during mitosis as the individual chromosomes separate to form daughter cells.

Researchers have discovered cytoplasmic filaments and/or microtubules associated with each of these forms of motility. Microtubules are long cylinders composed of proteins polymerized around a hollow core; they morphologically resemble the filaments of bacterial flagella. Microtubules are large enough to be easily seen in electron micrographs. The smaller cytoplasmic filaments permeate the cytoplasm of eucaryotic cells. They participate in movements within the cytoplasm and contribute to the architectural framework of the cell.

CELL MOTILITY: FLAGELLA AND CILIA

Cilia and flagella are external locomotor organelles of eucaryotic cells. Flagellated cells include sperm cells, flagellated protozoa, and numerous algae. These single cells usually contain one or two flagella that beat with a whip-like motion to propel the cell through liquid environments. Cilia (Figure 5.10) are morphologically similar to flagella; however, they are always arranged in groups and beat with a rhythmic motion that sets the surrounding liquid in motion. Cilia cover the surface of ciliated protozoa and ciliated cells line certain body cavities; for example, ciliated epithelial cells line the human trachea. The rhythmic, coordinated motion

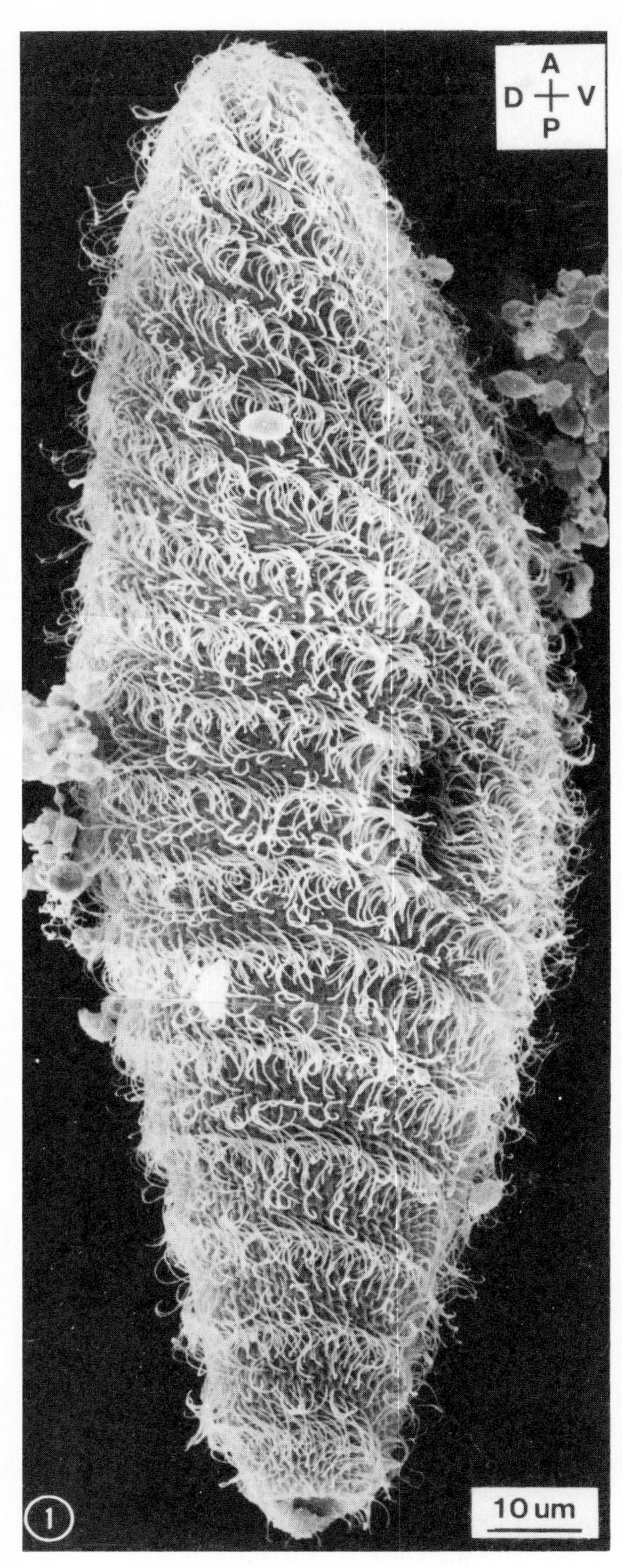

(a)

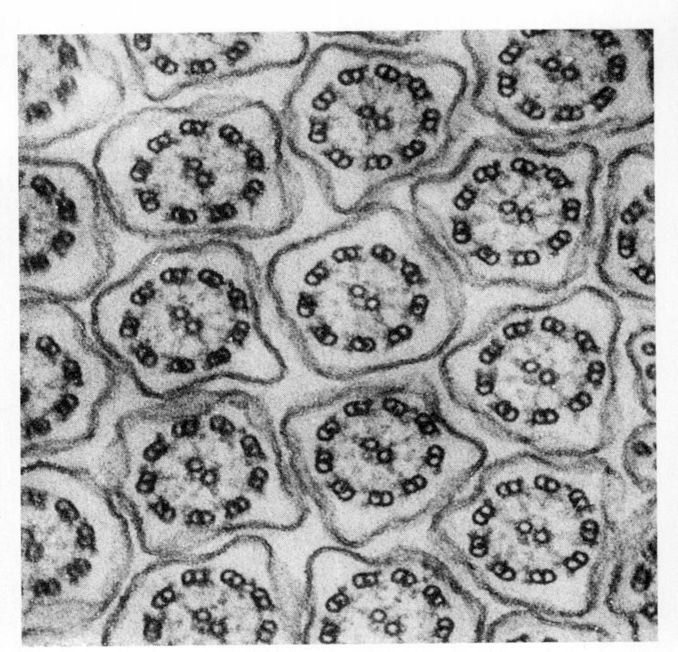

(b)

FIGURE 5.10 *(a)* Scanning electron micrograph of a *Paramecium,* showing cilia and its oral groove. *(b)* Transmission electron micrograph of a cross section through the cilia of a ciliate showing their symmetrical arrangement and internal structure. (Courtesy of *(a)* Dr. Sidney L. Tamm, *(b)* Dr. Peter Satir.)

of cilia, arranged in rows, is distinctly different from the independent whip-like motion of a eucaryotic flagellum.

Both cilia and flagella are protrusions from the plasma membrane that originate in a basal body lying beneath the plasma membrane. The distinctive morphology of the basal body is also seen in the cytoplasmic structures called centrioles, which are associated with the mitotic apparatus. The flagellar and ciliary basal bodies gives rise to the *axoneme,* the microtubule arrangement common to cilia and flagella (Figure 5.11), which protrudes from the cell's surface.

KEY POINT

Motility is one criterion for classifying single-celled animals, the protozoa. Movement in the protozoa is by flagella (flagellates), cilia (ciliates), and pseudopodia formed by ameba.

Microtubules and Motility

Microtubules are composed of proteins that polymerize around a central hollow core into long, thin cylinders. Microtubules have a constant diameter, between 25 and 30 nm, and an indeterminate length. In cross section, each microtubule is a circular array of 13 protein subunits arranged around the hollow core. The two main proteins present in isolated microtubules are α tubulin and β tubulin. These proteins exist as a heterodimer and self-assemble to form the basic tubular structure of the microtubule.

The microtubule doublet (Figure 5.11) is composed of an A tubule and a B tubule. The A tubule contains the normal 13 subunits of the heterodimer, whereas the B tubule has only 10 or 11 subunits. The semicircular hook structures attached to the pairs of microtubules in flagella and cilia are dynein (dy′neen) molecules. **Dynein** contains ATPase activity and releases energy for locomotion by hydrolyzing ATP.

The axonemes of flagella and cilia are composed of nine microtubule doublets arranged around two central microtubules (Figure 5.11). Surrounding the axoneme is a membrane that appears to be an extension of the plasma membrane. The movement of flagella and cilia occurs when the microtubule doublets slide over each other. The sliding, mediated by ATP hydrolysis, produces the forces necessary to bend the flagellum or cilium. This bending in turn creates the whip-like motion of the flagellum or the beating motion of cilia.

CYTOSKELETON

The protoplasm of the eucaryotic cell has an internal organization created by the extensive architectural network of microtubules and cytoplasmic filaments. This framework can be so extensive that it entraps organelles at specific cytoplasmic sites. In other cells it serves as a roadway on which organelles move about the cytoplasm. Microtubules are found in the spindle of the mitotic apparatus as an architectural support network in the cytoplasm of the cell (Figure 5.12*a*). In brain tissue, the high concentration of microtubules supports the elongated shape of brain and nerve cells. The architecture of the cell also depends on thin polymers of proteins collectively known as cytoplasmic filaments.

Cytoplasmic filaments are thin proteinaceous polymers that contribute to the cell's cytoskeleton.

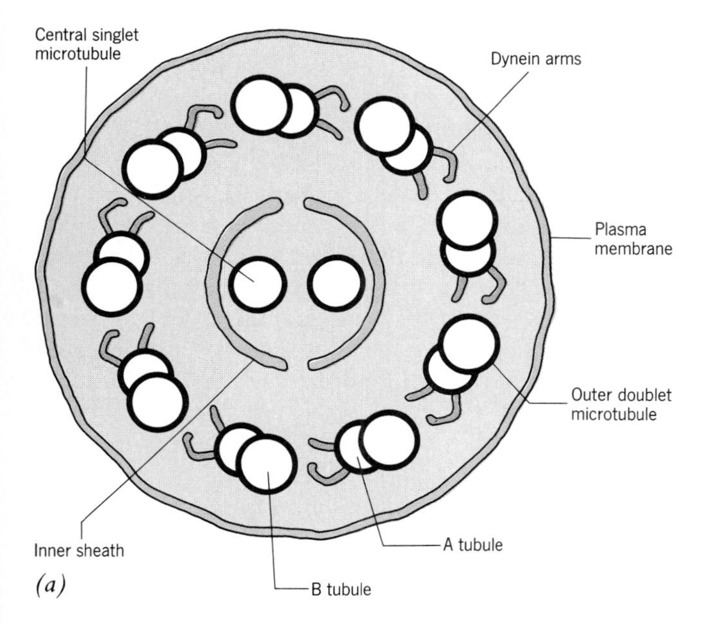

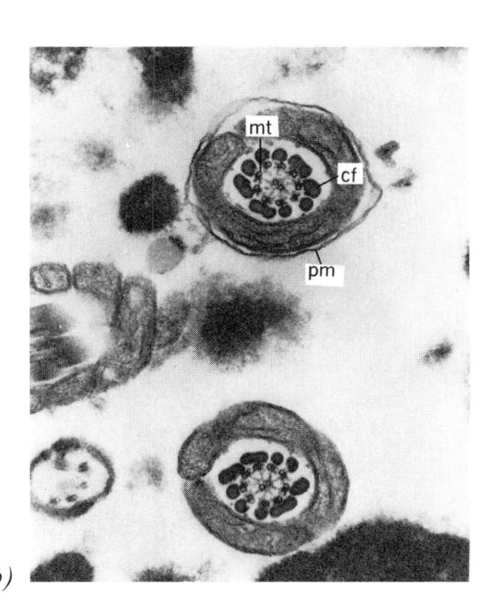

FIGURE 5.11 *(a)* Diagram of a cross section through a eucaryotic flagellum showing the "9 + 2" arrangement of microtubules, dynein arm, and surrounding plasma membrane. *(b)* Cross section through a typical eucaryotic flagellum. (Courtesy of Dr. C. Lindemann.)

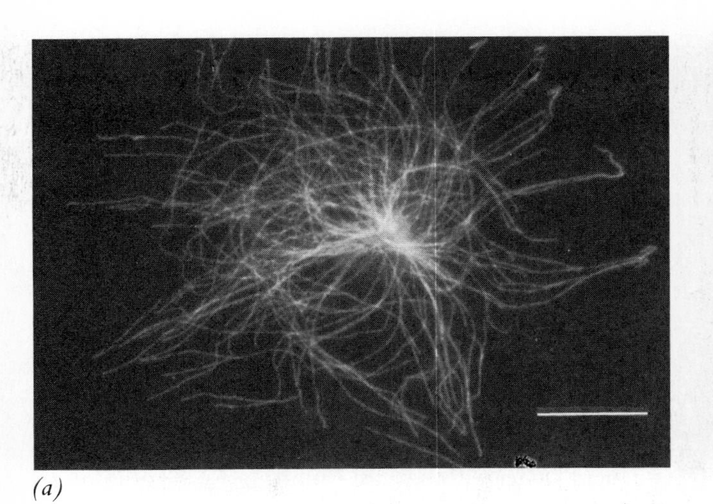

(a)

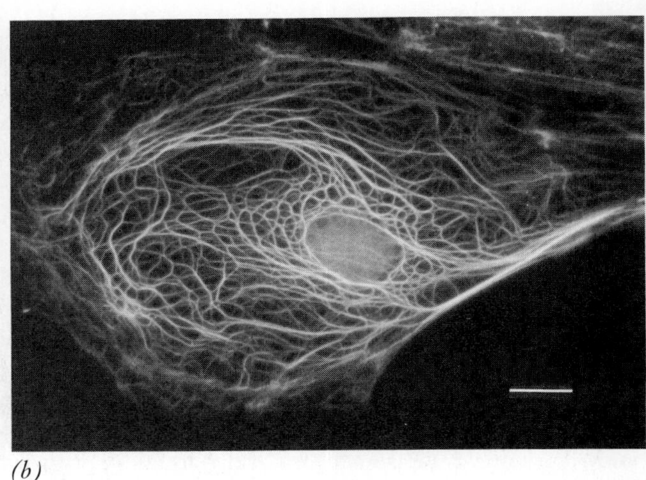

(b)

FIGURE 5.12 Animal cells have an internal architecture created by cytoplasmic filaments and microtubules that strengthens the cytoplasmic membrane and confers shape on the cell, and is involved in certain types of movement. The microtubules *(a)* of a sea urchin coelomocyte are visible when stained with fluorescent antibody directed against tubulin. *(b)* The distribution of intermediate filaments in cultured epithelial cells. The bar represents 10 μm. (Courtesy of Dr. Kenneth Edds, State University of New York at Buffalo.)

This type of filament, first observed in striated muscle cells, has also been found in many types of eucaryotic cells. The modern use of fluorescent-labeled antibodies enables microscopists to observe the cellular distribution of microfilaments. Although there are many types of these filaments, the most abundant type in the 5- to 6-nm class are the **actin** filaments, which are now commonly referred to as **microfilaments.** Microfilaments and microtubules provide the architectural framework that maintains the three-dimensional structure of many cells.

THE NUCLEUS AND HEREDITY

A prominent feature of the eucaryotic cytoplasm is the membrane-bound nucleus, which is the site of the cell's complement of DNA. The complexity of the eucaryotic cell is much greater than that of bacteria, so it follows that a eucaryotic cell possesses 100 times more DNA than a typical bacterial cell. The presence of the nucleus and many chromosomes necessitates that the process of eucaryotic cell division be significantly more complex than bacterial replication.

STRUCTURE OF THE NUCLEUS

The eucaryotic nucleus is separated from the cytoplasm by a double membrane known as the **nuclear envelope** (Figure 5.13). Both surfaces of the nuclear envelope are pockmarked by numerous nuclear pores through which proteins and RNA pass. Nuclear pores are holes in the double membrane formed where the two membranes fuse. Ribosomes may be attached to the cytoplasmic side of the outer nuclear membrane, which is continuous with the rough endoplasmic reticulum. The membrane-bound nucleus is always present in nondividing cells, but as we shall see, it disintegrates during mitosis only to reform after nuclear division.

DNA Is Organized into Chromosomes

In the nondividing nuclear karyoplasm, the cell's DNA is complexed with proteins that give it a fibrillar appearance (Figure 5.13). This negatively charged DNA is complexed with positively charged basic proteins called **histones** and other nonhistone proteins. The histones contribute to the structure of the chromosome and the nonhistone proteins are thought to be involved in regulating gene activation.

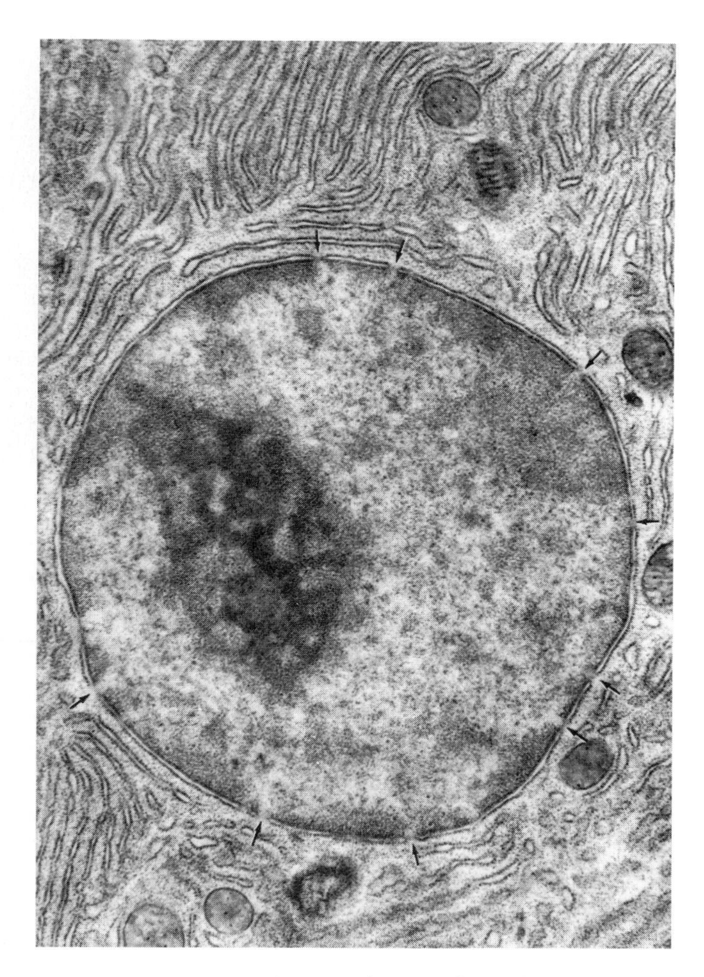

FIGURE 5.13 The interphase nucleus of an animal cell. The nuclear envelope surrounds the chromatin and the nucleolus. Material passes between the cytoplasm and the karyoplasm via the numerous nuclear pores (arrows). Numerous ribosomes are present on the outer membrane of the nuclear envelope, which has continuity with the endoplasmic reticulum. (Courtesy of Dr. Don W. Fawcett.)

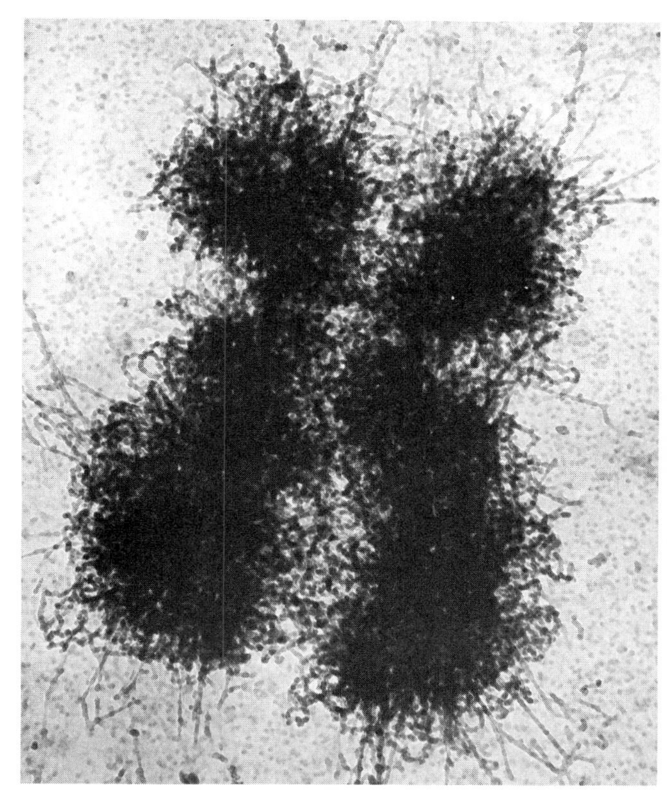

FIGURE 5.14 Electron micrograph of a human chromosome in metaphase showing its two distinct chromatids attached at the centrometer. (Courtesy of Dr. Gunther F. Bahr, Armed Forces Institute of Pathology, Washington, D.C.)

The eucaryotic cell's DNA is present as discrete large molecules called **chromosomes** (Figure 5.14). During most of the cell's reproductive cycle, the DNA exists in a relatively unfolded state known as **chromatin.** This changes during mitosis when the cell's individual chromosomes become visible as the individual DNA molecules coalesce into tightly packed DNA-protein structures. Based on the strict laws that govern the condensation of nuclear DNAs, each chromosome of a given cell has its own specific length and width that are maintained from generation to generation.

Eucaryotic cells of a given species always have a set number of chromosomes that is greater than one. Cells that possess one copy of each chromosome are haploid ($1n$). This applies to reproductive cells such as a sperm or an egg cell and the vegetative forms of certain fungi. Most other eucaryotic cells are diploid ($2n$) since they contain two copies of each chromosome, one from each parent.

Nucleolus Is the Site of Ribosome Formation

A major function of the nucleus is to synthesize RNA from nuclear DNA. The **nucleolus** (Figure 5.13) is the electron-dense area of the karyoplasm involved in

RNA synthesis. About 5 to 10 percent of the nucleolus is RNA, the rest is mainly protein. Very large RNA molecules are synthesized in the nucleolus before they are processed into the smaller RNA molecules found in ribosomes. Ribosomal proteins synthesized in the cytoplasm enter the nucleus through nuclear pores and then combine with the newly formed ribosomal RNA to form large and small subunits of the ribosomes. These subunits exit the karyoplasm via the pores in the nuclear envelope to become functional in the cytoplasm.

NUCLEAR DIVISION AND THE CELL CYCLE

Cell division in eucaryotic cells is more complex than in bacteria since eucaryotic cells must assure that each daughter cell receives an equal complement of chromosomes. This process is accomplished by replicating the cell nucleus before cell division. The chromosomes in the nucleus are separated to daughter cells by mitosis.

Eucaryotic cells intersperse periods of growth and DNA synthesis between nuclear divisions. The eucaryotic cell cycle (Figure 5.15) is divided into two parts: interphase and mitosis. **Interphase** is the period between nuclear divisions and consists of growth$_1$ (G_1), DNA synthesis (S), and growth$_2$ (G_2). During mitosis (M), the replicated chromosomes are equally divided between the two newly formed daughter cells. The duration of each stage can be determined by the rate of DNA synthesis or, in the case of mitosis, by microscopic observation. The period in which neither mitosis nor DNA synthesis occurs is referred to as G_1 when it follows mitosis and G_2 when it follows DNA synthesis. Interphase is the combined time of the G_1, S, and G_2 stages and it varies from one eucaryotic cell to another. Prior to cell division, two nuclei are formed

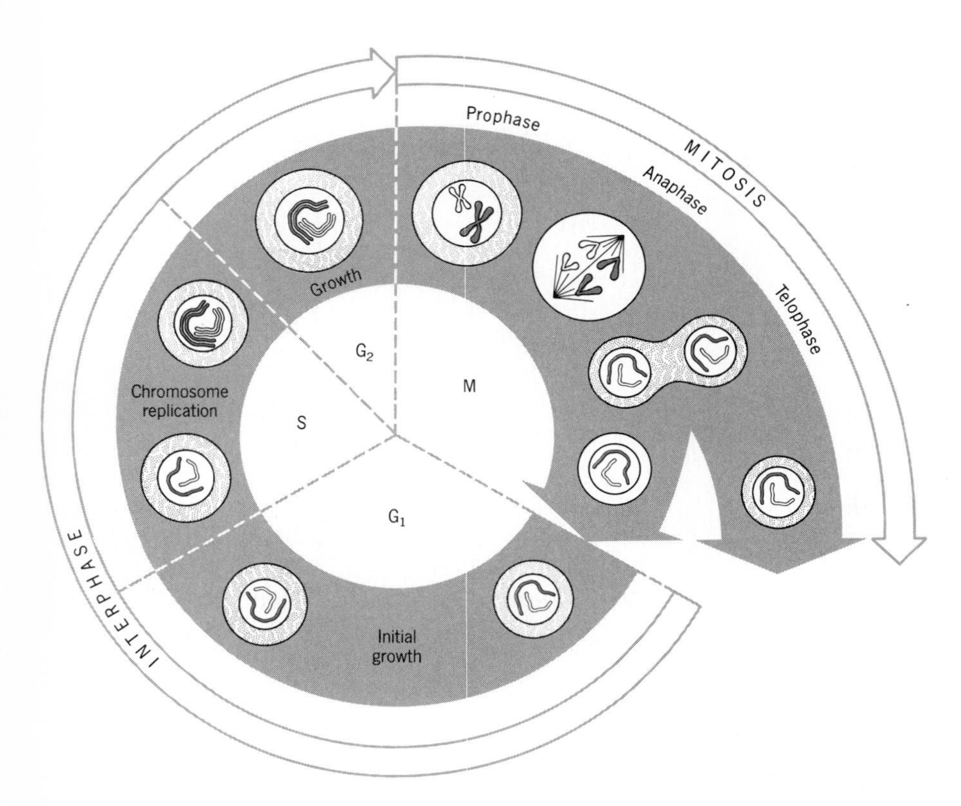

FIGURE 5.15 The cycle of growth and reproduction of a typical eucaryotic cell. (From P. Sheeler and D. Bianchi, *Cell Biology,* Wiley, New York, 1980.)

by mitosis. The cells of adult animals may then enter the nondividing stage (referred to as G_0 for zero growth), which can be of an indefinite duration.

Mitosis

The orderly division and separation of equal complements of nuclear DNA to daughter cells is accomplished by **mitosis** (Figure 5.16). Mitosis follows interphase and occurs as four distinct phases that can be observed with the light microscope (Figure 5.17). Between mitotic divisions, the chromosomes exist in an unfolded state bound to histone and nonhistone proteins and as such participate in cell maintenance and growth. The chromosomes are replicated during the S stage of interphase, so at the beginning of mitosis, each cell contains an extra copy of its genetic material ($4n$). This increased amount of nucleic acid first becomes evident to the microscopist during prophase.

Prophase, the first and longest stage of mitosis, occupies about 60 percent of the time required for mitosis. During early prophase, the chromatin coalesces into chromosomes. Each chromosome (Figure 5.14) contains two **chromatids,** which are duplicate copies of the cell's genetic information. Chromatids are joined at the region called the centromere, and remain joined until they are separated during anaphase. By late prophase, the condensed chromosomes have moved toward the equatorial plate, the nuclear envelope has disappeared, and the mitotic apparatus has formed.

Metaphase is the stage of mitosis in which the chromosomes line up in the equatorial plane of the spindle. This arrangement is governed by the mitotic apparatus, which is the organizer for the alignment of the chromosomes and their eventual allocation to daughter cells. The **mitotic apparatus** is composed of the spindle, two centrioles, two asters, and the chromosomes aligned in the equatorial plane of the spindle. The mitotic apparatus is a distinct cellular structure that has been isolated in an intact but nonfunctional form. For example, in Figure 5.18 the mitotic apparatus has been isolated from dividing sea urchin eggs. It originates from the centrioles (Figure 5.1) that are always present in animal cells; centrioles are not present in all plant cells.

The centrioles migrate to opposite poles of the cell during early prophase and at the same time serve as a growing point for the formation of the spindle and the asters. **Asters** are the array of fibers radiating from the centrioles in all directions that appear to merge with the spindle fibers. The spindle is composed of two types of microtubules (Figure 5.18): the "framework" microtubules connect the two centrioles at opposite poles of the cell, while the individual **kinetochore** microtubules extend from the centrioles and attach to the centromere of each chromatid.

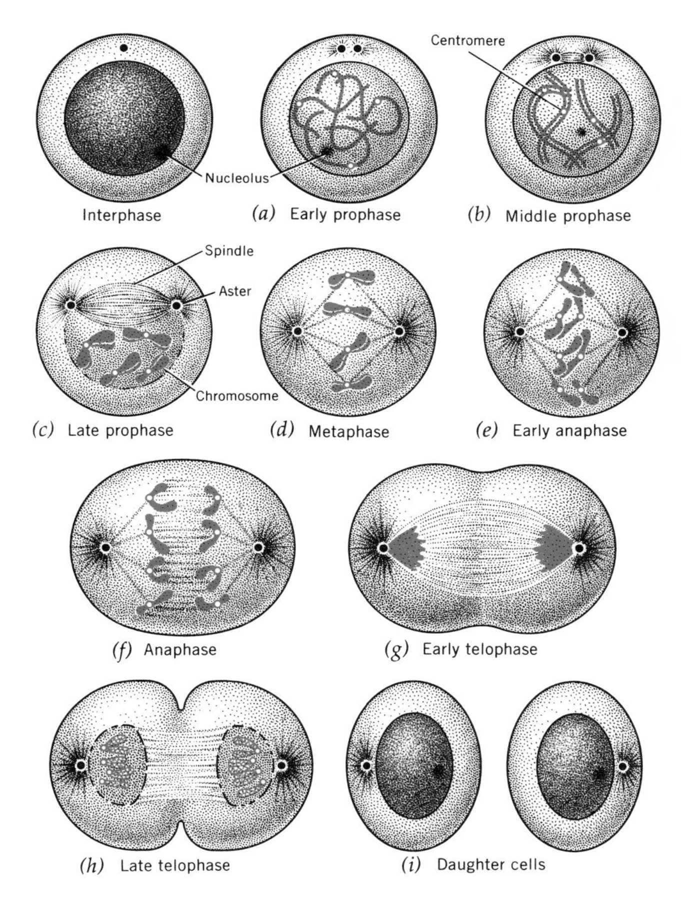

FIGURE 5.16 The major stages of mitosis—prophase, metaphase, anaphase, and telophase—are depicted sequentially. Photos of cells in the process of mitosis are shown in Figure 5.17. (From E. J. Gardner and D. P. Snustad, *Principles of Genetics,* 6th ed., Wiley, New York, 1981.)

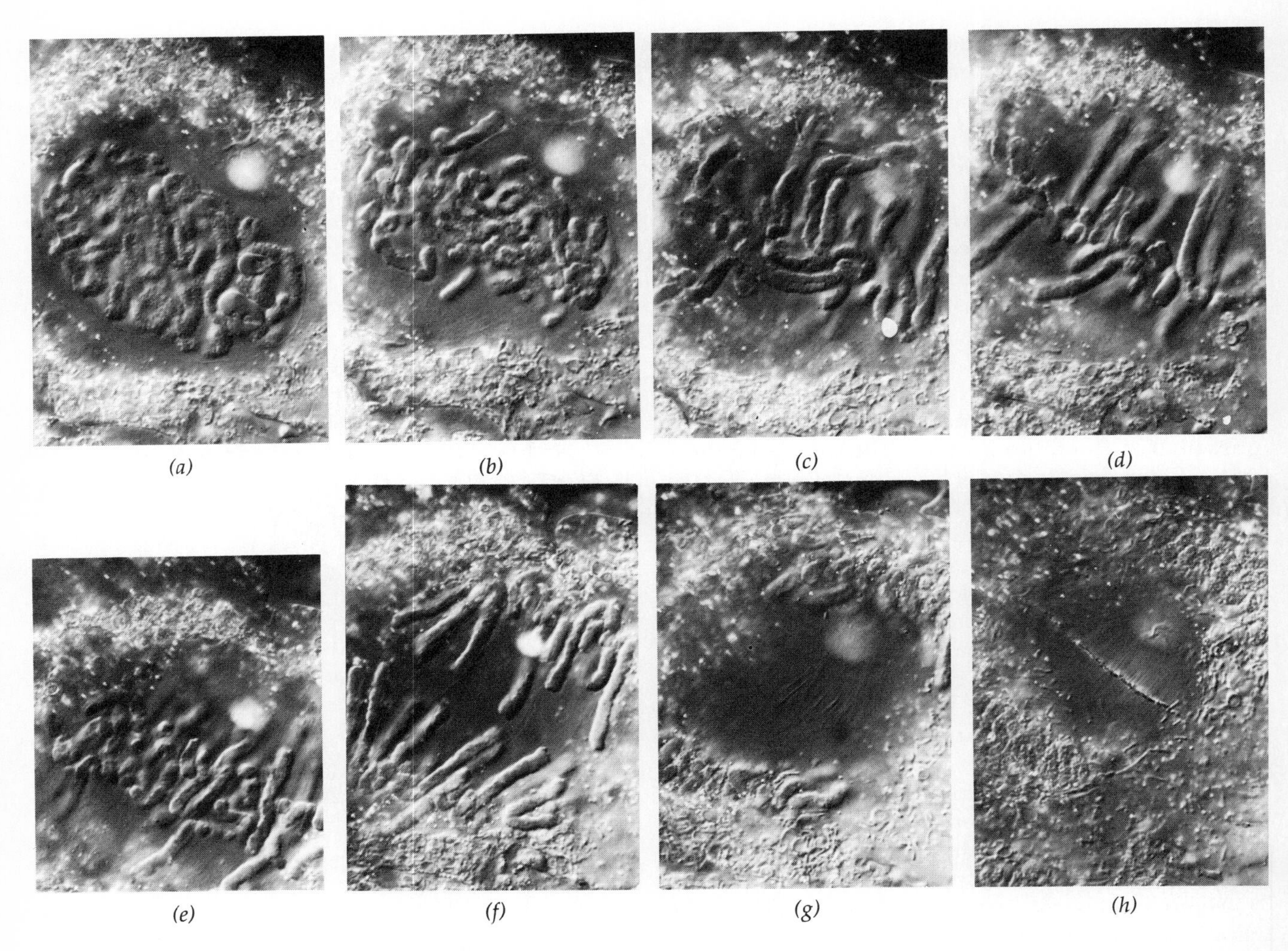

FIGURE 5.17 Photomicrographs from 16-mm time-lapse film taken of mitosis in the endosperm of *Haemanthus*. The time for the complete sequence was 160 minutes. The letters correspond to the stages of mitosis diagramed in Figure 5.16. (Courtesy of Dr. A. Bajer. From A. Bajer, *Chromosoma,* 24:384–417, 1968, copyright © 1968, Springer-Verlag, New York.)

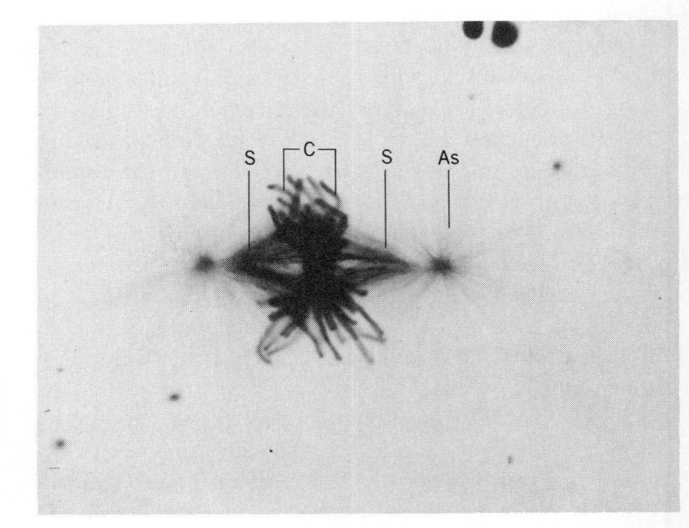

FIGURE 5.18 The mitotic apparatus isolated from Oregon newt showing the aster (As), the spindle (S), and the chromatids (C) lined up in the equatorial plane. (Courtesy of Dr. A. Bajer.)

During early **anaphase** (Figure 5.17), the chromatids take on an elongated appearance as they are pulled toward the centrioles. This movement is attributed to the shortening of the microtubules attached to the centromeres of the chromatids. Early **telophase** begins as the chromatids coalesce at the poles of the cell into a mass of chromatin consisting of indistinguishable chromatids. Next, the nuclear membranes are formed around the separated chromosomes and the equatorial plane of the spindle becomes the plane of cellular division. The cell divides (cytokinesis) such that the cytoplasm of each daughter contains one nucleus and becomes a diploid (2*n*) cell.

Eucaryotic cells need mitosis to replicate their chromosomes and to move a copy of each chromosome to the individual daughter cells. Eucaryotic cells need not replicate continuously, and often go through long resting stages between replications. This allows animal cells to grow rapidly during developmental periods, after which some cells, such as nerve cells, never again divide. In contrast, bacterial cells are constantly in the process of replicating their nuclear material.

Reductive Division by Meiosis

Eucaryotic organisms that reproduce sexually must produce gametes or cells that contain one half the DNA found in the adult. **Meiosis** is the process of reductive division, which forms haploid (1*n*) eucaryotic gametes. Meiosis is accomplished by two sequential nuclear divisions following a single replication of the cell's DNA. Haploid gametes (1*n*) from distinct parents unite during sexual reproduction to form a diploid individual.

EVOLUTION OF EUCARYOTIC AND PROCARYOTIC CELLS

Eucaryotic cells are distinct and different from the simpler procaryotic cell as summarized in Table 5.1. Both cell types perform the functions basic to life, but the eucaryotes are structurally much more complex. This leads one to wonder where the eucaryotes came from and if they developed after the procaryotic cell type was established.

The prebiotic earth was a very inhospitable place. The atmosphere consisted of carbon monoxide, carbon dioxide, water vapor, nitrogen gas, hydrogen gas, and hydrogen sulfide; there was no free oxygen (O_2). The first evidence of life is found in sedimentary rocks that were formed about 3.5 billion years ago. These rocks

TABLE 5.1 Procaryotic and eucaryotic cell characteristics

CHARACTERISTIC	PROCARYOTIC	EUCARYOTIC
	Structural Character	
Common cell size	0.3 to 2.5 μm	2 to 20 μm
DNA location	Nuclear region	Membrane-bound nucleus
Chromosome number	One	Generally more than one
Replication by mitosis	No	Yes
Cell Wall	Chemically complex, usually present	If present, less complex
Endoplasmic reticulum, lysosomes, Golgi, nucleolus	Absent	Usually present
Vacuoles	Absent or rare	Often present
Ribosome sedimentation	70S	80S
	Physiological Functions	
Site of photosynthesis	Cytoplasmic membrane extensions, thylakoids	Chloroplast grana
Site of respiration	Cytoplasmic membranes	Mitochondria
Movement	Flagella (simple)	Microtubules in "9 + 2" arrangement
Phagocytosis, pinocytosis	Absent	Often present

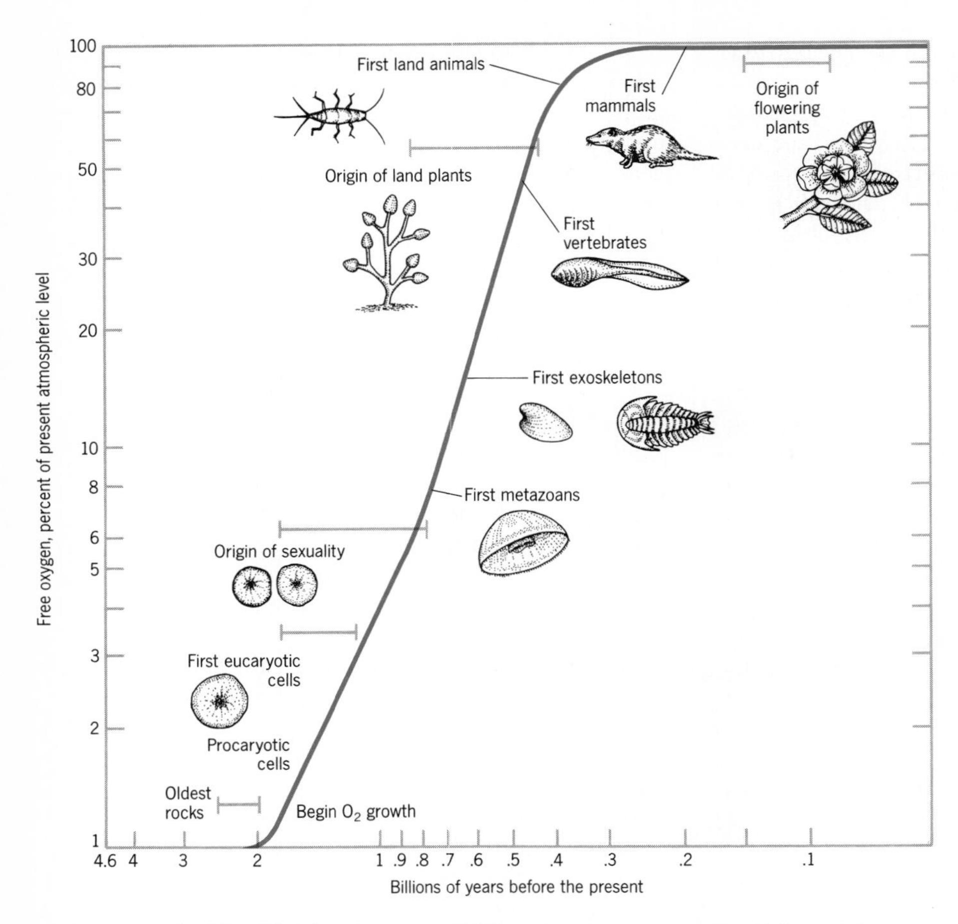

FIGURE 5.19 The development of life and oxygen on earth. (Adapted from Preston Cloud, "The Biosphere," *Sci. Am.*, 249:176–189, September 1983.)

contain evidence of photosynthetic activity. The oldest known fossils resemble procaryotic cells and were found in western Australia in sedimentary rocks 2.8 billion years old. In rocks that are only 2.0 billion years old, paleontologists have found evidence of cyanobacteria and other procaryotic cells (Figure 5.19). The cyanobacteria were probably responsible for increasing the concentration of free oxygen in the atmosphere beginning about 2.0 billion years ago. Eucaryotic cells date from about 1.4 billion years ago and their development and complexity coincide with the buildup of atmospheric oxygen. The overriding conclusion from these findings is that procaryotic cells preceded the emergence of eucaryotic cells.

A major problem for students of evolution is this: How did two distinct cell types come to exist in the earth's closed environment? Some biologists suggest that mutations in procaryotic cells, followed by natural selection, could have given rise to all the characteristics of eucaryotic cells. Others suggest that eucaryotic cells developed from more than one distinct cell type after they joined together as symbionts. A combination of these two lines of descent is probably closer to the truth.

ENDOSYMBIOTIC THEORY

Perhaps the following explanation of the endosymbiotic theory will help to clarify the picture for you. Proponents of the **endosymbiotic theory** argue that eucaryotic cells evolved from progenitor cells, which formed symbiotic relationships with cyanobacteria and aerobic bacteria. The progenitor cell is envisioned as a large cell capable of phagocytizing smaller procaryotic cells. An engulfed cyanobacterium presumably endowed the symbiont with the ability to photosynthesize, and eventually became the chloroplast of plant cells. Similarly, the progenitor cell engulfed aerobic bacteria that eventually evolved into the mitochondrion of the eucaryotic cell.

Current evidence supports the theory that chloroplasts were derived from free-living photosynthetic procaryotes (cyanobacteria and *Prochloron*). Marked similarities exist between chloroplast DNA and bacterial genomes; the ribosomes in bacteria and chloroplasts are very similar right down to the homology between their 16S ribosomal RNAs; and chloroplasts possess a protein-synthesizing system that is similar in structure and function to the systems in procaryotes.

The case for the endosymbiotic origin of mitochondria is much weaker, but still possible. The mitochondrial DNA varies greatly in its size and structure; mitochondrial ribosomes (55S to 70S) differ substantially from both cytoplasmic and procaryotic ribosomes; and the system for protein synthesis is markedly different from that of procaryotic cells. At one time, the genetic code was considered to be universal, but it is now known that some mitochondria have an alternative way to read the genetic code. None of the applicable biochemical evidence suggests that mitochondria had a common origin; however, it is still possible that mitochondria evolved at a much faster rate than did chloroplasts.

Currently there are no concrete answers to the questions of the origin and the meaning of the various cell types. Yet when we look at what cells must do in order to survive, we find a remarkable consistency among living things. This consistency is evident in the chemical mechanisms used by cells to perform cellular functions. If cell structure represents a divergence in living systems, then biochemistry must represent the continuity of life.

Summary Outline

OVERVIEW OF THE EUCARYOTIC CELL

1. Eucaryotic cells are large and have a structural sophistication significantly more complex than procaryotic cells.
2. Many of the biological functions of the eucaryotic cell are contained in cytoplasmic organelles.
3. Chromosomes located in the membrane-bound nucleus contain the cell's hereditary information. The nucleus divides by mitosis.

CELL WALLS AND THE PLASMA MEMBRANE

1. Cell walls of higher plants, fungi, and algae differ considerably in composition and physical structure from bacterial cell walls.
2. Cellulose, hemicellulose, and pectin are the major polysaccharides of plant cell walls.
3. Chitin and β glucans are major components of the cell walls of filamentous fungi; mannan is the major polysaccharide of yeast cell walls.
4. The protoplasm of each cell is surrounded by a plasma membrane that contains sterols and glycolipids.
5. The semipermeable plasma membrane is bridged by endocytotic mechanisms used to engulf large molecules (pinocytosis) and particles (phagocytosis).

CYTOPLASMIC ORGANELLES

1. Endoplasmic reticulum is a membrane network of flattened sacs and tubules that are often connected to the plasma and nuclear membranes.
2. The rough endoplasmic reticulum is studded with ribosomes that participate in protein synthesis.
3. The smooth endoplasmic reticulum (lacks ribosomes) is involved in glycogen, steroid, and lipid synthesis and in some cells in drug detoxification.
4. Cytoplasmic protein synthesis occurs on 80S ribosomes that are attached to membranes or are free in the cytoplasm. Eucaryotic ribosomes are structurally different from the 70S bacterial ribo-

somes and not affected by the antibacterial antibiotics that inhibit protein synthesis.

5. Mitochondria are discrete cytoplasmic organelles bounded by double membranes whose cristae are the site of ATP formation in aerobic respiration. The mitochondrial matrix is the cellular location of the citric acid (TCA) cycle enzymes.
6. Photosynthesis in eucaryotic cells occurs in chloroplasts. The photosynthetic pigments are localized in the thylakoids of the grana, whereas the stroma is the site of the dark reactions (CO_2 fixation) of photosynthesis.
7. Mitochondria and chloroplasts contain 70S ribosomes and closed circular DNA. Both organelles produce proteins; however, to function they also need proteins coded for by nuclear genes.
8. The flattened sacs of the Golgi apparatus are the site where the exportable polysaccharides and glycoproteins are packaged into vesicles. The membrane vesicles move to the plasma membrane, fuse, and then release their contents to the extracellular environment by exocytosis.
9. Hydrolytic enzymes are packaged by the Golgi apparatus into lysosomes, which serve a digestive role within the cell's cytoplasm.

MOVEMENT IN EUCARYOTIC CELLS

1. The many forms of movement in eucaryotic cells can be driven by external appendages known as cilia and flagella or by internal changes in the protoplasm that produce pseudopodia.
2. Flagella and cilia are locomotor organelles of eucaryotic cells that originate from a basal body lying beneath the plasma membrane. The axoneme protrudes from the cell surface and is surrounded by an extension of the plasma membrane.
3. Microtubules are long filamentous polymers made of repeating units of α and β tubulin heterodimers. The "9 + 2" arrangement of microtubules (axoneme) is the central structure of cilia and eucaryotic flagella.

CYTOSKELETON

1. The protoplasm of the eucaryotic cell has an internal organization created by the extensive architectural network of microtubules and cytoplasmic filaments.
2. Microtubules are present in the spindle of the mitotic apparatus and function as structural components of certain cells, such as nerve cells.

THE NUCLEUS AND HEREDITY

1. The double-layered nuclear envelope surrounds the karyoplasm to create the eucaryotic nucleus. The nucleus contains a heterogeneous array of fibrils associated with proteins known as chromatin, and the RNA synthesizing region called the nucleolus.
2. The eucaryotic cell cycle progresses through the stages of growth$_1$ (G_1), DNA synthesis (S), growth$_2$ (G_2), and mitosis.
3. The cell's genetic information is packaged into more than one DNA molecule (chromosome) and these are copied during DNA synthesis.
4. The mitotic apparatus, comprised of two centrioles, two asters, and spindle fibers, provides a structural orientation for segregating chromosomes to daughter nuclei.
5. The copied DNA condenses into chromosomes as the cell begins mitosis (prophase).
6. The chromosomes move to the equatorial plane of the mitotic apparatus (metaphase) as the nuclear envelope disappears.
7. The spindle fibers stretched between the centromeres of the chromosome and the centrioles pull the chromatids to separate poles of the cell during anaphase.
8. The daughter nuclei are formed during telophase as the new nuclear envelope forms around each new nucleus.
9. Meiosis is the process of reductive division to form haploid *(n)* cells, usually gametes.

EVOLUTION OF EUCARYOTIC AND PROCARYOTIC CELLS

1. There is biochemical support for the endosymbiotic origin of chloroplasts arising from cyanobacteria millions of years ago, but serious questions about the endosymbiotic origin of mitochondria exist.

Questions and Topics for Study and Discussion

QUESTIONS

1. What are the essential characteristics of the eucaryotic cell type that differentiate it from the procaryotic cell?
2. Describe the different types of endocytosis and give an example of each. Under what conditions do eucaryotic cells use exocytosis?
3. Why is mitosis essential to the eucaryotic cell and not to the procaryotic cell?
4. Name the four stages of mitosis and give a brief description of what occurs during each.
5. What is the function of the rough and the smooth endoplasmic reticulums in the eucaryotic cell?
6. Describe two functions of the Golgi apparatus. What structure in a procaryotic cell is comparable to the lysosome?
7. What internal structures would you expect to see in a plant cell, an animal cell, and both cell types?
8. What characteristics of chloroplasts suggest that they were derived originally from procaryotic cells by an endosymbiotic mechanism?
9. Compare the eucaryotic flagellum to the flagella found in procaryotic cells.
10. Define and explain the function of the following:

aster	kinetochore
chromatin	nuclear envelope
cytosol	nucleolus
histones	peroxisomes
karyoplasm	stroma

DISCUSSION TOPICS

1. What evidence is there to support the endosymbiotic theory? Postulate an argument against this theory.
2. How is the eucaryotic cell more adaptable to the varied functions required of cells in higher plants and animals?

Further Readings

BOOKS

Holtzmann, E., and A. B. Novikoff, *Cells and Organelles,* Saunders College Publishing, Philadelphia, 1984. A textbook in cell biology that emphasizes the experimental approach to this science.

Munn, E. A., *The Structure of Mitochondria,* Academic Press, New York, 1974. A major review of the structure and biochemical composition of the mitochondria.

ARTICLES AND REVIEWS

Dautry-Varsat, A., and H. F. Lodish, "How Receptors Bring Proteins and Particles into Cells," *Sci. Am.,* 250:52–58 (May 1984). This article describes how receptors on the plasma membrane in coated pits participate in endocytosis.

Dustin, P., "Microtubules," *Sci. Am.,* 243:66–76 (August 1980). A synopsis of the structure of microtubules and their function in nucleated cells.

Goodenough, U. W., and R. P. Levine, "The Genetic Activity of Mitochondria and Chloroplasts," *Sci. Am.,* 223:22–29 (November 1970). The influence of mitochondrial and chloroplast DNA on the structure and function of these organelles is discussed.

Gray, M. W., and W. F. Doolittle, "Has the Endosymbiont Hypothesis Been Proven? *Microbiol. Rev.,* 46:1–42(1982). A detailed review of the biochemical evidence both for and against the endosymbiotic theory.

Margulis, L., "Symbiosis and Evolution," *Sci. Am.,* 225:48–57 (August 1971). Evidence supporting the endosymbiotic theory, which is one explanation of the evolutionary development of eucaryotic cells, is presented.

Satin, P., "How Cilia Move," *Sci. Am.,* 231:44–52 (October 1974). A detailed look at the structure and function of eucaryotic flagella.

C H A P T E R

6

Microbial Physiology

OUTLINE

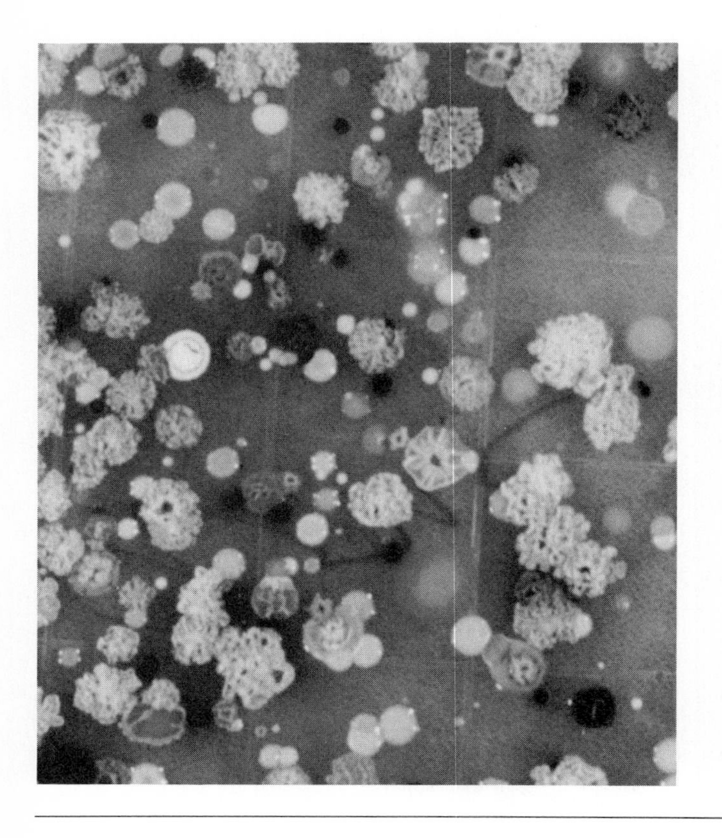

FOCUS OF THIS CHAPTER

Microorganisms use light or chemical energy to drive the chemical reactions that convert substrates and nutrients into cellular material. This chapter focuses on the biochemical machinery cells possess for extracting energy from chemical substrates and on the energy-dependent biosynthetic reactions they use to make new cellular components.

Microbial physiology is the study of the chemical reactions by which microbes grow and reproduce. These reactions are powered by energy from chemical sources or by the energy in sunlight. Most microorganisms use organic compounds as a source of energy, while a few have developed the specialized mechanisms needed to extract energy from inorganic compounds. Plants, algae, and the photosynthetic bacteria use the energy present in sunlight. Regardless of the energy's source, biological systems convert energy into a common currency, adenosine triphosphate (ATP), which they then use to power biosynthetic reactions. ATP is an essential link between energy-yielding and energy-utilizing reactions in biological systems.

The study of microbial physiology has broad implications because the biochemical reactions of microorganisms are common to many living systems. Since the basic building materials—carbohydrates, fatty acids, amino acids, nitrogenous bases—occur in all cells, it follows that the reactions required to manufacture those materials should be similar. For example, if we know how *Escherichia coli* synthesizes an amino acid, such as glutamic acid, we expect that a similar synthetic mechanism exists in humans, and indeed it does. The same principle applies to the way in which cells acquire and utilize energy.

BIOLOGICAL SOURCES OF ENERGY

Matter and energy are the two fundamental components of the earth. Matter is composed of atoms as described in Chapter 2, and **energy** is the capacity to do work. As was mentioned, biological systems can use chemical and/or light energy to perform the work necessary to produce new cellular material.

Sunlight is the ultimate source of biological energy. Sunlight provides heat to the earth's surface and is converted to chemical energy through the process of photosynthesis. Organisms that convert light energy into chemical energy are known as **phototrophs** (Figure 6.1). The chemical energy generated during photosynthesis is conserved in organic compounds, such as cellulose, glycogen, and the cellular matter of photosynthetic organisms. These compounds eventually serve as the renewable source of chemical energy for most nonphotosynthetic organisms.

Chemoorganotrophs (kem′oh-or-gan′oh-trohf) use the chemical energy in organic molecules to grow and reproduce, whereas **chemolithotrophs** (kem′oh-lith′oh-trohf) use the chemical energy in inorganic molecules for the same purpose (Figure 6.1). Most microbes are chemoorganotrophs that grow on sugars, amino acids, fatty acids, and other organic compounds. For example, yeasts are chemoorganotrophs that grow on the sugar in fruit juices. Chemolithotrophic growth is limited to a few bacteria that gain energy from inorganic compounds (lacking carbon) and are especially important to the natural recycling of nitrogen and sulfur compounds.

ENERGY FROM CHEMICAL REACTIONS

The amount of energy released or utilized during a chemical reaction is measured as an increase in heat and is expressed as the **Gibb's free energy change,** written as ΔG. ΔG represents the quantity of energy available to do useful work. When chemicals react or undergo a change from one form to another, energy is either released to the surroundings (usually as heat) or trapped in the chemical bonds of different compounds. Reactions that release energy have negative changes in free energy ($-\ \Delta G$) and are called **exergonic** reactions. Reactions that consume energy are termed **endergonic** reactions and have positive changes in free energy ($+\ \Delta G$). Examples of exergonic chemical reactions from which organisms gain energy are listed in Table 6.1.

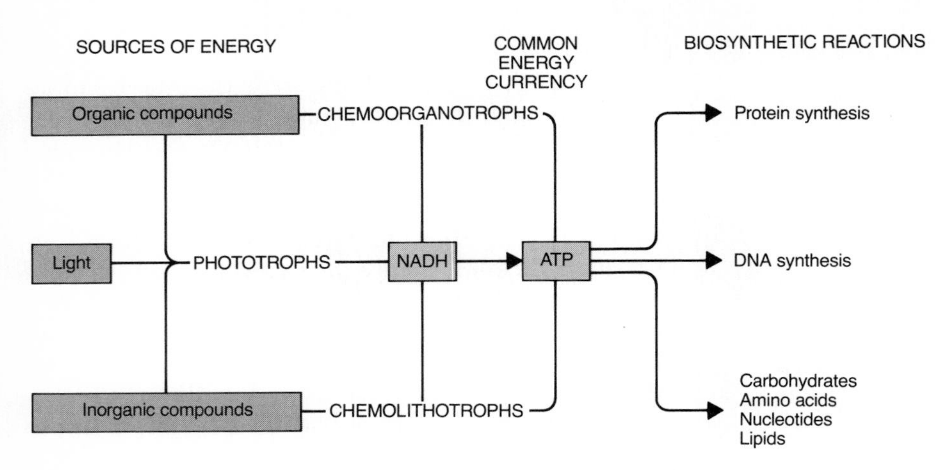

FIGURE 6.1 Sources of biological energy, energy conversion, and energy use.

Glucose is a common source of energy for biological systems. Under aerobic conditions, glucose is oxidized to carbon dioxide (CO_2) and water with the release of 686 kcal $mole^{-1}$. When this reaction occurs in cells at normal growth temperatures, the energy in the chemical bonds of glucose is trapped in biomolecules that in turn are used to do biological work. In contrast, when wood or paper is burned in a stove or fireplace, the glucose present (cellulose) is oxidized very rapidly and the chemical energy is released as heat.

Microorganisms can use chemicals other than glucose as sources of energy. Lactic acid and acetic acid are substrates for certain bacteria, but fewer kilocalories per mole are released during their oxidation than from the oxidation of glucose (Table 6.1). Inorganic compounds, such as ammonia, also produce heat when they are oxidized (Table 6.1), and some bacteria gain their energy from oxidizing inorganic compounds.

Endergonic reactions include most of the synthetic reactions of biological systems; for example, protein synthesis and DNA synthesis are endergonic reactions. The energy used in synthetic reactions can come directly from the breakdown of metabolites or indirectly from chemicals capable of energy storage. The amount of chemical energy contained in these compounds is conventionally expressed in kilocalories per mole and is measured by determining the heat loss (temperature change) during hydrolysis (Chapter 2). Chemists usually report the free-energy change of hydrolysis (or breakdown) because it is easier to determine this value than to determine the energy of formation. The ΔGs of hydrolysis of certain biologically important phosphate bonds are presented in Table 6.2.

KEY POINT

The energy in foods is measured by caloric content; for example, fats contain more calories per gram than do proteins. An active adult human needs approximately 2500 kcal of food every day.

TABLE 6.1 Chemical energy in molecules

COMPOUND	REACTION FOR YIELDING ENERGY	ENERGY, kcal $mole^{-1}$
Glucose	$C_6H_{12}O_6 + 6O_2 \rightarrow 6H_2O + 6CO_2$	−686
Lactic acid	$C_3H_6O_3 + 3O_2 \rightarrow 3H_2O + 3CO_2$	−318
Acetic acid	$C_2H_4O_2 + 2O_2 \rightarrow 2H_2O + 2CO_2$	−214
Ammonia	$NH_4^+ + 2O_2 \rightarrow NO_3^- + H_2O + 2H^+$	− 82.5

TABLE 6.2 Energy of hydrolysis of specific phosphate bonds in biologically important compounds[a]

COMPOUND	$\Delta G^{\circ\prime}$, kcal mole^{-1}
Glucose-6-phosphate + H_2O → glucose + phosphate	−3.3
AMP + H_2O → adenosine + phosphate	−3.4
ADP + H_2O → AMP + phosphate	−7.3
ATP + H_2O → ADP + phosphate	−7.3
GTP + H_2O → GDP + phosphate	−7.3
CTP + H_2O → CTP + phosphate	−7.3
UTP + H_2O → UDP + phosphate	−7.3
Acetyl phosphate + H_2O → acetic acid + phosphate	−10.1
1,3-Diphosphoglyceric acid + H_2O → 3-Phosphoglyceric acid + phosphate	−11.8
Phosphoenolpyruvic acid + H_2O → pyruvic acid + phosphate	−14.8

[a]These values were determined at standard conditions of 1 molar solutions, pH 7.0, and 30°C. Changes in these conditions of hydrolysis can change the ΔG for a given reaction.

ATP-MEDIATED ENERGY TRANSFER

Cells can capture the energy released from chemical reactions by forming energy-rich bonds in adenosine triphosphate (ATP) (Figure 6.2) and related compounds that serve as energy reservoirs (Table 6.2). These compounds transfer energy between exergonic and endergonic reactions as diagramed in Figure 6.1. ATP is a representative energy transfer molecule (Figure 6.2) composed of the nitrogenous base adenine, ribose, and three phosphate molecules. The high-energy phosphate

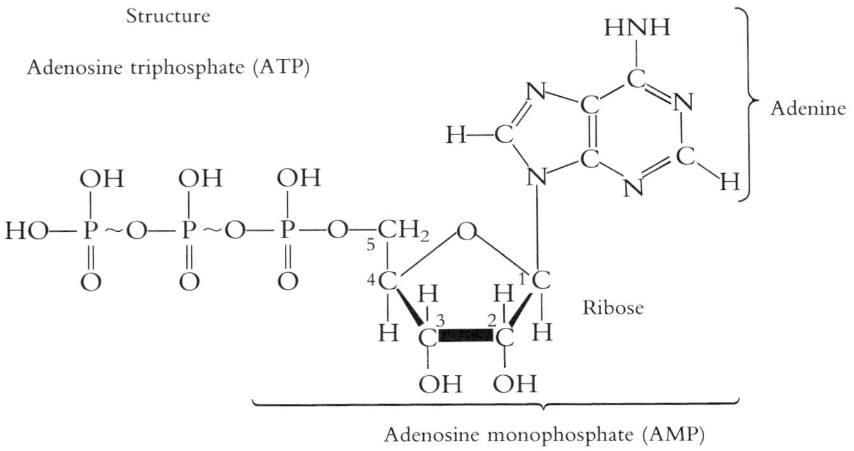

Structure

Adenosine triphosphate (ATP)

Adenine

Ribose

Adenosine monophosphate (AMP)

Reactions

$$ATP + H_2O \rightarrow ADP + H_3PO_4$$

$$ADP + H_2O \rightarrow AMP + H_3PO_4$$

FIGURE 6.2 Adenosine triphosphate contains energy-rich anhydride bonds (~) between the phosphate groups. Sequential hydrolysis of these two bonds yields adenosine diphosphage (ADP) and adenosine monophosphate (AMP) concomitantly with the release of 7.3 kcal mole^{-1} for each bond hydrolyzed.

bond in ATP is formed when adenosine diphosphate (ADP) and inorganic phosphate are joined in an endergonic reaction:

$$ADP + H_3PO_4 \rightarrow ATP + H_2O \qquad \Delta G = +7.3 \text{ kcal mole}^{-1}$$

The amount of energy in the terminal phosphate bond can be determined by measuring the heat released when the terminal phosphate is hydrolyzed from ATP. The value of 7.3 kcal $mole^{-1}$ of ATP hydrolyzed to ADP and phosphate under standard conditions will be used in our discussion of energetics. The terminal phosphate bond of adenosine diphosphate (ADP) is also a high-energy bond (Table 6.2).

ENERGY FROM REDUCED PYRIDINE NUCLEOTIDES

Another small molecule involved in energy reactions is the coenzyme nicotinamide adenine dinucleotide (NAD). NAD is made from adenine nucleotides, ribose, and the vitamin niacin (Figure 6.3). NAD is an electron carrier that exists in either a reduced form (NADH)* or an oxidized form (NAD^+). NAD^+ is converted to NADH when it gains two electrons and one proton:

$$\underset{\text{Oxidized}}{NAD^+} + 2e^- + 2H^+ \longrightarrow \underset{\text{Reduced}}{NADH} + H^+$$

*Reduced NAD is written as NADH. The second proton participating in the reduction is not bound to reduced NAD.

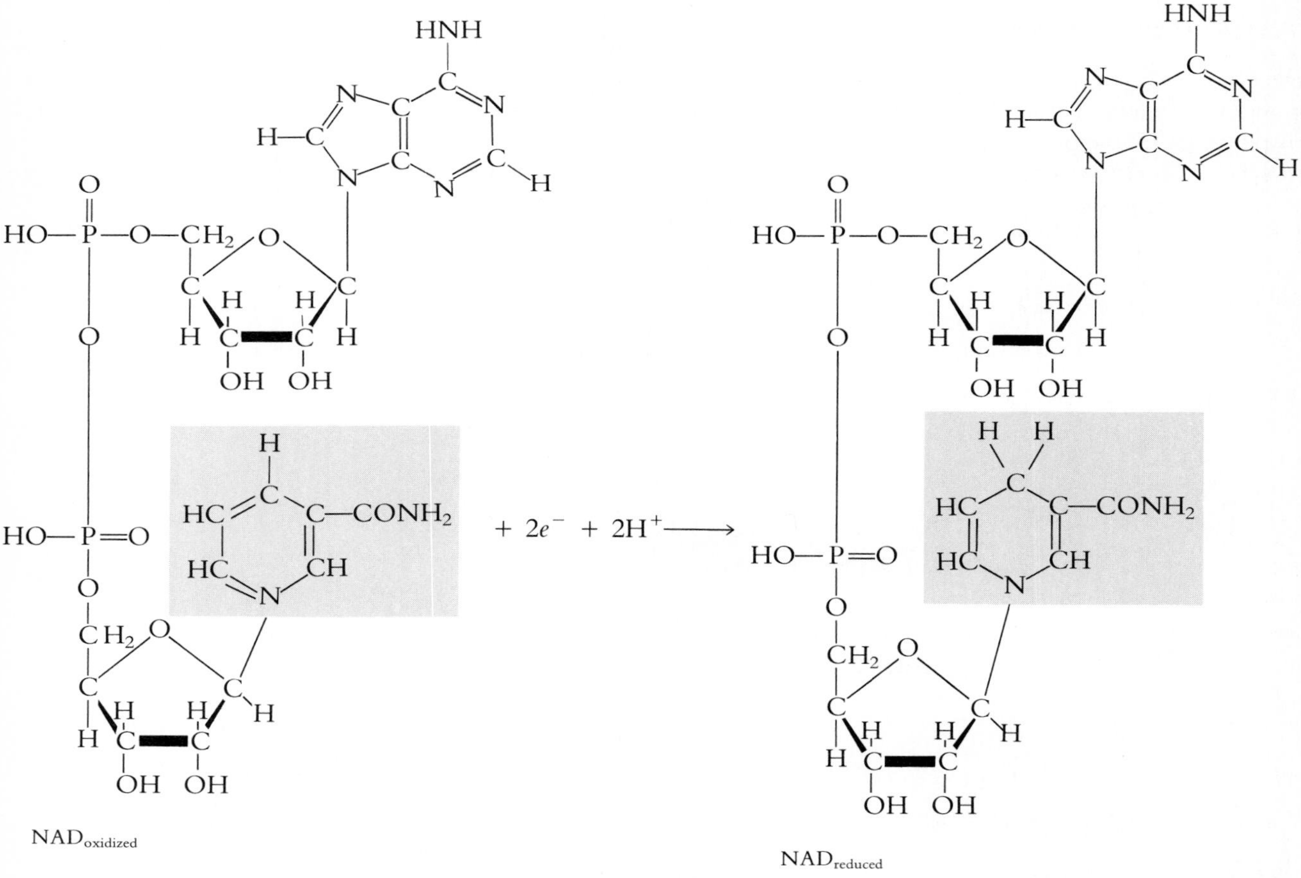

$NAD_{oxidized}$ $+ 2e^- + 2H^+ \longrightarrow$ $NAD_{reduced}$

FIGURE 6.3 Nicotinamide adenine dinucleotide (NAD^+) accepts two electrons and one proton when it becomes reduced. NAD is a coenzyme that participates in many enzyme-catalyzed oxidation–reduction reactions in cellular metabolism.

NADH is a strong reductant that participates in many enzyme-catalyzed oxidation–reduction reactions essential to both catabolism and biosynthesis.

Energy from Oxidation–Reduction Reactions

Although biological systems are not wired for electricity, they do take advantage of the energy released when electrons are transferred between compounds. Certain biomolecules such as NAD can act as electron-transfer agents by accepting electrons and then passing them on to other biomolecules. A compound that accepts an electron is said to be **reduced,** whereas a compound that gives up an electron is said to be **oxidized.** An example of an oxidation reaction is the conversion of lactate to pyruvate.

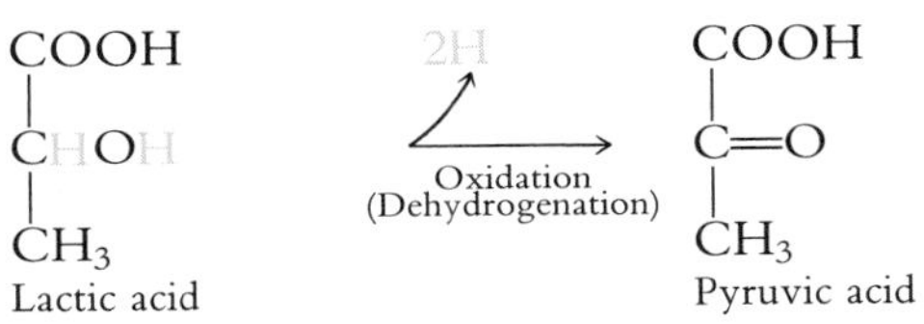

$$\underset{\text{Lactic acid}}{\begin{array}{c}\text{COOH}\\ |\\ \text{CHOH}\\ |\\ \text{CH}_3\end{array}} \xrightarrow[\text{Oxidation (Dehydrogenation)}]{2\text{H}} \underset{\text{Pyruvic acid}}{\begin{array}{c}\text{COOH}\\ |\\ \text{C=O}\\ |\\ \text{CH}_3\end{array}}$$

The lactic acid is oxidized when it gives up two hydrogens, a reaction that can be called a dehydrogenation. In the reverse reaction, pyruvic acid is reduced to lactic acid when it accepts two hydrogens.

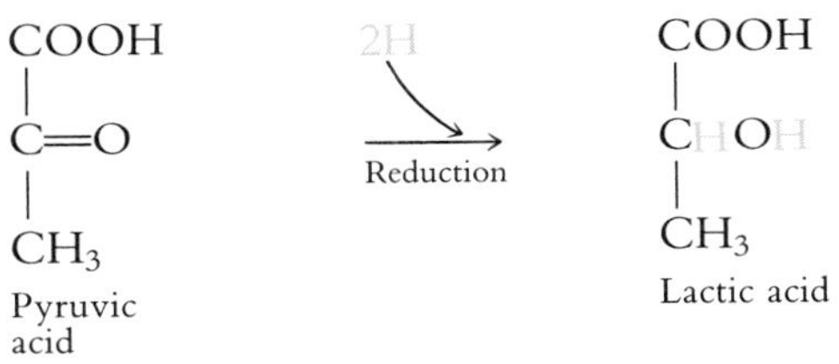

$$\underset{\text{Pyruvic acid}}{\begin{array}{c}\text{COOH}\\ |\\ \text{C=O}\\ |\\ \text{CH}_3\end{array}} \xrightarrow[\text{Reduction}]{2\text{H}} \underset{\text{Lactic acid}}{\begin{array}{c}\text{COOH}\\ |\\ \text{CHOH}\\ |\\ \text{CH}_3\end{array}}$$

In these reactions two protons (H^+) and two electrons are transferred—remember that a hydrogen atom is composed of a proton and an electron (Chapter 2). The hydrogen atoms in these reactions are transferred between compounds by electron carriers such as NAD. When there is a coupling of electron transfer between two compounds, it is called an *oxidation–reduction* reaction.

$$\begin{array}{c}\text{COOH}\\ |\\ \text{C=O}\\ |\\ \text{CH}_3\end{array} + \text{NADH} + \text{H}^+ \longrightarrow \begin{array}{c}\text{COOH}\\ |\\ \text{CHOH}\\ |\\ \text{H}_3\end{array} + \text{NAD}^+$$

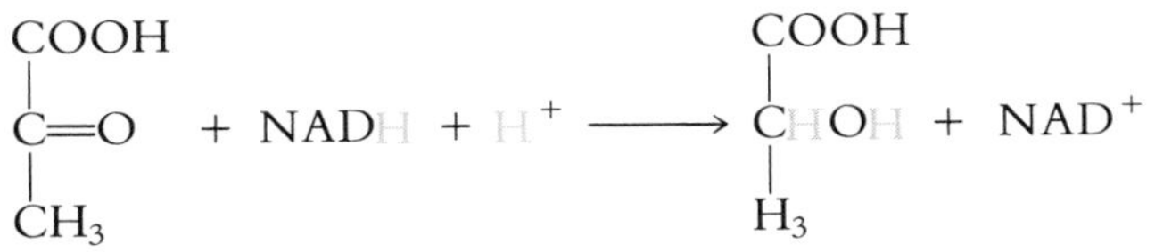

In this reaction pyruvic acid is reduced when it accepts two hydrogen atoms from NADH and NADH is oxidized to NAD^+.

The more reduced an electron carrier is, the greater is its potential to give up electrons and the more energy it contains. The ability of a reduced compound, such as NADH or molecular hydrogen (H_2), to give up electrons can be measured electrically as the **electromotive force** (**E'_0**) and is expressed in volts. The more negative the E'_0, the stronger is the compound's reducing power. When these molecules are arranged in a series, electrons travel from one molecule to the next and release energy in the process.

ENERGY SUMMARY

Almost all organisms depend on sunlight as their ultimate source of energy. Phototrophs convert light energy to chemical energy through the process of photosynthesis. The cellular matter produced during photosynthesis serves as the source of chemical energy for the growth and reproduction of chemoorganotrophs. Both chemoorganotrophs and chemolithotrophs employ oxidation–reduction reactions to release the energy stored in chemicals. The energy released is converted to a common energy currency such as ATP or NADH. Cells then use this chemical energy for cell synthesis.

ENZYMES CATALYZE CHEMICAL REACTIONS

Enzymes are specialized proteins that enable cells to perform chemical reactions under physiological conditions. Cells can be thought of as small chemical factories in which hundreds of chemical reactions occur simultaneously. The amounts and kinds of enzymes present in a cell determine its metabolic capability.

Enzymes are proteins that function as catalysts; that is, they speed up the rate of a reaction without being consumed in the process. These catalysts are active only within prescribed ranges of pH and temperature. The typical temperature optimum for an enzyme-catalyzed reaction falls between 30° and 37°C and the pH optimum is usually within the range of 4.5 to 8.5. As you would expect, enzymes have maximal catalytic activities at physiological temperatures and pH.

Most chemical reactions are temperature dependent, that is, the higher the temperature, the faster the reac-

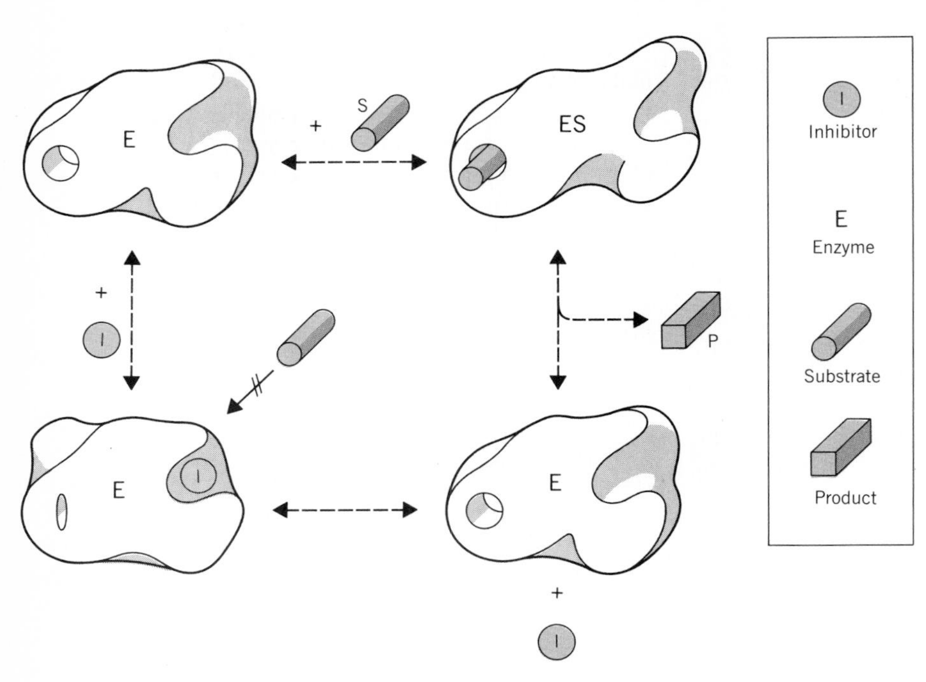

FIGURE 6.4 Diagramatic representation of enzyme action. Enzymes bind reversibly with their substrate to form an enzyme substrate (ES) complex. The reaction is then catalyzed, converting the substrate (S) to product (P). A noncompetitive inhibitor (I) can bind at a site other than the substrate site or active site. The inhibitor prevents the substrate from binding to the enzyme, so that no product is formed.

tion proceeds. For example, wood oxidizes very slowly at room temperature, but very quickly when burned in a fireplace. This is also true for enzyme-catalyzed reactions; enzymes, however, are inactivated by excessive heat, so these reactions have specific temperature maxima.

KEY POINT

Bacteria that grow in thermal springs above 70°C have specialized enzymes that require high temperatures for catalytic activity. Acidophilic bacteria grow only at a pH below 2.0. They survive in this environment by excreting protons so that the pH of their cytoplasm remains close to neutrality.

CHARACTERISTICS OF ENZYME REACTIONS

Each enzyme has one or more catalytic sites (or active sites) that interact with its substrate(s) (Figure 6.4). The substrate (S) combines with the enzyme (E) to form an enzyme–substrate complex, designated (ES). Because of its specialized structure, the enzyme exerts stress on the chemical bonds in the substrate molecule, causing either their breakage or rearrangement, and resulting in the formation of a new compound called the product (P). The equation for expressing the action of an enzyme on a single substrate is

$$S + E \rightleftharpoons ES \rightleftharpoons E + P$$

Enzyme-catalyzed chemical reactions are uniquely responsive to the concentration of their substrate, producing their product only after the enzyme substrate complex (ES) is formed. This binding is highly specific and is dependent on substrate concentration (Figure 6.5*a*). Thus when the substrate concentration is low, no enzyme–substrate complex is formed and no product is produced. As the substrate concentration is increased, the substrate molecules occupy all of the active sites on the enzyme so that each enzyme molecule is

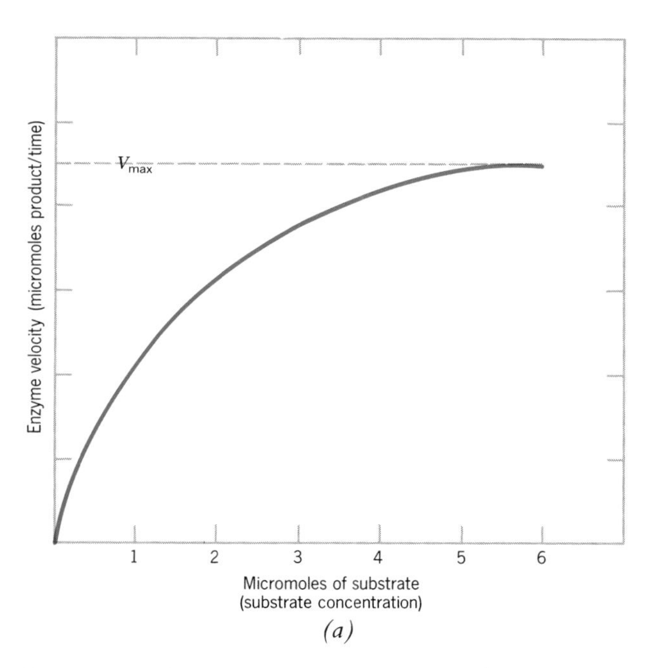

(a)

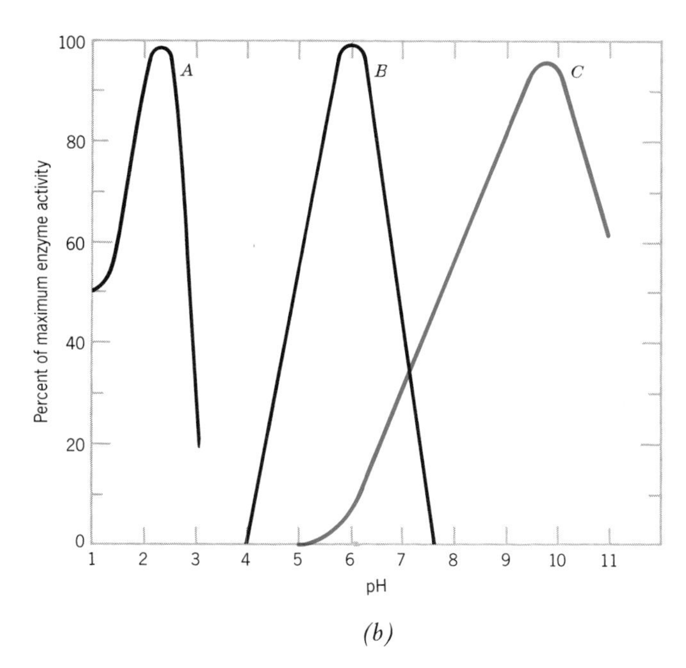

(b)

FIGURE 6.5 Enzyme reactions are substrate specific and dependent on substrate concentration *(a)*. The maximum enzyme velocity (V_{max}) is attained when each enzyme molecule is saturated with substrate (ES). *(b)* Effect of pH on enzyme activity: *A,* pepsin; *B,* glutamic acid; decarboxylase; and *C,* arginase.

complexed with substrate (ES). Only under these conditions does an enzyme-catalyzed reaction proceed at its maximal rate. Additional increases in the substrate concentration will not increase the rate of the reaction. Enzyme-catalyzed reactions are also dependent on pH (Figure 6.5*b*), temperature, and ionic strength.

ENZYME NOMENCLATURE

Each biological reaction is catalyzed by a unique enzyme, of which thousands are known. Each enzyme is named for its substrate and the type of chemical reaction it catalyzes. The names of most enzymes have the suffix *-ase*. Examples: **proteases** hydrolyze proteins, **nucleases** hydrolyze nucleic acids, and **lipases** hydrolyze lipids.

Any given bacterial cell will possess about 1000 different proteins, most of which will be enzymes. Each enzyme belongs to one of six major classes. **Oxidoreductases** catalyze oxidation–reduction reactions. **Transferases** take a chemical group from one molecule and transfer it to another. **Hydrolases** cleave covalent bonds through the addition of a water molecule. This is in contrast to **lyases** that cleave covalent bonds without the addition of water and do so in the absence of oxidation–reduction reactions. Pyruvic acid decarboxylase is an example of a lyase. **Isomerases** (eye-som′er-ase) catalyze structural changes within one molecule. Synthetic reactions are catalyzed by **ligases.** These enzymes catalyze the joining together of two molecules coupled to the hydrolysis of ATP.

COENZYMES

Coenzymes are organic molecules that participate as carriers of functional groups, specific atoms, or electrons in certain enzyme-catalyzed reactions. Every cell contains minute quantities of the essential coenzymes, which unlike the enzymes they serve are stable to heat. A cell that cannot synthesize these organic molecules must acquire them as growth factors from its growth medium. You know these growth factors as **vitamins,** which are a group of unrelated organic compounds, some or all of which are necessary in small quantities for normal metabolism. The important vitamins that function as coenzymes or as components of coenzymes

TABLE 6.3 Vitamins and their function

WATER-SOLUBLE VITAMIN	CHEMICAL NAME	COENZYME OR ACTIVE FORM	FUNCTION
Niacin	Nicotinic acid	Pyridine nucleotides (NAD, NADP)	Hydrogen atom (electron) transfer
Vitamin B_2	Riboflavin	Flavin nucleotides (FAD, FMN)	Hydrogen atom (electron) transfer
Vitamin B_1	Thiamine	Thiamine pyrophosphate (TPP)	Aldehyde group transfer
Vitamin B_6	Pyridoxine	Pyridoxal phosphate	Amino group transfer
Pantothenic acid	Pantothenic acid	Coenzyme A (CoA-SH)	Keto acid oxidation, fatty acid metabolism
Folic acid	Pteroylglutamic acid	Tetrahydrofolic acid (THF)	One-carbon group transfer
Vitamin H	Biotin	Biocytin	Carboxyl (CO_2) transfer
Vitamin B_{12}	Cyanocobalamin	Coenzyme B_{12}	Molecular rearrangements

are listed in Table 6.3. For example, flavin adenine dinucleotide (FAD) is the coenzyme made from vitamin B_2 (riboflavin), and NAD is the coenzyme made from the vitamin niacin.

The cell needs only a tiny quantity of these coenzymes since they are used repeatedly. A given coenzyme is not restricted to a single enzyme; instead it is recognized by many enzymes each of which competes for its availability. Coenzymes participate in oxidation–reduction reactions, transfer of chemical groups, and molecular rearrangement reactions (Table 6.3).

Enzymes and coenzymes are the workhorses of microbial physiology. The combined effect of all the enzyme-catalyzed reactions in a cell is to transform energy into cellular material, so that the cell can grow and reproduce.

AEROBIC RESPIRATION

Humans as well as many microorganisms require oxygen for growth and thus they are called aerobes. The energy for their growth is released from organic compounds during **aerobic respiration.** In this process organic molecules such as glucose are oxidized to carbon dioxide and water. The oxidant (O_2) is reduced to H_2O, while the carbon atoms of glucose are oxidized to CO_2. The amount of energy released by the complete oxidation of glucose to $6CO_2$ and $6H_2O$ is 686 kcal $mole^{-1}$ of glucose. Part of this energy is conserved as usable energy in the form of ATP, while the remainder is released as heat. The metabolism of glucose proceeds in a series of steps that include glycolysis, the tricarboxylic acid cycle, and the electron transport chain.

GLYCOLYSIS IS A PATHWAY FOR GLUCOSE METABOLISM

Glucose, one of the most abundant carbohydrates in nature, serves as a source of carbon and energy for many organisms. Cells assimilate glucose and then metabolize it in a series of enzyme-catalyzed reactions that we conceptualize as the glycolytic pathway (known also as the Embden–Meyerhof pathway). Although there are other pathways for metabolizing glucose, glycolysis is the most common one.

In glycolysis, glucose is converted to 2 moles of pyruvic acid, 2 moles of NADH, and 2 moles of ATP (Figure 6.6). This scheme illustrates the nine enzyme-catalyzed reactions and the intermediate metabolites of glycolysis. Four of the reactions are catalyzed by kinases, which are enzymes that transfer phosphate groups. By looking at the kinase reactions, one can see that there is a net gain of 2 molecules of ATP for each glucose metabolized in the glycolytic pathway.

Step 1 Metabolism of glucose requires an initial investment of energy. The first enzymatic reaction in glycolysis phosphorylates glucose by transferring a phosphate group from ATP, thus forming glucose 6-phosphate and ADP.

Step 2 Glucose 6-phosphate is converted to fructose 6-phosphate by an isomerase, an enzyme that catalyzes a geometric change within its substrate.

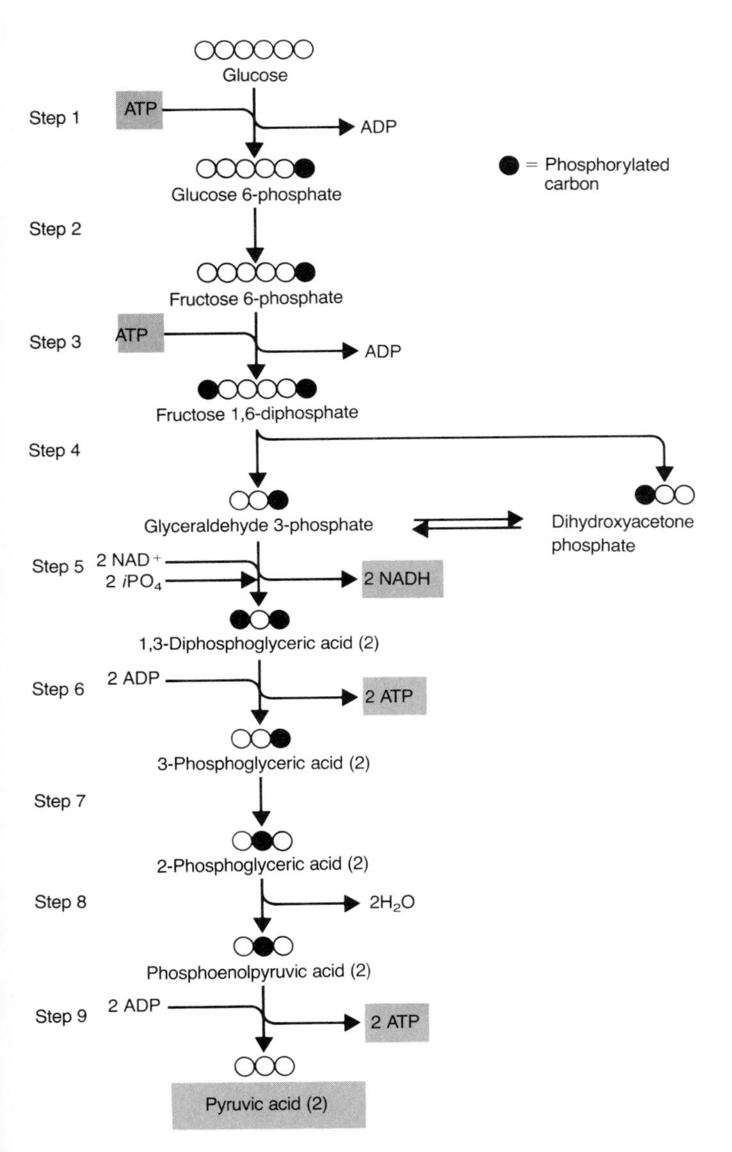

FIGURE 6.6 Glycolysis. One mole of glucose is metabolized by this metabolic pathway to 2 moles of pyruvic acid, 2 moles of ATP (net), and 2 moles of NADH. Note that fructose is split into two trioses, so each reaction from glyceraldehyde 3-phosphate to pyruvic acid takes place twice. The circles represent phosphorylated carbons.

Step 3 Fructose 6-phosphate is phosphorylated at carbon atom 1 at the expense of ATP to make fructose 1,6-diphosphate. At this point in the metabolic pathway there has been a net investment of 2 ATP molecules for each glucose molecule metabolized.

Step 4 Fructose 1,6-diphosphate is cleaved into two triose phosphates—dihydroxyacetone phosphate and glyceraldehyde 3-phosphate. These trioses are interconverted by an isomerase. The net result of step 4 is that 2 molecules of glyceraldehyde 3-phosphate are formed for each fructose molecule metabolized. The two glyceraldehyde 3-phosphates serve as the substrate for the remainder of the pathway, so all the remaining reactions occur twice.

Step 5 This is a key step in glycolysis because it results in the phosphorylation of glyceraldehyde 3-phosphate with inorganic phosphate and the reduction of NAD^+. There are three substrates for this enzymatic step: inorganic phosphate (iPO_4), NAD^+, and glyceraldehyde 3-phosphate. When the two glyceraldehyde 3-phosphates are oxidized in this reaction, $2NAD^+$ are reduced to 2NADH. Sufficient energy remains to phosphorylate the first carbon of glyceraldehyde 3-phosphate with iPO_4 to produce 1,3-diphosphoglyceric acid.

Step 6 The phosphate attached to the 1-carbon of 1,3-diphosphoglyceric acid is transferred to ADP to form ATP. The products of this reaction are ATP and 3-phosphoglyceric acid. At this point in the glycolytic pathway, the cell has regained the 2 ATP it invested to activate glucose.

Steps 7 and 8 The phosphate group on 3-phosphoglyceric acid is enzymatically moved from carbon atom number 3 to carbon atom number 2. Another enzyme removes water from this triose to form phosphoenolpyruvic acid.

Step 9 Pyruvic acid and ATP are formed from phosphoenolpyruvic acid. Since each of these reactions occurs twice, the net result is that 2 molecules of ATP and 2 molecules of pyruvic acid are produced for each molecule of glucose metabolized in the glycolytic pathway.

Summary of Glycolysis

Energy in the form of ATP and NADH is derived from glucose metabolized by the nine enzyme-catalyzed reactions of the glycolytic pathway (Figure 6.6). Energy must first be invested before the cell can capture the chemical energy present in the glucose molecule. The energy investment occurs when 2 molecules of ATP are utilized to produce fructose 1,6-diphosphate in the first half of the scheme. This energy is recaptured in step 6 where 2 molecules of ATP are formed. A net gain in energy occurs in step 5 where NADH is produced and in step 9 where 2 molecules of ATP are produced for each glucose molecule metabolized. The net overall reaction for glycolysis is

$$\text{Glucose} + i\text{PO}_4 + 2\text{ADP} + 2\text{NAD}^+ \rightarrow 2\text{Pyruvic acid} + 2\text{ATP} + 2\text{NADH} + 2\text{H}_2\text{O}$$

Note that a maximum of 2 ATPs (net) are produced in this complex pathway.

The cellular breakdown of glucose cannot stop at pyruvic acid. At this point in metabolism, a cell has two metabolic options, which depend on the cell's metabolic capabilities and on the availability of oxygen. In the absence of oxygen, the cell metabolizes pyruvic acid by a fermentation mechanism (see below). In the presence of oxygen, aerobic organisms completely oxidize pyruvic acid to CO_2 and H_2O. This continuation of aerobic respiration employs the tricarboxylic acid cycle and the electron transport chain.

TRICARBOXYLIC ACID CYCLE

Most aerobic organisms have the enzymes of the tricarboxylic acid (TCA) cycle, which is enzymatically linked with glycolysis by pyruvic acid dehydrogenase. This enzyme decarboxylates pyruvic acid in the presence of coenzyme A to form acetyl-CoA, carbon dioxide, and NADH (Figure 6.7).

$$\text{Pyruvic acid} + \text{NAD}^+ + \text{CoA} \xrightarrow[\text{dehydrogenase}]{\text{pyruvic acid}} \text{Acetyl-CoA} + \text{NADH} + \text{CO}_2$$

Carbon dioxide is excreted by the cell, while the acetyl-CoA is metabolized in the TCA cycle.

The **tricarboxylic acid cycle** (also called the Krebs cycle or the citric acid cycle) is a sequence of enzyme-catalyzed reactions that in concert oxidize acetic acid (as acetyl-CoA) to $2CO_2$ and capture energy in guanidine triphosphate (GTP) and reduced coenzymes (NADH and FADH). As the name implies, this pathway is a cyclic series of reactions that starts and ends with oxaloacetic acid (Figure 6.7).

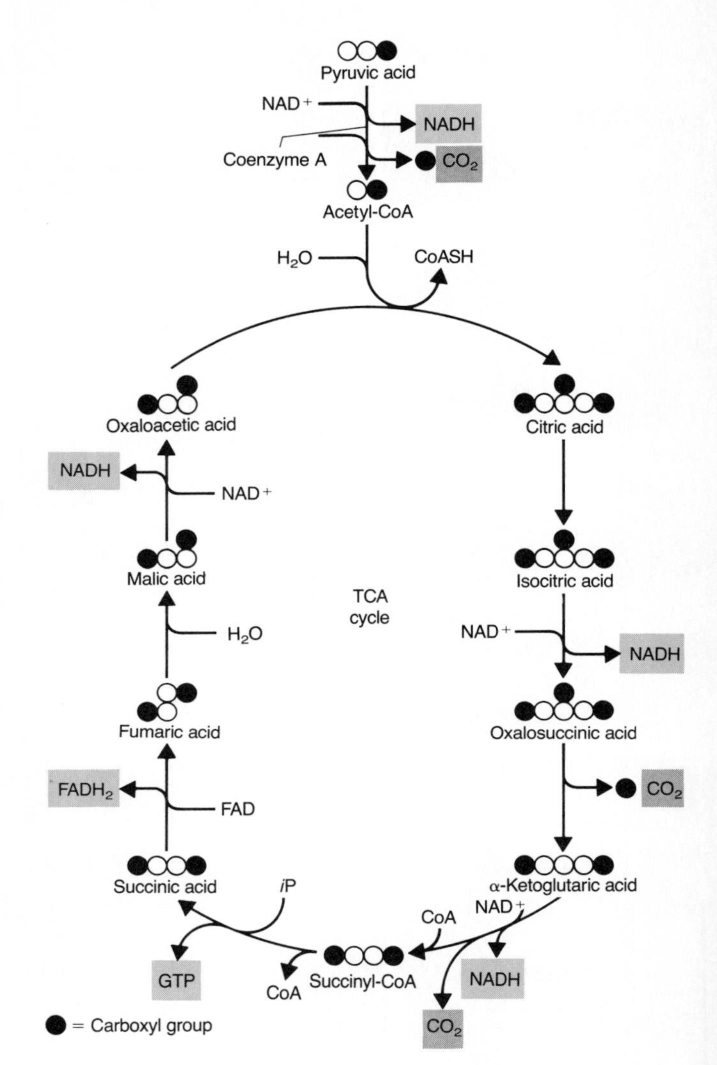

FIGURE 6.7 The tricarboxylic acid cycle (TCA) operates in aerobic cells to oxidize the carbon present in acetic acid. Acetyl-CoA, a product of the oxidative decarboxylation of pyruvic acid, is oxidized to CO_2 with the production of GTP, 3 NADH, and $FADH_2$. An additional NADH is produced during the oxidative decarboxylation of pyruvic acid. Many cells, both aerobic and anaerobic, use the reactions of the TCA cycle to generate metabolites for biosynthetic reactions.

Acetyl-CoA combines with oxaloacetic acid (Figure 6.7) to form citric acid, a tricarboxylic acid. During the cyclic reformation of oxaloacetic acid from citric acid by a series of enzyme reactions, 2 molecules of CO_2 are produced. In addition, 1 molecule of reduced flavin (FADH), 3 molecules of NADH, and 1 molecule of high-energy phosphate (GTP) are produced for each acetyl-CoA entering the cycle. The GTP can be used directly in synthetic reactions, or it can be converted to ATP without a loss in energy. The NADH and reduced flavins are oxidized by the electron transport chain (see below). In summary, the metabolism of the 2 molecules of pyruvic acid produced by the glycolytic pathway results in the overall reaction:

$$2\text{Pyruvic acid} + 8NAD^+ + 2FAD + 2ADP + iPO_4 \rightarrow 6CO_2 + 8NADH + 2FADH + 2ATP$$

The TCA cycle also provides carbon compounds for the synthesis of essential cellular metabolites. For example, α-ketoglutaric acid is a precursor for the synthesis of glutamic acid and glutamine, two essential amino acids. Oxaloacetic acid is the substrate for the synthesis of another essential amino acid, aspartic acid. In this way, the TCA cycle functions as a conduit between the catabolic (energy-generating) and the anabolic (synthetic or energy-utilizing) reactions of the cell.

As in glycolysis, one must envision all the reactions of the TCA cycle as occurring simultaneously, with many molecules of each intermediate and each enzyme in the cycle interacting together in a "closed sac." The enzymes of the TCA cycle are located in the cytoplasm of the procaryotic cell, whereas they are found in the innermost matrix of the mitochondria in eucaryotic cells.

REDUCED COENZYMES ARE OXIDIZED BY THE ELECTRON-TRANSPORT CHAIN

Every cell contains a limited amount of NAD. If all the NAD^+ in a cell became reduced, there would then be no oxidized NAD available to function as the substrate, and reactions requiring NAD^+ would not occur. So a cell must have a mechanism for oxidizing the NADH generated by glycolysis and the TCA cycle. Aerobic cells have an electron-transport chain that oxidizes reduced coenzymes and transfers the electrons to oxygen. This process occurs in cellular membranes and couples the oxidation–reduction reactions of the electron-transport chain to the generation of ATP.

The oxidation of NADH and reduced flavin ($FADH_2$) on the inside surface of the membrane (Figure 6.8) generates electrons that enter the membrane-bound transport chain. The electron transport chain is composed of flavoproteins, nonheme iron, quinones, and cytochromes that are involved in specific oxidation–reduction reactions. In eucaryotic cells, the electron-transport chain is located in the inner mitochondrial membrane, while in bacteria it is located either in the cytoplasmic membrane or in invaginated derivatives of the cytoplasmic membrane.

Each component of the electron-transport chain is oriented so it can participate in electron transfer or proton translocation. The arrangement depends on the inherent energy level of the compound, which is indicated in Figure 6.8*b*. A compound that gives up electrons to another is the stronger reducing compound, and the more reduced a compound is, the more energy it contains. Electrons generated by the oxidation of NADH are transferred down the electron-transport chain in a staircase fashion until they reach the oxygen, with which they react to form water.

The protons generated during the oxidation of NADH and reduced flavins are translocated across the membrane (Figure 6.8*a*). This creates a charge separation across the membrane with the outside being positively charged and rich in protons and the inside being negatively charged with hydroxyl ions (OH^-). The pH gradient and electrical potential across the membrane establish the energized membrane state and create the proton-motive force cells use to drive ATP synthesis, active transport, and flagellar movement.

CHEMIOSMOTIC COUPLING

One of the more difficult biochemical problems of the last 50 years has been to explain how cells couple electron transport to ATP formation. The complexity of membranes has made this a difficult problem to approach experimentally. The solution came from the theoretical model proposed by Peter Mitchell, who received a Nobel Prize in 1978 for his explanation of how **chemiosmotic coupling** drives the cellular synthesis of ATP.

The protons translocated across the membrane following the oxidation of reduced substrates on the interior surface (Figure 6.9) create a charge separation across the membrane. As protons accumulate outside the membrane, a pH potential is created between the inside and outside surfaces—similar to the electrical potential between the poles of a battery. The resulting proton-motive force drives the protons back into the cell at specific locations occupied by ATP synthetase.

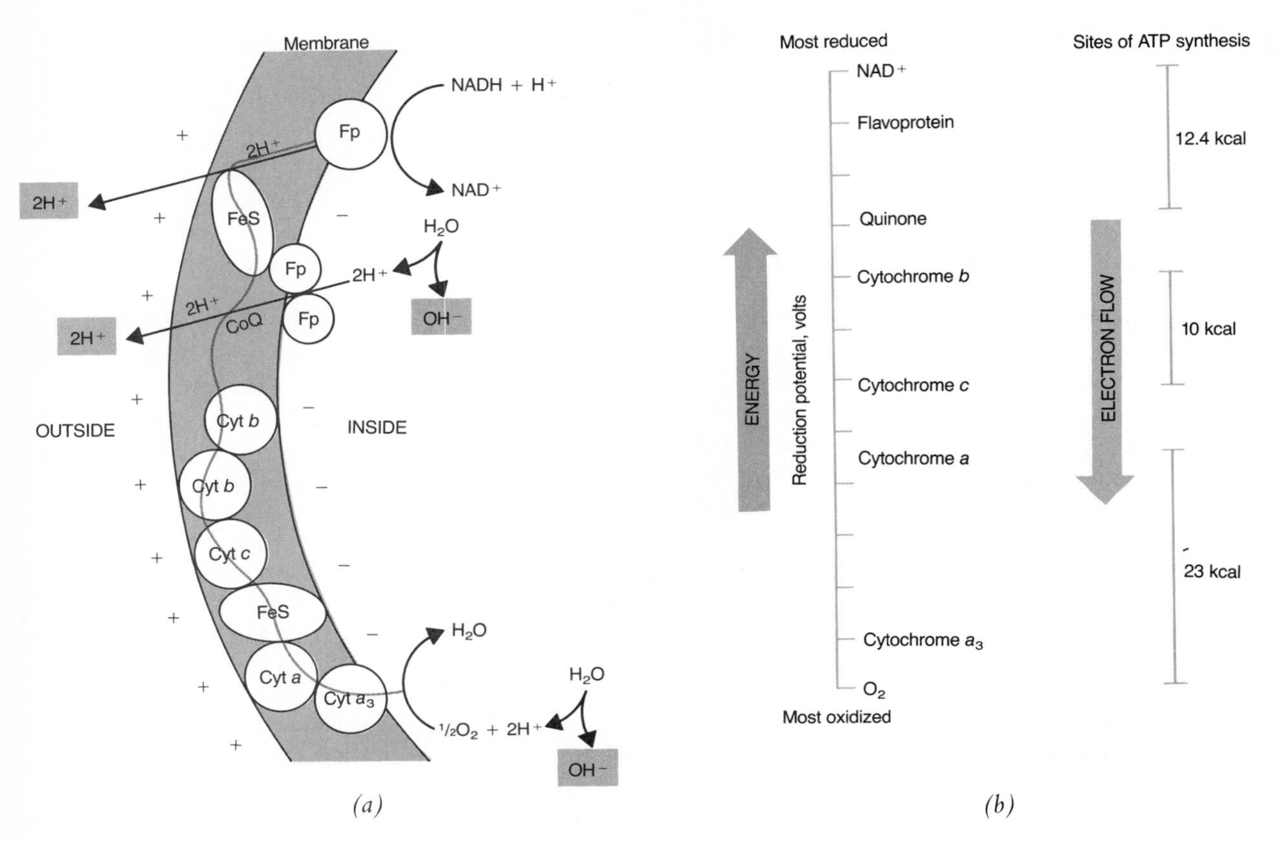

FIGURE 6.8 The electron transport chain found in membranes *(a)* transfers electrons from a reduced substrate (NADH + H^+) to oxygen. Flavoproteins (Fp), quinones (CoQ), iron sulfur compounds (FeS), and cytochromes are involved. These compounds possess different reduction potentials *(b)*. The more reduced a compound is the more energy it possesses. Energy is released as electrons flow from the more reduced to the more oxidized compounds.

Under these conditions, this enzyme chemically synthesizes ATP from ADP and inorganic phosphate (Figure 6.9).

Chemiosmotic coupling has been demonstrated in many microbes and explains how ATP is formed in such diverse groups as chemolithotrophs, phototrophs, and aerobic heterotrophs. Enough evidence has accumulated to accept chemiosmotic coupling as fact, even though details of proton translocation remain to be worked out.

OXIDATIVE PHOSPHORYLATION AND THE SITES OF ATP FORMATION

The coupling of electron transfer to ATP formation is referred to as electron-transfer phosphorylation, or when oxygen is the terminal electron acceptor, **oxidative phosphorylation** (Figure 6.10). In mitochondria, the electrons flowing down the electron-transport chain release sufficient energy to synthesize ATP at three specific sites containing ATP synthetase. The first ATP synthetase site is between NAD and flavoproteins; the second and third sites are located between cytochromes (Figure 6.10). When the electrons in NADH are transferred to oxygen, 3 molecules of ATP are formed per molecule of NADH oxidized. Only 2 molecules of ATP are formed when 1 molecule of reduced flavin is oxidized because electrons from reduced flavins enter the electron-transport chain after site 1.

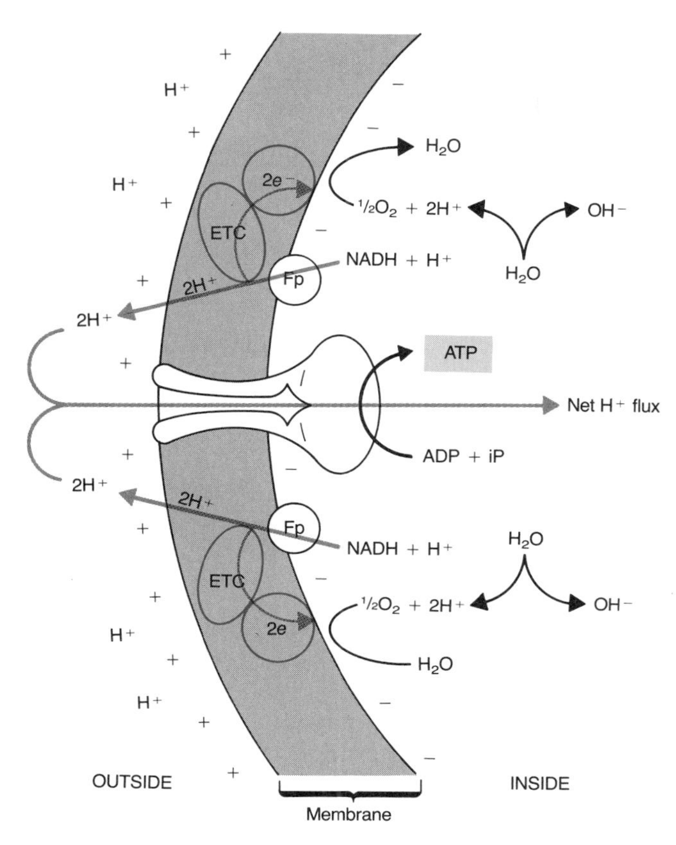

FIGURE 6.9 Chemiosmotic coupling explains how electron transfer and proton (H^+) translocation establish a pH and potential gradient across the membrane. This creates an energized state. When the protons flow back across the membrane, they create the proton motive force that drives the synthesis of ATP by ATP synthetase. Electron transport chain (ETC).

KEY POINT

To calculate the number of ATPs produced from the oxidation of NADH or FADH by the electron-transport system, remember that each NADH is equivalent to 3 molecules of ATP and each FADH is equivalent to 2 molecules of ATP.

OXYGEN AS THE TERMINAL OXIDANT

The electron-transport chain includes iron sulfur proteins, quinones, and heme-containing cytochromes. These proteins function as electron carriers through iron atom(s) located at their active sites. Cytochrome *a* **(cytochrome oxidase)** is the terminal cytochrome of the electron-transport chain, therefore it is directly oxidized by molecular oxygen. This reaction results in the formation of water together with some hydrogen peroxide (H_2O_2) and some superoxide (O_2^-). The superoxide radical is formed when oxygen is reduced by a single electron ($O_2 + e^- \rightarrow O_2^-$), whereas hydrogen peroxide is formed when oxygen is reduced by two electrons.

Hydrogen peroxide (H_2O_2) and superoxide (O_2^-) are toxic substances that react with and alter biomolecules. Aerobic cells generally have enzymes that metabolize these compounds to nontoxic products. The small amount of superoxide produced in cells is converted by **superoxide dismutase** to hydrogen peroxide, which in turn is decomposed by catalase. **Catalase** metabo-

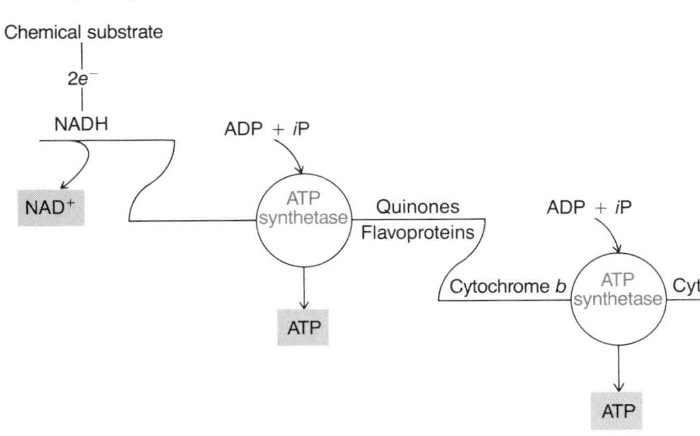

FIGURE 6.10 Electron-transport chain. This is an electrochemical gradient that operates in aerobic cells. Aerobic cells oxidize NADH by the membrane-bound electron-transport chain and use oxygen as the ultimate electron acceptor. NAD^+, water, and ATP are the products of this energy-yielding process.

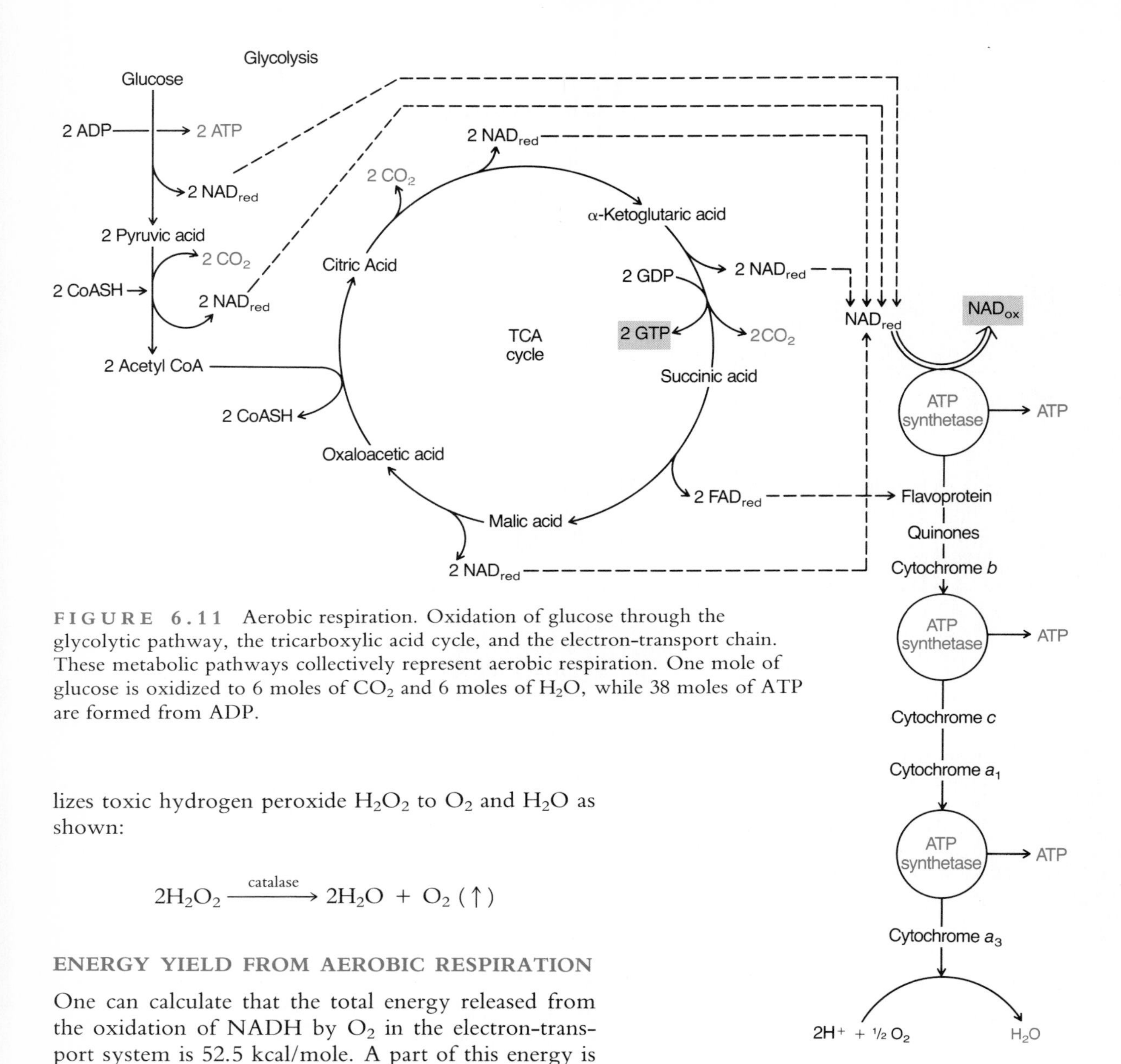

FIGURE 6.11 Aerobic respiration. Oxidation of glucose through the glycolytic pathway, the tricarboxylic acid cycle, and the electron-transport chain. These metabolic pathways collectively represent aerobic respiration. One mole of glucose is oxidized to 6 moles of CO_2 and 6 moles of H_2O, while 38 moles of ATP are formed from ADP.

lizes toxic hydrogen peroxide H_2O_2 to O_2 and H_2O as shown:

$$2H_2O_2 \xrightarrow{\text{catalase}} 2H_2O + O_2 (\uparrow)$$

ENERGY YIELD FROM AEROBIC RESPIRATION

One can calculate that the total energy released from the oxidation of NADH by O_2 in the electron-transport system is 52.5 kcal/mole. A part of this energy is captured in the chemical bonds of ATP, while the rest escapes as heat. The system is 42 percent efficient since 21.9 kcal (3×7.3 kcal mole^{-1} ATP) of the theoretical 52.5 kcal released from each mole of NADH oxidized is retained by the cell as chemical energy. This energy then becomes available for motility, cellular repair, biosynthesis, and reproduction.

Glycolysis, the TCA cycle, and the electron-transport chain are the means by which many aerobic cells gain energy from the oxidation of glucose. Figure 6.11 is a composite diagram of these processes and the energy generated from them. A total of 38 moles of ATP can be formed by bacteria from the complete oxidation of 1 mole of glucose to 6 moles of CO_2 and 6 moles of H_2O. When this theoretical yield is achieved, there are produced 277 kcal (38 moles ATP $\times$ 7.3 kcal mole^{-1} ATP = 277 kcal) per mole of glucose. Remember that the complete combustion of glucose yields 686 kcal mole^{-1} (Table 6.1). Therefore, the cell retains about 40 percent of the chemical energy in each glucose molecule (277/686 × 100 = 40 percent) metabolized during aerobic respiration.

ANAEROBIC ENERGY TRANSFORMATION

Recall that some biological systems can grow and reproduce in the absence of oxygen. Louis Pasteur (Chapter 1) was the first microbiologist to recognize this phenomenon, and he used the term anaerobic to describe growth in the absence of O_2. You remember that he studied diseases of wines and learned that wine is produced during anaerobic growth of yeast cells. We now know that there are numerous anaerobic environments throughout the biosphere and that an abundant variety of microbes thrive in these niches. Deep wells and the bottom of ponds are obvious examples of anaerobic environments. Less obvious is the fact that the rumen (stomach) of ruminates and the large intestine of humans are anaerobic environments where microorganisms grow.

The organisms of anaerobic environments are either obligate anaerobes or facultative anaerobes. **Obligate anaerobes** are intolerant of oxygen; they grow only in an oxygen-free environment. The **facultative anaerobes** grow by aerobic respiration in the presence of oxygen (aerobic environment) and by fermentation when oxygen is absent (anaerobic environment). The anaerobic microbes use the mechanisms of fermentation and/or anaerobic respiration for generating energy in the absence of oxygen.

ANAEROBIC RESPIRATION

Anaerobic respiration is the energy-yielding process involving an electron-transport system in which an inorganic compound other than oxygen serves as the terminal electron acceptor. The electron-transport chain may be similar to or identical to the one described in Figure 6.10 except that oxygen is not the terminal oxidant. Terminal electron acceptors for anaerobic respiration include nitrate (NO_3^-), nitrite (NO_2^-), elemental sulfur (S_0), and sulfate (SO_4^{2-}).

A number of aerobic bacteria will grow anaerobically on organic compounds when nitrate is available as a terminal electron acceptor. These organisms oxidize glucose or other organic substrates with the help of their electron-transport system, which transfers electrons to nitrate. The nitrate is reduced to nitrite (NO_2^-), which can be excreted into the environment. Denitrifying bacteria then reduce the nitrite to gaseous forms of nitrogen, including nitrous oxide (N_2O) and nitrogen gas (N_2). This process is an advantage for bacteria because it enables them to grow anaerobically; however, the nitrogen is lost to the atmosphere during anaerobic respiration, reducing crop yields among plants that depend on soil nitrogen for growth.

Other bacteria reduce sulfur compounds as the terminal electron acceptor in anaerobic respiration. *Desulfovibrio* and *Desulfotomaculum* oxidize organic compounds and reduce sulfate to elemental sulfur (S) and then to hydrogen sulfide (H_2S). The H_2S produced during sulfate reduction is indirectly responsible for the color of black mud and the color of the Black Sea—H_2S reacts with ferrous iron (Fe^{2+}) to form the black compound, ferrous sulfide (FeS). You may have smelled H_2S emanating from rotten eggs, sewage treatment facilities, and/or ponds.

FERMENTATION

Certain bacteria, fungi, and protozoa grow anaerobically by the process of fermentation (Table 6.4). A **fermentation** is an anaerobic energy-yielding process in which a metabolic intermediate (derived from the substrate) is the terminal electron acceptor. Fermentations are self-contained processes in which no outside electron acceptor (neither oxygen nor inorganic compounds) is involved. Instead, both electron donors and acceptors are organic compounds.

The glycolytic pathway becomes a fermentation when pyruvic acid is reduced to lactic acid. In glycolysis, 1 mole of glucose is metabolized to 2 moles of pyruvic acid, 2 moles of NADH, and 2 moles of ATP. When pyruvic acid accepts electrons from NADH, in the reaction catalyzed by lactic acid dehydrogenase, glycolysis becomes a fermentation. Pyruvic acid is the intermediate derived from glucose that serves as the terminal electron acceptor in this fermentation. This process is called the **lactic acid fermentation** since lactic acid is the major product.

TABLE 6.4 Microbial fermentations

FERMENTATION	SUBSTRATE	PRODUCTS	REPRESENTATIVE GENERA
Butyric acid	Glucose[a]	Butyric acid, acetone, butanol, isopropanol, acetic acid	*Clostridium* (saccharolytic)
Lactic acid[b]			
(a) homolactic	Glucose[a]	Lactic acid	*Streptococcus*
(b) heterolactic	Glucose	Lactic acid, CO_2, ethanol	*Leuconostoc*
Propionic acid	Glucose[a] or lactic acid	Propionic acid, CO_2, acetic acid	*Propionibacterium*
Mixed acid	Glucose[a]	Acetic acid, formic acid, succinic acid, lactic acid, CO_2, H_2, 2,3-butanediol	*Escherichia, Enterobacter*
Amino acid	Amino acids	CO_2, H_2, and organic compounds; products vary with amino acid(s) substrate	*Clostridium* (proteolytic)

[a]Most organisms will ferment a variety of carbohydrates in addition to glucose.
[b]Homolactic = approximately 80 percent of the products are lactic acid.
Heterolactic = no more than 50 percent of the fermentation products are lactic acid.

Usable energy is generated in fermentations by substrate level phosphorylation. In the lactic acid fermentation, a maximum net yield of 2 moles of ATP is generated from each mole of glucose metabolized (Figure 6.6). Most organisms that use this fermentation cannot metabolize lactic acid, so it is excreted into the environment.

KEY POINT

Aerobic organisms derive the major portion of their ATP from oxidative phosphorylation, which occurs when reduced coenzymes are oxidized by the electron-transport chain. The ATP produced during a fermentation is formed when a phosphate group on an organic molecule is transferred to ADP, forming ATP. This process is called substrate level phosphorylation.

Other fermentations either start with a different substrate or produce different products. Among the substrates that can be fermented are organic acids, amino acids, purines, and carbohydrates. Fermentation products include a mixture of gases, organic acids, alcohols, and some inorganic products such as ammonia. Fermentations are often named after their major product (Table 6.4), for example the butyric acid fermentation, the propionic acid fermentation, the lactic acid fermentation, and the mixed acid fermentation.

For centuries, people have used fermentations in the preparation of food and beverages. The propionic acid fermentation is important in the manufacture of Swiss cheese, and the lactic acid fermentation is used in the production of pickles, cheese, and yogurt. The yeasts naturally present on fruits ferment the sugar present in fruit juice to ethanol and carbon dioxide. This fermentation has been used for thousands of years to produce alcoholic beverages and is being used to produce ethanol as an additive to automobile fuel.

AUTOTROPHY

Microorganisms that derive all their nutritional needs from inorganic compounds are called **autotrophs.** These organisms are distinct from the **heterotrophs,** which gain their energy and nutritional needs from organic compounds. All autotrophs can use carbon dioxide as their sole source of carbon by combining it with a sugar molecule, in a process called the Calvin cycle. There are two subgroups of autotrophs: the **phototrophs** gain their energy from light; the **chemolithotrophs** gain their energy from the oxidation of inorganic compounds.

PHOTOTROPHS USE LIGHT ENERGY

Phototrophs use sunlight as their source of energy. They are distinct from other organisms because they contain photosynthetic pigments that participate in

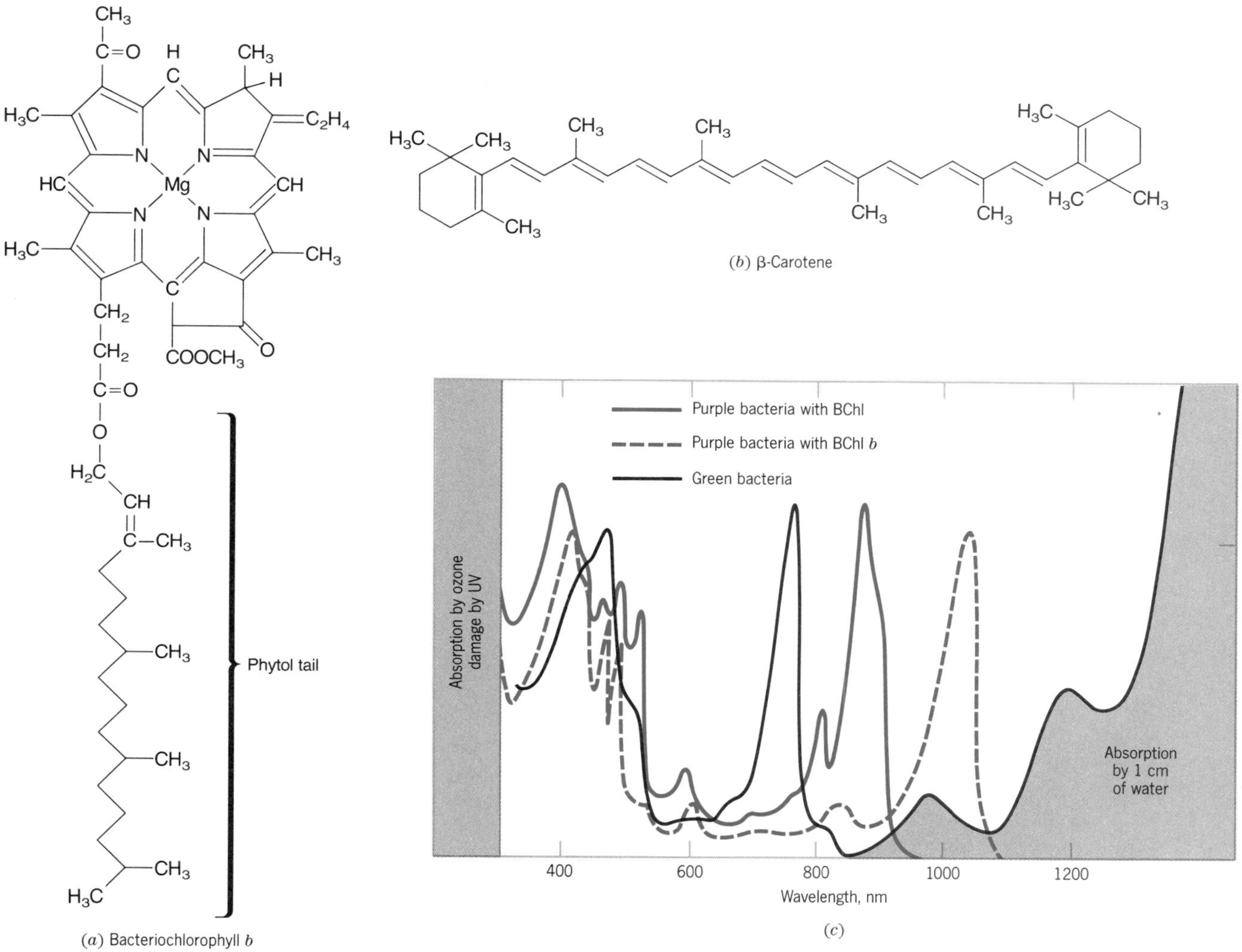

(*a*) Bacteriochlorophyll *b*

(*b*) β-Carotene

(*c*)

FIGURE 6.12 Chlorophyll *(a)* and carotenoids *(b)* are photosynthetic pigments that participate in converting light energy into chemical energy. *(c)* The absorption spectra of bacterial photosynthetic membranes. The usable part of the light spectrum is limited by the absorption by water and ozone in the atmosphere. (Courtesy of *(c)* Dr. R. Clayton. From *Light and Living Matter: The Biology Part*, McGraw-Hill, New York, 1971.)

harvesting light energy. The pigments, consisting of chlorophylls and carotenoids (Figure 6.12), are localized in photosynthetic membranes and absorb light in the visible spectrum. The chemical structures of these pigments vary slightly among different groups of photosynthetic organisms. These small chemical differences alter the wavelengths of light available for photosynthesis. This can be seen in the absorption spectra of the chromatophores from different photosynthetic bacteria (Figure 6.12*c*). The absorption peaks at the longer wavelengths are attributed to the bacteriochlorophylls. Researchers have discovered that a small proportion of the photosynthetic pigments function as reaction centers, where the light reactions of photosynthesis occur.

The eucaryotic green plants and algae grow by **oxygenic photosynthesis;** they produce O_2 from water by the following reaction:

$$CO_2 + 2H_2O \xrightarrow{\text{light}} (CH_2O)_n + O_2 + H_2O$$

Here the oxygen is released from water during the primary light reaction of photosynthesis. Absorption of light energy by the reaction center activates electrons. These electrons then participate in generating the ATP and NADPH needed to fix carbon dioxide into carbohydrate $(CH_2O)_n$. Oxygen-producing photosynthetic organisms contain two distinct reaction centers, which enable them to split water with the generation of oxygen. Cyanobacteria and *Prochloron* are the only procaryotes capable of oxygenic photosynthesis.

KEY POINT

Oxygenic photosynthesis is the major mechanism for returning oxygen to the biosphere. Oxygen is continually being consumed by the burning of fossil fuels and biological oxidations. We depend on green plants, algae, and cyanobacteria to maintain a balance in the world's supply of oxygen.

The other photosynthetic bacteria capture light energy only in anaerobic environments. This is called **anoxygenic photosynthesis** because the bacteria do not produce oxygen—in fact, oxygen inhibits their photosynthetic growth. These bacteria possess a single reaction center complex (Figure 6.13) and use it for the light-driven oxidation of reduced sulfur compounds such as H_2S or reduced carbon (R-H_2) compounds, which are their substrates for photosynthesis and sources of electrons.

The simplest mechanism for anoxygenic photosynthesis is found in the photosynthetic green sulfur bacteria. These bacteria convert light into chemical energy under anaerobic conditions, using hydrogen sulfide (H_2S) as a substrate. The formula for the oxidation of hydrogen sulfide and the fixation of carbon dioxide is

$$CO_2 + 2H_2S \xrightarrow{\text{light}} (CH_2O)_n + 2S + H_2O$$

All of the photosynthetic sulfur bacteria first produce elemental sulfur (S) from H_2S, and then oxidize it further to sulfate. One can actually see the culture turn chalky as elemental sulfur is deposited during the initial phase of photosynthetic growth.

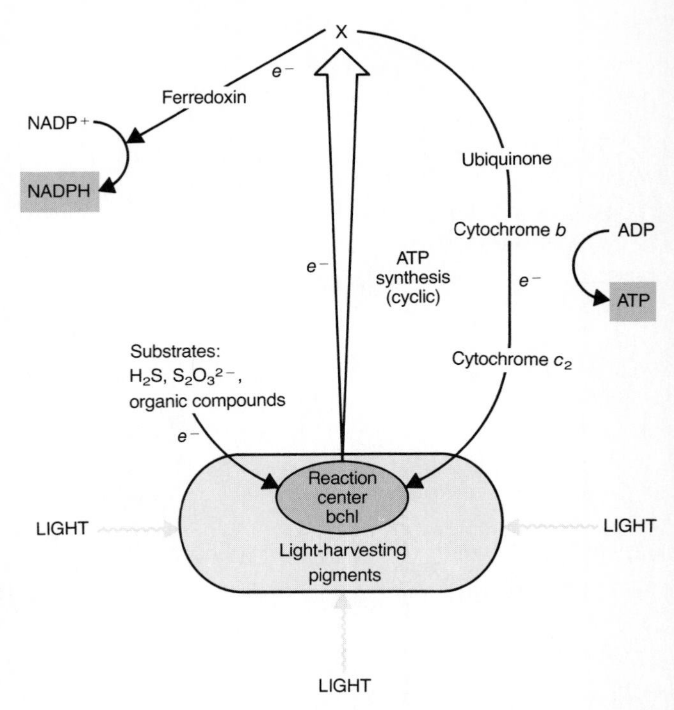

FIGURE 6.13 Bacterial photosynthesis. Electrons flow from the substrate to the reaction center bacteriochlorophyll (bchl). Light activates the electron in bchl, which either follows a noncyclic electron flow to reduce $NADP^+$ or enters the cyclic electron-transport chain that carries it back to the reaction center bchl. ATP is formed in the cyclic electron-transport chain by phosphorylation.

The primary act of photosynthesis occurs in bacterial membranes containing the photosynthetic pigments (Figure 6.13). These pigments are arranged in the photosynthetic membranes to function either as light-harvesting (antenna) pigments or as reaction centers (RC). The primary act of converting light energy into chemical energy occurs in the reaction center.

Photosynthetic reaction centers in green and purple bacteria contain bacteriochlorophyll, a carotenoid, ubiquinone, and proteins. The pigments in the reaction center absorb light energy that activates an electron present in bacteriochlorophyll. The light-activated electron reduces an electron acceptor(x); from here it can either reduce $NADP^+$ to form the NADPH needed for fixing CO_2 or it can enter the photosynthetic electron-transport chain used by the cell to generate ATP. As the light-activated electron is transferred back to bacteriochlorophyll, ATP is formed in the process called **cyclic photophosphorylation** (Figure 6.13). The source of electrons for initiating the process is either H_2S or R-H_2, substrates that are oxidized by the reaction center.

The ATP and NADPH generated by photosynthesis are used to "fix" carbon dioxide in the **Calvin cycle** (Figure 6.14). This is an enzymatic pathway in which CO_2 is combined with the 5-carbon sugar ribulose 1,5-diphosphate, after which two 3-carbon trioses are formed. The cell uses the trioses to produce hexoses, which are used to satisfy its need for cellular carbohydrates. The complex sequence of reactions in the Calvin cycle also regenerates ribulose 1,5-diphosphate, so the cycle can continue. The energy in ATP and the reducing power in NADPH that drive the Calvin cycle are generated by the light reactions of photosynthesis. For each hexose produced, the cycle uses 6 CO_2, 18 ATP, and 12 NADPH. Chemolithotrophs also use the Calvin cycle because they too depend on CO_2 as a source of carbon.

Through photosynthesis, chemoorganotrophs are constantly provided with organic matter. Photosynthesis by green plants also provides a continuing supply of oxygen, which these organisms generate from water. Both the supply of organic matter and the regeneration of molecular oxygen are essential to the survival of life on earth.

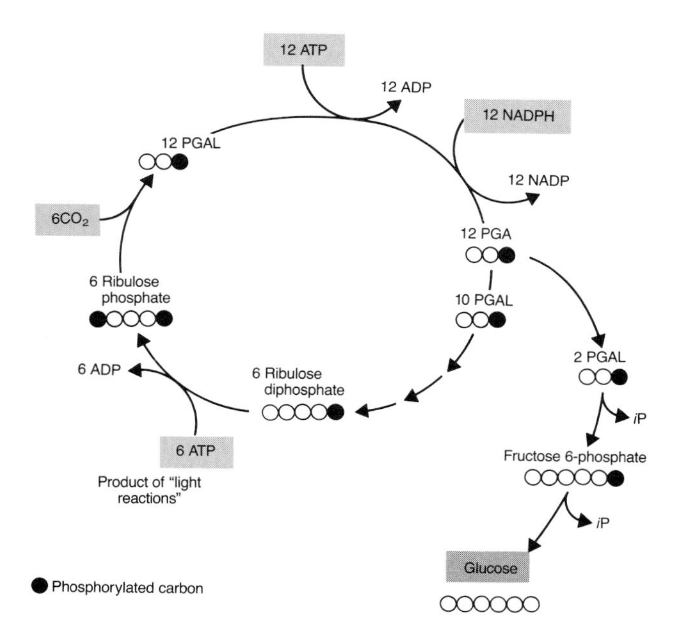

FIGURE 6.14 Phototrophs and chemolithotrophs use the Calvin cycle to fix carbon dioxide to carbohydrate. Overall, glucose is produced from 6 CO_2 at the expense of 18 ATP and 12 NADPH generated by the light reactions of photosynthesis. Phosphoglyceric acid = PGAL.

CHEMOLITHOTROPHS OXIDIZE INORGANIC COMPOUNDS

Chemolithotrophs are a small group of bacteria that meet all their energy needs from the aerobic oxidation of an inorganic compound and use CO_2 as their sole source of carbon. Many of the oxidations catalyzed by the chemolithotrophs are essential to the natural cycling of inorganic compounds (Table 6.5).

Nitrification describes the combined activities of the bacteria that oxidize ammonia to nitrate. Nitrification occurs in a stepwise fashion, in which ammonia (NH_3) is oxidized to nitrite (NO_2^-) by *Nitrosomonas* (nye-troh-so-moan'as), and nitrite is oxidized to nitrate (NO_3^-) by *Nitrobacter* (nye-troh-bak'ter). *Nitrosomonas* and *Nitrobacter* are representative of the many different nitrifying bacteria found in soil, marine, and aquatic environments, that are essential to the nitrogen cycle.

Reduced-sulfur compounds are oxidized aerobically by the sulfur-oxidizing bacteria of the genus *Thiobacillus* (thy'oh-bah-sill'us). These bacteria oxidize hydrogen sulfide (H_2S) to elemental sulfur, which is then oxidized in a series of reactions to sulfate (SO_4^{-2}). In addition to being able to oxidize reduced-sulfur compounds, *Thiobacillus ferrooxidans* can oxidize iron (II) to iron (III) as a source of energy (Table 6.5). Since iron (II) is autooxidized in air at biological pH, *Thiobacillus ferrooxidans* performs this oxidation only at a pH lower than 2. Such conditions exist in acid-mine waters, where these bacteria grow.

In summary, most microbes are chemoorganotrophs, so they gain energy from the aerobic oxidation of organic compounds. If it were not for photosynthesis, the supply of organic compounds for chemoorganotrophs would quickly be depleted from the earth. Fortunately, photosynthesis uses light energy to form organic materials that eventually resupply the chemoorganotrophs with their essential food source. Inorganic compounds have a parallel cycle on earth. Certain bacteria reduce sulfates to H_2S and nitrates to NO_2^-, N_2O, and N_2 when they grow by anaerobic respiration. Reduced inorganic compounds are sub-

TABLE 6.5 Oxidations by chemolithotrophs

NAME	REACTION[a]	CYCLE	REPRESENTATIVE GENERA
Ammonia oxidation	$NH_3 + 1\frac{1}{2}O_2 \rightarrow HNO_2 + H_2O$	Nitrogen	*Nitrosomonas*
Nitrite oxidation	$HNO_2 + \frac{1}{2}O_2 \rightarrow HNO_3$	Nitrogen	*Nitrobacter*
Sulfur oxidation	(*a*) $H_2S + \frac{1}{2}O_2 \rightarrow S + H_2O$	Sulfur	*Thiobacillus thiooxidans*
	(*b*) $S + 2O_2 \rightarrow SO_4^{2-}$	Sulfur	
Iron oxidation	$2Fe^{2+} + 2H^+ + \frac{1}{2}O_2 \rightarrow 2Fe^{+3} + H_2O$	Iron leaching	*Thiobacillus ferrooxidans*

[a]Chemolithotrophs gain their energy from the oxidation of inorganic compounds and make organic compounds by fixing CO_2. These bacteria are important in the cycling of inorganic nutrients, especially nitrogen and sulfur.

strates for the chemolithotrophs, which grow by oxidizing inorganic compounds and fixing CO_2. Each one of these processes fits neatly into the ecological scheme of the biosphere.

BIOSYNTHESIS OF CELLULAR MATERIALS

In the first part of this chapter we considered the mechanisms cells use to generate energy. Now we turn to the major use of that energy: the synthesis of cellular material. Cells need a great variety of biomolecules for manufacturing and maintaining cell structures. Even though some of the required biomolecules may be supplied by the growth medium, many cells can synthesize all the essential compounds they need to manufacture new cells. These cells use enzymatic pathways to form amino acids, purines, pyrimidines, carbohydrates, vitamins, coenzymes, and fatty acids. Here we present an overview of the mechanisms cells use to synthesize some of these biomolecules: carbohydrates, amino acids, lipids, purines, and pyrimidines.

RAW MATERIALS FOR CELLULAR SYNTHESIS

When we say a cell uses glucose as a carbon source, we mean that it can derive the organic biomolecules necessary for cell growth and reproduction from this compound. The pathways for the metabolism of glucose are interconnected with biosynthetic pathways for making essential biomolecules (Figure 6.15). The pentoses found in RNA and DNA are generated when glucose is metabolized through the pentose phosphate cycle. The glycerol found in simple lipids is also derived from the glycolytic pathway. Simple lipids also require fatty acids, which are synthesized from acetyl-CoA. Alternatively, acetyl-CoA enters the TCA cycle to generate the biosynthetic building blocks of oxaloacetic acid, succinyl-CoA, and α-ketoglutaric acid.

BIOSYNTHESIS OF CARBOHYDRATES

Cells need carbohydrates to make cell walls, glycoproteins, glycolipids, nucleic acids, starch, poly β-hydroxybutyric acid, slime layers, and capsules. The source of carbon for the various forms of carbohydrate is usually an organic nutrient, which also provides the cell with energy. As carbohydrates are catabolized, specific intermediates are shunted off from the degradative pathways to be used in biosynthetic pathways (Figure 6.16). Many carbohydrate polymers such as glycogen and the peptidoglycans of the bacterial cell wall are synthesized from activated sugars at the expense of ATP or uridine triphosphate (UTP). The sugars are derived from glucose as it is metabolized in the glycolytic pathway. Poly-β-hydroxybutyrate is a storage product of bacteria that is produced from acetate (Figure 6.16).

Production of Pentoses

Glucose is also the starting material for the synthesis of pentoses, the 5-carbon sugars that are essential com-

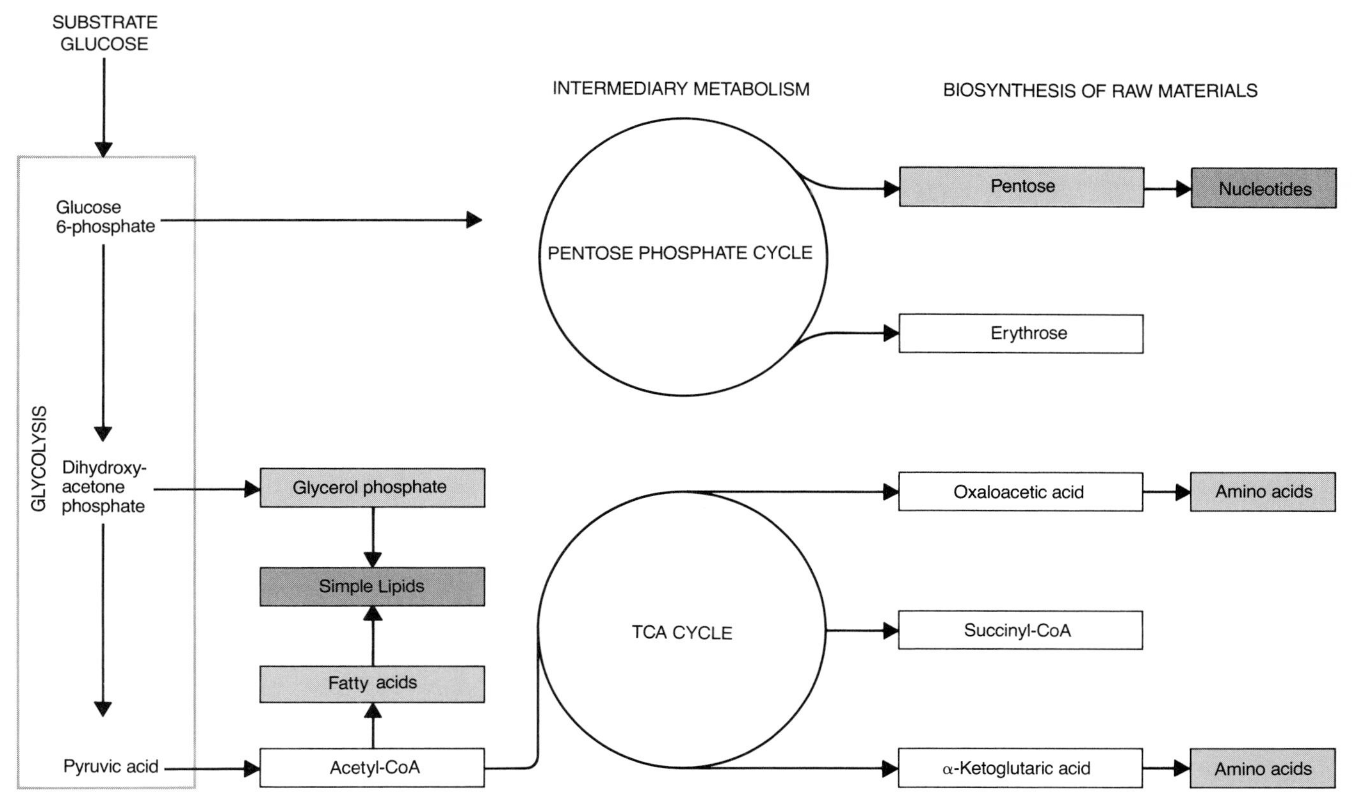

FIGURE 6.15 Raw materials for biosynthetic reactions. Glucose is metabolized by the pathways of intermediary metabolism (glycolysis, pentose phosphate cycle, and the TCA cycle) to produce the organic starting material used for biosynthetic reactions (simple lipids, nucleotides, amino acids).

ponents of the nucleotides in nucleic acids and coenzymes. Cells generate the pentoses they need through the initial steps of the pentose phosphate cycle (Figure 6.16). This series of reactions actually generates energy in the form of NADPH. The glucose 6-phosphate that starts this pathway is an intermediate of the glycolytic pathway. It is enzymatically converted to the pentose ribulose 5-phosphate with the production of CO_2 and 2 NADPH. Ribulose is a very important cellular metabolite because it is the precursor of both the ribose and the deoxyribose that form RNA and DNA.

BIOSYNTHESIS OF AMINO ACIDS

Biological systems need 20 amino acids for synthesizing proteins. α-Amino acids have the basic structure

$$H_2N{-}\overset{\displaystyle R}{\overset{|}{\underset{\underset{\displaystyle H}{|}}{C}}}{-}COOH$$

which consists of an amino group ($-NH_2$) and a carboxyl group ($-COOH$) attached to an alpha-carbon (α) atom. The R group can be any one of 20 structures (see Figure 2.10), which transform the basic amino acid structure into one of the 20 amino acids found in proteins.

The amino acids are synthesized by 20 different multienzyme pathways, some of which are very complex. Here we present the synthesis of two amino acids as examples. Amino acids derive their carbons from intermediates of the TCA cycle, glycolytic pathway,

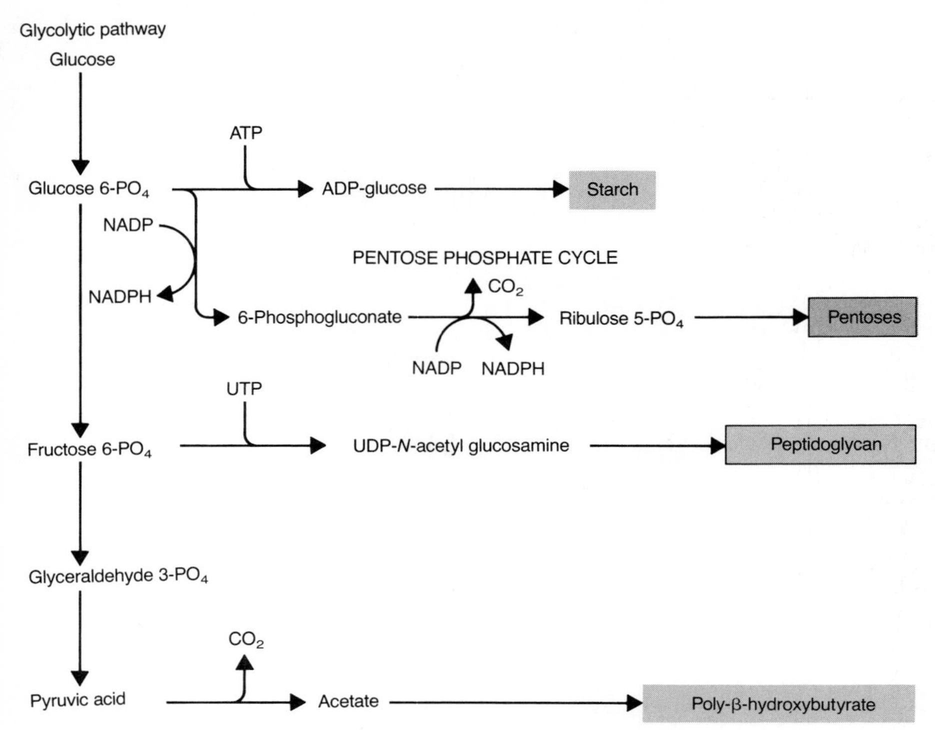

FIGURE 6.16 Cells synthesize complex carbohydrates from the phosphorylated intermediates of glucose metabolism. Pentoses (ribose and deoxyribose) needed to synthesize RNA and DNA are produced by the pentose phosphate cycle. Nucleoside diphosphate sugars such as ADP-glucose are the substrates for synthesizing large carbohydrate polymers. UDP-*N*-acetyl glucosamine is a precursor for synthesis of peptidoglycan.

and the pentose phosphate cycle, or from CO_2. Ammonia is assimilated into organic compounds by an amination reaction, such as the reactions used to synthesize glutamic acid (Figure 6.17). The carbon skeleton of glutamate is derived from α-ketoglutarate, an intermediate in the TCA cycle. Once formed, glutamate can donate its amino group to other molecules in transamination reactions, such as occurs in the formation of aspartic acid (Figure 6.17). Cells that contain the appropriate enzymes can synthesize all of the 20 amino acids from intermediates of cellular metabolism and ammonia. Cells lacking these enzymes must acquire the amino acid as a nutrient from the growth medium.

BIOSYNTHESIS OF LIPIDS

Simple lipids are composed of two components: a carbohydrate portion that is a derivative of glycerol, and one or more long-chain fatty acids. Glycerol is a 3-carbon alcohol that is synthesized from dihydroxyacetone phosphate (Figure 6.18), an intermediate in the glycolytic pathway. Fatty acids are synthesized from the acetyl-CoA by a complex series of reactions that require large amounts of NADPH and ATP. Phospholipids are made when fatty acids and glycerol-3-phosphate combine. Once formed, these simple lipids are incorporated into the cytoplasmic membranes of the cell or into the outer membrane of the gram-negative cell wall.

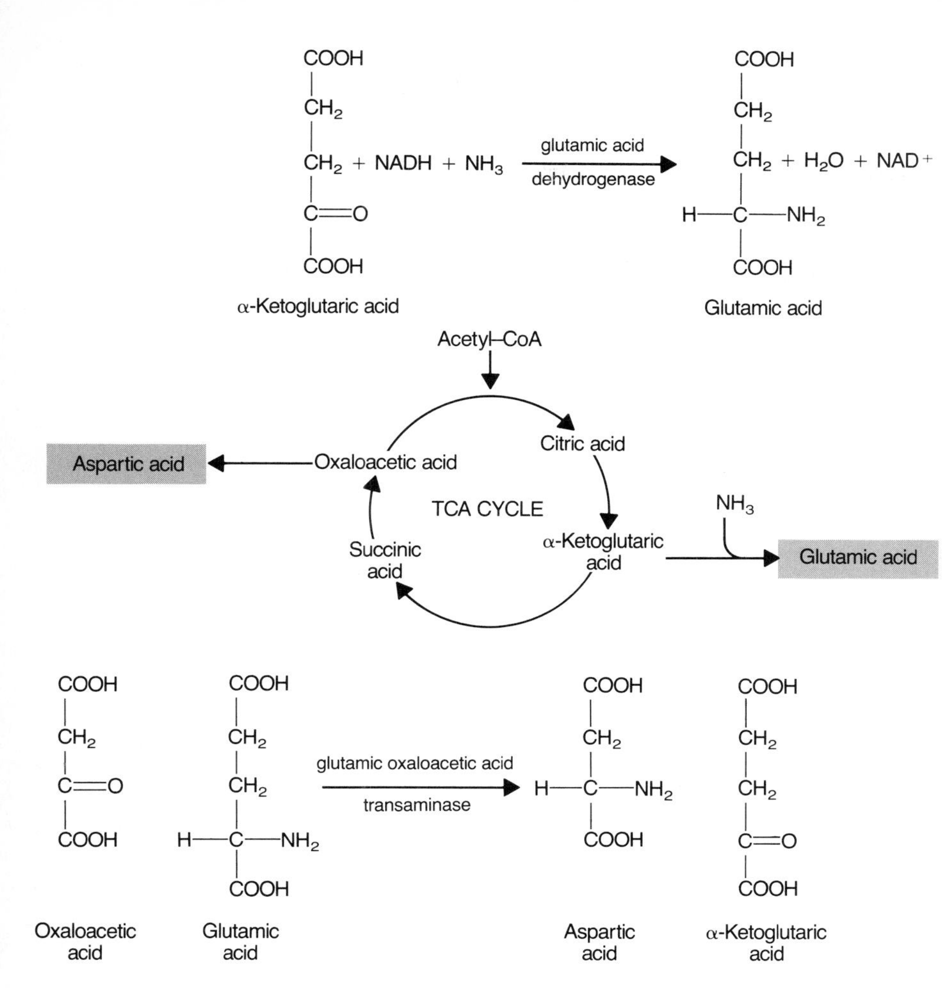

FIGURE 6.17 Amino acids are synthesized by many complex pathways; two examples are shown. Inorganic nitrogen (NH_3) is incorporated into α-ketoglutaric acid of the TCA cycle to form glutamic acid. Glutamic acid can then donate its amino group to oxaloacetic acid to form aspartic acid in a transaminase reaction.

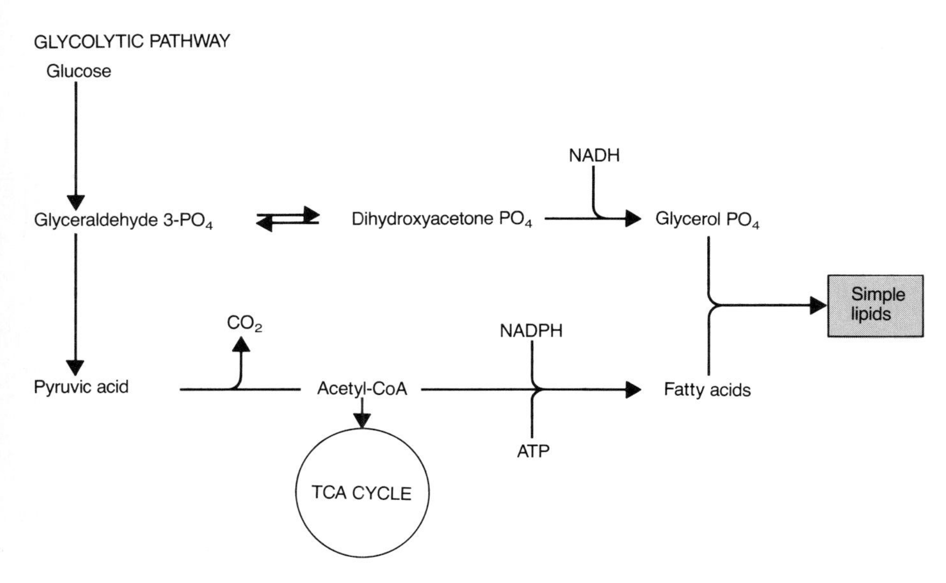

FIGURE 6.18 Lipids are synthesized from glycerol phosphate formed from the glycolytic pathway and from fatty acids produced from acetyl-CoA.

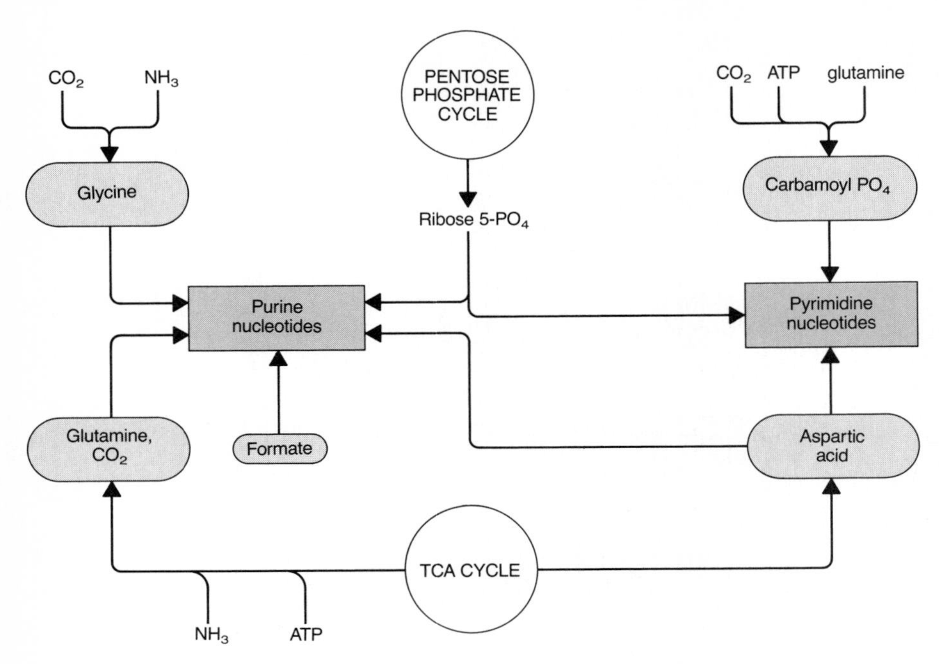

FIGURE 6.19 Nucleotides are built up from ribose 5-phosphate with carbons coming from the shaded compounds. These complex pathways make the purine and pyrimidine nucleotides that are substrates for DNA and RNA synthesis.

BIOSYNTHESIS OF NUCLEOTIDES

Nucleotides are composed of either a purine or a pyrimidine nitrogenous base, ribose, and phosphate. Synthesis of these complex molecules begins on ribose 5-phosphate generated in the pentose phosphate cycle (Figure 6.19). This ribose is the foundation on which the purine or pyrimidine base is built. The carbons in the nitrogenous bases come from aspartic acid, glycine, carbon dioxide, and formic acid, while the nitrogen atoms are derived from amino acids, specifically glutamic acid, glycine, and aspartic acid (Figure 6.19). Ribose 5-phosphate also contributes the first phosphate to the nucleotide.

Nucleotides are the building blocks for RNA and DNA, are involved in energy-transfer reactions, and are structural components of certain coenzymes. Cells use ATP and dATP (deoxyadenosine triphosphate) to synthesize respectively RNA and DNA.

The biosynthetic reactions of the cell are closely linked to the energy-generating mechanisms. Each of the synthetic pathways we have discussed either requires energy in the form of NADH or ATP, or utilizes intermediates of catabolic pathways that once used are unavailable for generating energy.

Summary Outline

BIOLOGICAL SOURCES OF ENERGY

1. Sunlight, the ultimate source of biological energy, is converted to chemical energy through the process of photosynthesis.

2. Chemoorganotrophs and chemolithotrophs employ oxidation–reduction reactions to release the chemical energy stored in organic or inorganic compounds.

3. Biological systems use the energy-rich phosphate bonds of adenosine triphosphate (ATP) and the reduced pyridine nucleotide (NADH) transfer chemical energy.
4. NADH contains chemical energy that is released upon oxidation.

ENZYMES CATALYZE CHEMICAL REACTIONS

1. All important chemical reactions in cells are catalyzed by enzymes at rates governed by temperature, pH, and substrate concentration.
2. Enzymes are specific for their substrate—bacterial cells have about 1000 different enzymes.
3. The names of most enzymes end in the suffix -ase and indicate the substrate acted upon.
4. Coenzymes are reusable organic molecules that contribute chemical groups to the enzymatic reaction. Many coenzymes are vitamins or contain a vitamin as a major structural component.

AEROBIC RESPIRATION

1. Glucose is oxidized in aerobic respiration through the glycolytic pathway and then through the TCA cycle.
2. Glycolysis is a sequence of nine enzymatic steps that converts glucose to 2 pyruvic acid, 2 NADH, and 2 ATP.
3. Pyruvic acid is decarboxylated to form acetyl-CoA, NADH, and CO_2. The tricarboxylic (TCA) acid cycle is a sequence of enzyme-catalyzed reactions that oxidize acetyl-CoA to 2 CO_2 and generate 1 GTP, 1 FADH, and 3 NADH.
4. The NADH and FADH produced in aerobic respiration are oxidized in the electron-transport chain. ATP formation occurs during electron-transfer phosphorylation, a process explained by chemiosmotic coupling.
5. The electron-transport system produces 3 ATP for each NADH and 2 ATP for each FADH oxidized by oxygen.
6. Aerobic respiration in bacteria generates 38 moles of ATP per mole of glucose oxidized to CO_2 and H_2O.

ANAEROBIC ENERGY TRANSFORMATION

1. Cells growing by anaerobic respiration oxidize an organic substrate and use an inorganic compound, such as sulfate or nitrate, as the terminal electron acceptor in place of oxygen.
2. Fermentation is an anaerobic energy-yielding process in which a metabolic intermediate (derived from the substrate) serves as the terminal electron acceptor. Energy in the form of ATP is generated by substrate-level phosphorylation.
3. Fermentations are named for their major products: lactic acid fermentation, butyric acid fermentation, and propionic acid fermentation.

AUTOTROPHY

1. Phototrophs use sunlight for energy and the Calvin cycle to fix CO_2 into carbohydrate.
2. Green plants, algae, and cyanobacteria grow by oxygenic photosynthesis, which is the major process for regenerating the earth's supply of molecular oxygen.
3. The purple and green photosynthetic bacteria grow by anoxygenic photosynthesis, using reduced organic or sulfur compounds as electron donors.
4. Chemolithotrophs gain energy from the oxidation of inorganic compounds and depend on CO_2 for their carbon. Specific groups of chemolithotrophs oxidize ammonia, nitrite, hydrogen sulfide, sulfur, or iron (II).

BIOSYNTHESIS OF CELLULAR MATERIALS

1. Cells convert their carbon source through a series of enzymatic reactions into the amino acids, nucleotides, carbohydrates, and lipids they need for growth and reproduction.
2. The pentose phosphate cycle produces ribulose 5-phosphate, which is a precursor of ribose in the synthesis of nucleotides.
3. Oxaloacetic acid and α-ketoglutaric acid are two intermediates of the TCA cycle that form amino acids by transamination reactions.
4. Fatty acids are synthesized from the acetyl-CoA produced by the decarboxylation of pyruvic acid.

The glycerol present in lipids is derived from the glycolytic pathway.

5. Nucleotides are synthesized by adding carbon and nitrogen groups from amino acids, formic acid, and carbon dioxide to ribose 5-phosphate in a complex series of reactions.

Questions and Topics for Study and Discussion

QUESTIONS

1. What properties does ATP have that enable it to function as the common energy currency?
2. Define or explain the following:

chemolithotroph	exergonic reaction
chemoorganotroph	fermentation
coenzyme	oxidation–reduction reaction
dehydrogenation	oxygenic photosynthesis
enzyme reaction	
endergonic reaction	

3. Write out one enzyme-catalyzed oxidation–reduction reaction and indicate which compounds are oxidized and reduced.
4. Diagram the key reactions of glycolysis. What is the maximum net yield of ATP from this pathway?
5. Describe how pyruvic acid is metabolized by aerobic bacteria. How do aerobic bacteria oxidize the NADH produced?
6. What is the maximum yield of ATP when glucose is metabolized by an aerobic bacterium? In eucaryotes, the NADH produced during glycolysis is transported into the mitochondrion as FADH. What is the yield of ATP from the metabolism of 1 mole of glucose by an aerobic eucaryote?
7. What are the major differences between anaerobic respiration and fermentation?
8. Give reasons why an obligate anaerobe could be killed by exposure to oxygen.
9. Explain how aspartic acid and glutamic acid are produced from inorganic nitrogen and intermediates of the TCA cycle.
10. What intermediates of the glycolytic pathway are used to synthesize carbohydrates, pentoses, and lipids?

DISCUSSION TOPICS

1. Why is it important to our environment that bacteria grow by diverse energy-yielding pathways?
2. What metabolic processes would you expect to encounter in disease-producing microbes? Which mechanism of energy metabolism would you not expect to find? Why?

Further Readings

BOOKS

Ingraham, J. L., O. Maaløe, and F. C. Neidhardt. *Growth of the Bacterial Cell,* Sinauer Associates, Sunderland, Mass., 1983. Contains a detailed presentation of the fueling reactions and energy costs for cellular synthesis in *Escherichia coli*.

Lehninger, A. L., *Principles of Biochemistry,* Worth, New York, 1982. An introductory textbook in biochemistry by an author whose writings are highly regarded for accuracy and style.

Stryer, L., *Biochemistry,* 2nd ed., Freeman, San Francisco, 1981. An introductory textbook in biochemistry that presents a foundation for microbial biochemistry.

ARTICLES AND REVIEWS

Hinkle, P. C., and R. E. McCarty, "How Cells Make

ATP," *Sci. Am.*, 238:104–117 (March 1978). A description of the biomolecules involved in photosynthetic and oxidative phosphorylation followed by an explanation of the "chemiosmotic" theory of ATP formation.

Lehninger, A. L., "How Cells Transform Energy," *Sci. Am.*, 205:62–73, (September 1961). An elementary discussion of energy transformations in mitochondria and chloroplasts.

Wood, W. A., "Fermentation of Carbohydrates and Related Compounds," in I. C. Gunsalus and R. Y. Stanier (eds.), *The Bacteria*, Vol. 2, pp. 59–149, Academic Press, New York, 1961. A classic paper on the microbial metabolism of carbohydrates by fermentation mechanisms.

CHAPTER

7

Nutrition, Cultivation, and Growth of Microorganisms

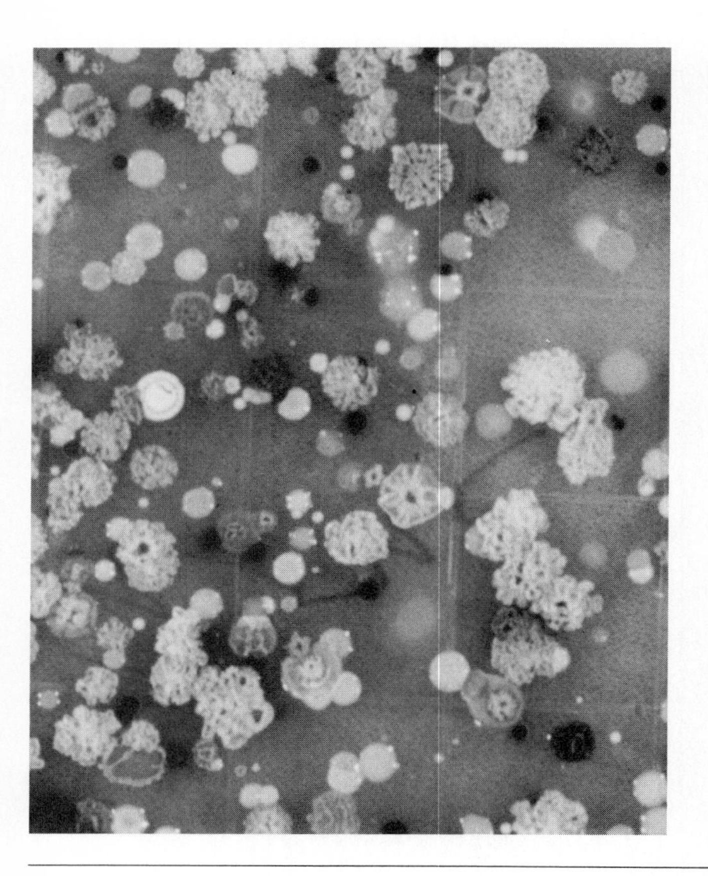

OUTLINE

FOCUS OF THIS CHAPTER

In this chapter we describe the techniques for growing pure cultures of microorganisms on media that provide their nutritional needs. The different culture media used in microbiology and the chemical and physical factors that affect microbial growth are discussed. This knowledge is applied to the study of microbial growth and the practical aspects of culturing microorganisms.

Microbial growth is the orderly increase in cellular constituents that ultimately results in the formation of new cells. Most bacteria divide by transverse binary fission to produce two daughter cells of equal size and composition. Before dividing, a cell must reproduce all of its essential constituents and double its mass. This process depends on the physiological capabilities of the species, the availability of nutrients, and the physical parameters of the environment, such as pH, temperature, hydrostatic pressure, and osmotic pressure.

The techniques used to isolate, grow, and maintain laboratory cultures of microorganisms are basic to microbiology. To grow microorganisms one must understand their nutritional requirements and be able to handle pure cultures.

NUTRITIONAL REQUIREMENTS OF MICROORGANISMS

The nutritional requirements of each microorganism depend on that organism's metabolic capabilities. A cell that is unable to synthesize an essential nutrient must be able to acquire that nutrient from its growth medium. When reduced to its basic ingredients, a microbiological medium must contain a source of energy, carbon, nitrogen, phosphorus, sulfur, certain inorganic salts, and trace elements.

MINIMUM NUTRITIONAL REQUIREMENTS

The bacterium *Escherichia coli* has minimal nutritional requirements since it can synthesize all of its cellular components from the simple glucose–salts medium (Table 7.1). *Escherichia coli* uses glucose as both the carbon and the energy source. It is able to convert glucose to the numerous organic metabolites required to make its lipids, proteins, nucleic acids, and other organic cellular constituents. The cell's phosphate requirement is satisfied by the potassium phosphate; the sulfur needed to form the sulfur-containing amino acids and vitamins is derived from magnesium sulfate; and the nitrogen in purines, pyrimidines, amino acids, and vitamins is supplied by ammonium chloride. Other essential components are supplied as salts (Mg^{2+}, K^{+}) or as trace elements.

SOURCES OF CARBON, NITROGEN, SULFUR, AND GROWTH FACTORS

As a group, chemoorganotrophs are able to metabolize a wide variety of carbon compounds, including organic acids, methane, amino acids, fats, oils, sugars, purines, pyrimidines, proteins, and polysaccharides. Some bacteria can use many organic compounds, while others are severely restricted. At one extreme are certain members of the *Pseudomonas* group that can metabolize over 90 different carbon sources. The other extreme is occupied by the autotrophs, for which carbon dioxide is the major source of carbon.

KEY POINT

For every natural organic compound, there exists a microorganism that is able to metabolize it. Unfortunately, many manufactured chemicals are toxic to birds and ani-

TABLE 7.1 Glucose–salts medium for *Escherichia coli*

INGREDIENT[a]	AMOUNT (g)
Glucose	5.0
Dipotassium phosphate (K_2HPO_4)	7.0
Monopotassium phosphate (KH_2PO_4)	2.0
Magnesium sulfate ($MgSO_4$)	0.008
Ammonium chloride (NH_4Cl)	1.0
Trace elements	1.0 ml
Distilled water	999.0 ml

[a]Solid medium can be prepared by adding 15 g of agar/1000 ml of medium.

mals and are not metabolized by microorganisms. Their accumulation in the environment presents a real hazard to many species. An example is DDT, an insecticide widely used in the United States until its hazardous effects on wildlife were discovered.

Reduced nitrogen (NH_3) is used by cells to form amino acids, purines, pyrimidines, and certain vitamins. Many microbes use organic forms of nitrogen that are added to media as amino acids, purines, or pyrimidines. Many microbes can also use inorganic nitrogen in the form of ammonia, and a few microbes use nitrate and/or nitrite as their nitrogen source. Molecular nitrogen (N_2) is the nitrogen source for a limited group of procaryotes called nitrogen fixers.

Almost all microbes can use sulfate in the biosynthesis of sulfur amino acids and vitamins. A few bacteria, including *Legionella* (le-gi-on-el'la), have an absolute requirement for the sulfur amino acid cysteine, but this is rare. Other bacteria can use reduced forms of sulfur such as hydrogen sulfide (H_2S), thiosulfate ($S_2O_3^{-2}$), or molecular sulfur (S_0).

Microbes that cannot grow on a chemically defined medium such as the glucose–salts medium usually require one or more growth factors—an amino acid, purine, pyrimidine, or vitamin. Species that require a growth factor are unable to synthesize it, and the factor must be supplied by the growth medium.

ALTERNATIVE SOURCES OF ENERGY

Phototrophs require light as their energy source either exclusively or in addition to chemical energy in organic compounds. Chemolithotrophs oxidize inorganic compounds including ammonia, nitrite, iron(II), and reduced sulfur compounds as their energy source. Specialized growth conditions and media are required to grow these microbes.

PURE CULTURES

Although some populations of mixed species are studied, most microbiological studies are done on pure cultures. A **pure,** or **axenic** (a-zee'nick), **culture** contains a single strain of microbe in which all the cells are derived from a single parent cell. All other cultures are mixed or contaminated. In nature, microbes exist in mixed populations that compete for substrates, nutrients, and metabolic products. For most purposes, microbiologists must work with pure cultures to ensure that the phenomena they observe are attributable to a given species, just as chemists must work with pure chemicals.

The techniques for growing pure cultures depend on physically separating microbes in an environment that enables them to grow. This was first attempted by Robert Koch when he streaked bacteria on the surfaces of boiled potato slices. Although these solid surfaces were ideally suited for physically separating microbes, the nutrients in the potato greatly restricted the type of bacteria that could grow.

The problem was solved by adding **agar** to the liquid medium as a solidifying agent. This polysaccharide is so widely used in microbiology that the term "agar" now refers to solidified media; for example, nutrient agar, glucose–salts agar. Certain properties of agar make it an ideal solidifying agent for microbiological media: (1) agar dissolves in boiling water to form solutions that remain liquid until they are cooled to about 44°C, at which temperature a solid gel is formed, (2) agar media remain solid until the agar melts at a temperature above 85°C, and (3) most bacteria are unable to metabolize agar. Agar can be incorporated into any liquid medium to make a solid vehicle for isolating microbes.

ISOLATION TECHNIQUES

A bacterium will grow into a distinct colony when it is physically isolated on a suitable solid growth medium. The **streak plate technique** is a common method for isolating pure colonies of bacteria and yeast as follows: a sterile* inoculating loop is streaked back-and-forth across the plate until the organisms are physically separated on the agar surface. The organisms grow into well-isolated colonies (Figure 7.1), which become visible to the eye when they contain about one million cells.

A culture is presumed to be pure if only one morphological type of colony grows on two successive streak plates. Bacterial colony morphologies can be described as small, large, mucoid, rough, raised, flat, filamentous, or spreading. In addition to requiring a single colony morphology, most microbiologists take the

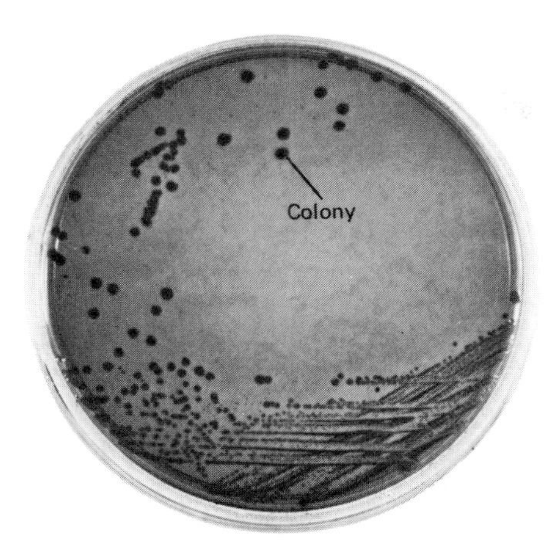

FIGURE 7.1 Colonies on streak plates are visible when they contain 1 million or more cells.

precaution of making a Gram stain of the culture as a second test of purity. Under ideal conditions, each pure culture is composed of organisms that have grown from a single cell, so the cells in a pure culture are genetically identical.

Pour plates can be made by first diluting a bacterial culture in a tube of melted agar at about 47° to 48°C. Dilutions of a bacterial culture are made by transferring 1 ml of the culture to 9 ml of sterile medium. This represents a 1:10 dilution since each milliliter of the new tube contains one-tenth the number of bacteria found in the original culture. This procedure is repeated until there are only a few bacteria per milliliter in the final dilution. Finally, the content of each tube is poured into a petri dish and allowed to solidify. Bacterial colonies will develop both on the agar surface and in the agar itself, depending on the cell's oxygen requirement.

Anaerobic bacteria can be isolated by a modification of this procedure known as the **shake-dilution tube** method (Figure 7.2). After the culture is diluted in a series of melted agar tubes, the medium is allowed to solidify. A layer of sterile paraffin wax/paraffin oil (called vaspar) is applied to seal the top of the tubes and prevent air from entering. After incubation, physically separated colonies will develop in the tubes of the greater dilutions. To obtain pure cultures, a new series of dilutions is made until only a single colony type is present.

Isolating colonies from these tubes is difficult. The vaspar plug is aseptically removed before the agar butt is blown out of the tube into a sterile petri dish with a gentle stream of gas directed between the tube wall and the agar butt. Colonies in the agar butt are then dis-

*Sterile means devoid of all living things.

FIGURE 7.2 Shake-dilution tubes are used to isolate bacteria under anaerobic growth conditions. In this example, a culture of a photosynthetic bacterium was serially diluted 1 to 10 in a liquefied agar medium before the medium solidified and the tubes were incubated in the light. A colony developed from each separate cell that was physically fixed in the solidified medium. The tube on the right has the highest dilution; therefore, it has the smallest number of colonies.

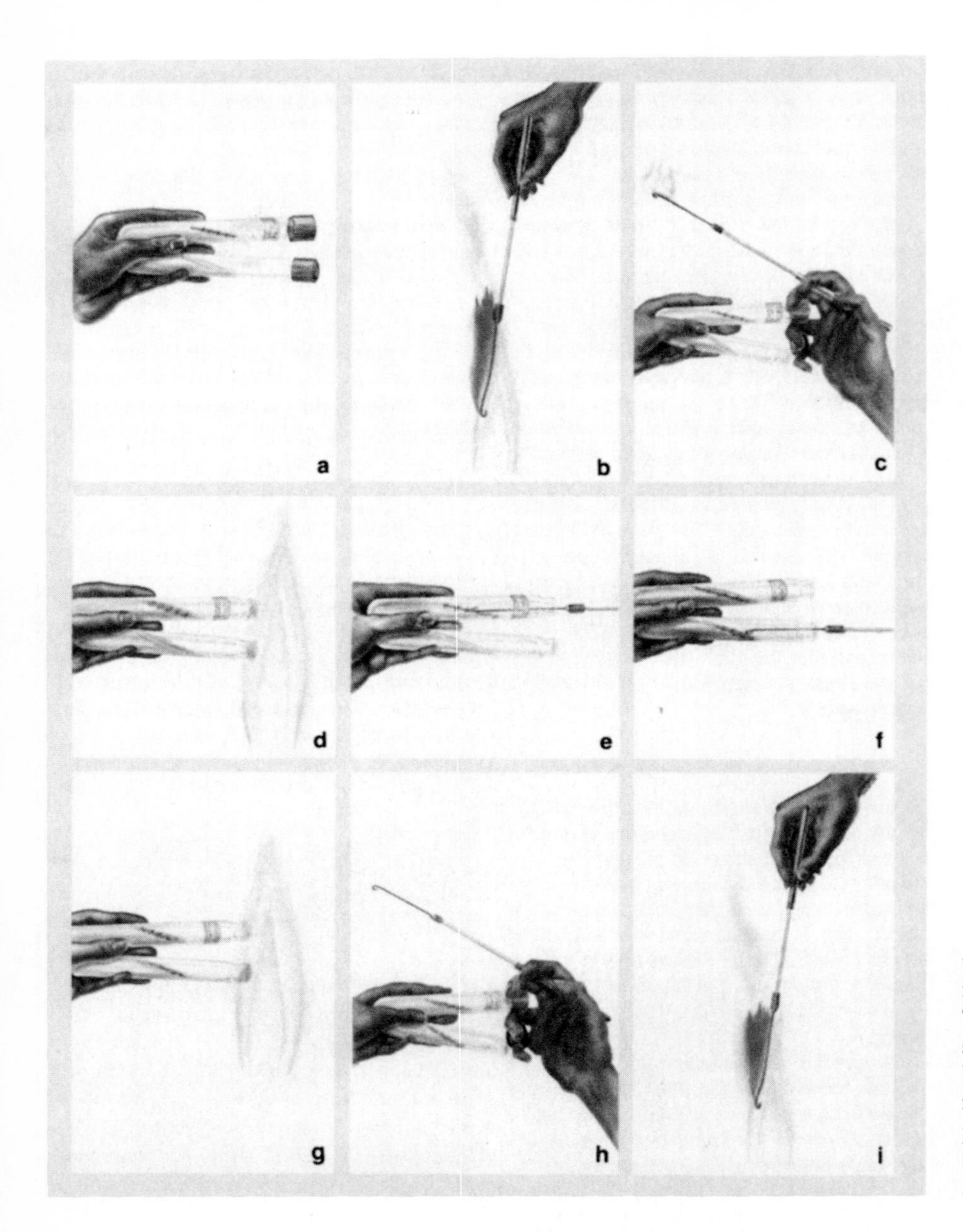

FIGURE 7.3 Aseptic technique is used to transfer bacteria from a culture to a fresh medium. The nickel–cadmium wire is sterilized by heating it in a Bunsen burner. The necks of the tubes are flamed to incinerate any dust particles that might contaminate the medium. (Courtesy of Carolina Biological Supply Company.)

sected with a knife or a sterile dissecting needle. Shake dilution tubes are used to isolate anoxygenic photosynthetic bacteria, *Propionibacterium* spp. (proh-pee-on-ee-bak-teer′ee-um), and other anaerobic bacteria.

ASEPTIC TRANSFER

Transfer of pure strains of bacteria for the purpose of inoculating new cultures (subculturing) must be done without allowing contaminating organisms to interfere. These transfers are carried out by **aseptic** (a-sep′tik) **technique**—without the introduction of new organisms. If the experimenter is not careful, contaminants from the air, desk top, hands, and so on may be introduced into the new culture.

A sterile loop or pipet is commonly used to inoculate sterile media aseptically (Figure 7.3). The loop is made from nickel–cadmium or platinum wire that is easily sterilized by flame incineration. Material is then transferred from a culture (solid or liquid) to a new medium with the air-cooled loop. A loop of a liquid culture contains about 0.005 ml of culture; larger volumes of liquid can be transferred with sterile pipets. With aseptic transfer techniques, pure cultures can be maintained in perpetuity.

STOCK CULTURES

Pure cultures of microorganisms can be maintained in stock cultures for periods varying from weeks to years depending on the organism and the stock-culturing technique employed. Preparing stock cultures is very important since many bacterial strains are difficult to obtain in pure culture, and painstaking experimentation is often required to isolate and characterize many mutants. Maintenance of bacterial strains is essential to the conduct of research; to medical, industrial, and agricultural microbiology; and to bacterial classification, now called systematics.

Stock cultures of bacteria, fungi, and algae are maintained on slants (aerobes) or in stabs (anaerobes). An **agar slant** (also called a slope) provides a large oxygen-exposed surface area for growth, whereas the agar in the bottom of a **stab** creates an oxygen-depleted environment for anaerobes. Some microbes remain viable in refrigerated stock cultures for weeks or months. Other organisms must be transferred periodically to new media. Layering sterile mineral oil over the growth on a slant prolongs the storage life of some organisms.

Stock cultures prepared for long-term storage and for culture collections are either freeze-dried or stored in liquid nitrogen at −196°C or at −75°C in a low-temperature freezer. The American Type Culture Collection (ATCC) located in Rockville, Maryland is a repository for thousands of bacterial, fungal, protozoan, and viral stock cultures that can be purchased by researchers and teachers.

FREEZE-DRIED CULTURES: LYOPHILIZATION

Freeze-dried cultures of microorganisms are prepared by **lyophilization** (ly-oh-fil-y-zay'shun), in which frozen cells are dried by evaporating the water in a vacuum. Once freeze-dried, cultures of bacteria, algae, and fungi will remain viable for years even when stored at room temperature. Before freezing, bacterial cultures are suspended in a medium containing milk products or lactose, which helps maintain their viability. The dried bacterial cultures can be regenerated by being suspended in an appropriate growth medium.

STORAGE OF CELLS IN LIQUID NITROGEN

Thermos flasks containing liquid nitrogen are used to store frozen cultures of bacteria and tissue culture cells at very low temperatures (Figure 7.4). Their viability

FIGURE 7.4 Bacterial cultures remain viable when stored at −196°C in liquid nitrogen or at −76°C in a low-temperature freezer. These vials containing stock cultures are stored in special Dewar flask (thermos bottle) containing liquid nitrogen.

is maintained during freezing by suspending them in a protective medium before the medium is slowly cooled at −1°C/minute. Glycerol and dimethyl sulfoxide protect cells against the harmful effects of freezing, and the frozen cultures remain viable for years. The culture is activated by thawing the vial quickly at room temperature, then transferring the culture to a suitable growth medium. Cultures of sperm, mammalian cells, and many bacteria are maintained by this means.

TABLE 7.2 Complex medium (MacConkey agar)

INGREDIENT	AMOUNT (g)
Peptone	17.0
Proteose peptone	3.0
Lactose	10.0
Bile salts	1.5
Sodium chloride	5.0
Agar	13.5
Neutral red	0.03
Crystal violet	0.001
Distilled water	1000 ml

Lactose degrading colonies (red) produce acid that changes the neutral red pH indicator from colorless to red. The bile salts and crystal violet inhibit growth of gram-positive bacteria.

MICROBIAL GROWTH MEDIA

CHEMICALLY DEFINED MEDIA

The glucose–salts medium presented in Table 7.1 is a **chemically defined** medium—that is, it is made from known quantities of pure chemicals. A defined medium is not changed by the addition of nutrients such as vitamins or amino acids, as long as known quantities of pure chemicals are included. Geneticists and physiologists routinely modify the glucose–salts medium (called minimal medium) in order to grow strains that have been altered (mutated) such that they require specific growth factors. Chemically defined media are also used in bioassays for measuring minute quantities of biomolecules.

CHEMICALLY UNDEFINED OR COMPLEX MEDIA

Fastidious microbes require an extensive array of nutrients for growth that ranges from common amino acids to complex vitamins. Most of these microbes grow best on a medium that contains a variety of nutrients such as are found in yeast extract, peptone, trypticase soy broth, casitone, and whole blood or blood-derived products. These are called **complex media** because their chemical composition is poorly defined. For example, MacConkey agar (Table 7.2) contains bile salts and two types of peptone, none of which is chemically defined. There are advantages in using complex media: they are easy to prepare and cost less than pure chemicals, and even more important, many microbes grow better on complex media.

The formula for MacConkey agar is given in grams liter; other media are described as percent solutions. In this context, percent means grams per 100 ml of distilled water. Therefore a medium that contains 5 percent glucose has 5 grams of glucose per 100 ml of medium.

LIVING CELLS AS CULTURE MEDIA

Animal viruses and some bacteria grow only in living cells. These microbes must be cultured in experimental animals, such as mice or chicken embryos, or in tissue cultures. Virologists use tissue cultures to isolate and cultivate animal viruses; the tissue cultures in turn are maintained in specially prepared culture media. Intracellular bacterial parasites, such as rickettsias (ri-ket′-see-ahs), chlamydias (klah-mid′dee-ahs), and some spirochetes are also grown in tissue cultures because they require a eucaryotic host cell for their multiplication.

ISOLATION AND CULTIVATION OF MICROORGANISMS

SELECTIVE ENRICHMENTS

Microbes can be isolated from natural environments or clinical specimens by means of **selective enrichment** techniques, which allow one microbe or group of microorganisms to outgrow all the other organisms present in the inoculum. The physiological characteristics of the desired organism dictate the composition of the special medium as well as the conditions of incubation. Specific organisms can be selected for by altering the composition of the medium with changes in carbon source, energy source, nitrogen source, and the presence of supplements such as vitamins and amino acids. Selective incubation conditions include different temperatures, presence or absence of oxygen, and presence or absence of light.

In some enrichments, **selective agents** are used to inhibit the growth of unwanted microbes. Chemical dyes, such as malachite green and crystal violet, are used to inhibit the growth of bacteria and yeast. Sodium azide is a metal-binding agent that inhibits the growth of aerobic bacteria, but does not affect the anaerobic lactic acid bacteria, which are fermentative.

Antibiotics in culture media prevent the growth of unwanted microorganisms. For example, penicillin and streptomycin are incorporated into tissue culture media to prevent bacterial contamination, and are also useful in preventing bacterial growth when the goal is to isolate certain protozoa. Nystatin is an antibiotic that prevents fungal contamination of bacterial cultures.

Heat is one selective agent that kills the vegetative cells without affecting endospores. The spore-forming bacteria can then be selected for by plating the heated solution on a suitable growth medium and incubating it under suitable growth conditions. For example, aerobic spore-forming *Bacillus* species are selected for by heating hay or soil samples before they are plated on an appropriate aerobic medium.

KEY POINT

The basis of selective enrichment techniques is the premise that each organism has a preferred ecological niche. When this niche is simulated in the laboratory, the organism outgrows the rest of the microbial population to become the predominant microbe. Standard isolation techniques can then be followed to obtain pure cultures.

SELECTIVE AND DIFFERENTIAL MEDIA

Selective enrichments are widely used in medical microbiology, for growing and isolating pathogens (disease-causing microbes) present in mixed populations of normal flora. Such selective media are commercially available (Table 7.3). **Differential media** distinguish between different species, usually through the physiological reactions unique to those bacteria. The most practical media are those that both select for and differentiate between common pathogens.

MacConkey agar (Table 7.2) is both a selective and a differential medium that contains lactose, peptone, neutral red, crystal violet, and bile salts. The bile salts and the crystal violet inhibit the growth of gram-positive bacteria so that they cannot compete with the enteric bacteria. Growth of the enteric bacteria result in visible changes characteristic of the species' ability to metabolize lactose.

Enteric bacteria that ferment lactose, such as *Escherichia coli,* cause a decrease in the pH, which turns the neutral red pH indicator red. The decrease in pH also

TABLE 7.3 Microbiological media for isolation and differentiation of medically important bacteria

MICROORGANISM OR GROUP	SELECTIVE MEDIA (ISOLATION)	DIFFERENTIAL MEDIA (DIFFERENTIATION)
Coliforms	MacConkey agar	MacConkey agar
	Levine EMB agar	Levine EMB agar
Mycobacterium spp.	Middlebrook media	Middlebrook media
Neisseria spp.	Chocolate agar	
	Thayer–Martin medium	
Salmonella and	Hektoen–enteric agar	
Shigella	SS agar	SS agar
	Xylose lysine (XL) agar	Xylose lysine (XL) agar
Staphylococcus spp.	Chapman–Stone medium	Chapman–Stone medium
	Staphylococcus 110 agar	Mannitol salts agar
Streptococcus spp.	Azide blood agar	Mitis salivarius agar

precipitates the bile salts surrounding the colony. Enteric bacteria that do not metabolize lactose, such as *Salmonella* and *Shigella,* grow as white colonies. MacConkey agar is a selective differential medium by virtue of its ability to select for and distinguish between various enteric bacteria.

CHEMICAL AND PHYSICAL FACTORS AFFECTING GROWTH

INFLUENCE OF pH ON GROWTH

The acidity or alkalinity of an aqueous environment has a major effect on the organisms that can live in it. Acidity or alkalinity of natural environments ranges from pH = 1.0 in acid mine waters to pH of approximately 11 in ammonia-rich soils such as feed lots for cattle.

Bacteria living in acid environments such as acid mine waters, volcanic soils, and volcanic waters are described as **acidophiles.** *Sulfolobus, Thermoplasma,* and some species of *Thiobacillus* are obligate acidophiles because they require acid environments. These bacteria must protect themselves against a low pH because many proteins and nucleic acids are inactivated at these low pHs. They maintain their cytoplasm near neutrality by preventing the entry of H^+ ions or by expelling H^+ ions.

The pH of marine environments is approximately 8.4, and, as you would expect, marine microbes prefer an alkaline environment. Certain soil bacteria also require alkaline environments. One such bacterium, *Bacillus pasteurii* (pas-teur′ee-eye), breaks down urea to CO_2 and NH_3, creating an alkaline soil that is selective for its growth. Though extremes of pH exist, the pH of most natural environments is close to neutrality, which is preferred by most microbes.

The pH of most natural environments is not significantly affected by the metabolism of microorganisms. This is not the case, however, in laboratory cultures, where the acidity or alkalinity resulting from metabolic activities of the microbes often changes the pH of the medium. For example, many fermenting bacteria produce organic acids, including lactic acid, formic acid, acetic acid, succinic acid, and propionic acid, and fermentation of amino acids releases ammonia, which raises the pH. When too much acid or alkali accumulates, there is a decrease in microbial growth.

KEY POINT

Knowledge that a low pH inhibits microbial growth is useful in food preservation. The microbial fermentations that are involved in the production of sauerkraut and pickles generate sufficient acid to prevent spoilage.

Buffers are chemicals that prevent extreme changes in pH (Chapter 2). Potassium phosphate (K_2HPO_4/KH_2PO_4) functions as a buffer between pH 6 and pH 8, while amino acids are effective buffers at a slightly acid pH. Many media are buffered by potassium phosphate, which also functions as the cell's source of phosphate and potassium.

ATMOSPHERIC CONDITIONS

Aerobic organisms use the oxygen (O_2), present in air for their metabolism. The oxygen content of a liquid culture can be increased by sparging (bubbling) it with compressed air or by shaking the culture on a mechanical shaker (Figure 7.5). Stock cultures of aerobes are often maintained on agar slants to provide a large surface area exposed to air.

Airborne microorganisms that fall into uncovered cultures are a constant problem. Cotton plugs prevent this type of contamination by filtering out particulate matter, while permitting the free passage of air. Synthetic sponge stoppers and plastic or metal caps that slip over the mouth of the test tube or flask also allow the free passage of air, while preventing the entry of dust particles and unwanted microbes.

There is a wide range of oxygen tolerance among anaerobic microorganisms: *Bacteroides* and *Fusobacterium* are obligate anaerobes, *Clostridium tetani* is moderately tolerant to oxygen, *Clostridium perfringens* is remarkably aerotolerant, and the lactic acid bacteria will grow slowly on aerobic plates. Microbes that grow best in environments that contain a low concentration of oxygen are termed **microaerotolerant** or **microaerophilic.** These organisms can also grow in aerobic environments, but their growth is slowed by high oxygen concentrations. **Anaerobes** grow in agar stab cultures or in growth chambers from which the oxy-

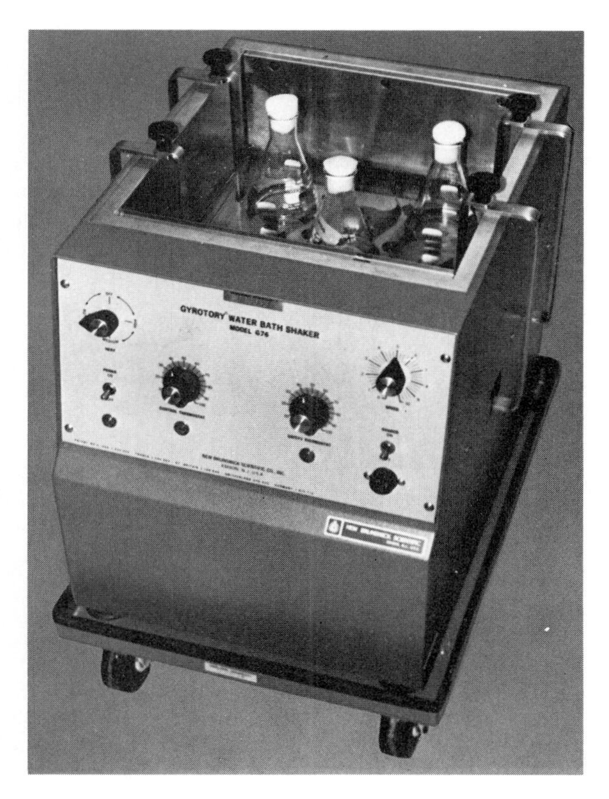

FIGURE 7.5 Temperature-controlled shakers are used to grow aerobic cultures.

gen has been removed, whereas **facultative anaerobes** are able to grow in either the presence or the absence of oxygen. Growth tests such as those illustrated in Figure 7.6 can be used to analyze an organism's ability to use oxygen.

Techniques for Culturing Anaerobes

Anaerobic cultures can be grown in **anaerobic growth chambers** from which all oxygen has been removed and replaced with an inert gas, such as molecular nitrogen, helium, or carbon dioxide. The cultures are placed in the growth chamber through an airlock door (Figure 7.7), which prevents air from reaching the interior. The cultures within the chamber are manipulated with airtight gloves mounted at the front of the chamber. Although such chambers are highly efficient, they are expensive both to purchase and to operate.

A much less expensive anaerobic environment can be created by burning a candle in a **candle jar.** Any jar large enough to hold a petri dish is suitable. The inoculated cultures are placed in the jar, a candle placed on top of the petri dishes is lit, and the jar is sealed with an airtight lid. As the candle burns it uses up most of the oxygen and produces carbon dioxide, which stimulates bacterial growth.

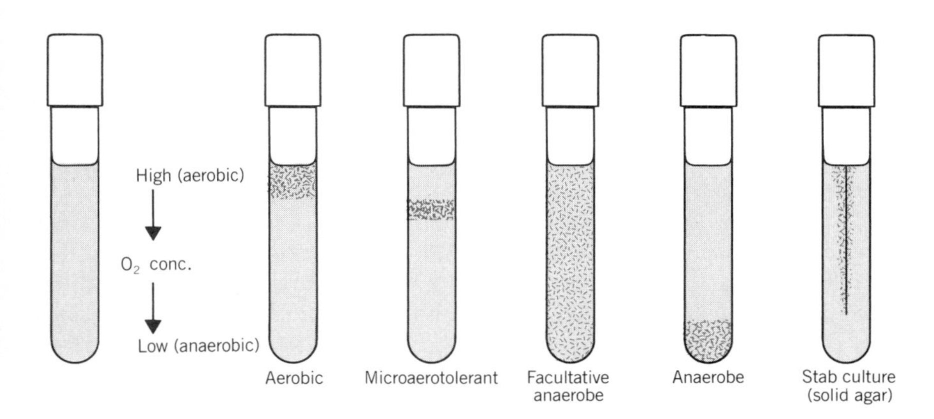

FIGURE 7.6 Growth response of different microbes to oxygen when grown in a semisolid medium or in a solid culture (1.5 percent agar). The highest concentration of oxygen is present at the top of the tube, whereas the lowest concentration of oxygen is at the bottom of the tube.

FIGURE 7.7 This anaerobic culture cabinet isolates the culture from both outside air and the worker. The anaerobic chamber is entered and exited through the air lock on the right. The worker can handle the material inside the cabinet through pliable, gas-tight gloves. (Courtesy of L. C. Edwards, Forma Scientific Company.)

The commercial **anaerobic jar** (Figure 7.8) is a refined version of the candle jar. The jar is equipped with a gas generator envelope, an anaerobic indicator strip, and a palladium catalyst. When water is added to the gas generator, H_2 and CO_2 are released into the jar. The oxygen is chemically removed as it reacts with H_2 (catalyzed by palladium) to form H_2O. When the oxygen is totally consumed, the indicator strip changes from blue to colorless. Anaerobic jars are routinely used in place of expensive growth chambers.

TEMPERATURE

Microorganisms grow in environments of extreme temperatures: in deep marine environments where the average temperature is 5°C; in hot springs where the temperature may be 90°C. The range for microbial growth appears to be limited by the availability of liquid water. The upper limit for growth would therefore be the temperature at which water boils and the lower limit would be the freezing point of water. Since the freezing point of water is lowered by increasing the concentration of solute,* the lower limit is probably determined by the cell's ability to withstand a high solute concentration (see "Osmotic Pressure" following). The upper limit is probably that recently discovered on the ocean floor near hydrothermal vents, where the water pressure is so great that the temperature is above 200°C. Since this water is still in liquid form, it may serve as an environment for thermophiles.

Each bacterium has an optimal growth temperature—a temperature at which its growth rate is maximal—and an upper temperature limit (Figure 7.9). Above its upper limit, the bacterium's growth rate falls off dramatically, apparently due to instability of its membranes and proteins. At lower temperatures, the drop in growth rate is not as precipitous (Figure 7.9).

Microorganisms are organized into three groups according to the temperature range in which they grow. Soil organisms and human pathogens usually grow at temperatures between 20° and 50°C and are called **mesophiles** (mes'oh-fyls). As a rule, human pathogens grow best at 37°C (the temperature of healthy humans), and soil microbes grow best at 30°C. Each mesophile has an optimum temperature range and many grow poorly or not at all at the extremes of the mesophilic range.

Thermophiles (therm'oh-fyls) require temperatures above 50°C for growth. Some eucaryotes grow at temperatures above 50°C, but only procaryotes grow at temperatures above 60°C. The upper limit for growth is not known; however, there are reports of bacteria growing in hot springs at temperatures approaching the boiling point of water (at sea level). The specialized enzymes, proteins, and membranes of the thermophiles are stable at these high temperatures.

Psychrophiles (sy'kroh-fyls) grow at temperatures below 20°C. An obligate psychrophile can usually grow at 0°C and grows optimally below 20°C. Psychrophiles are found in cold marine environments, glacial lakes, and in polar environments. The majority of psychrophiles are members of the genera *Pseudomonas, Flavobacterium,* and *Alcaligenes* (al-ka-li'je-nese).

*A *solute* is any substance dissolved in a liquid.

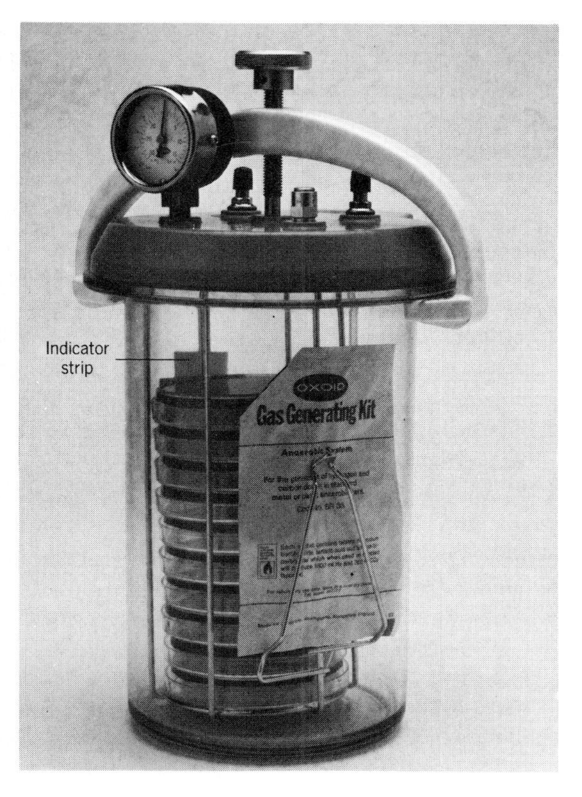

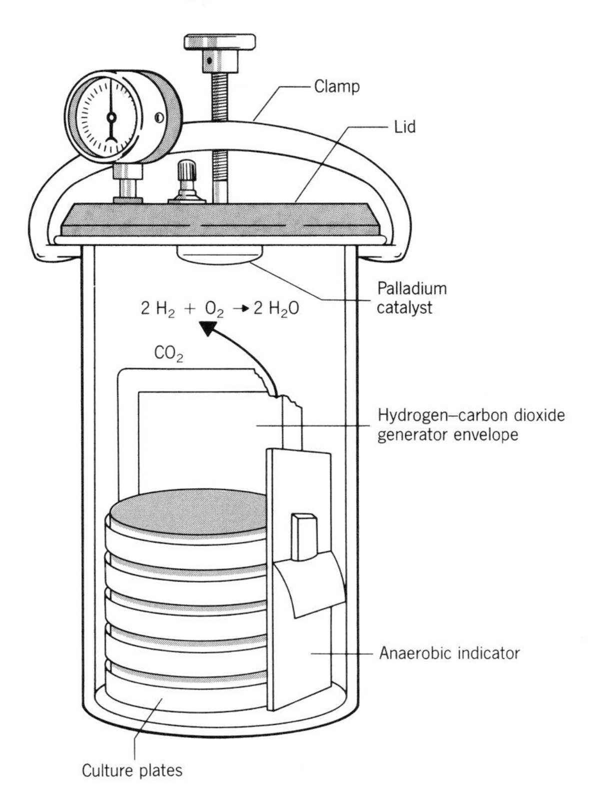

FIGURE 7.8 Anaerobic jars are used to incubate plates of anaerobic bacteria. Hydrogen and carbon dioxide gas are generated when water is added to the gas generator. Palladium catalyzes the reaction of oxygen and hydrogen to form water. The methylene blue strip turns colorless when the oxygen is removed. (Courtesy of Oxoid Ltd., Basingstoke, Hampshire, England.)

In general, temperature directly affects the rate of microbial growth (see Figure 7.9). A mesophile might grow into a visible colony in 12 hours at 37°C, but require 48 hours to produce a visible colony at 20°C. Being aware of this, microbiologists maintain their cultures in incubators at a prescribed temperature within ±0.5°C. Two or three incubators may be set at different temperatures to accommodate the requirements of the various microbes.

HYDROSTATIC PRESSURE

Barophilic organisms are those that grow optimally at high pressures. Until recently, the very existence of barophilic bacteria was controversial. Deep-sea bacteria live at extreme hydrostatic pressures exerted by the weight of the overlying water column. One marine bacterium, isolated from a depth of 5700 meters (about 19,000 feet), has been found to grow optimally at temperatures between 2° and 4°C. It is a true **barophile** (bair′oh-fyl) because it grows preferentially under high hydrostatic pressure. Great extremes in pressure are difficult to reproduce in the laboratory, so the study of barophiles is limited to specialized marine microbiology laboratories.

OSMOTIC PRESSURE

Cells possess a semipermeable membrane that separates the cytoplasm from the extracellular environment. This membrane is permeable to water, which enters the cell during **osmosis**—the tendency of a cell to

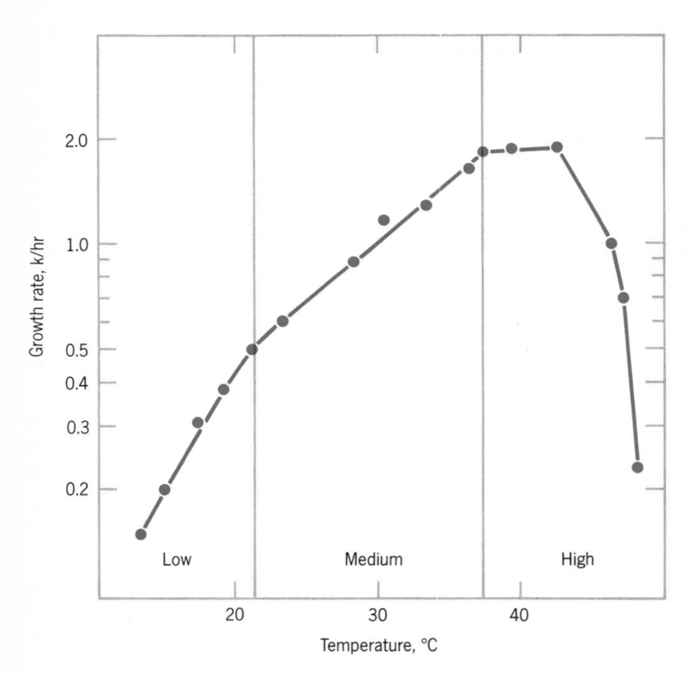

FIGURE 7.9 Bacterial growth rates are greatly affected by temperature. *Escherichia coli* is a mesophilic bacterium whose growth rate is maximal at 37°C. The growth rate of *E. coli* falls sharply below 20°C and above 45°C. (After S. L. Herendeen, R. A. VanBogelen, and F. C. Neidhardt, *J. Bacteriol.*, 139:185–194, 1979, with permission of Dr. Frederick Neidhardt and the American Society for Microbiology.)

accumulate water when the concentration of solutes inside the cell is greater than the concentration of solutes outside the cell. The migration of water across the membrane creates osmotic pressure inside the cell, as described earlier. This pressure is directly proportional to the concentration of solute particles and is independent of their nature.

Recall that most bacteria would burst under the osmotic pressure that builds up in their cytoplasm if it were not for their rigid cell wall. Protoplasts and cells that lack cell walls *(Mycoplasma)* must be protected against osmotic pressure to prevent them from bursting. In the laboratory, this protection is provided by maintaining the cells in a medium having a high solute concentration.

Plasmolysis occurs when the solute concentration of the environment is much greater than that in the cytoplasm. Under these conditions, water leaves the cell, causing the cytoplasm to collapse. This results in damage to the cell membrane and renders the cell incapable of growth.

Culturing Osmophiles and Halophiles

Osmophiles are tolerant of high osmotic pressure, and grow in environments with a high solute concentration. High salt concentrations are used to isolate osmophiles; *Staphylococcus aureus,* for example, is selected for in a medium that is 7.5 percent salt. **Halophiles** (hal′low-fyls) are obligate osmophiles because they require a sodium concentration of 12 percent or more. Halophiles grow in the Great Salt Lake in Utah and in food products preserved with salt, such as salted codfish.

Halophiles maintain an intracellular ionic concentration that is equal to the ionic concentration of their growth medium. They require this high intracellular ion concentration because many of their biochemical functions are dependent on high concentrations of Na^+ and K^+ ions. For example, enzymes isolated from halophiles require K^+ for activity, and their ribosomes require K^+ for stability. In addition, the integrity of the cell walls of certain halophiles, such as *Halobacterium,* depends on a high concentration of Na^+.

KEY POINT

Knowledge of the effect of osmosis on microorganisms has been used for hundreds of years to protect foods against decomposition. Examples include the use of syrup to preserve fruits and the salting of meats, fish, and vegetables, such as pickles.

DYNAMICS OF BACTERIAL GROWTH

Bacteria are ideal experimental entities since they grow rapidly and are so small that a population of billions of cells in a small flask can easily be studied. Microbiologists learn much about population dynamics from studying bacteria.

Bacterial growth is the orderly increase in cell constituents that results in an increase in the number of cells, as described earlier. Most bacteria divide into two equal daughter cells by the process of transverse binary fission; the daughter cells then double in mass before dividing. This sequence of growth and division continues as long as the cells are in an environment that provides the essential nutrients and physical conditions needed for growth.

GROWTH RATES

Most bacteria reproduce rapidly, and many divide more than once every hour. The rate of bacterial reproduction is measured as the **generation time**—the period between divisions of a growing microorganism or the period required for the number of cells in a culture to double. The shorter the generation time, the faster the cells will increase in number. Rapid bacterial growth does not last very long in a bacterial culture as is demonstrated by this example:

A bacterium growing unrestricted with a generation time of 20 minutes would produce 2.2×10^{43} cells in 48 hours. Since each bacterium weighs about 1×10^{-12} gram, the cells in the culture would weigh 2.2×10^{31} gram—about 4000 times the weight of the earth.

The duration of maximal growth is short, because bacterial growth depletes the available nutrients and generates waste products that may be toxic.

RESTRICTED GROWTH

Experiments in microbial growth are usually done on cell populations because it is too difficult to study individual cells. In a closed environment, bacterial growth is restricted by depletion of nutrients and production of toxic products. Under these conditions, the population follows a growth curve as depicted in Figure 7.10. The growth of *E. coli* in a closed environment can be followed by measuring the number of cells at timed intervals. Growth progresses from the lag phase through the exponential and stationary phases before it enters the death phase.

Lag Phase

Cells transferred to a new medium adjust to the new environment during the period called the **lag phase.**

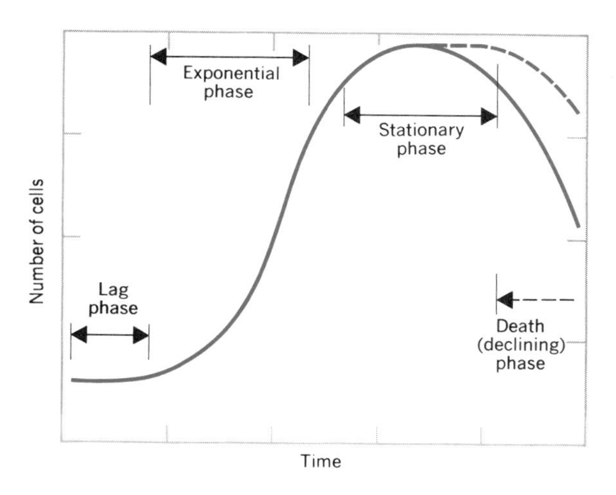

FIGURE 7.10 A typical bacterial growth curve for a liquid culture. The dashed line is the total number of cells, and the solid line is the number of viable cells.

The cells are metabolically active during this phase and may increase in size, but they do not divide.

The duration of the lag phase depends on the physiological condition of the transferred cells. Cells transferred from an exponentially growing culture to a new growth medium of the same composition will have a short lag phase (see "Exponential Phase"). A longer lag phase occurs when the inoculum is taken from a culture in stationary phase. The length of the lag phase will depend on the temperature, the nature of the nutrient(s) or toxic product(s) that limited growth in the original culture, and the length of time the culture was held in stationary phase. The lag phase is also lengthened when cells are transferred from a rich medium to a minimal medium.

During the lag phase, the transferred cells synthesize the enzymes and metabolites they need to grow at a maximal rate in their new environment. These enzymes and metabolites include the amino acids not present in the new medium and the enzyme(s) needed to metabolize new energy sources. At the same time, the toxic products are either metabolized by the cells or diluted by the new growth medium. The lag period ends when the cells begin growing exponentially with a constant generation time (Figure 7.10).

Exponential Phase

Once cells adjust to their new environment, they com-

TABLE 7.4 Mathematics of unrestricted bacterial growth

NUMBER OF GENERATIONS (*n*)	NUMBER OF BACTERIA (*B*)	$\text{Log}_2 B$	$\text{Log}_{10} B$
0	1	0	0.000
1	2	1	0.301
2	4	2	0.602
3	8	3	0.903
4	16	4	1.204
5	32	5	1.505
6	64	6	1.806
7	128	7	2.107
8	256	8	2.408
9	512	9	2.709
10	1024	10	3.010

mence to grow at the maximal rate. This rate persists until the composition of the medium is altered by the metabolic activities of the growing cells. The cells increase in mass by synthesizing new components until they double in size and divide.

Each time a bacterial cell divides it produces two daughter cells. When the cells in a culture are dividing at a constant rate, their number increases exponentially (2^n). The phase during which the growth rate is maximal (that is, minimum generation time) is termed the **exponential** or **logarithmic growth phase.** The growth rate of a given species during exponential growth will depend on the composition of the medium, its pH, ionic strength, carbon source, and the physiological capabilities of the organism.

The increase in the number of cells through 10 divisions is shown in Table 7.4. Plotting these data on an arithmetic graph gives a curved line (Figure 7.11*a*). Plotting the same data on a semilogarithmic graph (Figure 7.11*b*) gives a straight line, which is another way of demonstrating that bacterial growth during this phase is exponential.

During exponential growth, all the cells in the culture are physiologically identical for all practical purposes. Now all the individual components (proteins, DNA, and so on) are changing at a constant rate and the cells are said to be in **balanced growth.** The chemical composition of cells in balanced growth is virtually identical.

KEY POINT

Microbiologists need to know when a culture is in balanced growth since this is the easiest stage of growth to reproduce. Many physiological and cytological experiments must be done on cells from cultures in balanced growth in order to obtain reproducible results.

As we mentioned earlier, exponential growth cannot continue indefinitely. At some point the composition

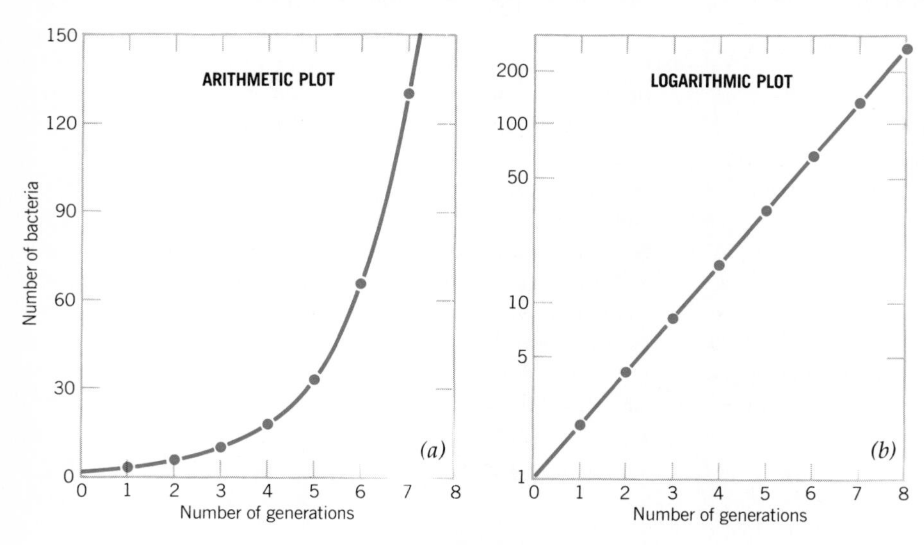

FIGURE 7.11 Arithmetic *(a)* and logarithmic *(b)* plots of the data in Table 7.5 showing the exponential increase in the bacterial population.

of the medium is altered because nutrients are used up, toxic products are produced, the pH changes, and the oxygen is depleted. The transition between exponential growth and the stationary phase is usually a gradual change caused by the slow accumulation of toxic products and depletion of nutrients. Growth becomes unbalanced during the transition when the synthesis of any cellular component ceases.

Stationary Phase

Bacterial cultures enter the **stationary phase** when the environment can no longer support an increase in cell mass. In the stationary phase, the number of cells in the culture remains constant because cells are dying at the same rate that new cells are being produced. The cells are uniformly smaller than during exponential growth since cell division continues as the rate of cellular synthesis slows.

Decline or Death Phase

The number of viable cells decreases as the culture enters the **death phase.** The cells die because they are unable to obtain the energy and nutrients needed to maintain essential functions. For example, cells that are unable to synthesize new cell walls lyse because they cannot manufacture cell-wall material fast enough to repair damaged sections. The cell mass decreases as the cells lyse. The number of living cells decreases faster than the total number of cells (Figure 7.10), because the total number of cells includes both living and unlysed dead cells.

Although most bacteria go through this normal growth cycle, some form spores, or cysts, in response to adverse environmental conditions. These bacteria enter a dormant state after the stationary phase.

MATHEMATICAL EQUATION FOR EXPONENTIAL GROWTH

The exponential phase of growth (Figure 7.10) can be expressed mathematically with logarithmic functions, as explained in Appendix 1. The following expression is applicable only to the exponential phase of growth. The symbols are as follows:

g = Generation time
B_0 = Number of bacteria at time zero
B_t = Number of bacteria at time t
t_0 = Zero time
t = Any time after t_0
n = Number of generations

Generation time (g) is the time required for the mass or the number of bacteria in the culture to double. Because bacteria divide by binary fission, the number of cells at any time after inoculation of the culture is equal to the number of cells at zero time (B_0) multiplied by 2 raised to the nth power, or

$$B_t = B_0 \cdot 2^n$$

where n is the number of generations (Table 7.4). When rearranged and solved for g (see Appendix 1), the formula for exponential bacterial growth is:

$$g = \frac{0.301\ (t - t_0)}{\log_{10}B_t - \log_{10}B_0}$$

This equation can be employed to determine any of the variables, g, time, B_t, or B_0 provided that the others are known.

METHODS OF MEASURING GROWTH

Bacterial growth can be measured on the basis of total number of cells, number of viable cells, or cell mass of the culture. Alternatively, one can measure the increase in a specific cellular constituent such as DNA or protein, but such measurements are difficult and are made only in special circumstances.

TOTAL CELL COUNT

The total number of cells in a liquid culture can be determined microscopically with a Petroff–Hausser counting chamber (Figure 7.12). The chamber consists of a glass slide thin enough to be used under a high-power oil-immersion lens. The chamber is marked off

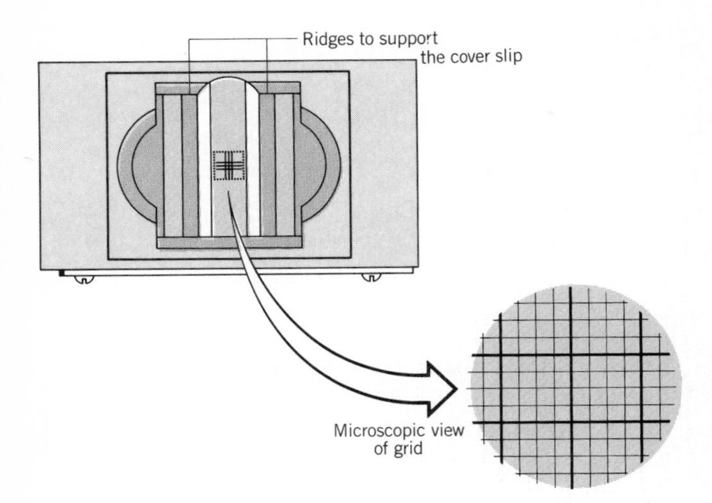

FIGURE 7.12 Diagram of a Petroff–Hausser bacterial counting chamber. The chamber is used to count the total number of bacteria in liquid samples.

in small squares of a predetermined size. Since the volume occupied by the liquid above each square is known, the number of bacteria per milliliter can be determined by dividing the number of bacteria in several squares by the volume of liquid above those squares. At least 10^6 cells per milliliter are required before they can be seen in the counting chamber. The primary advantage of this method is that it can be done quickly with a minimal amount of equipment.

Total cell counts of bacterial and mammalian cultures can be made with an electronic particle counter. The cells in the suspension are counted when they interrupt an electrical current as they individually pass through a small opening. The interruptions are registered as particles and are counted electronically. Electronic counters are used primarily for counting eucaryotic cells in research and clinical laboratories. Microscopic counts and electronic counters measure both viable and dead cells, so the total cell count is always higher than is the case with the colony-counting method.

VIABLE CELL COUNT

The number of viable cells in a culture can be ascertained by determining the number of colony-forming units (CFU) by the **colony-counting** technique. Between 30 and 250 CFU can be measured on a typical

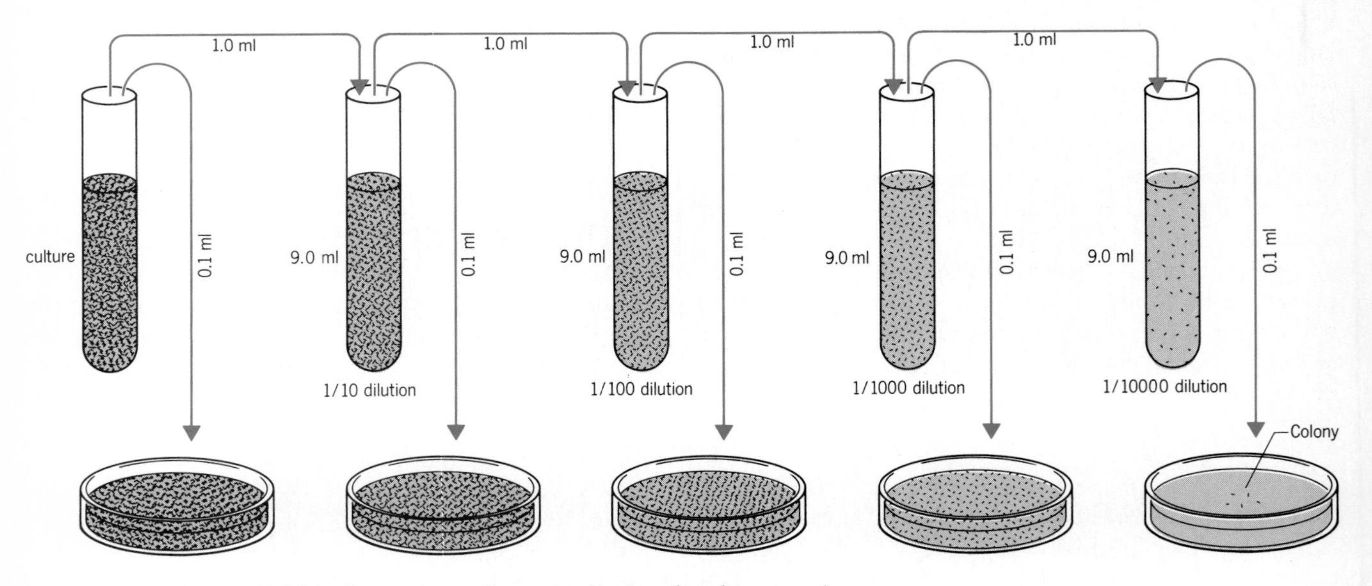

FIGURE 7.13 Viable cell counts are determined using the plate counting technique. The culture is diluted such that 0.1 ml of culture will contain between 30 and 250 colony forming units (CFU) when plated on the surface of an agar plate. Each bacterium that grows into an visible colony is counted as one CFU. The number of CFUs is a direct measure of the number of viable bacteria in the original culture.

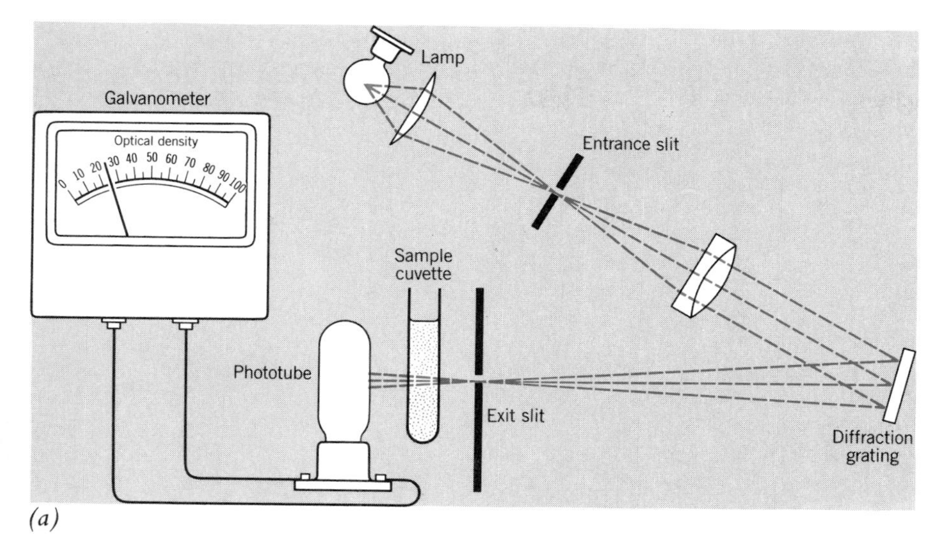

(a)

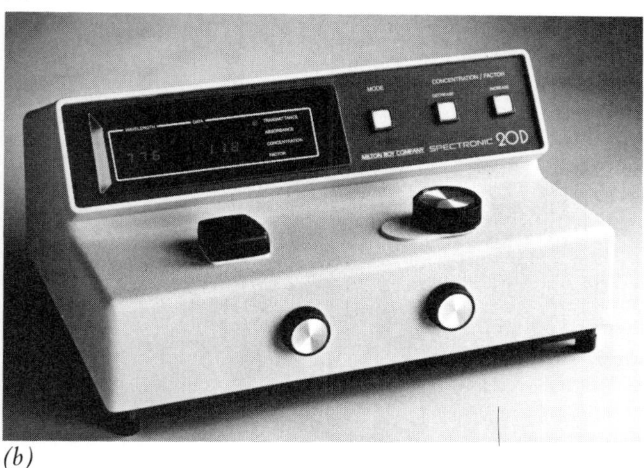

(b)

FIGURE 7.14 Spectrophotometers can be used to measure turbidity, which is directly related to the cell mass of a bacterial culture. *(a)* Diagram of a spectrophotometer, *(b)* photo of a Bausch & Lomb spectrophotometer.

petri plate—that is, having a diameter of 10 cm. Microbial cultures of a high cell density must be diluted before they are plated (Figure 7.13). A known volume of these dilutions is plated onto a suitable growth medium in the petri dish. After the plates have been incubated for 12 to 48 hours, the average number of colonies on duplicate plates is determined.

The number of viable bacteria per milliliter of the initial culture can be calculated from the average CFUs and the known dilution factor. The viable count is invariably less than the total cell count because it measures only cells capable of dividing. The major disadvantages of the colony plating technique are (1) the incubation period is lengthy, (2) sterile dilution media, pipets, and plates are required, and (3) sampling and dilution errors occur.

Cell populations in milk and polluted water can be estimated by a technique called most probable number. In this technique, the test solution is diluted so that samples added to the growth tubes have a low probability of containing a bacterium. Depending on the dilution, only a few of the inoculated tubes will exhibit growth. The **most probable number** (MPN) of bacteria in the original culture can be determined by analyzing the number of tubes with growth. This approach is often followed to test for bacterial contamination.

CELL MASS

The spectrophotometer (Figure 7.14) is an instrument that measures how much light passes through a solution or suspension. The wavelength of the light is controlled by a prism or a diffraction grating. Light passing through the sample strikes a phototube, which converts light energy into an electrical current. The current is registered on a scale of transmittance or absorbance (often called optical density [OD]). The more turbid the solution, the more it will scatter light. Since scattered light never strikes the phototube, the more turbid a solution the greater is its optical density. Light scattering is directly proportional to the size and number of particles, so light scattering is a measure of the cell mass in a culture sample.

Although optical density measurement is a rapid and accurate method of measuring growth, it presents some disadvantages: (1) large cells contribute more to the optical density than do small cells, (2) at least 10^6 cells per milliliter are required before a reliable reading can be obtained, (3) cells can aggregate and settle out before the measurement is made, and (4) both nonliving and living cells are counted. Problem 4 can be overcome by standardizing the optical density measurement with viable cell counts.

SPECIAL CONSIDERATIONS: BACTERIAL GROWTH

BIOASSAY

What happens when an essential nutrient is absent from the growth medium? By limiting the concentration of

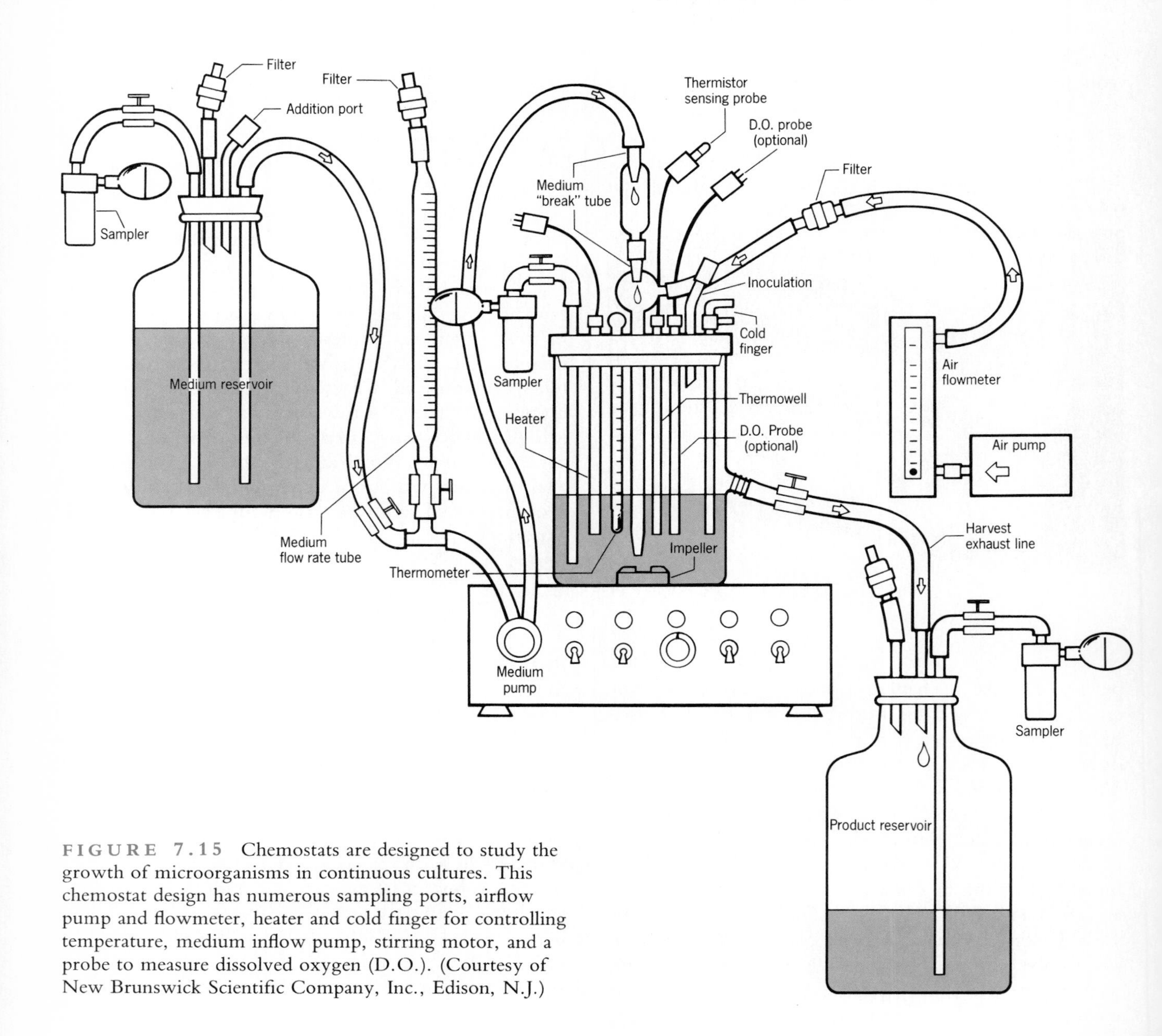

FIGURE 7.15 Chemostats are designed to study the growth of microorganisms in continuous cultures. This chemostat design has numerous sampling ports, airflow pump and flowmeter, heater and cold finger for controlling temperature, medium inflow pump, stirring motor, and a probe to measure dissolved oxygen (D.O.). (Courtesy of New Brunswick Scientific Company, Inc., Edison, N.J.)

a vitamin, a required amino acid, or a carbon and energy source, the experimenter can limit the final growth yield of a culture. Under these conditions, the final cell mass of the culture will be directly proportional to the concentration of the limiting nutrient. In the bioassay, this relationship measures the unknown concentrations of a required growth factor. Before the development of modern chemical instrumentation, bioassays were used to measure vitamins, such as folic acid, and certain amino acids. Now their function is to quantitate new growth factors and biomolecules that would otherwise be difficult to assay.

CONTINUOUS CULTURES

The *chemostat* makes it possible to maintain bacterial cultures in the exponential growth phase for extended periods of time. The chemostat shown in Figure 7.15 functions on the limiting-nutrient principle. As fresh medium enters the culture vessel from a reservoir of sterile medium, both cells and spent medium are carried off via the overflow.

The growth rate of the culture is limited by the availability of nutrients in the culture vessel. The faster the new medium enters the culture vessel, the higher the growth rate—up to a point. If the flow rate is too fast, the cell concentration decreases because the cells are diluted out of the chemostat by the incoming sterile medium faster than they can reproduce. Within these limits, the flow rate of new medium into the culture vessel establishes the growth rate of the culture.

A second approach to continuous culture employs cell density to regulate cell concentration. A **turbidostat** continuously measures the optical density of the culture and adds sterile medium when the optical density exceeds a predetermined setting. Addition of new medium dilutes the culture, causing the optical density to decrease as the cells are washed out in the overflow. Addition of new medium stops when the optical density reaches a preset lower limit. The culture then utilizes the nutrients in the new medium to increase its optical density until the process is repeated.

Chemostats maintain microbial cultures in exponential growth. Microbial ecologists use them to study mixed microbial populations and microbial growth at the very low nutrient levels found in natural environments. Small chemostats are used in industrial laboratories for studying natural product formation. Chemostats are especially important in pilot studies of fermentations done before a commitment is made to invest in large-scale commercial fermenters.

INDUSTRIAL FERMENTATION

Microorganisms are essential in the production of numerous commercially important products including ethanol, acetic acid (vinegar), beer and other alcoholic beverages, glutamic acid and other amino acids, and many antibiotics. These products are produced in huge fermenters, some of which are so tall (four and five stories) that they must be erected outside the plants in which their controls are housed.

Summary Outline

NUTRITIONAL REQUIREMENTS OF MICROORGANISMS

1. To satisfy the basic nutritional requirements of a microorganism, a growth medium must provide a source of carbon, nitrogen, sulfur, phosphorus, salts, trace elements, and growth factors (if required).
2. Some organisms grow on chemically defined media, such as glucose salts, whereas others require complex media containing vitamins, amino acids, and other growth factors.
3. Carbon is the energy source for chemoorganotrophs, light is the energy source for phototrophs, and inorganic compounds provide energy for chemolithotrophs.

PURE CULTURES

1. A pure or axenic culture contains a single strain of microbe in which all cells are derived from a single parent cell.
2. Streak plates are used to isolate cells that will

grow into visible colonies. Pour plates and shake-dilution tubes are also used to isolate pure cultures.

3. A culture is presumed to be pure only if one morphological type of colony grows on two successive streak plates.
4. Aseptic techniques are followed when pure cultures are transferred.

STOCK CULTURES

1. Pure cultures of many microbes can be maintained at 4°C as stabs (anaerobes) or slants (aerobes).
2. Cultures stored in liquid nitrogen and freeze-dried cultures retain viability for long periods. Microbial cultures, tissue cultures, and viruses are preserved by these means.

MICROBIAL GROWTH MEDIA

1. Chemically defined media are made from known quantities of pure chemicals.
2. Complex media are chemically undefined because they utilize complex nutrient sources such as peptone, yeast extract, and whole blood.
3. Living cells in tissue cultures are used to grow animal viruses and intracellular bacterial parasites.

ISOLATION AND CULTIVATION OF MICROORGANISMS

1. Selective enrichments permit one microbe or group of microorganisms to outgrow all others and are used to isolate microbes from a mixed inoculum.
2. The physical and chemical properties of a growth medium can be manipulated to selectively enrich for a specific species.
3. Chemical dyes, metal binding agents, and antibiotics inhibit the growth of unwanted microbes.
4. Selective and differential media are used to isolate and identify pathogens, especially in diagnostic microbiology.

CHEMICAL AND PHYSICAL FACTORS AFFECTING GROWTH

1. Chemical composition, pH, oxygen content, osmolarity, and temperature are important chemical and physical factors that affect the growth of microbial cultures.
2. Natural buffers such as phosphate salts and amino acids stabilize the pH of growth media. Acidophiles live in acidic environments by maintaining the pH of their cytoplasm near neutrality.
3. Air is the source of oxygen for most aerobes. Anaerobes are grown in anaerobic growth chambers, candle jars, or anaerobic jars from which the oxygen has been removed and replaced with CO_2 or an inert gas.
4. Microbes in general grow over a wide range of temperatures, but each species has both an optimum and a limiting growth temperature. Thermophiles have an optimum growth rate at temperatures above 50°C, psychrophiles at 20° or below, and mesophiles between 20° to 50°C.
5. Barophiles require high hydrostatic pressure for growth.
6. Most organisms are intolerant of high osmotic pressure and plasmolyze in media having a high concentration of solutes. Halophiles and osmophiles, however, require high concentrations of salt for growth.

DYNAMICS OF BACTERIAL GROWTH

1. Microbial growth is the orderly increase in cellular constituents that results in an exponential increase in the number of cells.
2. Bacteria cannot grow unrestricted for long without depleting the available nutrients and/or creating toxic by-products.
3. Bacterial cultures growing in a closed environment progress through a set sequence of phases: lag phase, exponential phase, stationary phase, and death phase.
4. The lag phase precedes the exponential phase, in which the cells attain their maximum growth rate.
5. As the culture depletes the available nutrients, it enters the stationary phase, characterized by no net change in the size of the population.
6. Cells die when they are unable to maintain essential physiological functions.

MATHEMATICAL EQUATION FOR EXPONENTIAL GROWTH

1. The mathematical expression for bacterial growth is used to calculate the generation time *(g)* from the number of bacteria present in the culture at two distinct times.

METHODS OF MEASURING GROWTH

1. The total number of cells in a liquid culture can be determined microscopically with a Petroff–Hausser counting chamber or electronically with an electronic particle counter.
2. The number of viable cells is determined by the colony counting technique. The total cell count of a culture is always greater than the viable cell count.
3. Cell mass can be measured by light scattering in a spectrophotometer.

SPECIAL CONSIDERATIONS: BACTERIAL GROWTH

1. Bioassays, used to determine the concentration of biomolecules, rely on the principle that the cell mass of a culture is directly proportional to the concentration of the limiting nutrient.
2. Chemostats and turbidostats are used to continuously culture microorganisms in the exponential growth phase. With chemostats, microbiolgoists study the effects of low nutrient concentrations on growth and run pilot studies of industrial fermentations.

Questions and Topics for Study and Discussion

QUESTIONS

1. Describe three techniques for obtaining pure cultures of bacteria using agar solidified media.
2. How do pH, salt concentration, and the temperature of a growth medium affect bacterial growth? How might a fever produced during a human infection affect the growth of an infectious microbe?
3. Define or explain the following:

barophile	microaerotolerant
defined medium	psychrophile
facultative anaerobe	selective enrichment
halophile	strict anaerobe
mesophile	thermophile

4. What techniques and equipment are used to culture obligate anaerobes? Describe the method each technique uses to replace oxygen.
5. Describe three methods of maintaining bacterial stock cultures.
6. What are the advantages/disadvantages of the viable cell count versus cell number or cell mass as an indicator of the bacterial population?
7. Draw a diagram of the normal bacterial growth curve and describe the phases of growth.
8. A bacterial culture containing 100 cells increases in population to 1 billion cells in 10 hours. What is the generation time of this culture in minutes?
9. A bacterial culture containing 100 cells has a generation time of 15 minutes. How long will it take for this culture to reach a population of 1 million cells?
10. A technician sequentially diluted a bacterial culture seven times by adding 1 ml of bacterial suspension to 9.0 ml of sterile medium. She took 0.05 ml of the final dilution and plated it onto a petri dish. After incubation, 97 colonies grew. How many bacteria per milliliter were in the original culture?
11. What will happen to the cell mass and the growth rate when the rate of adding fresh medium to a glucose-limited chemostat operating in a steady state is increased?
12. A technician plated 0.1 ml of the final dilution of

a culture onto a plate, incubated the plate, and counted 133 colonies. If he made five 1:10 dilutions of the original culture, how many bacteria were in the starting culture?

DISCUSSION TOPICS

1. What factors prevent continuous exponential growth of bacteria? Do similar factors influence the human population?
2. What everyday practices do you use to control the growth of bacteria and other microbes?

Further Readings

BOOKS

Brock, T. D., *Thermophilic Microorganisms and Life at High Temperatures,* Springer-Verlag, New York, 1978. This is a first-hand account of the important studies Brock and his group conducted on the hot springs at Yellowstone National Park.

Finegold, S. M., and W. J. Martin, *Bailey and Scott's Diagnostic Microbiology,* 6th ed., Mosby, St. Louis, 1982. A detailed reference source or advanced textbook that covers the techniques used in diagnostic microbiology.

Stanier, R. Y., J. Ingraham, M. L. Wheelis, and P. R. Painter, *The Microbial World,* 5th ed., Prentice-Hall, Englewood Cliffs, N. J., 1986. An advanced microbiology textbook that has two excellent chapters on the growth of microorganisms.

MANUALS

Difco Manual Dehydrated Culture Media and Reagents for Microbiology, 10th ed., Difco Laboratories, Detroit, 1984. This manual describes the composition of and uses of Difco's media and reagents.

ARTICLES AND REVIEWS

Brock, T. D., "Life at High Temperatures," *Science,* 230:132–138 (1985). This review of the thermophiles discusses the temperature limits of life, the possible origin of the thermophiles, and the potential application of their unique physiology to biotechnology.

Van Niel, C. B., "Natural Selection in the Microbial World," *J. Gen. Microbiol.,* 13:201–217 (1955). A classic article on the environmental influence over the growth and selection of microbial species.

C H A P T E R

8

Control of Microorganisms

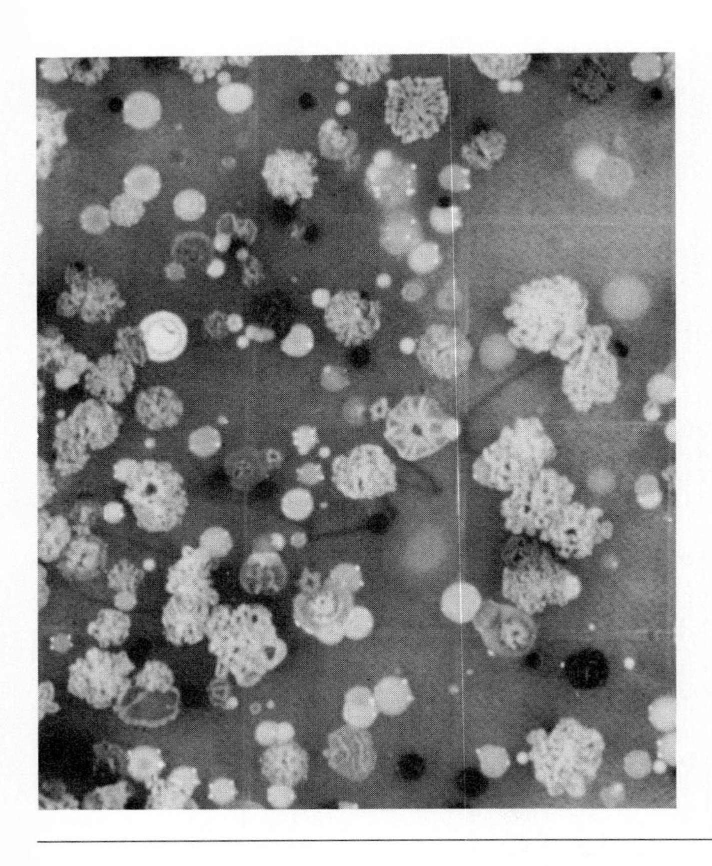

OUTLINE

FOCUS OF THIS CHAPTER

The major methods of protecting animals, persons, and food from the harmful effects of microorganisms are described in this chapter. We discuss the advantages and disadvantages of various agents for controlling microorganisms.

Protective clothing is worn by microbiologists, surgeons, nurses, and pharmaceutical workers to prevent dissemination of their natural microbial flora and to protect them from the infectious agents with which they are working. Microbiologists employ certain techniques for containing infectious agents. Microbes are grown in special laboratory glassware, and the transfer of cultures is performed with techniques that prevent the dissemination of infectious agents. When it is necessary to reduce or eliminate the microorganisms in a given situation, they can be killed by either physical or chemical methods.

The technology employed to control microorganisms is a necessary, integral part of a modern society. Without this technology, we would have tremendous problems related to the preparation, storage, and transportation of food; the modern practice of medicine would quickly become impossible; and we would be overwhelmed by contagious diseases.

PROTECTIVE CLOTHING

Every person is a potential source of microbial contamination. Microbes are associated with the human skin, hair, and mouth and are shed and disseminated in aerosols created by coughing, sneezing, and body movement, or by direct contact. Gowns, laboratory coats, and sterile trouser suits (Figure 8.1) are worn to curtail the transfer of microbes between the environment and the wearers.

The laboratory coat provides the maximum degree of freedom for the wearer, but it protects only against simple spills, splashes, and direct contact with infectious agents. The hospital gown, cap, and gloves afford greater protection than the laboratory coat, but microbes can escape and form aerosols from openings at the wrists, ankles, and face. The most restrictive protective clothing is the sterile trouser suit. This nylon suit has elastic cuffs at the wrists and ankles, overlapped gloves, overshoes, and headgear. A face mask prevents the dissemination of aerosol droplets and protects the wearer from inhaling infectious microbes (Figure 8.1). Sterile trouser suits are often worn by pharmaceutical workers during the bottling and packaging of drugs to prevent these products from being contaminated.

FIGURE 8.1 Protective outer clothing prevents normal human flora from becoming airborne. It is worn by attendants in hospital isolation wards to protect patients from infectious agents. Workers in drug companies wear protective clothing when preparing and bottling certain solutions and drugs. (Courtesy of Pfizer, Inc.)

The effectiveness of protective clothing has been measured by sampling the bacteria present in the air immediately surrounding individuals wearing different protective clothes (Table 8.1). The sterile trouser suit is the most effective protective clothing because there are fewer air leaks at the neck, wrists, and ankles of the wearer.

PRECAUTIONS FOR MEDICAL PERSONNEL

Medical personnel should take special precautions to control the spread of infectious diseases. Handwashing

TABLE 8.1 Dissemination of skin-borne organisms

ACTIVITY	AVERAGE VIABLE BACTERIAL COUNT/FT^3 OF AIR			
	NORMAL DRESS	LABORATORY COAT	STERILE[a] GOWN	STERILE[b] TROUSER SUIT
Sitting	6	3	1.5	0.5
Swinging arms	8	32	14.0	1.0
Walking	138	55	35.0	2.5
Running in place	–	140	53.0	4.0

Source: Adapted from M. C. Hooper, R. Smart, D. F. Spooner, and G. Sykes, "Sterility Testing and Assurance in the Pharmaceutical Industry," in D. A. Shapton and R. G. Board (eds.), *Safety in Microbiology,* Academic Press, New York, 1972.

[a]Surgical cotton gown with loose cotton trousers, overshoes, gloves, and headgear.

[b]Nylon suit with elastic at wrists and ankles, overshoes, overlapping gloves, and headgear.

is an essential practice since many infectious agents are transmitted by direct contact. Medical personnel should wash their hands after attending each patient; after handling contaminated objects such as bedpans, surgical dressings, and specimen containers; and before leaving the hospital. Scrubbing hands and forearms removes bacteria from the skin. Further protection is provided by gloves worn in operating rooms and delivery rooms, and when performing rectal and vaginal examinations, changing dressings, and handling specimens from patients with infectious diseases. Dentists and dental technicians should wear gloves when treating patients who have fever blisters or oral lesions indicative of a herpes infection.

Masks prevent the spread of disease by aerosol. They should be worn by hospital personnel who treat patients with contagious diseases. Masks are also worn in the delivery room to protect both the newborn and the mother. Gowns are worn in surgery, delivery rooms, and isolation units, and when contaminated specimens are being handled. Gowns should be sterilized before they are laundered.

CONTAINMENT OF MICROORGANISMS

Microbiology laboratories are specially designed to limit contamination and the spread of infectious organisms. Bench tops are swabbed routinely with a disinfectant to keep them clean and free of contaminating microbes. The circulation system should be designed to prevent excessive air currents from spreading aerosols that could contaminate cultures with dustborne microbes.

Microbiologists must follow other precautions when working with certain pathogens or when transferring sensitive cultures. Both the aseptic safety hood and the laminar-flow hood (Figure 8.2) have been designed to prevent airborne contamination. The aseptic safety hood is self-contained and completely separates the operator from the microbial material. The air that enters and exits this hood is sterilized either by heat or by passage through filters. The operator wears sealed gloves when manipulating the equipment inside the hood, so the operator has no contact with the infectious material.

The laminar-flow hood (Figure 8.2) is designed so that a curtain of filter-sterilized air continually covers the front opening. This prevents contaminants from entering the work space from outside and protects the user from infectious material in the hood. The air in the curtain is recirculated through the filters, which remove particulate material, including microbes. The main advantage of this design is that the operator's hands can penetrate the curtain and work freely within the hood. Laminar-flow hoods drastically reduce airborne contaminants and are used extensively for tissue culture transfer, work requiring bacteriological containment, and in the preparation of pharmaceutical products.

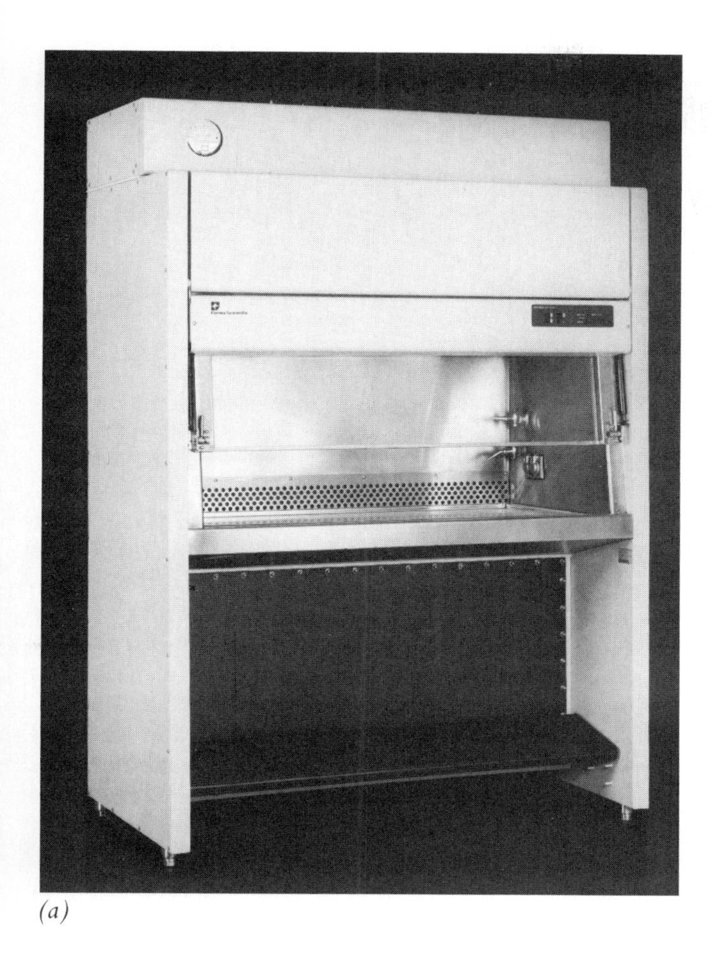

(*a*)

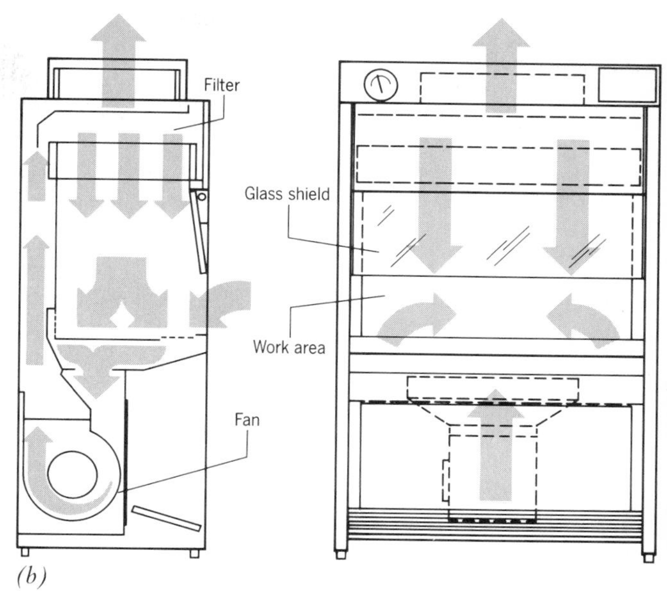

(*b*)

FIGURE 8.2 The laminar-flow hood (*a*) separates contaminants from the interior space by creating a curtain of filtered air across the front opening. The worker has complete access to the materials inside the hood. Air is circultated through filters (*b*) before it is recirculated or exhausted. Laminar-flow hoods are standard equipment in tissue culture laboratories and are required for containment of genetically engineered microorganisms. (Courtesy of L. C. Edwards, Forma Scientific.)

KILLING MICROORGANISMS

The destructive effects of microorganisms can be controlled by preventing their growth or by killing them. This is a never-ending problem, since potentially harmful microorganisms are ubiquitous in the human environment. Some microbes contaminate and spoil food, others infect livestock, and many are able to cause human disease. Physical approaches for controlling the destructive effects of microorganisms include the use of heat and irradiation, whereas antiseptics and disinfectants are chemical agents that control microorganisms.

Various situations require various levels of microbial control. Microbiologists use the word **sterile** to describe an object or place that is devoid of all life, including microbes, spores, and viruses. Prepared bacterial growth media and drugs must be sterile before they can be used. **Sterilization** is the process used to free a substance or object of all living things. Flasks and tubes of liquid media are sterilized routinely by heating them at an appropriate temperature for a specific time period. Aseptic techniques can then be used to inoculate this sterile medium. Microbiologists use the term **aseptic** to mean the absence of contaminating organisms and/or infectious agents.

The control of microorganisms, but not necessarily viruses, can be accomplished by removing them by filtration, by preventing their growth and reproduction, or by killing them. In **filtration,** special membrane filters remove microorganisms and some viruses from liquid solutions and gases. **Bacteriostatic** agents, such as the tetracycline antibiotics, prevent bacterial growth

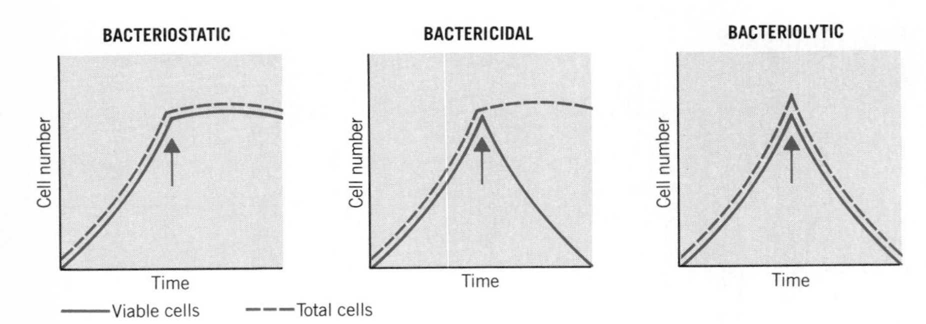

FIGURE 8.3 The growth of bacterial cultures can be inhibited by agents that prevent growth (bacteriostatic), kill cells (bactericidal), or lyse cells (bacteriolytic). The solid line indicates viable cells; the dashed line indicates total cells.

(Figure 8.3), but do not kill the bacteria. In contrast, penicillin causes lysis of growing bacterial cells so it is called a **bacteriolytic** antibiotic. Methods of killing bacteria other than cell lysis are referred to as **bactericidal** (Figure 8.3).

THEORY OF STERILIZATION

Practical sterilization techniques are devised to assure that the probability of a single cell surviving the process is infinitesimal. Cells die exponentially during sterilization (Figure 8.4); they do not all die at once.

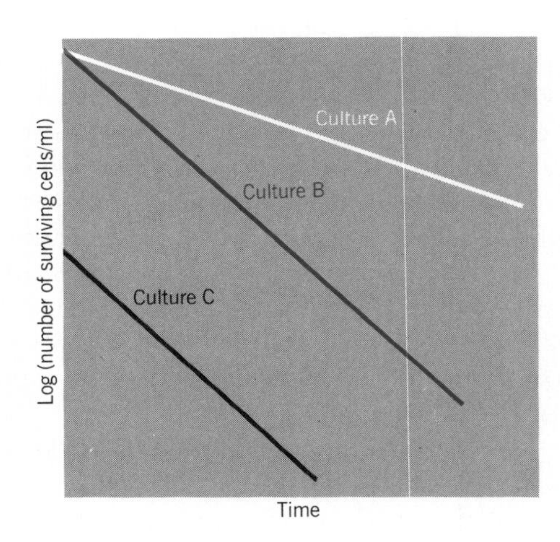

FIGURE 8.4 Death curves for three cultures exposed to a lethal agent. Culture B is more sensitive to the agent than culture A. Culture C is killed at the same rate as culture B, but it reaches zero survivors more quickly because there are fewer organisms at zero time.

The logarithm of the number of surviving bacteria per milliliter is a straight-line function of the duration of exposure to a lethal agent. This means that the majority of the cells die quickly, but that some cells are more resistant, so they survive. The explanation is that these cell populations are heterogeneous, and contain cells with varying levels of resistance to the lethal agent.

Species also vary in their resistance to a given lethal agent. Culture A in Figure 8.4 is more resistant to the lethal agent than is culture B. Culture B dies at a faster rate, so less time is required to sterilize this culture. The time required to sterilize a culture is dependent on the number of organisms in the initial culture. Cultures B and C are both killed at the same rate (the slopes of the lines are equal), but culture C starts with fewer organisms, so it reaches a value of zero survivors per milliliter in a shorter time than culture B.

PHYSICAL METHODS OF CONTROLLING MICROORGANISMS

Table 8.2 depicts several methods for controlling microbes. The heating of materials to be sterilized has many applications in microbiology laboratories and hospitals. In the food-processing industry, mild heat is used to preserve perishable foods. The hazards of radiation to humans restricts its use as a physical means of controlling microorgansisms. Filtration removes cellular organisms from filtered gas and liquid solutions and is used to sterilize heat-sensitive compounds (Table 8.2).

TABLE 8.2 Physical methods of control

TREATMENT	CONDITIONS	USES	DISADVANTAGES
Heat			
Dry heat	160°C, 2 hr	Sterilization of laboratory glassware	High temperatures; charring organic materials
Moist heat	121°C, 15 min, 15 lb pressure	Sterilization of aqueous solutions, liquid media	Limited volume; expensive equipment
Incineration		Destruction of contaminated animals, sterilizing loops, and inoculating needles	
Pasteurization	62°–66°C for 30 min, 71.7°C for 15 sec	Destruction of pathogens in milk, beer, wine; increases shelf life	Sterilization not possible
Filtration	Pore sizes from 0.05 to 0.2 μm	Sterilization of liquids containing heat-sensitive compounds	Viscosity; viruses pass through most filters
Radiation			
Ultraviolet		Decontamination of air, other gases, and surfaces	No penetration through regular glass; danger to eyes
Gamma rays		Food preservation	Danger to humans

KILLING WITH HEAT

Organisms and viruses can be killed by heat. Heat is the most widely used method of sterilization; however, its use is limited to heat-resistant materials. Plastics, clothing and gowns, liquids, and organic substances (including vaccines and drugs) cannot be sterilized by heat because they decompose at high temperatures. Laboratory glassware and most bacteriological media can withstand high temperatures and are routinely sterilized by heat. The process of heat sterilization depends on the temperature, duration of heating, and humidity.

Heat kills cells by disrupting their membrane functions and by denaturing proteins and nucleic acids. Membranes become more fluid at elevated temperatures, causing them to lose their selective permeability and disrupting their functions, including ATP formation by aerobes. High temperatures inactivate proteins and nucleic acids by breaking their hydrogen bonds, which unfolds proteins and separates double-stranded nucleic acids.

KEY POINT

The ultimate criterion for the effectiveness of heat sterilization is the survivability of *Clostridium* or *Bacillus* endospores. If a procedure kills endospores, among the most heat-resistant forms of life known, it will also kill other forms of life.

Sterilization

Dry heat generated by an electric or gas oven kills spores if they are treated for 2 hours at 160°C. The long heating period is necessary to kill all the spores, even though the vegetative cells of the organism will die within the first few minutes. One disadvantage of dry heat sterilization is that organic compounds char at 160°C. Moreover, liquid media cannot be heated above the temperature of boiling water (100°C) without undergoing excessive evaporation. Problems of this type are solved by sterilizing aqueous solutions of organic compounds with moist heat in a pressurized autoclave.

Moist heat is more effective than dry heat in killing bacterial endospores. In the presence of moisture, essential macromolecules are hydrolyzed, especially at the temperatures generated by pressurized steam. Steam sterilization is used to preserve foods in home pressure cookers during canning, and is extensively used in microbiology laboratories, hospitals, and industry. The **autoclave** (Figure 8.5), which is essentially a large pressure cooker, is specially designed for large-scale steam sterilization. Steam at 15 lb/in^2 in the

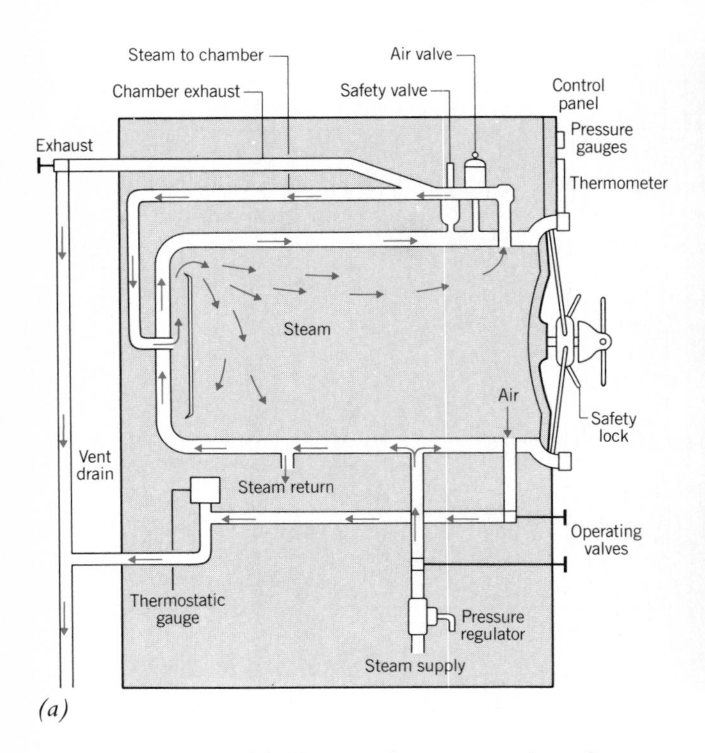

FIGURE 8.5 *(a)* Diagramatic representation of a steam autoclave, side view, and *(b)* photograph of an autoclave.

autoclave chamber reaches a temperature of 121°C, and this is sufficient to kill clostridial spores in 2 minutes. If the spores are contained in a liquid, the time of heating must be lengthened in order to enable the solution to reach 121°C. The normal sterilization time for an autoclave that is fully loaded with test tubes is 12 to 15 minutes at 121°C. The duration increases for sterilization of flasks containing 500 ml or more of liquid.

Microbiologists use the steam autoclave as a major method of sterilizing liquid media and decontaminating supplies and glassware. Aqueous solutions heated above 100°C boil over; however, the high pressure in the autoclave chamber counteracts vaporization so that solutions can be heated to 121°C under 15 lb pressure without boiling over. After sterilization, the superheated solutions must be slowly cooled to 100°C to prevent boiling over when the autoclave door is opened. Autoclaves have a slow exhaust cycle, which permits heated liquids to cool to 100°C as the pressure in the chamber is gradually reduced. Most autoclaves also have a fast exhaust and dry cycle for sterilizing and drying nonliquid materials, such as glassware. The usual cycle (pressure buildup, sterilization, exhaust) takes between 30 minutes and 1 hour.

Canning Foods with Moist Heat The sensitivity of bacterial endospores is affected by the pH of the material being sterilized, and this affects the procedures used in canning meats, fruits, and vegetables. Low-acid foods (pH above 4.5), such as meats, fish, vegetables, and eggs, require extensive heating so they are canned at temperatures between 115° and 121°C. Acidic foods, such as tomatoes, apricots, pineapples, berries, and certain fruits having a pH between 3 and 4.5, can be canned by boiling for 15 to 30 minutes. Acid environments make endospores much more sensitive to heat.

KEY POINT

In the laboratory, sterilization procedures are used to prepare the microbiological media needed to cultivate pure

cultures of microorgansims. Surgical instruments, bandages, and other surgical supplies are sterilized before use to prevent postsurgical infections.

Pasteurization of Perishable Foods

Food spoilage was a major problem before the widespread availability of refrigeration. Unrefrigerated food often spoils because of the activity of contaminating microorganisms. For example, unrefrigerated raw milk spoils within a day or two owing to contaminating microorganisms. The shelf life of perishable foods is now extended by pasteurization.

Louis Pasteur discovered the technique of heating a food just enough to preserve it without altering its composition. This is known as **pasteurization** (pas-tyoor-i-zay'shun), which is an ideal means of preserving milk, beer, and wine. Milk contains butterfat and proteins that cannot tolerate excessive heat. When milk is heated at 62° to 66°C for 30 minutes and then quickly cooled, most of the contaminating microbes are killed without precipitating the proteins. A newer technique, **flash pasteurization,** rapidly heats milk to 71.7°C, maintains it at this temperature for at least 15 seconds, and then rapidly cools it.

The U.S. Public Health Service establishes the conditions required for milk pasteurization based on the time and temperature required to kill *Coxiella burnetii* (kahks-ee-el'ah ber-net'ee-eye). This organism is the causative agent of Q fever and is the most heat-resistant pathogen found in milk. Other milk-borne diseases that are controlled by pasteurization include brucellosis (broo-sel-oh'sis) and certain types of tuberculosis. Pasteurization is designed to kill all the pathogens in milk and enough of the other microbes in the milk to extend its refrigeration shelf life. The shelf life of wine and beer can also be extended by pasteurization.

Incineration

Incineration converts the organic matter of cells to CO_2 and H_2O—this is the ultimate means of destroying cells. Small laboratory animals used in studies of infectious agents are routinely destroyed by incineration. This is also an appropriate means of destroying burnable laboratory wastes. Incineration is widely used to sterilize inoculating loops and needles (Figure 8.6).

FIGURE 8.6 Material on inoculating needles and loops is incinerated in the flame of a Bunsen burner. (Courtesy of BPS.)

PHYSICAL REMOVAL OF MICROORGANISMS BY FILTRATION

Bacteria, eucaryotic cells, and some viruses can be removed from liquids and gases by filtration. Filters are made from natural substances, such as cellulose and diatomaceous earth, or they are manufactured from chemical polymers. The cells (Figure 8.7) are adsorbed to the filter material by electrostatic charges or are trapped in the blind passageways of the filter. Membrane filters of nitrocellulose, cellulose acetate, and nylon are available with pore sizes ranging from 8.0 μm to less than 0.05 μm.

Membrane filters that remove bacteria are used to cold sterilize liquid solutions. Almost all bacteria are

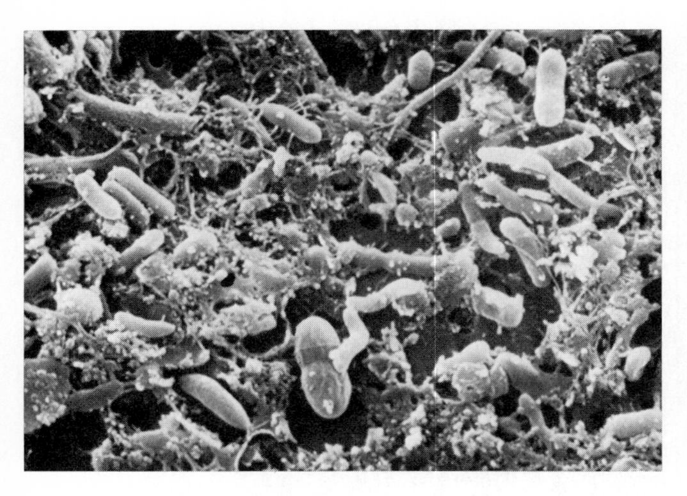

FIGURE 8.7 Scanning electron micrograph of microbes and debris in raw sewage collected on a membrane filter 10,000×. (Courtesy of Dr. A. M. Crundell.)

removed by filtering through 0.2 μm filters, which is the technique commonly employed to sterilize solutions of heat-sensitive compounds. A 0.2 μm filter does not retain the small viruses and, if allowed sufficient time (hours), some of the very thin spirochetes can squirm through these filters. Membrane filters with pore sizes of 0.05 μm are used to retain the larger viruses. The inability of most membrane filters to retain the smaller viruses must be considered when filtration is the method used to prepare bacterial-free solutions.

Filtration with sterile filters (Figure 8.8) is done with a vacuum to pull or with pressure to push liquids or gases through the filter. The filtrate is considered free of bacterial contaminants if a 0.2 μm filter is used and if the filtrate is collected in a sterile container. A practical problem of filtration is that the flow rate decreases as the viscosity of the solution increases and as the pore size of the filter decreases. Slow flow rates limit the volume of material that can be conveniently filtered and set a lower limit for the pore size for practical use.

Filtration techniques have been developed for the purpose of determining the presence of bacteria in large volumes of air and water. Bacteria collected on the filter will grow into colonies when the filter is placed on an appropriate growth medium (Figure 8.9). The number of colonies that grow is used to evaluate the microbes present in the sampled environment.

KILLING WITH RADIATION

The emission and propagation of energy through space or through a substance in the form of waves is called **radiation.** The energy content of different forms of radiation is dependent on the wave frequency (Figure 8.10). Each wave has electromagnetic properties and is composed either of particles or of light energy measured as photons.

Cosmic rays, alpha rays, and beta rays are particulate forms of radiant energy. Alpha and beta rays are emitted from radioactive atoms that are in the process of decaying. These two electromagnetic waves have sufficient energy to penetrate solid substances, so they must be contained to prevent damage to humans. Gamma rays and X rays (Figure 8.10) are electromagnetic waves that are generated as a pulse of electromagnetic energy called a **photon.** The amount of energy in an electromagnetic wave is a product of its frequency *(f)* times Planck's constant *(h)*. Gamma rays have higher frequencies than X rays and so they possess more energy. Gamma rays are generated by the decay of radioactive compounds such as cobalt 60; X rays are generated by a combination of heat and electrical energy.

Ultraviolet radiation is the spectrum of electromagnetic waves whose wavelengths are longer (shorter frequencies) than X rays, but shorter than visible light (Figure 8.10). Ultraviolet (UV) radiation can be generated by passing electrical current through vapors of mercury (black light). UV light is a low-energy, low-frequency form of radiation that is readily absorbed by ordinary glass and by water. Even though ultraviolet light is produced naturally by the sun, most of it is absorbed by the ozone layer of the earth's atmosphere and does not reach the earth's surface.

Gamma Radiation and X Rays Are Forms of Ionizing Radiation

Gamma radiation is useful in food preservation and sterilization of certain materials. Special precautions are necessary because gamma radiation can cause serious injury to humans. In cancer therapy, irradiation with gamma rays emitted from cobalt 60 is used to kill cancerous cells. Gamma radiation has applications in both experimental and therapeutic work.

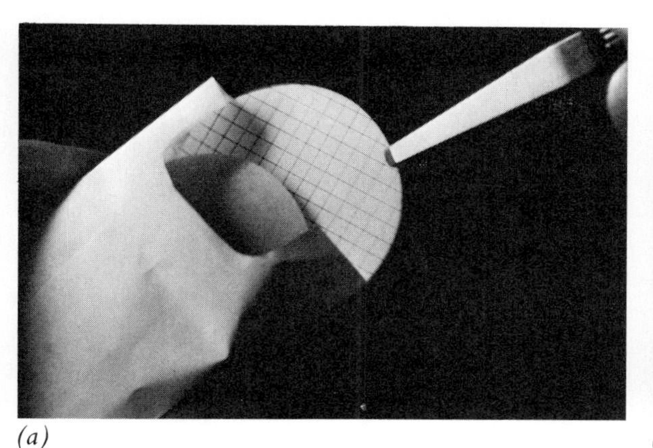
(a)

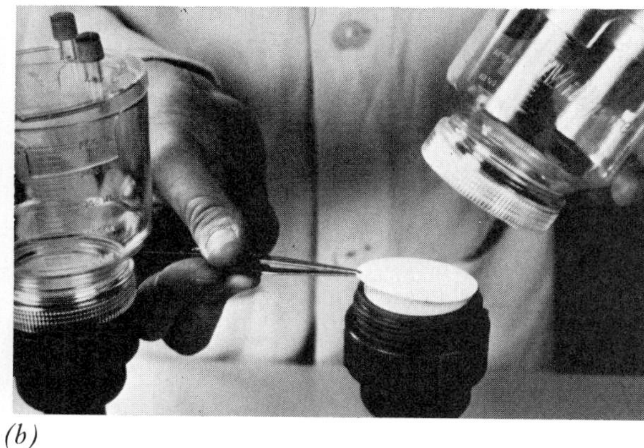
(b)

(c)

FIGURE 8.8 Membrane filtration can be used to remove microbes from liquids. The sterile membrane *(a)* is removed with sterile forceps and *(b)* is placed on the manifold. *(c)* The sample is filtered under vacuum. (Courtesy of Millipore AV Services, Bedford, Mass.)

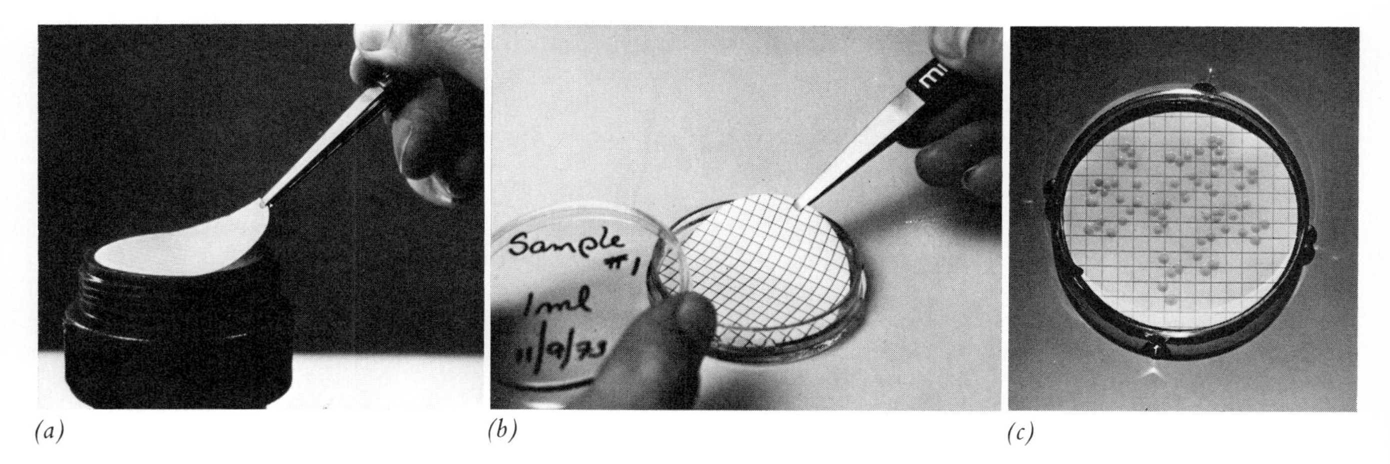

(a) (b) (c)

FIGURE 8.9 Bacteria retained on a membrane filter will grow into colonies when placed on a suitable growth medium. *(a)* Removal of the filter membrane, *(b)* laying the filter membrane on a petri dish, and *(c)* colonies growing on the membrane after incubation. (Courtesy of *(a, b)* Millipore AV Services, Bedford, Mass., *(c)* Gelman Scientific, photograph by Stanley Livingston.)

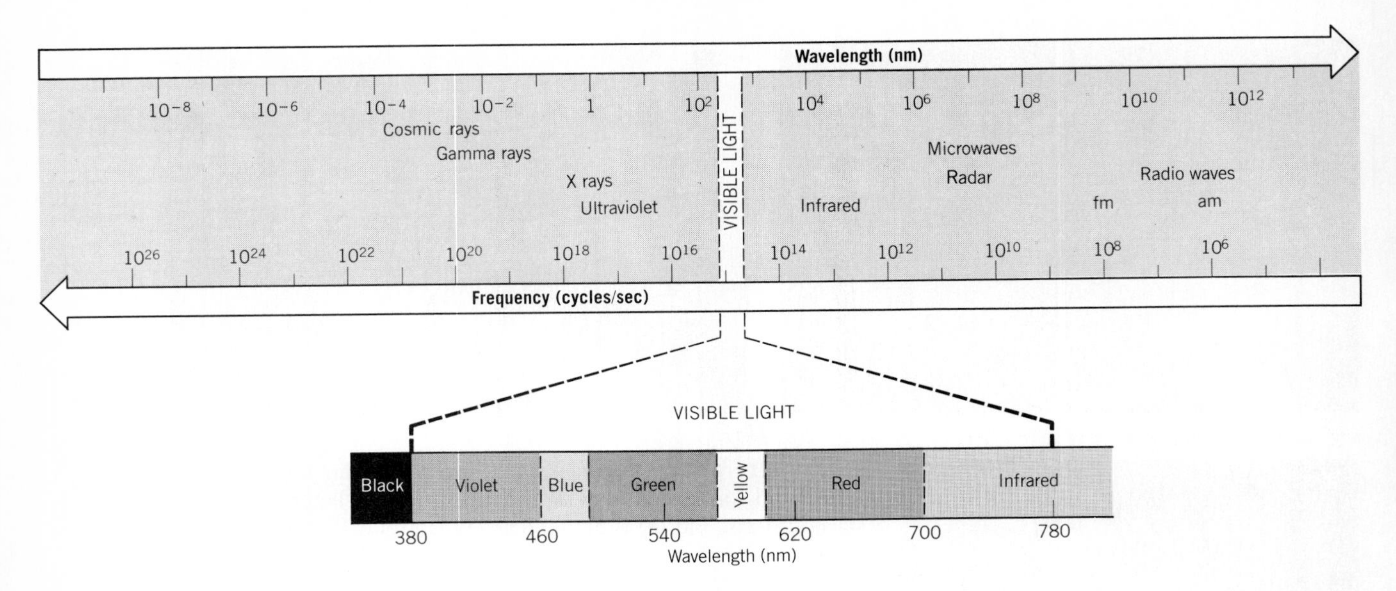

FIGURE 8.10 The electromagnetic spectrum. The amount of energy in a given electromagnetic wave is directly proportional to its frequency. Visible light is only a small part of the electromagnetic spectrum.

X rays are used in diagnostic medicine and dentistry for photographing bones, internal organs, and teeth. In X-ray photography the subject is positioned between a photographic emulsion and the X-ray source. The photons that penetrate the tissue expose the photographic film. The properties of body tissue affect the passage of X rays, and so a picture of the tissue is visible on the developed photographic plate.

Gamma rays and X rays ionize chemicals to form free radicals that kill cells. When gamma rays or X rays pass through body tissues, some of the energy is absorbed by atoms in the cells. This is called **ionizing radiation** because it causes atoms to gain or lose electrons (just as ions do) and become free radicals. The hydroxyl radical (·OH) is the predominant radical formed when aqueous solutions are exposed to ionizing radiations. These free radicals react with cellular proteins and nucleic acids to cause chemical alterations capable of killing the cell. Ionizing radiation usually kills a cell when a free radical reacts with and chemically alters the cell's DNA.

KEY POINT

Radiation therapy is a major treatment for certain types of human cancer. Healthy tissue is protected from the lethal rays by lead shielding and by focusing the radiation on the cancer tissue.

Endospores are much more resistant to ionizing radiation than are their vegetative counterparts. This resistance may be a direct result of the low water content in endospores. Among the nonsporing bacteria, pseudomonads are very sensitive to radiation, whereas some *Micrococcus* and *Streptococcus* species are more resistant. Resistance in this instance may be due to the presence of free-radical scavengers.

Gamma radiation and X-ray radiation have limited use in sterilization procedures because the free radicals produced in the materials irradiated pose a potential hazard to humans. Cosmic rays are natural forms of ionizing radiation, but the amount of cosmic radiation that reaches the earth's surface is so small that it does not appear to affect the biological world.

Ultraviolet Light

UV light from the sun or black lights (Figure 8.10) has wavelengths between 100 and 400 nm. This short wavelength light (250 to 270 nm) is absorbed by and damages nucleic acids by disrupting hydrogen bonding and causing chemical changes. Structural changes in

DNA lead to copying errors and thus increase the probability that a lethal mutation will occur. UV irradiation can be used to produce nonlethal mutations in bacteria or to kill them by varying the dose. Care is required when UV light is used because direct exposure can cause eye damage.

KEY POINT

UV light has limited application as a physical sterilizing agent because it is absorbed by most types of glass and by water. It is used to sterilize exposed surfaces and air, particularly in microbial transfer rooms and in operating rooms.

NONSELECTIVE CHEMICALS

Chemicals having diverse properties are used to control microorganisms. Chemicals that attack specific cells include both naturally produced antibiotics and synthetic (manufactured) drugs. Nonselective chemicals that affect cells are termed disinfectants. **Disinfectants** are substances used on inanimate objects to inhibit the growth of or destroy microorganisms (not necessarily effective against spores). The term is usually restricted to those compounds or preparations that are caustic to humans. **Antiseptics** are chemical agents that inhibit the growth of or kill microorganisms on human skin, mucous membranes, and other living tissue. Antiseptics usually disrupt the function of the plasma membrane.

Many disinfectants are active against vegetative bacteria, fungi, and viruses, but they do not affect spores. The most effective sporicidal chemicals include gluteraldehyde, formaldehyde, and ethylene oxide, but caution must be taken with these chemicals since they are hazardous to humans as well as to spores.

Every nonselective chemical developed for use against microorganisms has unique characteristics that affect its applicability. Criteria have been developed for the selection of these chemicals, as follows: (1) the compound must be an effective killer of microorganisms and, if possible, of viruses; (2) the compound should be soluble in water for ease of preparation and application; (3) the compound should have a low toxicity for humans; and (4) the compound should be available at a reasonable cost. Compounds that meet these criteria are listed in Table 8.3 according to the chemical group to which they belong.

ALCOHOLS

Ethanol, isopropanol, and benzyl alcohol are organic alcohols that are effective antiseptics when applied as 50 to 70 percent aqueous solutions. Alcohols precipitate proteins and solubilize lipids present in the cell walls and cell membranes of bacteria. They are used as antiseptics on human skin both individually and in combination with a halogen such as iodine. Alcohols effectively kill vegetative bacterial cells, and fungi, but they are not effective against spores and they do not inactivate all viruses.

HALOGENS

Iodine is a halogen that is both germicidal and antiseptic when dissolved in 70 percent alcohol (tincture of iodine) or in an aqueous detergent (povidone–iodine, known as Betadine). Solutions of iodine in water and alcohol (iodine tincture) are used as antiseptics to prophylactically prepare skin for injections and transfusions, presurgical disinfection of hands, preoperative preparation of skin, and the disinfection of the perineum prior to childbirth. These solutions are also used to treat cuts and abrasions of the human skin, and burns.

Iodine inactivates proteins and organic molecules by reacting with sulfhydryl groups, hydroxyl groups, and amines. The disadvantages of iodine are that it is caustic, causes pain when applied to human tissue, and stains skin and fabrics. Iodine has been used to disinfect drinking water and to treat swimming pools. **Iodophores** are loose complexes of elemental iodine or triiodide with a high molecular weight organic carrier. Iodophores are nonstinging and slowly release iodine at the site of application.

Chlorine gas is a halogen disinfectant that is used to decontaminate swimming pools, hot tubs, and community water supplies. Chlorine gas reacts with water to form hypochlorous acid (HClO),

$$Cl_2 + H_2O \longrightarrow HCl + HClO$$

which then reacts with water to form HCl and hydro-

TABLE 8.3 Antiseptics and disinfectants

CLASS OF COMPOUNDS	ACTION	USES	DISADVANTAGES
Alcohols			
Ethanol, isopropanol, benzyl alcohol (50–70% aqueous solutions)	Denature protein and solubilize lipids	Skin antiseptic	Not effective against spores and some viruses
Aldehydes			
Formaldehyde (8% solution), Glutaraldehyde (2% solution)	Alkylate, react with $-NH_2$, $-SH$, and $-COOH$	Effective disinfectants, sporicides	Toxic to humans, noxious fumes
Halogens			
Iodine (I_2) as tincture of iodine (2% in 70% alcohol), Betadine (I_2 in detergent)	Inactivate proteins (reacts with tyrosine)	Skin antiseptic, iodophores are nonstinging	Corrosive, stains skin and fabric
Chlorine (Cl_2) gas in neutral or acidic aqueous solutions	Reacts with water to make hypochlorous acid (HClO), strong oxidizing agent	Purify water, disinfectant in food industry	May react with water impurities to form toxic compounds
Heavy Metals			
Copper sulfate ($CuSO_4$)	Prevents growth of algae and fungi	Algicide and fungicide	
Silver nitrate ($AgNO_3$)	Precipitates proteins	Antiseptic, may be used in eyes of newborns	Neutralized by organic compounds, irritation
Mercuric chloride ($HgCl_2$)	Inactivates proteins (reacts with $-SH$)	Disinfectant	Poisonous in high concentrations, inactivated by organic compounds
Gases			
Sulfur dioxide (SO_2)	Reacts with many cell components	Food additive, especially fruit juices	Highly irritating to eyes, respiratory tract
Ethylene oxide (H_2C—CH_2, bridged by O)	Alkylates organic compounds	Sterilizes heat-sensitive objects, good penetration, sporicidal	Highly irritating to eyes, mucous membranes, highly flammable
β Propiolactone (H_2C—CH_2 / O—C=O, four-membered ring)	Alkylates organic compounds	Bactericidal and sporicidal	Potential human carcinogen, blisters on skin
Cationic Detergents			
Quaternary ammonium compounds with long-chain alkyl group	Disrupt cell membranes	Skin antiseptics and disinfectants	Inactivated by metal ions, low pH, organic material, and phospholipids
Phenols			
Phenol, carbolic acid and derivatives, Lysol, hexylresorcinol, hexachlorophene	Denature proteins and disrupt cell membranes	Disinfectant at high concentrations, used in soaps at low concentrations	Very toxic to human tissue

gen peroxide (H_2O_2). Both hypochlorous acid and hydrogen peroxide are strong oxidants that are toxic to biological systems.

Household bleach is a common source of hypochlorous acid; bleach contains 5.25 percent sodium hypochlorite (NaClO). A dilution (1/10) of household bleach in water is an effective disinfectant provided the surfaces to which it is applied can withstand its bleaching action. Bleach can also be used to control the growth of household mold, which grows on walls and household furnishings in damp climates.

KEY POINT

A weak solution of bleach and household detergent is an effective cleaning solution that inhibits mold growth. Bleach and household ammonia, however, should never be mixed together because they produce a highly poisonous gas.

ALDEHYDES

Formaldehyde and glutaraldehyde are alkylating agents that kill microbes and spores by reacting with the replaceable hydrogen atoms in organic compounds. Cellular targets include the amines ($—NH_2$), sulfhydryl (—SH), and carboxyl (—COOH) groups of proteins and small organic molecules. **Formaldehyde** is an effective disinfectant in an 8 percent solution, and **glutaraldehyde** is a useful disinfectant when present in a 2 percent solution. Both of these aldehydes have limited application because they give off noxious vapors.

HEAVY METALS

Copper, mercury, and silver are heavy metals that are effective in killing microbes. Eye drops containing **silver nitrate** were widely used to prevent infection by *Neisseria gonorrhoeae* in the eyes of newborns; other means are now used because silver nitrate is an irritant. **Copper sulfate** is toxic to the photosynthetic systems of algae and is an effective algicide; when added to marine paints, it prevents mussels, barnacles, and other fouling organisms from attaching to the bottom of boats. Mercuric compounds are both disinfectants and antiseptics. Although mercuric compounds have bacteriostatic or bactericidal activities, they are not widely used because of their cytotoxicity.

Precautions must be taken with lead, arsenic, and mercury as disinfectants because they are concentrated by the human body and can cause long-term disorders. Many young children living in homes decorated with lead paints have suffered mental disorders from consuming paint chips. Arsenic is a component of Salvarsan, which was the first effective drug for treating syphilis (penicillin is used now). Arsenic and lead are rarely used today. Mercury is added to ointments and solutions for treating skin infections. The organic mercurials such as merthiolate and mercurochrome are widely used as antiseptics.

GASES

Ancient Greeks burned sulfur (thereby producing sulfur dioxide) to disinfect their homes and their wine casks. Sulfur dioxide is still used as a food preservative, and to kill the natural flora in the grape juice used to manufacture wine.

Ethylene oxide is a simple organic molecule that vaporizes into a gas at temperatures above 10.8°C. Gaseous ethylene oxide is highly flammable unless it is mixed with freon or carbon dioxide. This unique sterilizing gas freely diffuses and penetrates paper, cellophane, cardboard, fabrics, and some plastics making it an ideal agent for sterilizing disposable medical laboratory supplies. Plastic petri dishes, tubes, pipets, and syringes are sterilized with ethylene oxide in pressurized chambers under conditions of controlled temperature and humidity. Ethylene oxide is both bactericidal and sporicidal. Caution must be exercised, however, because it irritates human eyes and mucous membranes, and can cause nausea and vomiting. Some evidence suggests that ethylene oxide may be a carcinogen.

Beta propiolactone, a nonflammable liquid that becomes a gas at 155°C, is active against a wide variety of microorganisms including spores, bacteria, and viruses. It is not widely used since it causes skin blisters and is a potential carcinogen.

DETERGENTS

Detergents are surface-cleansing compounds that act as wetting agents. These compounds dissociate in aqueous solutions to their ionic forms. **Anionic detergents,** such as sodium lauryl sulfate, dissociate into a negatively charged hydrophobic molecule (anion) and a sodium ion. Anionic detergents are good cleansers, but they have little effect on microorganisms.

$$\left[R{-}N(R_1)(R_3){-}R_2 \right] \cdot Cl^-$$

General formula

Cl (CH$_2$)$_{15}$—CH$_3$ N

Ceepryn®

$$\left[C_6H_5{-}CH_2{-}N(CH_3)_2{-}R \right]^+ Cl^-$$

R represents a mixture of alkyls from C_8H_{17} to $C_{18}H_{37}$

Roccol®

(a) Quaternary ammonium compounds

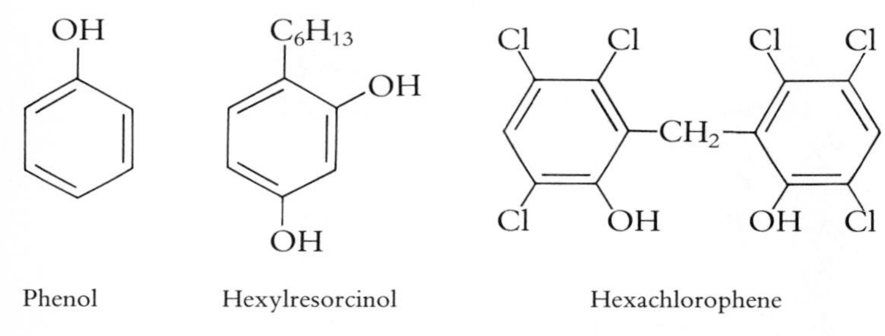

(b) Phenols

FIGURE 8.11 Quaternary ammonium compounds and phenols are used as disinfectants at high concentrations and as antiseptics at low concentrations.

Cationic Detergents

Cationic detergents, such as the quaternary ammonium compounds, produce positively charged hydrophobic molecules (cations) upon dissociation in solutions (Figure 8.11), and are effective disinfectants.

Quaternary ammonium compounds are bactericidal and have a low toxicity for mammalian tissue. They kill cells by disrupting their cytoplasmic membranes. This attribute makes them practical for preoperative preparation of the female vagina and other structures covered by sensitive mucous membrane. Cationic detergents are inactivated by low pH, phospholipids, organic compounds, and metal ions. Their use is limited because of their narrow antibacterial spectrum. Also, some species of *Pseudomonas* can grow in cationic detergents, such as benzalkonium chloride, used in hospitals for disinfection.

PHENOLS

Phenolic compounds and their substituted derivatives are extremely effective disinfectants. Carbolic acid was the disinfectant used by Joseph Lister in 1867 to prevent infection following surgery. Carbolic acid is quite caustic to human tissue, so it has been replaced by other phenol derivatives. Phenol derivatives in low concentrations are used as antiseptics and in higher concentrations as disinfectants. Phenols function by denaturing proteins and disrupting cell membranes.

Phenol itself is so caustic that it is no longer used as a disinfectant; however, the clear phenol derivatives such as Lysol, Clearsol, Stericol, and Sudol are very good disinfectants and are widely used. They are effective against many bacteria, are not readily inactivated by organic compounds, and are used in the laboratory when a disinfectant is needed.

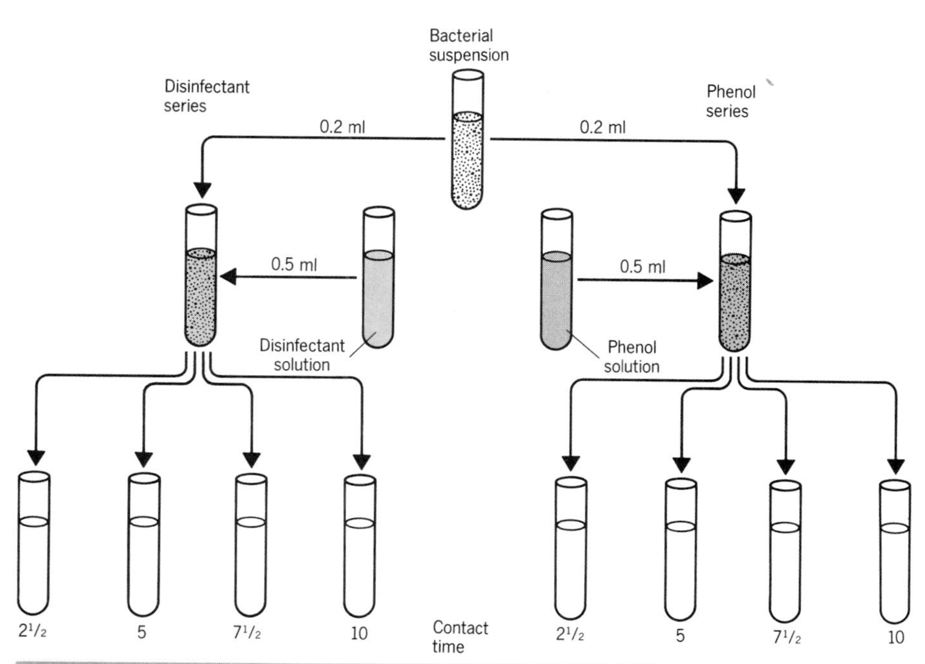

Disinfectant Concentration	Growth in Subculture After			
	2½ min	5 min	7½ min	10 min
1/700	–	–	–	–
1/800	+	–	–	–
1/900	+	+	–	–
1/1000	+	+	+	–
1/1100	+	+	+	+
Phenol control				
1/90	+	+	–	–

Phenol coefficient $= \frac{900}{90} = 10$

FIGURE 8.12 The phenol coefficient of a disinfectant is determined by comparing its antibacterial activity to that of phenol. The test organism, *Staphylococcus aureus,* was exposed to the disinfectant for different intervals (contact time) and then inoculated into subcultures. The phenol coefficient of **10** means that the disinfectant is ten times more potent than phenol.

Hexachlorophene is a chlorinated phenol that kills staphylococci at concentrations as low as one part per million. Hexachlorophene was once widely used as an antiseptic, but now it is available only on prescription because it is quickly absorbed through the skin and has been known to cause brain damage in laboratory animals.

Many disinfectants and antiseptics are sold commercially in the form of mouthwashes, soaps, creams, and facial washes. Others including hydrogen peroxide, mercurochrome, and merthiolate are used to treat cuts and abrasions. Disinfectants for the kitchen, bathroom, garbage can, and other potentially infested areas are also readily available. These conveniently packaged disinfectants and antiseptics have gained wide acceptance.

Phenol Coefficient

Disinfectants can be evaluated by standardized procedures such as the phenol-coefficient test. In this test, the antibacterial activity of the disinfectant is compared to the ability of phenol to inhibit bacterial growth. The **phenol coefficient** is determined by exposing the bacterial culture to different dilutions of the disinfectant over several periods of time, such as 2.5, 5, 7.5, and 10 minutes (Figure 8.12). A loop of each treated bacterial suspension is then transferred to a broth and incubated. Broth cultures that show growth are scored positive, whereas those with no growth are scored negative. The results are compared to those of an identical experiment done with phenol as the disinfectant. The phenol coefficient (determined as shown in Figure 8.12) indicates how effective the test disinfectant is in killing *Staphylococcus aureus* in comparison to phenol.

Summary Outline

PROTECTIVE CLOTHING

1. Protective clothing is worn by microbiologists, surgeons, nurses, and pharmaceutical workers to prevent dissemination of their natural microbial flora and to protect them from the infectious agents with which they are working.
2. Sterile trouser suits are worn by pharmaceutical workers to protect parenteral drugs during bottling and packaging.

PRECAUTIONS FOR MEDICAL PERSONNEL

1. Medical personnel wear protective clothing and follow procedures such as handwashing to protect their patients and themselves against contamination.

CONTAINMENT OF MICROORGANISMS

1. Dangerous infectious agents can be handled in aseptic safety hoods that completely separate the operator from the hazard.
2. Laminar-flow hoods reduce airborne contamination and are used to transfer tissue and microbial cultures.

KILLING MICROORGANISMS

1. Sterile describes a place or object that is devoid of all life. Sterilization is the process used to free a substance or object of all living things.
2. Aseptic means the absence of contaminating organisms and/or infectious agents.
3. Bacteriostatic agents prevent bacterial growth without killing them, bacteriolytic agents cause bacteria to lyse, and bactericidal agents kill bacteria.
4. Cells die exponentially during sterilization. Species vary in their resistance to lethal agents.

PHYSICAL METHODS OF CONTROLLING MICROORGANISMS

1. Dry heat (oven) and moist heat (autoclave) sterilization are used to kill all life in an area or contained space. Moist heat is more effective in killing bacterial spores than dry heat.
2. Endospores are used in biological sterility tests since endospores are the most heat-resistant forms of life known.
3. Pasteurization is a mild heat treatment that eliminates common pathogens from milk and other beverages. Pasteurization increases the shelf life of perishable foods and protects consumers from foodborne infectious diseases.
4. Incineration completely destroys cells and is used to sterilize inoculating loops and dispose of dead infected experimental animals.
5. Filtration is used to remove microorganisms from heat-sensitive solutions and to sterilize gases; it does not necessarily remove small viruses.
6. Gamma rays and X rays are penetrating forms of ionizing radiation that can kill cells by producing free radicals.
7. Ultraviolet light does not penetrate glass or water, but does decontaminate air and the surface of objects.

NONSELECTIVE CHEMICALS

1. Disinfectants are substances used on inanimate objects to inhibit the growth of or destroy microorganism (not necessarily effective against spores).

The term is usually restricted to those compounds or preparations that are caustic to humans.

2. Antiseptics are chemical agents that inhibit the growth of or kill microorganisms on human skin, mucous membranes, and other living tissue.
3. Common disinfectants include chlorine gas, formaldehyde, glutaraldehyde, ethylene oxide, sulfur dioxide, β propiolactone, and the higher concentrations of phenolic compounds. The potency of a disinfectant can be determined by measuring its phenol coefficient.
4. Antiseptics used to destroy microorganisms on humans include alcohol, iodine, iodine–alcohol mixtures, silver nitrate, mercuric compounds, cationic detergents, and low concentrations of phenolic compounds.

Questions and Topics for Study and Discussion

QUESTIONS

1. Where is protective clothing worn to prevent infections or contamination?
2. What are the advantages and disadvantages of dry heat versus moist heat in sterilization procedures?
3. What organisms are used as standards in determining the effectiveness of sterilization and pasteurization? Why are these organisms chosen?
4. What types of radiation will inhibit or control microbial growth? Describe a practical application of each type.
5. Explain the different meanings and uses of the words in the following pairs:
   antiseptic: disinfectant
   bactericidal: bacteriostatic
   aseptic: sterile
   pasteurization: sterilization
6. Name five nonselective chemicals that control microbial growth and describe a practical use for each one.
7. Describe how Cl_2 reacts with water to form bactericidal compounds.
8. What are the unique properties of ethylene oxide and what are the practical consequences of each?
9. Explain what is meant by the statement "A disinfectant has a phenol coefficient of 5."

DISCUSSION TOPIC

1. What household products do you use that can be classified as either disinfectants or antiseptics?

Further Readings

BOOKS

Block, S. S., *Disinfection, Sterilization, and Preservation,* 3rd ed., Lea & Febiger, Philadelphia, 1983. The sections of this compendium cover chemical and physical sterilization, antiseptics and disinfectants, medical and health-related applications, antimicrobial preservatives and protectants, mode of action, and methods of testing.

Borick, P. M., *Chemical Sterilization.* Dowden, Hutchinson & Ross, Stroudsburg, Pa., 1973. A collection of selected papers on the use of chemicals in sterilization.

Russell, A. D., *The Destruction of Bacterial Spores,* Academic Press, London, 1982. The effects of chemical and physical agents on bacterial endospores are expertly organized and discussed.

Russell, A. D., W. B. Hugo, and G. A. J. Ayliffe, *Principles and Practice of Disinfection, Preservation and Sterilisation,* Blackwell, Oxford, 1982. A current collection of papers on the control of microorganisms.

Shapton, D. A., and R. G. Board (eds.), *Safety In Microbiology,* Academic Press, New York, 1972. A series of articles describing the safe handling of microorganisms in industrial settings.

ARTICLES AND REVIEWS

Favero, M. S., "Sterilization, Disinfection, and Antisepsis in the Hospital," in E. H. Lennette, et al. (eds.), *Manual of Clinical Microbiology,* 4th ed., American Society for Microbiology, Washington, D.C., 1985. Control of microorganisms in hospitals is reviewed.

CHAPTER

9

Macromolecules

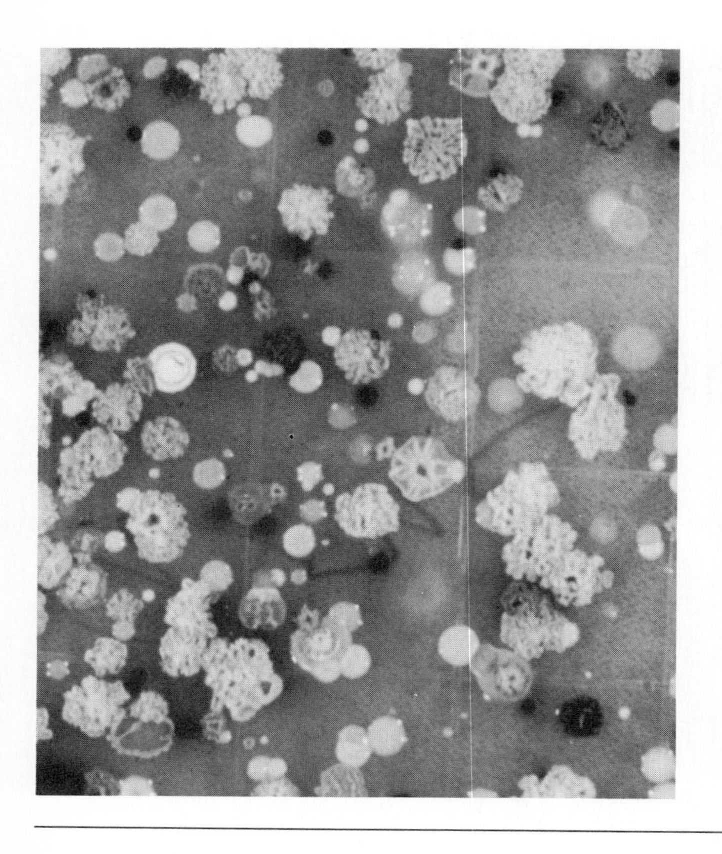

OUTLINE

FOCUS OF THIS CHAPTER

All the biological information the cell needs for growth and reproduction is stored in its DNA. This chapter describes how cells reproduce their DNA and how the information in DNA is converted to RNA and protein. The cellular mechanisms that regulate protein synthesis and enzyme activity are explained.

Beasts beget beasts, birds beget birds, and crawly things beget things that crawl. For hundreds of years, humans have applied this basic knowledge to improve animal breeds. Some dogs hunt, some tend sheep; thoroughbred race horses are capable of great speeds and draft horses pull heavy loads; some cattle provide beef and some produce milk. Animal breeding is possible because parental characteristics are passed on to their progeny. Similarly, microbes pass their traits on to their daughter cells and thereafter to their progeny for thousands of generations. DNA, RNA, and proteins are the cellular macromolecules through which traits are inherited and expressed. How can cells make exact replicas of themselves? They do so by replicating their deoxyribonucleic acid (DNA), the macromolecule that contains the code for the design of the organism.

DEOXYRIBONUCLEIC ACID AND HEREDITY

In 1869, Friederich Miescher discovered DNA while analyzing pus cells and salmon sperm. He showed that DNA was localized in the sperm cell's nucleus, so DNA was originally called "nuclein." However, the significance of DNA was not appreciated for almost a century, for it took that long to accumulate the circumstantial evidence that linked DNA and heredity.

DISCOVERY OF THE STRUCTURE OF DNA

Genetics, the science of heredity, was a full-fledged science long before scientists became interested in DNA. Even after DNA was known to be the hereditary material, the manner in which a macromolecule could carry all the hereditary information was imponderable. The problem was solved once the structure of DNA was understood. This discovery has had such a profound impact on biology that it will probably stand as the most important biological discovery of the twentieth century. Knowledge of the structure of DNA provides the basis for understanding biological inheritance and is the basis for the rapidly growing field of genetic engineering.

THE WATSON–CRICK DNA MODEL IS A DOUBLE HELIX

James Watson and Francis Crick were central figures in the discovery of the structure of DNA. These scientists worked together in the Cavendish Laboratory in Cambridge, England (Figure 9.1). They were convinced that the components of DNA would fit together like the pieces of a puzzle—once they had all the pieces. The following facts were known to Watson and Crick when they began to build their DNA model:

1. DNA is composed of the sugar deoxyribose, phosphate, and four nitrogenous bases.
2. The nitrogenous bases in DNA are adenine (A), thymine (T), guanine (G), and cytosine (C). These bases are attached to deoxyribose forming the four **nucleosides**★ that comprise DNA.
3. X-ray diffraction studies of DNA indicated that DNA had a double helical structure. It had previously been shown that protein chains can form a helical structure stabilized by hydrogen bonds.
4. In a double-stranded DNA molecule Erwin Chargaff had demonstrated that the quantity of adenine equaled that of thymine and the quantity of guanine equaled that of cytosine (A = T : G = C).

Using this information, Watson and Crick began building a model of DNA. The pieces of the puzzle fit together when (1) the sugar–phosphate backbone of the molecule took the shape of a helix and (2) the ni-

★A *nucleoside* is a nitrogenous base plus a sugar; a *nucleotide* is a nucleoside attached to one or more phosphate groups.

FIGURE 9.1 James Watson (left) and Francis Crick in front of the DNA Model. (From J. D. Watson, *The Double Helix,* Atheneum, New York, p. 215, copyright © 1968 by J. D. Watson. Photograph by A. C. Barrington Brown.)

trogenous bases were paired with their counterparts (adenine with thymine, and guanine with cytosine) located on the adjacent sugar–phosphate polymer running in the opposite direction. The purines (adenine and guanine) paired with their respective pyrimidines (thymine and cytosine) through the weak forces of hydrogen bonds.

Visualize this structure by thinking of a twisted ladder. The left sidepiece goes from the bottom to the top; the right sidepiece runs from the top to the bottom. The rungs of the ladder represent the nitrogenous bases bound together by hydrogen bonds. Now hold the bottom of the ladder and twist the top such that you create a **right-hand** twist in the ladder (Figure 9.2). This helical structure represents the structure of DNA.

The sidepieces of the ladder are made of sugar (deoxyribose) phosphate groups. Repeating units of the deoxyribose–phosphate joined by covalent bonds make DNA appear as a long thread. Extending from this sugar–phosphate backbone are nitrogenous bases (adenine, thymine, guanine, and cytosine). Each base is joined by hydrogen bonds to its complementary base on the adjacent sugar–phosphate backbone. Nitrogenous bases joined by hydrogen bonding are analogous to the rungs of our ladder.

Double-stranded DNA exists as a right-hand double helix that takes one complete turn every 34 Å (Figure 9.2). The nitrogenous bases are stacked next to each other perpendicular to the helical axis such that they occupy the central core of the helix. The sugar–phosphate backbone has directionality (see Figure 9.2a arrows) because of the way the phosphates are attached to the deoxyribose. We say that the sugar–phosphate backbones of the helix are **antiparallel** because they run in opposite directions.

Hydrogen Bonds Hold the Double Helix Together

The nitrogenous bases are neatly stacked so that they lie flat atop each other at right angles to the sugar backbone (Figure 9.2*b*). This structure is stabilized by hydrogen bonds between the carboxyl groups $\left(\rangle C{=}O\right)$ and the hydrogen atoms of the amino groups ($—NH_2$), and between the ring nitrogens and the hydrogens attached to nitrogen atoms on adjacent rings (Figure 9.3).

Because of the hydrogen bonding, adenine always pairs with thymine, and guanine always pairs with cytosine. This type of bonding is not permanent: hydrogen bonds are easier to break than are covalent bonds because hydrogen bonds contain less energy. When DNA is heated to temperatures between 60° and 90°C, large portions of the double helix separate into single strands. The exact temperature at which the DNA strands separate depends on the proportion of purines (A and G) to pyrimidines (T and C). The separated strands of DNA will rejoin upon cooling, a process called **annealing of DNA.**

Double-stranded DNA always contains equal molar concentrations of adenine and thymine and of guanine and cytosine. However, the amount of guanine plus cytosine compared to the amount of adenine plus thymine can vary from one organism to the next. This fact is used in bacterial systematics to determine relatedness between bacterial species.

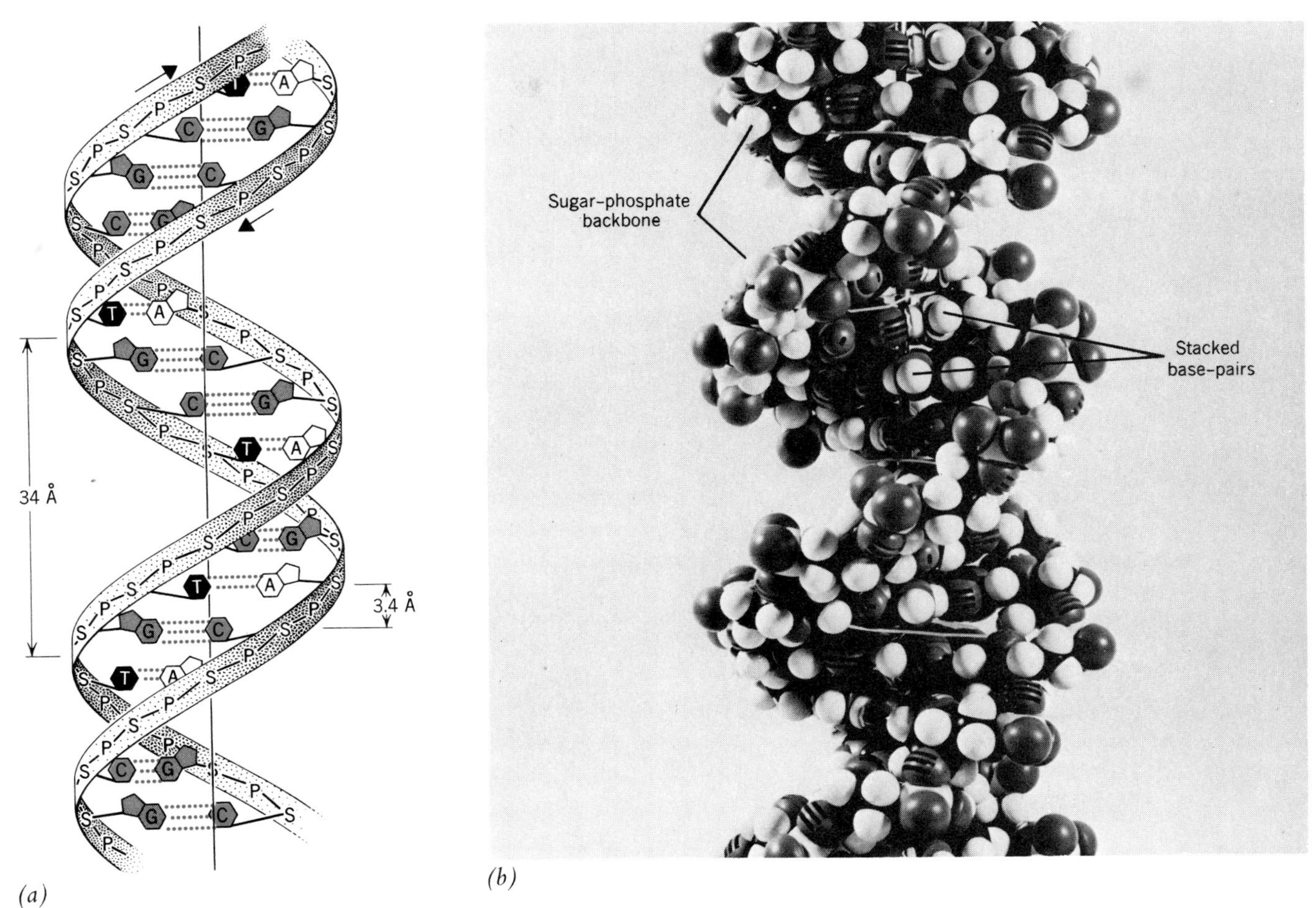

FIGURE 9.2 *(a)* Diagram of the Watson–Crick double-helix model of DNA. A, T, G, and C represent adenine, thymine, guanine, and cytosine; S represents the sugar (deoxyribose) and P represents phosphate. (From E. J. Gardner and D. P. Snustad, *Principles of Genetics,* 6th ed., Wiley, New York, 1981). *(b)* Photograph of a space-filling model of DNA showing the position of each atom. Notice how the bases are stacked in a horizontal position at right angles to the backbone. (Courtesy of Ealing Corporation.)

DNA REPLICATION

Cells transfer genetic information to their progeny by making exact copies of their DNA. DNA replication is a complex process that began to be understood only when Watson and Crick created their model and proposed how it could function in heredity. This model suggested a mechanism for self-replication. Suppose the two strands of DNA separate in the presence of the four nucleotides. One can envision a copy mechanism in which free nucleotides pair with bases on each of the separated DNA strands. Such a mechanism will preserve the nucleotide sequence in the DNA.

DNA viruses, procaryotes, and eucaryotes all replicate their DNA before producing progeny. Much of our current understanding of DNA replication is based on chromosomal replication in *Escherichia coli,* the system emphasized below.

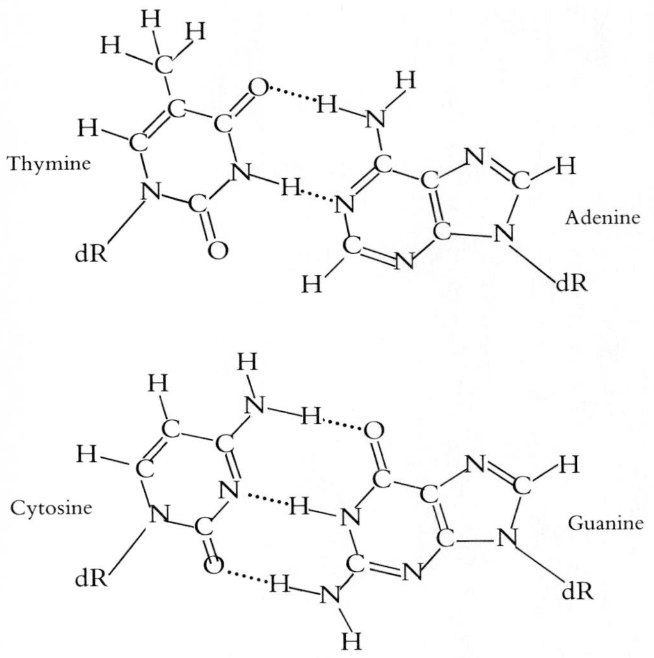

FIGURE 9.3 Base pairing in DNA: thymine pairs with adenine and cytosine pairs with guanine. The nucleotides are joined by hydrogen bonds. The sugar molecule is deoxyribose.

DNA Strand Separation

Double-stranded bacterial DNA separates during its replication to expose sections of single-stranded DNA that serve as the templates for making new DNA (Figure 9.4). Strand separation was demonstrated by labeling the DNA of growing bacteria and then determining the fate of the label as the cells divided in an unlabeled medium. After one generation, the cells contained a mixed DNA in which one strand (old) was labeled and the other strand (new) was unlabeled. This result requires separation of the original double-stranded DNA such that it can act as a single-stranded template for new DNA synthesis.

This method of replicating DNA is called **semiconservative replication**—each new double-stranded DNA contains one old strand (template) and one new strand. The semiconservative replication of double-stranded DNA and the double helical structure of DNA are now firmly established concepts.

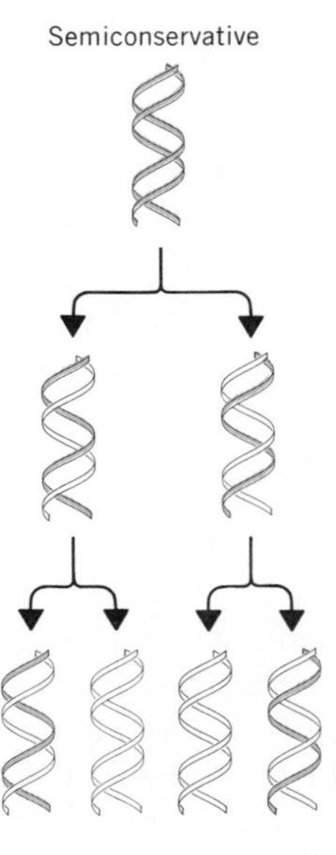

FIGURE 9.4 Semiconservative replication of DNA requires strand separation. Each new double-stranded DNA contains one old strand (template) and one new strand.

Replication Eye

Bacteria contain one circular chromosome that is continuously replicating. Labeling studies have shown that replication always begins at the same place (replication origin) and proceeds along the right and left replication forks as the DNA unwinds (Figure 9.5). The replication origin is a sequence of about 422 nucleotides designated *ori C*. This region of the chromosome probably folds back upon itself to form a clover-like structure that is recognized by the enzymes of DNA synthesis.

Escherichia coli replicates its DNA bidirectionally beginning at the *ori C* locus. Bidirectional replication means that there are two replication forks moving away from the replication origin. This process results in the **replication eye** as seen in electron micrographs of closed circular DNA molecules that are being replicated (Figure 9.6). It takes *E. coli* about 40 minutes to replicate its chromosome. Since this bacterium has a generation time of about 25 minutes, rapidly growing cells must be simultaneously making more than one

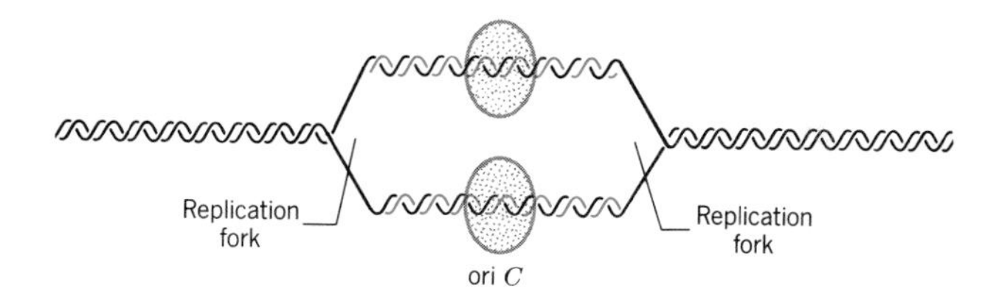

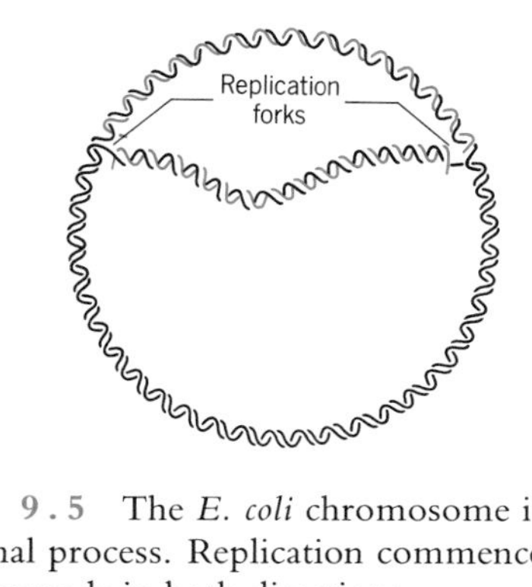

FIGURE 9.5 The *E. coli* chromosome is replicated by a bidirectional process. Replication commences at *ori C* locus and proceeds in both directions.

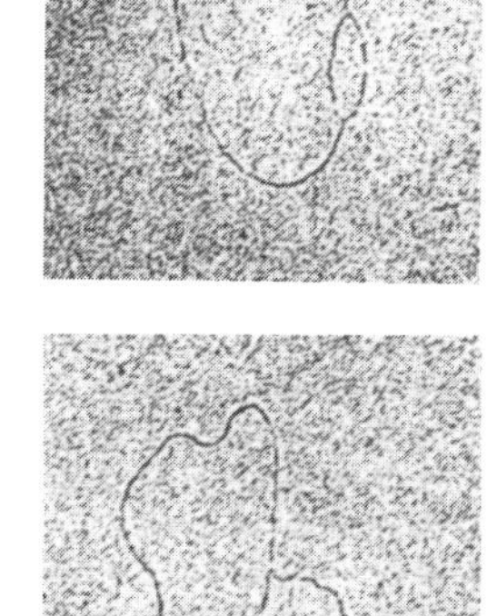

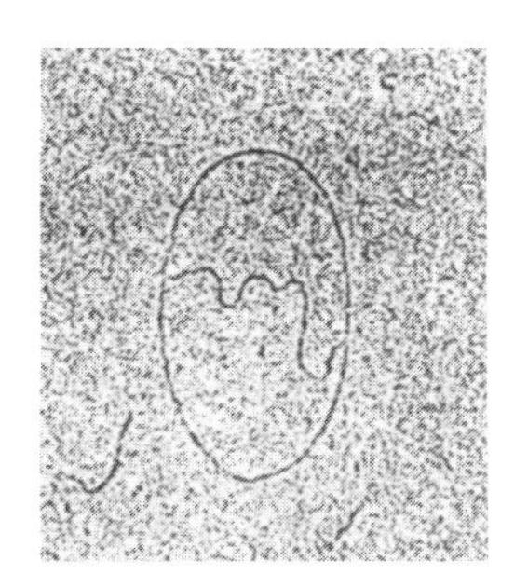

FIGURE 9.6 Bidirectional replication of closed circular DNA creates the replication eye seen in these electron micrographs. (From Lewin, *Genes II*, 2nd edition, John Wiley and Sons, 1985.)

replica. This is possible since the first sequences to be replicated are the *ori C* sequences and these new pieces of DNA provide two additional replication origins.

KEY POINT

The replication eye is formed when the circular bacterial chromosome is copied in both directions (bidirectional replication) from a single initiation site. DNA is replicated by a semiconservative mechanism; the old DNA strands separate to allow new DNA to be made on these DNA templates.

DNA Superstructure

Double-stranded cellular DNAs exist as covalently closed circular molecules. Because there are no free ends, circular DNAs are not degraded by exonucleases and do not combine with other DNAs. This structure, however, poses the perplexing problem of how to unwind its double helix during its replication. The answer is that DNA unwinding is accomplished by enzymes that nick the DNA, allow it to unwind, and then rejoin it, all without letting go of the free ends.

The untwisting that occurs when two strands of DNA are pulled apart to form stretches of linear, single-stranded DNA puts positive supercoils in the DNA. Cells avoid the problem of positive supercoiling by twisting the DNA in the opposite direction, thus forming negative supercoils. Such DNA is twisted back upon itself as can be seen in the micrograph of the bacterial nucleoid (see Figure 4.3).

Negative supercoils are formed in DNA by twisting it in the direction opposite to the twists of the DNA double helix. One supercoil is found for every 15 helical twists in native DNA. The negative supercoiled state is an energized conformation that tends to unwind the helix of double-stranded DNA. A nick in either of the strands releases the supercoiling and the DNA assumes its relaxed unwound state (Figure 9.7).

The relaxed uncoiled state is converted to supercoiled DNA by the enzymatic action of a **topoisomerase** (Figure 9.8). Gyrase is a bacterial topoisomerase that supercoils DNA at the expense of ATP hydrolysis. This enzyme nicks one of the strands, moves the end around the intact strand, and then rejoins the free ends. Negatively supercoiling of DNA is an essential step in the replication of DNA.

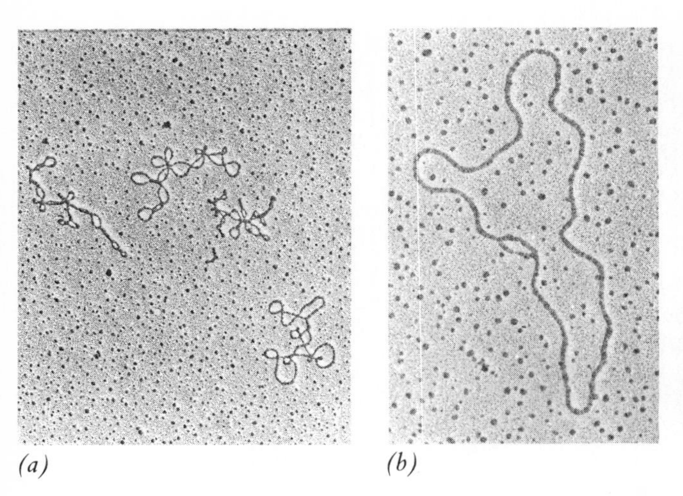
(a) (b)

FIGURE 9.7 An electron micrograph of supercoiled DNA *(a)* and relaxed DNA *(b)*. (Courtesy G. Glikin, G. Gargiulo, L. Rena-Descalzi, and A. Worcel, University of Rochester.)

Direction of DNA Replication

The independent strands of the DNA double helix are said to be antiparallel since they run in opposite directions. Each DNA strand has directionality because it has a 5′ end and a 3′ end (Figure 9.9). New strands of DNA always grow by the addition of a nucleotide to the 3′ end. The substrates for this reaction are the deoxynucleotide triphosphates, dATP, dGTP, dCTP, and dTTP, the three phosphates (P—P—P) being attached to the 5′ position of the deoxyribose. Pyrophosphate (P—P) is cleaved off as the nucleotide monophosphate is attached to the 3′ end of the growing DNA strand (Figure 9.9). The energy in the hydrolyzed phosphate bond is necessary to drive the synthesis of DNA.

KEY POINT

The energy for DNA synthesis is supplied by the nucleotide triphospate substrates whose phosphate bonds are hydrolyzed during DNA synthesis.

Proteins Involved in DNA Replication

DNA replication involves a number of proteins whose activities are summarized in Table 9.1. The single-strand binding proteins attach to the opened stretches of DNA, permitting them to be used as a template for replication. Topoisomerases put negative supercoils in DNA, which facilitate the unwinding of the double helix during replication, while unwinding at the replication fork is accomplished by **helicase.** Once unwound, short pieces of RNA, called RNA primers, are made by a special RNA polymerase called **primase.** Each RNA primer provides a 3′ end for the synthesis of new DNA (Figure 9.10) by DNA polymerase III. DNA polymerase I fills in the gaps between newly synthesized segments of DNA.

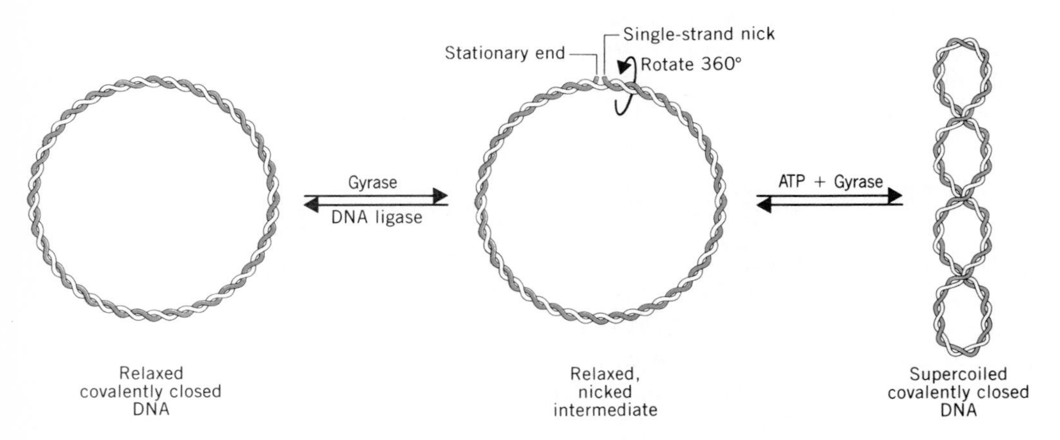

FIGURE 9.8 Relaxed covalently closed DNA is converted to supercoiled DNA through a relaxed nicked DNA intermediate. This reaction is catalyzed by a topoisomerase such as gyrase.

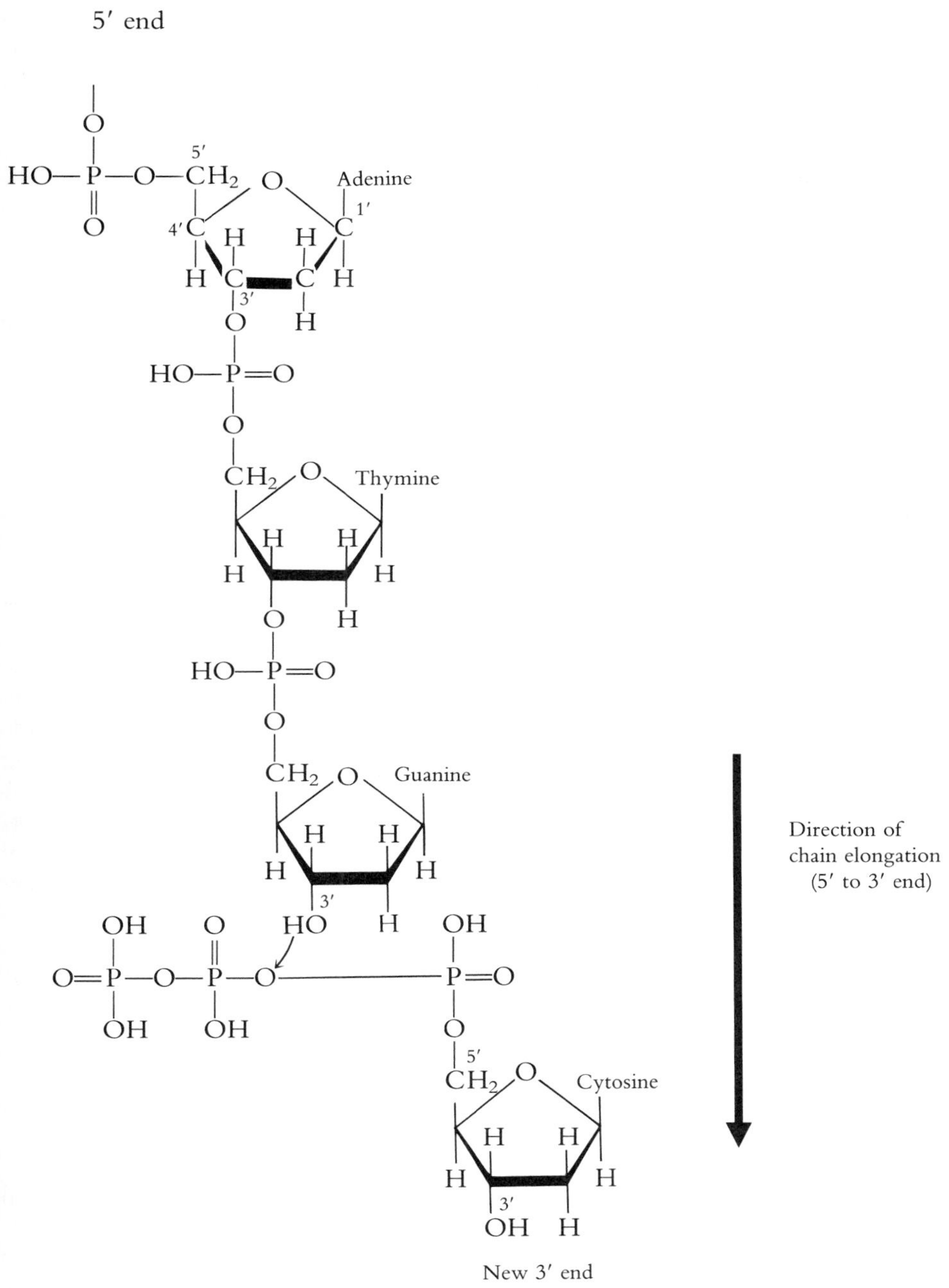

5′ end

Adenine

Thymine

Guanine

Cytosine

Direction of chain elongation (5′ to 3′ end)

New 3′ end

FIGURE 9.9 DNA is synthesized when nucleotides are added to the 3′ end of the growing nucleic acid polymer. In the process of elongation, pyrophosphate (P—P) is cleaved from the deoxyribonucleotide triphosphate.

TABLE 9.1 Major proteins involved in DNA synthesis

PROTEIN	SUBSTRATE(S)	FUNCTION
Single-strand binding proteins	Single-stranded DNA	Keeps strands separated
Topoisomerases (gyrase)	Double-stranded DNA, requires ATP	Negative supercoiling of DNA
Helicase	Double-stranded DNA	Unwinds DNA at replication fork
Primase (DNA-directed RNA polymerase)	Double-stranded DNA, ribonucleoside-5-triphosphates	Forms priming RNA
DNA polymerase III		
(1) synthetase	3′ end of priming RNA, deoxyribonucleoside-5-triphosphates	Forms new strand of DNA
(2) DNA exonuclease	Incorrect base pairing	Proofreading
DNA polymerase I		
(1) RNA exonuclease	RNA primer (5′→3′)	Removes RNA primer
(2) DNA exonuclease	Incorrect base pairing	Proofreading
(3) synthetase	Single-stranded DNA	Replaces RNA primer with DNA
DNA ligase	Single-stranded DNA fragments	Joins ends together

Bacterial DNA Polymerases Bacteria contain two DNA polymerases whose functions are understood, and DNA polymerase II whose function is unknown. Each cell contains between 10 and 20 molecules of **DNA polymerase III,** whose primary job is the synthesis of DNA. This enzyme reads the single-stranded DNA template and makes a new strand of DNA that is complementary to the old strand. For years it was thought that DNA polymerase I was a repair enzyme because mutants lacking detectable quantities of DNA polymerase I survived. Recently scientists found low qu^ntities of **DNA polymerase I** in these mutants, and now this enzyme is deemed essential for synthesizing the lagging strand of DNA. It is also involved in DNA repair and in synthesizing bacterial plasmids.

The Replication Fork To understand the roles of these specific enzymes, we must look at what happens at the replication fork. Replication occurs on both strands simultaneously with the replication fork moving away from the replication origin (Figure 9.10). In *Escherichia coli,* the **leading strand** is synthesized as a continuous polymer by DNA polymerase III, which adds nucleotides to the 3′ end of the growing DNA, the end closest to the replication fork. The second strand, called the **lagging strand,** is synthesized as discontinuous short pieces of DNA called **Okazaki fragments,** named after the scientist who discovered them. Synthesis of the lagging strand awaits the unwinding of the DNA and the formation of new RNA primers, so its formation lags the synthesis of the leading strand.

Helicases hydrolyze ATP and use the energy released to unwind the double helix at the replication fork. As the single-stranded DNA is exposed, primer RNA is laid down by primase (Figure 9.10). DNA polymerase III begins DNA synthesis by attaching new nucleotides, complementary to the template strand, to the 3′ end of the primer RNA. This process continues as long as a single-stranded template is available for the enzyme to work on. When the DNA polymerase III reaches the next primer RNA, it is replaced by DNA polymerase I, whose RNA exonuclease removes the RNA primer and replaces it with DNA. This leaves the Okazaki fragments lined up on the template strand. The Okazaki fragments then are joined by DNA ligase to form a continuous complementary DNA strand.

Proofreading DNA Both DNA polymerase I and III have proofreading functions. Before these polymerases move along the growing chain, they make sure that the last complementary match between nucleotide bases (Figure 9.11) is correct. If a mistake has occurred, the DNA polymerase excises the last nucleotide using its 3′–5′ DNA exonuclease and inserts the correct nucleotide. By this mechanism bacteria can reduce the DNA copying errors to about one error in every 1000 replication cycles.

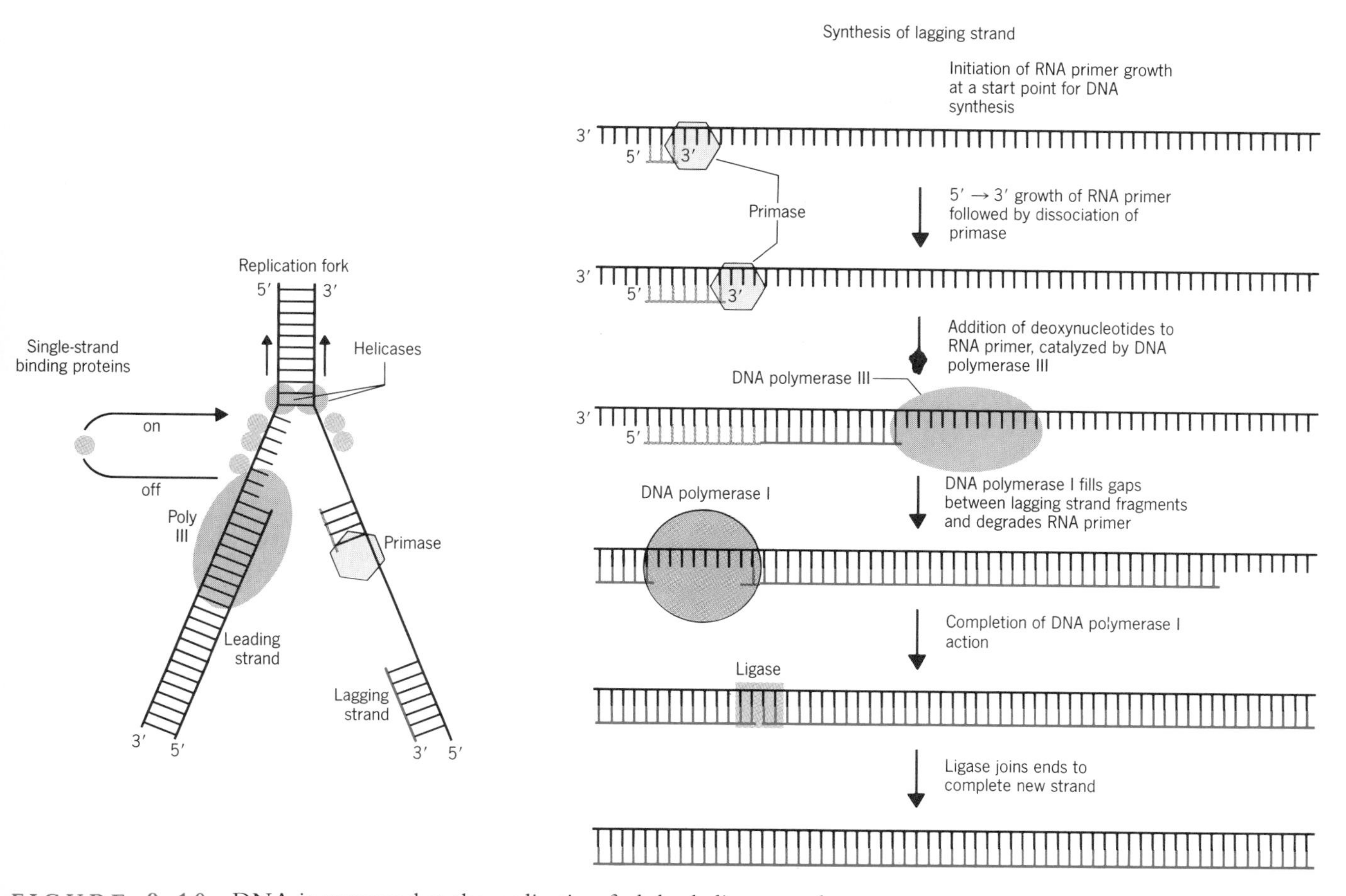

FIGURE 9.10 DNA is unwound at the replication fork by helicases at the expense of ATP. Single-strand binding proteins attach to the exposed stretches of single-stranded DNA, permitting DNA polymerase III to continuously synthesize the leading strand. The lagging strand is synthesized discontinuously on short pieces of priming RNA produced by primase. DNA polymerse III attaches deoxynucleotides to the 3′ end of the primer to synthesize a new DNA strand complementary to the template. Upon completion, DNA polymerase III is replaced by DNA polymerase I, which removes the RNA primer and replaces it with DNA. The ends of the new DNA strands are joined by ligase.

Rolling-Circle Mechanism of DNA Replication

The rolling-circle method of forming DNA results in the formation of a single-stranded DNA from a double-stranded template. The rolling-circle mechanism is used by certain double-stranded DNA bacteriophages and is probably used by bacteria to produce the single-stranded DNA involved in genetic transfer between bacteria. Replication of the double-stranded, closed-circular DNA molecule begins when a nick is put in one strand (Figure 9.12) exposing a 3′ end. DNA polymerase then adds nucleotides to the 3′ end, which effectively displaces the existing strand. The rolling-circle mechanism of DNA synthesis is used by the bacteriophage lambda, and current evidence suggests it is used for replicating the DNA transferred during bacterial conjugation.

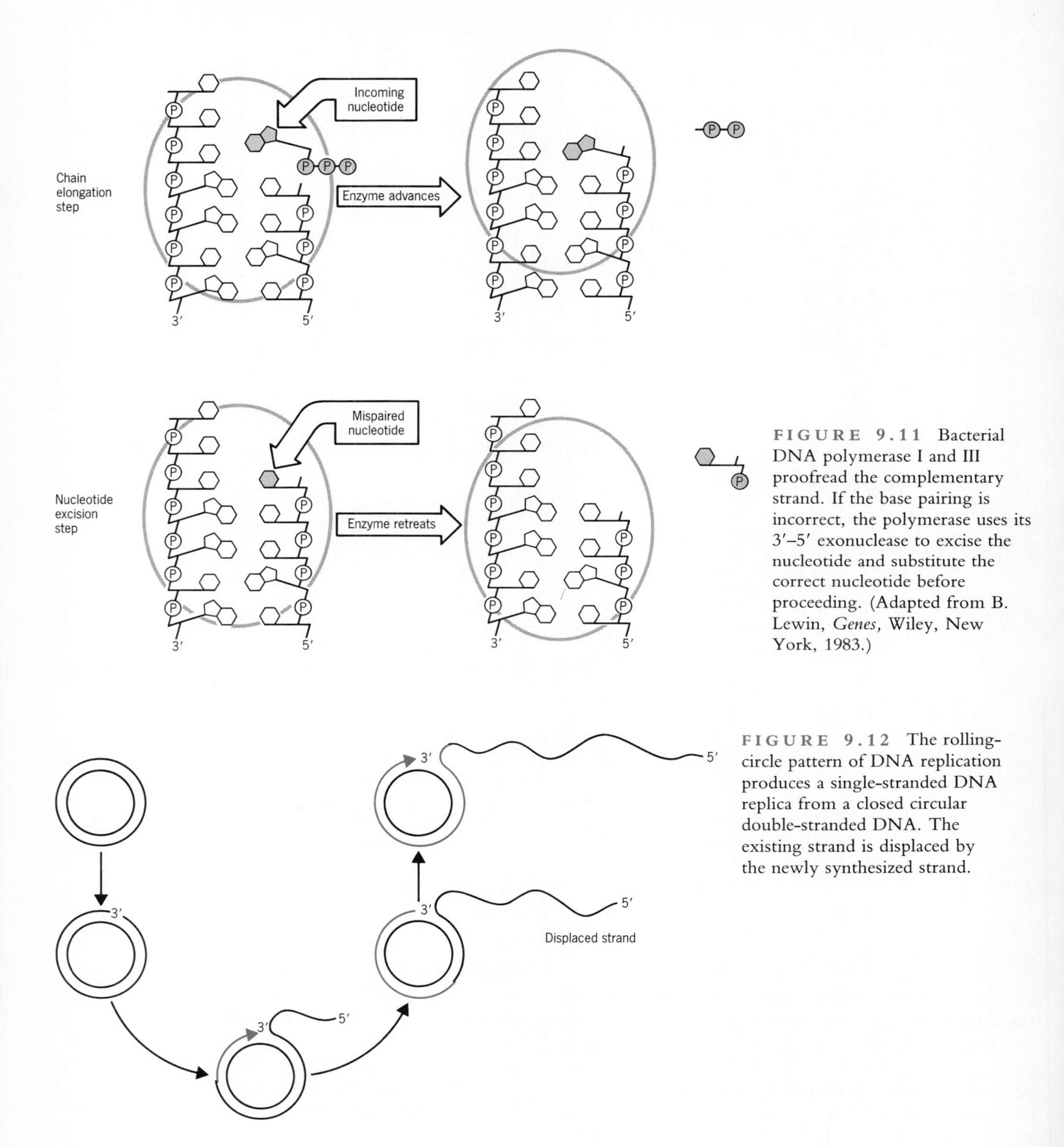

FIGURE 9.11 Bacterial DNA polymerase I and III proofread the complementary strand. If the base pairing is incorrect, the polymerase uses its 3′–5′ exonuclease to excise the nucleotide and substitute the correct nucleotide before proceeding. (Adapted from B. Lewin, *Genes,* Wiley, New York, 1983.)

FIGURE 9.12 The rolling-circle pattern of DNA replication produces a single-stranded DNA replica from a closed circular double-stranded DNA. The existing strand is displaced by the newly synthesized strand.

Heredity and DNA Synthesis

An organism's characteristics are passed from one generation to the next through the accurate replication of the cell's DNA. This DNA is organized into one or more chromosomes. One complete copy of a virus's or organism's essential DNA is referred to as its **genome.** The bacterial genome is a closed-circular DNA molecule (chromosome) that contains the essential genetic information needed by the cell to grow and reproduce. We identify genes through the traits the cell displays. For example, capsule formation, pigment production, and the ability to ferment lactose are bacterial traits. The observable traits of a cell are its **phenotype.** A cell's phenotype is encoded in its DNA and is expressed through protein synthesis.

PROTEIN SYNTHESIS

Biological information flows from DNA to RNA and from RNA to proteins (Figure 9.13). Proteins are polymers of amino acids joined by peptide bonds and arranged in an ordered sequence. The code for the sequence of amino acids in each protein is contained in DNA. This DNA code is first rewritten into a sequence of nucleotides in RNA by the process called **transcription** (Figure 9.13). The sequence of nucleotides in RNA is then decoded into a sequence of amino acids through the process of **translation,** which itself employs many different proteins and RNA molecules.

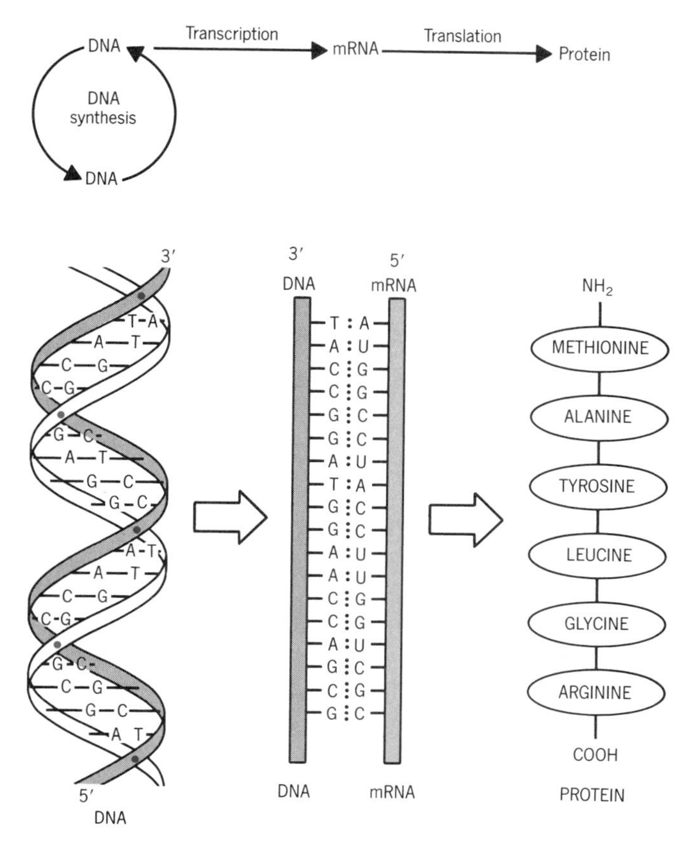

FIGURE 9.13 The flow of information from DNA to proteins

STRUCTURES OF RNAs

Ribonucleic acids (RNA) are polymers that contain adenine, uracil, guanine, and cytosine attached to a ribose–phosphate backbone. RNAs contain uracil and ribose in place of thymine and deoxyribose present in DNA. Cells contain three types of RNA (Table 9.2): ribosomal RNA (rRNA), transfer RNA (tRNA), and messenger RNA (mRNA). Each RNA is coded for by a sequence of nuleotides (gene*) in cellular DNA. The DNA is transcribed into RNA by RNA polymerase (Figure 9.13), an enzyme that acts on single-stranded DNA without requiring a primer nucleotide. This enzyme uses ribonucleotide triphosphates as its substrates and inserts a uracil into the growing RNA polymer wherever adenine appears in the DNA template.

*Genes are sequences of DNA; most code for proteins but some code for tRNAs and rRNAs.

RIBOSOMAL RNA

Ribosomal RNA plays a structural and a functional role in the bacterial ribosome. We know this because ribosomal RNAs are required for the assembly of ribosomes in test tube experiments. Procaryotic cells contain 70S ribosomes, which are composed of a 30S subunit and a 50S subunit. The 30S subunit contains a single 16S RNA molecule and 21 distinct polypeptides. The 50S ribosomal subunit contains a 5S RNA, a 23S RNA, and 31 distinct polypeptides. In addition to their structural role, ribosomal RNAs participate in the translation process: the 16S RNA is involved in binding mRNA and the 5S RNA is involved in binding tRNA.

KEY POINT

RNAs are nucleic acid polymers that differ from DNA because they contain uracil and ribose in place of thymine and deoxyribose. Each RNA is encoded by a gene in cellular DNA and is synthesized (transcribed) by RNA polymerase.

Messenger RNA

Messenger RNA (mRNA) is the link between the code contained in DNA and the sequence of amino acids in proteins. RNA polymerase makes RNA from the "sense" strand of the DNA (Figure 9.14) as the double helix sequentially unwinds. This RNA is complementary to the DNA and will specifically anneal to the single-stranded DNA from which it was formed. The RNAs that specifically code for the synthesis of proteins are called messenger RNAs because they are the transcripts of the DNA code used by the ribosomes to make proteins (Figure 9.13).

There is variation in the size of mRNA molecules, as you might expect from the variations in the molecular weights of the proteins for which they code. Once synthesized, bacterial mRNAs have a very short life span. The useful life of an mRNA in bacteria is on the order of 2 minutes, indicating that mRNAs are rapidly metabolized. This is not the case in eucaryotic cells where the mRNAs are relatively stable. Eucaryotic

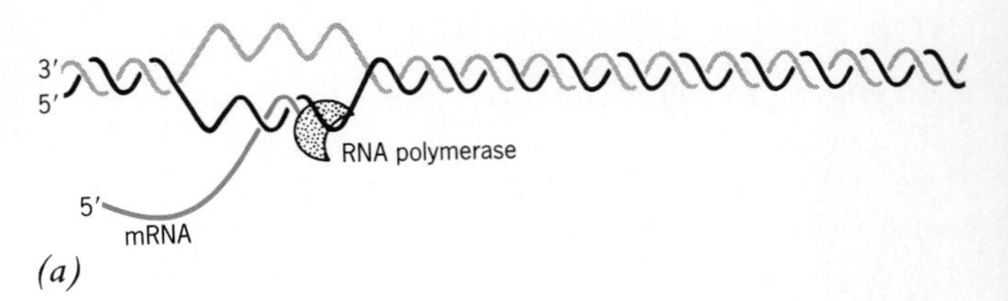

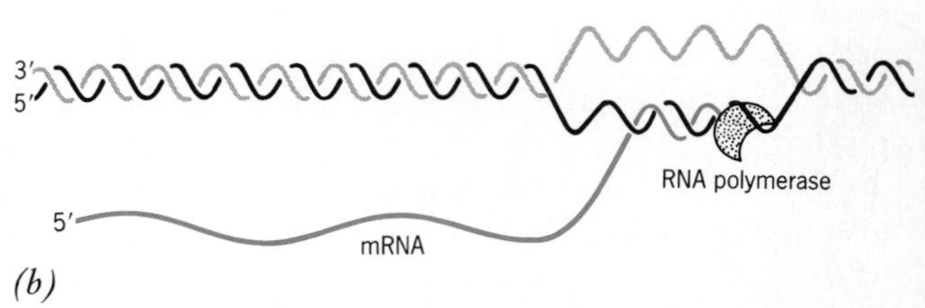

FIGURE 9.14 *(a)* Diagramatic representation of RNA-polymerase synthesizing RNA using one strand of the DNA as a template. *(b)* Electron micrograph of transcription (RNA formation) in *E. coli* of the large ribosomal RNA genes, 71,000×. The central strand is DNA; the dark objects are enzymes actively synthesizing RNA. (Courtesy of Dr. O. L. Miller. From O. L. Miller, B. A. Hamkalo, and C. A. Thomas, *Science,* 169:392–395, 1970, with permission of the American Association for the Advancement of Science.)

cells cleave their mRNAs from large precursor RNAs prior to translation so the process of transcription in eucaryotes is more complex than it is in procaryotes.

Transfer RNAs

Transfer RNAs are translator molecules because they deal with two languages: they read the nucleotide sequence of the mRNAs and in so doing arrange amino acids in the sequence directed by the mRNA. Every

TABLE 9.2 Properties of cellular RNAs

TYPE	SIZE (DALTONS)	HALF-LIFE	NUMBER OF KINDS IN CELL	FUNCTION
Ribosomal RNA (70S ribosomes)[a]	(5×10^5), 16S (4×10^4), 5S (1×10^6), 23S	Stable	3	Serves a structural and functional role in ribosomes
Messenger RNA	Variable	2 min in *E. coli*	1000	Carries code from DNA for making proteins, binds to ribosomes
Transfer RNA	(2.3 to 2.8×10^4)	Stable	Minimum of 31	Reads codon on mRNA and aligns amino acids for polypeptide synthesis

[a]The size and nature of the rRNAs in the eucaryotic 80S ribosomes are different.

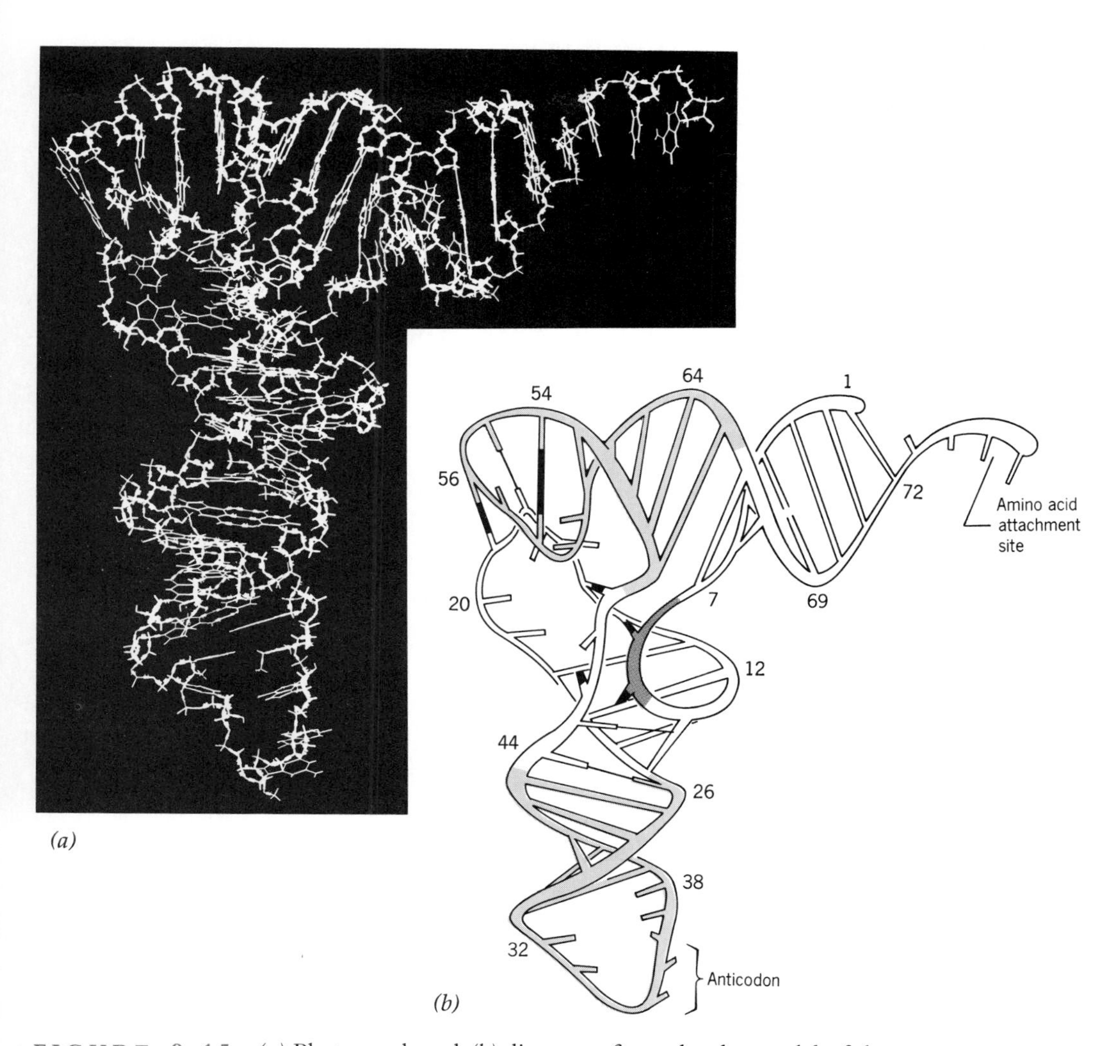

FIGURE 9.15 *(a)* Photograph and *(b)* diagram of a molecular model of the yeast phenylalanine tRNA showing its three-dimensional structure. The anticodon is at the bottom and the amino acid is attached to 3′ end of the tRNA. (Photograph reproduced with permission of Dr. S. H. Kim. From E. J. Gardner and D. P. Snustad, *Principles of Genetics,* 6th ed., Wiley, New York, 1981.)

cell has a complement of transfer RNA molecules. These small molecules contain between 73 and 93 nucleotide residues that fold back upon themselves to form a three-dimensional structure shaped like an inverted "L" (Figure 9.15). The form of the tRNAs is maintained by hydrogen bonding between nucleotide bases. Each transfer RNA binds a specific amino acid to the ribose portion of the adenosine located on the 3′ end of the tRNA. The rest of the tRNA is composed of a sequence of nucleotide residues, many of which are modified to make them chemically different from the nucleotides found in other RNAs. These unusual bases are modified by methylation after they are incorporated into the tRNA, for example, 5-methylcytidine and N^6 methyladenosine; or they are unique bases present in the nucleotide pool, for example, pseudouridine.

Each unique tRNA can carry a specific amino acid attached to its 3′ end. The amino acid is attached enzymatically to its tRNA by one of the 20 or more distinct **amino-acyl-tRNA synthetases** present in each cell. These enzymes are specific for one amino acid and its proper tRNA. The attachment of the amino acid to the tRNA to make a "charged" tRNA is an energy-dependent reaction driven by ATP hydrolysis.

The middle segment of the tRNA polymer contains a sequence of three nucleotides known as the **anticodon.** This is the interpretive part of the tRNA molecule since the anticodon recognizes a complementary sequence of nucleotides on the mRNA called the **codon.** Transfer RNAs are the link between the codons in mRNA and the amino-acid sequences of proteins.

The number of different tRNAs needed for protein synthesis is not known precisely. Since there are 20 different amino acids in proteins, there must be at least 20 different tRNAs. Theoretically, a cell requires a minimum of 31 tRNAs to recognize the 61 codons (exclusive of initiators). For a comparison of the properties of cellular RNAs, refer to Table 9.2.

THE GENETIC CODE IS UNIVERSAL

Somehow the sequence of nucleotides in DNA must encode messages for the protein-synthesizing apparatus. The information in DNA is transcribed into messenger RNA, and here is where we find the genetic code. Biological systems use a universal **triplicate code** composed of three-base codons written in mRNA.

The need for a triplicate code can be rationally deduced. Since there are 20 different amino acids in proteins, there must be at least 20 code words. There are four bases in messenger RNA* that can be used to write the code. If one base coded for one amino acid, there would be four possible code words, 4^1. If two bases coded for each amino acid, there would be 16 possible code words, 4^2. However, 16 is not enough to code for 20 amino acids. Therefore, *three* is the minimum sequence needed to provide codes for 20 amino acids when working with four nucleotide bases. This triplicate code provides 64 possible code words, 4^3, for cells to use in translating messenger RNA into proteins (Table 9.3).

*The genetic code of mRNA is described because this is the location of the codon.

The triplicate code is universal in almost all biological systems studied to date. This means that the genes of human cells use the same code bacterial cells use. This is very fortunate for it enables genetic engineers to do many experiments that would be impossible if multiple genetic codes existed.

The genetic code is **redundant;** that is, there is more than one code word for each amino acid—except for tryptophan and methionine, which have only one code word each (Table 9.3). The code has been deciphered, and each possible triplicate sequence codes for an amino acid except the code words UAA, UAG, and UGA. These code words are signals to terminate the message. The signal that initiates the message is also special; in bacteria it is AUG, which codes for methionine.

Number of tRNAs

How many tRNAs, or how many anticodons, are required for protein synthesis? Sixty-one of the code words code for amino acids, but that does not mean that there are 61 tRNAs. Protein synthesis requires a minimum of 32 tRNA anticodons to code for the 20 amino acids.

By studying the genetic code (Table 9.3), one can see that a change in the third base in a code word does not necessarily change the amino acid it codes for. For example, the first two bases (UC__) always code for serine regardless of the third base. This enables one tRNA molecule to read more than one code word for the same amino acid.

A complete analysis of how the genetic code operates has shown that cells need a minimum of 32 tRNAs. One of these is a specific initiator tRNA that recognizes the initiator sequence on the mRNA. The initiator tRNA, known as $tRNA_f^{Met}$, begins each polypeptide with the modified amino acid, *N*-formyl methionine.

In summary, the genetic code is written as a triplicate sequence of nucleotides in messenger RNA. The code uses uracil in place of thymine found in DNA. The triplicate code in mRNA is called the codon; it is recognized by the anticodon present on a tRNA molecule. The genetic code is redundant; that is, there is more than one codon for each amino acid. Some tRNAs read more than one codon, so a cell needs a minimum of 32 tRNAs to read the 61 codons.

TABLE 9.3 The genetic code

FIRST POSITION 5′ END	SECOND POSITION U		C		A		G		THIRD POSITION 3′ END
U	UUU	phe	UCU	ser	UAU	tyr	UGU	cys	U
	UUC	phe	UCC	ser	UAC	tyr	UGC	cys	C
	UUA	leu	UCA	ser	UAA	ter[a]	UGA	ter	A
	UUG	leu	UCG	ser	UAG	ter	UGG	trp	G
C	CUU	leu	CCU	pro	CAU	his	CGU	arg	U
	CUC	leu	CCC	pro	CAC	his	CGC	arg	C
	CUA	leu	CCA	pro	CAA	gln	CGA	arg	A
	CUG	leu	CCG	pro	CAG	gln	CGG	arg	G
A	AUU	ile	ACU	thr	AAU	asn	AGU	ser	U
	AUC	ile	ACC	thr	AAC	asn	AGC	ser	C
	AUA	ile	ACA	thr	AAA	lys	AGA	arg	A
	AUG	met	ACG	thr	AAG	lys	AGG	arg	G
G	GUU	val	GCU	ala	GAU	asp	GGU	gly	U
	GUC	val	GCC	ala	GAC	asp	GGC	gly	C
	GUA	val	GCA	ala	GAA	glu	GGA	gly	A
	GUG	val	GCG	ala	GAG	glu	GGG	gly	G

[a]Chain termination.

KEY POINT

The universality of the genetic code is a key argument for the common origin of both eucaryotes and procaryotes. The same genetic code is found in viruses, which use the biochemical machinery of cells to make their proteins.

BACTERIAL PROTEIN SYNTHESIS

Protein synthesis takes place on the ribosomes in the bacterial cytoplasm during translation, the process for converting the genetic code in mRNA to the correct amino acid sequence in proteins. A growing bacterium has about 20,000 ribosomes and 80 to 90 percent of these are actively synthesizing proteins at any given time. Single ribosomes can read an mRNA; however, most mRNAs are read by more than one ribosome (Figure 9.16) that sequentially attach to a single mRNA. These complexes are called **polyribosomes.** Protein synthesis is a rapid process in which 12 to 17 amino acids per second are added to the polypeptide chain. This means that a 300-residue polypeptide can be synthesized in about 20 seconds.

The process of protein synthesis can be separated into three separate phases. **Initiation** of protein synthesis occurs when the ribosome, mRNA, and the first amino-acyl tRNA join together. **Elongation** is the complex process in which amino acids are added to the growing polypeptide chain. **Termination** of protein synthesis results in the release of the polypeptide and recycling of the ribosome. The details of protein synthesis can best be understood by looking at each step independently.

Initiation Requires a Special tRNA

Initiation of bacterial protein synthesis involves protein initiation factors (IF), 70S ribosomes, *N*-formyl methionine tRNA, and GTP. The 70S bacterial ribosomes exist in a dynamic equilibrium with their 30S and 50S subunits (Figure 9.17), so that both subunits are available for initiation.

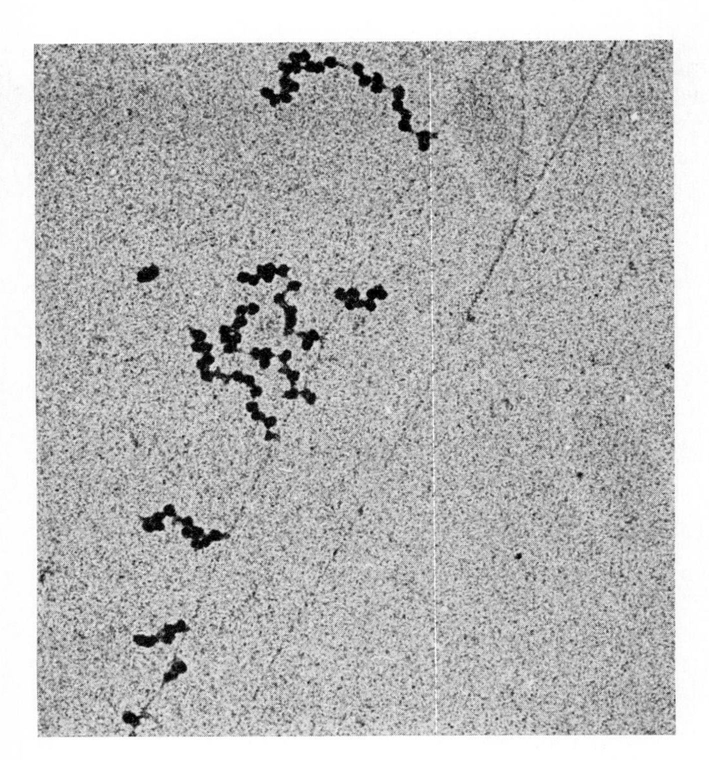

FIGURE 9.16 Electron micrograph of transcription and translation of *E. coli* genes, 165,000×. The mRNA is transcribed from one of the two DNA strands at the same time that polyribosomes translate this RNA message into protein. (Courtesy of Dr. O. L. Miller. From O. L. Miller, B. A. Hamkalo, and C. A. Thomas, *Science,* 169:392–295, 1970, with permission of the American Association for the Advancement of Science.)

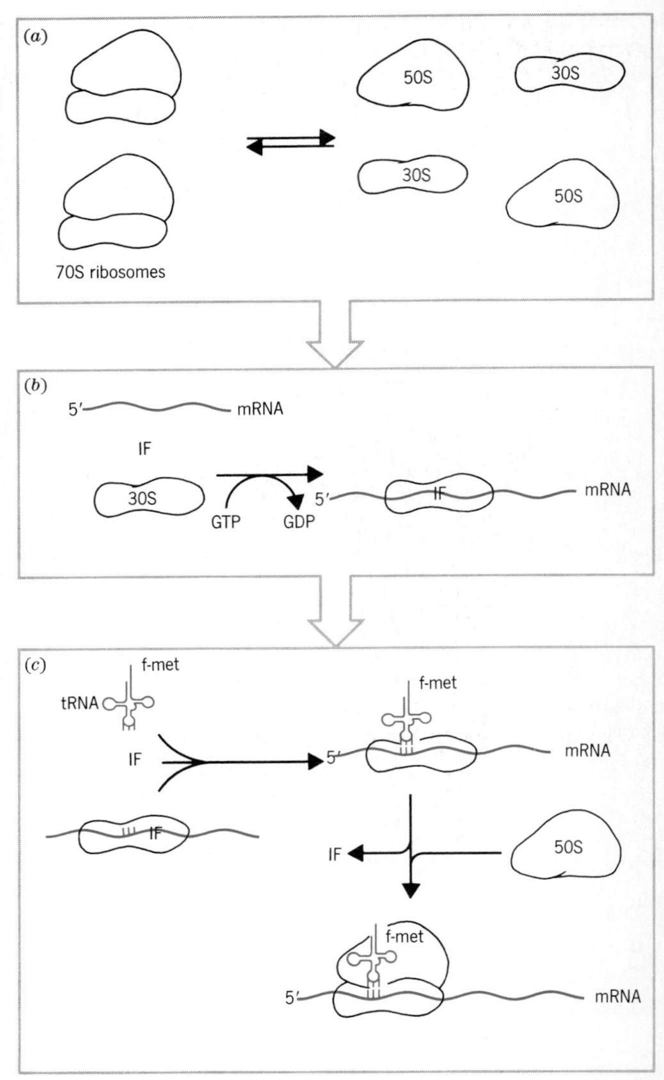

FIGURE 9.17 Initiation of protein synthesis. *(a)* Bacterial ribosomes (70S) dissociate to form an equilibrium with their two subunits. *(b)* The 30S subunit, an initiation factor (IF), and mRNA combine to form a complex. *(c)* The initiation codon on the mRNA is recognized by N-formyl methionine tRNA bound to an initiation factor. This tRNA binds to the 30S ribosomal-mRNA complex at the expense of GTP. The 50S ribosomal subunit joins this complex to begin protein synthesis. The initiation factors are released and recycled.

The 30S subunit binds to the 5′ end of mRNA by recognizing the codon (AUG) that codes for *N*-formyl methionine. *N*-formyl methionine is the amino acid that begins each bacterial polypeptide; however, either it is cleaved from the *N*-terminal end once the protein is formed or the formyl group is removed. The initiator tRNA, **$tRNA^{Met}_{f}$,** combines with an initiation factor and then binds to the 30S ribosomal subunit mRNA complex (Figure 9.17), at the expense of GTP hydrolysis. Initiation is completed when the 50S subunit joins the complex, after which the initiation factors are released and recycled.

Peptide Elongation Occurs on Ribosomes

Each protein is synthesized by sequentially attaching the correct amino acid to the growing polymer beginning at the amino-terminal end and progressing toward the carboxyl-terminal end. The entire process takes place on the 70S ribosome, which has two sites for binding tRNAs to the ribosomal–mRNA complex

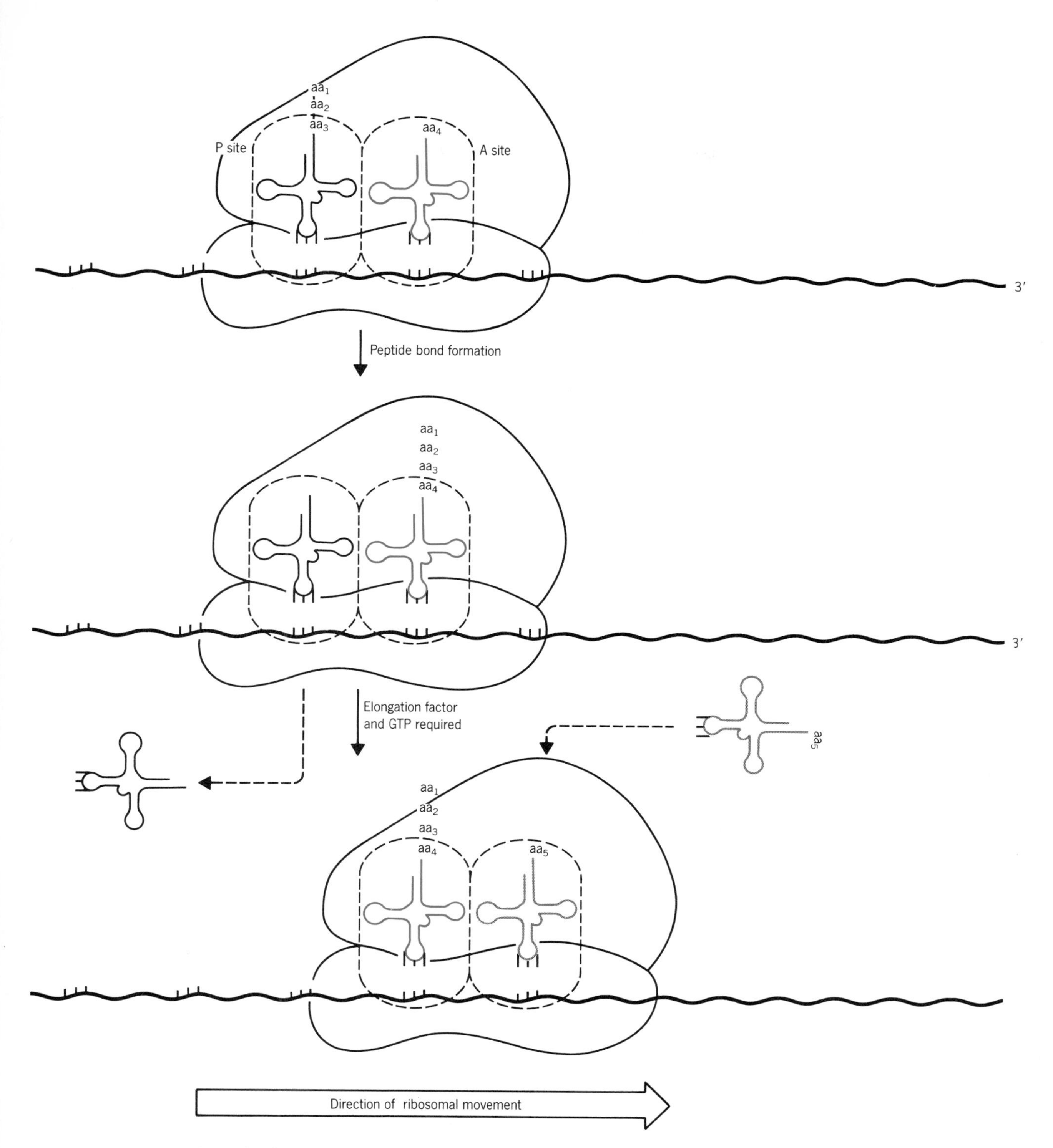

FIGURE 9.18 Peptide elongation occurs on the 70S ribosome. The P site holds the peptidyl tRNA, while the proper charged aminoacyl-tRNA is positioned at the A site. A peptide bond is formed as the peptide is transferred to the aminoacyl-tRNA aligned at the A site. The uncharged tRNA is released. The ribosome now moves along the mRNA one codon. This movement is driven by GTP hydrolysis and requires an elongation factor.

(Figure 9.18). The **A site** is the entry site that accepts charged amino-acyl tRNA whose anticodon recognizes the codon on the mRNA. One molecule of GTP is hydrolyzed every time an amino-acyl tRNA binds to the ribosome–mRNA complex. The **P site** carries the growing peptide chain (peptidyl–tRNA).

The ribosome moves on the mRNA toward the 3′ end so that each codon moves from the A site to the P site, making a new codon available at the A site. The movement occurs after the amino acid is attached to the growing polypeptide chain by **peptidyl transferase.** This enzymatic activity is a function of a protein of the 50S ribosomal subunit. Once the peptide bond is formed, an elongation factor and GTP participate in moving the ribosome to the next codon. As this translocation occurs, the used tRNA is released into the cytoplasm to be recharged with its specific amino acid. This sequential process continues until the ribosome complex reaches a termination sequence.

Peptide Termination

Any of three termination codons (Table 9.3) are read by protein **release factors** (RF) that recognize the nonsense codons (UAA, UAG, and UGA). These factors cause the release of the peptide from its tRNA, which is followed by the dissociation of the ribosomal subunits and the mRNA. The ribosomal subunits and the factors involved in protein synthesis are now free to initiate protein synthesis on another mRNA.

KEY POINT

The genetic information in DNA is transcribed into messenger RNA by RNA polymerase. The genetic code written in mRNA is read by tRNAs during translation, the process of converting the genetic code in mRNA to the correct amino acid sequence of proteins.

INHIBITION OF PROTEIN SYNTHESIS

Cells are dependent on the continual synthesis of proteins for growth and reproduction. Irreversible inhibition of protein synthesis, at the level of either transcription or translation, will kill a cell. **Antibiotics** are naturally produced microbial substances that inhibit or kill other organisms. Their amazing effectiveness in curing infectious diseases has had a tremendous impact on modern medicine. A number of antibiotics act by inhibiting protein synthesis.

Actinomycin D is an antibiotic produced by the bacterium *Streptomyces antibioticus* (strep-toh-my′sees). **Actinomycin D** binds to DNA and prevents mRNA synthesis. **Rifamycins** (ri′fah-my-sin) are antibiotics that bind to the bacterial RNA polymerase, preventing the initiation of mRNA formation. These antibiotics inhibit transcription.

Inhibitors of translation are useful drugs for treating animals with bacterial infections because these inhibitors often act on the 70S ribosomes of bacterial cells and do not inhibit protein synthesis on the 80S ribosomes of eucaryotic cells. **Streptomycin** binds to the 30S ribosomal subunit and prevents the dissociation of the 70S ribosome. **Tetracyclines** are antibiotics that prevent the binding of charged tRNAs to the A site on the 70S ribosome and thus stop peptide chain elongation. **Chloramphenicol** prevents protein synthesis by interfering with the peptidyl transferase reaction associated with the 50S ribosomal subunit. This is by no means a complete presentation of antimicrobial inhibitors of protein synthesis (see Chapter 21 for further details), but it indicates the importance of antibiotic inhibitors of bacterial protein synthesis in treating bacterial infections.

REGULATION

DNA contains all the information needed by a cell to grow and reproduce. Most cells contain additional information that is expressed only under certain conditions. This is obvious to us when we consider the fertilized egg (a zygote) of a frog. Even though a frog zygote appears round and egglike, we know that it will develop into a multicellular organism. All the DNA sequences present in the zygote are also present in each cell of the adult frog, but only certain information is expressed by these cells. We must conclude that regulatory processes control the expression of the genes contained in DNA. Since the code is expressed through the synthesis of proteins, one can conclude that there are regulatory systems controlling the synthesis of proteins, and indeed regulatory mechanisms for protein synthesis have been found in all cells studied.

REGULATION OF PROTEIN SYNTHESIS

Many proteins are formed at a constant rate in growing bacteria. These are the **constitutive proteins** always found in the cell, regardless of the conditions of growth. Enzymes of essential metabolic pathways such as glycolysis and structural proteins are synthesized constitutively and appear to be under little or no control other than the rate of cell growth. The synthesis of proteins that are needed occasionally is regulated by the cell to preserve metabolic resources. For example, (1) the enzymes needed to metabolize lactose are synthesized only when lactose is present in the medium and (2) the enzymes needed to produce amino acids are not produced when that amino acid is available in the cell's growth environment.

Cellular regulation of protein synthesis controls transcription and thereby acts at the level of DNA. These control mechanisms depend on the order of the genes on the chromosome, the products of regulatory genes, and on the activity of RNA polymerase.

Bacterial RNA Polymerase

Bacteria contain a single type of RNA polymerase, the enzyme responsible for synthesizing the three types of RNA. Bacterial RNA polymerase is composed of five polypeptides that function as two distinct units; the **core enzyme** and the **sigma factor** (Figure 9.19). The core enzyme does the actual transcription of the genes on DNA; however, by itself it cannot initiate transcription. Only when the core enzyme is bound to the sigma factor does RNA polymerase recognize and bind to a promoter sequence on the DNA. The sigma factor is released once the core enzyme begins to transcribe the DNA into RNA (Figure 9.19). When RNA polymerase reaches a terminator sequence on the DNA, both the core enzyme and the RNA dissociate from the DNA.

Each *Escherichia coli* cell contains about 7000 molecules of RNA polymerase. At any given time, a large proportion of these enzymes are synthesizing RNA at numerous locations on the cell's DNA.

The Operon Model for Controlling Protein Synthesis

Cells are selective in the types and quantities of proteins they manufacture, and thereby conserve their

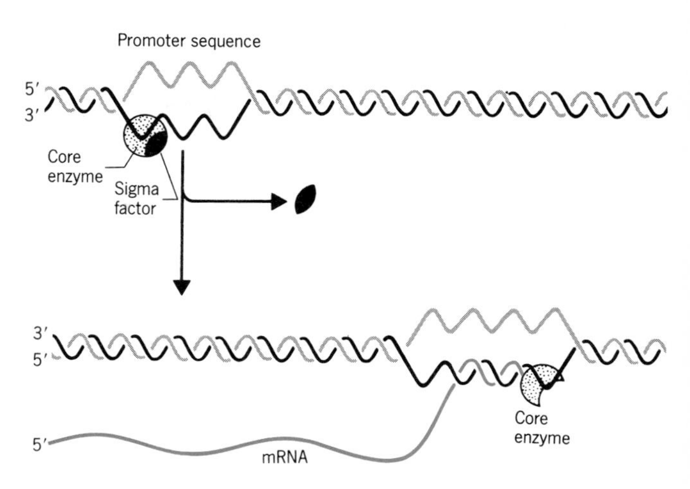

FIGURE 9.19 Initiation of RNA synthesis requires the sigma factor and RNA polymerase. The core enzyme–sigma factor complex recognizes the promoter sequence. The core enzyme then proceeds to transcribe the DNA.

limited energy resources. One level of control of protein synthesis occurs at the level of transcription (DNA → mRNA). Proteins synthesized under this control are encoded by genes organized in operons on the DNA. An **operon** is a sequence of genes dealing with a single cellular function that is controlled by a regulator gene (Figure 9.20). The operon consists of a **promoter** site and an **operator** site adjacent to the structural genes, and a regulator gene, which can be located anywhere on the chromosome. A regulator gene codes for the synthesis of a **repressor** protein that is synthesized constitutively—that is, it is always present in the cell.

Each repressor can exist in an active form that binds to the operator site on the DNA and an inactive form that doesn't bind to DNA. Binding of an active repressor to the DNA prevents RNA polymerase from transcribing the structural genes (Figure 9.20). In the absence of an active repressor, RNA polymerase binds to the promoter site on the operon and transcribes the structural genes into mRNA, thus initiating protein synthesis. The repressor is the switch that turns the operon on or off.

Bacterial operons can be functionally divided into two types. Enzymes involved in breaking down (catabolism) a substrate are encoded on operons regulated by enzyme induction. **Enzyme induction** increases the cellular concentration of specific enzymes in response to the presence of the substrate, which is called

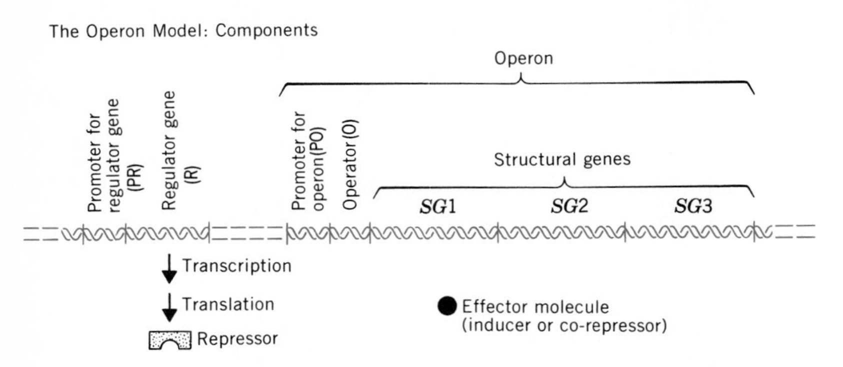

FIGURE 9.20 Diagramatic representation of the operon model for gene regulation. An operon is a group of one or more structural genes preceded by operator and promoter sequences. The promoter for the operon (PO) is the sequence to which RNA polymerase binds. The operator (O) is the sequence that binds the repressor protein. The regulator gene (R) and its promoter (RP) can be located anywhere on the genome. (From E. J. Gardner and D. P. Snustad, *Principles of Genetics,* 6th ed., Wiley, New York, 1981.)

the **inducer.** The cell makes these enzymes only when it needs to metabolize the substrate (inducer) as a source of energy and carbon. Biosynthetic (anabolic) pathways are controlled by enzyme repression and feedback inhibition (Figure 9.21). When the end product(s) of a biosynthetic pathway accumulate in the cell, it no longer needs to produce that substance. In **enzyme repression** the end product shuts down the operon and the cell stops producing the enzymes of the biosynthetic pathway. The end product also inhibits key enzymes of the biosynthetic pathway already present in the cell by **feedback inhibition.**

Enzyme Induction

Lactose metabolism is the classic example of enzyme induction (Figure 9.22). This regulatory system was discovered by two scientists who observed that *Escherichia coli* metabolized lactose only after all the glucose in the medium was consumed. Lactose is a disaccharide composed of galactose linked to glucose through a β 1—4 glycosidic bond, whose metabolism is governed by the *lac* operon. β-galactosidase breaks the β 1—4 glycosidic bond of lactose, producing 1 galactose and 1 glucose molecule. The other two enzymes of the *lac* operon are galactoside permease, which transports lactose into the cell, and galactoside transacetylase, the importance of which is unknown. These three enzymes are induced in *E. coli* when lactose is the substrate for growth. Lactose metabolism is controlled by the *lac* operon (Figure 9.22), which consists of a promoter and an operator gene that precede the structural genes for the three enzymes. When glucose and lactose are both present in the medium, cells preferentially use glucose.

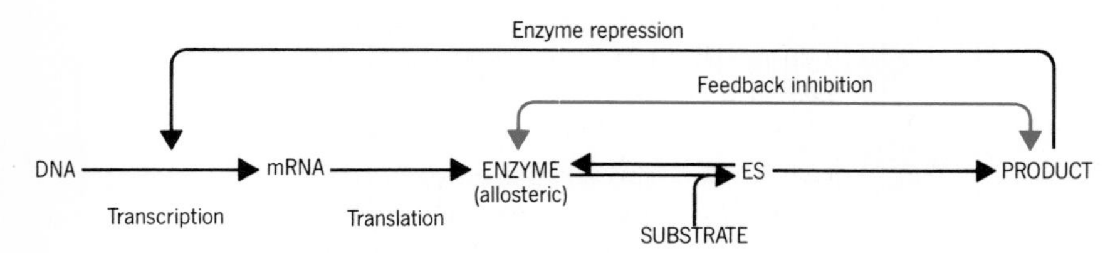

FIGURE 9.21 Enzyme repression acts at the level of transcription, whereas feedback inhibition affects the activity of an allosteric enzyme.

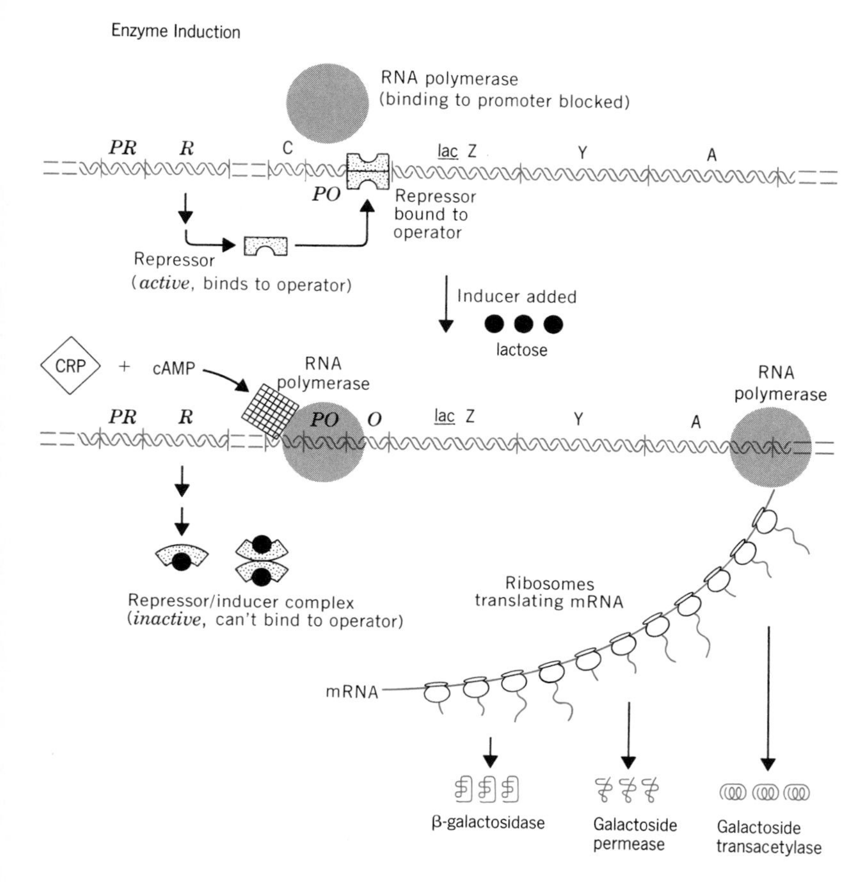

FIGURE 9.22 Enzyme induction. The repressor (protein) formed byt he lactose regulator (R) gene is *active;* when it binds to the operator (o) it prevents the trsncription of mRNA from the DNA. The inducer (lactose) combines with and inactivates the repressor to prevent it from binding to the operator. Under these conditions, RNA polymerase binds to the promoter (PO), ready to transcribe the structural genes into mRNA. Since the *lac* operon is controlled by catabolite repression, the cAMP receptor protein (CRP) must be bound to the C region before RNA polymerase can operate. CRP becomes active when it binds cAMP. Now mRNA for the *lac* Z, Y, and A genes is produced, and the enzymes needed to metabolize lactose are synthesized. (Modified from E. J. Gardner and D. P. Snustad, *Principles of Genetics,* 6th ed., Wiley, New York, 1981.)

The regulator gene (R), which is expressed continuously, codes for the synthesis of an *active* repressor protein, which binds to the operator gene and prevents the initiation of mRNA synthesis. When lactose is present, it binds to and thereby inactivates the repressor protein—now the repressor cannot bind to the operator. This permits RNA polymerase to bind at the promoter and initiate transcription of the structural genes. The resulting mRNAs are translated into β-galactosidase, galactoside permease, and galactoside transacetylase. The cell is now able to utilize lactose, the inducer of this operon.

Escherichia coli preferentially metabolizes glucose over other sugars, such as lactose, maltose, and arabinose, when these sugars and glucose are available in the medium. This is referred to as the **glucose effect** or **catabolic repression.** In the lactose system, glucose indirectly prevents the cell from synthesizing the enzymes of the lactose pathway.

Cyclic AMP Is the Secondary Message The glucose effect is mediated by the cellular concentration of cyclic AMP (cAMP), a secondary message. The cell makes a **cyclic AMP receptor protein** (CRP) that binds cAMP and acts as a switch to turn on those operons that metabolize secondary sugars. When the cell contains a high concentration of cAMP, CRP binds cAMP to form the cAMP–CRP complex. This complex then binds to the sequence of DNA before the promoter region and permits RNA polymerase to bind to the promoter (Figure 9.22). This switch must be on for the cell to transcribe the *lac* operon. The switch is turned off, even in the presence of lactose, when the

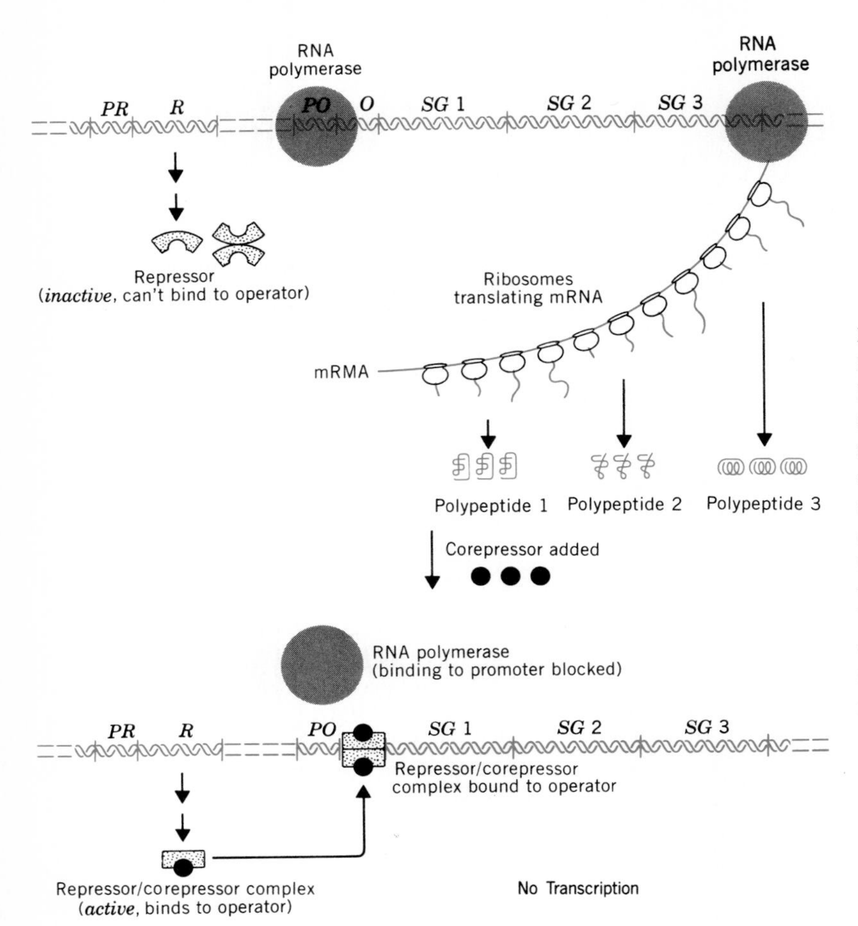

FIGURE 9.23 Enzyme Repression. Synthetic pathways are often controlled by enzyme repression. The repressor formed by the repressor gene (R) is *inactive*; i.e., it will not bind to the operator (O). Under these conditions, the enzymes of the synthetic pathway are synthesized. The operon is turned off in the presence of the corepressor, which is usually the end product of the biosynthetic pathway controlled by the operon. The corepressor binds with the repressor protein to form an *active* repressor. The active repressor binds to the operator, which prevents RNA polymerase from binding to the promoter (PO). No transcription of the structural genes (SG) occurs in the presence of the corepressor. (From E. J. Gardner and D. P. Snustad, *Principles of Genetics*, 6th ed., Wiley, New York, 1981.)

cellular concentration of cAMP is low because the cAMP–CRP complex is not formed.

The concentration of cAMP is low in cells growing on glucose, so glucose indirectly affects the induction of the *lac* operon. When both glucose and lactose are present, the glucose is preferentially metabolized as the substrate because there is insufficient cAMP to form the cAMP–CRP complex, and no β galactosidase is synthesized. As the glucose is utilized, the cell begins to make more cAMP and the *lac* operon is turned on. Now the cells make the enzymes of the *lac* operon, and lactose is utilized as the substrate.

Enzyme Repression

When a cell contains excess amounts of a metabolite, the synthesis of the biosynthetic enzymes of the pathway making that metabolite is stopped by **enzyme repression** (Figure 9.23). This mechanism controls the availability of amino acids, nucleotides, and other metabolites in cells. For example, cells growing in glucose–salts minimal medium contain the enzymes necessary to manufacture all the amino acids. If an excess of a given amino acid is added to the cell's environment, the cell can assimilate the amino acid directly, so it no longer needs to manufacture the enzymes required to produce it. An amino acid (end product) functions as a **corepressor** when it inhibits the synthesis of the enzymes used in its production.

Enzyme repression acts at the level of DNA transcription. Operons controlled by enzyme repression produce a repressor protein that is *inactive* until it combines with the corepressor (Figure 9.23). Reaction of the repressor with the corepressor activates the repressor, causing it to bind to the DNA at the operator site. This prevents the further transcription of the operon, so synthesis of the enzymes of the biosynthetic pathway ceases.

KEY POINT

The repressor protein that controls an inducible enzyme pathway is synthesized in an active form (inactivated by inducer), whereas the repressor protein involved in enzyme repression is inactive until it combines with its corepressor (usually an end product of the biosynthetic pathway).

Enzyme induction and enzyme repression are mechanisms that regulate protein synthesis in response to environmental factors. Cells also regulate their cytoplasmic inventory of small molecules by controlling the rate of enzyme activity. This control is exerted on the enzymes as they participate in chemical reactions.

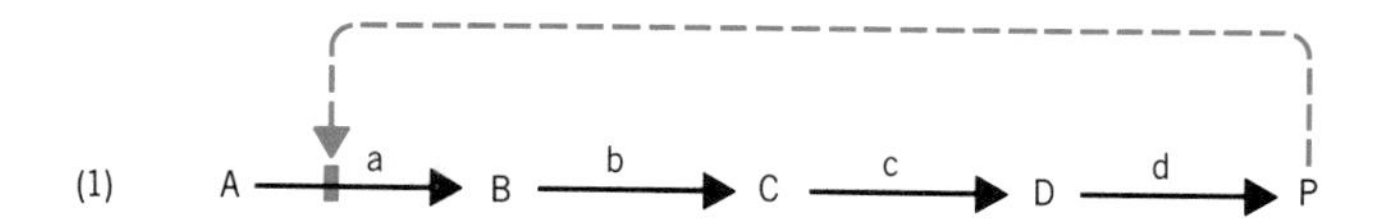

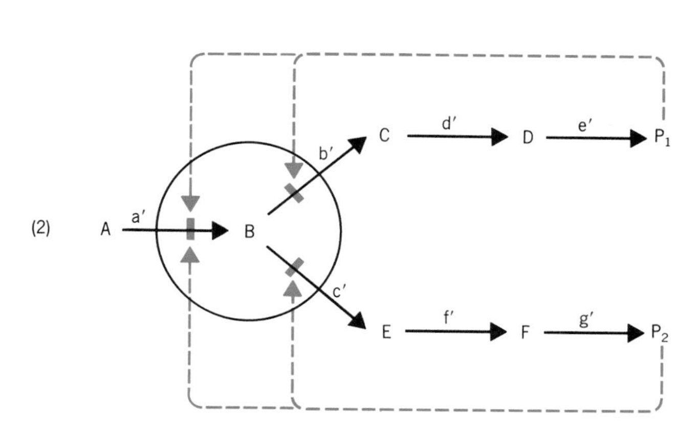

FIGURE 9.24 Feedback inhibition controls the production of small molecules. The first (allosteric) enzyme in a metabolic is inhibited by the end product(s) of the pathway. In the linear pathway (1), a would be the allosteric enzyme and P would be the allosteric effector. In the branched pathway (2), a′, b′, and c′ are allosteric enzymes differentially affected by P_1 and P_2.

REGULATION OF ENZYME ACTIVITY

Allosteric Enzymes

Enzymes strategically located in metabolic pathways often respond to the end product(s) of their pathway. This type of control is referred to as end-product or **feedback inhibition** (Figure 9.24). The enzyme affected is usually the first enzyme (a) in a linear pathway or the enzyme (a′) just prior to a branch point in a multiproduct pathway (Figure 9.24). These regulated enzymes possess one or more allosteric sites, distinct from their catalytic sites, to which a metabolite binds. **Allosteric enzymes** are controlled by feedback inhibition when the regulatory metabolite binds at an **allosteric site.** Allosteric enzymes control the economy of small molecules within a cell and can be catalysts in either linear or branched metabolic pathways.

Covalent Modification Controls Key Enzymes

Key enzymes positioned at a metabolic crossroad, or that utilize large amounts of energy, can be under additional control. These enzymes are covalently modified through the action of other enzymes that catalyze the covalent attachment of groups to the regulated enzyme. Covalent modification interconverts the enzyme between an active and an inactive state (Figure 9.25). This regulation is more permanent than allosteric regulation because reversal requires enzyme action.

An example of an enzyme controlled by covalent modification is glutamine synthase, a key enzyme in the assimilation of ammonia into organic compounds. Glutamine synthase (Figure 9.25) is a dodecamer (12 subunits) arranged as a double doughnut that can be covalently modified by an adenylating enzyme, which attaches one adenylate group to the exposed tyrosine residue on each of the twelve subunits. The completely adenylated enzyme [$AMP_{(12)} - GS$] is inactive. The enzyme is partially active when one or more of the subunits are not adenylated. Covalent modification thereby modulates the enzymatic activity of glutamine synthase.

KEY POINT

The commercial production of primary microbial end products, such as lysine, glutamic acid, and citric acid, is made possible by circumventing the biochemical mechanisms that control the metabolic pathways involved.

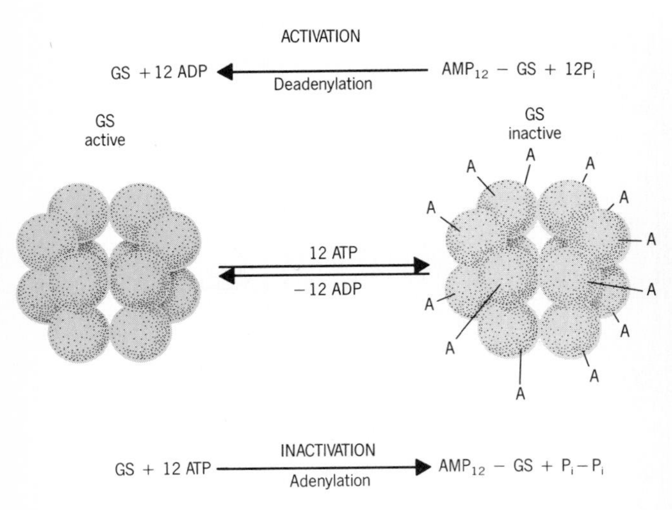

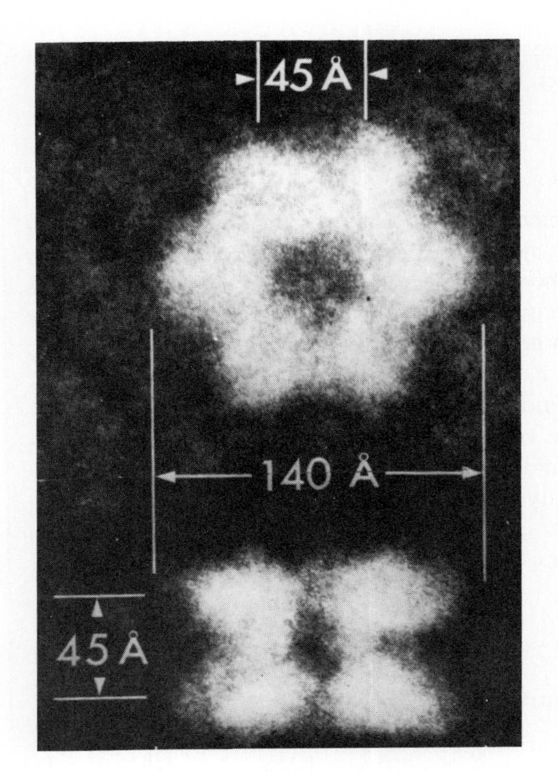

FIGURE 9.25 Glutamine synthase (GS) is a dodecamer that is interconverted between an active and an inactive form by covalent modification. An adenylating enzyme inactivates GS by adding adenyl groups (A) to tyrosine residues on each of the 12 subunits. The adenylated enzyme is activated by deadenylation. The insert is an electron micrograph of glutamine synthase. (Electron micrograph courtesy of Dr. Earl Stadtman.)

CONTROL AT THE CELLULAR LEVEL

Biologists now know that cell activities are intricately controlled. The examples of control presented here are multiplied many times over within each cell. Control over transcription, translation, and enzymatic activity is used by a cell to serve the immediate needs of that cell. Cells can respond almost immediately to their environment even when drastic changes occur.

Regulation in Eucaryotes

Multicellular organisms present more complex problems that require more elaborate regulatory systems. For example, these organisms possess hormones that simultaneously stimulate many cells to synthesize specific proteins. Multicellular organisms often go through complex developmental processes, many of which are poorly understood from the regulatory viewpoint. Experiments showing how eucaryotes process their genetic information after transcription open up many theoretical possibilities for explaining their complex regulatory phenomena.

The basic concepts of molecular biology have become firmly established during the years following the discovery of DNA. Much of our knowledge about heredity can now be explained by molecular biology. The application of this knowledge is manifest in the rapidly developing field of genetic engineering. Scientists use molecular techniques to modify DNA and the hereditary traits of an organism. We are now in an era when molecular biology and genetics can be combined to provide the world with new "modified" microorganisms capable of doing useful tasks for humanity.

Summary Outline

DEOXYRIBONUCLEIC ACID AND HEREDITY

1. Double-stranded DNA is the hereditary material of cells and certain viruses.
2. Cellular DNA exists as a double helix of two antiparallel polymers of deoxyribose nucleotides joined by the hydrogen bonding between adenine and thymine; guanine and cytosine.
3. The bacterial chromosome is a closed, circular DNA molecule that is replicated by a semiconservative mechanism.
4. Part of the energy needed to unwind DNA during replication comes from negative supercoils put in DNA by topoisomerases, such as gyrase.
5. Replication begins at an initiation site and proceeds in both directions.
6. Priming RNA synthesized on the single-stranded DNA template by RNA polymerase is the starting substrate for DNA polymerase III. This enzyme aligns the correct nucleotide and then joins it to the 3′ end of the growing macromolecule.
7. The priming RNAs are cleaved off by the RNA exonuclease activity of DNA polymerase III.
8. DNA ligase joins the two ends of the growing DNA molecule to complete the new double-stranded DNA, which contains one old strand and one new strand.
9. An alternative approach to DNA synthesis is the rolling-circle mechanism used by certain DNA bacteriophages to reproduce their DNA and by bacteria to synthesize DNA that is transferred by conjugation.

PROTEIN SYNTHESIS

1. Protein synthesis is a two-step process: the DNA is transcribed into messenger RNA by RNA polymerase (transcription) and then mRNA is translated into proteins (translation).
2. mRNA, the 30S ribosomal subunit, initiation factors, *N*-formyl methionine tRNA, and GTP interact to initiate protein synthesis. tRNAs with their attached amino acids are also required.
3. The genetic code is written in sequences of three bases (codon), is redundant, and is universal in biological systems.
4. Each tRNA specifically binds one amino acid and contains an anticodon that can pair with one or more codons in the mRNA.
5. The 50S subunit binds to the $tRNA^{Met}_f$-mRNA-30S ribosomal subunit complex. The peptidyl transferase associated with the 50S subunit catalyzes the elongation of the growing peptide chain as the ribosome moves along the mRNA.
6. When the ribosome reaches a termination codon at the 3′ end of the mRNA, the polypeptide is released and the ribosome dissociates from the mRNA.
7. Certain antibiotics interfere with bacterial protein synthesis by acting on RNA polymerase, preventing the dissociation of 70S ribosomes, or interfering with peptide chain elongation.

REGULATION

1. Cells regulate the transcription of genes that are organized into operons. Operons consist of a promoter, operator, and structural genes under the control of a regulator gene.
2. Enzyme induction controls the synthesis of the enzymes needed to metabolize a substrate. Transcription begins when the repressor protein is inactivated by an inducer, such as lactose.
3. cAMP levels in the cell control the cell's preference for metabolizing glucose over other sugar substrates (catabolic repression).
4. Enzyme repression prevents the synthesis of enzymes involved in forming a metabolite. The repressor protein for the operon is activated by a corepressor, usually the end product of the metabolic pathway.
5. The activities of key metabolic enzymes are regulated to control the inventory of cellular metabolites.
6. Allosteric enzymes contain one or more sites, separate from the enzyme catalytic site, that bind allosteric effectors, which are often end product(s)

of the metabolic pathway. This is called feedback inhibition when the binding decreases the enzymatic activity.

7. The activities of certain key enzymes are regulated when they are covalently modified by the action of a regulatory enzyme.

Questions and Topics for Study and Discussion

QUESTIONS

1. What scientific clues were available to Watson and Crick when they developed the double-helix model for DNA?
2. Define or explain the following:

antiparallel	genome
catabolic repression	priming RNA
codon–anticodon	semiconservative DNA
enzyme induction	replication
supercoiled DNA	feedback repression

3. Describe the functions of DNA polymerase III in DNA synthesis.
4. Describe the rolling-circle pattern of DNA replication. How does it differ from normal replication of the bacterial genome?
5. Describe the functions of the three types of RNA that participate in translation.
6. What are the energy requirements for protein synthesis?
7. Explain why the minimum number of tRNAs needed for protein synthesis is less than the 64 known codons.
8. How does CRP act as a switch in catabolic repression?
9. Name the components required for the initiation of protein synthesis and explain the function of each.
10. What mechanisms do cells use to regulate the intracellular concentrations of small molecules?

DISCUSSION TOPICS

1. What are the evolutionary implications of the universal genetic code with respect to procaryotes, eucaryotes, and viruses?
2. Explain how some antibiotics inhibit bacterial protein synthesis without affecting eucaryotic cells. Would you expect a similar group of antibiotics to be effective against viral replication in animal cells?

Further Readings

BOOKS

Freifelder, D., *Molecular Biology,* 2nd ed., Jones and Barlett, Boston, 1987. College-level introduction to molecular biology of bacteria, viruses, and eucaryotes.

Haynes, R., and P. Hanawalt, *The Chemical Basis of Life—Readings from Scientific American,* Freeman, San Francisco, 1973. A compendium of readings from *Scientific American* that traces the historical development of the concept of the gene through the early 1970s.

McCarty, M., *The Transforming Principle,* Norton, New York, 1985. The story of the discovery of DNA as the transforming principle in pneumococcus.

Stryer, L., *Biochemistry,* 2nd ed., Freeman, San Francisco, 1981. An excellent reference for the biochemical details concerning transcription, translation, and cellular regulation.

Watson, James D., *The Double Helix,* Atheneum, New

York, 1968. A personal account of the discovery of the double helix of DNA. This is probably the most significant discovery in biology during the twentieth century.

ARTICLES AND REVIEWS

Bauer, W. R., F. H. C. Crick, and J. H. White, "Supercoiled DNA," *Sci. Am.,* 243:118–133 (July 1980). Physical models are presented to explain the molecular biology of supercoiled DNA.

Miller, Jr., O. L., "The Visualization of Genes in Action," *Sci. Am.,* 228:34–42 (March 1973). An article on the use of the electron microscope in studies of translation and transcription.

C H A P T E R

10

Microbial Genetics

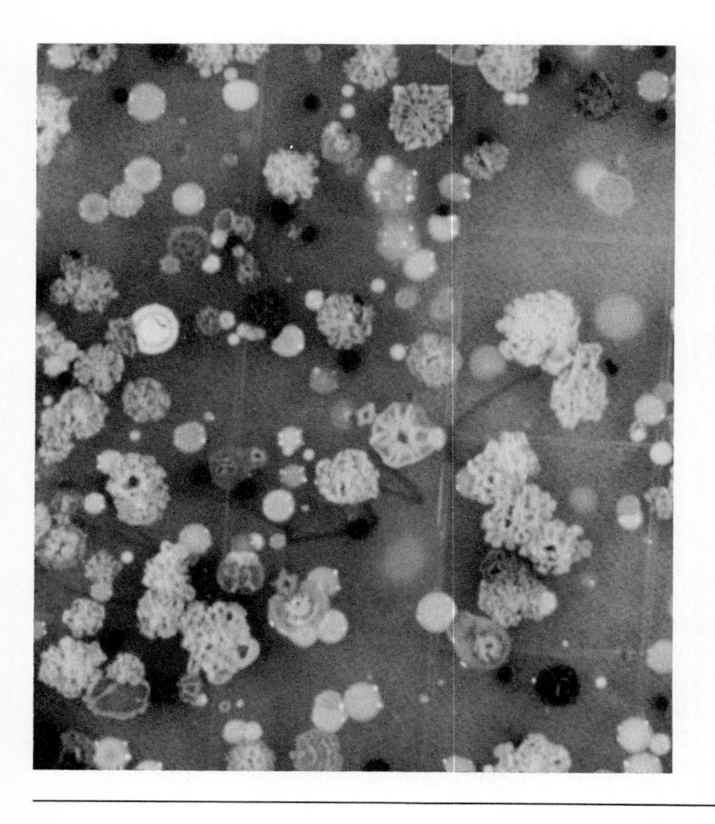

OUTLINE

FOCUS OF THIS CHAPTER

A cell's genetic information can be altered by mutations and by the acquisition of DNA from other organisms. Chapter 10 describes how DNA is changed by mutations and the mechanisms of transformation and conjugation by which microbes acqurie new DNA. These concepts are used to explain the significance of plasmids, genetic mapping of the bacterial chromosome, "jumping genes," and drug resistance.

Microbial diversity is maintained by inherited traits that are passed from generation to generation. These traits are encoded in genes, which are sequences of nucleotides in DNA. **Genetics** is the study of heredity, or more precisely, the study of how specific traits are passed from parents to their offspring.

Gregor Mendel (1822–1885) laid the foundations of classic genetics with his work on the inheritance of sweet pea traits. Mendel had entered an Austrian monastery in Brunn, Czechoslovakia (now Brno), where he worked in the gardens and performed detailed experiments on plant heredity. His experiments showed that the color of flowers, the shape of their seeds, and other traits are passed from parents to offspring in discrete units—now called genes. Mendel demonstrated that plants carry pairs of genes that separate and independently assort during gamete formation. Although Mendel published this work in 1865, it was far ahead of its time and scientists did not recognize its significance until the beginning of the twentieth century when it became the foundation for studies on heredity.

Modern genetics has developed concomitantly with molecular biology and both have depended heavily on studies of microorganisms. Microbes, and especially bacteria, are ideal organisms for genetic analysis because they grow rapidly and have a single chromosome. An understanding of genetics begins with a definition of the gene as the basic unit of inheritance.

CONCEPT OF THE GENE

Mendel had shown that traits such as flower color and plant height were inherited as discrete units. Every organism has many of these inheritable units, which are known as genes. We now know that genes are sequences of nucleotides in DNA, cumulatively known as the organism's **genotype.** The traits that an organism expresses are cumulatively known as its **phenotype.** The question geneticists posed in the early 1940s was, What is the link between the genes and an organism's phenotype?

This question was answered through experiments with amino acid-requiring mutants of *Neurospora crassa* (nur-ah'spor-ah cras'a). This pink bread mold was chosen because it can be grown in large quantities; and even though sexual reproduction occurs, *Neurospora* is haploid during most of its life cycle (Figure 10.1). **Haploid** *(n)* cells contain one copy of every gene so there is no problem with dominance of genes. Another advantage was that wild type *Neurospora* strains can synthesize all the amino acids required for growth on a sucrose minimal medium.

The experiments began with irradiation of the asexual spores (conidia) of *Neurospora* with X rays and isolation of amino acid-requiring nutritional mutants. After radiation, the conidia were mated (Figure 10.1), and the resulting ascospores were isolated and grown on a complete medium (one containing all the amino acids). The mating procedure assured the investigators that the mold cultures obtained from the haploid ascospores were genetically pure.

Next, pure cultures of *Neurospora* mutants that required one or more amino acid for growth were isolated. These nutritional mutants grew on complete medium, but did not grow on sucrose–salts medium (minimal medium). Mutants unable to synthesize a particular amino acid were identified by their ability to grow on minimal medium supplemented with that amino acid. For example, mutants able to grow on minimal plus tryptophan were called *trp*$^{-}$ mutants. Wild type *Neurospora* can grow on minimal medium because it can synthesize all of the amino acids from sucrose, ammonia, and salts.

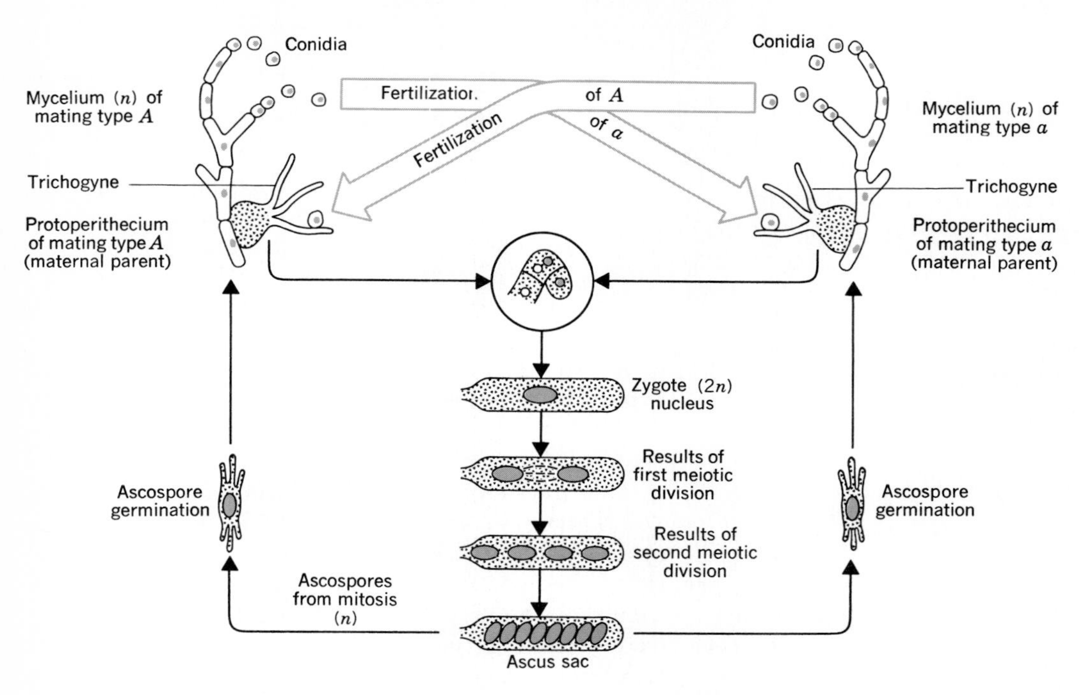

FIGURE 10.1 The life cycle of the pink bread mold, *Neurospora crassa*. Sexual reproduction occurs between conidia and protoperithecia of opposite mating types designated type *A* and type *a*. The resulting zygote (2*n*) undergoes two meiotic divisions to form eight ascospores (*n*) in an ascus. Each ascospore gives rise to a genetically pure strain of *Neurospora*. (From R. P. Wagner et al., *Introduction to Modern Genetics,* Wiley, New York, 1980.)

ONE GENE, ONE PROTEIN

Amino acid mutants were analyzed biochemically to determine why they could not form the amino acid they required for growth. It was found that each mutant lacked one of the enzymes needed to synthesize that amino acid. These results led to the conclusion that *one gene codes for one enzyme*.

This was a major contribution to the understanding of genetics, even though it was not completely correct. Contemporary biochemists knew that enzymes were proteins, but they did not know how complex certain enzymes are. We now know that many enzymes are composed of more than one polypeptide and that some polypeptides, such as the peptide hormones and ribosomal proteins, have no enzymatic activity. Therefore, the theory was modified as follows: one gene codes for one polypeptide. Genes also code for RNA, and some RNAs are not translated into polypeptides.

KEY POINT

A cellular gene is a segment of DNA that is transcribed into RNA: most genes are translated into a polypeptide; however, the genes for rRNA and tRNA make only RNA molecules.

MUTATIONS RESULT FROM PERMANENT CHANGES IN DNA

Microbiologists have exploited one of the basic phenomena of biology, genetic mutation. Instead of looking for natural or spontaneous mutants, they have produced mutations in *Neurospora* by exposing cultures to X rays, thus causing cellular **mutations,** which are inheritable changes in DNA. Most mutations are permanent so they are passed on to succeeding generations. These mutants are used as the basis of many genetic experiments.

Such experiments begin with a *wild type* strain. A wild type strain may require certain growth factors, like *Neurospora,* which requires biotin, or it may be like *Escherichia coli,* which grows on glucose–salts medium without any supplements. To make sure that all the cells in the culture are genetically identical, the experiment begins with an isolated colony of the desired strain. Isolated colonies are genetically pure because they arise from a single cell. If their genetic traits are stable, the strain will maintain its characteristics (phenotype) through many generations.

Mutants are isolated after exposure of a parent strain, called the **prototroph,** to a mutagenic agent. Many different mutations will be caused by the mutagen and not all of the cells will survive. **Lethal mutations** affect one or more essential cellular functions so they kill the cell. Mutations that prevent growth under certain environmental conditions are called **conditional lethal mutations.** Examples of such mutants are found among amino acid requiring strains that grow only when a specific amino-acid is added to the medium. Mutants that require nutritional supplements are known as **auxotrophs.**

KEY POINT

Alterations in cellular RNAs do not qualify as mutations since this information is not passed to succeeding generations. This does not apply to RNA viruses, since changes in their RNA genome are inherited.

RATE OF MUTATION

Physical and chemical forces on the genetic material and replication errors cause spontaneous mutations in microorganisms. Such mutations occur at different rates among the various genes and most occur independently of mutations in other genes. Some spontaneous mutations, such as mutations to antibiotic resistance, are very important in medical microbiology, as we shall see.

The rate of mutation of a given gene is a measure of the probability that permanent change will occur in the gene's DNA. If a mutation occurs once in 1 million divisions, the rate of mutation is 1×10^{-6} (1/1,000,000 divisions). It follows that the rate at which mutations occur in two separate genes is the product of the individual rates. For example: Assume that gene A normally mutates at a rate of 10^{-5} and gene B mutates at a rate of 10^{-8}. The formation of an organism containing mutations in both gene A and gene B will be one chance in 10 trillion divisions ($10^{-5} \times 10^{-8} = 10^{-13}$). Such double mutants occur only rarely in nature because the mutation rate is so low.

This concept is behind the use of two antibacterial agents for treating long-term infections such as tuberculosis. If the infectious agent develops resistance against one agent, there is a very low probability that it will become resistant to another agent, which will probably kill it.

TYPES OF MUTATIONS

DNA can be altered in a number of distinct ways that will result in a mutation (Figure 10.2). Addition of one or more nucleotides to the DNA will cause a mutation. Additions can occur when "jumping genes" (transposons) are inserted into the DNA (see "Transposons Are Mobile Genetic Elements," (see p. 240), or as copy errors. Similarly, deletions of one or two nucleotides

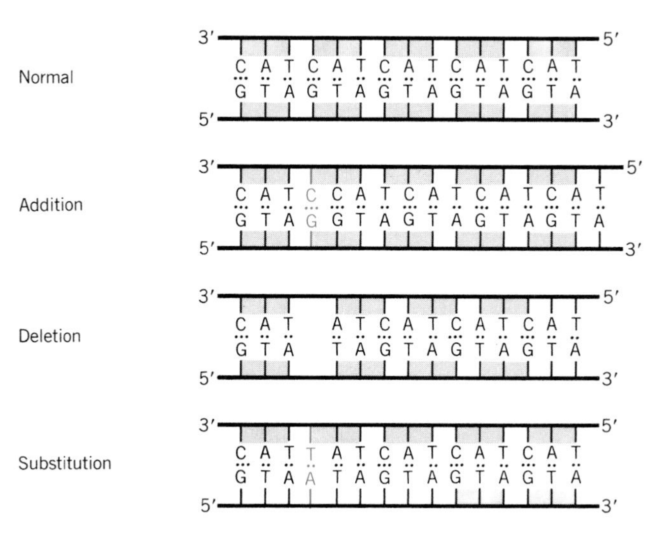

FIGURE 10.2 Mutations are inheritable alterations in DNA. Additions, deletions, and/or substitutions of nucleotide bases in DNA can cause mutations that alter the reading of the triplicate code.

from the DNA sequence will cause a mutation. Addition or deletion of a single nucleotide base will cause a **frameshift** in how the genetic code is read. Instead of reading THE FAT CAT ATE THE HAM, the message reads THE FAC ATA TET HEH, and so on.

Substitution of a single base in DNA will cause a mutation only when it alters the amino acid in the encoded polypeptide. The genetic code is said to be redundant because there is more than one codon for most amino acids. This means that certain base substitutions will not change the amino acid specified by that codon. For example, GCU, GCC, GCA, and GCG (see Table 9.3) all code for alanine. An alteration in the third position of this codon causes a **silent mutation** that will never be expressed.

Once a mutation occurs, it is not necessarily permanent. Many mutations are reversible by the process of back mutation or reversion. **Back mutations** can occur at the site of the original change in DNA or they may occur at a second site. For example, a frameshift mutation caused by deleting a single base can be reversed by deleting two additional bases. Reverse mutations caused by a deletion may lead to the restoration of the original phenotype, even though the base sequence in the DNA is permanently altered.

CHEMICAL AGENTS CAUSE MUTATIONS

Mutagens are chemical or physical agents that greatly increase the rate of mutation above the spontaneous level. Chemical mutagens directly alter the purine and/or pyrimidine bases in DNA. When the alteration changes the base pairing, a mutation can occur (Figure 10.3). Nitrous acid (HNO_2) is a good example of a chemical mutagen.

Nitrous acid alters the bases in DNA by replacing amino groups with a keto group (Figure 10.3). For example, adenine is changed to hypoxanthine in the presence of nitrous acid. Since hypoxanthine pairs with cytosine during the next cycle of replication (adenine would have paired with thymine), the mutation causes a replacement of an A—T pair in the DNA with a G—C pair. Likewise, cytosine is changed to uracil by nitrous acid. Since uracil pairs with adenine, there is a transition from G—C to A—T (Figure 10.3). No pairing change occurs when guanine is deaminated ($-NH_2$ removed) by nitrous acid, so guanine deamination does not result in a mutation.

Base analogues are compounds that resemble the natural bases found in DNA, but are chemically different. 5-bromouracil is an analogue of thymine (Figure 10.4). When 5-bromouracil is taken up by cells, it base pairs with adenine since it is structurally similar to thymine. This is fine; however, 5-bromouracil can also exist in an enol form (Figure 10.4), which base pairs with guanine. As a result, 5-bromouracil causes a replacement of a A—T pair in the DNA with an G—C pair.

KEY POINT

Base analogues that are incorporated into or inhibit the synthesis of viral DNA without affecting humans are constantly being sought as antiviral agents. Azidothymidine, acyclovir, and ribavirin are base analogues used effectively as antiviral drugs.

PHYSICAL CAUSES OF MUTATIONS

Microbes are constantly exposed to natural sources of radiation that can cause alterations in their DNA. Ultraviolet light (UV) from black lights or the sun has sufficient energy to chemically change the pyrimidine bases of DNA. In fact large doses of UV light will kill bacteria. This is why UV lights are used in sterile transfer rooms to decrease bacterial contamination on equipment surfaces and in the circulating air.

The formation of thymine dimers is the most common form of UV-caused DNA damage. Adjacent thymine residues absorb UV light between 250 and 270 nm and in turn form covalent bonds (T═T) that create dimers (Figure 10.3*b*, Figure 10.5). Thymine dimers prevent the strand from serving as a proper template for RNA and DNA synthesis. DNA-dependent polymerases cannot replicate the thymine dimer, so they "jump" over the affected section. The resulting deletion of two nucleotides (A—A) from the new strand usually results in a mutation.

Many cells have enzymatic mechanisms for reversing the damage caused by exposure to UV light. *Escherichia coli* possesses two mechanisms that deal with UV damage: a light-repair and a dark-repair mecha-

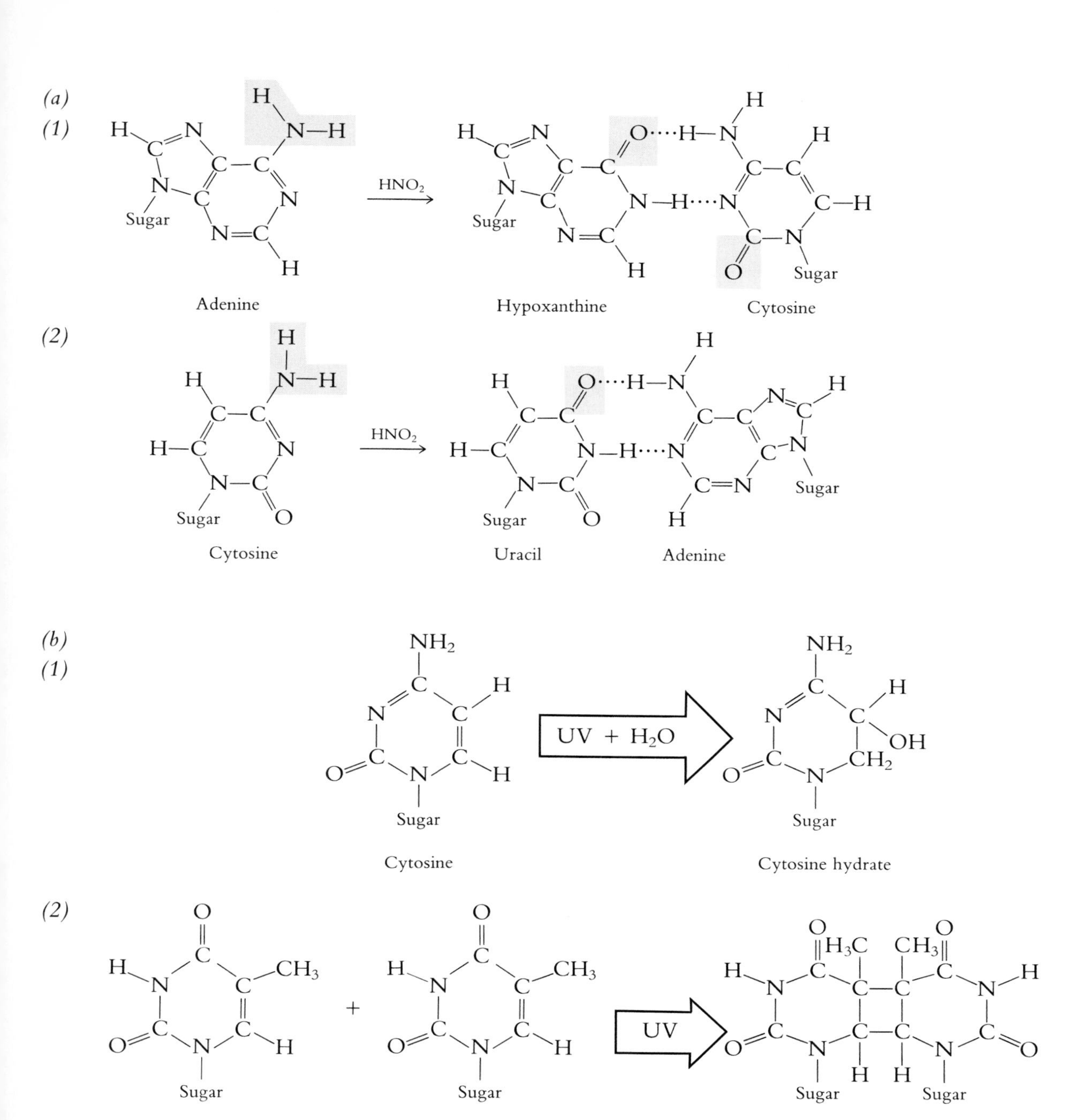

FIGURE 10.3 Chemical and physical mutagens. *(a)* Nitrous acid (HNO_2) chemically reacts with adenine (1) to form hypoxanthine, thus causing AT → GC transitions and with cytosine (2) to form uracil, thus causing GC → AT transitions. *(b)* Ultraviolet irradiation can cause (1) the hydrolysis of cytosine to form the hydrated form and (2) the formation of thymine dimers, which block replication. (Adapted from E. J. Gardner and D. P. Snustad, *Principles of Genetics,* 6th ed., Wiley, New York, 1981.)

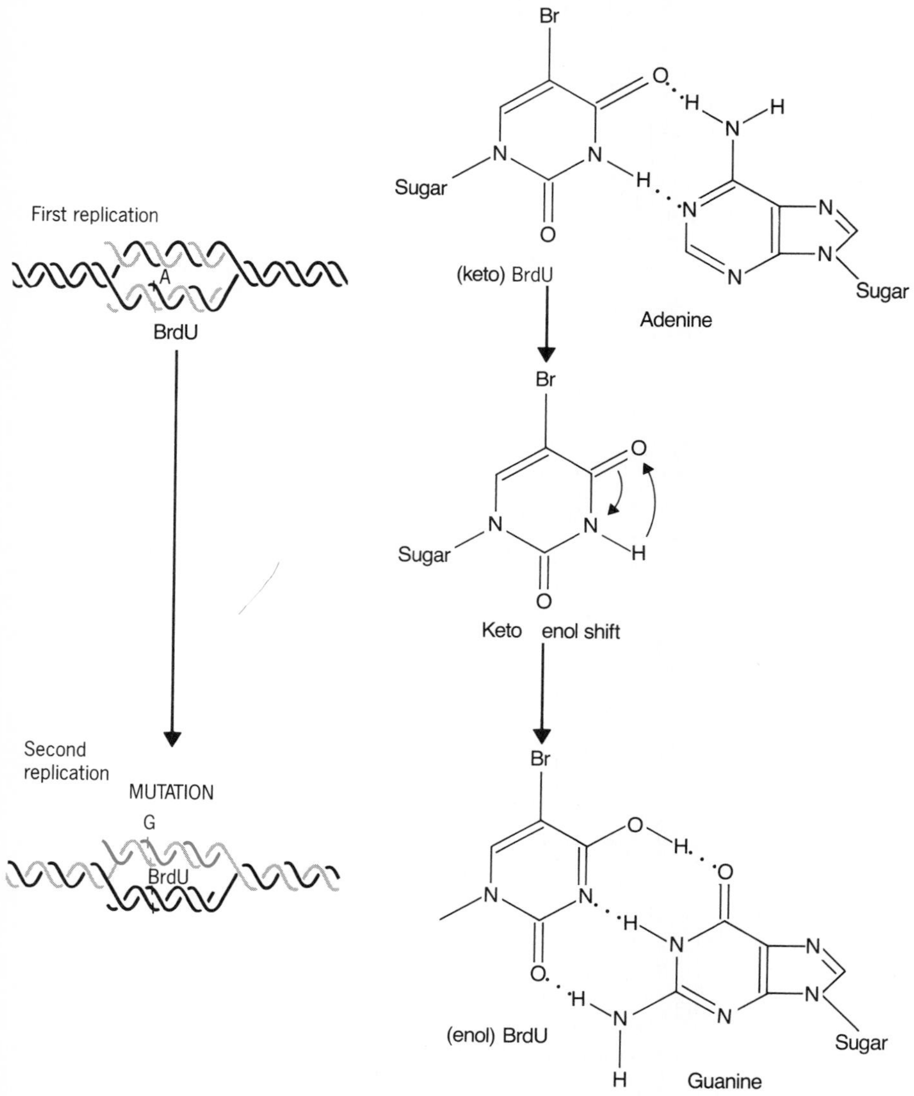

FIGURE 10.4 Mechanism of base analogue mutation. 5-Bromouracil is incorporated into DNA through its ability to base pair with adenine when it is in its natural keto form. This nucleotide can also exist in an enol form that base pairs with guanine. As a result, 5-bromouracil causes an AT → GC transition resulting in a mutation.

nism (Figure 10.5). The **light-repair** system depends on a photoactivated enzyme that breaks the bonds between thymine dimers. **Dark repair** is an enzyme-mediated excision-repair process that is independent of light. The dark-repair mechanism uses an endonuclease* to nick the 5′ side of the damaged DNA. This damaged DNA can then be excised by an exonuclease. The exposed single-stranded DNA is a suitable tem-

*Endonucleases are enzymes that nick double-stranded DNA at specific points in the center of the strand. Exonucleases are enzymes that act on the free 5′ or 3′ end of single-stranded or double-stranded DNA.

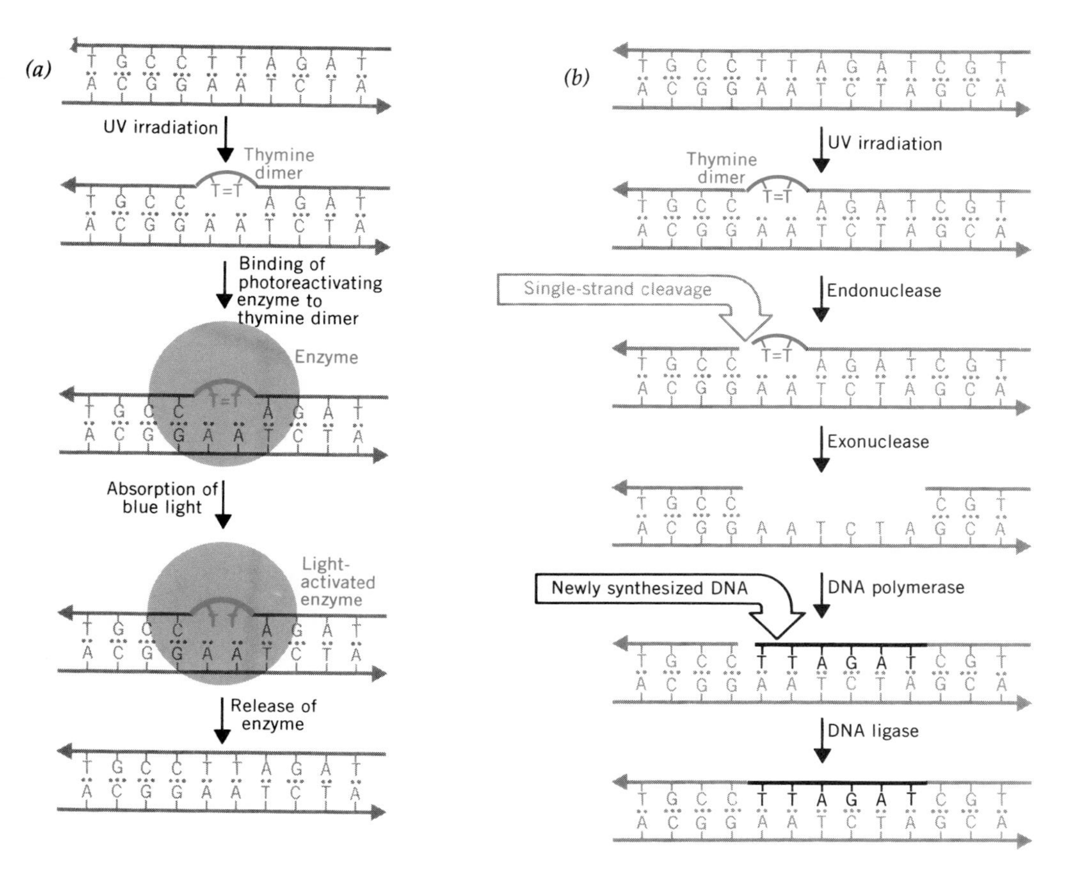

FIGURE 10.5 *(a)* The **Light-repair** mechanism of correcting thymine dimers in *E. coli*. Thymine dimers are formed by UV irradiation. The enzyme involved in the light-repair mechanism binds to the thymine dimer region of the DNA. In the presence of blue light, this enzyme cleaves the dimer cross-links. *(b)* The **dark-repair** mechanism involves a series of enzymatic steps. The damaged area is cut by a 5′ endonuclease and then excised by an exonuclease. DNA polymerase and DNA ligase act together to synthesize a new DNA strand complementary to the template. (From E. J. Gardner and D. P. Snustad, *Principles of Genetics,* 6th ed., John Wiley, New York, 1981.)

plate for DNA polymerase, which makes the complementary strand. DNA ligase then joins the free ends to complete the repair. In combination, the light- and the dark-repair mechanisms enable *E. coli* to repair UV damage to their DNA that otherwise would cause lethal mutations.

ISOLATION OF MUTANTS

Geneticists have devised ingenious approaches for isolating all types of bacterial mutants from large populations of cells. The selective enrichment techniques they use are successful even when the mutation is a very rare event. Bacteria are ideally suited for mutant isolations because they reproduce rapidly and grow to high cell densities—often attaining 10^{10} cells per milliliter.

The general approach to mutant isolation begins by exposing a large population of cells to a mutagen, such as UV light, X rays, or a mutagenic chemical. After exposure, it is desirable to select against the unmutated parent cells because they outnumber the mutants and

Mutations and Cancer

Many chemical mutagens, known as *carcinogens,* also cause cancer in animals. Ideally we would like to avoid contact with all carcinogens, and to this end the U.S. Congress has ordered the Federal Drug Administration to remove all carcinogens from consumable goods. To do this we must first determine what compounds are carcinogens. Traditional proof of carcinogenicity relies on animal testing at considerable cost in time and money. To facilitate testing, Bruce Ames devised a bacterial test that is quick and inexpensive. The Ames test determines whether a compound causes back mutations in an auxotroph of *Salmonella typhimurium* (ty-fi-mur′ee-um). This screening test indicates whether a compound has a carcinogenic potential; tests on animals are still required to prove carcinogenicity.

When histidine auxotrophs of *Salmonella typhimurium* are plated on minimal medium, they do not grow because they cannot make histidine. They grow into colonies only if they back-mutate to wild type, his^+. These cells have a very low rate of reversion to wild type except when they are grown in the presence of a mutagen. In the Ames test, cells of *Salmonella typhimurium* his^- are exposed to potential carcinogens present on filter disks. The compound diffuses away from the disk to establish a concentration gradient. Colonies of revertants occur around the filter disks containing mutagenic compounds. The potency of the carcinogen is indicated by the number of colonies around the particular disk.

There is a strong correlation between the mutagenicity of a compound determined by the Ames test and its known carcinogenic properties. A high percentage of the animal-tested carcinogens are bacterial mutagens. Similarly, a high percentage of the compounds that are noncarcinogenic in animals are nonmutagenic in the Ames test.

Although the Ames test is an important screening test for carcinogenic compounds, some concerns have been raised about its interpretation. For example, many common dietary substances, including black pepper, are mutagenic in the Ames test. Therefore, mutagens are more widely distributed in nature than we realized, and their importance to human cancer remains to be ascertained.

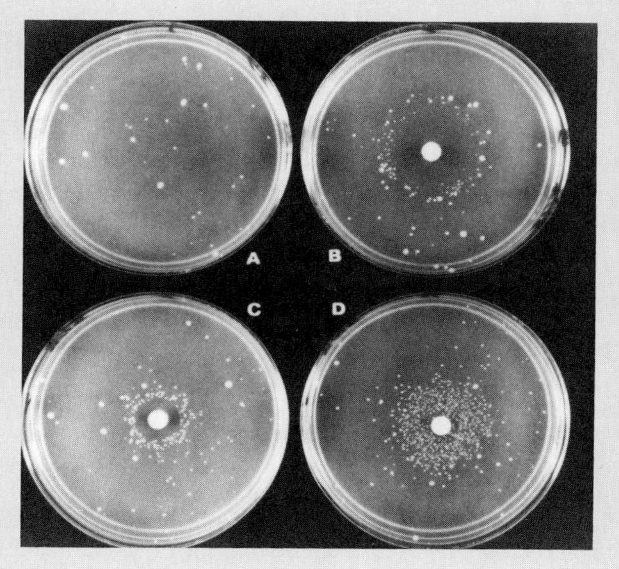

Ames test performed with a histidine auxotrophs of *Salmonella*. Plate *a* is the control, *b* furylfuramide (1 μg), *c* aflatoxin B_1 (1 μg), and *d* 2-aminofluorene (10 μg). (Courtesy of Dr. Bruce Ames. From B. Ames, J. McCann, and E. Yamasake, *Mutation Research,* 31:347–364, 1975, copyright © 1975, Elsevier Scientific Publishing Company.)

will interfere with the isolation. The mutants are then isolated by selective enrichments and characterized by nutritional or biochemical studies.

One method of selectively killing the parent cells is to expose the mutated culture to penicillin in a minimal medium. Penicillin is a bacteriolytic antibiotic that kills only *growing* cells. Since nutritional mutants require one or more growth factors, they are unaffected by the penicillin because they cannot grow on minimal medium. This same principle is used to isolate viral-resistant strains of bacteria. Many bacterial virus infections kill their host cell upon completion of their replication cycle. Bacterial cells plated in the presence of such viruses die unless they are resistant to the viral infection.

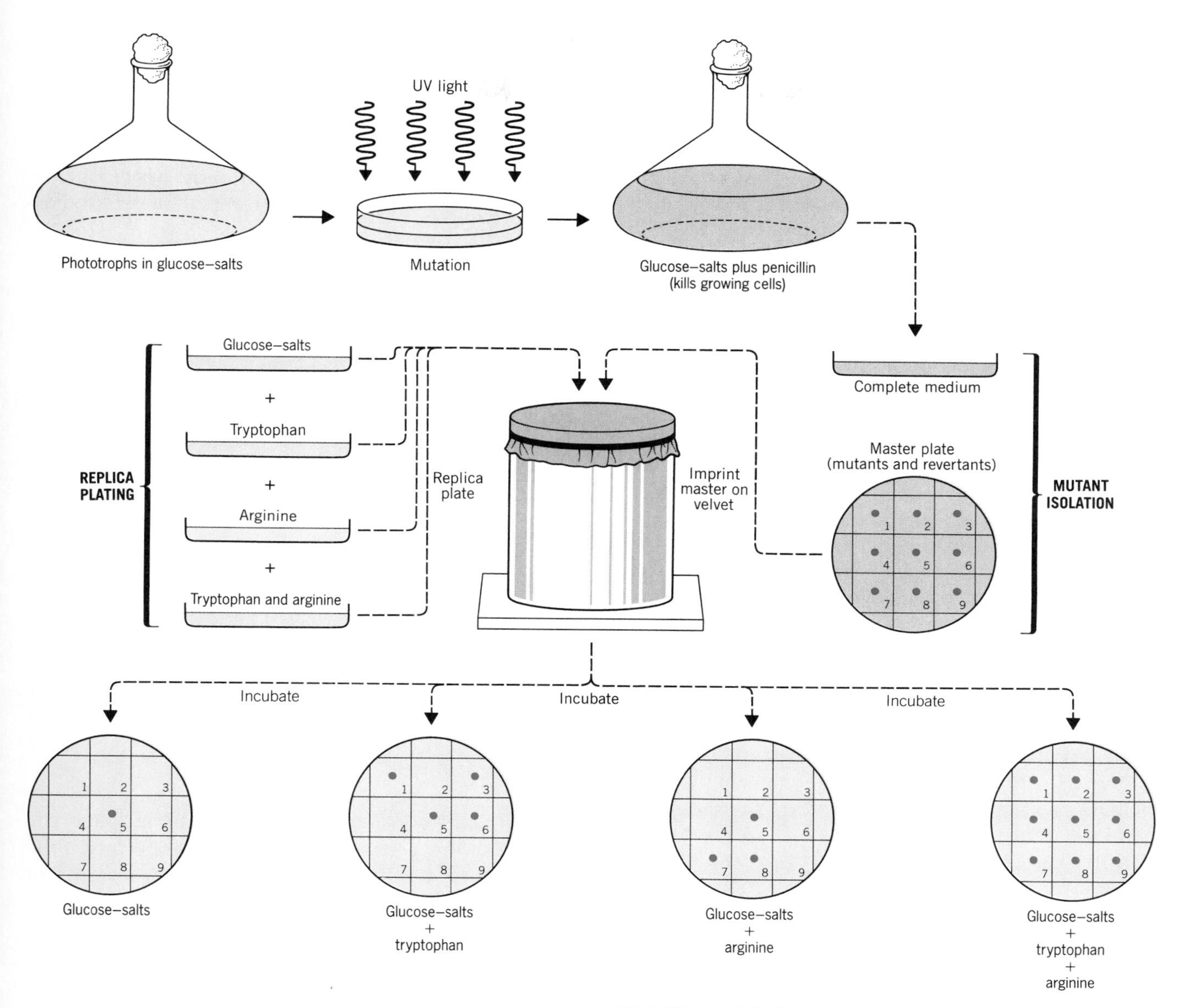

FIGURE 10.6 Nutritional mutants are isolated and characterized by replica plating. Mutants are isolated and grown on a master plate from a culture exposed to a mutagen. The master plate shown contains the nine mutant colonies that are replica-plated on four separate media. Isolate 5 is a revertant to wild type because it grows on the glucose–salts medium. Isolates 1, 3, and 6 are tryptophan auxotrophs (*trp*$^-$) because they grow on glucose–salts plus tryptophan, but not on glucose–salts alone. Similarly, isolates 7 and 8 are arginine auxotrophs (*arg*$^-$). Isolates 2, 4, and 9 are double auxotrophs (*trp*$^-$, *arg*$^-$) because they grow only in media containing both tryptophan and arginine.

Isolation of Nutritional Mutants

Studies of nutritional mutants have deepened our understanding of synthetic enzymatic pathways and the genetic controls over them. **Nutritional mutants** that have an altered ability to produce an essential metabolite, can be isolated from wild type strains, or strains with known genotypes, referred to as **prototrophs.** These nutritional mutants, called **auxotrophs,** are characterized by their growth response to different microbiological media. For example, wild type strains of

E. coli grow on glucose–salts medium without added nutritional supplements, whereas an auxotroph of *E. coli* grows on sucrose–salts medium only when it is supplemented with one or more essential nutrients.

The first step in isolating nutritional mutants (Figure 10.6) is to mutate the prototroph using one of the mutagenic procedures previously described. Many of the treated cells will acquire lethal mutations and die; however, a significant proportion of the mutant cells and prototrophs will survive. Two sequential selection steps are necessary to isolate the mutants from the surviving cell population.

Prototrophs usually outnumber the mutant cells, so the prototrophs must be eliminated. Penicillin-sensitive prototrophs are killed when grown on minimal medium containing penicillin. Nutritional mutants survive this treatment (even if they are sensitive to penicillin) because they cannot grow on minimal medium. These nutritional mutants are then isolated on complete medium.

Replica plating is an ingenious technique (Figure 10.6) developed to characterize nutritional mutants. It is based on the principle that a nutritional mutant such as *trp*$^-$★ grows on glucose–salts medium containing the nutritional supplement tryptophan, but not on minimal medium without the supplement. To identify these and more complex mutants, a master plate is prepared (Figure 10.6) that contains between 10 and 100 colonies. These colonies are grown on a complete agar medium in a petri dish. A sterile velvet pad on a transfer block is then used to inoculate each of the colonies from the master plate onto plates containing glucose-salts medium, plus one or more nutritional supplements (Figure 10.6).

After incubation, each strain is scored for growth on the supplements chosen by the investigator. Suppose a strain is growing on glucose–salts plus tryptophan and arginine, but not on glucose–salts plus tryptophan or on glucose–salts plus arginine (Figure 10.6, strains 2, 4, and 9). This strain is a double mutant with the genotype *trp*$^-$, *arg*$^-$. The study of bacterial genetics relies on the investigator's ability to isolate and analyze different mutants. The replica-plating technique greatly facilitates this process.

★Geneticists use ($^+$) to indicate that a gene is functional and ($^-$) to indicate that the gene is inoperative or absent.

BACTERIAL GENETICS: TRANSFORMATION

Genetic transfer of chromosomal genes between bacteria is a rare event that appears to occur in only a few bacterial species. Studies of these bacteria have greatly expanded our knowledge and understanding of heredity.

Genetic information is transferred between bacteria by three different mechanisms. When bacteria take up free DNA from the environment the process is called **transformation.** The transfer of genetic information from a donor cell in direct contact with a recipient cell is called **conjugation.** The third type of genetic transfer, called **transduction,** is mediated by bacterial viruses. An understanding of transduction requires a basic understanding of how bacterial viruses replicate, so transduction is described in Chapter 11 on bacterial viruses.

KEY POINT

Bacteria are ideal systems for studying genetic transfer because (1) genetic transfer between parents is not required for bacterial reproduction since bacteria divide by binary fission and (2) bacteria have one copy of each gene, meaning they are haploid and do not have dominance/recessive genes for each trait.

Genetic transfer is recognized when the recipient cell acquires a new phenotype. This is easily observed if the recipient cell is a mutant. For example, a tryptophan mutant *(trp*$^-$*)* acquires a new phenotype when it becomes *trp*$^+$. The *trp*$^+$ strain is easily detected because it grows on minimal medium, while a *trp*$^-$ strain does not. By constructing the correct mutants and using them in genetic crosses, geneticists have been able to demonstrate genetic transfer.

TRANSFORMATION INVOLVES UPTAKE OF FREE DNA

The earliest reported observations of genetic transfer in bacteria were made by Frederick Griffith in 1928 during his investigations of pneumococcal infections in mice (Figure 10.7). Disease-causing strains of *Streptococcus pneumoniae,* a bacterium that kills mice and causes

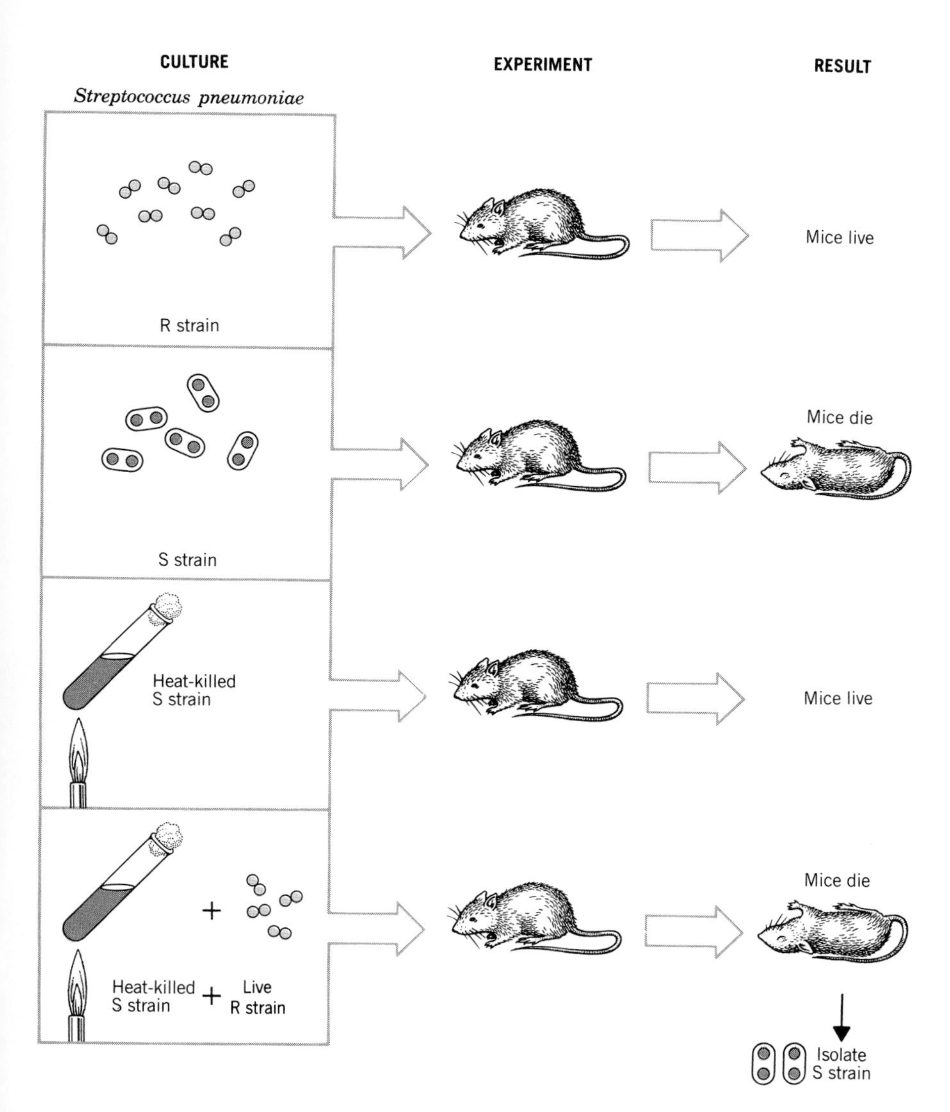

FIGURE 10.7 Griffith's transformation experiment with *Streptococcus pneumoniae*. Mice are able to survive injections of living *noncapsulated* (type R) pneumococci. However, mice suffer a serious bacterial infection and die when they are injected with capsulated pneumococci (type S). As expected, the mice survive when the type S pneumococci are heat-killed before being injected. The unexpected observation was that mice injected with a mixture of heat-killed type S and living type R died from a bacterial infection. From these dead mice, Griffith isolated only capsulated type S pneumococci, and concluded that noncapsulated (R) cells acquire a "transforming principle" from the heat-killed capsulated (S) cells, which converts them to the smooth type. (Modified from E. J. Gardner and D. P. Snustad, *Principles of Genetics,* 6th ed., Wiley, New York, 1981.)

pneumonia in humans, grow as smooth mucoid colonies because of the distinctive capsule they produce (called the S strain). The S strain spontaneously reverts to a noncapsulated strain that forms rough colonies (called the R strain). Only the capsulated S strain causes pneumonia and death in animals.

Mice were able to survive being injected with bacteria from cultures of the R strain and they survived injections of heat-killed cultures of the S strain (Figure 10.7). However, all the mice died when injected with living cultures of the S strain (Figure 10.7). Injections that combined cells from the R strain and heat-killed preparations of the S strain also killed the mice (Figure 10.7). From these dead mice Griffith isolated capsulated cells that produced smooth (S) colonies. Griffith postulated that a "transforming principle," released by the heat-killed S strain, was taken up by the viable cells of the R strain. This transforming principle converted the R strain into an S strain.

It is interesting to speculate what happened in this experiment. Presumably the phagocytic cells of the mice engulfed all the noncapsulated cells and killed them before they could cause disease. The smooth cells resisted phagocytosis so they were able to multiply and kill the mice. In essence, the mice used in Griffith's experiment acted as a selective enrichment for the disease-causing S strain.

The act of transformation can take place anywhere—in test tubes or in natural environments; animals are not required for transformation. Successful transformation experiments are done in test tubes when a procedure is available to select for the recombinant strains.

The true nature of the transforming principle eluded Griffith and his colleagues. The "transforming principle" was finally shown to be DNA in 1944 by Oswald Avery, Colin McCarty, and Maclyn MacLeod. This was the first direct evidence that genetic information was coded for by a macromolecule, and focused interest on DNA as the hereditary information.

The exchange of genetic information by transformation is inhibited by DNase, an enzyme that breaks down free DNA. DNase sensitivity distinguishes transformation from the other forms of genetic transfer between bacteria. Natural transformation has now been demonstrated in gram-positive and gram-negative bacteria (Table 10.1). In addition, it is possible to artificially transform *Escherichia coli, Salmonella typhimurium,* and *Pseudomonas aeruginosa.* These three species have no natural transformation mechanism, but they take up extracellular DNA after special laboratory treatments.

TABLE 10.1 Bacterial species capable of transformation

GRAM-POSITIVE BACTERIA	
Streptococcus pneumoniae	*Streptococcus sanguis*
Bacillus subtilis	*Bacillus cereus*
Bacillus stearothermophilus	*Bacillus licheniformis*
	Thermoactinomyces vulgaris
GRAM-NEGATIVE BACTERIA	
Natural Transformation	
Neisseria gonorrhoeae	*Acinetobacter calcoaceticus*
Moraxella osloensis	*Moraxella urethalis*
Azotobacter agilis	*Haemophilus influenzae*
Haemophilus parainfluenzae	*Pseudomonas stutzeri*
	Pseudomonas alcaligenes
Psychrobacter spp.	*Pseudomonas mendocina*
Artificial Transformation	
Escherichia coli	*Salmonella typhimurium*
Pseudomonas aeruginosa	

Natural transformation has been extensively studied in the gram-positive bacterium *Streptococcus pneumoniae,* and in the gram-negative *Haemophilus influenzae.* The mechanisms of transformation in these two bacteria are sufficiently different that we can describe them separately.

Transformation in Gram-Positive Bacteria

Transformation occurs when a recipient cell takes up short pieces of homologous DNA from its surroundings. The first step in transformation is the binding of the donor DNA to the surface of the recipient cell. Only **competent** cells participate in this binding. Competency depends on the synthesis of a low molecular weight protein—competency factor—that interacts with a receptor on the cell's surface (Figure 10.8). In turn, the cell is stimulated to synthesize 8 to 10 additional proteins involved in transformation. In *Streptococcus pneumoniae,* competency is acquired during exponential growth, while in other bacteria, competency depends on the phase of growth or on specific culture conditions.

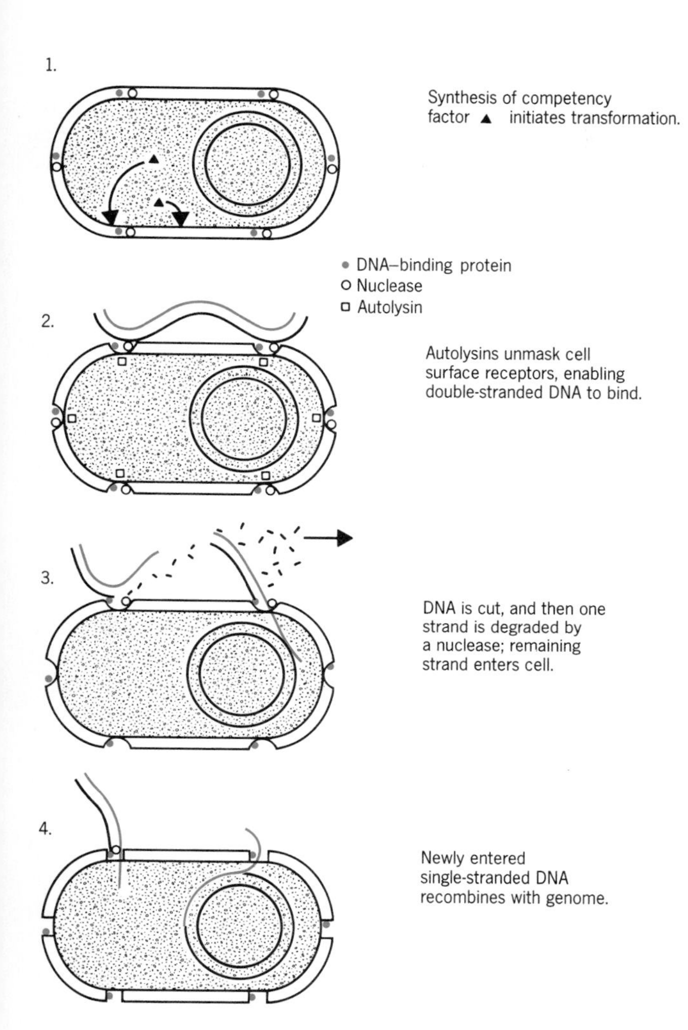

FIGURE 10.8 Transformation in a gram-positive bacterium is a complex process involving a DNA-binding protein on the cell's surface, enzymes, and a competency factor. Recombination occurs between the single-stranded transforming DNA and the cell's chromosome.

Each competent cell of *S. pneumoniae* has between 30 and 100 surface-receptor sites that bind double-stranded DNA. (Single-stranded DNA is inactive in transformation because it does not bind to the cell wall receptors.) While bound to the receptor, the double-stranded DNA is enzymatically cleaved into short pieces. These pieces are converted into single-stranded DNA by a cell-wall exonuclease that degrades one strand. A small polypeptide binds to the remaining single-stranded DNA, and the DNA-peptide complex enters the cell's cytoplasm. Once inside the cell, the DNA is incorporated into the host's genome by recombination.

Transformation in Gram-Negative Bacteria

Natural transformation has been studied extensively in *Haemophilus influenzae* (hee-mah′fi-lus in-floo-en′zee), which will serve as our model. Cultures of *Haemophilus* become 100 percent competent when they are transferred to a medium that permits protein synthesis, but does not permit cell division. Other gram-negative bacteria become competent when they enter a stationary growth phase.

Gram-negative bacteria can be transformed only by double-stranded DNA that comes from closely related strains. The DNA absorbed by *Haemophilus* must have an 11 base sequence (-5′-A-A-G-T-G-C-G-G-T-C-A-3′-) known as a recognition site, in order to be absorbed to the cell's surface. These DNA sequences are distributed throughout the cell's genome so most donor-cell DNA fragments will have at least one recognition site. DNA from unrelated bacteria lack the recognition sites and are not absorbed.

After the double-stranded DNA is absorbed, it is taken into the cell intact. Once in the cytoplasm, a single strand is incorporated into the host's genome. The mechanism of entry and recombination is not understood.

Artificial Transformation Many genetic engineering experiments depend on transferring modified DNA into a host organism. This is impossible to accomplish in most species because they lack a natural transformation system. Some bacteria, however, can be made to take up free DNA by a process called artificial transformation.

Escherichia coli lacks a natural transformation system, but will take up free double-stranded DNA during artificial transformation. The cells are treated first with Ca^{2+} at 0°C. The spherical cells formed during the cold treatment take up DNA when heated (90 seconds) at 40°C. The free DNA must be self-replicating (such as a plasmid, see following) in order for it to transform the cell. Short pieces of DNA will not transform the cell because they are quickly degraded in the cytoplasm of *E. coli* and never recombine with its genome. Artificial transformation is a key step in the gene cloning experiments discussed in Chapter 30.

BACTERIAL GENETICS: CONJUGATION

CONJUGATION REQUIRES CELL-TO-CELL CONTACT

Direct transfer of genetic information from a donor bacterium to a recipient bacterium is accomplished by the process of **conjugation** (Figure 10.9). Joshua Lederberg and Edward Tatum discovered conjugation in 1946, while working with *E. coli* K-12. Their choice of this bacterium was fortuitous since K-12 is a wild type strain of *E. coli* that contains the fertility factor *F*. Their experimental design was to mix two double auxotrophs of *E. coli* K-12, incubate the mixture, and then look for recombinants. The auxotrophs were bio^-, met^-, thr^+, leu^+ and bio^+, met^+, thr^-, leu^-. When 10^8 cells were plated on minimal medium, approximately 200 recombinants (bio^+, met^+, thr^+, leu^+) grew into colonies, indicating that genetic transfer had occurred. The parent strains used in this conjugation experiment were nutritional mutants that could not grow on minimal medium—only recombinants having the growth characteristics of wild type could grow.

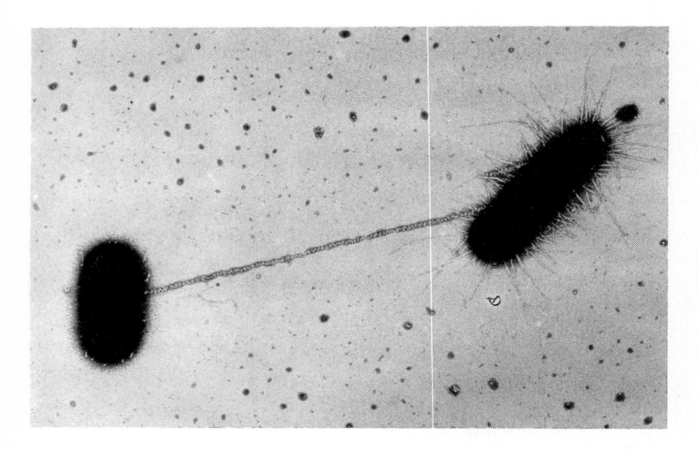

FIGURE 10.9 Bacterial conjugation between an Hfr and an F^- strain of *E. coli*. The donor bacterial cell with numerous appendages (on the right) attaches to the smooth-surfaced recipient cell (on the left) by means of a sex or F pilus. DNA from the donor cell is transferred to the recipient cell after the F pilus brings the cells into close contact. In this micrograph, the F pilus is identified by its ability to bind small virus particles, which are not involved in conjugation. (Courtesy of Dr. C. Brinton, Jr. and J. Carnahan.)

Genetic transfer requiring cell-to-cell contact is called bacterial conjugation. This mechanism is distinguished from other forms of gene transfer in bacteria by the following: (1) it is insensitive to DNase, a characteristic of transformation, and (2) it is not mediated by a bacteriophage (Table 10.2).

Further investigation revealed that the necessary cell-to-cell contact is mediated by **sex pili** (Figure 10.9) found on the surface of donor (male) cells. Each pilus is a hollow proteinaceous structure coded for by genes present only in donor cells. After coming in contact with a recipient cell, the pilus shortens and brings the two cells together. The transfer of genes probably occurs between the contact points joining the two cells and not through the hollow pilus.

Experiments with conjugation soon made it clear that extrachromosomal DNA was involved. In the early 1950s, Joshua Lederberg called this extrachromosomal DNA a plasmid and the name stuck. We now know that plasmids play a major role in bacterial genetics.

BACTERIAL PLASMIDS

Plasmids are self-replicating, extrachromosomal, circular, double-stranded DNA molecules that code for a wide variety of functions, ranging from genetic transfer to antibiotic resistance. Plasmids are present in many different bacteria and often have a major effect on the organism's phenotype.

Since plasmids are extrachromosomal, they usually code for nonessential, but not necessarily useless, information. This means a cell can lose, or be cured of, its plasmid and still survive. **Curing** is an irreversible process that can be expedited by chemical treatments (Figure 10.10) such as treatment with acridine orange. Acridine orange interferes with the replication of the F plasmid without interfering with chromosomal DNA synthesis and cell replication. As the cells continue to divide, the plasmids are diluted by the increased number of cells, and eventually the majority of the cells in the culture contain no plasmid. When a cell loses its plasmid, it also loses the genetic traits coded for by that plasmid. The irreversible loss of a trait or a group of traits under nonmutating conditions is genetic evidence for the location of those traits on a plasmid. The presence and loss of plasmids can be confirmed with physical techniques that separate and identify molecules of DNA of different sizes.

TABLE 10.2 Characteristics of DNA transfer in bacteria

	GENETIC PROCESS		
CHARACTERISTIC	TRANSFORMATION	TRANSDUCTION	CONJUGATION
DNase sensitive	+	−	−
Bacteriophage required	−	+	−
Cell-to-cell contact	−	−	+
Transfer of small DNA molecules	+	+	+
Transfer of large DNA molecules	−	−	+

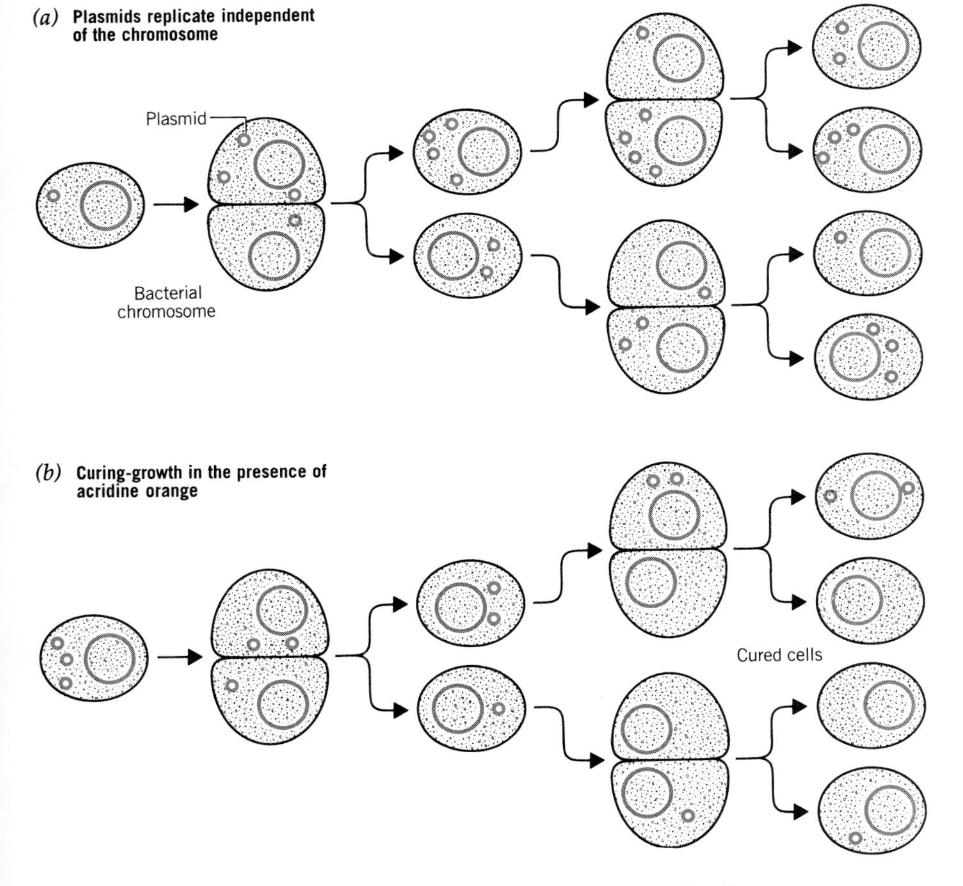

FIGURE 10.10 *(a)* Replication of plasmids in untreated cells occurs independently of chromosomal replication. *(b)* Acridine orange preferentially prevents plasmid replication so that after many replications the bacteria are cured of their plasmids.

Bacterial Plasmids Are Replicons

Bacterial plasmids replicate independently of the chromosome, and are called **replicons.** In some situations the numerous enzymes needed to replicate the bacterial chromosome (see Chapter 9) also participate in plasmid replication. This is true of the Col E1 plasmid of *E. coli*. Replication begins at a single origin on the Col E1 plasmid and progresses in one direction around the circular DNA molecule. Since the plasmid is replicated independently of the chromosome, each cell can have as many as 15 copies of the Col E1 plasmid. Other plasmids are present at one to two copies per chromosome and code for some of the enzymes involved in their replication.

Bacterial cells can contain different kinds of plasmids as long as the plasmids are compatible. For example, as many as seven compatible plasmids can be stably maintained in *E. coli*. Other plasmids are incompatible and cannot exist in the same cell.

Conjugative and Fertility Plasmids

Plasmids were first discovered when geneticists recognized that extrachromosomal genes were transferred between bacteria. Plasmids that can be transferred directly to another cell are called **conjugative plasmids.** The first conjugative plasmid to be discovered was the fertility or F plasmid, which is responsible for conjugation in *E. coli.* **F^+ cells** contain the F plasmid and are always the donor (male) cell in the transfer of genetic material by conjugation. The recipient is designated the **F^- cell** (female) because it lacks the fertility plasmid.

The F plasmid carries 13 transfer *(tra)* genes that code for the proteins needed to transfer the plasmid. These include the genes coding for the formation of the F pilus and for the enzymes used to replicate the DNA that is transferred. F plasmids do not normally code for chromosomal (essential) genes.

CONJUGATION, $F^+ \times F^-$ CROSS

When F^+ donor cells and F^- recipient cells of *E. coli* are mixed, the F plasmid is transferred with a high efficiency. A copy of the entire plasmid is transferred to the F^- cell so all the recipient cells are converted to F^+ cells. The genetic information is transferred on a single-stranded DNA copy of the F plasmid (Figure

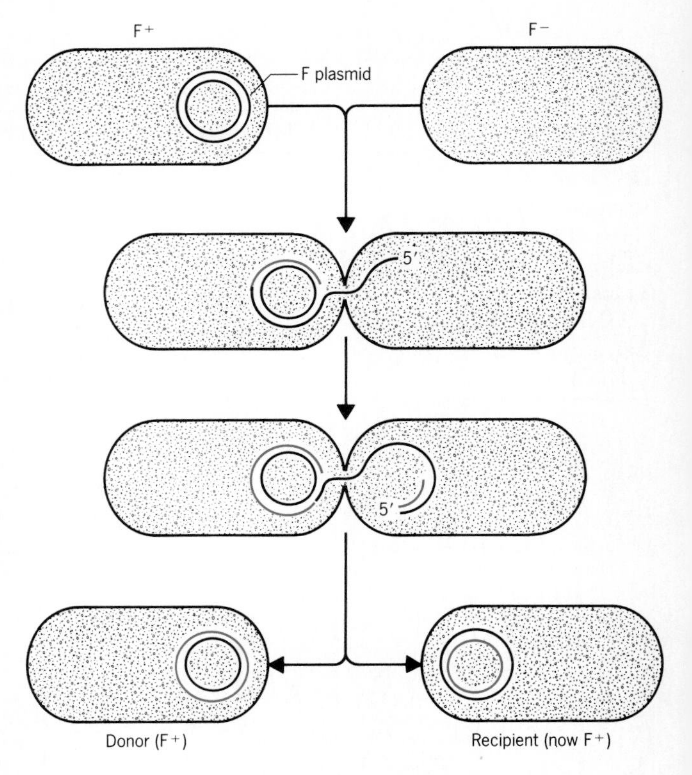

FIGURE 10.11 Conjugation between an F^+ and an F^- cell. The transferred DNA is replicated by the rolling-circle mechanism of DNA synthesis.

10.11) that is displaced during the replication of the F plasmid in the donor cell by the rolling-circle replication mechanism (see Chapter 9). The displaced strand is transferred to the recipient cell with the 5′ end entering first. Once transferred, recipient cell enzymes synthesize a complementary strand to form double-stranded DNA, which takes a closed circular form. This transferred F plasmid is now a replicon and the recipient is an F^+ cell. (The donor cell remains an F^+ cell since only a copy of the F plasmid is transferred.) Recombination with the host cell genome does not occur in this form of conjugation.

PLASMIDS CAN MOVE INTO THE CHROMOSOME

Fertility plasmids can be incorporated into the bacterial chromosome at DNA sequences identified as insertion sequences. The *E. coli* chromosome has at least 15 short sequences (700 to 2000 base pairs in length) of

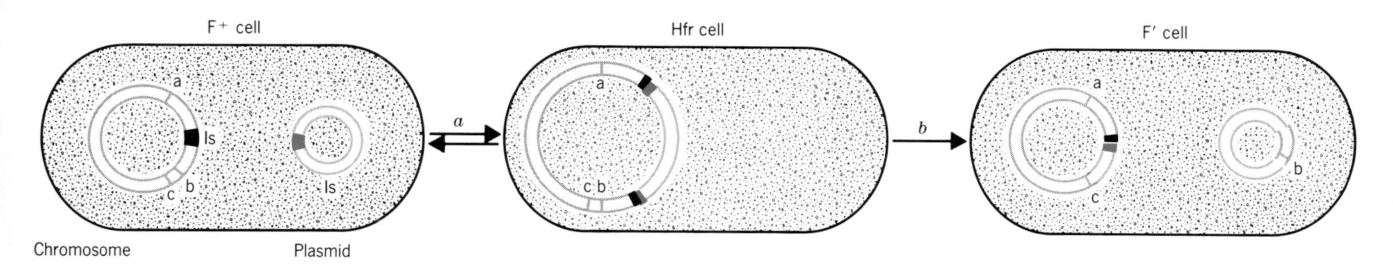

FIGURE 10.12 *(a)* The integration of an F plasmid into the host chromosome converts an F^+ cell into an Hfr cell. The Hfr cell readily reverts to an F^+ cell through the excision of the F plasmid. *(b)* When one or more chromosomal genes are excised with the F plasmid, the resulting cell is termed an F′ cell.

DNA known as **insertion sequences** (IS). Some of these IS sites match the three IS sites on the F plasmid. Each insertion sequence possesses repeat sequences at each end that enable it to insert into other segments of DNA.

Homologous insertion sequences present in the cell's chromosome and the F plasmid enable the F plasmid to be integrated into the chromosome. Insertion of an F plasmid into the host cell's genome (Figure 10.12) converts an F^+ cell to an *Hfr* cell. Hfr cells actively transfer genetic information to F^- recipient cells, resulting in a **high frequency of recombination** (Hfr). Plasmids that can exist in an autonomous, self-replicating state or integrated into the host cell's chromosome are termed **episomes.**

F Plasmids Can Acquire Chromosomal Genes

Hfr cells revert to F^+ cells with about the same frequency (10^{-4}) with which they are formed (Figure 10.12). Reversions occur when the homologous ends of the original insertion element join and again form a small closed-circular DNA molecule that becomes an independent replicon. Sometimes, instead of the original insertion elements, alternative insertion elements on either the plasmid or the chromosome are joined. When this happens, the excised plasmid will carry chromosomal genes. The specific chromosomal genes involved depend on the insertion site occupied by the F factor in the Hrf strain and on the presence and distribution of insertion elements in the host chromosome. Fertility factors that contain chromosomal genes are designated **F′ plasmids** and these genes are transferred in a F′ → F^- cross.

Conjugation with an Hfr Donor

The integrated fertility factor in the Hfr strain is capable of mediating genetic transfer to an F^- cell. When Hfr cells are mixed with F^- cells, they interact with each other through the F pilus contact point. Once contact is established, genetic transfer begins (Figure 10.13). At any given time after mixing, the conjugation can be stopped by physical disruption of cell-to-cell contact. This is accomplished by subjecting the paired cells to the shearing forces generated in a blender type of mixer.

Recombinants for different genetic loci form at different rates in the recipient cell because there is a sequential transfer of genetic information (Figure 10.13). The chromosome breaks in the middle of the F factor and is then replicated by the rolling-circle mechanism. The 5′ end of the displaced strand is transferred unidirectionally to the F^- cell. In *E. coli* at 37°C, complete transfer of the chromosome takes about 100 minutes. Complete transfer is a rare event, but when it does occur, the F^- recipient cell is converted to either an F^+ or an Hfr cell.

Bacterial conjugation is used to map the position of genes on the circular bacterial chromosome. By choosing different markers and by using different Hfr strains, we can attain a relatively complete map of the chromosome. The map of the *E. coli* chromosome (Figure 10.14) summarizes the data of many geneticists, including the experiments of Lederberg and Tatum.

KEY POINT

Many different Hfr strains exist because the F factor can integrate into the bacterial chromosome at numerous in-

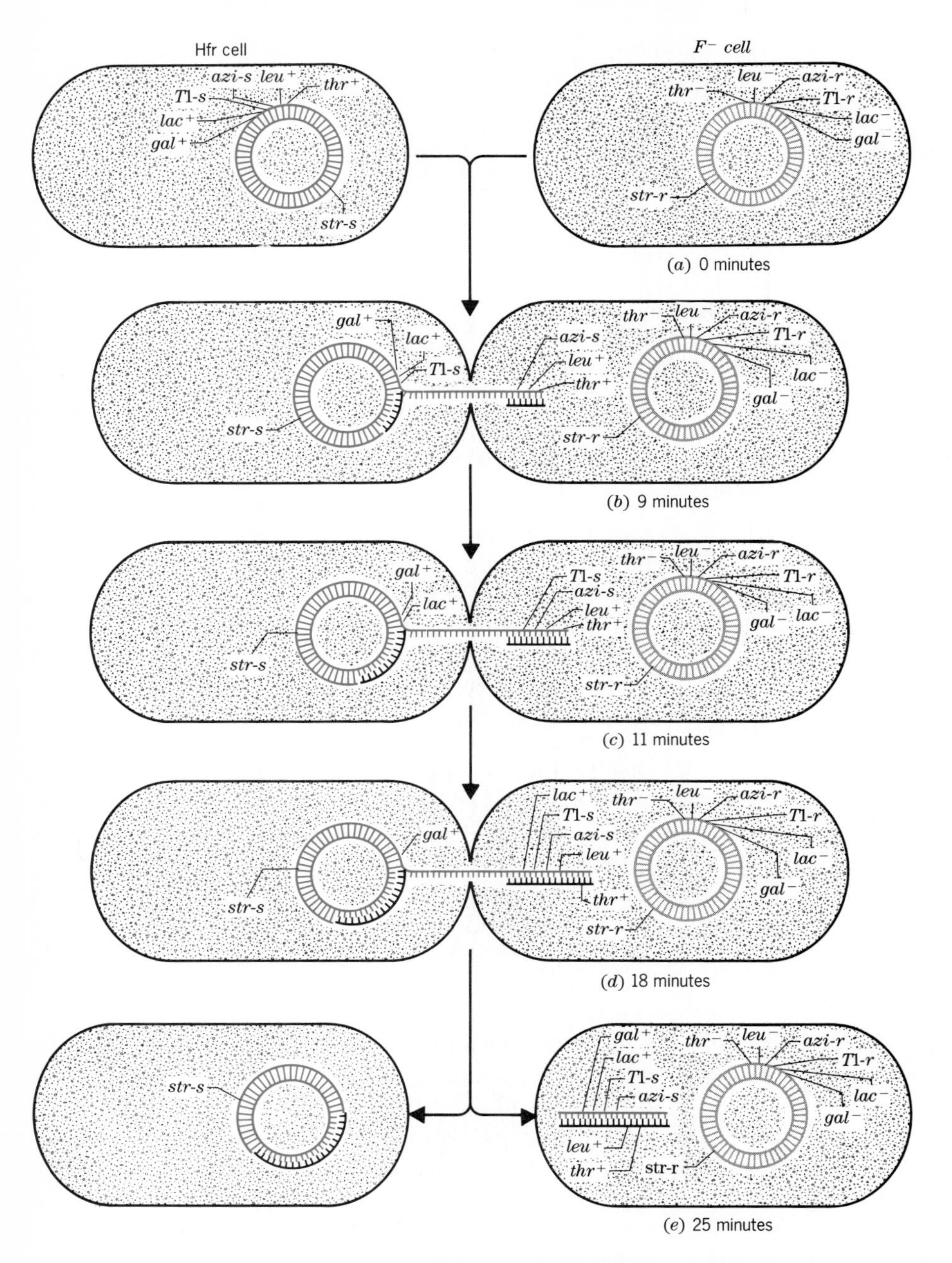

FIGURE 10.13 Conjugation between an Hfr and an F^- bacterium. Only one strand of the DNA is transferred into the recipient cell. The recipient cell forms the complementary strand of DNA to make the double helix. The minutes indicate the interval following initiation of conjugation. The genetic elements are not drawn to scale, only about one-fourth of the Hfr chromosome has been transferred during the 25 minutes of conjugation. (Adapted from E. J. Gardner and D. P. Snustad, *Principles of Genetics,* 6th ed., Wiley, New York, 1981.)

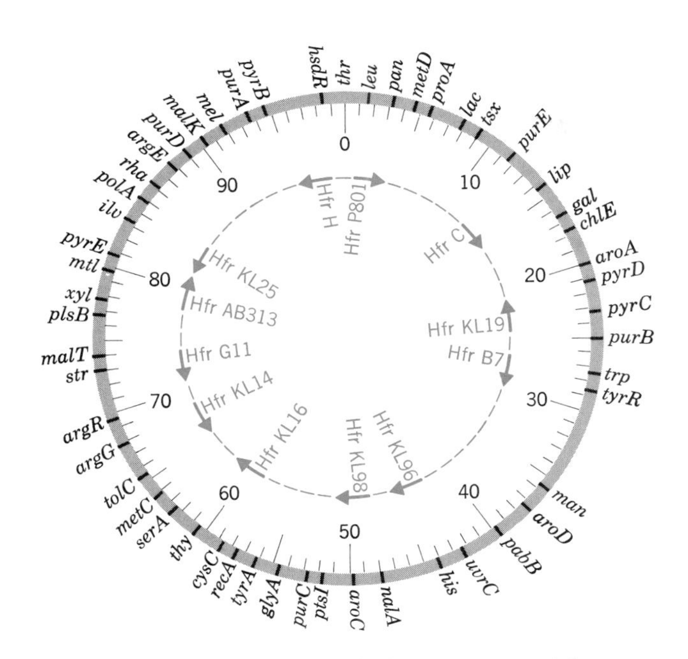

FIGURE 10.14 A simplified genetic map of the *Escherichia coli* K-12 genome. The map is divided into 100 minutes with the zero point arbitrarily set at the *thr* locus. The inner circle shows the F factor integration sites in various Hfr strains.

sertion sites. This changes how genes are transferred during conjugation. To correct for this, the genetic map is standardized to the transfer of the threonine marker, which is assigned a value of 0 time.

Conjugation with F′ Strains

Some F′ plasmids are stable and contain sufficient genetic information to be replicons. When such a plasmid is transferred to F$^-$ cells, the recipient contains two copies of the transferred chromosomal genes, one on the F′ plasmid and one on the chromosome. This creates a partial diploid or **merozygote** and is termed a **secondary F′.** These strains are useful for studying gene dominance in bacteria.

MAJOR GROUPS OF PLASMIDS

Fertility plasmids are conjugative plasmids that code for their own transfer. Bacteria contain other plasmids (Table 10.3) that may or may not be conjugative. These plasmids code for a variety of nonessential cellular functions.

DRUG-RESISTANT PLASMIDS

Plasmids have special significance for the control of bacterial diseases since they can code for drug resistance. This problem was first recognized in Japan in 1955 when a strain of *Shigella dysenteriae* (shi-gel′ah dis-en-tair′ee-eye), responsible for an epidemic of dysentery, was shown to be resistant to four antibacterial drugs: chloramphenicol (klor-am-fen′i-kol), streptomycin (strep-toh-my′sin), sulfanilamide (sul-fan-ill′am-ide), and tetracycline (tet-rah-sy′klyn). By 1956, 50 percent of all *S. dysenteriae* isolated at Japanese hospitals possessed resistance to multiple drugs. Genetic analysis showed that multiple-drug resistance was coded for by a plasmid that was readily transferred between strains.

TABLE 10.3 Representative bacterial plasmids

PLASMID	ORGANISM	SIZE (MOLECULAR WEIGHT)	PHENOTYPE
F	*Escherichia coli* K-12	63×10^6	Conjugative plasmid
Col E1-k30	*Escherichia coli*	4.3×10^6	Colicin E1
Δ	*Salmonella typhimurium*	61×10^6	Mobilizes nonconjugative R plasmids in some strains
R.1	*Salmonella paratyphi*	58×10^6	Resistance transfer factor and r-determinant (resistant to ampicillin, chloramphenicol, fusidic acid, kanamycin, streptomycin, sulfonamide)
Degradative	*Pseudomonas putida*	$50–200 \times 10^6$	Enzymes for metabolism of unusual or synthetic organic compounds
Ti	*Agrobacterium tumefaciens*	$90–160 \times 10^6$	Responsible for crown gall disease in dicotyledons

The many plasmids responsible for multiple-drug resistance are called **R plasmids.** They are found in both gram-negative and gram-positive bacteria and are quite common in the *Enterobacteriaceae* (en-ter-oh-bak′ter-i-ay′see-ee) and in staphylococci. The R plasmids found in gram-positive bacteria are nonconjugative, whereas R plasmids found in gram-negative bacteria are usually conjugative.

Nonconjugative R plasmids in gram-negative bacteria can be mobilized by conjugative plasmids. The genes encoding drug resistance are clustered in the **r-determinant** region of the R plasmid. This region is flanked by IS genes that can participate in its transposition. The r-determinant is mobilized when it is incorporated into a plasmid harboring the genes required for conjugation. The mechanism of mobilization could be the joining of two plasmids as shown in Figure 10.15, or the r-determinant could act as a transposon (see below) and move to a conjugative plasmid.

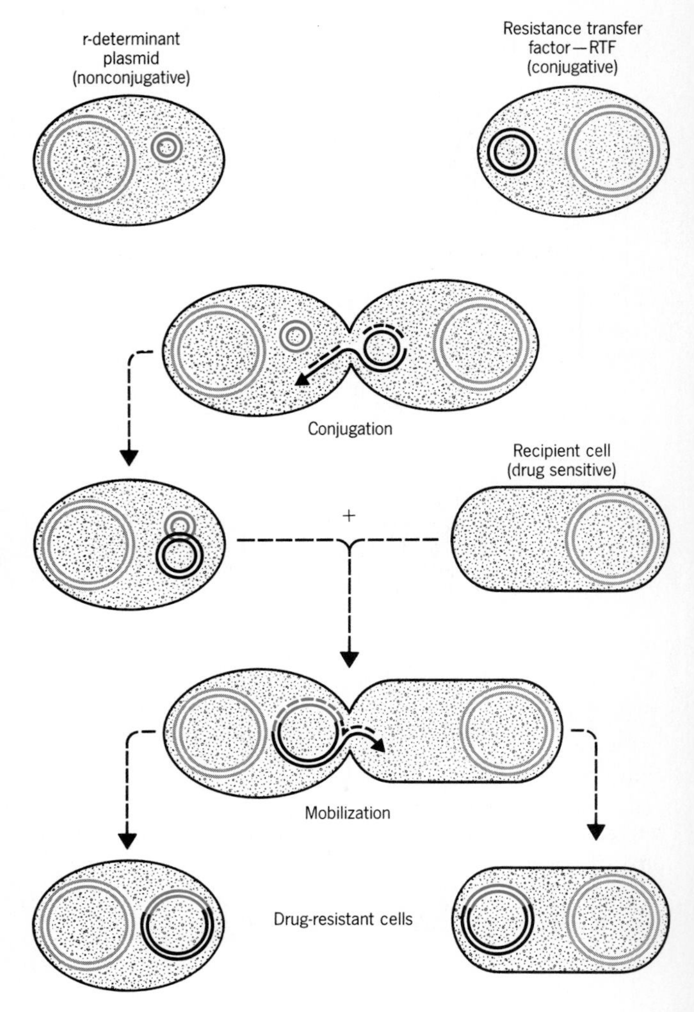

FIGURE 10.15 Plasmid mobilization. Multiple-drug resistance can be coded for by nonconjugative plasmids. These plasmids can be mobilized upon combination with a conjugative plasmid, such as the resistance transfer factor (RTF) and transferred to another cell.

BACTERIOCINS

Bacteria produce antibacterial substances called **bacteriocins** (bak-teer′ee-oh-sin), which kill or inhibit the growth of closely related bacteria. These proteins are produced by gram-positive and gram-negative bacteria. The colicins produced by *E. coli* and closely related species are encoded by col plasmids.

Colicins act in a variety of ways. Colicin E2 degrades DNA and colicin E3 cleaves 16S rRNA after they have penetrated the target cell's envelope. Other colicins bind to metabolite receptors on the surface of the target cell and block the uptake of essential nutrients. Most colicins prevent growth by disrupting the permeability of the cytoplasmic membrane to ions.

Every cell that synthesizes a colicin also makes immunity proteins to protect itself from the action of the colicin. Both the immunity protein and the colicin are coded for by col plasmids. Cells that produce bacteriocins are advantaged in nature because they can eliminate other microbes that would compete for common nutrients.

VIRULENCE PLASMIDS

Escherichia coli is a normal inhabitant of the human large intestine and is usually harmless; however, it will cause human diseases after acquiring virulence plasmids. One such plasmid codes for fimbriae that enable *E. coli* to colonize the human small intestine. Another plasmid codes for two toxins that upset the ionic balance across the intestinal mucosa, causing an outflow of water into the intestinal lumen. The result is a severe diarrheal disease known as traveller's diarrhea.

Other bacteria also acquire virulence plasmids. Cer-

Plasmids and Antibiotic Resistance

Which came first, drug-resistant plasmids or the medical use of antibiotics? Today a large percentage of clinical isolates are resistant to one or more antibiotics. Were pathogenic bacteria always drug-resistant or is this a recent development caused by widespread use of antibiotics? This question has been answered for a large group of enterobacteria studied by Veronica Hughes and Naomi Datta.

The enterobacteria studied were collected by the Canadian microbiologist E. D. G. Murray between 1917 and 1954 and stored in sealed vials until 1980. The collected strains were isolated from humans by Murray or were sent to him by colleagues who had isolated them from clinical specimens in various parts of the world. Of the 433 strains tested, antibiotic resistance occurred in only 11. Nine of these were tetracycline-resistant strains of *Proteus*, a resistance probably encoded by chromosomal genes, and two strains were resistant to ampicillin. Resistance to other antibiotics or to sulfonamides did not occur among any of the strains tested. Now antibiotic resistance is common in clinical isolates from humans and animals and is usually determined by plasmids. Plasmid-mediated antibiotic resistance is transferred between strains by conjugation. Hughes and Datta analyzed the Murray collection for the presence of conjugative plasmids and found them to be as common in the Murray collection as they are in collections of recently isolated drug-resistant enterobacteria. So the ability to transfer plasmids was not a factor in the low incidence of drug resistance in bacteria isolated prior to the medical use of antibiotics. This means the observed increase in antibiotic resistance results from the selection of resistant strains following the widespread use of antibiotics.

tain strains of *Staphylococcus aureus* possess the exfoliative (ex-foh'lee-a-tive) plasmid that encodes the exfoliative toxin responsible for the scalded skin syndrome. A plasmid found in *Agrobacterium tumefaciens* (ag-rho-bak-teer'ee-um too-meh-faysh'ee-enz) enables this bacterium to cause crown gall in fruit and other deciduous trees. *Agrobacterium tumefaciens* possesses the tumor inducer (Ti) plasmid, part of which is transferred to the plant cell where it causes tumor formation. The plasmids that code for virulence factors are able to change harmless bacteria into disease-causing strains.

METABOLIC PLASMIDS

The ability to degrade unusual or synthetic organic compounds is a nonessential function of certain bacteria that is often encoded on plasmid DNA. **Degradative plasmids** encode enzymes that metabolize unusual or synthetic organic compounds. Certain strains of *Pseudomonas putida* (pew'tid-ah) carry these degradative plasmids. Enzymes that degrade synthetic organic compounds, such as the herbicide 2,4-D (2,4-dichlorophenoxyacetic acid), are also coded for by degradative plasmids. Degradative plasmids are usually high molecular weight, conjugative plasmids.

Salmonella spp. are routinely characterized by their inability to metabolize lactose. However, recently isolated strains of *Salmonella* possess a plasmid that encodes the *lac* operon enabling these strains to metabolize lactose. This is a perplexing problem for microbiologists who have to identify them, because *Salmonella* are important causes of human disease.

Species of *Rhizobium* (rye-zoh'bee-um) fix molecular nitrogen while living in symbiosis with leguminous plants. The ability of *Rhizobium* to fix nitrogen and to stimulate nodule formation in plants has now been linked to the presence of a plasmid. Leguminous plants are extremely valuable to agriculture since their growth is not entirely dependent on the expensive application of chemical fertilizers (see Chapters 14 and 28).

KEY POINT

Plasmids are vehicles by which bacteria can acquire new genetic information. Plasmids can be transferred between members of the same species and to bacteria of closely related species. The acquisition of antibiotic resistance and virulence plasmids by bacteria is a major problem in medical microbiology.

TRANSPOSONS ARE MOBILE GENETIC ELEMENTS

Another mechanism of genetic change is the movement of genes from one place on the genome to another and between the genome and plasmids. These movements are mediated by mobile genetic elements whose existence has greatly altered our concepts of the fixed genome.

One of the more unexpected discoveries in genetics is that certain genes move from one part of the genome to another. Mobile genetic elements were discovered by Barbara McClintock in the early 1950s, during her studies of variegation in maize (corn). Her insight into this genetic phenomenon was not widely accepted during the 1950s and 1960s. After microbial geneticists discovered similar mobile elements in bacteria, the significance of Barbara McClintock's early contributions became universally recognized, and in 1983 she was awarded the Nobel Prize in Physiology or Medicine. The bacterial mobile elements are called **transposons.**

Transposons are segments of DNA that are composed of a core sequence flanked by IS-like sequences. The IS-like sequences at both ends of the transposon are identical, and enable copies of the transposon to move. The core sequence of many transposons codes for enzymes that inactivate antibiotics, so cells acquiring them become antibiotic resistant (Table 10.4). Transposons can also code for substrate catabolism (*lac* operon), and some code for toxins.

Movement of the transposon within the genome occurs when a copy of the transposon is inserted into a new region of the genome or into a plasmid. A few transposons move only to select sites, while most transposons can move to innumerable sites. A simple addition occurs when a transposon is inserted between genes; however, transposons can also be inserted into the middle of a gene. These transposons are mutagenic because they interrupt the continuity of the genetic code at the site of insertion, causing an altered message to be formed.

Cytoplasmic genetic elements can move back and forth between the bacterial chromosome and plasmids. A serious situation develops when transposons coding for antibiotic-drug resistance are transposed to a conjugative plasmid. Now the drug resistance can be transferred rapidly to all the cells in the population. This transfer process is called **infectious drug resistance** and has become a serious problem that is compounded by the transfer of plasmids between closely related genera of bacteria.

It was once thought that the genetic information of a cell changes slowly, and then only by mutation. Studies of classification and evolution have depended on the basic constancy of the cellular genome. We now know that genes can be transposed between chromosomal and plasmid DNA and that plasmids readily transfer genetic information between strains as well as to closely related genera. These facts have changed our concept of evolution and are having an effect on our approach to bacterial classification.

TABLE 10.4 Transposons

DESIGNATION	SIZE (BASE PAIRS)	CORE ENCODES
Tn1, Tn2, Tn3	About 5,000	Ampicillin resistance
Tn4	About 20,000	Ampicillin, streptomycin, sulfonamide resistance
Tn5	About 5,500	Kanamycin resistance
Tn9	2,638	Chloramphenicol resistance
Tn 10	About 9,300	Tetracycline resistance
Tn 501	About 8,200	Resistance to mercuric ion
Tn 951	About 16,600	Lactose operon *(Yersinia enterocolitica)*
Tn 1681	About 2,040	Stable toxin *(Escherichia coli)*

Summary Outline

CONCEPT OF THE GENE

1. Genetics is the study of the inheritance of biological traits, which are coded for by sequences of nucleotides in the DNA called genes.
2. Studies of *Neurospora* mutants showed that nutritional mutants lack an enzyme needed to synthesize the nutrient. It was proposed that one gene codes for one enzyme.
3. A cellular gene is a segment of DNA that is transcribed into RNA. Most are translated into a polypeptide; however, the genes for rRNA and tRNA make only RNA molecules.

MUTATIONS RESULT FROM PERMANENT CHANGES IN DNA

1. Mutations are inheritable changes in a cell's DNA that can be caused by chemical or physical agents.
2. Mutations in individual genes occur independently, so the rate of mutation in two genes is the product of the individual rates.
3. Additions and deletions of nucleotide bases cause frameshift mutations.
4. Chemical mutagens chemically alter existing nucleotide bases or they are incorporated into DNA as base analogues.
5. UV light is a physical mutagen that causes thymine dimer formation. Bacterial light-repair and dark-repair mechanisms correct UV damage before mutations become permanent.
6. Nutritional mutants are isolated by selecting against the prototrophs and then replica-plating the mutants on minimal medium containing nutritional supplements.

BACTERIAL GENETICS: TRANSFORMATION

1. Genetic information can be transferred by transformation, transduction, or conjugation among select bacteria.
2. Transformation is the transfer of free (naked) DNA from the environment to a competent recipient cell. DNase inhibits transformation since it hydrolyzes free DNA.
3. Competent gram-positive bacteria bind double-stranded DNA to surface receptors. One strand is degraded before the other strand is taken into the cell.
4. Gram-negative bacteria take up naked double-stranded DNA from closely related strains. Only DNA containing specific recognition sequences is able to transform the cell.
5. Artificial transformation is used in laboratory experiments to transfer naked, self-replicating, double-stranded DNA into gram-negative bacteria that lack a natural transformation system.

BACTERIAL GENETICS: CONJUGATION

1. Bacterial conjugation is the transfer of genes between cells in direct contact with each other.
2. Genetic transfer during conjugation is directional (donor to recipient) and is mediated by plasmids that are either free or integrated into the bacterial chromosome.
3. Plasmids are autonomous, self-replicating, extrachromosomal, circular DNA molecules. The F (fertility) plasmid codes for those structures and cellular activities needed for its transfer.
4. F^+ cells transfer a copy of the F plasmid to F^- recipient cells.
5. The F plasmid is integrated into the bacterial chromosome in Hfr cells, enabling chromosomal genes to be transferred at a high frequency.
6. Conjugation with Hfr donor cells is used to construct the circular chromosomal map of *E. coli* and other bacteria.
7. Excision of the F factor from the chromosome of Hfr cell creates an F' plasmid that can contain chromosomal genes.
8. Partial diploids are created when an F' plasmid containing chromosomal genes is transferred to an F^- cell.

MAJOR GROUPS OF PLASMIDS

1. R plasmids code for multiple-drug resistance and

create major medical problems when transferred to disease-causing bacteria.

2. Bacteriocins are bacterial products encoded on plasmids that kill closely related bacteria.
3. Certain strains of harmless bacteria cause disease after they acquire virulence plasmids.
4. Degradative plasmids code for enzymes that metabolize unusual organic compounds, synthetic organic chemicals, or common substrates such as lactose.

TRANSPOSONS ARE MOBILE GENETIC ELEMENTS

1. Transposons are mobile genetic elements composed of a core DNA sequence, often coding for drug resistance, flanked by IS-like sequences.
2. A copy of a transposon can be inserted at numerous places in the bacterial genome, or it can be transposed from a plasmid to the genome and vice versa.
3. Plasmids and transposons provide mechanisms by which bacteria acquire additional, nonessential genetic information.

Questions and Topics for Study and Discussion

QUESTIONS

1. Describe the current concept of the gene.
2. What is the molecular basis for mutations? How would a bacterial geneticist isolate a *trp*$^-$ mutant from a wild type *E. coli?*
3. Explain how the Ames test is used to detect potential carcinogens.
4. Define or explain the following:

auxotroph	Hfr cell
bacterial transformation	plasmid
bacteriocins	R plasmid
bacterial conjugation	silent mutation
F′ cell	transposon

5. Describe the mechanisms used by bacteria to correct mutations caused by UV irradiation.
6. How are plasmids involved in the transfer of genetic information between bacteria?
7. How is the rolling-circle mechanism of replication involved in the transfer of genetic information between bacteria?
8. Name four types of plasmids and explain the advantages or disadvantages they impart to the cell that possesses them.
9. Describe how conjugation is used to map the bacterial chromosome.
10. Explain how a transposon on a bacterial chromosome can cause infectious drug resistance.

DISCUSSION TOPICS

1. What impact do you think plasmids and transposons will have on our concepts of bacterial evolution and classification?
2. What is the molecular basis for a mutation occurring in an RNA virus? How does this relate to our definition of a gene?

Further Readings

BOOKS

Broda, P., *Plasmids,* Freeman, 1979. An excellent short book on bacterial plasmids.

Gardner, E. J., and D. P. Snustad, *Principles of Genetics,* 7th ed., Wiley, New York, 1984. A beginning college-level genetics textbook that is current and readable.

Hardy, K., *Bacterial Plasmids, Aspects of Microbiology 4,* 2nd ed., David Schlessinger (ed.), American Society for Microbiology, Washington, D.C., 1985. A short paperback that covers all aspects of bacterial plasmids.

Hayes, W., *The Genetics of Bacteria and Their Viruses,* Wiley, New York, 1968. This classic book offers a wealth of information on the experiments and conceptual development of microbial genetics.

Ingraham, J. L., O. Maaløe, and F. C. Neidhardt, *Growth of the Bacterial Cell,* Sinauer Associates, Sunderland, Mass. 1983. The authors do an excellent job of integrating microbial genetics with the molecular biology and physiology of the bacterial cell.

Watson, J. D., N. H. Hopkins, J. W. Roberts, J. A. Steitz, and A. M. Weiner, *Molecular Biology of the Gene,* Vol. I, Benjamin–Cummings, Menlo Park, Calif., 1987. Current material on molecular biology with one chapter devoted to the genetic systems of *E. coli* and its viruses.

ARTICLES AND REVIEWS

Brown, D. D., "The Isolation of Genes," *Sci. Am.,* 229:20–29 (August 1973). The application of molecular hybridization to the isolation of the ribosomal RNA genes from the frog *Xenopus.*

Clowes, R. C., "The Molecule of Infectious Drug Resistance," *Sci. Am.,* 228:18–27 (April 1973). An article on the role of bacterial plasmids in the transfer of antibiotic resistance between bacteria.

Cohen, S. N., and J. A. Shapiro, "Transposable Genetic Elements," *Sci. Am.,* 242:40–49 (February 1980). An introduction to the history of transposons followed by a detailed discussion of the significance of these genetic elements.

Novick, R. P., "Plasmids," *Sci. Am.,* 243:103–127 (December 1980). A brief review of the biochemistry and genetics of extrachromosomal bacterial DNA. The author raises penetrating questions concerning the ultimate biological role of bacterial plasmids.

Smith, H. O., D. B. Danner, and R. A. Deich, "Genetic Transformation," *Ann. Rev. Biochem.,* 50:40–68 (1981). The authors review the processes by which cells take up naked DNA from the surrounding medium.

C H A P T E R

11

Bacterial Virology

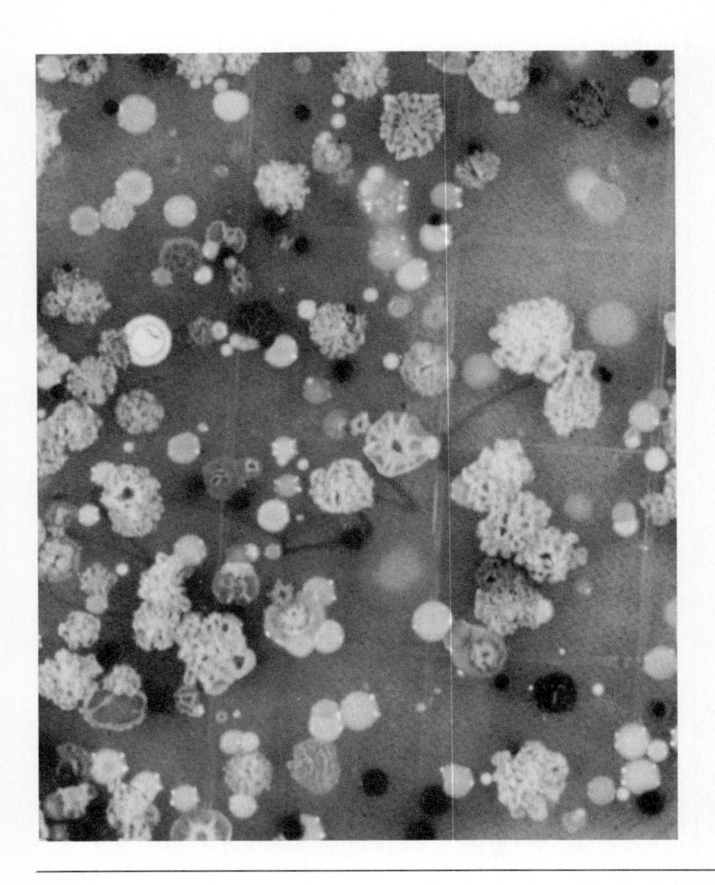

OUTLINE

FOCUS OF THIS CHAPTER

The basic principles of virology are developed in discussions of the isolation, assay, structure, classification, and replication of bacterial viruses. These principles are employed to explain how viruses transfer bacterial genes and alter bacterial phenotypes. The diversity of the bacterial viruses and their historical importance to the development of molecular biology are emphasized throughout the chapter.

Viruses are complex macromolecular entities that can replicate only after entry into specific types(s) of living cells, where they use their own genetic information, encoded in DNA or RNA, to replicate. They exploit the machinery of the host cell for their own reproductive purposes. Viruses are composed of one type of nucleic acid (either DNA or RNA but never both) surrounded by a protective protein coat. All viruses are noncellular and by necessity are smaller than the cells they infect.

The status of viruses as organisms has been debated by scientists for decades. Viruses are self-contained, reproductive entities that produce a continuous lineage, and as such they are similar to organisms. However, cells are the basic structural unit of all organisms, whereas viruses are noncellular. The evolutionary relationships among many organisms have been established, while the question of how and when viruses originated is still unanswered. These characteristics make it impossible to interrelate viruses with the evolutionary and taxonomic schemes of cellular organisms. Nevertheless, viruses are studied by microbiologists because of their small size, their unique composition, and their ability to replicate. Virology, the study of viruses, historically developed as an integral part of microbiology.

HISTORY OF VIROLOGY

Viruses were discovered over a period of years, more by inference than by conclusive experimentation. Attempts to control smallpox began over 2000 years ago in China where the process of variolation was developed to protect people against smallpox. Variolation involved taking material from a smallpox lesion and inoculating it into the skin of healthy people. This protected some patients against the dreaded outcomes of smallpox, but other patients became fatally ill after variolation.

Edward Jenner was the first physician to discover a safer and more effective approach for controlling smallpox (Table 11.1). His work led to the dramatic experiment that was initiated on May 14, 1796. Jenner inoculated a young boy with material taken from a cowpox lesion. The boy experienced a mild illness but recovered completely within ten days. On the first of July the boy was inoculated with material from a smallpox lesion. His reaction to the material was identical to that seen in patients who had recovered from smallpox. Even though Jenner did not know smallpox was caused by a virus, and had no knowledge of immunology, his procedure for vaccinating people with cowpox became the major medical technique for preventing smallpox.

Little progress was made with human virus diseases until Louis Pasteur showed that rabies could be passed between laboratory animals. From these infected animals he developed a rabies vaccine in 1885. This was quickly followed by discoveries concerning the transmissibility of plant diseases.

Plant virology began in 1886 when Adolf Mayer demonstrated the infectious nature of tobacco mosaic disease. Six years later Dimitri Ivanovsky reported that the agent for this plant disease passed through filters that retained bacteria. Then Martinus Beijerinck, a famous bacteriologist who was unaware of Ivanovsky's work, did similar experiments and proposed that tobacco mosaic disease was caused by a filterable infectious agent, smaller than bacteria, that replicated inside infected cells.

DISCOVERY OF ANIMAL VIRUSES

At the turn of the century foot-and-mouth disease was a serious illness in cattle. It has now been eliminated from the United States through the killing of the infected animals, but still causes enormous economic losses in other countries. In 1898, two scientists, Friedrich J. Loeffler and Paul Frosch, proved that foot-and-mouth disease is caused by an agent that is too small

TABLE 11.1 Historical events in virology

YEAR	SCIENTIST	DISCOVERY OR EVENT
1796	Edward Jenner	Smallpox vaccination
1885	Louis Pasteur	Rabies vaccine
1886	Adolf Mayer	Transmission of tobacco mosaic (TM) disease
1892	Dimitri I. Ivanovski	Filterability of TM virus
1898	Friedrich J. Loeffler and Paul Frosch	Filterability of foot-and-mouth disease (animal virus)
1898	Martinus W. Beijerinck	Intracellular reproduction of TMV
1901	Walter Reed and associates	Yellow fever virus
1903	P. Remlinger and Riffat-Bey	Rabies virus
1907	P. M. Asburn and C. F. Craig	Virus of dengue fever
1908	V. Ellerman and O. Bang	Transferred chicken leukemia
1909	S. Flexner and P. A. Lewis	Poliomyelitis virus
1911	J. Goldberger and J. F. Anderson	Measles virus
1911	Peyton Rous	Transferred chicken sarcoma (cancer)
1915	Frederick W. Twort	Bacterial viruses
1917	Felix d'Herelle	Rediscovered bacterial viruses, named them bacteriophages
1921	Jules Bordet and M. Ciucia	Described lysogeny
1934	C. D. Johnson and E. W. Goodpasture	Mumps virus
1935	Wendell M. Stanley	Crystallized tobacco mosaic virus
1936	Martin Schlesinger	Phage nucleic acids
1938	Y. Hiroals and S. Tasaka	Rubella virus
1939	G. A. Kausche, E. Pfankuch, and H. Ruska	Used electron microscopy to visualize virus
1949	John H. Enders, T. H. Weller, and F. C. Robbins	Cultivated poliovirus in nonneuronal tissue culture
1951	Alfred D. Hershey and Martha Case	Bacteriophage DNA enters cell
1952	Norton D. Zinder and Joshua Lederberg	Transduction in *Salmonella*
1953	James D. Watson and Francis H. C. Crick	Double-helix structure of DNA
1956	A. Gierer and G. Schramm	Infectious nature of TMV RNA
1970	David Baltimore, Howard Temin, and Satoshi Mizutaki	RNA-directed DNA polymerase (reverse transcriptase)
1983–1984	Luc Montagnier et al. and Robert Gallo et al.	Human immunodeficiency virus, cause of AIDS

to be seen through the microscope and that passes through bacteria-retaining filters. They transferred the agent into healthy calves by injecting small amounts of filtered lymph taken from a diseased cow. They proposed that other infectious diseases, including cowpox, smallpox, and measles are also caused by filterable agents.

By 1911, the filterable nature of the infectious agents responsible for yellow fever, rabies, dengue fever, poliomyelitis, measles, and chicken sarcoma had been reported (Table 11.1). Today we know there are hundreds of viruses that cause human diseases; there are over 100 viruses that cause the common cold. This knowledge combined with a basic understanding of animal virology is now applied to the problem of controlling viral diseases, resulting in the control of measles, mumps, rubella, and polio with immunizations. Unfortunately there are many viral diseases for which there is no effective prevention or treatment, so they run their course, causing illness and, in come cases, death.

DISCOVERY OF BACTERIAL VIRUSES

Frederick W. Twort spent the early years of his career studying the nutritional growth requirements of bacteria. As a logical extension of his work, Twort attempted to grow filter-passing viruses on artificial media. His attempts failed because viruses need living cells to grow; however, during these experiments he isolated a micrococcus that appeared to be diseased. Colonies of this bacterium became watery, and some of the colonies could not be subcultered. When kept for a long time the colonies turned glassy and transparent. Twort was able to transfer this diseased state to other micrococcal colonies, even after filtering the affected cultures through a bacteria-retaining filter. In 1915, Twort concluded that *this disease of the micrococcus was caused by a filterable infectious agent that multiplied within the micrococci and caused them to lyse.*

Felix d'Herelle, a Canadian microbiologist, rediscovered bacterial viruses in 1917. Unfortunately he was unaware of Twort's earlier publication, and for four years he was acknowledged as the discoverer of bacterial viruses.

D'Herelle was working in Paris when he discovered a filterable virus that lysed a dysentery bacillus. He isolated the bacillus from a patient's stools shortly after the patient afflicted with dysentery was admitted to the hospital of the Pasteur Institute. For four consecutive days d'Herelle filtered the patient's stools and added the filtrates to broth cultures of the dysentery bacillus. For three days the cultures grew normally. On the fourth day, however, the filtrate-inoculated culture of bacillus turned perfectly clear. D'Herelle soon found that this filterable agent could be transferred to other cultures of Shiga dysentery bacilli. He named these viruses **bacteriophages** (bak-teer′ee-oh-fayg-es), which means bacteria eating.

D'Herelle's discovery stimulated scientists to investigate the ability of bacteriophages (or phages as they are generally called) to cure infectious diseases. This discovery even caught the imagination of the popular American author, Sinclair Lewis.* The use of bacteriophages to treat infectious diseases never proved effective, but the concept led to a great amount of work in bacterial virology. In 1921, Jules Bordet reintroduced Twort's work and challenged the priority of d'Herelle's discovery. Indeed, there is still controversy over who deserves credit for discovering bacterial viruses.

VIRUSES AND THE DEVELOPMENT OF MOLECULAR BIOLOGY

Bacteriophages had been an enigma to biologists, who had a difficult time understanding the basic nature of these invisible, infectious agents. A major turning point occurred in 1935 when Wendell M. Stanley crystallized tobacco mosaic virus (TMV). For this purpose he used the techniques of protein purification and crystallization to purify the virus. The purified product took the form of needle-shaped crystals of TMV that were infectious even after a billionfold dilution. He concluded that tobacco mosaic virus was an "autocatalytic protein" that replicated in living tobacco cells.

Stanley's hypothesis was not entirely correct. It was soon demonstrated that purified bacteriophages contained nucleic acid in addition to protein and that TMV RNA was infectious in itself. Virologists then realized that all viruses are composed of nucleic acid (either RNA or DNA) surrounded by a protein capsid. Following the development of the electron microscope in the late 1930s, it became evident that the morphology of these protein capsids varies greatly (Figure 11.1).

KEY POINT

Viral genetic information can be contained in either RNA or DNA; viruses never contain both types of nucleic acids. The two major groups of viruses are therefore RNA viruses and DNA viruses.

Molecular biologists were attracted to the viruses because of their simple composition and structure. The bacteriophages were particularly appealing because the

*Sinclair Lewis's novel *Arrowsmith* portrays the excitement among scientists who hoped bacteriophages could cure infectious bacterial diseases.

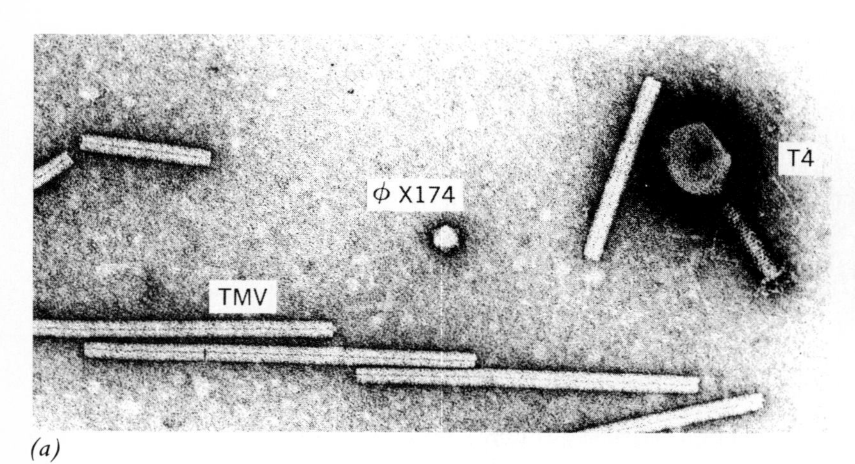

(a)

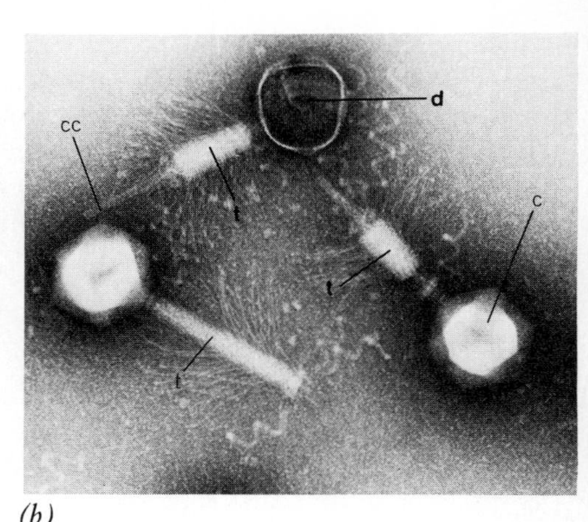

(b)

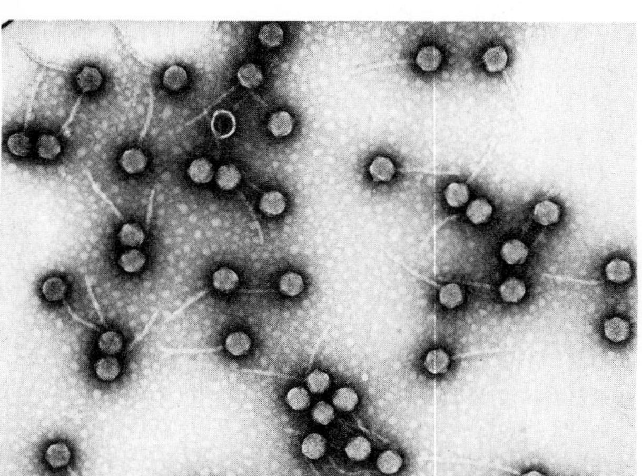

(c)

FIGURE 11.1 Viruses. *(a)* Transmission electron micrograph of a mixture of tobacco mosaic virus (TMV), T_4 bacteriophage, and ϕX174 bacteriophage. (Courtesy of Dr. F. A. Eiserling.) *(b)* Electron micrograph of *Bacillus* bacteriophages, 100,000×. The virus with the dark capsid *(d)* has lost its DNA. One virion has an extended tail, while the tails of two of the virions have contracted tails (t) and an exposed central core (cc). (Courtesy of Dr. R. K. Nauman.) *(c)* The lambda bacteriophage has a flexible tail that terminates in a single tail fiber. (Courtesy of Professor J. T. Finch, MRC Laboratory of Molecular Biology, Cambridge, Mass.)

bacteria they infect are easily cultured under controlled conditions. Experiments soon demonstrated that bacteriophages reproduce in assembly-line fashion—not by growth and binary fission. The component parts of a bacteriophage are produced inside the infected bacterium during a short latent period. These parts are then assembled into a complete infective **virion,** which is a fully developed virus particle. When enough virus particles are present inside the cell, the cell lyses and releases the bacteriophages into the environment.

Some early studies of DNA bacteriophages provided evidence that DNA was the molecule of heredity (Figure 11.2). In 1951 bacteriophages were cultured such that their DNA was labeled with ^{32}P and their protein was labeled with ^{35}S. When the labeled bacteriophages were used to infect host cells, the ^{35}S-labeled protein remained outside the cell, while the ^{32}P-labeled DNA entered the cell. This experiment firmly established that bacteriophages reproduce inside their host cell, and the information required for virus propagation is contained in bacteriophage DNA. Shortly after these experiments, James Watson and Francis Crick proposed their model of DNA (see Chapter 9), which ushered in the age of molecular biology.

FUNDAMENTALS OF BACTERIAL VIROLOGY

Much of modern virology is predicated on basic principles learned from experiments with bacteriophages. Since the discovery of bacteriophages, virologists have

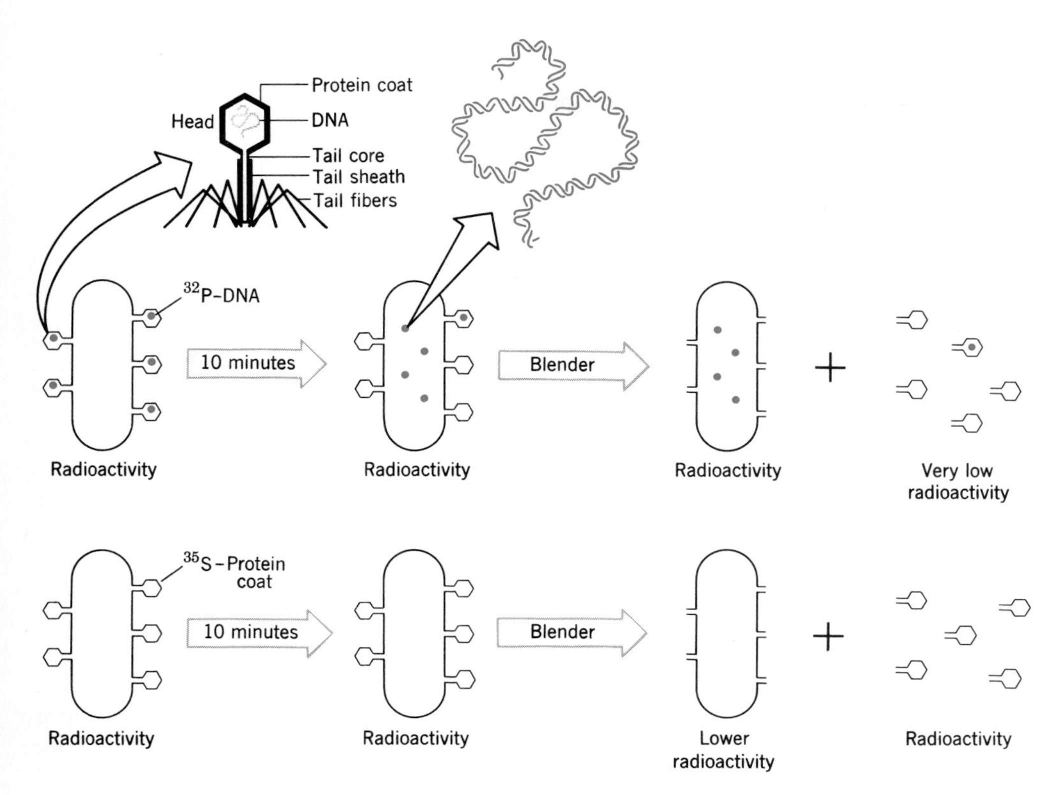

FIGURE 11.2 The Hershey–Chase experiment. T_4 bacteriophages whose **DNA** had been labeled with the isotope ^{32}P were used to infect *Escherichia coli*. Ten minutes later, the infected bacteria were placed in a blender and the bacteriophage capsids were severed from the bacterial cell wall by shearing forces. After centrifugation, the ^{32}P-labeled DNA was associated with bacterial cells found in the pellet. When T_4 bacteriophages whose **protein** had been labeled with the isotope ^{35}S were used to infect *E. coli*, the label was found in the supernatant solution. This experiment provided early evidence of the role of DNA in heredity. (From E. J. Gardner and D. P. Snustad, *Principles of Genetics*, 6th ed., Wiley, New York, 1981).

been able to study bacterial viruses in the laboratory because the bacteria they infect are easy to culture. Today, virologists study bacteriophages because they are interesting in their own right, they serve as model systems for molecular biology, and some transfer bacterial genes (transduction); and they have a major influence on bacterial infections through phage conversion.

CLASSIFICATION OF BACTERIOPHAGES

Chemical composition and morphology are the major criteria followed in bacteriophage classification. Viral nucleic acid is either DNA or RNA, arranged as a single-stranded, double-stranded, linear (two free ends), or closed molecule. Surrounding the nucleic acid is a protein capsid that can be symmetrical or helical. Additional proteins can be present in the bacteriophage tail assembly; however, not all bacteriophages have tails (Figure 11.3). Indeed, bacteriophage morphology can be described as cubic, filamentous, pleomorphic, or tailed. The major groups of bacteriophages are defined on the bases of viral morphology and nucleic acid composition (Table 11.2).

Some bacteriophage groups contain hundreds of individual viruses, whereas others contain only one. Therefore, chemical composition and morphology

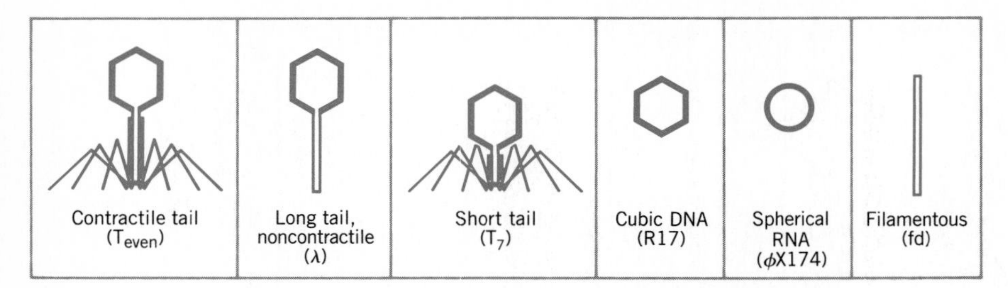

FIGURE 11.3 Morphological characteristics are used in the classification of bacterial viruses.

cannot be used as the sole criteria for identifying and classifying bacteriophages. Other criteria include the viral particle size, shape, structure, and weight, together with the type of nucleic acid and its molecular weight.

ISOLATION OF BACTERIOPHAGES

Almost all natural environments contain bacteriophages that can be isolated by filtration. Aqueous suspensions of natural samples containing bacteriophages are filtered through membrane filters that retain the debris and bacteria, but allow the viruses to pass. The filtrate is then mixed with a growing culture of the host bacterium. The bacteriophages reproduce in their host cells before they are released into the medium, usually following lysis of the host cell. Clones of individual bacteriophages can then be isolated from this medium by plating them on a lawn of host cells.

Plaque Assay

Bacteriophages produce plaques in a lawn of host bacteria growing over the surface of an agar plate (Figure 11.4). The turbid bacterial growth covers the surface of the plate except where the viruses have lysed the

TABLE 11.2 Classification of bacteriophages

SHAPE	NUCLEIC ACID[a]	EXAMPLE	CHARACTERISTICS	NUMBER KNOWN
Tailed	ds DNA, L	T_2	Contractile tail	450
	ds DNA, L	Lambda	Long, noncontractile tail	800
	ds DNA, L	T_7	Short tail	300
Cubic	ss DNA, C	φX174	Large, knoblike capsomeres	25
	ds DNA, C	PM2	Lipid-containing capsid	2?
	ds DNA, L	PRD1	Double coat, lipids, pseudotail	4?
	ss RNA, L	MS2, Qβ	Icosahedral shell	35
	ds RNA, L	φ6	Lipid-containing envelope	1
Filamentous	ss DNA, C	fd	Long rods	17
	ss DNA, C	MV-L1	Short rods	10
Pleomorphic	ds DNA, C	MV-L2	Lipid-containing envelope	3?

Source: Adapted from Hans-W. Akerman et al., "Guidelines for Bacteriophage Classification," in M. A. Lauffer, F. B. Bang, K. Maramorosch, and K. M. Smith (eds.), *Advances in Viral Research,* Academic Press, New York, 23:1–24 (1978).

[a]Key: ss, single-stranded; ds, double-stranded; C, circular; L, linear.

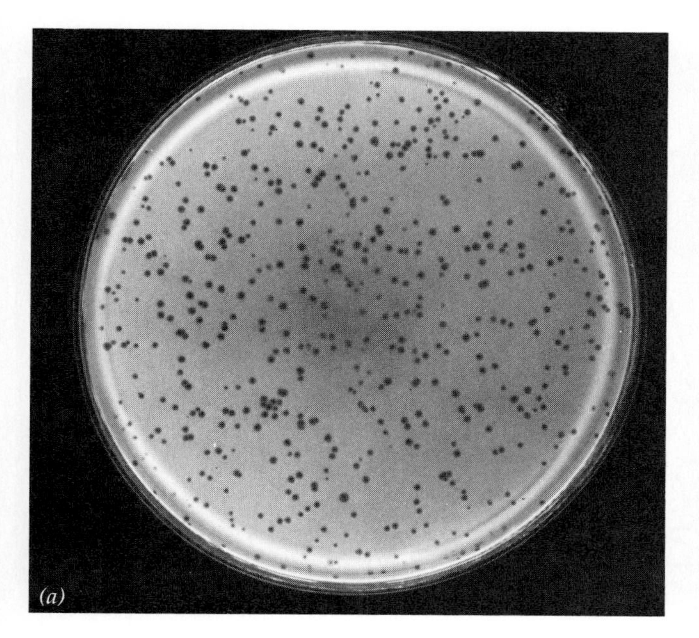

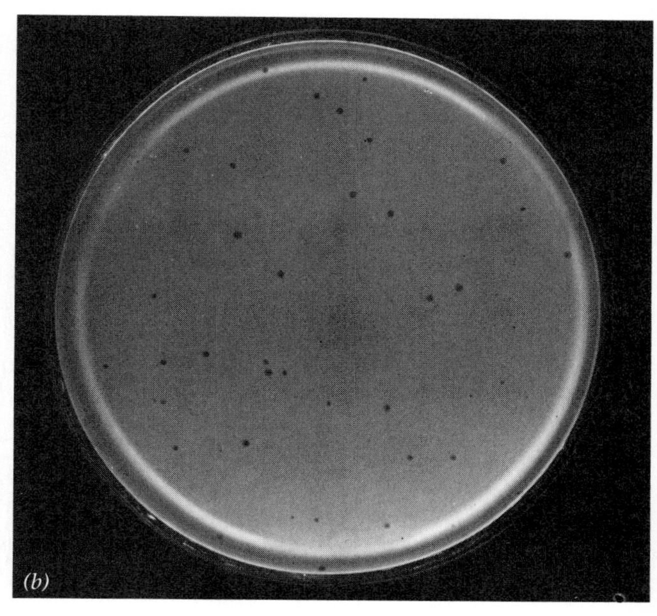

FIGURE 11.4 Plaques of T_4 bacteriophage growing on a lawn of *Escherichia coli* B. The number of plaques is directly proportional to the original concentration of phages in the sample plated on the lawn of bacteria. Petri dish *a* contains approximately 10 times more plaques than petri dish *b*. (Courtesy of Dr. Lee D. Simon, Waksman Institute of Microbiology, Rutgers University.)

bacteria. The clear spots in the bacterial lawn are called **plaques**, each containing thousands of bacteriophages. A pure clone of the virus can be obtained by replating the material in the center of the plaque on fresh host cells and isolating phage from a new plaque.

The plaque assay is a means of counting bacteriophages, since the number of plaques is directly proportional to the original number of infected bacteria. A plaque assay is performed by mixing a sample of bacteriophages containing 10 to 200 virions with approximately 10^8 cells of the host bacterium suspended in "top" agar (0.8 percent agar as opposed to the usual 1.5 percent). The mixture is then dispersed over a solid layer of nutrient agar. Each virus-infected bacterium produces progeny viruses, which in turn cause a plaque to form on the plate. A plaque arises from a single-virus-infected bacterium, so *the original number of plaques on a plate is equivalent to the number of bacteriophages mixed with the host bacterium.*

KEY POINT

Bacteriophages, released when the plated infected cell lyses, diffuse in a circular fashion and infect new cells. Lysis of these host cells enlarges the plaque. The size and morphology of the plaque vary depending on the species and physiological state of the host, strain of bacteriophage, and conditions of incubation.

DNA BACTERIOPHAGES

The T bacteriophages, designated T_1 through T_7, are double-stranded DNA viruses that infect *Escherichia coli*. The T-even bacteriophages (T_2, T_4, and T_6) are composed of a head, tail, tail plate, and tail fibers (Figure 11.5). The **capsid** or protein coat of a virus is composed of individual proteins known as **capsomeres**. The capsid or bacteriophage head surrounds a compartment that contains the nucleic acid (Figure 11.6),

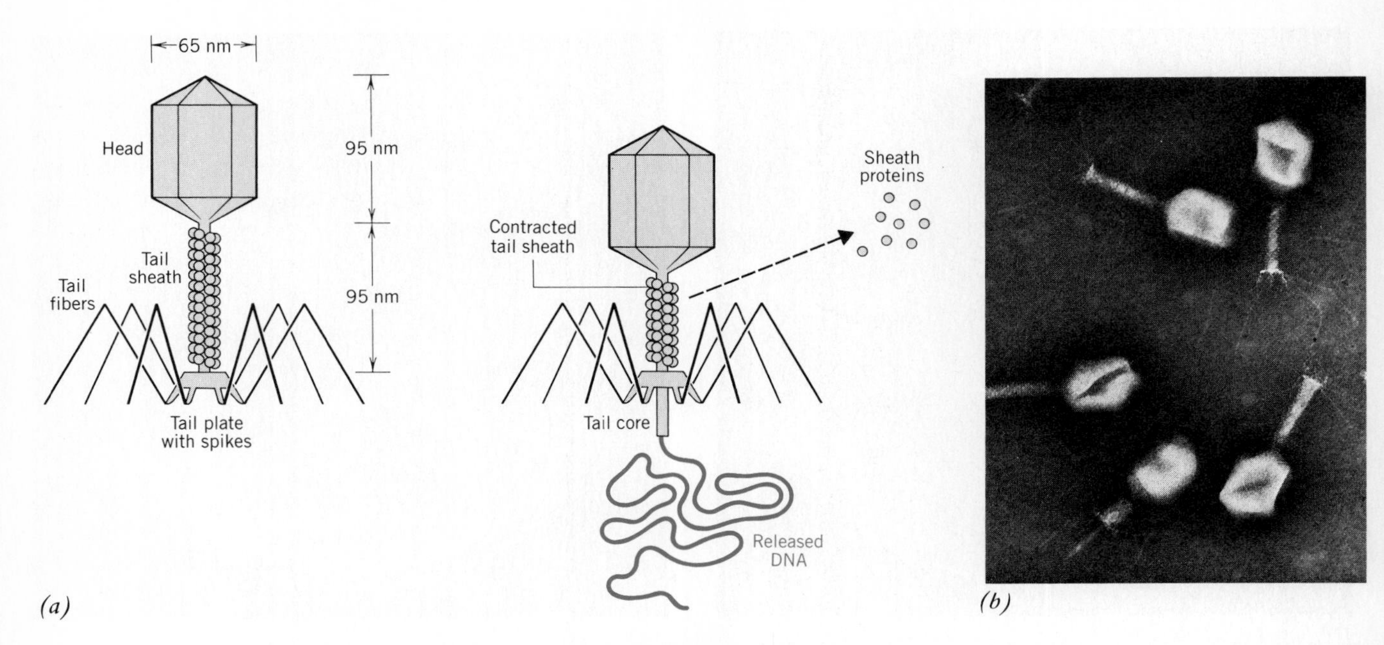

FIGURE 11.5 *(a)* Diagram of T-even bacteriophages, one with an extended tail sheath and one with a contracted tail sheath. Proteins are lost from the tail sheath during tail contraction. *(b)* Electron micrograph of negatively stained T_4 bacteriophages showing the complex morphology of the tail. (Courtesy of Dr. E. Boy de la Tour and E. Kellenberger.)

which in these phages is a linear DNA molecule. This DNA is protected against degradation by environmental nucleases because of its location inside the bacteriophage head.

The tail assembly is involved in attaching the phage to the host cell and injecting the phage DNA into the host's cytoplasm. The phage tail is composed of a core surrounded by a sheath and a terminal tail plate. The spikes and tail fibers emanating from the tail plate specifically bind to receptors on the outer membrane of the *E. coli* cell wall. Following adsorption to the host cell, the tail sheath loses proteins in a process that shortens the sheath and propels the tail core through the cell's outer layers (Figure 11.5). During this process, the DNA is extruded out of the bacteriophage head and into the cell's cytoplasm. This infection usually kills the host cell as the virus commandeers the cell's synthetic activities for its own replication.

Bacteriophage Growth Curve

The **lytic cycle** of bacteriophage replication is initiated by the injection of bacteriophage DNA into a host cell and culminates in the release of progeny viruses upon lysis of the host. The stages of the phage replication cycle are portrayed in the one-step growth experiment shown in Figure 11.7.

The one-step growth experiment follows the production of bacteriophages in a culture of host cells, all of which were infected simultaneously at zero time. The experiment depicted in Figure 11.7 was initiated when *Escherichia coli* was infected with bacteriophage T_1 such that the culture contained 10^5 virus-infected bacteria per milliliter. The number of infected units (infected bacteria plus newly produced T_1 bacteriophages) was measured by the plaque assay method and plotted as a function of time.

The results demonstrate that bacteriophage replication occurs in two distinct phases. The **latent period** is the time interval following infection (23 minutes), during which no increase in the number of infective units is observed. This is followed by a **rise period** (approximately 10 minutes), in which the number of infective units increases dramatically. In the experiment shown, the increase is a hundredfold. The difference between the initial and the final number of infected

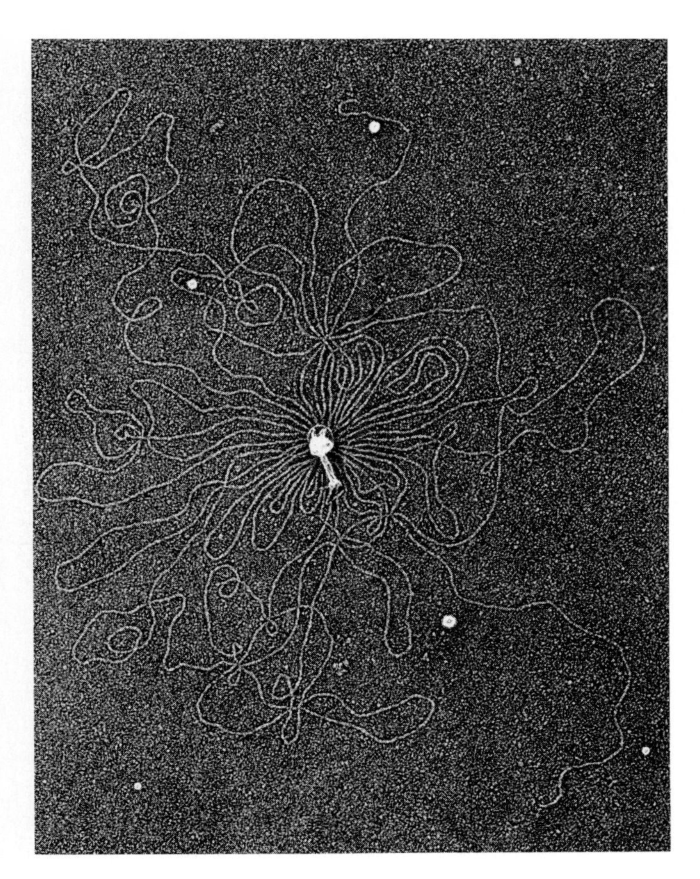

FIGURE 11.6 The T_2 bacteriophage capsid contains a single linear DNA molecule (note the two ends). The DNA has been released from the bacteriophage head (center) by gentle lysis. (Courtesy of Dr. A. K. Kleinschmidt, *Biochem. Biophys. Acta,* 61:861, 1962.)

units is called the **burst size**. The replicative process is called the lytic cycle because each infected bacterium is lysed by viral enzymes produced during phage replication.

Do bacteriophages divide inside the cytoplasm during the latent period? Experiments show that they do not. If infected cells are experimentally lysed during the first 11 minutes of the lytic cycle, no bacteriophages are recovered from the cell lysate. Following this 11-minute **eclipse period**, the number of infective bacteriophages continually increases inside the cell until the number of viruses reaches a plateau characteristic of the maximum burst size.

KEY POINT

Bacteriophage replication is similar to assembly-line manufacturing, where the individual parts are made before the desired product is assembled in a final form. In bacteriophage reproduction, the individual parts are produced during the eclipse period and are self-assembled to form infective viruses that are released during the rise period.

Replication of a T-Even Bacteriophage

The replication of bacteriophage T_4 in *Escherichia coli* was chosen to demonstrate the principles of viral replication. The intracellular development of this DNA virus in *E. coli* is shown in the sequence of electron micrographs presented in Figure 11.8. Each stage of the replicative process is diagramed in Figure 11.9.

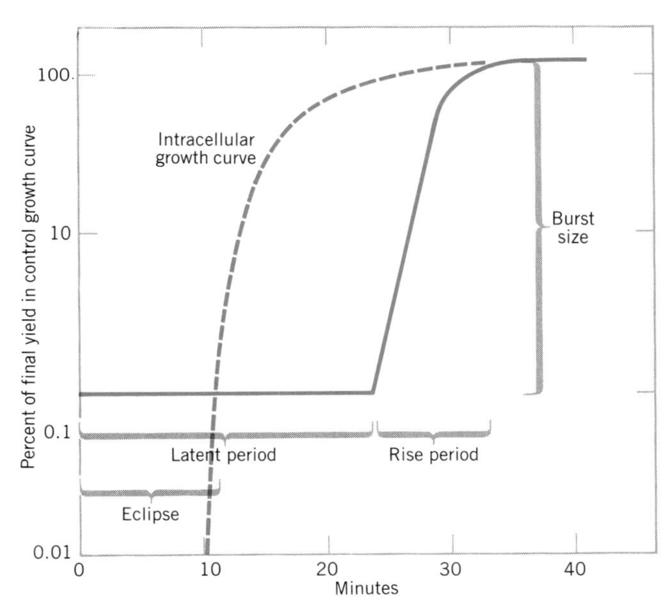

FIGURE 11.7 One-step growth curve of a bacteriophage (T_1) grown on *Escherichia coli*. The growth experiment was started with 10^5 infective units per milliter of culture. The dashed line represents the production of infective centers within the cell as observed following the artificial lysis of culture samples at sequential intervals. (Courtesy of Dr. A. H. Doermann. Reproduced from *J. Gen. Physiol.*, 35:645–656, 1952, by copyright permission of the Rockefeller University Press, New York.)

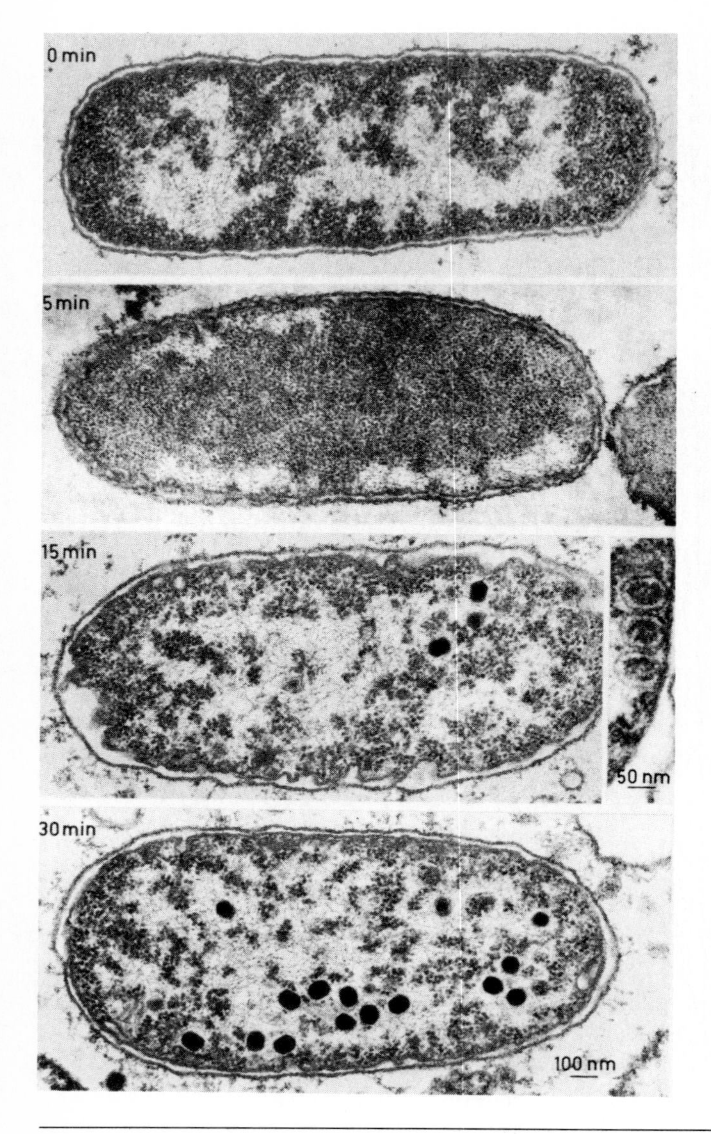

Adsorption Bacteriophage T_4 adsorbs to a specific lipopolysaccharide receptor on the outer membrane of the *E. coli* cell wall (Figure 11.10). The tail fibers help to position the bacteriophage over the cell's receptor in the initial steps of adsorption. Not all *E. coli* cells possess this lipopolysaccharide, and cells that do not are resistant to T_4 infection.

Other bacteriophages bind to different cell surface receptors. Bacteriophage receptors include the pili of *E. coli,* the Vi-antigen of *Salmonella,* glucosylated teichoic acid of *B. subtilis,* and portions of the outer membrane of the gram-negative bacterial cell wall. Because adsorption is a specific biochemical reaction, bacteriophages can infect only certain host cells. This adsorption is so specific that it is used to classify bacteria within a given species. Strains having the same bacteriophage receptors are classified together, just as bacteria possessing the same surface antigens are classified together.

Penetration Many of the tailed bacteriophages have a contractile mechanism that propels the central tail core through the cell wall and into the bacterium's cytoplasmic membrane. The T_4 tail sheath shortens when tail-sheath proteins are released into its surroundings. Tail "contraction" is thought to be powered by ATP associated with the tail rings. Penetration through the cell wall is assisted by lysozyme in the tail core, an enzyme that degrades cell-wall peptidoglycan. The T_4 DNA then passes through the central tail core into the host cell's cytoplasm, while the empty protein capsid remains attached to the cell surface (Figure 11.10). More than one phage can adsorb to a host cell and inject its chromosome. If enough bacteriophages

FIGURE 11.8 Intracellular development of bacteriophage T_4 in *Escherichia coli.*
Time 0′ This cell has the morphology of an uninfected cell with the typical nuclear region of cells in exponential growth. The infection starts after adsorption of the phage, partial penetration of the central tube of the tail, and injection of the DNA.
Time 5′ The bacteriophage infection results in "nuclear disruption" caused by the virus-directed hydrolysis of the host's DNA. The DNA and its breakdown products are found in marginal vacuoles. The nucleotides from the host DNA are later used by the bacteriophage to synthesize its own DNA.
Time 15′ About 8 minutes after infection, the synthesis of the structural proteins used for assembling the maturing bacteriophages begins. A few minutes later, the first infective viruses appear and continue to mature at a rate of five per minute. It is generally thought that a prehead with no or only a little DNA is made first (insert). The fine fibrillar material resembling the bacterial nucleus is the "vegetative bacteriophage."
Time 30′ A later stage of bacteriophage development showing increasing amounts of finished heads. The amount of virion particles per section is about 1/20 the total number per cell.
(Electron micrographs and legend provided by Beate Menge, Jacomina v. d. Broek, H. Wunderli, K. Lickfeld, M. Wurtz, and E. Kellenberger.)

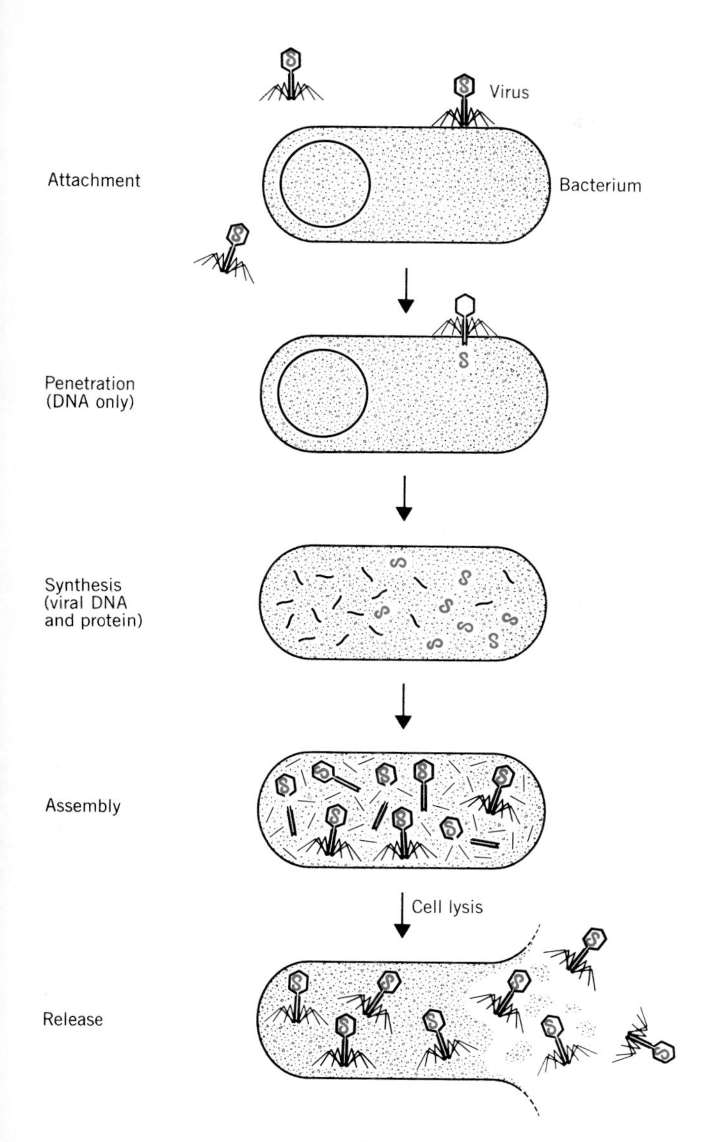

FIGURE 11.9 Steps in the reproduction of a DNA bacteriophage. (Modified from P. Sheeler and D. E. Bianchi, *Cell Biology*, Wiley, New York, 1980.)

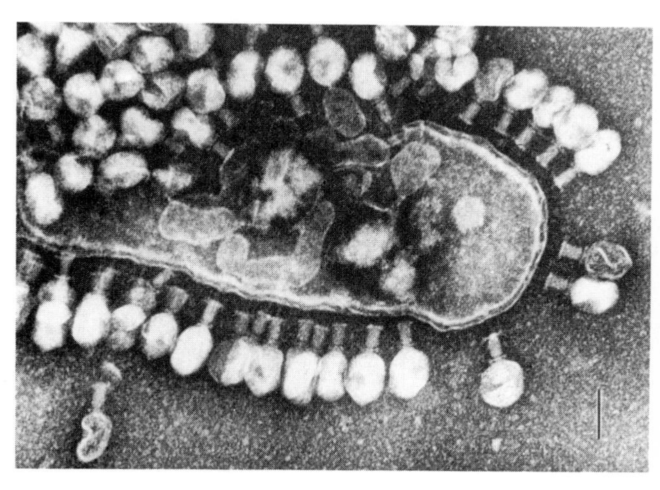

FIGURE 11.10 An electron micrograph of T_4 bacteriophages adsorbed to an *Escherichia coli* B cell. (Courtesy of Dr. Lee D. Simon. From L. D. Simon and T. F. Anderson, *Virology*, 32:279–295, 1967 and by permission of Academic Press, New York.)

adsorb, they punch so many holes in the cell that it dies, a phenomenon called lysis from without.

Early Protein Synthesis The T_4 bacteriophage genome is arranged in segments that are transcribed at different times during the replicative cycle. Early protein is made after the host cell's RNA polymerase transcribes a short segment of viral DNA into mRNA. The host's RNA polymerase recognizes the viral promoter that governs the transcription of the genes coding for early protein synthesis. One of the early proteins to be synthesized is the virus's own DNA-dependent RNA polymerase, which "recognizes" the other promoter regions on the viral DNA. The mRNAs coded for by the viral DNA are translated into proteins in the host cell's cytoplasm. The host cell's ribosomes, tRNAs, amino acids, ATP, other forms of energy, and enzymes are utilized for viral protein synthesis.

Early proteins are involved in commandeering the host's cellular machinery and in manufacturing progeny bacteriophages. Early proteins include nucleases, synthetic enzymes, and the RNA polymerase previously mentioned. The bacteriophage nucleases and protein kinases specifically prevent the cell from replicating its own DNA. Bacteriophages also make enzymes that are involved in the biosynthesis of precursors used to synthesize bacteriophage DNA. The actual synthesis of phage DNA is done by a viral DNA polymerase, formed early in the replication cycle, that makes the DNA used in the final assembly of the virions. Some bacteriophages also make an RNA polymerase that synthesizes the mRNA used in late protein synthesis.

Late Protein Synthesis Toward the end of the replication cycle, T_4 synthesizes *late proteins,* including the structural proteins necessary for virion self-assembly, the enzymes involved in maturation, and the proteins used in the release of bacteriophages from the cell. Bacteriophage DNA is synthesized in the cell's cytoplasm at the same time the viral structural proteins are being synthesized on the host's ribosomes. At this stage of the replication cycle, the cell becomes a repository for pools of all the bacteriophage proteins and copies of viral DNA.

Assembly Bacteriophages mature through a sequential process of self-assembly that is governed in part by the laws of quaternary protein structure (Chapter 2). The protein capsomeres join to form bacteriophage capsids at the same time they are filled with bacteriophage DNA (Figure 11.8). Concomitantly, the tails are formed from their component parts. The DNA-containing capsids combine with the tails before the tail fibers attach, to complete the process of phage assembly.

Release Infective T_4 bacteriophages are released into the surrounding environment when their host cells lyse. Release is mediated by a viral protein that damages the cell membrane and by bacteriophage lysozyme that damages the cell wall. The combined actions of these enzymes lyse the host cell and release the progeny virions.

TRANSFER OF BACTERIAL GENES BY TRANSDUCTION

On rare occasions, a bacteriophage acquires some of its host's DNA and transfers it to a recipient bacterium by the genetic process known as **transduction**. The recipient cell survives the process because the bacteriophages involved are unable to reproduce. The bacteriophages have no stake in the process—they accidentally acquire bacterial DNA and then act as a delivery system for the donor cell's DNA. Bacteriophages transfer bacterial genes by two basic mechanisms: (1) **generalized transduction**, in which random pieces of bacterial DNA are transferred; and (2) **specialized transduction**, which involves the process of lysogeny (see below).

GENERALIZED TRANSDUCTION

During the lytic cycle of phage infection, host-cell DNA is broken into small pieces by the viral nucleases synthesized during early protein synthesis. These pieces of host-cell DNA can be incorporated randomly into new phage particles in place of some or all of the phage DNA (Figure 11.11). The resulting phage particles are often defective because they lack a full complement of their own DNA. When a defective phage infects a bacterial cell, it injects bacterial DNA (from the donor cell) instead of or in addition to phage DNA. Once in the cell, the genes from the donor cell can recombine with the host cell's DNA to become a part of the host-cell genome. This process is known as **generalized transduction** because the phage can pick up any piece of host-cell DNA, so that any gene on the host chromosome can be transferred. Generalized transduction can be used to perform genetic mapping; however, its applicability is limited by the short pieces of DNA that are transferred.

KEY POINT

The volume of the phage head can contain only a limited amount of DNA. When the phage head mistakenly acquires bacterial DNA, it excludes segments of viral DNA required for viral reproduction. Such phages are defective.

LYSOGENY AND SPECIALIZED TRANSDUCTION

An unexpected quirk of the virus–host-cell relationship was the discovery that many host cells carry viral DNA as part of their genetic material. In many instances, the cell is not destroyed by the presence of these proviruses. This phenomenon was first discovered in bacteria isolated from nature and maintained in the laboratory for months before the presence of the virus was even suspected. Although the bacteria appeared normal, the cultures would occasionally lyse and release bacteriophages.

Any bacteriophage whose genome is an integral part

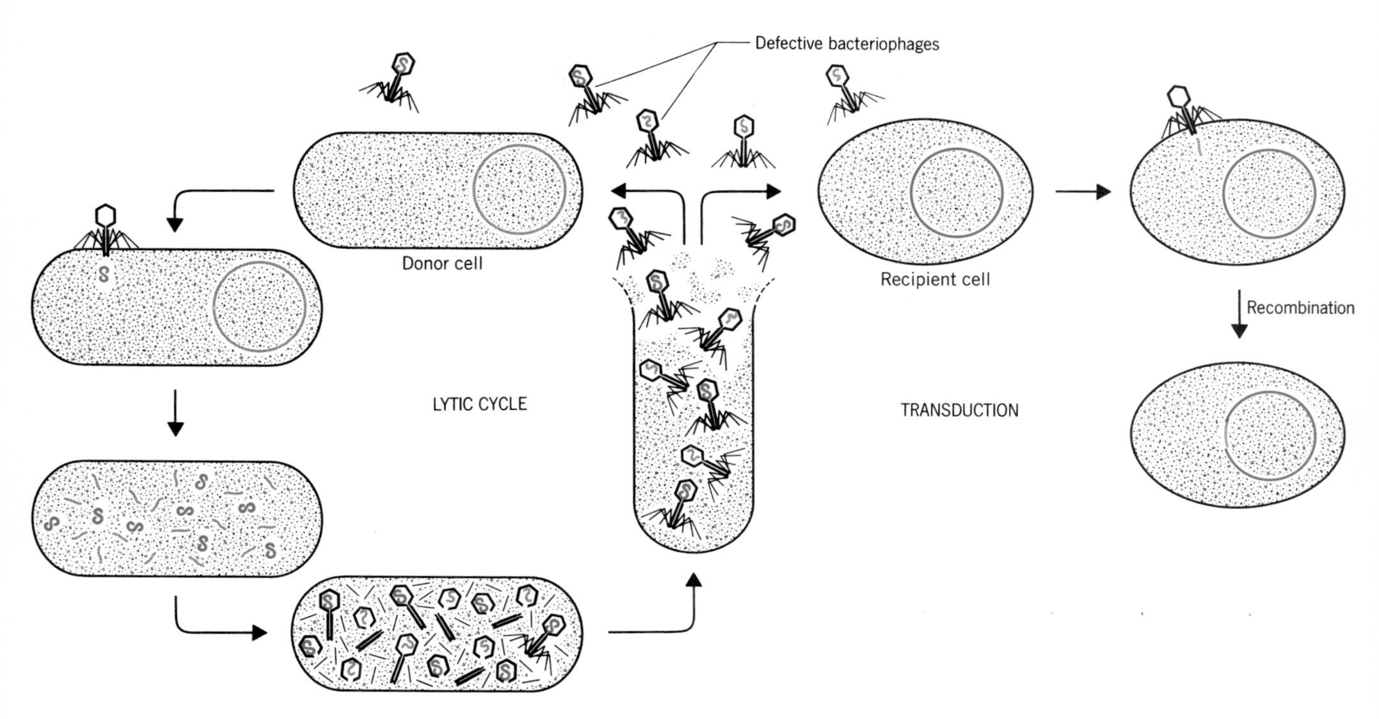

FIGURE 11.11 Generalized transduction is the bacteriophage-mediated exchange of bacterial DNA. During the normal lytic cycle of bacteriophage reproduction, defective bacteriophages are formed when empty heads take up host-cell DNA in place of phage DNA. Any piece of host-cell DNA can be incorporated by this process. These bacteriophages inject the bacterial DNA into the next cell they attach to.

of a host cell's chromosome is referred to as a **prophage** (or provirus). A bacterium that carries a prophage is a **lysogen** and the process of producing a lysogen is termed **lysogeny**. Bacterial strains can carry a prophage through thousands of generations without being adversely affected. The principles of lysogeny are explained below using bacteriophage lambda (λ) as an example.

Temperate bacteriophages are viruses that can exist as both a replicative form in the cell's cytoplasm, and as a prophage integrated into the host cell's chromosome. Lambda is a temperate DNA bacteriophage that infects *E. coli*. In its replicative form, lambda exists as a circular DNA molecule (physically similar to a plasmid) in the cytoplasm of its host. Its replication results in the destruction of the cell and the release of virulent lambda phages (Figure 11.12). Upon infection of another cell, the virulent lambda phages either reproduce via the lytic cycle or become prophages. The process that dictates the outcome of infection by a temperate bacteriophage—lytic cycle or lysogeny—in part depends on the formation of a repressor protein encoded by lambda DNA.

The Lambda Repressor Protein

Lambda DNA codes for a repressor protein that is formed as soon as the DNA enters the cell's cytoplasm. The lambda repressor binds to sites on the phage DNA (Figure 11.13) called the right operator. The genes to the right of this operator code for lambda virus production, while genes to the left code for the lambda repressor protein. When the lambda repressor binds to the right operator, it stimulates the transcription of the repressor gene (left) and prevents the transcription of lambda production genes (right).

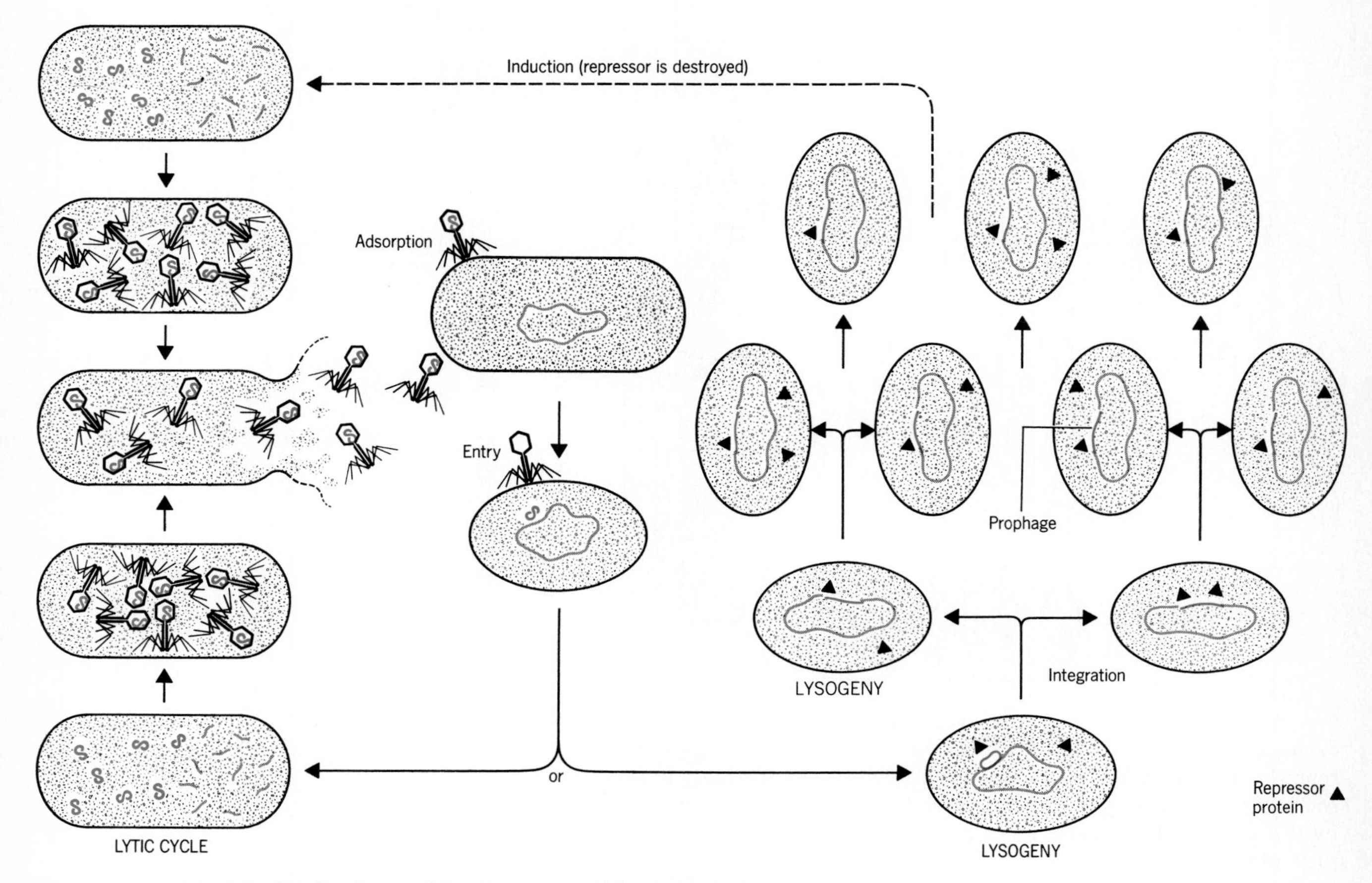

FIGURE 11.12 Replication and lysogeny caused by a temperate bacteriophage. After the genome of a temperate bacteriophage enters the host cell, the virus can progress through a lytic cycle to generate multiple progeny virions or it can be integrated into the bacterial genome as a prophage. Prophage genes code for the synthesis of a repressor protein that prevents transcription of the viral genome. Occasionally the repressor is inactivated and the prophage is induced to enter the lytic cycle that produces infective virions.

Virus production occurs only after the lambda repressor is inactivated. The destruction of the lambda repressor is mediated by the cell's **SOS response** to deleterious environmental effects. When the cell is exposed to UV light or mutagens, the SOS response is activated and certain proteins are formed. One of these proteins, the product of the **recA** gene, has a protease activity that attacks repressor proteins. The lambda repressor protein is inactivated by the recA protease, permitting transcription of the lambda genes to the right of the operator (Figure 11.13) and initiating the lytic cycle of lambda phage reproduction.

Integration of the Lambda Genome

Lambda DNA is double-stranded except for short 12-base sequences at its 5′ ends (Figure 11.14). These single-stranded tails are complementary, so lambda DNA can form a nicked ring structure. Integration is under control of the lambda *int* gene, which codes for **integrase,** and an integration factor synthesized by the host. Integrase recognizes specific sites on the genomes of both the bacteria and the phage genome called **attachment sites** *(att).* Because there is only one *att* site on the *E. coli* genome, lambda DNA always integrates at the same site on the bacterial chromosome (Figure

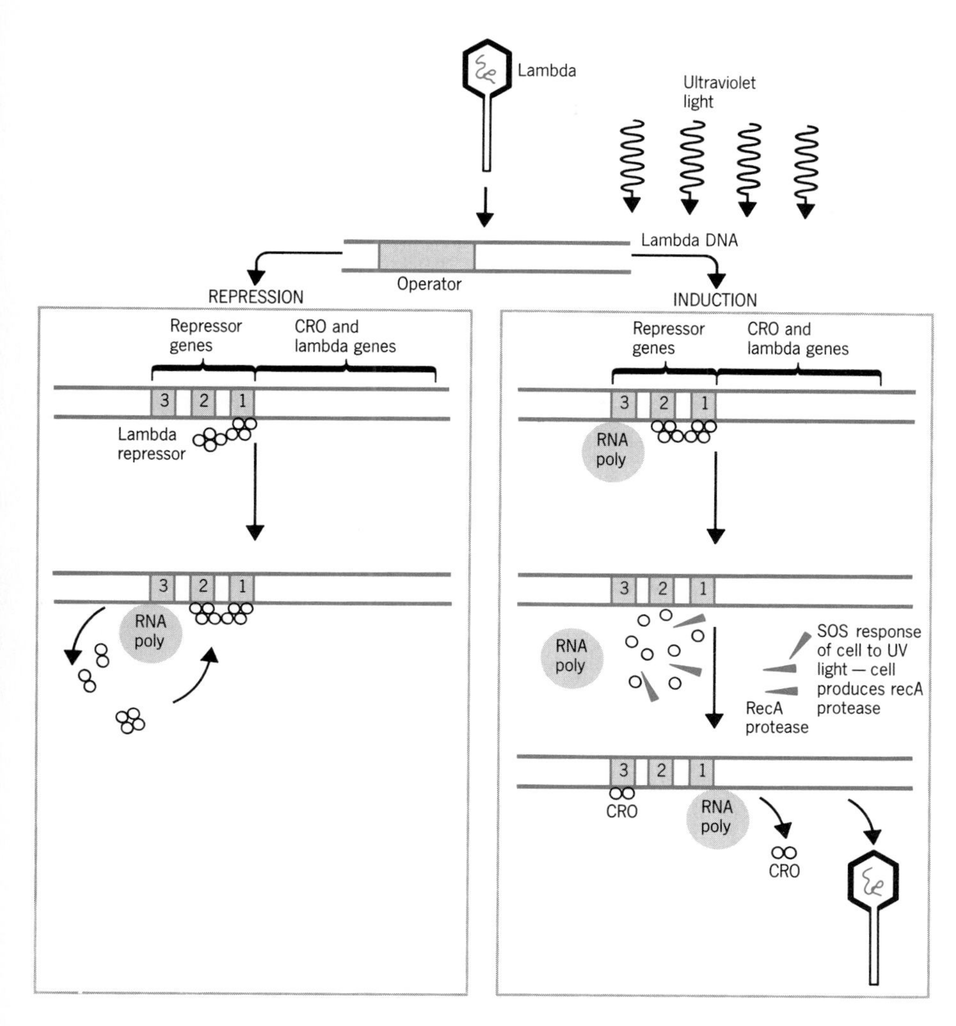

FIGURE 11.13 Lambda is a temperate bacteriophage that is maintained in the prophage state by a repressor protein. The lambda repressor binds to sites 1 and 2 on the right operator of lambda DNA. This activates the genes to the left of the operator, and lambda repressor protein is produced. These repressor proteins prevent transcription of the lambda genes located to the right of the operator. Progeny lambda phages are produced when the repressor protein is destroyed. Lysogenized cells exposed to UV light produce the recA protease, an enzyme that destroys lambda repressor. Now lambda and CRO genes are transcribed and progeny bacteriophages are produced. The CRO gene product prevents further repression by binding to site 3 on the right operator, which stops the synthesis of lambda repressor protein.

11.14). Therefore, prophage lambda is always present in the host genome between specific host genes.

KEY POINT

Three examples of how DNA segments integrate into the bacterial chromosome are: the F plasmid integrates into the chromosome of an Hfr cell; transposons move between plasmids and the bacterial chromosome; and prophage DNA integrates into the chromosome of a host cell to make it a lysogen.

Excision of the Lambda Genome

Excision is normally a precise process that results in a replicative form of lambda. Excision is catalyzed by the phage enzyme **excisionase** and results in a recombinational event to form the circular lambda DNA (Figure 11.15). A precise or legitimate recombination occurs when the entire lambda genome is excised. An illegitimate recombination generates an excised circular DNA molecule that is usually defective because essential phage genes have been replaced by short sequences of bacterial DNA.

Specialized Transduction

After excision, the circular lambda genome codes for the replication of progeny phages. A cell with a normal lambda genome goes through a lytic cycle and produces infectious lambda phages. Cells with defective

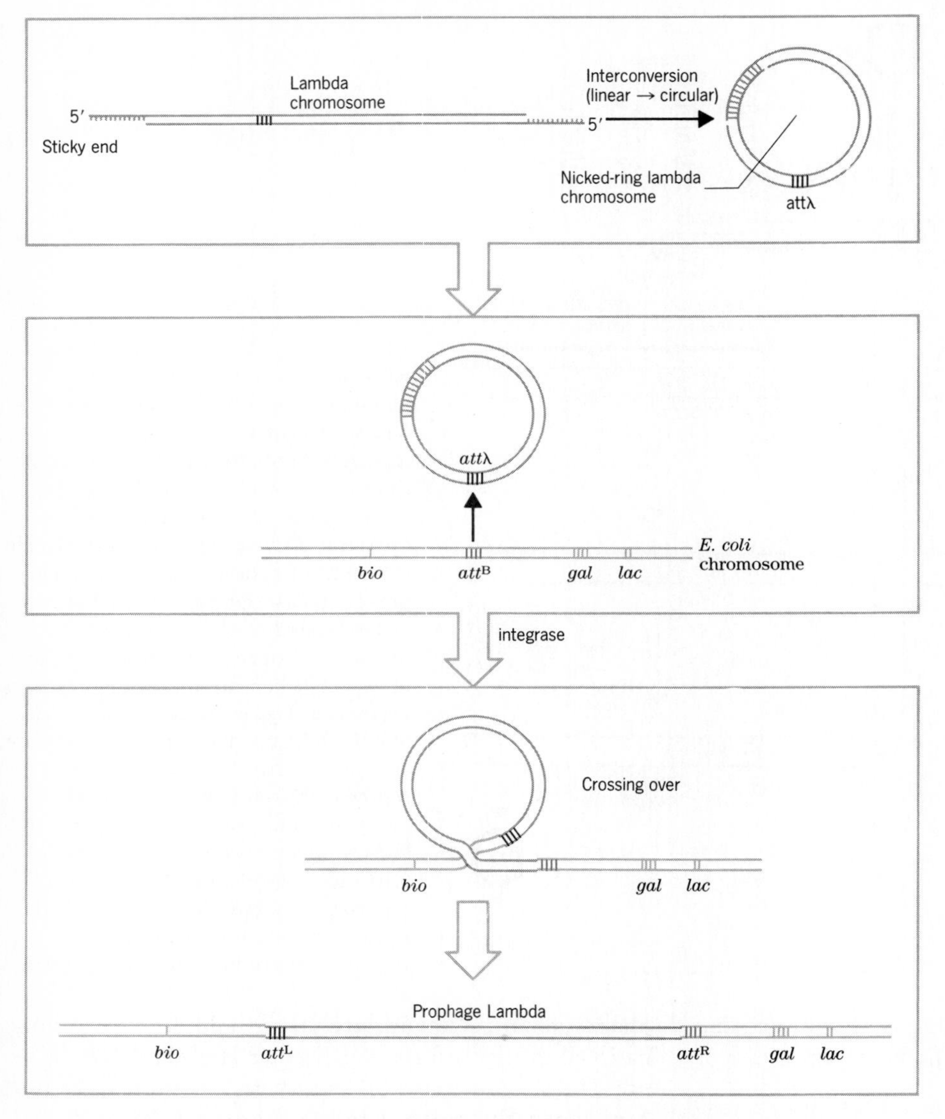

FIGURE 11.14 Integration of lambda bacteriophage into the *Escherichia coli* genome. Double-stranded lambda DNA spontaneously forms a closed nicked ring prior to being joined with the *E. coli* chromosome by integrase.

lambda genomes can produce only defective progeny viruses. Each defective lambda carries bacterial genes that were located adjacent to the attachment site in the bacterial chromosome. When these defective phages infect a recipient *E. coli* cell, it receives both the partial lambda genome and the accompanying bacterial genes. In essence, λ carries bacterial genes to the recipient cell. This type of bacteriophage-mediated genetic transfer is specialized because only the genes located adjacent to attachment sites in the donor bacterial chromosome can be transferred.

PHAGE CONVERSION

Many bacteria are lysogens; that is, they carry viral genetic information in their chromosome as a prophage.

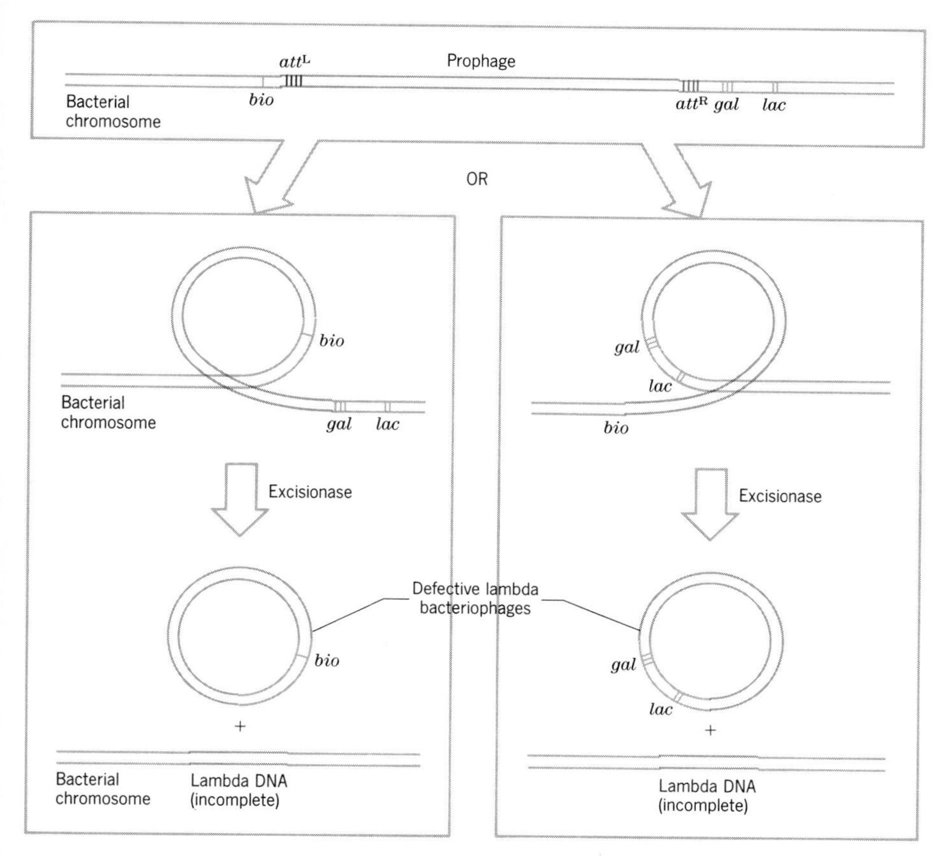

FIGURE 11.15 Lambda is a temperate bacteriophage that always integrates into the host chromosome at a specific point. Mistakes in excision can lead to the formation of incomplete lambda genomes that carry host-cell genes (*gal* or *bio*).

These viruses are taking a free ride since they are replicated each time the cell replicates. Normally the prophage's presence goes unnoticed since it does not affect the phenotype of the host cell. Some prophages, however, carry genes that markedly affect the host's phenotype. **Phage conversion** is an alteration of the phenotype of a lysogenized cell by the expression of a prophage gene. The prophage genes of special interest to medical microbiologists are those that code for toxins.

Lysogenized strains of *Corynebacterium diphtheriae* (kor-eye'nee-bak-teer'ee-um dif-theer'ee-eye) produce diphtheria toxin, which is directly responsible for the symptoms of diphtheria. This protein toxin is coded for by a β phage gene, so only cultures of *Corynebacterium diphtheriae* carrying the β prophage can cause diphtheria. Similarly, the erythrogenic toxins produced by *Streptococcus pyogenes* (py-ahj'en-eez) and the botulism toxins produced by some strains of *Clostridium botulinum* (klaw-stri'dee-um bot-choo-ly'num) are coded for by prophage genes. When the lysogenized cells lose their prophage, they also lose the ability to produce the toxin.

KEY POINT

Proviruses are not restricted to bacteria. Humans can carry retroviruses such as the human T cell lymphotropic virus (HTLV-I) and the human immunodeficiency virus (HIV). Other human viruses cause latent infections in which disease symptoms recur months or years after the primary infection. Human diseases attributed to latent viral infections include herpes cold sores, genital herpes, chickenpox, and infectious mononucleosis.

HOST-RANGE MODIFICATIONS

A constant war rages between viruses that possess the propensity to commandeer their host cell for replicative purposes and the host cells, which attempt to prevent viral infections. The offensive weapons of this war are nucleases that attack DNA, while the defensive mechanisms are enzymes that shield DNA from the attacking nucleases.

RESTRICTION ENDONUCLEASES

Bacteria produce restriction endonucleases that cleave double-stranded DNA at specific sites: **endonucleases** cleave bonds within the DNA molecules, whereas **exonucleases** act only on free ends of single- or double-stranded DNA. Restriction endonucleases limit the type of DNA that can be replicated within a cell. They recognize nucleotide sequences four to six bases long that are symmetrical about a central point (Table 11.3). EcoRI is an endonuclease that cleaves DNA in two places:

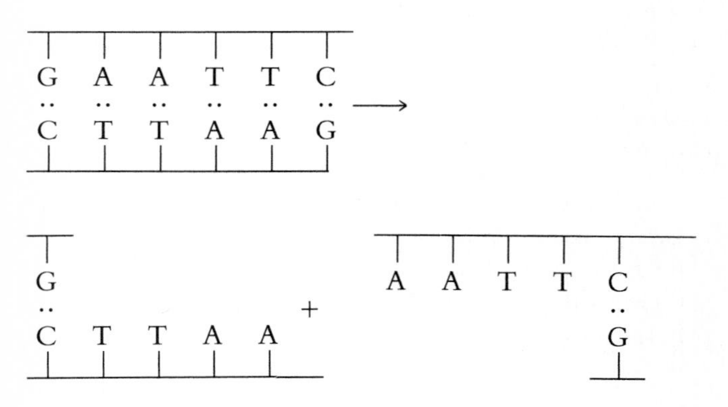

The resulting single-stranded ends are attacked by exonucleases that cleave off the —A—A—T—T ends, thus preventing reannealing. Other cellular nucleases then degrade the 3′ and 5′ free ends of the double-stranded DNA.

Since the genome of *Escherichia coli* has many —G—A—A—T—T—C sequences, *E. coli* must protect its own DNA from degradation by its own restriction endonuclease. Cells do this by enzymatically modifying their DNA. *Escherichia coli* protects itself from EcoRII (Table 11.3) by methylating the first cytosine present in the DNA sequence —mC—C—A—G—G—. EcoRII will not act on —mC—C—A—G—G—. Many bacteria have specific exonucleases and specific modifying enzymes that prevent foreign DNA from invading their cytoplasm.

Viruses defend themselves by modifying their own DNA. T_4 bacteriophages use hydroxymethyl cytosine and glucosylated hydroxymethyl cytosine in their

TABLE 11.3 Restriction endonucleases

ORGANISM	ENZYME	RECOGNITION SEQUENCE[a]
Escherichia coli	EcoRI	G↓A A T T C C T T A A↑G
	EcoRII	N↓C C A G G N N G G T C C↑N
Haemophilus influenzae	Hind II	G T Py↓Pu A C C A Py↑Pu T G
	Hind III	A↓A G C T T T T C G A↑A

[a]The arrows indicate the sites where the enzyme attacks DNA. Key: A, adenine; G, guanine; C, cytosine; T, thymine; Py, any pyrimidine; Pu, any purine; N, any nucleotide.

DNA. The presence of these unique nitrogenous bases renders their DNA resistant to most endonucleases. The host range of a given bacteriophage depends on the ability of its DNA to resist the cell's endonucleases.

Restriction endonucleases are essential for the procedures used in genetic engineering. Genetic engineers use restriction endonucleases to make fragments of DNA that can be incorporated into a plasmid or virus vector. Because restriction endonucleases make specific breaks in DNA, treatment of DNA with a specific endonuclease results in reproducible fragments of DNA. These DNA fragments are small enough to be sequenced chemically. These techniques have been used to determine the complete nucleotide sequence of ϕX174 (fee-X-174), λ, and other DNA bacteriophages.

REPLICATION OF SINGLE-STRANDED DNA BACTERIOPHAGES

Viruses with single-stranded DNA are either filamentous (Fl) or cubic (ϕX174) viruses. Since they lack a tail and tail fibers, their means of attachment and penetration differs from that of the tailed bacteriophages.

The filamentous bacteriophages attach to the hollow pili extending from the cell wall. How they enter the cell is still unknown: the entire virion could be pulled into the cytoplasm, or their DNA may travel down the pilus to the cytoplasm. ϕX174 is a cubic bacteriophage that attaches to the lipopolysaccharide of the outer membrane of *E. coli.* After attachment, its single-stranded DNA accompanied by a ϕX174 surface spike penetrates the cell's cytoplasm.

The entering single-stranded DNA serves as a template for synthesis of its complementary strand. The cell's DNA polymerase is used since this process occurs before transcription of mRNA occurs. The resulting double-stranded DNA is the replicative strand from which multiple copies of the complementary strand are synthesized.

REPLICATION OF RNA BACTERIOPHAGES

RNA bacteriophages are the simplest of the known viruses. Most RNA bacteriophages are single-stranded, plus RNA coliphages (plus RNA functions as mRNA). They infect *E. coli* after their A protein binds to the cell's F pilus (Figure 11.16). The A protein and the RNA enter the cell by a mechanism that may involve transfer through the pilus, and once inside the host's cytoplasm, replication commences. Between 5000 and 10,000 progeny bacteriophages are produced in a 30- to 60-minute replication cycle.

The male coliphages are single-stranded plus RNA viruses.* In the host cytoplasm the plus RNA functions as mRNA for the synthesis of A protein, coat protein, and RNA polymerase. This viral RNA polymerase makes a minus RNA strand (replicative strand) using the viral genome as a template (Figure 11.17). The minus strand is then used to form many plus RNA strands, which combine with coat protein and protein A to form infectious virions. RNA bacteriophages can be extruded from the cell without causing cell lysis.

*ϕ6 infects *Pseudomonas phaseolicola* and is the only known double-stranded RNA bacteriophage.

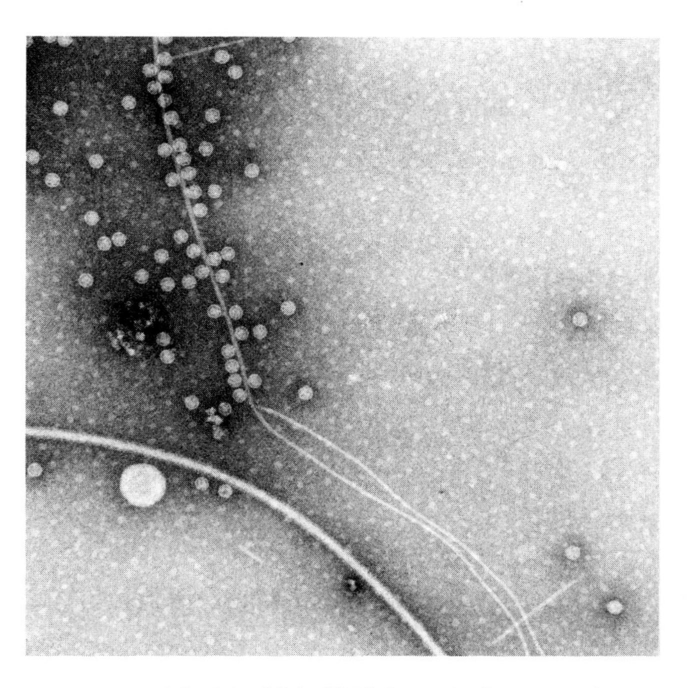

FIGURE 11.16 Male RNA bacteriophages are shown binding to the F pilus of this *E. coli* cell. (Courtesy of Professor Lucien Caro.)

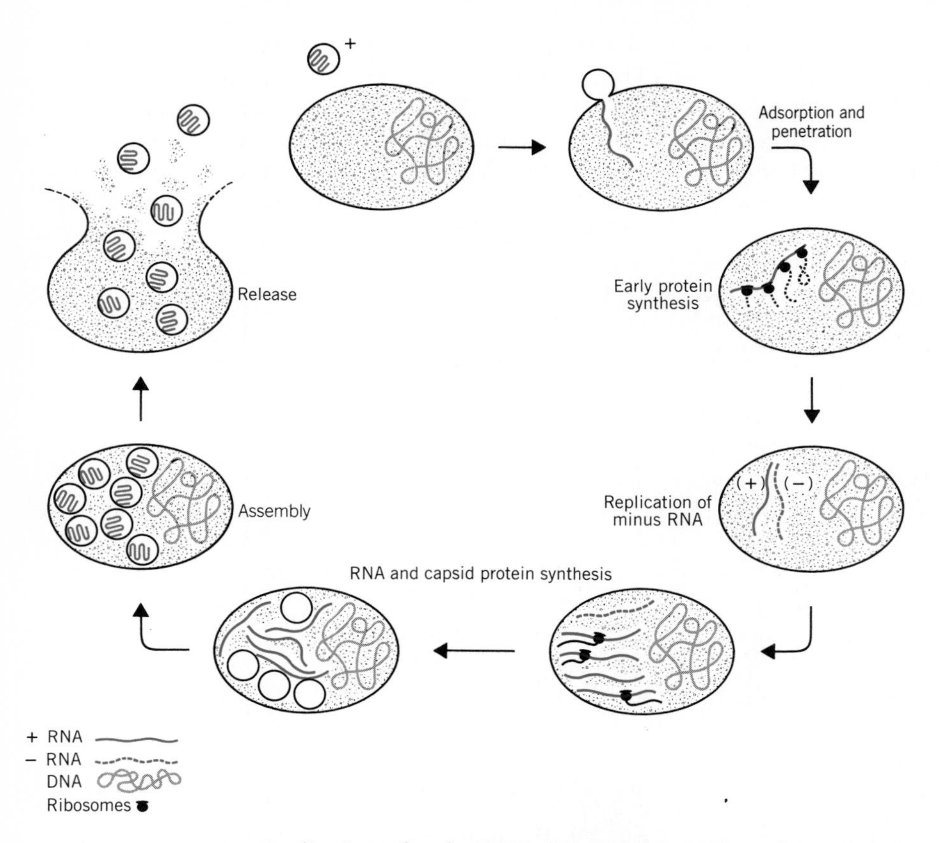

FIGURE 11.17 Replication of a plus RNA bacteriophage. The infective RNA strand (positive strand) is used as a message (mRNA) to make early proteins. These proteins make a replicative (minus RNA) strand that is used as a template to make viral plus RNA.

Summary Outline

BACTERIAL VIROLOGY

1. Viruses are complex macromolecular entities that can replicate only after entry into specific type(s) of cells, where they use their own genetic information encoded in either DNA or RNA to produce replicas of themselves.
2. Viruses are composed of one type of nucleic acid (either DNA or RNA) surrounded by a protective protein coat.

HISTORY OF VIROLOGY

1. Mayer, Ivanovski, and Beijerinck's work on TMV in the late 1800s contributed to the foundations of virology.
2. Pasteur investigated rabies and developed a rabies vaccine in 1885, but he failed to discover its viral nature.
3. Loeffler and Frosch described the agent for foot-and-mouth disease as being filterable. By 1911,

the filterable nature of the infectious agents responsible for yellow fever, rabies, dengue fever, poliomyelitis, measles, and chicken sarcoma had been reported.

4. Twort (1915) and d'Herelle (1917) discovered bacteriophages.
5. Stanley crystallized TMV in 1935 and suggested it was an autocatalytic protein.
6. The Hershey–Chase experiment with ^{32}P DNA from bacteriophage helped prove that DNA is the genetic material.

FUNDAMENTALS OF BACTERIAL VIROLOGY

1. Bacteriophages are classified according to their nucleic acid composition, structure, morphology, size, and shape. Viral nucleic acid is either single-stranded or double-stranded, linear or circular.
2. Almost all natural environments contain bacteriophages that can be isolated by plating filtrates of aqueous samples on a lawn of host bacteria.
3. Bacteriophages can be counted by determining the number of plaques produced on a lawn of host bacteria.
4. The T-even bacteriophages adsorb specifically and irreversibly to the outer membrane of *E. coli* before their DNA enters the cytoplasm, propelled by contraction of the phage tail.
5. Viral proteins are translated on the cell's ribosomes using mRNA transcribed from the bacteriophage DNA.
6. The T_4 bacteriophage DNA codes for nucleases, DNA polymerase, biosynthetic enzymes involved in DNA precursor formation, bacteriophage structural proteins, lysozyme, and a bacteriophage protein that destroys the cell's membrane.
7. Assembly of mature bacteriophages occurs through a sequential process of self-assembly using cellular pools of phage components.
8. Progeny bacteriophages are produced in assembly-line fashion during the lytic cycle, which ends with host-cell lysis and the release of viruses.

TRANSFER OF BACTERIAL GENES BY TRANSDUCTION

1. Generalized transduction is the phage-mediated transfer of any host-cell genes to a recipient bacterium. The transducing phage randomly acquires host DNA and transfers it to a recipient cell.
2. Specialized transduction is mediated by temperate bacteriophages that exist in a replicative form or are integrated into the host's chromosome as a prophage.
3. A prophage exists in a lysogen as a sequence of viral genes integrated into the host's genome.
4. Lambda synthesizes a repressor protein that maintains the lambda prophage in the cell genome. The repressor protein is inactivated by the SOS response.
5. Lambda acts as a specialized transducing phage when it transfers bacterial genes located adjacent to its integration site on the bacterial chromosome.
6. Phage conversion is the alteration of a lysogenized bacterium's phenotype caused by the expression of a prophage gene.

HOST-RANGE MODIFICATION

1. Restriction endonucleases are produced by bacterial cells to destroy foreign DNA.
2. Bacteriophages respond by chemically modifying their DNA to make it resistant to the endonucleases.
3. The host range of a given bacteriophage depends on the ability of its DNA to resist the endonucleases made by that host.
4. There are many applications of restriction endonucleases in genetics, genetic engineering, and nucleic acid biochemistry.

REPLICATION OF SINGLE-STRANDED DNA BACTERIOPHAGES

1. Filamentous DNA bacteriophages infect host cells by attaching to bacterial pili. Either the nucleic acid or the entire virion is transported into the cytoplasm, where it replicates.
2. The nucleic acid of single-stranded plus RNA viruses serves as mRNA for protein synthesis.
3. Using this plus RNA strand, a viral RNA polymerase makes a minus RNA strand that is used to replicate more viral RNA.

Questions and Topics for Study and Discussion

QUESTIONS

1. Explain why investigations on bacterial viruses were important to the development of virology.
2. How is plaque formation used to enumerate the concentrations of bacteriophages in liquid cultures? How can this technique be adapted to the counting of plant and animal viruses?
3. What mechanisms are used by bacteriophages to infect their host cells?
4. Compare the one-step growth curve to the normal growth curve for a bacterial culture.
5. Define or explain the following:

adsorption	lytic cycle
burst size	phage conversion
eclipse period	specialized transduction
generalized transduction	temperate bacteriophage
hydroxymethyl cytosine	
lysogeny	

6. Describe how the lambda repressor protein works *(a)* to maintain lysogeny and *(b)* to permit replication via the lytic cycle.
7. What are restriction endonucleases and what purpose do they serve in cell survival?
8. What special problems are encountered by single-stranded RNA and single-stranded DNA bacteriophages during their replication cycle?
9. How could you demonstrate that genetic transfer occurred by transduction as opposed to conjugation or transformation?
10. On what basis could you construct a genetic map using generalized transduction?

DISCUSSION TOPICS

1. Felix d'Herelle envisioned using bacteriophages to treat and cure patients with infectious bacterial diseases. Discuss why this has not occurred. Are there reasons why viruses were discovered when they were?
2. In what way could lysogeny have an impact on human health? What animal viruses would be involved and how can we deal with them?

Further Readings

BOOKS

Cairns, J., G. S. Stent, and J. Watson, *Phage and the Origins of Molecular Biology: Essays,* Cold Spring Harbor Symposium, Cold Spring Harbor, New York, 1966. A collection of articles about the beginnings of molecular biology as told by the scientists who participated in its development.

Luria, S. E., J. E. Darnell, Jr., D. Baltimore, and A. Campbell, *General Virology,* 3rd ed., Wiley, New York, 1978. An advanced, detailed textbook that covers all groups of viruses.

Mathews, C. K., E. M. Kutter, G. Mosig, and P. B. Berget, *Bacteriophage T_4,* American Society for Microbiology, Washington D.C., 1983. The papers in this compendium constitute a review of the molecular biology and genetics of bacteriophage T_4.

Stent, G. S., *Molecular Biology of Bacterial Viruses,* Freeman, San Francisco, 1963. A classic introduction to the bacteriophages emphasizing the early experiments in bacterial virology.

Watson, J. D., N. H. Hopkins, J. W. Roberts, J. A. Steitz, and A. M. Weiner, *Molecular Biology of the Gene,* Vol 1, 4th ed., Benjamin–Cummings, Menlo Park, Calif., 1987. One of the best textbooks on the molecular aspects of biological systems. This book covers the molecular biology of viruses, bacteria, and eucaryotic cells.

ARTICLES AND REVIEWS

Duckworth, D. H., "Who Discovered Bacteriophages?" *Bacteriol. Revs.,* 40:793–802 (1976). An interesting essay on the roles of Twort and d'Herelle in the discovery of bacteriophages.

Fiddes, J. C., "The Nucleotide Sequence of a Viral DNA," *Sci. Am.,* 237:54–67 (December 1977). The discovery of overlapping genes in the genome of φX174 is described together with the complete nucleotide sequence of this single-stranded DNA bacteriophage.

Hilts, P., "Mark Ptashne's Molecular Mission," *Science 82,* pp. 33–41 (December 1982). The story of the laborious work that went into the discovery of the lambda repressor system.

Maniatis, T., and M. Ptashne, "A DNA Operon-Repressor System," *Sci. Am.,* 234:64–76 (January 1976). A detailed description of the lambda bacteriophage repressor system.

CHAPTER

12

Animal and Plant Virology

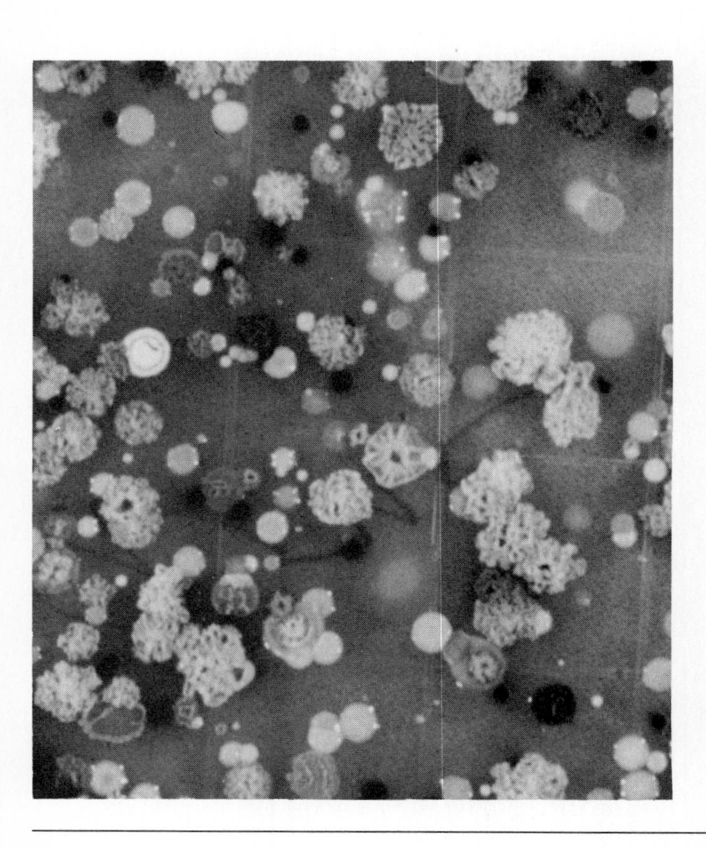

OUTLINE

FOCUS OF THIS CHAPTER

Animal and plant viruses are of great human and economic importance because of the diseases they cause. This chapter describes the structure, classification, and replication of animal viruses and their effects on animals and tissue cultures. General aspects of plant virology are presented, including material on transmission and control.

Animal and plant viruses are complex macromolecular entities that replicate in living eucaryotic cells by using their own genetic information, encoded in either DNA or RNA. They are dependent on their host's protein-synthesizing machinery and cellular enzymes for their replication. Although we recognize viral infections of animals by the symptoms they cause, in most cases there is no effective therapy for treating viral diseases. Our major hope for developing effective treatments for viral infections lies in virology research.

BASIC ANIMAL VIROLOGY

Humans and animals are susceptible to many different viral diseases. Some animal viruses cause mild or asymptomatic diseases, for example the common cold viruses, whereas others cause serious illness that can result in death. Even though the clinical descriptions of many viral diseases have been known for centuries, the techniques required to isolate and characterize the responsible viruses are relatively new. Modern animal virology could not flourish until tissue culture techniques were developed in the 1950s and scientists had a basic understanding of molecular biology. These two developments were essential to the rapid progress made in virology during the last three decades.

Animal viruses are self-replicating, intracellular parasites that are specific for the animals they infect. They differ from bacteriophages (see Chapter 11) in a number of significant ways. The entire animal virus, instead of just the nucleic acid, usually enters the host cell. This nucleic acid is released from the animal virus capsid by the process of decapsidation, also known as uncoating, which occurs within the cell before the virus replicates. Some animal viruses are surrounded by an envelope derived from a host-cell membrane, usually the cell's cytoplasmic membrane. Although there are differences, the basic principles learned from studying bacteriophages in Chapter 11 will help you to understand animal and plant virology.

CLASSIFICATION OF ANIMAL VIRUSES

Animal viruses are classified into families on the basis of their morphology and chemical composition. This classification scheme is constructed without regard to evolutionary relationships because the origin of viruses is unknown. The purpose of classifying viruses is to assign each virus a specific name and to describe each virus in scientific terms.

Classification Criteria

Before the molecular structures of animal viruses were known, the viruses were classified according to the hosts they infected and the diseases they caused. Now that many viruses have been isolated, they are characterized by their structure, size, and chemical composition. Using these data, virologists have constructed a modern viral classification scheme that assigns animal viruses into six families of DNA viruses and 11 families of RNA viruses (Table 12.1).

Viral Nomenclature

Viral family names always have the ending *-viridae.* The descriptive part of the family name sometimes connotes a characteristic of that viral family. For example, Picornaviridae is a viral family that contains small (pico) RNA (rna) viruses. Myxoviruses infect the respiratory tract and cause the formation of mucus (Gr. *myxa,* mucus). Other family names have historical origins.

The International Committee on Nomenclature of Viruses is responsible for naming viruses. This committee has recommended the use of Latinized generic

TABLE 12.1 Classification of animal viruses

FAMILY (REPRESENTATIVE GENUS)	GENERAL CHARACTERISTICS	TYPICAL AGENTS
	RNA Viruses	
Arenaviridae	Enveloped, helical nucleocapsid	Humans: lassa virus (rare, serious disease)
Bunyaviridae *(Bunyavirus)*	Enveloped, helical nucleocapsid, segmented, ss[a], minus RNA	Humans: bunyamwera viruses (arthropod borne)
Caliciviridae	Naked, icosahedral nucleocapsid	Humans: Norwalk virus
Coronaviridae	Enveloped, helical nucleocapsid, ss plus RNA, 80–120-nm particles, bulbous spikes	Humans: coronavirus strains cause common cold
Orthomyxoviridae *(Influenzavirus)*	Enveloped, helical nucleocapsid, segmented, ss, minus RNA	Humans: influenza A and B (infections in swine and birds)
Paramyxoviridae *(Paramyxovirus)*	Enveloped, helical nucleocapsid, 18–1000 nm, ss, minus RNA	Humans: mumps, measles, and respiratory syncytial viruses
Picornaviridae *(Enterovirus)*	Icosahedral, naked, ss, plus RNA, 27–35-nm capsid	Humans: poliovirus, coxsackievirus, rhinovirus
Reoviridae *(Rotavirus)*	Icosahedral, naked, double-shelled virions; segmented (10 or more), ds[a] RNAs	Humans: gastroenteritis (infections of mammals and birds)
Retroviridae *(Oncornavirus)*	Enveloped, round particles approx. 100-nm diameter, helical nucleocapsid, two identical plus RNAs, reverse transcriptase	Mammals, birds; human oncogenic viruses
Rhabdoviridae *(Rabdovirus)*	Enveloped, bullet-shaped, 70 × 175 nm, ss, minus RNA, helical nucleocapsid	Humans: rabies virus, vesicular stomatitis virus. Infections of fish, plants, insects, and animals
Togaviridae *(Alphivirus)*	Icosahedral, enveloped, ss, plus RNA, 50–70-nm capsid, 32 capsomeress	Humans: encephalitis, dengue, and yellow fever viruses (infections of insects, birds, and animals)
	DNA Viruses	
Adenoviridae *(Mastadenovirus)*	Icosahedral, naked, ds DNA (linear), 80-nm capsid, 252 capsomeres	Humans: adenovirus (causes upper respiratory diseases)
Herpetoviridae *(Herpesvirus)*	Enveloped, icosahedral, ds DNA (linear), 100-nm capsid, 162 capsomeres	Humans: herpes simplex viruses, Epstein–Barr virus, varicella–zoster virus, and cytomegalovirus
Iridoviridae	Naked or enveloped, icosahedral	Insects, swine, fish molluscs, protozoa
Papovaviridae *(Papillomavirus)*	Icosahedral, naked, ds DNA (closed circle), 45–55 nm capsid, 72 capsomeres	Humans: warts (SV_{40} and polyoma viruses cause tumors in animals)
Parvoviridae *(Parvovirus)*	Icosahedral, naked, ss DNA, 20-nm capsid, 32 capsomeres	Humans: adeno-associated virus (needs helper virus)
Poxviridae *(Orthopoxvirus)*	Enveloped (ether-resistant), 160 × 200-nm "brick shape," ds DNA (linear), enzymes associated with virion	Humans: variola, vaccinia

[a]Key: ss, single-stranded; ds, double-stranded.

names for each virus, although vernacular names are still widely used. Poliovirus type 1 and influenza virus are the vernacular names respectively for *Enterovirus h-polio-1* and *Influenzavirus*. Acceptance of these generic names is still in a state of flux.

Common viral names were often derived from the

name of the disease. Influenza is caused by the influenza virus; rabies is caused by rabies virus; and poliomyelitis is caused by poliovirus. Some viruses are described by their classic names. Smallpox is caused by variola virus (L. *variola,* speckled); German measles is caused by rubella virus (L. *rubella,* reddish); and the common cold is caused by rhinoviruses (Gr. *rhino,* nose). Although mixing classic names with Latinized viral nomenclature is confusing, it continues to be accepted because of its widespread use.

Chemical Composition of Animal Viruses

The simplest animal viruses are composed of one type of nucleic acid surrounded by a proteinaceous capsid. The more complex animal viruses can contain enzymes, segmented nucleic acids, simple or complex capsids, and/or a lipid-containing envelope. Animal viruses can be as small as a ribosome (20 to 30 nm) or as large as a very small bacterium (1000 nm). The chemical composition of animal viruses can vary considerably from one group to another depending on the presence or absence of an envelope and the type of nucleic acid present.

Structure of Animal Viruses

Most animal viruses can be either icosahedral or helical in shape (Figure 12.1); a few are pleomorphic, the poxviruses are brick-shaped, and the rhabdoviruses are bullet-shaped. Some animal viruses have envelopes, but they never have tails. Instead, they attach directly to receptors on the cytoplasmic membrane of their host cell (animal cells lack cell walls).

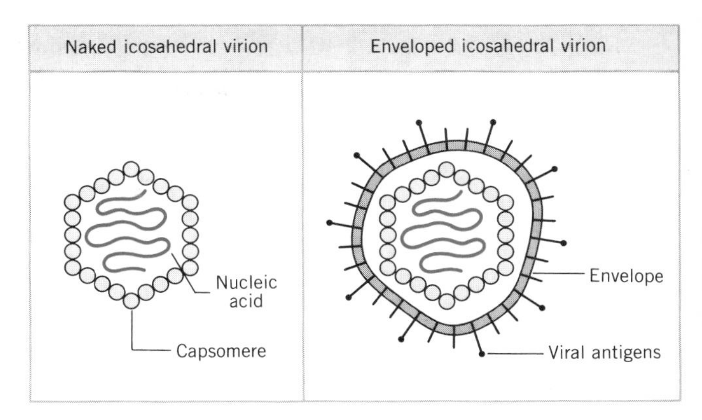

FIGURE 12.1 Enveloped and naked icosahedral and helical virions are the four basic structures of animal viruses.

Viral Capsids The viral capsid is composed of capsomeres that surround the viral nucleic acid. **Capsomeres** are the viral proteins (protomers) that self-assemble into either the helical or icosahedral structure of the capsid.

The capsid of an **icosahedral** virus is a symmetrical structure composed of 20 facets enclosing a central space (Figure 12.2). Each facet is an equilateral triangle constructed from the capsomeres. The size of an icosahedron depends on the size of the capsomeres, the number of capsomeres, and their arrangement. The number of capsomeres is variable within prescribed geometric constraints so an icosahedron can have 12, 32, 42, 72, 92, 162, 252, and so forth, capsomeres. The number of capsomeres in a given icosahedral virus is constant and is used in viral classification.

Helical animal viruses always contain RNA as their sole nucleic acid. The capsomeres of a helical virus self-assemble to form a cylinder around the RNA, resulting in the formation of naked helical virions (Figure 12.1). The self-assembly of a helical virus is not constrained by geometric form, but instead by the size of the viral genome.

Viral Envelopes The complex of the nucleic acid surrounded by capsid protein is known as a **nucleocapsid** (see Figure 12.1). The nucleocapsids of some helical and icosahedral viruses are surrounded by a lipid-containing membrane layer known as the **viral envelope.** The envelope is composed of protein (some are viral proteins), lipid, and carbohydrate and is de-

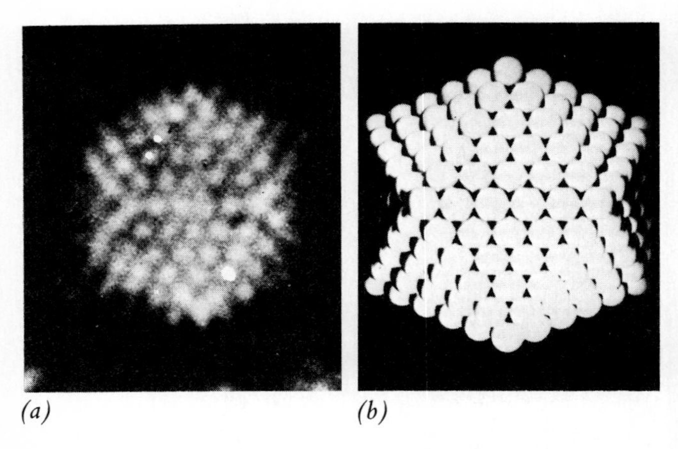

FIGURE 12.2 *(a)* The individual capsomeres are visible in this electron micrograph of adenovirus. *(b)* The model of adenovirus represents the three-dimensional orientation of an icosahedral capsid. (Courtesy of Dr. R. W. Horne. From R. W. Horne, S. Brenner, A. P. Waterson, and P. Wildy, *J. Mol. Biol.,* 1:84–86, 1959, copyright © 1959, Academic Press Inc., Ltd., London.)

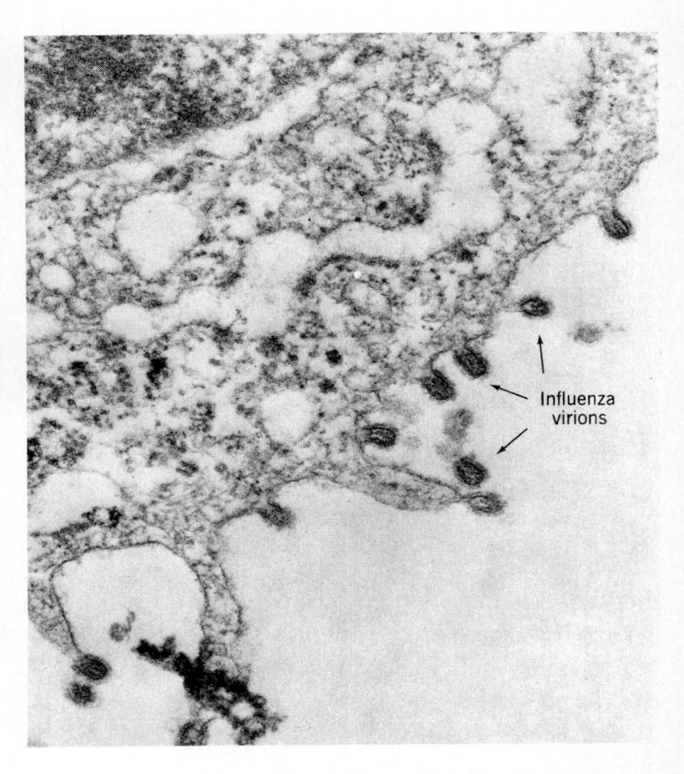

FIGURE 12.3 Release of influenza virus from the surface of a chicken embryo fibroblast. (Courtesy of Dr. R. W. Compans. From A. J. Dalton and F. Haguenau (eds.), *Ultrastructure of Animal Viruses and Bacteriophages: An Atlas,* copyright © 1973, Academic Press, New York.)

rived from a membrane of the host cell. In the simplest case, the nucleocapsid is surrounded by the cell's membrane (viral envelope) as it is extruded from the cell (Figure 12.3). Other viruses acquire their envelope from the cell's nuclear membrane or endoplasmic reticulum. Enveloped viruses (except poxviruses) are inactivated by lipid solvents, such as ether, which destroy the lipid structure of the viral envelope. The ether sensitivity of a virus is a test for demonstrating the presence of a viral envelope.

Viral Nucleic Acid Animal viruses possess a viral genome composed of either DNA or RNA. This nucleic acid can be single-stranded or double-stranded, linear or circular, monomolecular or segmented. By convention, single-stranded RNA that serves as a messenger RNA is designated plus RNA (+ RNA). Similarly, a single-stranded DNA molecule that serves as a template for transcribing mRNA is designated minus DNA (− DNA). The complementary macromolecules are designated minus RNA (− RNA) and plus DNA (+ DNA). Most viruses contain a single molecule of either DNA or RNA that can be either single- or double-stranded. A few RNA viruses contain more than one RNA molecule and are described as having a **segmented genome.** The nature and structure of the viral nucleic acid is a major criterion used to define viral families (Table 12.1).

ENUMERATION OF ANIMAL VIRUSES

Experimental virology depends on the investigator's ability to detect and count specific viruses. Animal virologists have adapted techniques from immunology, pathogenic bacteriology, and plaque assay to enumerate animal viruses. The most popular techniques involve the use of tissue cultures.

Tissue Culture Assays

John Enders and his colleagues were among the first researchers to propagate animal viruses in cell cultures, which in 1949 they developed for the growth of polio-

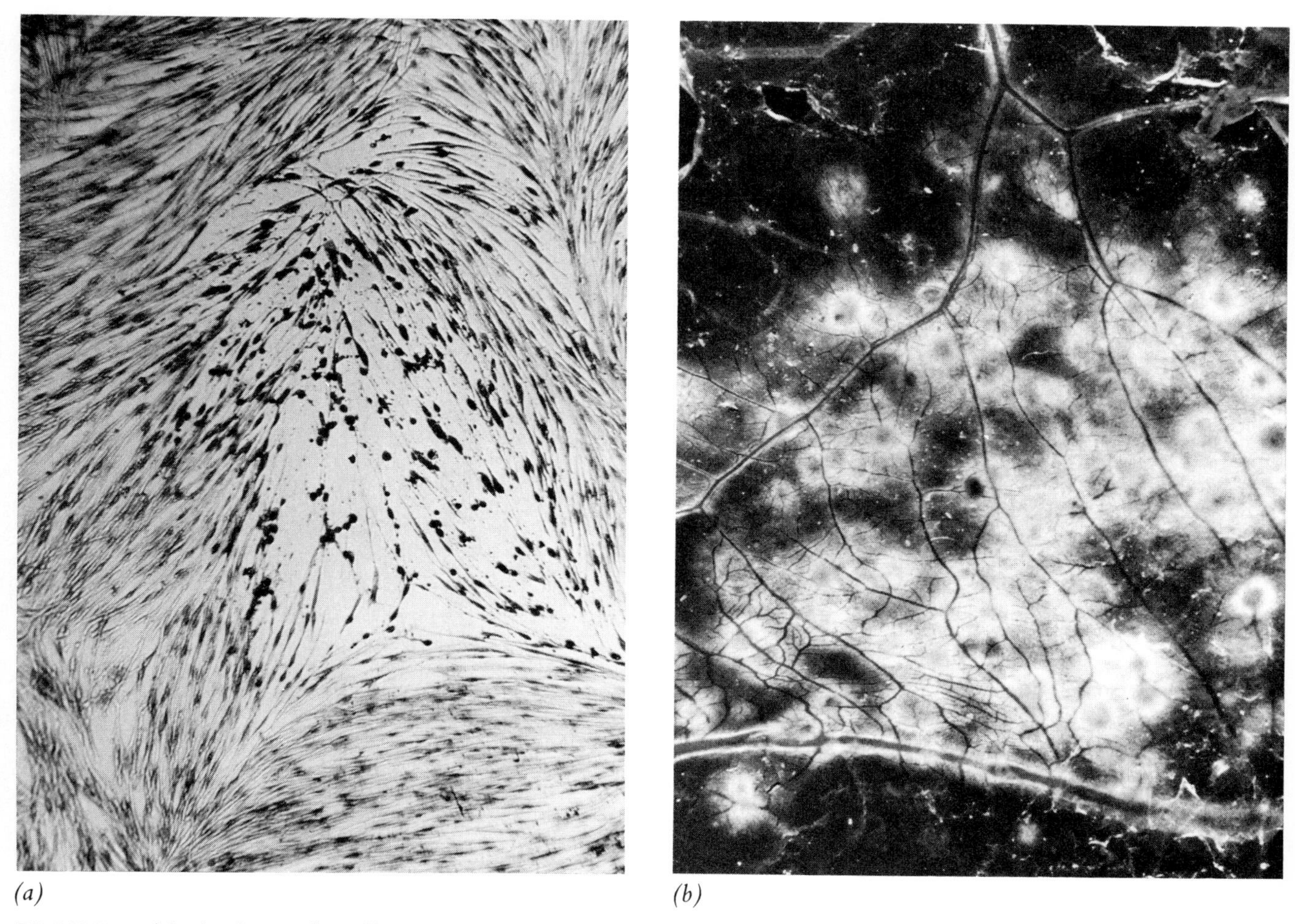

FIGURE 12.4 Cytopathic effects of animal-virus growth. *(a)* Herpes simplex growing on WI38 cells in tissue culture. *(b)* Vaccinia virus growing on a chicken embryo. (Courtesy of *(a)* Dr. N. Sharon, *(b)* Dr. L. LeBeau.)

viruses. Scientists can now grow many different animal cell lines suitable for propagating viruses in tissue cultures. These cell lines are maintained in complex media containing all the nutrients essential for cell growth and are transferred using the aseptic techniques developed by microbiologists.

Cultures of animal cells can be maintained through many aseptic transfers; however, these cells have an intrinsic age that prevents their perpetual reproduction. Cell cultures derived from normal tissues survive 35 to 120 transfers before they cease to propagate. In contrast, cell cultures derived from malignant (cancerous) animal tissue can be maintained indefinitely. Another property of cultured animal cells is **contact inhibition.** Animal cells grow in cultures as a sheet of cells, known as a monolayer (Figure 12.4). When the cells come in contact with their neighbors on all sides, they stop dividing.

Cytopathic effects (CPE) are visible changes in cell cultures resulting from an infection by a virus or other microbe (Figure 12.4*b*). Virus-caused CPE appear as foci of necrosis or as foci of transformed cell growth. The infection of a tissue culture cell by a cytopathic virus leads to a gradual and progressive destruction of the cells and local cell death (necrosis) that is readily visible. Other viruses cause animal cells to grow in an unrestricted manner. These **transformed** cells have lost their property of contact inhibition. So they grow as amultilayered mass of cells at the site of infection.

Monolayers of animal cells are grown in culture under a layer of liquid nutrient medium. These cultures support the growth of animal viruses when inoculated with stock viral suspensions or with clinical specimens. Their disadvantage as an assay system is that the liquid culture medium permits the viruses to spread and cause other foci of infection. This problem is prevented in the **plaque assay** by solidifying the culture medium with agar or agarose. The semisolid layer over the viral infected cells in the monolayer permits the development of distinct plaques, which can be counted. The plaque assay enables one to distinguish plaque morphologies caused by different viruses and to enumerate the viruses present in the inoculum.

KEY POINT

The plaque assay method assumes that each visible CPE results from the infection of a tissue culture cell with a single virion, so the number of CPE is equivalent to the number of virions in the inoculum. The plaque assay for animal viruses is analogous to the assay for enumerating bacteriophages.

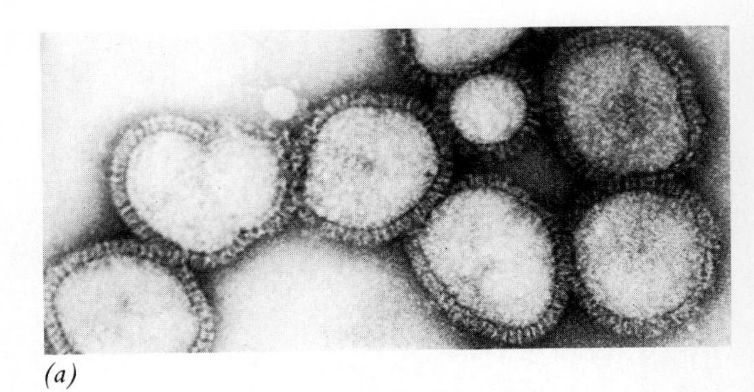

(a)

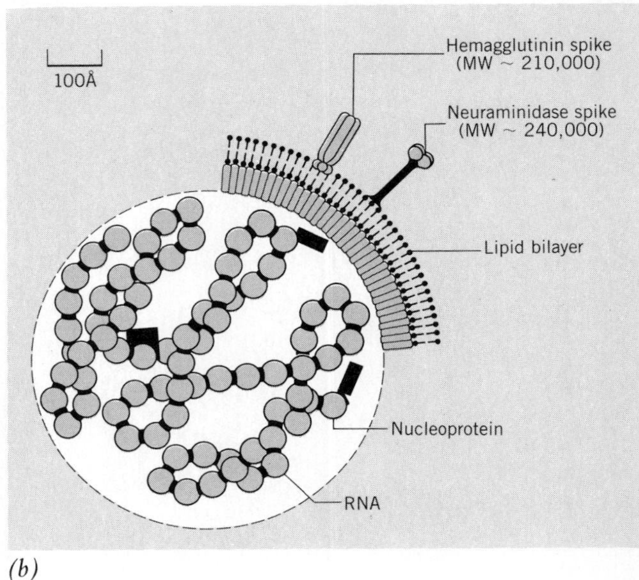

(b)

FIGURE 12.5 The surface of many enveloped viruses is covered by protein spikes that are involved in the adsorption of the virus to its host cell. *(a)* Electron micrograph of influenza A virus and *(b)* a diagram of the hemagglutinin and neuraminidase spikes attached to the viral envelope. (Courtesy of Dr. I. T. Schulze. From I. T. Schulze, *Virology,* 47:181–196, 1972, copyright © 1972, Academic Press, New York.)

Hemagglutination

Many animal viruses possess surface glycoproteins that react with receptors on the surfaces of animal cells. Hemagglutinins are viral proteins that react with receptors on red blood cells and cause hemagglutination when sufficient virions are added to cause the red blood cells to precipitate (agglutination). Precipitation results from the viral hemagglutinins (Figure 12.5) cross-linking the surface glycoproteins present on adjacent red blood cell membranes to form huge complexes. Each virus has many copies of the hemagglutinin so it can bind to two or more red blood cells, creating large complexes that precipitate.

Hemagglutination can be used for virus quantification by limiting the numbers of viruses present. This is done by adding different quantities of the virus to suspensions of red blood cells. The greatest dilution of virus that is able to agglutinate red blood cells is the end point, or titer, of the viral suspension.

Animal Assays

Current trends are away from killing animals for biological research; however, some viruses can be propagated and studied only after they have been grown in living animals. Often the end point of these infections is the death of the animal. The concentration of a virus in a suspension can be determined by measuring the effect of injecting dilutions of that viral suspension into susceptible animals. Laboratory animals such as mice and rats are used in animal assays for viral quantification. The viral dose that kills 50 percent of the animals (lethal dose 50 percent) is determined by statistical analysis from survival data. The availability of tissue cultures has greatly reduced the need to use animals for viral assays.

PROPAGATION OF ANIMAL VIRUSES

Vaccine manufacturers, viral diagnostic laboratories, and experimental virologists routinely grow animal viruses. Embryonated chicken eggs or one of a variety of tissue cultures is used as animal host cells for viral propagation. There are approximately 25 cell lines derived from humans, monkeys, hamsters, or mice that are commonly used to grow animal viruses. Although a virus may grow in more than one specific cell line, optimal growth is usually restricted to only a few.

Embryonated eggs containing a chicken embryo plus supportive tissues (see Figure 12.4*b*) provide virologists with a natural ready-made tissue culture. Embryonated chicken eggs are ideal culture environments for many different viruses. Viruses are inoculated into different parts of the egg after a hole is drilled in the egg's shell (Figure 12.6). The inoculated virus will then infect and grow in the cells of the embryo.

Diagnostic virology laboratories use commercially available cell lines maintained in tissue cultures. These cultures are inoculated with the patient's specimen and then incubated. Most human viruses grow at 36°C; however, some viruses such as the rhinoviruses, which cause human colds, have an optimum growth temperature of 33°C. The latent periods for animal viruses in tissue cultures vary from a few hours to many hours: many animal viruses require six days or longer to cause visible CPE. For this reason, diagnostic cultures are kept for two weeks before they are scored as being negative.

Vaccines are manufactured from viruses grown in tissue cultures or in embryonated eggs. For example, poliovirus is propagated in tissue cultures of monkey kidney cells for use in Salk and Sabin polio vaccines. Rabies virus was once grown in duck embryos, but it is now grown in human diploid tissue cultures, and the influenza viruses are grown in embryonated chicken eggs. These techniques are not applicable to all animal viruses, for some have never been grown outside their host.

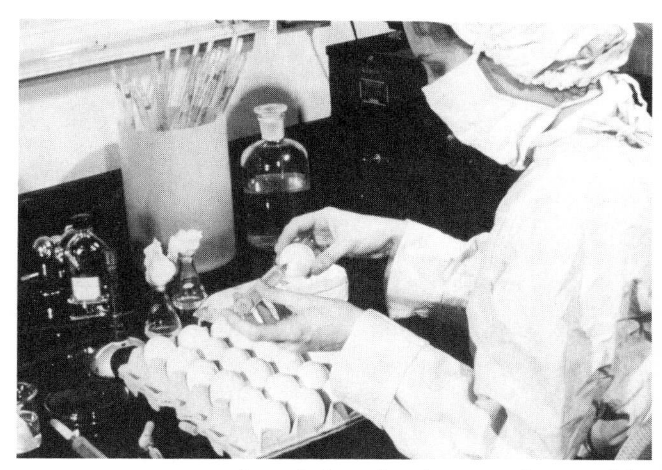

FIGURE 12.6 Inoculation of embryonated eggs with an animal virus. (Courtesy of the Centers for Disease Control.)

REPLICATION OF ANIMAL VIRUSES

Animal viruses have a limited host range because of the requirement for specific receptors on the surface of the cells they infect. Even within a given host, an animal virus will infect only certain cell types. Animal viruses infect cells in a two-step process involving adsorption and entry. **Adsorption**, the process by which a virion binds to the cell's surface, is followed by **entry** of the virion or its nucleic acid into the cell.

Viral Adsorption

Animal viruses have surface structures that specifically react with receptors on the surface of their host cell. Approximately 1 in every 10,000 collisions between the virus and a cell results in adsorption. Adsorption of the influenza virus to a host cell is mediated by the protein "spikes" extending from the envelope of the influenza virus (Figure 12.5). These spikes are the hemagglutinins that interact with cell-surface receptors on the host-cell membrane. Other viruses have different surface proteins that react specifically with cell-surface receptors.

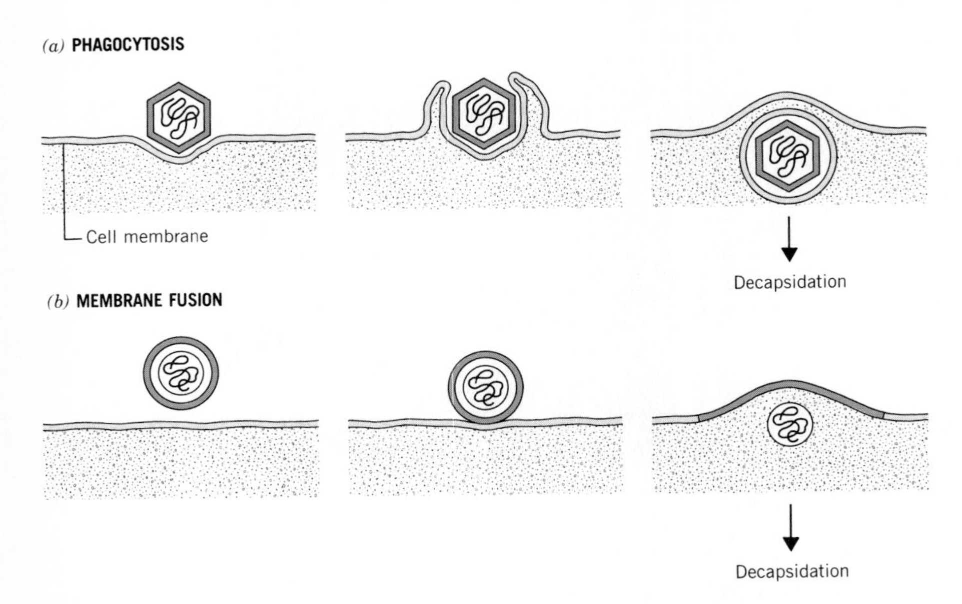

FIGURE 12.7 Mechanisms of viral entry: *(a)* phagocytosis of a naked virion and *(b)* membrane fusion mechanism for an enveloped virus entry.

The host cell's cytoplasmic membrane contains viral receptors, most of which appear to be glycoproteins. Only those cells that possess the correct receptor will be subject to infection by a given virus. Some viruses infect many cell types, whereas others are extremely limited. Measles virus infects many human cells, including those found in the nasopharynx, lung, lymphatics, testes, and nerves. Poliovirus infects a limited number of target cells. In humans, poliovirus infects cells of the nasopharynx, the intestinal tract, and the anterior horn cells of the spinal cord. Therefore, the nature of the viral binding proteins dictates, to a large extent, the cells that a virus can infect.

Viral Entry into Animal Cells

A virus, or at least the viral genome, must enter the cell's cytoplasm before it can establish an infection. Animal viruses enter their host cells by one of three poorly understood mechanisms. Most nonenveloped animal viruses enter a host cell by the process of phagocytosis (viropexis). The virus is taken into the cytoplasm in a vesicle whose outer layer is derived from the host cell's cytoplasmic membrane (Figure 12.7). Another mechanism of penetration requires the viral envelope to fuse with the cell membrane. Fusion is followed by a mechanical rearrangement that permits the virus to enter the cytoplasm of the cell. Still other viruses may enter their host cell as naked nucleic acid.

Once inside the cell, the viral nucleic acid becomes available to the biosynthetic machinery of the cell through the processs of **decapsidation** (uncoating). Some nonenveloped viruses are decapsidated via pH changes in endocytic vesicles or by the action of enzymes following fusion of the endocytic vesicle with a lysosome (Figure 12.8). The viral nucleic acid is then released from the vesicle, presumably by enzymatic action. Virologists do not yet know all the details of decapsidation, but they presume RNA viruses are uncoated in the cytoplasm since this is where they are replicated. Most DNA viruses replicate in the nucleus, so their DNA must be translocated to the cell's nucleus either before or after decapsidation.

Replication of the Viral Genome

DNA viruses (except the poxviruses) replicate their DNA in the nucleus of their host cell. They utilize the precursors for DNA synthesis present in the cell's nucleotide pool; however, they do not mediate the deg-

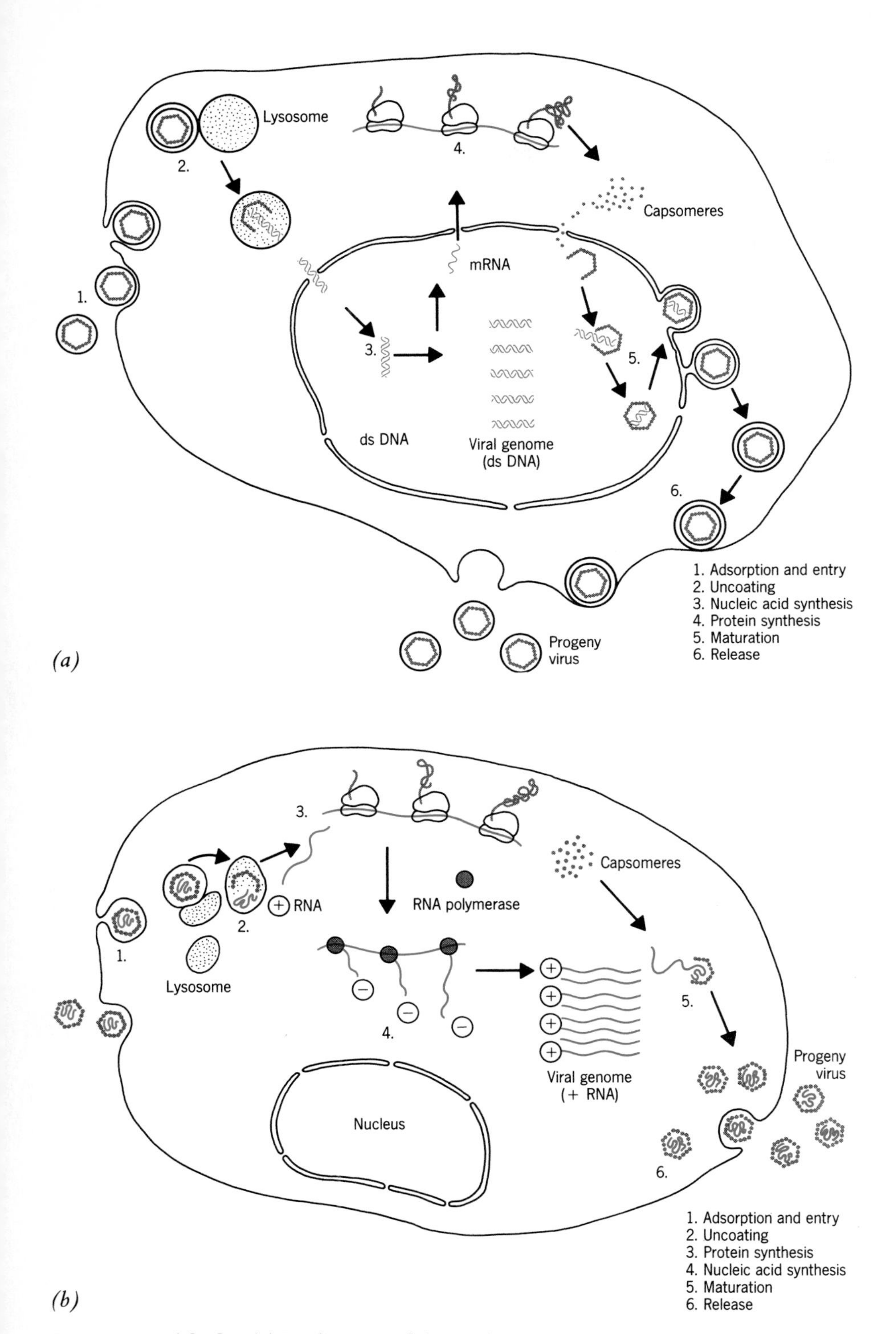

FIGURE 12.8 *(a)* Replication of the nucleic acid of most DNA viruses occurs in the nucleus. The mRNA formed in the nucleus moves to the cytoplasm where it is used to synthesize proteins. The viral proteins enter the nucleus where maturation occurs. This diagram represents the synthesis of a herpesvirus. *(b)* The entire replication of many RNA viruses occurs in the cytoplasm of the infected cell.

radation of the cell's DNA. The simpler DNA viruses utilize the cell's DNA polymerase(s) to synthesize viral DNA. The more complex viruses, such as the poxviruses and the herpesviruses, synthesize their own DNA polymerases, which then replicate viral DNA.

RNA viruses have a unique problem because cells *do not replicate RNA*. Therefore an RNA virus must supply the enzyme, **RNA transcriptase,** to make copies of its RNA genome. Some RNA viruses carry this enzyme with them into the cell during infection; others depend on the cell to make RNA transcriptase using the information in their RNA genome.

Single-stranded RNA viruses use this RNA transcriptase to make a **replicative RNA strand** from which they synthesize multiple copies of their genome. The replicative RNA strand never becomes a part of the progeny virions.

Retroviruses are a family of RNA viruses that replicate their RNA genome by first making a DNA intermediate. This is counter to the central dogma of molecular biology, which states that cellular information flows from DNA to RNA. The retroviruses have a **reverse transcriptase** (RNA dependent DNA polymerase), which makes the double-stranded DNA intermediate that functions as the template for synthesizing the viral RNA genome.

KEY POINT

Genetic information in cells flows from DNA to RNA. Retroviruses reverse this flow of information when their reverse transcriptase makes DNA from RNA. Reverse transcriptase has forced us to modify the central dogma of molecular biology and helped us explain retrovirus replication.

Synthesis of Viral mRNA

Since most DNA viruses replicate in the nucleus of their host cell, the nucleus is the site of viral mRNA synthesis (Figure 12.8*a*). Like the T-even bacteriophage described in Chapter 11, DNA viruses are able to form early and late messages that are translated into early and late proteins. The enzymes necessary for forming the early mRNA from the viral DNA genome are present in animal cells.

Animal RNA viruses are replicated in the cytoplasm of their host cells (Figure 12.8*b*). The simplest case is the single-stranded, plus RNA genomes since by definition, plus RNA functions as messenger RNA. Thus single-stranded plus RNA goes to the ribosome and is translated into proteins. Viruses with a single-stranded plus RNA genome require no special enzymes.

All other RNA virions that make RNA from RNA templates physically possess an RNA transcriptase in the virion. Animal cells do not contain such enzymes so the virus must provide it. RNA transcriptase makes mRNA from the viral RNA genome.

Retroviruses are a special group of RNA viruses. They contain two identical plus RNA strands and a reverse transcriptase. The reverse transcriptase makes double-stranded DNA, which functions as the template for mRNA formation.

Synthesis of Viral Proteins

All protein synthesis directed by viral nucleic acid takes place on ribosomes in the cell's cytoplasm. Although the actual mechanism of translation is essentially the same for all viruses, significant differences exist among the strategies of viral protein synthesis.

Poliovirus translates its single-stranded plus RNA genome into a giant polyprotein, which is then enzymatically processed into shorter polypeptides (Figure 12.9). The resulting viral proteins either have enzymatic activity or serve as capsomeres for the virus (VP_1, VP_2, VP_3, VP_4), or their function is unknown (P_x).

Another strategy is to synthesize multiple mRNAs from the viral genome. Multiple messages are synthesized from the segmented genome of the influenza virus by RNA transcriptase. The genome of the influenza virus is composed of eight different molecules of single-stranded minus RNA. Each of the eight RNAs is a template for one mRNA, which then codes for one of the eight viral polypeptides.

DNA viruses are replicated in the cell's nucleus. Messenger RNA is made from the viral DNA in the nucleus before it is translocated to the cytoplasm where protein synthesis takes place on the cell's ribosomes. This strategy is identical to the mechanisms used by eucaryotic cells: DNA synthesis and mRNA formation take place in the nucleus, whereas translation occurs in the cytoplasm.

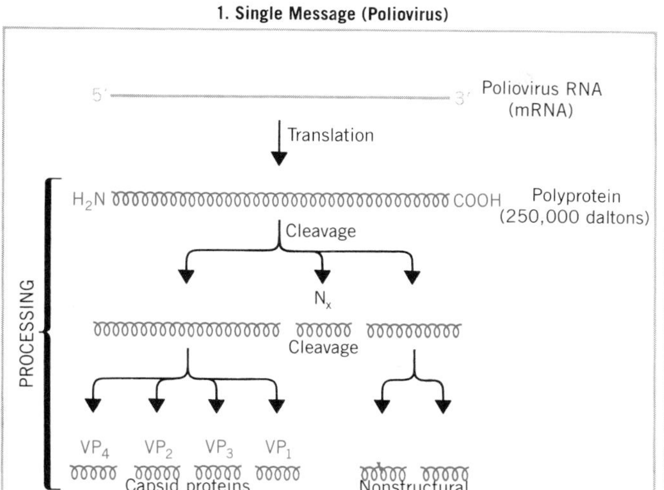

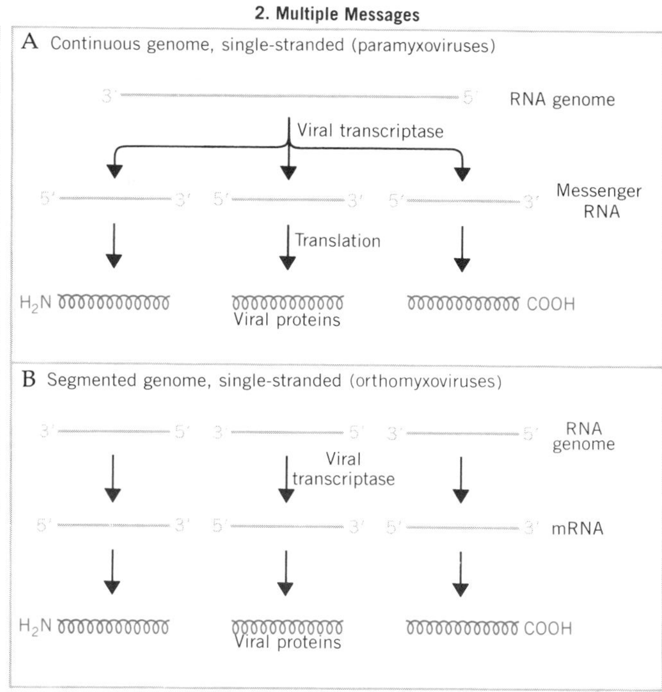

FIGURE 12.9 RNA viruses use various strategies for synthesizing proteins. (1) *Single message.* A single polyprotein is made directly from the single-stranded, plus RNA genome of the poliovirus. (2) *Multiple messages.* The paramyxoviruses transcribe multiple messages from their single-stranded, minus RNA genome. The orthomyxoviruses have a segmented, single-stranded, minus RNA genome that codes for individual mRNAs.

Maturation and Release

Complete virions are formed within the cell's nucleus or its cytoplasm, depending on the virus. Naked viruses self-assemble from preformed viral proteins and nucleic acids by a process governed by physiochemical forces. The aggregates of viral particles can be seen in electron micrographs (Figure 12.10) when they accumulate in the cell's cytoplasm. These aggregates are called **inclusion bodies** when they are visible in the light microscope. An example of this phenomenon occurs when the rabies virus accumulates in the cytoplasm of brain cells to form **Negri bodies.**

Some animal viruses are released into the extracellular environment when they destroy the integrity of

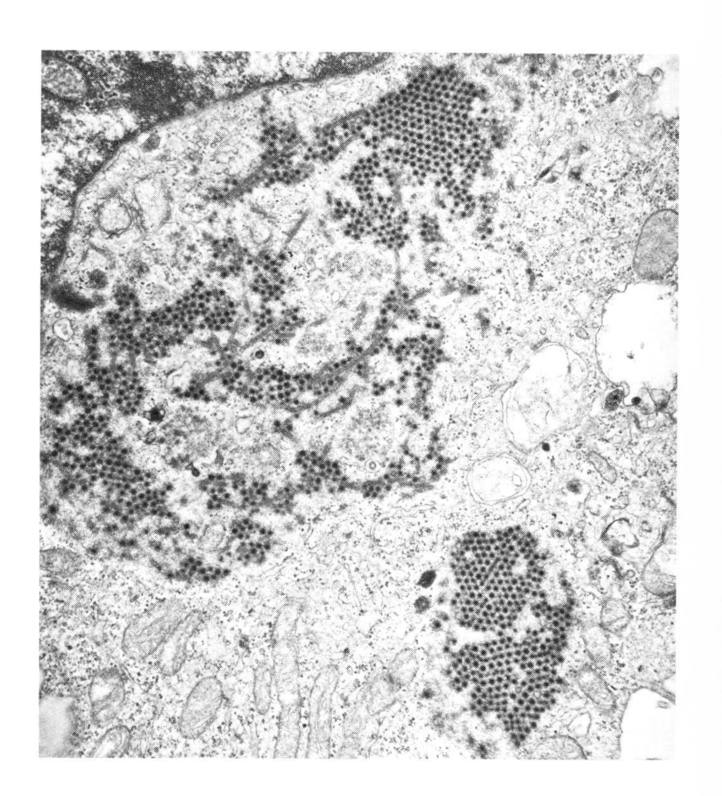

FIGURE 12.10 Cytoplasmic inclusions of reovirus 12 hours after inoculation. The inclusions are composed of virus particle aggregates. (Courtesy of Dr. S. Dales. From S. Dales, "The Structure and Replication of Poxviruses as Exemplified by Vaccinia," in A. J. Dalton and F. Haguenau (eds.), *Ultrastructure of Animal Viruses and Bacteriophages: An Atlas,* copyright © 1973, Academic Press, New York.)

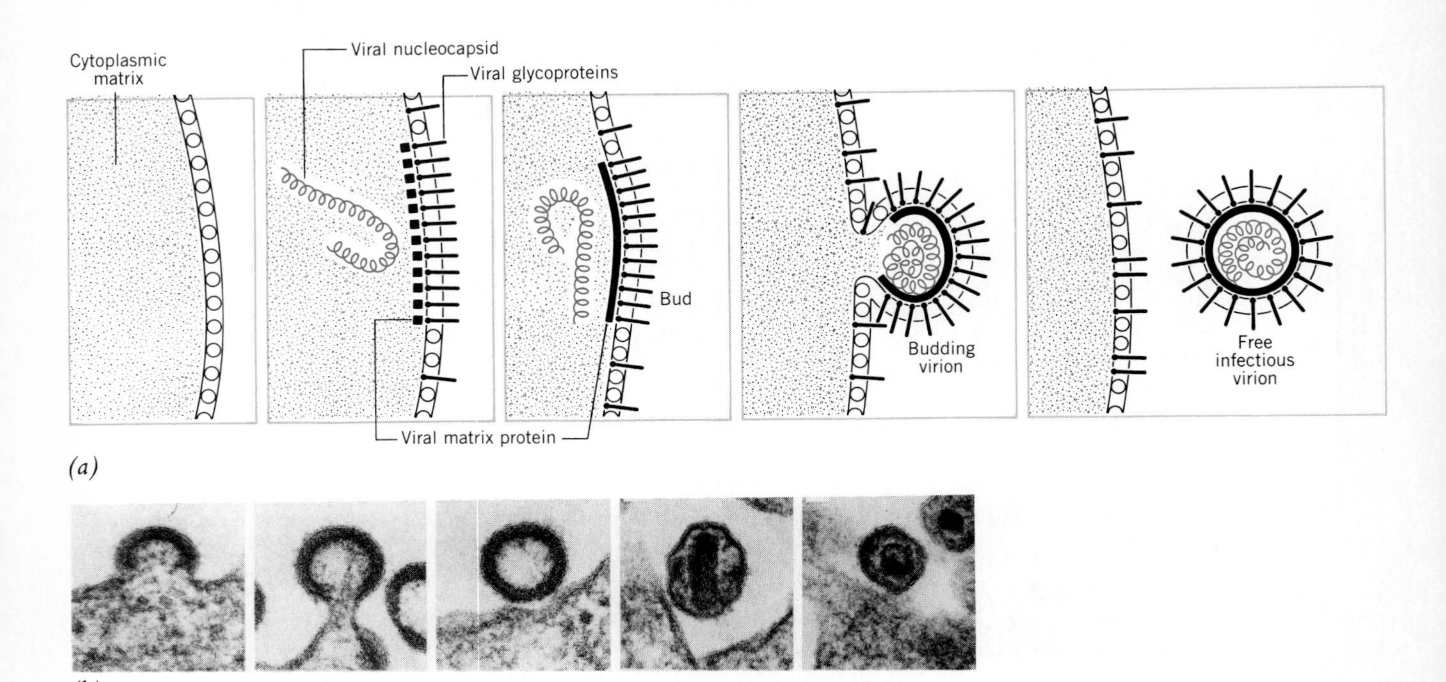

FIGURE 12.11 Sequential steps in the release of an enveloped virus by budding. *(a)* Diagram of the budding process, and *(b)* photos of the budding of visna virus. (Diagram reproduced with permission. From B. D. Davis, R. Delbecco, H. N. Eissen, and H. S. Ginsberg, *Microbiology,* 3rd ed., copyright © 1980, Harper & Row. Photographs courtesy of Dr. M. A. Gonda. From M. A. Gonda, F. Wong-Staal, and R. C. Gallo, *Science,* 227:173–177 (1985).)

their host-cell membrane. This process kills the cells and is one type of cytopathic effect attributed to infections by animal viruses.

Enveloped viruses usually acquire their envelope as they are extruded through the cell membrane (Figures 12.3 and 12.11). The cell membrane actually surrounds the virion to become its envelope. Cells infected with these viruses may produce and shed the virus over long periods of time, since the process of extrusion does not kill the cell.

The cell's membrane and the viral envelope are similar in chemical composition and structure, but they are not necessarily identical. Replicating viruses often produce viral proteins that are naturally incorporated into cellular membranes. This process adds unique antigens to sites in the membrane that will become the viral envelope (Figure 12.11). The nucleocapsids migrate to these membrane patches and acquire their envelope as they are extruded through the membrane.

EFFECTS OF ANIMAL VIRUS REPRODUCTION ON HOST CELLS

Animal viruses have a variety of effects on the cells they infect. These effects include killing the host cell (cytocidal effects), inducing the host to form specific proteins (inductive effects), and causing the cell to replicate in an unrestricted manner (transformation).

CYTOCIDAL EFFECTS

Cytocidal viruses cause extensive cell damage and eventually cell death by inhibiting the host cell's ability to synthesize cellular proteins and/or DNA. Soon after infection, cytocidal viruses prevent cell-directed protein synthesis by blocking the synthesis of ribosomal RNA and cellular mRNA. The synthesis of tRNA is not affected. Some viruses also prevent the synthesis of cellular DNA. The cytocidal animal viruses and the T-even bacteriophages employ similar biochemical mechanisms for commandeering the cell's synthetic machinery (see Chapter 11).

INDUCTIVE EFFECTS

Viruses induce host cells to manufacture viral proteins and other essential viral components required for virus propagation. Some of these viral proteins alter the host cell's membrane structure. These alterations can be detected by the human immune system, which is capable of destroying virus-infected cells. Another inductive effect occurs when the virus stimulates the cell to synthesize specific proteins that interfere with viral replication in other cells. Interferons are the best-understood class of viral-induced proteins.

Interferons

Many viruses induce animal cells to produce **interferons,** which are cellular proteins (often glycoproteins) that protect uninfected cells against viral infection. Their gross action can be measured by observing plaque reduction or virus-yield reduction when interferon-treated cell cultures are infected with a virus. Interferon is host-species specific, but not virus specific. Three types of human interferon are produced in response to a viral infection, predominantly by leukocytes and fibroblasts.

Human-fibroblast interferon is a glycoprotein produced when the cell encounters double-stranded RNA. Cells produce minute quantities of interferon (1 mg per 10 liters of tissue culture fluid), which is all that is needed to protect other cells from viral infection. The task of purifying interferon from tissue cultures is difficult, in large measure because of the small quantities produced. This problem has been surmounted by cloning the three human interferon genes and producing interferon by modern techniques of genetic engineering. Now sufficient quantities of human interferon are available for chemotherapeutic testing.

Interferon produced in a viral-infected cell leaves the producing cell and binds to the surface receptors on noninfected cells. This binding stimulates the cell to synthesize enzymes that are later activated during the early stages of viral infection. Once activated these enzymes inhibit viral replication. The protective state induced by interferon usually lasts for several days and then decays. Interferon is actively being investigated for its chemotherapeutic potential against viral infections and cancers.

VIRAL TRANSFORMATION

Oncogenic viruses induce tumor formation when they infect a host cell. This relationship between viruses and tumors is not a new concept. V. Ellerman and O. Bang (1908) and Peyton Rous (1911) were the first to report the transfer of cancer from one animal to another. Ellerman and Bang reported the transfer of chicken leukemia, and Rous demonstrated the transfer of chicken sarcoma (a tumor of connective tissue) from diseased to healthy chickens. The virus responsible for the chicken sarcoma was later identified and is now called Rous sarcoma virus (RSV). RSV is an RNA virus known for its ability to transform tissue culture cells. Oncogenic DNA viruses are also known; for example, the simian virus SV_{40} is an oncogenic DNA virus that was discovered in the rhesus monkey kidney tissue cultures used to grow poliovirus.

Rous Sarcoma Virus: An RNA Oncovirus

Domestic fowl can host a variety of viruses including the Rous sarcoma virus (RSV), a single-stranded RNA retrovirus. Each cell infected by RSV is transformed such that it loses control over replication and multiplies in an unrestricted manner, just as tumors do. The numerous daughter cells of a transformed cell line can each be stimulated to produce progeny RSV particles. Therefore, the transformed cells carry the RSV genome as a provirus just as the prophage is carried by lysogenized bacteria (see Chapter 11). Since RSV is an RNA virus, there must be a mechanism for maintaining an RNA genome in a host-cell line.

This poses a dilemma because we know that genetic information is passed from parent to daughter cells in DNA, not in RNA. The only way information in RNA can be passed to progeny cells is by converting that information to DNA (Figure 12.12). This indeed happens in the retrovirus infections, a fact confirmed experimentally by the isolation of viral RNA-directed DNA polymerase (reverse transcriptase) in the RSV virion. As soon as RSV infects its host cells, its reverse transcriptase makes a DNA using the viral RNA as a

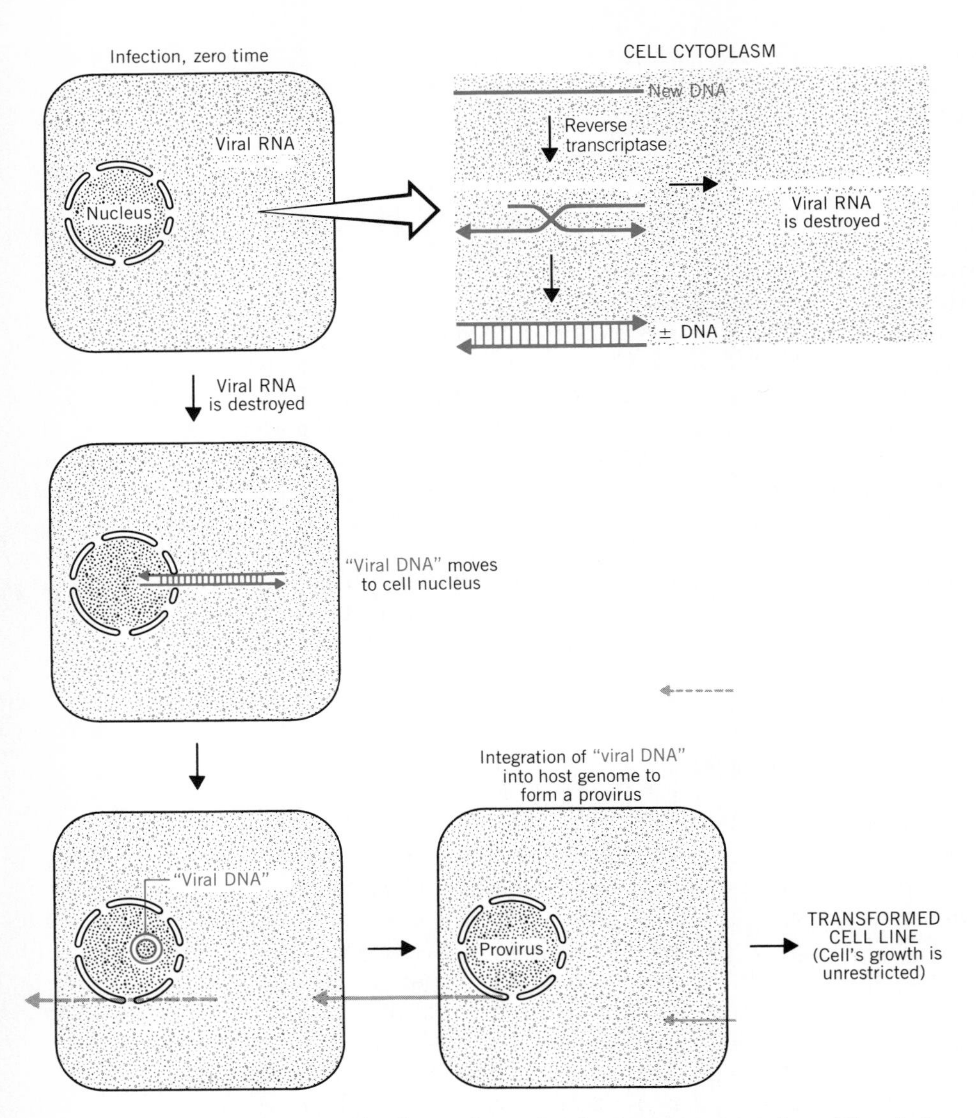

FIGURE 12.12 Animal cell transformation by Rous sarcoma virus. The viral RNA is copied into double-stranded DNA in the cytoplasm by reverse transcriptase. The double-stranded DNA replica then moves into the nucleus where it is incorporated into the cell's chromosomes as a provirus. This transformed cell now replicates in an unrestricted manner.

template. This DNA is then randomly incorporated into the host's genome as a provirus. Each daughter cell receives a copy of the provirus DNA and produces Rous sarcoma virions in a nonlethal replication process. The presence of the Rous sarcoma provirus causes the host cell to replicate in an unrestricted manner—that is, the cell is transformed. Other retroviruses do not transform host cells with such high efficiency.

Oncogenes Cause Tumors

Rous sarcoma virus carries a gene designated *src* for sarcoma. *Src* belongs to a family of genes called *oncogenes,* which are tumor-causing genes. Oncogenes carried by animal viruses have the potential to transform eucaryotic cells. The *src* gene is an altered version of a noncancerous (benign) gene found in healthy chicken tissue that somehow was picked up by the Rous sarcoma virus and transformed into an oncogene. Genes with the propensity to become oncogenes appear to exist in animal tissue and are designated "proto-oncogenes."

The *src* gene codes for a **protein kinase** that phosphorylates tyrosine residues on selected proteins. These phosphorylated cellular proteins are observed in RSV-transformed cells, but not in normal cells. Although some of the proteins phosphorylated by the *src* gene product have been identified, a clear explanation of the role of phosphorylation in cell transformation is still unavailable.

Great progress has been made in understanding the genetics and biochemistry of cell transformation following the identification of the RSV *src* oncogene. Similar approaches have uncovered oncogenes in other retroviruses and in cancerous tissue. The gene products of these oncogenes affect such cellular activities as growth factor receptors, DNA binding, and GTP binding. Studies of oncogenic viruses, oncogenes, and their gene products are providing significant insights into the complex problem of cancer.

SV_{40}: A DNA Oncovirus

Simian virus 40 was discovered as a passenger (unexpected) virus in monkey kidney tissue cultures used to cultivate poliovirus. This DNA virus produces tumors in baby hamsters and transforms human embryonic cells and cultured hamster cells at a low efficiency. During a normal infection process, only 1 out of every 10,000 infected cells is transformed. Early in their replication process the virus directs the synthesis of a new antigen that is incorporated into the cytoplasmic membrane. This antigen is the T antigen that is used as a marker to identify SV_{40}-transformed cells. SV_{40} is an example of a DNA virus that can transform nonpermissive mammalian cell lines; that is, they do not support viral multiplication.

There is a growing list of human viruses that are known or strongly suspected of being oncogenic viruses. These viruses include both RNA and DNA viruses. The Epstein–Barr virus infects lymph tissue and is known to cause Burkitt's lymphoma, and the human T-cell lymphotropic virus (HTLV-I) causes a rare human leukemia. Among the potential oncogenic human viruses are hepatitis B virus (liver cancer), papillomaviruses (penile and cervical carcinomas), and herpes simplex (cervical cancer).

ASPECTS OF HUMAN VIROLOGY

Even though most viral diseases cannot be cured by medical intervention, viral diagnosis is an important aspect of medical care. Some viral infections, such as rabies, can be treated by postinfection active immunization. Certain viral diseases in childbearing females (urogenital herpes, rubella) jeopardize the health of the fetus sufficiently to warrant taking precautions when the mother's disease is recognized. Other viral diseases are extremely contagious, and further illness can be decreased if proper preventive procedures are followed. Even when there are no medical procedures available, the mental outlook of the patient is usually improved by an accurate diagnosis and prognosis.

Viral diseases are diagnosed to enable the physician to administer treatment (if available), to provide an accurate prognosis to the patient, and to take steps to prevent the spread of the illness. Many large hospitals and almost all state public health laboratories have viral diagnostic facilities. A modern virology laboratory is established to isolate viruses from patients, to perform serodiagnostic tests, and to measure viral antigens. Many state public health laboratories are equipped to perform low-volume and specialized virology tests. The Virology Division of the Centers for Disease Control (CDC) tests specimens for the uncommon viral diseases, maintains a serological reference service for the identification of specific viruses, and conducts epidemiological investigations of major viral diseases.

ISOLATION OF HUMAN VIRUSES

Many human viruses can be isolated by inoculating specimens into tissue cultures. Tissue cultures of monkey kidney, human embryonic kidney, and human embryonic lung cells are commercially available for isolating human viruses. Inoculated cultures are incubated for an average of five days before 50 percent of the cultures are positive and an average of ten days before 90 percent of the cultures are positive. Some cultures show positive viral growth within one or two days. Serological and other tests are available for identifying a virus once it has been isolated.

CONTROL OF HUMAN VIRAL DISEASES

Most human viral diseases are diagnosed by the clinical symptoms they cause. Some of these symptoms can be

TABLE 12.2 Virus vaccines in clinical use

VACCINE	DESCRIPTION	ROUTE	FREQUENCY	ANTIBODY PERSISTENCE
Measles	Attenuated virus	Subcutaneous	Once	12 years
Mumps	Attenuated virus	Subcutaneous	Once	8 years
Rubella	Attenuated virus	Subcutaneous	Once	8 years
Poliomyelitis (Sabin)	Attenuated viruses, three types	Oral	Four	10 years
Poliomyelitis (Salk)	Killed viruses, three types	Subcutaneous	Four	2 years
Vaccinia (military personnel)	Poxvirus, protects against smallpox	Intradermal	Every 3 years	3 years
Yellow fever	Attenuated virus	Subcutaneous	Once	10 years
Influenza	Influenza A and B (prevalent strains)	Subcutaneous	Twice a year	1 year
Rabies	Human diploid cell rabies vaccine (HDCV)	Intramuscular	0, 7, and 28 days	2 years

treated, but most viral diseases run their course without medical intervention. A few viral diseases can be treated with antiviral substances, by vaccination, or by passive immunization. Vaccines are the major weapon available for preventing viral infections, but they are available for only a few viral diseases.

Viral Vaccines

Live-virus vaccines or killed-virus vaccines are used to immunize humans against specific viral diseases (Table 12.2). The vaccines are administered with a frequency and by a route that maximizes the human's immunological response. Measles, mumps, and rubella viruses are combined in a single vaccine (MMR vaccine), which is administered once as a subcutaneous injection. The polio vaccine (Sabin) contains three types of live attenuated polioviruses and is taken orally at four different ages. The Salk polio vaccine is a preparation of killed polioviruses that is injected intradermally.

The smallpox vaccine is no longer used, except by certain military personnel, because this viral disease has been eliminated by a worldwide vaccination and surveillance program. Today no country requires a smallpox vaccination certificate as a condition of entry. Only people who work with orthopox viruses need be vaccinated.

Yellow fever vaccine is given to persons in or traveling to areas where yellow fever is epidemic. This viral disease is transmitted by mosquitoes and is limited to tropical Africa, tropical South America, and the Caribbean. Influenza vaccine is produced from the type of influenza virus that is prevalent during a given influenza season. Influenza vaccine is administered to compromised patients and persons over 65 years. Preexposure rabies virus immunizations are given to high-risk groups such as animal handlers, certain laboratory workers, and field personnel. Postexposure rabies immunizations are also effective since the incubation period of the rabies virus is sufficiently long to permit immunization against this disease after infection.

Antiviral Compounds

Success in developing antiviral drugs has been seriously limited by the toxic effects antiviral agents have on humans. Therefore only a few antiviral compounds are available for chemoprophylaxis and chemotherapy (Table 12.3). **Amantadine** is licensed in the United States for the prophylaxis and therapy of influenza A (Figure 12.13). Chemically related to amantadine is **rimantadine,** a drug available and prescribed in the Soviet Union. These drugs prevent viral replication, apparently by interfering with the uncoating of the influenza A virus or the transcription of its viral RNA. Amantadine or rimantadine prophylaxis provides 55 to 80 percent protection against influenza-like illness during outbreaks of influenza A, but they are ineffective against influenza B and influenza C viral infections. Respiratory syncytial virus causes serious infections in

TABLE 12.3 Antiviral compounds effective against human viruses

SUBSTANCE	ANTIVIRAL ACTIVITY	APPLICATIONS	MODE OF ACTION
Amantadine (Symmetrel)	Influenza A	Oral	Inhibits uncoating of influenza A virus
Idoxuridine (Stoxil)	Herpes simplex, keratitis	Topical	Inhibits or interferes with viral DNA and protein synthesis
Acyclovir (Zovirax)	Herpes simples; genital, disseminated varicella–zoster	Topical, intravenous	Inhibits DNA synthesis by herpes viruses
AZT (azidothymidine)	AIDS		Interrupts elongation of DNA
Ribavirin	Respiratory syncytial virus	Inhaled aerosol	Interferes with guanosine synthesis and capping of viral mRNA

newborns and is a major problem in hospital nurseries. **Ribavirin** is a guanosine analog that is effective against respiratory syncytial virus when inhaled as an aerosol.

A number of antiviral compounds are available to treat the acute manifestations of herpes virus infections, both as topical agents and for systemic use. **Idoxuridine** is a halogenated analog of uridine (Figure 12.13) that is used for the topical treatment of mucocutaneous herpes infections, especially herpes simplex keratitis. It is not used systemically. **Acyclovir** (Figure 12.13) is active against herpes simplex virus-1, herpes simplex virus-2, and varicella–zoster virus. Acyclovir inhibits the cell's DNA polymerase after the drug is phosphorylated by enzymes in the virus-infected cell. This drug is used for topical treatment of mucocutaneous herpes and for intravenous applications in hospitalized patients.

An intensive effort is being made to develop and test drugs for treating acquired immune deficiency syndrome (AIDS). Partial success has been attained with **AZT** (azidothymidine), a drug that extends the life ex-

Azidothymidine (AZT)

Ribavirin

Idoxuridine

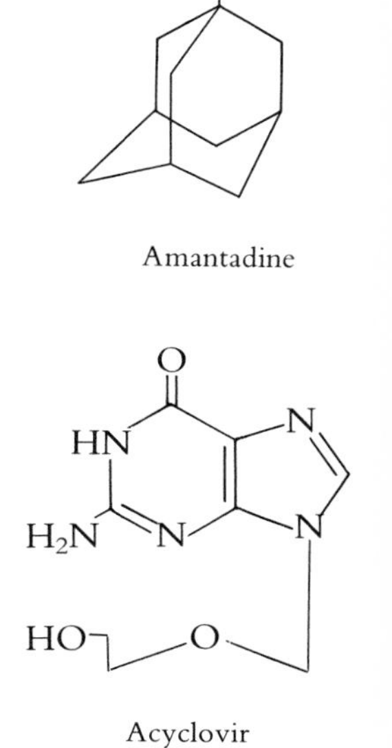

FIGURE 12.13 Amantadine is used to treat influenza A virus infections. Nucleoside analogs prescribed for treating viral infections include idoxuridine and acyclovir for herpesvirus infections, ribavirin for respiratory syncytial virus infections, and azidothymidine for AIDS.

pectancy of seriously ill AIDS patients, but is not a cure. AZT interferes with the formation of DNA by the AIDS virus reverse transcriptase.

KEY POINT

Human viral diseases that are prevented by immunizations include polio, measles, mumps, and rubella. Immunization was the approach used to eliminate smallpox from the world. Antiviral drugs have limited applications in viral chemotherapy in large measure because of their toxicity.

PRIONS CAUSE SLOW-DEVELOPING ANIMAL DISEASES

Prions are uncoventional viruses because they are resistant to inactivation by nucleic acid-modifying procedures and are extremely resistant to heating. Although infectivity of isolated prions is associated with a prion protein, the chemical nature of prions remains unknown. Prions infections are confined to the central nervous system and are associated with slow-developing diseases, diseases with incubation periods that range from months to decades. The clinical sign of a prion infection is a progressive degenerative neurological disease that eventually leads to death.

Prions were first discovered to cause scrapie, a neurological disease of sheep and goats. Researchers have succeeded in transferring the responsible agent to laboratory animals where it is being studied. Human diseases that may be caused by prions include Creutzfeldt–Jakob disease, kuru, and Gerstmann–Straussler syndrome. Decontamination of tissue infected with prions is a serious problem because these agents are more resistant to heating than endospores. To kill prions they must be autoclaved for 4.5 hours at 121°C or heated repeatedly at alkaline pHs (1*N* NaOH).

PLANT VIROLOGY

Most economically important crop plants can be damaged by viral infections. These plants fall into the categories of annuals grown yearly from seeds, perennials that have a yearly fruit crop, and vegetatively propagated plants. Many times the crop losses occur within a geographical locality; however, when infected seeds are the source of the disease the losses can be widespread. Serious economic losses from viral infections of rice, wheat, potatoes, tomatoes, lettuce, celery, cucumbers, pears, and tobacco have occurred. The symptoms of these diseases include stunting and dwarfing of the plant, necrotic lesions of the leaves, necrosis or distortion of the fruit, and lowering of seed or fruit yield.

Control of viral plant diseases requires an understanding of how the virus is spread and the techniques available for preventing infection. Many plant viruses are carried from generation to generation in seeds or in vegetatively propagated plants. Extensive efforts are made to produce virus-free seeds and stock plants as the major means of avoiding economic losses for the farmer.

CLASSIFICATION OF PLANT VIRUSES

Plant viruses are characterized by the type and organization of their nucleic acid, size of their nucleic acid, outline or particle shape, and the means by which they are transmitted. Most plant viruses contain RNA as their nucleic acid, and the majority of these have single-stranded RNA (Table 12.4). The shape of a plant virus is described as filamentous, spherical, or bacilliform. The categories used to describe the transmission of plant viruses include specific insects (e.g., aphids, leafhoppers, beetles, whiteflies), fungi, nematodes, without vector, and seed transmission. A compilation of this information has been used to construct a classification scheme for identifying and naming plant viruses.

Plant viruses are usually named after the plant they infect. Examples include the tomato spotted wilt virus, beet yellow virus, carnation latent virus, and tobacco mosaic virus. The name is often abbreviated with letters, TMV for tobacco mosaic virus. TMV is the single-stranded, filamentous, RNA virus that we will use as our model for studying plant virus replication (Figure 12.14).

REPLICATION OF PLANT RNA VIRUSES

The first filterable infectious agent ever demonstrated was the tobacco mosaic virus (TMV). This virus infects tobacco leaves and causes necrotic lesions at the site of infection. TMV is a helical RNA virus (Figure

TABLE 12.4 Characteristics of selected plant viruses

VIRUS	VIRION SIZE (nm)	MODE OF TRANSMISSION	REMARKS
	Single-Stranded RNA, Elongated Virions		
Tobacco mosaic	300 × 18	Mechanical wounds	First virus discovered
Potato X	515 × 13	Mechanical wounds	Latent in potatoes
Beet yellows	1250 × 10	Aphids	
Tobacco rattle	190 × 250 45–115 × 25	*Trichondorus* (nematodes)	Two particles[a]
	Single-Stranded RNA, Isometric Virions		
Cowpea mottle	28	Beetles	Two particles[a]
Cucumber mosaic	30	Aphids	Three particles
Barley yellow	22	Aphids	Transcapsidation
Tobacco ringspot	28	*Xiphinema* (trematodes)	Two particles
Turnip yellow mosaic	28	Flea beetles	
	Single-Stranded RNA, Bacilliform, Virions		
Potato yellow dwarf	380 × 75	Leafhopper	Multiplies in vector
	Double-Stranded RNA, Isometric, Virions		
Wound tumor	70	Leafhopper	Multiplies in vector
	Double-Stranded DNA, Isometric Virion		
Cauliflower mosaic	50	Aphids	Prominent inclusions in infected cells
	Single-Stranded DNA, Isometric Virion		
Maize streak	18–20	Leafhoppers or whiteflies	One molecule of plus circular DNA

[a]Specific segments of viral genome encapsulated in separate particles; all are needed to produce infection.

12.14) that infects various tissues of the tobacco plant, replicates in these tissues to produce infective progeny virions, and causes damage to the infected tissues. This virus is very prolific—a single infected root-hair cell can contain as many as 10^7 progeny tobacco mosaic virions.

The virus enters the cytoplasm of its host cell following the disruption of the plant cell wall by mechanical abrasion. Both the intact TMV and its plus RNA molecule are infectious. Natural infections occur when TMV is rubbed on the tobacco leaf, or when TMV is passed via intradermal bridges between plant cells. Virologists often study the replicative process of TMV in isolated tobacco cell protoplasts (a plant cell minus its cell wall) to avoid the complications involved in using entire plants.

Replication of TMV

Penetration of the TMV virion into the cell is followed by a rapid decapsidation that exposes the single-stranded RNA. This RNA functions as mRNA for the synthesis of viral proteins, one of which is an RNA transcriptase that synthesizes the minus RNA strand. This strand functions as the template for manufacturing more viral genome (plus RNA strands). TMV capsid proteins are among the other viral proteins formed in the infected cell from the plus RNA strand.

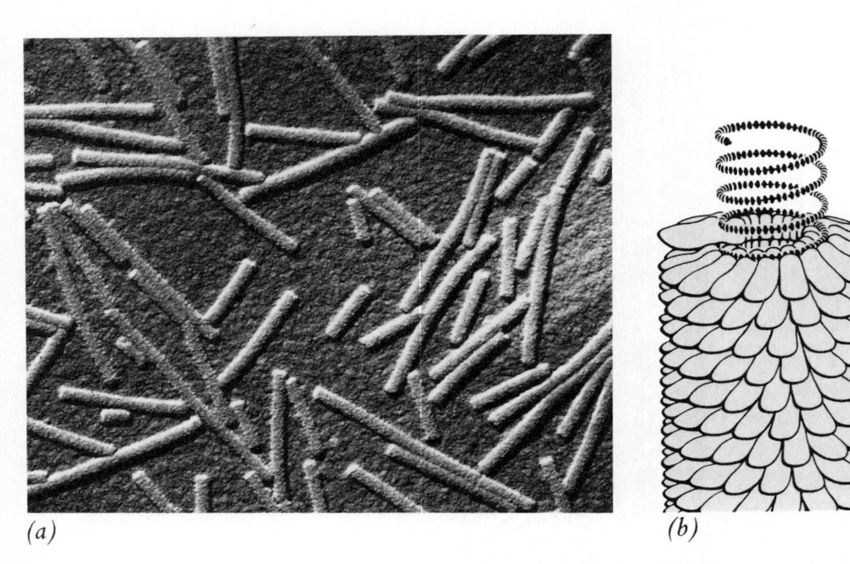

FIGURE 12.14 Tobacco mosaic virus. *(a)* Electron micrograph of TMV (Courtesy of Carl Zeiss Company.) *(b)* Diagram showing the relationship between the viral RNA and capsid proteins.

The long rod-shaped virus, 300 by 18 nm, is formed during viral maturation. Each of the identical capsid proteins binds to three nucleotides on the RNA genome, which becomes buried between the protein subunits as the helical capsid is formed. This process continues until the entire RNA is coated with protein subunits (Figure 12.14*b*).

VIROIDS ARE INFECTIOUS RNA MOLECULES

Viroids are infectious circular RNA molecules that so far have been shown to occur only in plants. They cause disease in a variety of plants, including some economically important crops such as tomato and potato. Viroids have no capsid or coat of any kind—they are simply single-stranded RNA molecules containing between 270 and 380 nucleotides. To date they have not been shown to code for any proteins, so they are completely dependent on their host's metabolic machinery for their replication. These RNA molecules infect and replicate autonomously in plant tissues. After the potato spindle tuber agent was isolated and shown to be a viroid, other viroids have been shown to cause other plant diseases.

VIRAL DISEASES OF PLANTS

Viral diseases in crop plants have a major effect on mankind. Perennial crops and crops that are propagated by vegetative material are particularly subject to viral disease. Raspberries, blueberries, strawberries, and fruit trees are perennial plants that bear fruit for many years. Once these plants become infected with a virus, there is no way to control the spread of the disease except to destroy the plant or to remove the infected growth. Vegetatively propagated plants include flowers propagated from cuttings, trees or bushes propagated by grafting, and plants grown from tubers, corms, or bulbs. To prevent viral disease in these plants, one must start with viral-free stock.

Plant diseases are recognized by the damage or alterations the viruses cause. Some of the most common symptoms of viral infections are local necrosis, a mottled appearance of green and yellow spots, yellowing of leaves, and leaf curl (Figure 12.15). Some viral infections cause stunted growth, while others cause alterations in flower color. The striping and mottling of once valuable tulips were caused by a tulip mosaic virus, which was perpetuated by cultivating infected tulip bulbs (Figure 12.16).

(a) (b)

(c)

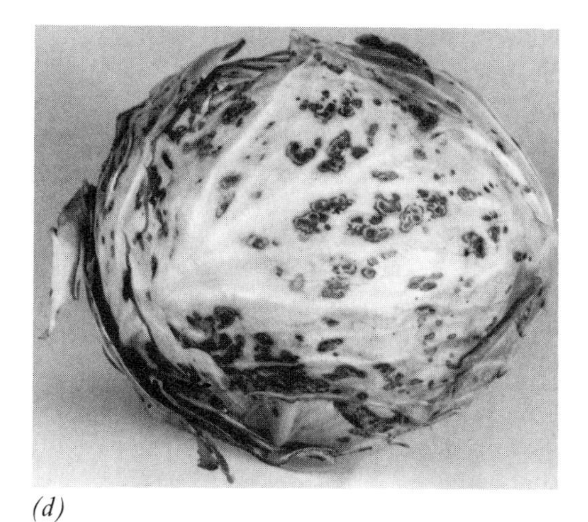

(d)

FIGURE 12.15 Plant viruses cause damage to the leaves, stems, and fruits of infected plants. Beetles transmit the cowpea chlorotic mottle virus, which *(a)* causes mottle in infected leaves (left)—healthy leaves (right). Turnip mosaic virus infects turnip leaves *(b)* to cause necrotic local lesions (NLL) and vein clearing (VC)—healthy leaves (H). Tomato mosaic virus *(c)* causes fruit necrosis and turnip mosaic virus causes black necrotic ring spots in cabbage *(d)*, cauliflower, and Brussels sprouts. (Courtesy of *(a)* and *(c)* Dr. Howard Scott, Dept. of Plant Pathology, University of Arkansas, *(b)* of Dr. John Hammond, USDA, and of *(d)* Dr. Thomas A. Zitter, Cornell University.)

FIGURE 12.16 Variegated color in tulip caused by a viral infection. (From Weier, et. al., *Botany: An Introduction to Plant Biology,* 6th ed., John Wiley & Sons, 1982.)

Transmission of Plant Viruses

Plant viruses can be transmitted between plants by mechanical transfer or via arthropod vectors. Although rare in nature, mechanical transfer of plant viruses occurs by the abrasive rubbing between plants caused by wind action. Viruses are also transferred by humans as they work with plants.

Natural transfer of plant viruses is often mediated by insects that acquire the virus by feeding on diseased plants and then transfer the viruses to healthy plants. Insects mechanically inject the virions into the host cell during their normal feeding process. Other insects serve both as intermediate hosts and as vectors for plant viruses. These insects do double damage to plant crops by first consuming plant material and then by transmitting viral diseases to healthy plants.

KEY POINT

Most plant viruses are single-stranded RNA viruses. Plant viruses infect their host through mechanical abrasions or are transmitted by insects. Once infected with a virus, the plant remains infected until it dies.

Control of Plant Viruses

Plant breeders have spent many years developing the virus-free plant strains available today. These strains are either naturally selected plants that have survived a viral infestation or plants that have been bred by plant geneticists. Plant suppliers are careful to supply only virus-free stocks of perennial plants to prevent the introduction of viruses into a new area or existing farm. Insect control is also necessary to prevent the spread of plant viruses.

Viruses and insects can overwinter in soil and in decaying plant material. Overwintering can be prevented by burning plant material or removing it from a garden. Crop rotation is another means of avoiding viral carryover from one season to the next.

Summary Outline

HISTORY OF ANIMAL AND PLANT VIROLOGY

1. Edward Jenner (1798) developed the first viral vaccination procedure, which protected people against smallpox.
2. Plant virology began when Adolf Mayer demonstrated (1886) the infectious nature of tobacco mosaic disease, which was followed by Ivanovsky's report (1892) that TMV passed through bacterial filters.
3. In 1898 Loeffler and Frosch discovered the viral cause of foot-and-mouth disease of cattle.

ANIMAL VIROLOGY

1. Humans and animals are susceptible to a wide variety of viral infections, ranging in symptoms from mild or asymptomatic to serious sickness that can cause death.

2. Animal viruses are self-replicating, intracellular parasites that are specific for the animals they infect.
3. Human and animal viruses are characterized by their structure, size, and chemical composition and are classified into six families of DNA viruses and 11 families of RNA viruses.
4. The names of viral families end in -viridae, whereas the name of a virus is often derived from the disease it causes, such as poliovirus.
5. Animal viruses are composed of one type of nucleic acid surrounded by a proteinaceous capsid. They can contain enzymes, segmented nucleic acids, simple or complex capsids, and/or a lipid-containing envelope.
6. Most animal viruses are either icosahedral or helical in shape; a few are pleomorphic, the poxviruses are brick-shaped, and the rhabdoviruses are bullet-shaped. Enveloped viruses have an outside lipid-containing membrane layer derived from a membrane of the host cell.
7. The viral capsid is composed of capsomeres arranged as a symmetrical icosahedron or an elongated helix. Helical animal viruses always contain RNA as their nucleic acid.
8. The animal virus genome is either single-stranded or double-stranded, linear or circular, monomolecular or segmented, DNA or RNA.
9. Single-stranded RNA that serves as a messenger RNA is designated plus RNA (+ RNA); the single-stranded DNA template for transcribing mRNA is designated minus DNA (− DNA).
10. Animal viruses are enumerated by the plaque assay using tissue cultures, animal assays, and by hemagglutination.
11. Embryonated chicken eggs and tissue cultures are used to propagate viruses for the manufacture of viral vaccine.

REPLICATION OF ANIMAL VIRUSES

1. After absorption, the complete virus enters the cell's cytoplasm and the viral nucleic acid is released by uncoating.
2. RNA viruses are replicated in the cytoplasm of the cell, whereas the DNA genome (except for the poxviruses) is transposed to and replicated in the cell's nucleus.
3. RNA viruses contain or manufacture specific enzymes for replicating their RNA. These include the reverse transcriptase of retroviruses and RNA transcriptase.
4. Viral proteins are synthesized on cytoplasmic ribosomes as a long polypeptide that is later cleaved into functional viral proteins or as individual proteins coded from multiple mRNAs.
5. DNA viruses produce mRNA in the cell's nucleus that is then translated on the cell's cytoplasmic ribosomes.
6. Infective virions accumulate in the cell, are released by extrusion through the cytoplasmic membrane, or are released during cell lysis.
7. Enveloped virions acquire their outer layer from a cell membrane, often as the virion is extruded through the cell's cytoplasmic membrane.

EFFECTS OF ANIMAL VIRUS REPRODUCTION ON HOST CELLS

1. Cytocidal viruses cause cell damage and death by inhibiting the host cell's ability to synthesize cellular proteins and/or DNA.
2. Virus-infected animal cells produce interferons, which are cellular proteins that protect uninfected cells against viral infection.
3. Oncogenic viruses such as RSV and SV_{40} induce tumor formation when they infect their host cells.
4. RSV is a single-stranded RNA oncogenic retrovirus that possesses reverse transcriptase and transforms each cell it infects. Each transformed cell contains an RSV provirus.
5. RSV carries the *src* oncogene, which codes for a protein kinase that phosphorylates selected proteins.
6. Simian virus 40 is a DNA oncovirus that produces tumors in baby hamsters and transforms human embryonic cells and cultured hamster cells at a low efficiency.
7. Human viruses that are known or strongly suspected of being oncogenic viruses include the Epstein–Barr virus, human T-cell lymphotropic virus, hepatitis B virus, pappillomaviruses, and herpes simplex virus.

ASPECTS OF HUMAN VIROLOGY

1. Human viruses are detected by inoculating speci-

mens into tissue cultures and identified by serological or other tests.

2. Only a few human viral diseases can be treated with antiviral substances, by vaccination, or by passive immunization.
3. Vaccines are routinely used to prevent measles, mumps, rubella, influenza, and polio.
4. The few antiviral compounds available for chemotherapy include amantadine and rimantadine (influenza A); idoxuridine, adenine arabinoside, and acyclovir (herpes infections); and AZT (AIDS).

PLANT VIROLOGY

1. Plant viruses are characterized by the type, size, and organization of their nucleic acid, outline or particle shape, and the means by which they are transmitted.
2. Most plant viruses contain RNA as their nucleic acid; the majority of these have single-stranded RNA.
3. Plant viruses are transmitted by insects, fungi, nematodes, infected seeds, and cuttings.
4. TMV is a helical RNA virus that infects tobacco plants following mechanical disruption of the plant cell wall.
5. The infectious single-stranded RNA functions as mRNA for the synthesis of viral proteins.
6. During viral maturation the RNA is buried between the protein subunits of the helical capsid.
7. Viroids are circular RNA molecules that are infectious only for plants—they have no capsid or coat.
8. Important food crops are protected from viral diseases by planting virus-free and/or virus-resistant strains, virus-free seeds, and controlling insects. Viruses able to overwinter in decaying plants can be destroyed by burning plant material.

Questions and Topics for Study and Discussion

QUESTIONS

1. What major criteria are used to classify animal viruses? Does this classification scheme indicate how viruses are related to each other? Explain.
2. Describe the organization of an enveloped virus by indicating the structure and chemical composition of its parts.
3. Define or explain the following:

capsid	icosahedron
capsomere	oncogene
contact inhibition	segmented genome
decapsidation	transformation
envelope	viral hemaglutination
helical capsid	viroid

4. Explain the importance of tissue cultures to modern animal virology.
5. Describe three methods of enumerating animal viruses.
6. Compare adsorption and entry by a typical animal virus to adsorption and entry by a T-even bacteriophage.
7. What protein-synthesizing strategies are available to a plus RNA virus that are not available to a minus RNA virus?
8. Describe how the RNA genome of poliovirus is used to synthesize individual proteins. How does this system limit viral control over protein synthesis?
9. What role do interferons play in resisting viral infections? How do interferons act in this process?
10. What is an oncogenic virus? Give an example of an oncogenic virus and explain how it transforms its host cell.
11. How are human viruses isolated? Are viruses isolated in the routine diagnosis of most viral infections? Explain.

12. What is/are the major method(s) used to control human viral diseases? Which diseases have been successfully controlled?

13. By what mechanisms do viruses infect plants? How can farmers protect their crops against viral infections?

14. What are viroids and how do they differ from other viruses?

DISCUSSION TOPICS

1. If oncogenic viruses cause tumors in animals, is it likely that they also cause human cancer?

2. Human virology developed about 50 years after the isolation of the majority of the bacteria that cause human diseases. What were the reasons for this delay?

3. Will interferons ever be as effective in treating viral diseases as antibiotics are against bacterial diseases?

4. What methods can be used to obtain virus-free plant seedlings?

Further Readings

BOOKS

Joklik, W. K. (ed.), *Virology,* 2nd ed., Appleton-Century-Crofts, Norwalk, Conn., 1985. One of the best virology textbooks available for its introduction to basic virology and comprehensive coverage of clinical virology.

Walkey, D. G. A., *Applied Plant Virology,* Wiley, New York, 1985. Practical information about plant virology in the format of an introductory text.

Zuckerman, A. J., J. E. Banatvala, and J. R. Pattison (eds.), *Principles and Practice of Clinical Virology,* Wiley, New York, 1987. Comprehensive articles on the medical aspects of human viruses.

ARTICLES AND REVIEWS

Bishop, J. M., "Oncogenes," *Sci. Am.,* 246:80–92 (March 1982). Recent research on Rous sarcoma virus has led to the identification of both viral and animal genes that are involved in oncogenesis.

Dolin, R., "Antiviral Chemotherapy and Chemoprophylaxis," *Science,* 227:1296–1303 (1985). A discussion of the current status of antiviral agents and the trials that support their use in treating influenza and herpesvirus infections.

Gallo, R. C. "The First Human Retrovirus," *Sci. Am.,* *255*:88–98 (December 1986). The story of how human T-cell lymphotrophic virus (HTLV I) was found to be the cause of a rare leukemia. This laid the scientific groundwork for discovering the AIDS virus.

Gordon, J., and M. A. Minks, "The Interferon Renaissance: Molecular Aspects of Induction and Action," *Microbiol. Revs.,* 45:244–266 (1981). An update on the molecular biology of the synthesis and activity of interferons.

Hirsch, M. S., and J. C. Kaplan, "Antiviral Therapy," *Sci. Am.,* 256:76–85 (April 1987). Describes how the more successful antiviral agents and interferon prevent viral replication.

Hunter, T., "The Proteins of Oncogenes," *Sci. Am.,* 251:70–79 (August 1984). The products of the known oncogenes provide insight into how these genes cause cancer in animals.

Simons, K., H. Garoff, and A. Helenius, "How an Animal Virus Gets into and out of Its Host Cell," *Sci. Am.,* 246:58–66 (February 1982). A detailed look at the entry, replication, and exit of an enveloped togavirus (Semliki Forest virus).

Spector, D. H., and D. Baltimore, "The Molecular Biology of Poliovirus," *Sci. Am.,* 232:24–32 (May 1975). The replication of this polycistronic, single-stranded, plus RNA virus is described and illustrated.

PART

2

The Microbial World

An important reason for studying microbiology is to heighten our awareness of the microorganisms in our environment and to understand how they affect our lives. Since most microorganisms go unseen by casual observation, few of us recognize their importance. To understand the great diversity of the microbial world, it is helpful to organize microorganisms into their natural biological groups. Our approach is to divide the microbial world into the bacteria (procaryotes) and the eucaryotic microorganisms, including the algae, fungi, protozoa; and multicellular parasites. Part 2 systematically presents examples of the microorganisms found in the microbial world and emphasizes their importance to human lives.

CHAPTER

13

Bacterial Systematics

OUTLINE

FOCUS OF THIS CHAPTER

This chapter explains how biologists systematically organize microorganisms into related groups, describe each organism so that new isolates can be recognized and identified, and assign each organism an acceptable scientific name. This chapter emphasizes the microbiological techniques employed in bacterial systematics.

Bacterial systematics is the scientific approach to describing, naming, and organizing bacteria into a classification scheme. The names and descriptions assigned to each organism are used by microbiologists to identify and name isolates obtained from natural sources. One goal of classification is to organize bacteria with related characteristics into groups. This facilitates the identification of new isolates and provides scientists with a scheme that reflects the natural relationships between bacteria.

Bacterial classification is a dynamic field because it continues to change as new discoveries modify past perceptions and modern techniques of molecular analysis reveal genetic relationships between the procaryotes. The long-range goal of bacterial systematics is to create a "natural" classification scheme for the bacteria that recognizes their common lines of descent.

SCIENTIFIC IMPORTANCE OF SYSTEMATICS

Systematics is the study of the diversity of organisms and how they relate to each other. This science encompasses the tasks of classification, identification, and nomenclature. The purpose of **classification** is to organize all organisms into groups whose members are interrelated. Bacteria are classified when they are assigned to one of the five kingdoms and then within a kingdom into divisions, families, genera, and species. **Identification** is the process by which the characteristics of new isolates are correlated with descriptions of known organisms. Newly discovered microorganisms are classified (as above) after they are shown to be distinct from those previously described. The processes of identification and classification require that each species of organism be given a name. Biological **nomenclature** attempts to follow a systematic process for assigning an internationally acceptable scientific name to each unique organism. Finally, systematists seek to understand the interrelationships between organisms by studying their evolutionary history—their *phylogeny*. Phylogeny of vertebrates depends heavily on the fossil records left by the ancestors of living animals. The scant fossil record of microorganisms shows that procaryotes predated higher plants and animals, but the fossil record of microbes is insufficient to exert a major effect on studies of their phylogeny.

THE FIVE KINGDOMS

Systematists devise kingdoms as the major framework for grouping organisms with similar characteristics and common lines of descent. Kingdoms reflect the evolutionary history (phylogeny) of each taxonomic group. Although there are other approaches, we will use the five-kingdom classification devised by Whittaker and presented in Chapter 1. The five-kingdom scheme places each organism in Animalia, Plantae, Fungi, Protista, or Procaryotae (see Figure 1.14).

The kingdom **Plantae** encompasses those multicellular eucaryotic organisms that obtain energy by photosynthesis. Multicellular eucaryotic organisms characterized by their motility and noted for ingesting their food belong to the kingdom **Animalia**. Animals lack the rigid cell walls found in plants. The kingdom **Fungi** contains nonphotosynthetic eucaryotic organisms that obtain their food by absorption. Fungi grow vegetatively as microscopic intertwined mycelia composed of hyphae or as single-celled yeasts (see Chapter 16).

The protozoa, single-celled algae, flagellated water molds, and slime molds are assigned to the kingdom **Protista**, the kingdom that bridges the gap between the eucaryotic plants, animals, fungi, and the procaryotes.

The diversity in structure and physiology among the protists is tremendous. Some have plant-like characteristics (algae) and others resemble animals (protozoa) because they are motile and lack rigid cell walls. The flagellated water molds are motile like protozoa, but in other respects they resemble the fungi. (The protists are described in more detail in Chapter 16).

THE KINGDOM PROCARYOTAE

Bacteria are phylogenetically distinct from other microorganisms because of their procaryotic cell structure. All bacteria are classified in the kingdom **Procaryotae**. Systematists have organized this kingdom into superfamilies or divisions* based primarily on the cytology of the layers external to the cytoplasmic membrane. The four divisions or superfamilies are the gram-negative bacteria (*Gracilicutes*), the gram-positive bacteria (*Firmicutes*), the bacteria without cell walls (*Tenericutes*), and the bacteria with widely varying cell-wall chemistries (*Mendosicutes*).

The higher taxa encompassed by this scheme represent the best current microbiological judgment as to how bacteria interrelate in a phylogenic sense. The archaebacteria are thought to be the most primitive procaryotes. Many live in extreme environmental conditions and on substrates that were present in the primordial earth. Separating bacteria with gram-negative cell walls from those with gram-positive cell walls is a sound taxonomic decision because cell-wall structures represent major differences between bacteria. Division III can be questioned as being distinctive because bacteria without cell walls may have evolved from gram-positive bacteria. Until further information becomes available, systematists will be unsure of how bacteria lacking cell walls relate to other procaryotes.

BACTERIAL CLASSIFICATION

Systematists attempt to describe and name each kind of organism (species) in a fashion that distinguishes it from all other known kinds. The description must be exact enough to enable an investigator to identify an unknown isolate. A name is attached to the description of each organism to facilitate communication about that species. Each species is then organized into an artificial classification scheme based on similarities and differences, or they are arranged into a phylogenetic or "natural" scheme based on evolutionary relationships and lines of common descent.

Each division (superfamily) of the kingdom Procaryotae is divided into classes and orders (not important for our discussion) and then into families, genera, species, and subspecies or types (Table 13.1). The four bacterial divisions are based on cell-wall structure and to some degree on 16S ribosomal RNA (see Chapter 9). Families are groups of similar genera, for example, all the genera in the *Enterobacteriaceae* are gram-negative facultative anaerobes. Genera are composed of groups of similar organisms known as species. For now we can define bacterial species as a collection of strains that share many features in common. Phenotypic variation within a species is permissible as long as the organisms are 70 percent or more related by total DNA and phenotypic criteria. An example of this is given in Table 13.1. Enterotoxic *E. coli* is a subspecies of *E. coli* because it produces a toxin (encoded on a plasmid not present in other strains) that causes diarrheal disease in humans.

SPECIES CONCEPT IN BACTERIAL CLASSIFICATION

Plant and animal systematists distinguish a species by grouping together those organisms that interbreed. Domestic cats mate with domestic cats, but are hardly able to mate with cougars. The cougar and the domestic cat are classified as distinct species and are given different species names. However, they share many characteristics, so they are given the same genus name, *Felis*. We even think of these animals as being in the same family and as having a common ancestry. Taxonomists have acknowledged the evolutionary relatedness of cats by placing them in the same genus, family, and kingdom.

Bacteria reproduce by binary fission or by budding; they do not breed. Bacteria *may* exchange genetic information by conjugation, but this is a rare event in nature and is not required for reproduction. Thus the concept that species are distinct breeding groups is not applicable to bacterial classification. Each bacterium is

*R. E. G. Murray, "Kingdom Procaryotae Murray, 1968, 252[AL]," pp. 35–36 in *Bergey's Manual of Systematic Bacteriology*, Vol. 1, Williams & Wilkins, Baltimore, 1984.

TABLE 13.1 Hierarchy of bacterial classification

CLASSIFICATION	EXAMPLE	BASIS FOR ASSIGNMENT	COMMENTS
Division (Superfamily)	*Gracilicutes*	Cell-wall structure, 16S ribosomal RNA	Two or more related families
Family	*Enterobacteriaceae*	Phenotype, 5S ribosomal RNA[a]	Group of similar genera, 15–40 percent related
Genus	*Escherichia*	Phenotype	Group of similar species, 45–65 percent related
Species	*Escherichia coli*	Total DNA, phenotype	70 percent or more related, less than 5 percent divergence in equivalent genome size
Subspecies	Enteropathic *E. coli*	Total DNA, including plasmids	Groups within species (for example, pathovars)

[a]This new technique may contribute to the definition of bacterial families.

given a species name, but in most cases this name is assigned on an artificial basis. The systematists must make a judgment as to the similarity of the organisms' characteristics and then make a decision about where to combine groups of bacteria into a single species and where to create new species.

KEY POINT

The establishment of bacterial species is based on available information and scientific judgment; there is no biological basis for establishing bacterial species on the basis of breeding. The artificial element in bacterial classification gives rise to ongoing changes in nomenclature and classification as new information becomes available. So species in bacteriology is an arbitrary concept devised as a convenience for scientists.

NAMES AND DESCRIPTIONS OF BACTERIA

BACTERIAL NOMENCLATURE

Each organism is given a name to facilitate the communication of ideas, concepts, and experimental results pertaining to that organism. Biologists use the **binominal system of nomenclature**, which assigns a two-part latinized name to each unique organism, called a **species**. The first name in the binomen is the **genus name**, which describes a group of closely related species. The first letter of the genus name is capitalized and the entire name is italicized. The second name is the **specific epithet** (ep′i-thet), which modifies the genus name and indicates the species of the genus. The specific epithet begins with a lowercase letter and is also italicized. *Escherichia coli* is an example of a binomen in which the genus name is *Escherichia* and the specific epithet is *coli*.

A bacterial species is a collection of strains that share essential characteristics and differ significantly from all other strains. Suppose a microbiologist isolates a number of different bacteria from 50 water samples, and each of these isolates has the same gross characteristics. This collection of isolates represents strains of the same species. A second microbiologist may have isolated the same species in another country, say China. This isolate could be designated the China strain. Each **strain** is composed of the descendants of a single colony or clone, so all cells of a strain are genetically homogeneous.

Some bacterial species are divided into two or more subspecies based on unique traits displayed by the strains. Officially, a subspecies is the lowest level of nomenclature that is recognized. A subspecies is a strain with one or more specific traits that make it unique. The subspecies name is a **biovar*** if it designates a specific biochemical or physiological property, a **serovar** if it designates a distinctive antigenic trait, a **pathovar** for pathogenic properties, a **phagovar** when the strain is lysed by a specific bacteriophage, and a **morphovar** when the strain possesses a morphological variation for the species.

*The "var" term is used interchangeably with biotype and serotype.

TABLE 13.2 Representative nomenclature changes for gram-negative rods reported in *Bergey's Manual of Systematic Bacteriology*[a]

LATEST CLASSIFICATION (1984)	*BERGEY'S MANUAL*, 8th ed. (1974)
Escherichia blattae	Not listed
Salmonella choleraesuis	*Salmonella cholerae-suis*
Klebsiella pneumoniae subsp.*pneumoniae*	*Klebsiella pneumoniae*
Klebsiella pneumoniae subsp. *ozaenae*	*Klebsiella ozaenae*
Klebsiella pneumoniae subsp. *rhinoscleromatis*	*Klebsiella rhinoscleromatis*
Klebsiella planticola	Not listed
Klebsiella terrigena	Not listed
Enterobacter agglomerans	*Erwinia herbicola*, *Erwinia stewartii*, and *Erwinia uredovora*
Serratia liquefaciens	Not listed
Serratia rubidaea	Not listed
Providencia alcalifaciens	*Proteus inconstans* biogroup A
Providencia stuartii	*Proteus inconstans* biogroup B
Providencia rettgeri	*Proteus rettgeri*
Morganella morganii	*Proteus morganii*
Gardnerella vaginalis	*Haemophilus vaginalis*

*Adapted in part from D. J. Brenner, "Facultatively Anaerobic Gram-Negative Rods," in *Bergey's Manual of Systematic Bacteriology,* Vol. 1, Williams & Wilkins, Baltimore, 1984.

Bacterial names are assigned according to the principle of publication priority. This sounds straightforward and simple, but it is not. As new information is revealed to the systematist, bacterial species can be combined, thus eliminating one species name; or species can be divided, which creates a new species. Table 13.2 gives examples of nomenclature changes that have occurred since the eighth edition of *Bergey's Manual* appeared in 1974. Examples of new species, combinations of three species into one new species, creation of subspecies, genus changes, and spelling changes are given. The rules governing the assignment of names to each species and for making changes in these names is regulated by the *International Code of Nomenclature of Bacteria*. The process of updating the list of approved bacterial names is a constant one. A revised list was published in 1980—*Approved List of Bacterial Names*—and serves as a basis for assigning names to new isolates. When a new species is isolated, described, and named, this information is published or announced in the *International Journal of Systematic Bacteriology* before the species is recognized under the international code.

DESCRIPTIONS OF BACTERIA

Descriptions of each bacterial species were compiled in the eighth edition of *Bergey's Manual of Determinative Bacteriology*, which was published as a single volume in 1974. Since that year, microbiologists have discovered many new bacteria and have uncovered new facts about known species. The new information has necessitated an extensive revision of bacterial classification as exemplified by the separation of the archaebacteria from the eubacteria. An updated version of *Bergey's Manual* is being published in a four-volume compendium entitled *Bergey's Manual of Systematic Bacteriology*. The first volume appeared in 1984 and covers the gram-negative bacteria of general, medical, and industrial importance. Volume II (published in 1986) covers the gram-positive bacteria of medical and commercial importance; volume III will cover the remaining gram-negative bacteria (archaebacteria, cyanobacteria, phototrophic, lithotrophic, sheathed, and appendaged bacteria); and volume IV will cover the filamentous actinomycetes and related bacteria. These volumes will represent the combined scientific knowledge of bacterial systematists, and will be the major reference work for bacterial classification in the foreseeable future.

KEY POINT

Bergey's Manual of Systematic Bacteriology serves as the most complete source of information for the identification, nomenclature, and classification of bacteria.

All bacterial species with official status at the time of publication are included in *Bergey's Manual of Systematic Bacteriology*. The criteria used to define each genus include morphological characteristics, colonial morphology (Figure 13.1), pigmentation, growth conditions, nutrition, physiology, metabolism, genetics, plasmids, bacteriophages, antigenic structure, pathogenicity, and ecology. These characteristics define the species that are classified under each genus name.

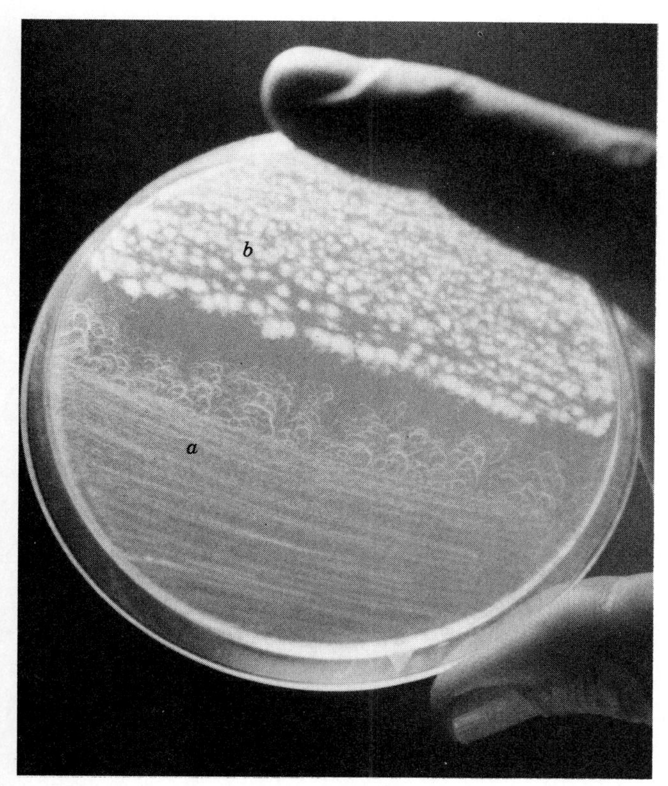

FIGURE 13.1 The bacteria known as *(a) Bacillus cereus* var. *mycoides* and *(b) Bacillus cereus* appear to be identical except for their colony morphology. *Bacillus cereus* var. *mycoides* forms rhizoid colonies because of the unequal nature of its cellular division.

BACTERIAL IDENTIFICATION: CLASSIC TECHNIQUES

Identification is such an important aspect of bacteriology that the criteria used to define bacteria must be easily determinable by tests that are highly reproducible. Emphasis should be placed on tests that can be reproduced in the bacteriology laboratory. Tests that depend on metabolic characteristics and growth parameters are preferable over tests requiring elaborate or expensive techniques such as the analysis of cell-wall chemistry or DNA homology. Each test should have a precise definition of what is a "positive" and what is a "negative" result. With these principles in mind, systematists choose the tests most appropriate for identifying each bacterial species. Bacterial identification is an acquired skill that requires technical competence, scientific judgment, and a broad understanding of bacterial systematics.

Characterization of a new bacterial isolate begins by describing the organism using information discovered with the tools and techniques available in the basic microbiology laboratory. These techniques include microscopy, bacterial growth, and the measurement of physiological traits.

CELL MORPHOLOGY

Microscopic observations of living and stained bacteria afford the earliest descriptive information about a new isolate. The differences between eucaryotic and procaryotic cells are ascertained by a search for the eucaryotic nucleus, which is usually visible upon microscopic observation. Living bacteria observed in wet mounts can be characterized morphologically as to their size, shape, presence of spores, and motility. The Gram-stain reaction, the acid-fast reaction, and the presence of capsules and flagella (Figure 13.2) are observed after specially staining the bacterial preparation.

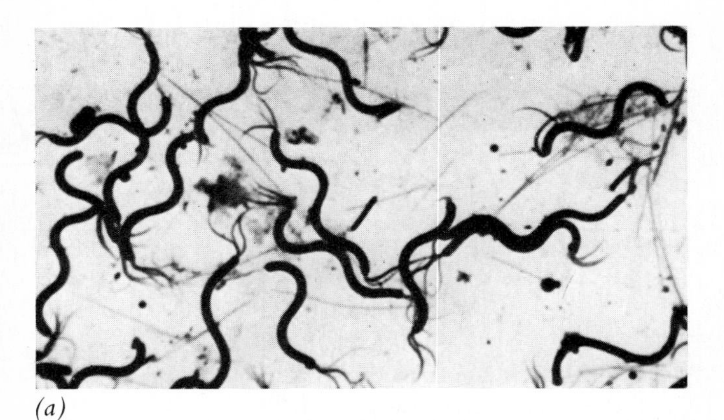

(a)

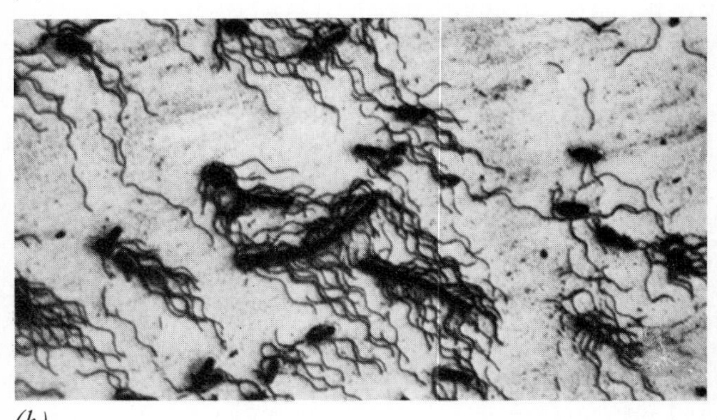

(b)

FIGURE 13.2 The presence and distribution of flagella on the surface of bacteria are inherited characteristics used in classification. Light micrographs of flagella-stained cells of *(a) Spirillum volutans* and *(b) Salmonella typhi.* (Courtesy of General Biological Supply House, Inc., Chicago.)

These morphological traits are the primary characteristics by which bacteria are identified.

CULTURE CHARACTERISTICS

Colony morphology and pigment production are evident when a bacterium grows on a solid medium (Figure 13.3). Colonies are colored when bacteria produce water-insoluble pigments during growth. The colors include shades of red, orange, yellow, and green. Some of these pigments are involved in photosynthesis, while others are metabolic by-products. When bacteria produce capsules or slime layers they usually form mucoid colonies, which may be soft and viscous or extremely hard when touched with an inoculating loop. Other colony characteristics include size, spreading (which implies motility), and general morphology (Figure 13.3). Although colony morphology varies with growth conditions, it is an important characteristic used in bacterial identification.

Information can also be deduced from the conditions under which the bacterium grows. Does the organism grow as an aerobe, an anaerobe, microaerophile, or a facultative anaerobe? What substrates can the bacterium use as a source of carbon and energy and what forms of nitrogen can it utilize? Are nutritional supplements required for growth and can the organism grow in high concentrations of salt? Is light required as an energy source? Can the bacterium grow on an inorganic medium? These questions can usually be answered by growth experiments done in an ordinary microbiology laboratory. Other bacteria, such as the obligate intracellular parasites, can be grown and characterized only with specialized laboratory equipment.

METABOLIC TRAITS

The metabolic by-products, enzymes, and toxins produced during bacterial growth are helpful in bacterial identification and classification. The metabolic by-products of fermentations can be used to group bacteria into genera or to differentiate between species belonging to a single genus. The production of propionic acid as the major fermentation product is one criterion used to assign bacteria to the genus *Propionibacterium* (proh-pee-on′ee-bak-teer′ee-um). The individual species belonging to the genus *Clostridium* are in part differentiated on the basis of their fermentation products, which can include acetone, isopropanol, butanol, butyric acid, and acetic acid. Fermentation products are a prime criterion in classification because they represent the end products of multienzyme pathways.

Cellular production of detectable enzymes and toxins can determine differences between species, or between closely related genera. Proteins that are easily detected include urease, coagulase, and hemolysins (Table 13.3). Cells containing urease hydrolyze urea; urease is present in most species of *Proteus*, but is absent in *Escherichia*, *Salmonella*, and *Shigella*. *Staphylococcus aureus* produces coagulase, an enzyme that coagulates serum plasma. The presence of coagulase distinguishes *S. aureus* from the other staphylococci.

Bacteria

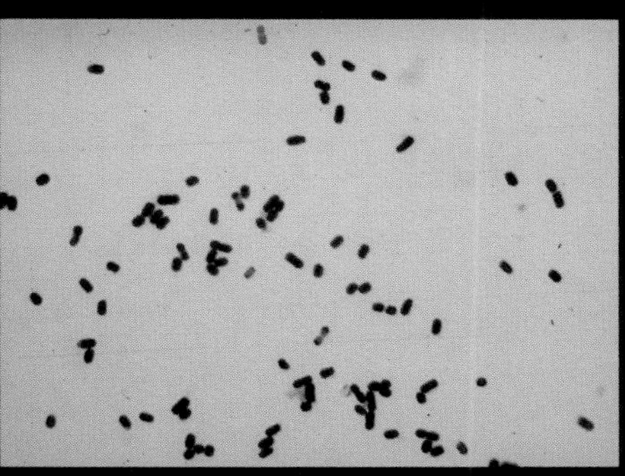

IA.1 Gram stain of *Clostridium perfingens*. (National Medical Audiovisual Center.)

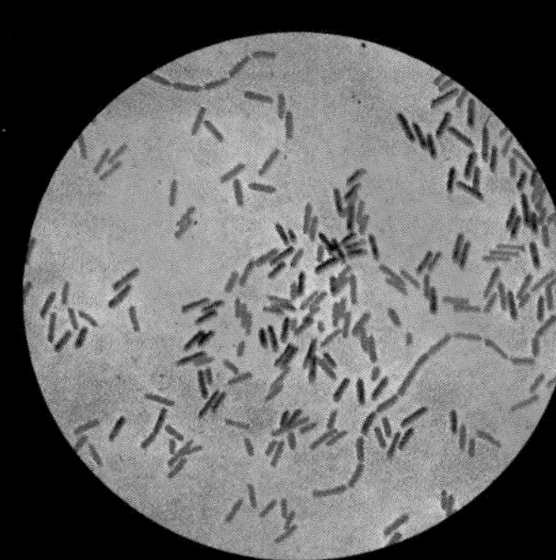

IA.2 Gram stain of *Salmonella typhi*. (National Medical Audiovisual Center.)

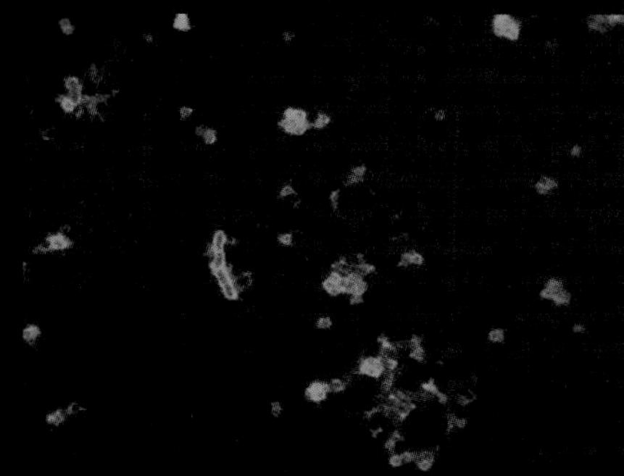

IA.3 Fluorescent antibody stain of *Yersini[a] pestis*. (National Medical Audiovisual Center.)

Algae and Protozoa

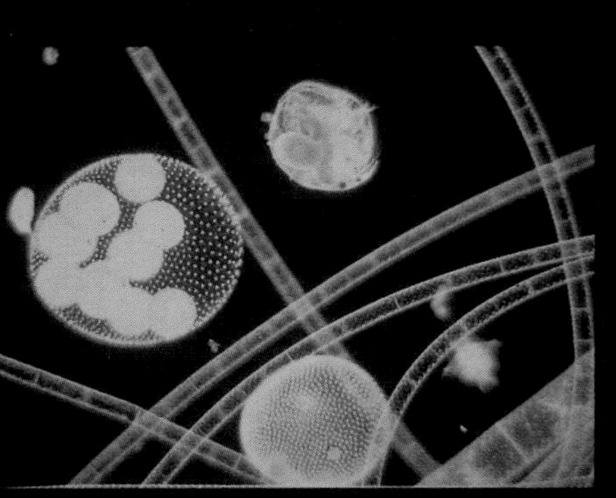

B.1 *Spirogyra*, *Volvox*, and *Chydorus* crustacean). (VU/T. E. Adams.)

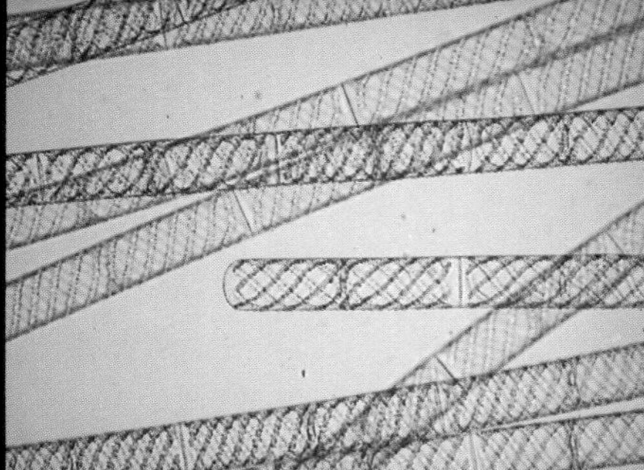

IB.2 Pure culture of *Spirogyra*, a filamentous green alga. (VU/T. E. Adams.)

IB.3 *Acetabularia*, a green alga. (VU/T. E. Adams.)

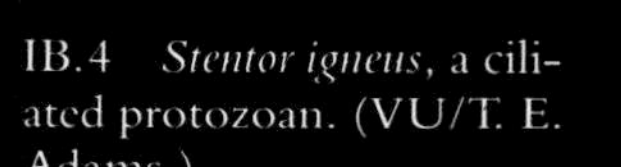

IB.4 *Stentor igneus*, a ciliated protozoan. (VU/T. E. Adams.)

Slime Molds and Fungi

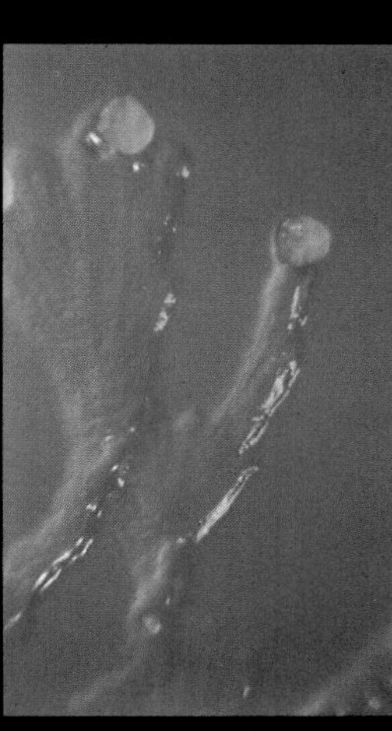

IC.1 Migrating colony of the slime mold *Polyangium* on agar. (R. H. D. McCurdy.)

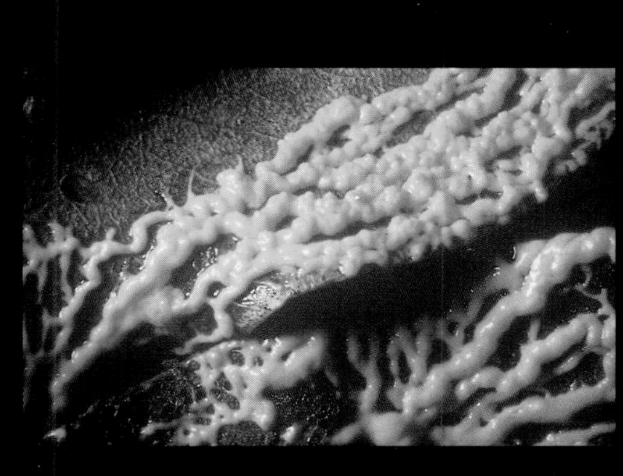

IC.2 Growth of the slime mold *Leocarpus fragilis*. (VU/V. Duran.)

IC.3 Fruiting body of the slime mold *Trichia sporangium*. (VU/V. Duran.)

IC.4 *Polyporus sulfuresus* growing on a tree trunk. (VU/John Serrao.)

IC.5 Coral fungus, *Clavaria*. (VU/William S. Ormerod, Jr.)

Mold and Yeast Cultures

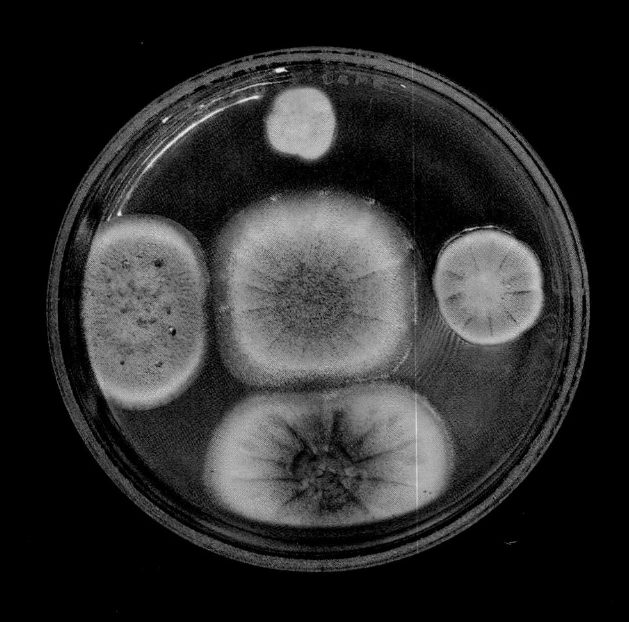

ID.1 Cultures of several commercially important molds after eight days of growth on nutrient agar. (Courtesy of Dr. H. J. Phaff.)

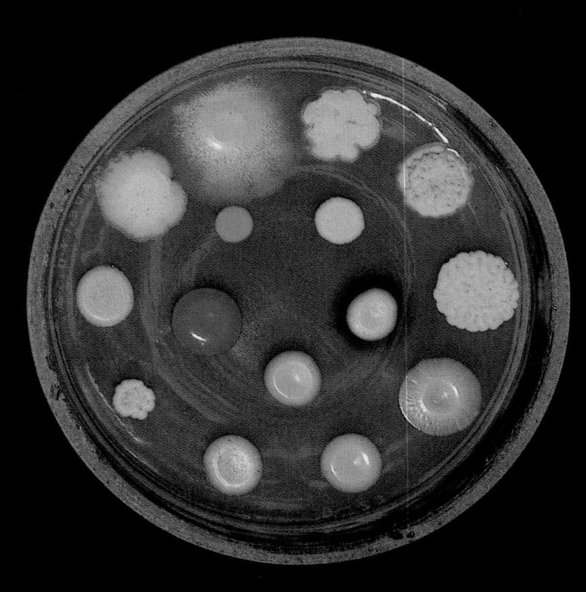

ID.2 Yeasts often grow as distinct colored colonies on nutrient agar. (Courtesy of Dr. H. J. Phaff.)

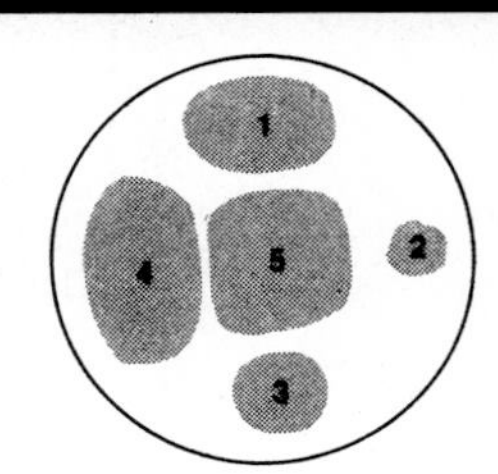

MOLDS
1 *Penicillium chrysogenum*
2 *Monascus purpurea*
3 *Penicillium notatum*
4 *Aspergillus niger*
5 *Aspergillus oryzae*

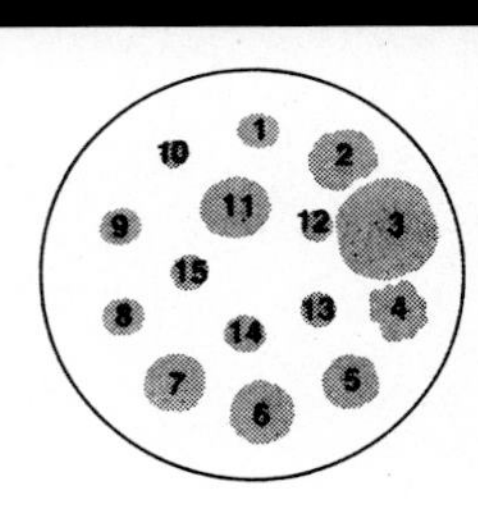

YEASTS
1 *Saccharomyces cerevisiae*
2 *Candida utilis*
3 *Aureobasidium pullulans*
4 *Trichosporon cutaneum*
5 *Saccharomycopsis capsularis*
6 *Saccharomycopsis lipolytica*
7 *Hanseniaspora guilliermondii*
8 *Hansenula capsulata*
9 *Saccharomyces carlsbergensis*
10 *Saccharomyces rouxii*
11 *Rhodotorula rubra*
12 *Phaffia rhodozyma*
13 *Cryptococcus laurentii*
14 *Metschnikowia pulcherrima*
15 *Rhodotorula pallida*

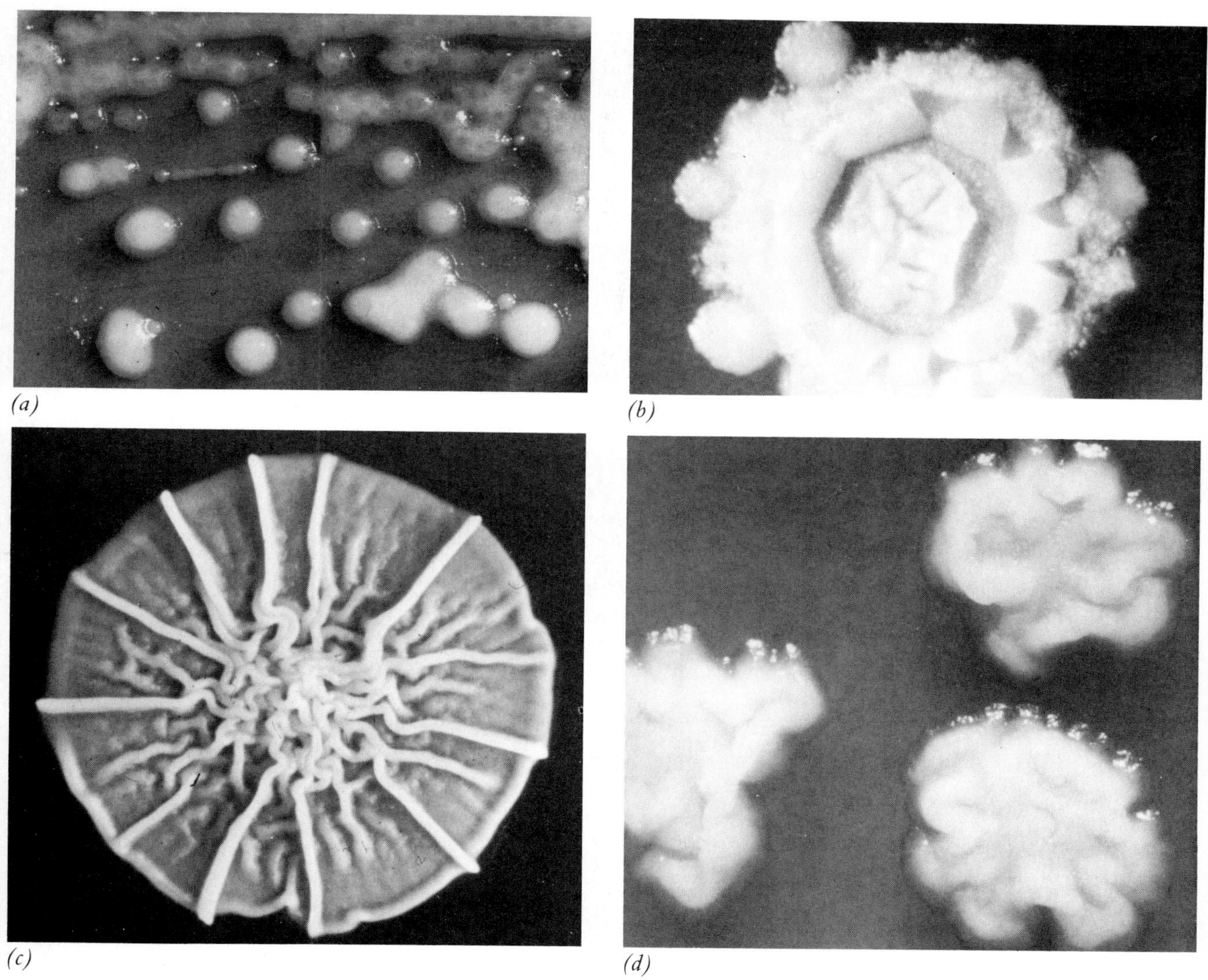

FIGURE 13.3 Colony morphologies are used to differentiate bacterial species. *(a)* Smooth colonies of *Klebsiella* on MacConkey agar, *(b) Nocardia asteroides* colony, *(c)* wrinkled colony of *Micrococcus luteus* on blood agar, and *(d)* rough colony of *Mycobacterium smegmatis* on Pennassay agar. (Courtesy of *(a, c)* Centers for Disease Control, *(b)* J. M. Slack and M. A. Gerencser, *(d)* R. J. Hawley and T. Imaeda, American Society for Microbiology/BPS.)

Toxins that lyse red blood cells (hemolysins) change the appearance of blood agar plates. The production of hemolysins is a means of differentiating certain species of *Streptococcus*. *Streptococcus pyogenes* produces hemolysins that lyse red blood cells, whereas *Streptococcus salivarius* does not produce hemolysins.

Metabolic tests are also used to differentiate between two closely related members of the *Enterobacteriaceae*. Members of the genus *Enterobacter* produce 2,3-butanediol from glucose, but they do not produce indole from tryptophan (Figure 13.4). Species of *Escherichia* produce indole from tryptophan, but they do not pro-

TABLE 13.3 Representative enzymes and toxins important in bacterial classification

	ASSAY RESULTS	GROUP OR ORGANISM
Enzymes		
Coagulase	Coagulation of plasma	*Staphylococcus aureus*
Tryptophanase	Indole production from tryptophan	Many bacteria
Urease	Alkaline medium from urea	*Proteus* spp.
Gelatinase	Protein hydrolysis	Many bacteria
Amylase	Starch hydrolysis	Many bacteria
Catalase	Oxygen evolution from H_2O_2	Aerobes
Oxidase	Ability to oxidize a dye	*Neisseria* spp. and *Pseudomonas* spp.
Toxins		
Hemolysins (β-hemolysis)	Clear lysis of red blood cells	*Streptococcus pyogenes* and other pathogens
Hemolysins (α-hemolysis)	Partial (green) lysis of red blood cells	*Streptococcus* spp. and other genera

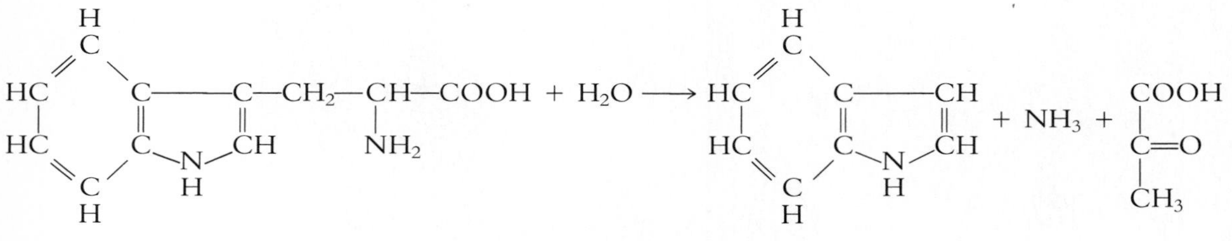

Tryptophan + H_2O → Indole + NH_3 + CH_3—C(=O)—COOH

Tryptophan — Indole — Ammonia — Pyruvic acid

(a) Indole production from tryptophan

2 CH_3—C(=O)—COOH → (CO_2) → HO—C(CH_3)(COOH)—C(=O)—CH_3 → (CO_2) → HO—CH(CH_3)—C(=O)—CH_3 → ($NADH_2$ → NAD^+) → CH_3—CH(OH)—CH(OH)—CH_3

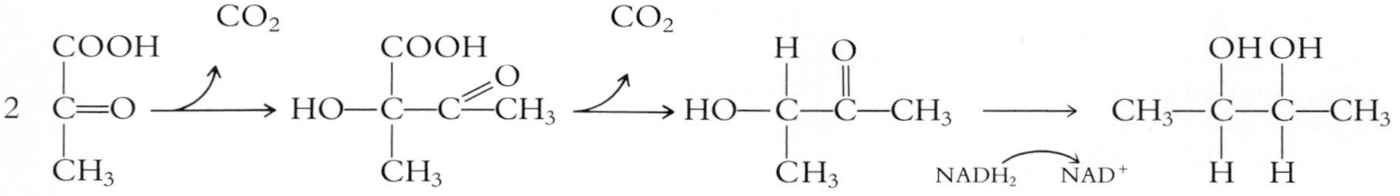

Pyruvic acid — Acetolactic acid — Acetoin — 2,3-Butanediol

(b) Formation of 2,3-butanediol

FIGURE 13.4 Metabolic by-products. *(a)* Indole is produced by the enzyme tryptophanase and is detected by adding Kovac's reagent to the medium. *(b)* Bacteria that possess the 2,3-butanediol pathway are identified by measuring the presence of acetoin in culture media, using the Voges–Proskauer test.

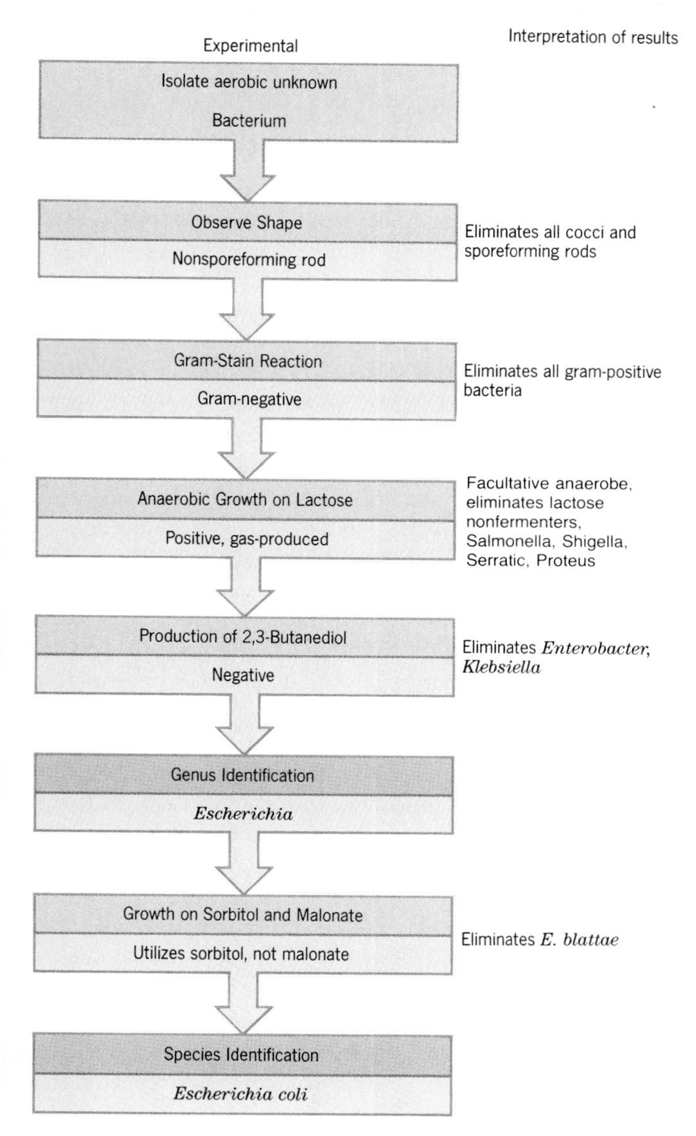

FIGURE 13.5 Sequential tests are used to identify unknown bacteria by eliminating groups and then individual bacteria. Bergey's Manual is the reference for the characteristics of each bacterial species.

duce 2,3-butanediol as a product of glucose fermentation. A synopsis of the biochemical, metabolic, and morphological traits, similar to those described here, is used to identify and classify bacterial isolates.

Everyday identification of bacterial isolates is accomplished by making a series of observations that progressively narrow down the possibilities. The flow chart presented in Figure 13.5 shows the steps necessary to identify an unknown bacterium isolated from a lactose broth. In general, an unknown bacterium is identified by determining the conditions required for growth, its colony morphology, cell morphology, motility, staining reactions, metabolic products, toxin production, and enzymatic activities. This information can be obtained by microscopic observations of living and stained preparations, growing the unknown in different media in the presence and absence of air, and by using chemical tests to detect metabolic products. The unknown in Figure 13.5 is identified as *Escherichia coli* by eliminating the other possible gram-negative rods from consideration.

CHEMOTAXONOMY

The most complete definition of an organism is written in its genetic code. Much of this information is revealed to the microbiologist indirectly through the cell's chemistry. Since physiological reactions are catalyzed by enzymes and enzymes are coded for by the cell's genome, the chemical composition of a cell directly reflects the cell's genotype. Chemical differences are found in bacterial cell walls, lipids, and quinones (kwin′ohns) and the protein profiles of both the total and the soluble protein fractions can reveal differences between cells. Chemotaxonomy can have a major impact on how systematists separate the major groups of microorganisms.

CHEMISTRY OF BACTERIAL CELL WALLS

The main divisions of bacteria are based on the characteristics of their cell walls. The major distinctions between the gram-negative and the gram-positive bacteria are based on the ultrastructure and chemical composition of their cell walls as described in Chapter 4.

The outer membranes of gram-negative bacteria contain lipopolysaccharide, which is a useful taxonomic marker. Differences in the lipopolysaccharide (O antigen) were first detected by immunological techniques. For example, the *Salmonella* genus contains over 1000 serovars differentiated by the antigenic structure of their lipopolysaccharide. These serovars can be important in tracing the source of salmonella infections, but they have little impact on the taxonomy of this genus. The lipopolysaccharide has other functional roles. The lipid A segment of the lipopolysaccharide is the endotoxin of gram-negative bacterial cell walls, while the polysaccharide component functions as a recognition site for bacteriophage binding.

Only the gram-positive bacteria contain teichoic acids in their cell walls. These polymers of ribitol phosphate, glycerol phosphate, or mannitol phosphate are bound covalently to the peptidoglycan of the cell wall or are attached to the cytoplasmic membrane (lipoteichoic acids). Immunological techniques detect differences in the teichoic acids and these differences define species within a genus, such as *Staphylococcus*.

Gram-positive cells can contain cell-wall polysaccharides in addition to the teichoic acids. The cell-surface polysaccharides of *Streptococcus* are used for classifying them into groups known as the Lancefield groups A to H and K to V. Group A and group B streptococci are each composed of a single organism, *Streptococcus pyogenes*, and *S. agalactiae* (a-ga-lac′ti-ae) respectively, and so are taxonomically significant. Other Lancefield groups have less taxonomic significance because they contain more than one species of streptococci.

Peptidoglycan is found in all procaryotes except wall-less bacteria, such as the mycoplasma, and the archaebacteria, which have cell walls lacking peptidoglycan. Both the presence of the peptidoglycan and its composition aid in classification. Variability occurs in the peptide portion of peptidoglycan. For example, the common diamino acid in gram-negative bacteria, meso-diaminopimelic acid, is replaced by ornithine in many spirochetes. The amino acid content and the presence of interpeptides add variability to the peptidoglycans of the gram-positive bacteria and these in turn are used in classification.

ESTER OR ETHER LINKAGES IN LIPIDS

The lipids of most living organisms are composed of fatty acids linked to glycerol by an ester linkage $\left(R\text{—}O\text{—}C\overset{O}{\underset{}{\diagup\!\!\!\diagup}}R_1\right)$. Ester-linked lipids occur in the membranes of all bacteria except the archaebacteria, which contain only ether-linked ($R\text{—}O\text{—}R_1$) lipids. The presence of unique fatty acids and/or lipids in a bacterium helps to identify and classify that species.

AMINO ACID SEQUENCES OF PROTEINS

Proteins are gene products and as such their presence reflects the genotype of the cell. An overview of a cell's proteins is provided by an electrophoretic "fingerprint" of the cell's soluble proteins. These fingerprints can reveal the presence of up to 200 proteins and can reveal differences between closely related strains.

A detailed view of this larger picture is revealed by looking at a specific class of proteins, such as a cell's cytochromes, the hemoproteins that participate in the redox reactions of the electron transport chain (see Chapter 6). Some anaerobes, such as the streptococci, contain no cytochromes, whereas other bacteria contain as many as six different cytochromes. The cytochrome profile is indicative of the cell's respiratory capability. Specialized cytochromes are observed in many bacteria, including the photosynthetic bacteria and the sulfur-reducing bacterium, *Desulfovibrio* (dee-sul-foh-vib′ree-oh).

It is possible to determine the amino acid sequence of small proteins isolated from phylogenetically different organisms. Proteins chosen for analysis include ferredoxins, lysozyme, and cytochrome *c*. When isolated from related organisms, these proteins have similar or identical amino acid sequences. When the amino acid sequences vary greatly, the organisms are considered to be unrelated. This follows from evolutionary theory, which assumes that all bacteria evolved from a common ancestor. Species that diverged in ancient times would have a greater opportunity to mutate, resulting in amino acid sequence differences, than species that diverged in recent millenia.

Amino acid sequences of cytochrome *c* isolated from many different organisms are known. The evolutionary relationships among these organisms have been deduced from the changes in the amino acid sequence of the organism's cytochrome *c*. Based on these data, a progression from the simplest bacterium to humans has been constructed. The importance of this approach to bacterial classification is limited because it is time-consuming and costly. However, it is reassuring to note that the sequence analysis of cytochrome *c* confirms the phylogenetic classification scheme established by other taxonomic approaches.

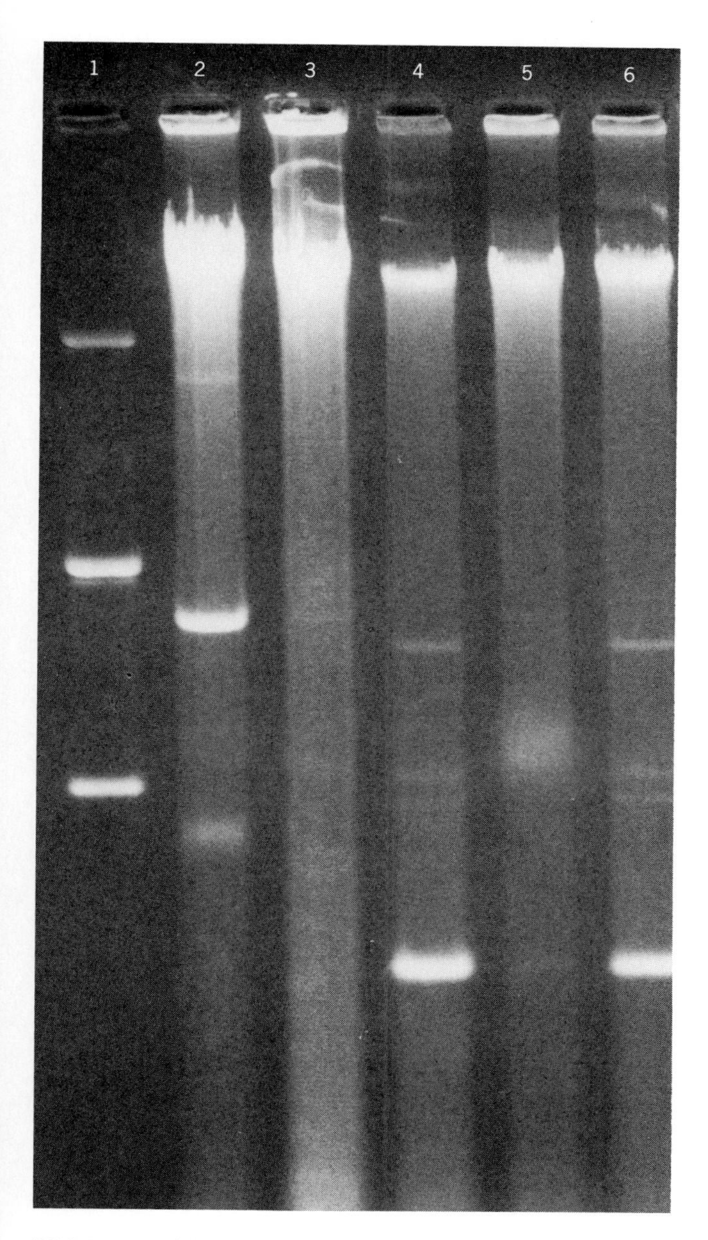

FIGURE 13.6 The plasmids present in different bacterial strains can be demonstrated after they are separated, using electrophoresis on an agarose gel, and stained with ethidium bromide. Lanes 2 through 6 are the plasmid profiles of five different clinical isolates of *Pseudomonas aeruginosa*. Lane 1 contains marker DNA. Strains analyzed in lanes 4 and 6 have the same plasmid profile, whereas strains analyzed in lanes 2, 3, and 5 possess different plasmid profiles and are therefore distinct. (Courtesy of Dr. Satish Walia.)

PLASMID PROFILES

In addition to their chromosomal DNA, bacteria can possess up to 10 or more different plasmids. These self-replicating, circular, extrachromosomal DNA molecules range in size from about 1×10^6 to 200×10^6 daltons. Plasmids are easily isolated from the cytosol of lysed cells and can be analyzed by their mobility in an electric field (gel electrophoresis, Figure 13.6). During electrophoresis, each plasmid migrates in a support medium (such as agarose) according to its size.

KEY POINT

All double-stranded DNA molecules have a net charge attributable to the phosphate groups in the DNA backbone. Since each plasmid has essentially the same charge per segment of DNA, they migrate in an electrical field according to size.

The plasmid profile of a bacterium is indicative of its past history and uniqueness and can be used to distinguish between different strains. A practical use of plasmid profiles is to trace the origins of disease-causing strains, as occurred in an outbreak of salmonellosis, when the causative organism, *Salmonella newport*, was traced by its plasmid profile.

NUMERICAL TAXONOMY

To the casual observer, some bacterial traits appear to be more important than others. For example, one may think that the ability to form spores is a more significant trait than the presence of flagellar motility. Systematists make subjective judgments when weighing the taxonomic importance of bacterial traits. The result is that the endospore-forming bacteria are classified together, whereas motile organisms are widely dispersed in the classification scheme.

Another approach to taxonomy is to weigh each trait equally and thereby avoid the biased judgment of the observer. **Numerical taxonomy** characterizes

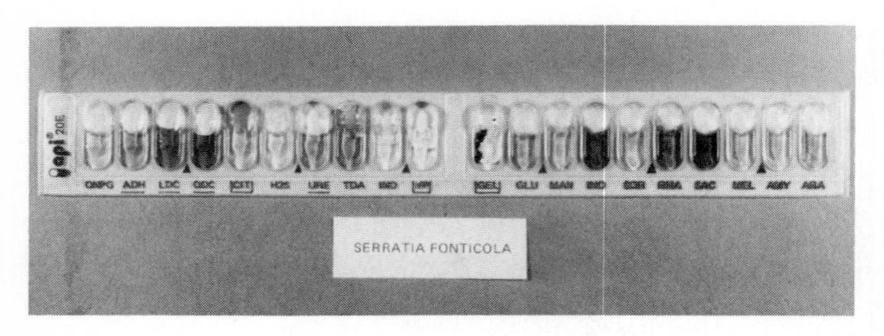

FIGURE 13.7 Each of the 20 media-containing cupules of the API 20-E kit is inoculated with the bacterial culture. After incubation, the API strip reveals the biochemical activities of the bacterium, including its use of different compounds for growth (acid and/or gas production), production of hydrogen sulfide and 2,3-butanediol, and the presence of five different enzyme activities. The numerically encoded results are used to identify the bacterium. (Courtesy of API.)

each strain by using as many traits as possible and assigning them all equal weight. Organisms that share similar traits are classified together; the more traits they share, the more tightly they are grouped.

The principles of numerical taxonomy are widely used in bacterial identification especially in diagnostic medical microbiology laboratories. For example, disposable multitest systems that provide the information needed to identify an isolate are available as test strips, plates, or tubes.

The API strip (**Analytical Profile Index**) is a multi-test strip containing sterile media in cupules (Figure 13.7). Each compartment contains a different medium that can be scored plus or minus for a phenotypic trait. The entire strip is inoculated with an unidentified isolate and then incubated. The results are read by the microbiologist and then converted into a numerical coding system. These numerical codes are analyzed by a computer or by reference to a chart listing the characteristics of all the potential isolates. The API system is read with the aid of the Analytical Profile Index, which correlates the test results with the characteristics of known organisms to provide an identification. Computers are now programmed to make these correlations and do statistical analysis of the results. These programs are used extensively in medical microbiology laboratories for identifying bacteria.

NUCLEIC ACIDS AND BACTERIAL CLASSIFICATION

Molecular biology has provided taxonomists with sophisticated tools for analyzing the genetic relationships between microorganisms. Since the traits of an organism are dictated by the genes of its cellular DNA, the degree of similarity between the base sequences in the DNAs of two cells is a measure of their evolutionary relatedness. Organisms sharing evolutionary ancestry will also share DNA base sequences.

MOLES PERCENT G + C

DNA isolated from bacteria varies widely in its nucleotide base composition. The nucleotide composition of DNA is expressed as the ratio of the moles of guanine plus cytosine to the moles of guanine plus cytosine plus adenine plus thymine ($\times$ 100). This value is the **G + C moles percent** or simply **mol% G + C**. Bacteria have values that vary from about 20 moles percent to more than 70 moles percent G + C (Table 13.4). In contrast, the mol% G + C values for animal DNAs fall within a narrow range with only slight variations between species.

The variation in the G + C content of DNA is related to genetic differences between organisms. If the G + C content of two species is significantly different, then these species must have different sequences of nucleotides in their DNA. Variation in the sequence of nucleotides in DNA is the basis of genetic differences between organisms, so mol% G + C values are used in classification.

If two strains have significantly different mol% G + C, systematists can conclude that they are not related to each other. In this sense, *mol% G + C is an exclusionary determinant.* Strains differing by more than 5 mol% should not be classified in the same species. Looking at the larger picture, the pseudomonads with a mol% G + C between 58 and 70 mol% are clearly different from humans with a mol% G + C of 42 (Table 13.4). The difference in these values is sufficient evidence to say that pseudomonads and humans are unrelated, on the basis of their DNA alone.

TABLE 13.4 Composition of DNA from different organisms

BACTERIA	MOL% G + C	EUCARYOTIC ORGANISM	MOL% G + C
Clostridia	21–28	*Salmo trutta* (brown trout)	43
Clostridium hemolyticum	21	*Gallus domesticus* (chicken)	42
Sarcina maxima	29	*Equus caballus* (horse)	43
Bacillus spp.	32–62	*Homo sapiens* (human)	42
Staphylococcus spp.	30–40	*Triticum aestivium* (wheat)	45
Cytophaga	34–43	*Saccharomyces cerevisiae* (yeast)	36
Treponema	32–50	*Aspergillus niger* (mold)	50
Proteus spp.	39–42	*Paracentrotus lividus* (sea urchin)	35
Escherichia spp.	50–53		
Bacteroides spp.	40–55		
Pseudomonas spp.	58–70		
Rhodospirillum spp.	62–66		
Micrococcus spp.	66–75		
Myxobacteria	68–71		

KEY POINT

No taxonomic conclusion can be drawn if two organisms have the same mol% G + C because both unrelated and closely related organisms can have the same mol% G + C. For example, numerous bacteria have a mol% G + C of 42, which is identical with that of human DNA. Obviously, bacteria are not closely related to humans.

Bacterial taxonomists use this technique to identify the evolutionary relationships between major groups of bacteria. At one time, both the fruiting myxobacteria and the nonfruiting myxobacteria were grouped together because they all have gliding motility. Now it is known that the nonfruiting group (Cytophaga) possess a low mol% G + C, whereas the fruiting myxobacteria (Myxobacteria) have a high mol% G + C (Table 13.4). These organisms are evolutionarily far apart and are now classified separately. The apparent relatedness of the genera *Micrococcus* and *Staphylococcus* has also been questioned on the basis of mol% G + C values. Members of these genera are aerobic, gram-positive, nonmotile, nonspore-forming cocci, yet they have vastly different mol% G + C values (Table 13.4). Based on their DNA composition these two groups of bacteria are unrelated. Nevertheless, *Micrococcus* and *Staphylococcus* remain classified in the same family for the purpose of identification.

HYBRIDIZATION OF DNA

Upon heating, double-stranded DNA dissociates into single strands. Under appropriate conditions of slow cooling, hydrogen bonding will be reestablished between the complementary strands to reform double-stranded DNA (Figure 13.8). This process is called **annealing of nucleic acids**. When single strands of DNA from different organisms are mixed together and allowed to anneal, the extent of double-stranded DNA formation depends on the homology of the base sequences between the two DNAs (Figure 13.8). Heterologous double-stranded DNA, containing one strand of DNA from two separate organisms, is called a **hybrid**, and the process of forming hybrids is called **DNA hybridization**. The formation of DNA/DNA hybrids or DNA/RNA hybrids is a measure of the relatedness between organisms and is used to compare closely related bacteria.

One common technique for measuring DNA hybrid formation involves the use of nitrocellulose filters. Denatured (single-stranded) DNA from the test organism is bound to a nitrocellulose filter. Next, the unused binding sites on the filter are covered so that additional DNA (single-stranded or double-stranded) will not bind unless it is complementary to the single-stranded DNA already bound to the filter (Figure 13.9). The filter is now incubated under standard conditions with denatured, radioactive-labeled DNA from

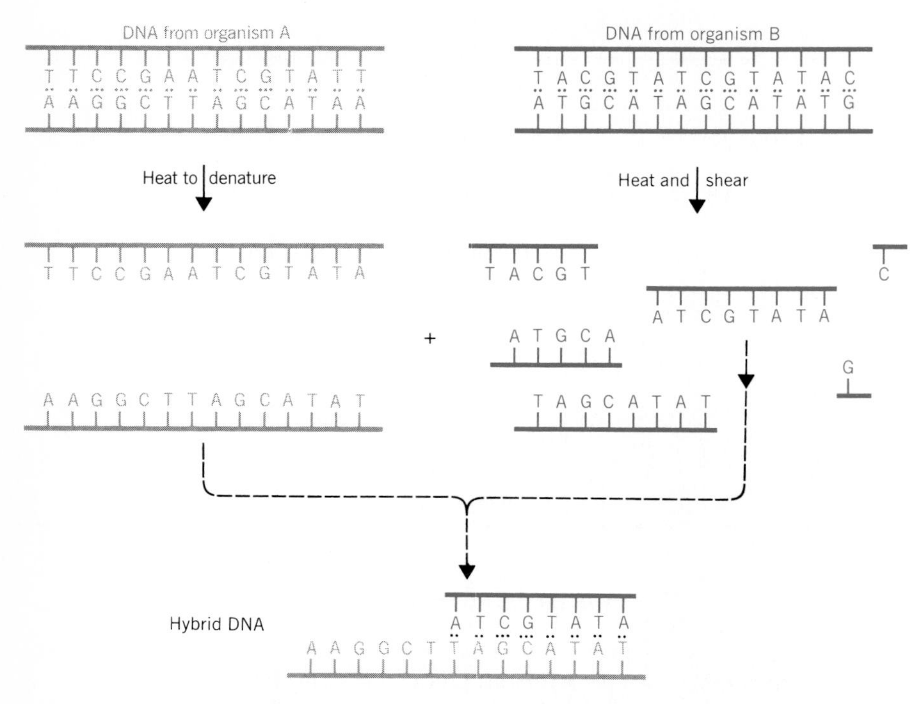

FIGURE 13.8 DNA hybrids are formed when complementary pieces of single-stranded DNA obtained from different organisms reanneal.

the control organism. After a prescribed time, unbound labeled DNA is washed off the filter, and the amount of labeled DNA bound to the filter is measured by radioactive counting.

Results of a DNA hybridization investigation of the *Enterobacteriaceae* are presented in Table 13.5. The control organism is *Escherichia coli*, which shows 88 to 100 percent binding with its own DNA. This value is established by incubating denatured DNA from *E. coli* bound to a filter with denatured ^{32}P-labeled DNA from the same strain of *E. coli*. The amount of radioactive DNA that binds to the filter under these conditions is equated to 100 percent homology and is the standard upon which the other hybridization results are measured.

Next, single-stranded DNA from another genus is prepared, bound to a filter, which then is incubated with the ^{32}P-labeled DNA from *E. coli*. The amount of radioactive *E. coli* DNA that binds to this filter is determined. This value is a measure of the base sequence

TABLE 13.5 Percent relatedness of DNA from bacterial sources[a]

SOURCE OF UNLABELED DNA	PERCENT RELATEDNESS[b]
Escherichia coli K12	88–100
Shigella flexneri	84–85
Citrobacter freundii	50
Salmonella typhimurium	45
Klebsiella pneumoniae	38
Enterobacter aerogenes	37
Serratia marcescens	24
Proteus mirabilis	6

Source: The data were derived from Kenneth E. Sanderson, "Genetic Relatedness in the Family *Enterobacteriaceae*," *Ann. Rev. Microbiol.*, 30:327–349 (1976).

[a]Source of labeled DNA is *Escherichia coli* K12.

[b]
$$\text{Percent relatedness} = \frac{E.\ coli\ \text{DNA}\ ^{32}\text{P that binds to unlabeled DNA} \times 100}{E.\ coli\ \text{DNA}\ ^{32}\text{P that binds to unlabeled}\ E.\ coli\ \text{DNA}}$$

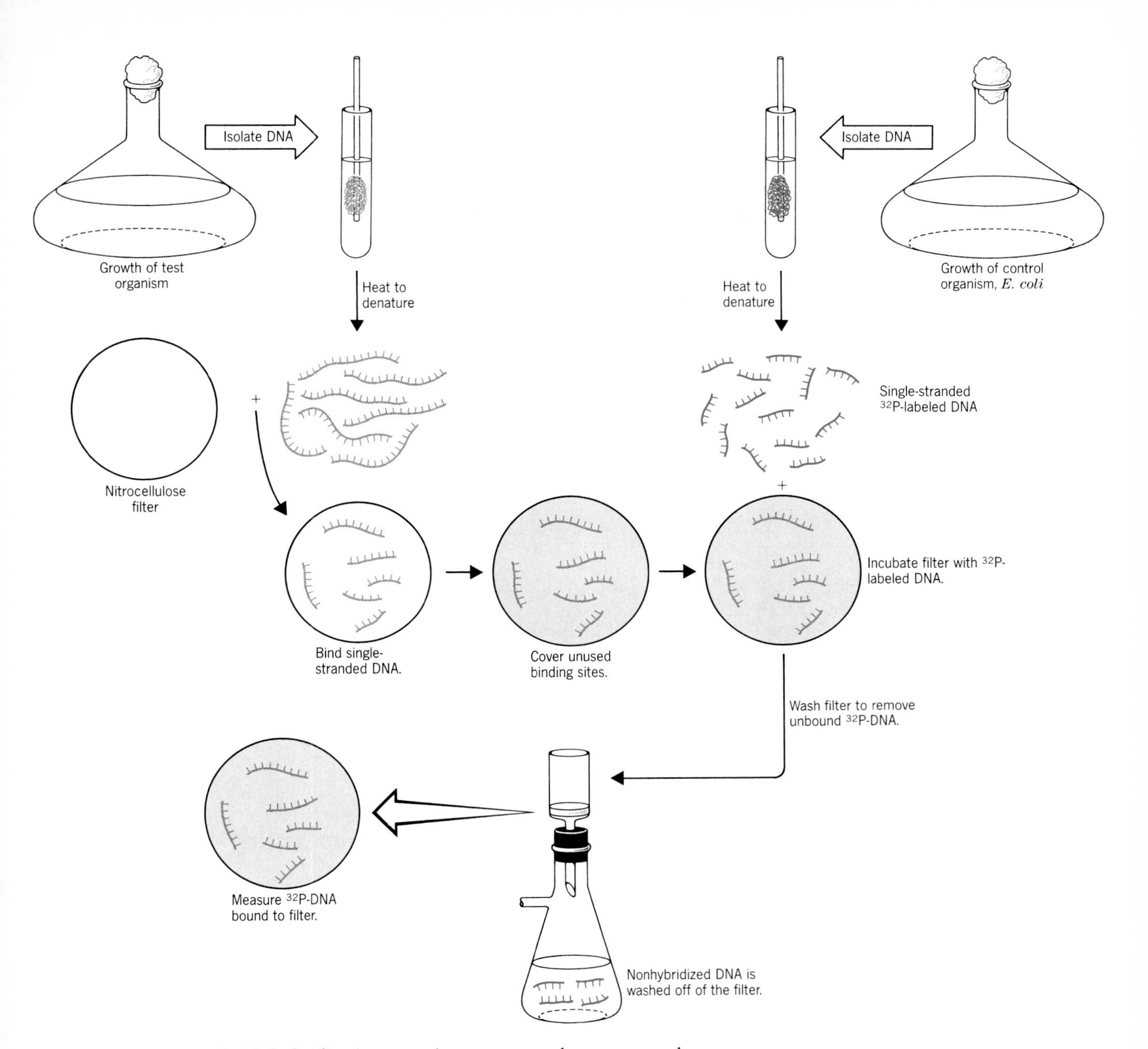

FIGURE 13.9 DNA hybridization experiments are used to measure the relatedness between the genomes of organisms. Double-stranded DNA is isolated from suspensions of lysed cells when the DNA is twirled onto a glass rod. The DNA is then purified and heat-denatured into single strands. Single-stranded DNA from the test organism is bound to a nitrocellulose filter. Additional binding sites are covered to prevent the binding of labeled DNA to the filter. The filter is then mixed with ^{32}P-labeled DNA from the control organism. Hybrids will form between the single-stranded DNA of the test organism (bound to the filter) and the labeled DNA from the control organism in direct relationship with their complementary nucleotide sequences. In the illustrated example, the DNA form *Escherichia coli* is labeled with ^{32}P, while the DNA of the test organism is unlabeled. The degree of homology between the DNAs from these organisms is measured as the amount of ^{32}P label retained on the filter.

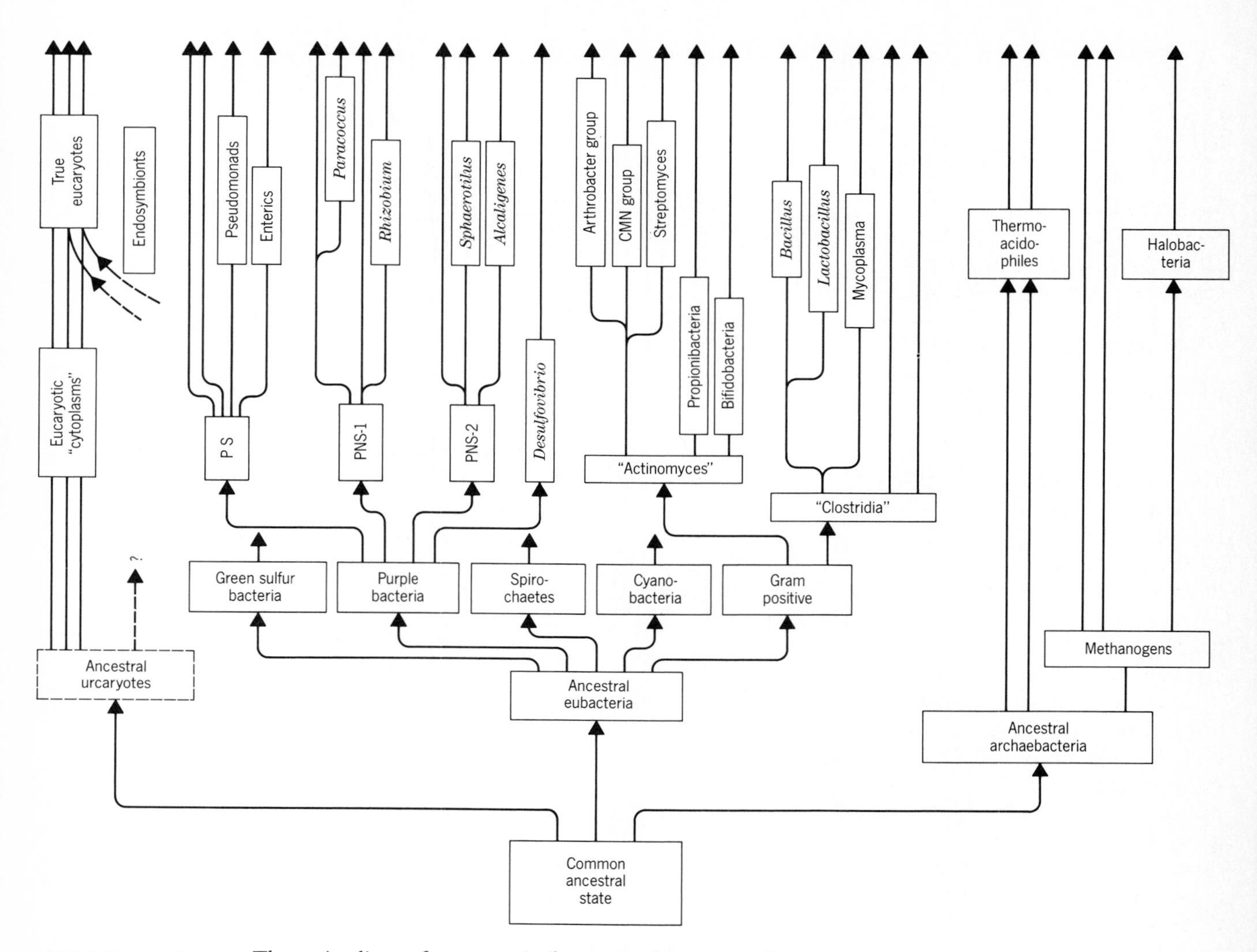

FIGURE 13.10 The major lines of procarytoic descent in this proposed scheme are based on the sequence analysis of 16S ribosomal RNAs isolated from the genera represented. (Courtesy of Dr. Carl Woese. From C. E. Woese et al., *Science,* 209:457–463, 1980, copyright © 1980 by the American Association for the Advancement of Science.)

similarity between the genera being tested and is always less than 100 percent. The results (Table 13.5) show that *Shigella flexneri* (shi-gel'ah fleks'ner-eye) shares more base sequences with *E. coli* than *Salmonella typhimurium. Proteus mirabilis* (proh'tee-us mi-rah'bi-lis) is even less related to *E. coli* than *S. typhimurium* on the basis of these DNA hybridization tests. This form of molecular analysis can provide important information on the taxonomy of closely related genera and species.

Hybridization of single-stranded DNA and isolated RNA is another measure of genetic relatedness. All cellular RNAs are single-stranded molecules coded for by DNA genes. Isolated mRNA and rRNA will form hybrids with complementary single-stranded DNA.

DNA/RNA hybridization experiments can detect genetic relatedness between species.

KEY POINT

Single-stranded DNAs and RNAs bind to complementary nucleotide sequences in DNA by low-energy hydrogen bonds that are broken by gentle heating. There is a high degree of specificity in the annealing reaction since short pieces of single-stranded DNA form hydrogen bonds only with complementary sequences of single-stranded DNA.

SEQUENCE ANALYSIS OF RIBOSOMAL RNA

The 16S ribosomal RNAs are universally present in bacteria as structural components of the 30S ribosomal subunit. This RNA is a direct transcript of the 16S rRNA gene present in the cell's DNA, hence, sequencing this RNA is equivalent to sequencing a chromosomal gene. Once obtained, the nucleotide sequences of the 16S rRNAs isolated from bacterial ribosomes can be used to classify members of the Procaryotae.

Sequence analyses of isolated 16S rRNAs (1541 bases in *E. coli*) have been done for many bacterial species. Analyses of the nucleotide sequences in over 170 bacterial 16S rRNAs were used to construct the scheme presented in Figure 13.10. Amazingly, the 16S rRNAs genes from eubacteria are more closely related to each other than to the archaebacteria. This suggests that the archaebacteria have a common line of descent that is distinct from the eubacteria. The distinctive nucleotide sequences present in the archaebacterial 16S rRNAs represent one of the criteria used to create a separate section for them in the kingdom Procaryotae (see Appendix 2).

The 50S subunit of the bacterial ribosome contains a 5S and a 23S rRNA (Chapter 9). The 5S rRNA (120 bases) from many different bacteria is being isolated and sequenced. The sequences of this gene product may contain useful information for grouping related bacterial genera into families (Table 13.1).

GENETICS AND THE FUTURE OF BACTERIAL CLASSIFICATION

Bacterial geneticists have elucidated the structure and functions of the bacterial genome in great detail. Until the 1950s, microbiologists believed that bacteria were genetically stable and evolved slowly over time. However, we now know that the genetics of a bacterium can be significantly altered by transposons, by plasmid acquisition, or by other methods of genetic transfer and recombination (Chapter 10).

Extrachromosomal DNA occurring as bacterial plasmids can be stably inherited. Moreover, certain plasmids can be transferred between bacteria by conjugation or artificial transformation. Bacteria can also acquire genetic information carried on bacteriophage genomes by the process of lysogeny. The acquisition of this genetic information can significantly alter a bacterium's role in nature. For example, *Escherichia coli* becomes a human pathogen when it acquires a virulence plasmid, and *Corynebacterium diphtheriae* becomes a toxin-producing strain when it is lysogenized by β-bacteriophage.

Transposons add another degree of complexity to bacterial systematics and our views on the mechanism of evolution. Transposons are genetic elements that move from plasmids to bacterial chromosomes and vice versa (see Chapter 10), and can be copied to new locations on the bacterial genome. How should systematists treat a species that is significantly altered by the acquisition of a plasmid? If the essential gene on a plasmid is transposed to the bacterial genome, should the bacterium become a new species?

These exciting developments in molecular genetics will compel bacterial systematists to deal with problems of gene acquisition by bacteria. Only when these possibilities for change are combined with the continued accumulation of data concerning known bacteria can the novice microbiologist appreciate the diversity of the bacterial world and the dynamic nature of bacterial systematics.

Summary Outline

SCIENTIFIC IMPORTANCE OF SYSTEMATICS

1. Bacterial systematics is the scientific approach to describing, naming, and organizing bacteria into a classification scheme.
2. Systematists describe and name each kind of bacterium so that investigators can identify their isolates.
3. Phylogeny is the study of the evolutionary relationships between organisms arrived at by studying their ancestory.

THE FIVE KINGDOMS

1. Organisms with similar characteristics and common lines of descent are assigned to one of five kingdoms: Plantae, Animalia, Fungi, Protista, and Procaryotae.

BACTERIAL CLASSIFICATION

1. Bacteria are assigned to one of four divisions (superfamilies) in the kingdom Procaryotae: gram-negative bacteria, *Gracilicutes;* gram-positive bacteria, *Firmicutes*; bacteria without cell walls, *Tenericutes*; and bacteria with widely varying cell-wall chemistries, *Mendosicutes*.
2. Within a division bacteria are assigned to sections then to a family, genus, species, and sometimes subspecies.

NAMES AND DESCRIPTIONS OF BACTERIA

1. A bacterial species is a collection of strains that share essential characteristics and differ significantly from all others.
2. The binominal system of nomenclature is used to give each bacterial species a binomen composed of a genus name and a specific epithet, which modifies the genus name and indicates the species.
3. The classification of all recognized bacterial species is compiled in the four-volume *Bergey's Manual of Systematic Bacteriology*.

BACTERIAL IDENTIFICATION: CLASSIC TECHNIQUES

1. The identification and classification of bacteria by classic techniques use information such as cell morphology, culture characteristics, enzymes, and metabolic traits.
2. Identifications are made by comparing observed traits to the description of each officially recognized bacterial species presented in *Bergey's Manual*.

CHEMOTAXONOMY

1. The chemical structure of bacterial cell walls, lipids, quinones, and the amino acid sequences of small proteins are used to classify bacteria into families.
2. Capsules, toxins, immunological surface structures, and plasmid profiles are used to identify bacterial subspecies.

NUMERICAL TAXONOMY

1. Numerical taxonomy characterizes each strain by using as many traits as possible and assigning them all equal weight.
2. This form of taxonomy is adaptable to computerized identification of bacterial isolates.

NUCLEIC ACIDS AND BACTERIAL CLASSIFICATION

1. Measurements of a cell's mol% G + C, DNA/DNA homology, rRNA sequencing, and DNA/RNA homology constitute the molecular basis for measuring bacterial diversity.
2. The mol% G + C is an exclusionary determinant: strains differing by more than 5 mol% are not recognized as the same species.
3. Hybridization between single-stranded DNAs or DNAs and RNAs reveals relationships between closely related bacteria.
4. Plasmid profiles can identify the similarity of strains, while more detailed information is ob-

tained through sequencing of 16S rRNA and 5S rRNAs.

5. The 16S rRNA gene is conserved in all bacteria except the archaebacteria—a fact that helped to establish a separate section of the Procaryotae for the archaebacteria.

GENETICS AND THE FUTURE OF BACTERIAL SYSTEMATICS

1. Bacterial taxonomy is based on the premise that species change slowly with time—speciation is a slow process. Recent discoveries about plasmids, transposons, and prophages have cast some doubt on the rate of speciation among the bacteria.

Questions and Topics for Study and Discussion

QUESTIONS

1. State the reasons why scientists classify microorganisms.
2. What is the concept of species in bacterial classification? How does this differ from the concept of an animal species?
3. Explain the basis for dividing the kingdom Procaryotae into four divisions.
4. Define the following:

API Test Strip	plasmid profiles
binominal system	serovar
division	16S rRNA
DNA hybridization	species
phylogeny	strain

5. Name three classic techniques used to classify bacteria. How can the data obtained from these techniques be applied to numerical taxonomy?
6. How is numerical taxonomy applied to diagnostic medical microbiology?
7. Explain how mol% G + C is used as an exclusionary determinant in bacterial classification.
8. On what principle of molecular biology is DNA hybridization based? How is DNA homology used as a criterion of classification?

DISCUSSION TOPICS

1. How would the acquisition of a plasmid, transposon, or a prophage affect a bacterium's classification?
2. Discuss the implications for students and the medical establishment of the continuing changes in the classification of bacteria.

Further Readings

BOOKS

Austin, B., and F. Priest, *Modern Bacterial Taxonomy,* Van Nostrand Reinhold Co. Ltd, Bershire, England, 1986. Concise paperback text on bacterial taxonomy.

Krieg, N. R., and J. G Holt (eds.), *Bergey's Manual of Systematic Bacteriology*, Vol. 1, Williams & Wilkins, Baltimore, 1984. The first volume of this standard reference for bacterial classification covers the gram-negative bacteria of general, medical and industrial importance. The introductory chapters on classification are excellent. Volume 2, see Sneath below. Volume 3 will cover the archaebacteria, cyanobacteria, and remaining gram-negative bacteria (J. T. Staley and J. G. Holt, eds.) and volume 4 will cover the actinomycetes (S. T. Williams and J. G. Holt, eds.).

Sneath, P. H. A., and J. G. Holt (eds.), *Bergey's Manual of Systematic Bacteriology*, Vol. 2, Williams & Wilkins, Baltimore, 1986. The second volume of this standard reference for bacterial classification covers the gram-positive bacteria other than actinomycetes.

Starr, M. P., H. Stolp, H. G. Truper, A. Balows, and H. G. Schlegel (eds.), *The Prokaryotes*, Springer-Verlag, New York, 1981. This excellent two-volume handbook on habitats, isolation, and identification of bacteria is highlighted by superb introductory chapters on procaryotic diversity, habitats, and isolation techniques.

MANUALS

Difco Manual of Dehydrated Cultured Media and Reagents, 10th ed., Difco Laboratories, Detroit, 1984. This is an important reference manual for the composition and use of the wide variety of bacteriological media, test, and reagents manufactured by Difco.

Lennette, E. H., A. Balows, W. J. Hausler, and H. J. Shadomy, *Manual of Clinical Microbiology*, 4th ed., American Society for Microbiology, Washington, D.C., 1985. This manual contains instructions and discussions of the major microbiological and immunological assays used to identify microorganisms in the clinical laboratory.

Skerman, V. B. D., *A Guide to the Identification of Genera of Bacteria*, Williams & Wilkins, Baltimore, 1967. An abridged form of *Bergey's Manual* that is a useful key for the identification of bacteria.

ARTICLES AND REVIEWS

Dickerson, R. E., "Cytochrome *c* and the Evolution of Energy Metabolism," *Sci. Am.*, 242:136–153 (March 1980). The amino-acid sequences of cytochrome *c*, isolated from many bacteria, are presented and analyzed from an evolutionary perspective. The author constructs a phylogenic tree for the bacteria based on the amino acid sequences in microbial cytochromes.

Schleifer, K. H., and E. Stackebrandt, "Molecular Systematics of Prokaryotes," *Ann. Rev. Microbiol.*, 37:143–187 (1983). A concise summary of the modern approaches to bacterial systematics.

Skerman, V. B. D., V. McGowan, and P. H. A. Sneath, "Approved lists of bacterial names," *Int. J. Syst. Bacteriol.*, 30:225–420 (1980). This is the official list of bacterial names and serves as a reference source for assigning new names to bacteria.

Staley, J. T., and N. R. Krieg, "Bacterial Classification I, Classification of Procaryotic Organisms: An Overview," in *Bergey's Manual of Systematic Bacteriology*, Vol. 1, pp. 1–4, N. R. Krieg and J. G. Holt (eds.), Williams & Wilkins, Baltimore, 1984. This article and the others that preface this compendium on bacterial systematics provide a concise overview of bacterial classification.

Stolp, H., and M. P. Starr, "Principles of Isolation, Cultivation, and Conservation of Bacteria," in M. P. Starr et al. (eds.), *The Prokaryotes*, Vol. 1, pp. 135–175, Springer-Verlag, New York, 1981. This chapter summarizes the salient techniques needed for isolating and cultivating bacteria.

C H A P T E R

14

Bacteria of Ecological, Industrial, and General Significance

OUTLINE

Photosynthetic Flexibacteria
Oxygenic Photosynthetic Bacteria
Cyanobacteria

ARCHAEBACTERIA
Methanogens
Sulfolobus, Thermoplasma, and *Halobacterium*

FOCUS OF THIS CHAPTER

In this chapter we introduce the biology of bacteria that are important in ecology, industry, and biology. The chapter is designed to bring out the great physiological diversity of bacteria by presenting descriptions of representative species. The impact of these bacteria on the environment and their industrial importance are amplified in later chapters.

Bacteria are ubiquitous in almost every environment on earth. They live in the extreme cold of the Arctic and Antarctic, in the heat of thermal springs, and in the depths of the oceans at thermal vent sites. Bacteria are also present in soil, air, and water that surround human habitats. The abundant distribution of bacteria in nature can be attributed to their amazing physiological diversity. Representative procaryotes grow by aerobic respiration, anoxygenic photosynthesis, oxygenic photosynthesis, fermentation, and/or anaerobic respiration.

The physiological activities of bacteria are the basis for their importance to ecology and their industrial applications. Many of the essential reactions of the carbon, sulfur, and nitrogen cycles are catalyzed by bacterial enzymes. Some bacterial products are commercially important and other bacteria are involved in the production of foodstuffs, as was described earlier. Our approach to presenting the bacteria of ecological, industrial, and general significance is to emphasize their physiological diversity, for only with this knowledge can one appreciate how bacteria contribute to the balance of life on earth and understand the practical applications of bacteria in industry.

FERMENTATIVE BACTERIA

Natural anaerobic environments, such as animal intestines, ponds, deep lakes, and mud, often support an abundant population of microorganisms. Bacteria capable of gaining energy from anoxygenic photosynthesis, anaerobic respiration, or fermentation grow in these anaerobic environments. In this section we will concentrate on the organisms that grow by fermentation.

Many bacteria metabolize carbohydrates, amino acids, purines, and/or other substrates by fermentation pathways that yield energy for growth. Fermentations are characterized by the substrates fermented and the products produced. For example, lactic acid is the major fermentation product produced in the homolactic fermentation, while equal quantities of ethanol and lactic acid are produced from glucose by the heterolactic fermentation (Chapter 6). Other bacteria ferment glucose and produce butyric acid or propionic acid as major products. Fermentation products are indicative of the bacterium's metabolic capabilities, so fermentation products are a useful characteristic in bacterial classification (Chapter 13).

LACTIC ACID BACTERIA

Souring of milk, curd production in cheese manufacturing, and the production of olives, pickles, and sauerkraut each involves one or more of the lactic acid bacteria. These nonspore-forming, microaerophilic, gram-positive rods or cocci produce lactic acid as a major product of sugar fermentation. They are unable to make heme, so they lack cytochromes and are catalase negative. The lactic acid bacteria are found in milk and dairy products, plants, vegetables, silage, wine, beer, animal intestines, and in humans. Lactic acid bacteria are assigned to the *Streptococcus, Leuconostoc* (leucoh-nos'tok), or *Lactobacillus* genera based on their fermentation (hetero- or homolactic acid fermentation) and their morphology (Table 14.1).

The streptococci are spherical cells that ferment glucose by the homolactic acid fermentation (Table 14.1). *Streptococcus lactis* and *S. cremoris* are nonpathogenic cocci found in milk and cheese. The lactic acid they

TABLE 14.1 Classification of the lactic acid bacteria

FERMENTATION	GENERA	
	SPHERICAL CELLS	ROD-SHAPED CELLS
Homolactic acid fermentation	*Streptococcus*	*Lactobacillus*
Heterolactic acid fermentation	*Leuconostoc*	*Lactobacillus*

produce during growth in milk decreases the pH sufficiently to turn the milk sour and eventually causes soft curd formation. Cottage and cream cheese are soft-curd food products produced by the growth of *Streptococcus lactis* and/or *S. cremoris* in milk. These species are also found in semisoft to hard, ripened cheeses such as cheddar, Monterey, Muenster, and Gouda.

Humans and other animals harbor streptococci of the enterococci group, which grow in their intestines and are excreted in feces. The presence of *Streptococcus faecalis* and *S. faecium* in water indicates contamination of water with human feces. Sometimes these bacteria cause hospital-acquired (nosocomial) infections, especially of the urinary tract.

Streptococci are also present in the human mouth, where they colonize the tooth surfaces and contribute to the formation of dental plaque and dental caries. *Streptococcus mutans* contributes to the formation of dental plaque by producing a sticky polymer of glucose known as glucan. This polymer binds *S. mutans* and *S. mitior* (my′ti-or) to the tooth surface. Here they grow and produce acidic products, which contribute to erosion of the tooth surface and the formation of caries.

Leuconostoc are found on plants and to a lesser extent in milk and milk products. They are spherical lactic acid bacteria that use the heterolactic acid fermentation to produce both lactic acid and ethanol. Dextran production by *Leuconostoc dextranicum* (dex-tra′ni-cum) and *L. mesenteroides* (mes-en-ter-oi′des) is easily detected by the mucoid appearance of the colonies they produce when grown on sucrose. They break down sucrose to fructose and glucose, converting the excess glucose into dextrans, which they secrete. Dextrans are commercially important as blood plasma extenders and are used in food production.

Lactobacilli occur in nature on plant surfaces and are important in the decay of plant material and in the preservation of certain foods, such as cucumbers (pickles) and cabbages (sauerkraut). These rod-shaped bacteria gain energy by fermenting glucose through either the homo- or the heterolactic acid fermentation (Table 14.1). Many species of *Lactobacillus* are commercially important in the manufacture of sourdough breads, and certain dairy products. *Lactobacillus bulgaricus* (bulga′ri-kus) is used in the production of buttermilk, *L. bulgaricus* and *Streptococcus thermophilus* are used in the manufacture of yogurt, *L. casei* (cay′se-eye) and *L. helveticus* (hel-ve′ti-kus) are used in the ripening of certain cheeses, and *L. plantarum* (plan-ta′rum) is used in the production of olives, pickles, and sauerkraut. Other lactobacilli are involved in the commercial production, especially in Europe, of the lactic acid used in food preservation, tanning animal hides, and textile manufacturing.

Although the lactose in milk has high nutritional value, some individuals are intolerant of milk lactose because they cannot digest it. The undigested lactose is fermented into gaseous products by the intestinal flora of lactose-intolerant people. Their solution is to drink a fermented milk containing low amounts of lactose. Acidophilus milk or sweet milk is produced by the action of *Lactobacillus acidophilus* (a-ci-do′phi-lus) on pasteurized milk. These bacteria convert almost all the lactose present to glucose and fructose, monosaccharides that are readily digested by all people.

PROPIONIC ACID BACTERIA

Propionibacteria are found in cheese and dairy products, on skin, and in the intestinal and respiratory tracts of animals. They ferment glucose and/or lactic acid to produce propionic acid, carbon dioxide, and acetic acid. The propionibacteria are nonmotile, nonspore forming, anaerobic, pleomorphic rods that grow best under strictly anaerobic conditions, although some are

FIGURE 14.1 Propionibacteria are involved in the ripening of Swiss cheese. The holes in this cheese are produced by the carbon dioxide generated during the propionic fermentation. The other cheese shown are ripened with species of *Lactobacillus* and *Streptococcus*. (Courtesy American Dairy Association.)

aerotolerant. They cannot compete with lactic acid bacteria for glucose, but they are able to ferment and grow on the lactic acid produced by lactic acid bacteria.

The propionibacteria are used in the ripening process of certain cheeses and often develop as colored colonies on cheese (or butter) that is left unrefrigerated. *Propionibacterium freudenreichii* (freu-den-reich'ee-eye) and *P. shermanii* (sher-man'ee-eye) are used in the manufacture of Swiss cheese. The large holes that characterize Swiss cheese (Figure 14.1) are created when carbon dioxide is produced by propionibacteria during the ripening process.

BUTYRIC ACID FERMENTATION: CLOSTRIDIA

Clostridia are the largest group of obligately anaerobic, endospore-forming, gram-positive rods (Figure 14.2). These bacteria are widely distributed in soil, sediments of marine and freshwater environments, and the intestines of humans and rumens of animals. They can be classified according to the substrates they ferment, which include sugars, amino acids, purines, and cellulose.

The saccharolytic clostridia ferment sugars to produce butyric acid, acetic acid, carbon dioxide, hydrogen gas, and various amounts of acetone, isopropanol, and butanol by the metabolic process called the butyric acid fermentation. A number of these products have commercial value as solvents. Clostridia have been used since the early 1900s to produce solvents for the production of synthetic rubber, gun powder, and nitrocellulose lacquers. Acetone produced by the butyric acid fermentation was used by the Allies for the production of chordite, a type of gunpowder, and later for the production of nitrocellulose lacquers.

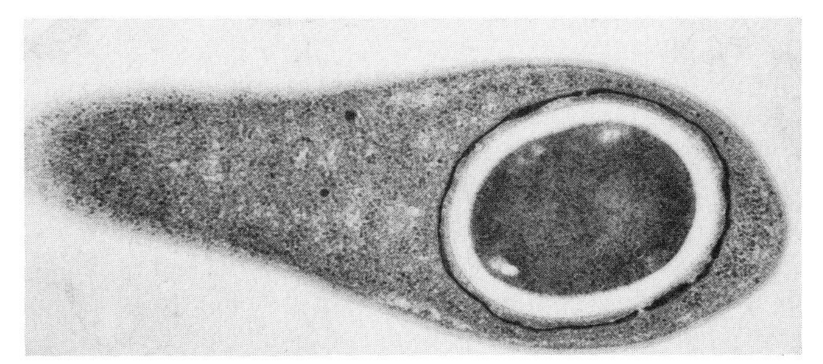

FIGURE 14.2 Electron micrograph of *Clostridium tetani* showing its terminal endospore in swollen sporangium. (Courtesy of Dr. T. J. Beveridge/BPS.)

Clostridium acetobutylicum (a-ce'toh-bu-ti'li-kum) produces commercial quantities of acetone when it ferments molasses and corn-steep liquor—two inexpensive and readily available feedstocks. The fermentation pathway of *Clostridium butylicum* is slightly different, so it produces butanol and isopropanol and essentially no acetone. Now the butyric acid fermentation is of little commercial importance because these solvents are produced more cheaply from petroleum.

Clostridia that ferment amino acids or purines are essential links in the natural recycling of nitrogenous compounds. *Clostridium tetani* (teh'tan-ee) and *Cl. sporogenes* (spo-rah'jen-ez) are examples of proteolytic clostridia that gain energy by fermenting amino acids. They contribute to the recycling of proteins in decaying vegetable and animal matter in anaerobic environments. *Clostridium acidiurici* (a-ci-di-u'ri-ci) ferments the purine uric acid as a source of energy and carbon and produces formic acid, glycine, carbon dioxide, and ammonia as fermentation products. Uric acid is the major nitrogenous excretory product of birds. You can imagine the problems that industrial-scale chicken and turkey farmers would have if there were no biological mechanism for metabolizing uric acid.

THE SULFUR-REDUCING BACTERIA

The sulfur-reducing bacteria are strict anaerobes that grow by oxidizing organic compounds and reducing sulfur or sulfate. In this form of anaerobic respiration, sulfate or elemental sulfur is used as the terminal electron acceptor. These bacteria are morphologically diverse, mostly gram-negative, obligate anaerobes found in sediments of both freshwater and marine environments, in deep wells, and in the intestinal tract of humans and animals.

Most naturally produced hydrogen sulfide (H_2S) is generated by the activities of sulfate-reducing bacteria. Hydrogen sulfide is often produced in deep (anaerobic) wells and taints the well water with a rotten-egg odor. This foul-smelling gas reacts with iron to form ferrous sulfide (FeS), which imparts a black color to anaerobic mud. The ferrous sulfide present in municipal water sources is a tremendous nuisance because it stains clothes and contributes to the corrosion of plumbing fixtures.

Cell morphologies of the sulfur-reducing bacteria include rods, vibrios, sarcinas, cocci, and filamentous gliding bacteria. The widely studied **sulfate-reducing** *Desulfovibrio* (dee-sul-foh-vib′ree-oh) are curved rods that oxidize lactic acid and fatty acids as they reduce sulfate to hydrogen sulfide. One species of sulfate-reducing bacteria forms true endospores; it is classified in the genus *Desulfotomaculum* (Figure 14.3). The other nonsporforming genera include *Desulfococcus, Desulfosarcina, Desulfobacter,* and *Desulfonema.*

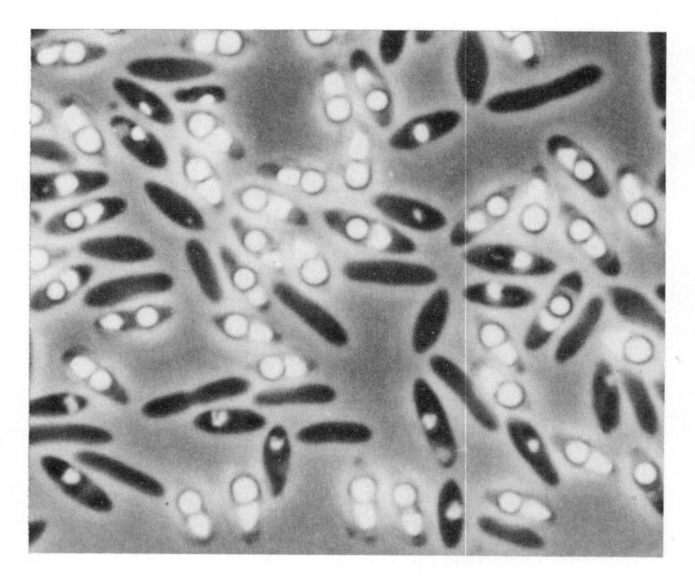

FIGURE 14.3 Cells of *Desulfotomaculum acetoxidans* containing spores and gas vacuoles. (Courtesy of Friedrich Widdel.)

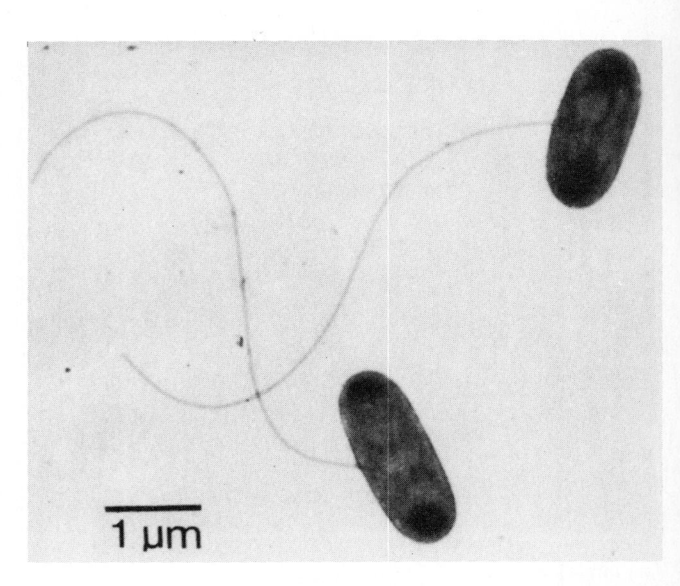

FIGURE 14.4 *Desulfuromonas acetoxidans* has a single flagellum attached to its side. (From N. Pfenning and H. Biebl, *Archives of Microbiology,* 110:3–12, 1976, copyright ©, Springer Verlag, Heidelberg.)

In contrast, *Desulfuromonas acetoxidans* (dee-sul-fur-oh-moh′nas a-cet-ox′i-dans) is a **sulfur-reducing bacterium** that uses molecular sulfur (S) as its terminal electron acceptor. This bacterium completely oxidizes acetate to carbon dioxide while reducing molecular sulfur to hydrogen sulfide. *Desulfuromonas acetoxidans* has a single flagellum that moves these cells with a propeller-like motion because of its lateral or subpolar insertion point (Figure 14.4).

INTERACTIONS BETWEEN PLANTS AND BACTERIA

Many bacteria live in close association with plants either in the soil surrounding plant roots **(rhizosphere),** or on the surfaces of plant stems and leaves **(phyllo-**

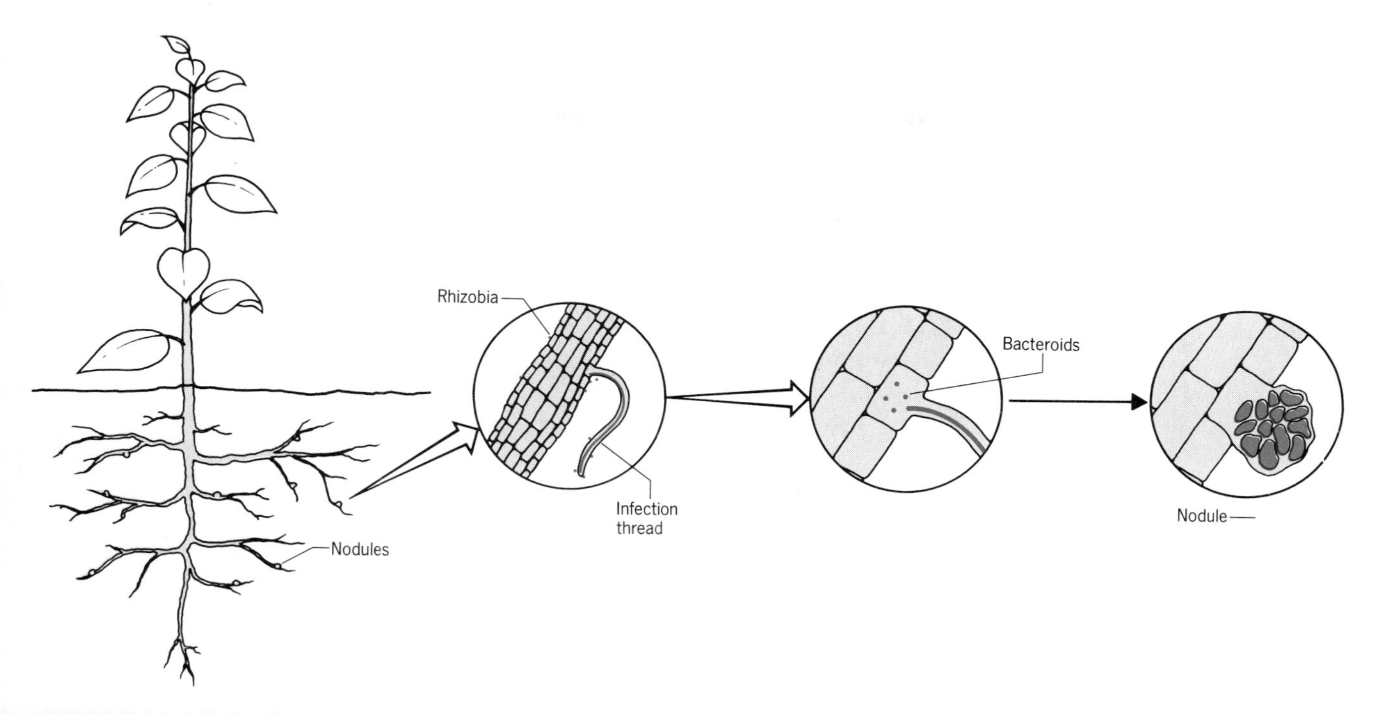

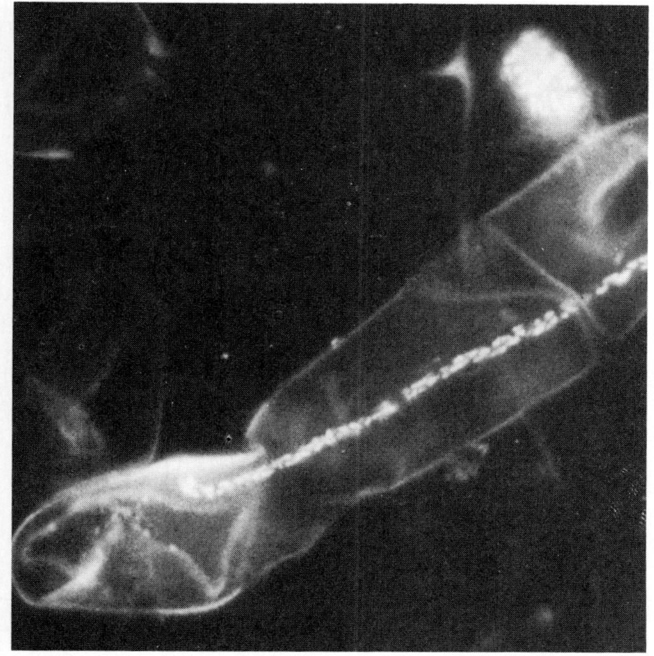

FIGURE 14.5 Root nodule formation in a leguminous plant being infected by a species of *Rhizobium*. Photo shows the fluorescent-stained cells that comprise the infection thread inside the root hair of a clover plant. (Courtesy of Ben Bohlool.)

sphere). Some soil organisms help the plant by (1) synthesizing vitamins and amino acids that the plant uses, (2) solubilizing minerals for plant use, and (3) removing toxic substances from the rhizosphere. In return, symbiotic bacteria gain nutrients from the plant, while other bacteria utilize plant nutrients and cause plant diseases.

RHIZOBIUM AND NITROGEN FIXATION

The nitrogen gas (N_2) present in the earth's atmosphere is the major reserve of nitrogen for biological systems. Nitrogen-fixing bacteria are the only organisms capable of utilizing this form of nitrogen. They possess an oxygen-sensitive enzyme called nitrogenase that reduces N_2 to ammonia (NH_3). The ammonia produced is converted into amino acids and other nitrogen compounds by plants, fungi, bacteria, and certain animals. The nitrogen-fixing bacteria are either free-living or fix nitrogen in symbiosis with plants (Figure 14.5).

Species of *Rhizobium* (ri-zoh′bee-um) are gram-negative, aerobic, rod-shaped bacteria that inhabit the rhizosphere of leguminous plants. The leguminous plants, such as peas, beans, vetch, clover, and alfalfa, form a symbiotic (mutually beneficial) relationship with species of *Rhizobium*. Alone, neither the plant nor the bacterium is able to fix N_2; together, however, they form a symbiotic relationship in which the bacterium fixes molecular nitrogen (N_2) and in the process supplies the plant with a usable source of reduced nitrogen.

The process by which rhizobia infect a plant root and stimulate nodule formation depends on complex interactions between the plant and the bacterium. The rhizobia are attracted to root hairs by amino acids secreted by the plant (Figure 14.5). Rhizobia swarm on the root-hair surface before they actively penetrate the root hair and stimulate the root hair to curl. Root-hair curling is probably due to bacterially produced plant growth hormones that cause unequal growth of the plant cell wall.

Each *Rhizobium* species infects a limited group of legumes through a recognition process by which the bacteria bind to specific root-hair lectins. **Lectins** are plant proteins (usually glycoproteins) on the root-hair surface that interact with sugars on the surface of the bacteria. After binding, the rhizobium enters the plant via the root-hair tip, and reproduces to form an **infection thread** that grows toward the root proper.

Formation of a root nodule begins when a rhizobium infects a root cell adjacent to the root hair. Only plant cells that are in the process of cell division (tetraploid cell) can give rise to a nodule; infected diploid plant cells usually die. The nodules that develop on the roots contain large spherical cells known as **bacteroids** (Figure 14.6), which are capable of fixing N_2. The nitrogenase present in bacteroids in the root nodule is protected against oxygen inactivation by the plant cell. By themselves, cultures of *Rhizobium* cannot fix N_2 unless the oxygen tension is kept extremely low.

Bacterial plasmids carry the genes necessary for both the infectious process and nitrogen fixation. These nitrogen-fixing genes code for nitrogenase, the molybdenum- and iron-containing enzyme that catalyzes the reduction of N_2 to ammonia. Nitrogenase is extremely sensitive to oxygen and would be inactivated if it were synthesized in free-living soil rhizobia. In the nodules, however, nitrogenase is protected by **leghaemoglobin** (leg-hee'moh-glob-in), a plant protein that binds oxygen and maintains a low oxygen tension in the nodule.

Farmers can ensure that the correct strain of *Rhizobium* will interact with the legumes they plant by sowing seeds inoculated with the correct species of *Rhizobium*. Preinoculated legume seeds assure the farmer that the correct strain of *Rhizobium* is present in the soil. Nitrogen fixation by leguminous plants is extremely important to agriculture since it reduces the fertilizer nitrogen needed for good crop yields (Figure 14.7). For example, rhizobia in symbiosis with clover will fix between 150 and 200 kg nitrogen per hectare per year, whereas only 3.5 kg nitrogen per hectare per year is fixed in fallow soils. Overall, biological nitrogen fixation accounts for approximately 70 percent of the total nitrogen fixed on the earth.

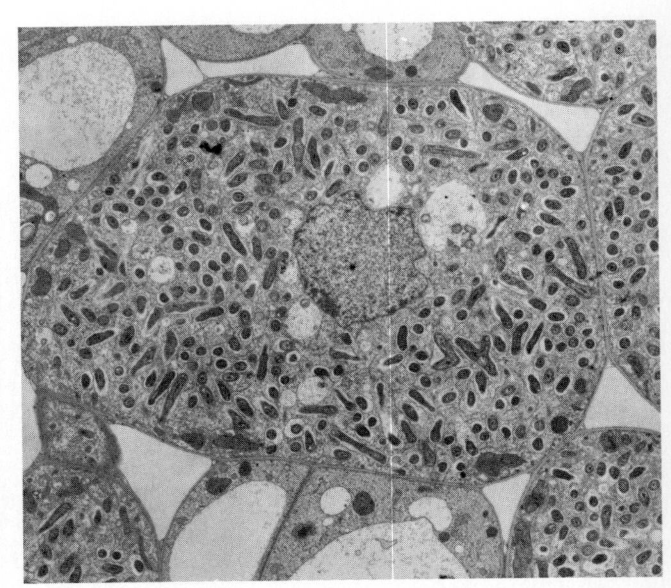

FIGURE 14.6 Bacteroides of *Rhizobium japonicum* in a soybean nodule. (Courtesy of Dr. L. Evans Roth.)

FREE-LIVING NITROGEN-FIXING BACTERIA

Many bacterial species, both aerobes and anaerobes, are able to fix N_2 as free-living organisms (Table 14.2). *Azotobacter* (ah-zoh'toh-bak-ter), *Beijerinckia* (by-jerink'ee-ah), and *Derxia* (derx'i-ah) are free-living bacteria that fix N_2 under aerobic conditions, whereas species of *Azospirillum* and *Aquaspirillum* fix N_2 only under microaerophilic conditions. Free-living anaerobic bacteria that fix nitrogen include the anoxygenic photosynthetic bacteria, the fermentative *Bacillus* spp., and *Clostridium pasteurianum* (pas-teu-ri-a'num). Many

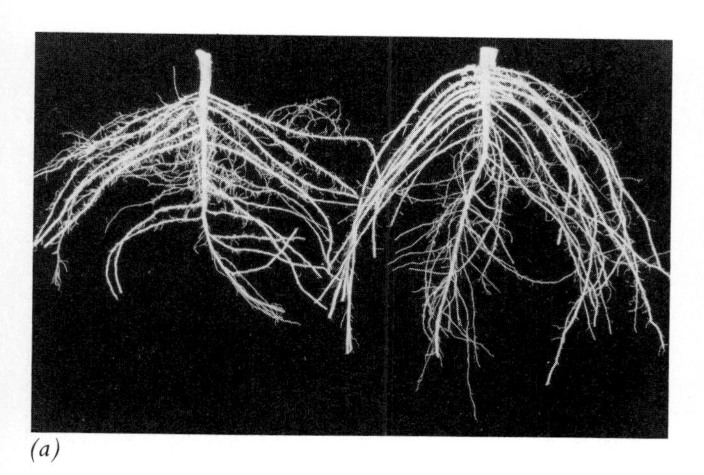

(a) (b)

FIGURE 14.7 Bare uninoculated roots of soybean plant *(a)* and nodulated roots of soybean plant *(b)* inoculated with *Rhizobium*. (Courtesy of Dr. Robert K. Howell, ARS, USDA.)

filamentous N_2-fixing cyanobacteria form heterocysts that protect nitrogenase against oxygen inactivation. The heterocysts are rounded cells arising from vegetative cells distributed along a filament or at one end of the filament. It is still unclear how nonheterocyst-forming cyanobacteria protect their nitrogenase from oxygen inactivation, especially since these bacteria produce oxygen during photosynthesis.

Azotobacteraceae

The *Azotobacteraceae* (a-zoh'toh-bak-te-ray'see-eye) are free-living aerobic soil bacteria that are noted for their ability to fix N_2. These bacteria are usually motile, large, gram-negative cells, and some species form cysts to carry them through droughts and other adverse conditions. Russian agriculturalists use these bacteria to in-

TABLE 14.2 Representative free-living nitrogen-fixing bacteria

CYANOBACTERIA	CHEMOTROPHIC BACTERIA
Heterocystous	
Anabaena	*Azospirillum* spp.
Nostoc	*Aquaspirillum fasciculus*
Scytonema	*Azotobacter* spp.
Gloeotrichia	*Azomonas* spp.
Stigonema	*Beijerinckia* spp.
Nonheterocystous	*Derxia gummosa*
Synechococcus	*Methylomonas methanitrificans*
Oscillatoria	*Methylococcus capsulatus*
Dermocarpa	*Klebsiella pneumoniae*
Myxosarcina	*Enterobacter aerogenes*
Pleurocapsa	*Bacillus polymyxa*
	Bacillus macerans
ANOXYGENIC PHOTOSYNTHETIC BACTERIA	*Desulfotomaculum ruminis*
Rhodospirillum rubrum	*Clostridium pasteurianum*
Rhodopseudomonas palustris	*Clostridium* spp.
Rhodomicrobium vannielii	*Corynebacterium autotrophicum*
Chromatium vinosum	*Desulfovibrio desulfuricans*
Chlorobium thiosulfatophilum	

crease crop yields by seeding fields with cultures of *Azotobacter*. The cost-effectiveness of this practice is still debated, even though there is no doubt about the ability of free-living nitrogen-fixing bacteria to enrich soil.

Species of *Azotobacter* are large gram-negative cocci found in alkaline soils; they are rarely found in soil with a pH below 6.0. Some species of *Azotobacter* survive in soil under adverse environmental conditions as cysts that are resistant to desiccation (Figure 14.8). Growing cells of *Azotobacter* have a high metabolic rate that rapidly consumes the available 0_2. This lowers the cytoplasmic oxygen concentration and effectively prevents the inactivation of nitrogenase. Studies of the physiology, biochemistry, and genetics of nitrogenase in *Azotobacter vinelandii* (vine-lan′di-eye) have contributed greatly to our understanding of nitrogen fixation.

Tropical acid soils are the natural habitat for species of *Beijerinckia*. These bacteria form bipolar lipid bodies (Figure 14.9) and produce a great amount of slime. They fix N_2 at pH values lower than 6.0 and have been isolated from tropical soils with a pH of 4.5. Another tropical nitrogen-fixing bacterium, *Derxia gummosa,* is widely distributed in neutral soils south of the equator.

BACTERIAL DISEASES OF PLANTS

Bacterial plant pathogens can cause widespread destruction of commercial crops. For example, *Spiroplasma citri* caused an outbreak of stubborn disease in Southern California in 1979 that led to the destruction of approximately 2 million citrus trees. Citrus canker caused a similar problem in the Florida citrus industry in 1984. Bacterial plant pathogens also cause plant diseases described as wilts, blights, rots, and galls.

Soft Rot, Wilt, and Blight

Plant diseases are recognized by the types of damage done to the plant tissue. **Soft rot** is caused by bacterial enzymes that hydrolyze the polysaccharides of the plant cell wall. This allows the water of the plant material to flow out of the cell and results in a soft slimy appearance to the damaged tissue. **Wilt** is caused by organisms that block the flow of water within the plant. The affected tissue loses its rigidity as the affected cells shrink in size. **Blight** is caused by organisms that destroy large portions of the plant by blocking its water-conducting system and by producing hydrolytic enzymes and toxins that inhibit metabolism.

Many of the bacteria that cause plant diseases are **epiphytes** (ep′i-fyts), that is they are bacteria that

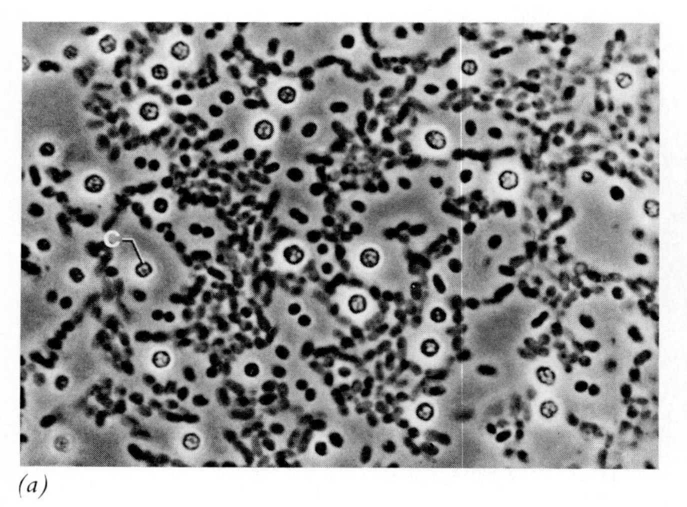

(a)

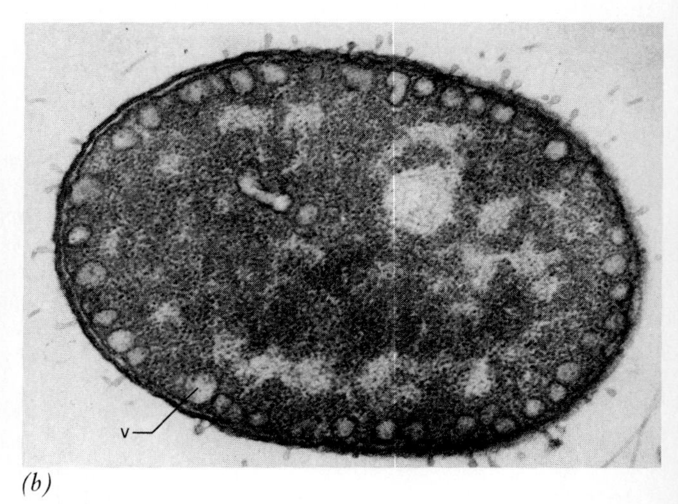

(b)

FIGURE 14.8 *(a)* Phase contrast micrograph of *Azotobacter vinelandii* containing cysts (c). *(b)* Thin section through the N_2 fixing cell of this large bacterium showing membrane vesicles (v) and poly-hydroxy butyric acid (pBA). (*(a)* Courtesy of Dr. J-H. Becking. From J-H. Becking, "The Family Azotobacteraceae," in M. P. Starr et al. (eds.), *The Prokaryotes,* with permission of Springer-Verlag, Heidelberg, 1981. *(b)* Courtesy of A. Ryter.)

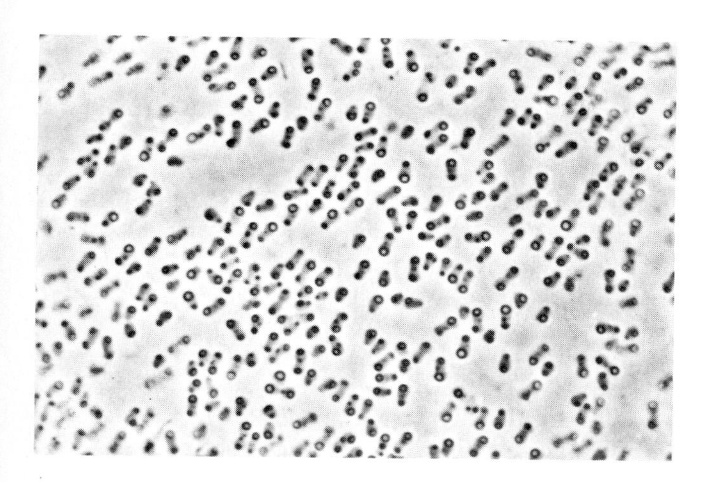

FIGURE 14.9 Photomicrograph of *Beijerinckia indica* showing their characteristic polar lipoid bodies. (Courtesy of Dr. J-H. Becking. From J-H. Becking, "The Family Azotobacteraceae," in M. P. Starr et al. (eds.), *The Prokaryotes,* with permission of Springer-Verlag, Heidelberg, 1981.)

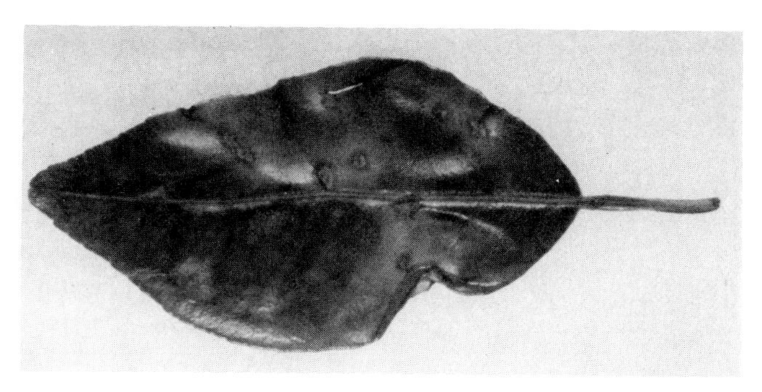

FIGURE 14.10 The lesion of citrus canker on the leaf of an orange tree caused by *Xanthomonas campestris* pathovar *citrii.* (Courtesy of V. Jane Windsor.)

grow on plants without receiving nourishment from the plants. Erwinias are a significant component of the epiphytic bacterial flora of plants. *Erwinia amylovora* (a-my-lo′vo-rah) causes fire blight in pears and apples, *E. stewartii* causes corn wilt, and *E. carotovora* (ca-roh-to′vo-rah) causes soft rot in fruit. These bacteria produce pectinases that break down the pectin in the cell walls of the fruit, which leads to soft rot. These bacteria are transferred between plants by insects, rain, and/or human contact.

Citrus Canker

Serious economic losses to the Florida citrus industry occurred in 1984 as a result of citrus canker, a bacterial disease caused by *Xanthomonas campestris* (cam-pes′tris) pathovar *citrii.* This infection is recognized by the presence of molted brown leaf lesions surrounded by a narrow yellow halo (canker) (Figure 14.10). The characteristic oozing of the citrus canker differentiates infections of *X. campestris* from similar disease caused by fungi. Once this canker develops on a tree, all the trees within a 125-foot radius must be destroyed to prevent the infection from spreading throughout the grove or nursery.

Citrus canker occurred in Florida citrus groves in the early 1900s, and at that time steps were taken to eradicate the causative agent. The outbreak of 1984 is thought to have been initiated by contaminated plant material brought into the United States from areas where citrus canker is epidemic, such as Asia and South America. The citrus farmers' only protection against such infectious diseases is a ban on the importation of all fruits and plant material that can carry plant pathogens.

Crown Gall Disease

Agrobacterium tumefaciens (too-meh-faysh′ee-enz) is a soil organism that infects wounded or damaged flowering plants and causes crown gall disease. A crown gall is a plant tumor composed of dividing plant cells (Figure 14.11). *Agrobacterium tumefaciens* enters the plant tissue via a wound, grows, and then spreads between the plant cells. Plant cells adjacent to the wound begin to proliferate and form a crown gall tumor. When the plant dies, the bacteria return to the soil where they overwinter before infecting the next generation of plants.

Bacteria that cause crown gall tumors carry the tumor-inducing (Ti) plasmid. The T portion of this plasmid is transferred to the nucleus of susceptible plant cells, where it is integrated into a plant chromosome. The presence of the T region of the plasmid causes the plant cell to divide continuously and independently of plant growth hormones. A discussion of how the Ti plasmid can be used to transfer genes between plants is presented in Chapter 30.

Other species of *Agrobacterium* infect plants via wounded tissues and cause crown gall, hairy root, or cane gall disease. These bacteria are motile, gram-negative, aerobic rods that live in soil. They are classified in this genus because of their ability to cause tumorous

FIGURE 14.11 Crown gall disease caused by *Agrobacterium tumefaciens* infection of dicotyledons. (Courtesy of Dr. Milton Gordon.)

plant diseases. Phytopathogenic* bacteria belonging to the genera *Pseudomonas, Corynebacterium,* and as we have discussed, *Rhizobium,* also cause plant hyperplasia (abnormal growth).

AEROBIC OXIDATIVE METABOLISM

Obligate aerobes gain energy by oxidative mechanisms and use oxygen as the terminal electron acceptor. The obligate areobes can be subdivided based on their substrates: the chemolithotrophs oxidize inorganic compounds as sources of energy; the chemoorganotrophs gain energy from the oxidation of organic compounds. Chemoorganotrophic bacteria play a major role in cycling organic carbon in nature. The special properties of a selected group of chemoorganotrophs are discussed below.

*Causing disease or abnormal growth in plants.

PSEUDOMONADS

The pseudomonads are a very large group of gram-negative, unicellular, motile rods that gain energy by respiratory metabolism. They are noted for their ability to oxidize a wide variety of organic compounds and for their inability to fix N_2. The pseudomonads are found in soil, fresh water, seawater, and assorted foods, fruits, flowers, and vegetables. Most are nonpathogenic, but some *Psuedomonas* cause disease in plants, humans, or domestic animals.

Phytopathogenic Pseudomonads

Phytopathogenic species of *Pseudomonas* infect plants to cause galls or scabs, rots, cankers, blights, or wilt diseases. Some are specific for the species of plants they infect, whereas others infect a wide variety of host plants. For example *Ps. solanacearum* (so-lah-nah-see-ah′rum) causes wilt in cultivated plants, including potato, tomato, tobacco, and peanut plants; and the pathovars of *Ps. syringae* (sy-rin′geye) are pathogenic to many unrelated plants. The phytopathogenic pseudomonads can be transmitted through infected soil and a few are transmitted by insects.

Nonpathogenic Pseudomonads

Soil, fresh water, salt water, and foods are major habitats for the nonpathogenic pseudomonads. As a group they oxidize a great variety of organic compounds, and some species can grow on over 100 organic compounds. The types of organic substrates they use vary from the long-chain hydrocarbons found in oil spills to ring-containing hydrocarbons, sugars, organic acids, and amino acids. Many of the enzymes involved in the metabolism of these compounds are encoded by plasmids. Although the pseudomonads are never fermentative, some species can grow by anaerobic respiration using nitrate or nitrite as a terminal electron acceptor.

Pseudomonas aeruginosa (ai-roo-jin-oh'sah) and *Ps. stutzeri* (stut'ze-reye) are denitrifiers that produce gaseous forms of nitrogen, including N_2, when they grow anaerobically on nitrite.

One group of pseudomonads is identified by the water-soluble fluorescent pigments they produce. *Pseudomonas aeruginosa* produces the pigments pyoverdin and pyocyanin, while *Ps. fluorescence* (flu-o-res'cens) produces only pyoverdin. The fluorescent pigments produced by pseudomonads are important in their classification.

Pseudomonads play a major role in spoilage of dairy products, meat, poultry, and eggs. They are especially important when the spoilage occurs at low temperatures. Growth of *Ps. fluorescens* on food stored at low temperatures is often associated with the emission of a sulfide-like odor.

ACETIC ACID BACTERIA

Acetic acid bacteria gain energy by oxidizing ethanol to acetic acid. They grow in slightly acid environments containing sugars and/or ethanol such as beer, wine, cider, souring fruit juices, honey, flowers, and fruits. The acetic acid bacteria occur naturally in the fermenting juices of pressed fruits and are involved in the production of vinegar from wine and cider.

Vinegar is a sour liquid consisting of dilute acetic acid that is manufactured by a microbial process using the acetic acid bacteria. Species of *Acetobacter* (a-ce-toh-bak'ter) and *Gluconobacter* (glu-con-oh-bak'ter) oxidize the ethanol in wine or fermented cider to acetic acid by the following equation:

$$CH_3CH_2OH + H_2O + 2NAD^+ \rightarrow CH_3COOH + 2NADH_2$$

The reduced NAD produced in this reaction is oxidized by the electron transport chain with the formation of ATP. Some acetic acid bacteria further oxidize the acetate to CO_2, others cannot.

Species of *Acetobacter* are gram-negative, acid tolerant, nonmotile or peritrichous flagellated rods that oxidize ethanol to acetic acid. They are called overoxidizers because they oxidize acetic acid to CO_2 and H_2O through their TCA cycle and respiratory chain. In contrast, species of *Gluconobacter* are polarly flagellated (if motile), gram-negative rods that are underoxidizers. These bacteria lack a functioning TCA cycle, so the acetic acid accumulates as the end product of ethanol oxidation. *Acetobacter aceti* (a'ce-tee), *A. pasteurianus* (pas-teur-i-ah'nus), and *Gluconobacter oxydans* (ox'y-dans) are used commercially to produce vinegar.

AEROBIC SPOREFORMING BACILLI

Members of the genus *Bacillus* (bah-cil'lus) are widely dispersed in soil, plant matter, and air. These bacteria are aerobic or facultative anaerobic, gram-positive, motile rods that grow readily on laboratory media. Under adverse conditions *Bacillus* spp. form true endospores that are extremely resistant to heat, desiccation, UV light, and chemical treatment. Subgroups of the *Bacillus* genus have been constructed based on endospore shape (oval, spherical), position of spore in the cell (central, terminal), and the species' ability to ferment sugars. Many species of *Bacillus* can be isolated from soil by plating a pasteurized soil suspension on a selective medium. The spores survive this heat treatment and grow into colonies on aerobic agar plates. These bacteria grow on a wide range of organic compounds including amino acids, organic acids, and sugars. Most *Bacillus* spp. are mesophilic; however, *Bacillus stearothermophilus* (ste-a-roh-ther-moh'phi-lus) is a thermophile that grows at 65°C.

Bacillus subtilis (sub'ti-lis) and *Bacillus polymyxa* (po-ly-my'xa) produce the peptide antibiotics bacitracin and polymyxin, respectively. Polymyxin destroys the cytoplasmic membrane of other bacteria, whereas bacitracin prevents bacterial cell-wall synthesis. The two important human pathogens in this genus are *B. cereus* (seer'ee-us), which causes gastroenteritis via contaminated food, and *B. anthracis,* which causes anthrax of sheep, cattle, and sometimes humans. Other species of *Bacillus* are pathogenic to insects.

Bacillus thuringiensis (thur-in-jee-en'sis) effectively kills the caterpillars of many moths and butterflies and the larvae of blackflies and mosquitoes. This bacterium produces a bipyramidal crystalline inclusion body during sporulation that is composed of a protein insect toxin. After ingestion, the protein toxin is solubilized by the alkaline environment in the insect's gut. The toxin then breaks down the gut epithelium, which permits the alkaline contents of the gut to spread into the body of the insect, causing instant paralysis. Insects that ingest sporulating *B. thuringiensis* cells present on plant material die within a short time. Commercial preparations of sporulating cells and of the isolated protein toxin are used in the biological control of insects.

BDELLOVIBRIO: THE PARASITIC BACTERIA

Bdellovibrio (del-loh-vib′ree-oh) are an integral component of the natural microbial flora of aerobic environments where they grow as parasites of gram-negative bacteria. The *Bdellovibrio* are the smallest known predators. These small, curved, gram-negative rods penetrate their host's cell wall and complete their complex life cycle in the host's periplasmic space. Each cell possesses a single polar flagellum that is sheathed by an extension of its cytoplasmic membrane (Figure 14.12). This flagellum propels the bdellovibrios rapidly through liquid media in search of host cells. Bdellovibrios infect only gram-negative bacteria—no bdellovibrio has been found that infects a gram-positive bacterium or cyanobacterium. Once the bdellovibrio smashes into a suitable living host cell, it drills a hole in the host's cell wall with the help of its cell wall-digesting enzymes. Eventually it works its way into the host's periplasmic space where its life cycle begins (Figure 14.13).

Infection of a host cell by bdellovibrio kills the host. Soon after cell wall penetration occurs, the host cell becomes immobile and loses its ability to perform essential metabolic activities. The bdellovibrios produce enzymes that break down the host's cytoplasmic components into the nutrients it utilizes for growth.

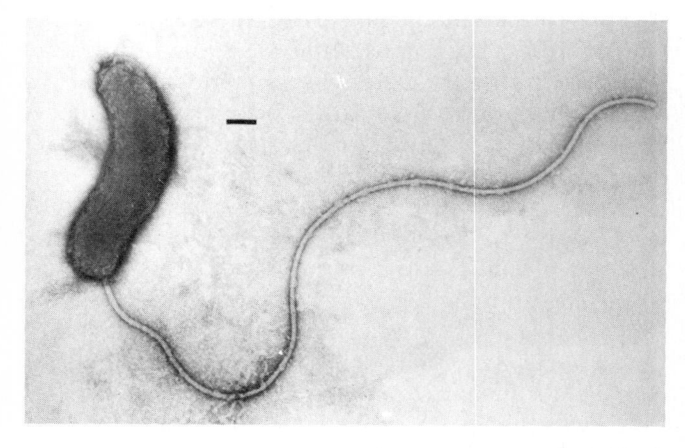

FIGURE 14.12 Electron micrograph of a *Bdellovibrio bacteriovorus* showing its large sheathed polar flagellum. (Courtesy of Dr. Jeffrey C. Burnham, Medical College of Ohio.)

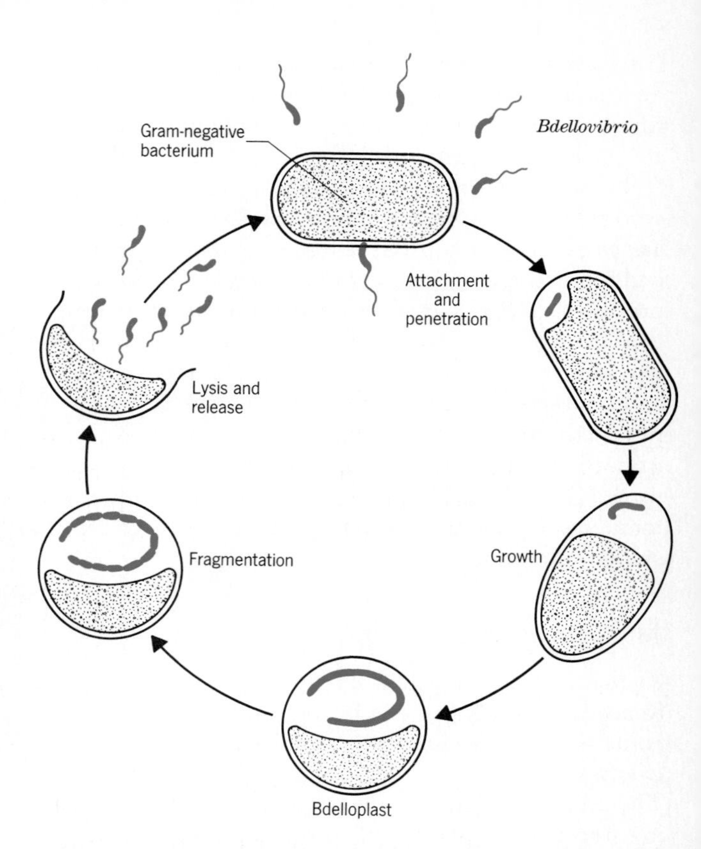

FIGURE 14.13 Development cycle of *Bdellovibrio bacteriovorus.* Motile *Bdellovibrio* penetrates the prey cell wall to enter its periplasmic space. Here the *Bdellovibrio* grows by cell elongation. Progeny *Bdellovibrio* are formed by fragmentation and released when the prey cell lyses.

The entire replicative cycle of *Bdellovibrio bacteriovorus* (bak-tee-ri-oh-vo′rus) occurs within the host cell's periplasmic space (Figure 14.13). After infection, the rod-shaped host cells become an osmotically stable spherical **bdelloplast.** The host's spherical shape enlarges the periplasmic space, which is now occupied by the growing predator. The *Bdellovibrio* cell elongates during growth to become up to 20 times longer than it was at the time of infection. Each elongated bdellovibrio eventually fragments into motile progeny cells, which are released when the host cell lyses.

Bdellovibrios can be isolated from natural environments, such as sewage and polluted waters, by the plaque isolation technique developed for enumerating bacteriophages (Chapter 11). All newly isolated bdellovibrios are dependent on a host cell for multiplication; however, saprophytic (growing on nonliving media) strains have been isolated from laboratory cultures and grown on complex media. Saprophytic strains appear to arise as variants within the larger population.

KEY POINT

Bdellovibrios are similar to bacteriophages because both replicate in living host cells, use host-cell nutrients for replication, and eventually kill their host. They are different in that bdellovibrios possess their own metabolic machinery for protein synthesis and some can replicate outside the host cell, whereas bacteriophages replicate only in living host cells.

Bdellovibrios that grow on eucaryotic cells have not been reported; however, a bdellovibrio-like bacterium has been identified that is parasitic on *Chlorella*. *Vampirovibrio chlorellavorus* (vam-pi-ro-vib′ri-oh chlo-rel-lah′vo-rus) attaches to the surface of the alga *Chlorella* and kills it, even though the parasite does not penetrate the algal cell wall.

GLIDING BACTERIA

Gliding bacteria move on the surfaces of decaying vegetable matter, dung, or agar plates by an unknown mechanism. Unlike the flagellated bacteria and the spirochetes, the gliding bacteria have no distinctive cytological features that explain their motility. Gliding is observed in a number of major groups, including the cyanobacteria, myxobacteria (myx-oh-bak-tee′ri-ah), and cytophagas (cy-toph′agah). The myxobacteria and the cytophagas are noted for their ability to grow on insoluble substrates, such as the cellulose in decaying plant material, chitin (ki′tin), and dead bacteria. Although both groups of bacteria have gliding motility, the myxobacteria and the cytophagas are phylogenetically unrelated based on moles percent G + C values (see Table 13.4).

FIGURE 14.14 Scanning electron micrograph of fruiting bodies of *Chondromyces crocatus* (magnification 820×). (Courtesy of P. L. Grilione. From P. L. Grilione and J. Pangborn, *J. Bacteriol.*, 124:1558-1565, 1975, with permission of the American Society for Microbiology.)

MYXOBACTERIA

Myxobacteria are unicellular gliding bacteria that grow on decaying vegetable matter and on the dead bacteria in the dung of animal herbivores, such as rabbits, hares, and moose. When they encounter unfavorable environmental conditions, they form fruiting bodies that contain the resting form of the bacterium called a myxospore (Figure 14.14).

Myxospores are highly resistant to desiccation, ultrasound, and UV irradiation, and they are slightly more resistant to heat (58° to 62°C for 10 to 60 minutes) than the vegetative cells from which they form. The fruiting body is a communal aggregation of bacteria in which individual cells assume different functions. The structure of the fruiting body varies greatly among species, from complex structures to relatively simple aggregates of cells. The more complex fruiting bodies possess **sporangioles** (spor-an′je-oles) filled with myxospores supported by a substantial stock (see Color Plate I).

Fruiting body formation is controlled by nutrient availability, pH, divalent cations, and temperature. Once induction has occurred, the cells in the commu-

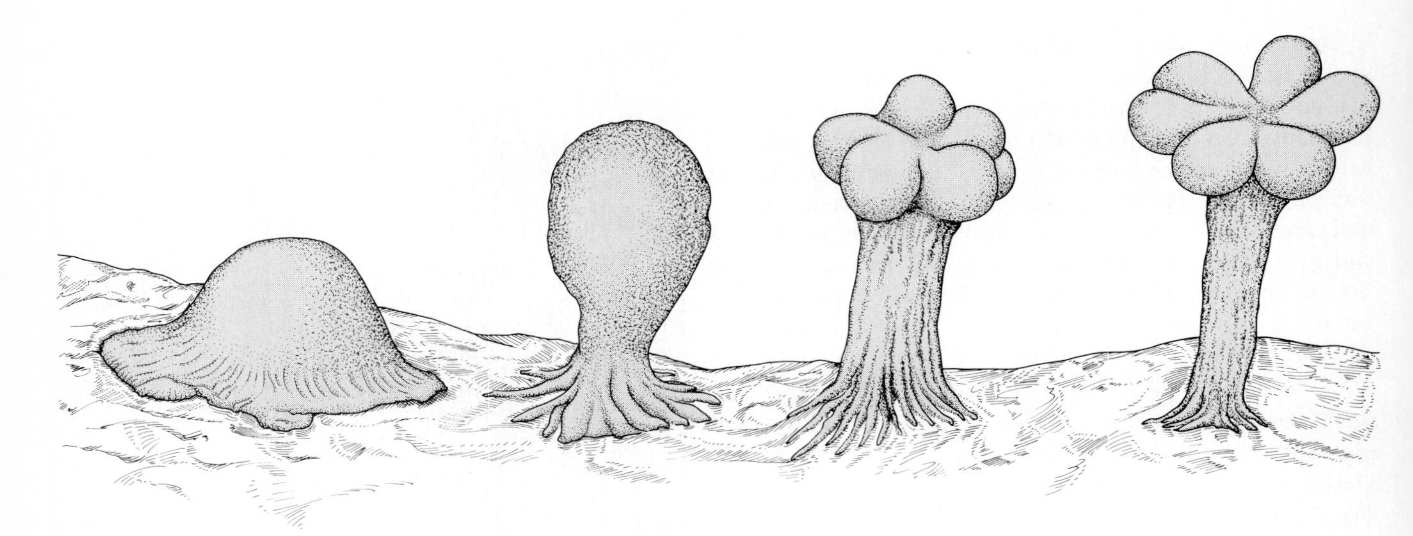

FIGURE 14.15 Development of the *Chondromyces* fruiting body.

nity aggregate, probably via a chemotactic mechanism. The cells forming the complex fruiting body rearrange themselves in clusters and produce slime and other structural components of the fruiting body (Figure 14.15). This stalk supports the myxospores formed by the cells at the top of the fruiting body.

Myxospore formation by *Myxococcus xanthus* (xan'thus) is stimulated by the presence of 0.5M glycerol or starvation for specific amino acids (tryptophan, or phenylalanine and tyrosine). Under these conditions, the long flexible vegetative rods of *M. xanthus* shorten into plump rods and then into spherical cells. The spherical cells become refractile as they make the final transition into myxospores. Myxospores exposed to favorable environmental conditions germinate into growing vegetative cells.

Myxobacteria play an essential role in cycling bioorganic polymers that cannot be attacked by other bacteria. They have been considered micropredators of bacteria and yeasts because of their role in breaking down the remains of dead cells. Myxobacteria grow only on dead bacteria and thus are different from the bdellovibrios.

CYTOPHAGAS

The cytophagas are gliding bacteria that do not form fruiting bodies or other complex morphological structures. They are gram-negative, usually long slender rods. Most species are aerobic; however, there are species that are facultative anaerobes. The cytophagas are found in fresh water and seawater, on the dung of herbivores, and in rich organic soils. Many of the aerobic cytophagas are able to decompose biopolymers, including agar, cellulose, chitin, pectin, keratin, and proteins.

Although most cytophagas are nonpathogenic, some, such as *Cytophaga columnaris,* are pathogens of freshwater and salt-water fish. Other pathogens of this group are the *Capnocytophaga* (kap'noh-cy-toph'agah) species found under the gums (gingival crevice) in the human mouth.

CHEMOLITHOTROPHS

The chemolithotrophs are bacteria that gain energy for cellular biosynthesis and maintenance from the aerobic oxidation of inorganic compounds. Many of the chemolithotrophs are also **autotrophs,** which means that they use carbon dioxide as their source of carbon. The chemolithotrophic autotrophs are unique because they can grow in the dark in completely inorganic media. Chemolithotrophs that can use either carbon dioxide or organic compounds as carbon sources are known as **mixotrophs.**

The classification of the chemolithotrophs is based on the substrates they oxidize, even though this physiological approach results in groups of bacteria with diverse morphologies. Chemolithotrophs are gram-negative, mostly aerobic rods, cocci, spirilla, or lobular organisms found in soil and aquatic environments. Selective enrichments for these bacteria utilize inorganic media containing the substrate the organism oxidizes for energy. The substrates oxidized by the chemolithotrophs include ammonia (NH_3), nitrite (NO_2^-), reduced sulfur compounds (H_2S, S_0, $S_2O_3^{2-}$), hydrogen (H_2), and ferrous iron (Fe^{2+}). Chemolithotrophs play an essential role in the natural cycling of inorganic compounds as described in Chapter 28.

THE NITRIFYING BACTERIA

Nitrification, an essential process of the nitrogen cycle, is the two-step bacterial oxidation of ammonia to nitrate by two distinct groups of bacteria. In the first step, the ammonia-oxidizing bacteria oxidize ammonia to nitrite. The second stage of nitrification is the oxidation of nitrite to nitrate by the nitrite-oxidizing bacteria.

Ammonia oxidizers include members of the genera *Nitrosomonas* (ni-tro-soh-moh'nas), *Nitrosococcus* (ni-troh-soh-kok'us), *Nitrosospira* (ni-troh-soh-spi'rah), and *Nitrosolobus* (ni-troh-soh-lob'us). These bacteria are difficult to study because of their long generation times, 8 to 24 hours, and their poor growth yields (0.5 gm per day in a 20-liter culture). During the same growth period one would expect to obtain 300 times more cell mass when growing a heterotroph. Nevertheless, they are important in natural environments and have been isolated from soil, compost, sewage, fresh water, and seawater.

The ammonia oxidizers are a morphologically diverse group of bacteria (Figure 14.16) that share the ability to gain energy from the oxidation of ammonia. These bacteria are rich in cytochromes and many have extensive intracytoplasmic membranes involved in generating ATP from ammonia oxidation. The extensive cytomembrane system of *Nitrosococcus oceanus* is of special interest because it appears to divide when the cell divides (Figure 14.16). The lobular external surface and the division of its cytoplasm into compartments by membranes makes *Nitrosolobus multiformis* (mul-ti-

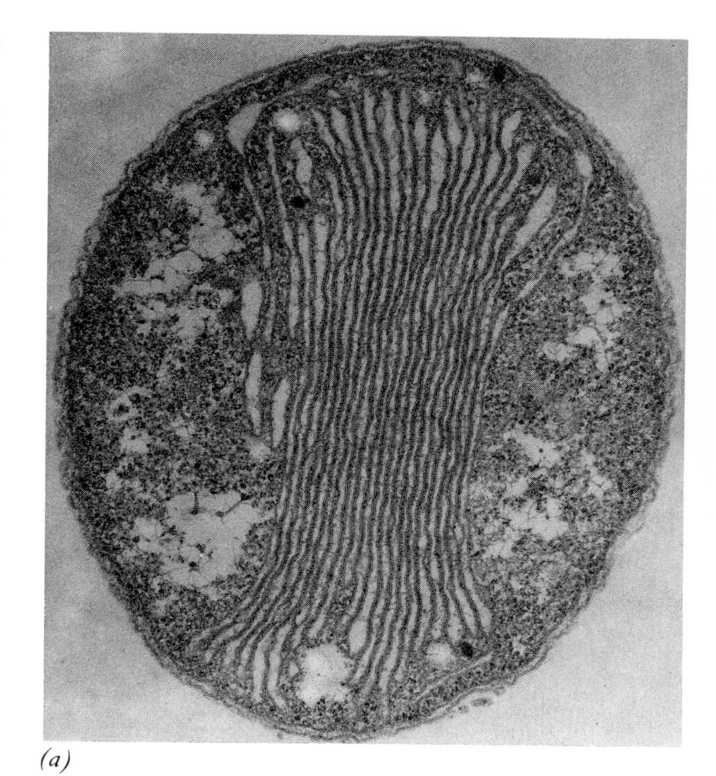
(*a*)

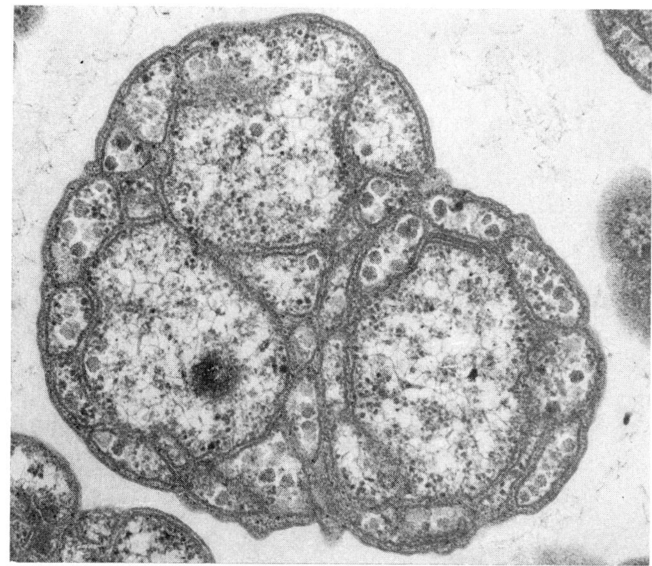
(*b*)

FIGURE 14.16 Representatives of the nitrifying bacteria that oxidize ammonia. Electron micrographs of (*a*) *Nitrosococcus oceanus*, and (*b*) *Nitrosolobus multiformis*, that show the morphologically different intracytoplasmic membranes. (Courtesy of Dr. Stanley W. Watson.)

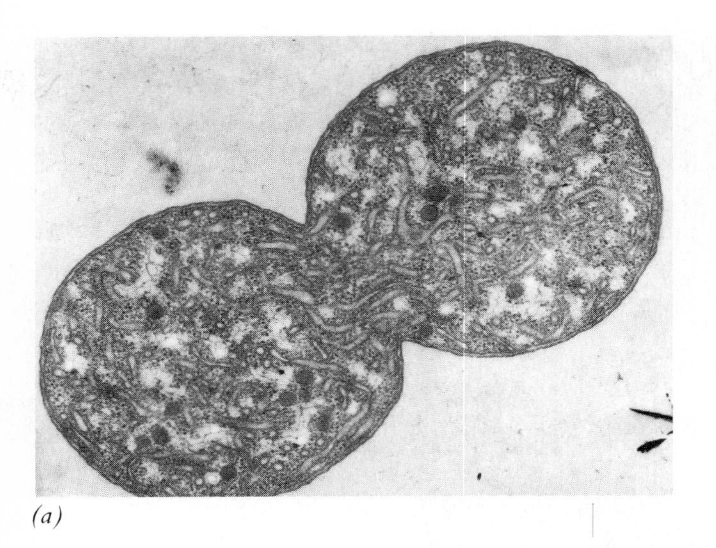

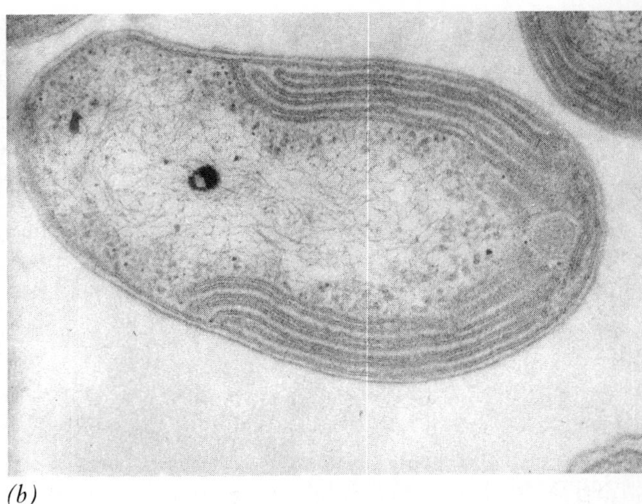

(a) *(b)*

FIGURE 14.17 Representatives of the nitrifying bacteria that oxidize nitrite to nitrate. Electron micrographs of *(a) Nitrococcus mobilis,* and *(b) Nitrobacter winogradskyi*. (Courtesy of Dr. Stanley W. Watson.)

for'mus) even more unusual (Figure 14.16). Although most ammonia oxidizers have intracytoplasmic membranes, such structures are not essential for ammonia oxidation since *Nitrosospira briensis* (bry-en'sis) has no apparent cytomembrane system.

The nitrite produced by the ammonia oxidizers is oxidized to nitrate by nitrite-oxidizing bacteria. Members of the genera *Nitrobacter, Nitrospina* (ni-troh-spi'nah), and *Nitrococcus* are found in the same habitats as ammonia oxidizers. These bacteria also have cytomembrane systems. *Nitrobacter winogradskyi* (vi-no-grad'sky-eye) has a polar cap of cytomembranes, whereas the membranes of *Nitrococcus mobilis* are tubular structures that permeate its cytoplasm (Figure 14.17). *Nitrospina gracilis* (grah'ci-lis) is a long slender rod that lacks an extensive cytomembrane system. The discovery of more nitrifying bacteria awaits patient investigators who are willing to wait the incubation period of 6 months or longer that is required to isolate them.

COLORLESS SULFUR BACTERIA

The colorless sulfur bacteria gain energy by oxidizing reduced forms of sulfur, eventually to sulfate. They are called colorless to differentiate them from the purple and green photosynthetic sulfur bacteria (see below).

Environmental sources of reduced sulfur come from the decomposition of sulfur-containing aminoacids and from the activities of sulfur-reducing bacteria such as the desulfovibrios. Much of this production occurs in anaerobic environments and results in the production of hydrogen sulfide (H_2S), which is a major substrate for sulfur-oxidizing bacteria. Other reduced forms of sulfur oxidized by these bacteria include molecular sulfur (S) and thiosulfate ($S_2O_3^{2-}$).

Chemolithotrophic Sulfur Oxidizers

Sulfur-oxidizing bacteria are found in a wide variety of environments where sulfur is present. They are present in the black mud of rivers and marine sediments, acid sulfate soils, hot-acid springs, and effluents from mines. The sulfur-oxidizing bacteria belonging to the genera *Thiobacillus* (thi'oh-bah-cil'us), *Thiomicrospira* (thi'oh-my-kroh-spi'rah), and *Sulfolobus* (sulf-oh-lo'bus) can gain all their energy from the oxidation of reduced sulfur compounds.

Species of *Thiobacillus* are small motile rods that grow over a wide range of pH by oxidizing reduced sulfur compounds and producing sulfuric acid (H_2SO_4) as a major product. In high concentrations, sulfuric acid is very corrosive and contributes to the acidification of ponds, lakes, rivers, and streams. This environmental problem is especially important in mining regions where pyrite (FeS_2) is found in the ore or coal

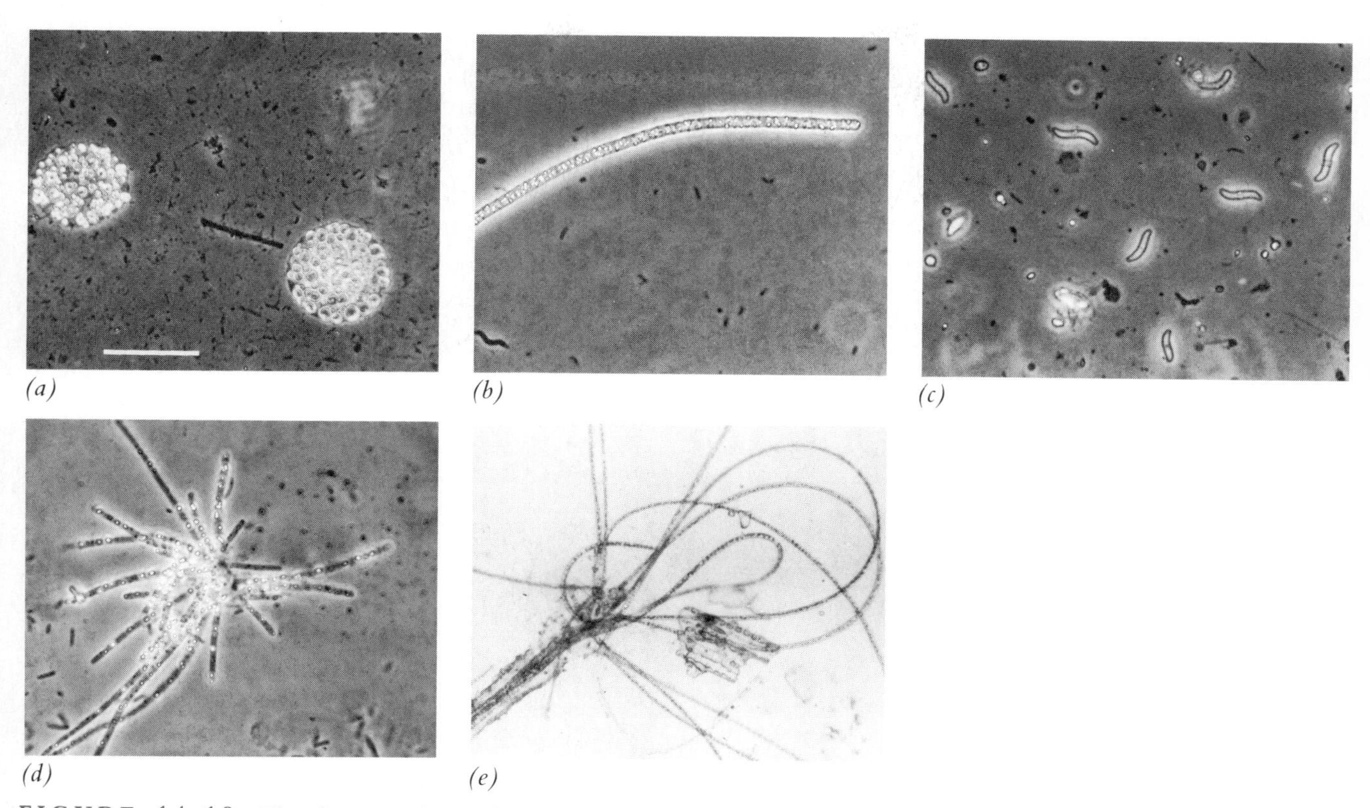

FIGURE 14.18 The chemotrophic sulfur-oxidizing bacteria. Photomicrographs of *(a) Achromatium volutans* packed with sulfur granules (bar = 20 μm), *(b)* a filamentous marine *Beggiatoa, (c)* the marine sulfur bacterium, *Thiospira bipunctata, (d) Thiothrix* isolated from activated sludge, and *(e)* cells of *Thioploca ingrica* emerging from the open end of its sheath. Note the different morphology of these bacteria and the refractile sulfur granules visible in the cells. (Courtesy of *(a)* Dr. K. Schmidt. From J. W. M. la Riviere and K. Schmidt, "Morphologically Conspicuous Sulfur-Oxidizing Eubacteria," in M. P. Starr et al. (eds.), *The Prokaryotes,* Springer-Verlag, Heidelberg, 1981. Courtesy of *(b, c),* Dr. Paul W. Johnson, *(d)* Dr. Michael Richard, *(e)* Dr. S. Maier.)

being mined. Pyrite is exposed to the activities of *Thiobacillus ferrooxidans* (fer-roh-ox'i-dans) and oxygen after the topsoil and overburden are removed. This bacterium then oxidizes the pyrite to ferric iron (Fe^{+3}) and sulfuric acid. The iron and sulfuric acid leach into the waterways and help to create the brown acid-mine waters of streams in coal-mining regions.

The production of strong acids by the colorless sulfur oxidizers also has beneficial effects. For example, sulfur is commonly added to basic soils to alleviate the high pH. The sulfur is then oxidized to sulfuric acid, which neutralizes the basic constituents present.

Chemotrophic Sulfur Oxidizers

A number of bacteria oxidize sulfur compounds as a source of energy while using organic sources of carbon. They are often found in microaerobic environments between an anaerobic source of H_2S and an aerobic environment. Many of the chemotrophic sulfur oxidizers have distinctive morphological characteristics that enable them to be identified under the microscope.

Achromatium (ay-kro-mah'ti-um) are very large (5 × 20 to 100 μm) spherical- to oval-shaped cocci (Figure 14.18) found in both freshwater and marine environments where H_2S is produced. These cells are full of sulfur granules and are easy to identify microscopically even though they have never been grown in pure culture. *Thiospira bipunctata* (by-punc-tah'tah) and *Macromonas mobilis* are two large colorless sulfur-oxidizing bacteria. They both contain sulfur granules and are motile by polar appendages that appear to be tufts of flagella. Under appropriate conditions, these bacteria can be maintained in the laboratory in mud enrichment cultures, but they have not been isolated in pure culture. Other chemotrophic sulfur oxidizers include *Thiobacterium, Thiovulum* (thi-oh'vu-lum), *Thiothrix* (thi'oh-thrix), *Thioploca* (thi-oh-plo'cah), and *Beggiatoa* (Figure 14.18).

PHOTOSYNTHETIC BACTERIA

Photosynthesis is the biological process by which light energy is converted to NADH and chemical energy in the form of ATP. The photosynthetic bacteria use this energy to synthesize carbohydrates from CO_2 according to the generalized equation for photosynthesis (Chapter 6). The primary light reaction of photosynthesis occurs in pigmented complexes situated in or on photosynthetic cytomembranes. Photosynthetic bacteria are classified according to their photosynthetic pigments, the electron donors they use, and their source of carbon (Table 14.3).

Anoxygenic photosynthetic bacteria use reduced sulfur or organic compounds as their photosynthetic electron donors. They photosynthesize only under anaerobic conditions and do not produce oxygen. The green and purple photosynthetic bacteria gain energy by anoxygenic photosynthesis. In contrast, **oxygenic** photosynthetic bacteria use water as their photosynthetic electron donor, have chlorophyll *a,* and generate oxygen during photosynthesis. Cyanobacteria are the largest group of oxygenic photosynthetic bacteria. A second type of oxygenic photosynthesis has recently been discovered in bacteria growing in symbiosis with didemnids, which are tunicates (invertebrate animals) found in tropical waters. *Prochloron didemni* is like an alga because it possess both chlorophyll *a* and *b;* however, since it lacks a nucleus it is classified as a bacterium.

KEY POINT

Oxygenic photosynthetic bacteria generate oxygen by splitting water in the primary reaction of photosynthesis. Anoxygenic photosynthetic bacteria cannot split water; instead, reduced sulfur or organic compounds donate electrons to the primary reaction of photosynthesis and no oxygen is generated.

ANOXYGENIC PHOTOSYNTHETIC BACTERIA

The photosynthetic bacteria that grow under anaerobic conditions are organized into four groups based on their coloration and primary photosynthetic electron donors.

TABLE 14.3 Characteristics of photosynthetic bacteria

	TAXONOMIC GROUP	PHOTOSYNTHETIC PIGMENTS	CARBON SOURCE(s)	PHOTOSYNTHETIC ELECTRON DONORS
Anoxygenic photosynthesis	Purple bacteria	Bacteriochlorophyll *a* and *b*	Organic and/or Co$_2$	H_2, H_2S, S, organic compounds
	Green bacteria	Bacteriochlorophyll *c, d,* or *e* and small amounts of *a*	CO_2	H_2, H_2S, S
Oxygenic photosynthesis	Cyanobacteria	Chlorophyll *a,* phycobiliproteins	CO_2	H_2O
	Prochlorophytes	Chlorophyll *a* and *b*	CO_2	H_2O

TABLE 14.4 Characteristics of the anoxygenic photosynthetic bacteria

DESCRIPTION (FAMILY)	REPRESENTATIVES	CHARACTERISTICS
Purple nonsulfur *(Rhodospirillaceae)*	*Rhodomicrobium, Rhodopseudomonas, Rhodospirillum*	Photoorganotrophs that use organic acids, amino acids, benzoate, and ethanol (anaerobic). Also grow as chemoorganotrophs in O_2.
Purple sulfur *(Chromatiaceae)*	*Chromatium, Ectothiorhodospira, Thiocapsa, Thiospirillum*	Photolithotrophs that use H_2S and S and fix CO_2 (anaerobic).
Green sulfur *(Chlorobiaceae)*	*Chlorobium, Pelodictyon*	Photolithotrophs that use H_2S and S and fix CO_2 (anaerobic). Will use simple organic compounds if H_2S is present (anaerobic).
Green flexibacteria *(Chloroflexaceae)*	*Chloroflexus*	Photoorganotrophs that use sugars, organic acids, or amino acids (anaerobic). Also grow as chemoorganotrophs in O_2.

Purple Nonsulfur Bacteria

The purple nonsulfur bacteria grow photoorganotrophically on a wide variety of organic compounds (Table 14.4). Organic acids, amino acids, benzoate, ethanol, and some sugars serve both as electron donors for photosynthesis and as sources of carbon. These bacteria grow photosynthetically under anaerobic conditions in the presence of light, or they grow chemoorganotrophically when oxygen is available. In the presence of oxygen they gradually lose their pigmentation since oxygen prevents the synthesis of the photosynthetic apparatus. The pigment containing photosynthetic membranes are again formed when the bacteria are placed under anaerobic conditions.

Rhodospirillum rubrum (rho-doh-spi-ril'lum rub'rum) is a photosynthetic spirillum found in environments rich in organic matter. It produces photosynthetic vesicles that arise as invaginations of the cytoplasmic membrane during anaerobic growth (see Figure 4.10*b*). These vesicles contain bacteriochlorophyll *a* and the carotenoid pigments involved in the light reactions of photosynthesis. Oxygen prevents the synthesis of both the photosynthetic pigments and photosynthetic vesicles, whereas anaerobiosis in the light or dark stimulates pigment and vesicle formation. Other purple nonsulfur photosynthetic bacteria include *Rhodomicrobium vannielii* (van-niel'ee-eye) and *Rhodopseudomonas palustris* (pa-lus'tris), which can divide by budding, and *Rhodocyclus purpureus* (rho-doh-cy'clus pur-pur'ee-us), which is an immobile, semicircular rod (Figure 14.19).

Purple Sulfur Bacteria

The purple sulfur bacteria are photolithotrophs that use reduced sulfur compounds as electron donors for photosynthetic growth. They oxidize hydrogen sulfide to molecular sulfur, which they deposit in granules. After all the hydrogen sulfide is oxidized to molecular sulfur, the molecular sulfur is oxidized to sulfate. Examples of purple sulfur bacteria include *Chromatium okenii* (kro-mah'ti-um oh-ken'ee-eye) and *Chromatium vinosum* (vi-noh'sum), two very large motile cells that often contain sulfur granules. The photosynthetic membranes of these bacteria are intracytoplasmic vesicles that contain bacteriochlorophyll *a*.

Species of *Ectothiorhodospira* (ec'toh-thi-oh-rhoh-doh-spi'rah) deposit sulfur outside their cells and possess a lamellar photomembrane system (Figure 14.20). Some species of *Ectothiorhodospira* are extreme halophiles that live in concentrated brines. The name of this genus is descriptive of its characteristics—a red spirillum (rhodospira) that deposits sulfur (thio) extracellularly (ecto).

Green Sulfur Bacteria

The green sulfur bacteria are obligate phototrophs that have bacteriochlorophyll *c, d,* or *e* as their predominant bacteriochlorophyll and small amounts of bacteriochlorophyll *a*. These anaerobic, nonmotile bacteria all have rigid cell walls and use reduced-sulfur compounds as phototrophic electron donors. Some are able to use

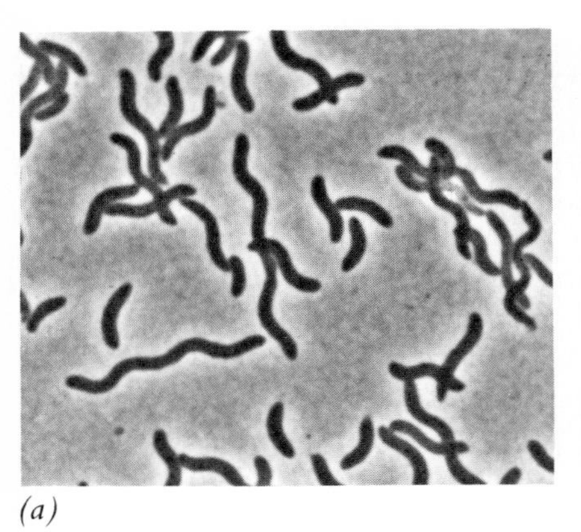

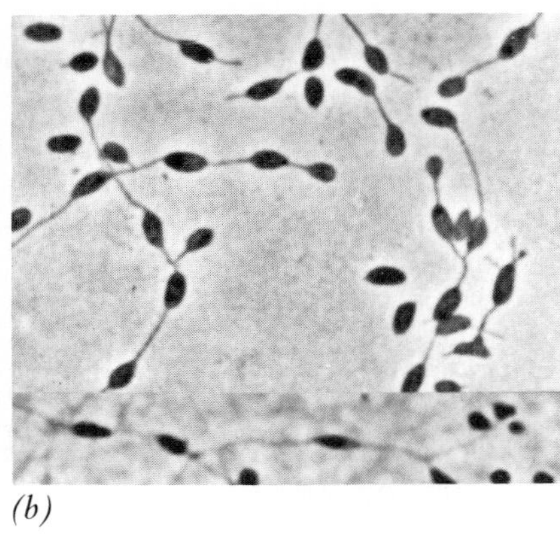

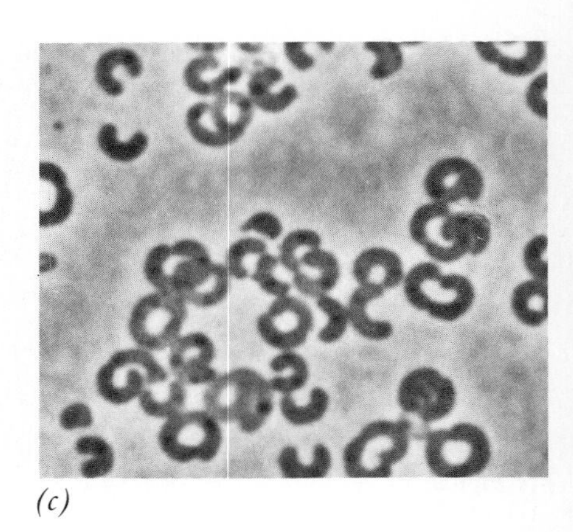

(a) *(b)* *(c)*

FIGURE 14.19 Morphological variations among the purple nonsulfur photosynthetic bacteria are *(a)* the spiral-shaped *Rhodospirillum rubrum, (b)* the budding *Rhodomicrobium vannielii,* and *(c)* curved cells of *Rhodocyclus purpureus.* (From H. Truper and N. Pfennig, "Characterization and Identification of the Anoxygenic Phototrophic Bacteria," in M. P. Starr et al. (eds.), *The Prokaryotes,* Springer-Verlag, Heidelberg, 1981.)

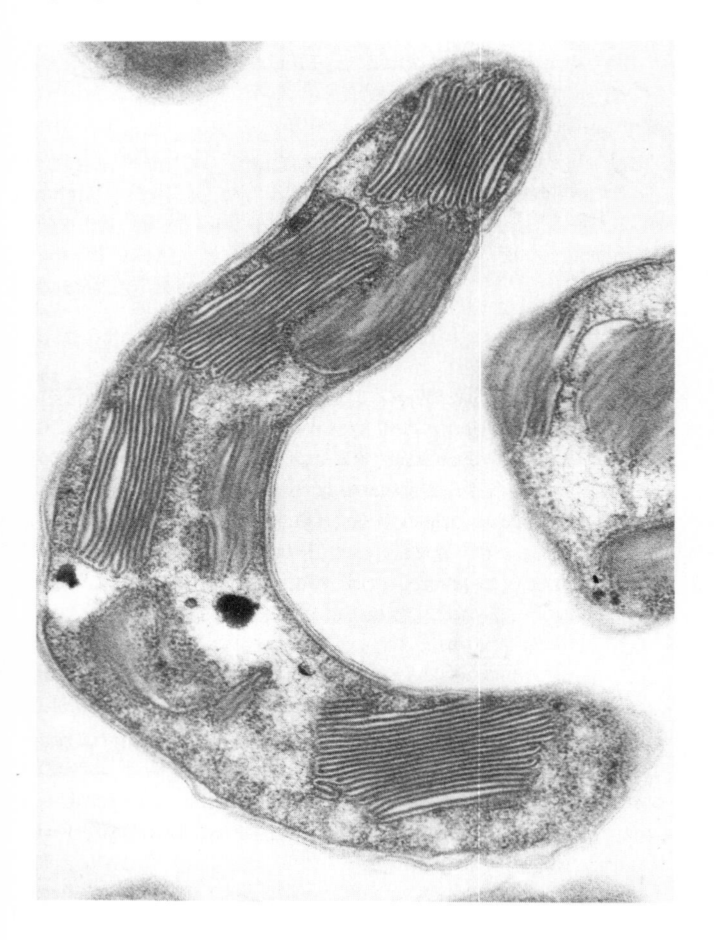

FIGURE 14.20 Electron micrograph of *Ectothiorhodospira mobilis* showing its lamellar photosynthetic membrane system. (Courtesy of Dr. Stanley W. Watson.)

simple organic compounds as electron donors provided that reduced sulfur is present as a sulfur source.

The green sulfur bacteria are morphologically diverse. *Chlorobium limicola* (chlo-roh'bi-um li-mi'coh-lah) is a short rod that deposits sulfur granules in the medium (Figure 14.21). Its photosynthetic apparatus is found in **chlorosomes** that lie just beneath the cytoplasmic membrane and appear to be independent of it. *Pelodictyon clathratiforme* (pe-loh-dic'ty-on clath-rah-ti-for'me) is a planktonic, green sulfur bacterium that forms a three-dimensional lattice network of branching cells. It has cytoplasmic gas vacuoles that enable it to move passively in the water column of lakes and ponds. The green sulfur bacteria live in anaerobic sulfide-containing environments exposed to light such as sulfur springs, the surfaces of mud, and the water column of freshwater ponds and lakes.

Photosynthetic Flexibacteria

Chloroflexus aurantiacus (chlo-roh-flex'us au-ran-ti'ah-cus) is a filamentous gliding photosynthetic bacterium that belongs to the family of green flexibacteria (Table

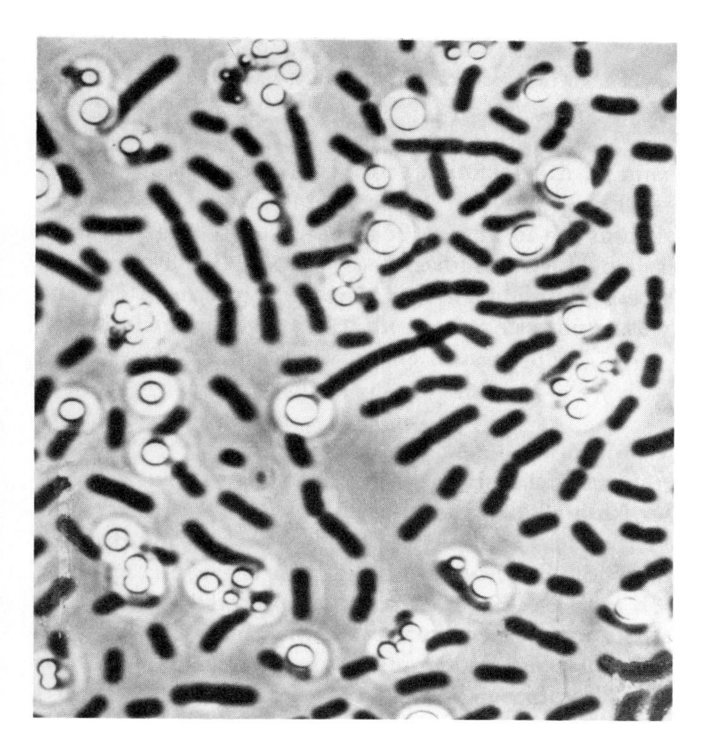

FIGURE 14.21 The green sulfur bacteria *Chlorobium limicola* deposits sulfur granules in the environment outside of the cell. (From H. Truper and N. Pfennig, "Characterization and Identification of the Anoxygenic Phototrophic Bacteria," in M. P. Starr et al. (eds.), *The Prokaryotes,* Springer-Verlag, Heidelberg, 1981.)

14.4). It has both bacteriochlorophyll c_s and small amounts of bacteriochlorophyll *a* that are found in chlorobium-like chlorosomes. *Chloroflexus aurantiacus* grows aerobically as a chemoorganotroph on a number of sugars, organic acids, or amino acids; as a photoheterotroph anaerobically in the light; or slowly as a photolithotroph in media containing H_2S. The photosynthetic metabolism of this organism is analogous to that of the purple nonsulfur bacteria, while its gross morphology makes it similar to the gliding cyanobacteria.

OXYGENIC PHOTOSYNTHETIC BACTERIA

Photosynthetic procaryotes that produce oxygen by splitting water during the primary act of photosynthesis belong to the cyanobacteria or the prochlorophytes (pro-chlo'roh-fytes). The cyanobacteria are by far the largest group of oxygenic photosynthetic bacteria. They occur in virtually every aquatic environment either attached to solid substrates or as truly planktonic forms suspended in the water column. Cyanobacteria can also be terrestrial, some of which form symbiotic relationships with fungi. Approximately 8 percent of known lichens are composed of cyanobacteria living in symbiosis with fungi.

Researchers have discovered a procaryotic photosynthetic organism that grows in symbiosis with tunicates. This bacterium is called a prochlorophyte because it resembles an alga by having both chlorophyll *a* and *b*. *Prochloron didemni* (proh-chlo'ron di-dem'nee) is thought to be a link between the photosynthetic procaryotes and the green algae.

Cyanobacteria

The cyanobacteria have two pigments in their photosynthetic membranes, chlorophyll *a* and phycobiliproteins. These pigments are localized on the photosynthetic membranes that occupy a significant portion of the bacterium's cytoplasm. The cyanobacteria use oxygenic photosynthesis to generate the ATP and reduced-pyridine nucleotides they need for the fixation of carbon dioxide. They depend on carbon dioxide as their carbon source since they have limited abilities to photoassimilate organic carbon.

About one-third of all cyanobacteria are able to fix molecular nitrogen. Many of the nitrogen-fixing cyanobacteria produce thick-walled heterocysts (Figure 14.22) that protect the nitrogenase within from being inactivated by oxygen. Heterocysts lack the photosystem needed to generate oxygen, which helps maintain their cytoplasm anaerobic. Heterocysts, however, are not essential for nitrogen fixation since many species of cyanobacteria fix N_2 even though they do not form heterocysts (see Table 14.2).

Blooms of cyanobacteria in flooded rice paddies are encouraged because their N_2-fixing ability provides reduced nitrogen compounds for the rice plants. Most blooms of cyanobacteria, however, are undesirable because of their negative impact on water quality, especially if the water is used for consumption or recreation.

Cyanobacteria can be classified by their morphological characteristics and their means of reproduction. The members of the genera *Synechococcus* (sen-e-cho-kok'us), *Gloeobacter* (glee-oh-bak'ter), *Gloeothece,* (glee-

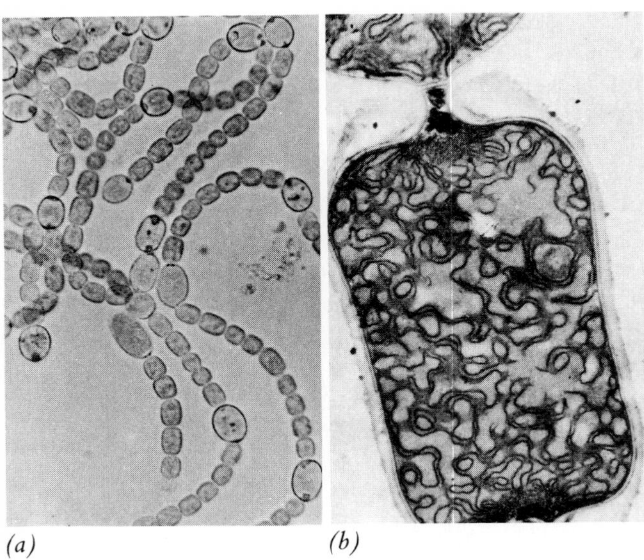

FIGURE 14.22 Ultrastructure of the heterocyst of the cyanobacterium *Anabaena: (a)* filaments of *A. azollae* showing enlarged heterocysts in different stages of development, *(b)* cross section through a heterocyst of *A. cylindrica* showing the contorted thylakoid membranes. (From Weier, et. al., *Botany: An Introduction to Plant Biology,* 6th edition, John Wiley and Sons, 1982.)

oh-thee'ka) and *Gloeocapsa* (glee-oh-cap'sah) are unicellular organisms that divide by binary fission (Figure 14.23). Another group of unicellular cyanobacteria, represented by members of the genera *Xenococcus* (zen-oh-kok'us) and *Pleurocapsa* divide by multiple fission. The filamentous cyanobacteria such as *Oscillatoria* (os-sil-la-tor'i-ah) and *Pseudoanabaena* (soo-doh-an-ah-bean'ah) grow as sheathed or unsheathed vegetative cells, whereas species of *Anabaena* and *Nostoc* form heterocysts when molecular nitrogen is the only available nitrogen source.

ARCHAEBACTERIA

The archaebacteria are so different from the true bacteria (eubacteria) that they are recognized as a separate division in the current classification of the Procaryotae. Microbiologists include the methanogens, and members of the genera *(Halobacterium* (hal-oh-bak-te'ree-um), *Sulfolobus* (sulf-oh-loh'bus), and *Thermoplasma* (ther-moh-plas'mah) in the archaebacteria. These bacteria are phylogenetically related through the conservation of their 16S ribosomal RNA genes. Other similarities: (1) they all lack peptidoglycan in their cell walls, (2) their lipids contain an ether linkage (C—O—C) between the glycerol and the fatty acid residues, (3) they are insensitive to chloramphenicol and kanamycin, and (4) translation is initiated by methionine-tRNA instead of *N*-formyl methionine-tRNA. They are referred to as the archaebacteria because their physiology is amenable to the extreme conditions thought to have existed during the development of life on earth.

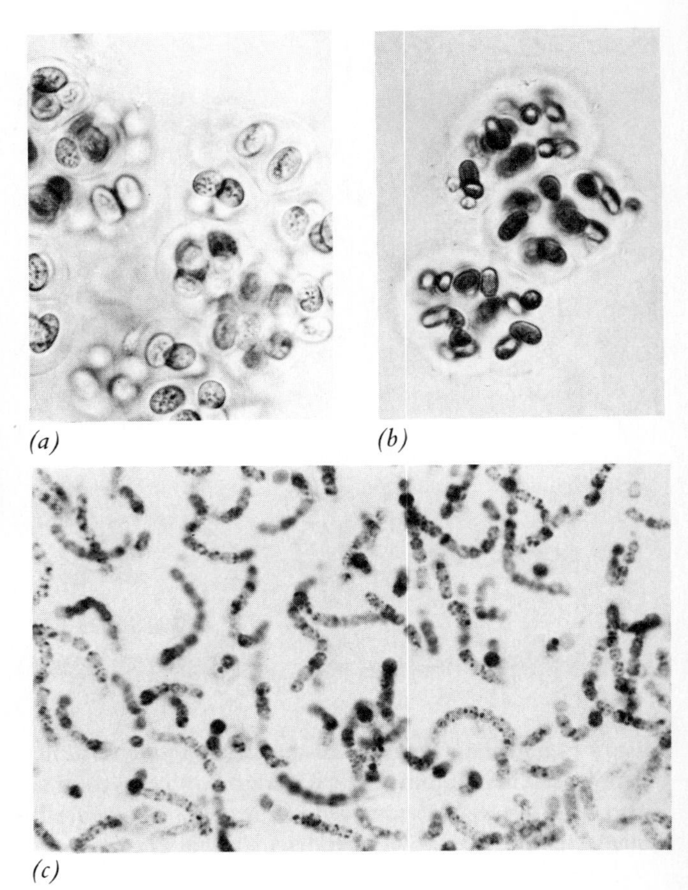

FIGURE 14.23 Diversity in cyanobacteria morphology. *(a)* Pairs of *Synechococcus* cells, *(b)* loosely organized colonies of *Gloeocapsa,* and *(c)* colony of the filamentous *Nostoc.* (From Weier, et. al., *Botany: An Introduction to Plant Biology,* 6th edition, John Wiley and Sons, 1982.)

TABLE 14.5 Selected examples of methanogens

BACTERIUM	MORPHOLOGY	GRAM STAIN	SPECIAL TRAITS
Methanobacterium bryantii	Long rod	+ or −	Nonmotile
Methanobrevibacteria ruminatium	Short rods	+	Requires coenzyme M
Methanomicrobium paynteri	Rods	−	Marine
Methanogenium thermophilicum	Small cocci	−	Grows at 55°C
Methanospirillum hungatei	Wavy filaments	−	Motile
Methanococcus vannielii	Irregular cocci	−	Selenium or tungsten required
Methanosarcina barkeri	Large cocci in packets	+	Metabolic versatility

METHANOGENS

Methane (CH_4) is produced in anaerobic environments from molecular hydrogen and carbon dioxide by bacteria collectively known as the **methanogens** (Figure 14.24). Methanogens can be long or short rods, spirilla, or cocci, and both gram-positive and gram-negative species are represented (Table 14.5). Their cell-wall chemistries are as diverse as their morphologies. Representatives have cell walls composed of different proteins, or of carbohydrate polymers other than peptidoglycan.

Methanogens are found in sewage sludge, aquatic sediments, the rumen of cattle, sheep, and goats, and the human intestine. As rumen bacteria in cattle, they can produce up to 200 liters of methane per animal per day. Most biologically produced methane is lost to the atmosphere; however, methane produced underground can be trapped in geological formations and then tapped as a source of heating gas. Methane can also seep into excavated tunnels where its propensity to explode poses a major hazard. Methane formation in ponds or swamps can be held by inversion layers that form in the air immediately above the water. Ignition of this "swamp gas" has been reported as unidentified flying objects (UFOs).

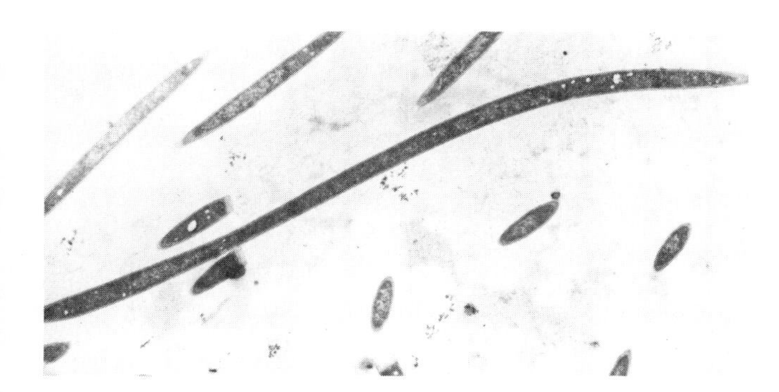

FIGURE 14.24 They wavy filament of the archaebacterium *Methanospirillum hungatei*. (Courtesy of Michigan Biotechnology Institute.)

The methanogens appear to have similar biochemical mechanisms for generating methane. Most methanogens can produce methane from CO_2 and H_2 gas, and all require **factor 420** and **coenzyme M** (2-mercaptoethane sulfonic acid) as cofactors for methane formation. These factors participate in the enzymatic reduction of carbon dioxide to methane. Methanogens apparently gain energy by an electron transport mechanism since substrate-level phosphorylation is not possible when they grow on CO_2 and H_2.

SULFOLOBUS, THERMOPLASMA, AND *HALOBACTERIUM*

Sulfolobus acidocaldarius (a-ci-doh-cal-dar′i-us) is a thermophilic acidophile that grows as a chemolithotroph by oxidizing reduced sulfur or as a chemoorganotroph by using amino acids or sugars. It grows optimally between 70° and 80°C at a pH between 2 and 3 and is found in hot-acid, sulfur springs, and in hot-acid soils. *Sulfolobus* has a specialized cell membrane composed of a monolayer of long-chain hydrocarbons connected at both ends to glycerol by ether linkages. This specialized cell membrane may enable it to grow in hot-acid environments.

Thermoplasma is a thermophilic acidophile that lacks a cell wall and is therefore similar to species of *Mycoplasma*. However, the fact that its cell membrane is similar to *Sulfolobus* and its 16S ribosomal RNA gene is conserved with the methanogens places *Thermoplasma* with the archaebacteria.

Halobacterium is an extreme halophile that is prevalent in the Great Salt Lake where high concentrations of salt—4 to 5 molar salt—provide an ideal growth environment. When sufficient oxygen is present, *Halobacterium* grows by aerobic respiration; however, in high-salt solutions the oxygen content is usually low since 0_2 solubility decreases with increasing salt concentration. This archaebacterium adapts to anaerobic environments by making purple membranes that enable it to generate ATP by a light-driven reaction. *Halobacterium* makes the purple pigment, **bacteriorhodopsin,** which accumulates as patches in the cytoplasmic membrane. The absorption of light by bacteriorhodopsin in the purple membranes provides the energy necessary to establish a proton gradient across the cell membrane. This gradient drives the formation of ATP and provides *Halobacterium* with a source of energy in anaerobic or microaerophilic environments.

Summary Outline

FERMENTATIVE BACTERIA

1. The lactic acid bacteria are assigned to the genera *Streptococcus, Leuconostoc,* and *Lactobacillus* based on their morphology and the pathways they use to ferment glucose. These bacteria are used commercially in the production of pickles, sauerkraut, yogurt, and cheese.
2. The propionic acid bacteria ferment glucose and/or lactate and produce propionic acid, acetic acid, and carbon dioxide. Propionibacteria are found in Swiss cheese and other dairy products.
3. The clostridia are spore-forming rods that ferment amino acids, purines, cellulose, or sugars.
4. *Clostridium acetobutylicum* grows by the butyric acid fermentation and produces commercially important solvents. Other clostridia are important to the natural recycling of nitrogenous compounds.

THE SULFUR-REDUCING BACTERIA

1. The sulfur-reducing bacteria are strict anaerobes that grow by oxidizing organic compounds and reducing sulfur or sulfate to hydrogen sulfide during anaerobic respiration. Cell morphologies of the sulfur-reducing bacteria include rods, vibrios, sarcinas, cocci, and filamentous gliding bacteria.
2. *Desulfovibrio* are sulfate-reducing bacteria, *Desulfuromonas* reduces sulfur, not sulfate, and *Desulfotomaculum* forms true endospores.

INTERACTIONS BETWEEN PLANTS AND BACTERIA

1. Nitrogen fixation is a bacterial process by which atmospheric nitrogen is reduced to ammonia by nitrogenase.
2. Species of *Rhizobium* fix N_2 in symbiosis with leguminous plants. They specifically infect the root hairs of their host and migrate to root cells where they participate in nodule formation.
3. The *Rhizobium* bacteroids in the nodule fix nitrogen and the plant leghaemoglobin binds oxygen to protect the bacterial nitrogenase.
4. *Azotobacter, Beijerinckia,* and *Derxia* are examples of free-living nitrogen-fixing bacteria.
5. Bacterial plant pathogens are widely distributed in natural environments and cause diseases described as wilts, rots, blights, and galls.
6. Species of *Agrobacterium, Corynebacterium, Erwinia, Mycoplasma, Pseudomonas, Spiroplasma, Streptomyces,* and *Xanthomonas* cause plant diseases. *Erwinia* spp. cause fire blight, corn wilt, and soft rot.

7. *Agrobacterium tumefaciens* causes crown gall disease when a portion of the Ti plasmid is transferred to the nucleus of the susceptible plant cell. Here it causes the cell to divide continuously.
8. Citrus canker is caused by *Xanthomonas campestris.*

AEROBIC OXIDATIVE METABOLISM

1. Pseudomonads are noted for their ability to oxidize a wide variety of organic compounds. Some cause disease in plants, whereas others live in soil, fresh water, salt water, and foods.
2. Pseudomonads play an important role in food spoilage, especially in foods stored at low temperatures.
3. The acetic acid bacteria oxidize ethanol in wine and fermented cider to acetic acid during vinegar production. Species of *Acetobacter* are overoxidizers, whereas species of *Gluconobacter* are underoxidizers that produce only acetic acid.
4. The aerobic or facultative anaerobic, gram-positive, motile, spore-forming bacteria belong to the genus *Bacillus* and are ubiquitous in nature.
5. *Bacillus cereus* and *B. anthracis* cause human disease, whereas *B. thuringiensis* forms a protein during sporulation that is toxic to insects. Commercial preparations of *B. thuringiensis* are used for the biological control of insects.

BDELLOVIBRIO: THE PARASITIC BACTERIA

1. Bdellovibrios, the smallest known predators, penetrate the cell wall of gram-positive hosts and multiply in the periplasmic space.
2. They kill their host cell and break down the host's cytoplasmic components into nutrients that are utilized for growth.
3. The Bdellovibrio cell elongates and eventually fragments into motile progeny cells, which are released when the host cell lyses.

GLIDING BACTERIA

1. Myxobacteria are unicellular gliding bacteria that form fruiting bodies and myxospores under adverse environmental conditions.
2. The fruiting bodies may be complex structures or simple aggregates of cells that support myxospores, which are resistant to ultrasound, UV light, and heat.
3. The cytophagas glide on solid surfaces, do not form fruiting bodies, but may be pathogenic to fishes.
4. Both the myxobacteria and the cytophagas are noted for their ability to break down cellulose, peptides, and biopolymers of complex carbohydrates.

CHEMOLITHOTROPHS CAN GROW ON INORGANIC MEDIA

1. The chemolithotrophic autotrophs oxidize inorganic compounds in the absence of light as their source of energy and fix carbon dioxide as their source of carbon.
2. The nitrifying bacteria catalyze the two-step process in which ammonia is oxidized to nitrite and nitrite is oxidized to nitrate. *Nitrosomonas* is an ammonia oxidizer and *Nitrobacter* is a nitrite oxidizer.
3. The colorless sulfur bacteria gain energy from the oxidation of reduced-sulfur compounds and are essential to the sulfur cycle.
4. *Thiobacillus ferrooxidans* can oxidize either ferrous iron or reduced sulfur. Other colorless sulfur bacteria use organic compounds as carbon sources.

PHOTOSYNTHETIC BACTERIA

1. Anoxygenic photosynthetic bacteria are organized into four groups based on their pigments and the nature of their photosynthetic electron donors.
2. The purple nonsulfur bacteria contain bacteriochlorophyll *a* and use organic compounds as both electron donors and carbon source.
3. The purple sulfur bacteria are photolithotrophs that use reduced sulfur as electron donors and fix carbon dioxide as their source of carbon. Most purple sulfur bacteria possess bacteriochlorophyll *a* in membrane vesicles.
4. Green sulfur bacteria are phototrophs that contain either bacteriochlorophyll *c, d,* or *e.* They use reduced sulfur as electon donors; some can use simple organic compounds.
5. *Chloroflexus* is a green flexibacterium that can

grow on a variety of organic compounds as a chemoorganotroph in air or as a photoorganotroph in anaerobic-light environments.

6. The cyanobacteria are oxygenic photosynthetic bacteria that contain chlorophyll *a* and phycobiliproteins in thylakoid membranes. They are widely dispersed in aerobic aquatic environments and about one third of them can fix N_2.
7. The procaryote *Prochloron didemni* has chlorophyll *a* and *b* and grows by oxygenic photosynthesis in a tunicate.

ARCHAEBACTERIA

1. All archaebacteria lack peptidogylcan, contain ether linkages in their lipids, are insensitive to chloramphenicol and kanamycin, and have conserved 16S ribosomal RNA genes.
2. The methanogens are morphologically diverse, strictly anaerobic archaebacteria, most of which grow on carbon dioxide and hydrogen gas. They produce methane through a biochemical process that requires coenzyme M and factor 420.
3. Species of *Sulfolobus* and *Thermoplasma* are thermophilic acidophiles that are aerobic archaebacteria.
4. *Halobacterium* is an aerobic extreme halophile that can grow anaerobically by making bacteriorhodopsin, which in the presence of light establishes a proton gradient for ATP synthesis.

Questions and Topics for Study and Discussion

QUESTIONS

1. Name three bacterial fermentations and describe the organisms able to perform them. What products of bacterial fermentations are commercially important and how are they used?
2. Explain anaerobic respiration as it applies to the sulfur-reducing bacteria.
3. Describe the symbiotic relationship between *Rhizobium* and its legume host.
4. Describe the cause of citrus canker and explain what steps should be taken to control its spread.
5. Why is *Gluconobacter* called an underoxidizer, while *Acetobacter* is described as an overoxidizer?
6. Explain how *Bacillus thuringiensis* is used as an insecticide.
7. Describe the life cycle of the bacteria that are parasitic on other bacteria.
8. Under what conditions do myxobacteria form fruiting bodies? What is the advantage of fruiting body formation to the myxobacteria?
9. What is a chemolithotrophic autotroph? Give three examples of chemolithotrophic autotrophs and explain the role they play in the environment.
10. Describe three natural environments where sulfur oxidation would occur and name the sulfur-metabolizing organisms you would find there.
11. What is meant by anoxygenic photosynthesis? What groups of organisms grow by anoxygenic photosynthesis?
12. How do cyanobacteria differ from other photosynthetic bacteria?
13. What distinguishing characteristics separate the archaebacteria from other bacteria?

DISCUSSION TOPICS

1. What problems would the biological world encounter if there were no nitrifying bacteria?
2. How do bdellovibrios differ from bacteriophages?
3. What evidence would you need to prove that *Prochloron didemni* is a bacterium and not just an enucleated alga?
4. Are dark reactions of photosynthesis (CO_2 fixation) necessary for bacterial photosynthesis?

Further Readings

BOOKS

Postgate, J. R., *The Fundamentals of Nitrogen Fixation,* Cambridge University Press, Cambridge, 1982. This book discusses the enzymology, physiology, and ecology of the nitrogen-fixing bacteria.

Postgate, J. R., *The Sulfate Reducing Bacteria,* 2nd ed., Cambridge University Press, London, 1984. The author covers the biochemistry and physiology of the sulfate-reducing bacteria, including material on recently isolated species.

Starr, M. P., H. Stolp, H. G. Truper, A. Balows, and H. G. Schlegel, *The Prokaryotes, A Handbook on Habitats, Isolation, and Identification of Bacteria,* Vols. 1, 2, Springer Verlag, Berlin, 1981. A comprehensive two-volume work that contains valuable information on the characteristics, ecology, and methods of growing bacteria.

REVIEWS AND ARTICLES

Burnham, J. C., T. Hashimoto, and S. F. Conti, "Ultrastructure and Cell Division of a Facultatively Parasitic Strain of *Bdellovibrio bacteriovorus," J. Bacteriol.,* 101:997–1004 (1970). Describes the life cycle of Bdellovibrio with excellent electron micrographs.

Dworkin, M., and D. Kaiser, "Cell Interactions in Myxobacterial Growth and Develoment," *Science,* 320:18–24 (1985). The contact-mediated interactions and extracellular signals involved in the life cycle of myxobacteria are described.

Larkin, J. M., and W. R. Strohl, *"Beggiatoa, Thiothrix,* and *Thioploca," Ann. Revs. Microbiol.,* 37:341–67 (1983). This is a review of the current understanding of the gliding sulfur bacteria.

Pfenning, N., "Microbial Behavior in Natural Environments," in D. P. Kelly and N. G. Carr (eds.), *The Microbe 1984, Part II Prokaryotes and Eukaryotes,* Thirty Sixth Symposium of the Society for General Microbiology, pp. 23–50, Cambridge University Press, Cambridge, 1984. A stimulating essay on the behavior of microorganisms in their efforts to survive and flourish in their natural habitats.

Reichenbach, H., "Taxonomy of the Gliding Bacteria," *Ann. Rev. Microbiol.,* 35:339–64 (1981). This overview points out the diversity of the gliding bacteria by discussing the relationships between the *Cyanobacteria, Choloflexaceae,* and *Flexibacteriae.*

Schlegel, H. G., and H. W. Jannasch, "Prokaryotes and Their Habitats," in M. P. Starr, H. Stolp, H. G. Truper, A. Balows, and H. G. Schlegel (eds.), *The Prokaryotes,* Vol. 1, pp. 43–82, Springer-Verlag, Berlin, 1981. An excellent introduction to the ecology of microorganisms.

CHAPTER

15

Bacteria of Medical Importance

OUTLINE

- GRAM-NEGATIVE BACTERIA
  - Spirochetes
  - Aerobic Curved Bacteria
  - Aerobic Rods and Cocci
    - Pseudomonads
    - Legionella
    - Neisseria
    - *Brucella, Bordetella,* and *Francisella*
- FACULTATIVELY ANAEROBIC GRAM-NEGATIVE RODS
  - *Enterobacteriaceae*
    - Classification
    - *Escherichia*
    - *Shigella*
    - *Salmonella*
    - *Klebsiella*
    - Opportunistic Pathogens
    - *Yersinia*
  - Vibrios Pathogenic to Frogs, Fish, and Humans
    - Cholera
  - Parasites of Mammals and Birds
    - *Pasteurella*
    - *Haemophilus*
- OBLIGATE ANAEROBIC GRAM-NEGATIVE BACTERIA
  - *Bacteroides* and *Fusobacterium*
- RICKETTSIAS AND CHLAMYDIAS
- MYCOPLASMAS
- GRAM-POSITIVE BACTERIA
  - The Pyogenic Cocci
    - Streptococci
    - Staphylococci
  - Endospore-Forming Rods
    - *Bacillus* and *Clostridium*
- BRANCHING BACTERIA
- *CORYNEBACTERIUM* AND DIPHTHERIA
- THE MYCOBACTERIA

FOCUS OF THIS CHAPTER

In this chapter we introduce students to the biological characteristics of the bacteria that live in and cause damage to humans and other animals. The focus is on the biology of the pathogenic bacteria; the medical aspects of the diseases they cause are discussed in Part 4. The diversity among these pathogens is systematically presented through discussions of the major bacterial groups that include disease-causing species.

Microbes that live on or within the tissue of another organism are called **parasites.** Many parasites are infectious agents since they can be transmitted from one host to another. If these parasites also harm or cause damage to their host, they are called **pathogens.** The subject of this chapter is the biology and characteristics of the major bacterial pathogens.

Virtually everyone has suffered from a sore throat, diarrhea, or other symptoms caused by pathogenic bacteria. The few species of bacteria that cause disease can make a person feel very sick, and some can cause death. The ability of a bacterium to cause disease often depends on the physiological health of the host. **Opportunistic pathogens** cause disease in compromised hosts—persons who are malnourished, diabetic, immunosuppressed, catheterized, burned, elderly, or suffering from AIDS, cancer, or respiratory distress. Opportunistic pathogens may be part of the normal human bacterial flora or they may be found in natural environments and cause disease when compromised persons come in contact with them. In contrast, **obligate pathogens** require a host for their survival and cause disease in susceptible hosts.

GRAM-NEGATIVE BACTERIA

As you recall from Chapter 13, cell-wall structure is a major classification criterion used by bacterial systematists. For convenience we have divided the medically important bacteria according to their response to the Gram stain, which closely follows the organization of *Bergey's Manual of Systematic Bacteriology.*

SPIROCHETES

Spirochetes are responsible for syphilis, yaws, pinta, relapsing fever, and leptospirosis. They are readily identified by their unique morphology, and are thereby classified as a distinct group.

Helical bacteria that are motile by means of periplasmic flagella are called **spirochetes** (Figure 15.1). These organisms are composed of a **protoplasmic cylinder** surrounded by a multilayered membrane structure known as the **outer sheath.** Beneath the outer sheath is a typical gram-negative cell-wall cytoplasmic–membrane complex, which surrounds the cell's cytoplasm. Between the cell wall and the outer sheath are two to 100 periplasmic flagella (also called axial filaments or endoflagella). One end of each flagellum is inserted into the cell wall at the distal end of the protoplasmic cylinder; the other end extends toward the middle of the bacterium where the periplasmic flagella overlap (Figure 15.1). By rotating in place, these flagella create a torque that spins the protoplasmic cylinder. This spinning propels the spirochete with a corkscrew motion through liquid or semisolid media.

Spirochetes are differentiated by their size, morphology, and physiology. Spirochetes that cause human diseases are classified in the genera *Treponema* (trepo-nee'mah), *Borrelia* (bor-rel'ee-ah), and *Leptospira* (lep'toh-spy'rah).

Treponemes are tightly coiled anaerobic or microaerophilic spirochetes that live in association with humans and animals. Some cause serious human illness, such as syphilis, whereas others occur as normal flora of the human oral cavity.

Syphilis is a serious human sexually transmitted (venereal) disease caused by *Treponema pallidum* (pal'lidum) subspecies *pallidum.* Transmission occurs when individuals come in contact with the treponemes present in the lesions of an infected partner. After transfer, the treponemes enter the host through the mucous membranes of the urogenital tract, anus, or mouth. The first sign of syphilis is an open lesion or chancre

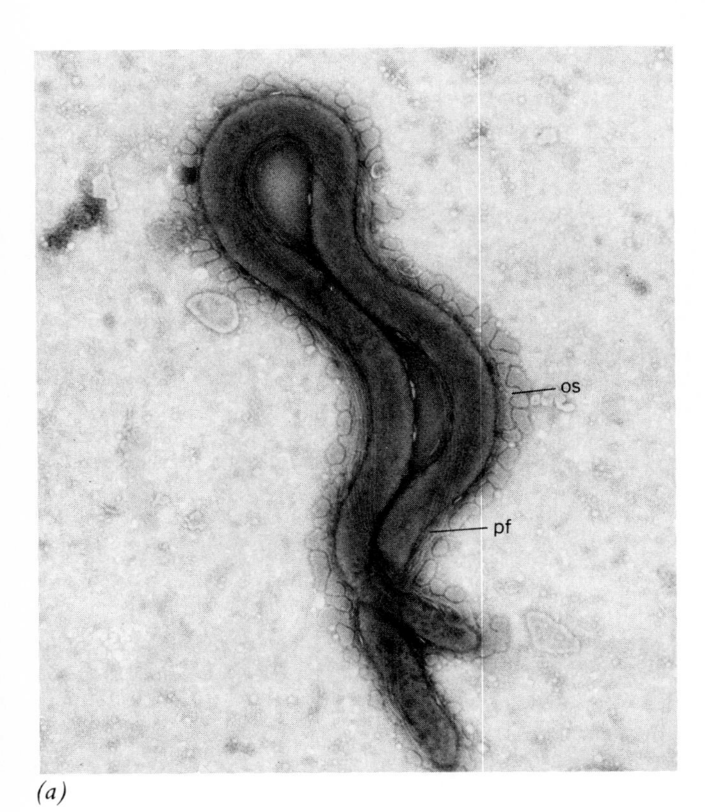

(a)

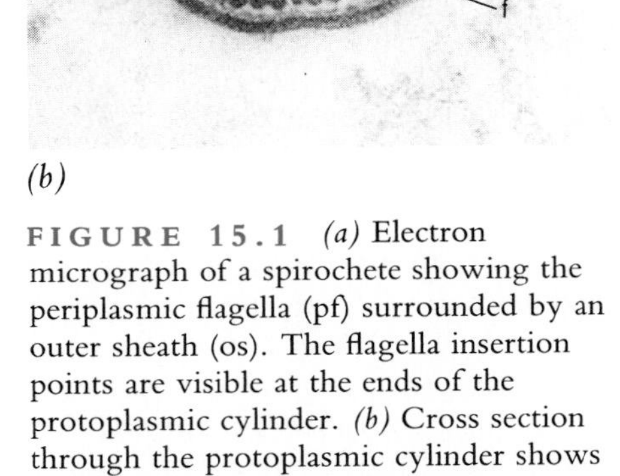

(b)

FIGURE 15.1 *(a)* Electron micrograph of a spirochete showing the periplasmic flagella (pf) surrounded by an outer sheath (os). The flagella insertion points are visible at the ends of the protoplasmic cylinder. *(b)* Cross section through the protoplasmic cylinder shows the location of the periplasmic flagella (f). (Courtesy of Dr. Max A. Listgarten.)

(shang′ker) that develops at the site of infection. Syphilis can be treated at this stage with penicillin; however, if not treated, syphilis can progress to the more serious secondary stage. Tertiary syphilis can develop in untreated patients years after the symptoms of secondary syphilis disappear. Tertiary syphilis can involve the heart and central nervous system and eventually contributes to the death of the patient.

Syphilis is one of the few diseases that can be passed *in utero* from an infected mother to her fetus. The disease in the fetus is known as **congenital syphilis,** and this often results in a stillbirth. The greatest risk of congenital syphilis occurs when the mother's disease has progressed to the secondary stage.

Other treponemal diseases include **yaws** (yoz) and **pinta,** infections of the skin that occur in tropical countries. Fortunately, all of the treponemal diseases can be treated by readily available antibiotics such as penicillin.

The treponemes that cause disease in humans have not been grown in artificial media *(in vitro),* despite much effort to do so. Although researchers have had some success in growing *T. pallidum* in tissue cultures, they have been unable to grow treponemes in ordinary bacteriological media. The most effective method of producing *T. pallidum* for experimental and diagnostic purposes is to grow it in rabbit testicles.

Borrelias are spirochetes that have coarse, uneven, or irregular coils. They are visible upon microscopic examination of blood samples taken from naturally infected mammals, insects, and birds. The borrelias are transmitted between cattle, horses, birds, and humans by ticks or lice. Humans are susceptible to **louse-borne** (epidemic) **relapsing fever** caused by *Borrelia recurrentis* (ree-kur-ren′tis), and to **tick-borne relapsing fever** caused by other species of *Borrelia. Borrelia burgdorferii* (burg-dor-fer′ee-eye) causes another tick-borne borrelial disease, called **Lyme disease.** The symptoms of Lyme disease were first observed in children living in the vicinity of Lyme, Connecticut; however, now the disease is known to be widespread.

Leptospiras are free-living or parasitic, aerobic spirochetes with morphologically distinctive hooked ends (Figure 15.2). They are found in ponds, streams, and

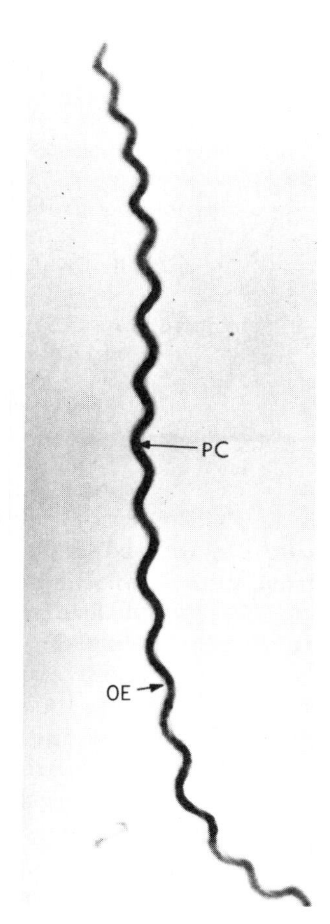

FIGURE 15.2 Electron micrograph of tightly wound leptospiras, showing the outer envelope (OE) and the protoplasmic cylinder (PC), 36,000×. (Courtesy of Dr. Robert K. Nauman. From R. K. Nauman, S. C. Holt, and C. D. Cox, *J. Bacteriol.*, 98: 264–280, 1969, with permission of the American Society of Microbiology.)

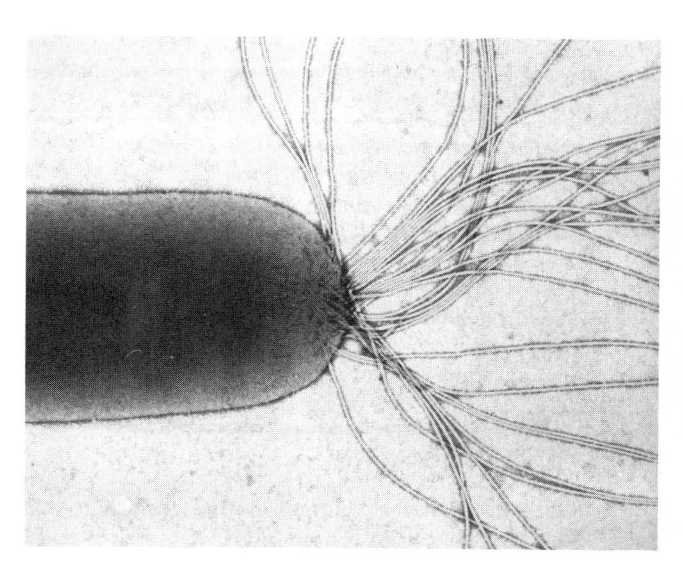

FIGURE 15.3 The negatively stained preparation of *Aquaspirillum graniferum* shows its polar bundles of flagella and their multiple insertion points. (Courtesy of Roche Diagnostics.)

infected animals. Leptospirosis is primarily a zoonosis* (zoh-no′sis) of domestic and wild animals, but humans can contract it via skin abrasions or mucus membranes, or by drinking contaminated water. Leptospirosis in humans is a very serious disease that must be diagnosed early if the patient is to survive.

AEROBIC CURVED BACTERIA

A **spirillum** is a curved or helical motile bacterium whose flagella are arranged as bipolar clusters (Figure 15.3). Spirilla are found in fresh water, marine habitats, and associated with plants. Spirilla that cause animal infections are aerobes belonging to the genus *Campylobacter* (cam-py-lo-bak′ter) or the species *Spirillum minus,* the agent of rat-bite fever.

Campylobacters are small, curved, gram-negative rods that move corkscrew-like as they propel themselves by a single polar flagellum or by flagella at either end. *Campylobacter* are found worldwide in the intestinal tract of domestic animals including cattle, sheep, cats and dogs, as well as in rodents, chickens, and wild birds. They are transmitted to other animals in contaminated food and water, and unpasteurized milk. *Campylobacter jejuni* (je-ju′ni) infects the human intestinal tract by the fecal–oral route and is a major cause of acute gastroenteritis (gas′troh-en-te-ry′tis). *Campylobacter fetus* (fe′tus) is apt to infect the blood and cause a bacteremia* leading to infections of the central nervous system and the heart.

AEROBIC RODS AND COCCI

Of the eight families of aerobic rods and cocci, three contain species that cause human disease: Pseudomonads, Legionellas, and Neisseria. Other aerobic rods and

*Zoonosis is an infectious disease that occurs primarily in animals; humans are occasionally infected.

*Bacteremia indicates the presence of nonmultiplying bacteria in the blood.

cocci play a significant ecological role as described in Chapter 14.

Pseudomonads Are Aerobic Rods

The pseudomonads (soo-doh-moh'nads) are straight or curved gram-negative, polar-flagellated bacteria that have the ability to grow in simple media on a wide variety of organic compounds. They are strict aerobes found widely dispersed in nature. These bacteria often contain plasmids that code for degradative enzymes, fertility, antibiotic resistance, and/or resistance to heavy metals.

Pseudomonas aeruginosa (soo-doh-moh'nas ai-roo-jin-oh'sah) is widely distributed in nature and human habitats. It is an important opportunistic pathogen that causes urinary tract infections and is a major problem in the treatment of burn patients. *Pseudomonas aeruginosa* produces the blue pigment (pyocyanin) responsible for the color of "blue pus" seen in infected wounds. The pervasive existence of *P. aeruginosa* in the environment and its resistance to antibiotics make it a significant cause of urinary tract infections, especially in catheterized patients.

Legionella and Legionellosis

In the summer of 1976, an outbreak of febrile pneumonia occurred among legionnaires attending the American Legion Convention in Philadelphia. This serious disease killed many patients before the appropriate antibiotic therapy was devised. The causative agent was finally isolated by growing it in guinea pigs and then in the yolk sac of embryonated eggs (Figure 15.4). The isolated aerobic rod was a new bacterial species and was named *Legionella pneumophila* (lee-jun-nel'ah noo-moh'fill-ah). Pneumonia caused by this bacterium is commonly referred to as **legionnaires' disease.**

Legionella are widely dispersed in the environment and contaminate the water-cooling towers of air-conditioning systems. The aerosols made by these cooling towers carry the *Legionella* to susceptible people who become infected when they inhale the bacteria. Infections caused by *Legionella* can be effectively treated with antibiotics.

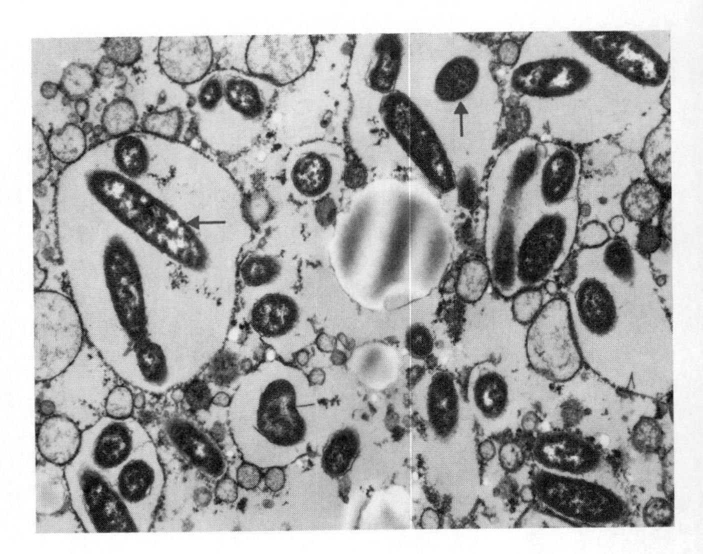

FIGURE 15.4 Electron micrograph of *Legionella pneumophila* (↑) growing in the yolk sac of an infected embryonated egg. (Courtesy of Dr. Francis W. Chandler, Centers for Disease Control.)

Neisseria Are Gram-Negative Cocci

The neisseria (nye-see'ree-ah) are gram-negative, nonmotile, aerobic diplococci that inhabit the mucous membranes of animals. Some species are part of the normal flora of the human throat, others can cause sexually transmitted diseases, eye infections, and meningitis.

Neisseria gonorrhoeae (gon-or-ree'eye) is the causative agent of the sexually transmitted human disease, **gonorrhea.** Albert Neisser in 1889 first observed this bacterium in the pus cells from the genital exudate of an infected patient. *Neisseria gonorrhoeae* grows in the mucous membrane of the human urogenital tract where it can cause sufficient tissue damage to result in sterility of the male or female patient. Urogenital *N. gonorrhoeae* infections of females are predominantly asymptomatic and consequently difficult to control.

Neisseria meningitidis (men-in-jy'ti-dis) inhabits the nasopharynx of human carriers. A small percentage of these carriers develop cerebrospinal **meningitis** (an inflammation of the membranes surrounding the brain and spinal column) after the bacteria are transmitted by the blood to the central nervous system. Both species of *Neisseria* are readily grown on laboratory media and are susceptible to antibiotics. Rapid direct co-agglutination tests for quickly identifying neisserial infections are available.

Two genera that are closely related to the *Neisseria* are *Moraxella* and *Acinetobacter*. *Moraxella lacunata* (morax-el'lah la-ku-nah'tah) is a short gram-negative rod found in the human conjunctiva. In the past, this bacterium was considered to be a cause of human conjunctivitis (kon-junk-tiv-eye'tis), but now it is rarely isolated from inflamed membranes of the eye (conjunctivae). Species of *Acinetobacter* (ah-si-nee'toh-bak-ter) occur naturally in soil and water and are considered to be nonpathogenic. They do, however, cause nosocomial (hospital associated) infections, possibly by contaminating surgical instruments.

Brucella, Bordetella, and *Francisella*

Sir David Bruce discovered *Brucella melitensis* (broo-sel'ah meh'lih-ten'sis), on the island of Malta, and it was found to be the cause of Malta fever (now called **undulant fever**). Goats and sheep are natural hosts for this bacterium, which infects humans who drink contaminated goat's milk. The brucellas are obligate parasites that cause a variety of diseases in sheep, cattle, dogs, and laboratory animals. Brucellosis also occurs among veterinarians and workers in the meat-packing and livestock industries who come in contact with infected animals.

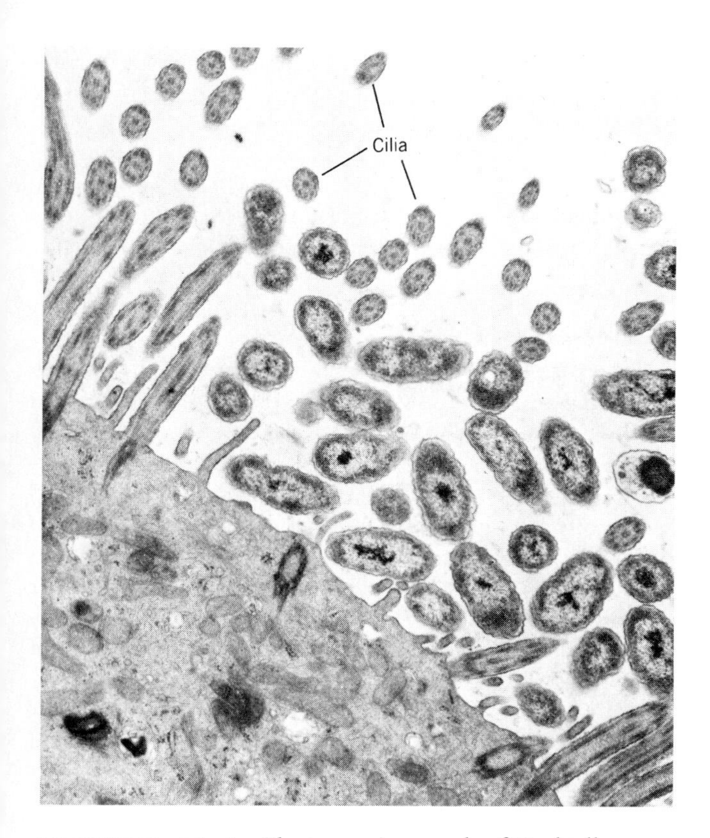

FIGURE 15.5 Electron micrograph of *Bordetella pertussis* associated with cultured hamster tracheal cells. Note the 9 pairs plus 2 structure of the tracheal cilia. (Reprinted from A. M. Collier, L. P. Peterson, and J. B. Baseman, *J. Infect. Dis. Suppl.*, 136S:196–203, 1977, by permission of The University of Chicago Press.)

Bordetellas (bor-deh-tell'as) are mammalian parasites named after Jules Bordet, who first isolated the bacterium responsible for pertussis (per-tus'sis). **Pertussis** is a serious childhood disease also known as **whooping cough.** Humans are the only host for *Bordetella pertussis,* which grows among the cilia of the respiratory epithelium (Figure 15.5) and produces the toxin responsible for the symptoms of pertussis. Most youngsters are immunized against pertussis when they receive the DTP (diphtheria, tetanus, pertussis) vaccine.

Francisella tularensis (fran-sis-el'ah too-lah-ren'sis) is an extremely small rod that causes tularemia in animals and humans. It naturally infects wild rabbits and fur-bearing animals such as muskrats, rabbits, beavers, and squirrels. Humans are infected directly when they handle or ingest contaminated animal carcasses or indirectly through the bite of ticks or deerflies. This bacterium is named after Edward Francis, who did extensive field, laboratory, and clinical studies on tularemia; and after Tulare County, California, where the disease was first observed.

FACULTATIVELY ANAEROBIC GRAM-NEGATIVE RODS

Facultative anaerobes grow in the presence of air by aerobic respiration, and by fermentation of carbohydrates in the absence of oxygen. There are so many gram-negative facultatively anaerobic rods that their description and identification is beyond the scope of an introductory microbiology course, therefore we will confine our discussion to selected examples.

ENTEROBACTERIACEAE

The *Enterobacteriaceae* (en'te-roh-bak-te-ree-ay'see-eye) are found in soil, water, animals, fruits, vegetables,

TABLE 15.1 Disease capabilities of the *Enterobacteriaceae*

BACTERIUM	DISEASES
Escherichia coli	Opportunistic pathogen, travelers' and infantile diarrhea, shigella-like syndrome
Enterobacter spp., *Proteus* spp., and *Serratia* spp.	Opportunistic pathogens
Erwinia	Diseases of plants and vegetables
Klebsiella pneumoniae	Pneumonia; opportunistic pathogen
Salmonella spp.	Gastroenteritis, salmonella septicemia, enteric fevers
Salmonella typhi	Typhoid fever
Shigella spp.	Dysentery
Yersinia pestis	Plague
Yersinia ruckeri	Diseases of tropical fish

grains, flowering plants, and trees throughout the world. Some species are harmless; however, many cause disease of human, animals, or plants. Table 15.1 is a summary of the key members of this family that cause disease.

Classification of the *Enterobacteriaceae*

The large number of organisms in this family make their classification a very difficult task. In many cases, the assignment of known strains to species and even to genera poses a difficult problem for systematists. In these situations, DNA-relatedness data help microbiologists understand the relationships between the bacterial species (Figure 15.6). Some genera are clearly distinguished by DNA-relatedness; however, others are not. For example, the 85 to 100 percent DNA-relatedness between certain strains of *Escherichia* and *Shigella* (shi-gel'ah) indicates that these genera are identical, yet these bacteria are assigned to separate genera because their reclassification would seriously disrupt the practice of medical microbiology.

Practical techniques have been developed for identifying the gram-negative rods, which are the most commonly encountered bacteria in clinical isolates. New isolates are identified by scoring each organism's response to a complex set of biochemical tests, which are conveniently done using self-contained biochemical assay systems such as API Test strips (Color Plate II) or Enterotube (Figure 15.7). Some medically important *Enterobacteriaceae* are described below.

Escherichia Inhabits the Human Intestine

One of the most widely studied organisms in the history of science is *Escherichia coli*. It has been the object of extensive biochemical, genetic, viral, environmental, and cytological studies. Many of the experiments in genetic engineering directly involve *E. coli,* or use plasmids or genetic information derived from *E. coli*. Numerous basic concepts of intermediary metabolism, biosynthesis, molecular biology, regulation, bacteriophage, replication, and genetics were developed from experiments on *E. coli*.

Escherichia coli is a normal inhabitant of the human intestinal tract where it grows and is excreted in human feces. Since it survives well in natural environments, *E. coli* is used as an indicator of human fecal contamination. The presence of *E. coli* in municipal water supplies, swimming water, wells, or other water sources indicates that human sewage is contaminating the water. Common reasons for contamination are improper sewage treatment, improper location of a septic system, and overflow from an existing sewage treatment facility.

Although *E. coli* is not normally a pathogen, ordinary strains can cause human disease in compromised patients, and they are common causes of urinary tract infections, especially in women. The same bacterium can acquire virulence plasmids that enable it to cause disease. Enterotoxigenic *E. coli* (ETEC) has a plasmid that codes for enterotoxin, which in turn causes a diarrheal disease in humans. ETEC strains are prevalent in many developing countries and are one cause of trav-

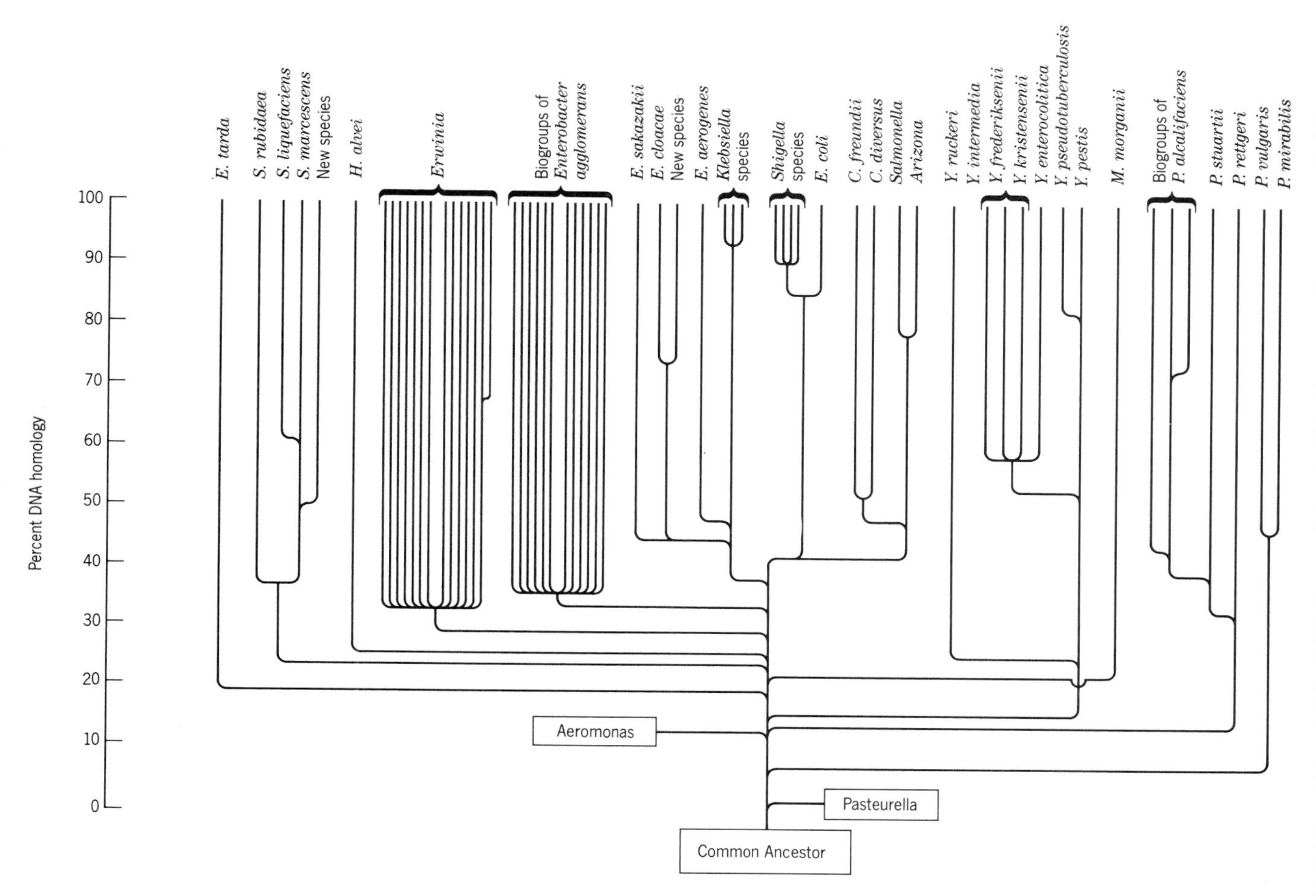

FIGURE 15.6 The relatedness between species in the family *Enterobacteriaceae* is depicted using percent DNA homology on the ordinant. The scheme assumes a common ancestor from which all *Enterobacteriaceae* have diverged. (Courtesy of Dr. Don J. Brenner. From Dr. D. J. Brenner, "Facultatively Anaerobic Gram-Negative Rods," in *Bergey's Manual of Systematic Bacteriology,* Vol. 1, 1984, Williams and Wilkins, Baltimore.)

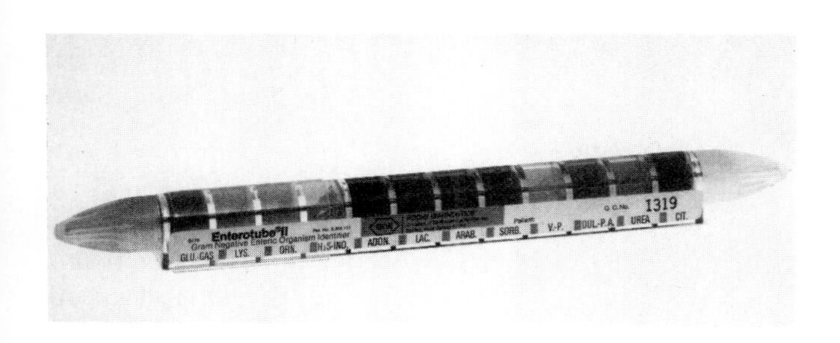

FIGURE 15.7 The Enterotube is used to determine biochemical characteristics of isolated enterics.

elers' diarrhea. Other serovars include enteropathogenic *E. coli* (EPEC), which is responsible for infantile diarrhea, and enteroinvasive *E. coli* (EIEC), which causes a shigella-like syndrome.

Shigella Cause Dysentery

The shigella are nonmotile, facultatively anaerobic rods that do not usually produce gas during sugar fermentation. Their natural host range is limited to humans (though laboratory animals have been infected with shigella for experimental purposes). All four species of *Shigella* infect humans by the oral–fecal route and cause shigellosis (bacillary dysentery). These pathogens invade the large intestine and produce lesions in the intestinal wall and rectum. **Shigellosis** is an acute diarrheal illness characterized by the presence of blood and mucus in stools. Most of these illnesses are self-limiting and of short duration, and the patient usually recovers without requiring antimicrobial therapy.

Salmonella

Salmonella, which live in the intestines of cold-blooded and warm-blooded animals, are a major cause of foodborne and waterborne bacterial disease. Some *Salmonella* (sal-mohn-el′ah) species are restricted to a single host, while others are not restricted. Classification of the *Salmonella* is complicated by the description in the literature of over 1400 serovars. These serovars are important in tracing the sources of foodborne and waterborne salmonella outbreaks.

Species of *Salmonella* are responsible for enteric fevers, salmonella septicemia,* and salmonella gastroenteritis in humans. The most serious disease is typhoid fever, caused by *Salmonella typhi*. This bacterium is found only in humans who are ill or who have become carriers after recovering from typhoid fever. Other *Salmonella* cause gastroenteritis, the major foodborne bacterial disease in the United States. Contamination of the foodstuffs is usually from an animal reservoir such as poultry or cattle. Direct fecal–oral spread between humans is possible, but is uncommon because high numbers of salmonella are required for infection.

Other species of *Salmonella* are carried by cold-blooded animals such as the green turtles that are kept in small turtle bowls as pets. Salmonella excreted in the feces of infected turtles contaminate the water of the turtle bowl. The fecal–oral route of infection is completed when children play with the turtle, then put their fingers in their mouths.

Klebsiella and Pneumonia

Species of *Klebsiella* (kleb-see-ell′ah) are nonmotile facultative anaerobic gram-negative rods that are identified by their biochemical characteristics. The klebsiella are widely distributed in soils, water, and grains and are normal inhabitants of the intestinal tract of warm-blooded animals. They are difficult to isolate from human fecal matter because of their low numbers in relation to *Escherichia coli*.

Klebsiella pneumoniae (noo-moh′nee-eye) is an important cause of human pneumonia (inflammation of the lungs). The strains responsible produce polysaccharide capsules (Figure 15.8) that protect them against phagocytosis. *Klebsiella* species also cause bacteremia and urinary tract infections and are implicated increasingly in nosocomial infections. The strains involved in nosocomial infections often contain antibiotic resistance plasmids, which complicate the chemotherapy used to treat infected patients.

*Septicemia is the presence of multiplying microorganisms in an animal's blood.

Opportunistic Pathogens

Enterobacter, Proteus, and *Serratia* are common inhabitants of soil and water and some species are found on plants or in animal intestines. Species of each genus cause disease in compromised patients. *Enterobacter cloacae* (en-te-roh-bak′ter clo-ay′see) and *E. aerogenes* (ai-rah′jen-eez) have been isolated from urine, the respiratory tract, and occasionally from the blood and spinal fluid of humans, usually as one component of a mixed infection. *Serratia marcescens* (ser-rah′tee-ah mahr-ses′ens) and *S. liquefaciens* (li-que-fay′shenz) are

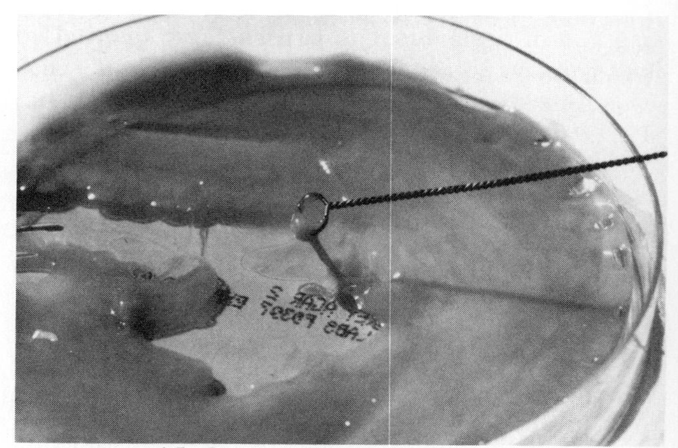

FIGURE 15.8

Capsule-producing strains of *Klebsiella pneumoniae* grow as mucoid colonies. Antiphagocytic capsules contribute to this species' virulence.

occasionally encountered as opportunistic human pathogens, whereas *Proteus mirabilis* (proh'tee-us mi-rah'bi-lis) and *P. vulgaris* (vul-gah'ris) cause urinary tract infections in compromised humans.

Yersinia Infections of Fish, Animals, and Humans

Yersinia pestis causes enzootic disease in over 100 different rodents, including rats and ground squirrels. It causes **bubonic plague** in humans, a disease that devastated the human population in past centuries. Today, patients with bubonic plague are treated successfully with antibiotics; however, this disease has a high fatality rate if undiagnosed and untreated. Fleas living on infected rodents transmit *Y. pestis* to other rodents and to humans. Rodents can also be infected by burrowing in contaminated soil, which is the natural reservoir for *Y. pestis*. Human plague occurs at a low rate in geographical regions harboring infected rodent populations.

Other pathogenic yersinia cause disease in wild and domestic animals and in fish. Humans who come in contact with diseased animals can be infected, probably by ingesting the bacteria. These bacteria cause intestinal tract infections that can lead to infected lymph glands and septicemia. Fish are also susceptible to infection, for example *Y. ruckeri* (ruck'er-eye) occurs in North America where it causes **red-mouth disease** in rainbow trout.

VIBRIOS PATHOGENIC TO HUMANS

The family *Vibrionaceae* (vib-ri-oh-na'see-eye) is the second major group of gram-negative, facultatively anaerobic rods. Members are curved or straight rods that are motile by flagella attached to the pole of the cell (Figure 15.9). The most important human pathogens of the *Vibrionaceae* are *Vibrio cholerae* (kol'er-eye) and *V. parahaemolyticus* (para-hae-moh-ly'ti-kus).

Cholera and Other Human Diseases

Vibrios live in aquatic environments of various salinities (salt concentration) and cause disease when humans ingest contaminated water or contaminated seafood. *Vibrio cholerae* is the causative agent of epidemic or **Asiatic cholera** that is a major waterborne bacterial disease in developing countries, especially in Asia. This organism infects the human intestine and produces the cholera toxin, which is responsible for the purging diarrhea characteristic of the disease.

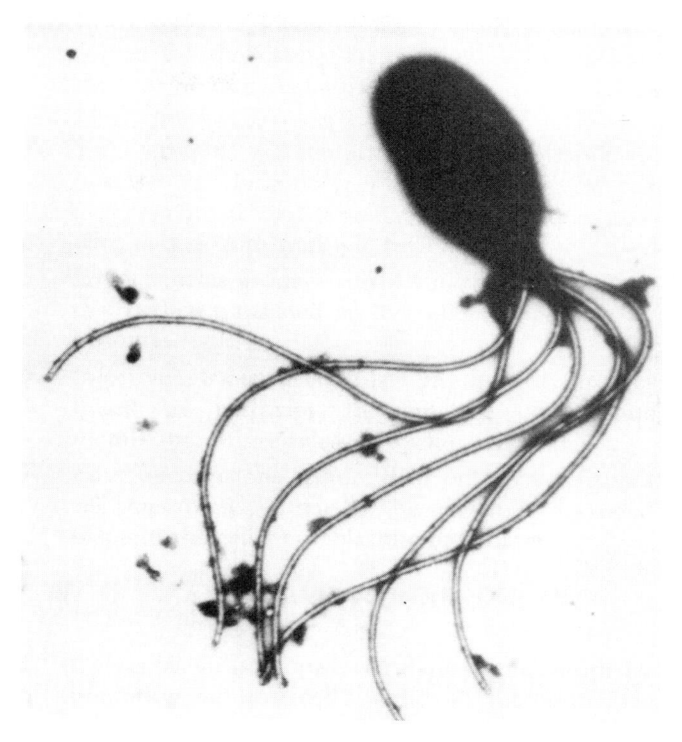

FIGURE 15.9 Electron micrograph of *Vibrio fischeri* showing its tuft of sheathed flagella. The individual flagella are larger than normal bacterial flagella because they possess an outer sheath that completely surrounds them. (Courtesy of Dr. Kenneth H. Nealson, The University of Wisconsin-Milwaukee.)

Vibrio parahaemolyticus appears to be a common inhabitant of coastal marine waters and estuaries. This bacterium is associated with seafoods and causes gastroenteritis when contaminated food is consumed. Other marine vibrios form symbiotic relationships with fish. For example, *Vibrio fischeri* is a luminescent bacterium present in the luminous organ of fishes (Figure 15.10) and squids.

PARASITES OF MAMMALS AND BIRDS

The third family of gram-negative, facultatively anaerobic rods is the *Pasteurellaceae* (pas-teu-rel-la'see-eye). Members of this family are small, straight, nonmotile coccoid-to-rod-shaped bacteria that require organic

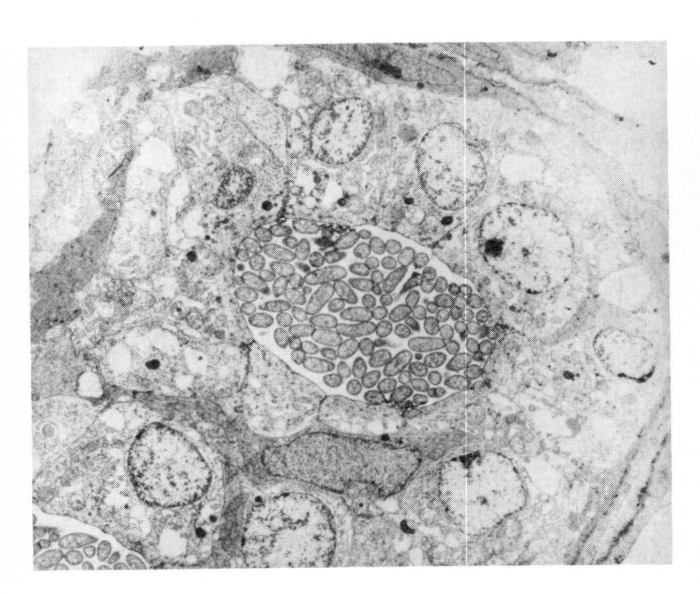

FIGURE 15.10 Electron micrograph of the luminous organ of *Monocentris japonicus* showing the presence of *Vibrio fischeri*. (From K. H. Nelson and B. M. Tebo.)

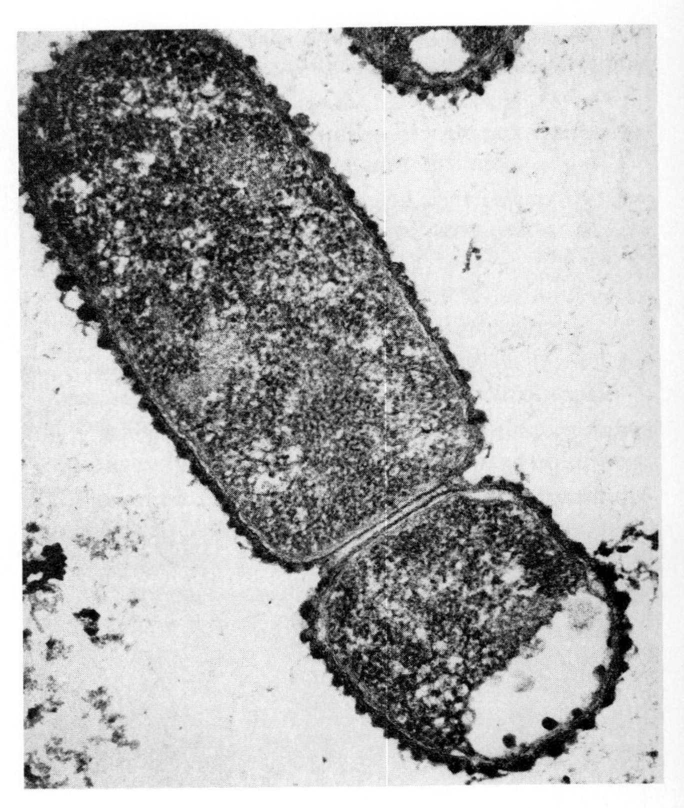

FIGURE 15.11 These cells of *Haemophilus* are surrounded by capsular material. (Courtesy of Dr. T. J. Beveridge.)

sources of nitrogen. These bacteria are all parasites, some of which cause serious diseases in domestic animals and birds.

Pasteurella Infections of Mammals and Birds

Species of *Pasteurella* infect the mucous membranes of the upper respiratory tract of mammals (rarely humans) and birds. *Pasteurella multocida* (pas-teu-rel′lah mul-to-ci′dah) causes hemorrhagic septicemia of cattle and buffaloes and causes fowl cholera in chickens, turkeys, ducks, geese, and wild fowl. Fowl cholera is responsible for serious economic losses to North American turkey farmers. Humans are not a major host, but can be infected with *P. multocida* from the scratch or bite of domestic dogs and cats. *Pasteurella haemolytica* has been associated with pneumonia of domestic animals and with "shipping fever" of cattle.

Haemophilus and Human Disease

All species of *Haemophilus* (hee-mof′ill-us) are obligate parasites of the mucous membranes of animals (Figure 15.11) and numerous species cause human disease. The bacteria were named *Haemophilus* (blood loving) because they require two growth factors present in blood—the X and/or Y factor. The X factor is haematin and the Y factor is NAD or NADP.

Haemophilus influenzae (in-floo-en′zee) is a major cause of meningitis in young infants. In older patients, it is a secondary invader of the lower respiratory tract, especially following viral infections. *Haemophilus influenzae* also causes inflammation of the epiglottis and ear infections, and is often isolated from patients with conjunctivitis (inflammation of the eye). *Haemophilus aegyptius* (ee-jip′tee-us) is responsible for a very contagious form of conjunctivitis known as **"pink eye."** *Haemophilus influenzae* and *H. aegyptius* are so similar that their identification is sometimes confused—both cause conjunctivitis. A third species, *Haemophilus ducreyi* (du-kray′ee), appears to be responsible for **soft chancre** or **chancroid** (shang′kroid), a sexually transmitted disease characterized by nonsyphilitic lesions in human genitals. Although *H. ducreyi* and *H. aegyptius* are isolated only from patients with disease symptoms, other species of *Haemophilus* are carried by apparently healthy individuals.

OBLIGATE ANAEROBIC GRAM-NEGATIVE BACTERIA

Obligate anaerobes are found in mud, sewage, the rumen of ruminates, and in body cavities of humans and other animals. The obligately anaerobic, gram-negative rods associated with humans grow by fermentation.

BACTEROIDES AND *FUSOBACTERIUM*

Obligate anaerobes make up more than 90 percent of the bacteria present in the human lower intestine. *Bacteroides fragilis* (bak-ter-oid′eez fra′jil-is) is a component of this bacterial flora and it is the anaerobe most often isolated from soft-tissue infections. Soft-tissue infections are associated with appendicitis, peritonitis (inflammation of the visceral cavity), heart-valve infections, septicemia, rectal abscesses, and surgical wounds of the urogenital tract. Occasionally *B. fragilis* is isolated from the human mouth and vagina.

Species of *Fusobacterium* (fyoo-soh-bak-teer′ee-um) are obligately anaerobic, nonmotile rods that are isolated from the oral cavity and upper respiratory tract of humans and warm-blooded animals. The pathogenic species cause infections of soft tissue and are secondary invaders of gangrenous tissue. Fusobacteria produce butyrate as a major fermentation product, which differentiates them from species of *Bacteroides*.

RICKETTSIAS AND CHLAMYDIAS

Some bacteria are obligate intracellular parasites and as such they must infect an animal or insect host to survive. Within an animal population they are often spread by arthropod or insect vectors. The geographical distribution of the hosts and vectors restricts the incidence of these diseases in humans.

The rickettsias (ri-ket′see-ahs) and chlamydias (klah-mid′ee-ahs) are obligate intracellular parasites that grow in animals and insect tissue. These bacteria can cause serious or even fatal human diseases. Because of their small size and intracellular growth, these agents were once thought to be viruses; however, they are bacteria that divide intracellularly by binary fission. The rickettsias and the chlamydias are distinguished by their metabolism and their mechanism of intracellular replication (Table 15.2).

TABLE 15.2 Classification of the human rickettsial pathogens

FAMILY, GENERA	MAJOR CHARACTERISTICS
Rickettsiaceae	
Rickettsia	Multiply in cytoplasm (rarely in nucleus) of host, most are never cultivated in a cell-free medium, glutamate is metabolized, glucose is not metabolized
Coxiella	Multiply in host cytoplasmic vacuoles, resistant to physical and chemical agents, endospore-like structure seen in electron micrographs
Chlamydiaceae	
Chlamydia	Unique growth cycle, multiply within vesicle, host cell provides ATP (energy parasites), glucose metabolized

RICKETTSIACEAE ARE OBLIGATE INTRACELLULAR PARASITES

Rickettsias are dependent on their host cell to supplement their own limited metabolic machinery. The rickettsias are unable to synthesize adenosine monophosphate (AMP), and so they depend on the AMP produced by their host cell. Their host cell also provides them with glutamate, which the rickettsias metabolize through the TCA cycle as their major source of carbon and energy. These bacteria are unable to grow on hexoses because they lack the enzymes necessary to metabolize either glucose or glucose-6-phosphate. Rickettsias infect and grow in the cytoplasm of their host cell (Figure 15.12). Occasionally they are found in the cell's nucleus, but never in phagocytic vacuoles (see discussion of chlamydias below).

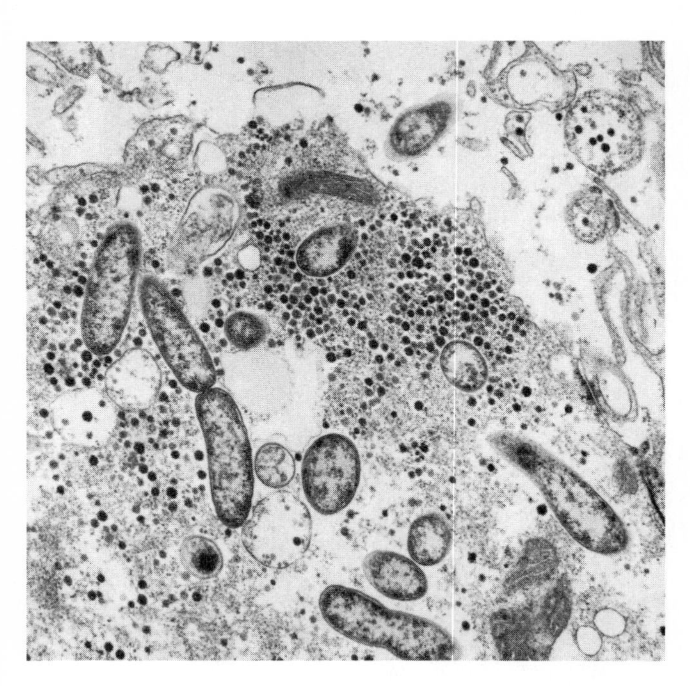

FIGURE 15.12 The spotted fever agent, *Rickettsia rickettsii,* in the salivary gland tissue of the vector tick, *Dermacentor andersoni,* 24,750×. (Courtesy of Drs. S. F. Hayes and W. Burgdorfer, Laboratory of Pathobiology, Rocky Mountain Laboratories, Hamilton, Mont.)

Because they are obligate parasites they need an animal reservoir in order to survive. Rats, mice, other rodents, and insects serve as reservoirs for the rickettsias, while humans are usually incidental rickettsial hosts. Fleas, lice, mites, and ticks function as vectors for transmitting the rickettsias between animal reservoirs and humans. Ticks can serve as both vector and reservoir for certain rickettsias. In this case infected tick eggs carry the bacteria from parent to offspring, a process known as **transovarian passage.**

Epidemic typhus is unique among the rickettsial diseases because humans are the major reservoir for the causative agent, *Rickettsia prowazekii* (prow-ah-zee′kee-eye). Epidemic typhus is treated with antibiotics, which eliminate the causative agent from its reservoir. Large outbreaks of epidemic typhus occurred during wars and in crowded urban populations infested with body lice, the vectors of *R. prowazekii.* Patients who recovered from epidemic typhus became carriers. *Rickettsia prowazekii* also causes **Brill Zinsser disease,** a recurrence of primary typhus that can erupt years after a patient's recovery. Antibiotic chemotherapy has eliminated the carrier state so the occurrence of these two diseases is very rare.

Rodents are the reservoir for *Rickettsia typhi,* which causes **flea-borne typhus fever** in humans. Fleas carry the rickettsias from the infected rodent to other rodents or incidentally to humans. Flea-borne typhus fever occurs at a low but persistent rate in areas of the world inhabited by infected rodents.

KEY POINT

An epidemic is an illness, caused by a specific agent, that affects many persons in the general population. An endemic disease is an illness affecting a population at a low but persistent rate.

Rocky Mountain spotted fever or tick-borne typhus fever occurs in the western hemisphere where *Rickettsia rickettsii* (ri-ket′see-eye) is found in wild rodents and

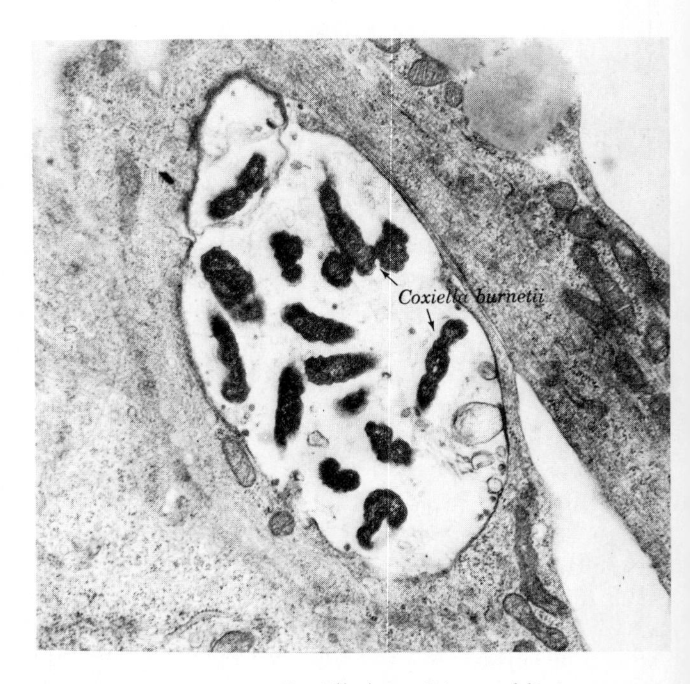

FIGURE 15.13 *Coxiella burnetii* is an obligate intracellular parasite that grows inside cellular phagosomes. (Courtesy of Dr. F. Eb. From F. Eb, J. Orfila, and J-F. Lefebvre, *J. Microscopie Biol. Cell.,* 25:107–210, 1976.)

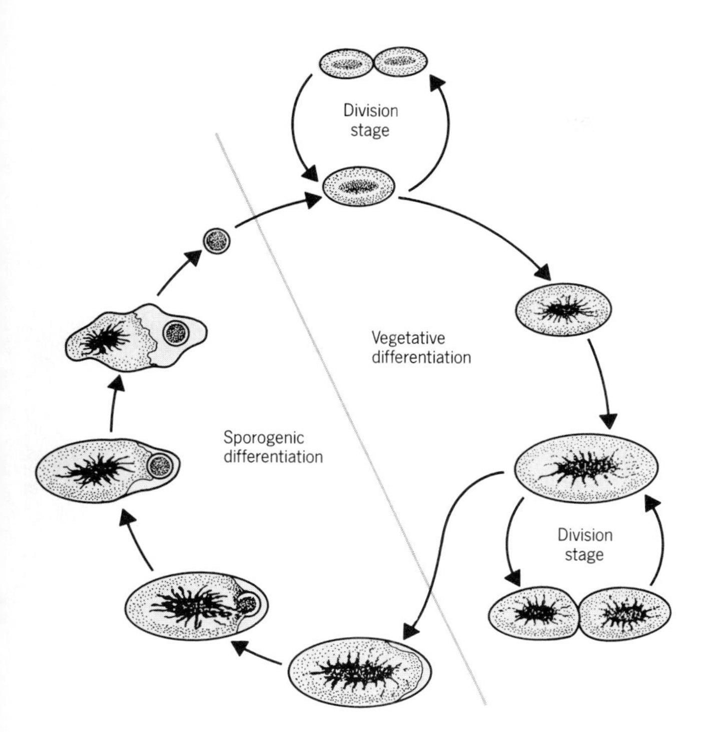

FIGURE 15.14 The developmental cycle of *Coxiella burnetti* suggests the formation of a spore-like structure, which can germinate into a vegetative cell. (Courtesy of Dr. J. C. Williams. From T. F. McCaul and J. C. Williams, *J. Bacteriol.*, 147:1063–1076, 1981, with permission of the American Society for Microbiology.)

ticks. It is the most significant rickettsial disease seen in the United States. People contract Rocky Mountain spotted fever during the summer season in areas where ticks reside.

Coxiella burnetii (kahks-ee-el'ah bur-net'ee-eye) is the cause of **Q fever** in humans. This bacterium grows intracellularly in phagocytic vacuoles (Figure 15.13) in both ticks and animal tissue. It reproduces by a distinctive developmental cycle (Figure 15.14) that includes the formation of an endospore-like body. These spore-like bodies may explain why *C. burnetii* is more resistant to chemicals and heat than other rickettsias.

Coxiella burnetii infects domestic animals such as cows, sheep, and goats and is found in their milk and placentas. Most patients with Q fever were infected when they inhaled aerosols of *C. burnetii* generated from infected animal matter. Q fever occurs predominantly in sheep country during lambing season.

CHLAMYDIACEAE ARE OBLIGATE ENERGY PARASITES

The chlamydias are obligate intracellular parasites that grow within membrane-bound cytoplasmic vesicles (Figure 15.15). They reproduce through a distinctive developmental cycle, which progresses from the small dark-staining elementary body, through an intermediate body, to a reticulate body. When the reticulate bodies divide, they give rise to elementary bodies, the only growth stage of a chlamydia that is infectious. Chlamydias are energy parasites because they depend completely on their host cell for ATP.

Humans are the major reservoir for *Chlamydia trachomatis* (trah-koh-ma'tis), which causes infections of the human eye and the urogenital tract. Chlamydia in-

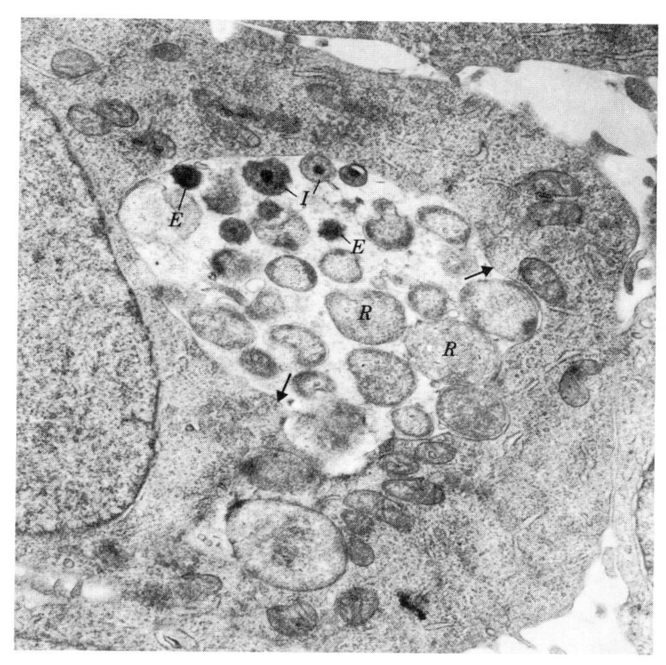

FIGURE 15.15 Electron micrograph of an HeLa cell 24 hours after being infected with *Chlamydia psittaci* showing the three stages of the chlamydial reproduction cycle. The larger light bodies are the reticulate (R) cells, the large cells with dark centers are the intermediate (I) forms, and the smaller dark-staining cells are the elementary bodies (E). Chlamydia grow within a membrane-(↑) limited phagosome. 21,000×. (Courtesy of Dr. F. Eb. From F. Eb, J. Orfila, and J-F Lefebvre, *J. Microscopie Biol. Cell.*, 25:107-210, 1976.)

fections of the female vagina cause one of the most widespread sexually transmitted diseases. These infections can lead to sterility in the female. Infection of the eyes of an infant born of an infected mother results in inclusion conjunctivitis. The same bacterium causes **trachoma,** a severe inflammation of the eye that is the leading cause of blindness in the world. Trachoma is common in the Middle East where the eyes of children are infected by contaminated towels and other inanimate objects.

Many species of birds are reservoirs for the chlamydias that cause **psittacosis** (sit-ah-koh'sis), which is primarily a disease of parrots and other psittacine birds. Humans are infected by inhaling *C. psittaci* present in the dried discharges of infected birds, especially parrots.

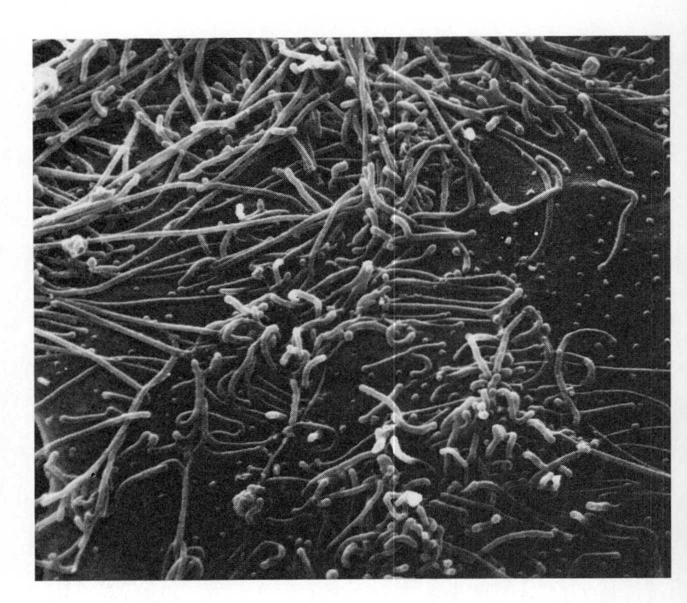

FIGURE 15.16 Scanning electron micrograph of *Mycoplasma* showing the pleomorphic nature of this bacterium that is unable to make a cell wall. (Photograph by Dr. David M. Philips, The Population Council.)

MYCOPLASMAS LACK CELL WALLS

Mycoplasmas (my-koh-plaz'mahs) are truly distinctive because they are biochemically incapable of making a cell wall. Their cytoplasmic membrane is the outermost layer of these small bacteria. The absence of a cell wall makes them very sensitive to osmotic pressures and quite pliable. To compensate, mycoplasmas strengthen their cytoplasmic membranes with sterols derived from their growth environment. The pliability of the mycoplasmas enables them to pass through filters that retain almost all other procaryotes.

L-phase variants lack cell walls when they are grown in the presence of inhibitors of cell-wall synthesis. L-phase variants or L-forms probably develop as laboratory artifacts because they can revert to cells with intact cell walls once the inhibitors of cell-wall synthesis are removed. This is in contrast to the mycoplasmas which are permanently unable to form cell walls.

Most of the mycoplasmas live in association with animals, plants, or insects. Species of *Mycoplasmsa* grow as branching filaments or elongated cells (Figure 15.16) and are found in animals, including sheep, goats, rats, and humans. They reproduce by simple binary fission, or their filaments form chains of individual cocci that eventually break apart to become independent cells.

Mycoplasmas cause disease in domestic sheep, goats, cattle, and humans. In cattle they cause pleuropneumonia, mastitis (inflammation of the mammary glands), and eye infections. *Mycoplasma pneumoniae* is the bacterium responsible for **atypical pneumonia** or "walking pneumonia" in humans. This lung infection is atypical because the causative agent is difficult to culture from infected patients. Another mycoplasma, *Ureaplasma urealyticum* (u-re-ah-plas'mah u-re-ah-ly'ticum), infects the human urogenital tract and causes nongonococcal urethritis.

KEY POINT

Because mycoplasmas lack cell walls, they are resistant to antibiotics that inhibit cell-wall synthesis and are unaffected by lysozyme, an enzyme that degrades the peptidoglycan of bacterial cell walls. Mycoplasmal infections are treated with broad-spectrum antibiotics that inhibit protein synthesis.

GRAM-POSITIVE BACTERIA

Classification of the gram-positive bacteria at the genus level relies on their morphological and physiological

TABLE 15.3 Morphological and physical characteristics of selected medically important gram-positive bacteria

GENUS	MORPHOLOGICAL AND PHYSIOLOGICAL CHARACTERISTICS
Streptococcus	Cocci in pairs or chains; catalase negative, fermentative
Staphylococcus	Cocci in clusters; catalase positive, facultatively anaerobic
Bacillus	Endospore-forming rods; aerobic or facultatively anaerobic
Clostridium	Endospore-forming rods; fermentative
Corynebacterium	Slightly curved rods with irregular stained segments; aerobic to facultatively anaerobic
Propionibacterium	Pleomorphic short rods; anaerobes that produce propionic acid and CO_2 as fermentation products
Nocardia	Branching mycelia; aerobes
Actinomyces	Branching mycelia; anaerobic to facultatively anaerobic
Mycobacterium	Acid-fast rods (difficult to stain with Gram stain); aerobes

characteristics (Table 15.3). The gram-positive cocci can be differentiated by patterns of cell division that result in the formation of diplococci, tetrads, streptococci (chains of cocci), clusters, and sarcina. The gram-positive rods are divided into those that form spores *(Bacillus, Clostridium)* and the asporogenous bacteria such as *Lactobacillus*. Branching gram-positive bacteria are assigned to separate genera such as *Nocardia* (no-car'di-ah) and *Actinomyces* (ak-tin-oh-my'sees). The unusual lipid content in the cell walls of *Mycobacterium* (my-koh-bak-teer'ee-um) species gives them the acid-fast staining properties that provide a morphological approach to their classification and identification (see Chapter 3).

Some gram-positive bacteria cause human infections that vary in severity from a simple skin boil to anthrax, diphtheria, and tuberculosis. Among the most prevalent human diseases are those caused by the gram-positive pyogenic (pus producing) cocci.

THE PYOGENIC COCCI

Species of *Streptococcus* and *Staphylococcus* (staf-il-oh-kok'us) exist as part of the normal flora of the human skin, mouth, nasopharynx, and vagina. When these bacteria cause disease they attract white blood cells to the infection site, so they are referred to as the **pyogenic cocci.** The streptococci and the staphylococci cause pyogenic diseases ranging in severity from boils to generalized septicemias.

Pyogenic Streptococci

Although many streptococci are harmless to humans (Chapter 14), certain species produce toxins and capsules that enable them to infect humans and cause disease. Streptococci are easily isolated on laboratory media and characterized as gram-positive, spherical cells that grow in pairs or chains (Figure 15.17).

Streptococcus pyogenes (py-ah'jen-eez) is the pathogen that causes **streptococcal pharyngitis** (fa-rin-jy'tis) or "strep throat" in humans. The bacterium infects the back of the throat causing swelling, redness, and other symptoms of a sore throat. Strep throat is diagnosed by isolating *S. pyogenes* or by using a rapid ten-minute immunological test (co-agglutination) specific for this species. Positive cases are treated with antibiotics to prevent late sequelae owing to this bacterium. Certain strains of *S. pyogenes* produce an erythrogenic toxin that causes **scarlet fever** in humans. *Streptococcus pyogenes* can also infect the human skin to cause the localized purulent dermatitis known as **streptococcal impetigo** (im-pah-ty'goh).

Streptococcus pneumoniae is the encapsulated diplococcus responsible for most cases of bacterial pneumonia in humans. It is a normal inhabitant of the upper respiratory tract from where it infects the lungs of compromised individuals. A few patients, especially those over 70 years old, die despite antibiotic chemotherapy. Other streptococci cause disease in newborns, nosocomial infections, and participate in tooth decay.

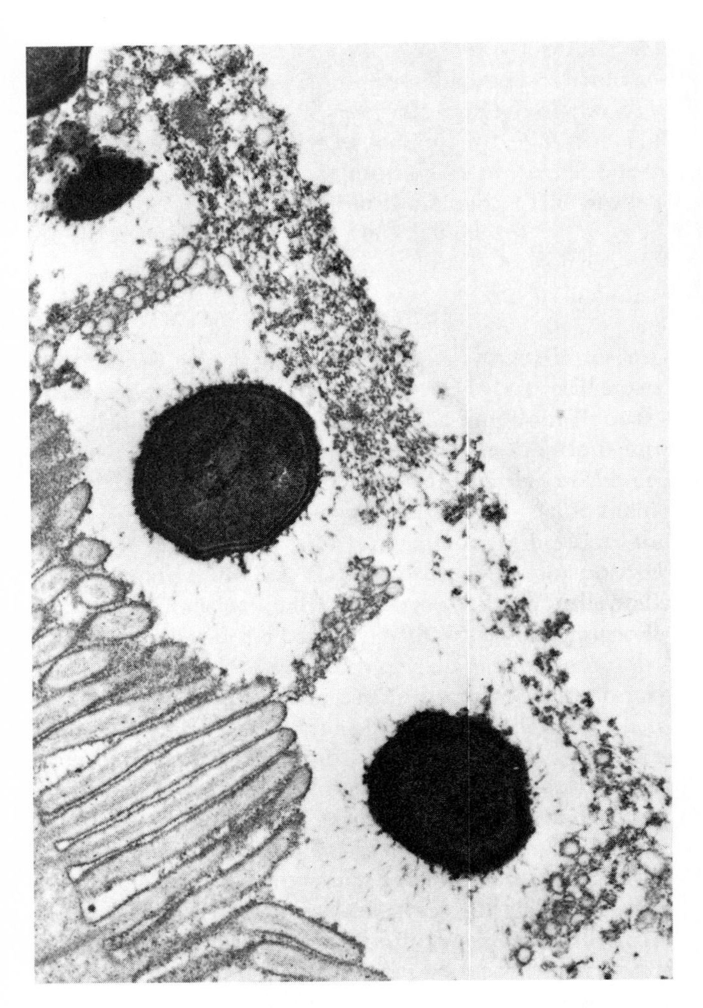

FIGURE 15.17 Transmission electron micrograph of streptococci attached to the duodenal brush border in a chicken infected with *Streptococcus faecium*. (Courtesy of R. Fuller. From R. Fuller, S. B. Houghton, and B. E. Booker, *Applied Environ. Microbiol.*, 41:1433–1441, 1981.)

Staphylococci

Staphylococci are facultatively anaerobic, gram-positive cocci that grow in clusters (Figure 15.18). *Staphylococcus aureus* (aw'ree-us) is a common inhabitant of the human skin, nasal cavity, mucous membranes, and intestine. *Staphylococcus aureus* is responsible for the majority of staphylococcal infections in humans. Among the diseases it causes are localized pyogenic infections of the skin such as boils, wound infections, impetigo, and toxic shock syndrome. Staphylococci

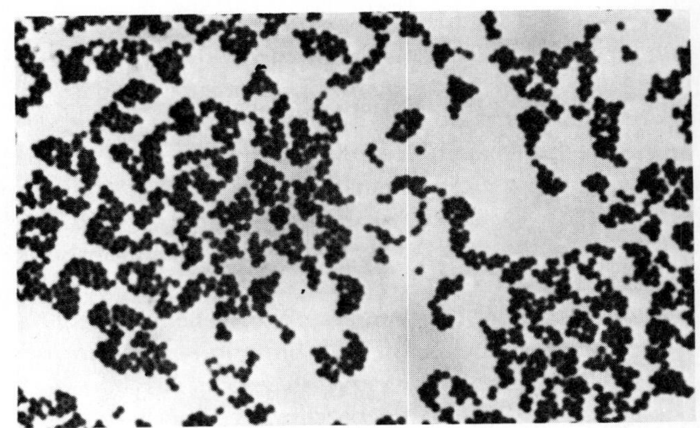

FIGURE 15.18 Gram stain of staphylococci showing their typical cell clusters.

also cause systemic infections that can lead to septicemia, pneumonia, and bone infections (osteomyelitis).

About one half the *Staph. aureus* strains produce a heat-stable enterotoxin that is responsible for **staphylococcal food poisoning.** The toxin is produced while the bacteria are growing in custards, stuffing, and creamed foods that become contaminated during preparation. Unfortunately, warming contaminated foods before serving does not destroy the enterotoxin responsible for the symptoms of staphylococcal food poisoning.

ENDOSPORE-FORMING RODS

Of the six bacterial genera known to produce endospores, only a few species of *Bacillus* and a few species of *Clostridium* (klaw-stri'dee-um) cause disease in humans. Most sporeforming bacteria are widely dispersed in nature and do not cause disease.

Bacillus and *Clostridium*

Species of *Bacillus* are either aerobic or facultatively anaerobic gram-positive rods that form endospores. *Bacillus anthracis* (an-thray'sis) is the organism responsible for anthrax, which is often a fatal disease of cattle and sheep. Its spores persist for a long time in contaminated fields and on hides and wool derived from infected animals. Humans who work with contaminated animal hides or wool are the most likely people to contract anthrax. In humans, *Bacillus anthracis* can cause

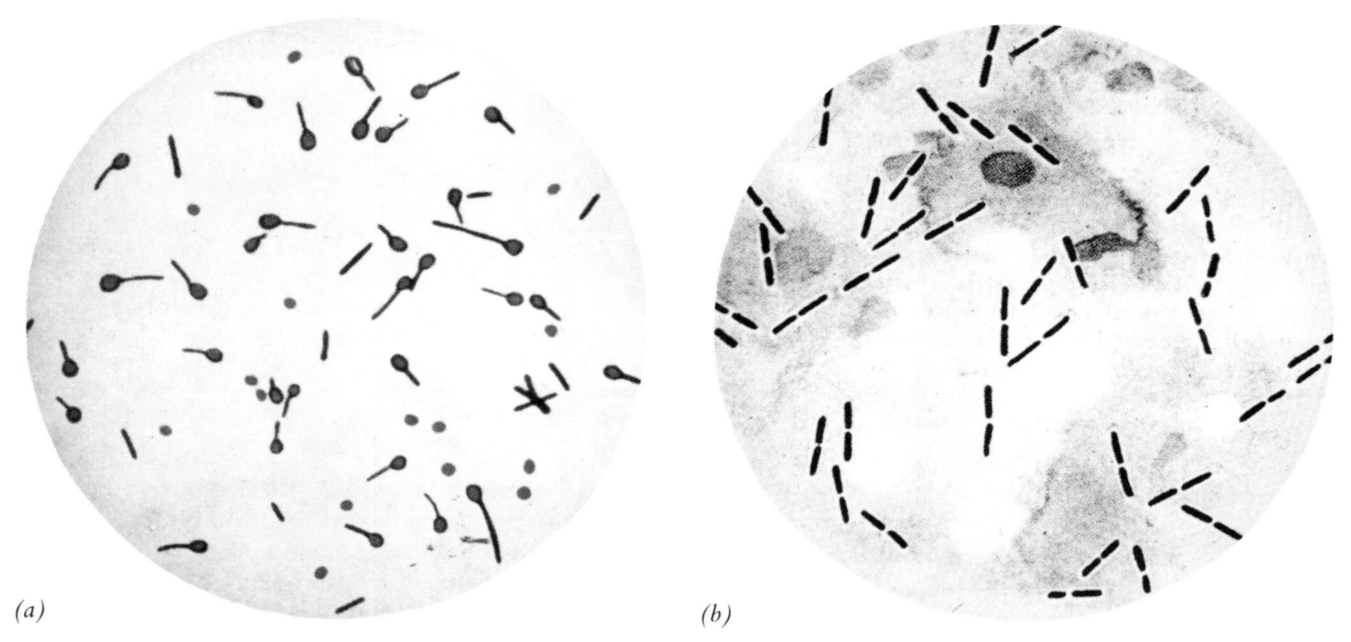

FIGURE 15.19 *(a)* Spore stain of *Clostridium tetani* showing the terminal position of the spores. (Courtesy of the Center for Disease Control, Atlanta, Ga.) *(b)* Gram stain of *Clostridium perfringens* from gas gangrene specimen. (Courtesy of the National Medical Audiovisual Center.)

pulmonary anthrax if the spores are inhaled, or cutaneous anthrax if the spores enter via abrasions on the skin. Human anthrax is a very rare disease.

Bacillus food poisoning is caused by endotoxin-producing strains of *Bacillus cereus* (ce′re-us). Consumption of the endotoxin, often in contaminated rice or rice dishes, causes abdominal pain, diarrhea, and vomiting. The toxin is produced when the bacterium grows in food. This type of food poisoning can be prevented by refrigerating prepared food and/or by heating the food to 56°C for 5 minutes to destroy the enterotoxin.

Two species of the anaerobic, spore-forming, rod-shaped clostridia produce powerful toxins that cause serious human intoxications. *Clostridium botulinum* (bot-choo-ly′num) produces the botulism neurotoxins that cause **botulism,** a potentially lethal intoxication of humans, animals, and birds. Humans consume the botulism toxin produced by this bacterium in canned foods, such as fish and vegetables. The toxin can be lethal if the patient is not treated quickly with the appropriate antitoxin. Another clostridia, *Cl. perfringens* (per-frin′jens), is a major cause of bacterial food poisoning in the United States; fortunately this type of food poisoning is nonlethal.

Clostridium tetani (te′tan-eye) is a common soil organism infecting humans who sustain deep tissue wounds (war, accidental, or puncture wounds) that become contaminated with soil. These clostridia (Figure 15.19) are able to grow in deep anaerobic wounds where they produce the tetanus toxin. Humans are immunized against tetanus with the DTP vaccine. Other soil clostridia enter tissue wounds and cause gaseous gangrene.

BRANCHING BACTERIA

Nocardias (no-kar′dee-ahs) and actinomyces (ak-tin-oh-my′sees) grow as long, continuous protoplasmic cylinders known as mycelia (Figure 15.20). Nocardias

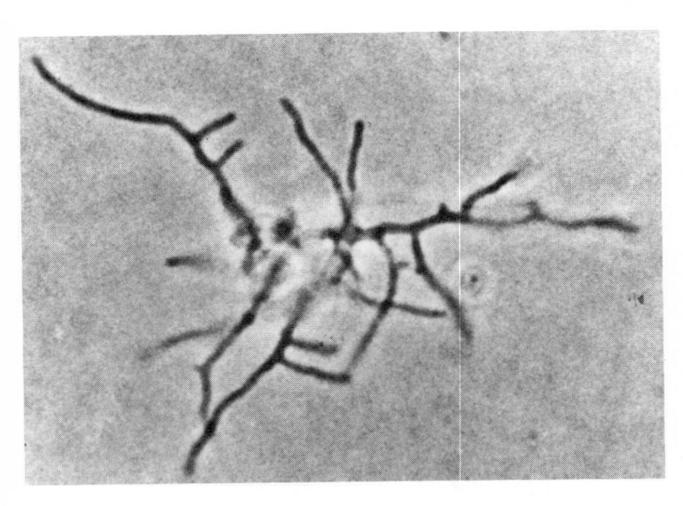

FIGURE 15.20 Photomicrograph of *Nocardia asteroides* illustrating its branching mycelial growth. (Courtesy of Drs. R. J. Hawley and T. Imaeda.)

are gram-positive, aerobic soil bacteria that are somewhat acid-fast. *Nocardia asteroides* (as-ter-oid'eez) is the cause of human lung infections known as **nocardiosis.** This disease begins when *N. asteroides* in soil dust is inhaled into the lungs where it grows into a mass of intertwining mycelia. The infection can spread to the central nervous system and cause death. Nocardias can also cause chronic abscesses under the skin known as **mycetomas.**

Actinomyces are gram-positive, anaerobic, to facultatively anaerobic branching bacteria. *Actinomyces israelii* (is-ray'li-eye) causes chronic destructive abscesses in human connective tissue, referred to as actinomycosis. This bacterium is a normal inhabitant of human tonsils and gums, that causes disease in compromised or traumatized patients, such as the tissue trauma resulting from tooth extractions.

CORYNEBACTERIUM AND DIPHTHERIA

The corynebacteria either are parasites of humans, animals, or plants or occur in soil, water, and air. One species, *Corynebacterium diphtheriae* (kor-eye'nee-bak-teer'ee-um dif-theer'ee-eye), causes the serious human disease called diphtheria.

Corynebacterium diphtheriae grows in the human nasopharynx and is spread by aerosols generated when the patient coughs. Diphtheria is caused only by lysogenized strains of *C. diphtheriae*. These strains carry the β phage that has the genes coding for the diphtheria toxin. This protein toxin interferes with protein synthesis in eucaryotic cells and causes tissue damage throughout the human body. Diphtheria is prevented by immunizing young children with the DTP vaccine and is treated by the administration of antitoxin.

THE MYCOBACTERIA

Mycobacteria are responsible for tuberculosis and leprosy, two serious human diseases. The infectious agents are identified in specimens by their positive acid-fast staining reaction. The tubercle bacillus can be grown on bacteriological media; however, the leprosy bacillus has not been grown in an artificial medium.

Mycobacterium tuberculosis (my-koh-bak-teer'ee-um too-ber-kyoo-loh'sis) is an aerobic rod that is the major cause of human **tuberculosis.** Tuberculosis is spread from active tubercular patients in sputum or aerosols generated by coughing. The white blood cells in the lungs resist the infection by phagocytizing the bacteria. Many are killed by phagocytosis; however, some survive and grow inside the white blood cells. This intracellular growth together with certain immunological reactions of the host bring about the formation of tubercles, which are walled-off structures found in lung tissue. The tubercle protects *M. tuberculosis* from being destroyed and makes chemotherapeutic treatment of tuberculosis both prolonged and difficult.

Leprosy is caused by *Mycobacterium leprae* (lep'ree), a bacterium that has never been grown on artificial media. *Mycobacterium leprae* grows very slowly in the cold parts of the human body, including the extremities (toes and fingers), the face, and male testicles. Nobody knows how humans are contaminated with *M. leprae,* but once a person is infected, it takes 3 to 5 years and sometimes 40 years for the symptoms of leprosy to develop. The progressively destructive effects of leprosy can be prevented by treating patients with a sulfonamide called dapsone.

Summary Outline

MEDICALLY IMPORTANT GRAM-NEGATIVE BACTERIA

1. Spirochetes are helical bacteria that are motile by means of periplasmic flagella.
2. Treponemes are tightly coiled, anaerobic or microaerophilic spirochetes found in association with humans or animals. Syphilis, yaws, and pinta are caused by species of *Treponema*.
3. Borrelias have coarse, uneven, or irregular coils and live in association with mammals, insects, and birds. *Borrelia recurrentis* causes louse-borne relapsing fever in humans.
4. Species of *Campylobacter* are motile, microaerophilic, curved bacteria that are found in the intestinal tract of birds and animals, and contaminate food, water, and milk. They cause fever and enteritis in humans.
5. The pseudomonads are rods known for their ability to metabolize many different organic compounds. *Pseudomonas aeruginosa* infects wounds and causes urinary tract infections.
6. Species of *Legionella* infect human lungs and cause pneumonia. *Legionella pneumophila* causes legionnaires' disease.
7. Neisseria are diplococci that inhabit the mucous membranes of animals. *Neisseria gonorrhoeae* causes the sexually transmitted human disease, gonorrhea, and *N. meningitidis* can cause human meningitis.
8. *Brucella* spp. infect domestic animals and cause brucellosis or undulant fever in humans, especially in persons who work in the meat-packing and livestock industries.
9. *Bordetella pertussis* grows in the human respiratory tract and causes whooping cough.
10. *Francisella tularensis* causes tularemia in animals and humans, who become infected when they handle or ingest contaminated animal carcasses.

FACULTATIVELY ANAEROBIC GRAM-NEGATIVE RODS

1. Bacteria of the *Enterobacteriaceae* family are widely distributed in natural environments and some cause diseases of plants and animals.
2. *Escherichia coli* is a normal inhabitant of the human intestinal tract. It is a major cause of urinary tract infections and certain strains (EPEC, ETEC, and EIEC) cause diarrheal diseases in humans.
3. *Salmonella typhi* causes typhoid fever, while other species of *Salmonella* and *Shigella* cause diseases of the human intestinal tract.
4. *Serratia, Proteus*, and *Enterobacter* are opportunistic pathogens known to cause infections of the urogenital tract, especially nosocomial infections.
5. Capsulated strains of *Klebsiella pneumoniae* are a major cause of human pneumonia.
6. The cause of bubonic plague in humans, *Yersinia pestis,* exists as an enzootic infection in rodents and is transmitted to humans by fleas.
7. Cholera is a serious waterborne infection of the intestinal tract caused by *Vibrio cholerae. Vibrio parahaemolyticus* in contaminated seafood causes foodborne gastroenteritis.
8. The *Pasteurellaceae* are nonmotile rods that are parasites of warm-blooded animals. *Pasteurella* infect domestic animals, domestic birds, and wild fowl.
9. *Haemophilus influenzae* and *H. aegyptius* infect the conjunctiva of the human eye. *Haemophilus influenzae* is a major cause of meningitis in young children.

OBLIGATE ANAEROBIC GRAM-NEGATIVE BACTERIA

1. *Bacteroides* and *Fusobacterium* are genera of obligate anaerobes found in the human intestine, oral cavity, and upper respiratory tract that cause infections of soft tissues.

RICKETTSIAS AND CHLAMYDIAS

1. The rickettsias and chlamydias are obligate intracellular bacterial parasites that grow in animal reservoirs and are transmitted between hosts by ticks, fleas, lice, or mites.
2. Rickettsias multiply in their host cell's cytoplasm using glutamate as a source of energy. Species of *Rickettsia* are responsible for epidemic typhus, flea-

borne typhus fever, and tick-borne typhus fever (Rocky Mountain spotted fever).

3. *Coxiella burnetii* infects domestic animals and causes Q fever predominantly in sheep country during lambing season.
4. Chlamydias are energy parasites that reproduce with a unique growth cycle within cytoplasmic vesicles. They cause trachoma and psittacosis in humans.

MYCOPLASMAS LACK CELL WALLS

1. The mycoplasmas, which are permanently devoid of cell walls, cause disease in domestic animals and humans.
2. *Mycoplasma pneumoniae* is the causative agent of primary atypical pneumonia in humans, a disease treated with broad-spectrum antibiotics.
3. L-phase variants are eubacteria that have temporarily lost their ability to form a cell wall.

THE GRAM-POSITIVE COCCI

1. The streptococci and the staphylococci cause pyogenic diseases ranging in severity from boils to generalized septicemias.
2. Streptococci cause "strep" throat, streptococcal impetigo, nosocomial infections, and pneumonia.
3. *Staphylococcus aureus* is a common inhabitant of human nasal passages and skin that causes boils, wound infections, impetigo, and toxic shock syndrome. Certain strains of *Staph. aureus* cause staphylococcal food poisoning.

ENDOSPORE-FORMING RODS

1. *Bacillus anthracis* and *B. cereus* are aerobic endospore-forming rods that cause respectively anthrax and bacillus food poisoning.
2. *Clostridium tetani* and *Cl. botulinum* are obligately anaerobic, spore-forming rods that cause respectively tetanus and botulism.

THE BRANCHING BACTERIA

1. Nocardias and actinomyces grow as mycelia. *Nocardia asteroides* can cause pulmonary nocardiosis and *Actinomyces israelii* causes chronic abscesses of connective tissue.

CORYNEBACTERIUM AND DIPTHERIA

1. Diphtheria is caused by lysogenized strains of *C. diphtheriae* that carry the β phage coding for the diphtheria toxin.
2. Diphtheria toxin interferes with protein synthesis in eucaryotic cells and causes tissue damage in humans.

THE MYCOBACTERIA

1. Tuberculosis is caused by *Mycobacterium tuberculosis* infections of the human lung; the bacterium grows intracellularly and forms the tubercles characteristic of this disease.
2. Leprosy is caused by *Mycobacterium leprae,* a bacterium that has never been grown on an artificial medium.
3. Leprosy is characterized by progressive tissue destruction of the cold parts of the human body, including the extremities (toes and fingers), face, and male testicles.

Questions and Topics for Study and Discussion

QUESTIONS

1. Describe the different forms of motility observed among the gram-negative bacteria. Name three major bacterial groups that are distinguished on the basis of their mechanism of motility.
2. Describe the biology of two bacteria that cause sexually transmitted disease in humans.
3. Name three bacteria that can be isolated only from animal tissue. Which ones are obligate parasites and why?
4. Immunizations with the DTP vaccine protect patients against bacterial diseases. Name these bacterial diseases and the organisms that cause them.
5. Describe a gram-negative rod, a gram-positive

coccus, and a cell wall-less bacterium that are able to cause human pneumonia.

6. Compare the life history of the bacterium responsible for flea-borne typhus to the life history of the bacterium responsible for bubonic plague.

7. Name and compare the distinguishing characteristics of two bacteria that cause bacterial meningitis.

8. Describe the unique growth cycles of the *Chlamydia* and explain why these bacteria are obligate parasites. How do they differ from the *Rickettsia?*

9. What are L-phase variants and how do they differ from the mycoplasmas?

10. What cytological test is used to detect the mycobacteria? What is the basis for this test?

11. Name an opportunistic human pathogen from three different families of bacteria. What is the natural source of each and what diseases do they cause?

DISCUSSION TOPICS

1. Why are *Escherichia* and *Shigella* classified in distinct genera, when certain members of these genera are 100 percent related according to DNA homology?

2. What biochemical and cytological changes would an energy parasite have to make to become energy self-sufficient?

Further Readings

BOOKS

Konenman, E. W., and S. D. Allen, *Color Atlas and Textbook of Diagnostic Microbiology,* 2nd ed., J. B. Lippincott, Philadelphia, 1983. This textbook for clinical microbiology students contains a practical introduction to the laboratory identification of microbial agents associated with infectious diseases. The book is extensively illustrated with color plates of important microbial reactions.

Krieg, N. R., and J. G Holt (eds.), *Bergey's Manual of Systematic Bacteriology,* Vol. 1, Williams & Wilkins, Baltimore, 1984. This first volume of this standard reference source for bacterial classification covers the gram-negative bacteria of general, medical, or industrial importance.

Lennette, E. H., A. Ballows, W. J. Hausler Jr., and H. J. Shadomy, *Manual of Clinical Microbiology,* 4th ed., American Society for Microbiology, Washington, D.C., 1985. A valuable reference source for the test procedures used to identify microorganisms of clinical importance.

Skinner, F. A., and L. B. Quesnel (eds.), *Streptococci,* Academic Press, New York, 1978. This symposium covers a wide variety of topics concerning the streptococci.

Sneath, P. H. A., and J. G. Holt (eds.), *Bergey's Manual of Systematic Bacteriology,* Vol. 2, Williams & Wilkins, Baltimore, 1986. The second volume of this standard reference source for bacterial classification covers the gram-positive bacteria other than actinomycetes.

Volk, W. A., D. C. Benjamin, R. J. Kadner, and J. T. Parsons, *Essentials of Medical Microbiology,* 3rd ed., J. B. Lippincott, Philadelphia, 1986. This textbook introduces advanced students to medical microbiology.

ARTICLES AND REVIEWS

Barksdale, L., "*Corynebacterium diphtheriae* and Its Relatives," *Bacteriol. Revs.,* 34:378–422 (1970). An extensive, well-illustrated review of the bacteriology of *C. diphtheriae* that includes a section on the corynebacteriophages.

Barksdale, L., and K-S. Kim, "*Mycobacterium,*" *Bacteriol. Revs.,* 41:217–372 (1977). An extensive review of the mycobacteria and the mycobacteriophages.

Fraser, D. W., and J. E. McDade, "Legionellosis," *Sci. Am.,* 241:82–99 (October 1979). The story of the discovery of *Legionella pneumophila* as the causative agent of legionellosis.

Holt, S. C., "Anatomy and Chemistry of Spirochetes," *Microbiol. Revs.,* 42:114–160 (1978). The anatomy of the spirochetes is reviewed with extensive use of superb electron micrographs.

Razin, S., "The Mycoplasmas," *Microbiol. Revs.,* 42:414–470 (1978). An extensive review of the bacteriology of mycoplasmas and mycoplasmal viruses.

Van Ness, G. B., "Ecology of Anthrax," *Science,* 172:1303–1307 (1971). This article describes the soil ecology of *Bacillus anthracis* in relation to known outbreaks of anthrax in the United States.

C H A P T E R

16

Algae, Fungi, Protozoa, and Multicellular Parasites

OUTLINE

FOCUS OF THIS CHAPTER

The biological principles for organizing the eucaryotic microorganisms into the algae, fungi, and protozoa include their morphology, method of food acquisition, reproduction, and motility. This chapter emphasizes the biological properties, classification, and significance of the major groups of eucaryotic microorganisms. Selected multicellular parasites, including representative helminths and parasitic arthropods, are introduced.

Eucaryotic microorganisms include algae, fungi, and protozoa. Most members of these biological groups are unicellular microbes; however, certain algae and fungi exist as multicellular organisms large enough to be seen with the naked eye. All possess the eucaryotic cell structure in which physiological functions are organized in subcellular organelles. Hereditary information is contained in more than one chromosome located in a nucleus; respiration occurs in mitochondria; chloroplasts are the site of photosynthesis; and motility is mediated by cilia and flagella (see Chapter 5). Helminths and ectoparasites are included in this chapter because of their ability to cause infectious diseases or their involvement in disease transmission.

OVERVIEW OF EUCARYOTIC MICROORGANISMS

Biologists have intuitively thought of organisms as being either animals or plants—animals being the motile heterotrophs and plants being the sessile phototrophs (Chapter 13). The fungi are nonphotosynthetic eucaryotic organisms that obtain their food by adsorption, and are classified in their own kingdom, Fungi. They are unlike animals because they have distinctive cell walls and they are nonmotile, with the exception of the aquatic fungi and cellular slime molds. Even though we are most familiar with the visible fruiting bodies of the mushrooms, all of the fungi spend most of their life cycles as microscopic organisms.

Other eucaryotic microorganisms do not fit easily into the kingdom of Plantae, Animalia, or Fungi. Protozoa are similar to animals because they are motile; however, some have plantlike cell walls. Algae are studied by botanists because algae are phototrophs; however, some algae grow heterotrophically, and many are motile or have motile stages during their life cycle. The algae and protozoa occupy the middle ground of the living world and have been classified in the kingdom Protista (see Figure 1.14).

PHOTOSYNTHETIC ALGAE

Algae are the simplest photosynthetic eucaryotes. They are assumed to be photoautotrophic, even though many species of algae have not been grown in culture. Photoautotrophic algae produce oxygen during photosynthesis and grow on inorganic media with carbon dioxide as the sole source of carbon.

Algae may be unicellular, loosely organized clumps of cells (colonial), intertwining filaments (Figure 16.1), or organisms containing structures that resemble roots, stems, and leaves. Unlike higher plants, algae are nonvascular. They do not require a vascular transport system since most algal cells photosynthesize and assimilate water and nutrients directly from their surroundings. When favorable growth conditions prevail, algae tend to reproduce asexually. Many algae also reproduce sexually, a process that enriches their own kind with genetic variation.

Algae are ubiquitous in natural environments, including streams, ponds, lakes, and moist soil. Some grow in symbiosis with fungi or animals, and others are epiphytic (grow on another plant). In nature, algae are found growing on the substratum or on other plants (benthic) and as free-floating organisms (planktonic) in the upper 75 meters of the water column. The algae contribute greatly to primary productivity* in

*Conversion of CO_2 to organic carbon.

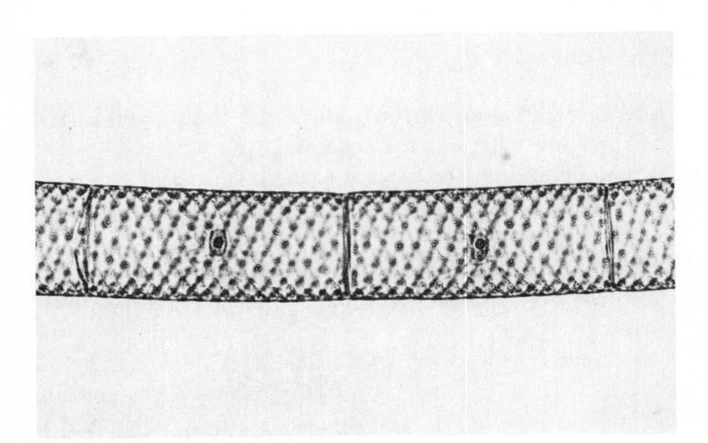

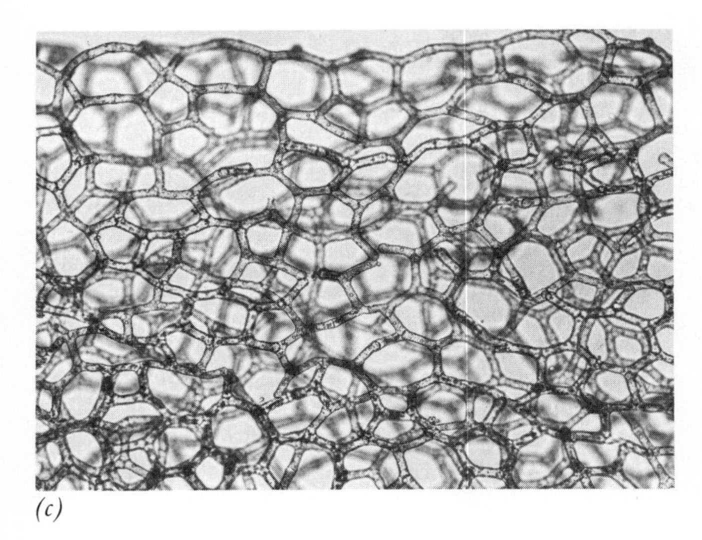

(c)

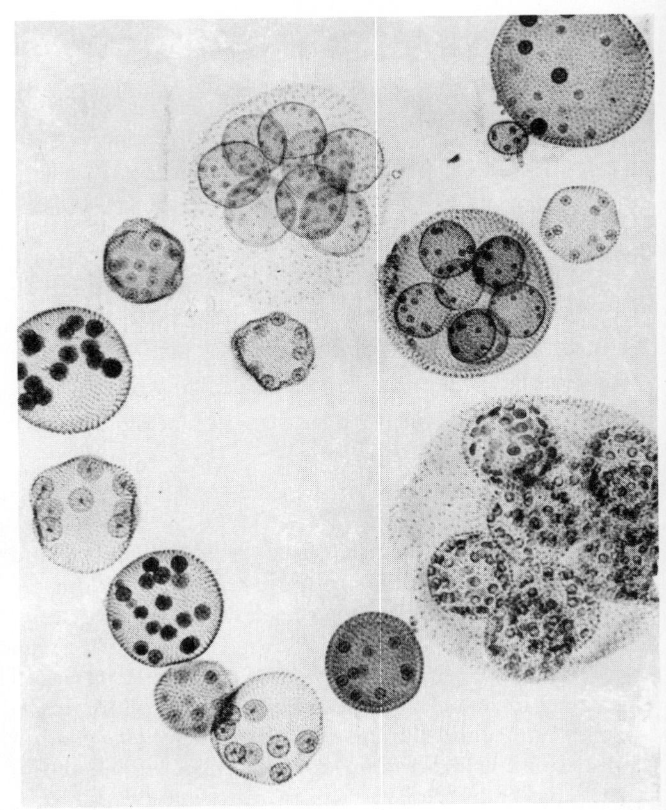

(b)

FIGURE 16.1 Morphological diversity of the green algae. *(a)* The filamentous alga, *Spirogyra, (b) Volvox,* a colonial alga, and *(c)* a nonseptate filamentous alga, *Hydrodictyon*. [*(c)* Courtesy of Dr. J. D. Cunningham.]

the oceans and other aquatic environments and are essential participants in the oxygen cycle. Worldwide estimates of the importance of algal photosynthesis suggest that they produce 50 percent of the oxygen released into the atmosphere by photosynthesis.

CLASSIFICATION OF ALGAE

Major divisions of the algae are based on their photosynthetic pigments, cell-wall composition and structure, storage products, and motility (Table 16.1). All algae contain chlorophyll *a* as their primary photosynthetic pigment. The **Rhodophyta** (rho-doh-fye'tah) resemble the cyanobacteria since they contain chlorophyll *a* and phycobilins (fie-koh-bill'ins). The other algae contain chlorophyll *a* and either chlorophyll *b* or *c*. Fucoxanthin (few-koh-zan'thin) is the accessory pigment present in brown algae, diatoms, and golden algae, whereas α, β, and/or γ carotenes occur as accessory pigments in other algae. The animal-like **Euglenophyta** (you-gleen-oh-fye'tah) and **Pyrrhophyta** (pye-roh-fye'tah) do not have a cell wall, whereas other algae possess cell walls composed of silica, cellulose, other polysaccharides, or organic acids. Some algal cell-wall components, such as agar, carrageenan (kar-a-geen'an), and algin are commercially important. Algae are able to store energy in organic compounds such as fats, oils, and carbohydrates. Many of the algae are motile by means of eucaryotic flagella, which vary in their basic structure, number, and point of insertion. All these characteristics are used to assign the algae to divisions (Table 16.1).

TABLE 16.1 Classification of the algae

DIVISION	CHLOROPHYLLS AND ACCESSORY PIGMENTS	CELL-WALL CHEMISTRY	STORAGE PRODUCTS	FLAGELLA	SPECIAL CHARACTERISTICS
Rhodophyta (red algae)	Chlorophyll *a* (+*d* in some) phycobilins	Cellulose-agar or carrageenan	Floridean starch	None	Mostly seaweeds
Phaeophyta (brown algae)	Chlorophyll *a* + *c* fucoxanthin	Cellulose + algin	Laminaran, mannitol	2, unequal, lateral	No unicellular forms, kelps
Chrysophyta (golden algae)	Chlorophyll *a* + *c* β carotene	Cellulose	Fats, oils, chrysolaminaran	2, unequal, anterior	Mainly fresh water
Bacillariophyta (diatoms)	Chlorophyll *a* + *c* fucoxanthin	Silica + pectin	Fats, oils, chrysolaminaran	Generally none	Prominent phytoplankton
Chlorophyta (green algae)	Chlorophyll *a* + *b*	Cellulose	Starch	2, equal, anterior	Possible precursors of higher plants
Euglenophyta (euglenas)	Chlorophyll *a* + *b*	None	Paramylon	1–3, equal, anterior	Protozoa-like
Pyrrhophyta (dinoflagellates)	Chlorophyll *a* + *c*	None or plates of cellulose	Starch (?), oil in some	2, unequal, lateral	Prominent in phytoplankton

Rhodophyta: Red Algae

The red algae are morphologically diverse seaweeds usually found in warm marine water, with only a few species found in fresh water. The red algae have chlorophyll *a* and phycobilins, produce cellulose as a major component of their cell wall, and store energy as floridean starch.

Agar and carrageenan are commercially important polysaccharides found in cell walls of red algae. **Agar** is a solidifying agent used in microbiological media. It is extracted from *Gelidium* (jel-id′ee-um) and *Gracilaria* (gras-ill-ar′i-ah), which are seaweeds that grow to a depth of 3 to 12 meters in warm coastal waters. In the United States, these red algae are harvested off the coasts of California and certain southeastern states. **Carrageenan** is a cell-wall polysaccharide found in the red alga, *Chondrus crispus* (kon′drus kris′pus), commercially known as Irish moss (Figure 16.2). It takes its name from the town of Carragheen, County Cork, Ireland, where the properties of this polysaccharide were discovered. Carrageenan is a thickening/smoothing agent that reacts with milk proteins and is used in ice cream, whipped cream, custards, evaporated milk, and other foods. Irish moss is commercially harvested in Maine as a source of carrageenan.

Phaeophyta: Brown Algae

The large kelps that grow off the coast of California (Figure 16.3) are characteristic of brown algae. All brown algae are multicellular organisms; there are no unicellular members of this division. They contain so much brown pigment (fucoxanthin) that it masks the presence of chlorophyll *a* and *c,* and makes these algae brown–black in color. The structural organization of the brown algae is more complex than that of other algae, with differentiated structures serving specific functions. For example, *Macrocystis pyrifera* (mak-row-

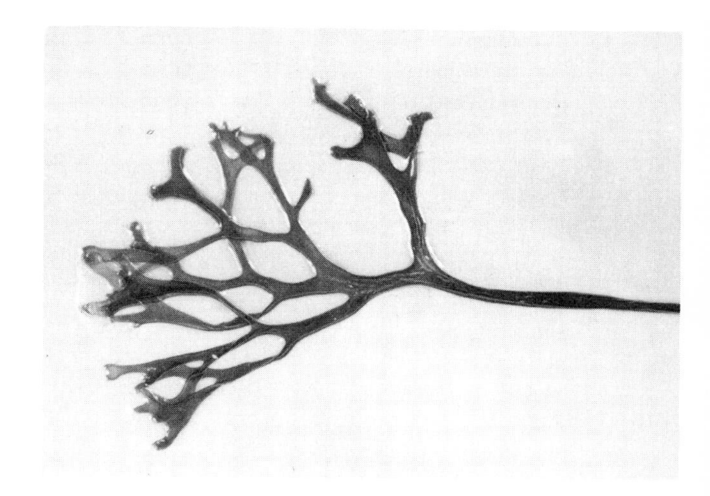

FIGURE 16.2 The polysaccharide thickening agent carrageenan is extracted from the red-alga seaweed known as Irish moss, *Chondrus crispus.* (Courtesy of Carolina Biological Supply Company.)

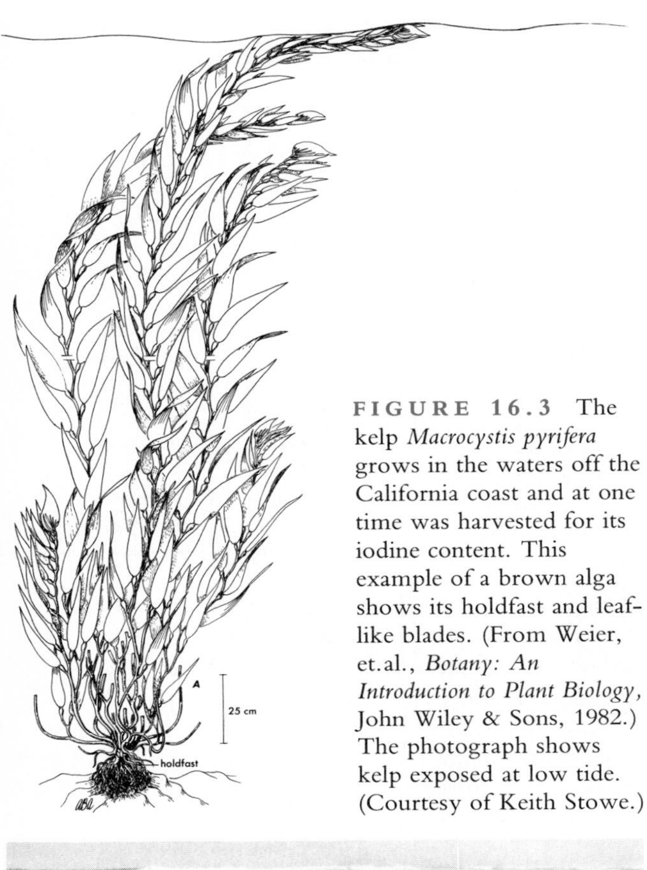

FIGURE 16.3 The kelp *Macrocystis pyrifera* grows in the waters off the California coast and at one time was harvested for its iodine content. This example of a brown alga shows its holdfast and leaf-like blades. (From Weier, et.al., *Botany: An Introduction to Plant Biology,* John Wiley & Sons, 1982.) The photograph shows kelp exposed at low tide. (Courtesy of Keith Stowe.)

sis'tiss pye-rif'er-ah) anchors itself to the substrata with its holdfast and performs photosynthesis in its leaf-like blades that move with the tidal currents (Figure 16.3).

Chrysophyta: Golden Algae

The morphologically diversified **Chrysophyta** (kris-oh-fye'tah) found in both freshwater and marine environments are the golden algae. Their golden color is caused by the presence of carotenoids, which predominate over their chlorophyll *a* and *c*. β carotene is present throughout this division, while some classes also have fucoxanthin. Many Chrysophyta are flagellated by two unequal flagella attached to the anterior end, while others are nonmotile. Storage products in this group include chrysolaminaran, fats, and oils.

Bacillariophyta: Diatoms

Diatoms are golden unicellular algae with a hard shell-like covering or **frustule** (frus'tewle) composed of silica covered by an organic layer. The frustule is arranged as two overlapping halves or valves (Figure 16.4*a*). The larger valve fits over the edges of the smaller valve to make a compartment that encloses the algal cell. The photosynthetic pigments and storage materials in diatoms are similar to those in the Chrysophyta, and for these reasons some phycologists classify diatoms as a subgroup of the Chrysophyta. Diatoms contain chlorophyll *a* and *c,* β carotene, and fucoxanthin and store energy as fats, oils, or chrysolaminaran (Table 16.1).

Diatoms are ubiquitous in nature, being found in most soils, and in marine and freshwater environments. The 200-plus genera of diatoms are identified by the intricate structure and ornamentation of their frustules (Figure 16.4*b*). These intricate designs are obvious in scanning electron micrographs especially after the cellular material has been removed from the frustule. When you look down on a diatom, you see its characteristic shape, which is either circular or elongated, corresponding to the two main orders of **Bacillariophyta** (ba'sill-air-ee-oh-fye'tah), Centrales and Pennales (Figure 16.4*b*).

Fossil diatoms are present in many terrestrial deposits over which water once stood. As the water receded or evaporated, large deposits of their cell-wall parts remained. These deposits are known as **diatomite** or diatomaceous earth. The rich deposit of diatomite near Lompoc, California is actively mined for use as a building material additive that provides bulk, a filtering agent, and an abrasive.

Diatoms have an interesting way of dealing with their rigid silica cell wall during division. When diatoms reproduce asexually by vegetative cell division, the valves of the frustule move apart to accommodate the new protoplasm. As the cell divides, it forms new

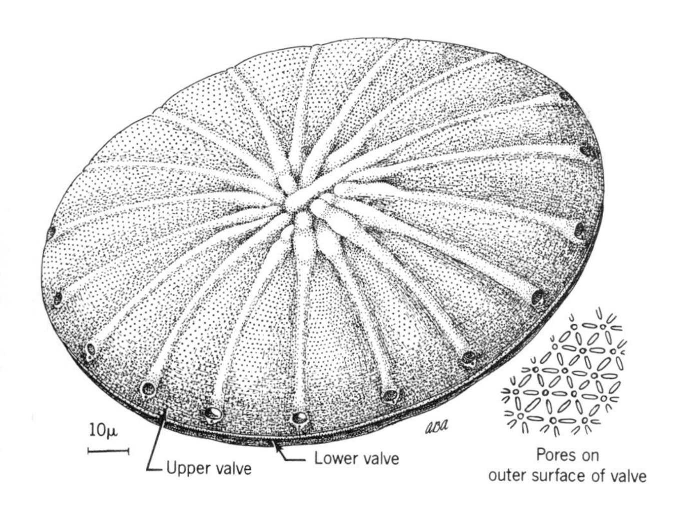

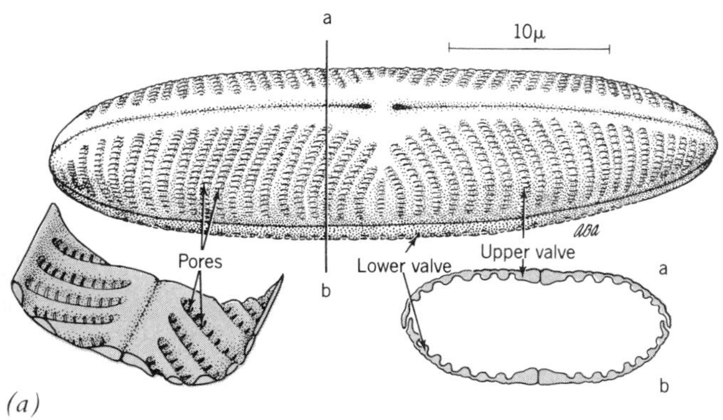

(a)

(b)

FIGURE 16.4 *(a)* A drawing of two diatoms showing their symmetry and component parts; *(b)* scanning electron micrographs of diatoms with (top) circular symmetry, *Stephanodiscus,* and (bottom) elongated symmetry, *Cymbella.* Note the small and large cell of *Stephanodiscus.* (Photographs courtesy of F. E. Round.)

valves inside the existing valves (Figure 16.5). In many species, this necessitates that the progeny cells occupy a smaller and smaller frustule. The diminution in size continues until the diatom reproduces sexually. The diploid algal cells formed from sexual reproduction construct a completely new frustule that is the maximal size for the species.

Chlorophyta: Green Algae

Green algae are widely distributed in both freshwater and marine environments where both benthic (bottom) and planktonic forms exist. They also grow in snow, in thermal springs, in moist soil, and as epiphytes. The photosynthetic pigments in the green algae are very similar to those of green plants, and some evolutionists think higher plants evolved from the green algae.

The morphological diversity among the green algae extends from simple unicellular forms to the complex colonial filamentous, and tubular forms (Figure 16.1). *Chlamydomonas* (klah-mi-doh-mon′as) is a representative unicellular green alga (Figure 16.6). Most motile green algae have two smooth flagella of equal length

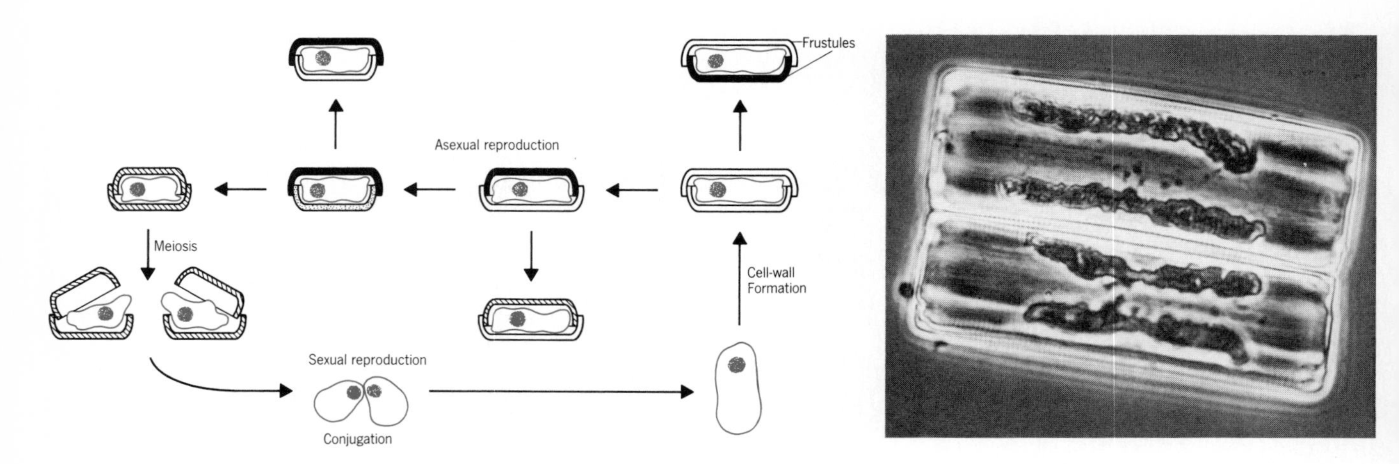

FIGURE 16.5 A diatom becomes progressively smaller during asexual reproduction. The regular size is regained following sexual reproduction. Insert: living marine diatoms in the process of asexual cell division, 2400X. (Courtesy of Walter Dawn.)

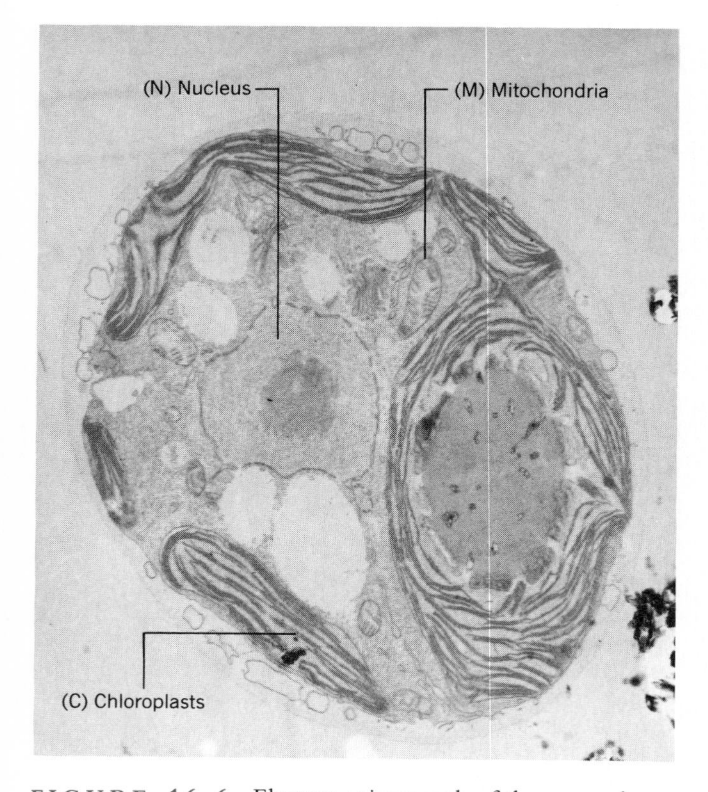

FIGURE 16.6 Electron micrograph of the green alga, *Chlamydomonas,* showing chloroplasts (c), mitochondria (m), and nucleus (n). (Courtesy Dr. Norma J. Lang, Department of Botany, University of California, Davis.)

attached at the anterior end, just like *Chlamydomonas.* The cytoplasm of *Chlamydomonas* is dominated by its nucleus and extensive chloroplast. The chloroplast is intimately associated with the **pyrenoid,** which is involved in manufacturing and/or storage of starch. Chlorophyta store starch as a reserve food in chloroplast starch granules. Mitochondria are found associated with the basal bodies of the flagella and in other areas of the cytoplasm (Figure 16.6).

Sexual reproduction occurs in the Chlorophyta with a simplified alternation of generations. *Chlamydomonas* (Figure 16.7) spends most of its life as either + or − haploid flagellated cells that reproduce by mitosis. Conjugation between these + and − algal cells results in a diploid zygote, the only diploid cell in the life cycle of *Chlamydomonas.* The zygote divides by meiosis to form two + and two − gametes. This form of algal reproduction is called a zygotic life cycle.

Euglenophyta: Euglena

Euglena occupy a transitional position between the algae and protozoa and are claimed by both phycologists and protozoologists. *Euglena granulata* (you-gleen'ah gran-u-lah'tah) is a flagellate that contains chlorophylls *a* and *b* in its chloroplast (Figure 16.8). Euglena have essentially no cell wall; instead they are bounded by a serrated plasmalemma. Their cytoplasmic organelles

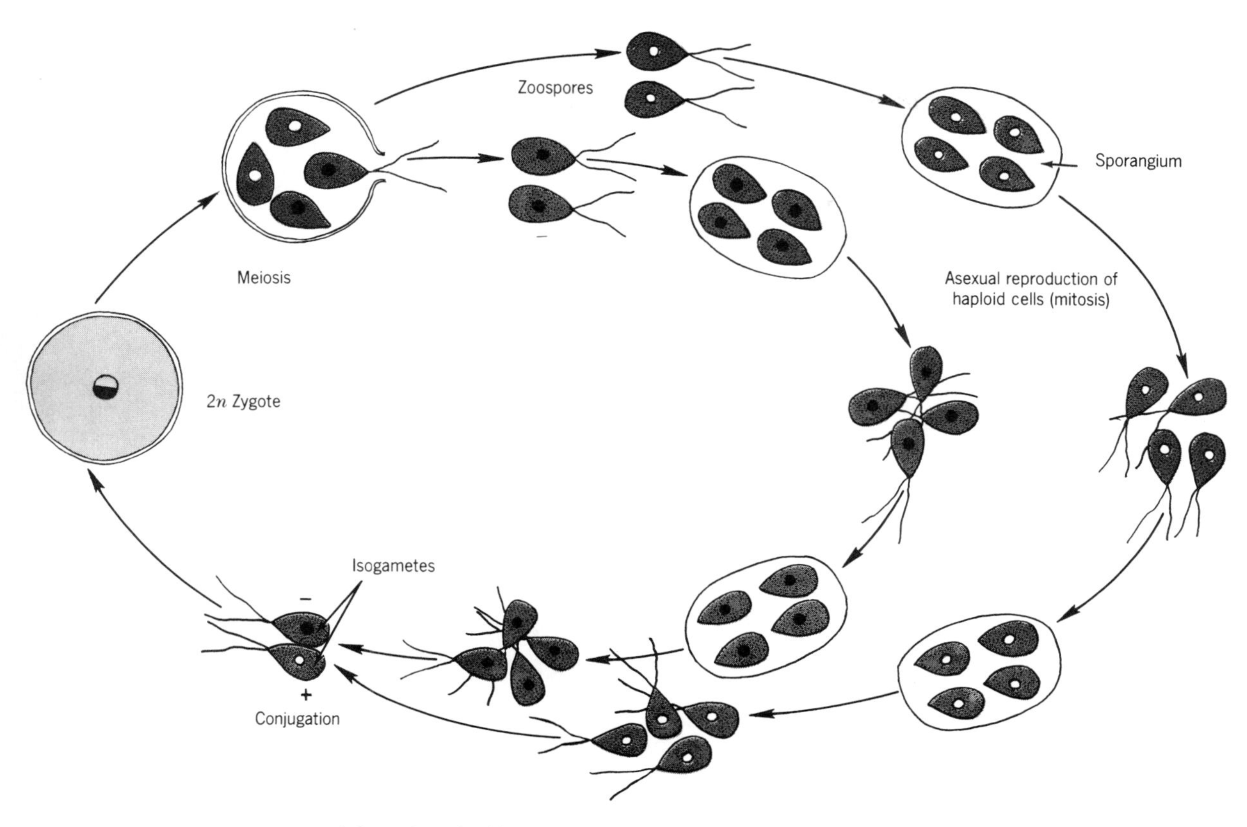

FIGURE 16.7 Zygotic life cycle of *Chlamydomonas*. (Redrawn after F. Moewns, from Weier, et. al., *Botany: An Introduction to Plant Biology*, 6th edition, John Wiley & Sons, 1982.)

include contractile vacuoles, a nucleus, chloroplasts, mitochondria, paramylon (a glucose polymer) granules, and an eyespot or stigma. Euglena deposit paramylon in cytoplasmic granules or in the dark-staining chloroplast regions called pyrenoids.

Chlorophyll containing euglena can be bleached to colorless forms by growing them at 32° to 35°C, exposing them to UV light, or by treating them with streptomycin. When the euglena's chloroplasts fail to divide as fast as the cell divides, the cell loses its chloroplast and becomes colorless. These cells cannot regain their chloroplasts since they are unable to reproduce sexually. A colorless euglena meets the definition of a protozoan, since it is motile and cannot photosynthesize.

Euglena are found in fresh water, brackish water, marine environments, moist soils, and in mud. Like some of the protozoa, certain genera of euglena can encyst to survive adverse environmental conditions. A few species have been grown in pure cultures, but most cultures are contaminated with other microbes. *Euglena gracilis* (gra′sil-is) is often used in biology laboratories to show the unique properties of this interesting microorganism.

Pyrrhophyta: Dinoflagellates

Dinoflagellates (dine-oh-fla′gell-ates) are biflagellated unicellular algae found in fresh water, brackish water, and marine environments. These morphologically diverse algae grow heterotrophically in the dark or photosynthetically in light, using chlorophylls *a* and *c*, β carotene, and xanthophylls. The heterotrophs can be saprophytic, symbiotic, parasitic, or free-living.

Certain marine dinoflagellates produce blooms in

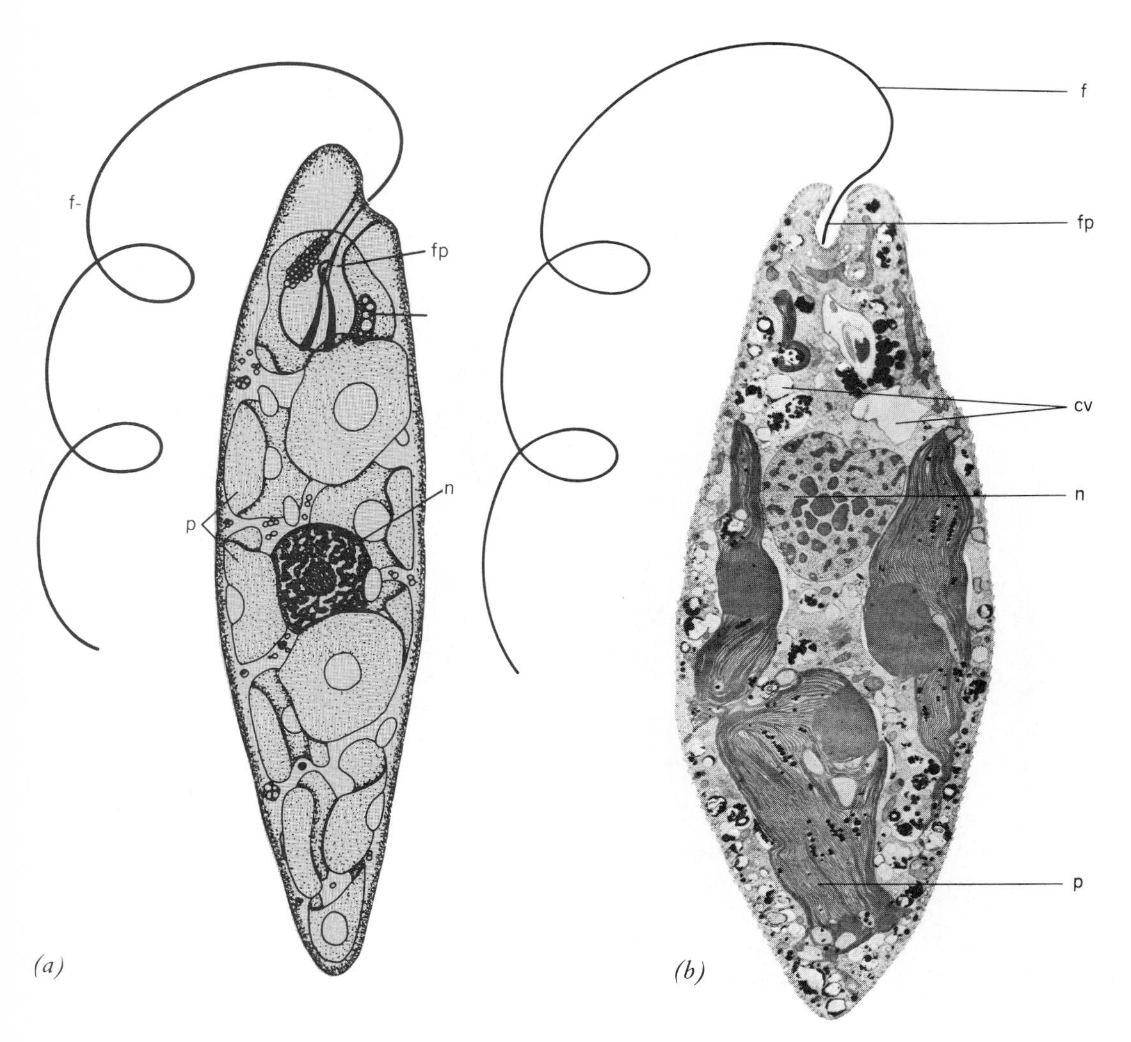

FIGURE 16.8 Drawing of *Euglena gracilis* (50 μm) showing flagellum (f), plastids (p), nucleus (n), flagellar pocket (fp). (From M. A. Sleigh, *The Biology of Protozoa,* 1973, "Contemporary Biology Series" published by Edward Arnold (Publishers) Ltd. Mainly based on information from G. F. Leedale, *Euglenoid Flagellates,* Prentice-Hall, Englewood Cliffs, N.J., 1967) *(b)* Electron micrograph of *Euglena granulatum.* (Courtesy of P. L. Walne and H. J. Arnott, *Planta,* 77:325–354, 1967.)

surface waters that color the water red. These red tides are sometimes associated with massive fish or invertebrate deaths that have been attributed to the toxins produced by the dinoflagellates. **Saxitoxin** is a neurotoxin produced by *Gonyaulax catenella* (gon-yawl'ox cat-en-ell'ah), which is found in waters off the Pacific coast. This toxin is concentrated by filter-feeding bivalve molluscs (clams, mussels, oysters, and scallops) and causes paralytic shellfish poisoning in humans who consume the shellfish. Another species, *Gonyaulax excavata* (ex-ca-vat'ah), has caused paralytic shellfish poisoning on the East coast (see Color Plate III). Shellfish should not be harvested from waters containing large concentrations of these dinoflagellates.

MYCOLOGY IS THE STUDY OF FUNGI

Mycologists are scientists who study the fungi. **Fungi** (sing. fungus) are spore-bearing, achlorophyllous organisms most of which possess a cell wall and are nonmotile. Their morphologies range in complexity from the single-celled microscopic yeasts to the large visible fruiting bodies we know as mushrooms (Figure 16.9). The part of the mushroom we see grows from hyphae (Figure 16.10) that spread as protoplasmic filaments under the ground. All fungi reproduce asexually by budding, fission, or spore formation, and most fungi also reproduce sexually. Although many fungi depend on wind and air currents to move their spores about, some of the lower fungi are ameboid and others move by their own flagella.

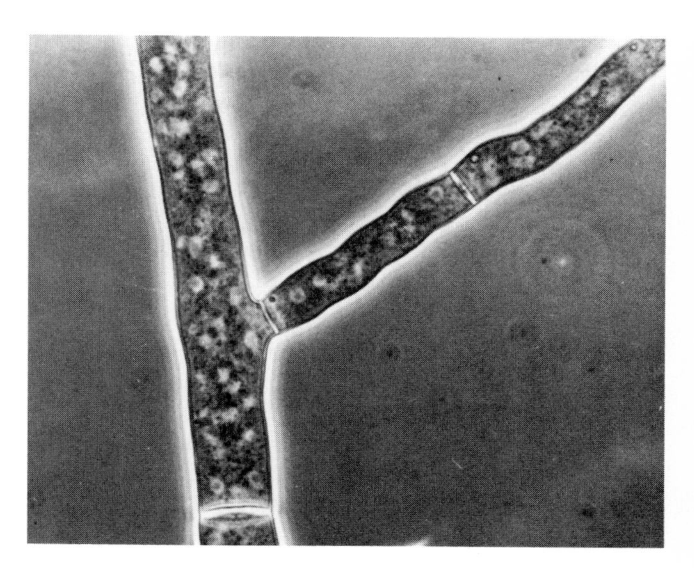

FIGURE 16.10 Light micrograph of a branching hypha in a myceliated fungus that possesses septae. (Courtesy of Dr. James F. Wilson.)

THE IMPORTANCE OF FUNGI

Fungi are ubiquitous in soil, decaying plant material, and aquatic environments. They assimilate organic compounds and other nutrients from their environment to support their heterotrophic growth. Most fungi are called **saprobes** (Gr. *sapros,* rotten; and *bios,* life) because they grow on dead organic matter. Fungi are a key factor in recycling dead vegetation matter such as fallen trees, composted garden material, and leaf litter. In addition, fungi are important because they can destroy food, fabrics, and leather; they can produce potent animal toxins; and some are parasites of plants and animals. About 100 fungal species infect humans and cause disease.

FIGURE 16.9 Stages in the growth of the poisonous mushroom *Amanita.* The fruiting bodies of mushrooms grow from the nutrients acquired through their hyphae, which penetrate soil, rotten wood, or other nutritious substrata. (Courtesy of Carolina Biological Supply Company.)

The fungal plant pathogens grow on leaves, stems, or roots causing serious damage to the plant. There are over 4000 species of **rusts** that cause tremendous economic loses to cultivated food crops. The black-stem rust of cereals, white-pine blister rust, and rusts that attack coffee plants, asparagus, beans, and flowering plants each cause special problems. The smuts, all of which are plant pathogens, form black dusty spore masses that resemble soot or smut. They cause diseases such as corn smut, loose smut of oats, onion smut, and stinking smut of wheat. Fungal infections result in millions of dollars' worth of damage to food crops and ornamentals each year.

Mushrooms and puffballs are eaten by humans and animals. (I have often watched grey squirrels eat mushrooms in the late summer before the nut crop is ready.) Morels and truffles are rare mushrooms—difficult or impossible to cultivate—that are highly prized by chefs who cook gourmet meals. Both fresh and canned cultivated mushrooms are commercially available.

One must be careful when collecting edible mushrooms since many are highly poisonous. *Amanita phalloides* (am-ah-ni′tah fal-loi′dez) produces mycotoxins that act as potent neurotoxins in humans (Figure 16.9). This mushroom is commonly called the "death angel" because of its toxicity. *Claviceps purpurea* (kla′vi-seps pur-pu-ree′ah) is a plant pathogen responsible for ergotism. It produces an alkaloid toxin known as ergot that causes hallucinations in humans who ingest the toxin or eat contaminated rye. Under controlled conditions, ergot is used as a drug to contract the involuntary muscles and to control hemorrhaging.

A number of fungi are important to humans because of the compounds they produce. For example, *Penicillium chrysogenum* (pen-i-sill′ee-um kry-soh′jee-num) produces the antibiotic penicillin; *Saccharomyces cerivisiae* (sak-ah-ro-my′seez sara-vis′ee-eye) is used in the manufacture of both alcoholic beverages and bread; and certain fungi are involved in the manufacture of cheese.

Production of Ethanol by Yeasts

Ethanol is one of the most important products commercially produced by microorganisms. The ethanol produced by the yeast sugar fermentation is used as a gasoline additive, an organic solvent, a chemical feedstock, and for human consumption. The yeast *Saccharomyces cerevisiae* (Figure 16.11) is used commercially to produce ethanol from the sugars present in corn, roots, tubers, molasses, and wood wastes. This yeast is also used to produce beer, wine, and distilled spirits.

FUNGAL MORPHOLOGY

The term **fungus** (fung′gus) encompasses both the filamentous tubular molds and the single-celled yeasts. The tubular protoplasmic structure of the molds is known as a **hypha,** the structure that differentiates molds from the single-celled yeasts. Each fungus has a predominant morphology enabling it to be characterized as a yeast or a mold. However, intermediate structures such as pseudohyphae and the ability of a fungus to grow as either a mold or a yeast often confuse beginning mycologists.

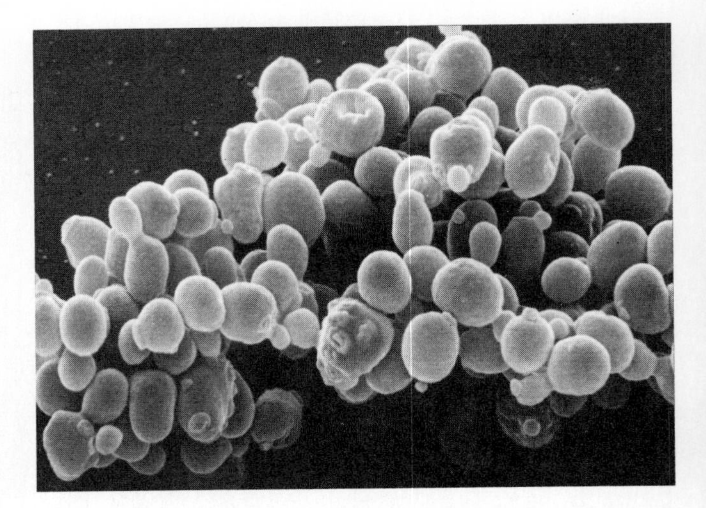

FIGURE 16.11 Scanning electron micrograph of budding yeast cells of *Saccharomyces cerevisiae,* which is employed in the manufacture of beers, bread, wines, and industrial ethanol. (Courtesy of Dr. Thorner.)

Unicellular Morphology

The term **yeast** is used to describe a single, oval, or spherical fungus that reproduces asexually by budding

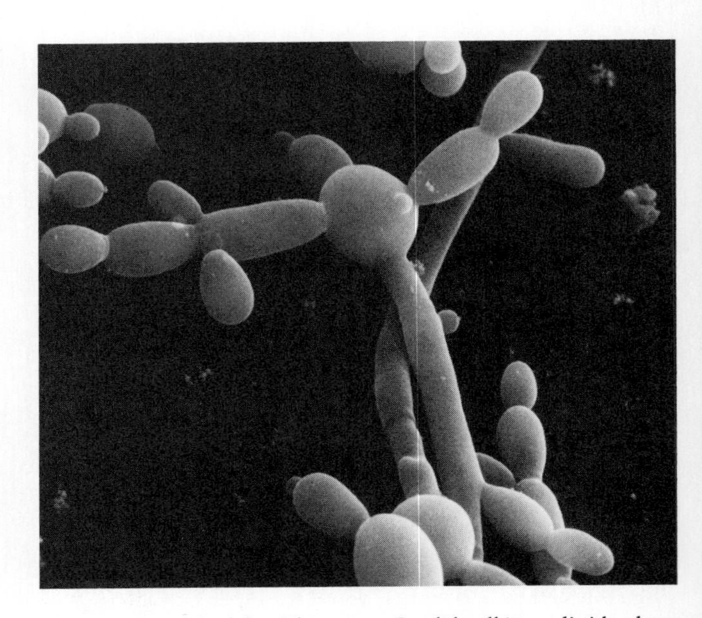

FIGURE 16.12 The yeast *Candida albicans* divides by forming buds at the tips of the pseudohyphae. (Courtesy of Dr. Garry T. Cole/BPS.)

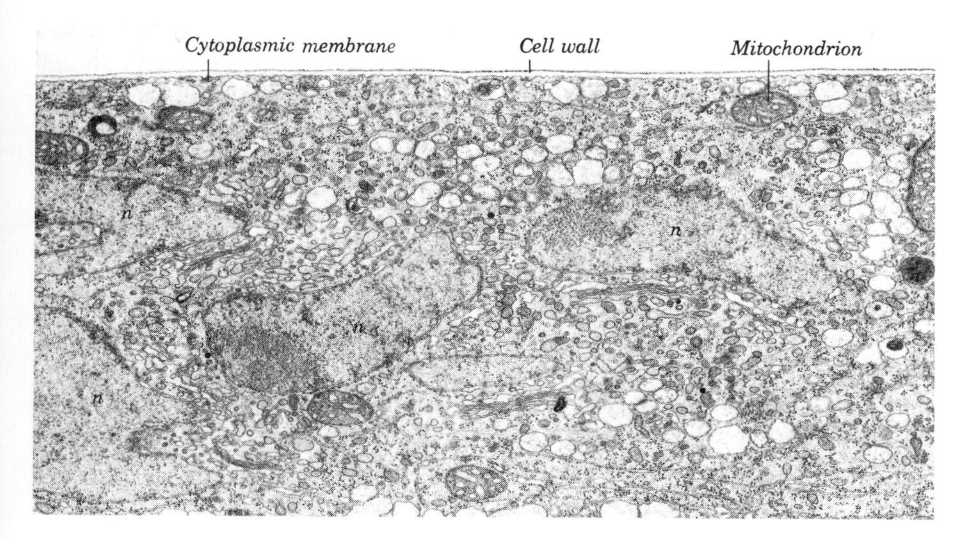

FIGURE 16.13 Electron micrograph of a longitudinal section of a hypha of *Pythium aphanidermatum*, showing the eucaryotic cell structure. This section of the coenocytic hypha contains numerous nuclei (n). (Courtesy of Dr. C. E. Bracker. From S. N. Grove and C. E. Bracker, *J. Bacteriol.*, 104:989–1009, 1970. With permission of the American Society for Microbiology.)

or fission. Microscopically, a budding yeast cell is characterized by the presence of attached buds and by the unequal size of the daughter cells. Some yeasts grow as elongated, attached cells known as **pseudohyphae** (soo-do-hi′fe), which are intermediate forms between the morphologies of yeasts and molds (Figure 16.12). You may be familiar with the bakers yeast, *Saccharomyces cerevisiae,* that is available in grocery stores in either dried packets or cakes.

Filamentous Morphology

Many fungi grow as an extensive, multibranched, interwinding hyphal mats known as a **mycelium** (mi-see′le-um). The hyphae are protoplasmic cylinders surrounded by a fungal cell wall (Figure 16.13). Some hyphae are divided into compartments by cross-wall structures called **septa** (sing. septum); while others are nonseptate. The septa divide the hyphae into compartments (cells) containing one to many nuclei. There are pores in the septa, which are usually large enough to permit the movement of nuclei from one compartment to the next. Fungi possessing a multinucleated protoplasm are **cenocytic** (se-no-sit′ik), which literally means "common cell."

MOLDS, YEASTS, AND DIMORPHISM

Colonial morphology aids in fungal identification, especially when one is dealing with human pathogens. **Molds** grow as dry, cotton-like masses of protoplasm (Figure 16.14). These filamentous fungi produce aerial hyphae that may support terminal spores, surface hyphae that laterally extend the colony, and submerged hyphae that extract nutrients from the substratum. In contrast, yeasts grow as moist, opaque, pasty, or creamy colonies (Color Plate I). Fungi are further characterized by microscopic determinations of their size, presence of capsules, cell-wall thickness, presence of pseudohyphae, and by the types of spores they form.

The **dimorphic** fungi grow in two distinct forms: as single cells (yeasts) or as hyphae (molds), depending on the growth environment. Many pathogenic fungi are dimorphic and grow as yeast cells in infected tissue at 37°C, but as a mycelium when they are cultured at room temperature. Clinical mycologists must consider fungal dimorphism when identifying clinical isolates.

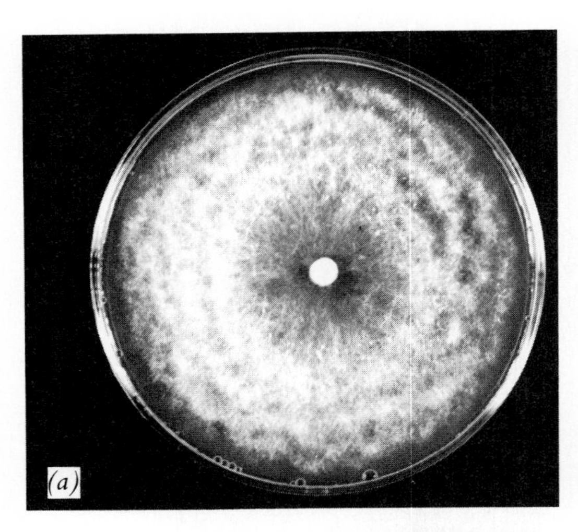

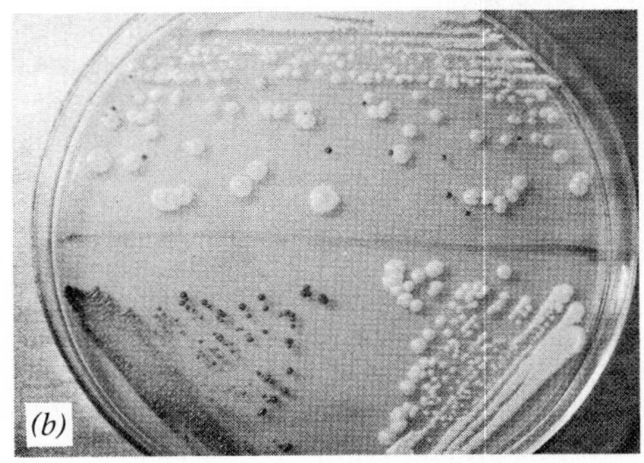

FIGURE 16.14 Growth characteristics of fungi and yeasts. *(a)* Fungi grow as cottony filamentous colonies that can rapidly spread on agar plates. (Photographs by R. W. Scheetz. Courtesy of C. J. Alexopoulos and C. W. Mims, *Introductory Mycology,* Wiley, 1979.) *(b)* Yeasts grow as distinct colonies on agar media. (Photograph by Phoebe Rich. From M. C. Campbell and J. L. Stewart, *The Medical Mycology Handbook,* Wiley, New York, 1980, copyright © 1980, by John Wiley & Sons, Inc.)

FUNGAL GROWTH AND REPRODUCTION

Yeast cells grow by increasing the size of individual cells and reproduce asexually by budding or fission to form daughter cells of unequal size. In contrast, a mold grows by increasing the mass of its mycelium by extension of its hyphae. The diameter of the hypha remains essentially constant as the mycelium increases in mass through elongation and branching.

Many fungi can be grown in the laboratory on defined media. Fungi grow as aerobic heterotrophs on defined media containing salts, NH_4^+ or NO_3^- as their source of nitrogen, vitamins (biotin and/or thiamine), and a sugar. Because many fungi prefer an acid pH (4.0 to 6.0) for growth, mycologists can grow them selectively in the presence of bacteria, most of which do not grow below pH 5.0. Antibiotics, such as chloramphenicol or penicillin, can be added to a fungal isolation medium to prevent bacterial growth.

KEY POINT

Certain precautions should be strictly observed when pathogenic fungi are grown in the microbiology laboratory. Since inhaled fungal spores can cause infections, fungal cultures should never be smelled, and petri dishes containing fungal cultures should be opened only in a bio-safety hood.

Asexual Reproduction

Fungi reproduce asexually by budding (yeasts) or by producing asexual spores. The diverse morphologies of these spores give them a prominent role in fungal classification. Asexual spores produced in a sporangium are called **sporangiospores** (spo-ran′jee-oh-spores), whereas asexual spores produced at the end of a hypha are referred to as **conidia** (ko-nid′e-ah). Macroconidia are large, multicellular conidia; small, unicellular conidia are known as microconidia. **Chlamydospores** (klam′i-do-spores) are surrounded by a thick, double-walled spore coat and are formed either on the hyphal tip or within the hypha (Figure 16.15). **Arthrospores** (ahr′thro-spores) are produced within the hyphae by septation, whereas **blastospores** (blas′to-spores) are bud-like projections formed by yeast cells. These spores function to disperse the species to new locations by air currents.

Sexual Reproduction

Sexual reproduction occurs when haploid gametes fuse to form a zygote (zi′goht). The haploid gametes *(n)* come either from reproductive structures on the same fungus, or from reproductive structures produced by different mating types of the same species (Figure 16.16). After fusion, the zygote (2*n*) divides by meiosis (reductive division) with the accompanying reassortment of chromosomes to form haploid sexual spores.

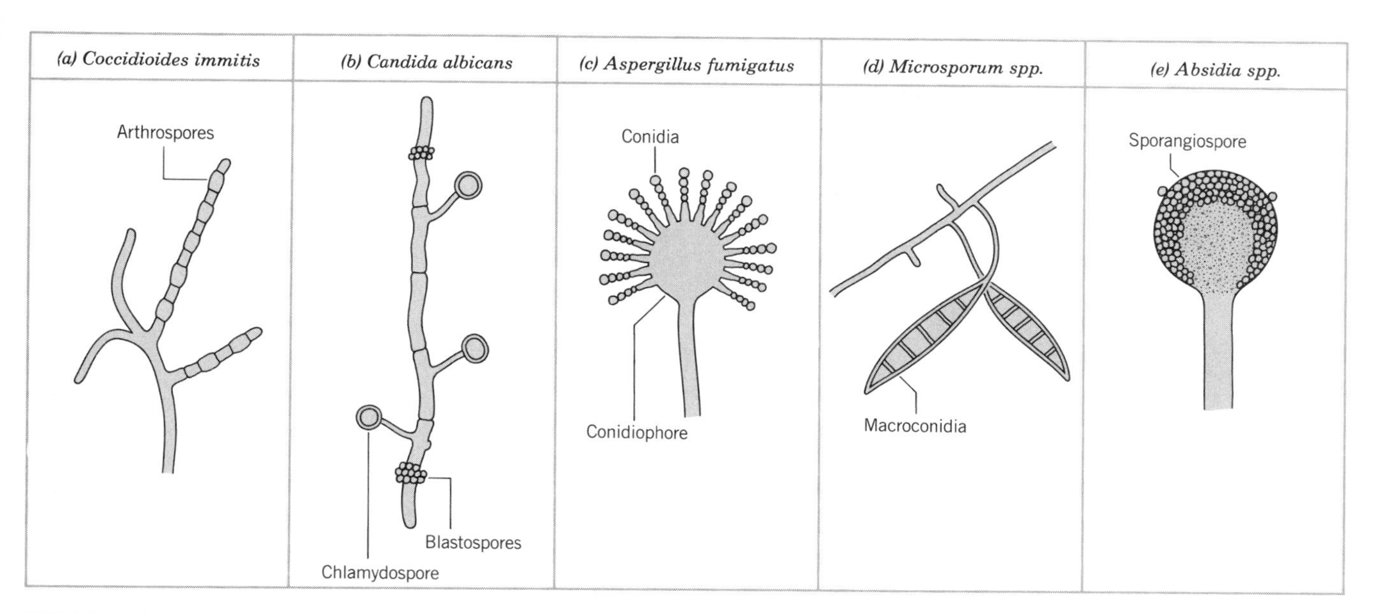

FIGURE 16.15 Asexual fungal spores. *(a)* Arthrospore formation in *Coccidioides immitis,* *(b)* chlamydospores and blastospores in the pseudohyphae of *Candida albicans,* *(c)* conidia of *Aspergillus fumigatus,* *(d)* macroconidia of *Microsporum,* and *(e)* sporangiospore of *Absidia.* *(f)* This scanning electron micrograph of *Aspergillus flavus* conidia provides a three-dimensional perspective of asexual spore formation. (Photograph courtesy of Dr. Garry T. Cole/BPS.)

The vegetative fungi that grow from these spores are haploid.

The morphology of the sexual spores (Figure 16.16) and their supporting structures are important in identifying and classifying the fungi. **Zygospores** (zi′go-spores) are sexually produced by *Zygomycetes* (zi-go-mi-se′tes), **ascospores** (as′ko-spores) are sexually produced by *Ascomycetes* (as-ko-mi-se′tes), and **basidiospores** (bah-sid′e-oh-spore) are sexually produced by *Basidiomycetes* (bah-sid′e-oh-mi-se′tez). Sexual spores are blown on the wind to new locations and function as a major means of dispersing fungal species. Once in an appropriate environment, the sexual spores germinate to form a vegetative mycelium.

CLASSIFICATION

Fungi are separated into three divisions (Table 16.2): slime molds, Gymnomycota (jim′no-mi-kot′ah); flagellated lower fungi, Mastigomycota (mas′ti-go-mi-cot′ah); and terrestrial fungi, Amastigomycota (ah-mas′ti-go-mi-kot′ah). Because they are nonmotile and possess cell walls, terrestrial fungi are distinguished from slime molds, whose vegetative cells lack a cell wall, and from motile aquatic fungi.

The cellular slime molds (Gymnomycota) exist as vegatative amebas when environmental conditions permit them to grow and replicate. These amebas lack cell walls, so they are able to ingest particulate nutrients

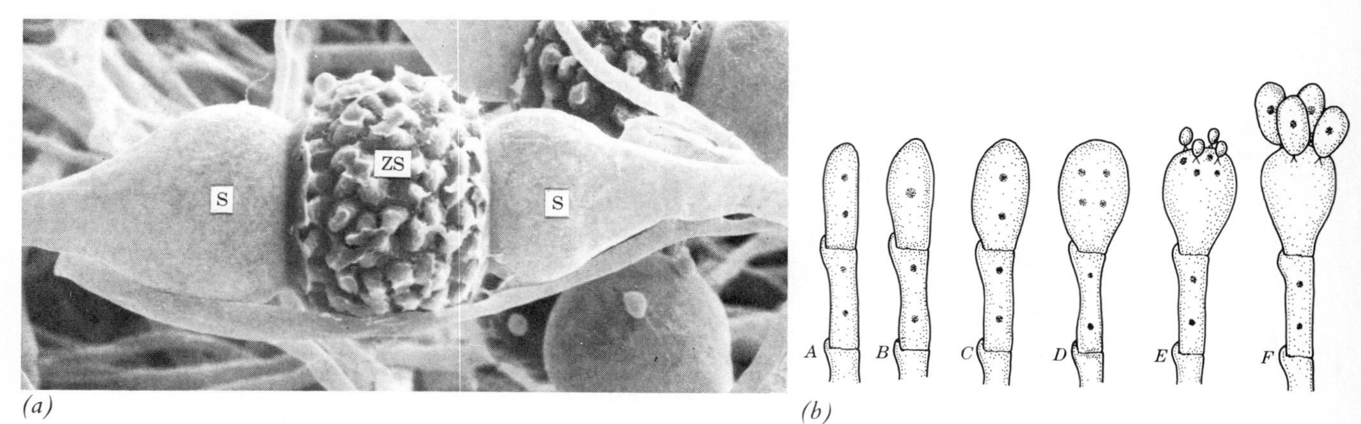

(a) (b)

(c)

FIGURE 16.16 Sexual reproduction in fungi. *(a)* Scanning electron micrograph of a mature zygosporangium (zs) of *Gilbertella persicaria.* (Courtesy of Dr. K. L. O'Donnell. From K. L. O'Donnell et al., *Can. J. Bot.,* 55:662–675, 1977.) *(b)* Six successive stages in the formation of a basidium and *(c)* scanning electron micrograph of a basidium bearing four basidiospores. (Photographs by S. L. Flegler. Courtesy of Dr. C. J. Alexopoulos and C. W. Mims, *Introductory Mycology,* Wiley, New York, 1979.)

and to move by ameboid motion. During adverse times, the amebas aggregate and form fruiting bodies (see Figure 1.16) and produce resting spores that help the species survive. The growth cycle begins again when favorable conditions exist.

The Mastigomycota are aquatic fungi that produce motile flagellated cells. Some are obligate parasites on higher plants, others are parasites of fish and their eggs. The flagellated cells of these fungi enable them to move in their aquatic environment and find new host cells to infect.

The terrestrial fungi are grouped into four classes of the division Amastigomycota. The are distinguished by sexual and asexual reproductive processes and by the presence or absence of septa. Flagella and ameboid movement are absent. Morels and truffles, *Saccharomyces,* and *Neurospora* produce sexual spores in a sac-like structure called an **ascus.** These fungi produce asexual conidia and all belong to the class *Ascomycetes.* Mushrooms and puffballs, rust, and smuts produce sexual spores on club-like cells called basidia and are classified in the class *Basidiomycetes.* Fungi that do not demonstrate a sexual form of reproduction are classified as imperfect fungi in the class *Deuteromycetes* (doo'ter-o-mi-se'tez). *Rhizopus* (ri-zo'pus) and *Mucor* produce thick-walled zygospores (Figure 16.16) during sexual reproduction and form multispores in a sporangium during asexual reproduction, characteristics that place them in the class *Zygomycetes.*

Sexual reproduction in many of the pathogenic fungi either is unknown or has only recently been dis-

TABLE 16.2 Major groups of the kingdom Fungi

DIVISION/ CLASS	DISTINGUISHING CHARACTERISTICS	REPRESENTATIVES
Gymnomycota (slime molds)	Ingest particulate nutrients, vegetative cells lack cell walls	
Acrasiomycetes (cellular)	Vegetative stage: free-living amebas that aggregate to form fruiting body	*Dictyostelium discoideum*
Myxomycetes (acellular)	Vegetative stage: multicellular, wall-less plasmodium, forms organized sporangia	*Physarum polycephalum*
Mastigomycota (flagellated lower fungi)	Aquatic fungi producing motile, flagellated cells	*Allomyces macrogynus*
Amastigomycota	Nonmotile terrestrial fungi	
Ascomycetes	Septate mycelium, sexual spores in sac-like ascus, asexual conidiospores at ends or sides of hyphae	*Neurospora crassa*
Basidiomycetes	Septate mycelium, sexual spores on surface of basidium, asexual conidiospores at ends or sides of hyphae	*Filobasidiella*
Zygomycetes	Nonseptate mycelium, zygospores formed by gamete fusion, asexual mitospores	*Absidia, Mucor, Rhizopus*
Deuteromycetes (Fungi Imperfecti)	Unicellular or septate, sexual reproduction absent, asexual conidiospores at ends or sides of hyphae	*Candida, Coccidioides, Sporothrix*

covered. Demonstrating sexual reproduction in a given isolate is not always easy; it often necessitates that the investigator isolate both mating types of the fungus. When sexual reproduction of a member of the Deuteromycetes is discovered, by definition the organism must be reclassified.

PROTOZOA ARE MICROSCOPIC ANIMALS

Protozoa—literally meaning the "first animals"—are the simplest forms of animal life. Protozoa are either free-living, residing in oceans, fresh water, and soil, or they are symbionts, living on or in other organisms. The prevalence of free-living protozoa in an aquatic environment depends on the available nutrients, temperature, pH, and, for some protozoa, on light. Many protozoa assimilate chemical nutrients, whereas the **holozoic** (eat other organisms) protozoa feed on bacteria and other protozoa. *Didinium* is an example of a holozoic protozoan as is apparent from the micrograph of *Didinium* engulfing a *Paramecium* (Figure 16.17). Of the more than 65,000 protozoan species that have been named, over half are fossil and about 10,000 are parasitic.

Symbiotic protozoa coexist on or in another organism in an intimate association that is either commensal, beneficial, or harmful. A **commensal** (ko-men'sal) interaction is one in which the protozoan gains and the other organism neither benefits nor loses. Examples of commensalism include the stalked ciliates that attach themselves to plant material and or to the gills of bi-

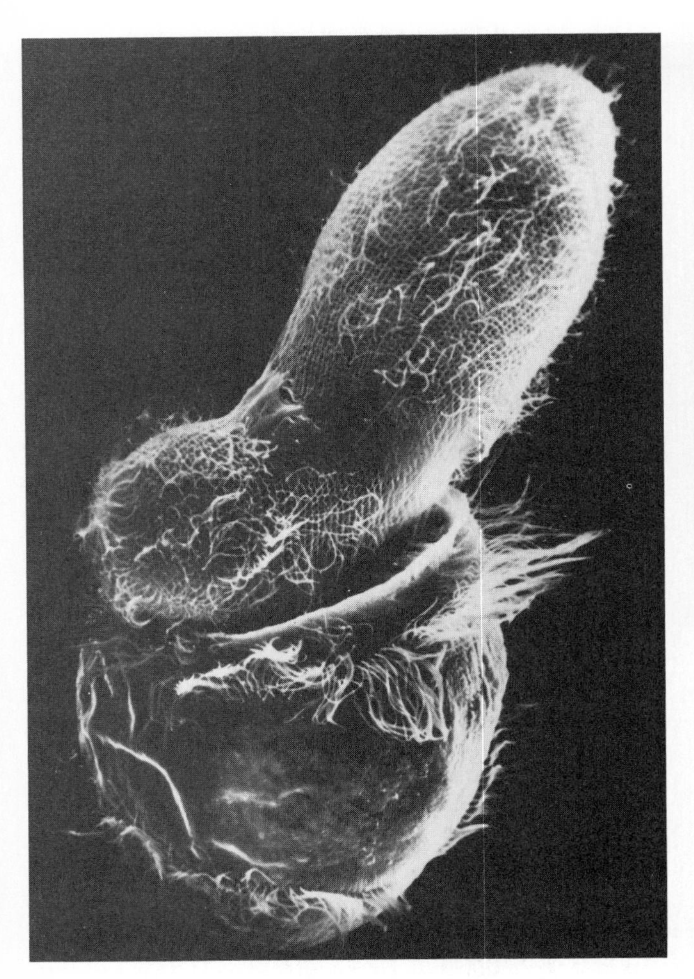

FIGURE 16.17 Scanning electron micrograph of a *Paramecium* being engulfed by a *Didinium*. (From H. S. Wessenberg and G. Atipa, *J. Protozoology,* 17:250–270, 1970.)

FIGURE 16.18 *Giardia* trophozoites present on the microvillous surface of mouse jejunum villi are visible in this scanning electron micrograph. The ventral surface of *Giardia* contains an adhesive disk for adherence to host tissue. (Courtesy of Dr. R. L. Owen. From R. L. Owen, P. C. Nemanic, and D. P. Stevens, *Gastroenterology,* 76:757–769, 1979.)

valve molluscs or fish. From these vantage points the protozoan is better able to feed on detritus, while the plant or animal serves only as a support. In the **mutually beneficial** relationship between flagellated protozoa in the termite intestine and the termite, the protozoa partially digest woody material eaten by the termite. These digests provide the termite with nutrients otherwise unavailable to it. Neither the termites nor the flagellates can survive independently. **Parasitic** protozoa benefit at the expense of their host and are responsible for numerous animal diseases. The first such parasite to be described was *Giardia lamblia* (je-ar'de-ha lam'ble-ah), seen by Anton van Leeuwenhoek in his own stools (Figure 16.18). The few protozoa that infect humans are a major cause of morbidity and mortality in the world, especially in tropical countries.

BIOLOGY OF THE PROTOZOA

Protozoa are single-celled eucaryotic animals that range in size from 1 μm to 50 mm or more with most being between 5 and 250 μm. Many of the differences between the protozoa are attributed to their specialized cellular organelles, which function in feeding, locomotion, osmoregulation, and reproduction. Most protozoa have a single nucleus, but some are multinuclear and the Ciliophora (sil-ee-of'or-ah) have nuclei of unequal size and responsibilities. The diversity of the protozoa is reflected in the structures of their subcellular organelles and their methods of sexual and asexual reproduction.

Locomotor Organelles

Motility of protozoa involves four basic organelles: cilia, flagella, undulating ridges and membranes, and pseudopodia. In many protozoa the entire flagellum is free to whip about in the medium, whereas in others the axoneme is part of an undulating membrane that lies along the outside of the animal. Gliding motility in the sporozoa (spo-ro-zo′ah) is attributed to undulating ridges beneath their outer membrane. Finally, amebas form pseudopodia that function in movement and food acquisition.

Flagella and Cilia Flagella are the locomotor organelles of the flagellates. The axoneme of a flagellate originates in the basal body, which in trypanosomes is located in the flagellar pocket (Figure 16.19). The free end of the flagellum extends into the extracellular medium and beats with a whip-like motion that imparts movement to propel the flagellate. Protozoa can have one or more flagella arranged in a variety of conformations. The flagellum of the trypanosomes is attached to an undulating membrane (see below).

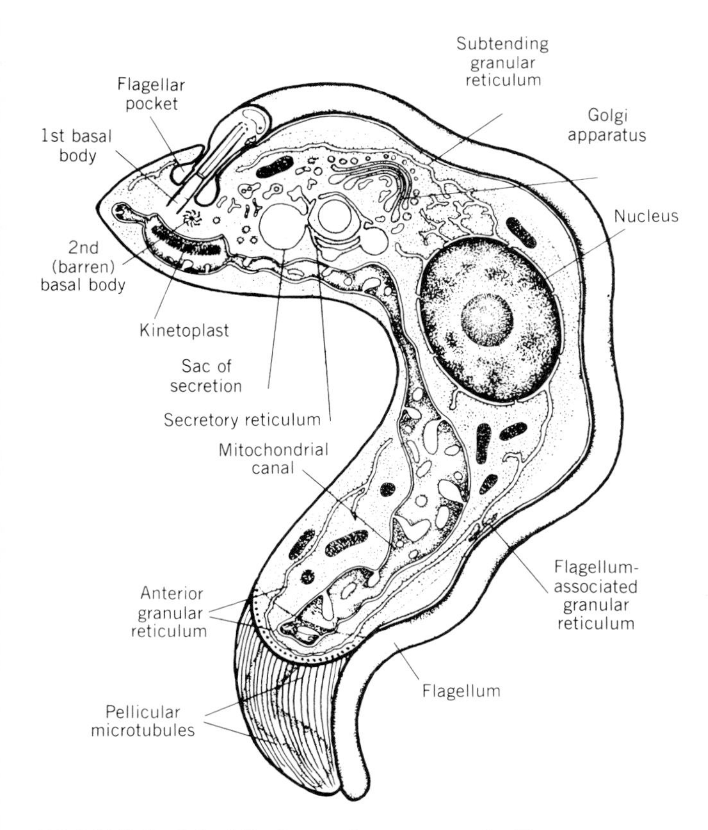

FIGURE 16.19 Internal structure of a *Trypanosoma* showing the DNA-rich kinetoplast, mitochondria, flagellar pocket, and the flagellum attached to an undulating membrane. (Courtesy of Dr. K. Vickerman. From K. Vickerman, *J. Protozoology,* 16:54–69, 1969.)

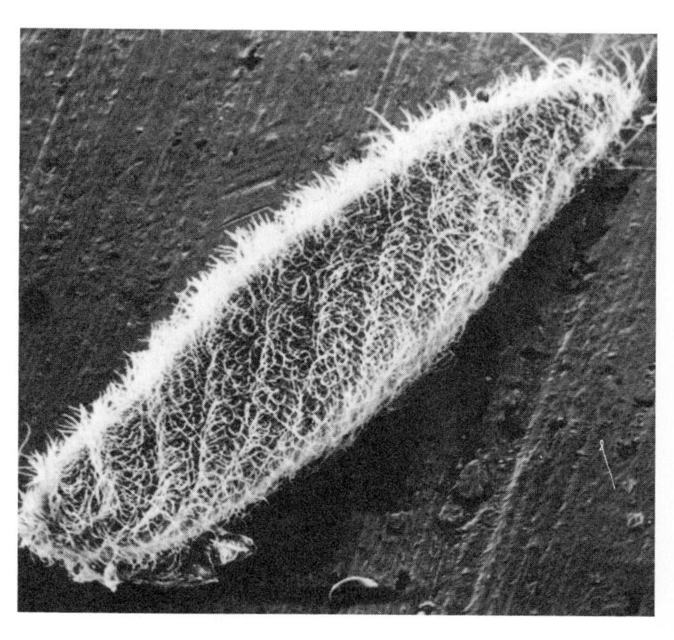

FIGURE 16.20 Scanning electron micrograph of the ciliate *Paramecium.* (Photograph by E. Vivier, from W. J. Wagtendonk (ed.), *Paramecium,* Elsevier North Holland Bio-Medical Press, New York, 1974.)

The cilia of protozoa are similar in structure to flagella; however, they beat with a rhythmic coordinated motion. Each cilium has an axoneme, a basal body, and a surrounding membrane. Unlike flagella, cilia are always found in groups or rows (Figure 16.20). The basal bodies of cilia are joined by interconnecting fibers known as **kinetodesma** (ki-ne-to-des′mah), responsible for coordinating their characteristic rhythmic motion. Cilia function as organelles of locomotion in food gathering and as tactile organelles.

Pseudopodia Amebas are noted for their ability to form pseudopodia, which are temporary protoplasmic extensions of their plasmalemma (cell membrane) that function in locomotion and food gathering (Figure

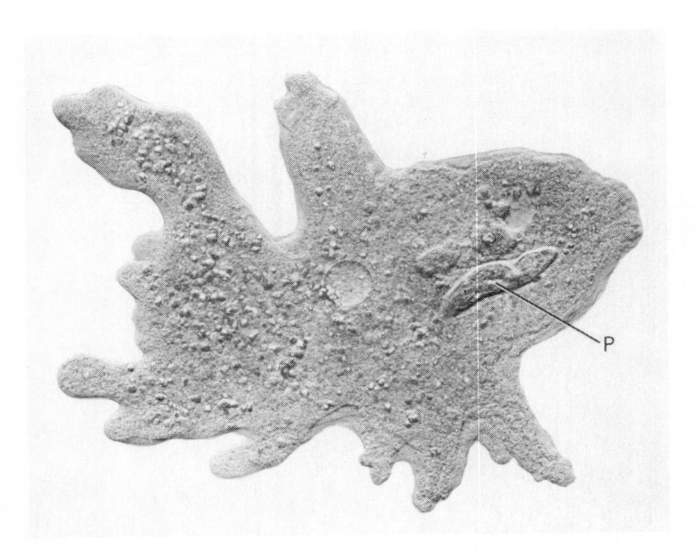

FIGURE 16.21 Sarcodina are protozoa that move by pseudopodia. Light micrograph of an *Amoeba* that has engulfed a *Paramecium* (p). The nucleus and numerous lophopodia are evident. (Courtesy of Carolina Biological Supply Company.)

16.21). The ameba's surface is composed of a very flexible plasmalemma, which is easily distorted by the streaming of the internal protoplasm. When the protoplasm streams in an outward direction, a pseudopodium is formed. This extension of the animal moves it toward its food or away from negative stimuli. Once in contact with a food particle, the pseudopodium surrounds the food and forms a phagocytic vacuole that is taken into the animal's cytoplasm.

Undulating Membranes and Ridges Some protozoa are moved by the wave motion generated in a fin-like **undulating membrane.** The undulating membrane of trypanosomes extends along the length of the animal. The basal body of the flagellum is closely associated with a mitochondrion that supplies ATP for the axoneme's motion. Mitochondria are present in many flagellates such as the trypanosomes, but are noticeably absent from anaerobic protozoa, including the flagellates *Giardia* and *Trichomonas* (trik-o-mo′nas).

Certain sporozoans contain **undulating ridges** lying just beneath the plasmalemma. These easily overlooked ridges are formed by submicroscopic contractile structures whose composition and mechanism of action are unknown. Apparently, their contractions propagate a wave motion on the animal's surface that enables the organism to glide.

Food Acquisition and Nutrition

Protozoa are heterotrophs that generate energy from simple carbohydrates by aerobic respiration or by anaerobic metabolism. Many protozoa have been grown and studied in laboratory cultures, and some, such as *Tetrahymena,* are used as experimental animals in student biology laboratories.

Protozoa assimilate food by a variety of mechanisms, including permanent cytostomes (mouths), phagocytosis, pinocytosis, and membrane transport. Ciliates have a permanent cytostome through which large particles or entire microorganisms can be engulfed (Figure 16.17). The cilia surrounding the cytostome set up local currents that move food into the oral cavity. Once inside the animal, the food is digested in a food vacuole. Amebas use their pseudopodia to capture food and to form food vacuoles by the mechanism of phagocytosis. Protozoa also take up nutrients directly from the environment by pinocytosis, diffusion, and active transport.

Parasitic protozoa exist within their hosts as extracellular or intracellular parasites. The extracellular parasites are found in the intestine, blood, and other organs of infected animals. Intracellular parasitic protozoa actively penetrate their host cell's membrane using a boring action or are phagocytized by specialized host cells. They derive their nutritional needs from their host cell and often destroy the host cell in the process.

Reproduction

Among protozoa, asexual reproduction is more common than sexual reproduction. In some species, only one form of reproduction occurs, whereas others have distinct life cycles in which asexual and sexual processes alternate. As in other eucaryotic cells, copies of the cell's genetic information are partitioned to daughter cells during mitosis. Protozoa use many variations on the classic process of mitosis, and sometimes not all the classic steps are represented.

Asexual Reproduction Asexual reproduction occurs by binary fission (Figure 16.22), budding, or multiple fission, depending on the species of protozoa. Bi-

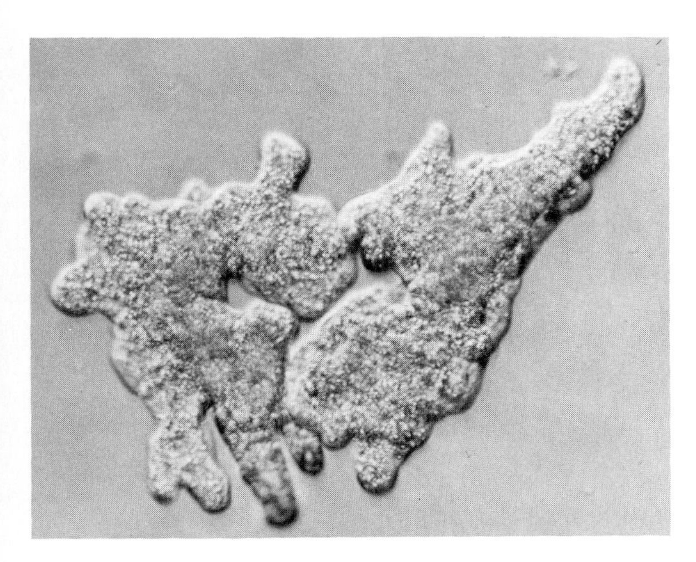

FIGURE 16.22 Binary fission in *Amoeba proteus*. (Courtesy of Carolina Biological Supply Company.)

nary fission is transverse in ciliates, whereas flagellates divide along their long axis by longitudinal fission. Protozoa that reproduce by budding form two daughter cells of unequal size.

Multiple fission is an asexual process of reproduction that results in many progeny of equal size. It is called **schizogony** (ski-zog′o-ne) when it occurs in sporozoans. The dividing mother cell, called a **schizont** (skiz′ont), divides its nucleus and other organelles repeatedly prior to cell division. The daughter cells that result from schizogony are known as **merozoites** (mer-o-zo′ytes). They eventually break away from the parent cell to begin another cycle of schizogony or sexual reproduction. Schizogony is an essential reproductive stage in the life cycle of *Plasmodium* (plaz-mo′de-um), the sporozoan that causes malaria.

Sexual Reproduction The first stage of sexual reproduction is the formation of gametes during **gametogenesis** (gam-e-to-jen′e-sis). The animal undergoes meiosis (reductive division) to form haploid cells called gametes or to form haploid nuclei. The gametes produced by a protozoan can be morphologically similar isogametes or of different sizes designated macrogametes and microgametes. When two gametes fuse during sexual reproduction they form a diploid zygote (zi′goht). A zygote can be formed from the union of two isogametes from different mating types, or the union of a microgamete and a macrogamete. When the zygote undergoes multiple fission, the process is called **sporogony** (spo-rog′o-ne) and the daughter cells are known as **sporozoites** (spo-ro-zo′ytes).

Protozoan **conjugation** is a form of sexual reproduction seen only in ciliates. It involves exchange of nuclei between two ciliates in direct contact with each other.

Encystment

Vegetative forms or **trophozoites** (troph-o-zoh′ytes) of some protozoa are able to protect themselves against adverse environments by forming cysts. The cysts are resting forms that can survive adverse environments such as desiccation, low nutrient supply, or anaerobiosis. When the cyst encounters favorable conditions, the organism excysts to form trophozoites that feed, grow, and reproduce asexually.

Many parasitic protozoa of the intestine form cysts as a necessary part of their life cycle. The cysts are formed in the intestinal tract and excreted in the animal's feces. In many cases, the cyst is the only form of the organism able to survive outside the host. The life cycle continues when a cyst is ingested by an animal host. Most protozoa form cysts directly from trophozoites; however, sporozoans form cysts from zygotes.

Classification

The protozoa are a diversified group of microorganisms classified in a subkingdom of the Protista. Most of the protozoa have animal-like characteristics, including locomotion, ingestion of food, and absence of a rigid cell wall. Other protozoa have characteristics that overlap those of the algae or the fungi. For example, the photosynthetic euglena and dinoflagellates are classified as phytoflagellates by protozoologists, and as algae by botanists. The ameboid cellular slime molds are claimed by both the mycologists and the protozoologists.

Early schemes for classifying the protozoa were based on organelles of locomotion. New information about the protozoa, much of it derived from electron micrographic studies, necessitated the development of a more elaborate scheme. The current classification scheme, highlighted in Table 16.3, assigns the protozoa

TABLE 16.3 Classification of the protozoa: highlights

TAXONOMIC GROUP	CHARACTERISTICS	REPRESENTATIVE GENERA
Phylum I. *Sarcomastigophora*		
Subphylum *Mastigophora*	Single nucleus, trophozoites have one or more flagella, longitudinal binary fission	
Class *Phytomastigophorea*	Plant-like flagellates typically with chloroplasts, mainly free-living	*Euglena*, *Cryptomonas*
Class *Zoomastigophorea*	Animal-like flagellates, chloroplasts absent, one to many flagella, most are parasitic	*Giardia*, *Trichomonas*
Subphylum *Sarcodina*	Pseudopodia or locomotive protoplasmic flow, binary fission, sexual reproduction in some species, most are free-living	*Amoeba*, *Entamoeba*, *Naeglera*
Phylum II. *Labyrinthomorpha*	Network of spindle-shaped cells, ameboid cells move within network, parasitic on algae, mostly in marine and estuarine waters	*Labyrinthula*
Phylum III. *Apicomplexa*	Organisms that form spores during their life history, apical complex of polar rings, subpellicular microtubules, sexual reproduction, all species parasitic	
Class *Sporozoea*	Trophozoites small, sexual and asexual reproduction, intercellular growth, some cause important human diseases	*Plasmodium*, *Toxoplasma*
Phylum IV. *Microspora*	Unicellular spores with polar tube and capsule, no mitochondria, obligate intracellular parasites of all major animal groups	*Hessea*, *Nosema*, *Glugea*
Phylum V. *Ascetospora*	Multicellular spores, without polar capsule or polar filaments, all parasitic	*Marteilia*, *Haplosporidium*
Phylum VI. *Myxozoa*	Spores with multicellular origin, with one or more polar capsules, all parasitic	*Ceratomyxa*, *Myxidium*
Phylum VII. *Ciliophora*	Cilia or ciliary organelles typical in one stage of life cycle, two types of nuclei, typically macronucleus and micronucleus, transverse binary fission, but budding and multiple fission occur, sexual conjugation, most are free-living, some commensalistic or parasitic.	
Class *Kinetofragminophorea*	Cilia surround oral cytostome on surface of body, cytopharyngeal apparatus prominent	*Didinium*, *Balantidium*
Class *Oligohymenophorea*	Oral cilia distinct from body cilia, oral apparatus located ventrally or near anterior end, cysts common	*Tetrahymena*, *Paramecium*, *Pleuronema*

Source: N.D. Levine, et al., "A Newly Revised Classification of the Protozoa," *J. Protozool.*, *27*:37–58 (1980).

into one of seven phyla. Here we discuss selected representatives of four protozoan phyla.

Amebas (Subphylum Sarcodina) Amebas are single-celled organisms that produce pseudopodia (Figure 16.21) for locomotion and food gathering, and divide by transverse binary fission. They are classified in the subphylum Sarcodina (sar-ko-di'nah). *Amoeba proteus* contains a nucleus and numerous cytoplasmic vacuoles for processing food, wastes, and water. Surrounding the cytoplasm is the plasmalemma, a malleable membrane that enables the cell to produce pseupodia in response to chemical and physical stimuli.

Amebas are common in polluted waters and in the intestinal tract of amphibians and other animals. *Entamoeba coli* and *E. gingivalis* (jin-ji-vahl'is) are common inhabitants of humans that do not cause disease; however, *Entamoeba histolytica* (his-to-lit'ik-ah) is a cyst-forming pathogen that causes amebic dysentery in humans.

Foraminifera (fo-ram'i-nif'er-ah) are free-living amebas that produce a chalky compartmentalized shell. The shell is perforated by holes through which the animal extends its pseupodia for food gathering and locomotion. Fossils of the foraminifera are important to geologists since their presence is indicative of oil-bearing strata. The white cliffs of Dover, England and the chalk deposits of Mississippi contain deposits of extinct fossilized foraminifera.

Flagellates (Subphylum Mastigophora) Unicellular flagellates comprise a large group of organisms that are subdivided into phytoflagellates and zooflagellates. The **phytoflagellates** are unicellular free-living motile organisms that contain a chloroplast, which enables them to grow photosynthetically. They have all the characteristics of a protozoan, except for their chloroplast. Although the protozoologists lay claim to the dinoflagellates and the euglena, we have classified them with the algae.

The **zooflagellates** contain one or more flagella, lack chloroplasts, and are clearly protozoa. These unicellular heterotrophs vary greatly in size, shape, position, and form of flagella, and in their internal organelles (Figure 16.23), and are identified by their distinctive morphologies. They divide by transverse longitudinal fission; sexual reproduction is rare.

Many zooflagellates are parasitic on plants and animals. Among the serious human diseases caused by flagellates are trichomoniasis (trik-oh-mon-eye'ah-sis), leishmaniasis (leesh-mah-ni'ah-sis), giardiasis (ji-ar-di'ah-sis), and trypanosomiasis (tri-pan-o-so-mi'ah-sis).

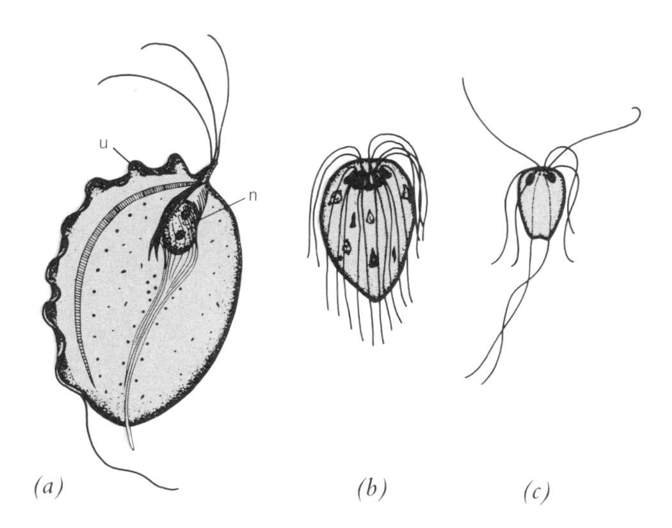

FIGURE 16.23 Representative zooflagellates. *(a) Trichomonas muris* (15 μm) showing an undulating membrane (u) and flagella, *(b) Giardia muris* (10 μm), *(c) Hexamita intestinalis* (10 μm). (Courtesy of M. A. Sleigh, *The Biology of Protozoa*, 1973, "Contemporary Biology Series" published by Edward Arnold (Publishers) Ltd.)

Sporozoa (Phylum Apicomplexa) All sporozoa (spohr-oh-zoh'ah) are parasitic on one or more animals. Their small trophozoites grow intracellularly in the animals they infect and have complex life cycles that often alternate between asexual and sexual reproduction. They form spores or cysts during their life cycle.

Two sporozoans, *Plasmodium* spp. and *Toxoplasma gondii* (toks-o-plaz'mah gon'dee-eye) cause major diseases in humans. Malaria is caused by *Plasmodium,* which is transmitted to humans by the bite of a mosquito. This sporozoan infects the liver before replicating asexually in human erythrocytes. Release of merozoites from the erythrocytes is responsible for the cycles of chills and fever characteristic of malaria. Toxoplasmosis is caused by *Toxoplasma gondii,* which replicates sexually only in the intestinal cells of a member of the cat family, such as domestic cats. The symptoms of toxoplasmosis vary greatly depending on the tissue and organs infected and on the age of the host at the time of infection.

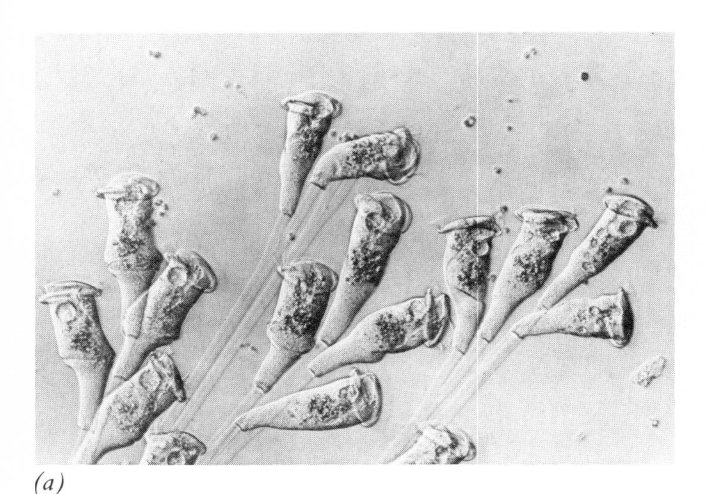

(a)

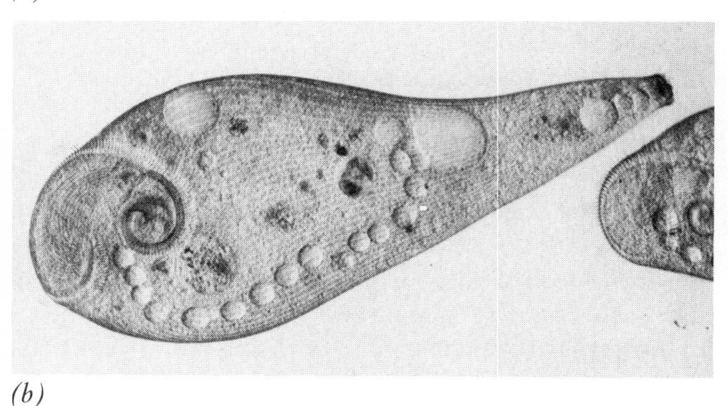

(b)

FIGURE 16.24 The diversity among the ciliates is demonstrated by micrographs of *(a) Vorticella* and (b) *Stentor*. (Courtesy of Carolina Biological Supply Company.)

Ciliates (Phylum Ciliophora) All unicellular ciliates are classified as Ciliophora (sil-ee-of-o'rah). These protozoa are readily identified by their rows of body cilia or by the cilia around their cytostome (Figure 16.24). Ciliates reproduce asexually by transverse fission and sexually by conjugation.

Two types of nuclei are found in the ciliates: the macronucleus governs metabolic activities and regeneration and the micronucleus is involved in reproduction. Conjugation occurs when unlike mating strains come in contact and align themselves so their oral grooves are together. The micronuclei divide to form haploid nuclei, which are then exchanged. The transferred nucleus fuses with the existing one to form a diploid nucleus. These animals then proceed to reproduce asexually.

Paramecium (par-ah-me'she-um) is a common ciliate whose oral groove is surrounded by cilia that move food particles into its cytostome (Figure 16.20). These cilia are distinct from the body cilia involved in locomotion. *Paramecium* divide by binary fission and occasionally by conjugation. Most ciliates are free living; however, *Balantidium coli* (bal-an-tid'ee-um koh'lee) infects the human intestine and is the only ciliate known to cause human disease.

MULTICELLULAR PARASITES

Helminths (hel'minths) are macroscopic parasitic worms, many of which have complex life cycles that involve humans as a host. Helminths cause widespread human disease, especially in tropical and developing countries. They are considered in microbiology courses because of their infectious nature and because microbiologists are often called upon to identify their larvae and eggs.

Certain arthropods are **ectoparasites,** which either live outside of, or burrow into the skin of their host. Ectoparasites depend on their host for part of their life cycle and many are vectors that transmit infectious microorganisms or viruses.

HELMINTHS: FLATWORMS AND ROUNDWORMS

The parasitic worms range in complexity from simple animals with a single opening for their digestive and excretory systems to animals with complex digestive, excretory, nervous, and circulatory systems. Based on their evolutionary development, the worms are classified in three phyla: Platyhelminthes (flatworms), Aschelminthes (roundworms), and Annelida (segmented worms). The parasitic worms are either roundworms or flatworms.

Platyhelminthes: Flatworms

Flatworms have a primitive bilateral symmetry resulting from their flattened dorsal–ventral shape (Figure 16.25). They have a gastrovascular cavity with a single opening that functions as both a mouth and an anus. They are **hermaphrodites** (her-maf'ro-dit) since both

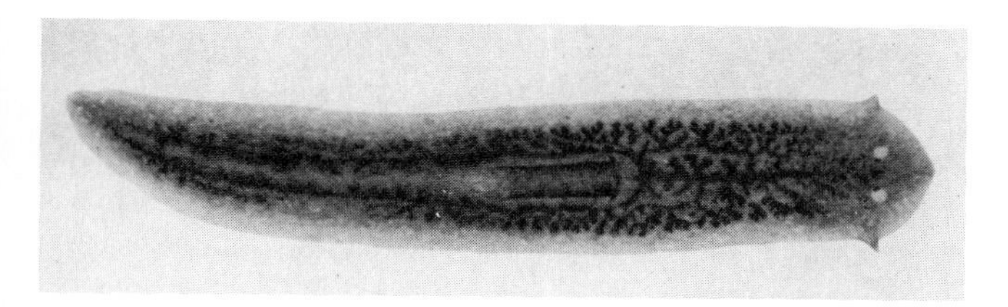

FIGURE 16.25 *Planaria* are common freshwater flatworms. This worm has bilateral symmetry and a single opening for its alimentary tract. (Courtesy of Dr. John D. Cunningham.)

male and female organs coexist in the same animal. Human parasites in this phylum belong to the class Cestoda (tapeworms) or Trematoda (flukes).

Class Cestoda: Tapeworms Tapeworms are long, flat, ribbon-like worms that are usually composed of many segments (Figure 16.26) known as proglottids (pro-glot′ids). Adult tapeworms inhabit the intestinal tract of vertebrates, where they attach to the intestinal mucosa with the suckers and hooks found on the specialized anterior end called the **scolex** (sko′leks). Tapeworms reside in the intestine of infected animals where they absorb nourishment from the partially digested food of the host. Depending on the species, adult tapeworms can live in their hosts for up to 25 years, attain lengths of 10 meters, and possess 4000 proglottids.

The common tapeworms of humans require one or more intermediate hosts to complete their life cycle (Figure 16.26). The intermediate hosts can be simple invertebrates or higher animals such as dogs, hogs, and cattle. The intermediate host is infected by ingesting an embryonated egg produced in the human intestine and excreted in human feces. The egg develops into a larval stage in the intermediate host and can encyst in the host's muscle. New human infections occur when the tapeworm larvae present in contaminated meat of the intermediate host are ingested or the larvae are acquired by direct contact with an intermediate host or infected human.

Humans are the definitive host for the beef tapeworm, *Taenia saginata* (te′ne-ah sa-ji-nah′tah), and the pork tapeworm, *Taenia solium* (sol′i-um). Beef tapeworm infections result from eating undercooked beef from cattle that grazed on land contaminated with human feces. Pork tapeworm infections begin with ingestion of infected pork (Figure 16.26). Both of these infections, known by the general term **taeniasis** (te-ni′ah-sis), are rare diseases in countries where hogs and cattle never have access to human feces. Tapeworm infections are diagnosed by the finding of eggs or gravid proglottids in the feces. They can be prevented by proper sanitary practices, by careful inspection of food, and by thorough cooking of meats.

Class Trematoda: Flukes Adult trematodes (trem′-ah-tohd) are usually elongated, flat, leaf-shaped worms ranging in size from 1 mm to several centimeters. The animal is enveloped by a noncellular tegument (skin) through which it absorbs nutrients. It uses its cup-shaped muscular suckers situated at its anterior end to attach to host tissue. These worms have a primitive, single-opening digestive system, a primitive nervous system, and a rudimentary excretory system.

Most human trematodes are hermaphrodites in which sexual reproduction is followed by asexual multiplication of the larvae in freshwater snails, which are the primary intermediary host. After infecting and developing in the snail, secondary hosts such as fish, crustaceans, other snails, or aquatic plants are infected. Humans are infected when they ingest trematodes in contaminated foods or when the waterborne larvae burrow through their skin (Figure 16.27).

Schistosomiasis (shis′to-so-mi′ah-sis) is an important human disease caused by blood flukes in the genus *Schistosoma* (Figure 16.27). Schistosomiasis, also called **Bilharziasis** (bil′harz-eye′a-sis), is prevalent in parts of Africa, South America, and the Far East. The occurrence of schistosomiasis depends on the distribution of snails that can serve as the intermediate host. Humans entering contaminated water to swim, bathe, wash clothes, or fish are infected by waterborne **cercariae** (larval stage), which burrow through the skin. The burrowing cercariae (ser-ka′re-ee) enter the capillaries, are transported throughout the circulatory system, and develop in the venules. Adult blood flukes can live for 30 years in humans.

Aschelminthes: Roundworms

Roundworms or nematodes have elongated bodies that often taper at both ends. Many are free-living inhabitants of soil and water, while others are parasites of plants, molluscs, annelids, arthropods, and vertebrates.

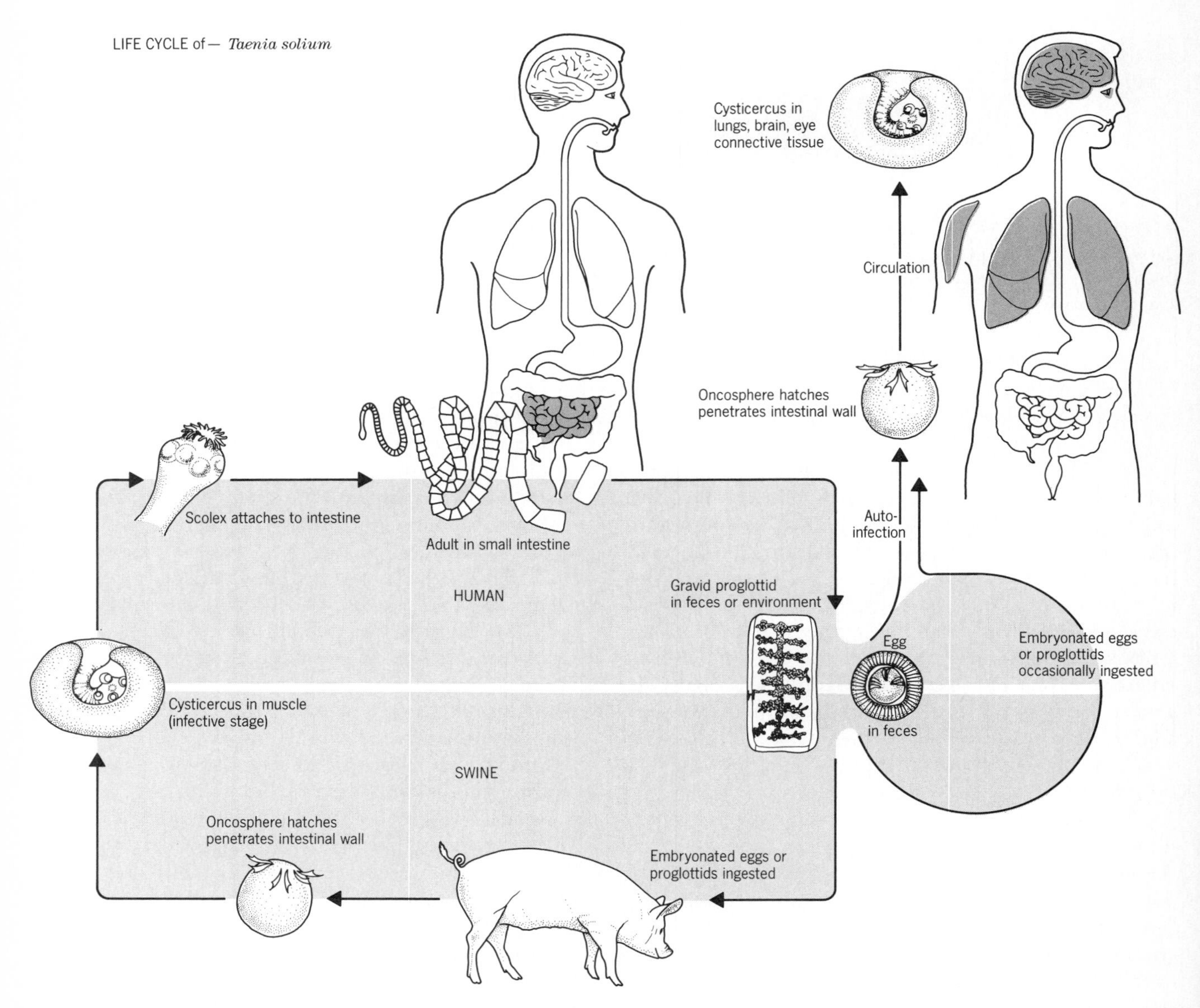

FIGURE 16.26 Life cycle of *Taenia solium*. The tapeworm in the human small intestine produces gravid segments containing about 80,000 eggs that pass out of the anus. The eggs survive in soil for several weeks or months. After they are ingested by pigs, the larva penetrate the intestinal wall and reach muscle tissue via the circulatory system. Here they develop into cysticercus. People are infected when they eat undercooked contaminated meat.

Nematodes have a well-developed digestive tract with an anterior mouth and a posterior anus. The male worms are separate and smaller than the females (Figure 16.28). They have a primitive nervous system and longitudinal muscles that produce their sinuous movements. Nematodes have no circulatory system.

Most parasitic nematodes have only one host in which they produce eggs or larvae. The larvae may exist in a free-living state in soil or water before the host is infected. Arthropods function as intermediate

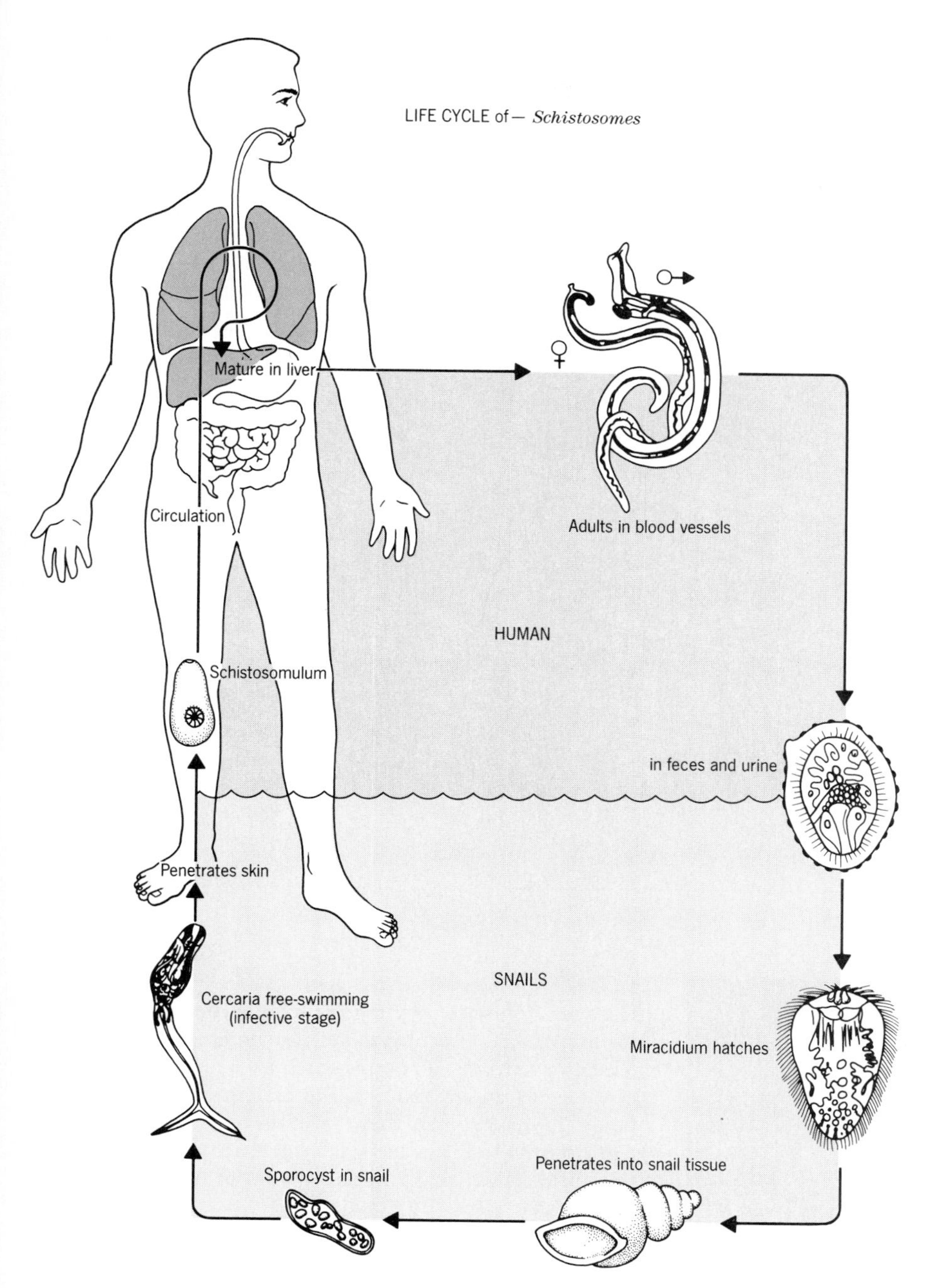

FIGURE 16.27 Freshwater snails are the intermediate host for the *Schistosoma* that cause schistosomiasis. Infected snails release cercariae into water. These motile forms of the *Schistosoma* penetrate human skin and enter the circulatory system. The male and female worms mature and mate in the blood vessels. The female produces eggs that infect target organs and are excreted in the feces. In water the miracidium emerges from the egg and infects a snail to complete the cycle.

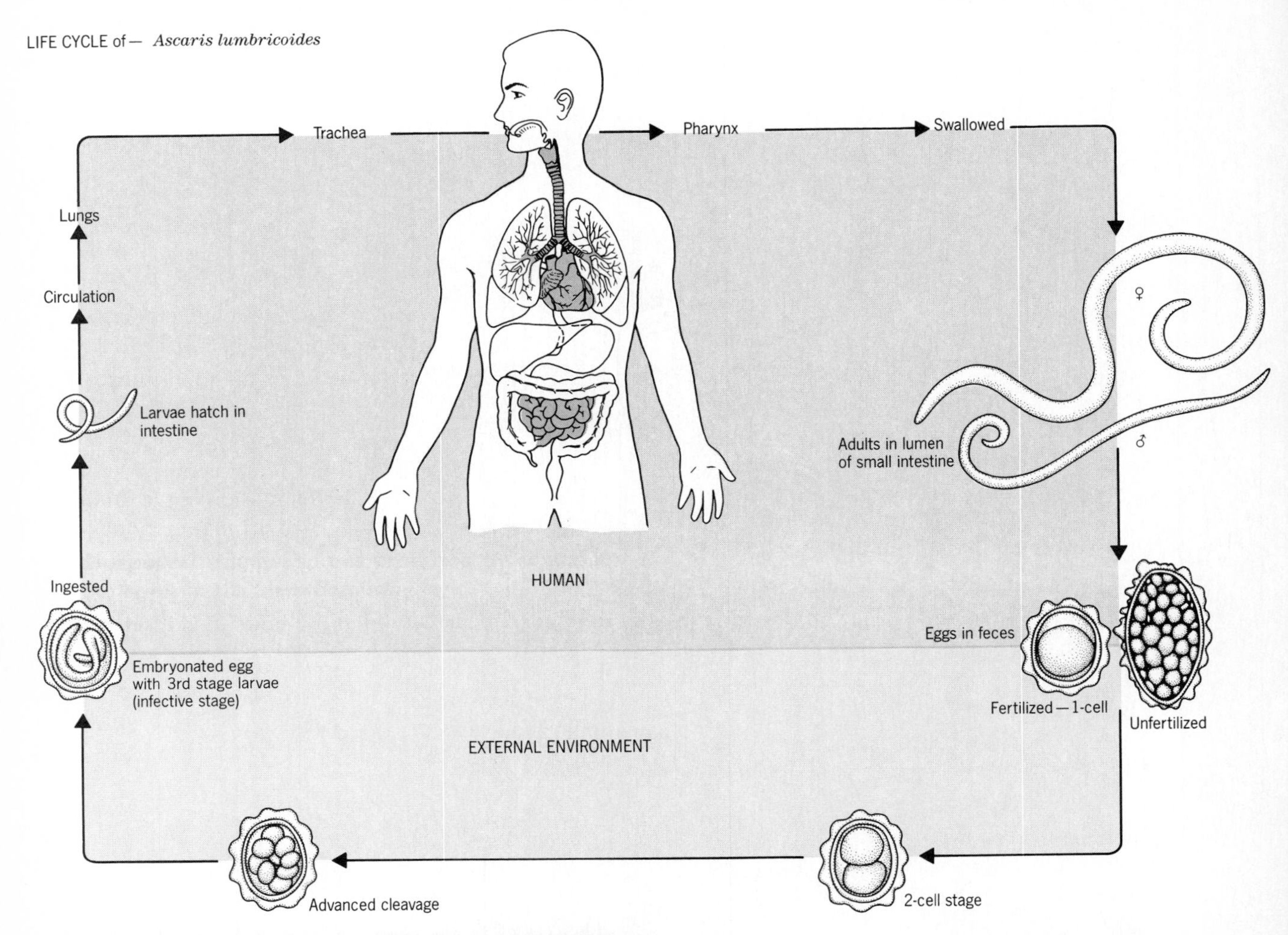

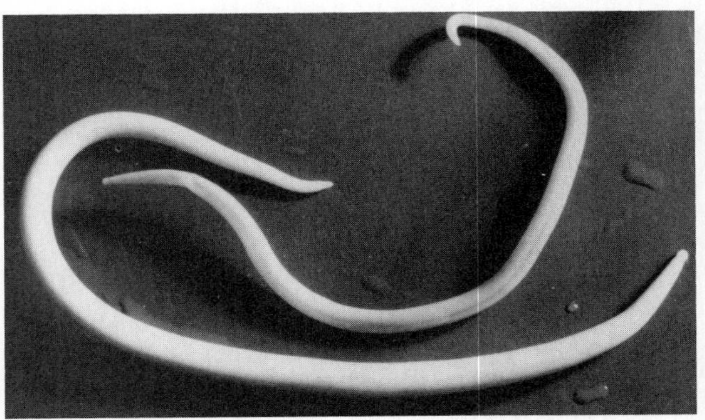

FIGURE 16.28 Life cycle of *Ascaris lumbricoides*. Humans are infected by ingesting embryonate eggs on inanimate objects or vegetables contaminated with nightsoil used as a fertilizer. The eggs hatch in the jejunum; the larvae penetrate the mucosa, and are carried to the heart and lungs. From the lungs they travel through the esophagus and stomach to infect the small intestine, where eggs are produced and excreted in feces. Photo of Ascaris worms.

hosts for some nematodes and consequently are vectors. Humans are the hosts for nematodes that infect the intestinal tract, blood, and tissues.

Class Nematoda *Trichinosis* (trik-i-no′sis) is a disease humans acquire by eating undercooked pork contaminated with *Trichinella spiralis* (trik-i-nel′ah spy-rah′lis). Trichinosis is rare in the United States, but does occur in farming communities and among immigrants who prefer lightly cooked pork dishes. The source of contamination is the muscles of animals infected with larval cysts. These larvae remain viable for years and are transmitted to carnivores that consume infected meat, or to humans who eat lightly cooked contaminated pork. Animals remain free of *Trichinella* if they do not eat garbage or other sources of infected meat.

Human **hookworm** infections are a serious health problem, especially in the tropics and subtropics. *Ancylostoma duodenale* (an-si-los′to-mah du-o-de-na′le) and *Necator americanus* (ne-ka′tor ah-me-ri-ka′nus) infect humans almost exclusively. In tropical soils, these worms produce eggs that develop into larvae, which are especially abundant around dwellings that have no sanitation facilities. Humans are infected when the larvae enter through hair follicles, pores, or even unbroken skin—especially the interdigital spaces of the toes and fingers of barefoot individuals and the hands of farmers. The infecting larvae enter the lymphatics or venules and are carried by the blood to the alveoli. They are released into the lumen of the lung, ascend the bronchi and trachea to the esophagus, and are swallowed. In the intestine they mature into adults capable of reproducing and the cycle begins again.

Pinworm is the human nematode infection that has the widest geographical distribution. *Enterobius vermicularis* (en′ter-o′be-us ver′mik-u-lar′is) is the small nematode that causes pinworm infections, also known as **enterobiasis** (en′ter-o-bi′ah-sis). The female worm is 8 to 13 mm long, while the male is only 2 to 5 mm long. They live in the human cecum and adjacent portions of the large and small intestines. Gravid females migrate to the perianal region where their eggs are released en masse. The life cycle is completed when a human ingests the eggs either as airborne particles or following hand-to-mouth transfer. In households with heavily infected patients, *Enterobius* eggs can be widely dispersed on bed linens, clothes, furniture, and other fomites. Even the fur of domestic animals is a source of *Enterobius* eggs, although cats and dogs do not har-

Controlling River Blindness

A young boy walks across the fertile fields of a West African valley, leading a middle-aged man who has lost his sight to river blindness. Onchocerciasis (on′ko-ser-ki′ah-sis), a helminth disease endemic in western Africa, is associated with rivers in which the blackfly vectors reproduce. This incapacitating disease has prompted entire communities to abandon fertile river valleys and relocate in less productive areas uninhabited by blackflies. Blackflies are found close to oxygenated, usually fast-moving rivers in which they reproduce. The female blackfly infects a human with *Onchocerca* larvae when it takes a blood meal required for its reproductive cycle. The larvae mature into worms in human skin and exist in skin nodules characteristic of onchocerciasis. Here they breed and produce millions of microfilariae. During this prelarval stage, the worm lives and dies in the skin, causing tormenting, painful itching. The symptoms of onchocerciasis are dependent on the cumulative presence of microfilariae in the body and often takes years to develop. The accumulation of microfilariae in the eye causes blindness, which can occur in 5 to 10 percent of the population in endemic areas.

The ONCHO control program mounted in West Africa attacks the life cycle of the blackfly with a variety of pesticides. Temephos, a biodegradable orthophosphate, and chlorphoxim were effectively used until resistance to these pesticides was reported in blackfly larvae. Resistance to suitable pesticides almost scuttled the program in the early 1980s, but the development of *Bacillus thuringiensis* H-14 as an effective biological control agent against blackfly larvae brought new support for the control program. Now river breeding areas are being sprayed from planes and helicopters with new formulations of insecticides and with *B. thuringiensis* H-14 to control the blackflies.

Extended control of the blackfly populations must be accomplished to reduce onchocerciasis significantly, for the adult worms can live for 11 to 12 years in an infected patient. Spraying may have to continue for 15 to 20 years in the target areas to break the life cycle of *Onchocerca volvulus;* however, this period could be shortened if drugs were available for treating infected humans. Recent research indicates that the veterinary antiparasitic drug, ivermectin, reduces the microfilariae in infected humans. A combination of pesticide spraying and effective antiparasitic drug therapy for onchocerciasis may bring this debilitating helminth disease under control and enable Africans to again cultivate the fertile valleys of West Africa.

bor this worm. The symptoms associated with pinworm include perianal itching, insomnia, and restlessness. It is diagnosed by the finding of eggs or adult worms in feces or in the perianal region.

Ascariasis (as'kah-ri'ah-sis) is caused by *Ascaris lumbricoides* (as'kah-ris lum-bri-koi'dez), a nematode that grows to a length of 10 to 31 cm (male) and 22 to 35 cm (female) long (Figure 16.28). They live in the lumen of the human intestine where, after mating, the female produces an average of 200,000 eggs per day. The eggs are excreted in human feces and develop in soil to an infectious eight-cell stage. The infectious egg is ingested, develops into a worm, and establishes itself in the intestinal lumen.

Ascariasis is a common parasite in temperate and tropical regions where sanitary facilities are not used. Epidemiologists have estimated that 900 million people worldwide are infected by this worm. The infection becomes serious if the worm migrates out of the intestinal lumen and infects vital organs such as the lungs, bile ducts, or liver. *Ascaris* worms have been known to penetrate the intestinal wall, enter the peritoneal cavity, and exit through the umbilicus in children. Prevention of ascariasis depends on proper disposal of human feces, good hygienic practices, and health education.

ECTOPARASITES

Ectoparasites, including lice, bedbugs, flies, fleas, mites, and ticks, cause disease even though they live outside the body. Some cause disease directly (scabies) whereas others serve as vectors for bacterial, viral, rickettsial, protozoan, or helminthic diseases. Ectoparasites are no longer a serious public health problem in the United States; however, ectoparasitic diseases and the infectious agents they transmit are responsible for tremendous human and economic losses in tropical countries.

Lice Cause Pediculosis

Lice are insects that live on the scalp or body of humans and function as vectors for certain rickettsias and spirochetes that cause human diseases. The human body louse, *Pediculus humanus* var. *corporis* (pe-dik'u-lus hu-man'us), grows on the scalp and deposits eggs on body hairs or clothing fibers (Figure 16.29). These eggs appear as oval protrusions called nits, which develop into nymphs after a period of seven to ten days. The nymphs feed on the scalp by biting. Bacteria carried by the lice infect the host either during the bite, or they are excreted in the feces and infect the host through abrasions. Relapsing fever is a spirochetal disease transmitted by body lice.

Fleas, Flies, and Mosquitoes

Fleas feed on warm-blooded animals and serve as the vector for the bacteria that cause flea-borne typhus fever, bubonic plague, and endemic typhus. These diseases occur in geographical regions where the reservoirs (rats, mice, ground squirrels) live. Humans are infected during the bite of an infected flea.

Numerous species of dipterous flies cause disease in humans when their larvae infect open wounds or penetrate unbroken skin. The eggs deposited by adult flies in tissue develop into larvae commonly known as maggots. The larvae grow in open wounds and in the

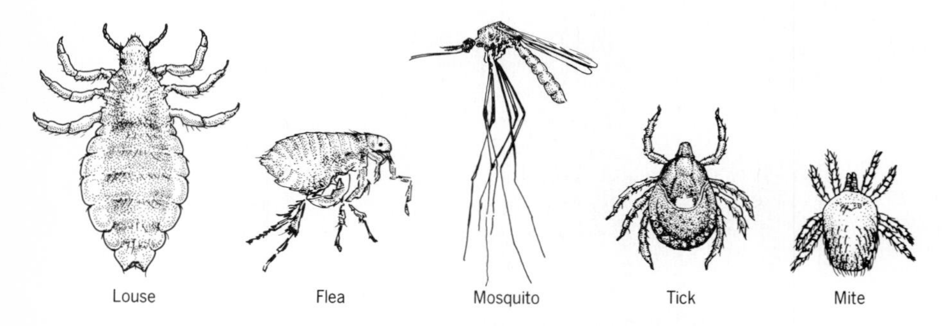

FIGURE 16.29 Representative ectoparasites. Insects: *(a)* louse, *(b)* flea, *(c)* mosquito; arachnida; *(a)* tick, *(b)* mite.

tissues of dead animals before they mature into adult flies.

Mosquitoes are vectors for bacterial and protozoan diseases, and for viruses that cause encephalitis, dengue, and yellow fever. Many of these diseases are prevalent in the tropics where the appropriate mosquito species and reservoirs exist. Mosquito-borne diseases in temperate climates occur seasonally during the summer mosquito season.

Ticks and Mites

Ticks and mites are arthropods assigned to the class Arachnida, which also includes the spiders. Ticks have a life cycle that depends on animal blood for their development. The tick attaches itself to a human or warm-blooded animal to take its blood meal. Rocky Mountain spotted fever and Lyme disease are caused by bacteria transmitted by ticks.

Mites are voracious creatures that cause itching and an inflammatory reaction at the bite site. Larvae of the mite hatch from eggs laid in soil or foodstuffs. The larvae (chigger) of the common mite bite humans on the lower legs or exposed skin to cause an inflamed papule. The reaction depends on the individual's sensitivity to the oral secretions of the mite. Some mites are the vectors for rickettsial diseases.

Human scabies is caused by the itch mite, *Sarcoptes scabiei* var. *hominis* (sar-kop′tes). The fertilized female burrows into the skin to a depth of several millimeters and lays her eggs. The eggs hatch into nymphs that molt several times before they become adults, which mate to initiate a new cycle. Scabies is characterized by intense itching, usually worst at night.

Summary Outline

OVERVIEW OF EUCARYOTIC MICROORGANISMS

1. Eucaryotic microorganisms are classified as algae, fungi, and protozoa. Algae are phototrophs, fungi are nonphotosynthetic heterotrophs (most are nonmotile), and the protozoa are heterotrophs (most are motile).
2. The multicellular parasites include the helminths and the ectoparasites, which either cause human diseases directly or are vectors for infectious microorganisms and viruses.

PHOTOSYNTHETIC ALGAE

1. Algae are nonvascular photosynthetic eucaryotes that are unicellular, colonial, filamentous, or complex organisms with structures resembling roots, stems, and leaves. They reproduce both asexually and sexually.
2. Major divisions of the algae are based on their photosynthetic pigments, cell-wall structure, storage products, and motility.
3. Red algae (Rhodophyta) are morphologically diverse seaweeds usually found in warm marine waters. Agar and carrageenan are commercially important polysaccharides produced from red algae.
4. The brown algae (Phaeophyta) are multicellular structurally complex organisms. The brown kelp has a life cycle that involves both a sporophyte and gametophyte plant.
5. The golden algae (Chrysophyta) are found in both freshwater and marine environments. Many golden algae are flagellated by two unequal flagella attached to the anterior end of the cell.
6. Diatoms (Bacillariophyta) are golden unicellular algae with a hard shell-like covering or frustule arranged as two overlapping valves. Fossil diatoms in diatomite are actively mined for commercial uses.
7. Green algae (Chlorophyta) have photosynthetic pigments similar to the green plants. These morphologically diverse algae exist as unicellular, colonial, filamentous, and tubular forms and are widely dispersed in nature.
8. Flagellated euglena (Euglenophyta) are claimed by both the botanists and the protozoologists because they are motile and grow both photosynthetically and heterotrophically. These cells are surrounded by a serrated plasmalemma instead of a cell wall.
9. Dinoflagellates (Pyrrhophyta) are biflagellated unicellular algae that grow photosynthetically or as

heterotrophs. Certain marine dinoflagellates produce toxins in surface waters that are concentrated by filter-feeding bivalve molluscs and cause paralysis in humans who consume the shellfish.

MYCOLOGY IS THE STUDY OF FUNGI

1. Fungi are spore-bearing, achlorophyllous, organisms most of which possess a cell wall and are nonmotile.
2. Most fungi are heterotrophs that grow on dead material (saprobes), others attack living plants (rusts and smuts) and cause tremendous damage to food crops. Mushrooms and puffballs are eaten by humans and animals.
3. Certain fungi are important producers of antibiotics, and the yeast *Saccharomyces cerevisiae* is used in making bread and alcoholic beverages.
4. Fungi grow as either unicellular yeasts that reproduce asexually by budding or fission or as extensive multibranched intertwining hyphae known as mycelia. The dimorphic fungi grow as single cells or as hyphae, depending on growth conditions.
5. Yeast cells divide by budding, whereas myceliated fungi grow by increasing the mass of their hyphae through elongation and branching.
6. Asexual spores include chlamydospores, arthrospores, blastospores, and conidia. Sexual spores include zygospores, ascospores, and basidiospores.
7. Fungi are separated into three divisions: Gymnomycota (slime molds), Mastigomycota (flagellated lower fungi), and Amastigomycota (terrestrial fungi).
8. The terrestrial fungi have cell walls and are nonmotile; the vegetative cellular slime molds are cell-wall-lacking amebas; the aquatic fungi produce motile flagellated cells.
9. The terrestrial fungi are grouped into four classes (Ascomycotina, Basidiomycotina, Mastigomycotina, or Zygomycotina) on the basis of sexual and asexual reproductive processes and by the presence or absence of septa. When sexual reproduction is not known, the fungus is placed in the subdivision Deuteromycotina.

PROTOZOA ARE MICROSCOPIC ANIMALS

1. Protozoa are the single-celled eucaryotic animals that assimilate chemical nutrients or feed on bacteria and other protozoa (holozoic protozoa).
2. Symbiotic protozoa coexist with other organisms in either a commensal, beneficial, or harmful relationship. Parasitic protozoa benefit at the expense of their host and are responsible for numerous animal diseases.
3. Most protozoa have a single nucleus, but some are multinuclear and the Ciliophora have nuclei of unequal size and responsibilities.
4. Protozoa reproduce asexually by binary fission, budding, and multiple fission and sexually by conjugation or gamete fusion.
5. Motility of protozoa involves four basic organelles: cilia, flagella, undulating ridges and membranes, and pseudopodia.
6. Protozoa are heterotrophs that gain food by a variety of mechanisms, including permanent cytostomes (mouths), phagocytosis, pinocytosis, and membrane transport.
7. Amebas (subphylum Sarcodina) are single-cell organisms that produce pseudopodia for locomotion and food gathering and divide by transverse binary fission. Most are free-living; however, some are human pathogens such as *Entamoeba histolytica*.
8. Flagellates (subphylum Mastigophora) are divided into the phytoflagellates and the zooflagellates. Zooflagellates cause human diseases including trichomoniasis, leishmaniasis, giardiasis, and trypanosomiasis.
9. Sporozoa (class Sporozoea) grow intracellularly in the animals they infect and have complex life cycles—all are parasitic for one or more animals. *Plasmodium* spp. (malaria) and *Toxoplasma gondii* (toxoplasmosis) cause major diseases in humans.
10. Ciliates (phylum Ciliophora) have cilia that function in locomotion and/or oral cilia that function in food gathering. Asexual reproduction is by transverse fission, whereas sexual reproduction is accomplished by conjugation.

MULTICELLULAR PARASITES: HELMINTHS

1. Helminths range in complexity from simple animals with a single opening for their digestive and excretory systems to animals with complex digestive, excretory, nervous, and circulatory systems.

2. Flatworms (Platyhelminthes) have bilateral symmetry with a gastrovascular cavity and a single opening that functions as a mouth and an anus. Both male and female organs are found in the same animal.
3. Tapeworms (class Cestoda) inhabit the intestinal tract of vertebrates by attaching to the intestinal mucosa by the suckers and hooks on their scolex. Humans are the definitive host for the beef tapeworm, *Taenia saginata,* and the pork tapeworm, *Taenia solium.*
4. Flukes (class Trematoda) are usually elongated, flat, leaf-shaped worms having a single-opening digestive system, a primitive nervous system, and a rudimentary excretory system. Schistosomiasis is an important human disease caused by blood flukes in the genus *Schistosoma.*
5. Roundworms or nematodes (Aschelminthes) have elongated bodies that often taper at both ends. Nematodes cause trichinosis, human hookworm infections, enterobiasis, and ascariasis.

MULTICELLULAR PARASITES: ECTOPARASITES

1. Ectoparasites, including lice, bedbugs, flies, fleas, mites, and ticks, cause disease even though they live outside the body. Some cause disease directly (scabies) whereas others are vectors for microbial agents of disease.
2. Lice are insects that live on the scalp or body of humans and function as vectors for the spirochetes of relapsing fever.
3. Fleas and mosquitoes are vectors for viral (encephalitis), bacterial (plague), and protozoan (malaria) diseases of humans. Mosquito-born diseases occur seasonally in correspondence with the mosquito season.
4. Ticks and mites (class Arachnida) transmit diseases when they take a blood meal. Mites are voracious creatures that cause an inflammatory reaction accompanied by itching at the bite site.

Questions and Topics for Study and Discussion

QUESTIONS

1. Describe the commercial importance of two algae and explain how they benefit humanity.
2. Explain why most diatoms get smaller and smaller as they reproduce asexually. How do they surmount this problem?
3. What algae are involved in food poisoning? Explain when and how these algae cause human illness.
4. Describe and compare the morphologies of a filamentous fungus and a yeast cell. What mechanisms do these cells use to reproduce?
5. Define or explain the following:

cenocytic	mycelium	saprobe
conidia	parasite	septa
hyphae	Protista	

6. What is meant by fungal dimorphism, and what impact does dimorphism have on medical mycology?
7. Name the methods of motility used by the protozoa and give an example of each. What role does locomotion play in protozoa classification?
8. How do pseudopodia function in motility and food acquisition?
9. Define or explain the following:

axoneme	merozoite	trophozoite
cytostome	schizont	vector
encystment	sporozoite	zygote

10. Describe the reproduction cycle of a human tapeworm.
11. What role do snails play in the life cycle of the blood fluke, *Schistosoma?*
12. Describe the transmission of human pinworm.

DISCUSSION TOPICS

1. Are euglena algae or protozoa? Defend your response.
2. What steps could be taken to stop the spread of schistomiasis within a native population?

3. What characteristics separate the algae from green plants?
4. How do ectoparasites affect the incidence and spread of infectious diseases? What steps can be taken to stop their spread?

Further Readings

BOOKS

Ainsworth, G. C., and A. S. Sussman (eds.), *The Fungi, An Advanced Treatise,* Vols. 1–4, Academic Press, New York, 1965. A recognized authoritative series on the biology and classification of the Fungi.

Alexopoulos, C. J., and C. W. Mims, *Introductory Mycology,* 3rd ed., Wiley, New York, 1979. An advanced textbook that is an excellent reference for the biology of the nonpathogenic fungi.

Beck, J. W., and J. E. Davies, *Medical Parasitology,* 3rd ed., Mosby, St. Louis, 1981. A well-illustrated textbook that covers protozoa, helminths, and the medically important arthropods involved in disease transmission.

Beneke, E. S., and A. L. Rogers, *Medical Mycology Manual with Human Mycoses Monograph,* 4th ed., Burgess, Minneapolis, 1980. A manual that can be used in the clinical laboratory and in medical mycology courses.

Bold, H. C., and M. J. Wynne, *Introduction to the Algae, Structure and Reproduction,* Prentice-Hall, Englewood Cliffs, N.J., 1985. This introductory textbook of phycology is profusely illustrated with micrographs, photographs, and line drawings. The authors use a taxonomic approach to present the algae to beginning students.

Brown, H. W., and F. A. Neva, *Basic Clinical Parasitology,* 5th ed., Appleton-Century-Crofts, Norwalk, Conn., 1983. This is an excellent short textbook for introducing students to the biology, epidemiology, diagnosis, and treatment of the parasitic protozoa and helminths.

Emmons, C. W., C. H. Binford, and J. P. Otz, *Medical Mycology,* 3rd ed., Lea & Febiger, Philadelphia, 1977. A classic approach to the biology and laboratory diagnosis of the pathogenic fungi.

Garraway, M. O. and R. C. Evans, *Fungal Nutrition and Physiology,* Wiley, New York, 1984. This book presents a rigorous treatment of the principles of nutrition and physiology of fungi.

Schmidt, G. D., and L. S. Roberts, *Foundations of Parasitology,* 2nd ed., Mosby, St. Louis, 1981. A comprehensive introductory college textbook in parasitology that emphasizes the relationships between the biology, morphology, and ecology of eucaryotic parasites and the diseases they cause.

Sleigh, M., *The Biology of Protozoa,* Arnold, London, 1973. This paperback emphasizes cell function and ultrastructure as it is reflected in the diversity of the protozoa.

ARTICLES

Levine, N. D., et al., "A Newly Revised Classification of the Protozoa," *J. Protozool., 27*:37–58 (1980). This new scheme for the classification of the protozoa was created by the Committee on Systematics and Evolution of the Society of Protozoologists.

PART

3

Host-Parasite Relationships

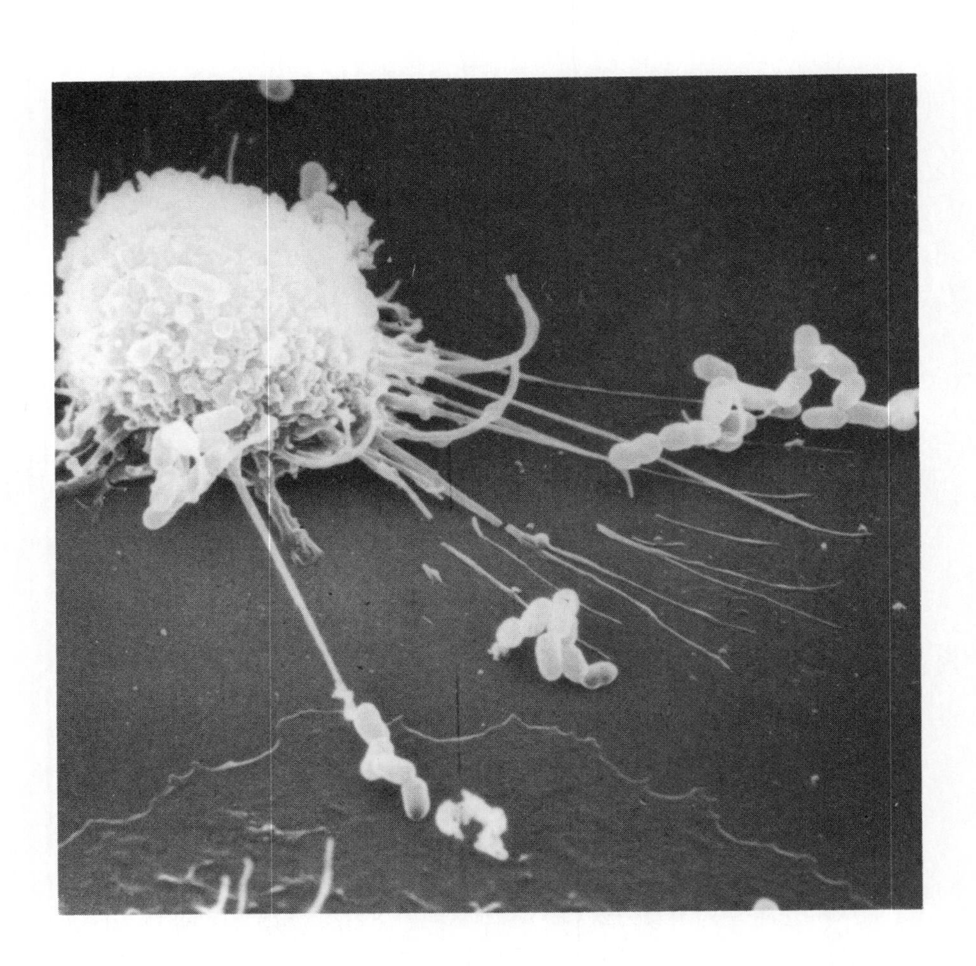

Humans cannot escape a dynamic ongoing association with microorganisms. For the most part we live peacefully with the majority of microbes, which do not cause disease. There are, however, a few species that are harmful, and they evoke swift reactions. We defend ourselves against these infectious agents at three levels: our bodies possess physical mechanisms that resist infection; society takes measures to prevent diseases; and medical researchers and professionals develop procedures and drugs that protect against infection or treat it.

C H A P T E R

17

Host–Parasite Relationships

OUTLINE

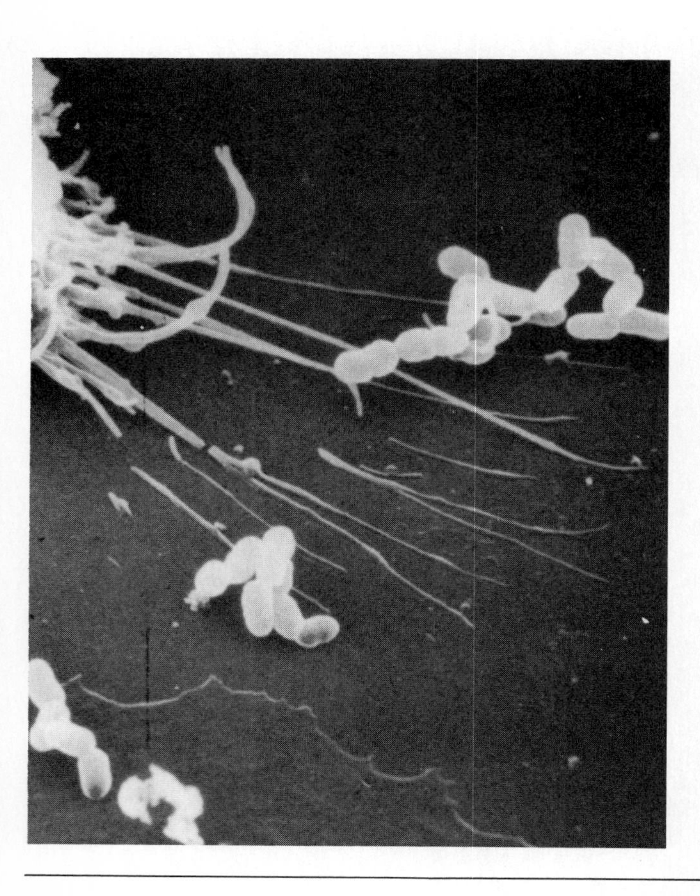

FOCUS OF THIS CHAPTER

Humans live in a dynamic relationship with microbes, some of which have the potential to cause disease. This chapter describes the unique characteristics that enable the pathogens to cause disease and the defense mechanisms humans invoke to resist them.

We are constantly exposed to microorganisms—in the water we drink, in the food we eat, on the hands of people who touch us, and in the air we breathe. Most microbes are harmless and cause no ill effects, in large part because they are unable to grow in the body. A **pathogen** (path'o-jen), on the other hand, is a microbe that can cause disease in a susceptible host. How a person deals with pathogens often determines whether illness or health will prevail.

NATURE OF THE DISEASE STATE

Any continuing disturbance in structure or function of a bodily organ that causes damage is a **disease.** Diseases can be classified as hereditary, degenerative, metabolic, neoplastic (cancer), and infectious. **Infectious diseases** are caused by agents that have the potential to be transferred between susceptible hosts. This description applies to certain bacteria, fungi, protozoa, helminths, and viruses. Any organism capable of causing an infectious disease possesses characteristics that enable it to colonize and invade the host.

An organism that causes damage to the host is a **parasite** and the relationship thereby established is **parasitism.** Some parasites are **obligate:** they are unable to reproduce outside the host, and invariably cause disease. Viruses and rickettsias, for example, are dependent on host cells in order to replicate. Other microbes are **opportunistic parasites,** causing disease only under specific conditions. Such organisms normally exist either separately or symbiotically with the host; however, disease can develop following a significant change in host resistance or following a change within the parasite itself.

Infectious diseases are characterized by the interaction between the microbe's ability to persist and the host's ability to resist. The arsenal used by pathogenic microbes includes capsules, potent toxins, and enzymes—collectively referred to as **virulence factors** (vir'u-lence). The host retaliates through the first line of defense—mechanical barriers including the skin, tears, mucous membranes, and ciliary epithelia of the respiratory tract. If the microbe penetrates these barriers, the host invokes the second line of defense—phagocytic cells, which interact with the third line of defense—the immune response.

MICROBIAL FLORA OF THE HEALTHY HUMAN

Over 100 microbial species are regarded as normal flora of the adult human. Most are bacteria, and a few are fungi or protozoa. Among the benefits of having this normal flora is the constant stimulation of the immune system by foreign antigens, the participation of intestinal bacteria in digestion, and the exclusion of less desirable microbes by the occupation of ecological space. Included among the normal flora are those microbes having the potential to cause disease, in persons who are malnourished; those whose defense mechanisms are compromised, as by diabetes, lymphoma, and leukemia; and in older persons whose ability to resist infections has diminished.

ORIGIN OF NORMAL FLORA

A healthy fetus is free of foreign organisms until the birth membranes break and it is exposed to the flora of the mother's vagina. The newborn is then exposed to the flora of its mother's breasts and skin, and to the flora on the hands of those who care for it. With its first breath, the infant comes in contact with airborne microbes, many of which originated in the respiratory tracts of other humans. The multitude of microbes

TABLE 17.1 Predominant flora according to site[a]

BODY SITE	NORMAL FLORA
Respiratory Tract	
Tongue and buccal mucosa	Streptococci (viridans group), *Neisseria* spp., *Branhamella* spp., *Staphylococcus* spp.
Teeth and gingival crevices	*Bacteroides* spp., *Fusobacterium* spp., *Eikenella, Streptococcus* spp., *Actinomyces* spp.
Nasopharynx	Carrier state: *Streptococcus pneumoniae, Neisseria meningitidis, Haemophilus* spp., *Staphylococcus* spp.
Bronchi and lungs	Sterile
Gastrointestinal Tract	
Esophagus	Transient mouth flora
Stomach	Rapidly becomes sterile after ingestion of a meal
Small intestine	Variable and ill defined
Large intestine (colon)	
During breast feeding	*Lactobacillus* spp., streptococci, *Bifidobacterium* spp.
After weaning	*Bacteroides* spp., *Fusobacterium* spp., fecal streptococci, *Clostridium* spp.; *E. coli, Proteus* spp. and other *Enterobacteriaceae; Pseudomonas aeruginosa* and numerous other microbes
Vagina	
Prepubertal and postmenopausal women	Skin and some colonic organisms
Childbearing women	Lactobacilli, streptococci, staphylococci, peptostreptococci, clostridia, and yeasts
Blood and Cerebrospinal fluid	Sterile in healthy humans

[a]Modified from J. C. Sherris (ed.), *Medical Microbiology, An Introduction to Infectious Disease,* Elsevier, New York, 1984.

having access to the newborn all compete for favorable growth environments. Thus begins the lifelong process by which the microbial flora establishes itself through colonization, competition, and succession.

A fetus can be infected by pathogens during the birth process. Congenital herpes and infections of the eyes by *Neisseria gonorrhoeae* can occur during birth if the mother carries these agents in her vagina. Very few diseases are transmitted to the fetus *in utero,* but those that are can have serious consequences. For example, congenital rubella, congenital syphilis, and AIDS can be acquired by the fetus.

NORMAL FLORA OF THE ADULT

Microorganisms colonize the skin, mouth, nasopharynx, eyes, ears, intestinal tract, and anterior urogenital tract soon after an infant's birth. Colonization of a specific site depends on the host's condition, and varies greatly from site to site. Age and sex also influence the normal flora, as we shall see. The important sites of colonization are summarized in Table 17.1. Note, however, that many organs are normally free of microbes: blood, cerebrospinal fluid, urinary bladder, uterus, fallopian tubes, middle ear, paranasal sinuses, and kidneys.

Colonization of the Respiratory Tract

The respiratory tract includes the mouth, tonsils, nasopharynx, throat, trachea, bronchi, and lungs (Figure 17.1), and microorganisms are normal inhabitants of the mouth, nasopharynx, teeth, tonsils, and throat. The trachea, bronchi, and lungs in both infants and

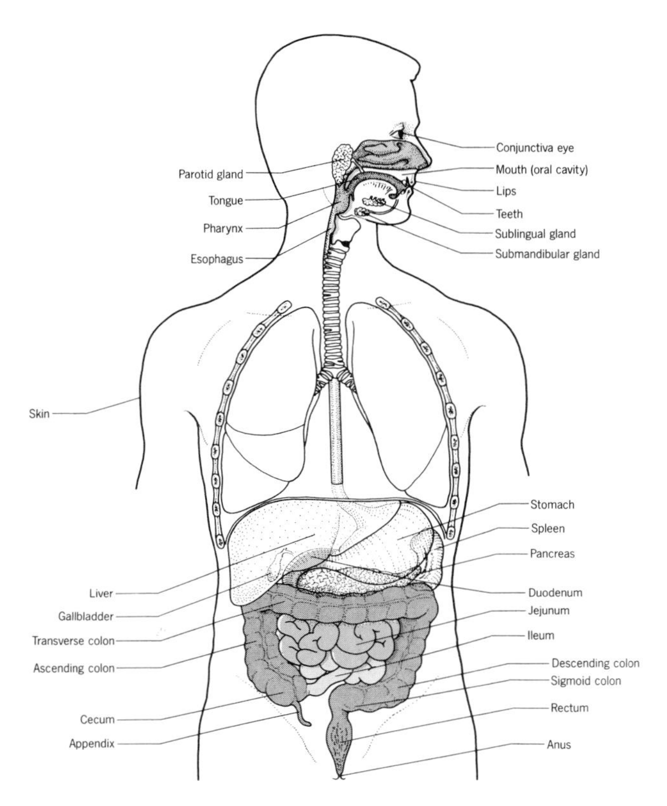

FIGURE 17.1 Normal microbial flora of the adult human. Tissues that are colonized by bacteria are shown in color.

adults are protected by the action of the ciliary epithelia and by the movement of mucus, thus these tissues are free of microorganisms (except those that are transiently inhaled). Bacteria or viruses able to penetrate these defenses and infect the lungs cause **pneumonia.**

The normal flora of the mouth includes many facultative anaerobes and some strict anaerobes. They colonize the surfaces of the teeth, the spaces between the gums and the teeth called the gingival (jin'ji-val) crevice, the tongue mucosa, and the nasopharynx. Bacterial communities on the tooth surface are thought to be the major cause of dental caries (tooth decay) and periodontal (per'e-oh-don'tal) disease. *Streptococcus mutans* and lactobacilli play a significant role in tooth decay. In addition, the nasopharynx is host to transient bacteria, including *Streptococcus pneumoniae, Haemophilus* spp., and *Neisseria meningitidis,* which, under appropriate conditions, cause serious illnesses.

Dental Caries and Periodontal Disease Teeth are coated by an organic film derived from saliva proteins called the **dental cuticle**, which enables bacteria to adhere to the tooth surface. A large variety of microorganisms colonize tooth surfaces, including *Streptococcus mutans,* lactobacilli, diphtheroids, and members of the genera *Fusobacterium* (fu'zo-bak-te're-um) and *Leptotrichia* (lep'to-trik'i-ah) (Figure 17.2).

Dental plaque is formed on the tooth surfaces when bacteria deposit a film of polysaccharides. This film entraps bacteria that produce lactic acid and other organic acids during their fermentation of glucose. These acids contribute to the demineralization of the tooth surface that leads to the formation of dental caries.

Periodontal disease, the progressive loss of supporting tissue for teeth, is in part due to bacteria growing in the dental plaque formed on the tooth surfaces under the gums. This condition is the major cause of tooth loss in persons 40 years of age and older.

Gastrointestinal Tract

The esophagus, stomach, gallbladder, duodenum, small intestine, and large intestine (colon) comprise the human gastrointestinal tract (see Figure 17.1). Microbes usually do not multiply in the esophagus and stomach, but are present in ingested food and as tran-

FIGURE 17.2 Human plaque near the gingival margin showing clusters of bacteria in polysaccharide matrix. (Courtesy of Dr. Z. Skobe, Forsyth Dental Center, Boston/BPS.)

sient flora derived from the mouth. Those that enter the stomach either are destroyed by the stomach's inhospitable environment or pass through the stomach enmeshed in food; they are unable to grow in the stomach because of its digestive enzymes and low pH, so the stomach becomes sterile soon after a meal.

Bacteria are prevalent in the lower ileum and the large intestine, where populations of between 10^8 and 10^{11} bacteria per gram of solids are normal. Obligate anaerobes such as *Bacteroides* and *Fusobacterium* species make up more than 90 percent of the microbial population of the large intestine, and the anaerobe *Clostridium perfringens* (per-frin′jens) is invariably present. Other organisms resident in the large intestine include *Escherichia coli, Proteus* spp., the fecal streptococci (enterococci), *Pseudomonas aeruginosa,* and yeasts.

Newborns lack an intestinal microbial flora, but very soon after birth the infant's intestine is colonized by microorganisms. The predominant organisms in the gastrointestinal tract of breast-fed infants are *Lactobacillus acidophilus, Bacillus* spp., *Clostridium* spp., coliforms,* and enterococci. Children develop normal adult intestinal flora after they begin to eat solid foods.

Microorganisms in the intestine contribute to the bulk of feces, participate in the breakdown of food, and manufacture some of the vitamins assimilated by their host. Their presence usually does not give rise to disease; if, however, they gain entrance to sterile tissues and organs, disease may develop.

Gastrointestinal diseases include peritonitis, cholecystitis, and gastroenteritis. **Peritonitis** (per′i-to-ni′-tis), an inflammation of the peritoneum (the membrane lining the abdominal cavity), is a very serious illness. It is caused by microbes from the bowel that have invaded the peritoneum following surgery or rupture of the appendix. **Cholecystitis** (ko′le-sis-ti′tis), an inflammation of the gallbladder, may be caused by intestinal microbes. **Gastroenteritis** (gas′tro-en-ter-i′tis) is an inflammation of the mucous membranes of the stomach and the small intestine. Numerous bacteria that produce enterotoxins can be the cause of this illness. Gastroenteritis can also result from ingestion of enterotoxins in contaminated food.

Escherichia coli is an inhabitant of the lower ileum and the large intestine of almost all humans, and occurs in the lower bowel of other warm-blooded animals. Its presence in water and food is often used as an indicator of contamination by human feces. *Streptococcus faecalis* and *S. faecium* are also present in the human intestine and accordingly are referred to as the enterococci. Their presence in food and water is also a useful indicator of fecal contamination; however, since enterococci are also found in vegetable matter, they are not necessarily an indicator of fecal contamination.

*Coliforms are lactose-fermenting gram-negative rods that inhabit the intestinal tract.

Genitourinary Tract

In healthy persons, the kidneys, bladder, and the fallopian tubes are free of microorganisms. The urethra (u-re′thrah) is free of microorganisms over most of its distance from the bladder to the body's exterior; however, the anterior segment of the urethra is colonized by microbes (Figure 17.3), derived from the skin flora and the flora of the female perineum (per′i-ne′um). The microbial population of the urethra is kept low by the flushing action of urination.

The urethra is a susceptible site for nosocomial (hospital acquired) infections that often result from the insertion of tubes called catheters. **Urethritis** (u′re-thri′tis) is an infection of the urethra, **cystitis** (sis-ti′-tis) is an infection of the bladder, and **pyelonephritis** (pi′e-lo-ne-fri′tis) is an inflammation of the kidney. These infections develop from surgical procedures or catheterization, when the organisms normally associated with the external genitalia infect the urogenital tract. Males are less susceptible to urinary tract infections than females, who are more susceptible to urethral infections caused by the indigenous flora of the vagina and perineum.

Prior to puberty and in postmenopausal women, the microbial flora of the vagina is scanty and composed of organisms resident on the skin and in the colon. During childbearing years, the predominant microbes in the vagina are species of *Lactobacillus,* which produce acid during the lactic acid fermentation of glucose. This acid contributes to the acidic pH of the vagina and curtails the growth of nonacidophilic organisms. Organisms that are able to survive the low pH of the vagina include *Staphylococcus epidermidis,* small numbers of anaerobic gram-negative rods, and some yeasts. During childbearing years, the acidic pH is maintained in part by the deposition of glycogen in vaginal epithelial cells. The glycogen is metabolized by the lactic acid bacteria to produce acid by-products.

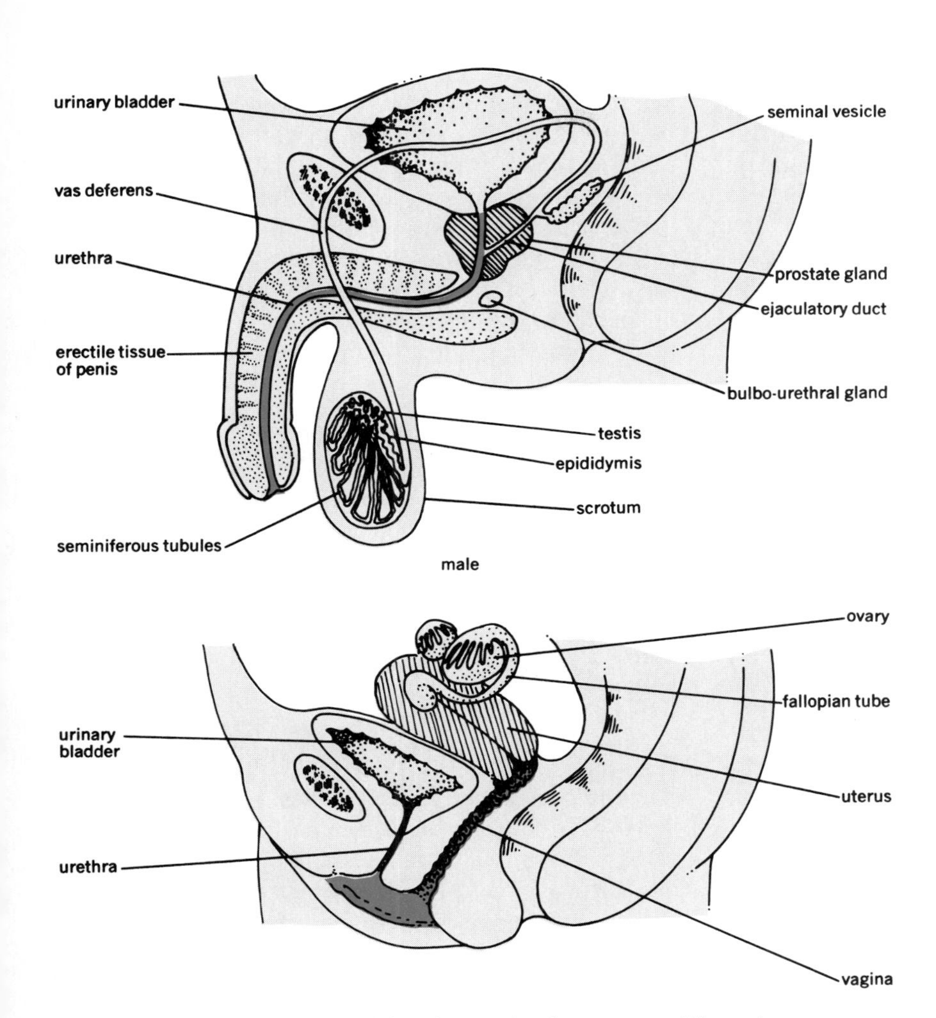

FIGURE 17.3 Male and femal reproductive systems. The color represents areas that contain normal microbial flora. (From J. R. McClintic, *Physiology of the Human Body,* Wiley, New York, 1978, copyright © 1978, by John Wiley & Sons, Inc.)

Antibiotic chemotherapy can disrupt the normal flora of the vagina and thereby change the vaginal environment. The pH of the vagina rises if an antibiotic significantly affects the acid-producing bacteria. Organisms that can multiply in this altered environment often cause an inflammation known as **vaginitis.** The yeast *Candida albicans* (kan'di-dah al'bi-cans) is part of the normal vaginal flora that can cause vaginitis. Its growth is stimulated by the high pH and by the lack of competition from the bacteria killed by the antibiotic chemotherapy, with the result that a woman can contract *Candida* vaginitis while being treated for a bacterial infection. The protozoan *Trichomonas vaginalis,* another common inhabitant of the vagina, can also cause vaginitis.

Normal Flora of the Skin, Ears, and Eyes

Although skin is the outermost barrier to infectious agents, it harbors many microorganisms, some of which are also present in the external auditory canal and in the healthy conjunctiva (the membrane covering the eyelids and the exposed surface of the sclera). *Staphylococcus aureus* is the major organism of the skin that has the potential to cause disease, including boils, impetigo, and mastitis. **Boils,** common infections of hair follicles or sebaceous glands, are characterized by the accumulation of pus. **Impetigo** (im'pe-ti'go) is an infection of the skin that is usually caused by *Staphylococcus aureus* or *Streptococcus pyogenes*. **Acne** (ak'ne), well known to many teenagers, is characterized by the development of pimples. *Staphylococcus epidermidis* (e-pi-der'mi-dis) and *Propionibacterium acnes* (pro'pe-on'e-bak-te're-um ak'neez) are indigenous microflora of the skin often associated with acne.

Healthy conjunctiva often contains microorganisms derived from the skin. The external canal of the ear contains a microbial flora derived from resident skin organisms, while the middle and inner ears are usually sterile. An infection of the inner ear **(otitis media)** is often caused by the normal microbial flora of the nasopharynx and not by the normal skin flora.

KEY POINT

Although microorganisms live and multiply in many parts of the body, most do not cause disease. Opportunistic pathogens cause disease in compromised patients or when introduced into parts of the body where they do not belong. The blood, cerebrospinal fluid, urinary bladder, uterus, fallopian tubes, middle ear, paranasal sinuses, and kidneys are normally free of microorganisms.

MICROBIAL VIRULENCE

The diversity among microorganisms capable of inducing a disease state is paralleled by the variety of diseases to which they may give rise. The discussion that follows centers on bacterial pathogens since there is a extensive body of information available relating to their mechanism of virulence.

Pathogens are species that have caused disease in the past and therefore are likely to cause disease in the future. Not all isolates of a given pathogenic species will be equal in their ability to cause disease; the measure of the degree to which a given isolate or culture is able to cause disease is termed its **virulence.** It would require only a few cells of a highly virulent strain to cause disease in a susceptible host, whereas many cells of a strain with low virulence would be required to cause a similar disease state.

Bacteria cause damage by producing toxins, invading tissue, eliciting adverse immune reactions, or by some combination of these mechanisms. Four attributes of an infectious agent are: (1) The organism must be able to enter a host. Some pathogens are highly invasive and accomplish entry on their own; other pathogens, however, gain entrance to the host as a result of a bite from an arthropod vector such as a tick or mosquito. (2) The pathogen must be able to multiply once inside the host. Some organisms can multiply in various host tissues, whereas others are able to multiply in only one or a few local tissues. (3) Most successful pathogens can interfere with or bypass the host's defense mechanisms. For example, capsulated bacteria are less susceptible than noncapsulated ones to phagocytosis by white blood cells. A few bacteria survive phagocytosis and then grow within the macrophages that engulfed them. (4) By definition, pathogens must cause damage to their host. A pathogen may cause either immediate damage or a long-term debilitating disease.

INTOXICATION VERSUS INVASIVE DISEASES

Bacteria damage the host by growing in infected tissue, destroying tissue through chemical means, or by stimulating the inflammatory response. Those that invade tissue usually cause localized tissue damage and stimulate the inflammatory response. During this process, they often produce chemicals, known as **toxins,** that cause damage to cells or tissues. A disease attributable to the effects of toxins is described as an **intoxication** (in-tok-si-ka'shun).

An intoxication can be caused by the effects of the toxin itself; the toxin-producing organism need not be present in the host. Botulism food poisoning is an example of this kind of intoxication. It is due to *Clostridium botulinum,* which produces a neurotoxin during its growth in improperly prepared canned food. Persons who consume the neurotoxin will thus exhibit the clinical symptoms of botulism.

KEY POINT

Botulism food poisoning is an intoxication caused by the botulism neurotoxin alone: infection of the host by the toxin producing *Clostridium* is not necessary. In contrast, diphtheria is caused by an invasive bacterium that produces the diphtheria toxin after infecting the host.

TABLE 17.2 Properties of exotoxins and endotoxins

EXOTOXINS	ENDOTOXINS
Protein in composition	Complexes containing lipopolysaccharide
Heat labile	Heat stable
Soluble products of bacterial metabolism	Low solubility, components of cell-wall structures
Usually very high toxicity, specific in effect	Low toxicity, nonspecific
Produced by certain gram-positive and gram-negative bacteria	Produced by gram-negative bacteria
Detected by biological effect and laboratory methods	Detected by *Limulus* amebocyte lysate assay and other methods
Toxoids formed by chemical alteration	Toxoids usually cannot be formed

For an invasive disease to develop, the microorganism must gain entry into and be able to multiply in the host. Syphilis, for example, is due to the *Treponema pallidum,* which penetrates the mucous membranes of the genitourinary tract, enters the circulatory system, then invades the tissues.

Many pathogens cause disease through a combination of toxigenic and invasive mechanisms. For example, *Streptococcus pyogenes* produces the toxins responsible for scarlet fever after colonizing the throat. The invasive characteristics of *Strep. pyogenes* enable it to multiply in the throat and cause streptococcal pharyngitis (far'in-ji'tis). While growing here, certain strains of *S. pyogenes* produce the toxins responsible for the red scarlet fever rash. Most bacterial diseases result from similar combinations of toxigenic and invasive characteristics.

TOXINS

Toxins are natural substances that chemically damage cells or tissues and thereby cause disease. Toxins that affect humans are produced by snakes, jellyfish, plants, fungi, and bacteria. The bacterial toxins are of two major types: exotoxins and endotoxins (Table 17.2).

Exotoxins

Bacterial **exotoxins** are highly toxic (usually), heat-labile proteins produced and released into the environment during bacterial growth (Table 17.3). Their effects on human tissues are usually highly specific and they can be produced by either gram-negative or gram-positive bacteria. Most exotoxins can be destroyed by chemical or physical means, while chemically altered exotoxins are used to immunize individuals against the toxigenic diseases they cause.

Exotoxins usually "recognize" a specific single target tissue to which they are able to cause damage. Many exotoxins, including the diphtheria toxin, shigella cytotoxin, cholera toxin, and tetanus toxin are composed of two polypeptide subunits; the B subunit specifically binds to the target cell before the A subunit is released to enter the cell. These toxins are synthesized as inactive proteins that are activated by proteolytic cleavage, usually after binding to the target tissue. Upon cleavage, the A polypeptide is transported into the cell where it exerts its toxic effect (Figure 17.4).

Each of the toxins listed in Table 17.3 has a specific effect on the host tissue. The toxins of botulism and tetanus are neurotoxins; diphtheria and shigella cytotoxin are inhibitors of protein synthesis; and certain enterotoxins produced by *Escherichia coli* and *Vibrio cholerae* alter retention of fluid by the small intestine. Being proteins, most exotoxins are destroyed by heat at cooking temperatures; staphylococcal enterotoxin is a notable exception, however, since it is heat-stable.

Botulism toxin is the most toxic naturally produced substance known. It causes paralysis and can lead to death from respiratory failure. Humans are not immunized against botulism because the incidence of the disease is low (the total number of cases averaged 109 in the United States from 1980 to 1984). In a diagnosed case, the patient is treated with antitoxins, which neutralize the toxin.

TABLE 17.3 Properties of selected bacterial toxins

TOXIN (ORGANISM)	PROPERTIES	ACTION
	Exotoxins	
Botulism toxins *(Clostridium botulinum)*	Heat-labile proten, Seven antigenic types	Neurotoxin, muscle paralysis
Tetanus toxin *(Clostridium tetani)*	Heat-labile protein	Neurotoxin, spasmodic muscle contraction
Diphtheria toxin *(Cory. diphtheriae)*	Heat-labile protein	Inhibition of protein synthesis in eucaryotes
Enterotoxins *(Staphylococcus)* (Enterotoxic *E. coli*) *(V. cholerae)* cholera toxin	Proteins, some are heat-stable	Emetic (causes vomiting), net loss of fluid from small intestine
Erythrogenic toxins *(Streptococcus)*	Protein types A, B, and C	Scarlet fever rash
Whooping cough toxin *(Bordetella pertussis)*	Heat-labile protein	Dermal necrosis (slows ciliary action)
Plague (murine) toxins *(Yersinia pestis)*	Protein types A and B	Necrosis, hemorrhage
Shigella cytotoxin *(Shigella dysenteriae)*	Heat-labile protein	Inhibits protein synthesis—60S ribosomal subunit
	Endotoxins	
Pyrogen (gram-negative bacteria)	Toxic lipid A, heat-stable protein antigen	Diarrhea, fever, shock, intestinal hemorrhage, necrosis

Enterotoxins (en′ter-o-tok′sin) are bacterial exotoxins that act on the intestine (Gr. *enteron,* intestine). Enterotoxic *Escherichia coli* and *Vibrio cholerae* infect the intestinal tract, producing enterotoxins that cause diarrhea. The cholera toxin and the heat-labile enterotoxin of *E. coli* are structurally similar and have the same effects. Both act by increasing the intracellular concentration of cyclic AMP (cAMP) in the cells of the intes-

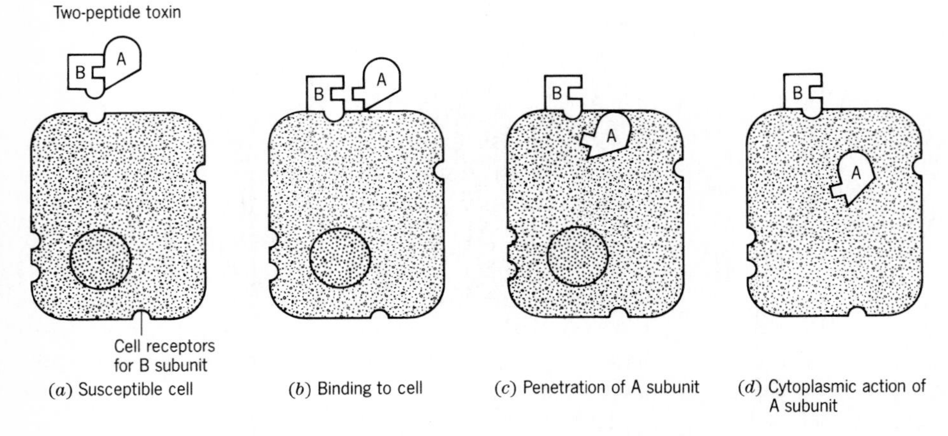

FIGURE 17.4 Mechanism of action of a two-peptide toxin.

Endotoxin and the LAL Assay

Many modern drugs are administered parenterally, that is, by routes other than oral. For the FDA to certify parenteral drugs and intravenous solutions, they must be free of bacterial contamination and free of endotoxins. Endotoxins, the lipid A portion of the outer membrane of gram-negative cell walls, are best avoided since they cause diarrhea, shock, and fever in humans and other animals. The fever response in rabbits is a major method for detecting endotoxins. A more sensitive test for endotoxins has been developed using blood components of the primitive horseshoe crab.

The *Limulus* amebocyte lysate (LAL) test evolved from studies of the immunological response of the horseshoe crab, *Limulus polyphemus,* to the presence of gram-negative bacteria. *Limulus* blood forms a gel or clot when it comes in contact with endotoxins either in a purified form or as components of gram-negative bacterial cell walls. This clotting reaction is presumed to benefit the crab by localizing the infection in a walled-off area. It also prevents blood loss from damaged appendages since gram-negative bacteria live on the surface of the horseshoe crab.

The horseshoe crab has a rudimentary circulatory system, containing blue blood and a single type of white blood cell called an amebocyte. Each amebocyte is packed with granules, which contain the proteins involved in the clotting reaction. When the cell contacts endotoxin, the granules migrate to the cell surface and release their contents of soluble coagulogen and clotting enzymes. The clotting enzymes are stimulated by the endotoxin to cleave coagulogen into coagulin, the protein that forms the insoluble gel.

Assays that measure very low quantities (picograms) of endotoxin have been developed using the *Limulus* gel-forming reaction. LAL is made from amebocytes harvested from the blood of horseshoe crabs, packaged under guidelines established by the Food and Drug Administration, and used by drug companies and medical laboratories to detect endotoxins. After bleeding, the horseshoe crabs are returned to the waters from which they were collected.

Limulus polymixia is the Atlantic Coast horseshoe crab. Its primitive circulatory system contains amebocytes which react with the endotoxin of gram-negative bacteria. (Courtesy of Dr. Thomas J. Novitsky and Associates of Cape Cod, Inc.) Collecting Limulus blood (below).

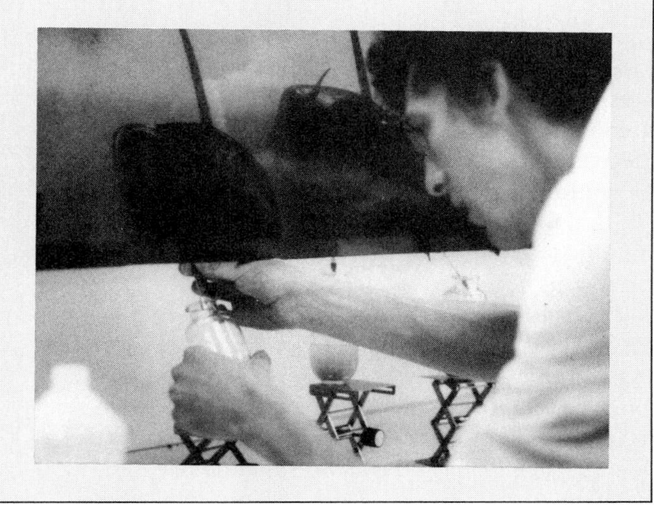

tinal mucosa. High cAMP stimulates these cells to secrete chlorine ions (Cl^-) and prevents them from taking up potassium ions (K^+). These changes trigger an outpouring of fluid into the intestinal lumen, resulting in the clinical symptoms of diarrhea.

Certain strains of *Staphylococcus aureus* produce **staphylococcal enterotoxins,** which, unlike other exotoxins, are relatively heat stable. These toxins are globular proteins, resistant to the acidity and enzymes of the stomach as well as to boiling for 30 minutes.

When ingested, staphylococcal enterotoxins cause food poisoning with the symptoms of vomiting, nausea, cramps, and diarrhea.

Another group of exotoxins causes necrosis (tissue destruction) or hemorrhage. *Bordetella pertussis* produces **whooping cough toxin,** which destroys the ciliary epithelial cells in the trachea. The **plague toxin** (murine toxin) produced by *Yersinia pestis* causes necrosis and also damages the vascular system, resulting in subcutaneous hemorrhages. *Shigella dysenteriae* produces shigella cytotoxin, which inhibits protein synthesis in mammalian cells.

Toxoids are modified bacterial exotoxins that have lost their toxicity, but retain their ability to stimulate the formation of antibodies. Toxoids can be created from most exotoxins. The diphtheria toxoid and the tetanus toxoid are used in the United States to immunize persons against tetanus and diphtheria. As we will see, endotoxins are chemically very different from exotoxins and cannot be made into toxoids.

Endotoxins

Endotoxins (pyrogens) are heat-stable, poisonous substances that are structural components of the outer membrane of gram-negative bacterial cell walls. They cause fever, so are also called **pyrogens** (pi′ro-jens). They are lipopolysaccharide in nature and can be released as subcomponents of the cell only after the cell lyses. Endotoxins are effective poisons both in the bound state and as products of cell lysates. Their toxic effects are associated with the lipid A portion of the gram-negative cell wall lipopolysaccharide (Table 17.3). Endotoxins are stable to boiling, have low immunogenic characteristics, and have low nonspecific toxicity in humans.

The presence of endotoxins in animals causes diarrhea, shock, and fever—the fever being due to the **endogenous pyrogen** released by white blood cells upon contact with the endotoxin. Exposure to endotoxins occurs during infections by gram-negative bacteria, accidental wounds to the abdominal cavity, and situations that result in the release of intestinal flora (such as a ruptured appendix).

HYDROLYTIC ENZYMES

Bacteria produce certain enzymes that hydrolyze tissue components. These enzymes are classed as toxins when they cause cellular damage. Hydrolysis (as described in Chapter 2), is the process of cleaving a covalent bond with the concomitant addition of water to the reactants (Figure 17.5). All cells produce hydrolytic enzymes for the purpose of recycling their essential components and utilizing storage products. Cells also excrete hydrolytic enzymes that break down large molecules, some of which are structural components of tissue. Pathogens produce various hydrolytic enzymes, as depicted in Table 17.4.

Bacterial **collagenase** and **hyaluronidase** (hi-ah-lu-ron′i-dace) are enzymes that are able to hydrolyze components of human connective tissue. Collagen, the most abundant protein in connective tissue, is hydrolyzed by bacterial collagenase; hyaluronic acid, which also is a component of connective tissue, is hydrolyzed by bacterial hyaluronidase. Hyaluronidase is produced by invasive staphylococci, enabling them to spread rapidly through the skin and invade tissues. Production of these hydrolytic enzymes contributes to bacterial invasiveness.

Lecithinase (les′i-thin-ace) hydrolyzes lecithin, a phospholipid of animal cell membraness. This enzyme can destroy the integrity of the cell's plasma membrane and cause the cell to lyse.

Animals localize infections by forming a fibrin clot surrounding an injury. Some bacteria can negate this defense mechanism by producing enzymes that destroy the clot. **Streptokinase** is such a bacterial enzyme; it converts the plasminogen in an animal's serum to plasmin, which in turn hydrolyzes the fibrin clot. Once the clot is destroyed, the bacteria can invade adjacent tissue thereby spreading the infection.

Proteinases, which hydrolyze proteins to peptides, are abundant in microorganisms that cause severe damage to tissue such as skin and muscle. These proteinases are diverse in their action: some hydrolyze many different peptide bonds, others hydrolyze only bonds formed between two specific amino acids.

The types of hydrolytic enzymes, their specificity, and the quantity produced by the microorganism all contribute to the organism's virulence. The hydrolytic enzymes produced by invasive pathogens enable them to spread more deeply into the infected tissue and to invade adjacent tissues.

ENZYME	SUBSTRATE	PRODUCTS	REACTION
Collagenase	Helical regions of collagen	Polypeptides	Hydrolysis of peptide bonds
Fibrinolysin	Fibrin	Soluble products	Hydrolysis of peptide bonds (esp. arg and lys)
Hyaluronidase	Hyaluronic acid 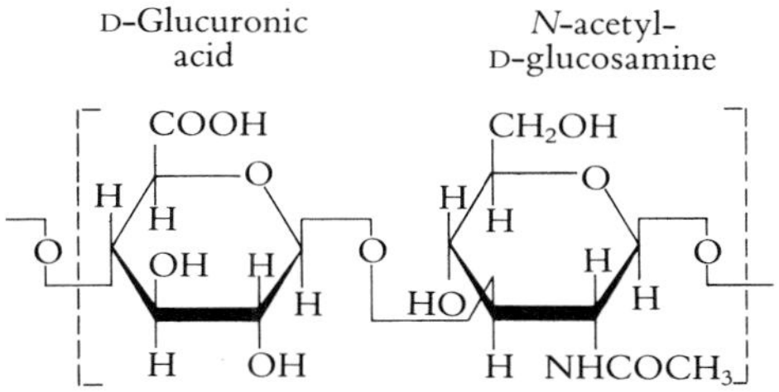	D-glucuronic acid N-acetyl-D-glucosamine	Hydrolysis β1-4 glycosidic bond
Lecithinase (phospholipase A)	Lecithin 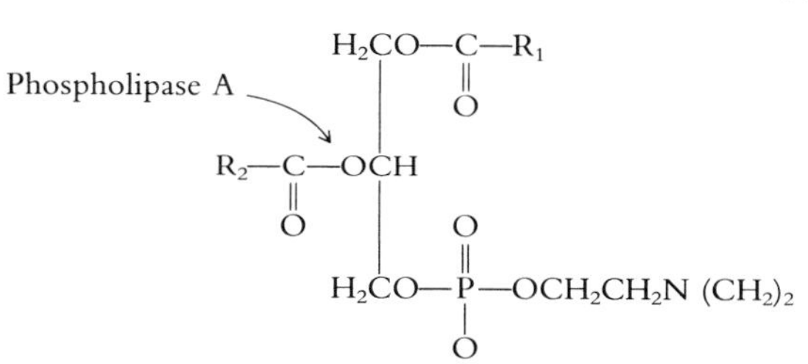	1-Acylglycerophosphocholine + a fatty acid anion	Lipid hydrolysis

FIGURE 17.5 Reactions of hydrolytic enzymes produced by pathogenic bacteria.

CAPSULES

Some bacteria produce extracellular polymers of amino acids or carbohydrates that surround the cell with a protective capsule. The capsular material is deposited outward of the bacterial cell wall and lends a smooth viscous appearance to the producing colonies. The capsule can be recognized by the use of simple staining techniques (Figure 17.6).

Capsulated bacteria are much less likely to be destroyed by phagocytes. This attribute generally makes

TABLE 17.4 Bacterial enzymes involved in virulence

ENZYME	ACTION	ROLE IN DISEASE
Coagulase	Activation of fibrin, clot formation	Unclear function: may coat bacteria, useful in diagnosis
Collagenase	Hydrolysis of collagen	A factor in invasiveness
Hyaluronidase	Hydrolysis of hyaluronic acid in tissue	A factor in invasiveness
Lecithinase (alpha toxin)	Hydrolysis of membrane lipids	Lysis of cells, destroys cell membranes
Streptokinase (plasmin[a])	Hydrolysis of fibrin into soluble products	A factor in invasiveness, used in diagnosis
Proteinases	Hydrolysis of proteins	A factor in invasiveness

[a]Streptokinase converts plasminogen in plasma to plasmin, which has the proteolytic activity that acts on fibrin.

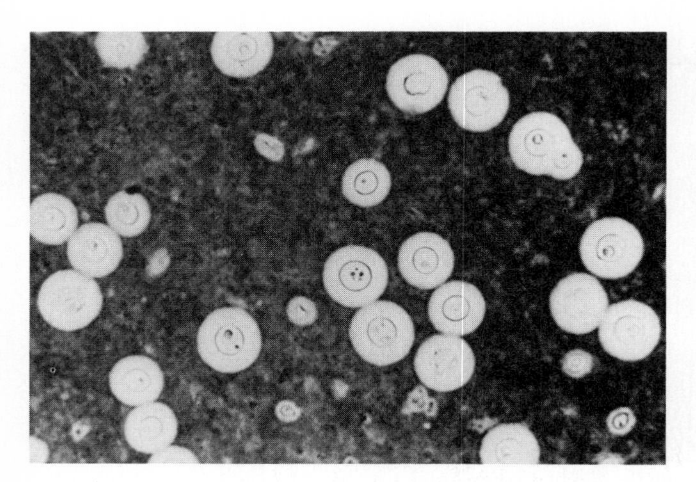

FIGURE 17.6 Microbial capsules can be observed in the light microscope with simple staining procedures. The capsule of *Cryptococcus neoformans* is clearly visible when the background is stained with india ink. (Courtesy of Dr. N. L. Goodman.)

these strains more virulent than the noncapsulated ones, as was demonstrated by Griffith's transformation experiments (see Chapter 10). The host response is to produce specific antibodies, called opsonins, that aid in overcoming the effects of the capsules. In fact, vaccines have been developed from capsular material and used to protect persons against infections by capsulated bacteria.

ADHERENCE OR COLONIZATION

Bacteria that colonize specific regions of the body appear to have surface antigens that facilitate adherence to that particular tissue. (Recall that colonization of tooth surfaces by species of streptococci plays a role in dental caries.) A second example is the presence of pili on the surface of gonococci, which appears to facilitate bacterial adherence to urethral epithelium. Once adherence is accomplished, the bacteria are able to colonize that tissue.

MEASURES OF VIRULENCE

Pathogens can possess one or many virulence factors. Some virulence factors, such as hydrolytic enzymes, can be measured by simple biochemical techniques, whereas others represent a combination of properties detectable only by a biological assay. For example, the ability to multiply inside a phagocytic white blood cell can be measured only by biological assay. The **lethal dose 50 percent**—LD_{50}—is a biological assay that measures virulence or toxicity by determining the amount of toxin or the number of bacteria that will cause death in 50 percent of a population of susceptible animals.

The LD_{50} is perhaps the only way to demonstrate the extreme toxicity of certain bacterial products, such as botulism toxin. Botulism type A toxin in the mouse has an LD_{50} of 1.2 nanograms (ng) per kg of body weight. From these results, experts estimate that 1 ng per kg of human body weight is a lethal quantity for a human. Obviously, botulism toxin should be avoided and cultures of *Cl. botulinum* handled with care.

These procedures sacrifice animals and should not be used when scientifically valid and legal alternatives are available. Alternative tests can be used in drug testing only after they are approved by the U.S. Food and Drug Administration. The use of bioassays that cause the death of laboratory animals and drug testing procedures that harm animals are being challenged by various organizations in the United States and abroad. Bills have been introduced in the U.S. Congress that would curtail the use of animal tests for certifying biological products. The issues surrounding this controversy are being debated.

NONSPECIFIC RESISTANCE TO INFECTION

Establishment of an infectious disease depends on the route of infection, the strength of the infecting dose, and the health of the host. Under normal circumstances, the host's defense mechanisms will act quickly to ward off the infectious agent. However, several factors come into play here, including race, age, and the overall state of the host. These factors form the basis of the next discussion.

SPECIES, GENETICS, AND RACE

Some bacterial pathogens and viruses are species specific and have a very limited host range. *Yersinia pestis* can be carried by ground squirrels and transmitted to humans by fleas to cause plague. Human plague is usually fatal if untreated, yet in ground squirrels *Y. pestis* causes no apparent ill effects. To cite another example: poliovirus infects humans and monkeys, but not quadrupeds or other species unrelated to the human. A species that never succumbs to a particular infectious disease is said to have **species resistance** to that disease.

Most pathogens exhibit species specificity and will cause disease only in a limited number of hosts. This is beneficial to us since there are only a few infections that we can contract from domestic animals. On the other hand, species resistance is a major obstacle to scientists because diseases that cannot be reproduced in laboratory animals are difficult to investigate. Syphilis and leprosy are two such diseases since neither can be produced experimentally in animal models.

There is historical evidence that the inherited characteristics of a host population, represented by its gene pool, can confer either resistance or susceptibility to an infectious agent. For example, the population of the Massachusetts Bay Indians prior to colonization by Europeans was estimated to be about 30,000 individuals; by the mid-1600s, however, fewer than 1000 had survived the severe smallpox epidemics that were introduced by European settlers. A similar devastation occurred when the Plains Indians of North America lost two-thirds of their population to smallpox and tuberculosis. Their resistance to these diseases was very low, because they had not previously been exposed to them, whereas the European settlers, the progeny of ancestors who had been so exposed, survived because they were genetically more resistant to the diseases. More recently, during the 1970s, an Amazonian indian tribe was devastated by Asian flu following contact with a team of government workers. Thus, inhabitants of remote regions who have been isolated from various pathogens are super susceptible to the diseases they cause because neither the inhabitants nor their ancestors ever developed resistance to them.

A population possessing a common set of genetically determined physical characteristics is designated a **race.** American Indians are a race, as are the Caucasian and the Negro peoples. Racial resistance and susceptibility to a disease are perpetuated in the gene pool of that race. For example, members of the Negro race have a high resistance to erysipelas, a streptococcal skin infection, and they respond readily to treatment for gonorrhea. Subgroups of Caucasians also demonstrate differences in resistance to and susceptibility to various diseases. People of Irish descent have a low resistance to tuberculosis, whereas people of Eastern European Jewish descent have a very low mortality from tuberculosis.

The facts about malaria provide an excellent example of one type of resistance. Persons with sickle-cell anemia are highly resistant to malaria. The basis for this resistance is as follows: In sickle-cell anemia, the hemoglobin is abnormal, causing the red cells to assume a sickle shape. Plasmodias, the causative agents of malaria, replicate in red cells during an essential stage in their life cycle; however, they are unable to replicate in the sickled cells of persons with sickle-cell anemia. Sickle-cell anemia is prevalent in Negroes living in tropical regions where malaria is endemic. Here we have a clear-cut case of inherited resistance to a disease.

RESISTANCE AND AGE

After the first few months of life, resistance to infectious disease increases during one's lifetime, largely as a consequence of the development of the immune response. An infant acquires from the mother a short-lived immunity to infectious diseases, which may last for six months to one year. As the child grows, its immune response develops, either through immunizations or as a result of contracting and successfully surviving sublethal cases of infectious disease.

Tuberculosis and pneumonia appear to be exceptions to the general rule of increased resistance with age. The incidence of tuberculosis increases with age, beginning with the group 5 to 14 years of age, and is highest among persons older than 25 years of age (Figure 17.7). A similar pattern is true of pneumonia. The gradual deterioration that usually accompanies advanced age is believed to be responsible for this lowered resistance.

RESISTANCE AND THE HOST

Resistance to infections is directly related to the host's physiological well-being. Surgery, physical wounds, organic disease, nutritional deficiencies, stress exhaustion, and fatigue all lead to diminished resistance of various sorts.

Poor nutrition and, worse, outright starvation will

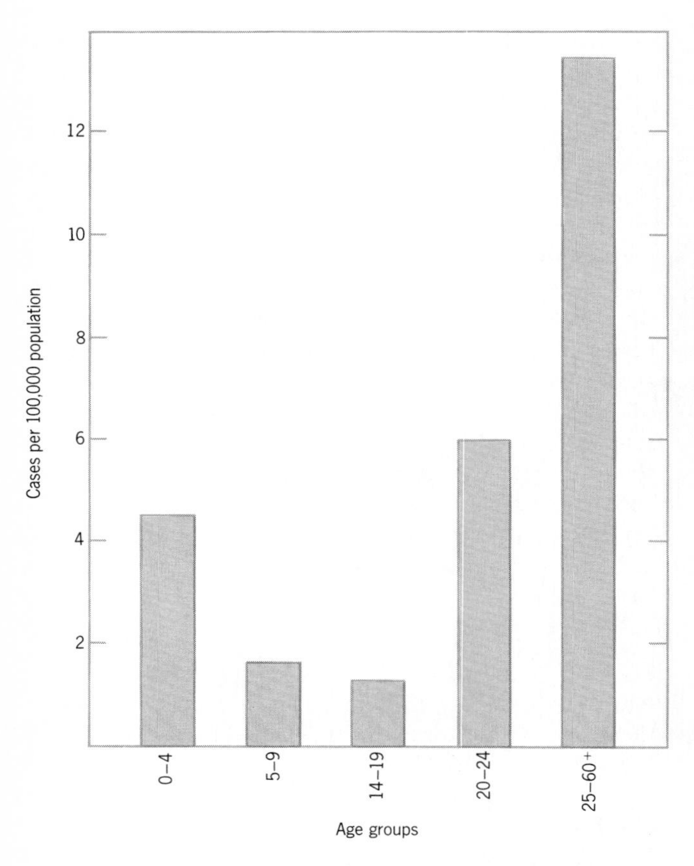

FIGURE 17.7 Cases of tuberculosis by age group in the United States during 1984, plotted as cases per 100,000 population.

FIGURE 17.8 Starvation and other physical stresses reduce the ability to resist infectious diseases. A father and his children at a Korem camp hoping to survive another night in the open at near-freezing temperatures. (UNICEF 88/84/Ethiopia, photograph by Bert Demmers.)

quickly exacerbate susceptibility to infection. Persons living in refugee camps are particularly vulnerable because they are more likely to suffer fatigue, nutritional deficiencies, or starvation (Figure 17.8). Fatigue and crowding also increase susceptibility, as has been observed during outbreaks of meningococcal meningitis among military recruits. In this situation the rigorous basic training induces fatigue, and the crowded living conditions in barracks facilitate the transmission of meningococcus, the causative agent. All these factors taken together—species, genetics, race, age, and health—play an overriding role in susceptibility to infection. This brings us to the topic of the defense mechanisms that are activated when host and microbe come together.

FIRST-LINE DEFENSE

As was previously stated, the first line of defense consists of the skin; the conjunctiva and tears; mucous membranes in the nasopharynx; the ciliary epithelium of the trachea; and the mucous membranes of the genital and urinary tracts. Together, these anatomical features function as a barrier against microbes that have the ability to colonize tissues and cause disease.

SKIN AS A BARRIER TO INFECTION

Healthy skin is an effective barrier against the invasiveness of most infectious agents, though many bacteria normally reside there. Additionally, the outer layer of skin is continually being sloughed off, carrying with it large numbers of normal flora. And skin contains sebaceous glands, which produce fatty acids that are bactericidal. However, should bacteria be able to penetrate this barrier and infect for example the hair follicles or sebaceous glands, the infection will remain localized as a result of fibrin deposition around the infection site.

THE EYE

The **conjunctiva** (kon′junk-ti′vah) is the mucous membrane lining the eyelid and the exposed surface of the sclera. Underneath the eyelid is the lacrimal fluid (tears) that bathes the cornea and the sclera (white) (Figure 17.9). Tear ducts carry this fluid from the lacrimal glands to the conjunctiva. Tears contain lysozyme, which, as we have learned, is an enzyme that hydrolyzes peptidoglycan in bacterial cell walls, and secretory immunoglobulins that inactivate specific infectious agents. Thus the eye is well protected by the conjunctiva and tears. Tears not only lubricate the eyeball to facilitate its movement, but wash away foreign matter from its surface and so protect the eye from infectious agents.

RESPIRATORY TRACT

The respiratory tract is protected by nasal hairs, ciliary epithelium, and mucous membranes. Mucus secreted by the membranes lining the nasopharynx moves to the oropharynx on a regular basis and is either expectorated or swallowed. Foreign particles and agents that are inhaled are either filtered out by hairs in the anterior nares or trapped in the flow of mucus. The mucus itself is not bactericidal, but its viscosity causes bacteria to be entrapped within it so they cannot gain entrance into the lungs.

The trachea is also lined with mucous membranes interspersed with ciliary epithelium. Mucus containing trapped foreign matter is propelled upward toward the oropharynx either by the ciliary epithelium or by coughing. This material is then expectorated or swallowed.

These mechanisms normally protect the bronchioles from infectious agents. As a matter of fact, air expired from healthy lungs is normally sterile. When microorganisms are present in the expired air they are exhaled in water droplets generated in the nasopharynx region by a cough or sneeze.

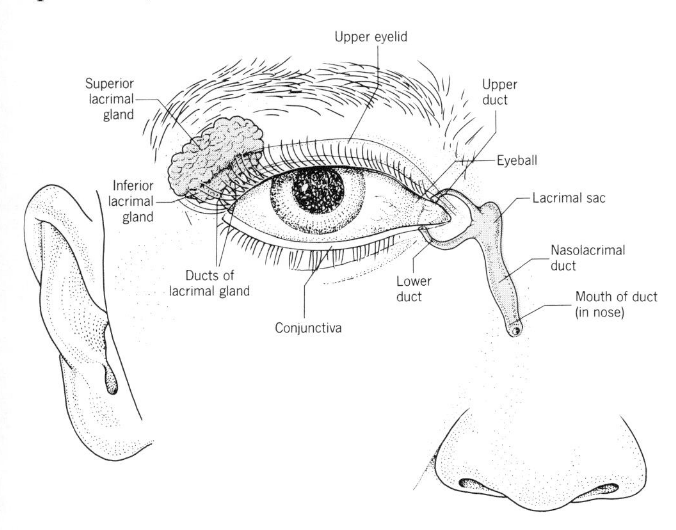

FIGURE 17.9 The human eye.

DIGESTIVE TRACT

The bacteria normally resident in the mouth—on the tongue, tooth surfaces, and gums—are washed from surfaces by saliva and swallowed in liquids or food. When the bacteria reach the stomach, its inhospitable environment usually prevents them from gaining entrance to the intestine. In contrast, bacterial spores, the cysts of microbes, and microbes enmeshed in food are able to resist the acidity and enzymes of the stomach, and will reach the intestine, where they may become part of the intestinal flora.

URINARY AND GENITAL TRACT

Voiding of urine cleanses the urethra and normally maintains the urinary tract in a bacteria-free state. The female vagina is colonized during childhood and remains colonized by bacteria throughout life. Vaginal mucus and the acidity of the vagina are to some degree bactericidal.

KEY POINT

The first lines of defense against infectious agents are the mechanical and chemical barriers that include the conjunc-

tiva, tears, mucous membranes of the respiratory and urogenital tract, ciliary epithelium of the trachea, acids and enzymes of the stomach, secretions of the sebaceous glands, and the skin.

SECOND-LINE DEFENSE

CELLULAR DEFENSE MECHANISMS

The **second-line defense** mechanisms consist of the body's array of phagocytic cells and the inflammatory response. The phagocytes are capable of destroying foreign agents that have penetrated the anatomical barriers.

Leukocytes

White cells in blood, lymph, and tissues are collectively called **leukocytes** (Table 17.5). Leukocytes are identified by their function, anatomical location, morphology (Table 17.6), and their staining reactions (Figure 17.10, Color Plate II). Leukocytes are derived from stem cells that reside in the bone marrow. These stem cells differentiate into functional white cells that occur in various parts of the body. Leukocytes are divided into the granulocyte and the monocyte series.

TABLE 17.5 Leukocytes in normal human blood

CELL TYPE	PERCENTAGE RANGE
Granulocyte Series	
Neutrophils	40–80
Eosinophils	1–7
Basophils	0–1
Monocyte Series	
Monocytes	2–11
Lymphocytes	15–50

The granulocyte series contains cells possessing a lobed or irregularly shaped nucleus and a granular cytoplasm. **Neutrophils** (also known as polymorphonuclear neutrophils or PMN) are identified by their lobed nucleus and by the pink-to-violet staining granules present in their cytoplasm. Neutrophils are the major phagocytes in the circulatory system. **Eosinophils** (e-o-sin′o-fil) comprise 1 to 7 percent of the blood's white cells. They are morphologically similar

TABLE 17.6 Properties of leukocytes

LEUKOCYTE	CELL TYPE	MORPHOLOGY	BODY LOCATION	FUNCTIONS
Granulocytes	Neutrophil (polymorphonuclear)	Lobed nucleus	Circulatory system, few in tissue	Phagocytosis
	Basophil	Lobed nucleus, granules in cytoplasm	Circulatory system	Release of histamine (serotonin in mice)
	Eosinophil	Lobed nucleus, red–yellow staining granules	Circulatory system	Control of inflammatory response; helminthic infections
Monocytes	Monocytes	Single nucleus, abundant cytoplasm	Circulatory system	Phagocytosis, precursors of macrophages
	Macrophages (free) Histiocytes (fixed in tisssue)	Single nucleus, abundant cytoplasm description applies to both macrophages and Histiocytes	All tissues	Phagocytosis and destruction of cell debris
Lymphocytes	Lymphocytes	Single nucleus, little cytoplasm	Lymphoid tissue and circulation	Participation in immunological response
	Plasma cells	Single nucleus, ovoid cell	Lymphoid tissue	Antibody synthesis

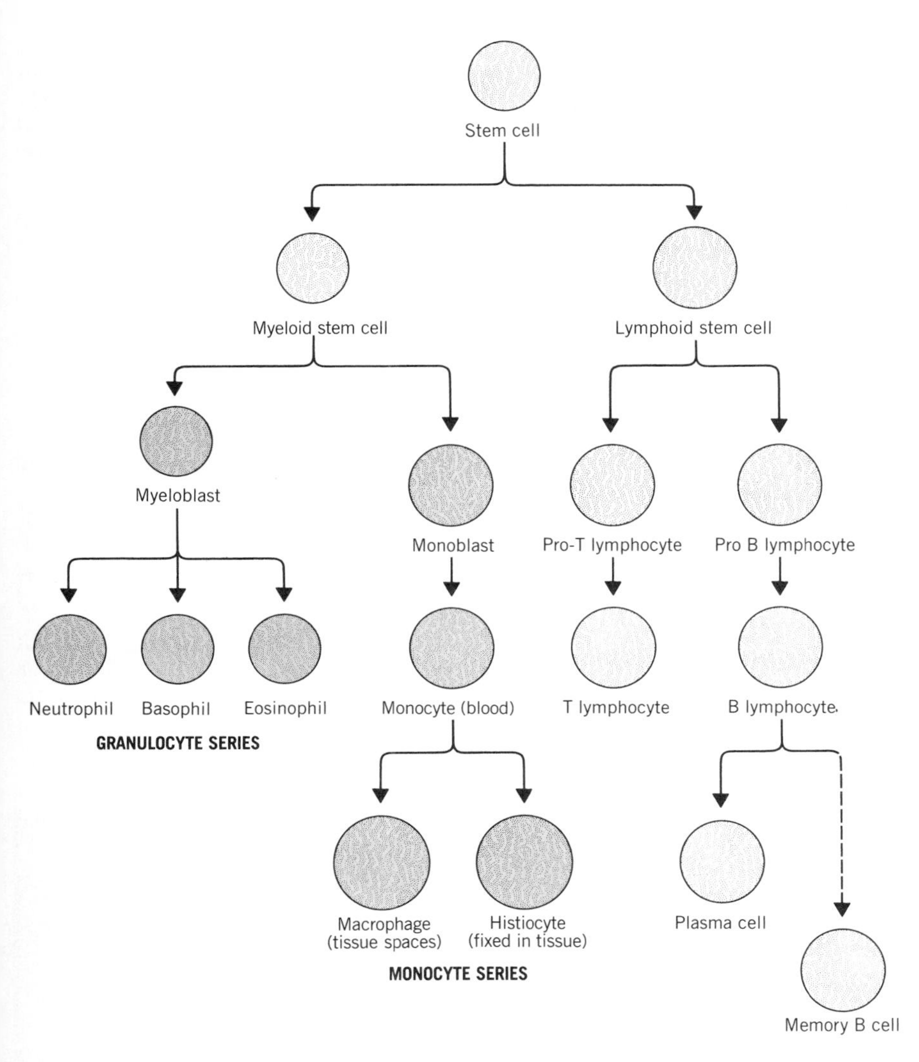

FIGURE 17.10 Leukocytes are all derived from stem cells through the process of differentiation.

to neutrophils; however, they take up the acid eosin (hence their name), which stains their cytoplasmic granules red. Eosinophils are involved in resisting infections by helminths. **Basophils** have an irregularly shaped nucleus and possess large cytoplasmic granules. These granules contain heparin, slow reacting substance A, and eosinophil chemotactic factor, which are released on degranulation, usually induced by an allergenic stimulus. Basophils are present in the blood and contribute about 0.2 percent of the leukocytes. Mast cells are morphologically similar cells present in tissues.

The monocyte series (Table 17.5) includes the lymphocytes, which are involved in the immune response, and the monocytes, which are circulating forms of differentiated stem cells that eventually become phagocytic tissue cells called macrophages (see below). Though monocytes have phagocytic activity in the circulation, blood monocytes probably represent a transitional stage in macrophage development.

Phagocytosis Phagocytosis is the mechanism by which cellular debris, foreign matter, and infectious agents are removed from tissue. Phagocytes present in blood, tissues, and tissue spaces seek out foreign matter, ingest it, and in most cases destroy it.

Neutrophils are the major phagocytes in the blood, comprising 40 to 80 percent of the normal white cell population. They are the terminal stage in stem cell differentiation and have a circulating half-life of only 7 hours. Neutrophils are attracted to the site of infection by the presence of bacteria and by certain products of inflammation. Their job is to phagocytize and kill the invading microorganisms.

Monocytes are mononuclear phagocytes, making up 2 to 11 percent of the circulating white cell population. They remain in the blood for only a few days before migrating into the tissue spaces, where they increase in size (by as much as tenfold) and differentiate into macrophages (Figure 17.11). **Macrophages** are present in most tissues and function as scavengers, by removing and destroying foreign matter and breakdown products of tissue. Macrophages move freely in the tissue spaces or become fixed in the reticulum of the spleen, lymph nodes, and liver. The fixed macrophages are known as **histiocytes** (his′te-o-syt). The dispersed array of phagocytes associated with the connective tissue framework of the liver, spleen, and lymph nodes is known as the **reticuloendothelial system** (re-tik′u-lo-en′do-the′le-al), whose job is to remove particulate matter.

The first step in phagocytosis is attachment of the particle to the phagocytic cell. The particle is then engulfed (Figure 17.12) by the plasma membrane, which invaginates and pinches off to form a vacuole called the **phagosome.** When the phagosome collides with a lysosome (see Chapter 5), the two structures fuse to form a phagolysosome. This joining is accompanied by the explosive rupture of the lysosome and the exposure of the bacteria within the phagolysosome to the degradative effects of the lysosome. Lysosomes contain many bactericidal substances, including hydrogen peroxide and superoxide radical. They also contain degradative enzymes such as lysozyme, proteinases, acid hydrolases, and methylperoxidase, all active in the acidic interior of the phagolysosome. Few bacteria survive this hostile environment.

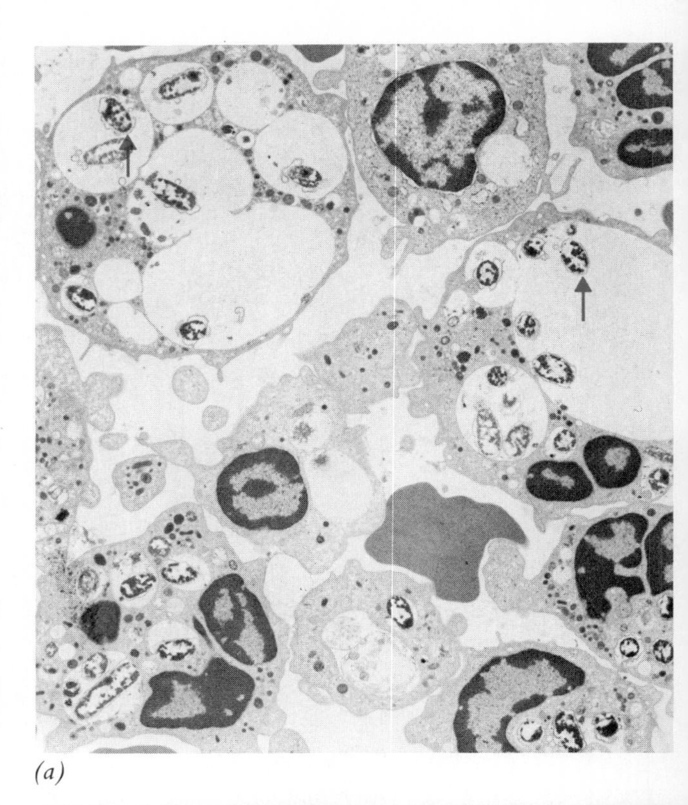

(*a*)

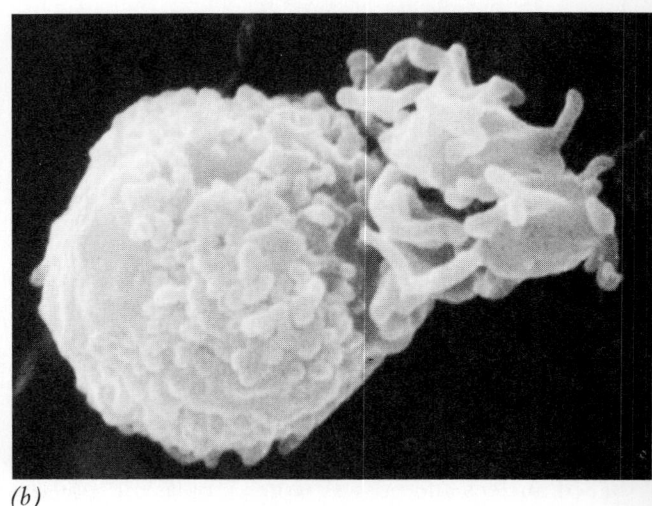

(*b*)

FIGURE 17.11 (*a*) Transmission electron micrograph of phagocytic leukocytes containing ingested bacteria in phagocytic vesicles (Courtesy of Dr. D. M. Philips, Visuals Unlimited.) (*b*) A scanning electron micrograph of a lymphocyte associated with platelets and red blood cells. (Courtesy of Dr. R. Chao.)

Inflammation

Inflammation represents a mobilization of defense mechanisms where tissue damage has occurred. It is

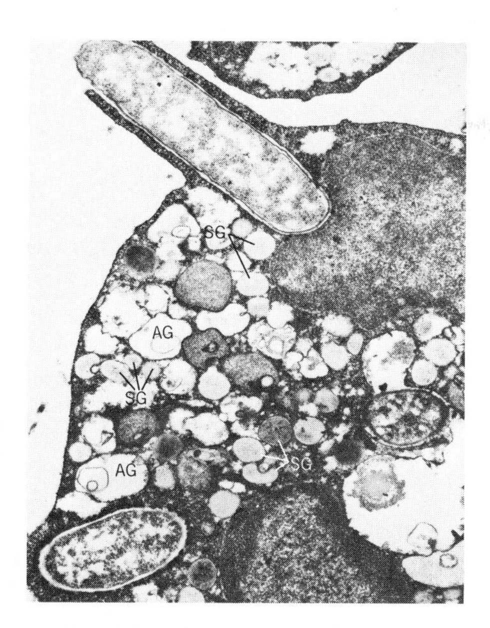

FIGURE 17.12 Phagocytosis of *E. coli* cells by a rabbit polymorphonuclear neutrophil. The bacteria are degraded by enzymes released into the phagocytic vesicles. (Courtesy of Dr. Dorothy F. Bainton.)

marked by rednesss, swelling, pain, heat, and tenderness, which result from an increase in the flow of blood and tissue fluid to the site of infection. Some infectious agents stimulate the white cells to accumulate at the site, which we observe as pus. The swelling is due to the local dilation of blood vessels, and this allows fluid and white cells to penetrate the area. Redness results from the increased flow of blood to the area. Neutrophils gain entrance to the inflamed area through the dilated capillaries, whereas macrophages migrate to the inflammation site through tissue spaces.

Lymphokines (lim′fo-kyn) are released into the fluid surrounding sensitized lymphocytes. The lymphokines are a large group of proteins that exert effects on other cells, and they are involved in the inflammatory reaction at the infection site. They are named and described on the basis of their effects on other cells: migration-inhibiting factor, macrophage-chemotactic factor, macrophage-activating factor, pyrogen-releasing substance, T-cell growth factors (called interleukins), and lymphocyte-chemotactic factor. Although most lymphokines are produced by activated T lymphocytes, some are produced by B lymphocytes, macrophages, and virus-infected fibroblasts.

KEY POINT

The second line of defense against infectious agents is the complement of phagocytic cells located in the circulatory system, tissues, and tissue spaces. The phagocytes assume specific defensive roles after differentiating from a common stem cell.

In the next chapter we discuss the immune response—the third and final line of defense, called into play against those bacteria and viruses as they are being processed by phagocytic cells.

Summary Outline

NATURE OF THE DISEASE STATE

1. Damage or disturbance in structure or function of a bodily organ caused by microbes is classified as infectious disease.
2. Obligate parasites are unable to reproduce outside their host and always cause disease, whereas opportunistic parasites cause disease only under specific conditions.
3. Infectious diseases are characterized by the dynamic interaction between the microbe's virulence and the host's defenses against the agent.

MICROBIAL FLORA OF THE HEALTHY HUMAN

1. Microorganisms coexist with humans, residing in the mouth, nasopharynx, eyes, external ears, re-

spiratory tract, gastrointestinal tract, anterior urethra, and the vagina.

2. Some of these organisms are opportunistic pathogens that cause disease in compromised hosts, whereas others are not known to cause disease in humans.
3. Blood, cerebrospinal fluid, kidneys, urinary bladder, fallopian tubes, lungs, and urethra (except the anterior urethra) of healthy persons are free of microbial contaminants.
4. Microbes growing on the surface of teeth and in the gingival crevices contribute to the development of dental caries and periodontal disease.
5. The lower ileum and the large intestine contain large populations of anaerobic bacteria. Diseases of the intestinal tract caused by microbes include peritonitis, cholecystitis, and gastroenteritis.
6. The skin is colonized by streptococci and staphylococci that can cause boils and impetigo. Resident skin organisms can also cause infections of the eyes and ears.

MICROBIAL VIRULENCE

1. Pathogens are microorganisms capable of causing disease. The measure of a pathogen's immediate ability to cause disease is termed virulence.
2. Invasive diseases are caused by microbes able to penetrate the host's defense mechanisms and colonize tissues within the host.
3. Toxin-producing microbes cause toxigenic diseases.
4. Exotoxins are highly toxic, heat-labile proteins produced during the growth of certain gram-negative or gram-positive bacteria. Exotoxins include the botulism toxin, tetanus toxin, diphtheria toxin, enterotoxin, erythrogenic toxin, whooping cough toxin, and plague toxin.
5. Endotoxins are heat-stable lipopolysaccharide components of the outer membrane of gram-negative bacteria. They cause diarrhea, shock, and fever.
6. Bacterial collagenase, hyaluronidase, lecithinase, proteinases, and streptokinase are hydrolytic enzymes that contribute to bacterial invasiveness.
7. Bacterial capsules protect the cell against phagocytosis.
8. The toxicity of a toxin and the virulence of a bacterial strain can be measured by means of the LD_{50} test.

NONSPECIFIC RESISTANCE TO INFECTION

1. Animals are not all subject to the same infectious agents because susceptibility to an infectious agent depends on the species, gene pools within the population, race, age, and general physiological state of the animal.
2. Isolated human populations have been devastated by infectious diseases introduced by explorers and government officials.
3. Resistance to most infectious diseases increases during one's lifetime, but decreases with a decline in physiological well-being.

FIRST-LINE DEFENSE

1. The first lines of defense against infectious agents are anatomic barriers, including the skin; the conjunctiva and tears, which protect the eyes; mucous membranes in the nasopharynx; ciliary epithelium of the trachea; and the mucous membranes of the genital and urinary tracts.
2. The eye is protected by tears that contain lysozyme and secretory immunoglobulins.
3. The respiratory tract is protected by hairs, ciliary epithelium, and mucous membranes.
4. The acidity and enzymes of the stomach are a barrier to infectious agents reaching the intestine.
5. Voiding of urine cleanses the urethra. The mucus and acidity of the vagina are to some degree bactericidal.

SECOND-LINE DEFENSE

1. The second-line defense mechanisms consist of the body's array of phagocytic cells and the inflammatory response.
2. Invading microorganisms are phagocytized by circulating neutrophils and by macrophages and histiocytes present in the tissue and tissue spaces.
3. Infectious agents cause tissue damage that initiates inflammation resulting in redness, swelling, pain, heat, and tenderness at the damaged site.
4. Leukocytes produce lymphokines that stimulate macrophages and other white blood cells to participate in removing the infectious agents.

Questions and Topics for Study and Discussion

QUESTIONS

1. What changes occur in the intestinal flora as a breast-fed infant grows into childhood?
2. How does the normal flora of the vagina change during sexual maturation? Which microorganisms in the vagina are potential pathogens?
3. Define and explain the following:

basophil	intoxication	pyrogen
cystitis	otitis media	streptokinase
disease	pathogen	urethritis
gastroenteritis	pyelonephritis	virulence

4. What anatomical parts of a healthy human are free of microorganisms?
5. Compare the characteristics of an invasive pathogen with those of a pathogen that causes a toxigenic disease.
6. Explain the mechanism employed by the two subunit exotoxins to recognize their target tissue.
7. What enzymes contribute to bacterial invasiveness and how do they act on host tissue?
8. How does a bacterial capsule contribute to the virulence of the bacterium?
9. Are all human races equally susceptible to all infectious diseases? Give examples to support your answer.
10. How does age influence a person's susceptibility to disease?
12. What anatomical barriers do humans possess that protect them against infectious agents?
13. Which cells participate in phagocytosis and where are they located?
14. What are lymphokines and how do they participate in host resistance to infectious agents?

DISCUSSION TOPICS

1. Consider the defensive mechanisms we possess to resist infectious agents. Which of these mechanisms are also present in plants? Do plants have alternative mechanisms to resist disease? Explain.
2. Should the human race be concerned about introducing infectious agents into extraterrestrial environments? Discuss the historical evidence that supports your answer.

Further Readings

BOOKS

Lennette, E. H., A. Balows, W. J. Hausler, Jr., and H. J Shadomy, *Manual of Clinical Microbiology,* 4th ed., American Society for Microbiology, Washington, D.C., 1985. A comprehensive reference for the identifying of infectious agents. Contains a chapter on the microorganisms of the human.

Mims, C. A., *The Pathogenesis of Infectious Diseases,* 2nd ed., Academic Press, New York, 1982. A short, comprehensive description of the mechanisms of infectious disease.

ARTICLES AND REVIEWS

Costerton, J. W., R. T. Irvin, and K.-J. Cheng, "The Bacterial Glycocalyx in Nature and Disease," *Ann. Rev. Microbiol.,* 35:299–324 (1981). This review emphasizes the importance of extracellular polysaccharides in virulence and colonizing inert surfaces.

Gill, M. D., "Bacterial Toxins: A Table of Lethal Amounts," *Microbiol. Revs.,* 46:86–94 (1982). The toxicities of bacterial toxins are related to the guidelines for cloning bacteria toxin genes.

Loesche, W., "Role of *Streptococcus mutans* in Human Dental Decay," *Microbiol. Revs.,* 50:353–380 (1986). Reviews how the accumulation of bacterial communities on the tooth surface causes both decay and periodontal disease.

McNabb, P. C., and T. B. Tomasi, "Host Defense Mechanisms at Mucosal Surfaces," *Ann. Rev. Microbiol.,* 35:477–496 (1981). Reviews the role of secretory IgA and mucins.

Middlebrook, J. L., and R. B. Dorland, "Bacterial Toxins: Cellular Mechanisms of Action," *Microbiol. Revs.,* 48:199–221 (1984). The authors relate the action of toxin binding to target tissue to the hormone receptor system.

C H A P T E R

18

Immunology: Host Protection

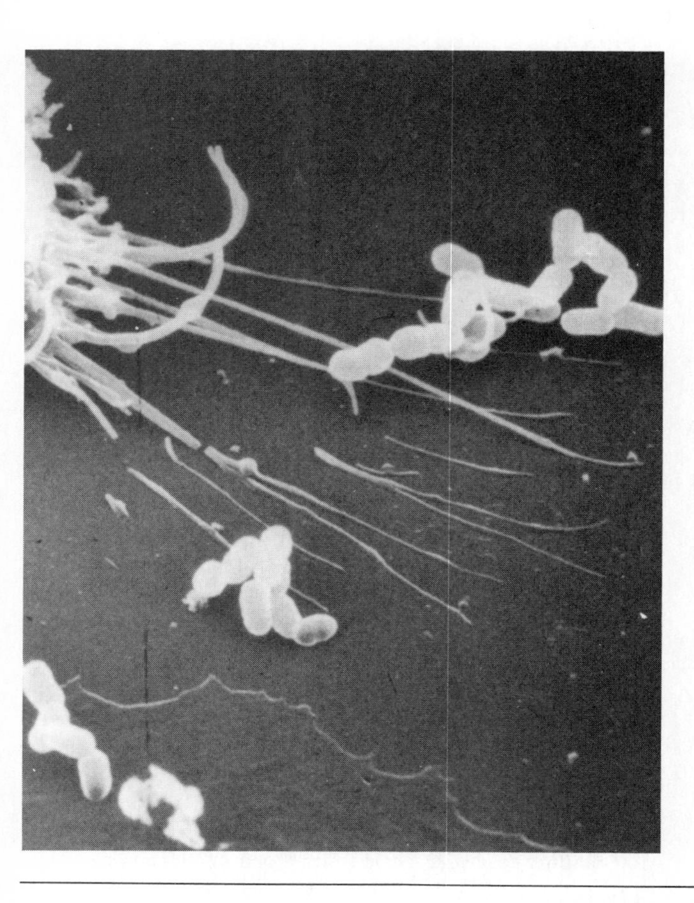

OUTLINE

FOCUS OF THIS CHAPTER

In this chapter we focus on the mammalian immune system and how it protects against infectious agents. The discussion centers on the cells involved in the immune response, their interactions, and their products. The utility of immunological reactions in medical diagnosis and biology is demonstrated, with examples of antibody–antigen reactions.

All vertebrates possess an immune system that affords protection against infections by viruses and microorganisms. These immune reactions are mediated by specialized cells present in the lymphoid tissue and the blood. The cells either are directly involved in immune reactions or produce products that circulate in the blood and fluids that bathe the tissues. The immune system is of paramount importance to survival. A person born with a major immune deficiency or one whose immune system is destroyed by viruses (such as AIDS victims) will soon die from an infectious disease.

OVERVIEW OF THE IMMUNE SYSTEM

The immune system affords protection against specific diseases, has a memory, and distinguishes between foreign substances and self. After recovering from the ill effects of an infectious disease, the patient usually develops lifelong protection against a recurrence of the same illness. This type of resistance is developed against each infectious agent to which s(he) has been exposed and is known as immunity. The immune system has the remarkable ability to recognize, attack, and destroy infectious agents that might otherwise cause disease.

Immunoglobulins, complement, and lymphocytes are the central components of the immune system. **Immunoglobulins** are proteins (antibodies) that specifically react with foreign substances called antigens, which are usually large proteins or polysaccharides. Immunoglobulins are synthesized by a special class of lymphocytes present in the bone marrow and lymph tissue. **Complement** is a series of blood proteins that participate in the destruction of cells and the elimination of foreign substances from the body. Together these components destroy infectious bacteria, virus-infected cells, and foreign vertebrate cells in tissue grafts. The significance of immunology extends far beyond microbiology and is now a full-fledged science unto itself. Our emphasis will be on those aspects of immunology that have a direct impact on microbiology.

TYPES OF IMMUNITY

Immunity is the increased resistance of a person or a population against a disease caused by an infectious agent or against the toxic effects of an antigenic substance. This is not an absolute protection since a large enough dose of a toxin or infectious agent can cause illness. The importance of immunity is that it enables a healthy person to resist diseases resulting from natural contact with toxins or infectious agents.

The human population possesses different kinds of immunity (Table 18.1). **Innate immunity** is the inborn resistance to disease possessed by a given population or race. For example, humans do not become ill with dog distemper because they are innately immune to this disease of domestic dogs. Innate immunity is inborn, has no dpendency on previous exposure to the infectious agent, and largely depends on the nonspecific resistance factors discussed in Chapter 17.

Each person develops a unique spectrum of acquired immunity that changes during one's lifetime. A person acquires immunity from exposure to foreign substances that elicit an immune response or from the injection of preformed immune substances. This type of immunity can be acquired in four different ways: (1) naturally acquired by an active or (2) a passive means or (3) artificially acquired by an active or (4) a passive means (Table 18.1).

Naturally acquired active immunity develops when a patient has a sublethal case of an infectious disease, for example measles. The host makes specific antibodies against the microbe while it is causing the dis-

TABLE 18.1 Acquired and innate immunity

IMMUNITY	MEANS OF ACQUISITION	HOST RESPONSE
Naturally Acquired		
Active	Recovery from sublethal or transient infection	Antibodies formed against agent, long-lasting
Passive	Transfer of antibodies to fetus through the placenta, to infants through mother's colostrum	Antibodies (IgG), short duration
Artificially Acquired		
Active	Immunization	Antibodies against toxoids or attenuated infectious agents, long-lasting
Passive	Injection of antibodies (antibodies formed in another animal)	Host receives antibodies from another animal; temporary, specific resistance
Innate Immunity		
Innate immunity	Inborn	Natural resistance to infectious agents

ease. **Naturally acquired passive** immunity is the resistance to disease passed from mothers to their newborns. The term passive indicates that the newborn did not form the antibodies, instead they were formed by its mother. For several months after birth, newborns are protected by the antibodies they acquired when the mother's serum crossed the placenta into the fetal circulation. Breast-fed infants also acquire antibodies from the mother's colostrum—the thin, yellow, milky fluid secreted by the mammary gland for a few days following birth.

A person can be artificially immunized in either an active or a passive way. The vaccines and toxoids used to protect persons against specific diseases stimulate the formation of protective antibodies. The protective response thus produced is called **artificially acquired active** immunity—the patient produces the antibody. If a patient is diagnosed as having a disease to which s(he) lacks immunity, a physician can passively immunize the patient by administering antibodies that were formed in an animal or another person. This creates **artificially acquired passive** immunity. Passive immunity is always of short duration (a few months), whereas all types of active immunity are longer lasting and some may last a lifetime.

ANTIBODIES AND ANTIGENS

An **antibody** is a protein produced in response to a foreign substance (antigen) that will react specifically with that substance. Unattached antibodies are present in blood and in secretions of the exocrine tissues, while other antibodies attach to the surface of cells. Antibodies were first discovered in **serum,** the fluid part of blood after cells and clotting factors have been removed. Many diseases can be diagnosed by detecting the presence of specific antibodies in serum, a branch of immunology that is called **serology.**

An **antigen** is a substance that stimulates the animal to synthesize antibodies that can specifically react with that antigen. Antigens are usually complex macromolecules that are foreign to the animal—that is, self recognizes self and does not make antibodies against its own complex molecules. Some substances such as large foreign polysaccharides and proteins are excellent antigens, whereas other foreign substances do not stimulate antibody formation. As a rule, substances that are degraded by metabolic activities are good antigens, whereas nondegradable organic polymers are poor antigens.

ANTIBODY STRUCTURE

Antibodies are high molecular weight proteins present in serum and in other body fluids. They were first identified as a distinct group of serum proteins by electrophoresis experiments done in 1939. Like all proteins, antibodies migrate in an electrical field depending on size and the electrical charge on their surface. Antibodies were discovered in the gamma globulin fraction of serum, which is now called the immunoglobulin fraction.

Immunoglobulins (Ig) are a class of globular proteins that function in immune reactions. Within this class there are literally thousands of different molecules, each of which reacts specifically with a different antigen. Immunoglobulins have been studied extensively by biochemists to determine why their reactions with antigen are so specific and how a person is able to make so many of these proteins. An understanding of the complexity of antibodies came from studying the largest class of immunoglobulins, the IgGs.

The Heavy and Light Chains of the Immunoglobulins

Single immunoglobulin molecules have the basic form of a Y, consisting of two identical light polypeptide chains and two identical heavy polypeptide chains (Figure 18.1). The two heavy chains are connected by disulfide bridges (-S—S-) and each of the light chains is attached to a heavy chain by other disulfide bridges. This conformation places the amino-terminal ends of the light and heavy chains at the tip of the "arms" of the Y and the carboxyl-terminal ends of the heavy

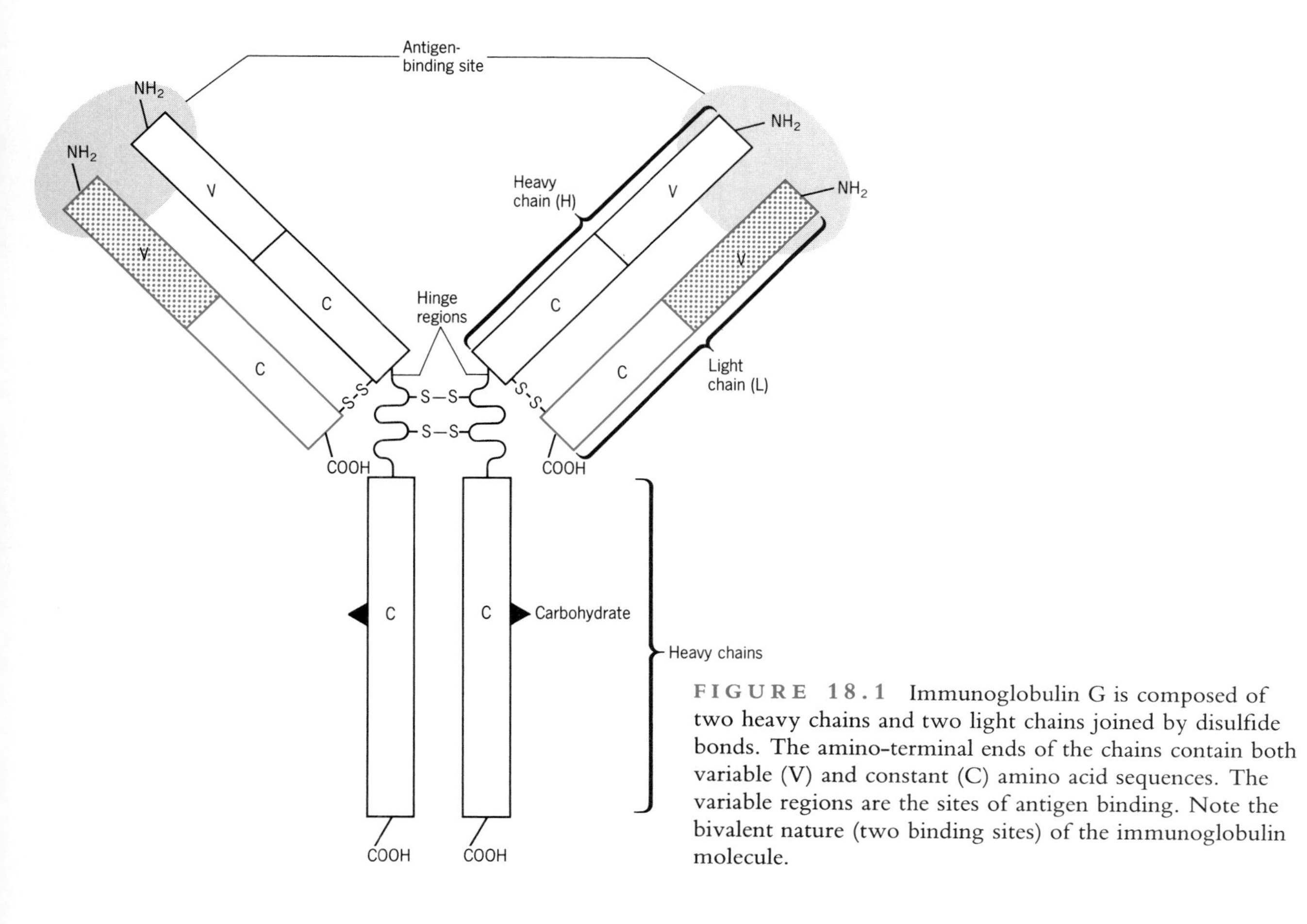

FIGURE 18.1 Immunoglobulin G is composed of two heavy chains and two light chains joined by disulfide bonds. The amino-terminal ends of the chains contain both variable (V) and constant (C) amino acid sequences. The variable regions are the sites of antigen binding. Note the bivalent nature (two binding sites) of the immunoglobulin molecule.

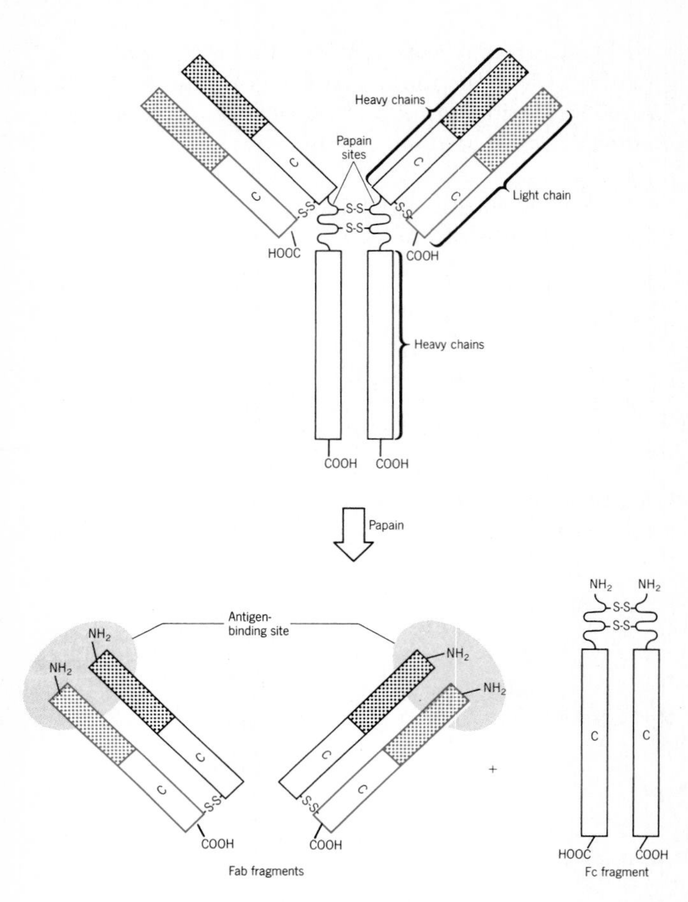

FIGURE 18.2 Immunoglobulin G is cleaved into one Fc fragment and two Fab fragments by the enzyme papain. The Fab fragments contain the antigen-binding sites.

chains in the "tail" of the Y (Figure 18.1). The arms of the immunoglobulin molecule are free to move because of the flexible hinge region in the middle of the heavy chain.

The antigen-binding sites have been identified by taking immunoglobulin molecules apart and analyzing the pieces as to structure and function (Figure 18.2). Treating IgG with the proteoloytic enzyme papain cleaves it at the hinge region into three components, two antibody-binding fragments (Fab) and one Fc fragment (c for crystallizability). The amino-terminal ends on the Fab pieces bind antigen. Since each immunoglobulin molecule contains two Fab pieces, immunoglobulins are bivalent and bind antigen at two separate sites (with the same specificity). This bivalence is important because it permits large antigen–antibody complexes to form when the bivalent antibody reacts with its antigen. The Fc piece of the immunoglobulin is also important in that many immunoglobulins possess Fc sites for binding complement and for binding to cell surfaces.

We are able to respond to the thousands of different antigens to which we are exposed by making an enormous variety of distinctive antibody molecules. Although each antibody molecule has the same basic two light and two heavy chain structure, there are variable regions within each chain that make the antibody unique. Between these variable segments are constant regions (having the same amino acid sequence) in both the heavy and light chains. The variable regions of the light and heavy chains are found toward the amino-terminal ends of the polypeptide chains (Figure 18.1), where they contribute to the specificity of the antigen binding site.

CLASSES OF IMMUNOGLOBULINS

Humans produce five distinct classes of immunoglobulins designated IgA, IgD, IgE, IgG, and IgM. Each class of immunoglobulin is functionally and structurally different, although each possesses two light and two heavy chains. The molecular weights, serum concentration, half-life in serum, and characteristics of the immunoglobulin classes are summarized in Table 18.2

Immunoglobulin G

Immunoglobulins of the G class (IgG) make up the greater percentage of antibody molecules in blood. This class is composed of four subclasses (IgG_1, IgG_2, IgG_3, IgG_4) each having a distinct heavy chain, $\gamma 1$, $\gamma 2$, $\gamma 3$, and $\gamma 4$. These immunoglobulins combine with small antigens (precipitins), combine with and neutralize toxins (antitoxins), participate in complement fixation (except IgG_4), are the class of antibodies formed during the secondary response (see below), and are able to cross the placenta (except IgG_2). IgGs have half-lives of 23 days, except for IgG_3, whose half-life is only 8 to 9 days.

Immunoglobulin M

IgM (M = macroglobulin) is synthesized in almost all cases during the primary resonse to an antigen. This

Diagnostic Bacteriology

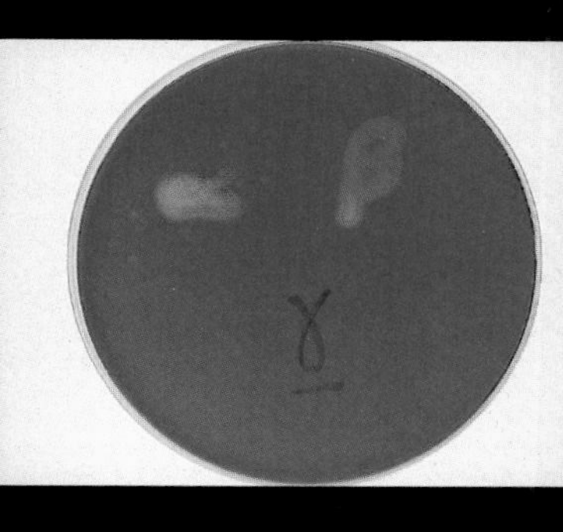

IIA.1 Alpha, beta, and gamma hemolysis caused by different streptococci grown on sheep blood agar.*

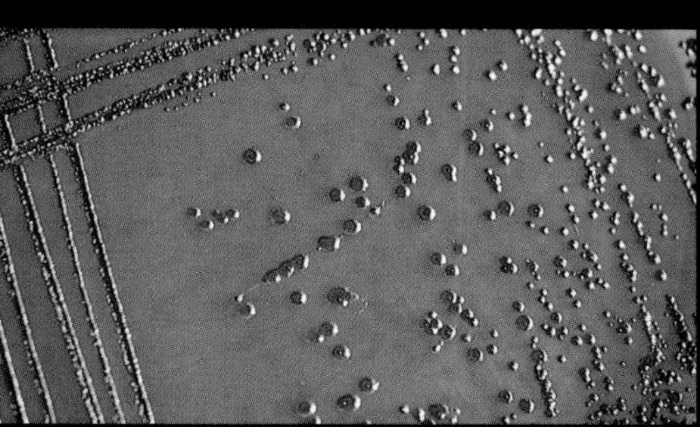

IIA.2 *Escherichia coli* produces colonies with a green sheen when it grows on EMB agar. (Nancy Rosenthal.)

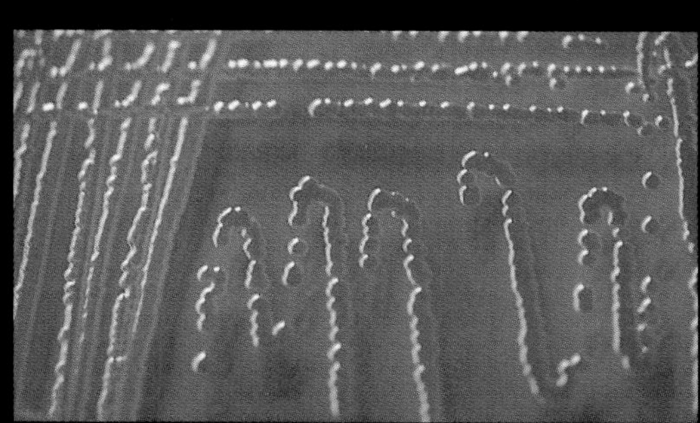

IIA.3 *Enterobacter aerogenes* produces pink mucoid colonies when it grows on EMB agar. (Nancy Rosenthal.)

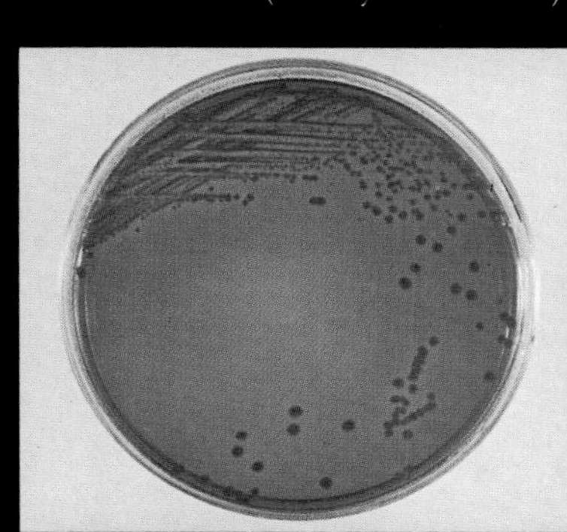

IIA.4 Lactose fermenters grow as red colonies on McConkey agar.

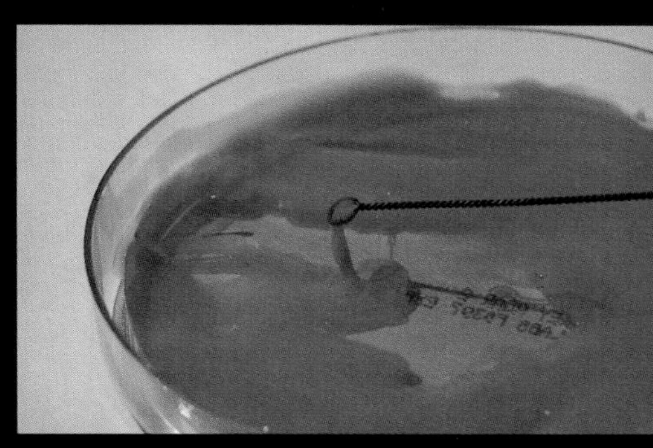

IIA.5 Mucoid growth of *Klebsiella pneumoniae*. (Virginia Ellis.)

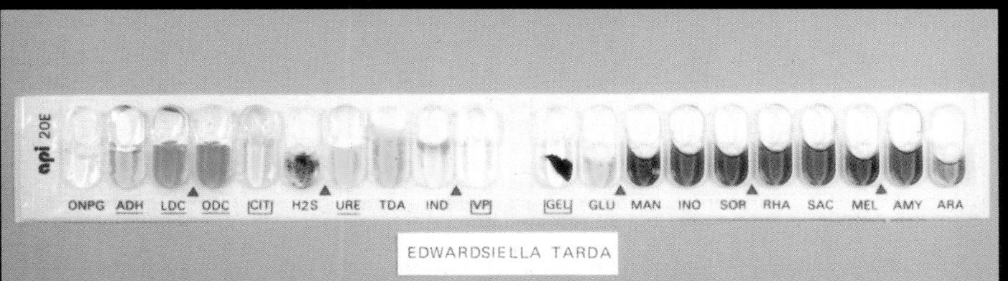

IIA.6 API strip used to identify gram-negative bacteria.

Mycroscopy of Pathogens

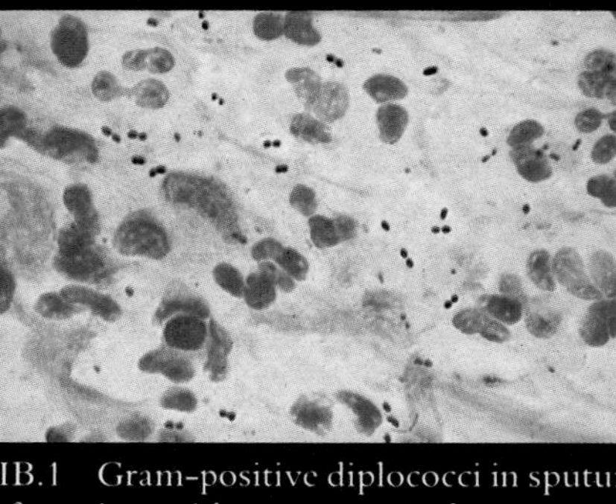

IB.1 Gram-positive diplococci in sputum of a patient with pneumococcal pneumonia.*

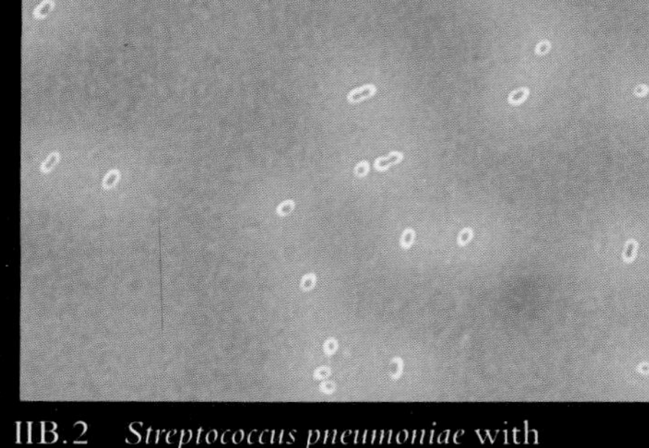

IIB.2 *Streptococcus pneumoniae* with capsules. (VU/E. Chan.)

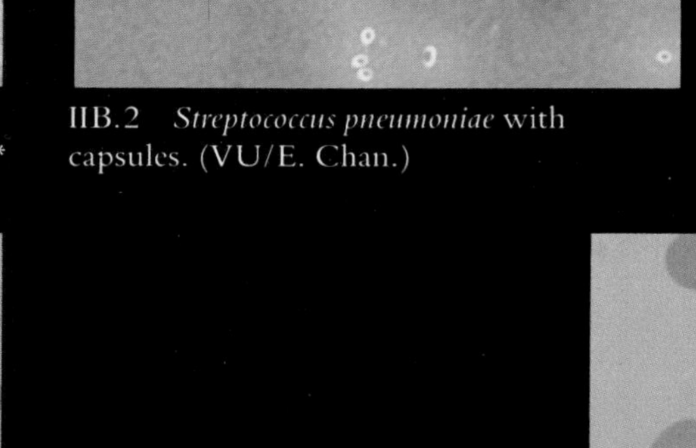

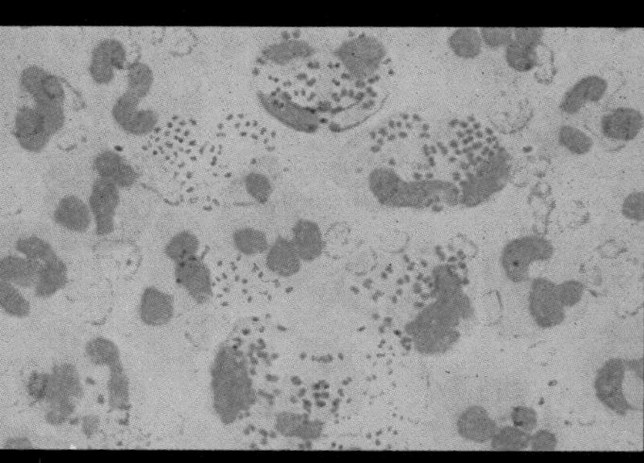

IB.3 Gram-negative diplococci inside eukocytes from the urethral exudate of a nan with gonorrhea.*

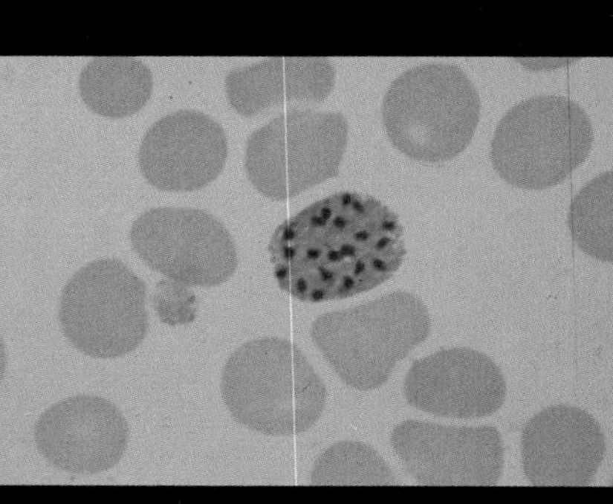

IIB.4 Schizont of *Plasmodium vivax*.*

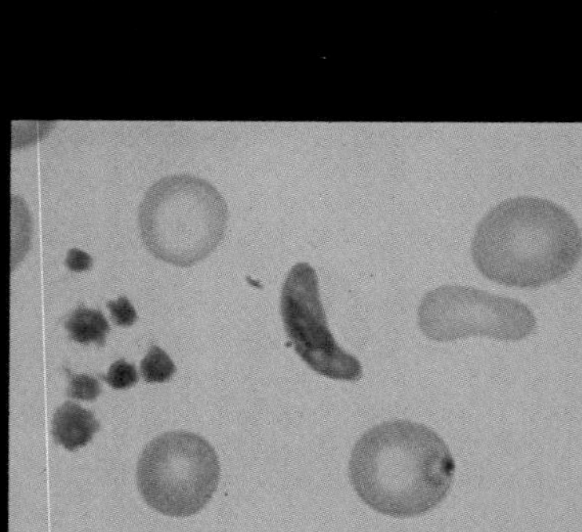

IIB.5 Gamont of *Plasmodium falciparum*.*

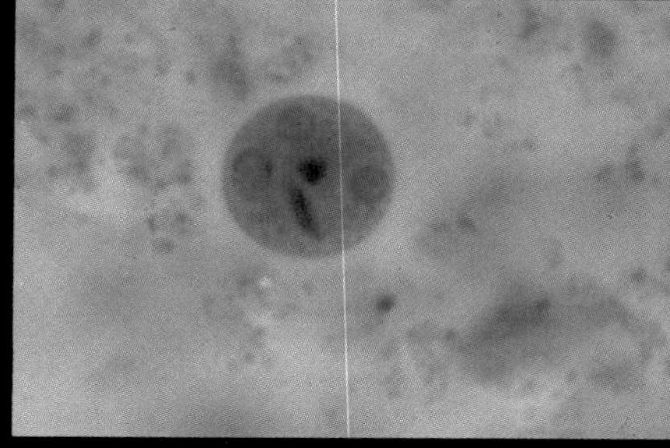

IIB.6 Mature cyst of *Entamoeba histolytica*. Three of the four nuclei are seen in the plane of focus. (X200)*

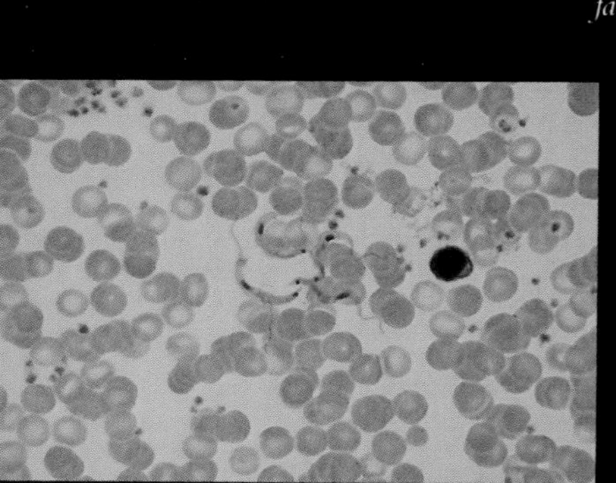

*From G. L. Mandell, J. E. Bennett, and R. G. Douglas, Jr., *Principles and Practice of Infectious Diseases*, 2nd ed., John Wiley and Sons, 1985.

Clinical Signs of Infection

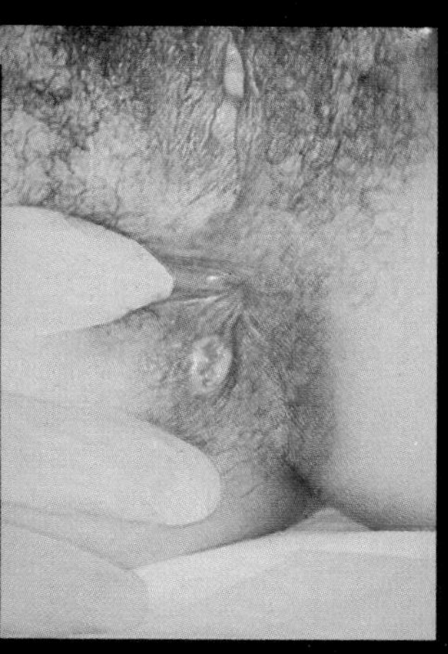

IC. 1 Primary syphilitic hancre of perineum.*

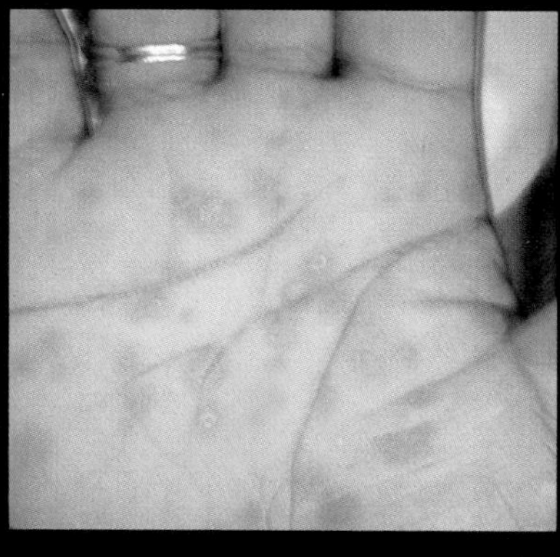

IIC. 2 Mucous patch lesions of secondary syphilis.*

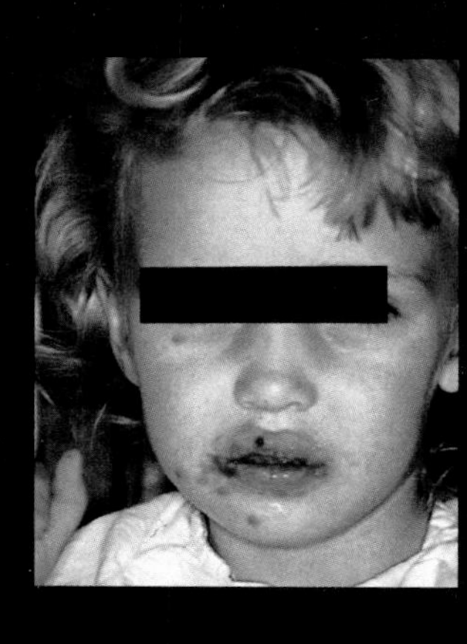

IIC. 3 Primary herpes gingivostomatitis in a child, involving the cheek, chin, and periocular skin.*

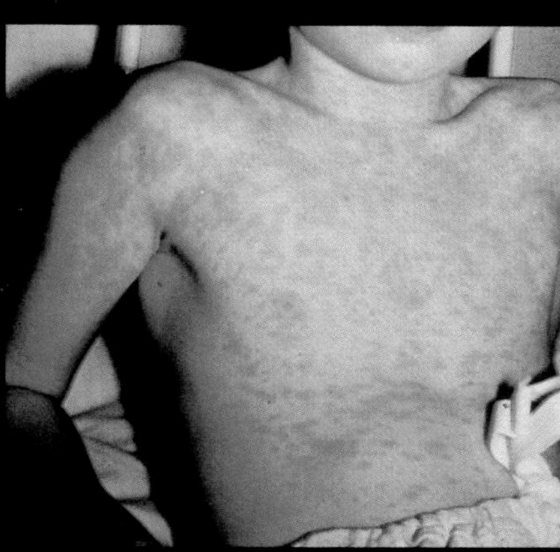

IIC. 4 Typical measles rash on trunk.*

IIC. 5 Herpes zoster rash in a 67-year-old woman.*

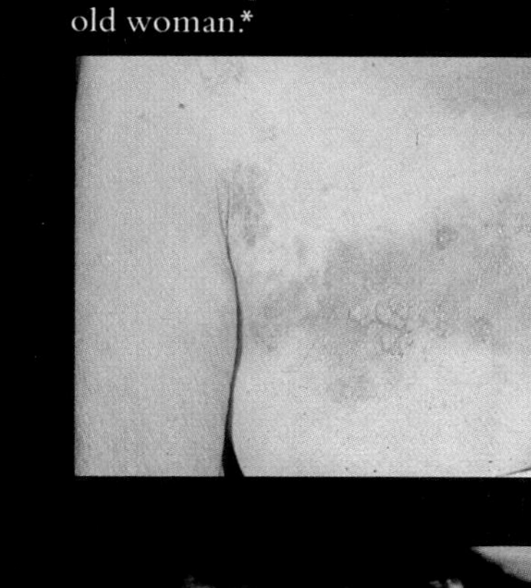

IIC. 6 Severe case of smallpox.*

IC. 7 Desquamation of soles of feet during recovery from toxic shock syndrome.*

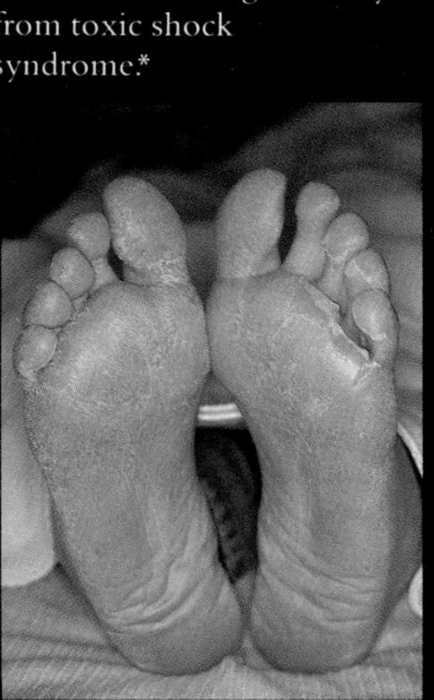

IIC. 8 Lymphocutaneous sporotrichosis.*

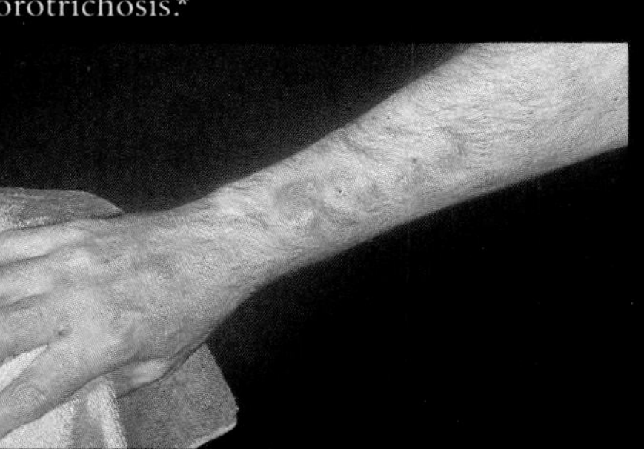

Morphology of Blood Elements

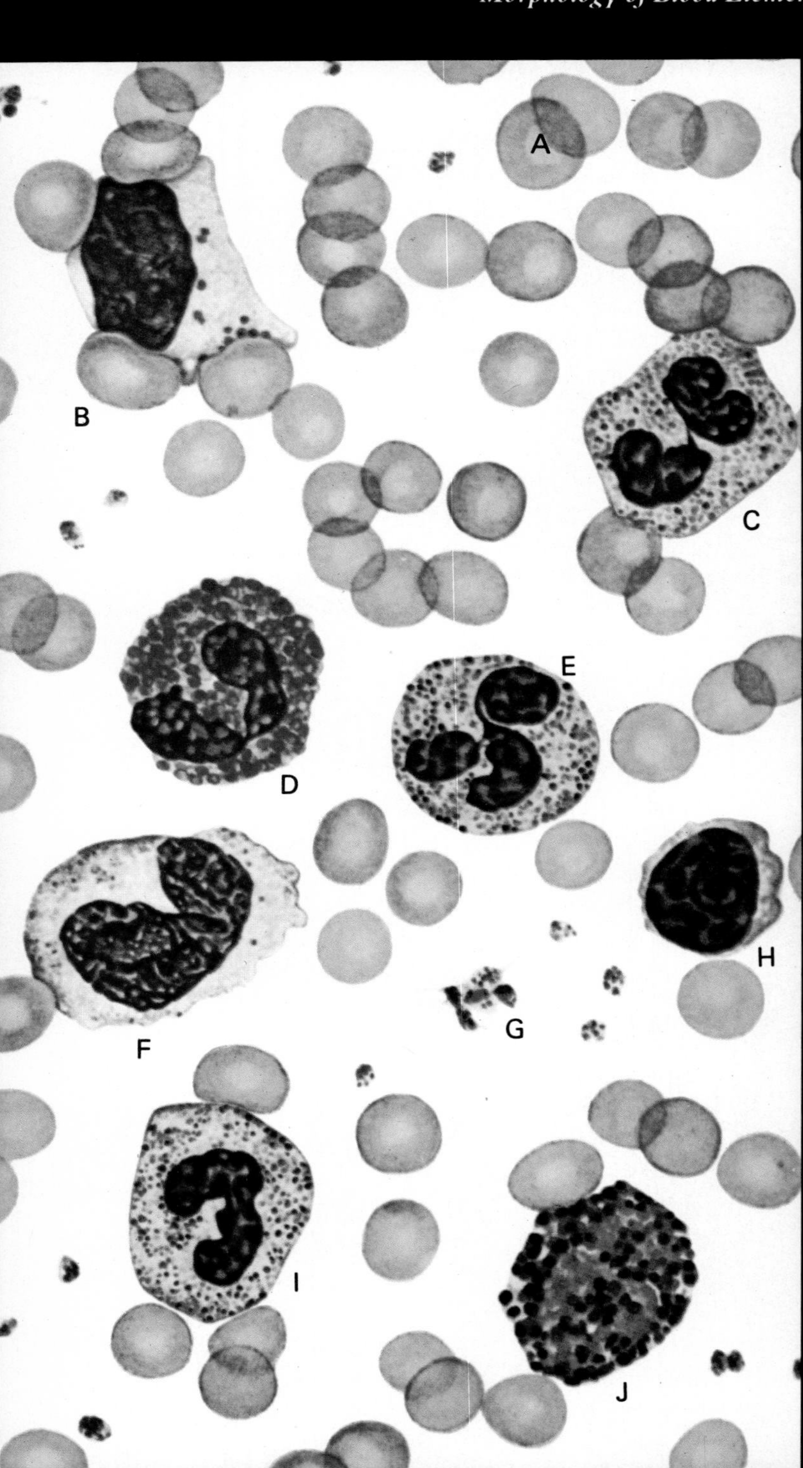

IID.1 Morphology of blood elements. (Courtesy of L. W. Diggs, M.D. From L. W. Diggs, L. W. Strum, and A. Bell, *The Morphology of Human Blood Elements*, 3rd ed., Abbott Laboratories, North Chicago, Ill., 1954.)

Legend Key. Cell Types Found in Smears of Peripheral Blood from Normal Humans. The arrangement is arbitrary, and the number of leukocytes in relation to erythrocytes and thrombocytes is greater than would occur in an actual microscopic field.

- A Erythrocytes
- B Large lymphocyte with azurophilic granules
- C Neutrophil
- D Eosinophil
- E Neutrophil segmented
- F Monocyte with blue-gray cytoplasm, coarse linear chromatin, and blunt pseudopods
- G Thrombocytes
- H Lymphocytes
- I Neutrophilic band
- J Basophil

TABLE 18.2 Properties of antibody molecules

IMMUNOGLOBULIN CLASS	SERUM CONCENTRATION (mg/100ml)	MOLECULAR WEIGHT	HALF-LIFE (days)	CHARACTERISTIC PROPERTIES
IgG (total)	range (800–1600)			Precipitins, antitoxins, complement fixation (except IgG_4), placental transfer (except IgG_2), secondary response
IgG_1	900	146,000	23	
IgG_2	300	146,000	23	
IgG_3	100	165,000	8–9	
IgG_4	50	146,000	23	
IgM	50–200	900,000	5	Agglutinins, opsonins, primary response
IgA	150–400	160,000	6–8	Protection of mucous membrane surfaces
Secretory IgA		385,000		
IgD	0.3–4.0	184,000	2.8	On surface of B lymphocyte
IgE	0.0001–0.0007	200,000	2.5	Antibody to allergens

immunoglobulin is a pentamer composed of five monomeric untis held together by disulfide bonds between the heavy chains of adjacent molecules (Figure 18.3). The J chain joins two of the monomeric units and either initiates the assembly of the pentamer or completes the circular structure (it is not known which). IgM is the first immunoglobulin synthesized following initial exposure to an antigen. It binds to the surface of the B lymphocyte and acts as an antigen receptor. IgMs can bind complement and participate in agglutination (clumping of cells) and opsonization (rendering bacteria susceptible to phagocytosis) reactions. Because of its large size, IgM is confined to the blood vascular system; it has essentially no role in tissue spaces and does not cross the placenta.

Immunoglobulin A

Immunoglobulins of the A class are known as the secretory antibodies because they are found in breast milk, respiratory and intestinal mucin, saliva, tears, vaginal secretions, and prostatic fluid. IgAs protect these parts of the body from infectious agents.

Monomeric IgAs have a molecular weight of 160,000; however, functional IgAs usually occur as dimers joined by one J chain and enveloped (Figure 18.4) by a secretory polypeptide, having a molecular weight of 71,000. The IgA dimer combines with the secretory polypeptide produced by the epithelial cells lining the intestine, bronchi, and other exocrine tissue. After leaving the vascular system, the dimer binds with the secretory polypeptide on the surface of the epithelial cells. This large complex is transported through the

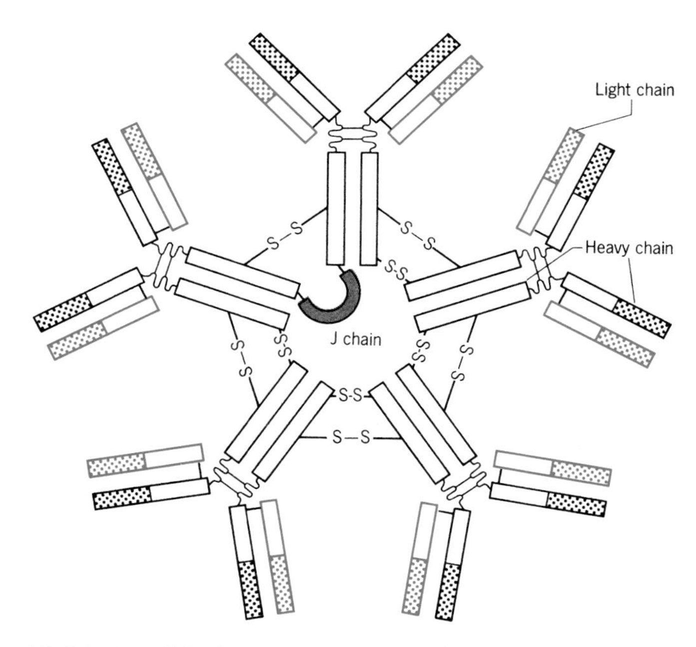

FIGURE 18.3 Immunoglobulin M is a pentamer of five monomeric subunits that individually are the size of an IgG molecule. Each of the monomers is joined by disulfide bonds -S—S-. The J chain joins two of the monomers and either initiates or completes pentamer formation. Immunoglobulin M is formed early during the immune response.

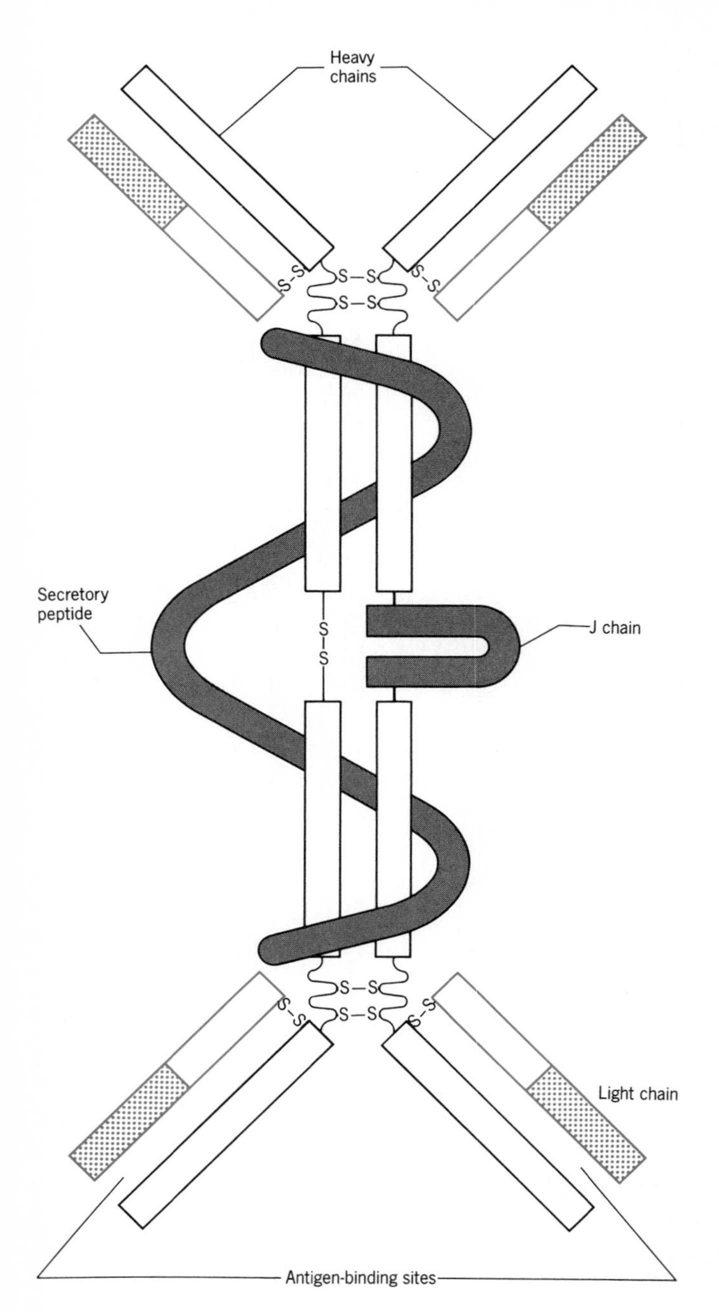

FIGURE 18.4 A diagramatic representation of the secretory form of IgA. Two IgAs are joined by a disulfide bond and a J chain. IgAs are secreted by mucous membranes after the secretory peptide is wrapped around the heavy chains.

epithelial cell before it is excreted into the secretory lumen of the tissue.

Humans normally produce equal amounts of IgA and IgG in a given time period; however this is not apparent in the serum data given in Table 18.2 because significant quantities of IgA are lost in secretions. The serum level of IgA is actually about one-fifth the level of serum IgG (Table 18.2).

Immunoglobulin D

Immunoglobulins of the D class are very difficult to study because of their low serum concentration and their sensitivity to proteolytic enzymes. Immunoglobulin D has no disulfide bonds between its heavy chains, which makes it vulnerable to denaturation by mild heating and probably makes it susceptible to proteolysis. IgD is found on the surface of B lymphocytes where it functions as a specific antigen receptor in conjunction with surface-bound IgM.

Immunoglobulin E

Immunologists didn't detect IgE until the 1970s because its concentration in the serum of normal humans is very low (Table 18.2). Higher levels of IgE are found in persons with type I hypersensitivities (allergies and anaphylaxis) and those who have been infected by parasites. The activities of IgE are involved in both the resistance to parasitic infections and type I hypersensitivities. The heavy chains of IgE possess a unique structure through which it binds to receptors on the surface of mast cells. Allergens specifically bind to these cell-bound IgEs. This binding triggers degranulation, releasing cellular substances active in the allergic response.

KEY POINT

Each class of immunoglobulins has constant and variable regions within the light and heavy chains. The variable regions confer the uniqueness on individual antibody molecules that enables the antibody to react specifically with its antigen.

CHARACTERISTICS OF ANTIGENS

As previously described antigens are foreign substances that stimulate the formation of antibodies. Antigens are usually complex macromolecules such as proteins or polysaccharides that can be degraded by animal cells, whereas inert molecules such as plastics are poor anti-

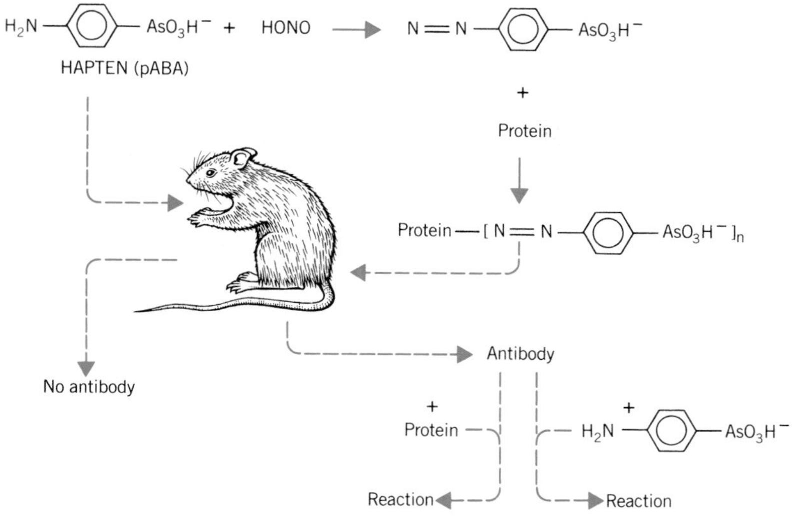

$H_2N-C_6H_4-AsO_3H^-$ + HONO → $N{=}N-C_6H_4-AsO_3H^-$

HAPTEN (pABA)

\+ Protein

Protein—[N=N—C_6H_4—AsO_3H^-]$_n$

No antibody

Antibody

\+ Protein

\+ $H_2N-C_6H_4-AsO_3H^-$

Reaction

Reaction

FIGURE 18.5 Hapten molecules are too small to elicit an immunological response. Para-aminobenzene arsenate (*p*ABA) is a hapten that is easily attached to a large protein. In this state, *p*ABA can stimulate the production of specific antibodies against both the protein and pABA, when injected into an animal.

gens. Lipids and nucleic acids can react with antibodies, but are poor antigens unless they are coupled first with proteins.

Most antigens are large molecules that contain many **antigenic determinants,** the chemical structures that actually bind to the antibody. The size of an antigenic determinant is small in comparison to the antigen; antigenic determinants can be as small as six hexose residues or five amino acids.

A microbe contains numerous antigenic determinants, each being able to elicit the formation of a specific antibody. Some of the antibodies formed against a bacterium participate in killing it, either through the action of complement (see below) or by making it susceptible to phagocytosis. Since antibodies react specifically with their antigens, slight differences in closely related strains of microorganisms can be detected using antibody–antigen reactions.

Antigenic Cross Reactivity

Different species of microorganisms may contain similar or identical antigenic determinants. Antibodies formed against these antigenic determinants will **cross react** with the same antigenic determinants on other cells. A good example of cross reactivity involves two bacteria: *Proteus vulgaris* and the obligate intracellular parasite, *Rickettsia rickettsii,* which causes Rocky Mountain spotted fever. Persons with Rocky Mountain spotted fever make antibodies that cross react with *Proteus vulgaris* OX-2—a fact that was useful in diagnosing rickettsial infections by the Weil–Felix assay before better immunological tests were available.

Haptens

An antigenic determinant is the smallest chemical unit that will react specifically with an antibody. **Haptens** are low-molecular weight foreign compounds that are too small to elicit antibody formation, yet they function as antigenic determinants. Certain common drugs become haptens when by-products of their metabolic breakdown bind to large protein molecules.

The experiment depicted in Figure 18.5 demonstrates how haptens can be tricked into eliciting an immune response. The hapten *p*-aminobenzene arsenate (*p*ABA) is a small molecule that can be chemically conjugated with a protein, such as serum albumin. By itself, *p*ABA is unable to elicit an antibody response; however, the protein–hapten complex elicits the formation of antibodies when injected into an animal. The antibodies formed in response to the protein–hapten complex can be made to react with the protein–hapten complex, the protein alone, and with *p*-aminobenzene arsenate. These experiments demonstrate that small molecules, such as *p*ABA, can serve as antigenic determinants even though they do not elicit antibody formation (are not antigens) by themselves.

KEY POINT

Antibodies react specifically with antigens by recognizing a determinant on the antigen. Large antigens have many determinants, each being defined by its chemical structure.

ANAMNESTIC RESPONSE

Up to this point, we have viewed antibodies as biochemical entities and have described them in chemical terms. Let us return to humans in order to understand the immunological mechanism by which antibodies are formed. As described, antibodies are formed following exposure to antigens, which by definition are foreign substances. Once exposed, the person remembers the antigens and responds differently upon a second or subsequent exposure. This phenomenon is called the anamnestic (Gr. *anamnesko,* memory) or memory response.

A person is said to be immunologically competent when he synthesizes antibodies in response to an antigen challenge (Figure 18.6). On first exposure, there is a delay or latent period (usually three to four days), during which no circulating antibody can be detected. Following the latent period, the concentration of detectable antibody in the serum increases, reaches a peak level, and then declines. IgM antibodies are the first to be produced, followed by antibodies of the IgG class. These two classes of immunoglobulins constitute the **primary response** (Figure 18.6). As the antibody molecules combine with and remove the antigenic stimulation, there is a gradual decrease in the level of circulating antibody. This decrease is caused in part by the short half-life of circulating antibodies (see Table 18.2) and in part by the absence of continued antigenic stimulation of the antibody-producing cells.

Upon a second or subsequent exposure to the same antigen, the level of circulating antibody in the serum dramatically increases, with a much shorter latent period. The person behaves as if s(he) remembers the previous experience with the antigen. The rapid increase in the level of circulating antibody, predominated by IgGs, is known as the **secondary** or **anamnestic response.** The ability to produce high levels

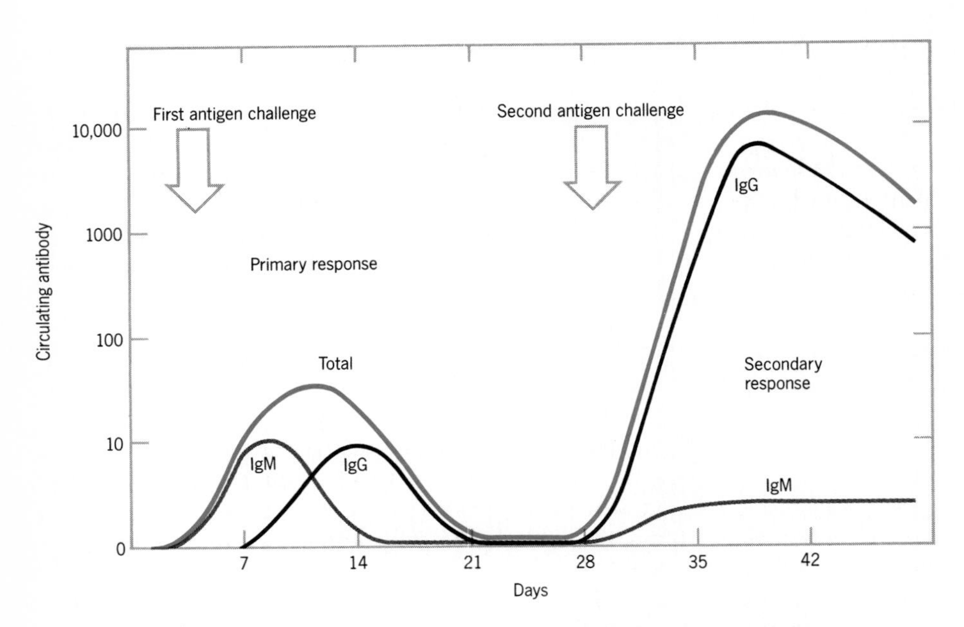

FIGURE 18.6 The primary response to a foreign agent is the formation of IgM. The early IgM antibody is followed by the synthesis of IgG. Upon a second or subsequent exposure to the antigen, the level of circulating IgG increases dramatically. This secondary response is known as the anamnestic response.

TABLE 18.3 Active immunization schedule for normal infants and children

IMMUNIZATION	VACCINE	SCHEDULE OF IMMUNIZATIONS					ROUTE
		1st	2nd	3rd	4th	5th	
Diphtheria	DTP	2 mo	4 mo	6 mo	1.5 yr	4–6 yr	Intramuscular
Pertussis	DTP	2 mo	4 mo	6 mo	1.5 yr	4–6 yr	Intramuscular
Tetanus	DTP	2 mo	4 mo	6 mo	1.5 yr	4–6 yr	Intramuscular
Poliovirus	TOPV	2 mo	4 mo	1.5 yr	4–6 yr		Oral
Haemophilus b	Hib or Hflu	2 years					Intramuscular
Measles	MMR	15 mo					Intramuscular
Mumps	MMR	15 mo					Intramuscular
Rubella	MMR	15 mo					Intramuscular

DTP = diphtheria and tetanus toxoids combined with pertussis vaccine.
TPOV = trivalent oral poliovirus vaccine.
MMR = measles, mumps, and rubella vaccine.

of circulating antibodies against microorganisms and viruses is an important mechanism by which infectious agents are resisted.

The level of circulating antibody continues to rise during the secondary response as long as the antigen is present in the system. When the antigen is removed, the level of circulating antibody again decreases because of the short half-life of serum immunoglobulins and the absence of continued antigenic stimulation. The secondary response can be repeated many times to increase the level of circulating antibody. This phenomenon is the scientific basis for **booster "shots,"** which are given to assure continued immunity against diseases such as tetanus, diphtheria, and whooping cough.

IMMUNIZATIONS PROTECT AGAINST INFECTIOUS AGENTS

The absolute amount of antibody that is formed will depend on (1) the nature of the antigen, (2) the route of exposure, and (3) the dose and sequence of exposure. The best procedures for routine immunizations have been established through clinical trials. For maximal protection, children should be immunized against bacterial and viral diseases on the recommended schedule (Table 18.3).

These immunizations are administered orally or by an intramuscular injection of the antigen. Polio vaccine is taken orally (TOPV = trivalent oral poliovirus vaccine), whereas the DTP (diphtheria, tetanus, pertussis) "shots" are given intramuscularly. The nature of the antigens also varies. Immunization against polio is accomplished by having the person ingest attenuated strains of the poliovirus. A formaldehyde-inactivated preparation of diphtheria toxin is given to immunize against diphtheria, whereas killed cultures of *Bordetella pertussis* are used to immunize against pertussis (whooping cough). Clinical trials have established which antigens are best, which route is most effective, and how to sequence the administration of the "shots." The newest immunization is the *Haemophilus* b vaccine (sometimes called Hib or Hflu), which is recommended for children 18 to 24 months old to prevent *Haemophilus influenzae* infections.

THE IMMUNE SYSTEM

An explanation of the immune response must take into account the foreign nature of antigens, the specificity of antibodies, and the anamnestic response. In part, the immunolgoical response can be explained through the role of the white cells known as **lymphocytes.** These specialized cells are found in blood and lymph and in the bone marrow, thymus, spleen, appendix, and lymph nodes (Figure 18.7). Lymphocytes differentiate from the embryonic stem cells, which in adults are found primarily in the bone marrow and in lymph.

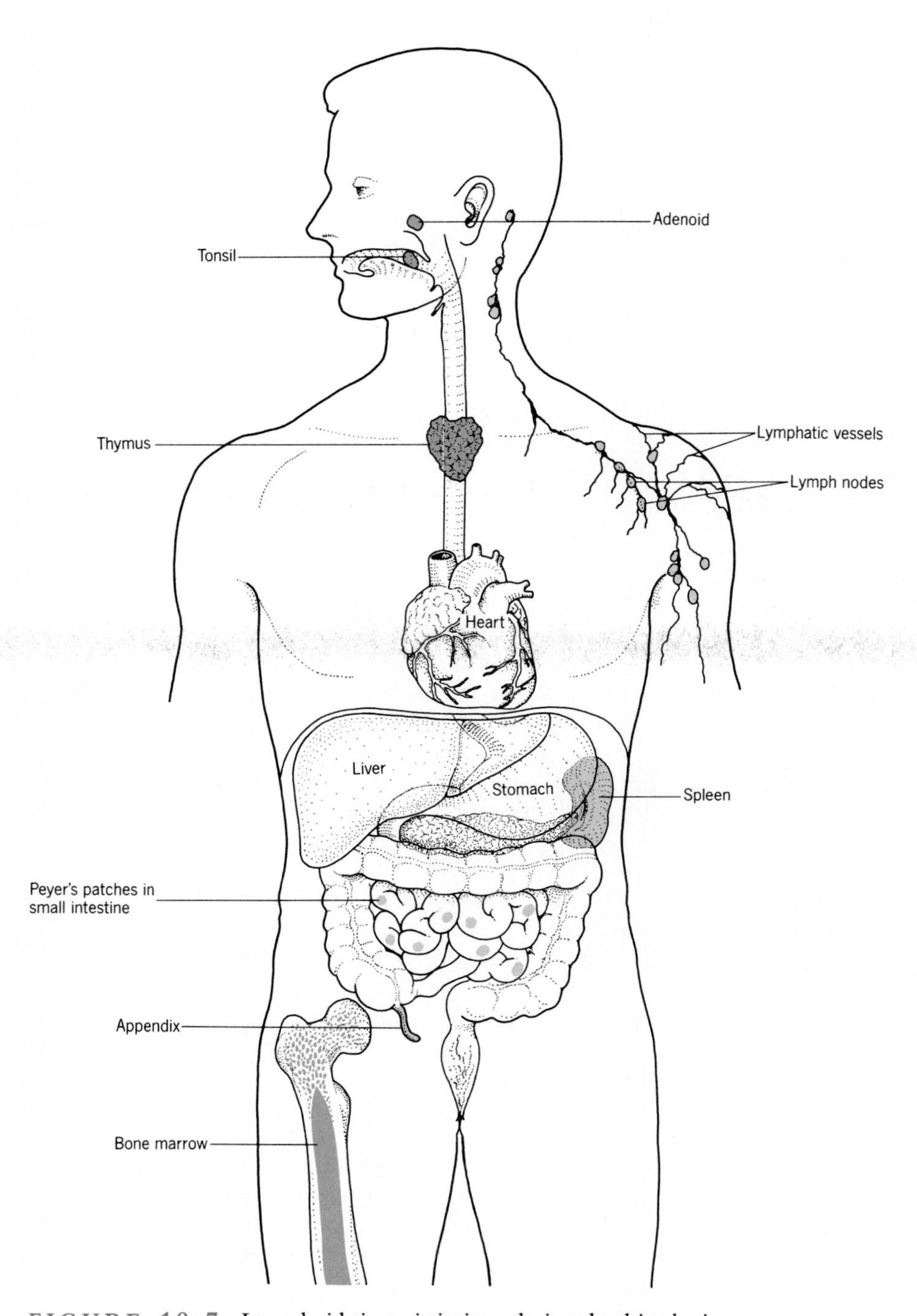

FIGURE 18.7 Lymphoid tissue is intimately involved in the immune response. Lymphocytes develop in the central lymphoid tissue (dark color), which includes the thymus and bone marrow. The peripheral lymphoid tissue (light color) includes the spleen, lymph nodes, Peyer's patches in the small intestine, and the appendix.

FORMATION AND CHARACTERISTICS OF T AND B LYMPHOCYTES

Lymphocytes are spherical cells ranging in size from 7 to 15 μm in diameter that belong to the mononuclear series of white cells (see Chapter 17). The lymphocytes involved in the immune reaction are subdivided into two types of cells (Figure 18.8): the T lymphocytes and the B lymphocytes (Table 18.4). The B lymphocytes are named for the bursa of Fabricius (fa-bris′e-us), which is lymphoid tissue found in chickens, but not in mammals. (The equivalent tissue in mammals is probably the intestinal lymph tissue, spleen, and bone marrow.) The B lymphocytes contain IgD and IgM antibodies on their surface that mark them for their role as the antibody-producing cells.

The lymphocytes that develop in the thymus gland are designated thymus-derived (T) lymphocytes. The T lymphocytes comprise the majority of the leukocytes in the peripheral blood (80 percent) and are identified by the *Thy-1* antigen on their surface. These cells have an average life span of six months, but may live as long as ten years (Table 18.4). The functions of the T lymphocytes are (1) killing of specific foreign or virus-infected host cells, (2) helping B lymphocytes respond to antigen, and (3) participating in regulation of the immune response. Each function is performed by a

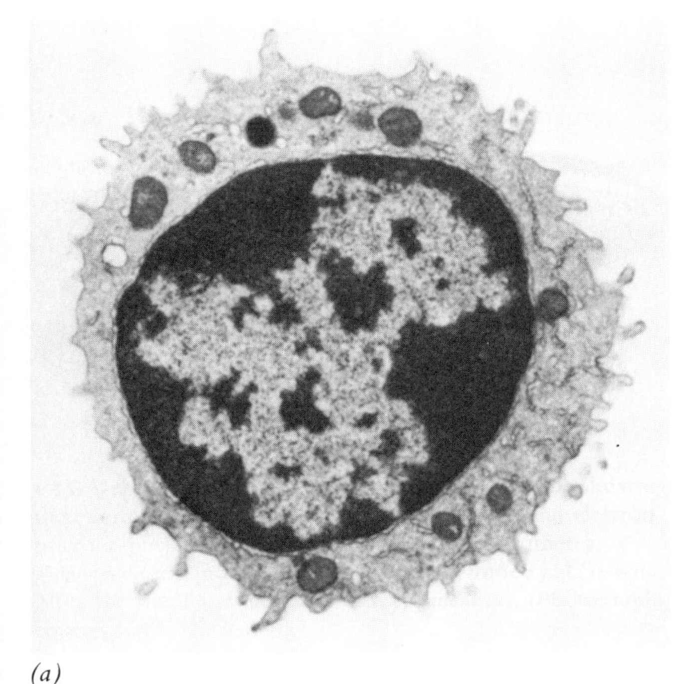

(a)

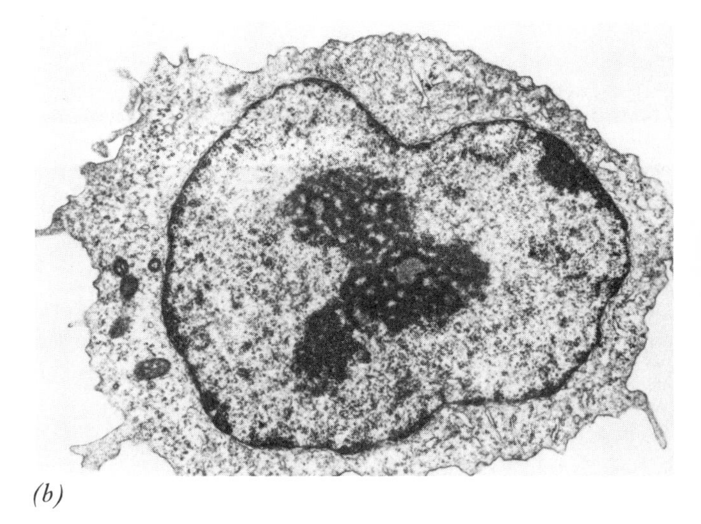

(b)

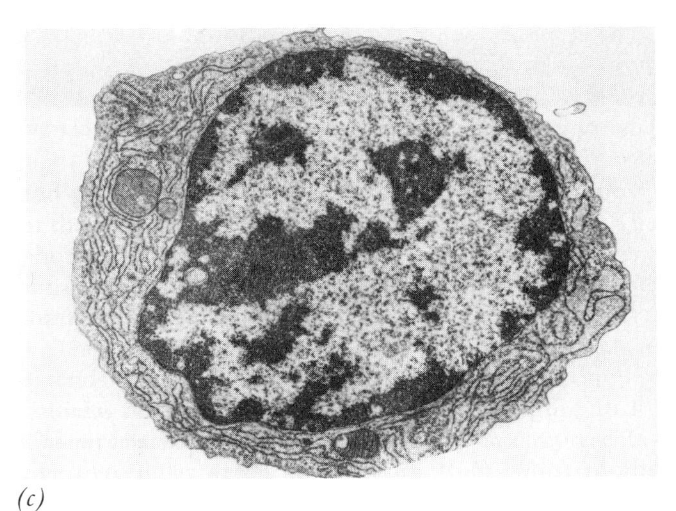

(c)

FIGURE 18.8 Electron micrographs of (*a*) an unstimulated lymphocyte, (*b*) an activated T lymphocyte, and (*c*) an activated B lymphocyte. A small lymphocyte can give rise to either a B or a T lymphocyte, depending on its developmental history. This T lymphocyte has greatly increased in size following activation. The B lymphocyte is maturing into a plasma cell whose cytoplasm will increase in volume and will be filled with endoplasmic reticulum, upon which antibodies will be synthesized. (Photographs courtesy of Dr. Don Fawcett. From W. Bloom and D. W. Fawcett, *A Textbook of Histology,* 11th ed., Philadelphia, W. B. Saunders, 1986.)

TABLE 18.4 Characteristics of human B and T lymphocytes

CHARACTERISTICS	B LYMPHOCYTE	T LYMPHOCYTE
Derivation	Stem cells that mature in intestinal lymph tissue, spleen, bone marrow (named for bursa of Fabricius found in chickens)	Stem cells that mature in thymus (T) gland
Body location	Bone marrow and lymph tissue	Peripheral blood and lymph tissue
Identification	IgM and IgD on surface, converted to plasma cells containing endoplasmic reticulum	*Thy-1* antigen on surface
Function	Antigen converts B cells to antibody-producing plasma cells, other B cells become memory cells	Cell-mediated immunity, help in B cell response, regulation of immune response
Life span	6 to 8 weeks, memory B cells much longer	6 months, memory T cells live up to 10 years

specific subpopulation of T lymphocytes: cytotoxic T cells, helper T cells, and suppressor T cells. T lymphocytes cooperate in the production of immunoglobulins, but they do not produce them.

KEY POINT

Antibody synthesis is accomplished by B lymphocytes in cooperation with helper T lymphocytes. Another function of T lymphocytes is to destroy infected host cells such as bacteria- and virus-infected cells and to reject foreign tissue, such as tissue grafts.

THEORY OF ANTIBODY FORMATION

The complexities of the immune system are difficult to understand in the absence of a working hypothesis or theory. The **clonal selection theory** of antibody formation, proposed in 1957 by Sir MacFarlane Burnet, provides a framework for understanding the major phenomena of the immune response. The main points of Burnet's theory are these.

1. Every person possesses millions of lymphoid cells, each of which carries a different antibody on its surface capable of reacting with a single antigenic determinant.
2. Combination of the antigen with one particular cell-bound antibody stimulates these cells to proliferate (to clone themselves) and to differentiate into antibody-producing cells and memory cells.
3. The antibody that is thus produced reacts specifically with the antigen that elicited the response.

The dividing lymphoid cells constitute a **clone**—all the progeny of a single cell—that produces a specific antibody as long as the antigen is present in the system. After the antigen has been destroyed by the defense mechanisms, descendants of the clone remain in the body as memory cells. Upon subsequent exposure to the antigen, the memory cells are stimulated to divide and to synthesize antibody molecules. The clonal selection theory explains the anamnestic immune response.

B LYMPHOCYTES ARE THE ANTIBODY-PRODUCING CELLS

Our understanding of the cellular basis of the immune system began with the identification of the B and T lymphocyte populations. Studies of B lymphocytes showed that they carry specific immunoglobulin molecules (IgD and IgM) bound to their cell membrane. The 20,000 to 200,000 antibody molecules on the surface of each B lymphocyte function as antennas for antigens (Figure 18.9). Each of the antibody molecules on a given B lymphocyte reacts with the same antigenic determinant. When an antigen enters the system, it seeks out that B lymphocyte carrying the surface antibodies with which it can react. There are millions of different B lymphocytes in the system each having different antenna antibodies.

The B lymphocyte is stimulated to differentiate and divide after it binds with its antigen, a process that

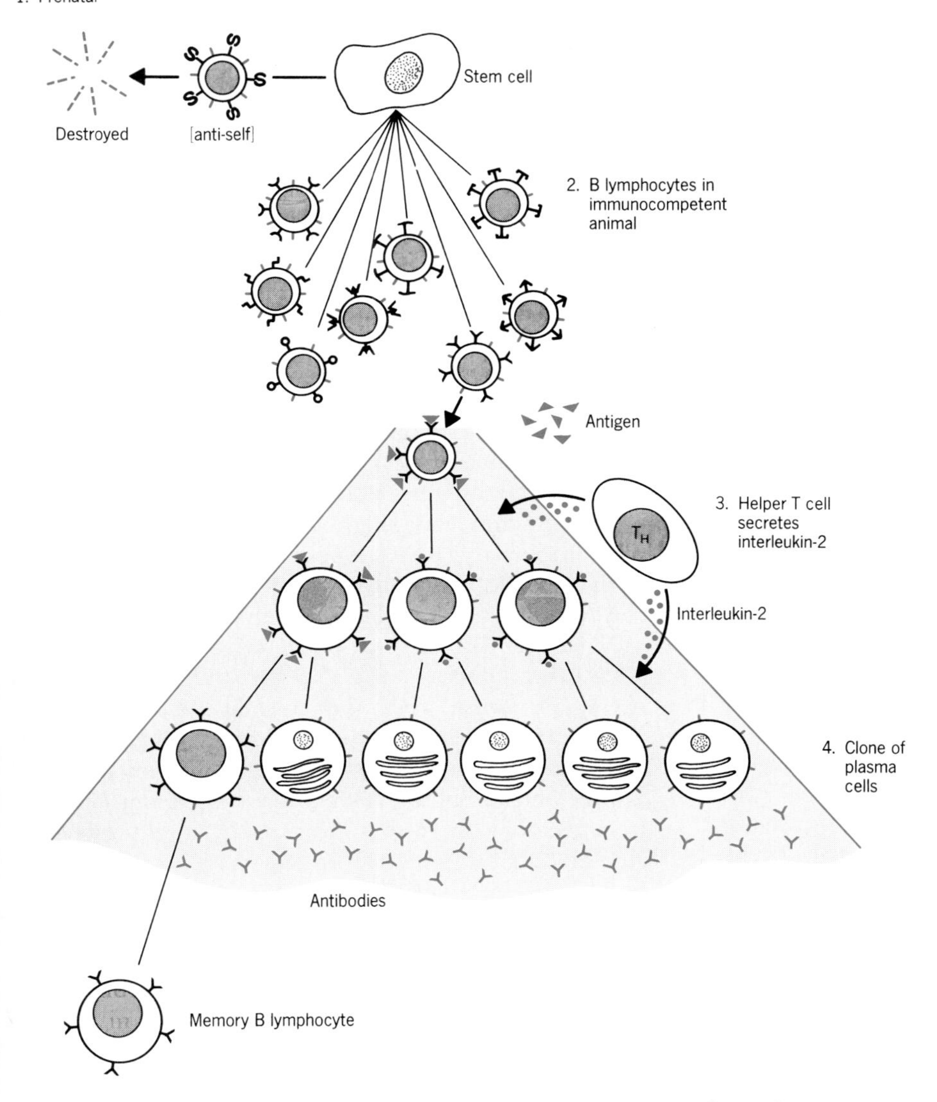

FIGURE 18.9 The B lymphocytes, each containing a specific surface antibody, are differentiated from stem cells. Cells producing antibody against self (1) are destroyed. Antigens select a specific B lymphocyte (2) by binding to its surface antibody. Mature helper T cells interact with the activated B lymphocyte and produce interleukin-2 (3), which stimulates the B lymphocytes to divide and differentiate into a clone of antibody-producing plasma cells (4).

depends on cooperation of the helper T cells as described below. The B lymphocyte differentiates into antibody-producing plasma cells (Figure 18.9) and long-lived memory cells, which remain in the system ready to respond upon a subsequent exposure to the same antigen.

Each plasma cell produces a single kind of antibody and continues to produce antibody as long as the antigen is present. Plasma cells are protein-synthesizing "factories" that possess an extensive rough endoplasmic reticulum for this purpose (Figure 18.8). A given plasma cell synthesizes and secretes about 2000 identical antibody molecules every second during its short life span of a few days.

Although B lymphocytes can circulate in the peripheral blood, they are found mostly in the bone marrow and the lymphoid tissue, which are the important sites of antibody formation. Once synthesized, the antibodies make their way into the peripheral blood and then into other tissues. Since humans produce many different antibodies, they possess many different clones of plasma cells.

IMMUNOLOGICAL TOLERANCE AND THE FOREIGN NATURE OF ANTIGENS

Immunological tolerance provides one explanation of how self is recognized and antibodies made only against foreign substances. We lack immunocompetence during some or all of the period of fetal development. We know this because a fetus exposed to foreign substances during its immuno-incompetent period will recognize these substances as self and in adulthood will be immunotolerant to them. In humans the period of immunological tolerance is the first trimester of gestation. This period varies among animals, for example immunocompetence in laboratory rodents develops only after birth.

One explanation of immunological tolerance involves the destruction of B lymphocytes that react with antigens before immunocompetency develops. The B lymphocytes in young persons have little IgD bound on their surface. Any of these cells that react with antigen (self) are destroyed so that in adulthood there are no B lymphocytes capable of making antibodies against self. Nevertheless, the system is not foolproof—there are diseases in which antibodies are made against self. To compensate, surveillance goes on into adult life. This is accomplished in part by the **suppressor T cells,** which help to regulate the immune response.

KEY POINT

Animals establish what is foreign by destroying their own antibody-producing cells that recognize self. The remaining cells recognize only foreign antigens. This process occurs mostly during fetal development before immunocompetency prevails.

MONOCLONAL ANTIBODIES AND HYBRIDOMAS

The clonal selection theory could be proven if one could isolate a clone of plasma cells and show that it produced a single type of antibody. Immunologists have not done this experiment because the life span of a plasma cell is too short. However, they have been able to beat this problem by fusing antibody-producing plasma cells with cancerous cells to form long-lived hybrid cells called hybridomas. As postulated by the clonal selection theory, each hybridoma produces a single type of antibody, now called a **monoclonal antibody.** Because monoclonal antibodies all have the same structure and single antigenic specificity, they have important applications in biological analysis and in the study of immunoglobulin structure.

Monoclonal antibodies were first produced in 1975 by George Kohler and Cesar Milstein. To culture antibody-producing cells, Kohler and Milstein fused myeloma (cancer) cells with normal plasma cells (Figure 18.10). By selecting specific hybridomas and growing them in culture, they created clones that produced monoclonal antibodies. For this work Kohler and Milstein received the Nobel Prize in Physiology or Medicine in 1982.

The method of making monoclonal antibodies against a specific antigen has become a routine procedure (Figure 18.10): A mouse is immunized with the antigen of interest. Cells from its spleen are prepared and mixed with cultured mouse myeloma cells. (These cancer cells replicate continuously in culture, but do not produce immunoglobulins.) Fusion between the two cell types is facilitated by the presence of polyethylene glycol. About one out of 200,000 spleen cells forms a viable hybrid with a myeloma cell.

The challenge is to select out the hybridomas from all the other cells. This can be accomplished by taking advantage of a genetic defect in the myeloma cells that prevents them from replicating in a special medium designated HAT. The hybridomas can multiply in HAT medium because the defect is corrected by genes present in the spleen cells. Hence only the hybrid cells multiply in HAT medium. These cells are now dispersed into separate culture chambers where they mul-

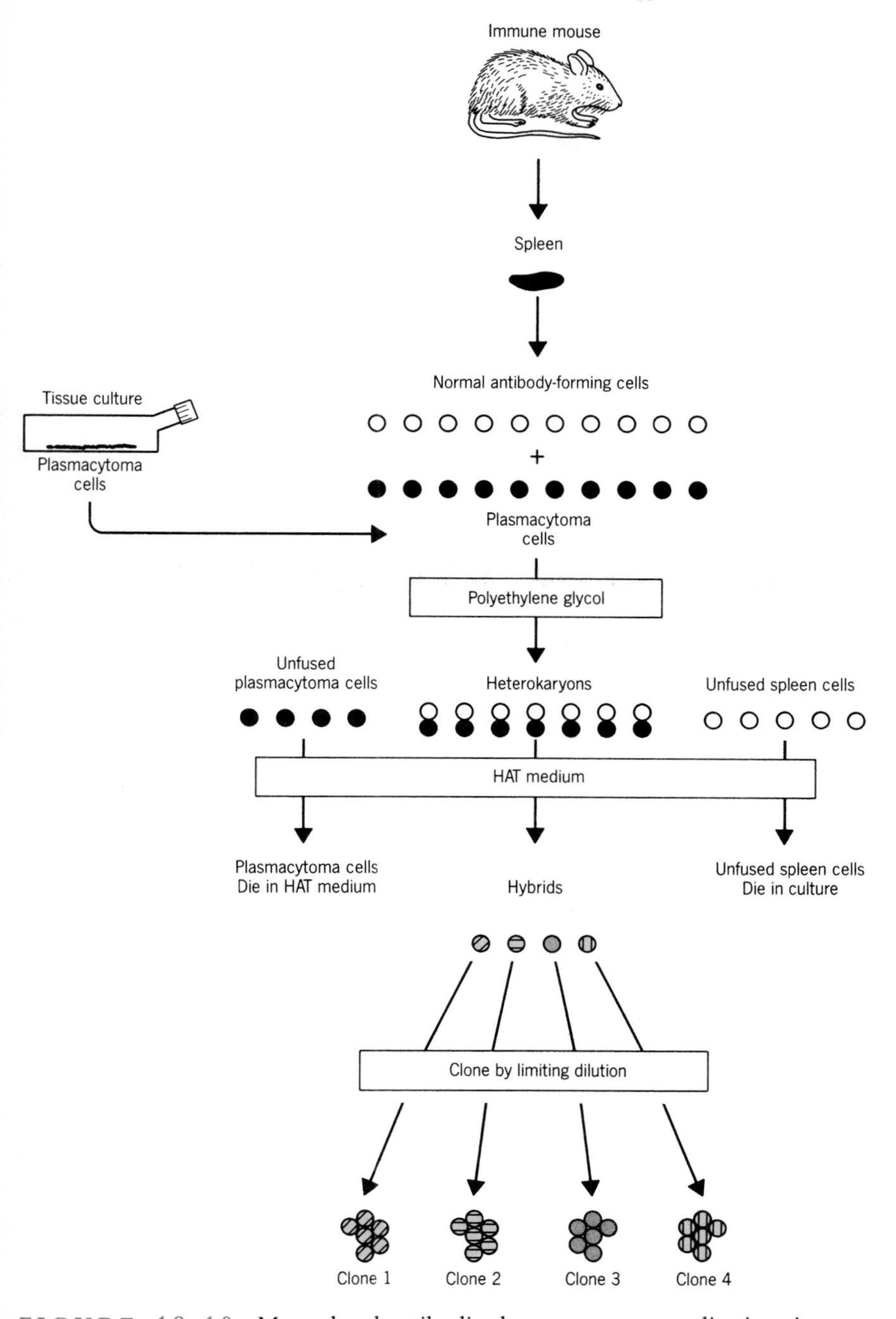

FIGURE 18.10 Monoclonal antibodies have numerous applications in biology. To produce a monoclonal antibody, a mouse is immunized with a specific antigen. Antibody-producing cells from the mouse spleen are isolated and mixed with myeloma cells (plasmacytoma). Fusion takes place between these two cell types, resulting in hybrid cells. The hybrids are selected for in a special medium (HAT) and then cloned. Clones producing the desired antibody are detected by immunoassays. (Modified from J. W. Golding, *Monoclonal Antibodies: Principles and Practice,* Academic Press, London, 1983.)

tiply into a clone. Clones that produce antibody against the desired antigen are identified by antibody–antigen reactions, recloned, and then used to produce monoclonal antibodies.

KEY POINT

Production of monoclonal antibodies by hybridomas proves a major point of the clonal selection theory. In the immune system the B lymphocytes are activated to divide and produce a clone of antibody-producing cells when they react specifically with their antigen.

FUNCTIONS OF T LYMPHOCYTES

Immunity to some infectious diseases cannot be transferred from an immune to a nonimmune individual by transferring antibodies in serum. Immunity to those diseases is transferred only by certain lymphocytes, so it is called cell-mediated immunity. This type of immunity is directed against those infectious microbes with the capacity to live and multiply *within* the host's cells and is mediated by T lymphocytes independent of B lymphocytes. This system polices the body for virus-infected cells, cells infected with bacteria, grafts of foreign tissues, and some types of tumors.

The cells involved in this type of immunity are the T lymphocytes, often called T cells. There are many subsets of T cells that are recognized by their surface antigens and functions in immunity (Table 18.5). Though T cells do not synthesize immunoglobulins, they are involved in immune reactions that specifically recognize antigens and have a memory response. Their ability to recognize antigens is mediated by the T-cell receptor present on their surface.

MAJOR HISTOCOMPATIBILITY ANTIGENS (MHC)

The T lymphocytes develop in the thymus gland where they are imprinted with the ability to recognize self. Each person possesses genes that encode the proteins marking our tissue as unique. These genes are found in a region of the cell's DNA called the **major histocompatibility complex** (MHC). A person produces a class I MHC-encoded protein present on most cells within the body and class II MHC-encoded protein found on T lymphocytes, B lymphocytes, and macrophages. There are millions of variants of the MHC-encoded proteins among humans. T cells are imprinted with their own MHC-encoded proteins in the thymus gland, after which they recognize other MHC-encoded proteins as foreign. This is the reason why grafts of skin or organs are recognized as foreign and more likely than not rejected by the immunological response.

THE T LYMPHOCYTE AND ITS RECEPTOR

Each T cell possesses one kind of protein receptor molecule, similar to an immunoglobulin, on its surface. The T-cell receptor is composed of two polypeptide chains called the alpha chain and the beta chain, each containing constant and variable regions. As is true of immunoglobulins, there are probably millions of possible variants of the T-cell receptor. This receptor rec-

TABLE 18.5 Functions of T lymphocytes

SUBCLASS OF T LYMPHOCYTES	FUNCTION
Helper T cell	Recognize class II MHC-encoded proteins on B lymphocytes, stimulate proliferation of activated B lymphocytes, secrete interleukins
Cytotoxic T cells	Recognize class I MHC-encoded proteins (all tissues except lymphocytes), bind to antigen presenting cells and kill them; act against virus-infected cells, intracellular bacterial growth, and some types of tumors
Suppressor T cells	Help to regulate the immune response, produce lymphocyte inhibitory factor
Delayed hypersensitivity T cells	Release lymphokines that modulate the behavior of monocytes and marcrophages

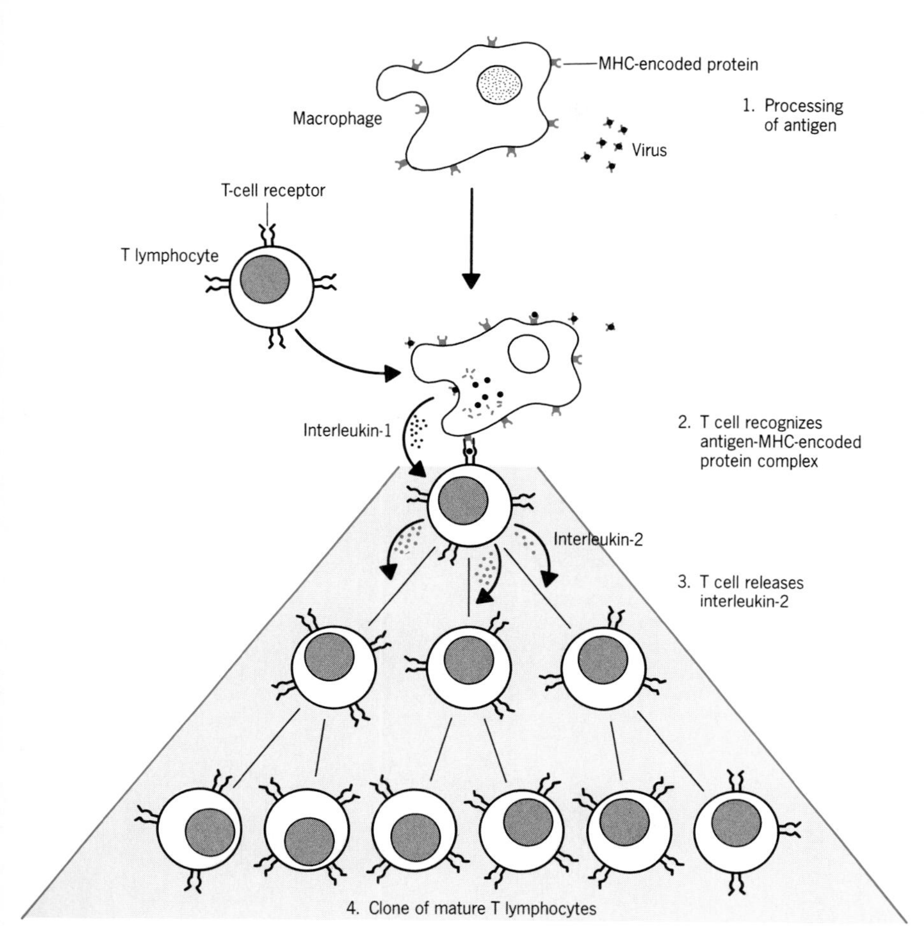

FIGURE 18.11 Immature T lymphocytes form clones of mature helper T cells or cytotoxic T cells in the presence of antigen. The antigen is processed by a macrophage (1) that contains MHC-encoded surface protein. A specific T cell recognizes the antigen MHC-encoded protein complex (2) and binds to the macrophage. The macrophage excretes interleukin-1, which then stimulates the T cell to divide and release interleukin-2 (3). The T cells form a clone (4) of mature T lymphocytes, which can act as either helper T cells or cytotoxic T cells.

ognizes a specific foreign antigen, but only when the antigen is bound to the surface of an antigen-processing cell, such as a macrophage, in association with the host's MHC-encoded protein.

ACTIVATION OF IMMATURE T LYMPHOCYTES

Foreign substances are processed by macrophages after which the antigens are bound to the MHC-encoded protein on the surface of the macrophage (Figure 18.11). The immature T cell recognizes this antigen complex and binds to the surface of the macrophage, which stimulates the macrophage to release **interleukin-1,** a small peptide that stimulates T cells to divide. In the process the T cells are stimulated to produce interleukin-2. **Interleukin-2** stimulates the T cells to differentiate into mature T cells and to divide as long as the antigen is present. Thus a clone of mature T cells is created that recognizes a specific antigen. This mech-

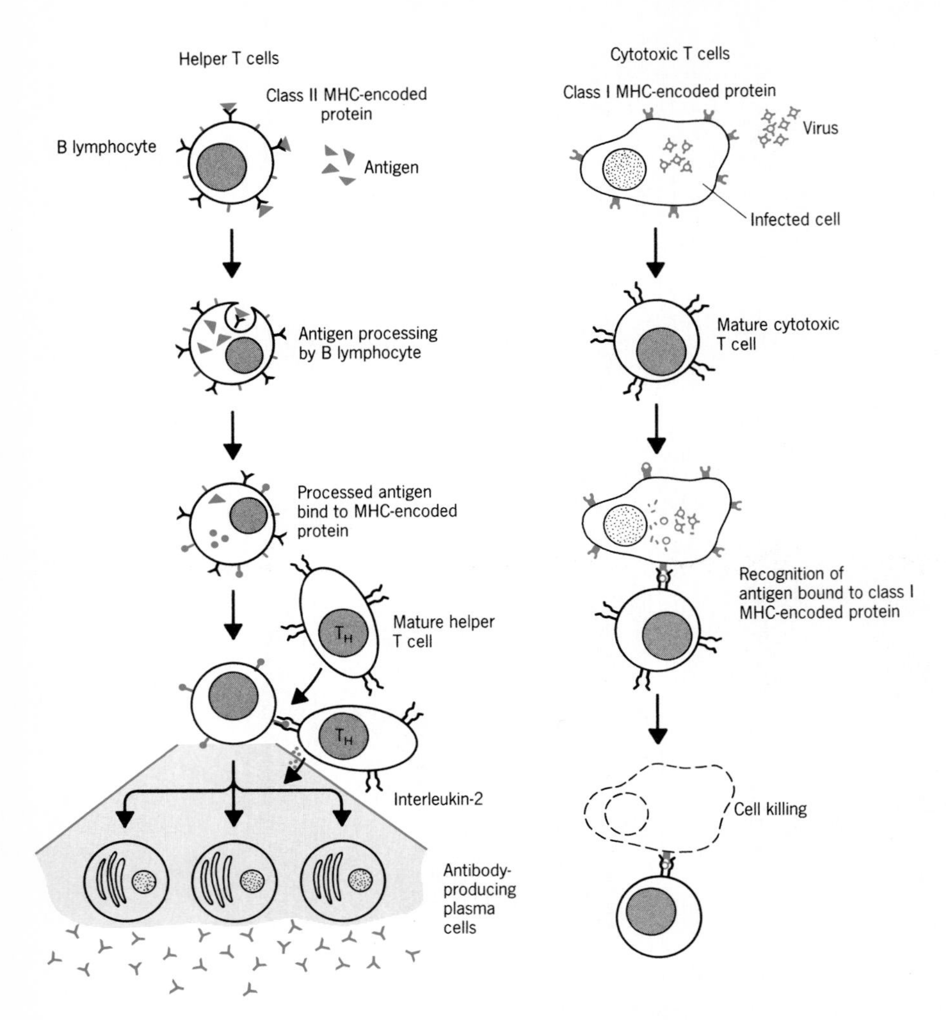

FIGURE 18.12 Mature helper T cells (1) recognize class II MHC-encoded proteins and attached antigen on the surface of an activated B lymphocyte. The helper T cell binds and releases interleukin-2, which stimulates the B lymphocyte to divide and differentiate into antibody-producing plasma cells. Mature cytotoxic T cells (2) recognize class I MHC-encoded protein attached to antigen on the surface of an infected host cell. The T receptor binds to the cell and releases substances that kill the infected host cell.

anism functions in the production of clones of both cytotoxic T cells and helper T cells.

CYTOTOXIC T LYMPHOCYTES KILL INFECTED HOST CELLS

Cytotoxic T cells recognize virus- and microbe-infected host cells that contain class I MHC-encoded proteins (Figure 18.12). The cytotoxic T cell is postulated to have a single kind of T-cell receptor that recognizes antigen(s) of the infectious agent when it is bound to the MHC-encoded protein on the surface of the infected (target) cell. The cytotoxic T cell binds to its target cell, then quickly destroys it; this mechanism enables the host to "police" his cells and destroy the infected ones.

HELPER T LYMPHOCYTES

Helper T cells produce the interleukin-2 required for a B lymphocyte to form an antibody-producing clone of plasma cells (Figure 18.12). Mature helper T cells possess receptors that recognize specific antigens when they are bound to a class II MHC-encoded protein present on a B lymphocyte. The receptor on the helper T cell programs it to interact only with B lymphocytes and macrophages.

When the receptor on the mature helper T cell binds to the processed antigen on the B lymphocyte (Figure 18.12), the helper T cell is stimulated to release interleukin-2. This protein stimulates the B cell to divide and differentiate into plasma cells and to begin secreting antibody. Some of the mature helper T cells become long-lived memory cells.

ACTIVATION OF T LYMPHOCYTES AND CELL-MEDIATED IMMUNITY

Resistance to certain infections can be transferred between persons only by transferring T lymphocytes—this is the classic definition of cell-mediated immunity. This form of immunity requires prior exposure to the antigen and has a demonstrable accelerated response upon second exposure to the antigen; it has a memory. The cells involved in this response are activated cytotoxic T cells and T lymphocytes that produce a variety

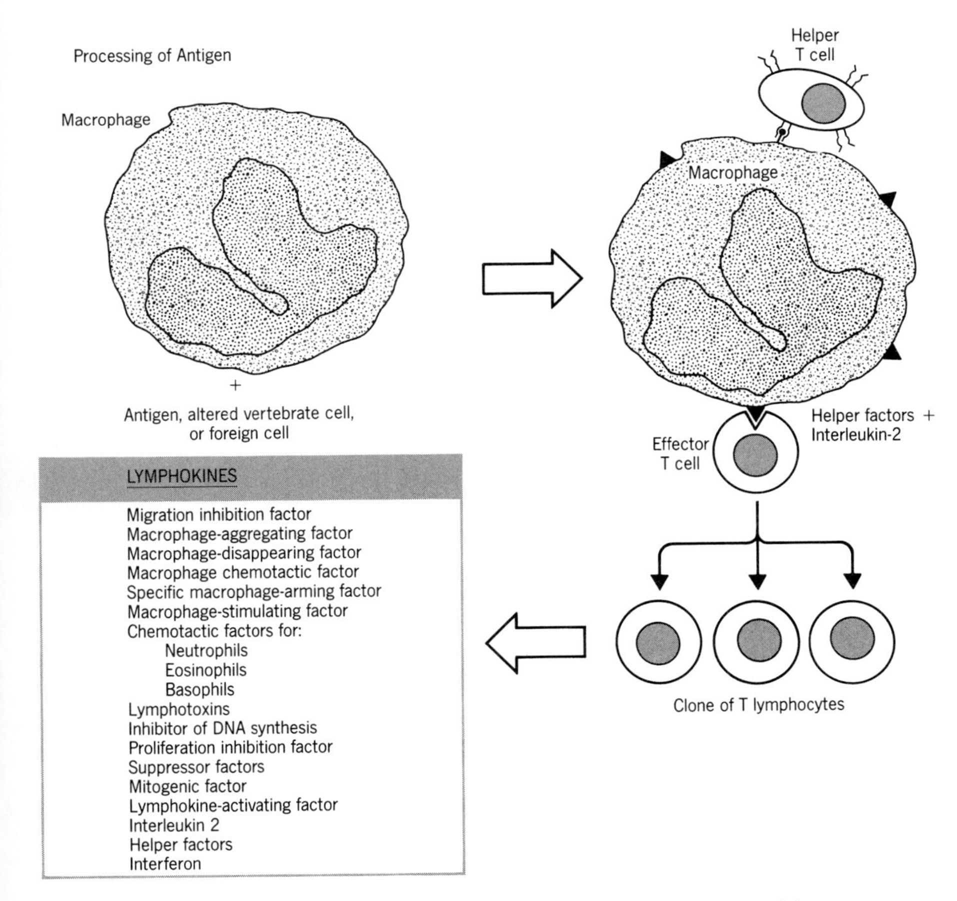

FIGURE 18.13 Macrophages processing antigens are recognized by T lymphocytes, which are then stimulated to divide and produce a clone of sensitized T lymphocytes. These cells release lymphokines that participate in mobilizing the cellular defenses.

of substances collectively known as lymphokines (Figure 18.13).

The **lymphokines** (1) affect macrophages, (2) act as chemotactic factors that attract neutrophils, basophils, and eosinophils, (3) possess cytotoxic and growth inhibitory effects, or (4) stimulate growth of lymphocytes. Some lymphokines attract macrophages to the site of infection, fixing them, causing them to aggregate, and activating them. These macrophage-activating factors enhance macrophage bactericidal activity, especially against bacteria capable of intracellular growth such as *Mycobacterium tuberculosis* and *Legionella pneumophila*. Interferons are lymphokines that prevent the replication of viruses in virus-infected cells (see Chapter 17). Interleukins stimulate the multiplication and differentiation of lymphocytes. The release of lymphokines mobilizes the host's cell-mediated immune system against the foreign material, and this, combined with the killing ability of the cytotoxic T cells, protects the host against intracellular bacterial and viral infections.

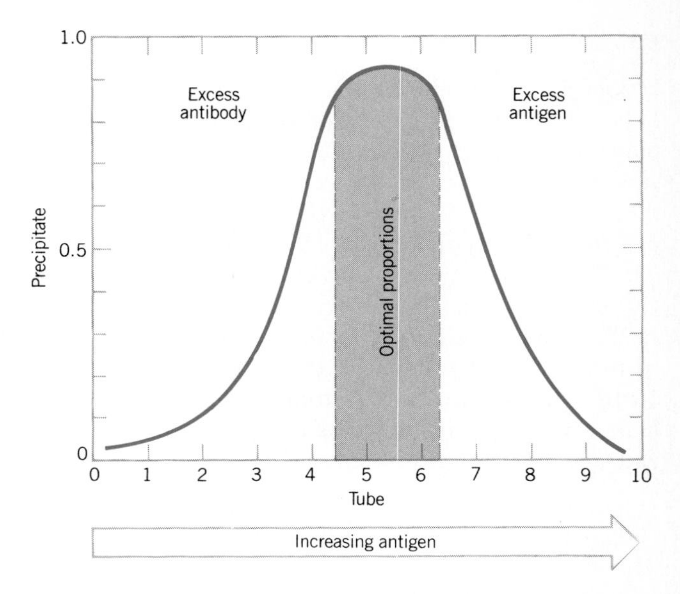

FIGURE 18.14 Precipitin reactions occur between small (noncellular) antigens and antibodies. A visible precipitate is formed when the concentration of antigen and antibody is optimal. No precipitate is formed when either excess antibody or excess antigen is present.

ANTIBODY–ANTIGEN REACTIONS

We return now to the laboratory procedures for measuring the reactions of the circulating antibodies present in the serum of immunized persons. Specific antibody–antigen reactions are widely used in biology as a sensitive and accurate measurement tool.

Serology is the branch of immunology concerned with the reactions between antigens and the antibodies present in blood sera. Blood drawn from patients can be analyzed to detect salts, cholesterol, white and red cells, hormones, and a myriad of physiological components indicative of the person's state of health. Blood serum is the straw-colored solution remaining after the cells and clotting factors are removed. Immunologists have devised many tests for detecting antibodies in sera.

PRECIPITIN REACTIONS

The precipitin reaction provides us with a useful example of an antibody–antigen reaction. This reaction occurs when a multivalent antibody reacts with its soluble multivalent antigen. Under the right conditions it produces a large visible aggregate that precipitates. The antibody—called a **precipitin**—can be detected by any of several precipitin reactions.

The **tube precipitin** reaction is performed in a series of tubes each of which contains an identical amount of serum (Figure 18.14). As antigen at varying concentrations is added to the tubes, a precipitate forms in those tubes containing the correct proportions of antigen and antibody molecules. The results depicted in Figure 18.14 show a visible precipitate in tubes 4, 5, and 6. Tubes 1 and 2 contained *excess antibody,* so there was not enough antigen to form large visible aggregates. Tubes 8 through 10 contained *excess antigen,* so only small, nonprecipitating aggregates of antigen and antibody were formed (Figure 18.15). The **optimal proportion** between antigens and antibody molecules occurred in tubes 5 and 6. These antibody–antigen aggregates are large because both the antigen and the antibody molecules are multivalent (Figure 18.15).

Modifications of the precipitin reaction have been devised to enable the precipitate to be seen. In the **Ouchterlony double-diffusion plate,** the antigens

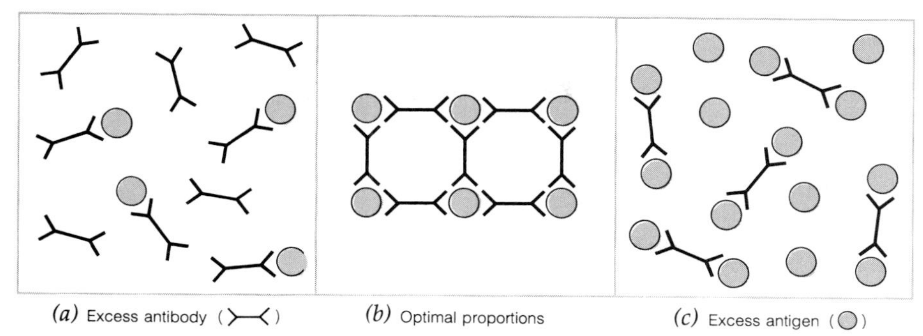

FIGURE 18.15 Schematic representation of the zone of (*a*) excess antibody, (*b*) optimal proportions for a precipitin reaction, and (*c*) the zone of excess antigen.

and antibodies form a concentration gradient as they diffuse out from their point of application (Figure 18.16). Symmetrical holes are cut in an agar matrix in a petri dish, and each well is filled with either an antigen or an antibody. The assay is called a double-diffusion test since both antigen and antibody diffuse out from their respective locations in the plate. A precipitate forms at the position in the agar where optimal proportions between the antigen and the antibody occur.

The structure of the precipitate band indicates whether the antigens are identical, partially identical, or nonidentical. Identity is demonstrated when the precipitin bands between two antibody wells join smoothly at the apex and show no spurs. Partial identity is indicated by the presence of a single spur at the apex, and nonidentity is indicated by the presence of two spurs at the apex (Figure 18.16). The Ouchterlony precipitin reaction is used to demonstrate the degree of similarity between antigens or antisera.

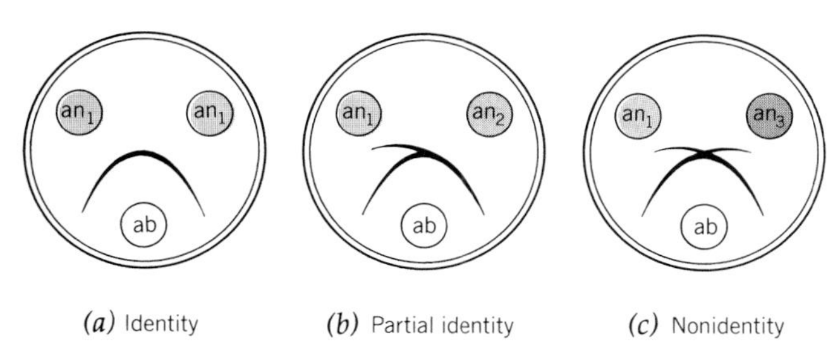

FIGURE 18.16 Ouchterlony precipitin reaction. Wells in the agar are filled with immune serum or antigen. After diffusion, precipitin bands form where the reactant concentrations are at optimal proportions. The wells contain different antigens (an_1 an_2, an_3) or immune sera prepared against all three antigens. A smooth continuous precipitin band (*a*) represents identity, a single spur (*b*) represents partial identity, and a double spur (*c*) represents nonidentity.

AGGLUTINATION REACTIONS

Agglutination occurs when antibodies react with large antigens, usually cells, causing them to clump into large aggregates. Bacterial cells and red cells are examples of large multivariant antigens. These large antigens form clumps—agglutinate—when multivalent antibodies bring them together by binding to antigens on their surface. Agglutination reactions are quantified by diluting the serum in saline and adding a sample of each dilution to a constant concentration of the antigen. Typical results of this procedure are diagramed in Figure 18.17.

Greater consistency in the agglutination reaction is obtained by coating microscopic latex particles with soluble antigen or antibody. Serum antibodies are detected with this technique by adding antigen coated latex beads to the serum. Clumping of the beads, observed in a test tube or on a microscopic slide, indicates the presence of the antibody. The reverse experiment is done by coating latex beads with antibody and using these beads to detect antigen. These assays are called **latex agglutination** tests.

One measure of the amount of antibody in a serum

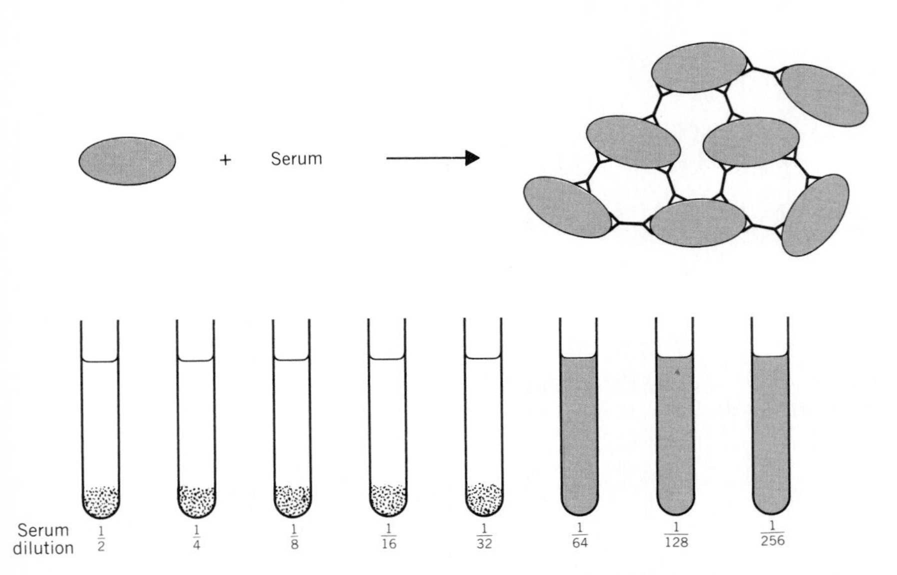

FIGURE 18.17 Agglutination reactions occur when antibodies react with antigens on the surface of cells, causing them to clump. The serum is diluted out in tubes, then a constant amount of antigen is added. The titer of this serum is the reciprocal of the dilution in the last tube (1/32) in which agglutination occurred (a titer of 32 results from the experiment shown).

is its titer. **Titer** is defined as the reciprocal of the highest dilution that will cause clumping of the antigen. If clumping occurs in the tube with a dilution of 1/32, but not in the tube with a dilution of 1/64, the titer would be the reciprocal of 1/32 or 32. The titer tells us how well the animal responded to antigenic stimulation and this often correlates with the degree of resistance against a particular infectious agent. Agglutination tests have their place in typing of bacterial species, in determining prior exposure to an infectious agent, and as a means of diagnosing certain infectious diseases.

ENZYME-LINKED-IMMUNOSORBENT ASSAY (ELISA)

An enzyme-linked-immunosorbent assay, commonly known by its acronym ELISA, can be used to detect either antibody (Figure 18.18) or antigen. Polystyrene tubes (dishes) are used to take advantage of the fact that proteins bind to polystyrene. To detect antibody against a protein antigen, the antigen is bound to the surface of the tube. The excess antigen is washed out with saline, and the unbound sites are covered with another protein such as serum albumin. The patient's serum is now added to the tube. If specific antibodies are present, they will bind to the antigens coating the tube before the excess antibodies are washed out with saline.

A special enzyme-linked reagent is used to detect the presence of the patient's antibodies in the tube. The reagent, called an enzyme-linked antiglobulin, is made by chemically attaching an enzyme to antihuman immunoglobulin (produced in a goat). When this reagent is added to the tube it will bind to any human antibodies present (Figure 18.18). After unbound reagent is washed out of the tubes with saline, the amount of bound reagent is determined by measuring the amount of enzyme remaining in the tube. This value is a direct measure of the patient's antibody in the tube. ELISAs are very sensitive assays, and have been developed for important serological tests including a screening test for antibodies against the AIDS virus.

SPECIAL SEROLOGICAL TESTS

It is not always possible to obtain large quantities of the antigen one wishes to work with. This is true in

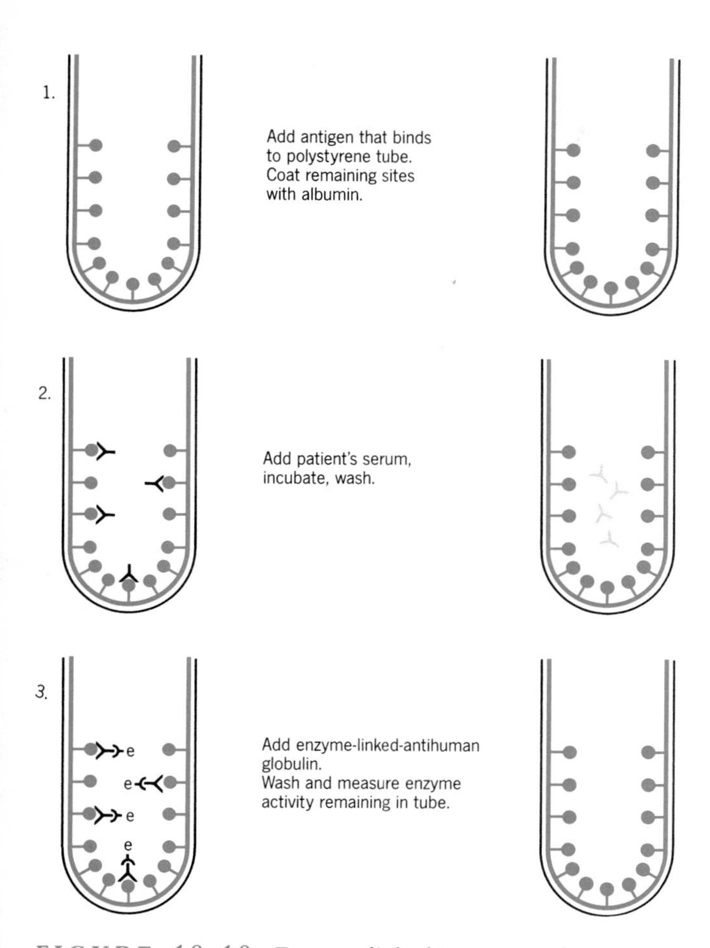

FIGURE 18.18 Enzyme-linked immunosorbent assay (ELISA) uses an enzyme assay to detect a patient's antibody. The protein antigen is bound to the surface of a polystyrene tube (1). The excess is removed and any unused protein-binding sites are covered with albumin. The patient's serum is added (2). In a positive test (left), the patient's antibody binds to the immobilized antigen and remains attached during washing. Enzyme-linked antiglobulin is added (3), incubated, and the excess is removed by washing. The amount of bound enzyme is determined by measuring the enzyme activity in the positive test, and by the absence of enzyme activity in the negative test (right).

the case of the most common sexually transmitted agent, *Chlamydia trachomatis,* and for many human viral infections. One approach that surmounts this problem is the use of fluorescent antibody tests.

Fluorescent Antibody Assays

Although individual antibody molecules are much too small to be seen in a light microscope, concentrations of antibodies chemically labeled with fluorescent compounds can be seen in a fluorescent microscope (see Chapter 3). Binding of the fluorescent antibody to its antigen concentrates the fluorescence sufficiently to outline the microbe or virus being stained. Specific microbes and viruses can be detected with this technique by using fluorescent-labeled monoclonal antibodies that react with specific strains or species. Because these antibodies are specific, fluorescent-antibody assays can be used to detect infectious agents even in contaminated specimens taken directly from a patient. Fluorescent-antibody assays are available for the laboratory detection of herpesvirus types I and II, and in diagnosing infections caused by *Chlamydia trachomatis.*

Toxin Neutralization

Antigens that affect a biological system can be assayed by the ability of antibodies to neutralize the effect. Very small quantities (microgram and even nanogram) of certain bacterial toxins kill laboratory animals. Premixing a toxin (antigen) with its antitoxin neutralizes the toxin's biological effect: an animal injected with such a mixture survives, even though the toxin by itself would be lethal. This type of assay is called a **toxin neutralization test.**

Hemagglutination and Hemagglutination Inhibition Tests

Certain viruses contain hemagglutinins, which are surface antigens that cause red cells to agglutinate. These viral antigens react with receptors on red cells causing them to form large aggregates (Figure 18.19), a process called **hemagglutination.** Hemagglutination tests are used to measure the presence of virus particles. Animals make antibodies against viral hemagglutinins. **Hemagglutination inhibition** tests (Figure 18.19) detect the presence of these antibodies, which inhibit hemagglutination.

COMPLEMENT-MEDIATED CELL LYSIS

Normal serum contains proteins collectively known as **complement.** There are 17 components of the human

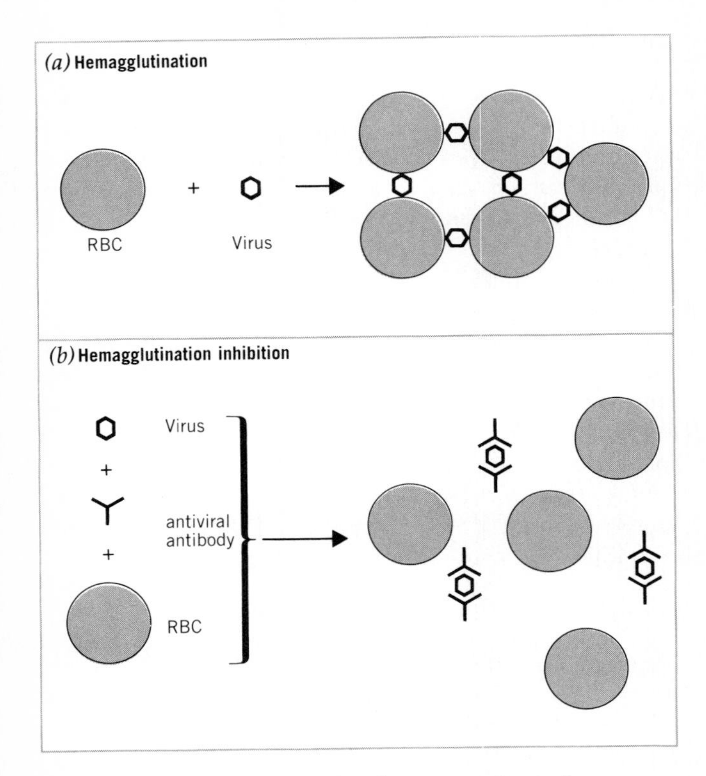

FIGURE 18.19 Certain viruses contain surface antigens that agglutinate red blood cells (RBC). (*a*) Hemagglutination is used to detect these viruses and (*b*) hemagglutination inhibition is used to detect antibodies against the hemagglutinating viruses.

complement system, most of which are β globulins. These proteins can have individual functions involving inflammation, opsonization, and phagocytosis, or can act in concert to cause cell lysis. Complement is heat labile since it can be destroyed by heating at 56°C for 30 minutes.

The classic complement pathway is activated by an antigen–antibody complex (Figure 18.20). Reaction of the antigen with its antibody exposes sites on the heavy chains of the IgM or IgG (except IgG_4) immunoglobulins that bind the first component of complement (C1q). This initiates a sequential cascade, in which one activated complement component enzymatically activates the next one, and so on. Since each protein can activate many subtrate molecules, each step has a multiplier effect on the next sequential reaction until all 17 components are involved. The complete C1 complex (c1q, C1r, C1s) activates C2 and C4, which in turn activate C3. C3 is converted to C3b, which enhances the phogocytosis of bacteria. C3b also initiates the formation of the membrane attack complex from the aggregation of components C5–C9. This complex forms a hole in the cell membrane that brings about cell lysis. The alternative pathway is activated by microorganisms alone and doesn't require antibody. Again the C3 is converted to C3b and the cytolytic complex (C5–C9) forms and causes cell lysis.

The reactions discussed in this chapter benefit the

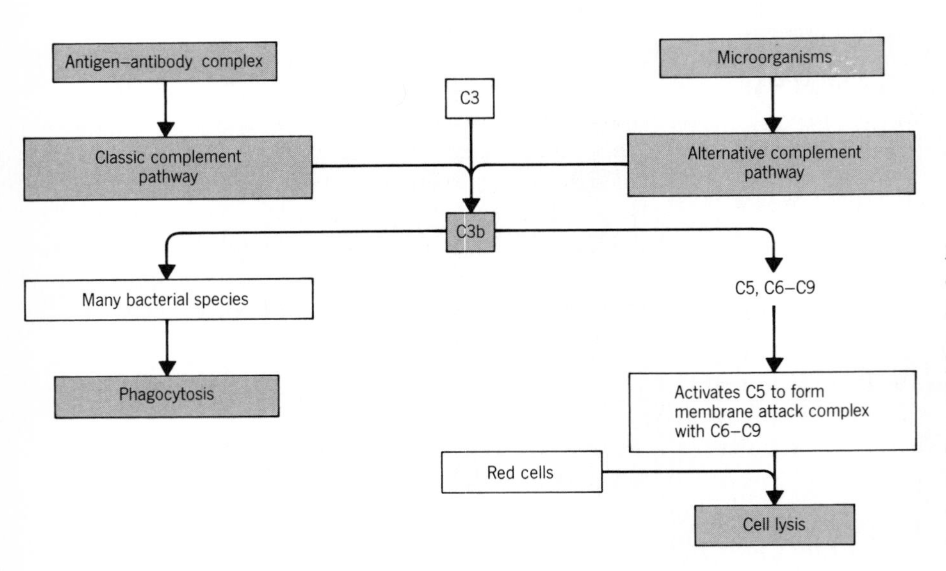

FIGURE 18.20 The classic complement pathway is initiated by an antigen–antibody complex. The alternative pathway is initiated by the lipopolysaccharides of microorganisms. Both pathways activate C3 by converting it to C3b. This component coats many bacterial cells, making them subject to phagocytosis. C3b also activates C5 to form the membrane attack complex that causes cell lysis.

animal, but not all immune reactions are beneficial. Those reactions that cause damage to the host are referred to as hypersensitivity. Atopic allergies, anaphylaxis, immune complex disorders, delayed hypersensitivity, and contact dermatitis are examples of adverse immune reactions discussed in Chapter 19.

Summary Outline

OVERVIEW OF THE IMMUNE SYSTEM

1. The immune system protects against infectious diseases, has a memory, and distinguishes between foreign substances and self.
2. Immunoglobulins, complement, and lymphocytes are the central components of the immune system.

TYPES OF IMMUNITY

1. Immunity is the increased resistance of an individual or population against infectious agents or against the toxic effects of an antigenic substance.
2. Innate immunity is the inborn resistance to disease possessed by a given population or race.
3. The four types of acquired immunity are natural active, natural passive, artificial active, and artificial passive.

ANTIBODIES AND ANTIGENS

1. Antibodies are immunoglobulins formed by cells in response to a foreign substance (antigen) that react specifically with that substance.
2. Antigens are substances that stimulate a person to synthesize antibodies. Antigens are usually complex foreign macromolecules that contain many antigenic determinants.
3. Immunoglobulin monomers are composed of two light and two heavy polypeptide chains joined in a Y configuration. Each chain has variable and constant regions.
4. The largest class of immunoglobulins (IgGs) function as precipitins and antitoxins, participate in complement fixation (except IgG_4), and cross the placenta (except IgG_2).
5. Immunoglobulin M, a pentamer that is confined to the blood vascular system, is active in agglutination. This antibody is present on B lymphocytes and is formed early in the immune response.
6. IgA is the secretory immunoglobulin, IgD is found on the surface of B lymphocytes, and IgE participates in resisting parasitic infections and type I hypersensitivities.
7. Haptens are too small to elicit antibody formation by themselves, but function as antigenic determinants.

ANAMNESTIC RESPONSE

1. Upon first exposure to an antigen, the host forms specific antibodies (IgM and IgG) against that antigen, following a brief lag.
2. The host responds to a second or subsequent exposure to an antigen by repidly making high levels of circulating antibodies (mainly IgG), in the secondary or anamnestic response.
3. Children are actively immunized against the bacterial diseases of diphtheria, tetanus, and whooping cough, and the viral diseases of rubella, mumps, polio, and measles. Booster "shots" stimulate the secondary response.

THE IMMUNE SYSTEM

1. B lymphocytes develop from stem cells in bone marrow and contain surface immunoglobulins that interact with one specific antigen.
2. T lymphocytes develop in the thymus gland and are the major lymphocytes in the peripheral blood. Subclasses of T lymphocytes include cytotoxic T cells, helper T cells, and suppressor T cells.
3. The clonal slection theory explains the anamnestic response, formation of specific antibodies, and the ability to distinguish between self and nonself.
4. Immunological tolerance is created during a period of fetal development by destruction of B lymphocytes capable of making antibodies against self.

5. Activated B lymphocytes differentiate into clones of antibody-producing plasma cells with the help of mature interleukin-2-producing helper T cells. Some of the B lymphocytes and mature helper T cells become memory cells.
6. Monoclonal antibodies are produced by hybridomas formed upon fusion of a plasma cell and a myeloma cell.

FUNCTIONS OF T LYMPHOCYTES

1. T lymphocytes develop in the thymus gland where they are imprinted to recognize their own MHC-encoded proteins.
2. Each T cell possesses a distinctive T-cell receptor on its surface that recognizes antigens bound to MHC-encoded proteins.
3. Clones of T cells form after interacting with antigen-processing macrophages. The macrophages release interleukin-1, which stimulates the T cells to divide.
4. Mature cytotoxic T cells recognize and kill virus- and microbe-infected host cells that contain class I MHC-encoded proteins.
5. Mature helper T cells interact with B lymphocytes (possess class II MHC-encoded protein) and produce interleukin-2, which stimulates B lymphocytes to divide.
6. Other T cells are involved in the cell-mediated immune response by producing lymphokines.

ANTIBODY–ANTIGEN REACTIONS

1. Serology is the branch of immunology concerned with analysis and study of blood to detect the presence of antigens and antibodies.
2. Precipitin reactions occur when soluble multivalent antigens react with their multivalent antibody. Precipitins can be detected by the Ouchterlony double diffusion assay.
3. Bacteria are multivalent large antigens that normally agglutinate upon reaction with antibodies directed against them. Specialized tests include hemagglutination and latex agglutination.
4. An enzyme-linked-immunosorbent assay (ELISA) is a very sensitive assay for detecting either antigens or antibodies.
5. Direct fluorescent antibody tests with monoclonal antibodies are valuable for detecting viruses and microbes.
6. Complement is a group of blood proteins that act in opsonization, bacterial cell lysis, lysis of foreign cells, and the inflammatory response.

Questions and Topics for Study and Discussion

QUESTIONS

1. Describe the four types of acquired immunity and explain how they protect against infectious diseases.
2. Describe the physical and functional properties of the different classes of immunoglobulins. Which ones can cross the placental barrier?
3. Define or explain the following:

anamnestic response	hemagglutination	plasma
antigen	interleukin-2	serology
complement	monoclonal antibody	serum
ELISA	lymphokines	titer
hapten	optimal proportions	

4. What is the basic structure of an antibody? How does this structure explain the ability of an antibody molecule to react with two separate antigenic determinants (bivalence)?
5. What are the functions of the interleukins in the development of an immune response?
6. How are virus-infected cells killed by the immune system?
7. Explain the cellular biology of the anamnestic response to a soluble antigen.
8. What is the clonal selection theory of antibody formaion and what evidence is available to support it?

9. What are the differences between a precipitation reaction and an agglutination reaction?
10. What protocol would you use to detect an antigen with an ELISA?

DISCUSSION TOPICS

1. Smallpox has been eliminated from the human population. Can other infectious agents be eliminated from nature using the principles of immunology? Explain.
2. What are the potential uses of monoclonal antibodies in biology and medicine? How might monoclonal antibodies be used to treat cancer?
3. Interleukin-2 has been produced by genetic engineering techniques. What potential medical uses of this natural product do you envision?

Further Readings

BOOKS

Alberts, B., D. Bray, J. Lewis, M. Raff, K. Roberts, and J. D. Watson, *Molecular Biology of the Cell,* Garland, New York, 1983. Chapter 17 on the immune system is an excellent introduction to this topic from the viewpoint of a molecular cell biologist.

Barrett, J. T., *Basic Immunology and Its Medical Application,* 2nd ed., Mosby, St. Louis, 1980. An introduction to the fundamentals of immunology for allied health students. The clinical applications of immunology are nicely brought out in the patients' case histories presented at the end of each chapter.

Roitt,I., *Essential Immunology,* 5th ed., Blackwell, Oxford, England, 1984. An introductory textbook in immunology for the undergraduate.

Tizard, I. R., *Immunology: An Introduction,* Saunders, Philadelphia, 1984. An introductory textbook for the undergraduate.

ARTICLES AND REVIEWS

Ada, G. L. and Sir G. Nossal, "The Clonal-Selection Theory," *Sci. Am.,* 257:62–69 (August 1987). Historic development of the clonal-selection theory of antibody formation and descriptions of the experiments that led to its acceptance.

Cooper, M. D., and A. R. Lawton, III, "The Development of the Immune System," *Sci. Am.,* 231:58–72 (November 1974). An introduction to the development of B and T lymphocytes and their role in the immune response.

Edelson, R. L., and J. M. Fink, "The Immunologic Function of Skin," *Sci. Am.,* 252:46–53 (June 1985). The body's largest organ is described as an active element in the immune system.

International Conference, "Can Infectious Diseases Be Eradicated?" *Rev. Infect Dis.,* 4:912–984 (1983). An authoritative debate on eradicating infectious diseases.

Marrack, P. and J. Kappler, "The T Cell and Its Receptor," *Sci. Am.,* 254:36–45 (Febuary 1986). An explanation of how T cells function in the immune system based on their T-cell receptors.

Milstein, C., "Monoclonal Antibodies," *Sci. Am.,* 243:66–74 (August 1980). The technique for creating hybridomas for the production of monoclonal antibodies is described.

Tonegawa, S., "The Molecules of the Immnue System," *Sci. Am.,* 253:122–131 (October 1985). An excellent description of how diverse antibody structures are made from a limited number of genes.

CHAPTER

19

Hypersensitivity and the Immune System

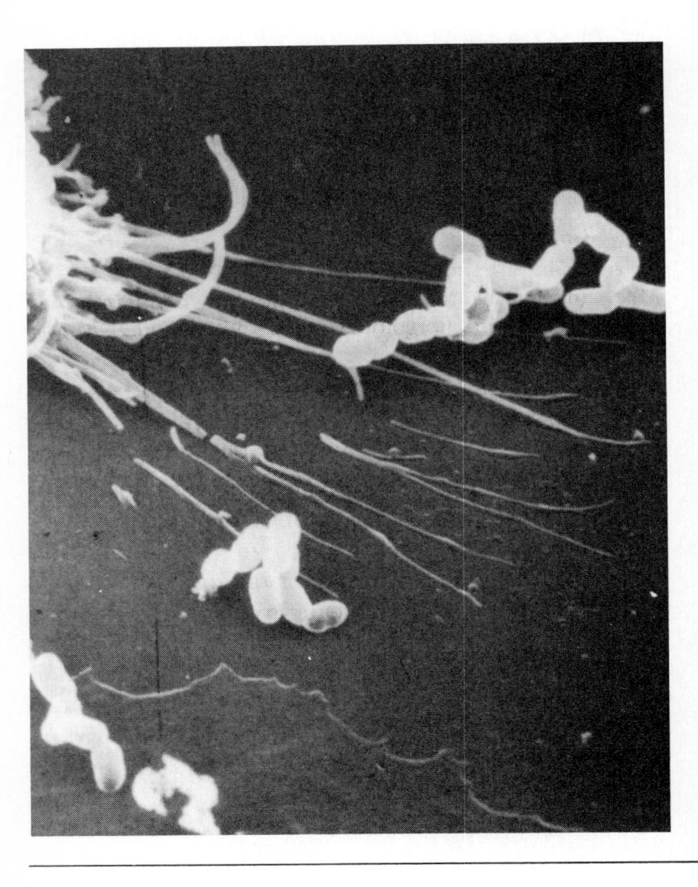

OUTLINE

FOCUS OF THIS CHAPTER

The adverse effects of the immune system—hypersensitivity—are the focus of this chapter.

The physiological and cellular bases of the four types of hypersensitivity reactions are discussed in relation to their importance to infectious diseases.

Those who have experienced asthma or hay fever, or who react violently to bee stings, have experienced an adverse effect of the immune system. These conditions are manifestations of immunological reactions that provoke one or more of the five signs of inflammation: heat, redness, swelling, pain, loss of function. Inflammation and the immune response protect us against infectious agents and participate in healing; but sometimes these mechanisms go awry and cause gross tissue damage, especially if the antigen is present in excessive amounts and the individual is super (hyper) sensitized. These adverse effects are known as **hypersensitivity reactions,** which are exaggerated immune responses to foreign antigens that adversely affect the person.

TYPES OF HYPERSENSITIVITY

There are many ways to induce hypersensitivity, and an enormous variability in the response. One approach to explaining these differences is to describe the four classes of hypersensitivity that were devised by Philip Gell and Robin Coombs (Table 19.1). **Immediate** (type I) hypersensitivities are inflammatory reactions mediated by the release of pharmacologically active substances from mast cells and basophils coated with IgE immunoglobulin molecules. **Cytotoxic** (type II) hypersensitivities result in the destruction of cells through the action of antibody and complement or by cytotoxic (T cells, macrophages, and neutrophils). **Immune complex** (type III) hypersensitivities are the complement-mediated consequences of immune complex deposition in tissue. **Delayed** (type IV) hypersensitivities are cell-mediated immune reactions mediated by the release of lymphokines from activated T lymphocytes.

Because hypersensitivity reactions are a component of the immune response, the person must be sensitized through exposure to the antigen or allergen responsible for the reaction. Sensitized individuals experience type I reactions immediately upon exposure to the antigen, whereas the delayed (type IV) hypersensitivity response develops over a period of one or more days. Types II and III reactions involve serum antibodies and complement. There are other differences also, having to do with the cells involved and the mediators of the inflammatory reaction.

INFLAMMATION

Inflammation, the host's response to injury, can be triggered by an invading microbe, a wound, or an immune response that results in tissue damage. A local

TABLE 19.1 The four classes of hypersensitivity reactions

CLASS	DESCRIPTION	EXAMPLES
Type I (immediate)	IgE bound to mast cells and basophils, release of pharmacologically active substances	Asthma, hay fever, anaphylaxis
Type II (cytotoxic)	Antibody and complement, cytotoxic cells	Blood groups, transfusions, Rh problem
Type III (immune complex)	Immune complex deposition, complement	Farmer's lung, Arthus reaction
Type IV (delayed)	Cell-mediated, activated T lymphocytes	Tuberculin reaction, contact dermatitis

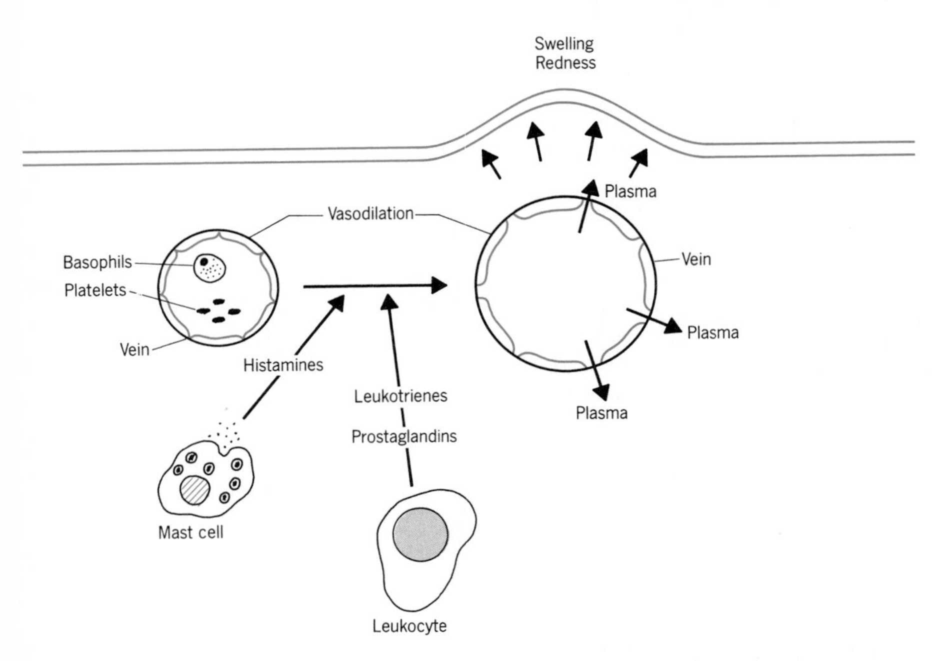

FIGURE 19.1 Vasodilators released by mast cells, basophils, and platelets are mediators of inflammation. Plasma and leukocytes enter the tissue from the dilated blood vessels to cause swelling, while increased blood flow causes the tissue to redden.

inflammatory response involves a complex sequence of events that bring phagocytes to the damaged area so they can destroy foreign material and remove cell debris preliminary to tissue repair. The stages of the inflammatory response are as follows:

1. Injury by invading microbe, physical wound, or immune complex.
2. Increase in blood flow from vasodilation of local vessels.
3. Plasma enters site following a marked increased in vascular permeability.
4. Fluids accumulate causing local swelling.
5. Neutrophits (PMN) and later monocytes and macrophages enter the tissue.
6. Foreign matter and cell debris are removed by phagocytosis prior to healing.

These events are mediated by the release of cellular substances at the site of inflammation. Many of the gross manifestations of inflammation—swelling (edema), redness, heat, pain—can be explained by the release of vasodilators (vas'o-di'lah-tor), which increase vascular permeability and bring an influx of protein-rich plasma to the site (Figure 19.1).

Numerous pharmacologically active compounds produced within various cells cause vasodilation (Table 19.2). Basophils and mast cells are sources of **histamine,** which causes vasodilation in type I hypersensitivities. Histamine also causes contraction of bronchial smooth muscles, which is a major cause of the symptoms of asthma. Leukocytes synthesize **leukotrienes** and **prostaglandins** (pros'tah-glan'din), two families of small molecules that are made from the fatty acid arachidonic (ah-rak'i-don'ic) acid (Figure 19.2), after its release from membrane phospholipids. Prostaglandins PGE_2 and PGI_2 and leukotrienes LTC_4 and LTD_4 play significant roles as vasodilators. **Anaphylatoxins** (an'ah-fi-lah-tok'sin) are vasoactive polypeptides derived from the cleavage of the C3 and C5 components of complement; they act indirectly as vasodilators by

TABLE 19.2 Mediators of inflammation and their sources

MEDIATORS	SOURCE
Vascular Permeability Factors	
Histamine	Mast cells, basophils
Leukotrienes LTC_4, LTD_4	Leukocytes (synthesized from arachidonic acid)
Prostaglandin PGE_2, PGI_2	Leukocytes (synthesized from arachidonic acid)
Anaphylatoxins	Cleavage products of C3 and C5 components of complement
Lymphokines	Lymphocytes
Kinins (bradykinin)	α globulin
Smooth Muscle Contractors	
Histamine	Mast cells, basophils
Leukotrienes LTC_4, LTD_4, LTE_4	Leukocytes
Chemotactic Factors	
C5a (attracts neutrophils)	Breakdown product of C5 component of complement
Bacterial products	Infecting bacteria
Leukotriene LTB_4	Leukocytes (from arachidonic acid)
Lymphokines	Activated T lymphocytes

stimulating mast cells to release histamines. **Bradykinin** (brad'e-ki'nin) is a short peptide derived from α globulin that is a powerful vasodilator and increases capillary permeability.

As the vessels dilate, more blood is moved to the affected site, bringing with it neutrophils and later monocytes and macrophages. The increased capillary permeability enables cells to squeeze through the vessel walls and enter the tissue spaces where phagocytosis occurs. Cells are attracted to the site by chemotaxis and fixed there by lymphokines as explained in Chapter 17. Plasma also enters the site through the dilated capillaries and with it come immunoglobulins and complement, which participate in the inactivation and destruction of microbes by phagocytosis. The gross manifestations of this process are redness due to the increased blood flow and swelling from the buildup of fluid in the inflamed tissue. The pain associated with inflammation is in part caused by this swelling.

The phagocytes involved in inflammatory reactions initiated by invading microbes are neutrophils, monocytes, and macrophages. These cells ultimately are responsible for removing invading microbes from the host. Eosinophils are the cells in great preponderance when the inflammation is triggered by certain allergies and helminthic infections. Eosinophils help to control inflammation by releasing enzymes that destroy the mediators of inflammation. An acute inflammation clears up in a matter of days, whereas a chronic inflammation can last for months or years.

KEY POINT

Inflammation reactions mobilize the defense mechanisms at the site of infection by stimulating the release of pharmacologically active substances. Mediators of inflammation include vasodilators and chemotactic factors that draw leukocytes to the site.

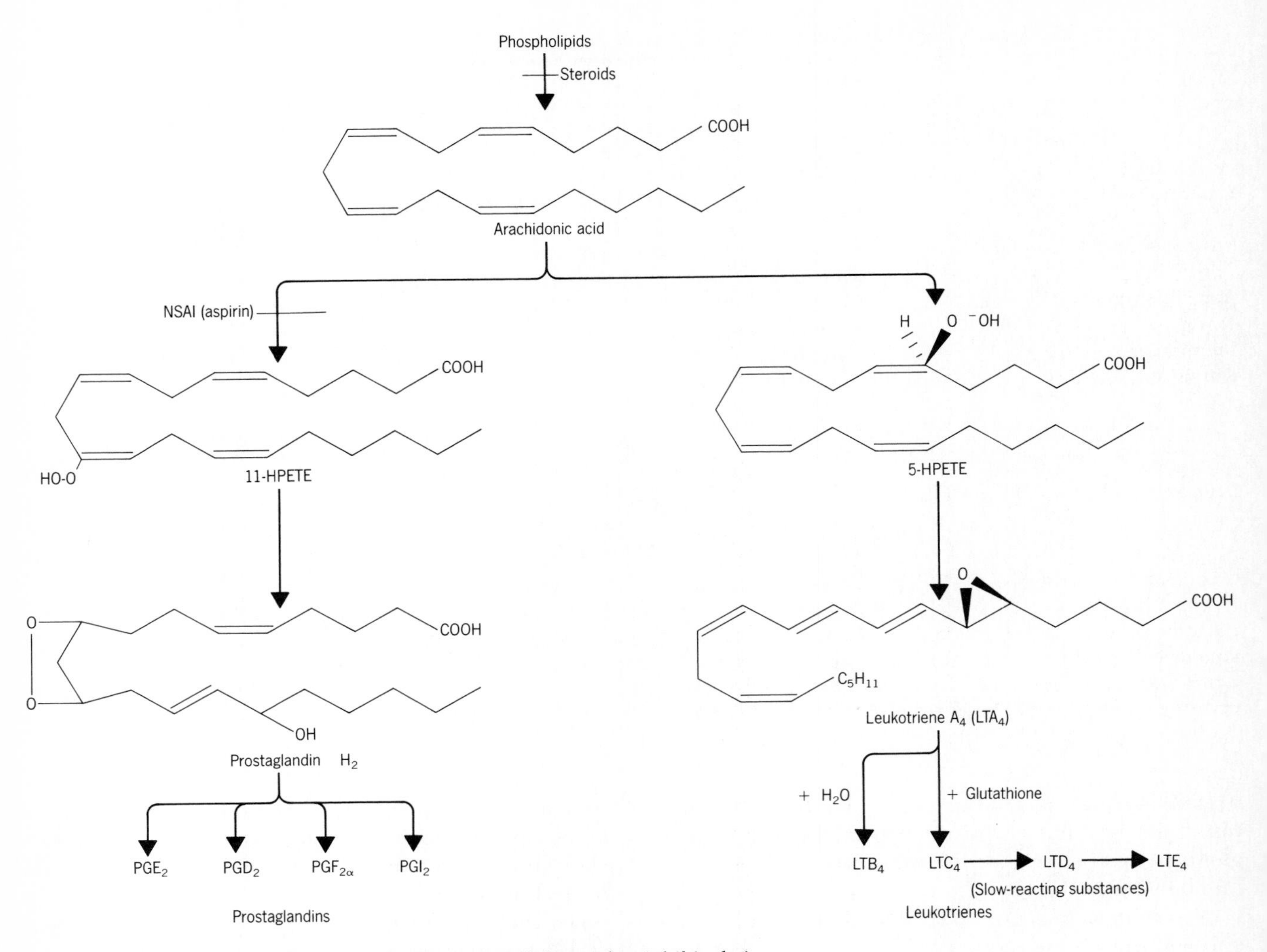

FIGURE 19.2 Mediators of inflammation and drugs that inhibit their function. Leukocytes synthesize prostaglandins and leukotrienes. Steroids inhibit the production of arachidonic acid from membrane phospholipid by the action of phospholipase A2. Arachidonic acid is the substrate for the pathway leading to the formation of the leukotrienes and the prostaglandins. The prostaglandin pathway is inhibited by nonsteroid anti-inflammatory compounds (NSAI) such as aspirin. Leukotrienes LTC_4, LTD_4, and LTE_4 are the slow-reacting substances involved in allergic reactions.

TYPE I HYPERSENSITIVITIES: ALLERGIES AND ANAPHYLAXIS

Immediate hypersensitivities such as atopic allergies and anaphylaxis (an'ah-fi-lak'sis) cause great discomfort. Early investigators used the term **atopy** (Gr. *atopy*, no place) to describe hay fever, asthma, and eczema because there was no obvious cause of these afflictions. Unlike infectious diseases, atopic allergies are chronic conditions that can last many years or even a lifetime. The symptoms can be treated, but there is no known cure. **Anaphylaxis** is an exaggerated reaction to a foreign substance and is the most severe im-

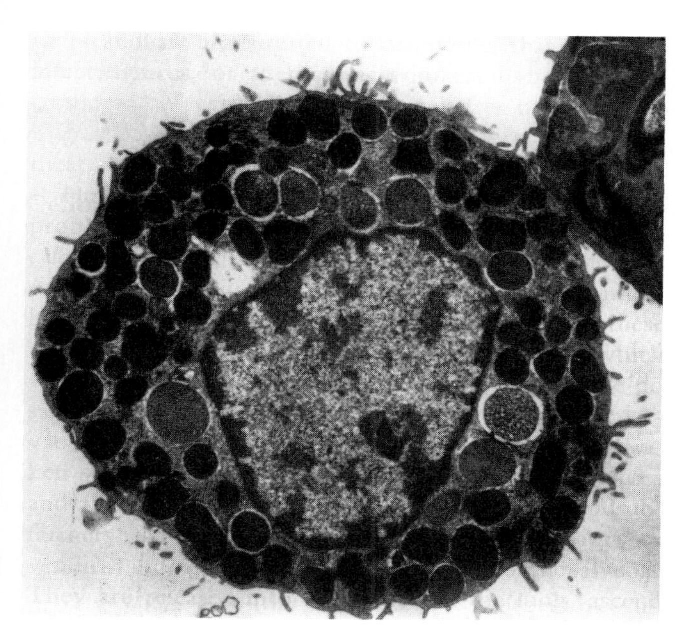

FIGURE 19.3 Electron micrograph of a slightly immature basophilic cell showing its numerous cytoplasmic granules. These granules are the source of histamine and other chemicals involved in immediate hypersensitivity reactions. (Courtesy of Dr. Ernesto O. Hoffman.)

mediate form of hypersensitivity. It is seen in persons who are sensitized to bee venom and certain drugs, and can cause death if preventive steps are not followed immediately.

Atopic allergies and anaphylaxis are mediated by IgEs bound to the surface of basophils in the circulation and mast cells, which are basophilic cells found in tissues (Figure 19.3). The heavy chains of the IgE bind to the cell surface and leave the antigen-binding sites free to act as antennas.

Reactions involving IgE, such as those that give rise to asthma, are obviously detrimental; but some reactions involving IgE are protective. For example, resistance to helminthic infections is associated with an IgE response. In parts of the world where large segments of the population are infected with helminths, the serum level of IgE is exceptionally high, and this apparently facilitates the destruction of the helminths. IgEs may also play a part in resistance to some forms of cancer. Some data are available suggesting that persons who manifest atopic àllergies have a greater chance of recovering from cancer following diagnosis and treatment than nonallergic people. The reason for these findings is not understood—but certainly the allergy sufferer will not readily by persuaded that an allergy is a desirable affliction.

ATOPIC ALLERGIES

Allergic asthma (az′mah) is marked by recurrent attacks of difficulty in breathing—called paroxysmal dyspnea—with wheezing, due to spasmodic contraction of the bronchi. Asthma begins when a sensitized individual inhales an allergen. The reaction between the allergen and the IgE-carrying mast cells in the lungs causes release of cellular substances that bring on the symptoms. **Hay fever** is an acute inflammation of the mucous membranes of the respiratory tract and the conjunctiva caused by an allergy to pollen. Hay fever is stimulated by airborne allergens that stimulate mucus secretions and cause inflammation of the conjunctiva. **Eczema** (ek′ze-mah) is a superficial inflammation of the epidermis characterized by dry itching skin, redness, and the appearance of minute vesicles, usually in folds or joints, for example behind the knees. All atopic allergies involve previous sensitization to the allergen and are dependent on IgE immunoglobulins.

Predisposition to Atopic Allergies

Children born of parents who are affected by atopic allergies have a significantly greater risk than the general population (one chance in two) of experiencing allergy (Figure 19.4). The reasons for this are not entirely clear; however, as we have noted, their serum IgE level is higher than that of normals. In the overall population, the allergic response peaks in persons 20 to 24 years of age, then slowly declines. Fortunately, many young patients outgrow the allergic state.

Prevention of atopic allergies is difficult; however, there is good evidence that breast-fed babies are less likely than bottle-fed babies to become asthmatic. This is one reason why pediatricians favor breast feeding, especially for newborns of parents who have a history of allergies.

Mechanism of the Allergic Response in Asthma

Initial exposure to an allergen stimulates B lymphocytes to synthesize IgE, though why some B lympho-

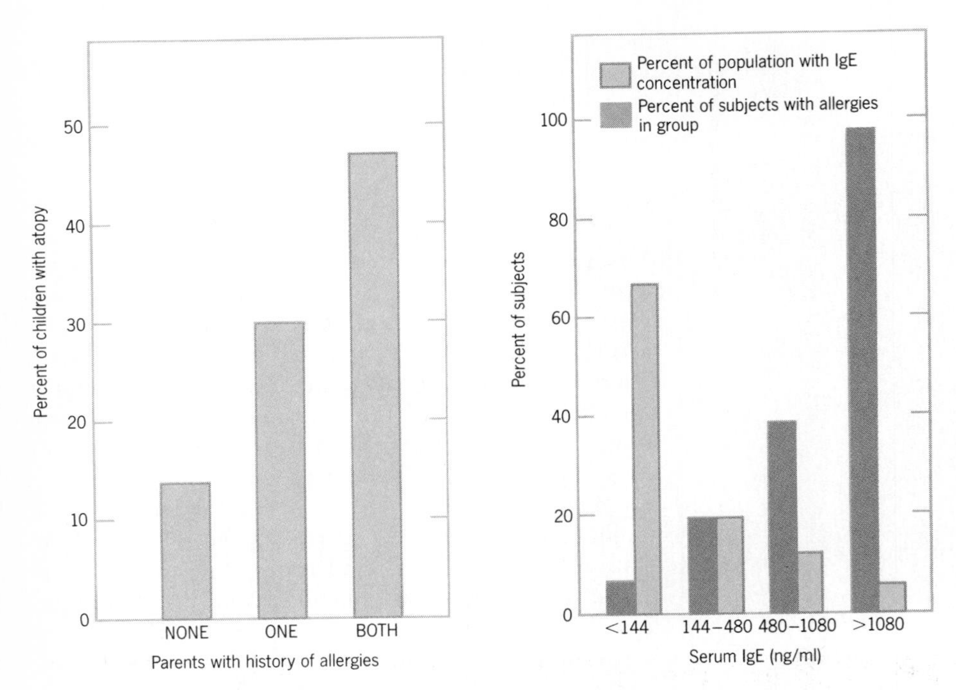

FIGURE 19.4 Predisposition to atopy is correlated with parental history of allergies and high serum concentrations of IgE. (With permission, from I. M. Roitt, *Essential Immunology*, 5th ed., Blackwell Scientific Publications Limited, Oxford, 1984.)

cytes make IgE instead of other immunoglobulins is not known. The IgEs are specific for the allergen, so genetically predisposed persons sensitized to cat dander will be allergic to cats, but not necessarily to tree pollen. The IgEs enter the circulation and find their way to mast cells in lung tissue. Each IgE has cytotrophic regions on its heavy chains through which it binds to mast cells and basophils (Figure 19.5). The surface of a sensitized mast cell is coated with between 100,000 and 500,000 IgE molecules. People are sensitive to the specific allergens that will react with the IgEs bound to the surface of their mast cells.

Asthma develops when the inhaled allergens come in contact with the sensitized mast cells in the bronchioles. The allergen binds with the IgE molecules, initiating a series of cytoplasmic changes that trigger the release of the substances in the basophil's cytoplasmic granules (Figure 19.5). These granules migrate to the membrane surface, fuse with the plasma membrane, and release their contents to the surroundings in the process of cell **degranulation.** Histamine, heparin, tryptase, eosinophil chemotactic factor, and platelet activating factor are released. Degranulation also activates the mast cell phospholipase A2, which releases arachidonic acid needed to begin the synthesis of leukotrienes and prostaglandins (Figure 19.2).

The immediate symptoms are caused by pharmacologically active substances that initiate bronchiole constriction and stimulate mucus secretion. The released histamine immediately stimulates smooth-muscle contraction and vasodilation. The effects of histamine and those of the slow-reacting substances (SRS) combine to cause the symptoms of asthma. The SRS act over extended periods and stimulate the contraction of bronchial smooth muscles (bronchoconstrictor). SRS are a composite of leukotrienes LTC_4, LTD_4, and LTE_4, which are part of the inflammatory response (Figure 19.2). These leukotrienes are 100 to 1000 times more potent than histamine or prostaglandins in their ability to constrict the smallest bronchial airways. Although the leukotrienes are the major pharmacological agents responsible for asthma, other agents are also involved.

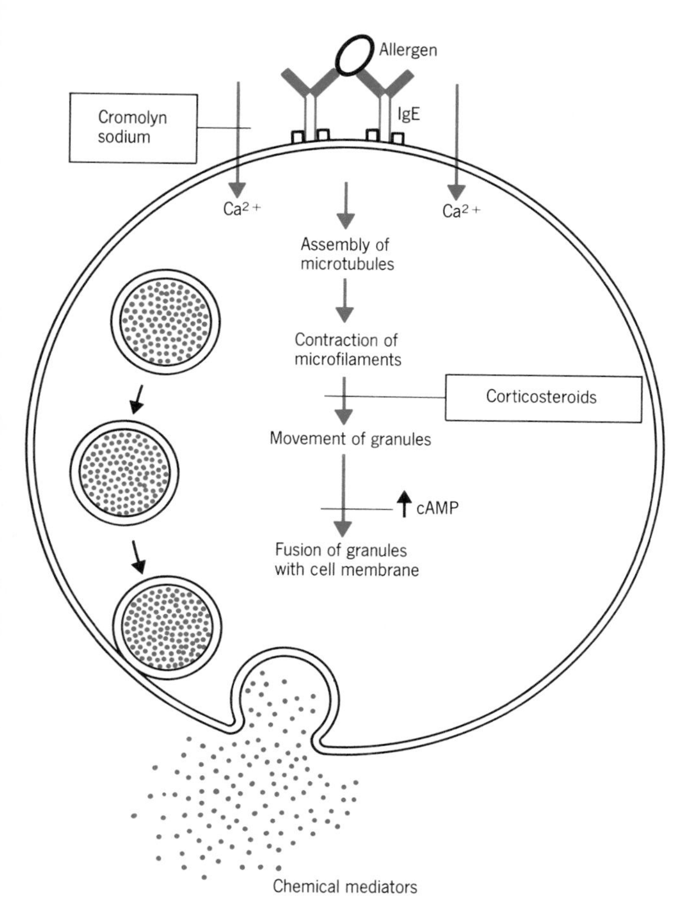

FIGURE 19.5 Degranulation of mast cells is triggered by the cross-linkage of two surface IgEs. This causes an influx of Ca^{2+} that initiates the movement of the granules to the cell membrane, where they fuse and degranulate. The sites of action of corticosteroids and cromolyn sodium are shown.

Treating Asthmatics

Current understanding of the physiological basis of asthma and the availability of specific drugs enables physicians to treat asthmatics. Allergen avoidance is the most effective approach to preventing allergies. Avoidance is possible with food and pet allergies, but it is almost impossible to avoid house-dust mites and pollens. Allergists treat the symptoms of asthma with drugs that inhibit mast cell degranulation and in severe cases with anti-inflammatory corticosteroids. **Hyposensitization,** the process of desensitizing a person against one or more allergens by injecting increasing doses of those antigens, is another medical approach to preventing asthma.

Hay Fever

Hay fever (allergic rhinitis) sufferers are usually sensitized to airborne pollens and plant spores that are seasonally released by plants during their reproductive cycles. These allergens elicit an immediate hypersensitivity response when they come in contact with the mucous membranes of sensitized persons, causing a runny nose, sneezing, lacrimation, and inflammation of the eyes. Antihistamines, which are not effective in treating asthma, help to relieve the symptoms of hay fever.

Food Allergies

Food allergies cause local reactions in the intestinal mucosa, manifested by abdominal pain and diarrhea. Some sufferers also develop hives after ingesting an offending food. Chocolate, cow's milk, nuts, wheat and wheat products, strawberries, and fish are common causes of distress.

Interestingly, aspirin prevents the intestinal symptoms of food allergies in certain patients. Allergic patients who take aspirin before ingesting the food experience reduced abdominal pain and do not produce increased levels of prostaglandins. Aspirin inhibits the pathway leading to the synthesis of prostaglandins (Figure 19.2).

ANAPHYLACTIC SHOCK

Anaphylactic shock, the most severe form of immediate hypersensitivity, develops when a significant quantity of allergen goes directly into the circulatory system of a sensitized person. It is rare, but is a potential problem in persons sensitized to insect venom (bees and wasps) or to various drugs. Anaphylactic shock can rapidly lead to death. What happens is that the allergic reaction releases large amounts of histamine throughout the body. Within minutes, the bronchioles constrict and the person begins to wheeze, and the smooth muscles of the intestinal tract and urinary bladder contract, causing involuntary urination and defecation. Finally the person is unable to breathe, and dies from asphyxiation.

Anaphylactic shock results from the explosive degranulation of mast cells, releasing histamine, heparin, eosinophil and neutrophil chemotactic substances, and platelet-activating factor. The mast cell phospholipase A2 is activated and begins the enzymatic release of arachidonic acid from phospholipids, thus initiating the synthesis of prostaglandins and leukotrienes.

Fortunately, anaphylactic shock is rare in humans; however, people do become sensitized to insect venom and drugs. Hyposensitization against insect venom is usually recommended for patients who react severely to bee or wasp stings. Anaphylactic shock is treated by the immediate administration of epinephrine and oxygen. As a precaution, patients being hyposensitized to atopic allergens or insect venoms should always receive their allergy injections in a medical facility where epinephrine and oxygen are available.

TYPE II HYPERSENSITIVITIES: CYTOTOXIC

Antibody-dependent cytotoxicity is an immune process for destroying infecting organisms, cancer cells, and virus-infected cells. Many cytotoxic mechanisms require the target cell to be sensitized with an antibody (IgG, IgM, or both) bound to a surface antigen. Opsonins are antibodies or other body components that render bacteria and other cells susceptible to phagocytosis. Natural killer cells attack virus-infected cells, and cells sensitized with bound surface antibodies are lysed by the complement cascade system. Antibody-dependent cytotoxicity is beneficial because it is a means of destroying bacteria, virus-infected cells, and probably tumor cells. As we shall see, however, it can be destructive in some situations.

BLOOD GROUPS

There are at least 15 genes that code for the erythrocyte surface antigens used to classify erythrocytes. The existence of these antigens is the basis of blood typing. The main blood groups are designated by letters—A, B, AB, and O. A, B, and AB indicate the presence of antigen A or B or both (Figure 19.6); type O indicates their absence. The ABO blood group antigens are glycolipid complexes present on the surface of red cell membranes. About 75 percent of persons who produce A or B also possess these glycolipids in their serum, urine, and saliva.

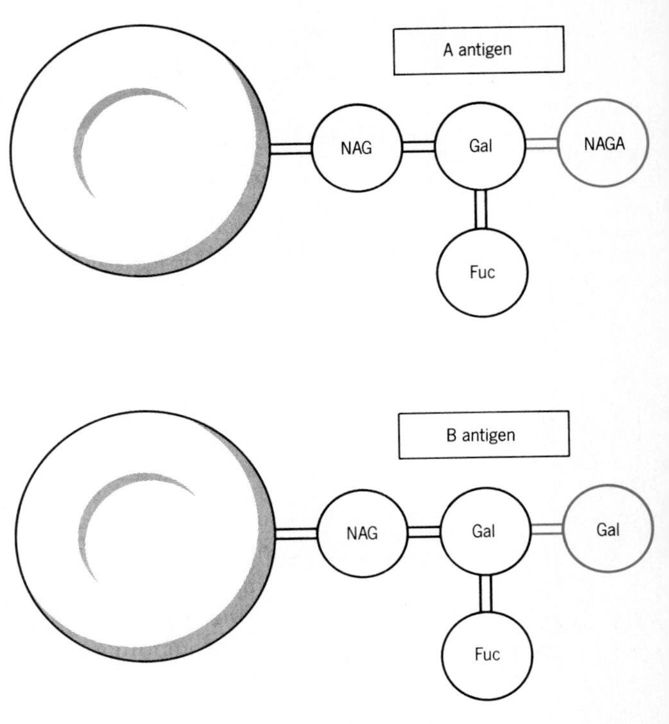

FIGURE 19.6 The A and B blood group antigens differ by a single carbohydrate residue on a glycoprotein, which is found on the red blood cell's surface. *N*-acetyl glucosamine (NAG), galactose (Gal), fucose (Fuc), *N*-acetyl galactosamine (NAGA).

Antibodies against the ABO blood group antigens are called isohemagglutinins. Persons with O blood have anti-A and anti-B isohemagglutinins, whereas those with type AB blood lack isohemagglutinins against A or B. Those with type B blood have anti-A, and those with type A blood have anti-B isohemagglutinins (Table 19.3). Both of them agglutinate red cells possessing the complementary antigen.

Presumably, the isohemagglutinins are formed upon exposure to blood group antigens that the person does not produce. For example, an O type person would recognize A and B antigens as foreign and would produce antibodies against them (Table 19.3). There are many natural sources of these antigens, including

TABLE 19.3 Human blood groups

BLOOD GROUP	FREQUENCY IN POPULATION (per hundred)	SERUM ISOHEMAGGLUTININS	ANTIGENS ON RBC
O	43	Anti-A; anti-B	Neither A nor B
A	39	Anti-B	A
B	13	Anti-A	B
AB	5	Neither anti-A nor anti-B	A and B

plants, bacteria, and other microorganisms. Presumably, isohemagglutinins are formed following normal exposure to naturally occurring A and B antigens.

Blood Transfusions

Often during surgery, or following a severe accident, a person requires a blood transfusion. Blood from his or her ABO type should be selected for the transfusion because an incompatible transfusion can cause massive hemagglutination with erythrocyte lysis. Symptoms of incompatibility include fever, hypotension, lower back pain, feeling of chest compression, nausea, vomiting, and ultimately shock.

Donor red cells are tested to determine their ABO group with anti-A and anti-B serum. (The Rh factor is also tested for; see below). Donor red cells are of primary importance because they carry the antigens that could react with the recipient's isohemagglutinins. The antibodies in the donated blood (0.5 liter) are unlikely to cause harm to the recipient because they are rapidly diluted when they enter the recipient's circulation (4.7 to 5.7 liters). Hence isohemagglutinins in donor blood do not react with recipient red cells to cause hemagglutination.

Blood type can be determined on a microscope slide or in a rapid tube agglutination test. Before transfusion, the donor's red cells are mixed with the recipient's serum. If hemagglutination occurs, the blood types are incompatible and the transfusion is not done. Screening of donor blood for the presence of antibodies to red cell antigens is also done to ensure that the donor hasn't been exposed to red cells other than his or her own.

Hemolytic Disease of the Newborn (Erythroblastosis Fetalis)

There is a second way in which sensitization to blood can occur, and it involves the Rh factor. Approximately 85 percent of the population is Rh-positive; that is, their red cells manifest the D antigen. Rh-negative persons, who number some 15 percent of the population, lack the D antigen and normally do not possess anti-D immunoglobulins.

When an Rh-negative woman and an Rh-positive man produce an Rh-positive child, the infant will produce the D antigen. The antigen then enters the maternal circulation when the placenta ruptures during the birth process (Figure 19.7), or (in some situations) during an abortion. The Rh-negative mother then manufactures anti-D antibodies that are able to cross the placenta. In a second or later pregnancy, she may have enough anti-D antibodies to overwhelm the Rh-positive fetus and cause its death.

In the past, hemolytic disease of the newborn, or **erythroblastosis fetalis,** was a dreaded disease. Today, however, it is common practice for the Rh-negative woman to be passively immunized with anti-D antibody, called RhoGAM, within 72 hours of giving birth. The anti-D antibody reacts with her child's D antigen and therefore she does not produce her own anti-D antibody. Since passive immunization is of short duration, subsequent pregnancies are similar to the first pregnancy and RhoGAM should be given again. Since 1969, when passive immunization with anti-D antibody was introduced, hemolytic disease of the newborn has become a rare occurrence.

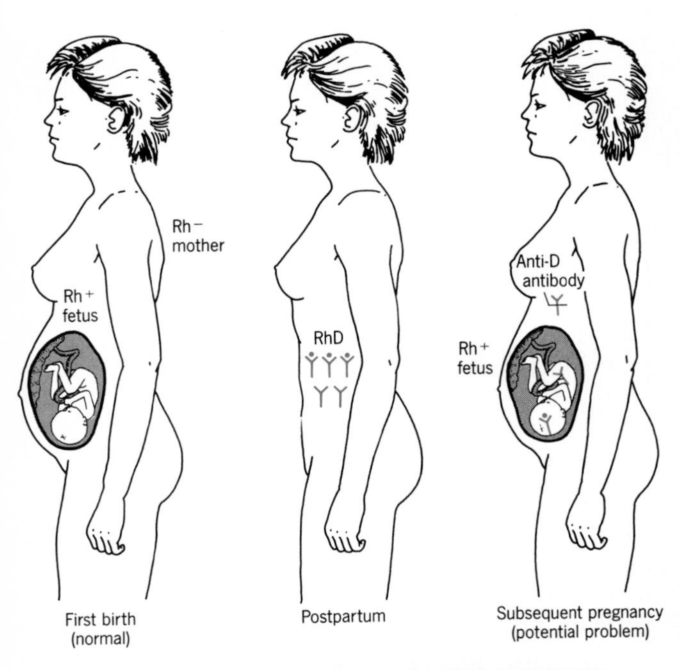

			MOTHER		
			Rh−	Rh+	
			dd	DD	Dd
FATHER	Rh+	DD	Rh+	Rh+	Rh+
		Dd	Rh+ / Rh−	Rh+	Rh+ / Rh−
	Rh−	dd	Rh−	Rh+ / Rh−	Rh−

FIGURE 19.7 Hemolytic disease of the newborn can occur if the mother is Rh-negative and the child is Rh-positive (shaded areas in chart). The mother makes antibodies against infant's D antigen, which enters the mother during birth or during an abortion. The mother's anti-D antibodies can cause fetal hemolytic disease during a subsequent pregnancy.

KEY POINT

To prevent red cell lysis by their isohemagglutinins, a person requiring a blood transfusion receives only blood of his or her own ABO and Rh type.

NATURAL KILLER (NK) CELLS

All of us have a small population of large granular lymphocytes, called **natural killer** (NK) cells, which are capable of destroying tumor cells, fetal cells, and virus-infected cells *in vitro*. They do this by binding to the target cell and causing it to lyse. Though the role of natural killer cells in resistance to cancer is still controversial, it is known that they have the ability to destroy leukemia, myeloma, and certain sarcoma cells. Their functioning is enhanced by interferon (see Chapter 16), which is synthesized by the target cells.

TYPE III HYPERSENSITIVITIES: IMMUNE COMPLEXES

Under normal conditions, precipitins react with antigens to form immune complexes that are destroyed by the reticuloendothelial system. Excessive antigen or antibody in a person can stimulate the formation of large quantities of immune complexes that cause type III hypersensitivity. Either a localized Arthus reaction or a systemic reaction, such as serum sickness, can result.

THE ARTHUS REACTION

A sensitized animal will experience a localized inflammation in the presence of excessive sensitizing antigen. This response is called the Arthus reaction, and it can be demonstrated by injecting the antigen into the skin of a sensitized animal. The circulating antibodies react with the antigen to form localized immune complexes (Figure 19.8) that bind complement, resulting in the release of C5a and C5-7 (recall the complement cascade, Chapter 18). Neutrophils, attracted to the site, trigger the inflammatory response. There is local swelling and redness owing to the release of vasodilators; and hemorrhaging is possible if the reaction progresses. In a severe case, the involved tissue may necrose and be sloughed off, as the hydrolytic enzymes present in the neutrophil lysosomes are released.

ACUTE SERUM SICKNESS

Immune complexes in the circulation cause the systemic symptoms of acute serum sickness, a type III hypersensitivity that can occur in a person who has been passively immunized with animal gamma globulin. Before toxoid vaccines were widely employed in im-

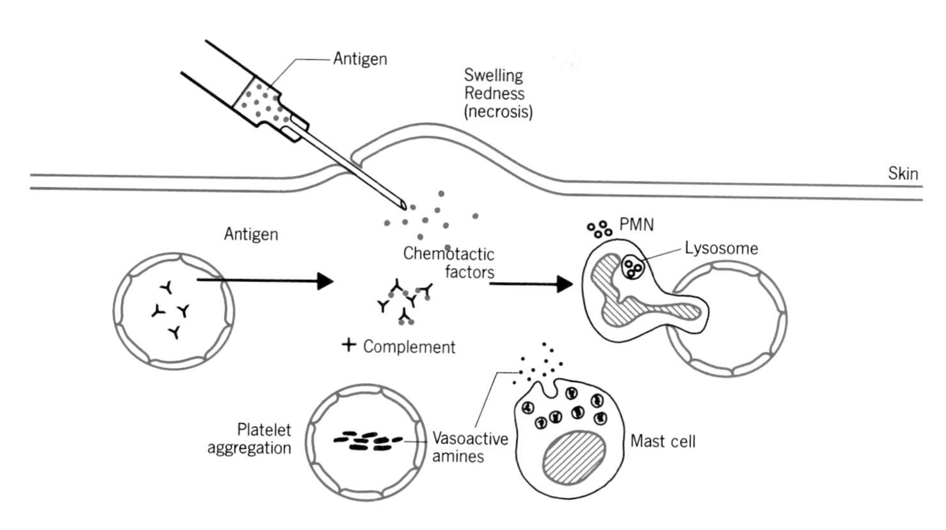

FIGURE 19.8 An Arthus reaction occurs when an antigen is injected into the tissue of a sensitized animal. The binding of complement to immune complexes in turn causes the characteristic swelling and redness. Necrosis may follow due to the release of enzymes from the lysosomes of polymorphonuclear neutrophils.

munization, a person who had been exposed to diphtheria or tetanus was passively immunized with antitoxins (gamma globulin) from domestic animals such as the horse. The person became sensitized to the horse's gamma globulin, and upon subsequent passive immunization with horse gamma globulin, developed immune complexes in his circulatory system.

These complexes tend to be deposited on blood vessel walls, where local complement attachment leads to the inflammatory response. The symptoms of acute serum sickness include fever, joint pain, and a rash, and are usually followed within a few days by spontaneous recovery. IgGs are the antibodies involved in an immune complex syndrome such as serum sickness.

IMMUNE COMPLEX PNEUMONITIS

A farmer exposed to the spores of microorganisms that multiply in damp hay may suffer a type III hypersensitivity known as **farmer's lung.** This is one type of immune complex pneumonitis; others include pigeon breeder's lungs and cheese washer's disease (their names are good descriptions). Farmer's lung is caused by the actinomycete, *Micropolyspora faeni,* which multiplies in wet, stored hay. Heat generated by the decomposition of the hay creates an environment favorable to the sporulation of *M. faeni.* When the farmer inhales the spores, his or her immune system is stimulated to develop a high titer of precipitating antibody. The antibodies react with the spores inside the alveolar walls, where they form immune complexes. Complement then binds to the immune complexes, neutrophils accumulate, and tissue damage occurs from the resulting inflammation. A farmer so sensitized will experience difficulty in breathing within 5 to 10 hours after being exposed.

KEY POINT

In the presence of excessive antigen, immune complexes may appear in serum and cause serum sickness; in tissue these complexes cause a local Arthus reaction. The complexes bind complement, which contributes to the inflammatory response that is characteristic of type III hypersensitivity reactions.

CELL-MEDIATED TYPE IV HYPERSENSITIVITIES

A cell-mediated (delayed) hypersensitivity reaction is characterized by a delayed response of between 24 and 48 hours following exposure to the antigen. This type of hypersensitivity is dependent on the presence of sensitized T lymphocytes. The sensitized state requires prior exposure to the antigen and can be transferred between individuals only by transferring lymphocytes.

The presence of intracellular parasites poses a special challenge to the immune system because it necessitates destruction of the infected cells without doing harm to the healthy ones. The body accomplishes this through a two-phase approach involving lymphokine-producing leukocytes and cytotoxic T lymphocytes, collectively known as cell-mediated immunity. A good example of this type of immunity is the body's response to *Mycobacterium tuberculosis,* the causative agent of tuberculosis, which multiplies intracellularly. Cell-mediated immunity protects against intracellular infectious agents; yet it also is responsible for the occurrence of some disorders, including contact dermatitis.

THE TUBERCULIN REACTION

Cell-mediated immunity reactions were first described in 1891 by Robert Koch. **Tuberculin,** an extract of *Mycobacterium tuberculosis,* causes a delayed hypersensitivity reaction when injected into the skin of a sensitized person. A local area of inflammation denoted by a red, firm (indurated) swelling develops approximately 24 hours later (Figure 19.9) in persons who have been exposed to *M. tuberculosis.*

Type IV hypersensitivity reactions involve sensitized T lymphocytes and not immunoglobulins. The delayed hypersensitivity to tuberculin can be passively transferred by injecting washed live white cells from an exposed person into a nonexposed one.

The sensitized T lymphocytes bind to the tuberculin after it becomes fixed to tissue cells at the site of injection. This reaction stimulates the T lymphocytes to divide and produce lymphokines that attract macrophages, basophils, and mast cells. The macrophages become fixed at the site and together with cytotoxic T cells participate in the destruction of the offending material. The mast cells and basophils release the vasoactive substances that cause the redness and swelling of the inflammatory response.

The tuberculin test is used extensively to screen for exposure to tuberculosis. In this test, tuberculin or its purified-protein derivative (PPD) is injected into the skin. A person who has been sensitized to tuberculin reacts by manifesting an induration (hard swelling) of 10 mm or greater at the site of injection 24 to 48 hours after exposure. This hypersensitivity means that the person has an active case of tuberculosis; has recovered

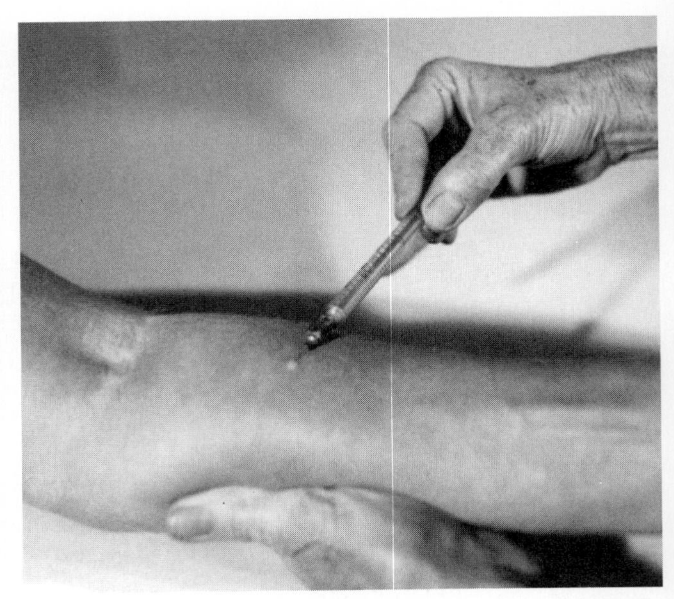

(a)

(b)

FIGURE 19.9 The tuberculin test is an example of delayed (type IV) hypersensitivity. *(a)* The antigen is injected intradermally and *(b)* a positive reaction results in a red, firm (indurated) swelling; it is read after 48 hours. (Courtesy of the Centers for Disease Control.)

from tuberculosis; or has been immunized with bacillus Calmette–Guérin (BCG) vaccine.

CONTACT DERMATITIS

Contact dermatitis results from skin contact with reactive chemicals that form complexes with cellular proteins. Chemicals known to elicit contact dermatitis include formaldehyde, picric acid, aniline dyes, plant resins (such as uroshiol [u-roo′she-ol] from poison ivy), organophosphates, and salts of certain metals. These chemical protein complexes are recognized as foreign and stimulate a cell-mediated immune response. Upon subsequent contact, there is local swelling and redness that reaches a maximum intensity 24 to 48 hours later. In a severe case large blisters filled with serous fluids and leukocytes form (Figure 19.10).

KEY POINT

Cell-mediated hypersensitivity reactions are characterized by a delayed response and dependence on sensitized T lymphocytes. These reactions play an important role in resistance against intracellular parasites, graft rejection, and contact dermatitis.

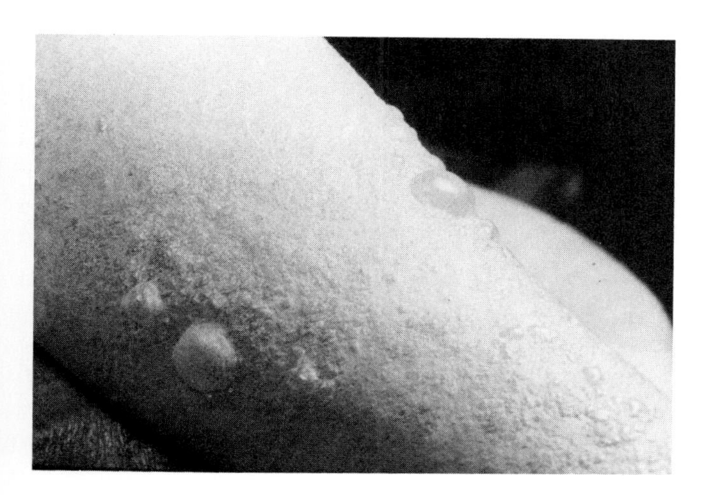

FIGURE 19.10 Acute allergic contact dermatitis is characterized by the appearance of large blisters filled with serous fluid and leukocytes. (Courtesy of David Bronstein, Academy Professional Information Services, New York.)

DESTRUCTION OF FOREIGN AND ABNORMAL CELLS

Advances in scientific knowledge, along with modern surgical techniques, have made it possible to transplant organs between unrelated persons. The heart, liver, kidney, bone marrow, and cornea have been transplanted, with varying degrees of success. The transplant may be an isograft, or it may be an allograft.

An **isograft** is a transplant between two genetically identical persons, such as identical twins; an **allograft** is a transplant between two genetically dissimilar persons. An isograft is generally accepted by the recipient, whereas an allograft is rejected by the recipient, unless both parties are histocompatible and the recipient is treated with drugs to suppress the immune response (called immunosuppressive drugs).

HISTOCOMPATIBILITY ANTIGENS

Within the cells of every person is a segment of DNA called the major histocompatibility complex (MHC), which encodes protein antigens that are inherited according to Mendelian genetics. The MHC-encoded antigens of Class I, Class II, and the ABO group antigens are involved in transplant rejection. Class I MHC-encoded antigens are detected by the use of specific sera that react with them. Class II antigens are found on lymphocytes, macrophages, epidermal cells, and sperm and are detected by means of a cellular assay system.

GRAFT REJECTION AND ACCEPTANCE

Allograft rejection is based on an immunological mechanism as was determined in experimental animals. The first kidney transplanted between unrelated animals is rejected after the graft survives for about one week in its new host. When a second kidney is transplanted to the same recipient, it will be rejected within one to two days. This accelerated rate of graft rejection is characteristic of a secondary immune response. Graft rejection is mediated by lymphocytes and antibodies specific for the tissue antigens of the grafted tissue. Matching the MHC-encoded antigens between donor and recipient and administering immunosuppressive drugs greatly increases the likelihood of allograft acceptance. **Cyclosporin A** is an important drug in this effort, as it appears to suppress the type of cell-mediated immunity that causes graft rejection.

MALIGNANT DISEASE AND IMMUNITY

Cancer results when eucaryotic cells grow unrestricted. Immunologists at one time hoped that a better understanding of the immune system would explain the basis of cancerous growths. The immune surveillance theory proposed that cancer cells develop in a persistent fashion and are eliminated by cell-mediated immune responses and natural killer cells. As long as the immune system is functioning properly, the cancer cells would be destroyed by the immune cytotoxic response. However, counter evidence was obtained in experiments involving some 15,000 nude mice* that lacked T-cell immunity. In this population, not a single mouse developed a tumor, suggesting that T-cell immunity is not involved in preventing tumors, at least in mice.

This is not to say that immunity plays no role in cancer prevention. Many tumor cells contain unique antigens on their surface, and humans respond immunologically to them. Patients born with immunodeficiencies and AIDS patients have unusual cancers or have a higher incidence of cancer than normal, indicating that immunological responses help in resistance to cancer. Researchers are attempting to understand the role of tumor antigens and to apply the information thus gained to control of tumor growth and spread.

*A furless breed of mice that lacks the thymus gland.

Summary Outline

TYPES OF HYPERSENSITIVITY

1. A hypersensitivity reaction is an immune-mediated exaggerated response to a foreign antigen that adversely affects the individual.
2. Hypersensitivity reactions are divided into four types: immediate, cytotoxic, immune complex, and delayed.

INFLAMMATION

1. The signs of inflammation—swelling, redness, heat, pain—are mediated by pharmacologically active substances that cause vasodilation and smooth muscle contraction.
2. Histamines, leukotrienes, prostaglandins, anaphylatoxins, lymphokines, and bradykinin stimulate increased vascular permeability and act as mediators of inflammation.

TYPE I HYPERSENSITIVITIES: ALLERGIES AND ANAPHYLAXIS

1. Atopic allergies are mediated by IgEs bound to the surface of basophils and mast cells.
2. Patients with asthma, hay fever, and ezcema have been sensitized to the allergens. Predisposition to asthma is genetically based, but can be influenced by environmental factors.
3. Asthmatics inhale the allergen that reacts with IgEs on the surface of mast cells in the lungs. The release of histamine and the production of SRS (LTC_4, LTD_4, and LTE_4) cause the symptoms of asthma.
4. Hay fever—allergic rhinitis—is a seasonal allergy caused by airborne pollens and plant spores. It gives rise to localized response in the mucous membranes of the eyes and nasal passages.
5. Allergies to certain foods cause abdominal pain, diarrhea, and/or hives.
6. Anaphylactic shock, the most severe type I hypersensitivity, causes the explosive degranulation of mast cells throughout the body following injection of an antigen into the circulatory system of a sensitized person.
7. Anaphylaxis occurs in persons sensitized to certain drugs and insect venom. It is treated with oxygen and epinephrine.

TYPE II HYPERSENSITIVITIES: CYTOTOXIC

1. Antibody-dependent cytotoxicity is an immune process for destroying foreign cells, cancer cells, and virus-infected cells.
2. The main blood groups are designated by letters—A, B, AB, and O. A, B, and AB indicate the presence of antigen A, B, or both. Persons with the D antigen on their cells are Rh positive.
3. Hemolytic disease of the newborn (erythroblastosis fetalis) occurs when an Rh^- mother carries an Rh^+ fetus. It is prevented by passively immunizing the mother with anti-D immunoglobulin (RhoGAM).

TYPE III HYPERSENSITIVITIES: IMMUNE COMPLEXES

1. Local injection of an antigen into a sensitized animal causes an Arthus reaction due to the formation of immune complexes in tissue. The ensuing inflammatory response results from the binding of complement to the immune complexes.
2. Immune complex pneumonitis is initiated by inhalation of an antigen. Farmer's lung and pigeon breeder's lung are examples.
3. Acute serum sickness occurs when large quantities of immune complexes form in the serum and bind complement after being deposited on blood vessel walls.

CELL-MEDIATED (TYPE IV) HYPERSENSITIVITIES

1. Cell-mediated hypersensitivity reactions are characterized by delayed onset and are dependent on sensitized T lymphocytes. Contact dermatitis and the tuberculin reaction are examples.
2. Cell-mediated immunity helps the host resist infection by intracellular parasites.
3. The tuberculin test is a cell-mediated immunity reaction that is diagnostic for exposure to *M. tuberculosis* or the BCG vaccine.

DESTRUCTION OF FOREIGN AND ABNORMAL CELLS

1. An allograft is rejected by a cell-mediated immune response and the second transplant to the same individual is rejected faster than the first.
2. Acceptance of an allograft depends on a match of the MHC-encoded antigens between donor and recipient, and the use of immunosuppressive drugs such as cyclosporin A.

Questions and Topics for Study and Discussion

QUESTIONS

1. Describe the role of mast cells, lymphocytes, and complement in the inflammatory response.
2. Define or explain the following:

allograft	histamine
anaphylaxis	hyposensitization
atopy	isograft
cell-mediated immunity	leukotrienes
contact dermatitis	prostaglandins
farmer's lung	Rh factor
hay fever	RhoGAM

3. Describe two forms of atopic allergies. What steps can be taken to prevent these diseases? What drugs are used to treat asthma and hay fever and how do they work?
4. How would you know if you are sensitized to bee stings?
5. How would you prove that a certain inflammatory response is a type IV hypersensitivity?
6. How does the tuberculin test work and why is it used to screen patients for tuberculosis?
7. What would happen if type A blood were transfused into a person with type B blood?

8. Under what circumstances does the D antigen elicit an immune response in a woman who has just given birth, and what are the consequences?

DISCUSSION TOPICS

1. Design a questionnaire to determine which members of your class have a predisposition to atopic allergies. Significant data could include present age, age of onset, type of allergies, whether breast- or bottle-fed as infants, allergens to which they are sensitive, and the treatment. Discuss the conclusions that might be drawn from this survey.

2. Scientists are now able to identify histocompatible tissues and successfully transplant organs. What ethical issues has this raised?

Further Readings

BOOKS

Barrett, J. T., *Basic Immunology and its Medical Application,* 2nd ed., Mosby, St. Louis, 1980. Introduction to the fundamentals of immunology with clinical applications of immunology explained in case histories.

Bryant, N. J., *An Introduction to Immunohematology,* 2nd ed., Saunders, Philadelphia, 1982. Good coverage of blood immunology as it applies to blood transfusion.

Hyde, R. M., and R. A. Patnode, *Immunology,* Reston, Reston, Va., 1978. Fundamentals of immunology presented in lecture note form. The extensive questions at the end of each chapter make it useful in self-study.

Roitt, I., *Essential Immunology,* 5th ed., Blackwell, Oxford, England, 1984. An introductory text.

Tizard, I. R., *Immunology: An Introduction,* Saunders, Philadelphia 1984. This concise text includes chapters on the four types of hypersensitivity.

ARTICLES AND REVIEWS

Buissert, P. D., "Allergy," *Sci. Am.,* 247:86–95 (August 1982). Recent advances that have led to a greater understanding of allergy.

Samuelsson, B., "Leukotrienes: Mediators of Immediate Hypersensitivity Reactions and Inflammation," *Science,* 220:568–575 (1983). The author reviews his work on leukotrienes for which he shared the Nobel Prize in Physiology or Medicine in 1982.

CHAPTER

20

Epidemiology

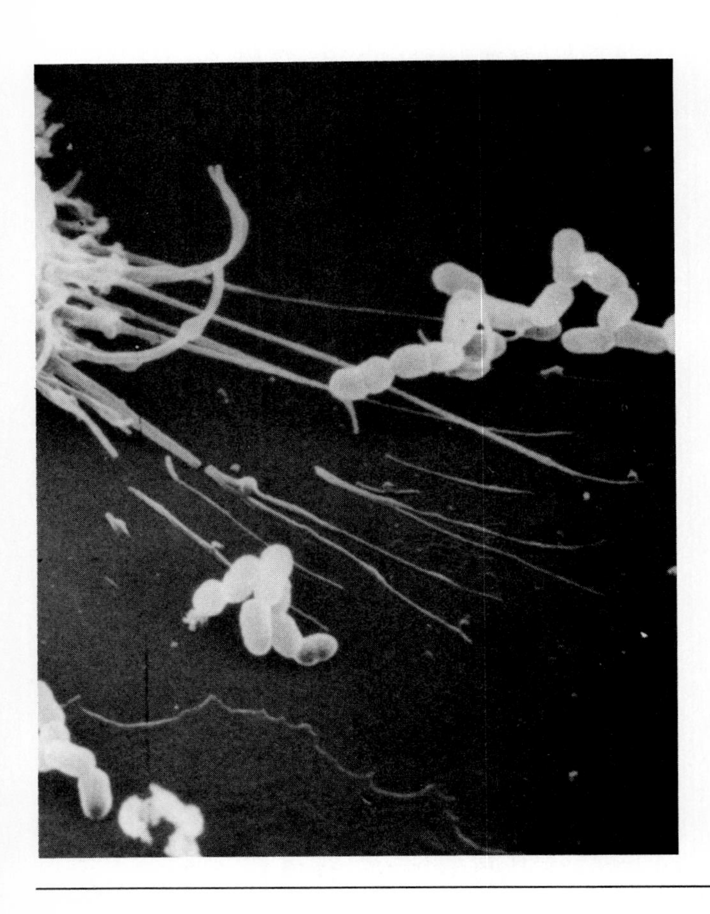

OUTLINE

FOCUS OF THIS CHAPTER

This chapter focuses on the nature of epidemics, on how infectious diseases are spread within a population of susceptible persons, and on the public health measures society uses to control the occurrence of infectious disease.

Epidemiology is the study of the factors that determine the occurrence of disease in a population. In its broad sense, epidemiology is concerned with the occurrence of all forms of illness within a population, including mental illness, accidents, alcoholism, and cancer and other organic diseases. As microbiologists, we are concerned with the spread of infectious diseases within populations of humans, animals, and plants. Infectious diseases are those caused by agents that can be transmitted from one susceptible host to another. Epidemiologists study the factors that influence the spread of infectious agents and devise strategies to prevent the illnesses they cause.

Epidemics of infectious diseases occur when there is a sudden outbreak of illness that affects many persons. An infectious **endemic** disease is an illness in a localized population that occurs at a low but persistent rate. Many endemic diseases are contained within the habitat range of the reservoir(s) and vector(s) that harbor and transmit the infectious agent.

Though twentieth century science has made great advances in the treatment and prevention of disease, infectious diseases continue to cause widespread illness. Major epidemics of childhood diseases occur yearly; and in many geographic areas endemic diseases persist. The story of earlier epidemics is one we need to heed so we can apply this lore to current problems.

HISTORICAL PERSPECTIVE

During these times there was a pestilence, by which the whole human race came near to being annihilated . . .
Procopius of Caesarea (written about 550 A.D.)

Infectious diseases have altered the course of civilization by depopulating entire countries. Pestilences that occurred in China and Egypt between 1500 and 1000 B.C. were remembered through the centuries because of the suffering and death they brought, and with it the disruption to the countries' political and social life. Centuries later, toward the end of the Parthean War, 161 to 166 A.D., the Roman Empire suffered a major epidemic (probably smallpox). So many Romans died that Emperor Marcus Aurelius was forced to recruit conquered peoples to serve in his armies and to settle his lands. The Roman Empire was again devastated by an epidemic (probably plague) that began in 251 A.D. and lasted until 266 A.D. So terrible was this epidemic that Alexandria, the most populous city of the Roman Empire, lost two-thirds of its population.

The 52-year plague (referred to in the opening quotation) that almost annihilated the human race was documented by Procopius of Caesarea, the historian for Justinian (the Byzantine emperor who ruled from 527 to 565 A.D.). Procopius accurately described the clinical symptoms of plague, reporting that those who became ill usually died within the week. There was no cure, and only rarely did a person survive. The plague so ravaged the Byzantine Empire that its social structure never recovered. Leaders became powerless as they lost the work force required to maintain their empires. There were no farmers to cultivate the land, no produce to be taxed, and no armies to defend the outlying territories. The disintegration of the Byzantine Empire marked the beginning of the Middle Ages and the end of the glory and magnificence that was the Graeco–Roman culture.

Some years later came the fabled Black Death—The Plague. This devastating epidemic began in Central Asia and by the 1340s had spread to Italy (Figure 20.1). One account of this period tells of the siege to the fortified cathedral town of Kaffa (now Feodosiya in the Ukraine) by a rival army. When the plague struck, the attacking army quickly sickened and began to die in great numbers. So many soldiers died that the disposal of bodies posed a serious problem; to deal with it, the commanding officer of the attacking army ordered the soldiers' bodies to be catapulted by hurling machines into the fortified city. The plague rendered Kaffa uninhabitable, and the survivors fled in their galleys into the Mediterranean Sea. In their search for refuge, they carried the plague to many port cities, and soon the disease spread throughout Europe.

FIGURE 20.1 Bubonic plague terrorized people of the seventeenth century. Sammlung Brettauer made this etching of the devastation of the plague.

Colonial America battled epidemics during much of its history. By some accounts, fewer than 40 percent of the early settlers survived the Atlantic passage and the hardships that accompanied the colonizing of North America. The population of the colonies grew very slowly in large part because of infectious diseases. American Indians suffered and died along with the colonists, and American Indian tribes were ravaged by the infectious diseases brought by the Europeans.

Diphtheria and yellow fever were rampant among the colonists. Diphtheria caused many deaths during the middle 1700s, especially among children under 10 years of age; since no cure was known, it was not unusual for a large family to lose four or five children to diphtheria in a single year. Epidemics of yellow fever occurred in colonial port cities, especially in those located south of Philadelphia. Yellow fever was carried from the tropics to the northern port cities by ships plying the West Indies trade. Its spread was curtailed only when the colonial authorities began to quarantine all ships arriving from the West Indies.

JOHN SNOW AND CHOLERA

Out of this tragic history of suffering, death, and social upheaval came epidemiology. John Snow (1813–1858), with his famed report on cholera, was the first person to publish a paper that presented a logical approach for controlling epidemics. Snow, an English physician, studied the effects of cholera in a section of London that was served by two separate water companies, both of which obtained water from the Thames River. The Lambeth Company took water from the river upstream of London; the Southwark and Vauxhall Company took water from the populated London section of the river. Snow discovered that the number of deaths from cholera in houses supplied with water from the Southwark and Vauxhall Company was significantly higher than that in houses supplied by the Lambeth Company. He concluded that cholera was carried in the water supply that originated in the sewage-polluted basin of the Thames River. He also proposed that cholera was caused by a microorganism—well before Robert Koch proved the infectious nature of anthrax.

John Snow followed a similar line of reasoning to solve the mystery of the Broad Street cholera epidemic of 1854. The population living within walking distance of a water pump on Broadstreet suffered a very high incidence of cholera (Figure 20.2). Snow surmised that the well was the source of the cholera epidemic. He removed the pump handle from the well to prevent the well water from being used—and this ended the epidemic. These early studies led to **epidemiology.**

RECENTLY DESCRIBED INFECTIOUS DISEASES

Modern societies still must contend with newly described diseases caused by unknown infectious agents. Two examples will suffice to make the point.

Since 1970, two critical infectious diseases have appeared in the United States. The first one made its appearance in Philadelphia in the summer of 1976, among American Legionnaires attending a convention. The culprit—which at the time could not be identified—caused a serious respiratory infection that hospitalized many Legionnaires, and was fatal to a number of them. Enormous effort was brought to bear on the situation, involving health agencies at the city, state, and federal levels. Eventually (January 1977) the causative bacterium was isolated and named *Legionella pneumophila*—the bacterial cause of what has become known as Legionnaires' disease.

The story of AIDS, our second example, is as tantalizing as any mystery, but is too lengthy to recount

FIGURE 20.2 Asiatic cholera and the Broad Street pump. (From J. Snow, *Snow on Cholera,* The Commonwealth Fund, New York, 1936.)

here in detail. Acquired immunodeficiency syndrome, or AIDS as it is called, was initially diagnosed in the United States in 1979. Since that time, diagnosed cases of AIDS have increased alarmingly fast (Figure 20.3). The victims are predominantly homosexuals or bisexuals with many sex partners (72 percent of patients). Intravenous drug abusers account for 17 percent; Haitian entrants to the United States for 5 percent, and hemophiliacs for 0.8 percent.

These data and numerous investigations have led health authorities to conclude that AIDS is caused by a virus transmitted by semen and other body fluids between sexual partners and by contaminated needles. AIDS is the foremost epidemiological problem in the

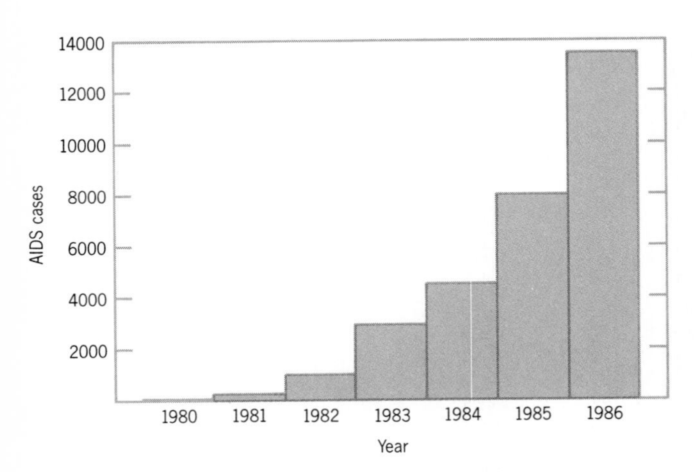

FIGURE 20.3 Reported cases of acquired immunodeficiency syndrome (AIDS) in the United States through December 1986.

United States because it is fatal in almost all cases; and perhaps even more significantly, no cure or effective treatment is at hand (as of the publication of this book).

KEY POINT

The ability to limit the devastating effects of infectious diseases through immunizations, antimicrobial agents, and modern microbiological practice has been available for only a short segment of human history. The presence of AIDS demonstrates that there remain serious threats to the human population from infectious diseases.

EPIDEMIOLOGY OF INFECTIOUS DISEASES

Comprehension of the etiological basis of infectious diseases, their progressive effects, and their mode of transmission constitute the heart of epidemiology. Control of infectious diseases is exercised in hospitals and health care facilities, in schools, at the county level through local and state health departments, and nationally through the Centers for Disease Control (CDC). The World Health Organization (WHO) is heavily involved in research and programs to contain infectious diseases. Through the work of these organizations, morbidity and mortality owing to infectious diseases have been sharply reduced.

Morbidity, the measure of the incidence of all diseases, both fatal and nonfatal, in a population, tells us much about the health of that population—its diseases, the number of workdays lost to illness, the nature of the illnesses, the number of persons affected, and so on.

Mortality is the proportion of deaths within the population attributable to a specific cause. In the United States, cancer, heart disease, and stroke, along with several other illnesses, are more frequent causes of death than infectious diseases. Although many people die from infectious diseases, a number of infectious diseases are preventable through the widespread use of immunization procedures and the use of suitable drugs.

THE NATURE OF EPIDEMICS

An epidemic occurs when a significant proportion of a given population contracts an infectious disease. The baseline on which the epidemic is determined varies according to the specific disease. Figure 20.4 presents an example related to pneumonia. Between 400 and 650 deaths from pneumonia associated with influenza occur annually in 121 large U.S. cities (Figure 20.4). Only when the mortality exceeds these expected levels is influenza considered to be an epidemic.

In an epidemic there must be an origin and a mode of transmission. We need to learn where and how the disease begins and how it is carried from one person to another. Diseases transmitted from one person to another often have their onset during winter months, when people tend to stay indoors. Person-to-person transmission is characteristic of a propagated epidemic. The incidence of disease in a **propagated epidemic** (Figure 20.5) increases gradually and then falls as the patients either recover or die. **Common-source epidemics** originate from a single comtaminated source, have a rapid onset, quickly reach a maximum incidence of new cases, and then decline (Figure 20.5). This type of epidemic can be caused by infectious agents found in contaminated swimming water or in food and water consumed by a large number of people. Common-source epidemics can be prevented by eliminating the source of contamination, but propagated epidemics are far more difficult to control.

An endemic disease is, as we have noted, one that is present in a population at a low but persistent rate. For example, flea-borne typhus fever (Figure 20.6) is caused by a bacterium whose natural reservoir is wild rats and field mice. It is transmitted sporadically to hu-

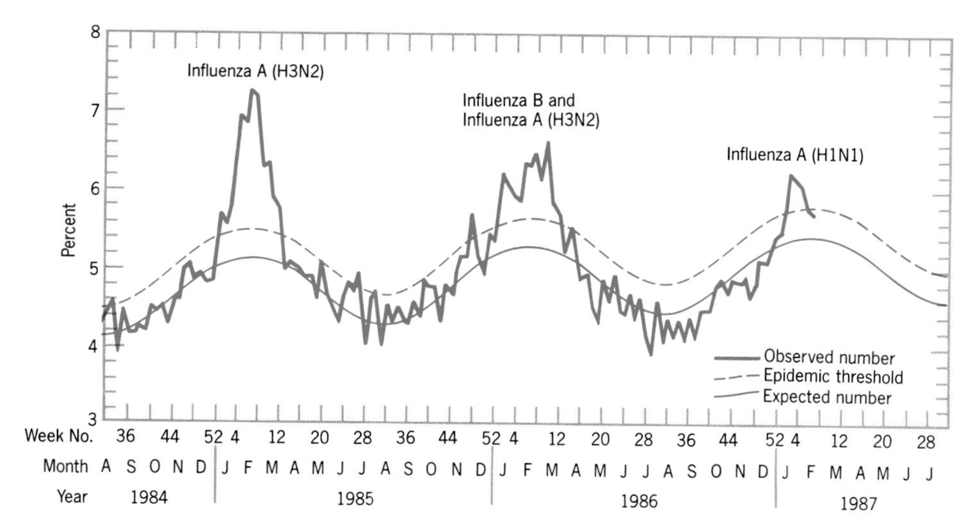

FIGURE 20.4 Reported deaths in 121 U.S. cities from pneumonia–influenza, 1984–February 1987. There is an expected number of cases at any given time of the year in the U.S. population. Epidemiologists know an influenza epidemic is in progress when the number of deaths from pneumonia–influenza exceed the epidemic threshold. (Source: *Morbidity and Mortality Weekly Report.*)

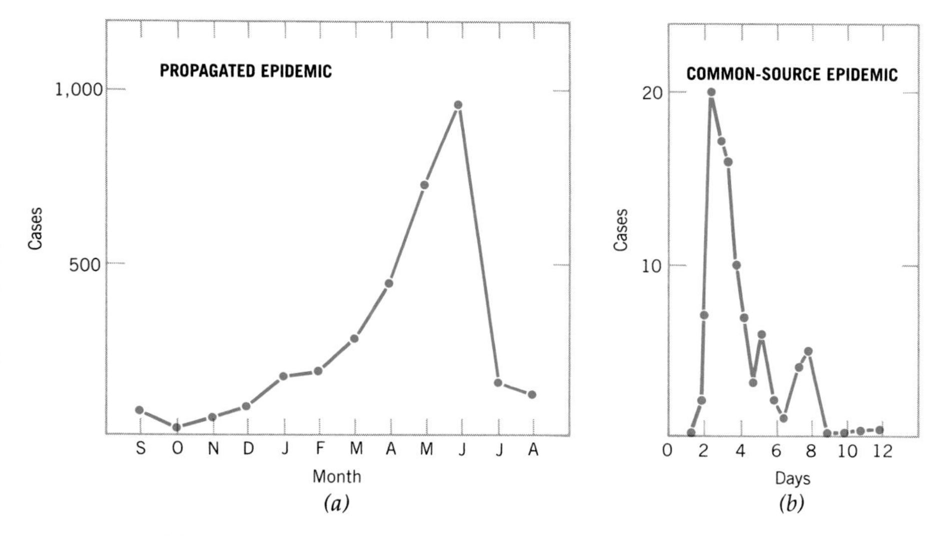

FIGURE 20.5 *(a)* A propagated epidemic of German measles (rubella) in Michigan (1978). Only a few measles cases occurred during the summer months while children were out of school. There was a gradual increase in the number of cases during the winter and then an abrupt decline as the summer recess began. These are characteristics of a propagated epidemic involving person-to-person transmission. *(b)* A common-source epidemic caused by *Vibrio parahaemolyticus* aboard a cruise ship. Ninety-one of the 1059 persons on board were afflicted by gastroenteritis traced to a seafood salad. This type of epidemic is characterized by a rapid increase in the number of cases. (Centers for Disease Control, "Annual Summary 1979: Reported Morbidity and Morality in the United States," *Morbidity and Mortality Weekly Report,* Vol. 28, No. 54, 1980.)

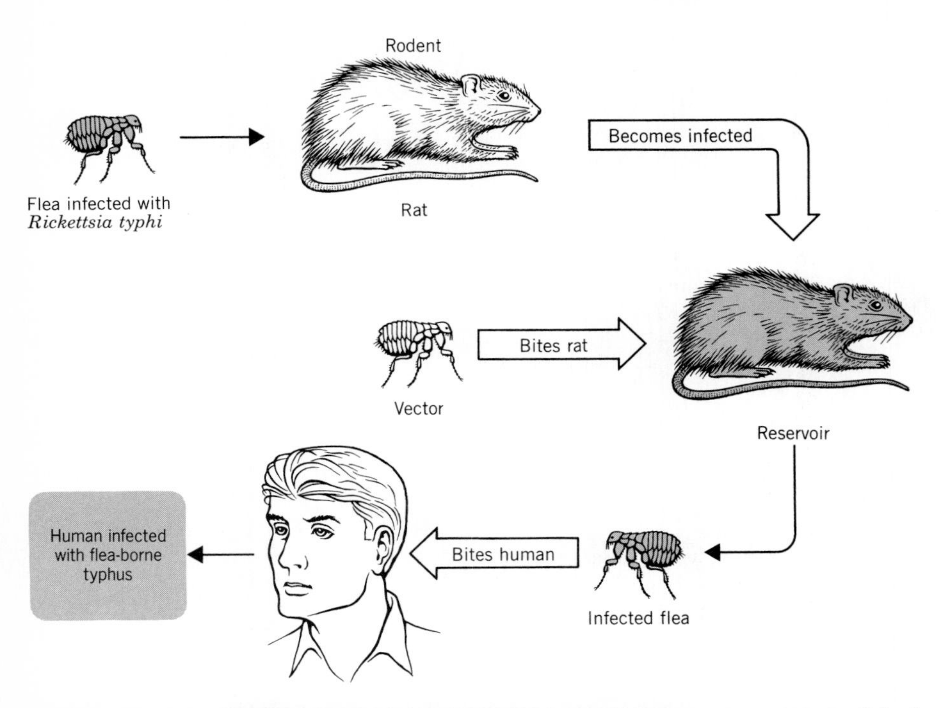

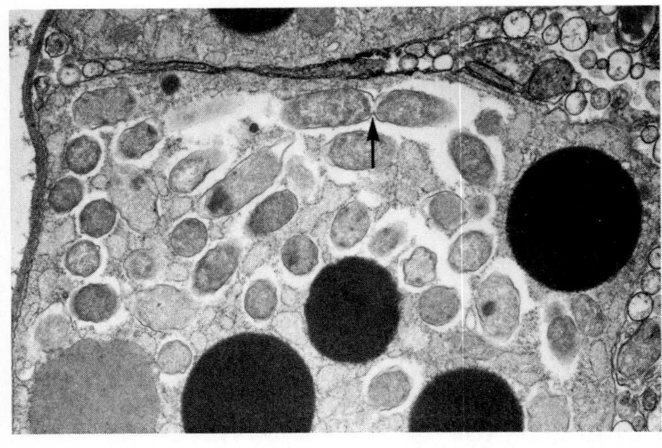

FIGURE 20.6 *Rickettsia typhi* lives in rodent reservoirs until fleas transfer this bacterium to humans. *Rickettsia typhi* causes flea-borne typhus in humans. *Rickettsia* in salivary gland of *D. andersoni*. (Courtesy of Dr. W. Burgdorfer, Rocky Mountain Laboratories, Hamilton, Mont.)

mans by the bite of fleas, while its natural reservoir provides a constant source of the infectious agent. Flea-borne typhus fever is seen in the susceptible population in whom natural immunity has not developed. It is geographically restricted because its natural reservoir is restricted (Figure 20.6).

A **pandemic** disease is one that spreads throughout the world, causing extensive morbidity. Influenza can be (and has been, at times) pandemic, because it is easily transmitted between persons and the virus has the ability to alter its antigenic makeup. New antigenic types of influenza viruses arise that are able to bypass the immunological resistance that developed in infected persons during earlier epidemics. Thus influenza remains a serious viral disease worldwide, with high morbidity and often mortality.

FACTORS INFLUENCING EPIDEMICS

Climate, seasons, geography, and social practices are factors that affect the occurrence of infectious disease within a population of susceptible individuals.

Vectors Transmit Infectious Agents

A living agent that is capable of transferring a pathogen from one person to another is a **vector.** Ticks, lice,

TABLE 20.1 Selected arthropod-borne diseases of humans

Viral Diseases	Animals Affected	Reservoir	Vector
St. Louis encephalitis	Humans	Perching birds	Mosquito (*Culex* spp.)
Western equine encephalitis	Horses and humans	Wild birds	Mosquito (*Culex* and *Culiseta* spp.)
Venezuelan equine encephalitis	Horses (rare in humans)	Rodents and horses	Mosquito (*Culex* spp.)
Bacterial Diseases	**Etiological Agents**	**Reservoir**	**Vector**
Tick-borne typhus fever (Rocky Mountain spotted fever)	*Rickettsia rickettsii*	Rodents, dogs, and foxes	Wood tick (*Dermacentor* spp.)
Flea-borne typhus fever	*Rickettsia typhi*	Rats and field mice	Rat flea (*Xenopsylla cheopis*) Rat louse (*Polyplax spinulosa*)
Bubonic plague	*Yersinia pestis*	Rats and ground squirrels	Flea (*Xenopsylla cheopis*)
Relapsing fever	*Borrelia recurrentis*	Rodents, ticks, humans	Louse (*Pediculus humanus*), tick (*Ornithodoros*)
Protozoan Diseases	**Etiological Agents**	**Reservoir**	**Vector**
Malaria	*Plasmodium* spp.	Humans, monkeys	Mosquito (*Anopheles* spp.)
African sleeping sickness	*Trypanosoma brucei*	Humans, antelope, hogs	Tsetse fly (*Glossina morsitans*)

fleas, and mosquitoes are arthropod vectors (Table 20.1) that transfer specific viral and bacterial diseases.

Mosquitoes transfer encephalitis viruses from animal and bird reservoirs to humans. The mosquitoes pick up the infectious agent as they take a blood meal from an infected reservoir (Table 20.1), and transfer the virus to other reservoirs and/or humans. Rickettsias, obligate intracellular parasitic bacteria (see Chapter 14), are transferred between reservoir and human host by the arthropod vectors.

Seasonal death of the vector due to cold or desiccation before it transfers the pathogen to a person or other reservoir usually results in the death of the pathogen. In some vector–parasite relationships, however, the pathogen can be passed to the vector's progeny by egg infection. For example, *Rickettsia rickettsii,* the pathogen responsible for tick-borne typhus fever (Rocky Mountain spotted fever) is transmitted from one generation of ticks to the next by **transovarian** (across the ovum) **passage.**

Reservoirs of Infectious Agents

Animals that harbor pathogens without being adversely affected are **reservoirs.** The reservoirs of the arthropod-borne diseases (Table 20.1) are animals with which the pathogen can coexist. Perching birds, rodents, and horses are reservoirs of the encephalitis viruses. Rickettsial disease and nonpulmonary bubonic plague are arthropod-borne bacterial diseases that are maintained in animal reservoirs. Rats, ground squirrels, dogs, foxes, and humans are reservoirs for these arthropod-borne bacterial diseases. In most cases, the pathogen replicates in the vector or reservoir without causing damage to this intermediate host. If a specific disease is not passed between humans, the human is said to be a definitive host.

Geographic Considerations

The survival of many infectious agents depends on the climate (temperature, humidity, rainfall) and the local fauna of a region. Arthropod vectors live in distinct geographic areas, thus the incidence of arthropod-borne diseases is directly related to the boundaries of their reservoirs and vectors. For example, Rocky Mountain spotted fever has a restricted distribution (Figure 20.7) because its transmission depends on the bite of an infected tick. *Rickettsia rickettsii* is carried by the wood tick in the Rocky Mountain states and by the

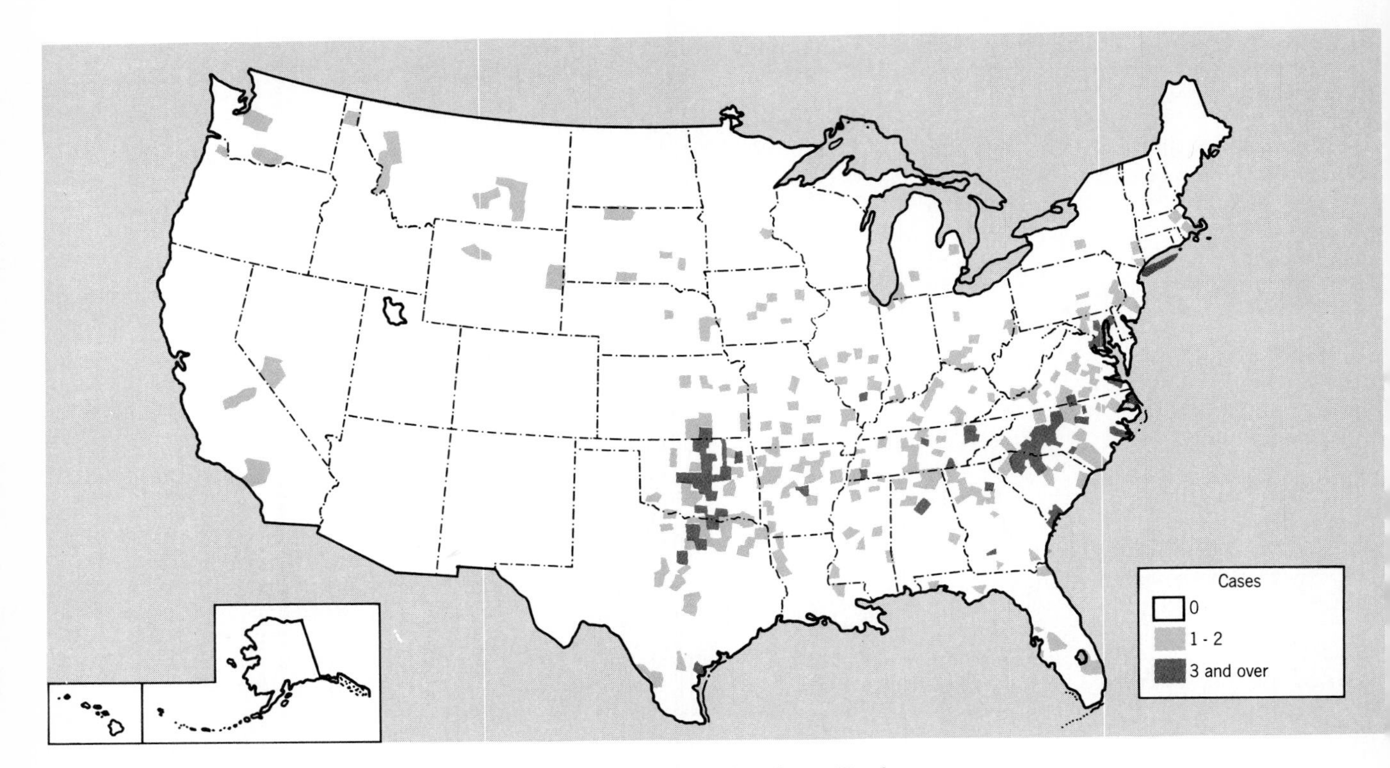

FIGURE 20.7 Incidence and distribution of tick-borne typhus fever (Rocky Mountain spotted fever) in the United States, reported by county in 1984. Of the total reported cases, 47 percent were reported from the South Atlantic states. More cases than usual (24 percent) were reported from the west central states. (Centers for Disease Control, "Annual Summary 1984: Reported Morbidity and Mortality in the United States," *Morbidity and Mortality Weekly Report,* Vol. 32, No. 54, 1986.)

dog tick in the eastern United States. The disease is most frequent in the east central United States, though in 1983 more cases than usual developed in the west and south central United States. Very few cases are reported in the northern, western, or Rocky Mountain states (Figure 20.7).

Seasonal variations in climate also restrict the distribution of arthropod-borne diseases. Many of the viruses that cause human encephalitis, for example, are transmitted by mosquitoes that hatch during the summer months, so the incidence of encephalitis peaks between July and September (Figure 20.8). The last major epidemic of St. Louis equine encephalitis, in which 429 cases were reported in Illinois, 416 in Ohio, and 290 in Indiana, peaked in September 1975, and declined during October and November. The normal decline in encephalitis cases coincides with the first frost, which kills mosquitoes.

The viruses of herpes simplex, varicella zoster (chickenpox), mumps, and measles can cause encephalitis even though they are not anthropod-borne agents. These viruses contribute to the cases of encephalitis seen during the remaining months of the year (Figure 20.8).

Some fungal diseases are prevalent in distinct climatic regions. Coccidioidomycosis (kok-sid-e-oi'do-mi-ko'sis) is a fungal disease caused by *Coccidioides immitis,* which is found in the soil of arid regions of the southwestern United States and Mexico. Human infections result from the inhalation of airborne fungal spores prevalent in dry desert regions and can progress to serious systemic infections.

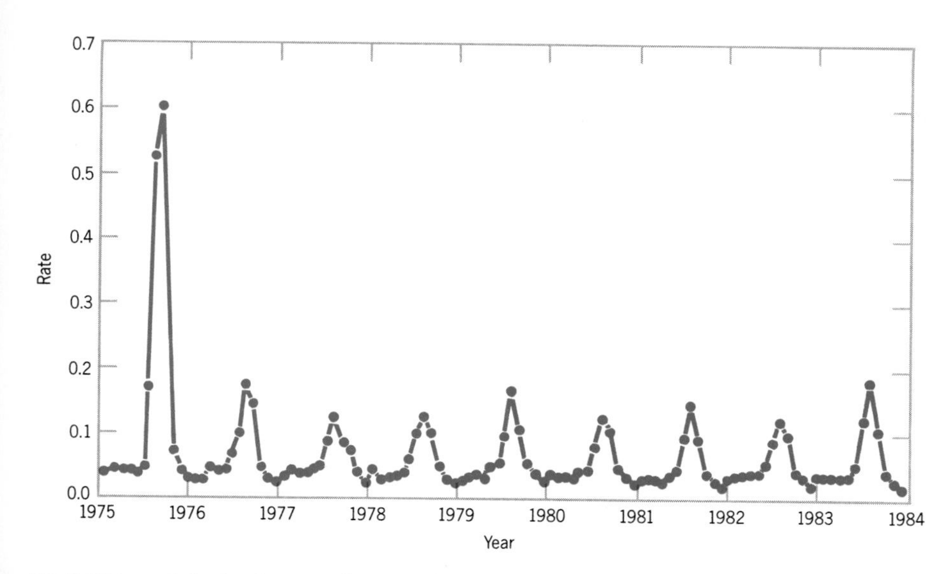

FIGURE 20.8 Reported cases of encephalitis per 100,000 population by month of onset in the United States, 1975–1984. Epidemics occurred in the summer months and coincided with the peak of the mosquito season. A major outbreak of St. Louis encephalitis occurred in 1975. (Centers of Disease Control, "Annual Summary 1984: Reported Morbidity and Mortality in the United States," *Morbidity and Mortality Weekly Report,* Vol. 32, No. 54, 1986.)

Social Practices That Contribute to Disease Transmission

Some of our everyday practices can facilitate transmission of disease. Even so simple an act as shaking hands in greeting readily transfers bacteria and viruses from one person to another. Sexual promiscuity increases the risk of contracting sexually transmitted diseases (Figure 20.9). The congregation of children in day-care centers and schools promotes infectious diseases, as seen with chickenpox, which annually reaches epidemic proportions in the late winter and early spring (Figure 20.10), when children are in school.

An interesting case in point is kuru, a viral disease that at one time was perpetuated by an isolated social practice. Kuru, a degenerative disease of the central nervous system, causes death four to five years after infection. It was discovered among the young males and the women of a cannibalistic New Guinea tribe that followed the custom of consuming their dead tribesmen. The adult male members were honored with the choicer portions of muscle, while the women and the young males were given the entrails and the brains. Kuru is caused by an agent that infects the brain—hence the disease was prevalent in the women and young males of the tribe. When Western doctors discovered how kuru was transmitted, they convinced the tribe to cease their cannibalistic customs and thus brought an end to the transmission of kuru.

KEY POINT

Infectious agents that depend on a reservoir or vector for survival are restricted to those geographic regions of the world where the vector and/or reservoir exist. Climate and seasonal variations affect the survival of the infectious agents, reservoirs, and vectors. Social practices influence human behaviors that can cause changes in the incidence of, or susceptibility to, infectious diseases.

DISEASE TRANSMISSION

Epidemics occur only when there is a significant population of susceptible persons and a means of transferring the infectious agent among them. There are five mechanisms by which transfer can occur: (1) person-

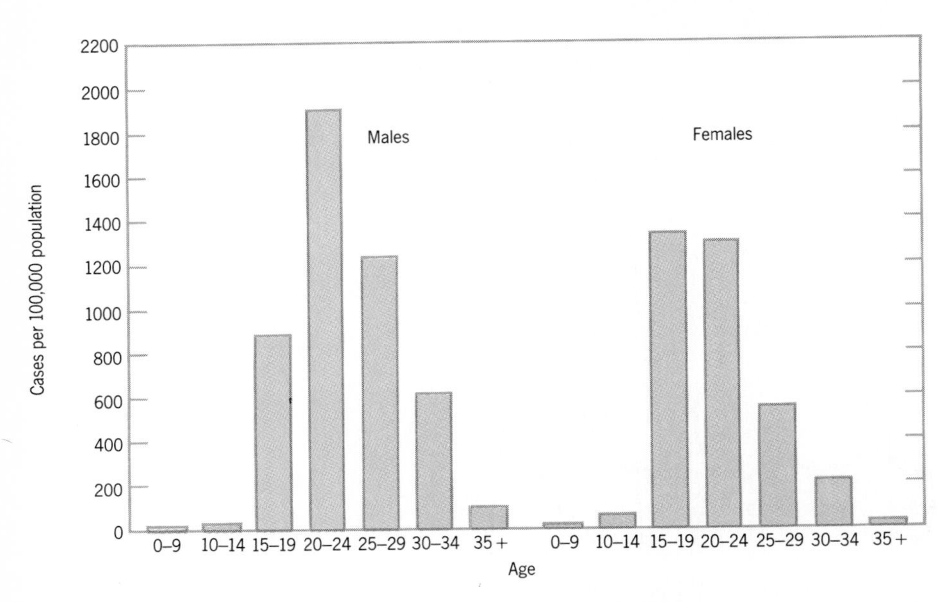

FIGURE 20.9 Cases of gonorrhea per 100,000 population reported by age and sex in the United States in 1983. The incidence of sexually transmitted diseases is related to the promiscuity of the population. The substantially higher morbidity for younger females places them at higher risk for sequelae of gonococcal infections such as pelvic inflammatory disease and infertility. (Centers for Disease Control, "Annual Summary 1983: Reported Morbidity and Mortality in the United States," *Morbidity and Mortality Weekly Report,* Vol. 32, No. 54, 1985.)

to-person, (2) airborne, (3) foodborne and waterborne, (4) arthropod vectors, and (5) zoonoses. Since most infectious agents are transferred by only one of these mechanisms, epidemiologists can control the spread of infectious disease by interrupting its transfer.

Person-to-Person

Sexually transmitted diseases (venereal diseases) are transferred by direct contact between persons during sexual intercourse. Syphilis, gonorrhea, chlamydial infections, genital herpes, and AIDS are sexually transmitted diseases. The causative agents are transferred by direct contact with mucous membranes or body fluids (semen, vaginal fluid) and not by casual contact with infected persons. To prevent sexually transmitted diseases, people need to be educated to take precautions both before and during sexual intercourse and seek immediate medical help should an infection occur. Correct use of condoms, treatment, and avoidance of infection by scrupulous practices all have a part in prevention and treatment.

Nonsexual touching is an important means of transmitting infectious agents. Common colds, influenza, and infectious mononucleosis are viral diseases transmitted by nonsexual direct contact, including shaking hands and kissing.

Objects such as drinking glasses, needles, toothbrushes, combs, and towels, called fomites (fo'miteez), are effective vehicles for indirect person-to-person transmission. It is difficult to interrupt this type of transfer. Preventive measures include isolation (as required) or, in extreme cases, quarantine of the infected person.

Airborne Transmission

Respiratory diseases including diphtheria, influenza, and the common cold are readily transferred through air. Air exhaled from the lungs is normally free of bac-

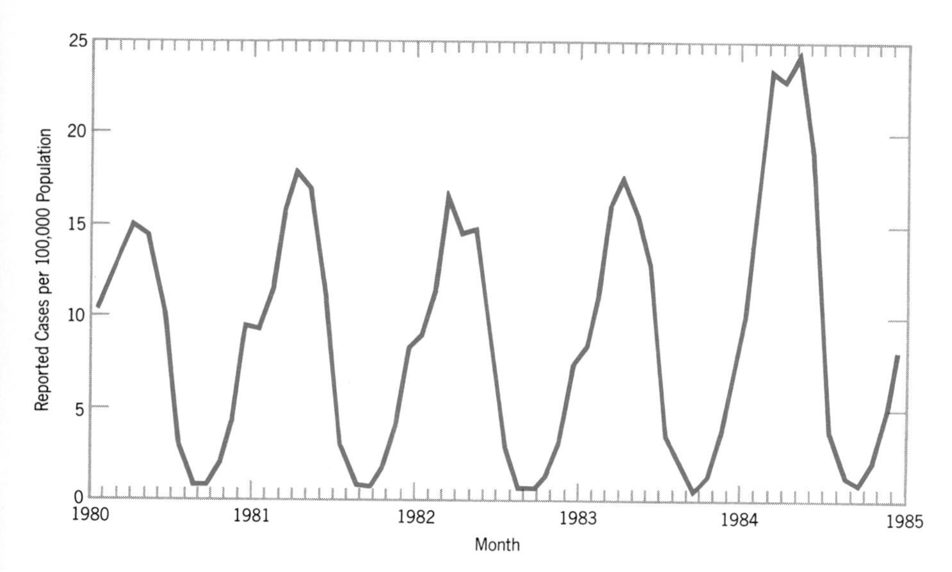

FIGURE 20.10 Reported case rates of chickenpox by month in the United States between 1980 and 1985. The incidence of chickenpox peaked between March and May and occurred primarily among school age children. (Source: *Morbidity and Mortality Weekly Report*.)

teria and viruses. When a person coughs or sneezes, however, the exhaled air picks up water droplets that contain viruses and microorganisms from the mouth and the nasopharynx (Figure 20.11). Once outside the body, the water quickly evaporates, leaving the infectious agent suspended in the aerosol. As these suspensions move in air currents, they transmit the infectious agent to susceptible persons.

Airborne transmission is greatest among congregations of people in enclosed quarters, especially in winter. Wearing a face mask may prevent transmission, but outside of hospitals, this practice is not widespread.

Foodborne and Waterborne Transmission

Viruses, protozoa, and bacteria that cause gastrointestinal diseases can be carried in contaminated water or food. Once inside the host, the microbes grow in the intestinal tract before they are excreted in the feces. Controlling the spread of these diseases is dealt with through improved hygiene, public sanitation, and sewage treatment facilities.

Drinking water may be contaminated when inadequately treated or untreated human sewage enters the water source. People become infected by consuming contaminated water or swimming in it. Waterborne outbreaks of infectious disease usually occur as common-source epidemics (Table 10.2).

FIGURE 20.11 Violent, unstifled sneeze. (Courtesy of American Society for Microbiology.)

TABLE 20.2 Etiology of waterborne disease outbreaks reported to the Centers for Disease Control, 1983

ETIOLOGY	OUTBREAKS	CASES	PERCENTAGE OF TOTAL CASES
AGI[a]	15	16498	78.92
Giardia lamblia	17	2207	10.55
Hepatitis A	3	167	0.78
Salmonella	2	1150	5.50
Shigella	1	12	0.06
Campylobacter	1	871	4.10
Chemical	1	3	0.01
		20908	99.92

[a]Acute gastrointestinal illness of unknown etiology.

Food becomes infected through improper handling, unhygienic practices, and contamination with human feces. *Clostridium* spp., *Salmonella,* and *Staphylococcus aureus* are the major causes of food poisoning (Table 20.3). *Clostridium botulinum* from the environment contaminates fish, meats, and garden vegetables and multiplies in improperly prepared canned foods. *Salmonella* enter the food supply through contaminated meat and poultry and the unhygienic handling of food by infected persons, whereas people carry *Staphylococcus aureus* as part of their skin flora and transfer it to food during food preparation. Outbreaks of food poisoning occur as common-source epidemics when infected food is served to large numbers of people.

Certain helminths infect animal muscle and are transmitted to people who eat raw or undercooked meat. For example, the trematode *Trichinella spiralis* infects the muscle of pigs and wild game animals and causes trichinosis in people who consume contaminated meat. Major outbreaks of trichinosis have been associated with the eating of raw or partially cooked pork in ethnic dishes prepared with pork purchased directly from local farmers.

Hepatitis A has been traced to sewage-contaminated saltwater from which the virus is taken in by filter-feeding shellfish. Oysters harvested from these waters are one source of foodborne hepatitis. There are no known cures for trichinosis and hepatitis, so they are among the more serious of the foodborne diseases.

Arthropod Vectors

Trying to monitor and reduce the population of fleas, ticks, lice, and mosquitoes is the chief means of controlling the spread of arthropod-borne diseases. These diseases usually are seasonal because they depend on the life-cycle of the vector. Some infectious agents replicate in the vector, so the latter also serves as a reservoir; some vectors merely transfer the infectious agent from an infected reservoir to a susceptible host. Insecticides help to control the vector population once it is identified.

Zoonoses

Zoonoses (zo′o-no′sez) are diseases of animals that can be transmitted to humans under natural conditions. Tularemia, rabies, Q fever, and brucellosis are examples of zoonotic diseases. Tularemia, found in wild rabbits and other small game animals, is passed to humans through direct contact with an animal's entrails when it is being dressed in the field. Rabies is most prevalent in skunks, bats, raccoons, and foxes; but it also occurs in domestic dogs, cats, and cattle. The rabies virus is present in the saliva of infected animals and is transmitted to humans who are bitten by a rabid animal. Domestic dogs and cats in the United States are now immunized against rabies, greatly reducing the possibility of human contact with a rabid pet. Brucel-

TABLE 20.3 Major foodborne disease outbreaks by etiology

ETIOLOGY	1977	1978	1979	1980	1981
Bacterial					
Bacillus cereus	–	6	–	9	8
Campylobacter jejuni	–	–	–	5	10
Clostridium botulinum	20	12	7	14	10
Clostridium perfringens	6	9	20	25	28
Salmonella	41	45	44	39	66
Shigella	5	4	7	11	9
Staphylococcus aureus	25	23	34	27	44
Vibrio parahaemolyticus	2	2	2	4	2
Parasitic					
Giardia lamblia	–	–	–	–	1
Trichinella spiralis	14	7	11	5	7
Viral					
Hepatitis A	4	5	5	10	6

Source: Centers for Disease Control: *Foodborne Disease Outbreaks Annual Summary*, 1981, issued June 1983.

losis, a disease of domestic cattle, can be passed to humans through milk, though modern pasteurization has eliminated milk as a major source of this disease. The few widely scattered cases of brucellosis in the United States occur primarily among workers in the meat-packing and livestock industries.

PUBLIC HEALTH AND THE CONTROL OF DISEASE

Effective collection of data on morbidity and mortality is a prerequisite for the control of infectious diseases. The attending physician is responsible for diagnosing the illness, and the clinical microbiologist is responsible for identifying the infectious agent. In compliance with federal law, this information is transmitted via each state's department of public health to the Centers for Disease Control (CDC) in Atlanta, where it is compiled and statistically analyzed. Special attention is given to emerging antibiotic-resistant strains of bacteria and to the increased virulence of infectious agents that has been noted in recent years. The CDC compiles statistics on both morbidity and mortality and publishes these data in a weekly report that is available to health care professionals.

Controlling the spread of infectious diseases falls under the purview of public health agencies, which include the federal Food and Drug Administration, the U.S. Department of Health and Human Resources, and state health departments. These agencies establish and enforce regulations governing the preparation and distribution of food, water quality, and immunization programs.

WATER QUALITY AND SEWAGE TREATMENT

Infectious agents enter the water supply when they are excreted in the feces of infected hosts. *Salmonella, Shigella,* and *Vibrio cholerae* are examples of bacterial pathogens that are transmitted by water contaminated with human feces, and poliomyelitis and infectious hepatitis are examples of waterborne viral diseases. The chief approach to reducing the spread of waterborne infections is by preventing raw sewage from entering the water supply.

Large numbers of bacteria that are naturally present

Characteristics of a Measles Outbreak—New Hampshire

Between April 9 and June 1, 1984, 37 cases of measles were reported from the Hanover, New Hampshire area to the state Division of Public Health Services (DPHS). Twenty-one cases were serologically confirmed. Twenty-nine infections occurred among Dartmouth College students and their family contacts, one in a bookstore employee, and one in a resident of a nearby town. The six remaining cases occurred in four employees, one outpatient, and one neurosurgical inpatient at the Mary Hitchcock Memorial Hospital, a teaching facility of Dartmouth Medical School.

Symptoms were first diagnosed in a 27-year-old male graduate student on April 3. The source of his infection was unknown since he had traveled to Michigan, Massachusetts, New Jersey, and North Carolina for job interviews during the likely period of exposure. Indigenous measles had not been reported in New Hampshire during the previous 22 months. New measles cases occurred from April 13 to April 16 (see the accompanying figure) in the unimmunized 2 1/2-year-old son of the index patient and in six additional graduate students. Groups of new measles cases continued to appear during the following month as shown in the diagram.

The Dartmouth College Student Health Service, assisted by DPHS, instituted outbreak-control measures following the

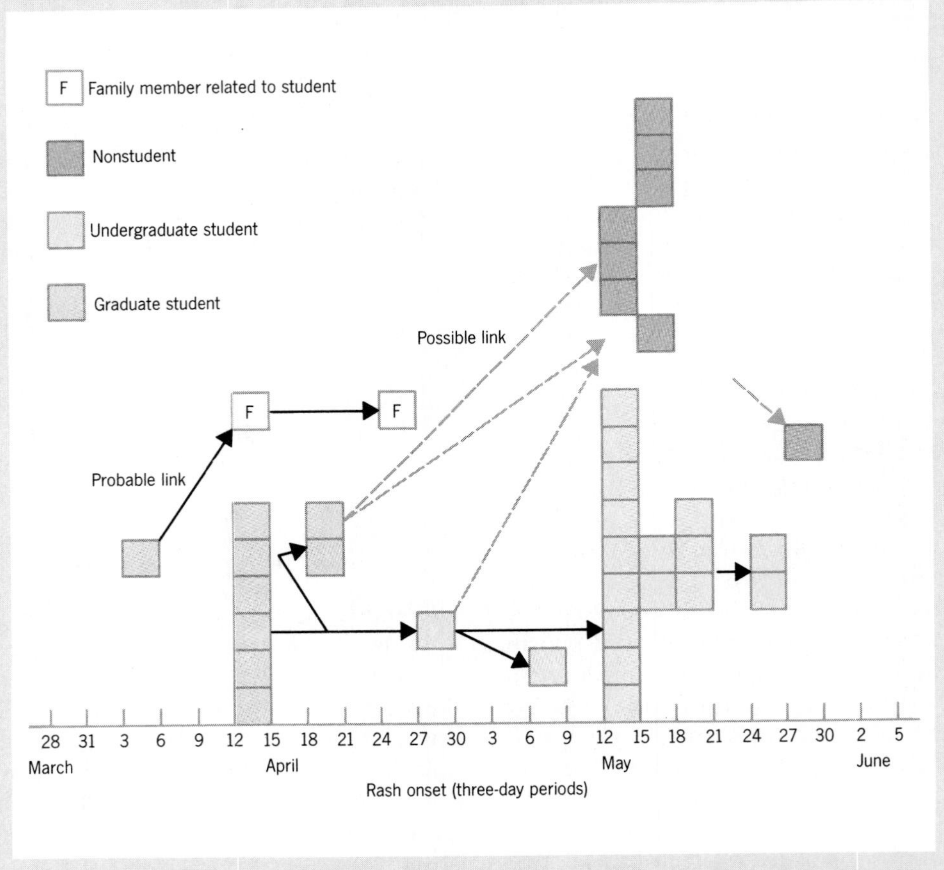

Chain of transmission, measles outbreak, Dartmouth College and Mary Hitchcock Memorial Hospital, Hanover, New Hampshire.

report of the index patient on April 9. These measures included determining the immune status of students at Dartmouth's Tuck graduate school and vaccinating on a voluntary basis susceptible Tuck graduate students, susceptible family members, and neighbors exposed to the index patient. On April 13, DPHS recommended that the entire student body (4,903 persons) be notified that measles had occurred on campus and that all student health immunization records be reviewed to identify susceptible individuals. Seventy student health aides assisted in surveillance and education. Steps were taken to identify and immunize susceptible athletes, to prohibit them from participating in off-campus events, and to advise teams from other colleges that measles had occurred on campus.

An audit of the 4,903 student health records revealed that 2,923 did not have adequate documentation of measles immunity (live measles vaccine on or after the first birthday, physician-diagnosed measles, or laboratory evidence of measles immunity). In most cases dates of vaccination were lacking. Letters were sent to each of these possible susceptible students explaining that they should be vaccinated unless they could provide documentation of immunity. The 375 employees of the college under 28 years of age received similar letters.

Vaccination clinics staffed by DPHS were opened on campus April 13 and maintained through April 27. The Student Health Service also established a walk-in vaccination clinic from April 16 throughout the outbreak.

Comments

Measles transmission among college students may be sustained by several factors: (1) children growing up in the mid-1960s may have missed measles vaccination in the first years following the licensure of measles vaccine; (2) students may not have been immunized under comprehensive school laws now in effect in many states covering students in kindergarten through grade 12; (3) many colleges and universities do not have requirements concerning immunization; (4) students may have escaped natural measles infection because of decreasing transmission; (5) some students may have been vaccinated with the ineffective formalin-inactivated ("killed") measles vaccine, which was administered to 600,000 to 900,000 individuals from 1963 to 1967; (6) students may have been vaccinated with live virus vaccine before their first birthday, when measles vaccine is known to be less effective; and (7) the tendency of college students to congregate in large numbers contributes to transmission. These factors contribute to the persistence of measles within the population in the face of efforts to eliminate this viral disease through effective immunization procedures.

Adapted from *Morbidity and Mortality Weekly Report*, Vol. 33, No. 39, October 5, 1984.

in the lower bowel are expelled in feces. The majority are obligate anaerobes along with smaller but significant populations of facultative anaerobes, including *Escherichia coli*. This bacterium can easily be detected by laboratory methods, so it is used as an indicator of human fecal contamination. Water free of *E. coli* is considered to be free of fecal contamination.

Modern sewage treatment facilities are designed for the individual home or an entire community (Figure 20.12). Their purpose is to remove contaminants from sewage or render them harmless. Methods for treating human waste are described in Part 5.

CONTROLLING FOODBORNE DISEASES

Foodborne diseases are controlled by regulating and inspecting facilities where food is prepared and served to the public–such as restaurants, hospitals, school cafeterias. Inspection of these facilities and enforcement of public health regulations rests with local governing bodies. Nevertheless, foodborne outbreaks do occur and these are monitored by the CDC, as part of its ongoing effort to control and prevent epidemics.

The case of Mary Mallon is a classic example of food being contaminated by an infected human. Mary Mallon, more commonly called Typhoid Mary, was a cook who worked in the New York City area during the early part of the twentieth century. She recovered from typhoid fever only to become a chronic carrier of the causative agent, *Salmonella typhi*. Public health officials discovered her by tracing the source of a typhoid outbreak to a kitchen in which she was the cook. Microbiologists proved that she was a carrier of *S. typhi*. Since she refused to give up her work, health officials had her imprisoned to prevent her from spreading typhoid. (If this case had occurred today, Mary Mallon would have been treated with antibiotics instead of being imprisoned.)

(a) (b)

FIGURE 20.12 *(a)* Sewage treatment plant showing a sprinkling filter in the foreground and a sludge drying bed in the background. *(b)* Waste water treatment facility composed of an aerobic digester (foreground), a circular sludge thickener, and a clorination building. (Both Runk/Schoenberger, Grant Heilman.)

TABLE 20.4 Immunizations recommended according to occupation, life-style, environmental situation, travel, and for foreign students and immigrants

INDICATION	IMMUNIZATIONS
Occupation	
Hospital, laboratory, other health care personnel	Hepatitis B, polio, influenza
Staff of institutions for mentally retarded	Hepatitis B
Veterinarians and animal handlers	Rabies
Selected field workers	Plague
Life-style	
Homosexual males	Hepatitis B
Illicit drug users	Hepatitis B
Environmental situation	
Inmates of long-term correctional facilities	Hepatitis B
Residents of institutions for mentally retarded	Hepatitis B
Travel	Measles, rubella, polio, rabies, yellow fever, hepatitis B, meningococcal polysaccharide, typhoid, cholera, plague
Foreign students, immigrants, refugees	Measles, rubella, diphtheria, tetanus

Source: Morbidity and Mortality Weekly Report Supplement, "Adult Immunizations," *Vol. 33,* September 1984.

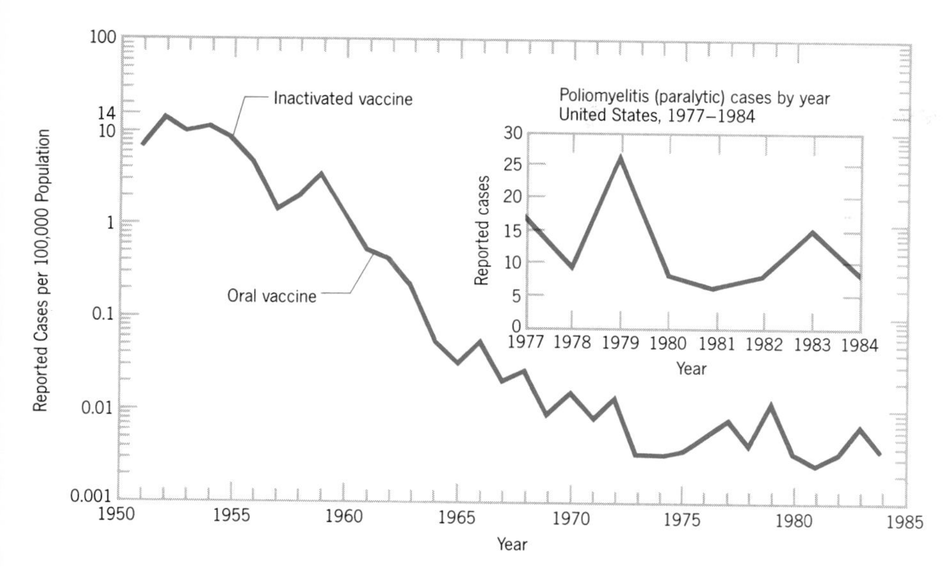

FIGURE 20.13 The incidence of poliomyelitis (paralytic) in the United States dropped dramatically after the introduction of the polio vaccines. After 1976, the number of polio cases in the United States was fewer than 30 per year. (Centers for Disease Control, "Annual Summary 1984: Reported Morbidity and Mortality in the United States," *Morbidity and Mortality Weekly Report,* Vol. 32, No. 54, 1986.)

IMMUNIZATION PROGRAMS

The purpose of large-scale immunization programs is to reduce the numbers of susceptible people and thereby limit the spread of infectious agents. Infants are immunized during the first two years of life. Babies younger than six months have some degree of maternally acquired immunity against infectious diseases, but this passive immunity is short-lived, so early immunization is required. In the United States babies are routinely immunized against tetanus, diphtheria, whooping cough, mumps, measles, rubella, and polio. Certain life-styles, occupations, and environmental situations call for various types of immunizations, as may be seen in Table 20.4. Adults, especially those who are vulnerable to influenza, may be immunized against the "flu" in preparation for its expected appearance in winter months.

The effect of a broad immunization program on infectious disease can be dramatic, as indicated for poliomyelitis in Figure 20.13. Observe that the decline in cases continued following the introduction of oral (Sabin) vaccine in 1961. The introduction of the rubella (German measles) vaccine in 1969 was followed by a corresponding dramatic reduction. Rubella continues to occur at a fairly high rate in the United States, mainly among high school students and young adults who as a group were inadequately immunized.

International Travel

Efforts to prevent the spread of disease worldwide are detailed in the International Health Regulations, generated by the World Health Organization (WHO) in Geneva, Switzerland. WHO recommends the immunization certificates required of international travelers who enter foreign countries, designates quarantinable diseases, and distributes a weekly "Blue Sheet" listing the areas infected with quarantinable diseases. Vaccinations against yellow fever, typhoid, and cholera are required for travel to countries where these diseases are endemic.

NOSOCOMIAL INFECTIONS

The hospital environment is uniquely suited to the spread of infections, as here are housed both susceptible patients and patients with difficult-to-treat infec-

TABLE 20.5 The 15 most frequently isolated pathogens from hospital patients and their percentage distribution for each site of infection, 1983

PATHOGEN	URINARY TRACT	SURGICAL WOUNDS	LOWER RESPIRATORY	PRIMARY BACTEREMIA	CUTANEOUS	OTHER	TOTAL ISOLATES
Escherichia coli	31.7	11.4	7.1	9.5	7.7	7.3	5779
Staphylococcus aureus	1.6	19.0	12.8	12.8	33.3	16.4	3356
Streptococcus (Enterococcus)	14.9	11.4	1.6	7.3	9.5	7.3	3308
Pseudomonas aeruginosa	12.5	8.1	15.1	6.1	7.2	6.2	3286
Klebsiella spp.	7.6	4.8	12.8	9.1	4.4	4.0	2288
Staphylococcus epidermidis	3.7	8.4	1.1	14.2	9.5	11.1	1892
Enterobacter spp.	4.4	6.9	10.0	6.9	4.1	4.1	1811
Proteus spp.	7.3	5.0	4.4	1.7	3.5	3.2	1667
Candida spp.	5.1	1.4	4.2	5.6	4.5	13.4	1570
Serratia spp.	1.2	2.0	5.6	2.8	1.8	1.8	691
Other fungi	2.0	0.4	1.8	1.0	0.4	2.3	471
Bacteroides spp.	0.0	4.4	0.2	3.4	1.0	1.9	439
Streptococcus Group B	1.0	1.8	0.7	2.8	1.3	1.7	418
Citrobacter spp.	1.4	1.3	1.7	1.1	1.1	0.9	403
Other anaerobes	0.0	2.3	0.0	2.1	0.9	4.2	337
All others[a]	5.6	11.4	20.9	13.6	9.8	14.2	3296
Number of isolates	13165	6163	4490	2292	1798	3104	31012

Source: Centers for Disease Control: "Nosocomial Infection Surveillance, 1983," In CDC Surveillance Summaries, 1984; 33 (No. 2SS) pp. 9SS–21SS.

[a]No other pathogen accounted for more than 3 percent of the isolates at any site.

tious diseases. **Nosocomial infections** are diseases acquired during the course of a hospital stay. It is estimated that between 5 and 6 percent of all patients admitted to general hospitals acquire infections while there.

INCIDENCE OF NOSOCOMIAL INFECTIONS

Hospital personnel routinely compile records on the number and types of infectious agents isolated from their patients. One six-month summary report (January to June 1983) of nosocomial infections in 54 United States hospitals is shown in Table 20.5. This study involved 865,282 discharged hospital patients, of whom 28,248 (3.3 percent*) were identified as having acquired a nosocomial infection. The most frequent site of infection was the urinary tract with the second highest rate of infection involving surgical wounds. *Escherichia coli* was the most frequent pathogen isolated (18 percent of all isolates) and was the predominant organism isolated from infections of the urinary tract (Table 20.5). *Streptococcus* of the enterococcus group was the second most prominent pathogen isolated from the urinary tract. *Staphylococcus aureus* was the most common pathogen isolated from surgical wounds.

*Experts believe this figure to be an underestimate.

SUSCEPTIBLE PATIENTS

Studies indicate that bacterial infections are most frequent on the surgical and medical services, which have more high-risk patients, and least frequent on the pediatric and newborn services.

The reasons for this are not hard to fathom. A patient with cancer or one who has just received or is about to receive a transplant, and persons recovering from surgery, are at considerably higher risk than healthy individuals. Patients receiving immunosuppressive drugs or cytotoxic drugs experience a suppressed immune response. The use of various devices that must be inserted into a body orifice, such as catheters, promotes infections. Finally, organs that are exposed and handled during surgical procedures are highly vulnerable to infection. At the same time, viral nosocomial infections are more common in children than in adults.

KEY POINT

Many hospital patients acquire an infection during the course of their stay, and these infections constitute an ever-present problem to the health professions and an ever-present threat to the patient.

MONITORING AND CONTROL IN THE HOSPITAL

Nurses serve as the first line of control against nosocomial infections. They need to be ever alert to the signs of infection such as fever, chills, shock, presence of pus, and diarrhea, and the information gained should be recorded and acted on as soon as possible. The nurse also has the responsibility of following hygienic practices at all times and of protecting the patient from those who do not.

The hospital environment must be monitored to detect potential sources of pathogens. This involves taking samples of the surfaces and the ambient air in operating and delivery rooms, testing the effectiveness of autoclaves and other sterilizing equipment, testing antiseptics and disinfectants, and monitoring to detect contamination of food including infant formulas. Hospitals establish standards regarding levels of contamination, and take corrective steps when measured levels exceed those standards. Some units, such as burn treatment facilities, require more frequent and extensive monitoring.

Summary Outline

HISTORICAL PERSPECTIVE

1. Epidemics of infectious diseases have caused extensive death, morbidity, and suffering, and have even altered the course of ancient civlizations.
2. Plague—The Black Death—devastated the societies of the Roman and Byzantine Empires and marked the beginning of the Middle Ages. The Black Death later ravaged Central Asia before spreading to the European continent.
3. John Snow is regarded as the first epidemiologist, because of his studies on cholera. AIDS is a current epidemic that poses grave dangers to humans.

EPIDEMIOLOGY OF INFECTIOUS DISEASES

1. Epidemiology is the study of factors that influence the occurrence of disease in a population.
2. Propagated epidemics are caused by diseases transmitted by person-to-person contact, whereas common-source epidemics originate from a single contaminated source, have a rapid onset, quickly reach a maximum incidence of new cases, and then decline.
3. Endemic diseases are illnesses in a localized population that occur at a low but persistent rate. Pandemic disease is one that spreads worldwide causing extensive morbidity.
4. Infectious agents that depend on a reservoir or vector for survival are restricted to those geographical regions of the world where the vector and/or reservoir exist.
5. Climate and seasonal variations affect the survival of the infectious agents, reservoirs, and vectors.
6. Social practices influence human behavior and can alter the incidence of, or susceptibility to, infectious diseases.
7. Infectious agents may be airborne, foodborne, or waterborne, or can be transferred by person-to-person contact, by indirect contact with fomites, by arthropod vectors, or by animals.
8. Drinking water supplies are contaminated when improperly treated or untreated human waste enters the water source. Waterborne diseases usually occur as common-source epidemics when people drink contaminated water or use it for recreational purposes.
9. Food becomes infected through improper han-

dling of contaminated meats and poultry, unhygienic practices, and contamination with human feces.

10. Zoonoses are animal diseases that may be transmitted to humans under natural conditions, such as tularemia, rabies, Q fever, and brucellosis.

PUBLIC HEALTH AND THE CONTROL OF DISEASE

1. Preventing an epidemic can be accomplished by interrupting the mode of disease transmission. Public health measures that have been established to prevent disease transmission include sewage treatment, data collection, surveillance of and investigation of disease outbreaks, monitoring to ensure proper handling of food, and immunization programs.
2. Immunization programs decrease the number of susceptible people which in turn limits the spread of specific infectious agents. Infants in the United States are immunized against tetanus, diphtheria, whooping cough, mumps, measles, rubella, and poliomyelitis.
3. The World Health organization establishes the quarantinable diseases and advises travelers of the necessary immunizations for entry into foreign countries.

NOSOCOMIAL INFECTIONS

1. Nosocomial infections are diseases acquired during the course of a hospital stay, and occur in 5 to 6 percent of all patients admitted to general hospitals.
2. The urinary tract is the most commonly infected site and *E. coli* is the most common cause of nosocomial infections. *Staphylococcus aureus* is the pathogen most often isolated from surgical wounds.
3. Nosocomial infections are controlled by monitoring the hospital environment to detect potential sources of pathogens.

Questions and Topics for Study and Discussion

QUESTIONS

1. What are the characteristics of endemic, epidemic, enzootic, and pandemic disease?
2. What characteristics distinguish a propagated epidemic from a common-source epidemic? Name two infectious agents that cause each type of epidemic.
3. What role do vectors and reservoirs play in infectious diseases? What reservoirs and vectors of infectious diseases exist in your area?
4. What social practices in your culture may contribute to the spread of infectious disease? Should these practices be stopped or curtailed?
5. Name five modes of disease transmission and describe an infectious disease that exemplifies each. What are some practical means of preventing the transmission of these diseases?
6. What are nosocomial infections and how serious a problem are they? What procedures and practices contribute to nosocomial infections?

DISCUSSION TOPICS

1. What effect has our ability to control infectious diseases had on life expectancy and culture? Has there been a different effect on other countries attributable to modern techniques?
2. How would a major disaster, such as earthquake, hurricane, tornado, or war affect the spread of infectious agents within your community? What infectious agents would be most likely to cause disease and how could their transmission be prevented?

Further Readings

BOOKS

Benson, A. S. (ed.), *Control of Communicable Disease in Man,* 12th ed., American Public Health Association, Washington, D.C., 1975. A handbook of the essential facts on infectious diseases of humans including symptoms, occurrence, infectious agents, reservoirs, transmission, incubation period, susceptibility, and methods of control.

Burnett-Conners, E. B., H. J. Simon, and D. C. Dechairo (eds.), *Epidemiology for the Infection Control Nurse,* C. V. Mosby Company, St. Louis, 1978. A collection of essays on topics pertaining to the control of nosocomial infections.

Linton, A. H., *Microbes, Man and Animals: The Natural History of Microbial Interactions,* Wiley, New York, 1982. The principles of pathogenicity and epidemiology are brought together, with chapters on the modes of disease transmission.

Roueche, B., *The Medical Detectives,* Washington Square Press, New York, 1980. The epidemiological investigations of numerous disease outbreaks are recounted in a delightful style of this renowned medical journalist.

Roueche, B., *The Medical Detectives,* Vol. 2, Dutton, New York, 1985. Further adventures in epidemiology.

ARTICLES

Cliff, A., and P. Haggett, "Island Epidemics, " *Sci. Am.,* 250:138–147 (May 1984). An analysis of the public health records of measles in Iceland that led to general principles of epidemiology.

Henderson, D. A., "The Eradication of Smallpox," *Sci. Am.,* 235:25–33 (October 1976). An epidemiological report on the total, worldwide elimination of a serious viral disease.

Kaplan, M. M., and R. G. Webster, "The Epidemiology of Influenza," *Sci. Am.,* 237:88–106 (December 1977). A review of influenza that covers its history, epidemiology, and virology.

Merson, M. H., et al., "Traveler's Diarrhea in Mexico," *N. Engl. J. Med.,* 294:1299–1305 (1976). A case study of diarrhea among 73 physicians and 48 family members who visited Mexico City to attend a medical conference.

Wishnow, R. M., and J. L. Steinfeld, "The Conquest of the Major Infectious Diseases in the United States: A Bicentennial Retrospect," *Ann. Rev. Microbiol., 30:427–450 (1976).* A review of the progress made in the United States to control many of the major infectious diseases of humans.

C H A P T E R

21

Antimicrobial Agents and Chemotherapy

OUTLINE

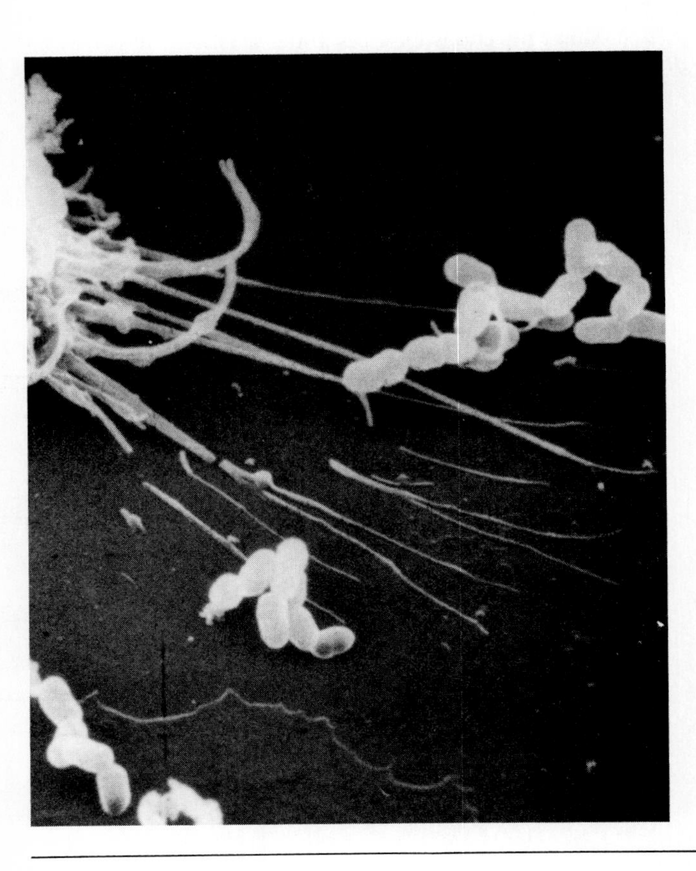

FOCUS OF THIS CHAPTER

Natural and synthetic antimicrobial compounds are the chief weapons with which society tries to control microbial infections. In this chapter we discuss the major antimicrobial agents, their modes of action, the types of resistance bacteria develop to them, and the importance of antimicrobial chemotherapy in treating bacterial, fungal, and protozoan diseases.

The discovery and development of antimicrobial drugs is one of the greatest medical stories of the twentieth century. These drugs have saved countless human lives and have facilitated the rapid progress of medicine.

Alexander Fleming, Howard Florey, and Ernst Chain are the three scientists credited with discovering penicillin and demonstrating its effectiveness as an antibiotic drug. Fleming discovered penicillin in 1929, but was unable to obtain enough of it to proceed with clinical trials. The clinical use of antibiotics began in 1941 when Florey and Chain administered penicillin to a patient suffering from a staphylococcal infection. The remarkable response of the patient gave them reasons to be elated, even though the supply of penicillin ran out and the patient died. With further efforts Florey and Chain obtained sufficient penicillin to successfully treat other patients. Their success stimulated an international effort to develop methods for producing commercial quantities of penicillin. Fleming, Florey, and Chain received the Nobel Prize in Physiology or Medicine in 1945 for introducing antibiotics to the world.

Compelled by the medical success of penicillin, many drug companies and independent scientist began a worldwide search for microbes that produced other antibiotics. The resulting array of compounds that are effective against bacteria, fungi, and protozoa has helped us overcome the devastating effects of many infectious diseases, and has permanently changed the practice of medicine. The success of these drugs lies in the selective way they kill microorganisms without harming people.

SELECTIVITY OF ANTIMICROBIAL AGENTS

Achieving antimicrobial selective toxicity is the goal of those charged with developing and testing antimicrobials; that is, the agent should either kill the microbe or prevent its multiplication, without harming the host. This selectivity is the major difference between chemotherapeutic agents and disinfectants. Selective toxicity is based on the principle that there is a unique structure or process of the microbe that is absent in the host. The research efforts of many years paid off by revealing the uniqueness residing in bacterial cell walls, ribosomes, and the protein synthesis mechanism.

The chemicals employed to treat infectious disease are called **chemotherapeutic drugs,** and they fall into two main classes—antibiotics and chemosynthetic agents. **Antibiotics** are substances produced by one organism that kill or inhibit the growth of another. They are naturally produced so they probably enhance the producing organism's ability to survive. Of all the antibiotics known, only a small group are highly effective chemotherapeutic drugs, the others are too toxic for human use. The **chemosynthetic antimicrobials** are chemically synthesized organic compounds that are selectively toxic to microorganisms. (The distinction between chemosynthetic antimicrobial drugs and antibiotics has become clouded because some natural antibiotics, such as chloramphenicol, are now produced by chemical synthesis, and the semisynthetic antibiotics are chemically modified natural products.) The term antimicrobial is used here to describe all agents that are selectively toxic to microorganisms.

To be an effective drug, an antimicrobial must exhibit other properties besides its ability to inhibit or kill microbes. Its **efficacy** is a measure of the combination of characteristics that contribute to its use as a drug. Its stability *in vivo,* rate of absorption, rate of elimination, and penetration of the infection site all contribute. The agent may rank high in one characteristic and lower in another, and it is often possible to circumvent a drug's deficiencies by chemical modification or by administering the drug by a designated route (e.g., by injection rather than by mouth).

A drug's **therapeutic index** is the minimum dose of the drug that is toxic to the host divided by its effective therapeutic dose. The higher the index, the greater is the efficacy of the drug. In part, the index depends on the host's ability to absorb the drug. The blood level of the drug depends on the dose administered, the host's body weight, the route of administration, the schedule of medication, and the rate of drug elimination. Drug efficacy also depends on the complex interaction between host, drug, and invading microbe.

KEY POINT

Antimicrobials exhibit selective toxicity toward the microbes, sparing the host. This characteristic distinguishes antimicrobials from disinfectants.

FIGURE 21.1 Alexander Fleming (1881–1955) discovered the first antibiotic, penicillin. (Courtesy of National Library of Medicine.)

ANTIBACTERIALS

The natural antibiotics have provided chemists with the key they needed to synthesize additional and perhaps better antibacterial agents. The accidental discovery of penicillin opened an era that has had a dramatic impact on everyone's life.

DISCOVERY OF ANTIBIOTICS

Alexander Fleming (Figure 21.1) was a Scottish physician trained in microbiology who before discovering penicillin was known for his work on host-parasite resistance. After World War I, he studied the phenomenon of phagocytosis and discovered a substance in tears, mucus secretions, and other body fluids that destroyed bacteria. A colleague of Fleming's named this substance lysozyme, an enzyme that attacks the peptidoglycan of bacterial cell walls, as explained in Chapter 4.

At a later point, Fleming discovered penicillin while working with staphylococcus variants. Fleming wrote:

While working with staphylococcus variants a number of culture-plates were set aside on the laboratory bench and examined from time to time. In the examination these plates were necessarily exposed to the air and they became contaminated with various micro-organisms. It was noticed that around a large colony of a contaminating mould the staphylococcus colonies became transparent and were obviously undergoing lysis.

[BR. J. EXP. PATHOL. 10:226–236 (1929).]

Fleming identified his contaminant as a species of *Penicillium* (Figure 21.2) and gave the name penicillin to its active principle. He cultured the strain of *Penicillium* and discovered that the active substance present in the culture filtrates was bactericidal against many pathogenic bacteria. Fleming measured the stability of penicillin to heat and to pH, and demonstrated that its toxicity was very low. Although he advocated the use of penicillin in treating infectious diseases, his attempts to purify penicillin were unsuccessful, and this greatly delayed its clinical use.

FIGURE 21.2 This symmetrical mold colony is the strain of *Penicillium chrysogenum* that is used to produce most of the world's commercial penicillin. (Courtesy of Pfizer Inc.)

For lack of purified penicillin, it was not clinically tested until the early 1940s. Howard Florey and Ernst Chain of Oxford University (England) were finally able to purify a sufficient quantity of penicillin, and with it they demonstrated its effectiveness against streptococcal infections in mice. Florey and Chain then began in earnest to purify sufficient quantities of penicillin from *Penicillium* cultures to treat a human patient.

By February 1941 they had collected all the penicillin then available, and administered it to a person who was dying of staphylococcal osteomyelitis. For a period of five days the patient improved; but the penicillin supply ran out, and the patient died. A decision was made to purify more penicillin and to administer it only to children, who needed smaller doses of the scarce drug. The results of these trials created excitement on both sides of the Atlantic Ocean; however, production problems still hampered the availability of penicillin. Moreover, England's energies and labor were heavily engaged in World War II.

Large-scale production of penicillin was first accomplished at the Northern Regional Research Laboratory in Peoria, Illinois. Both American and British scientists contributed to developing the process by which large quantities of penicillin could be produced. By the end of World War II, penicillin was being marketed by major drug companies and used to treat numerous infectious diseases.

Penicillin research made another big advance in 1959 when 6-amino penicillanic acid (6-APA) was produced in Britain. At about the same time, John C. Sheehan, a chemist at the Massachusetts Institute of Technology, succeeded in synthesizing penicillin. The chemical techniques developed by Sheehan and the availability of 6-APA led to the production in the late 1950s of semisynthetic penicillins. Penicillins are now among the most widely used drugs for treating infectious disease.

PRODUCTION OF ANTIBIOTICS BY MICROORGANISMS

Natural antibiotics are produced by microorganisms living in soil or aquatic environments. Following the success with penicillin, microbiologists developed techniques for screening soil and water samples from around the world for antibiotic-producing microorganisms.

These microorganisms are limited to a few genera of bacteria and fungi. Penicillin is produced by a few species of *Penicillium,* and the prototype of the cephalosporins is produced by another fungus, *Cephalosporium.* Members of the genus *Streptomyces* are by far the most important bacterial producers of antibiotics, accounting for more than 50 percent of the known antibiotics. Clinically important antibiotics produced by a *Streptomyces* species include streptomycin, the tetracyclines, vancomycin, kanamycin, rifampin, and chloramphenicol. A few antibiotics are produced by *Bacillus* species isolated from soil, but most of them have limited clinical applications because of their low therapeutic indices. Although the search continues, experts believe that most of the organisms capable of producing antibiotics have already been discovered. Efforts in antibiotic research now concentrate on improving the characteristics of identified antibiotics through chemical means.

MICROBIAL SUSCEPTIBILITY TO ANTIBIOTICS

The range of microbes affected by an antimicrobial is called its spectrum. Each antibiotic is active against a limited number of genera and species of bacteria. **Broad-spectrum** antibiotics inhibit a wide range of gram-negative and gram-positive bacteria. The tetracyclines are broad-spectrum antibiotics that affect gram-positive, gram-negative, aerobic and anaerobic bacteria, spirochetes, mycoplasmas, rickettsias, chlamydias, and some protozoa. Other broad-spectrum antibiotics are chloramphenicol and the cephalosporins. **Narrow-spectrum** antibiotics act against only a few groups of bacteria. For example, benzyl penicillin is effective against gram-positive cocci, gram-negative cocci, and spirochetes and ineffective against gram-negative rods. The spectrum of some antibiotics is so limited that they are effective against only one or two organisms; for example, rifampin is given to treat tuberculosis, and spectinomycin to treat gonorrhea caused by penicillin-resistant *Neisseria gonorrhoeae*.

Each antibiotic has its special mode of action which can be broadly characterized as inhibiting protein synthesis, nucleic acid synthesis, cell-wall synthesis, or interfering with membrane function (Figure 21.3). **Bactericidal** antibiotics, such as streptomycin and penicillin, kill the bacterial cell without intervention of the

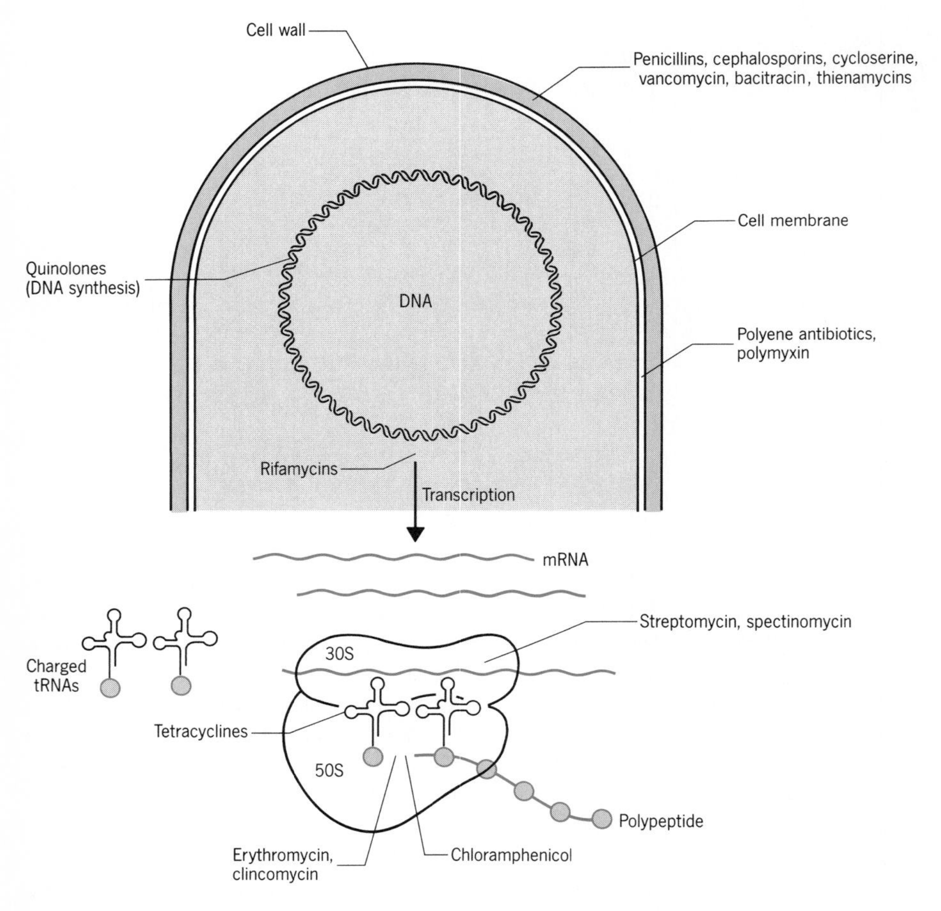

FIGURE 21.3 The sites and physiological functions inhibited by antibacterial antibiotics include cell-wall synthesis, protein synthesis, DNA synthesis, and membrane function.

TABLE 21.1 True antibiotics: their source and mode of action

ANTIBIOTIC	SOURCE	MODE OF ACTION	YEAR DISCOVERED
Antibacterial Antibiotics		*Inhibits*	
Bacitracin	*Bacillus licheniformis*	Cell-wall synthesis	1945
Cephalosporin	*Cephalosporium* spp.	Cell-wall synthesis	1948
Chloramphenicol	*Streptomyces venezuelae*	Protein synthesis	1947
Erythromycin	*Streptomyces erythraeus*	Protein synthesis	1952
Gentamicin	*Micromonospora purpurea*	Protein synthesis	1963
Imipenem	*Streptomyces cattleya*	Cell-wall synthesis	1970
Kanamycin	*Streptomyces kanamyceticus*	Protein synthesis	1957
Lincomycin[a]	*Streptomyces lincolnensis*	Protein synthesis	1962
Penicillin	*Penicillium* spp.	Cell-wall synthesis	1929
Polymyxin	*Bacillus polymyxa*	Cell-membrane function	1947
Rifampin	*Streptomyces mediterranei*	DNA transcription	1959
Streptomycin	*Streptomyces griseus*	Protein synthesis	1944
Tetracycline	*Streptomyces aureofaciens*	Protein synthesis	1948
Vancomycin	*Streptomyces orientalis*	Cell-wall synthesis	1956
Antifungal Antibiotics			
Amphotericin B	*Streptomyces nodosus*	Membrane function	1955
Griseofulvin	*Penicillium griseofulvum*	Cell wall, microtubules	1939
Nystatin	*Streptomyces noursei*	Cell-membrane function	1950
Antiprotozoan Antibiotics			
Fumagillin	*Aspergillus fumigatus*	Protein synthesis	1951
Quinine	Cinchona bark	Unknown	1927[b]
Cyclosporin A	Fungus	Unknown	1970–72

[a]The derivative clindamycin is more active and is now used in place of lincomycin.
[b]Year extraction procedure developed.

host's immune system. Penicillin kills susceptible dividing bacteria by inhibiting their ability to synthesize cell walls. The absence of an intact cell wall makes the bacterium osmotically sensitive, and it eventually lyses. **Bacteriostatic** antibiotics prevent the cell from multiplying by inhibiting an essential function such as protein or nucleic acid synthesis. Bacteria so affected are eventually destroyed by the host's cellular defense system. The mode of action of the major antibiotics is listed in Table 21.1.

Bacteria that possess innate resistance to a given antibiotic will be unaffected by it. Innate resistance can result from the inability of the antibiotic to get into the cell, the absence of the affected target (e.g., mycoplasma lack a cell wall), or the ability of the cell to circumvent the metabolic pathway affected.

The spectrum of activity of a given antibiotic is not absolute in every instance since an individual strain of bacteria may develop resistance to it. This being the case, most organisms isolated from ill patients (clinical isolate) must be tested to determine their antibiotic sensitivity so that the appropriate drug can be administered.

INHIBITORS OF BACTERIAL CELL-WALL SYNTHESIS

Substances that interfere with formation of the bacterial cell wall either inhibit the enzymes involved in cell-wall synthesis or activate autolysins. Then, having no cell wall to protect it, the bacterium dies (lyses). See Table 21.1 for a list of antibiotics that inhibit bacterial cell-wall synthesis.

Penicillins

Penicillin G, the antibiotic naturally produced by *Penicillium notatum* and *P. chrysogenum* (kris-o'jen-um), inhibits the synthesis of the peptidoglycan of the bacterial cell wall. Most penicillins are narrow-spectrum antibiotics

Penicillin G	β-lactam ring	A natural product
	Semisynthetic Penicillins	**Characteristics**
Penicillin V		Acid-resistant
Ampicillin		Broad spectrum, acid-resistant
Methicillin		Resistant to penicillinase
Oxacillin		Resistant to acid and penicillinase

effective against gram-positive bacteria, spirochetes, and a few gram-negative cocci. They are largely ineffective against gram-negative bacilli because they do not penetrate the outer membrane of the cell wall. The high therapeutic indices of the penicillins make them the drugs of choice for treating the majority of cases of gonococcal, treponemal, and streptococcal infections.

The Penicillin Family *Penicillium chrysogenum* produces different penicillins depending on its growth medium. Although penicillin G (benzylpenicillin) is a very effective antibacterial agent (Figure 21.4), it has the disadvantage of being inactivated by penicillinase (β lactamase) and it slowly decomposes in the acid environment of the stomach. To surmount this problem, chemists make semisynthetic penicillins that possess improved chemotherapeutic properties (Figure 21.4).

The manufacture of the semisynthetic penicillins was helped tremendously by the isolation in 1959 of 6-aminopenicillanic acid from cultures of *P. chrysogenum*. Armed with this inexpensive source of starting material, chemists were able to produce penicillin derivatives that were resistant to gastric acids and/or penicillinase. The semisynthetic penicillins include penicillin V, oxacillin, and ampicillin, which are acid-resistant; and methicillin and oxacillin, which are resistant to penicillinase (Figure 21.4). In addition, ampicillin has a wide spectrum because it is able to penetrate the outer membrane of the gram-negative bacterial cell wall.

Penicillin-Binding Proteins Penicillins inhibit bacterial cell-wall synthesis by interacting with penicillin-binding proteins on or near the cytoplasmic membrane. The penicillin-binding proteins vary among bacterial species in number, kind, and probably function. Some of these proteins are involved in peptidoglycan synthesis and in cell division; however, the function of most remains unknown. We do know that the cellular function of key penicillin-binding proteins is inhibited by penicillin.

Resistance to Penicillins Resistance to the penicillins comes about when bacteria produce penicillinase (or β lactamase), which hydrolyzes the β-lactam ring of the penicillin (Figure 21.4). This inactivates the antibiotic before it is able to inhibit cell-wall synthesis. Penicillin-resistant strains of staphylococci produce an extracellular penicillinase, while the penicillinases produced by gram-negative bacteria occur in the periplasmic space. Penicillinase-producing bacteria can be killed with other antibiotics (for example, spectinomycin for penicillin-resistant gonococcus) or with the penicillin derivatives that are modified to be penicillinase resistant (such as methicillin).

Bacteria can become resistant to penicillins by mutational events. These mutations probably affect the penicillin-binding proteins or the autolytic enzymes that cause the cell to lyse.

Pharmacology of Penicillins The pharmacological advantage of the penicillins is their high therapeutic index. They have a low toxicity except in the case of a few people who are allergic to them and experience an anaphylactic reaction or serum sickness. A person who is allergic to one penicillin should be considered allergic to all penicillins.

KEY POINT

The penicillins are a family of antibiotics many of which are made chemically by attaching different R groups to 6-aminopenicillanic acid isolated from *Penicillium* cultures. The semisynthetic penicillins have increased resistance to gastric acids and/or are resistant to penicillinases.

Cephalosporins

The *Cephalosporium* fungus produces the antibiotic cephalosporin C as an extracellular product. Cephalosporin C is chemically similar to the penicillins by virtue of its β-lactam ring (Figure 21.5) and its action as a cell-wall inhibitor. The family of cephalosporin drugs includes the naturally produced cephalosporin C and the semisynthetic cephalosporins that are grouped as "generations" based on their antibacterial spectra. The "first-generation" cephalosporins are active against gram-positive cocci including penicillinase-producing *Staphylococcus aureus* and some of the enterics. The "second generation" cephalosporins have an expanded spectrum that includes species of *Proteus, Enterobacter,* and *Citrobacter*. The "third-generation" cephalosporins are particularly active against gram-negative bacilli and are very important in the control of nosocomial infections.

Cephalosporins are similar in their mode of action to the penicillins. Although the biochemical reactions

◀FIGURE 21.4 Structure of the natural penicillin and selected semisynthetic penicillins with indications of their unique properties.

Cephalosporin group

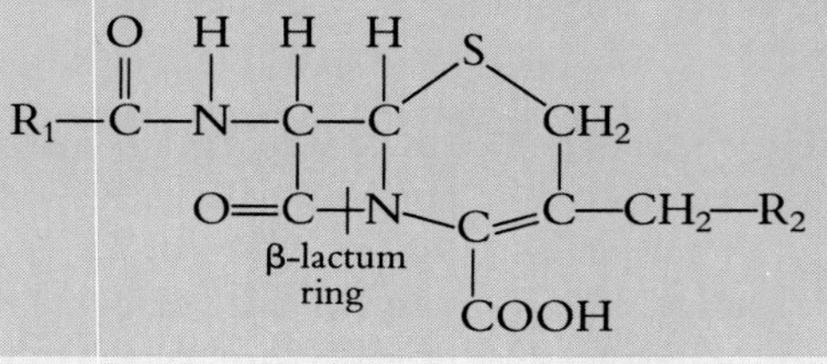

$$R_1—\overset{O}{\overset{\|}{C}}—\overset{H}{\overset{|}{N}}—\overset{H}{\overset{|}{C}}—\overset{H}{\overset{|}{C}}—S—CH_2—C(—CH_2—R_2)=C(COOH)—N(—C=O)$$

β-lactum ring

ROUTE OF ADMINISTRATION

	R_1		R_2	PARENTERAL	ORAL
Natural product	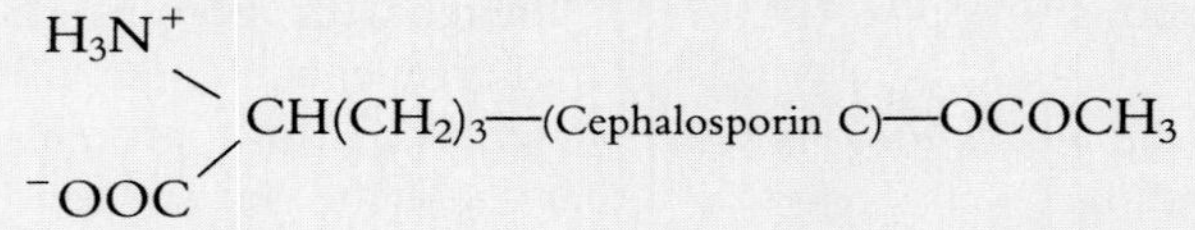 H_3N^+ / ^-OOC >CH$(CH_2)_3$—	(Cephalosporin C)	—$OCOCH_3$		
First generation	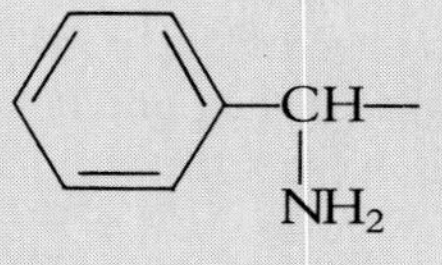 phenyl—CH(NH_2)—	(Cephalexin)	—CH_3	−	+
	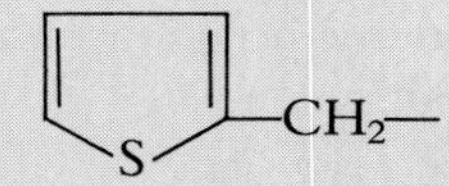 2-thienyl—CH_2—	(Cephalothin)	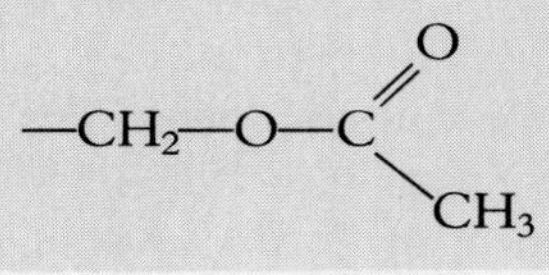 —CH_2—O—C(=O)—CH_3	+	−
Second generation	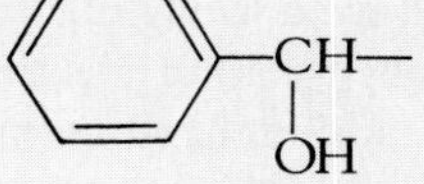 phenyl—CH(OH)—	(Cefamandole)	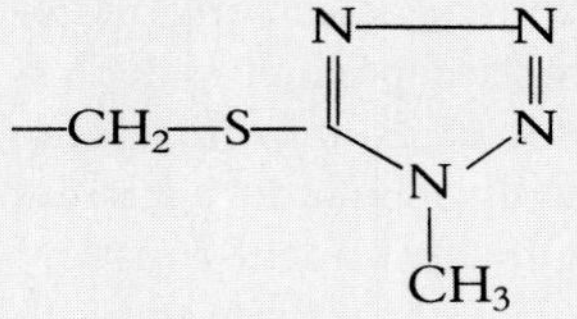 —CH_2—S—(1-CH_3-tetrazolyl: N—N, =N, N—CH_3)	+	−
	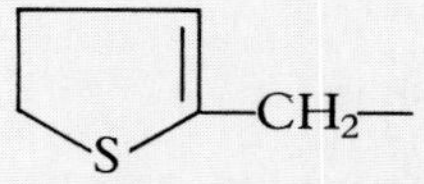 2-thienyl—CH_2—	(Cefoxitin)	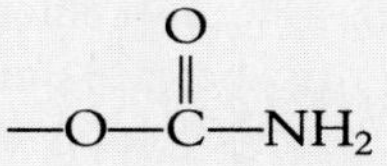 —O—C(=O)—NH_2	+	−
Third generation	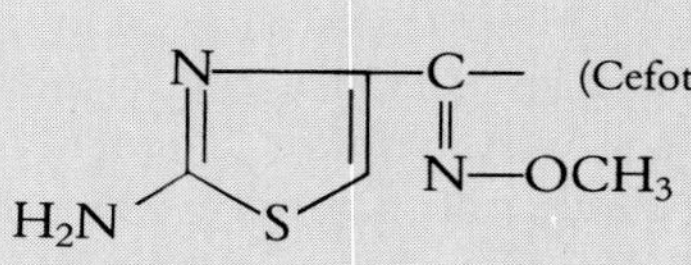 (H_2N-thiazolyl: N, S)—C(=N—OCH_3)—	(Cefotaxime)	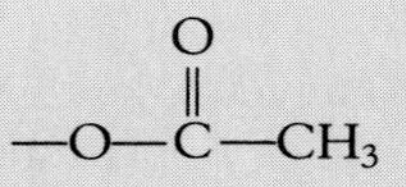 —O—C(=O)—CH_3	+	−
	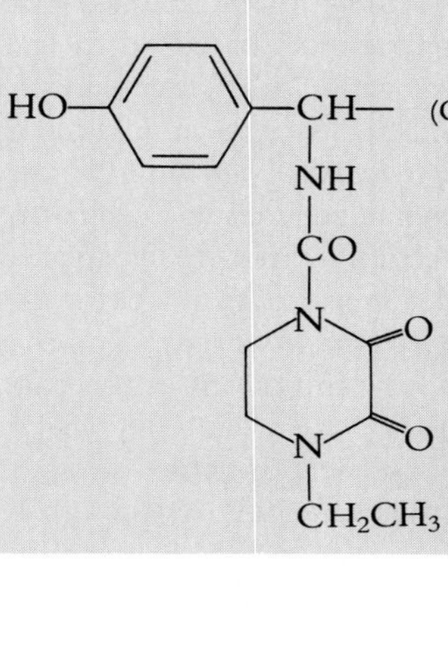 HO—(4-HO-C_6H_4)—CH(NH—CO—N(dioxopiperazine, =O, =O, N—CH_2CH_3))—	(Cefoperazone)	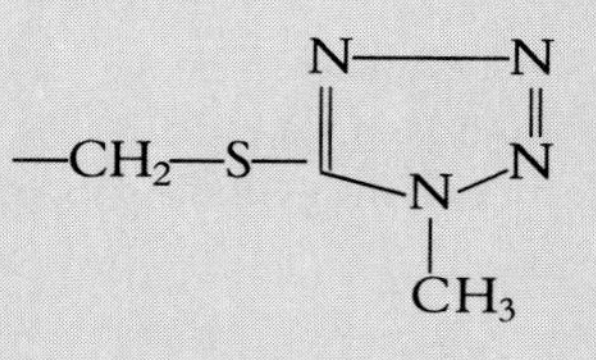 —CH_2—S—(1-CH_3-tetrazolyl: N—N, =N, N—CH_3)	+	−

FIGURE 21.5 The β-lactam structure of cephalosporin C and selected first, second, and third generations semisynthetic cephalosporins.

they inhibit are not entirely understood, it is known that they bind to enzymes and membrane proteins that appear to be involved in cell-wall synthesis. Cephalosporins are inactivated by **cephalosporinase,** which is produced by resistant strains. The first generation cephalosporins are sensitive to the β lactamases produced by gram-negative bacilli; however, this problem has been overcome with the third generation cephalosporins.

The cephalosporins are widely used and relatively safe. Some patients are allergic to the cephalosporins; however, of these only a few (5 to 10 percent) are also allergic to penicillin. The third-generation cephalosporins reach the cerebrospinal fluid and are effective in treating meningitis caused by gram-negative bacteria.

Thienamycins: Imipenem

A class of β-lactam antibiotics called thienamycins (thi-en'ah-my'cin) was discovered in the 1970s as natural products of *Streptomyces cattleya.* The prototype drug is **imipenem** (im-i'pen-em), which possesses novel stereochemical features that differentiate it from the penicillins and cephalosporins. Imipenem inhibits bacterial cell-wall synthesis by interacting with penicillin-binding proteins, and is effective against a broad spectrum of bacteria, including the gram-positive cocci, the enteric gram-negative bacilli, most anaerobic species, and *Pseudomonas aeruginosa.* Another advantage of this drug is its resistance to hydrolysis by most of the known β lactamases, penicillinases, and cephalosporinases. Imipenem is given by intravenous infusion because it is hydrolyzed by gastric acids.

β-Lactamase Inhibitors

Specific inhibitors of the β lactamases produced by bacteria are now used in combination with penicillins in chemotherapy. *Streptomyces clavuligerus* produces **clavulanic acid,** which inhibits β lactamases from a number of gram-positive and gram-negative organisms. This β-lactam analog is strongly bound by β lactamases, thus initiating the destruction of the enzyme. Clavulanic acid is combined with amoxicillin as an effective treatment against certain penicillin-resistant bacteria.

Other Antibiotics That Inhibit Cell-Wall Synthesis

Vancomycin is a glycopolypeptide antibiotic produced by *Streptomyces* spp. that is chemically unrelated to other known antibiotics. It inhibits the transfer of the peptide to the peptidoglycan and therefore inhibits cell-wall synthesis. Vancomycin is a narrow-spectrum antibiotic with activity against the gram-positive cocci. It must be injected because it is poorly absorbed from the gastrointestinal tract. Vancomycin is used to treat staphylococcal infections caused by methicillin-resistant staphylococci.

Bacitracin is a peptide antibiotic produced by *Bacillus subtilis* and *B. licheniformis.* The first antibiotic-producing strain of *B. subtilis* was isolated from the dirt in a wound suffered by a girl named Tracy and was thus named bacitracin. This antibiotic prevents peptidoglycan synthesis and is effective against gram-positive bacteria. It is not used to treat systemic infections because of its high toxicity, but is applied to wounds and skin infections, often in combination with polymyxin and neomycin, as a topical ointment.

INHIBITORS OF BACTERIAL PROTEIN SYNTHESIS

Bacterial ribosomes are composed of 50S and 30S subunits, which join with messenger RNA to form a 70S ribosome mRNA complex. Protein synthesis (translation) begins with the binding of tRNA to this complex. At the termination of translation, the ribosomes revert to their subunit state and remain as 50S and 30S subunits until they form another complex with mRNA (see Chapter 9). Some antibiotics inhibit translation on the 70S bacterial ribosomes (Figure 21.3). These antibiotics also inhibit the functioning of the 70S ribosomes found in mitochondria and chloroplasts, but this inhibition is not lethal to eucaryotic cells.

Chloramphenicol, Erythromycin, and Clindamycin

These three antibiotics are produced by different species of *Streptomyces* (Table 21.1). They are structurally diverse antibiotics (Figure 21.6) that inhibit protein synthesis by binding to the 50S bacterial ribosomal subunit. Their probable mechanism of inhibition is to

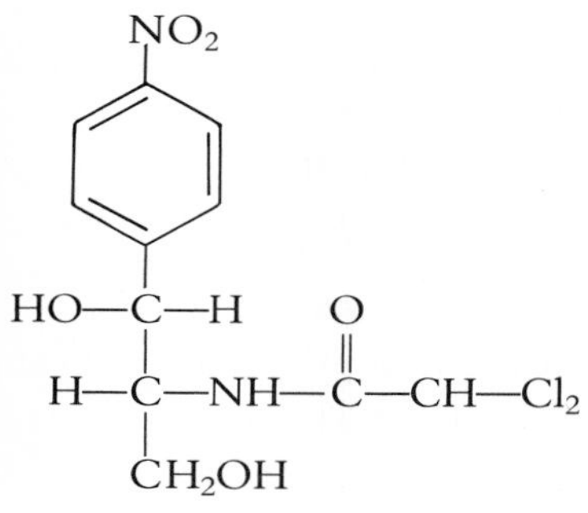

Chloramphenicol

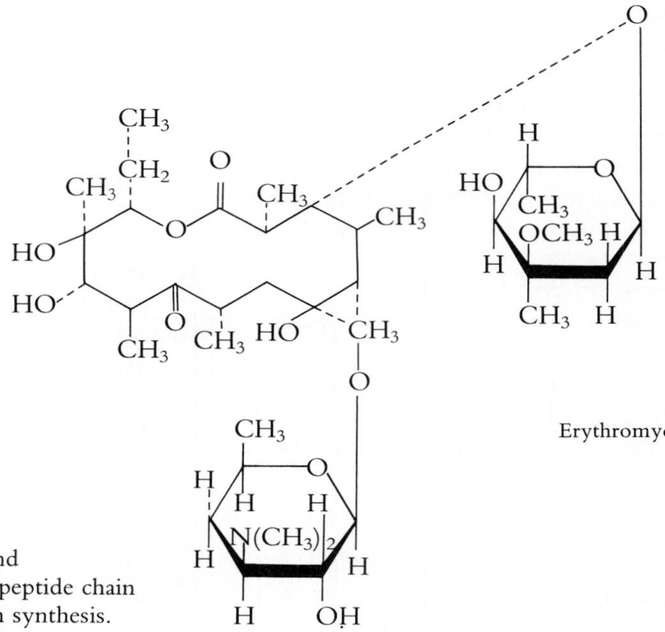

Erythromycin

FIGURE 21.6 Chloramphenicol and erythromycin are antibiotics that inhibit peptide chain elongation to prevent procaryotic protein synthesis.

prevent peptide-bond formation by binding to the peptidyl transferase center on the ribosome.

Chloramphenicol is a broad-spectrum antibiotic originally derived from a culture of *Streptomyces;* it is now produced by chemical synthesis. This bacteriostatic antibiotic binds to the 50S ribosomal subunit and blocks the peptidyl transferase reaction necessary for protein synthesis. Chloramphenicol penetrates the blood–brain barrier and quickly enters the cerebrospinal fluid, a characteristic that makes it useful in the treatment of bacterial meningitis.

Chloramphenicol carries certain risks. It disrupts normal development of red blood cells in one of every 50,000 persons, causing an extremely severe and sometimes fatal condition called aplastic anemia. Aplastic anemia also occurs in persons who are not taking chloramphenicol, so it is clear that other factors are involved in causing this disorder.

Chloramphenicol is the drug of choice in typhoid fever and serious infections such as meningitis and epiglottitis when no alternative antibiotic is appropriate. Long-term therapy with the drug is, obviously, contraindicated.

Erythromycin is the most commonly prescribed member of the macrolide family of antibiotics (Figure 21.6). It is produced by *Streptomyces erythraeus* (e-ryth′rae-us) and it inhibits protein synthesis by binding to the 50S ribosomal subunit and blocking translocation. It is effective against gram-positive bacteria, *Legionella,* mycoplasma, and chlamydias. It is the drug of choice in mycoplasmal pneumonia, pertussis, *Chlamydia trachomatis, Legionella pneumophila,* and streptococcal infections of the respiratory tract in patients allergic to penicillin. Erythromycin is administered orally, is readily absorbed, and has a high therapeutic index.

Clindamycin (klin′dah-mi′sin) is a derivative of lincomycin produced by *Streptomyces lincolnensis* (lin-cuhn-en′sis). Clindamycin is active against both gram-positive bacteria and anaerobic gram-negative bacilli. It is the drug of choice in *Bacteroides* infections occurring outside the central nervous system. Lincomycin is rarely indicated since clindamycin is more active. These antibiotics inhibit protein synthesis by binding to the 50S ribosomal subunit.

Aminoglycosides

Streptomycin, gentamicin, kanamycin, tobramycin, and amikacin are aminoglycoside antibiotics. They are bactericidal, probably because they prevent the synthesis of essential proteins. Kanamycin and streptomycin are produced by *Streptomyces* spp., while gentamicin is produced by *Micromonospora purpurea* (mi′kro-mo-nos′po-rah pur-pu-re′ah). Amikacin (am′i-ka′sin) is a semisynthetic derivative of kanamycin.

Streptomycin is rarely prescribed except to treat tuberculosis and, in combination with penicillins, to treat bacterial endocarditis. Streptomycin was the first effective aminoglycoside antibiotic to be discovered and is important historically for its effectiveness against *Mycobacterium tuberculosis.* In therapy, it is combined with other drugs because bacteria readily develop resistance to it. Combinations of streptomycin and isoniazid are credited with significantly reducing the incidence of tuberculosis during the 1950s. Streptomycin must be injected and is toxic to the auditory and vestibular branches of the eighth cranial nerve, so it can cause loss of hearing and balance.

Streptomycin binds to one protein on the 30S ribosomal subunit and thereby interferes with protein synthesis. Cells become resistant to streptomycin when the chromosomal gene for that protein is mutated such that the streptomycin no longer binds to the ribosome. Resistant cells can also produce enzymes encoded by plasmid-borne genes that enzymatically inactivate streptomycin.

Gentamicin is active against *Pseudomonas aeruginosa* and most of the *Enterobacteriaceae.* Gentamicin is given in nosocomial infections caused by gram-negative bacilli and with a second antibiotic in severe systemic infections. **Tobramycin** is an aminoglycoside with an antimicrobial spectrum similar to gentamicin, but it is more active against *Pseudomonas aeruginosa.* **Amikacin** is a semisynthetic antibiotic that has the broadest spectrum of all the aminoglycosides. It is not inactivated by many of the aminoglycoside-inactivating enzymes and therefore can be used against gentamycin- or tobramycin-resistant bacteria.

Spectinomycin

This antibiotic is an aminocyclitol that was isolated in 1960 from *Streptomyces spectabilis* (spec-ta′bi-lis). Spectinomycin prevents protein synthesis by binding to the 30S ribosomal subunit. Its primary use is to treat gonorrhea caused by penicillin-resistant *Neisseria gonorrhoeae;* other bacteria readily develop resistance to spectinomycin. It is not effective in sexually transmitted diseases other than gonorrhea; *Treponema pallidum* (syphilis) is insensitive to it, and most strains of *Chlamydia trachomatis* are resistant.

Tetracycline

The tetracyclines are a family of broad-spectrum antibiotics that inhibit the initiation of protein synthesis. They are effective against many bacteria and are indicated in brucellosis, chancroid, chlamydial and rickettsial infections, cholera, relapsing fever, and mycoplasmal pneumonia. Tetracyclines have a low therapeutic index in large part because they bind to proteins in food and to serum proteins in blood. To counteract this binding, relatively high oral or parenteral* (pah-ren′ter-al) doses are given.

The tetracyclines have bothersome side effects. Gastrointestinal disturbances including diarrhea is one. They cause staining and deformity of developing teeth, which is especially pronounced in young children whose permanent teeth are still developing. For this reason, children under 8 years of age and pregnant women are not usually treated with tetracyclines. Resistance to a tetracycline occurs in some bacteria, which apparently prevent the antibiotic's entry into their cytoplasm.

ANTIBIOTICS THAT ACT ON BACTERIAL CELL MEMBRANES

The **polymyxins** are cyclic polypeptide antibiotics produced by *Bacillus polymyxa* that disrupt the function of the cytoplasmic membrane. Polymyxin B and polymyxin E are active against gram-negative bacteria and may be indicated for certain *Pseudomonas aeruginosa* infections. They are rarely given parenterally because of their toxicity. Topical ointments containing polymyxin B are used in treating skin, ear, and eye infections. The polymyxins bind to phospholipids in the cell membrane and prevent membrane transport. They kill resting cells without causing lysis.

RIFAMPIN INHIBITS mRNA SYNTHESIS

Rifampin selectively binds to the bacterial RNA polymerase and inhibits the initiation of mRNA synthesis. It does not interfere with the activity of RNA polymerases of eucaryotes. Rifampin is a semisynthetic derivative of rifamycin, and is active against the gram-

*Not through the alimentary canal, but rather by injection.

positive cocci as well as *Mycobacterium tuberculosis*. Rifampin is used in combination with isoniazid to treat tuberculosis since *M. tuberculosis* can become resistant to rifampin during the required chemotherapy period.

ACQUIRED RESISTANCE TO ANTIBIOTICS

From the beginnings of antibiotic therapy, scientists have realized that species of normally sensitive bacteria become resistant to antibiotics. To understand how this happens, we have to know the nature of this resistance. Mutations that affect a cellular structure, such as a ribosome, can alter that structure so it no longer binds the antibiotic. Such mutations occur in the bacterial chromosome and are not readily transferred to other cells. A second mechanism is mediated by the acquisition of a gene encoding an antibiotic-inactivating enzyme, such as penicillinase. These genes are often carried on plasmids and can be transferred between bacterial strains and closely related species.

Streptomycin and erythromycin prevent bacterial translation of mRNA by attaching to proteins of the bacterial ribosome. If mutations alter these proteins such that the antibiotic can no longer attach, the bacterium becomes resistant to that antibiotic. Streptomycin resistance has been traced to alterations of one protein in the 30S ribosomal subunit, and mutational resistance to erythromycin has been traced to an alteration in a protein of the 50S ribosomal subunit.

Enzyme inactivation of an antibiotic can confer resistance on the producing cell. For example, cells that produce a β lactamase can cleave the β-lactam ring of penicillin and cephalosporin (Figures 21.5, 21.6) and inactivate these antibiotics. Antibiotic-inactivating enzymes are encoded by genes usually located on R plasmids or transposons. The R plasmids are readily transferred between cells, and genes on transposons can jump between plasmids and the chromosome (see Chapter 10). The substrate specificity of the β lactamases varies among organisms so not all β-lactam antibiotics are destroyed by a given enzyme. Some of the semisynthetic penicillins and cephalosporins have been modified to be resistant to β lactamase, and imipenem, a new β-lactam antibiotic, is resistant to attack by many β lactamases.

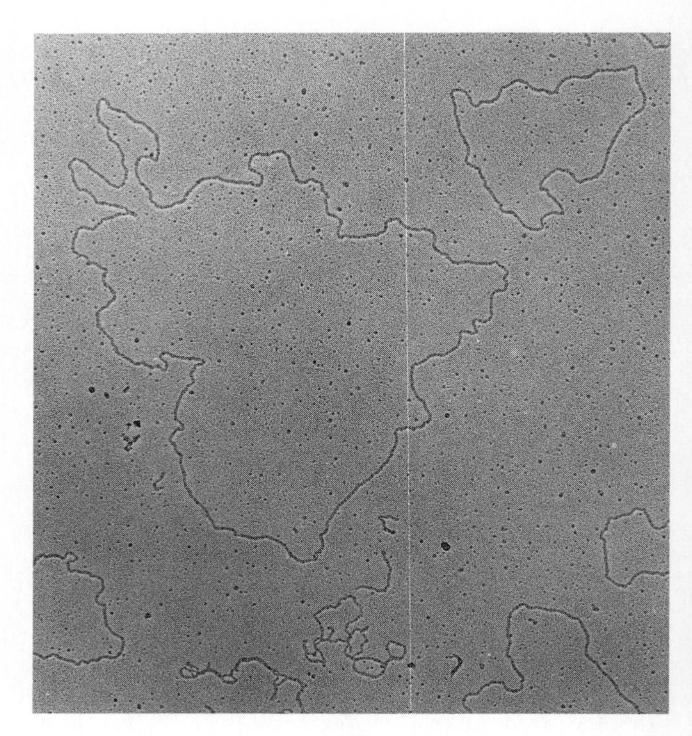

FIGURE 21.7 Electron micrograph of two bacterial plasmids that encode antibiotic resistance. The large plasmid pBF4 from *Bacteroides fragilis* encodes resistance to clindamycin; the small plasmid pSC101 from *E. coli* encodes resistance to tetracycline. 30,000 ×. (Courtesy of Dr. R. Welch, University of Wisconsin Medical School/BPS.)

Other antibiotics are inactivated by covalent modification—the addition of an acetyl group, an adenyl group, or a phosphate group to the antibiotic. The aminoglycosides, such as kanamycin, gentamicin, and streptomycin, are inactivated by covalent modification, and the responsible enzymes are found on R plasmids (Figure 21.7) and transposons. Finally, some cells acquire resistance by becoming impermeable to an antibiotic, a mechanism known to result in resistance to erythromycin and tetracyclines.

PROBLEMS ASSOCIATED WITH ANTIBIOTIC RESISTANCE

The use of antibiotics in animal feeds is an international scientific concern of increasing importance. It had been discovered, rather accidentally, that animals whose feed included large quantities of penicillin or tetracycline (Figure 21.8) fattened much faster and so were ready for slaughter sooner than animals not fed antibiotics. The growth-promoting effects of antibiotics are attributed to changes in the microflora of the animal's gastrointestinal tract. In the United States a major part of the market for antibiotics is for animal feeds, with annual sales of 270 million dollars in 1983. This large-scale use of antibiotics is believed to select for antibiotic-resistant bacteria, and the practice has been challenged by scientists who fear the consequences.

In less-developed countries of the world, antibiotics for human use can be obtained over-the-counter, that is, without a prescription. This is what happened in the Philippines, where a penicillin-resistant strain of *Neisseria gonorrhoeae* emerged. The strain was then carried to the United States by an infected patient, with the end result that penicillin-resistant strains of *N. gonorrhoeae* are being found with some frequency.

Hospitals have been notorious overusers of antibiotics—both to prepare patients for surgery and to treat

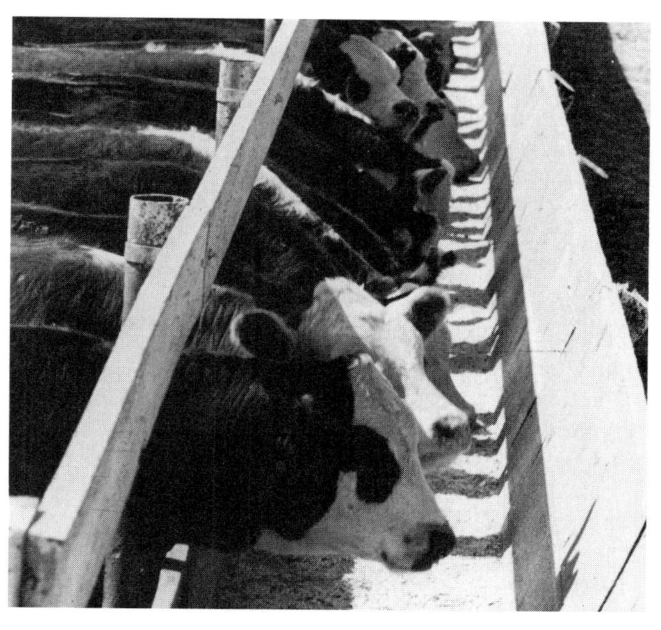

FIGURE 21.8 Antibiotic-resistant strains of bacteria are selected for when cattle are fed antibiotics as growth-promoting substances. (Photograph courtesy of USDA.)

Monobactams—A New Family of Antibiotics

New antibiotics are increasingly difficult to find since many of the compounds identified resemble known antibiotics. Nevertheless, some drug companies pursue the search for new antibiotics. One study, begun in 1978, sought microbes that produced compounds with a β-lactam ring, the chemical structure common to penicillins and cephalosporins. While screening over one million isolates, the investigators isolated a strain of *Chromobacterium violaceum* from the Wadding River in the coastal plain of New Jersey that produced an unstable compound with the desired ring structure. This compound attracted their attention because it was resistant to the β lactamases that inactivate so many penicillins and cephalosporins. Chemists were able to stabilize the compound by chemically converting it to a compound they called aztreonam.

NH_3^+ — CH — CH_2 ; C — N ; C=O ; N — SO_3^-

Aztreonam

This compound proved to be an effective antibiotic, especially against aerobic, gram-negative bacteria.

Aztreonam is characteristic of a new family of antibiotics called monobactams. It has a high therapeutic index and has been effectively used to treat infections of *E. coli*, *K. pneumoniae*, *P. mirabilis*, *Serratia marcescans*, *Ps. aeruginosa*, *Enterobacter*, and *Providencia*. In addition to being resistant to the beta lactamases, aztreonam does not cause allergic reactions in people allergic to penicillins. Aztreonam is now approved for chemical use and is an example of how microbiologists and chemists work together to develop new antibacterial drugs.

them after surgery, when the risks of infection can increase. And here again, the end result in increased resistance to antibiotics.

How can these events be explained? How is it that such beneficial and life-saving drugs as penicillin and tetracycline are proving to be less potent and less useful than in the past? The explanation appears to be that the injudicious use of powerful antibiotics has promoted an increase in R factors (antibiotic-resistance plasmids) in the flora of both food animals and people. To put it another way, the microorganisms that normally inhabit the intestinal tract (food animals and humans) are overgrown by the antibiotic-resistant strains as soon as the person or animal consumes or is treated with antibiotics.

The agricultural and pharmaceutical industries have supported the use of antibiotic supplemented animal feeds in part because there was little scientific evidence against this practice. In 1984, however, a scientist named Scott Holmberg reported epidemiological evidence showing that an antibiotic-resistant strain of *Salmonella newport* isolated from human cases of salmonellosis originated in cattle grown on antibiotic-supplemented feed. This evidence reinforces the hypothesis that antibiotics should be reserved for the use for which they were originally developed—the treatment of infectious diseases.

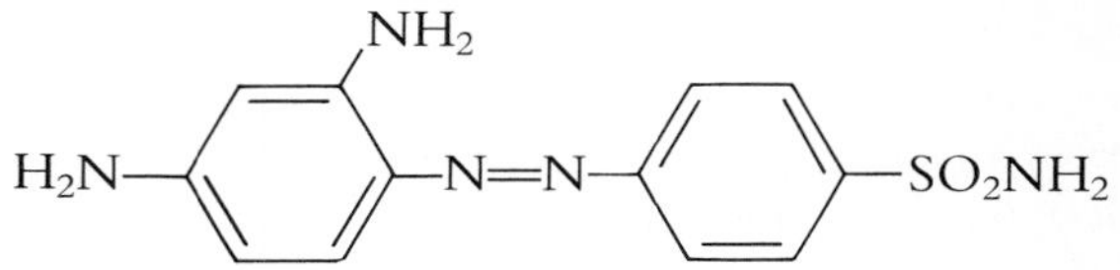

KEY POINT

Enzymes that inactivate antibiotics and confer resistance to microorganisms encoded on R plasmids can be genetically transferred between closely related bacteria. Structural changes in a cell that result in antibiotic resistance usually occur when a chromosomal gene is mutated, and such changes are not readily transferred.

Prontosil

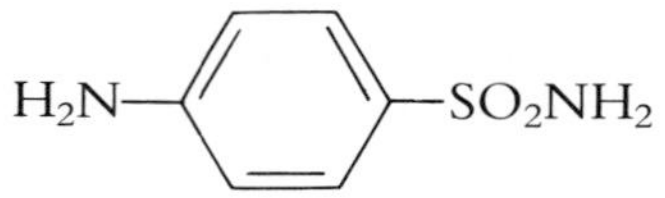

Sulfanilamide

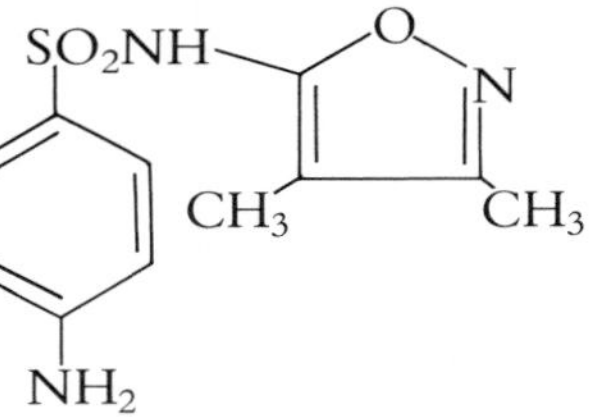

Sulfisoxazole

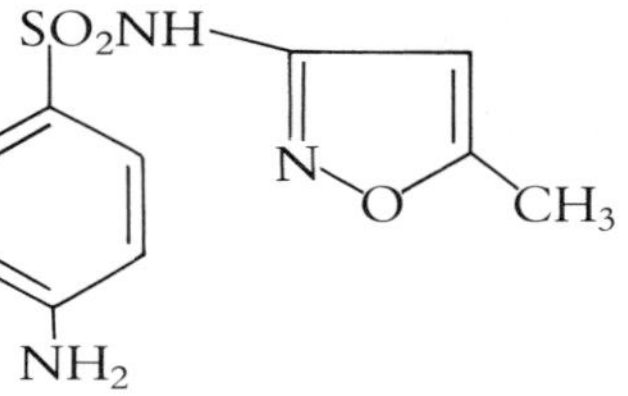

Sulfamethoxazole

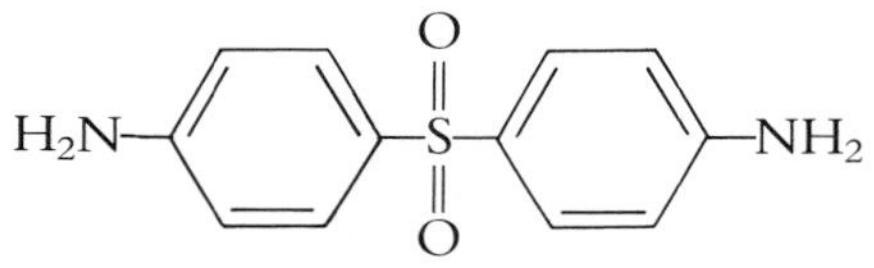

Dapsone

FIGURE 21.9 Representative sulfonamide drugs. Prontosil, one of the first sulfa drugs, breaks down in the body to sulfanilamide, which is the active agent. Dapsone is used to treat leprosy; sulfisoxazole and sulfamethoxazole are used to treat urinary tract infections.

CHEMOSYNTHETIC ANTIBACTERIALS

Paul Ehrlich (1854–1915) devoted much of his career to the development of chemicals that would selectively destroy infectious microorganisms. His search began when he tested the inhibitory effects of aniline dyes. Ehrlich reasoned that for a chemotherapeutic agent to be effective, it would have to react with a chemoreceptor on the infectious agent and he knew of dyes that reacted selectively with bacteria. This reasoning led Erhlich to study dyes from the standpoint of their potential use as chemosynthetic agents. His most successful discovery was an arsenical called **Salvarsan,** which was effectively used to treat syphilis before penicillin was developed. Ehrlich formulated a scientific approach to the discovery of chemicals that could cure infectious diseases, and he coined the word **chemotherapy** to describe the treatment of disease by chemical agents.

Chemotherapy developed slowly during the early 1900s because many of the compounds that killed bacteria were also toxic to the host. This was a critical problem, and it led scientists to conduct systematic studies of organic compounds. One result of these intensive investigations was the discovery of the drug named prontosil by Gerhard Domagk in 1935. Even though **prontosil** (Figure 21.9) had no effect on the multiplication of bacteria in laboratory cultures, it was nontoxic to mice and was an effective antistreptococcal agent when given to humans.

The active component of prontosil was **sulfanilamide** (Figure 21.9), a product of prontosil metabolism. Sulfanilamide was the forerunner of the sulfonamide drugs and served as the model for their synthesis and manufacture.

The mechanism of sulfanilamide inhibition was clarified in 1940 as follows: Its inhibitory action resulted from its structural similarity to para-amino benzoic acid (Figure 21.10), a metabolite needed for the biosynthesis of folic acid. In the presence of sulfanilamide, the bacteria are unable to make folic acid because sulfanilamide competitively inhibits a necessary enzyme in the pathway of folic acid synthesis. In effect, this inhibition poisons the bacteria. Humans are not affected by sulfanilamide because they obtain folic acid from their diet and do not synthesize it. Ultimately, all of the sulfonamide drugs were shown to act by inhibiting folic acid formation or by interfering with its action as a coenzyme in the metabolic pathways leading to synthesis of some amino acids, purines, and thymine.

SULFONAMIDES (SULFA DRUGS)

By modifying sulfanilamide, chemists have developed a family of sulfur-containing antimicrobials known as the sulfa drugs. These drugs are mainly used in the treatment of uncomplicated urinary tract infections. They are also incorporated into topical ointments, and are administered before intestinal surgery to suppress the susceptible bowel flora. Bacteria can become resistant to the sulfa drugs by overproducing para-aminobenzoic acid or by altering the enzymes involved in folic acid synthesis by mutation. Thus, emergence of resistant strains has limited the use of the sulfa drugs.

Sulfisoxazole (sul′fi-sok′sah-zol) is the preferred sulfa drug for treating urinary tract infections (Figure 21.11). Given orally, sulfisoxazole is readily absorbed into the blood and appears in the urine. It is effective against *Escherichia coli,* which is the causative agent in most initial episodes of unobstructed urinary tract infections.

Sulfamethoxazole (sul′fah-meth-oks′ah-zol) in combination with trimethoprim (tri-meth′o-prim) has a broad spectrum of activity against gram-negative bacteria and is used to treat urinary tract infections.

Folic Acid

COOH, CH_2, CH_2, H_2N, C—OH, O, OH, N, CH_2—N—, H, C—N—CH, COOH

Para-aminobenzoic acid

2-Amino-4-hydroxy-6-methylpteridine

Para-aminobenzoic acid

Glutamic acid

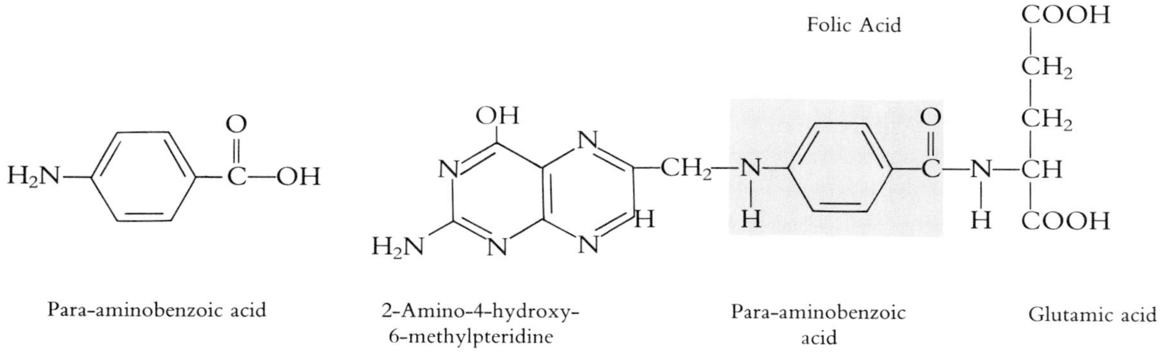

FIGURE 21.10 Para-aminobenzoic acid is a substrate for the bacterial synthesis of folic acid. Sulfanilamide, an analog of para-aminobenzoic acid, acts as a competitive inhibitor to block folic acid synthesis.

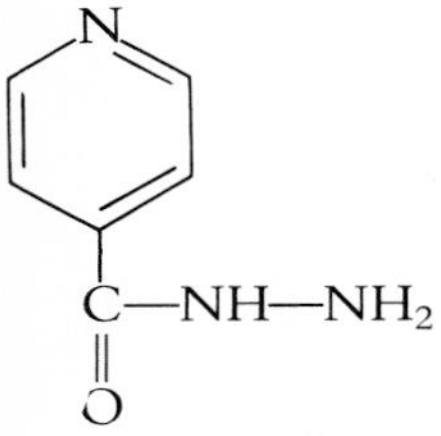

Isoniazid

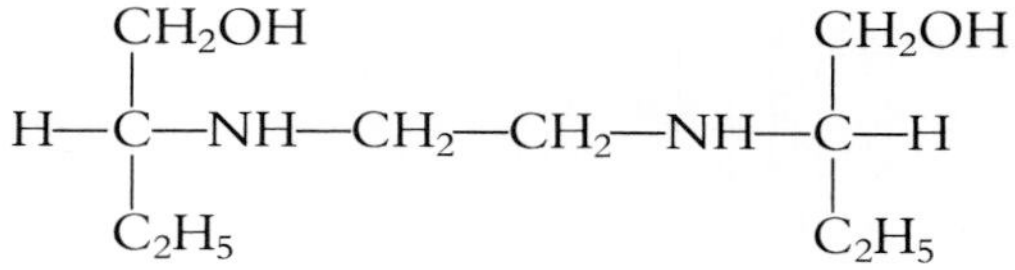

Ethambutol

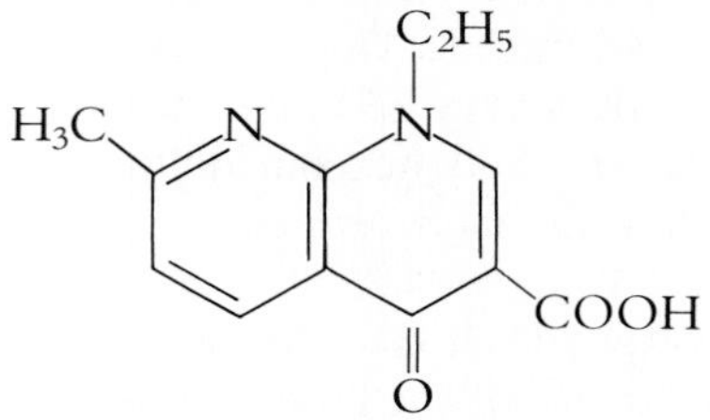

Nalidixic acid

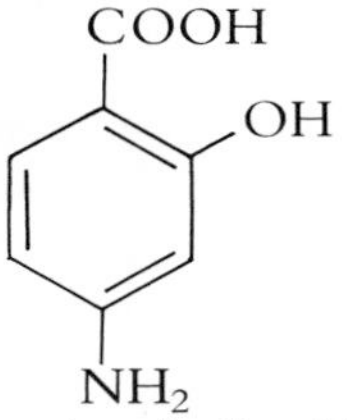

Para-aminosalicylic acid

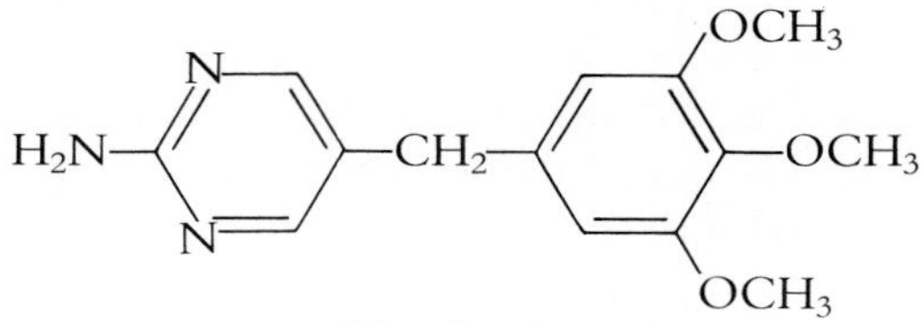

Trimethoprim

FIGURE 21.11 Representative chemotherapeutic drugs.

Trimethoprim inhibits the reduction of folic acid to its active form. In combination, trimethoprim and sulfamethoxazole are given in recurrent or chronic urinary tract infections caused by sensitive bacteria. This combination is also effective against *Pneumocystis carinii,* the protozoan that is a frequent cause of pneumonia in patients with AIDS.

Dapsone, a sulfone, is the principal drug given in leprosy. Its mechanism is presumed to parallel that of the sulfonamides.

QUINOLONES INHIBIT BACTERIAL DNA SYNTHESIS

Quinolones are antibacterials that concentrate in the urinary tract. Hence, they are prescribed in acute and recurrent uncomplicated urinary tract infections caused by susceptible bacteria. The quinolones include nalidixic acid, oxolinic acid, cinoxacin, and nitrofurantoin. **Nalidixic acid** (nal-i-diks′ik) is taken orally to treat urinary tract infections caused by some gram-negative rods. It prevents DNA synthesis by inhibiting the enzyme (gyrase) that puts supercoils in bacterial DNA prior to its replication (see Figure 9.8). Cinoxacin and oxolinic acid inhibit DNA synthesis apparently by a similar mechanism.

ANTIMYCOBACTERIAL CHEMOSYNTHETIC AGENTS

Isoniazid (i′so-ni′ah-zid) and ethambutol (e-tham′bu-tawl) are first-line antituberculosis drugs. **Isoniazid** (Figure 21.11) inhibits the synthesis of mycolic acid, a component of the cell wall of *Mycobacterium tuberculosis,* the causative agent of tuberculosis. Isoniazid is bacteriostatic even against the intracellular growth of this bacterium. However, the bacterium can become resistant to isoniazid, so the drug must be given in combination with either rifampin or streptomycin. **Ethambutol** is also effective against *M. tuberculosis,* though its mode of action is unknown.

ANTIFUNGALS

The fungi, which were described in Chapter 16, are unaffected by most antibacterials, so the infections they cause must be treated with other drugs.

Nystatin (nis′tah-tin), amphotericin B (am′fo-ter′i-sin), and griseofulvin (gris′e-o-ful′vin) are antifungals that are produced by *Streptomyces* spp. and *Penicillium griseofulvin* (Table 21.1). **Nystatin,** named for New

York State where it was first isolated, is given to treat superficial yeast infections of the mouth, skin, and vagina. It is toxic, so it cannot be administered parenterally; its use is restricted to oral and topical medications.

Amphotericin B, the most effective drug in systemic fungal infections, can cause serious adverse reactions in humans, including renal damage, anorexia, nausea, weight loss, chills, fever, and vomiting. It is, nevertheless, the drug of choice in most life-threatening systemic fungal infections.

Griseofulvin is active against fungi that infect hair, nails, and skin. It is administered orally over long periods so that it can reach a therapeutic level in these tissues.

There are a number of chemosynthetic antifungal agents that are incorporated into creams and powders for topical application and some that are given parenterally to treat systemic mycoses. **Flucytosine** (flu-si'to-seen') is an analog of cytosine that is used in combination with amphotericin B to treat systemic infections of *Candida* and *Cryptococcus*.

Miconazole (mi-kon'ah-zol) is a synthetic antifungal that is incorporated in topical creams and lotions to treat cutaneous candidiasis and fungal skin infections such as athlete's foot. Miconazole can be given intravenously as a treatment for coccidioidomycosis and paracoccidioidomycosis.

Ketoconazole (ke'to-kon'ah-zol) is a synthetic that shows promise as an antifungal, being the first orally administered antifungal that is active against a broad spectrum of pathogenic fungi. It has therapeutic activity in paracoccidioidomycosis, chronic mucocutaneous candidiasis, and fungal skin infections. (Amphotericin B is still the drug of choice in severe systemic fungal infections, since most patients recover slowly from the infection when treated with ketoconazole.)

ANTIPARASITICS

Chemotherapy against parasitic infections has not matched the success of antibacterial chemotherapy, mainly because many antiparasitics are too toxic for human use. However, a number of antiparasitics are under development. Some are approved for investigational use only and others are available in the United States only from the Centers for Disease Control.

ANTIPROTOZOANS

The most effective drugs against protozoan infections are chemosynthetics. **Metronidazole** (me'tro-ni'dah-zol) is administered orally in amebiasis or giardiasis, and orally or intravaginally in trichomoniasis. The drug is absorbed well and kills protozoa both within and outside the intestine.

Quinine occurs in cinchona bark and has been used for centuries by the natives of Peru to treat malaria. It interferes with the life cycle of *Plasmodium falciparum* by inhibiting the schizont stage. **Quinacrine hydrochloride,** originally used as an antimalarial, is now given to treat giardiasis. **Chloroquine** phosphate and **primaquine** phosphate are given prophylactically in malarious regions. Chloroquine phosphate is used to treat all forms of malaria, though some strains of *Plasmodium* are now resistant to it. Chloroquine acts on the asexual erythrocytic forms of *Plasmodium*. Other antiparasitics are mentioned in the discussion of specific diseases.

SENSITIVITY TO ANTIMICROBIALS

Besides isolating and identifying infectious agents, clinical microbiologists are responsible for advising physicians of their isolates' susceptibility to antimicrobials. This is of paramount importance now that antibiotic resistance has become commonplace.

Sensitivity to antimicrobials is determined by *in vitro* testing of isolates against an array of chemotherapeutic drugs. The basic measure of this susceptibility is the minimum inhibitory concentration (MIC). The **MIC** is the least amount of antimicrobial agent that prevents visible growth of the microbe under standard conditions. Another measure is the minimun lethal concentration (MLC)—the least amount of antimicrobial agent required to kill a portion (usually 99.9 percent) of the inoculum within a specified period. Two basic techniques for determining the MIC of an antimicrobial agent are (1) the tube dilution technique and (2) the diffusion sensitivity disk method.

TUBE DILUTION METHOD

Dilution susceptibility tests afford the most direct method of measuring susceptibility to an antimicrobial.

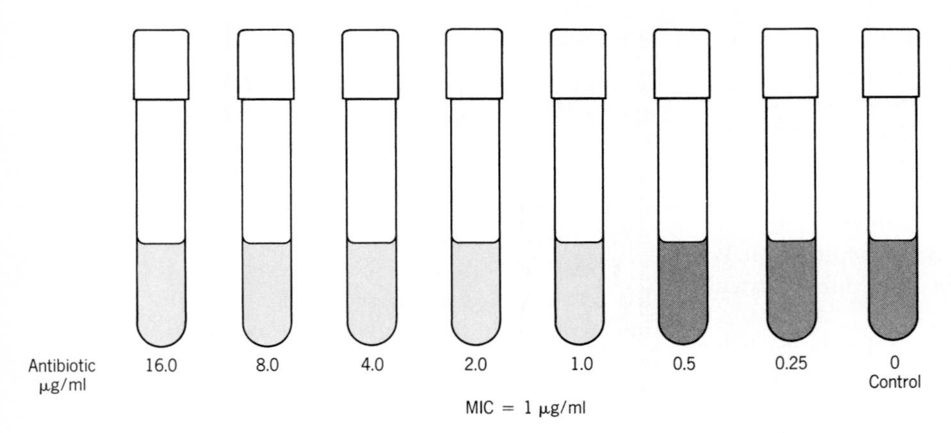

FIGURE 21.12 Dilution susceptibility test for determining minimum inhibitory concentration (MIC) of an antibiotic. The lowest antibiotic concentration that inhibits growth is the MIC.

In a typical macrodilution test, as it is called, twofold dilutions of the antimicrobial are prepared in growth medium such that the concentration of the drug covers its clinically significant range. An equal volume of broth containing 10^5 to 10^6 bacteria per milliliter is added to each tube and to a control tube that contains no antimicrobial (Figure 21.12). The tubes are read to determine visible turbidity after overnight incubation.

The macrodilution test is the method for determining antimicrobial susceptibility in liquid media; however, it is cumbersome and has given way in the clinical laboratory to microdilution tests and automated testing procedures. All are used to determine the MIC, upon which appropriate drug therapy is based. In most clinical situations the MIC determined *in vitro* is substantially below the level of drug actually needed at the infection site. Thus, the MIC serves as a guide to an organism's relative susceptibility.

DIFFUSION SENSITIVITY DISK METHOD

The diffusion sensitivity disk method is a simple and economical substitute for the more direct dilution test. In this method, a culture of known concentration is spread on an appropriate agar medium (Figure 21.13). Mueller–Hinton medium is widely used because it is an appropriate growth medium for most bacterial pathogens and it contains very few substances that inactivate antibiotics.

Filter paper disks containing predetermined concentrations of an antimicrobial are placed on the seeded agar plate in a set pattern, with equal spacing between disks. Automated dispensers apportion eight or more different antimicrobial agents simultaneously (Figure 21.13). Each disk is identified by the antimicrobial and

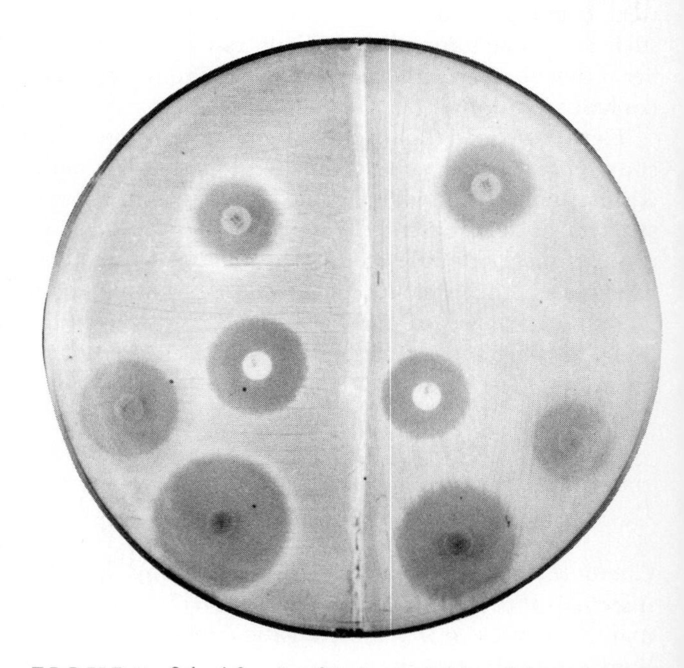

FIGURE 21.13 Antibiotic sensitivity disk testing. Each filter-paper disk contains a set concentration of an antibiotic. The zone of growth inhibition indicates the culture's sensitivity to that antibiotic. (Centers for Disease Control, Atlanta, GA.)

its concentration. During incubation, the agent diffuses out of the disk, creating a concentration gradient that decreases according to distance away from the disk. The plate is incubated for 16 to 18 hours at 35°C, after which the organism's sensitivity is measured on the basis of the size of the zone of inhibition (no growth) around each disk. This measure (recorded in millimeters) is compared to values on a standard chart, and it indicates whether the organism is resistant, intermediate, or sensitive to the agent. By sensitive is meant that the organism is inhibited by the antimicrobial at clinically attainable concentrations; by resistant is meant that the organism is not inhibited; and by intermediate is meant that special considerations need to be followed if the agent is to be used.

The Kirby–Bauer method for diffusion susceptibility testing has standardized many of the variables of the diffusion sensitivity disk method. In this procedure the concentration of antibiotics to be tested, the composition of the test medium, the incubation time, the nature of the inoculum, and the measurement of the inhibition zone are each standardized. The Kirby–Bauer method has been accepted as the standard operating procedure by many clinical microbiology laboratories.

Summary Outline

SELECTIVITY OF ANTIMICROBIAL AGENTS

1. Clinically effective antimicrobial agents are selectively toxic against microbes without adversely affecting the host.
2. Chemotherapeutic drugs are chemicals used to treat infectious diseases. The true antibiotics are substances produced by one organism that kills or inhibits the growth of another, whereas other antimicrobials are chemically synthesized.
3. The higher the therapeutic index of an antimicrobial, the greater its efficacy as a drug.

ANTIBACTERIALS

1. Alexander Fleming discovered penicillin in 1929; however, it was not tested in a human until 1941 when Florey and Chain purified enough penicillin to carry out clinical trials.
2. Penicillin G became widely available after World War II. The development of the semisynthetic penicillins awaited the microbial production of 6-aminopenicillanic acid in the late 1950s.
3. Members of the genus *Streptomyces* are the most important producers of antibiotics. Some *Bacillus* species and a few fungi also produce clinically important antibiotics.
4. The range of microbes affected by an antimicrobial agent is its spectrum: antibiotics are either broad-spectrum or narrow-spectrum antimicrobials.
5. Antibiotics inhibit the growth of (bacteriostatic) or kill (bactericidal) microorganisms by interfering with protein synthesis, nucleic acid synthesis, cell-wall synthesis, or cell-membrane function.
6. Penicillins are bactericidal antibiotics that inhibit bacterial cell-wall synthesis. Semisynthetic penicillins are made to be acid-resistant and/or penicillinase-resistant.
7. Penicillins are widely used against gram-positive and gram-negative cocci. Resistant cells acquire a penicillinase or mutate their penicillin-binding proteins.
8. The semisynthetic cephalosporins have a wide spectrum of antibacterial activities and inhibit cell-wall synthesis.
9. Imipenem, the prototype drug of the thienamycins, inhibits bacterial cell-wall synthesis, is effective against a broad spectrum of bacteria, and resists hydrolysis by most of the known β lactamases, penicillinases, and cephalosporinases.
10. Clavulanic acid, a β-lactam analog that inhibits β lactamases, is combined with amoxicillin as an effective treatment against certain penicillin-resistant bacteria.
11. Vancomycin and bacitracin are polypeptide antibiotics that inhibit bacterial cell-wall synthesis.
12. Chloramphenicol, erythromycin, and clindamycin inhibit procaryotic protein synthesis by binding to the 50S ribosomal subunit.

13. Chloramphenicol is the drug of choice for treating typhoid fever and gram-negative bacterial meningitis. The risk that aplastic anemia will develop may contraindicate its use.
14. Erythromycin is the drug of choice for treating pertussis, *Legionella pneumophila,* and mycoplasmal and chlamydial infections.
15. Clindamycin is used in *Bacteroides* infections outside the central nervous system.
16. The aminoglycosides, including streptomycin, gentamicin, kanamycin, tobramycin, and amikacin, act by inhibiting bacterial protein synthesis by binding to the 30S ribosomal subunit.
17. The aminoglycosides can be inactivated (covalently modified) by attaching adenyl, phosphate, or acetyl groups.
18. Streptomycin was used at one time to treat tuberculosis, but now it is rarely prescribed and only in combination with other drugs.
19. Gentamicin and tobramycin are active against *Pseudomonas aeruginosa* and most of the *Enterobacteriaceae.*
20. Spectinomycin is used to treat infections of penicillin-resistant *Neisseria gonorrhoeae.*
21. Tetracyclines are a family of broad-spectrum antibiotics that inhibit protein synthesis; however, they have a low therapeutic index and cause discoloration of developing teeth.
22. Polymyxins are polypeptide antibiotics that disrupt the function of the cytoplasmic membrane.
23. Rifampin binds to bacterial RNA polymerase and inhibits the initiation of mRNA synthesis.

ACQUIRED RESISTANCE TO ANTIBIOTICS

1. Bacteria acquire resistance to antibiotics when a structure is altered (mutation) so as to prevent antibiotic binding, or the cell acquires genes (plasmid) encoding enzymes that inactivate the antibiotic.
2. Selection for antibiotic resistance through the indiscriminate use and prescription of antibiotics is a major problem.

CHEMOSYNTHETIC ANTIBACTERIALS

1. Sulfonamides (sulfa drugs) interfere with the synthesis of folic acid or with its reduction to an active cofactor.
2. Sulfisoxazole is the preferred sulfa drug for treating urinary tract infections. Combinations of sulfamethoxazole and trimethoprim are also used.
3. Nalidixic acid is a quinolone given in urinary tract infections, dapsone is a sulfone used to treat leprosy, and isoniazid and ethambutol are antituberculosis drugs.

ANTIFUNGALS

1. Griseofulvin is an antifungal antibiotic that interferes with the mitotic apparatus of eucaryotes. Nystatin and amphotericin B are antifungal antibiotics that bind to membrane sterols and disrupt membrane permeability.

ANTIPARASITICS

1. Miconazole and ketoconazole are antifungal imidazoles and flucytosine is an antifungal analog of cytosine.
2. Quinine is an antibiotic given in protozoan infections. Metronidazole, quinacrine hydrochloride, and chloroquine are chemosynthetic antiprotozoan agents.

SENSITIVITY TO ANTIMICROBIALS

1. Clinical microbiologists are responsible for advising physicians of the drug susceptibility of microorganisms.
2. The minimum inhibitory concentration (MIC) is the least amount of antimicrobial that prevents growth of the microbe under standard conditions.
3. Tube dilution susceptibility tests are the most direct method of measuring a strain's susceptibility to an antimicrobial.
4. The diffusion sensitivity disk method is a simple and economical way for testing the susceptibility of a strain to many different antimicrobials.

Questions and Topics for Study and Discussion

QUESTIONS

1. What groups of organisms produce antibiotics? What is the function of antibiotics in nature?
2. Describe the mode of action of two antibiotics that inhibit cell wall formation and two that inhibit protein synthesis.
3. What are broad-spectrum antibiotics?
4. Describe two ways bacteria become resistant to antibiotics. Explain the genetic basis for each.
5. How is drug resistance transferred between bacteria? What modern uses of antibiotics contribute to the natural selection of antibiotic-resistant bacteria?
6. What are sulfa drugs and how do they work?
7. Describe a practical method for determining antibiotic sensitivity.
8. How could you experimentally determine if an antibiotic is bactericidal or bacteriostatic?

DISCUSSION TOPICS

1. Why are fewer antibiotics available for use against eucaryotic cells than against procaryotic cells?
2. Are there biological reasons for science's inability to develop effective broad-spectrum antiviral drugs?

Further Readings

BOOKS

Crueger, W., and A. Crueger, *Biotechnology: A Textbook of Industrial Microbiology,* Sinauer Associates, Sunderland, Mass., 1984. The commercial approaches to manufacturing antibiotics are discussed in this textbook.

Garrod, L. P., H. P. Lambert, and F. O'Grady, *Antibiotic and Chemotherapy,* Churchill Livingstone, New York, 1981. An authoritative reference work on the biology, biochemistry, and pharmacology of chemotherapeutic agents.

Sheehan, J. C., *The Enchanted Ring, The Untold Story of Penicillin,* The MIT Press, Cambridge, Mass., 1982. The author's personal account of the development of penicillin from its early production problems in the United States through his successful synthesis of penicillin at MIT in 1957.

Stewart, G. T., *The Penicillin Group of Drugs,* Elsevier, Amsterdam, 1965. The history, chemistry, and impact of penicillin are the topics of this monograph.

ARTICLES AND REVIEWS

Abraham, E. P., "The Beta-Lactam Antibiotics," *Sci. Am.,* 244:76–86 (June 1981). An excellent review of the history and biology of the penicillins and cephalosporins.

Benveniste, R., and J. Davis, "Mechanism of Antibiotic Resistance in Bacteria," *Ann. Rev. Biochem.,* 42:471–506 (1973). A review of the genetic and biochemical mechanisms through which bacteria develop antibiotic resistance.

Blumberg, P. M., and J. L. Strominger, "Interaction of Penicillin with the Bacterial Cell: Penicillin-Binding Proteins and Penicillin-Sensitive Enzymes," *Bacteriol. Rev.,* 38:291–335 (1974). A biochemical review dealing with the interaction of penicillin with the components of bacterial cells.

Goldstein, G. W., and A. L. Betz, "The Blood–Brain Barrier," *Sci. Am.,* 255:74–83 (September 1986). The authors explain why the lipid solubility of an anitibotic is the key to its ability to cross the blood–brain barrier.

Kobayashi, G. S., and G. Medoff, "Antifungal Agents: Recent Developments," *Ann. Rev. Microbiol.,* 31:291–308 (1977). A review that covers the structure and function of antifungal agents.

PART

4

Infectious Diseases

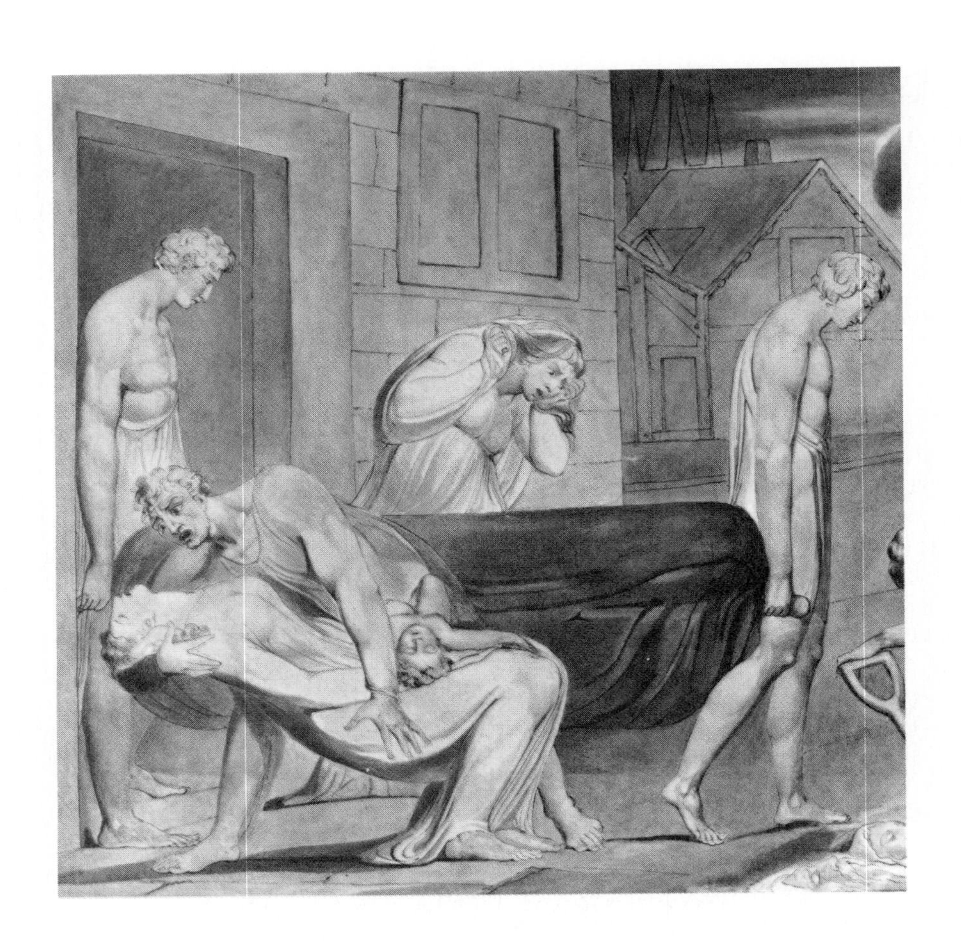

Though many infectious diseases have been eliminated or reduced in frequency and severity, at least in the developed countries of the world, there are others that remain a threat to all of society or to some of its population groups. This is especially true of those diseases that sometimes reach epidemic proportions and spread rapidly through a susceptible population. To control, reduce, or eliminate these diseases, we need to learn their underlying causes and modes of transmission.

Part 4 is devoted to infectious diseases of body systems, since it is at the system level that the diseases are recognized and, when possible, treated. Within each main discussion, the text follows the taxonomic grouping of the pathogens in order to emphasize their distinctive attributes.

C H A P T E R

22

Infectious Diseases of the Skin and Eyes

OUTLINE

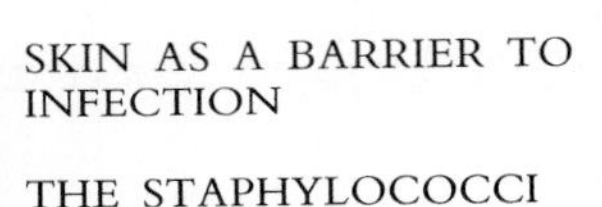

- SKIN AS A BARRIER TO INFECTION
- THE STAPHYLOCOCCI
  - Extracellular Products
  - Staphylococcal Diseases
    - Toxic Shock Syndrome
    - Epidemiology
- THE STREPTOCOCCI
  - *Streptococcus pyogenes*
    - Toxins and Virulence Factors
    - Diseases Caused by *Streptococcus pyogenes*
    - Epidemiology
- *BACILLUS:* ANTHRAX
- SPIROCHETE: TREPONEMA
  - Pinta
  - Yaws
- WOUND INFECTIONS
  - *Pseudomonas aeruginosa* and Burns
  - Infectious Gangrene
    - The Histotoxic Clostridia
    - Clinical Symptoms of Gas Gangrene
- INFECTIONS OF THE EYE
  - Conjunctivitis
  - Ophthalmia Neonatorum
  - Inclusion Conjunctivitis
  - Trachoma
- VIRAL DISEASES
  - Measles
    - Measles Virus
    - Symptoms
    - Epidemiology
  - Rubella
    - Rubella Virus
    - Symptoms
    - Epidemiology
  - The Herpesviruses
    - Orofacial Herpes
    - Varicella–Zoster Virus
    - Prevention and Treatment
  - Poxviruses
    - Smallpox
    - Eradication of Smallpox
  - Warts: Papillomaviruses

FUNGAL DISEASES
- Cutaneous Mycoses
  - Tinea Corporis
  - Tinea Cruris
  - Tinea Pedia
  - Tinea Capitis
- Subcutaneous Mycoses
  - Sporotrichosis
  - Chromomycosis
  - Mycetomas

PROTOZOAN DISEASES
- Kala-azar
  - Cutaneous and Mucocutaneous Leishmaniasis

FOCUS OF THIS CHAPTER

Skin, the first line of defense against pathogens, is usually a highly effective barrier against invasion by them. But under some circumstances pathogens can penetrate the skin, with resulting infection. This chapter is concerned with the ways pathogens gain access to the body through skin penetration, the problems that can ensue, and preventive measures and treatment.

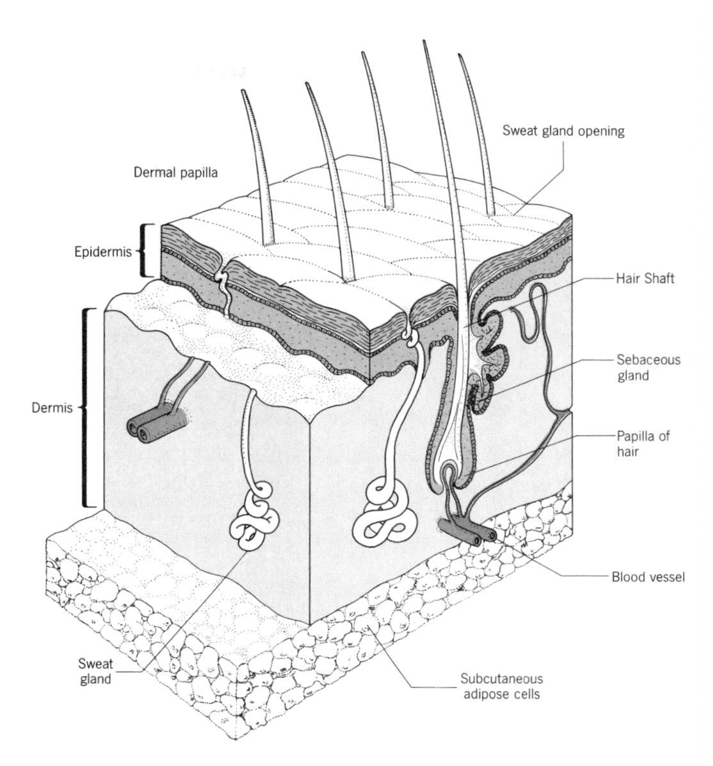

FIGURE 22.1 The skin is the first line of defense against infectious agents.

SKIN AS A BARRIER TO INFECTION

Skin is made up of the epidermis, which is the outer layer, and dermis, the underlying layer (Figure 22.1). The stratum corneum, which is the outermost layer of epidermis, is constantly being sloughed off, carrying with it bacteria and other normal inhabitants (species of *Corynebacterium, Haemophilus, Micrococcus, Neisseria, Propionibacterium, Staphylococcus,* and *Streptococcus).* Their growth is restrained by the presence of fatty acids, lactic acid, and salts of perspiration, all of which are bacteriostatic.

Skin bacteria are transmitted between individuals by direct contact. Among the most common skin diseases are those due to cocci that naturally inhabit the skin, but that can cause infection when host defenses weaken. Species of *Staphylococcus* and *Streptococcus* are gram-positive cocci and the diseases due to them range in severity from boils to generalized bacteremia and rheumatic fever.

THE STAPHYLOCOCCI

Staphylococci, as described in Chapter 15, are catalase-positive, gram-positive, facultative-anaerobic bacteria that grow as clusters of spherical cells (Figure 22.2). These bacteria are highly resistant to drying, withstand heat of 60°C for 30 minutes, and grow in the presence of 7.5 percent NaCl. *Staphylococcus aureus* and *S. epidermidis* are common inhabitants of the skin, respiratory tract, mucous membranes, and intestine.

EXTRACELLULAR PRODUCTS

Staphylococcus aureus is the most important species of the genus *Staphylococcus. Staphylococcus aureus* is differentiated from other staphylococci by means of the coa-

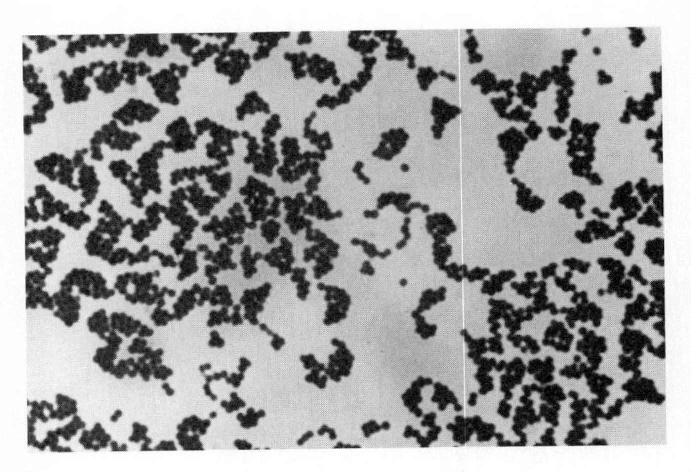

FIGURE 22.2 Gram stain of *Staphylococcus aureus*. The spherical cells are arranged in clusters. 1000 X. (Courtesy Dr. L. LeBeva.)

gulase test. Coagulase is an extracellular enzyme that reacts with citrated plasma to form a clot. Its role in virulence is unclear, though most pathogenic staphylococci produce the enzyme. Coagulase-positive cultures also produce other extracellular virulence factors including hemolysins, lipase, hyaluronidase, DNase, and fibrinolysin (Table 22.1).

In addition, some 50 percent of the coagulase-positive strains produce heat-resistant enterotoxins (see Chapter 17). There are five staphylococcal enterotoxins coded for by prophage genes that cause the acute gastrointestinal upset associated with staphylococcal food poisoning.

STAPHYLOCOCCAL DISEASE

Staphylococcus aureus is found in the nasal cavity of most adults and on the skin of most children. It establishes a local infection by invading a sebaceous gland, a hair follicle, or a wound in the epidermis. The resulting inflammatory reaction attracts lymphocytes to the site and increases the capillary permeability, causing localized swelling and the formation of a pimple. Plasma then enters the area, bringing its clotting factors, which surround the site with a fibrin barrier to prevent the bacteria from spreading into the circulatory system. The pimple becomes suppurative (produces pus) as the lymphocytes and bacteria accumulate. The pus-filled vesicle near the epidermal surface is a **boil** or a **furuncle** (fu'rung-k'l), depending on the degree of penetration. This is the genesis of boils and furuncles—the least severe forms of "staph" infection.

Impetigo (im'pe-ti'go) is a localized purulent dermatitis that commonly affects the face and hands. Between 30 and 80 percent of the cases of impetigo are

TABLE 22.1 Extracellular products of staphylococci

EXTRACELLULAR PRODUCT	MODE OF ACTION	PROPERTIES
Alpha toxin	Lysis of rabbit RBC	Heat-labile protein
Beta lysin	Lysis of sheep RBC	Inactive at 37°C
Gamma lysin	Lysis of human and rabbit RBC	Heat-stable toxin
Delta lysin	Toxic to leukocytes	
Leukocidins	Kill leukocytes	Three different types
Exfoliative toxin	Desquamation in scalded-skin syndrome	Coded for by plasmid
Enterotoxin (A, B, C, D, and E)	Acute gastrointestinal upset (food poisoning)	Relatively heat-stable toxin
Coagulase	Clots plasma	High correlation with virulence
Hyaluronidase	Breaks down hyaluronic acid in connective tissue	Invasive factor

caused by *S. aureus,* many of the remaining cases are caused by *Streptococcus pyogenes* (see following). Impetigo is spread from one site to another by towels and touching. It is treated with antmicrobial ointments.

In hospital nurseries staph infections are a serious problem. *Staphylococcus aureus* carried on the skin of mothers and hospital personnel is readily transmitted to newborns by direct contact. As a result, the newborn's umbilical cord and eyes are often colonized by staphylococci very quickly. How severe these infections will become depends on the strain involved. Infecting strains that produce the exfoliative toxin can cause the **scalded-skin syndrome** characterized by desquamation (des'kwah-ma'shun) of surface skin layers (Figure 22.3). Staph infections in the nursery cause serious problems when antibiotic-resistant strains are associated with high infection rates. Hexachlorophene and chlorhexidine are antiseptics used in hospitals to control staph infections.

Surgical procedures, accidents, burns, and debilitating diseases such as cancer can so severely weaken the patient that (s)he is rendered vulnerable to the critical problems of pneumonia, osteomyelitis, deep tissue abscesses, endocarditis, meningitis, and suppurative arthritis caused by staphylococci. Intensive treatment with drugs is required in these cases.

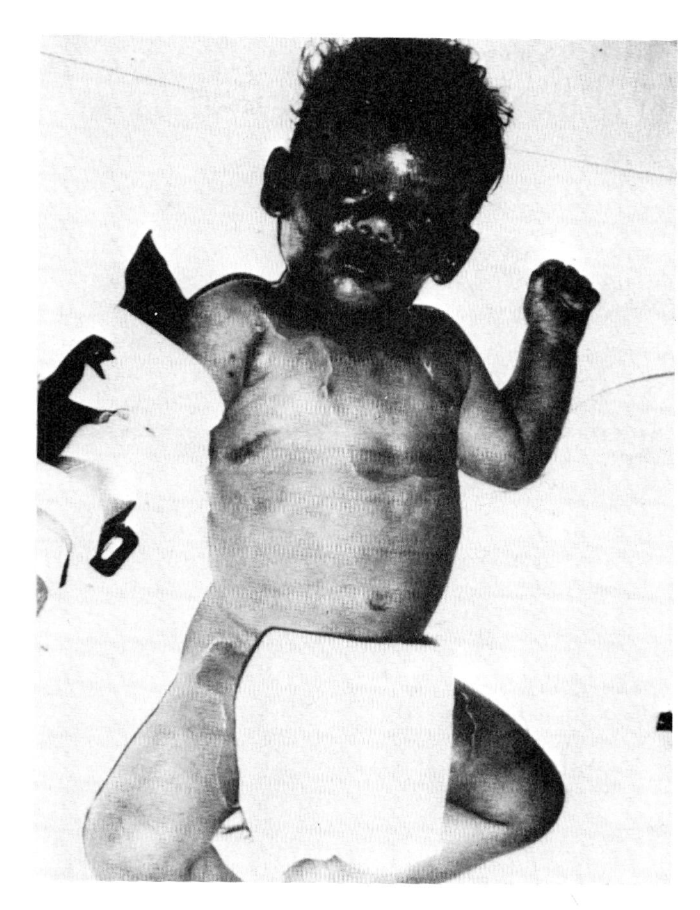

FIGURE 22.3 Scalded-skin syndrome. (From G. Mandell, R. G. Douglas, and J. E. Bennett (eds.), *Principles and Practice of Infectious Diseases,* 2nd edition, John Wiley & Sons, 1985.)

Toxic Shock Syndrome

Toxic shock syndrome (TSS) was first reported in the United States in 1975 as an occurrence unique to menstruating women. By 1980, the year in which 299 TSS cases and 25 TSS deaths had occurred between January and September, the public was alarmed. Microbiologists studying this newly reported and dangerous illness were able to link *Staphylococcus aureus* and the newly introduced superabsorbent tampons with the occurrence of TSS; 95 percent of the TSS cases occurred in menstruating women and almost all were using the new tampon. The superabsorbent tampon was involved in 71 percent of TSS cases, as determined by a careful survey. (It may be of interest to note that the increase in TSS paralleled the increase in market share of this product—which suffered a steep drop in sales before it was taken off the market in late 1980.)

Symptoms of TSS include the sudden onset of high fever, vomiting, diarrhea, and muscle cramps. A rash similar in appearance to sunburn develops about ten days after the symptoms first appear. In a severe case the patient goes into shock; in some 8 percent of reported cases the patients died. TSS in men and in nonmenstruating women was also correlated with *S. aureus* present in a focal lesion of skin, bone, or lung.

From all that has been learned, it appears that TSS is caused by exotoxin (TSS-1) producing strains of *Staphylococcus aureus.* Both toxin production and growth of the responsible strains are controlled by the availability of magnesium. With a low concentration of magnesium, the bacterium ceases to multiply and begins to produce the extracellular toxins. The superabsorbent tampon directly affects the availability of this metal because the tampon binds ten times as much magnesium as other tampons. Thus the tampon effectively decreases the magnesium concentration to a point where the indigenous *S. aureus* begin to produce the toxin.

Epidemiology

Staphylococci are normal flora of the skin, mucous membranes of the respiratory tract, and the gastrointestinal tract, and are transmitted by direct contact. Adults carry *Staphylococcus aureus* in the nose, and spread it to skin and clothing with their hands. With this direct route of transmission, one can easily see why the carrier rate is higher among medical personnel than among the general population. Patients with severe furuncles often reinfect themselves from contaminated bandages, clothes, and bed linens, and can serve as a significant source of contamination in a hospital.

Staphylococcal infections require rapid and effective treatment because the bacteria readily become resistant to penicillins. Their resistance to penicillin is carried on a plasmid and mediated by an extracellular penicillinase. Infections caused by penicillinase-resistant *Staphylococcus* can be treated with methicillin, which is not hydrolyzed by penicillinase.

THE STREPTOCOCCI

Streptococci are gram-positive, spherical, nonmotile cells that grow in pairs or chains (Figure 22.4) by fermenting glucose under anaerobic or microaerophilic conditions. The streptococci are one of the few groups of air-tolerant bacteria unable to make catalase. The absence of catalase is a key marker for differentiating between the streptococci and the catalase-positive staphylococci.

Only a few of the many species of *Streptococcus* cause disease in humans. Most of the pathogenic species produce either an α or β hemolysin that is easily detected when isolates grow on blood agar plates. Other characteristics by which the pathogenic streptococci can be differentiated include the antigenic serotype, sensitivity to bacitracin or optochin (an inhibitory chemical), presence of capsules, and bile solubility. Many nonpathogenic species are found in natural environments and some have industrial applications.

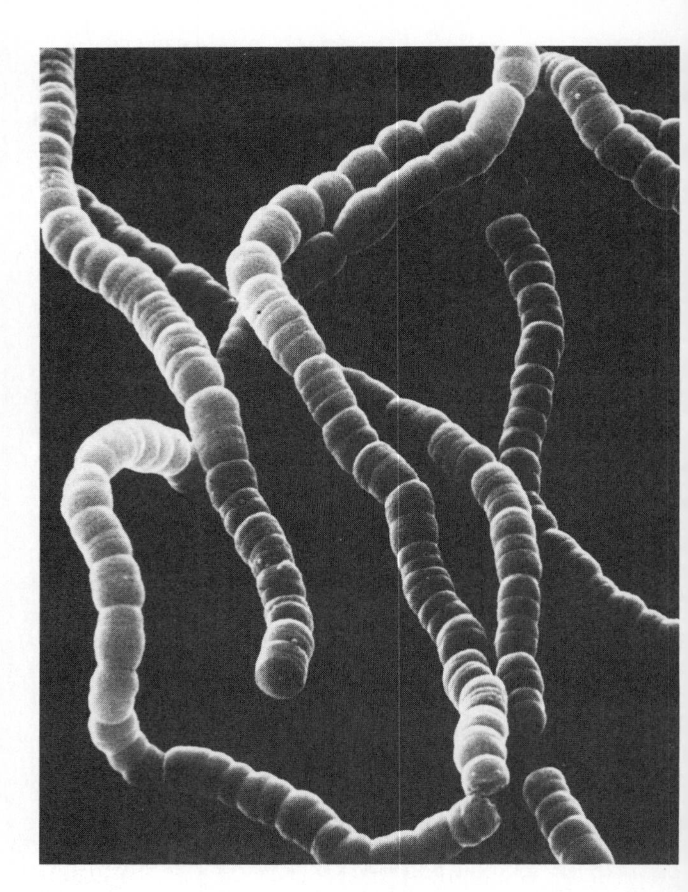

FIGURE 22.4 Scanning electron micrograph of *Streptococcus*. Note that the bacteria divide in a single plane to form long chains of cells. (Courtesy of Dr. D. M. Philips.)

STREPTOCOCCUS PYOGENES

Streptococcus pyogenes is medically the most important species of the genus and includes all of the group A, β-hemolytic streptococci. *Streptococcus pyogenes* is an opportunistic pathogen associated with humans that causes both cutaneous and systemic infections. It is the chief cause of acute bacterial pharyngitis.

Toxins and Virulence Factors

Many extracellular toxins and antiphagocytic substances contribute to the virulence of *S. pyogenes*. This bacterium produce capsules of hyaluronic acid that retards phagocytosis by polymorphonuclear leukocytes and macrophages. The capsule is not immunogenic in humans since hyaluronic acid is a component of human connective tissue. **M protein** (there are more than 60 serotypes) is another antiphagocytic virulence factor of these streptococci. M protein is found attached to the fimbria-like structures that protrude from the cell surface.

Streptolysins are streptococcal toxins that cause β hemolysis (Color Plate II). Streptolysin *S* is stable in air and lyses erythrocytes on aerobic plates, whereas streptolysin *O* is inactivated by oxygen, so it is able to lyse erythrocytes only under anaerobic conditions. Antibodies against streptolysin *O* are found in the serum of patients who have recovered from streptococcal infections. Since both streptolysins *S* and *O* have the ability to lyse cells, both destroy the phagocytic leukocytes.

Streptococcus pyogenes also produces enzymes that contribute to its virulence. One of them, **streptokinase,** dissolves fibrin clots by breaking down plasminogen to plasmin. (This enzyme was used at one time to dissolve clots in persons who had experienced a myocardial infarction; better products produced by genetic engineering are now available.) NADase breaks down NAD, and hyaluronidase breaks down hyaluronic acid in connective tissue. Since hyaluronidase can also attack the capsular material of streptococci, its role in pathogenicity is unclear. *Streptococcus pyogenes* also produces DNase, which decreases the viscosity in tissue spaces and promotes bacterial invasiveness.

Diseases Caused by *Streptococcus Pyogenes*

Streptococcal pharyngitis, or "strep throat" as it is usually called, is the most common disease caused by *Streptococcus pyogenes*. It develops when *S. pyogenes* colonizes the mucous membranes of both the pharyngeal region and the tonsils. Symptoms include fever, purulence, and swelling and inflammation of the throat. The disease rarely leads to pneumonia or meningitis, though it may involve the middle ear (otitis media) or the sinuses (sinusitis).

A diagnosis of streptococcal pharyngitis is indicated when throat cultures reveal a β-hemolytic *Streptococcus* that is sensitive to bacitracin. The group A streptococci are sensitive to low concentrations (0.02 unit) of bacitracin, whereas most other β-hemolytic streptococci are resistant to it. Streptococcal pharyngitis is one of the most common bacterial diseases of children between the ages of 5 and 15. Untreated, streptococcal pharyngitis can lead to rheumatic fever and other complications. Treatment is with penicillin (or erythromycin in hypersensitive patients). Penicillin resistance is not a problem because resistant to this antibiotic has not been observed in *S. pyogenes*.

Scarlet fever is caused by lysogenic strains of *Streptococcus pyogenes* that produce an erythrogenic toxin. The three **erythrogenic toxins** (A, B, and C) are extracellular proteins produced by lysogenized strains and are responsible for the rash of scarlet fever. Streptococci present in the throat, and occasionally in wounds, release the toxin into the circulation. A red rash develops, appearing initially on the chest, then spreading to the rest of the trunk and the extremities. Scarlet fever is diagnosed on the basis of the clinical symptoms. Immunity to one of the erythrogenic toxins does not confer immunity to the other two, so it is possible for a person to contract scarlet fever more than once.

Group A streptococci can cause impetigo and erysipelas (er′i-sip′e-las). **Streptococcal impetigo** is a localized purulent dermatitis that commonly affects the face and hands. Some 5 to 10 percent of cases of this highly contagious disease are caused by *Streptococcus pyogenes*. Group A streptococci also cause **erysipelas,** an acute febrile disease characterized by a red rash (Figure 22.5). Both impetigo and erysipelas can be treated with antibiotic ointments or with ammoniated mercury ointments. Proper hygienic practices normally prevent the spread of these skin infections.

Epidemiology

Almost all streptococci are transferred from person to person by direct contact or by a cough, though streptococci can survive on inanimate objects such as bed linens. Young children who have not built up a resistance to these bacteria are the most susceptible during an epidemic. Epidemics usually occur during the colder months and are more prevalent in regions with a dry climate. The carrier rate for group A streptococci is usually below 10 percent of the population, but it often increases just before an epidemic. Prevention of strep-

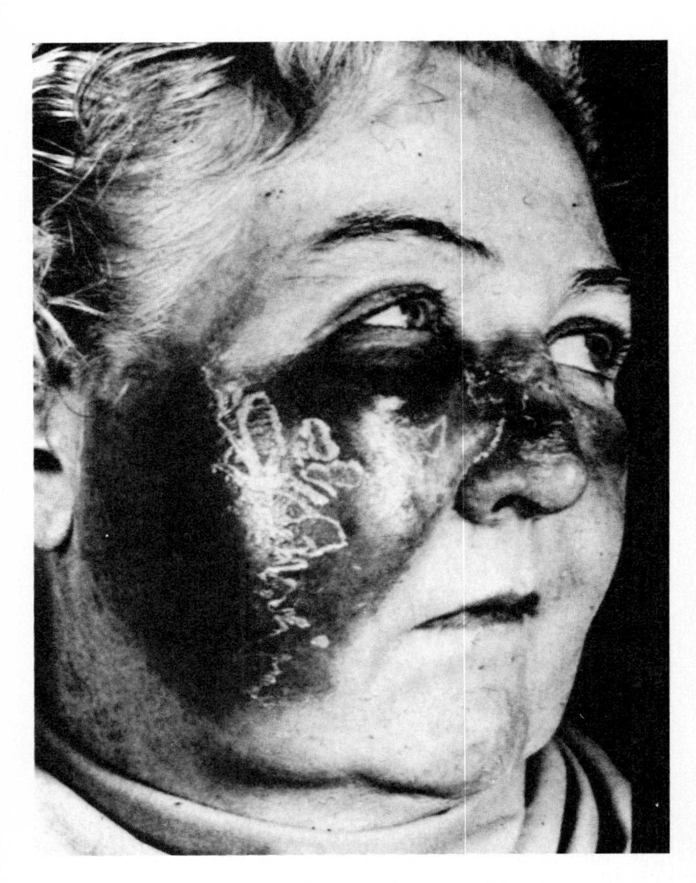

FIGURE 22.5 Facial erysipelas caused by *Streptococcus pyogenes*. There is a sharp demarcation between the inflamed areas of both cheeks and the normal surrounding skin. (Courtesy of T. B. Fitzpatrick et al.)

tococcal infections depends on early detection of the causative agent followed by antibiotic therapy.

BACILLUS: ANTHRAX

Robert Koch gave anthrax a prominent place in the history of microbiology when he proved that it was caused by an infectious bacterium (Chapter 1). Anthrax is a zoonosis of cattle, sheep, and goats that is occasionally transmitted to humans. The incidence of human anthrax in the United States is very low; only 29 cases were reported between 1969 and 1984, and only two cases were reported in the five years preceding the writing of this book. Sporadic epidemics still occur among cattle and sheep, but most are controlled by quarantine, herd immunizations, and destruction of infected herds. Cases of anthrax in the United States have been traced to contaminated imported animal materials such as hides.

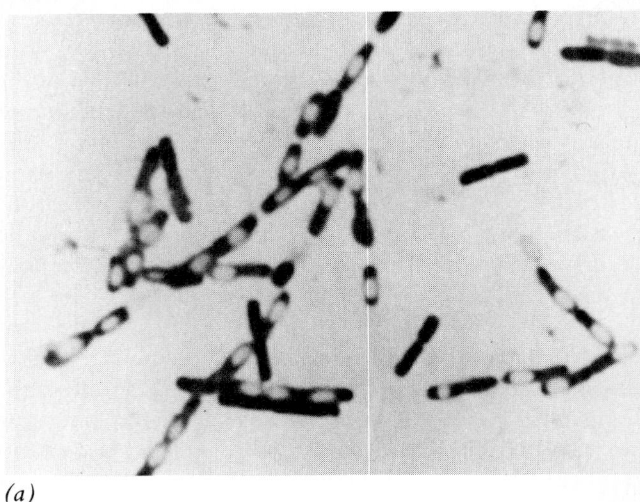

(*a*)

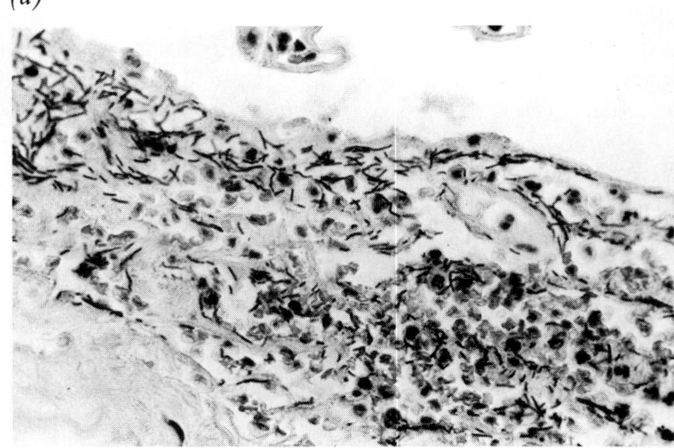

(*b*)

FIGURE 22.6 (*a*) Photomicrograph of *Bacillus anthracis* after it was stained with the fuchsin-methylene blue spore stain. The clear areas in the cell's cytoplasm are ellipsoidal endospores. (Reproduced by permission from E. J. Bottone and Abbott Laboratories. From R. Girolami and J. M. Stamm (eds.), *Shneierson's Atlas of Diagnostic Microbiology,* 8th ed. Abbott Laboratories, North Chicago, Ill., 1982. Copyright © 1982, Abbott Laboratories.) (*b*) *Bacillus anthracis* in tissue taken from a patient with fatal inhalation anthrax. (Courtesy of the Centers for Disease Control.)

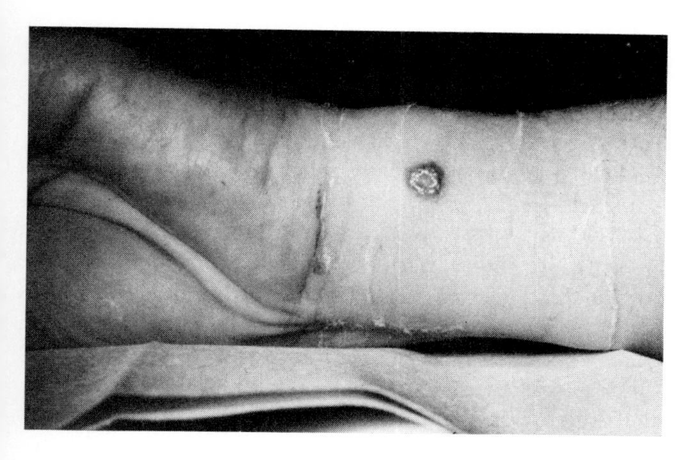

FIGURE 22.7 A cutaneous anthrax lesion on the arm of a 50-year-old woman who had been a carder in a wool factory. (Courtesy of the Centers for Disease Control.)

Anthrax is caused by the aerobic endospore-forming bacterium *Bacillus anthracis* (Figure 22.6). Infected animals contaminate the fields with *B. anthracis* spores, which have long survival times in natural environments. The spores infect horses, sheep, and cattle when animal wounds become contaminated with soil or when the animal ingests them. The spores germinate and cause localized or systemic infections, depending on the animal and mode of infection. Virulent strains of *B. anthracis* produce D-glutamyl polypeptide capsules and three protein toxins.

Anthrax in humans can occur as either a skin or lung infection. **Cutaneous anthrax** is a localized infection (Figure 22.7) that develops when *B. anthracis* contaminates a wound or abrasion. This form of anthrax is often seen on the hands of persons who work with animal skins. Patients treated with penicillin will usually recover if the infection does not become systemic. **Inhalation anthrax** is a much more serious disease because the vegetative *B. anthracis* cells often enter the blood and cause a septicemia, from which most patients die.

SPIROCHETE: TREPONEMA

Treponemes produce rashes, open sores, or lesions on the skin, from which the spirochetes are transmitted by person-to-person contact. Syphilis, a sexually transmitted disease, is discussed later. Yaws and pinta are nonvenereal diseases that are common in tropical countries. All three diseases are caused by subspecies of *Treponema pallidum* (see Chapter 15).

PINTA: *TREPONEMA PALLIDUM* SUBSPECIES *CARATEUM*

Pinta (peen′ta), characterized by discolored skin lesions, is typically seen in rural populations living in the tropical climates of Central and South America. The infection begins as a papule and, after four to five weeks, spreads and reddens, becoming about 1 cm in diameter. Secondary lesions develop about five months later both at the site of the initial papule and elsewhere (Figure 22.8). These secondary lesions become darkly pigmented and then undergo gradual depigmentation, over two years. Penicillin is effective in treating pinta.

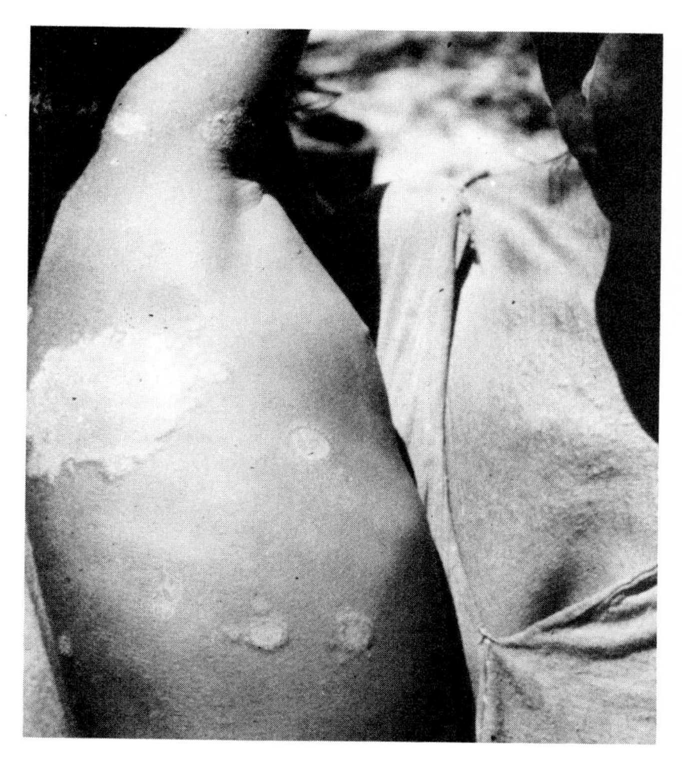

FIGURE 22.8 Large primary and smaller secondary lesions of pinta (AFIP 75-5536-2) (Courtesy of Dr. D. H. Connor, Armed Forces Institute of Pathology, Washington, D.C.)

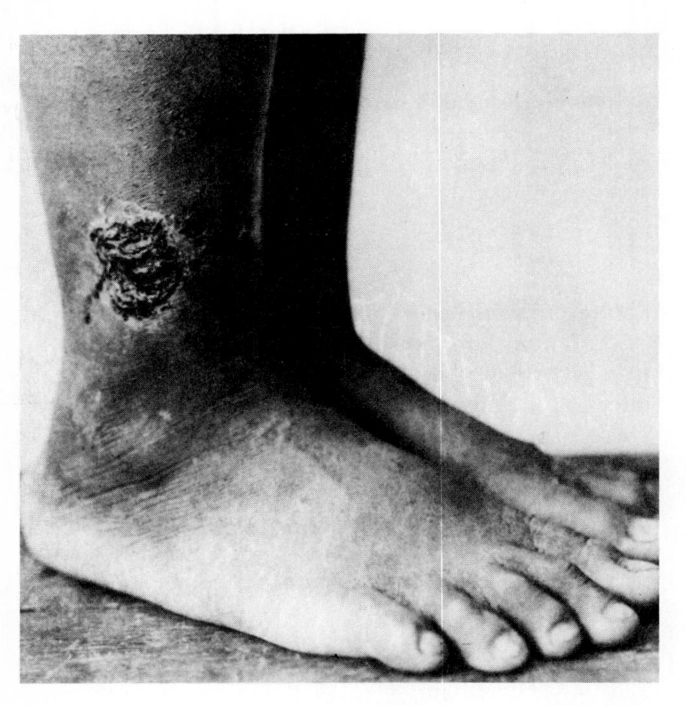

FIGURE 22.9 Open lesions of yaws (AFIP 39207) (Courtesy of Dr. D. H. Connor, Armed Forces Institute of Pathology, Washington, D.C.)

YAWS: *TREPONEMA PALLIDUM* SUBSPECIES *PERTENUE*

Yaws occurs among natives of tropical countries, where little protective clothing is worn and where the high humidity promotes persistence of open skin lesions. The responsible spirochete is transmitted by person-to-person contact and possibly by flies that feed on the open lesions.

The bacteria infect humans through skin abrasions that usually occur on the legs or feet. The primary lesion appears three to four weeks after infection and progressively develops into an open lesion called a **yaw,** which contains the infectious spirochetes (Figure 22.9). A scab eventually forms, with healing. Six weeks to three months after the development of the yaw, secondary eruptions occur on the neck, the extremities, and at the juncture of the skin and mucous membranes (nose, mouth, anus). A single injection of long-lasting penicillin is an effective treatment.

WOUND INFECTIONS

Wounds can result from trauma, surgery, or physiological phenomena. Most **surgical wounds** are clean and are not infected if proper care is exercised, especially during the first 3 hours following surgery, which is the most critical period for preventing infection. The most common cause of surgical wound infections is *Staphylococcus aureus,* a common inhabitant of the skin that can cause serious systemic infections (see above) once it penetrates the body's first line of defense.

Traumatic wounds, such as cuts, compound fractures, frostbite necrosis, and burns are often contaminated by environmental microbes. Cleaning wounds and application of antiseptics is a major approach to preventing infections. Contamination of traumatic wounds can lead to wound botulism or tetanus (as discussed in Chapter 27), pseudomonad infections, and gas gangrene.

PSEUDOMONAS AERUGINOSA AND BURN WOUND INFECTIONS

Pseudomonads are aerobic, gram-negative rods found in water, soil, and on fomites in both natural and domestic environments. *Pseudomonas aeruginosa* appears harmless to healthy adults; however, it is a significant cause of infections of burned tissue and the urinary tract. Avoiding infection is difficult because of its ubiquitousness in the environment and its resistance to antimicrobials.

After *P. aeruginosa* infects a burn wound, it multiplies and produces exotoxin A and other toxins that prevent healing. **Exotoxin A** inhibits protein synthesis by inactivating the eucaryotic cell's elongation factor—very similar to the action of diphtheria toxin (see Chapter 23). This kills host cells in the wound, resulting in tissue necrosis.

Hospital burn centers are greatly concerned with the control of pseudomonad infections. Their personnel go to great lengths to decrease the environmental incidence of *P. aeruginosa* so that burn wounds will not become infected. The third-generation cephalosporins have notable activity against *P. aeruginosa;* however, recent isolates are resistant to these drugs.

Pseudomonas aeruginosa is also involved in **cystic fi-**

brosis, a hereditary chronic disease that progressively affects digestion and respiration. Patients are susceptible to lung infections by *P. aeruginosa,* especially strains that make capsules of alginic acid. With the help of antibiotic chemotherapy, patients recover from the initial infection, but they are soon reinfected. Each succeeding infecting strain generates a greater quantity of alginic acid, and eventually the patient cannot clear the bacteria from the lungs and dies. Currently no solutions exist for preventing this infection cycle.

INFECTIOUS GANGRENE

Many bacteria are able to invade damaged tissue and cause local tissue necrosis known as **infectious gangrene.** Genera of the bacteria involved include *Streptococcus, Staphylococcus, Bacteroides,* and *Clostridium.* Patients affected are usually elderly, have poor circulation, are recovering from surgery or cancer, or have some other predisposing condition. After infecting the tissue, these bacteria cause tissue necrosis. Some forms of gangrene are treated with antimicrobials; however, the poor blood circulation in the infected tissue complicates therapy. Excision of the necrotic tissue combined with antibiotic treatment is usually required.

The Histotoxic Clostridia

Clostridial species from soil, mud, and the intestine cause gas gangrene when they multiply in anaerobic tissue. Anoxia at the wound site is created by an insufficient blood supply. These bacteria are called **histotoxic clostridia** because they produce toxins that destroy tissue.

Clostridium perfringens—the most common isolate—*C. novyi,* and *C. septicum* all cause gas gangrene. In addition to gangrene, *C. perfringens* can cause uterine infections and is the causative agent of clostridial food poisoning (see Chapter 24). The enzyme toxins produced by *Clostridium perfringens* include the α toxin, which is a β lecithinase that disrupts mammalian cell membranes; collagenase; proteinases; hyaluronidase; and deoxyribonuclease.

Clinical Symptoms of Gas Gangrene

Gas gangrene caused by histotoxic clostridia is characterized by progressive myonecrosis and gas production. Local tissue death perpetuates the anoxic character of the infected area by cutting off the blood supply. Gas gangrene is often a multiple infection involving more than one species of *Clostridium.* Symptoms include swelling, a dark yellow appearance of the skin, and the presence of a thin, dark, watery exudate. The gaseous fermentation products, temporarily trapped in the tissue spaces, escape by bubbling up through the exudate of the wound. Symptoms can begin 6 to 72 hours after a wound is infected.

Quick removal of dead tissue, called debridement, is recommended for controlling the progression of gas gangrene. Systemic antimicrobial chemotherapy is given when a sufficient blood supply exists, otherwise local irrigation with bacitracin is called for. Gas gangrene develops in patients recovering from serious automobile accidents and in people with circulatory problems, especially circulation to the limbs.

KEY POINT

Wound infections can seriously affect healing and may lead to life-threatening systemic diseases. Most traumatic wounds are treated as unique occurrences because the responsible pathogens are not transmitted to others. Burns are easily infected by *P. aeruginosa* and require special care.

INFECTIONS OF THE EYE

The eye (Figure 22.10) is protected against infectious diseases by chemical and mechanical means. As discussed in Chapter 17, tears contain secretory immunoglobulins (IgA), which inactivate infectious agents, and lysozyme, which attacks the bacterial peptidoglycan. Dust and other foreign particles are carried away by the cleansing action of tear flow and the mechanical action of the eyelid.

The **conjunctiva** (Figure 22.10) is a vascular membrane that covers the posterior side of the eyelid and the surface of the globe (the white sclera), ending at the corneal epithelium. The cornea and the underlying lens are avascular tissues that permit light to pass. The conjunctiva is the source of the lymphocytes that phago-

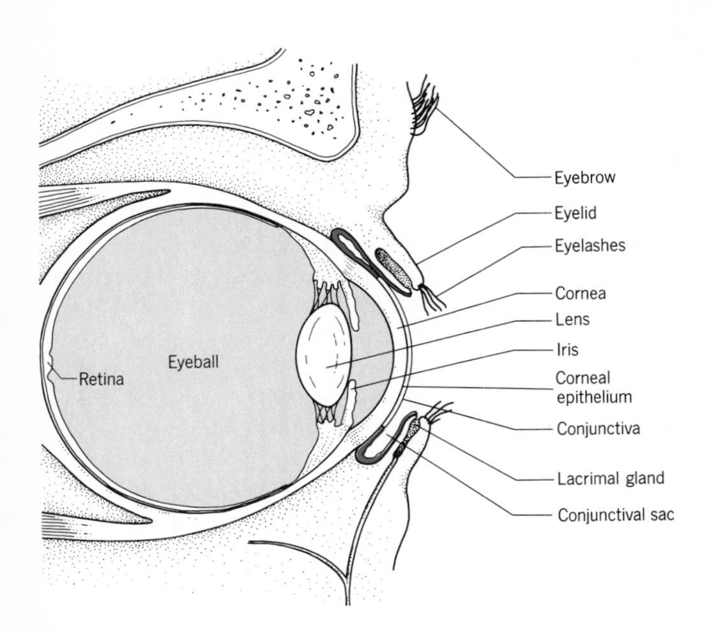

FIGURE 22.10 The eye is bathed by tears that wash away foreign matter and carry IgAs and lysozyme. Most eye infections result in inflammation of the conjuctiva called conjunctivitis.

cytize pathogens and is the site of the inflammatory response. **Conjunctivitis** (kon-junk′ti-vi′tis) is an inflammation of the conjunctiva; it may be caused by infectious agents, foreign matter, or allergic reactions (see Chapter 19). **Keratitis** (ker′ah-ti′tis), an inflammation of the cornea, is usually more serious than conjunctivitis since the passage of light to the lens may be obstructed. **Keratoconjunctivitis** describes an infection that involves both the cornea and the conjunctiva.

CONJUNCTIVITIS

Infections of the conjunctiva stimulate an inflammatory response, characterized by a swelling of the eyelid and redness of the conjunctiva and sclera. Infections caused by *Haemophilus aegyptius* (he-mof′i-lus ae-gyp′ti-us) are popularly called "pinkeye" because of the intense red color imparted to the sclera. Other bacteria cause suppurative infections of the eyelid. The pus may accumulate in the corners of the eyes, and in severe cases the eyelids may become stuck together during sleep. The array of responsible bacteria includes *Streptococcus pneumoniae, S. pyogenes, Haemophilus influenzae,* and *H. aegyptius. Staphylococcus aureus* is a principal cause of infections of the eyelid and cornea, and of sty at the base of the eyelash. These infections can be treated with topical antimicrobials. Adenovirus, herpes simplex, measles, and varicella–zoster virus also cause conjunctivitis.

OPHTHALMIA NEONATORUM

The eyes of newborns can be infected by the vaginal flora during passage through the vagina. A woman with gonorrhea (gon′on-re′ah) transmits *Neisseria gonorrhoeae* (nys-se′ri-ah gon-or-re′eye) to the newborn's eyes. If uncontrolled, the ensuing infection can cause blindness, so the eyes of newborns are treated with antimicrobial ointments or drops soon after birth. Herpes simplex can also be transmitted to the eyes of newborns during the birth process and cause herpetic keratitis. Treatment is with antiviral ointments.

INCLUSION CONJUNCTIVITIS

Both trachoma (trah-ko′mah) and inclusion conjunctivitis are eye diseases caused by *Chlamydia trachomatis* (klah-mid′ee-ah trah-ko′mah′tis). Inclusion conjunctivitis is clinically different from trachoma though both diseases are caused by serologically related bacteria.

Asymptomatic *C. trachomatis* infections of the adult cervix appear to be epidemic in developed countries. When the eyes of newborns are infected with *C. trachomatis* during the birth process, the infection appears 2 to 25 days later as an inflamed conjunctiva with a purulent exudate. Inclusions of *C. trachomatis* in host cells are visible in scrapings from the conjunctival surface (Figure 22.11) when stained with fluorescent antibodies. The infection usually clears spontaneously after several months without causing permanent damage. Newborns can be treated with systemic antimicrobials such as sulfonamides or erythromycin.

In adults, inclusion conjunctivitis is transmitted by towels or fingers contaminated with *C. trachomatis* from exudates of the eye or the genitourinary tract. Swimming in contaminated, unchlorinated pools is another source of infection.

The vagina is the natural reservoir of *C. trachomatis,* so sexual contact can be the route of spread. Inclusion conjunctivitis is characterized by an inflammation of the conjunctiva and a purulent exudate. Blindness does not follow because the cornea does not become vascu-

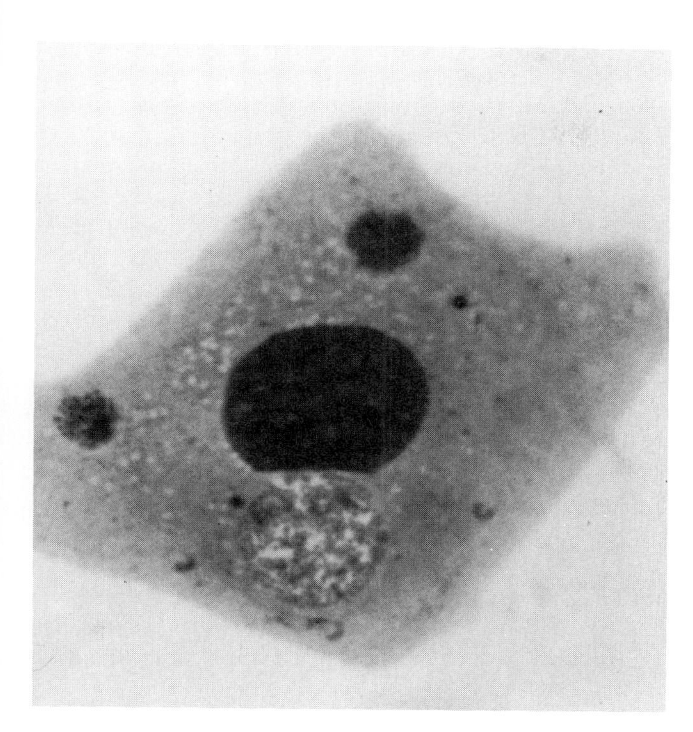

FIGURE 22.11 The Giemsa stain of a cell from a conjunctival scraping reveals the chlamydial inclusion bodies in the cell's cytoplasm. 1000×. (Courtesy of Dr. N. M. Jacobs.)

larized. The disease is self-limiting, but can be treated by systemic administration of tetracyclines or sulfonamides. It is difficult to prevent inclusion conjunctivitis because of the widespread incidence of asymptomatic *C. trachomatis* infections. Control is augmented by chlorination of swimming pools and good personal hygiene.

TRACHOMA

Trachoma is the world's leading cause of preventable blindness. Experts estimate that 360 million people are afflicted with trachoma and approximately 2 million people have been rendered permanently blind. Trachoma is a serious illness in the Middle East, where it is so prevalent that most young children are infected. The agent is transmitted by direct contact or by towels and other inanimate objects.

Chlamydia trachomatis infects the cornea as well as the conjunctiva, causing a severe inflammation that stimulates lymphocytes, polymorphonuclear neutrophils, and macrophages to enter the area and form follicles (minute cavities) beneath the conjunctival surface. These changes stimulate corneal vascularization, which can lead to permanent corneal scarring and finally blindness. Trachoma can be treated with topical applications of sulfonamides and by systemic administration of tetracyclines. The host responds to the infection by secreting IgAs against *C. trachomatis* in tears, and by developing a cell-mediated immune response. Reinfections or relapses do occur because immunity to trachoma is not long lasting.

KEY POINT

It is remarkable how rarely eye infections occur, when you consider how often you rub your eyes or subject them to windblown dust. Tears and the inflammatory response protect the human eye from infectious agents.

VIRAL DISEASES OF THE SKIN

Dermatotrophic viruses infect humans via various routes to cause the skin disorders described either as a maculopapular rash, a vesicular lesion, or a wart. A maculopapular rash is a spotty skin blemish with a raised, inflamed center such as is seen in measles and rubella. Before the use of vaccines in the 1970s, these diseases were common among children, but now are largely prevented. DNA viruses of the Herpetoviridae and the Poxviridae family cause vesicular lesions. Herpes simplex viruses cause cold sores on the lips (herpes labialis) and genital herpes, which is a sexually transmitted disease. Another member of the Herpetoviridae, the varicella–zoster virus, causes both chickenpox and zoster. Smallpox, a devastating disease that frequently ended in the patient's death, has been conquered, as described below. Warts, skin growths that are usually benign, are caused by papillomaviruses.

MEASLES

Measles, a highly contagious viral disease, is recognized by the presence of Koplik spots on the inside the cheek and a maculopapular rash. Measles vaccine has greatly reduced its incidence in the United States; however, sporadic outbreaks still occur and in 1986 there was a dramatic increase in the incidence of measles, especially among persons under 20 years of age.

The Measles Virus

The measles agent is a single-stranded, RNA virus that has hemolysin and hemagglutinin associated with its envelope. The hemagglutinin binds the virus to host cells, and the hemolysin functions in the fusion process of viral entry (see Chapter 12). Hemagglutination inhibition, the immunological assay described earlier, is a standard test for measuring the presence of measles antibodies in serum. Though measles is diagnosed on the basis of the symptoms, the responsible virus can be isolated in tissue cultures of human and monkey cells. Now that there are so few cases of measles in the United States, health authorities recommend that each case be confirmed by isolating the virus.

Symptoms

The measles virus is transmitted by aerosols or direct contact with secretions of the respiratory tract, eye, and urine from infectious patients. Susceptible patients are infected by way of the mucous membranes of the respiratory tract and conjunctiva. They are infectious for three days before the onset of symptoms, and remain infectious until the rash clears.

The incubation period is 12 to 14 days (Figure 22.12), during which the virus replicates inside the respiratory epithelium, then infects the reticuloendothelial leukocytes. Here it replicates again before causing a generalized viremia (Figure 22.12). The viremia coincides in appearance with the early symptoms, which include a cold, fever, cough, malaise, loss of appetite, and conjunctivitis.

The telltale **Koplik spots** are bluish-gray specks on an inflamed background; they last for a few days, then disappear as the rash develops. The rash (Color Plate II) appears at the hairline, then spreads to the face,

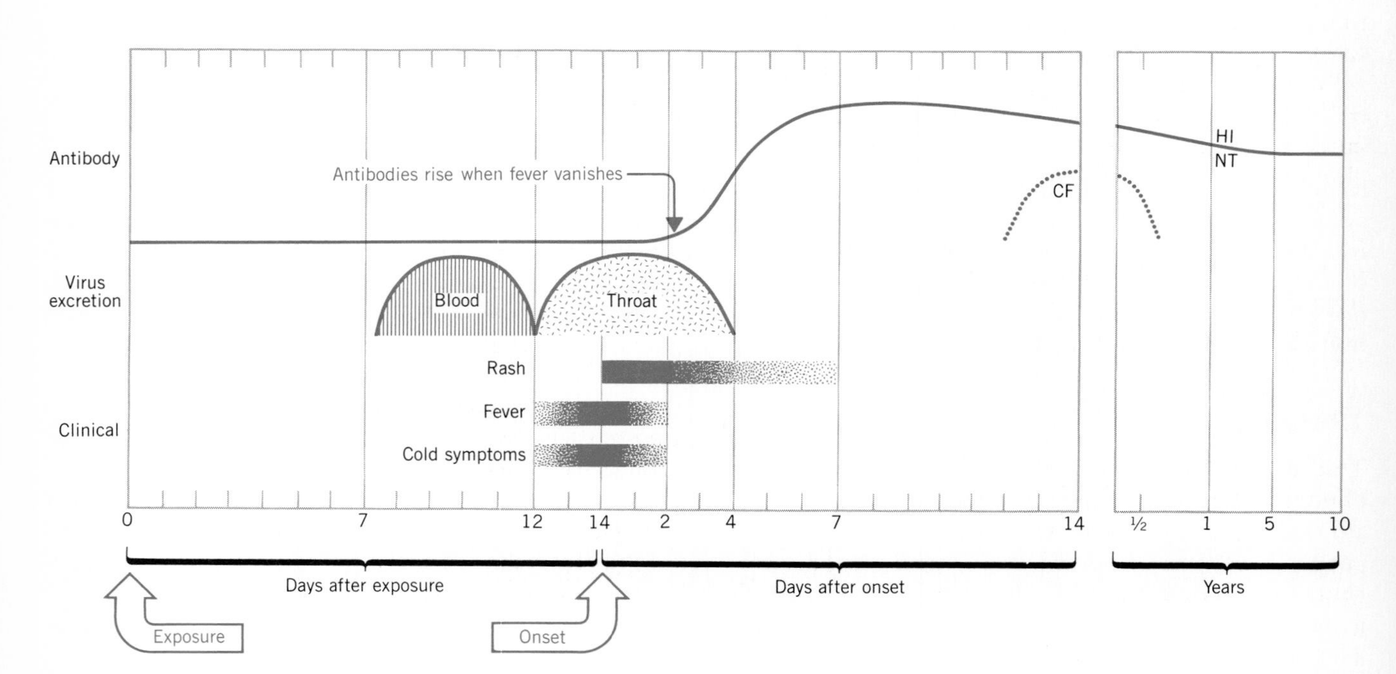

FIGURE 22.12 Clinical symptoms of measles. CF, complement–fixation antigen; HI, hemagglutination inhibition; NT, neutralizing antibody. (Reproduced by permission of Dr. D. M. McLean. From D. M. McLean, *Virology in Health Care,* p. 123, Williams and Wilkins, Baltimore, copyright © 1980, Williams and Wilkins.)

trunk, and extremities. The rash initially is maculopapular, then coalesces over the neck and face. In some patients the skin desquamates as the rash clears. Most of the discomfort is felt on or about the second day of the rash. On the fifth day, the rash subsides, the virus disappears from the throat, and the level of circulating antiviral antibody increases.

Epidemiology, Prevention, and Treatment

Before the vaccine was introduced in 1963, measles occurred cyclically every two to five years among young children. The epidemics were more severe in winter months, possibly because the virus is more stable in dry air.

The vaccine is usually administered in combination with live mumps and live rubella virus vaccine (MMR) to children 15 months old. Whereas measles cases were fewer than 1500 in 1983, they increased to 2700 cases in 1985 and to over 6000 cases in 1986. Although there are comprehensive school laws requiring vaccination, many children growing up in the 1960s escaped natural measles, were never vaccinated, were vaccinated with the ineffective killed vaccine between 1963 and 1967, or were vaccinated before their first birthday. These individuals are susceptible to measles. The observed increase in cases belies the public health commitment to eliminate indigenous measles from the United States.

There is no specific treatment for measles other than keeping the patient comfortable. Should an epidemic occur, great effort is made to control its spread by immunizing susceptible contacts. Even exposed persons can be immunized since they will respond to the vaccine in about 7 days, well before symptoms develop from the natural infection (Figure 22.12).

RUBELLA (GERMAN MEASLES)

Early investigators confused rubella with other diseases characterized by the red rash, until a group of German physicians demonstrated it to be a distinct disease. As a result of their work, rubella became popularly known as German measles.

Rubella is a mild illness, especially in children. It is important because a pregnant woman who contracts the disease can transmit it to her fetus, with dire consequences (described below).

The Rubella Virus

The rubella virus is an RNA virus classified as the sole member of the *Rubivirus* genus. It infects primates and certain laboratory animals, but only humans develop the typical symptoms. The virus is transmitted in aerosol droplets, and infects the respiratory tract. Rubella virions can be detected by hemagglutination; hemagglutination inhibition (HI) activity is a measure of a person's immunity to rubella.

Symptoms

The severity of rubella is dependent on the person's age. Infants and young children experience a subclinical to mild illness, but adults usually experience fever, rash, and swelling of lymph nodes (lymphadenopathy), especially those in the back of the neck. The incubation period ranges from 12 to 22 days, with an average of 18 days. The virus is present in the blood and the throat (Figure 22.13) before symptoms are manifest. Fever, rash, and lymphadenopathy last for 4 to 5 days after onset. The virus is shed from the throat before the onset of symptoms and for 10 to 15 days thereafter. An increase in serum antibodies coincides with fever abatement. Lifelong immunity develops in most patients following recovery.

Congenital Rubella Syndrome Congenital cataract (clouding of the lens) was first recognized in 1941 to be caused by rubella infection (Figure 22.14). Now an array of problems, including premature delivery, stillbirth, and permanent defects are known to result from *in utero* infections: hearing loss, cataract, severe myopia, myocardial abnormalities, mental retardation, and diabetes mellitus. These problems are known collectively as the **congenital rubella syndrome** (CRS), which is obviously the most serious complication of rubella infections.

The possibility that a fetus will contract CRS is greatest when the mother becomes infected during the first two months of gestation (40 to 60 percent possibility), and it diminishes progressively thereafter. Fetal rubella during the first trimester often results in spontaneous abortion. CRS developing during the third or fourth month of gestation usually has a single congenital defect as its outcome.

Affected infants continue to shed the virus for many months; however, how these viruses are associated

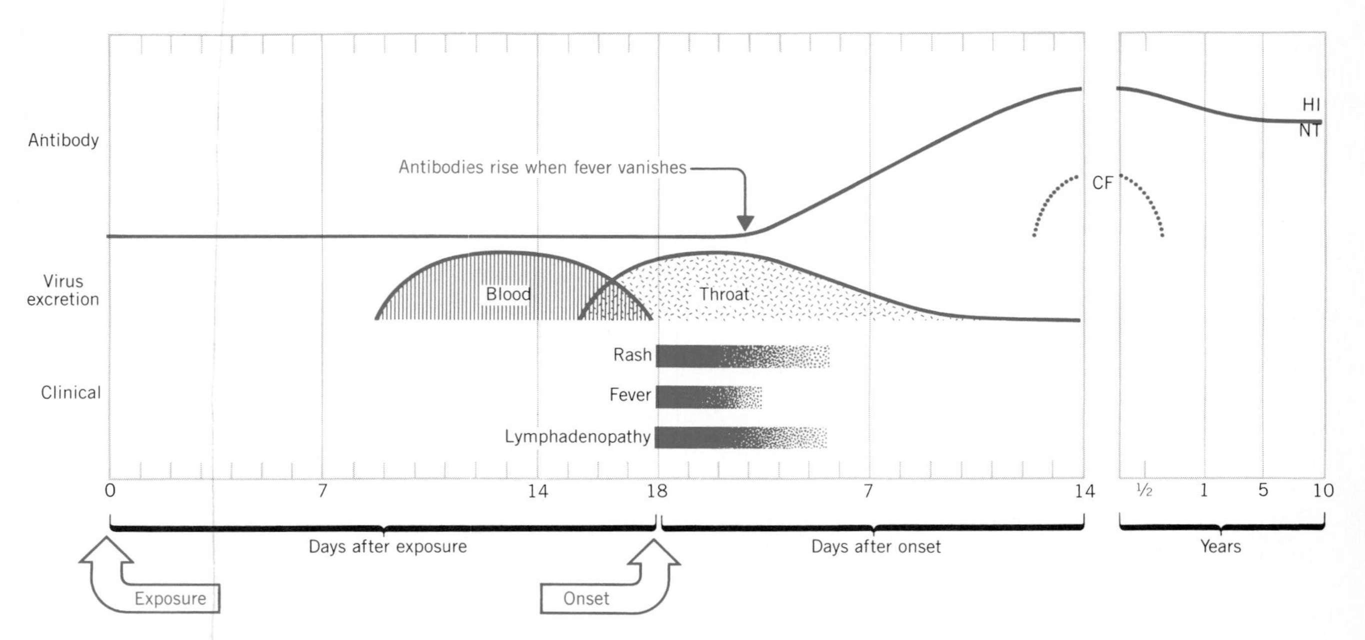

FIGURE 22.13 Rubella (German measles). CF, complement fixation antigen; HI, hemagglutination inhibition; NT, neutralizing antibody. (Reproduced by permission of Dr. D. M. McLean. From D. M. McLean, *Virology in Health Care,* p. 140, Williams and Wilkins, Baltimore, copyright © 1980, Williams and Wilkins.)

with the symptoms of CRS remains a puzzle. The potential defects are so serious to the health of the child that mothers should avoid exposure to rubella during pregnancy.

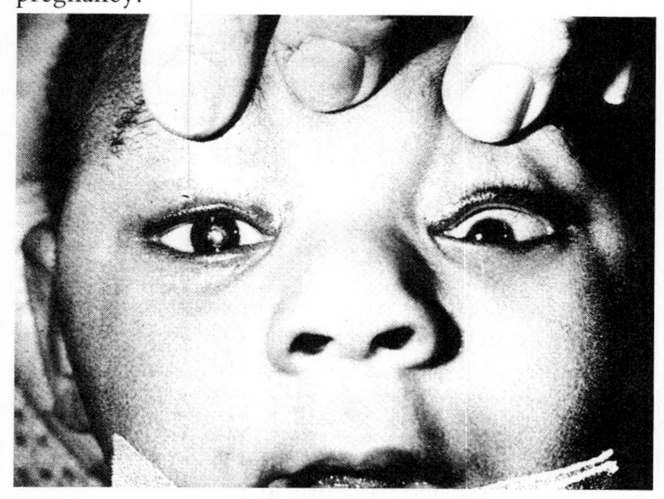

FIGURE 22.14 Congenital rubella cataract in a 9-month old infant. (Reproduced with permission of Wilson et al.)

Epidemiology

Before the vaccine was introduced in 1969, rubella was most prevalent in children ages 6 to 9. The frequency of rubella is now much lower (Figure 22.15), and most cases occur in persons 15 years of age or older.

The vaccine, which is part of the MMR vaccine, is recommended for all prepubertal children older than 15 months of age, and it has sharply reduced the number of CRS cases. During the most recent epidemic, in 1964, 30,000 infants were born with congenital rubella syndrome; by 1984, only 752 cases of rubella and only five cases of CRS were reported nationally.

KEY POINT

Since humans are the only reservoir for the measles and rubella viruses, we could eliminate these diseases completely if everyone were effectively immunized with the MMR vaccine.

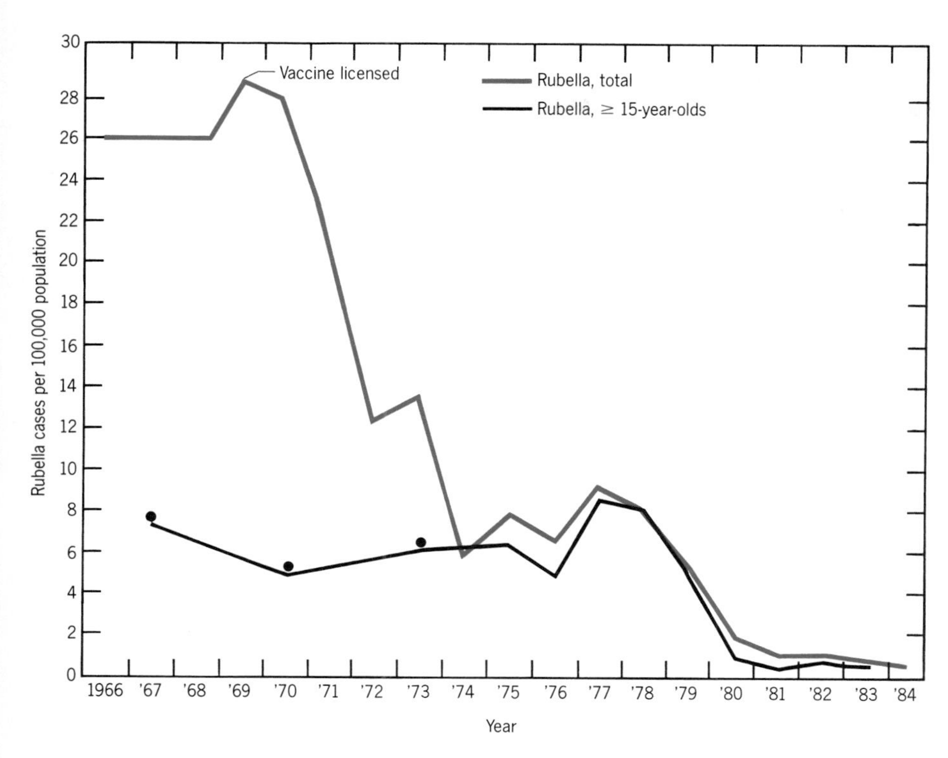

FIGURE 22.15 The effect of the licensing of the rubella vaccine on the incidence of rubella in the United States. The decline in the incidence of rubella in persons 15 years of age and older has had a dramatic effect on the incidence of congenital rubella syndrome. Can you explain why? (Centers for Disease Control, "Annual Summary 1984: Reported Morbidity and Mortality in the United States," *Morbidity and Mortality Weekly Report,* Vol. 32, No. 54, 1986.)

THE HERPESVIRUSES

Herpesviruses are large, double-stranded DNA viruses (Figure 22.16) that infect humans and animals. They are ubiquitous in the human population, in which they cause disease ranging in severity from asymptomatic infections to fatal encephalitis and cancer. Each of the herpesviruses causes an overt disease (usually) upon primary infection after which the virus remains latent within the host's neuronal cells or lymphocytes (Table 22.2). Infections characterized by vesicular skin lesions include zoster, chickenpox, genital herpes, and orofacial herpes.

The two herpes simplex viruses, HSV-1 and HSV-2, share about 50 percent of their genomes, but are antigenically different and cause distinctive illnesses (Table 22.2). HSV-1 causes orofacial infections (infec-

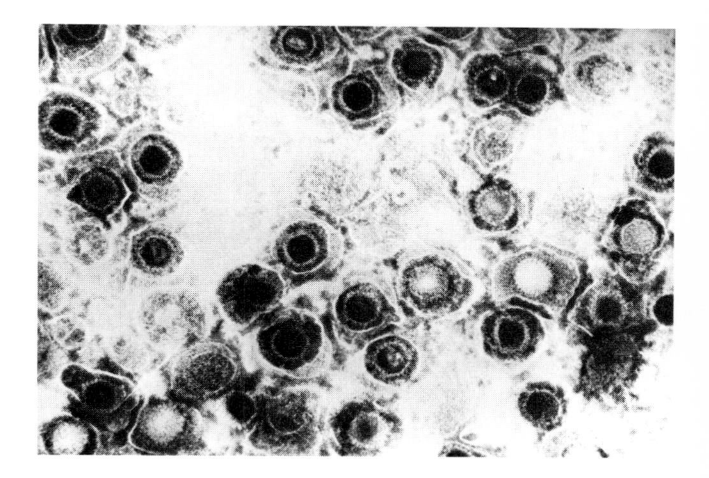

FIGURE 22.16 Electron micrograph of herpes simplex virus. (Courtesy of the Centers for Disease Control.)

TABLE 22.2 Diseases caused by the herpetoviridae

VIRUS	DISEASES	SITE OF LATENT INFECTION
Herpes simplex virus type 1	Gingivostomatitis in children, recurrent orofacial infections (cold sores), keratitis, herpetic encephalitis	Trigeminal nerve root ganglion and autonomic ganglia of superior cervical and vagus nerves
Herpes simplex virus type 2	Genital herpes, neonatal herpes	Sacral nerve root ganglia
Varicella–zoster virus	Chickenpox (primary), zoster or shingles (recurrent)	Thoracic nerve root ganglia
Epstein–Barr virus	Infectious mononucleosis (heterophile-positive)	B lymphocytes
Cytomegalovirus	Infections of neonates and transplant patients, infectious mononucleosis (heterophile-negative)	Leukocytes (neutrophils and lymphocytes)

[a]Reprinted by permission of the publisher from J. C. Sherris, (ed.), *Medical Microbiology: An Introduction to Infectious Diseases,* Elsevier, New York, 1984. Copyright 1984 by Elsevier Science Publishing Co., Inc.

tions above the waist) while HSV-2 is the predominant cause of genital herpes (see Chapter 24). Primary infection by the varicella–zoster virus causes chickenpox. The virus remains dormant in the thoracic root nerve ganglia (Table 22.2) and can erupt to cause the recurrent infection of zoster. Cytomegalovirus (si'toh-meg'ah-lo-vi'rus) and Epstein–Barr virus infect lymphocytes (see Chapter 26); Epstein–Barr virus is the cause of most cases of infectious mononucleosis.

Orofacial Herpes Infections

Primary infections by HSV-1 may be asymptomatic or may cause vesicular lesions of the lips (Color Plate II) and mouth, commonly known as cold sores or fever blisters. The lesions occur singly or in groups that may coalesce. On dry skin, scabs form and healing takes place; in the mouth, lesions remain open as new epithelium is laid down during the healing process. The primary infection of the gums and oral mucosa known as **gingivostomatitis** (jin'ji-vo-sto'mah-ti'tis) is most frequent in 3- to 5-year-old children (Figure 22.17) and may be accompanied by fever, malaise, irritability, and swelling of the regional lymph glands. The lesions heal in 5 to 12 days, after which HSV may become latent in the nerve root ganglion of the fifth cranial nerve. Reactivation can occur under the stimulus of exposure to the sun, menstruation, or emotional excitement.

HSV-1 is transferred by direct contact between persons and by autoinoculation to other parts of the body, including the eyes and genitals. Mucocutaneous HSV-1 infections occur in dentists, dental hygienists, and others who have direct contact with the lesions. The virus is also transmitted by direct contact among wrestlers and other participants in contact sports. HSV-1 infection of the cornea and the genitals is probably a result of autoinoculation. Infection of the cornea causes **herpetic keratitis,** which can develop into blindness. Most sexually transmitted cases of genital herpes are caused by HSV-2, though in HSV-1 infections, autoinoculation may be the source. Occasionally, HSV-1 reaches the brain, resulting in **herpetic encephalitis,** a serious, life-threatening infection.

Topical treatment of vesicular lesions with acyclovir or idoxuridine (Chapter 12) lessens both the symptoms and their duration, but is not curative. Herpetic en-

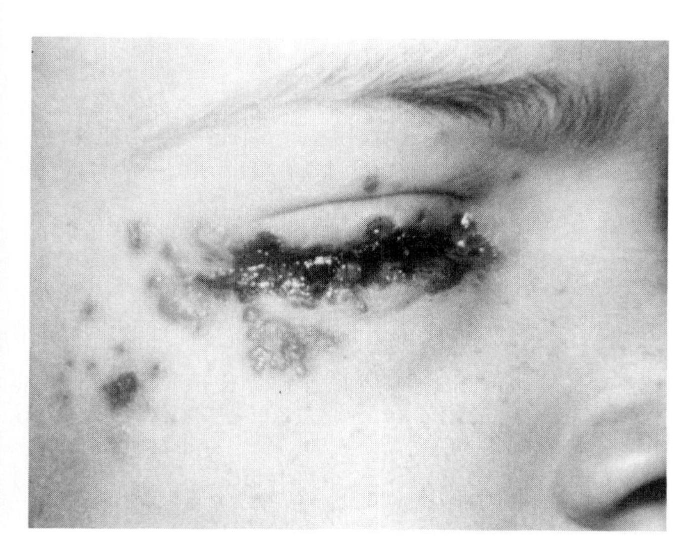

FIGURE 22.17 Acute herpes simplex virus infection of the periocular region. (Courtesy of Centers for Disease Control.)

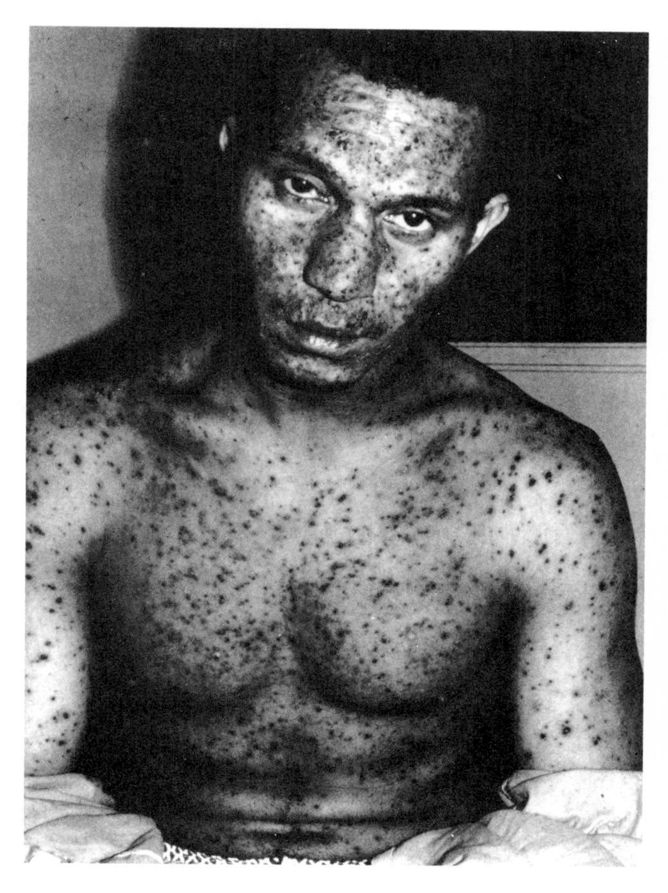

FIGURE 22.18 Adults can become severely ill with chickenpox. This adult patient has vesicular varicella lesions over the face, trunk, extremities, and the conjunctiva of the right eye. (From G. Mandell, J. E. Bennett, and R. G. Douglas (eds.), *Principles and Practice of Infectious Diseases,* 2nd ed., Wiley, New York, 1985.)

cephalitis is treated by intravenous administration of acyclovir or adenine arabinoside.

Avoidance of direct contact with the herpes simplex viruses is the chief means of preventing the spread of herpes. For example, medical and dental personnel should wear gloves when they treat a patient with herpes.

Varicella–Zoster Virus: Chickenpox and Zoster

Varicella (chickenpox) and zoster are two clinical diseases caused by a single virus, the varicella–zoster virus (Table 22.2). The common etiology of these diseases was first recognized in 1892 when it was observed that varicella occurred in households where one or more zoster patients resided. **Chickenpox** is due to an initial infection by the varicella–zoster virus, whereas **zoster,** or shingles, results from the activation of latent infections of the same virus.

The varicella–zoster virus is an icosahedral, DNA virus that is morphologically indistinguishable from the herpes simplex viruses. It is present in the fluid of the vesicular lesions on skin (Figure 22.18) and can be grown in human cell tissue cultures. The viruses isolated from chickenpox lesions are identical to those isolated from lesions of zoster patients.

Chickenpox (Varicella) Chickenpox is an annoying childhood disease characterized by an itching pock. This extremely infectious viral disease is caused by the varicella–zoster virus after a 14- to 15-day incubation period, during which the virus is present in the blood and throat. The first clinical signs are fever and erupting vesicles on the scalp or trunk. The lesions spread outwardly to the extremities and can eventually involve much of the body surface (Figure 22.18). The fever lasts for two to three days, after which new lesions are rarely observed.

The lesions begin as red spots that gradually become vesicular. With healing, a crust forms and intense itching follows. As the patient scratches, bacterial contamination can be introduced, and when this happens scar tissue forms. Circulating antibody against the varicella–zoster virus appears in the serum as healing begins. The antibodies do not destroy the virus; the virus merely enters a latent state in the thoracic nerve root ganglia.

Chickenpox is transmitted in aerosols of respiratory tract secretions and by direct contact with vesicular lesions. Yearly epidemics occur during the late winter and early spring with a seasonal peak between April and May. Only a few cases are seen in July and August when the susceptible population of school children is on summer vacation.

Zoster Activation of the latent varicella–zoster virus results in the disease called zoster. Zoster occurs sporadically throughout the year in persons of any age, but is more prevalent in older persons. The lesions are manifest in crops that erupt, usually on the trunk (Color Plate II), but also on the face, neck, and back. Most remain localized, although a disseminated zoster can develop in which fresh lesions appear over the entire body about a week after the initial eruption. Recovery is uneventful in the absence of exacerbating physical problems. Irradiation and immunosuppressive therapy increase susceptibility to zoster.

Prevention and Treatment

Varicella in children is a mild disease. Recovery brings total immunity, which, however, does not prevent the activation of a latent varicella–zoster infection that causes zoster. Development of a vaccine against this virus, therefore, is theoretically possible, and in fact a vaccine is in the final stages of testing as of this writing. Its advantages and disadvantages are still being debated.

Compromised and immunodeficient patients can be passively immunized with zoster-immune globulin, a therapy that lessens the symptoms in this high-risk group. No special treatment is called for in most cases, and the intense itching can be relieved with calamine lotion and oral trimeprazine.

KEY POINT

Not only varicella, but all the herpesviruses cause latent infections. The body is unable to eliminate these viruses, which can remain in cells for years and cause recurrent disease.

POXVIRUSES

Smallpox is the only viral disease to be controlled by elimination of the virus from the human population. Smallpox was known in China and India more than 2000 years ago. It was introduced into Europe between the fifth and seventh centuries A.D. and caused major epidemics in Europe during the Middle Ages.

Smallpox: Variola Virus

The poxviruses are large, brick-shaped, DNA viruses (Figure 22.19) that replicate in the host's cytoplasm,

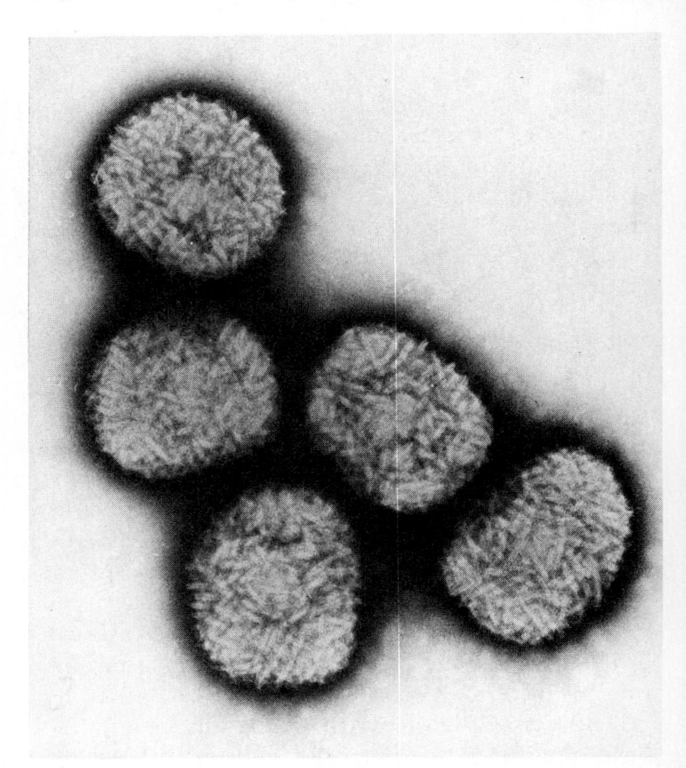

FIGURE 22.19 Purified vaccinia virus showing its brick-shaped structure and its convoluted surface. (Courtesy of Dr. S. Dales. From P. Gold and S. Dales, *Proc. Natl. Acad. Sci.* USA, 60:845–850, 1968.)

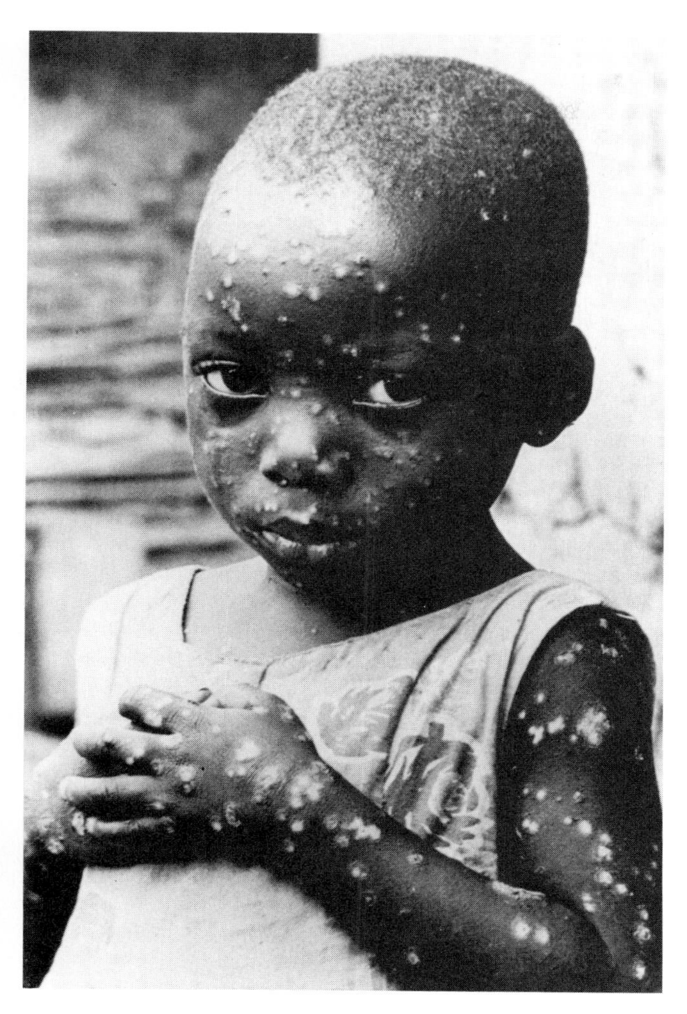

FIGURE 22.20 The scars that cover this youngsters face and arms developed from smallpox lesions. (Courtesy of WHO/UN.)

whereas most DNA viruses replicate in the cell's nucleus. At one time, two variola virus strains caused smallpox: the more virulent Asian strain, **variola major,** caused death in 20 to 40 percent of patients, while **variola minor** present in South America and southern Africa caused a mortality of less than 1 percent.

If the patient survived, (s)he was left permanently scarred. The rash appeared first on the scalp and face then spread to the back, chest, arms, and legs (Figure 22.20). It changed progressively into firm, distinct papules, then into vesicles, over a 24- to 48-hour period (Color Plate II). Finally, the vesicles became pearly white pustules that dried up, with scab formation, in the second week. The scabs then dropped off, leaving pitted scars.

Eradication of Smallpox

The World Health Organization began its effort to eradicate smallpox worldwide in 1967. Smallpox was an appropriate target for eradication because the vaccine is effective and because humans are the only known host for the virus. The WHO mounted a massive surveillance program to identify and then vaccinate every contact of every known smallpox patient. No attempt was made to vaccinate the world population. These public health measures resulted in the eradication of the virus from all of humankind.

The last case of variola major on the American continent occurred in Brazil in 1971 and the last reported case in the world was recorded in October 1975. Persons in modern societies, with the exception of certain military personnel, are no longer vaccinated against smallpox.

WARTS: PAPILLOMAVIRUSES

Warts are benign tumors caused by papillomaviruses (pap′i-lo′mah-vi′rus). They are seen principally in children and young adults and are classified according to their location and appearance. **Common warts** are flesh-colored to brown, 2 to 10 mm in diameter (Figure 22.21). **Flat warts** have smooth, rounded surfaces and are found on the face, hands, and legs. **Plantar warts** grow from the skin surface inward, often on the feet. Plantar warts are more painful and more difficult to control than the others. **Condyloma acuminatum** (kon′di-lo′mah ah-ku′mi-nat′um) occurs on the genitalia or perineum in (Figure 22.21) pink to brown clusters, and is sexually transmitted.

Warts are annoying and sometimes painful afflictions, but most pose no long-term threat to health. The exception is condyloma acuminatum, which can become malignant. Condyloma acuminatum in females may be associated with squamous carcinomas of the cervix and vulva. It precedes squamous carcinomas of the penis in 15 percent of males who contract these carcinomas.

Warts are caused by small icosahedral DNA **papillomaviruses** transmitted by direct contact with an in-

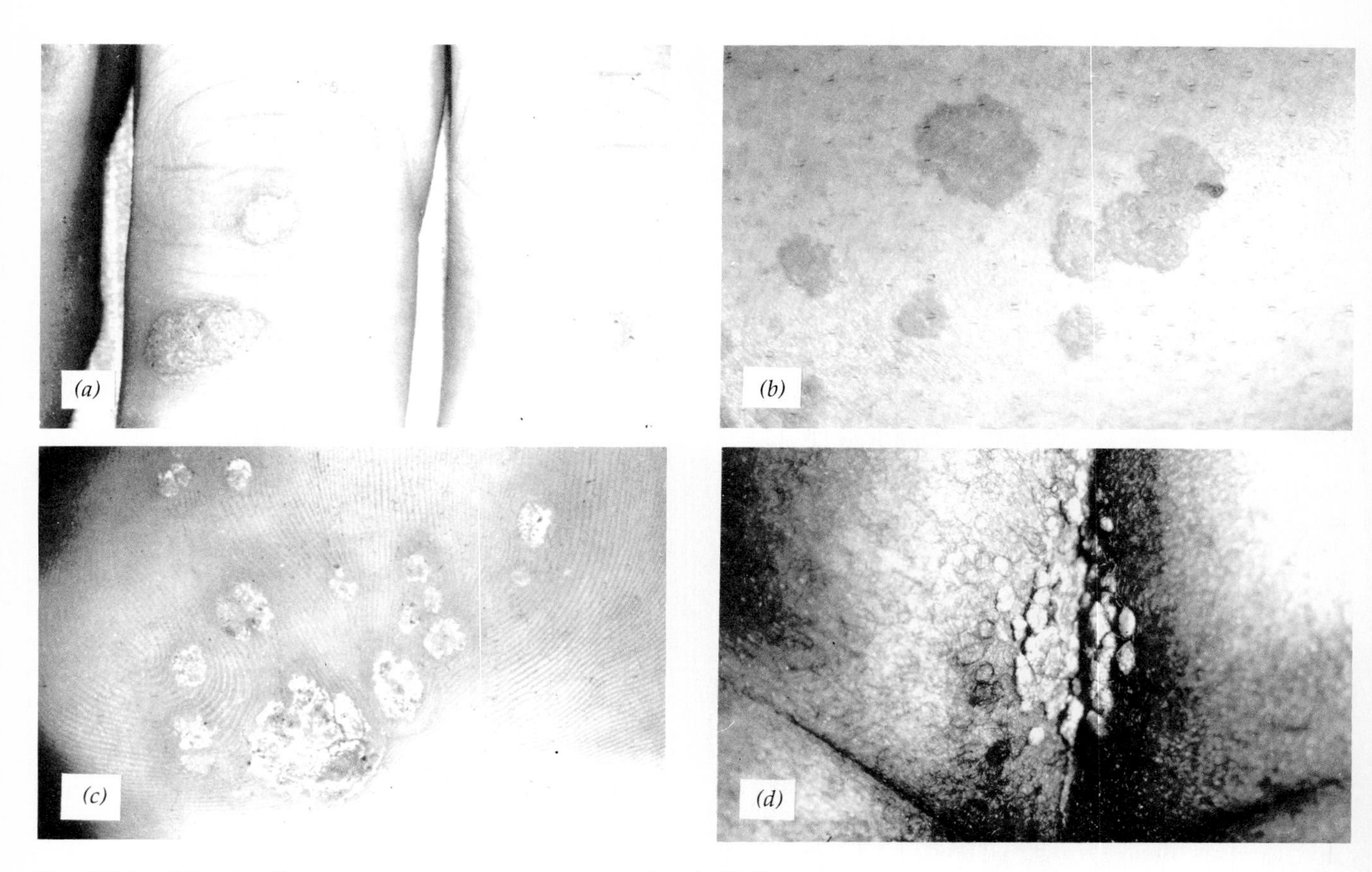

FIGURE 22.21 Human warts: (*a*) common wart on hand, (*b*) flat warts on thigh, (*c*) plantar warts, and (*d*) perianal condylomata acuminata. (Courtesy of Dr. K. E. Greer. From K. E. Greer, "Papovaviridae," in G. L. Mandell, R. G. Douglas Jr., and J. E. Bennett (eds.), *Principles and Practice of Infectious Diseases,* 6th edition, John Wiley & Sons, 1985.)

fected object or person. The virus penetrates minor abrasions to infect a single epithelial cell. The infection stimulates the cell to divide so that it develops into a visible wart after an incubation period of one to six months.

Treatment

Although many warts spontaneously disappear, leaving no scar, others persist and should be removed by surgical or chemical treatments. Warts can be excised by electrosurgery, or the affected tissue can be destroyed by cryosurgery; or topical application of salicylic acid or 5-fluorouracil may be administered. Spontaneous regression is common.

KEY POINT

Most papillomaviruses are entirely benign and are not associated with cancer. Nevertheless, some papillomaviruses precede carcinoma in a large enough percentage of cancer cases to arouse considerable investigative efforts.

FUNGAL DISEASES OF SKIN

Of the more than 50,000 species of fungi, only about 100 cause disease. Most of the pathogenic fungi are incidental or opportunistic pathogens whose normal habitat is soil or vegetable matter. The most common fungal infections, such as athlete's foot, are superficial infections easily treatable. Other fungi cause systemic infections that often originate in the respiratory tract (see Chapter 23).

CUTANEOUS MYCOSES (DERMATOMYCOSES)

Cutaneous mycoses are fungal infections of the keratinous tissues—nails, hair, and the stratum corneum. The keratinous material is usually dead (sloughed off) except for a germinating zone in the nail plate or the hair follicle, and it is within the dead tissue that the **dermatophytes** (der′mah-to-fyt′) develop. These fungi are classified on the basis of their asexual spores (Chapter 16) and belong to the genus *Epidermophyton* (ep′i-der-mof′i-ton), *Microsporum* (mi-kros′po-rum), or *Trichophyton* (tri-kof′i-ton). Dermatophytes are often grouped according to their primary natural reservoirs, which include humans (anthropophilic), animals (zoophilic), and soil (geophilic).

Cutaneous mycoses often appear ring shaped as the dermatophyte grows outward from the infection site. These infections were originally thought to be caused by worms and are still called **ringworm** or *tinea* (tin′e-ah). More than one fungus is able to infect a given anatomical location, so the diseases have been given anatomical designations. Ringworm of the body is **tinea corporis;** of the groin, **tinea cruris** ("jock itch"); of the head, **tinea capitis;** and of the foot, **tinea pedis** ("athlete's foot").

Tinea Corporis

This dermatophytic infection of the body (excluding the head, scalp, feet, beard, and groin) occurs worldwide, but is most common in tropical climates. The causative fungi include *Trichophyton mentagrophytes* (Figure 22.22), *T. rubrum,* and *Microsporum canis.* The infection usually begins as a papule, which grows to form an annular lesion. Many other diseases have similar clinical manifestations, and diagnosis is difficult in the absence of microscopic examination of scrapings. Tinea corporis is treated with topical antifungal ointments, such as miconazole and clotrimazole.

Tinea Cruris ("Jock Itch")

Ringworm of the groin is caused by *Epidermophyton floccosum* (Figure 22.23), and *T. mentagrophytes.* It is most common in men, but is seen also in women. Moisture from a wet bathing suit, an athletic supporter, tight-fitting slacks or pantyhose, and obesity encourage the infection. Tinea cruris is effectively treated with medicated powders.

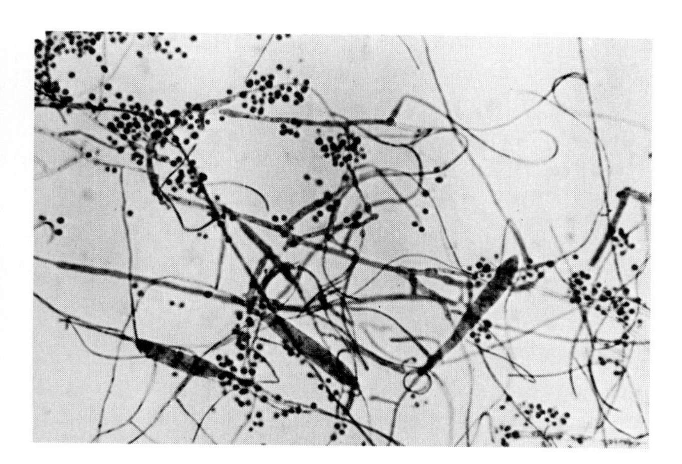

FIGURE 22.22 Micro- and macroconidia of *Trichophyton mentagrophytes.* 552×. (Courtesy of Dr. L. Ajello, Mycology Division, Centers for Disease Control. With permission of the American Society for Microbiology.)

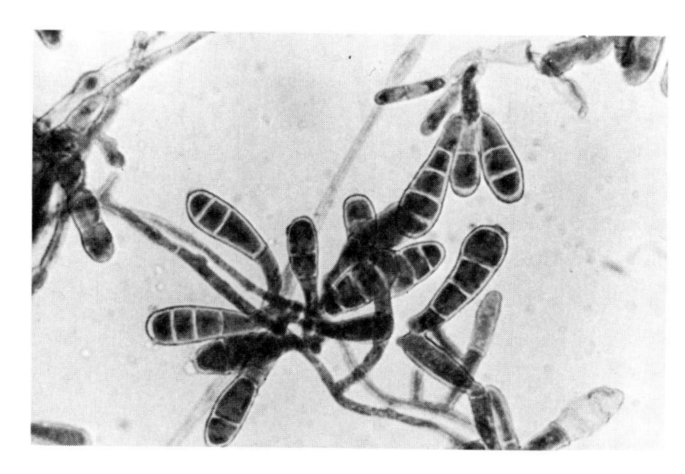

FIGURE 22.23 Clusters of smooth macroconidia of *Epidermophyton floccosum.* 552×. (Courtesy of Dr. L. Ajello, Mycology Division, Centers for Disease Control. With permission of the American Society for Microbiology.)

Tinea Pedis ("Athlete's Foot")

Tinea pedis occurs worldwide, though it is mainly seen in persons who wear shoes. The causative agent grows between the toes, giving rise to inflammation and desquamation. The infection is commonly spread through the use of shared shower facilities. The pathogens are all anthropophilic and include the same species as those responsible for tinea cruris. Treatment is as for tinea corporis.

Tinea Capitis

Dermatophytes of the genus *Microsporum* spp. (Figure 22.24) and *Trichophyton* spp. cause infection of scalp hair. The infection begins in the stratum corneum with hyphae extending into the opening of the hair follicle. The hyphae penetrate the hair near the bulb and extend into the hair shaft, generating spores near the surface of the hair shaft. The fungus weakens the hair; a plucked hair will break off next to the scalp. **Alopecia** (al′o-pee′sh-ia) (hair loss) is a common consequence especially when associated with an inflammation that destroys the follicle. The alopecia may be localized and spotty or may involve large areas of the scalp. This form of tinea capitis can be treated with oral griseofulvin.

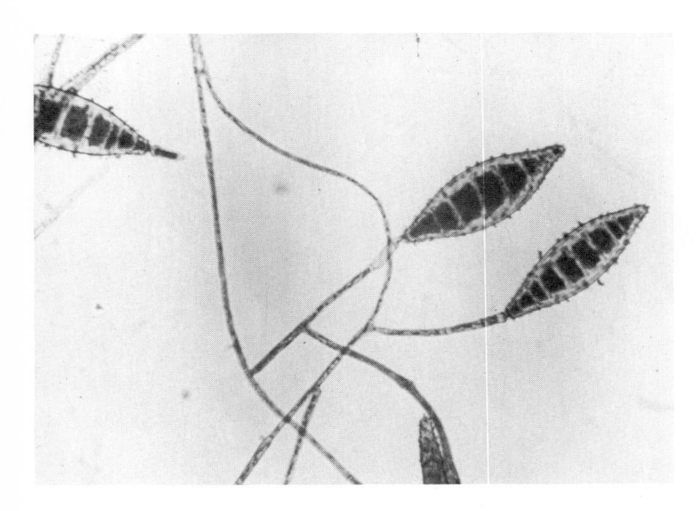

FIGURE 22.24 Rough-walled macroconidia of *Microsporum canis*. (Courtesy of Dr. L. Ajello, Mycology Division, Centers for Disease Control. With permission of the American Society for Microbiology.)

A few species of *Trichophyton* cause an infection of the hair shaft, but do not destroy the follicle. These fungi generate arthrospores in the shaft that greatly weaken the hair, causing it to break off just below the follicular surface. The resulting cavity soon fills with bacteria and debris, imparting a characteristic black-dot appearance to the scalp. Griseofulvin taken orally is prescribed.

SUBCUTANEOUS MYCOSES

Fungal diseases of the epidermis usually result from infected abrasions and wounds. These disorders are most prevalent in the tropics, where people wear a minimum of protective clothing. Sporotrichosis (spo′ro-tri-ko′sis), chromomycosis (kro′mo-mi-ko′sis), and maduromycosis (mah-du′ro-mi-ko′sis) are three major subcutaneous mycotic diseases.

Sporotrichosis

Sporotrichosis is caused by *Sporothrix schenckii* (Figure 22.25), which begins as an ulcerous lesion. The epidermis becomes infected when a contaminated splinter or thorn penetrates the outer layer. After an incubation period of one to two weeks, a painless papule develops at the site, and enlarges to become an ulcerated lesion. **Lymphocutaneous sporotrichosis** occurs when the fungus forms multiple nodules after being spread by the lymph channels. These nodules also can ulcerate. Untreated lesions remain for years and spontaneous recovery is rare, though complete recovery is the rule with proper iodide therapy. **Extracutaneous sporotrichosis** is more serious and requires amphotericin B administration. The prognosis for these patients is still poor.

Chromomycosis

A number of darkly pigmented fungi cause *chromomycoses* (kro′mo-mi-ko′sis), which are chronic, localized fungal infections of the skin. The causative agents are found in the soil and decaying vegetation in tropical and subtropical climates. The fungi enter a skin wound, developing into a wart-like, scaly nodule or a small, firm tumor. They are unsightly, and patients usually seek treatment for cosmetic reasons or because

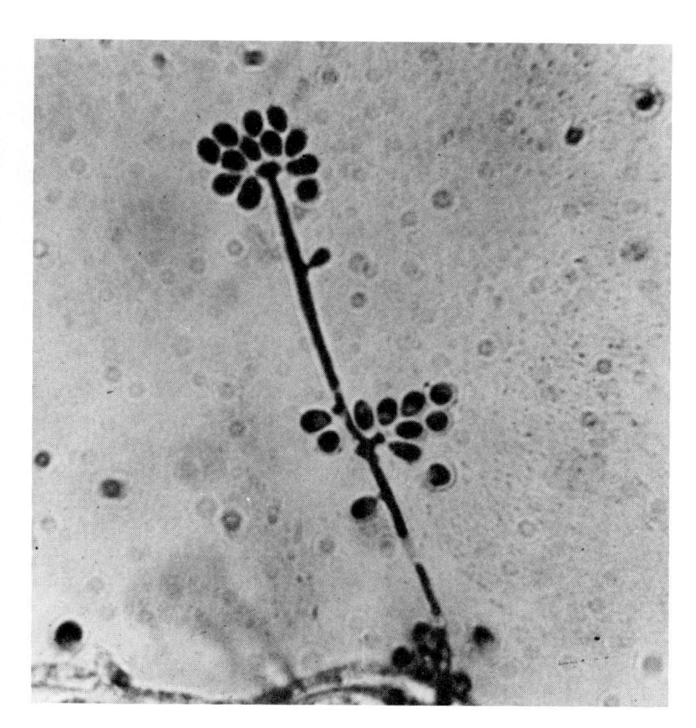

FIGURE 22.25 Typical morphology of conidiophores of *Sporothrix schenckii*. (Courtesy of Dr. C. Emmons.)

of secondary infections. Most infections can be treated by surgical excision. Drugs are ineffective, presumably because the lesions are inaccessible.

Mycetomas

Mycetomas (mi-see-toh'mah) were initially reported in the district of Madura in India, and were called maduromycoses. A widely used name is Madura foot. Mycetoma is a chronic, local, subcutaneous infection of the foot and, occasionally, of the hand. It is progressively destructive, destroying muscle, bone, and connective tissue.

Mycetomas occur most often in barefoot farmers in rural parts of the tropics, where *Madurella mycetomi* is found. The initial infection is usually associated with a penetrating wound of a foot often caused by a thorn or splinter. The pathogens are visible as dark grains within the exudate of the wound. Scar tissue forms and causes local disfigurement. There are no spontaneous recoveries and no dependable therapy.

PROTOZOAN DISEASES AFFECTING THE SKIN

The *Leishmania* (lesh-ma'ne-ah) are transmitted between humans or to humans from animal reservoirs by blood-sucking insects. They are most prevalent in the tropical climates. Species of *Leishmania* cause kala-azar (kah'lah-ah-zar'), cutaneous leishmaniasis (lesh'mah-ni'ah-sis), and mucocutaneous leishmaniasis.

KALA-AZAR

The most important form of leishmaniasis in India is kala-azar, which means "black fever." The protozoan *Leishmania donovani* is an obligate parasite whose prominent reservoirs are humans, dogs, and foxes.

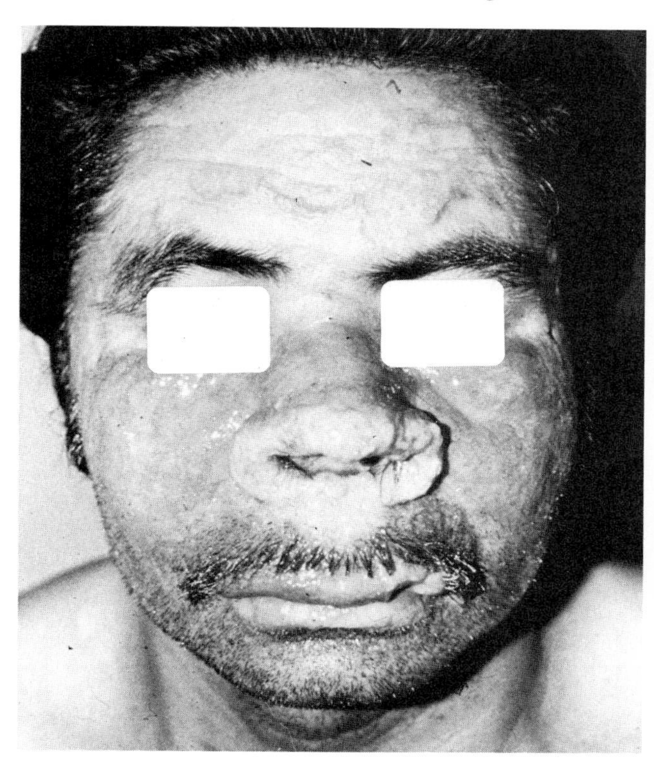

FIGURE 22.26 Mucocutaneous leishmaniasis with extensive involvement of the nose and upper lip. The nasal septum has been completely destroyed. (From G. Mandell, J. E. Bennett, and R. G. Douglas (eds.), *Principles and Practice of Infectious Diseases,* 2nd ed., Wiley, New York, 1985.]

Sandflies of the genus *Phlebotomus* (fle-bot′o-mus) are the parasite's intermediate host and vector.

Kala-azar is an insidious disease. Symptoms usually develop 2 weeks to 15 months after a bite by an infected sandfly. There is vague discomfort, diarrhea or constipation, low-grade fever, and anorexia; or there may be an acute onset characterized by a high fever and chills. In the later stages, the abdomen becomes bloated because the liver and spleen are swollen. The skin at this stage takes on a characteristic gray hue, the sign from which the disease derived its name. The patient deteriorates gradually due to the prolonged fever, weight loss, and weakness, and may succumb to secondary bacterial infections.

CUTANEOUS AND MUCOCUTANEOUS LEISHMANIASIS

Dogs, cats, and rodents are the reservoir for *Leishmania tropica,* which causes cutaneous leishmaniasis after being transmitted to humans by infected sandflies. A papule forms at the infection site following an incubation period of two to six months, and becomes ulcerous. The ulcer usually heals spontaneously over a period of several months; however, it leaves the patient disfigured if the ulcer was located in the oral or nasal mucosa (Figure 22.26). Immunity is permanent following recovery. The infection can be treated with antimony compounds.

Summary Outline

BARRIERS AGAINST INFECTIONS

1. The first line of defense against infectious agents is the anatomical barrier, including the skin, conjunctiva and tears, mucous membranes, and the ciliary epithelia of the trachea.
2. The outermost layer of epidermis (stratum corneum) is constantly being sloughed off, carrying with it bacteria and other microbes that are normal inhabitants of the skin.
3. The body orifices are also protected against infectious agents; mucus secreted by mucous membranes entraps microbes, and the secretory antibodies (IgAs) and lysozyme produced in mucus and tears help resist invading microbes.

BACTERIAL INFECTIONS OF THE SKIN

1. Pathogenic staphylococci are coagulase-producing, gram-positive cocci that infect the skin, causing boils, furuncles, scalded-skin syndrome, and impetigo.
2. Certain strains of *Staphylococcus aureus* produce enterotoxin(s) that cause food poisoning; strains that produce TSS-1 exotoxin are responsible for toxic shock syndrome.
3. *Streptococcus pyogenes* is a gram-positive, catalase-negative, β-hemolytic pathogenic coccus that can cause streptococcal pharyngitis, scarlet fever, impetigo, and puerperal sepsis.
4. Streptococcal pharyngitis can lead to rheumatic fever if not treated with an appropriate antibiotic such as penicillin or erythromycin.
5. Anthrax is a zoonosis of sheep and cattle caused by the endospore-forming bacterium *Bacillus anthracis*.
6. Humans contract cutaneous anthrax when spores infect an open sore on the skin, or pulmonary anthrax by inhaling the *B. anthracis* spores.
7. Yaws and pinta are tropical skin diseases caused by two subspecies of *Treponema pallidum* that are transmitted by direct contact. Pinta is characterized by discolored skin lesions; yaws by open lesions that contain infectious treponemes.
8. Surgical wounds, physiological wounds, and traumatic wounds disrupt the first line of defense and this enables opportunistic pathogens to cause infections.
9. *Pseudomonas aeruginosa* infects burned tissue as an opportunistic pathogen. This ubiquitous aerobe grows in burned tissue and produces toxins that prevent healing.
10. Alginic acid-producing strains of *P. aeruginosa* in-

fect the lungs of cystic fibrosis patients and contribute to the patient's death when s(he) cannot clear the bacteria from the lungs.

11. *Streptococcus, Staphylococcus, Bacteroides,* and *Clostridium* invade damaged tissue and cause local tissue necrosis known as infectious gangrene.
12. *Clostridium perfringens,* the most common isolate, *C. novyi,* and *C. septicum* all cause gas gangrene. The symptoms include swelling, a dark yellow appearance of the skin, and the presence of a thin, dark, watery exudate.

INFECTIONS OF THE EYE

1. Bacterial conjunctivitis is an inflammation of the conjunctiva caused by *Haemophilus aegyptius, H. influenzae, Streptococcus pneumoniae,* or *S. pyogenes.* Keratitis is an inflammation of the cornea.
2. The eyes of newborns can become infected during birth by *Neisseria gonorrhoeae* present in the mother's vagina, and this can result in ophthalmia neonatorum.
3. *Chlamydia trachomatis* from the vagina can cause inclusion conjunctivitis in newborns. Adults contract inclusion conjunctivitis through contaminated towels, swimming pool water, or autoinoculation.
4. *Chlamydia trachomatis* can also cause trachoma, with permanent corneal scarring leading to blindness.

VIRAL DISEASES OF THE SKIN

1. Measles is a highly contagious childhood disease caused by an RNA virus and diagnosed by the presence of Koplik spots inside the cheek and by a maculopapular rash.
2. The incidence of measles is on the rise although efforts to eliminate the virus by vaccination have been renewed.
3. Rubella (German measles) is usually a mild childhood disease caused by an enveloped RNA virus.
4. Congenital rubella syndrome can cause abortion or serious birth defects if the virus infects the fetus.
5. Measles and rubella are controlled by immunizing children with live attenuated viral vaccines, which were introduced in 1963 and 1969, respectively.
6. The herpes simplex viruses and the varicella–zoster virus are DNA viruses (Herpetoviridae) that cause vesicular lesions and latent infections.
7. Herpes simplex virus type 1 is primarily associated with orofacial herpes (cold sores or gingivostomatitis). These infections are transmitted by direct contact and recur when triggered by stress, emotional upset, too much sun, or menstruation.
8. HSV-1 can be transmitted from one part of the body to another by self-infection.
9. Initial infection by the varicella–zoster virus causes chickenpox (varicella), a highly contagious febrile childhood disease characterized by itching vesicular lesions.
10. Reactivation of a latent varicella–zoster infection results in zoster, which can occur months or years after recovery from chickenpox.
11. Immunity against the herpes simplex viruses or the varicella–zoster virus is incomplete at best.
12. Smallpox was the first viral disease to be eradicated from the human population. This was accomplished by a worldwide immunization and surveillance program.
13. Papillomaviruses are DNA viruses that cause benign skin tumors known as warts. Many warts spontaneously regress, while persistent warts can be treated chemically or by electro- or cryosurgery.
14. Condyloma acuminatum in females may be associated with squamous carcinoma of the cervix and vulva, and in males with carcinoma of the penis. It is sexually transmitted.

FUNGAL DISEASES OF THE SKIN

1. *Epidermophyton, Microsporum,* and *Trichophyton* are dermatophytic fungi that infect keratinous tissue, including hair, nails, and stratum corneum from their reservoirs in humans, animals, or soil.
2. Cutaneous mycoses (dermatomycoses) are collectively known as ringworm (tinea), and include tinea corporis, tinea cruris, tinea capitis, and tinea pedis.

3. Sporotrichosis and chromomycosis are subcutaneous mycoses initiated by penetration of skin by splinters, thorns, and dirt contaminated by the causative soil fungus.
4. *Madurella mycetomi* causes progressive, destructive infections in the feet and hands, called mycetomas, of barefoot farmers in the tropics.

PROTOZOAN DISEASES AFFECTING THE SKIN

1. *Leishmania donovani,* an obligate parasite protozoan, causes kala-azar or "black fever" in humans. Sandflies *(Phlebotomus)* are its intermediate host and vector and its reservoirs are humans, dogs, and foxes.
2. Cutaneous leishmaniasis is caused by *L. tropica* transmitted from dogs, cats, or rodents to humans by sandflies. Facial disfigurement is a result of the ulcers that develop in the oral or nasal mucosa.

Questions and Topics for Study and Discussion

QUESTIONS

1. How does the skin function as a barrier against infectious diseases? What microbes discussed in this chapter can cause disease only after the skin is punctured or broken?
2. What kinds of diseases are caused by the pathogenic staphylococci? How are the staphylococci differentiated from the streptococci?
3. Describe the virulence factors produced by *Staphylococcus aureus.* How do these characteristics relate to this bacterium's ability to cause skin infections?
4. What links the superabsorbent tampon and *Staphylococcus aureus* as being responsible for toxic shock syndrome?
5. Describe the relationship between "strep throat" and scarlet fever. What protection do patients develop against scarlet fever?
6. Why is *Pseudomonas aeruginosa* a constant threat to burn patients? What steps can be taken to prevent *P. aeruginosa* infections of burn wounds?
7. Describe the symptoms and causes of gangrene. What group of patients would you expect to suffer from a histotoxic infection?
8. Which bacteria and viruses cause eye infections in newborns and how are their eyes infected? Can any of these lead to blindness?
9. What mechanisms does the eye use to protect itself against infectious agents?
10. How has the measles vaccine altered the incidence of measles? Is postinfection vaccination with the measles vaccine an effective treatment? Explain.
11. What are the most serious consequences of rubella infections and how can these complications be prevented?
12. Which viruses cause recurrent skin infections in humans? What effect do recurrent infections have on the transmission of these viral diseases?
13. Explain how smallpox was eradicated from the human population. Can we do the same thing with other viruses and, if so, which viral diseases are prime candidates?
14. Are the papillomaviruses oncogenic viruses? What type of warts have the potential to become cancerous and how are they transferred?
15. Describe the clinical signs of four common dermatomycoses. How are these infections prevented and treated?

DISCUSSION TOPICS

1. How are humans protected against the recurrence of chickenpox, while at the same time they remain susceptible to zoster?
2. What are the arguments against using a live, attenuated strain of the varicella–zoster virus to immu-

nize the population against varicella (chickenpox)?

3. Rubella occurs seasonally, in the winter. In which months would an infant with CRS be most likely to be born?

Further Readings

BOOKS

Beneke, E. S., and A. L. Rogers, *Medical Mycology Manual with Human Mycoses Monograph,* 4th ed., Burgess, Minneapolis, 1980. A manual that can be used in the clinical laboratory and in medical mycology courses.

Easmon, C. S. F., and C. Adlam (eds.), *Staphylococci and Staphylococcal Infections,* Vols. 1 and 2, Academic Press, New York, 1983. These two volumes bring together various aspects of staphylococcal research and their application to clinical and veterinary practice.

Fraser, K. B., and S. J. Martin, *Measles Virus and Its Biology,* Academic Press, London, 1978. A monograph on the biology of the measles virus and the disease it causes in humans.

Sherris, J. C. (ed.) *Medical Microbiology: An Introduction to Infectious Diseases,* Elsevier, New York, 1984. This comprehensive textbook contains chapters on staphylococci, streptococci, and the genus *Bacillus*. An excellent source for detailed information on infectious diseases.

ARTICLES AND REVIEWS

Behbehani, A. M., "The Smallpox Story: Life and Death of an Old Disease," *Microbiol. Rev.,* 47:455–509 (1983). An extensive history of smallpox from antiquity to present.

Brunell, P. A., "Protection Against Varicella," *Pediatrics,* 59: 1–2 (1977). A commentary on the use of a vaccine against chickenpox.

Chesney, P. J., M. S. Bergdoll, J. P. Davis, and J. M. Vergeront, "The Disease Spectrum, Epidemiology, and Etiology of Toxic Shock Syndrome," *Ann. Rev. Microbiol.,* 38:315–338 (1984). This review covers the history, epidemiology, toxicology, and clinical manifestations of toxic shock syndrome.

Henderson, D. A., "The Eradication of Smallpox," *Sci. Am.,* 235:25–33 (October 1976). An epidemiological report on the successful elimination of this serious human viral disease.

Langer, W. L., "Immunization Against Smallpox Before Jenner," *Sci. Am.,* 234:112–117 (January 1976). The history of variolation in England.

Wolontis, S., and S. Jeansson, "Correlation of Herpes Simplex Virus Type 1 and 2 with Clinical Features of Infection," *J. Infect. Dis.,* 135:28–33 (1977). A research report on the symptoms associated with more than 300 herpes simplex viruses isolated from human patients.

CHAPTER

23

Infectious Diseases of the Respiratory Tract

OUTLINE

FOCUS OF THIS CHAPTER

In this chapter the focus is on the bacteria, viruses, and fungi that cause respiratory diseases. These diseases are grouped according to the causative agents and the type of damage they cause. Symptoms, modes of transmission, and prevention and treatment are discussed.

The upper respiratory tract is constantly exposed to infectious microbes. The microbes gain direct entrance to the tract when inhaled with dust or droplets generated by coughing, or indirect entrance viá the mucous membranes of the nares or the conjunctiva. They then cause the symptoms we commonly associate with respiratory infections. The illness may be as mild as a cold or as debilitating as tuberculosis.

THE RESPIRATORY TRACT

The upper respiratory tract includes the mouth, tonsils, nasal cavity, nasopharynx, and throat (Figure 23.1). It is protected against infectious agents by mucus secretions and saliva, which contain IgAs and lysozyme as previously described (Chapter 17). Nonetheless, some microbes colonize the nares, gums, and teeth to cause dental caries and periodontal disease; others are opportunistic pathogens found in carriers.

The lower respiratory tract, consisting of the trachea, bronchi, and lungs, is free of microbes in healthy persons, because in these organs the mucous membranes and ciliated epithelial cells eliminate pathogens present in inspired air. The mucus secreted by the membranes entraps microbes, and the IgAs of mucus inactivate them. Should any pathogens escape the guards and reach the bronchi, they will be swept toward the esophagus by the ciliary action of the epithelium and expectorated in sputum, or be swallowed.

BACTERIAL INFECTIONS

Bacterial pneumonia, diphtheria, tuberculosis, pertussis (whooping cough), Q fever, Legionnaires' disease, and ornithosis (psittacosis) are all due to pathogenic bacteria. Their virulence factors vary from potent exotoxins to a combination of factors that enable them to develop as intracellular parasites. Most bacterial diseases of the respiratory tract can be treated with antimicrobials and many are preventable through immunizations.

DIPHTHERIA

Diphtheria is of historic importance, because this once-dreaded killer of children was the first bacterial disease to be treated and controlled by immunological procedures. Diphtheria derives its name from the whitish-gray veil or membrane that forms on the tonsils and pharynx.

Isolation of the diphtheria toxin in the late 1880s led to the practical protective procedures with which we are familiar. Emil von Behring and Shibasaburo Kitasato (1890) made an inactive toxoid from the diphtheria toxin and with it immunized laboratory animals. They discovered that the animal's blood contained a substance (an antitoxin) that neutralized the diphtheria toxin, and they demonstrated that the antitoxin protected unimmunized laboratory animals against the toxin. In other words, they were the first to passively immunize susceptible subjects with an antitoxin. This approach was adopted in 1891 by a group of German physicians who used it to immunize human subjects, a practice that continues today.

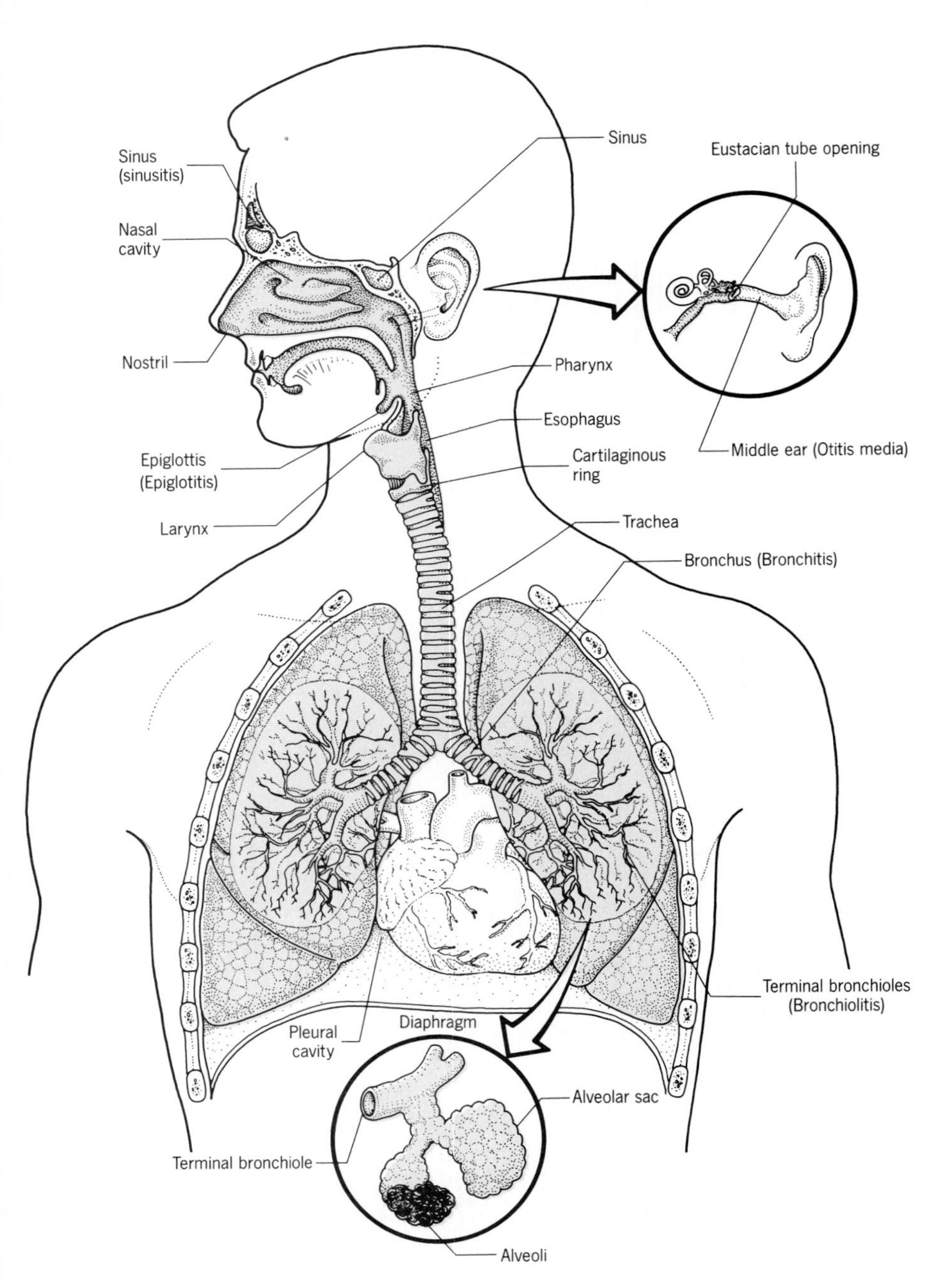

FIGURE 23.1 Upper torso, showing the anatomy of the upper and lower respiratory tract. (From A. Nason, *Modern Biology,* Wiley, New York, 1965, copyright © 1965, by John Wiley & Sons, Inc.)

Corynebacterium diphtheriae

The causative agent of diphtheria, *Corynebacterium diphtheriae,* colonizes the nasopharynx where it forms a whitish-gray membrane over the tonsils and nasopharynx and produces the toxin that is carried by the circulatory system to all parts of the body. Symptoms then occur, including a slight fever, fatigue, malaise, and a sore throat that is often accompanied by a dramatic swelling of the neck. An untreated patient's health can deteriorate rapidly and death from progressive organ failure due to the toxin can follow.

Corynebacterium diphtheriae is a gram-positive straight or slightly curved rod that may have a club-like appearance. It can be cultured on tellurite blood agar (Figure 23.2). Diagnosis is made on the basis of symptoms, since treatment cannot await bacteriological confirmation.

Virulence, Lysogeny, and the Diphtheria Toxin

All toxin-producing (virulent) strains of *C. diphtheriae* are lysogenic for β phage (see Lysogeny, Chapter 11), whose genome carries the *tox* gene coding for diphtheria toxin. Strains cured of their prophage lose their ability to produce the toxin and become avirulent. The process is reversed during phage conversion; nonvirulent strains of *C. diphtheriae* become virulent when the β phage establishes itself as a prophage in a host cell (Figure 23.3).

The diphtheria toxin is released as a single 62,000 MW protein. The toxin molecule is then nicked by an extracellular protease to produce the A and B subunit that remain bound until they reach a target cell. The B subunit defines which cells are targets by recognizing and binding to cell surface receptors (Figure 23.4). Once bound, the A subunit is released and translocated into the cell's cytoplasm, where its toxic effects are expressed.

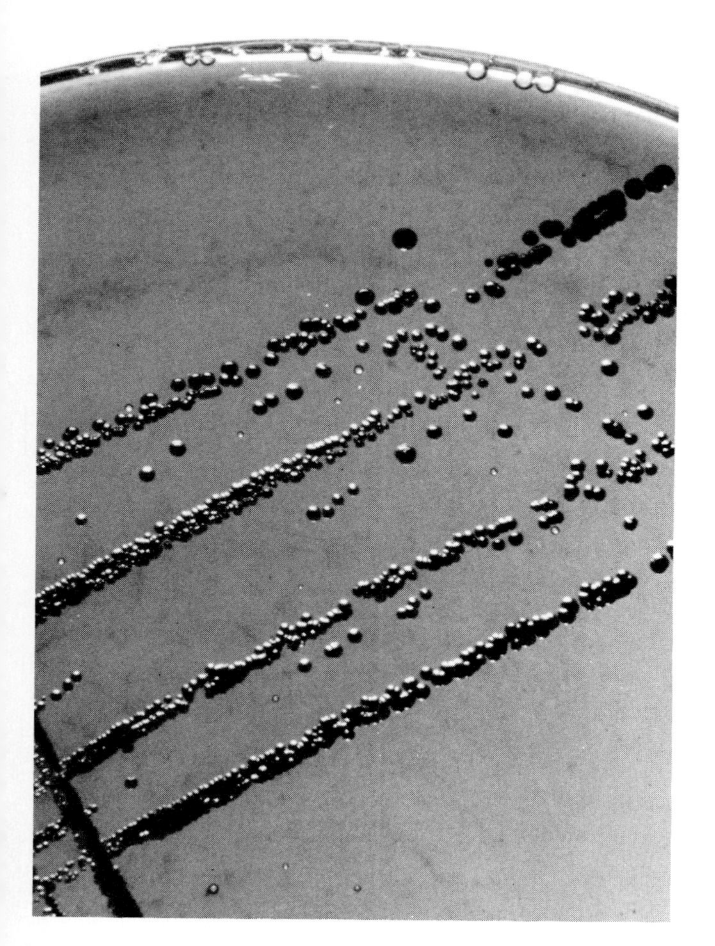

FIGURE 23.2 Colonies of *Corynebacterium diphtheriae* growing on tellurite medium. (Courtesy of Dr. I. S. Snyder.)

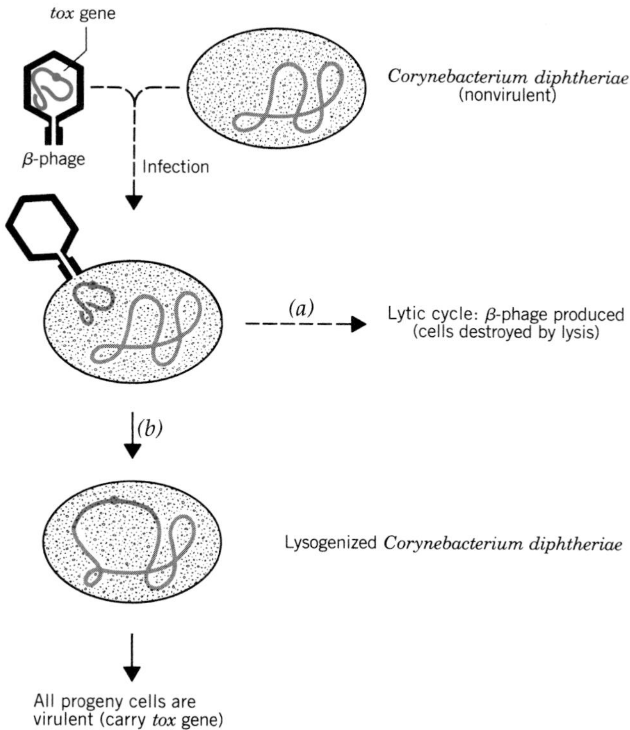

FIGURE 23.3 β-phage infection of nonvirulent *Corynebacterium diphtheriae* can convert the bacterium to a toxin-producing, virulent strain. This temperate bacteriophage can *(a)* reproduce itself through a replication cycle that kills the host cell or *(b)* be incorporated into the host cell's genome as a prophage. The prophage carries a *tox* gene that codes for diphtheria toxin.

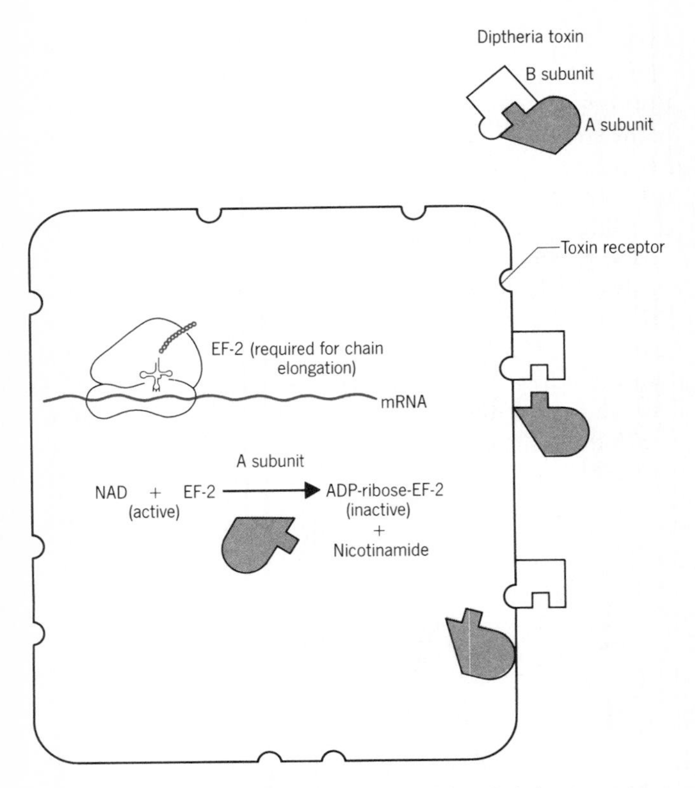

FIGURE 23.4 The B subunit of the diphtheria toxin recognizes receptors on the surface of eucaryotic target cells. After binding, the A subunit enters the cell's cytoplasm where it inhibits protein synthesis by inactivating EF-2.

The A subunit inhibits eucaryotic cell protein synthesis by enzymatically inactivating the elongation factor (EF-2). EF-2 is necessary for movement of the tRNA polypeptide–ribosomal complex along messenger RNA (Figure 23.4). EF-2 is inactivated when the A subunit covalently attaches an ADP-ribose to it. When all of the EF-2 is inactivated, protein synthesis stops and the cell dies. Procaryotic cells contain an elongation factor (EF-G) that is not affected by diphtheria toxin.

Immunobiology of Diphtheria Toxin

Diphtheria can be controlled by active or passive immunization. An unimmunized person who has been exposed should be passively immunized with antitoxin and treated with erythromycin. The antitoxin inactivates unbound toxin, while erythromycin inhibits toxin production and prevents the person from becoming a carrier of *C. diphtheriae*. Passive immunization is accomplished with equine diphtheria antitoxin, which neutralizes diphtheria toxin before the A subunit enters the tissues.

Formalin-inactivated diphtheria toxoid is part of the DTP vaccine administered to babies to protect them against diphtheria, tetanus, and pertussis. It is a highly effective preventive. In countries where active immunization is required, the disease is close to nonexistent—only 16 cases were reported in the United States between 1980 and 1984.

Epidemiology

Unimmunized individuals continue to be the victims of diphtheria. Because the bacteria can be so readily carried, both asymptomatic carriers and sick persons can spread them in cough-produced droplets and by direct contact. Though antibiotic therapy is helpful, as previously described, the chief method of control is immunization.

PERTUSSIS (WHOOPING COUGH)

Pertussis (per-tus'is), popularly known as whooping cough, is caused by *Bordetella pertussis* and less commonly by *Bordetella parapertussis*. The genus derives its name from Jules Bordet, a scientist who in 1900, aided by another scientist, Octave Gengou, first identified the organism. Pertussis is a serious disease in newborns and in older patients, despite the widespread use of the DTP vaccine.

Bordetella pertussis is a small, nonmotile, gram-negative bacterium that grows aerobically on media containing charcoal, starch, blood, or blood products. These substances protect the bacteria by reacting with unsaturated fatty acids, which inhibit bacterial growth. Strains of *B. pertussis* newly isolated from the human respiratory tract, the bacterium's only natural reservoir, grow as smooth colonies on these media.

Symptoms

Bordetella pertussis infects the trachea, especially in children less than 1 year old, and grows preferentially on the ciliated epithelial cells of the bronchi (Figure 23.5). After a ten-day incubation period, symptoms de-

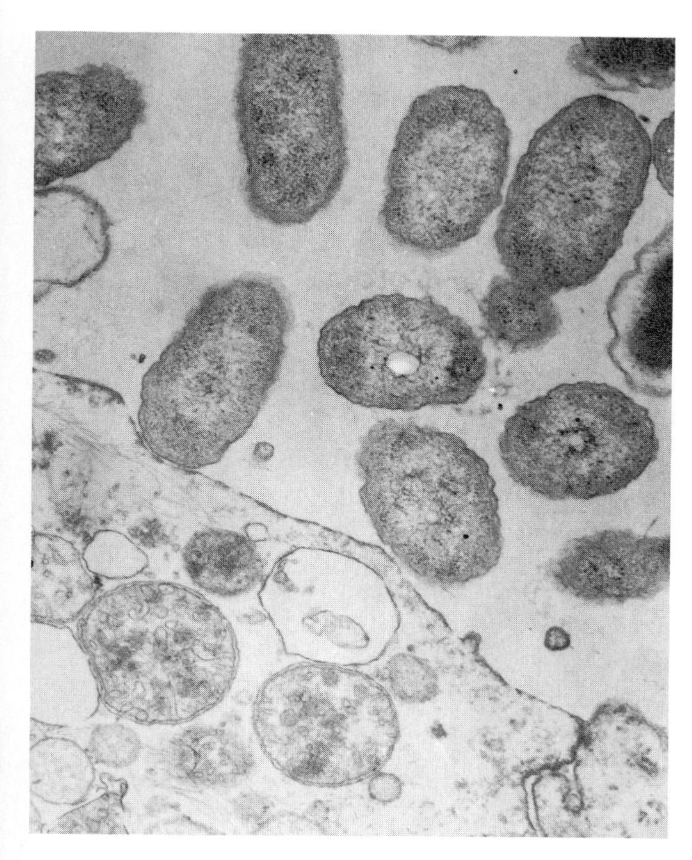

FIGURE 23.5 Electron micrograph of *Bordetella pertussis* growing in association with cultured hamster tracheal cells. (Reprinted from A. M. Collier, L. P. Peterson, and J. B. Baseman, *J. Infect. Dis. Suppl.*, 136S:S196–S203, 1977, by permission of The University of Chicago Press.)

velop—a mild cough, sneezing, and inflammation of the nasal mucous membranes. During the next 10 to 14 days, the infection spreads to the lower respiratory tract; the cough is now severe, characterized by whooping as air is inspired. Vomiting is common. As mucus accumulates heavily in the bronchi, there may be anoxia, sometimes with convulsions. The whooping aspect of the cough lasts for approximately two weeks and is followed by a milder cough, for another two to three weeks.

Immunity and the Pertussis Vaccine

Recovery from whooping cough confers resistance to subsequent infections, but does not provide complete protection. A second episode would be less severe. Active immunization of all babies at 2 months of age with the DTP vaccine is advocated because they have little or no immunity against *B. pertussis*.

The vaccine, made from inactivated whole cells of *B. pertussis*, can have mild to serious side-effects. A few babies experience redness and swelling at the injection site, and, in an extremely small number of babies, permanent brain damage occurs after vaccination.

Epidemiology and Treatment

Pertussis is a highly contagious disease that is worldwide in occurrence. Most cases of pertussis develop in unimmunized babies. The preferred drug for chemotherapy is an enylhromycin. There has been a steady decline in the incidence of whooping cough (Figure 23.6) since widespread use of the vaccine began in the 1950s. Even so, the median number of reported cases in the United States from 1981 to 1985 was 2197 and in 1986 it surpassed 4000 cases.

TUBERCULOSIS

There was a time when tuberculosis caused one-fourth of all adult deaths in Europe. Despite modern methods of therapy, tuberculosis remains a significant respiratory disease suffered annually by more than 20,000 persons in the United States. Early diagnosis and lengthy antimicrobial therapy are the cornerstones of public health efforts to control tuberculosis.

The Infection Process and Development of the Tubercle

Mycobacterium tuberculosis (mi-ko-bak-te′re-um too-ber-ku-lo′sis) is the major cause of tuberculosis in developed countries. Its ability to resist destruction by phagocytosis and multiply intracellularly contributes to the unique characteristics of tuberculosis.

Mycobacterium tuberculosis enters the alveoli upon being inhaled in droplet nuclei generated by the coughing of an infected person. A nonspecific inflammatory reaction develops at the infection site, and the tubercle bacilli (another term for *M. tuberculosis*) are engulfed by phagocytes. Many bacilli are killed, but some survive and multiply as intracellular parasites. Infection spreads by means of these infected leukocytes to the regional lymph nodes, and from there to other parts of the body, where additional foci of infection are established.

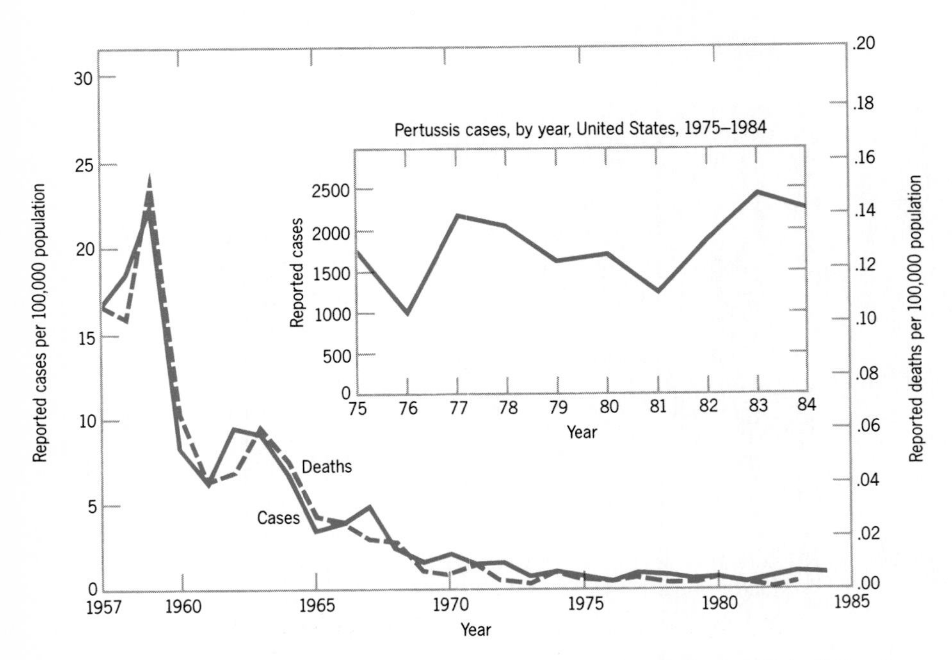

FIGURE 23.6 Incidence of pertussis (whooping cough) in the United States, 1957–1984. The dashed line depicts the death rate from whooping cough. (Centers for Disease Control, "Annual Summary 1984: Reported Morbidity and Mortality in the United States," *Morbidity and Mortality Weekly Report,* Vol. 32, No. 54, 1986.)

Cell-Mediated Immunity Two to six weeks after the primary infection, there is a cell-mediated immune response and the person becomes hypersensitive to the proteins of *M. tuberculosis*. The immune response is necessary for the development of tubercles, which are small rounded nodules surrounding leukocytes infected by bacilli. Antigens of the tubercle bacilli react with activated T lymphocytes, causing them to multiply into a clone of cells and to secrete lymphokines (Figure 23.7*a*). The lymphokines attract macrophages to the area, activate the macrophages, and immobilize them at the site. These phagocytic cells have numerous lysosomes and vacuoles (Figure 23.7*b*) that participate in destroying the bacilli.

Not all the bacilli are killed; and some of the infected macrophages fuse to form giant cells that become the foci of the tubercle. Intracellular replication of the tubercle bacilli slows down and eventually stops as the macrophages in the center of the nodule die, leaving a cheesy residue. The tubercular lesion slowly heals, often with the deposition of fibrin and calcium, making the tubercles visible in X-ray films. The lesions may heal completely; however, a number of patients never destroy all the bacilli, and the surviving bacteria can remain in dormant tubercles for years or decades.

Active cases of tuberculosis are diagnosed by demonstrating the presence of acid-fast rod-shaped bacteria in a patient's sputum (Figure 23.8). The acid-fast staining (see Chapter 3) is attributed to the high-lipid content (up to 60 percent) in the bacterial wall. Diagnosis is confirmed by growing the organism aerobically on complex laboratory media. The slow-growing cells adhere to each other to form flowing waves of aggregated cells, which organize into colonies within three to six weeks.

Immunology of *M. tuberculosis*

The tuberculin screening test detects a delayed cell-mediated hypersensitive state that develops in persons who have been exposed to *M. tuberculosis* or its products. Hypersensitivity develops as early as 1 month after infection and may last a lifetime.

The **tuberculin test** is done by intradermally inject-

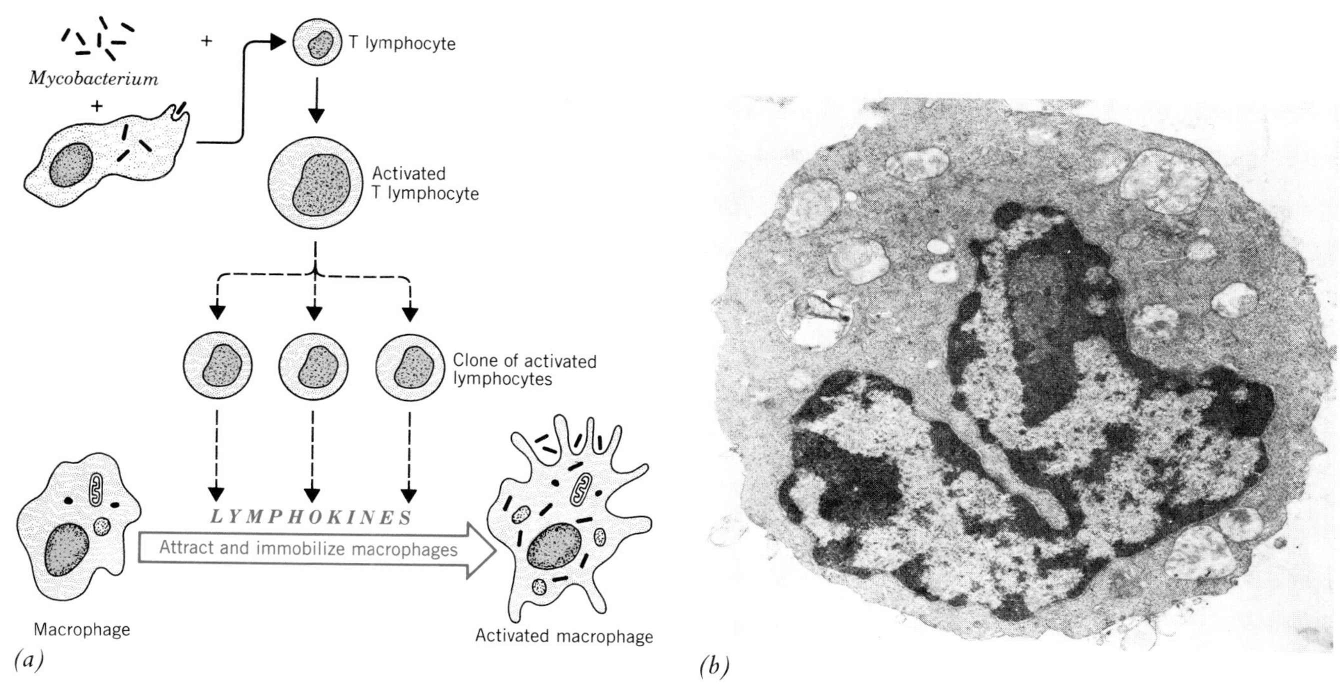

FIGURE 23.7 *(a)* Cell-mediated immune response to *Mycobacterium tuberculosis*. The T lymphocytes are activated by the bacteria to form a clone of lymphokine-producing cells. The lymphokines attract macrophages to the infection site, where they engulf the tubercle bacilli. *(b)* Electron micrograph of an activated T lymphocyte. (Courtesy of Dr. T. E. Mandel. From T. E. Mandel, "The Ultrastructure of Mammalian Lymphocytes and Their Progeny," in J. J. Marchalonis (ed.), *The Lymphocyte: Structure and Function,* Dekker, New York, 1977, copyright © 1977, Marcel Dekker.)

ing a purified protein derivative (PPD) taken from culture filtrates of *M. tuberculosis*. In a person who tests positive, a red, hardened area will appear at the site of injection in about 48 hours. A positive reaction indicates that the person either has an active case of tuberculosis, was previously infected, or has been immunized. A positive chest film, the presence of acid-fast bacteria in sputum or biopsy material, and isolation and speciation of *Mycobacterium* confirm the diagnosis.

The BCG Vaccine The **bacille Calmette Guérin** (BCG) vaccine is made with an attenuated strain of *Mycobacterium* and is administered to persons living in Scandinavia and some other parts of Europe. The BCG immunization is effective in preventing childhood tuberculosis, though its efficacy in preventing adult tuberculosis is still in question. Most immunized persons become tuberculin positive for about four years, commencing two months after they receive the BCG vaccine. Reversion to a tuberculin-negative state is construed to mean that reimmunization is necessary. Americans are not routinely immunized against tuberculosis; active cases are detected through routine tuberculin-screening procedures, and treated with drugs.

Epidemiology

New cases of tuberculosis originate when susceptible persons are infected by a person with active tuberculosis, or by the reactivation of dormant tubercles in previously infected persons. Epidemiologists estimate that some 75 percent of cases arise though the latter means. Crowded living conditions facilitate the transmission of tuberculosis from these active cases to others. Drug treatment was initiated in this country in the early 1950s, and has brought a dramatic decrease in the

Iditarod

In the winter of 1925, the city of Nome, Alaska, was threatened with a diphtheria epidemic. The only available diphtheria antiserum was 1000 miles away in Anchorage, and the only means of transporting it was by dog sled.

Libby Riddles or Susan Butcher with her dog sled team. (From Susan Butcher, "A woman's struggle: Thousand-mile Race to Nome," *National Geographic,* March 1983.)

Leonhart Seppala, one of the Yukon's most famous dog handlers, responded to the call for help. He left Nome by dog sled to meet the serum dispatched by relay teams from Anchorage. After four days on the trail covered with ice and snow, he met a dog team carrying the serum. After picking up the serum, he headed back to Nome. Soon Seppala met a fresh dog sled team, which carried the serum to Nome in time to prevent a serious diphtheria epidemic.

In 1973 the Iditarod was begun as a 1,049-mile dog sled race from Anchorage to Nome commemorating the epic diphtheria serum run of 1925.

Alaskans look forward to three great events in the long Alaskan winter: Christmas, New Year's, and the Iditarod. For years this race was dominated by men, even though women have competed in the Iditarod every year since 1976. In 1985 Libby Riddles became the first woman to win the $50,000 first-place prize, after 18 days on the trail. Susan Butcher, a close second in the 1982 and 1984 races, won the Iditarod in 1987 to become a heroine in her own right.

number of cases seen annually—from 88 per 100,000 persons in 1935 to fewer than 10 per 100,000 in 1984.

Prevention and Treatment

Combinations of isoniazid, rifampin, and ethambutol are used to treat tuberculosis. Isoniazid, or simply INH, as it is known, is widely used because it is effective, is only slightly toxic, and is inexpensive. The combination drugs are given to control the *Mycobacterium* if by chance it becomes resistant to one of the drugs. Drug resistance is a special problem in tuberculosis because ample time is available to select for resistant strains during the 9 to 12 months of therapy needed to effect a cure.

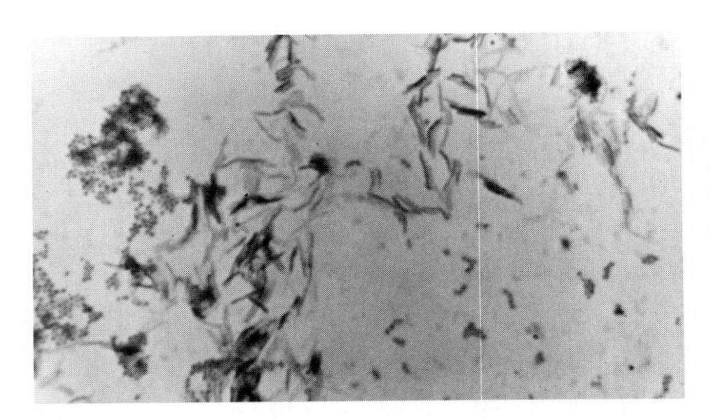

FIGURE 23.8 Acid-fast stain of *Mycobacterium smegmatis*. 1200×. (Courtesy of Dr. L. LeBeau.)

KEY POINT

If a person has once been infected with *M. tuberculosis,* (s)he will have to be continually vigilant against a recurrence because the tubercle bacilli survive intracellularly in dormant tubercles.

BACTERIAL PNEUMONIA

Several bacteria are capable of causing pneumonia, although *Streptococcus pneumoniae* and *Mycoplasma pneu-*

TABLE 23.1 Bacterial infections of the human lungs

ORGANISM	SOURCES	DISEASES	COMMENTS
Streptococcus pneumoniae	Human respiratory tract	Pneumonia	About 70 percent of bacterial pneumonias
Mycoplasma pneumoniae	Upper and lower respiratory tract	Atypical or walking pneumonia	About 20 percent of bacterial pneumonias
Klebsiella pneumoniae	Human respiratory tract	Pneumonia	About 3 percent of bacterial pneumonias
Haemophilus influenzae	Nasopharynx	Pneumonia, obstructive epiglottitis	Affects young children and viral-compromised patients
Bordetella pertussis	Respiratory tract of humans	Whooping cough	Prevent with pertussis vaccine (DTP shots)
Legionella pneumophila	Widely dispersed in nature	Legionnaires' disease	Spread in dust and by air-conditioning systems
Coxiella burnetii	Domestic animals	Q fever	Rickettsia, transmitted by aerosols
Chlamydia psittaci	Domestic and psittacine birds	Ornithosis	Spread in aerosols of parrot droppings

moniae together account for about 90 percent of the bacterial pneumonia cases.

STREPTOCOCCAL PNEUMONIA

As indicated in Table 23.1, approximately 70 percent of bacterial pneumonias are caused by *Streptococcus pneumoniae,* a normal inhabitant of the upper respiratory tract in many people. Encapsulated strains (Figure 23.9) grow on blood agar plates in the form of round glistening colonies surrounded by a zone of α hemolysis. They produce smooth mucoid colonies and the capsule protects them against phagocytosis.

Streptococcus pneumoniae gain entrance to the alveoli after being inspired in respiratory secretions. A compromised host, especially an older one, one who is ill or infirm, or one weakened by a viral infection, is a likely victim. The lung infection leads to edema, and this makes breathing difficult. With edema comes a profuse outpouring of fluid; in turn, the fluid inhibits normal gas exchange. Some 70 percent of untreated patients spontaneously recover after five to six days, while others recover following treatment with either penicillin or erythromycin. A few patients, especially those over 70 years old, die despite antibiotic therapy.

A vaccine called **pneumococcal polysaccharide vaccine** is available and is effective against most strains of *S. pneumoniae.* Its administration is recommended for adults with chronic illnesses and healthy adults above the age of 65.

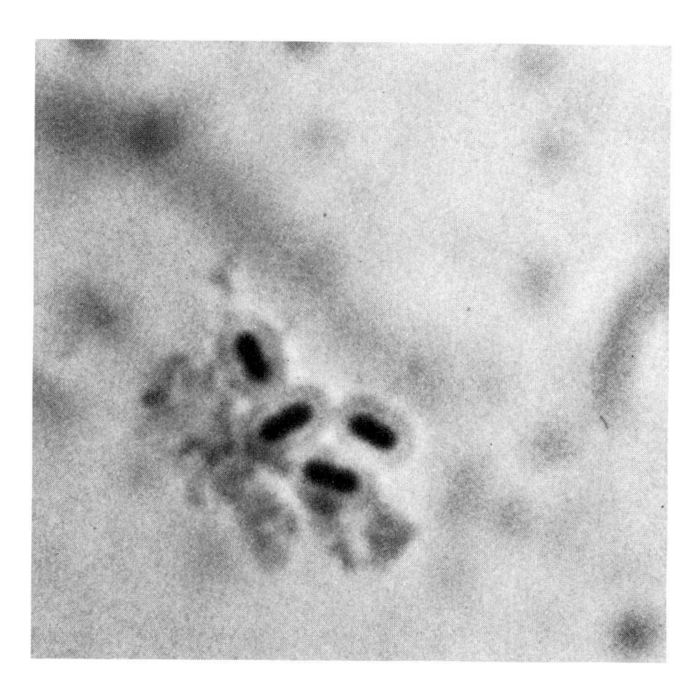

FIGURE 23.9 Capsules of *Streptococcus pneumoniae* as demonstrated by the Quellung reaction. (Courtesy of Dr. G. Goodhart.)

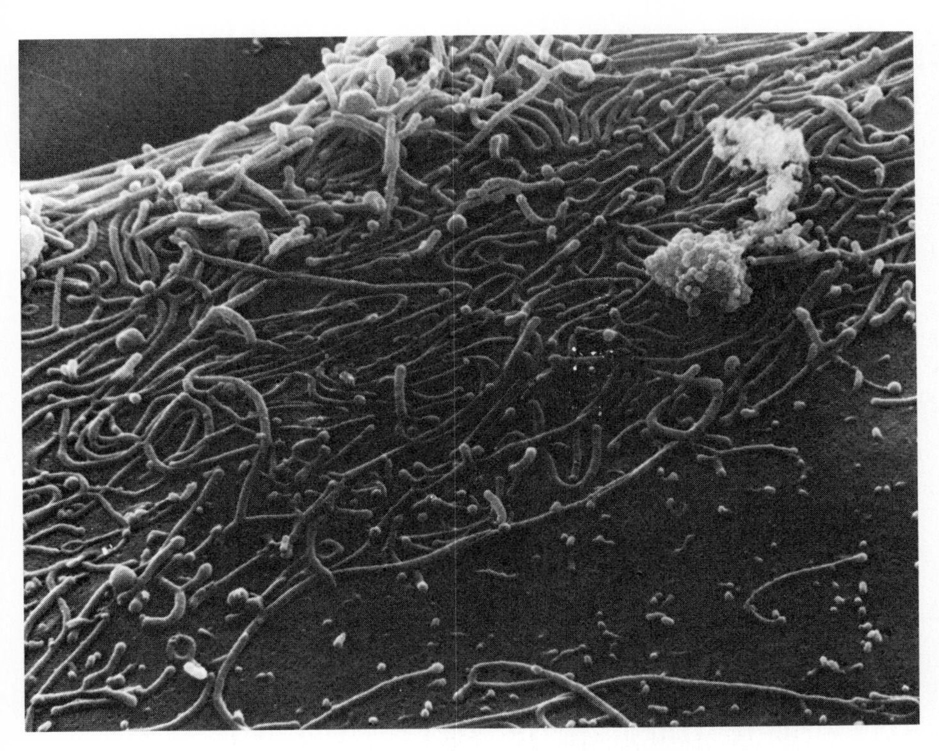

FIGURE 23.10 The pleomorphic structure of bovine *Mycoplasma* is shown in this scanning electron micrograph. (Photograph by Dr. David M. Philips, The Population Council.)

THE MYCOPLASMAS

The mycoplasmas (mi'ko-plaz'mah) are among the smallest free-living bacteria known (Figure 23.10). They were initially identified as the causative agent of pleuropneumonia in cattle. Their small size enables them to pass through membrane filters, thus they were first thought to be viral in nature; however, mycoplasmas possess all the normal features of bacteria—except that *they lack a cell wall.* Mycoplasmas possess DNA and RNA; they have bacterial ribosomes; and they are able to grow on special laboratory media. They are insensitive to antibiotics that inhibit cell-wall synthesis, such as the penicillins and cephalosporins.

Mycoplasmas are widely found in nature living in association with plants, insects, and animals. Humans are the major reservoir for *Mycoplasma pneumoniae,* which infects the lungs and causes pneumonia. *Mycoplasma pneumoniae* is a pleomorphic organism that can grow as filaments (Figure 23.10) or as spherical cells that average 0.3 μm in diameter. They readily pass through small (0.22 μm) membrane filters because they lack a rigid cell shape. Mycoplasmas grow on beef heart infusion agar as a flattened colony with a raised center, described as a "fried-egg" colony. Unlike other bacteria, the cytoplasmic membranes of *Mycoplasma* contain sterols, obtained from their growth medium. The presence of sterols appears to strengthen the cytoplasmic membrane enabling it to resists the osmotic forces that would otherwise destroy the cell's structural integrity.

Atypical or Walking Pneumonia

Mycoplasma pneumoniae infects the lung, causing *atypical* or *walking pneumonia.* The incubation period is two to three weeks, and infections are most prevalent among the 5- to 15-year age group. Outbreaks occur among close social groups, such as the family, in which the rate can reach 60 percent. A family outbreak usually begins in a school-age child and is transmitted by respiratory secretions. The bacterium infects the surface epithelium in the trachea and bronchi, and inhibits ciliary motility. The patient has a fever, cough, and malaise. Although the infection is self-limiting, immunity is incomplete and reinfections are well documented.

Diagnosis and Treatment

Mycoplasmas can be isolated in laboratory media, but the five- to ten-day incubation period is too long to make isolation a practical diagnostic procedure. Infection can also be diagnosed by demonstrating a rise in serum antibody with standard laboratory procedures, though treatment of symptoms with tetracyclines or erythromycins is begun before the test is completed. The association of Q fever, ornithosis, Legionnaires' disease, and viral pneumonia with atypical pneumonia is discussed later.

PNEUMONIA: *KLEBSIELLA*

Klebsiella (kleb'se-el'lah) is a gram-negative, nonmotile, encapsulated rod that grows readily on selective laboratory media. *Klebsiella pneumoniae,* which is widely distributed in soil, water, vegetable matter, and the human intestinal and respiratory tracts, is an important human pathogen.

Some 5 to 10 percent of healthy Americans carry *K. pneumoniae* in their respiratory tract. Capsulated strains are isolated from sick individuals since the virulence of *K. pneumoniae* depends on the presence of a polysaccharide capsule. The organism is also responsible for nosocomial infections in surgical wounds and the urinary tract (see Table 20.5). Treatment is with antibiotics and sulfonamides.

PNEUMONIA: *HAEMOPHILUS*

Species of *Haemophilus* (he-mof'i-lus) are small, gram-negative, coccoid to rod-shaped bacteria that cause disease only in humans. Many healthy individuals carry *H. influenzae* (in-flu-en'ze) as a symbiont in the nasopharynx. It causes lung infections in compromised persons and is a significant cause of meningitis in young children.

Haemophilus influenzae causes bacterial pneumonia in children 2 to 5 years old and in patients recovering from viral infections, who are especially vulnerable to secondary bacterial infections. *Haemophilus influenzae* is one of the secondary bacterial invaders that contribute to the increased mortality among elderly patients with viral pneumonia.

Haemophilus influenzae is the pathogen responsible for the rare disease of *obstructive epiglottitis* (ep'i-glot-ti'tis). In this disease, the bacteria spread from the pharynx to the epiglottis, and thence to the larynx. Laryngeal swelling can so obstruct the trachea that an emergency tracheotomy is required to save the patient's life. *Haemophilus* infections are treated with chloramphenicol or combinations of chloramphenicol and ampicillin. Public health officials recommend that children be immunized with the *Haemophilus* b vaccine between 18 and 24 months of age to prevent illnesses caused by *H. influenzae.*

Q FEVER: *COXIELLA BURNETII*

We can demystify Q fever at the outset by mentioning that "Q" merely stands for "query"—because the nature of the illness was not understood at first. Q fever is caused by *Coxiella burnetii* (kok-se-el'lah bur-net'e-i), an obligate intracellular parasite that is enzootic in cattle, sheep, and goats and is found throughout the world wherever these animals are domesticated. The genus *Coxiella* consists of a single strain of bacterium that is distinct from the other rickettsias.

Coxiella grows in cells inside specialized vacuoles known as phagolysosomes (fag'o-li'so-sohm), which enlarge as the bacteria multiply (Figure 23.11). *Coxiella* is released from infected animals in saliva, nasal secretions, as well as afterbirth, which is why outbreaks of Q fever on sheep ranches correspond with the lambing season.

The bacterium is distinct from the other rickettsias in being amazingly resistant to desiccation, heat, and many chemical agents, and can survive in dried blood or on wool for more than six months. These unique characteristics may be explained by the observation that it forms endospore-like structures during its intracellular growth.

The bacterium is transmitted between its animal reservoirs by ticks, and is transmitted to human lungs by aerosols. It is manifest in a sudden onset of fever and chills, with symptoms resembling those of atypical pneumonia. In humans the infection is readily cured with tetracyclines.

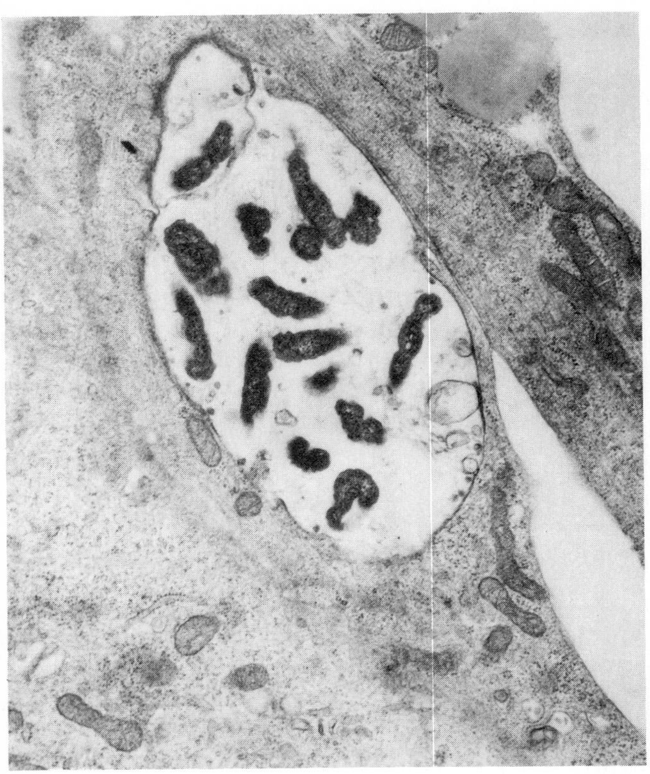

(a)

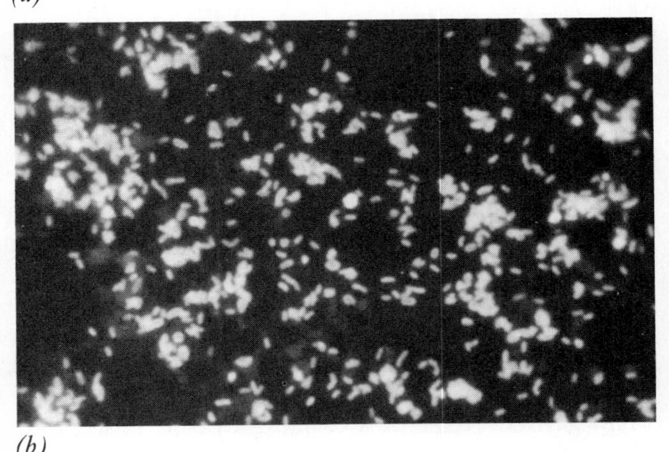

(b)

FIGURE 23.11 *(a) Coxiella burnetii* growing inside a vesicle of an infected L cell. Courtesy of Dr. F. Eb. From F. Eb, J. Orfila, and J-F Lefebvre, *J. Microscopie Biol. Cell., 25*:107–210, 1976. *(b)* Fluorescent antibody staining of *Coxiella burnetii* in a smear of yolk sac tissue. (Courtesy of Dr. W. Burgdorfer and Rocky Mountain Laboratory/BPS.)

CHLAMYDIAS

Chlamydias (klah-mid′e-ah) are nonmotile, spherical, obligate intracellular bacterial parasites (Figure 23.12). They grow in the cytoplasmic vesicles of animal cells and have a unique reproductive cycle. These bacteria are energy parasites that depend on the host cell to provide them with ATP. One of the two recognized species, *Chlamydia psittaci,* is responsible for the infection associated with birds called ornithosis (or′ni-tho′sis) or psittacosis (sit-ah-ko′sis).

ORNITHOSIS (PSITTACOSIS)

Many species of birds are reservoirs for chlamydias, which cause ornithosis in flocks of domestic birds such as turkeys and in wild birds existing under stressful or crowded conditions. Ornithosis was first observed in parrots; hence the causative agent was named *C. psittaci* (Gr., parrot) (sit′ah-si). Diseased domestic and wild birds discharge the bacteria in their feces and in exudates from the nose and eyes. Infection in domestic flocks can run unchecked, with mortality rates as high as 30 percent of the flock.

Humans become infected by handling domestic fowl or pet birds or by inhaling *C. psittaci* from dried exudates or feces. The disease in humans may be asymptomatic, or may develop into a severe, sometimes fatal, pneumonia following a one- to three-week incubation period. The severe form begins abruptly with chills, fever, headache, and the symptoms of atypical pneumonia.

Ornithosis is diagnosed by the demonstration of a high antibody titer against antigens of the *Chlamydia* family. Diagnosis is confirmed when chlamydias from blood or sputum samples are grown in embryonated chicken eggs or tissue cultures. The symptoms can be relieved by tetracycline therapy.

KEY POINT

The major causes of typical bacterial pneumonia are *Streptococcus pneumoniae, Klebsiella pneumoniae,* and *Haemophilus influenzae,* organisms that can be isolated readily on laboratory media and identified. *Mycoplasma pneumoniae,*

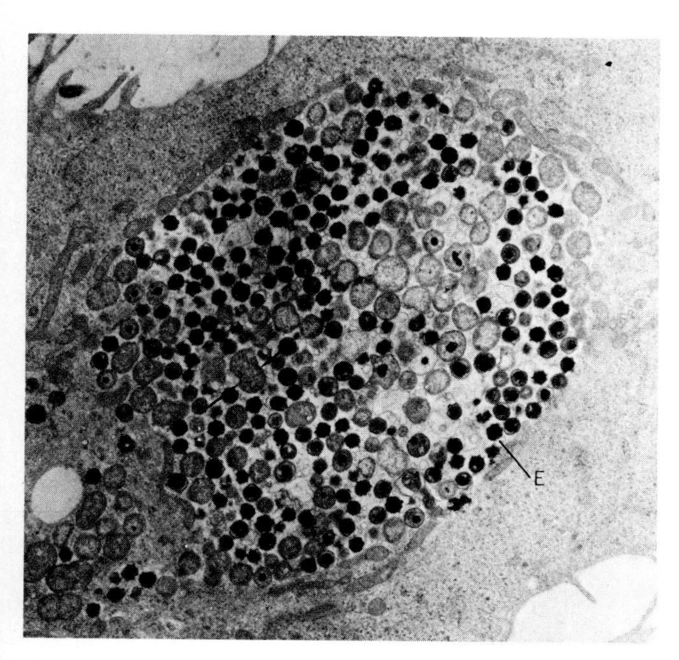

FIGURE 23.12 Electron micrograph of a HeLa cell 24 hours after being infected with *Chlamydia psittaci* showing the three stages of their reproduction cycle. Note the two dividing reticulate cells. Elementary bodies (E) are more prevalent at this stage of infection. (Courtesy of Dr. F. Eb. From F. Eb, J. Orfila, and J-F. Lefebvre, *J. Microscopie Biol. Cell.,* 25:107–210, 1976.)

***Coxiella burnetii*, and *Chlamydia psittaci* cause atypical pneumonia in humans, which is diagnosed by clinical symptoms or immunological techniques.**

LEGIONELLA

The description of Legionnaires' disease (Chapter 20) may be recalled in preparation for the following discussion.

The causative agent of what came to be known as Legionnaires' disease was isolated in January 1977 by investigators from the Centers for Disease Control. Bearing in mind the near frenzy that was generated by the many deaths due to the "new" microorganism, microbiologists made an all-out effort to unravel what was, at the time, an enigma. They inoculated guinea pigs with suspensions of lung tissue taken from persons who had died from the yet-unidentified disease (Figure 23.13). Bacteria were observed in stained tissue samples from the spleen of those guinea pigs that became febrile. The spleens were homogenized, then were inoculated into embryonated eggs, and finally a pure culture of a rod-shaped bacterium (Figure 23.14) was isolated. By means of indirect fluorescent antibodies, the investigators were able to demonstrate that patients with the disease had a high antibody titer against this bacterium. Since the bacterium was closely associated with an American Legion convention, it was given the name *Legionella pneumophila* (le-gi-on-el′la pneu-mo′phi-la).

Legionella pneumophila can be isolated directly from natural sources on several media, including a charcoal yeast extract agar and a modified Mueller–Hinton agar enriched with iron and the amino acid cysteine. *Legionella pneumophila* requires cysteine for growth and utilizes other amino acids as its major source of carbon and energy. It is a gram-negative, short to filamentous rod (Figure 23.14), motile by a single flagellum.

Legionella pneumophila is naturally found in aquatic ecosystems worldwide. The reservoir usually is contaminated air-conditioning cooling water or a contaminated potable water system. Since the discovery of *L. pneumophila,* microbiologists have described more than 22 other species of *Legionella* that cause febrile illnesses. These new isolates have not diminished the importance of *L. pneumophila,* which remains the most important *Legionella* pathogen. The general term for the symptoms produced by *Legionella* is **legionellosis.**

LEGIONNAIRES' DISEASE

Two to ten days after exposure, the person experiences a malaise, with muscle ache and headache. The body temperature rises rapidly to between 38.9° and 40.5°C and there is a cough, chest and abdominal pain, diarrhea, and shortness of breath. Chest films reveal pneumonia in 90 percent of patients. Untreated patients experience progressive deterioration and between 15 and 30 percent of these patients die from pneumonia followed by shock. With treatment, the patient gradually recovers over a period of one to two weeks. Most patients need to be hospitalized, so that treatment can be closely monitored. Erythromycin is the drug of choice.

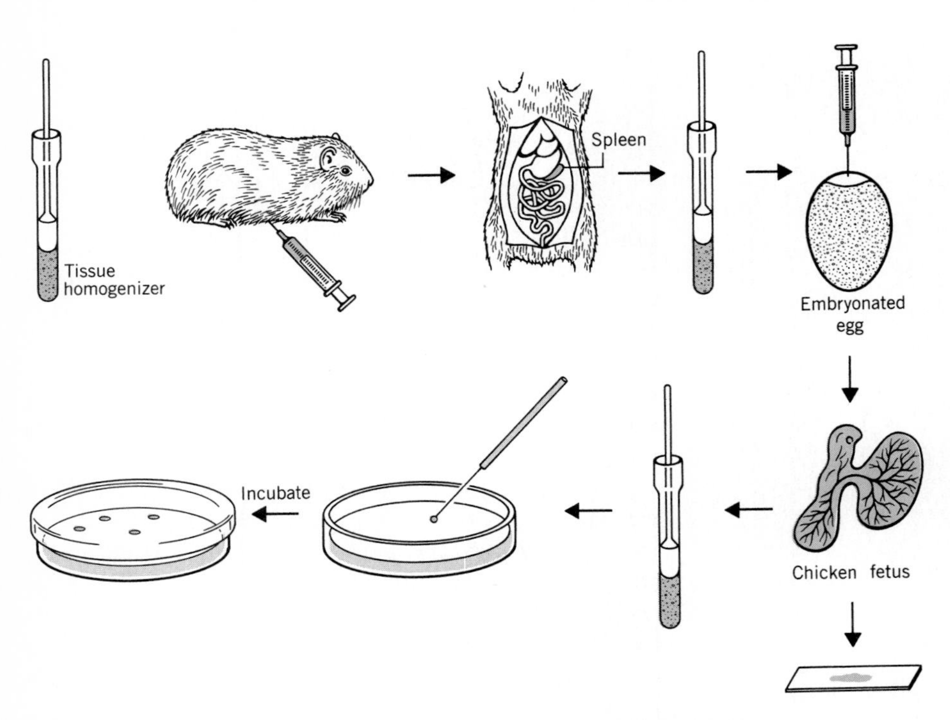

FIGURE 23.13 The initial isolation of *Legionella pneumophila* was accomplished by injecting lung tissue from an infected patient into a guinea pig. Embryonated eggs were inoculated with material from the spleen of a sick animal. Colonies of *Legionella pneumophila* were grown on modified Mueller–Hinton agar as the final step in the isolation.

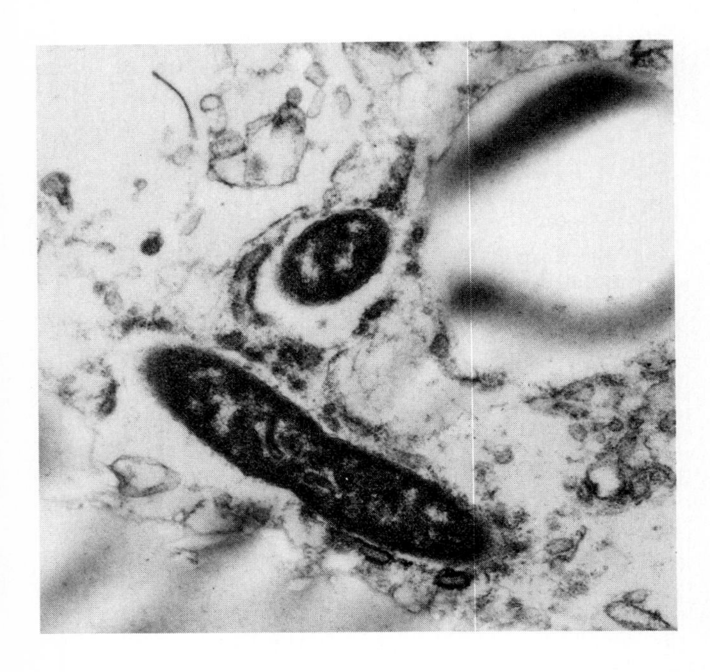

◀FIGURE 23.14 Electron micrograph of *Legionella pneumophila* growing in the yolk sac of an infected embryonated egg. (Courtesy of Dr. Francis W. Chandler, Centers for Disease Control.)

Epidemiology

A question originally posed was, Why did the epidemic occur during one particular convention? From all that was learned, it is believed that the disease originated as a common-source epidemic, spread by aerosol. The bacterium prefers the moist environment of water-cooling towers where the refrigerant of air-conditioning systems is cooled. This environment was rarely subjected to decontamination measures. And such was the environment of the air-conditioning system in the Legion's convention headquarters hotel.

Legionnaires' disease is not transmitted by person-to-person contact. This fact was apparent early in the

CDC investigation because family members of infected patients did not become ill. The bacterium can survive for more than a year in tap water, and has been isolated from many cooling towers and numerous mud samples—ideal environments for *L. pneumophila*. Fortunately only 1 percent of the population exposed to natural sources of *L. pneumophila* becomes ill.

Prevention

Prevention of Legionnaires' disease is a complex problem because the bacterium is widely distributed. Procedures have been developed for cleaning and decontaminating air-conditioning water towers. Should an outbreak occur, it is necessary to discover the source of the pathogen and institute appropriate remedial action.

VIRAL DISEASES OF THE RESPIRATORY TRACT

Viral infections of the upper respiratory tract may be asymptomatic or may be debilitating, yet they rarely cause serious complications. A few viruses, including influenza and mumps viruses, cause viremias that can lead to serious complications involving the central nervous system or to secondary bacterial infections. Viral respiratory diseases are among the most common afflictions known to humankind.

TABLE 23.2 Viruses associated with the common cold

VIRUS	ANTIGENIC TYPES	PERCENT OF COLDS
Rhinoviruses	>100	25–30
Coronaviruses	3 or more	5–15
Parainfluenza viruses	4	
Respiratory syncytial virus	1	10–15
Influenza viruses	3	
Adenoviruses	33	
Other viruses (known)		5
Other viruses (presumed)		30–40

Source: Adapted with permission from J. M. Gwaltney, Jr., "The Common Cold," in G. L. Mandell, R. G. Douglas, Jr., and J. E. Bennett (eds.), *Principles and Practice of Infectious Diseases*, 2nd ed., Wiley, New York, 1985.

THE COMMON COLD

Common head colds are one of the most annoying among viral infections. **Coryza** (ko-ri'zah), an acute inflammation of the nasal mucous membranes associated with a profuse discharge, is the hallmark of the common cold. Many viruses can infect the nasal membranes, bringing on mucus secretion, chills, headache, sneezing, sore throat, and cough. A head cold is hardly life-threatening, but it does cause significant discomfort and absenteeism from work and school.

Common Cold Viruses

The head cold is caused by any of more than 100 types of rhinoviruses (ri'no-vi'rus), at least three types of coronaviruses (kor'o-nah-vi'rus), four types of parainfluenza viruses, and 33 types of adenoviruses (ad'e-no-vi'rus) (Table 23.2). Indeed, the common cold viruses are so ubiquitous that Americans and Britons as a group have an average of one head cold per year per person. Children are the chief source of infection for adults, and direct contact with an infected person is the main mode of transmission.

The small, naked, RNA **rhinoviruses** comprise the largest group of viruses responsible for the common cold (Table 23.2). These viruses were first isolated from affected persons in 1959 by virologists who succeeded in growing them in tissue cultures. The rhinoviruses require a low pH and have an optimum growth temperature between 33° and 35°C—conditions that simulate the pH and temperature of the human nasal passages.

Over a lifetime a person has many head colds, each

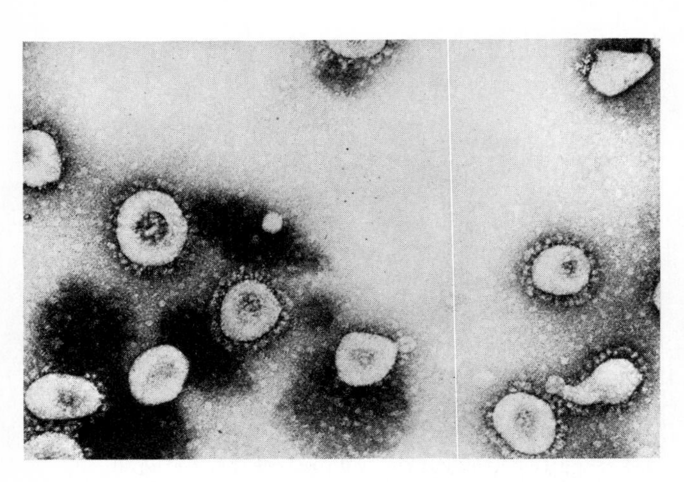

FIGURE 23.15 Human coronavirus from infected human embryonic tissue culture cells (approximately 144,000×). The club-shaped projections from the viral envelope have been likened to the sun's corona. (Courtesy of the Centers for Disease Control.)

caused by a distinct virus. During an infection the body responds immunologically and produces antiviral IgAs. These antibodies protect against reinfection, but not against other rhinoviruses, so a person can experience multiple head colds during a single cold season. The immunological protection is not long lasting and diminishes after 18 months.

Coronaviruses are pleomorphic, enveloped, RNA viruses that possess petal-shaped structures projecting from their viral envelope (Figure 23.15). Each virus particle looks like a sun surrounded by a corona—hence the name coronavirus. Three coronaviruses cause mild upper respiratory infections in both children and adults and probably are responsible for up to 15 percent of common colds.

Cold symptoms also accompany viral infections of the lower respiratory tract due to parainfluenza virus, respiratory syncytial (sin-sish'al) virus, influenza viruses, and adenoviruses (Table 23.2). Virologists estimate that up to 40 percent of the cold viruses have not been isolated.

KEY POINT

More than 100 rhinoviruses cause the common cold, and this diversity is what has (at least until this writing) made it impossible for virologists to develop an effective vaccine.

Symptoms

The incubation period is two to three days. The symptoms are readily recognized: sneezing; nasal discharge; a sore, dry, scratchy throat; headache; cough; malaise; and chills. Initially, the nasal discharge is colorless and watery, but later it becomes heavy and tan-colored.

Most patients are afebrile and recover completely. A few patients, however, develop an infection of the sinuses called sinusitis (si'nu-si'tis) or a middle ear infection called otitis media (o-ti'tis me'di-ah) (Figure 23.1). Their symptoms are evident upon physical examination. Respiratory syncytial virus and parainfluenza viruses (see below) infect young children and may lead to pneumonia, croup, and bronchiolitis (brong'ke-o-li'tis). In the absence of these complications, symptoms of the common cold, whether due to rhinoviruses, coronavirus, parainfluenza virus, or respiratory syncytial virus, are indistinguishable.

Epidemiology

Incidence rises beginning in September, peaks through February, then gradually falls in the spring to a low point in summer. There is no scientific explanation for the greater number of colds during the colder months. At one time it was thought that one "caught a cold" by becoming chilled, but experiments have cast doubt on this notion. Susceptibility does not increase with chilling. What is perhaps more significant is that the viruses themselves have seasonal fluctuations. Coronaviruses cause more colds in the winter months, while the rhinoviruses are doing their mischief in the early fall and the spring.

The viruses are transmitted by direct contact and to a lesser extent by aerosols. They are excreted in mucus and transferred to the hands by a tissue or handkerchief used to clear the nose. Handshaking and other forms of direct contact then transfer the viruses to other persons. The susceptible person then infects their conjunctival mucosa or nasal mucosa by direct contact. Transmission in aerosols, or as large particles of respiratory excretions suspended in air, probably occurs only over short distances, and is not a significant means of transmission.

Prevention and Treatment

The only way to prevent a cold is to interrupt transmission of the virus. One should take extra care in washing one's hands after touching clothes, facial tissues, or any part of the body of a person who has a cold. Symptomatic treatment will abate the annoying manifestations of a cold. Because there is an immunological reaction to each cold virus, a person may be able to resist one virus, yet experience several colds during a single season—each due to a different virus.

Phenylepherine (fen′il-ef′rin) or **ephedrine** (e-fed′rin) may be given to decrease swelling and allow drainage of secretions. Postnasal drip is relieved by nasal decongestants, which lessen the discharge of mucus into the pharyngeal region. Aspirin is often taken to relieve the accompanying headache and muscle pains.

RESPIRATORY INFECTIONS DUE TO ADENOVIRUSES

Human adenoviruses are icosahedral (Figure 23.16), double-stranded DNA viruses that cause mild to asymptomatic infections and have the propensity to remain in lymphoid tissue after recovery. Such latent adenovirus infections are usually asymptomatic in humans, but adenoviruses do cause tumor formation in hamsters and other animals.

Approximately 50 percent of the 35 known serotypes of adenovirus cause disease in humans, with symptoms of the common cold, pharyngitis, acute respiratory disease, pneumonia, and epidemic keratoconjunctivitis. In infants, the adenoviruses cause a mild head cold and pharyngitis. All of the adenoviral infections are self-limiting and rarely go on to cause serious illnesses.

INFLUENZA

Influenza, or flu, is a serious viral infection of the lower respiratory tract. The influenza A virus is noted for its ability to change its antigenic makeup, a characteristic that enables it to cause pandemics. The most severe pandemic occurred in 1918–1920, when 21 million people died worldwide, 549,000 in the United States. Since 1920 there have been five pandemics, the most recent one occurring in 1977. Persons who recovered from influenza during a previous epidemic are not necessarily immune when a new influenza virus emerges with altered antigens.

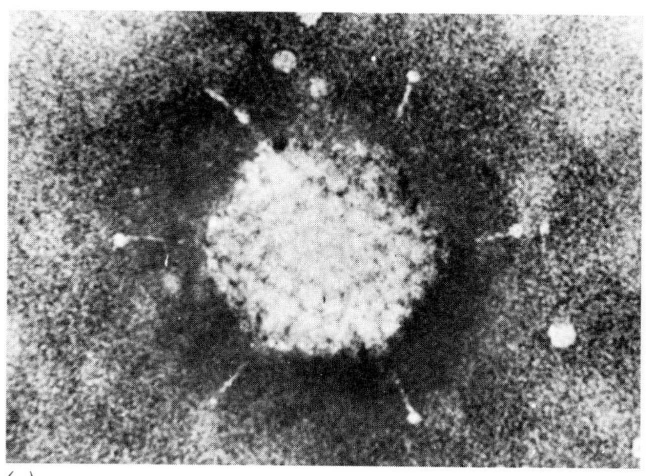

(a)

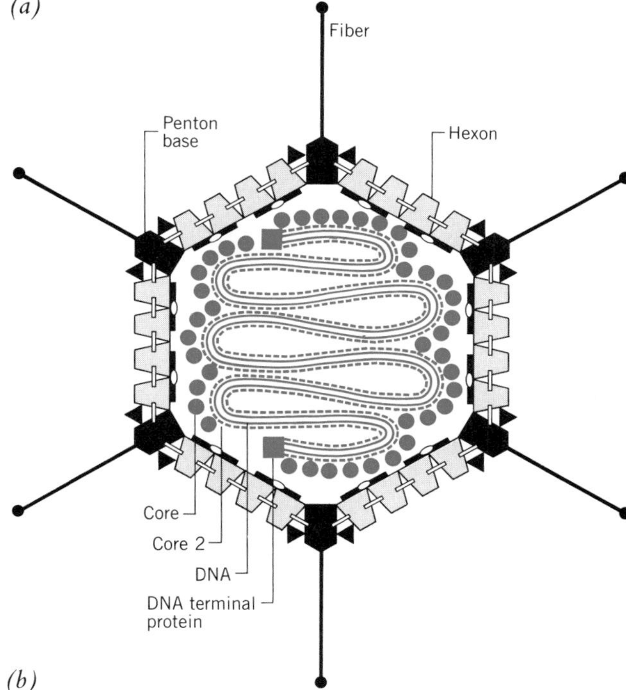

(b)

FIGURE 23.16 *(a)* Electron micrograph of adenovirus type 5 showing the icosahedral structure of the capsid and the projecting fibers. (Courtesy of Dr. H. G. Pereira. From R. C. Valentine and H. G. Pereira, *J. Mol. Biol.*, 13:13–20, 1965, copyright © by Academic Press Inc., Ltd., London.) *(b)* Diagram of human adenovirus illustrating the structure of the capsid and the positioning of the three capsid antigens. (Courtesy of Dr. H. S. Ginsberg. From H. Fraenkel-Conrat and R. Wagner (eds.), *Comprehensive Virology*, 13:409–457, 1979, Dover Publications.)

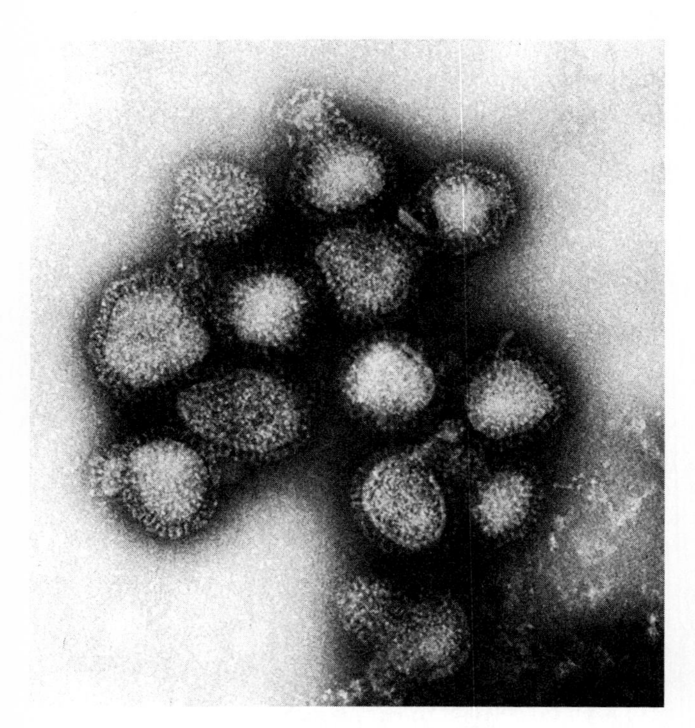

FIGURE 23.17 Electron micrograph of the spherical forms of influenza virions. (Courtesy of Omikron/Photo Researchers.)

The Influenza Viruses

The three distinct influenza viruses are differentiated by their matrix protein and nucleoprotein into types A, B, and C. Influenza A and influenza B viruses infect the lower respiratory tract of humans and cause similar symptoms. Influenza A also infects horses, pigs, and birds and is responsible for pandemic human infleunza. Infleunza C infects humans, but with milder symptoms than the illnesses caused by influenza A or B.

The viruses are pleomorphic, enveloped RNA viruses that replicate as spherical or filamentous virions in tissue culture (Figure 23.17). Their viral genome is composed of eight single-stranded RNA molecules, each coding for a specific protein (Table 23.3). Among these proteins are hemagglutinin, neuraminidase, and nucleoprotein, which can be detected by immunological reactions.

Hemagglutinin and Neuraminidase Spikes

Hemagglutinin, neuraminidase, and the matrix proteins accumulate at the host's cytoplasmic membrane where they will become part of the viral envelope (Figure 23.18). The mature virion is formed by budding at this membrane site. Budding occurs such that the hemagglutinin (HA) and neuraminidase (NA) protrude outside the viral envelope and serve as surface antigens.

The function of the hemagglutinins is to recognize and bind to receptors on the surface of the eucaryotic target cell. Binding is the first step in the process of viral infection. The role proposed for neuraminidase is to prevent viral aggregation.

How the Influenza Virus Changes Its Coat

Over the years, influenza type A has demonstrated the ability to alter its hemagglutinin and neuraminidase, its

TABLE 23.3 Type A infleunza viral genome

RNA SEGMENT (NUCLEOTIDE LENGTH)	GENE PRODUCT (DESIGNATION)	PROPOSED FUNCTION OF VIRAL PRODUCT
1 (2341)	Polymerase (P_3)	Plus RNA synthesis
2 (2341)	Polymerase (P_1)	Plus RNA synthesis
3 (2233)	Polymerase (P_2)	Minus RNA synthesis
4 (1778)	Hemagglutinin (HA)	Attachment to cell membrane
5 (1565)	Nucleoprotein (NP)	Structural component of nucleocapsid
6 (1413)	Neuraminidase (NA)	Release from membrane: prevents viral aggregation
7 (1027)	Matrix proteins (M1,M2)	Structure of envelope
8 (890)	Nonstructural (NS1, NS2)	RNA synthesis?

Source: Adapted with permission from R. Gordon Douglas, Jr. and R. F. Betts, "Influenza Virus," in G. L. Mandell, R. G. Douglas, Jr., and J. E. Bennett (eds.). *Principles and Practice of Infectious Diseases,* Wiley, New York, 1985.

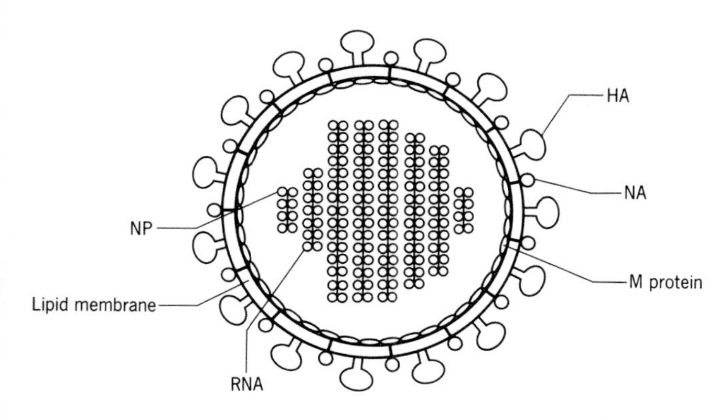

FIGURE 23.18 Structure of influenza virus showing lipid envelope, hemagglutinins (HA), neuraminidase (NA), nuclear proteins (NP), segmented RNA, and matrix (M) proteins.

chief surface antigens. To date, three distinct hemagglutinins (H_1, H_2, and H_3) and two distinct neuraminidases (N_1 and N_2) have been identified (Table 23.4).

Specific antibodies are developed against each viral antigen. The 1918 influenza pandemic was caused by the H_1N_1 subtype of influenza A. This strain was prevalent until 1957, when the Asian strain with its unique antigenic structure (H_2N_2) appeared and caused a severe pandemic. Another antigenic change occurred in 1968 when the Hong Kong virus (H_3N_3) appeared. Changes in the antigenic structure of influenza A virus enable the new strain to infect persons who have survived a previous influenza infection by a different strain. There are two theories to explain how influenza A virus changes its antigenic coat: antigenic drift and antigenic shift.

Antigenic Drift Minor antigenic variations in either the hemagglutinin or the neuraminidase gene probably result from mutation. RNA viral genomes are known to mutate at a higher rate than DNA genomes, and mutations in the influenza A genome have been confirmed by sequencing the genes and the resulting proteins. **Antigenic drift** describes this slow, sequential change in the antigenic structure of viral proteins (Figure 23.19). A change from one antigenic type to another occurs in nature with new antigenic types appearing over a period of years. The mutant virus is able to replicate because its antigenic structure is sufficiently different so as not to react with the host antibodies formed against the "old" influenza A antigens.

Antigenic Shift Major antigenic changes have occurred in both the hemagglutinin and neuraminidase of influenza A to create "new" viruses. Such changes are referred to as **antigenic shift,** and cause the most severe pandemics (Table 23.4). The appearance in 1957 of the Asian virus (H_2N_2) is an example of a major change in the structure of the virus.

Antigenic shift probably occurs during the production of virions in cells infected with more than one strain. Since each of the surface antigens is coded for by an independent segment of RNA (Table 23.4), the reassortment of viral RNA segments during assembly in multiple-infected cells (Figure 23.19) provides a reasonable explanation for the observed antigenic shifts. The "new" virus will then have a combination of surface antigens distinctly different from either of its parent virions. Experiments with mixed influenza infections of swine and turkeys have provided support for this theory.

Epidemic Influenza

Influenza epidemics appear as respiratory illnesses most frequently in the winter months among school-age

TABLE 23.4 Major antigenic subtypes of influenza A virus associated with pandemic influenza

YEAR OF PREVALENCE	INFLUENZA A VERNACULAR NAME	ANTIGENIC STRUCTURE	DISEASE CHARACTERISTICS
1918?	Swine	H_1N_1	Severe
1933–1946	PR8	H_1N_1	No pandemic
1947–1957	FM1	H_1N_1	Mild
1957–1967	Asian	H_2N_2	Severe
1968–1979	Hong Kong	H_3N_2	Moderate
1977–1979	Russian	H_1N_1	Mild

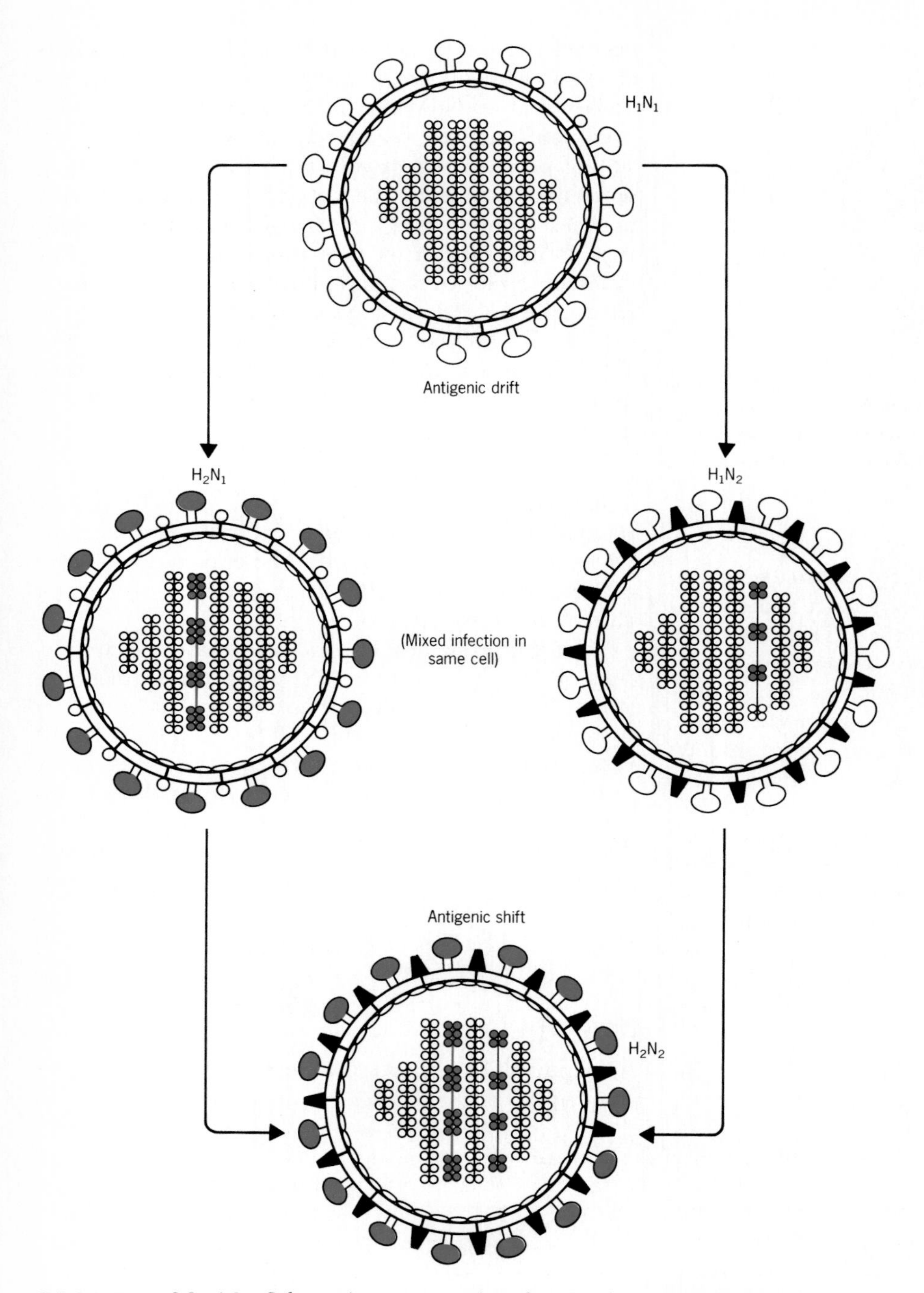

FIGURE 23.19 Schematic representation showing how antigenic drift produces changes in the antigenic structure of the influenza virus. Major changes in both HA and NA could occur by the reassortment of RNA segments when a single cell is coinfected with two strains of influenza A.

children, then in adults and preschool children. The illness usually affects between 10 and 20 percent of a country's population. Between pandemics, the population builds up immunity to the extant strains, but they remain susceptible to "new" influenza A strains, which cause epidemics every 9 to 17 years (Table 23.4).

The incubation period is between 18 and 72 hours following person-to-person transmission by direct contact or aerosols. The viruses infect the respiratory tract and replicate in the epithelial cells of the tracheobronchial mucosa and the nasopharynx. At the onset of the symptoms, the viruses are present in nasal discharges, whence they become suspended in aerosols by coughing, sneezing, or talking. Infection is by direct contact with an infected person and by contact with contaminated fomites; the rate of infection is much higher when the virus is inhaled.

KEY POINT

Lasting immunity against the viruses of mumps, measles, and rubella is possible because these viruses have stable surface antigens. But lasting immunity against the influenza virus cannot occur because it is able to create multiple antigenic subtypes.

Symptoms

Influenza is an acute febrile illness. The mucous membranes of the nose and mouth are smooth and glossy; there is severe muscle pain and weakness; fever (39° to 41°C); a dry, hacking cough; headache; and chills. The fever lasts only two to three days, but the weakness it induces may persist for up to two weeks.

Some patients experience a primary viral pneumonia following the onset of typical influenza symptoms. The pneumonia is characterized by labored breathing (called dyspnea), and persistent coughing. It can be fatal, especially in compromised patients. Patients are also vulnerable to secondary bacterial lung infections, commonly due to *Streptococcus pneumoniae, Staphylococcus aureus,* and *H. influenzae.* The increased death rate (see Figure 20.4) above the expected number is attributed to the viral pneumonia or to influenza-related bacterial pneumonia.

Diagnosis, Treatment, and Prevention

Influenza is diagnosed by isolating the virus in tissue cultures or in embryonated chicken eggs during the initial outbreak. Once the existence of an epidemic is established, influenza diagnosis is based on symptoms.

The drugs **amantadine** and **rimantadine** are effective therapies for treating uncomplicated influenza A (see Chapter 12). Amantadine inhibits the replication of the influenza A virus by preventing uncoating (see Chapter 12), and reduces the duration and the symptoms of influenza by approximately 50 percent. Neither amantadine nor rimantadine is effective against influenza B infections. The associated cold symptoms can be treated with common cold remedies such as aspirin and phenylephrine. Pulmonary complications caused by secondary bacterial infections are treated with the appropriate antibiotics.

Influenza can be prevented by administering the appropriate inactivated viral vaccine. The influenza vaccines used in the United States are composed of inactivated type A and type B influenza viruses of the antigenic serotype currently causing disease. The vaccines are produced from inactivated whole or disrupted ("split") influenza virus isolated from infected embryonated chicken eggs. Influenza viral vaccines are administered preferentially to persons with chronic illness and to persons over 64 years of age.

PARAMYXOVIRUSES

Parainfluenza viruses, respiratory syncytial virus, and the mumps virus belong to the Paramyxoviridae (par′ah-mik′so-vi′ri-de) family. They cause respiratory infections that lead to viral pneumonia, croup, bronchitis, bronchiolitis, and mumps (Table 23.5). The parainfluenza viruses and the respiratory syncytial virus cause important respiratory infections in infants.

Croup Is Caused by Parainfluenza Viruses

The three types of parainfluenza virus cause respiratory illnessess in young children (Table 23.5) that range in severity from bronchitis with cold symptoms to croup. All three types of parainfluenza virus can cause croup in youngsters under 30 months of age. The symptoms of croup include a barking cough and hoarseness that may be accompanied by respiratory blockage. Its severity depends on the specific causative virus and the patient's immunologic history. Most older persons have some degree of resistance to the viruses.

The viruses are transmitted by direct contact and by

TABLE 23.5 Diseases caused by paramyxoviruses

VIRAL AGENT	NUMBER OF SEROTYPES	DISEASES	AGE GROUP AFFECTED
Parainfluenza viruses	Type 1 and 2	Croup, bronchitis	8 to 30 months
	Type 3	Croup, pneumonia, bronchiolitis	1 to 24 months
Respiratory syncytial virus	One	Bronchiolitis, pneumonia	Peaks at 2 months Under 7 years
Mumps	One	Parotitis	5 to 9 years old

large droplets discharged from the respiratory tract. Type 1 is most frequently isolated in the fall and winter months, whereas type 3 is present at a low but persistent rate throughout the year. Young children are most often infected by type 1 and type 2 before age 6. Older persons are reinfected with all three types, but the symptoms are usually restricted to a cold and bronchitis. The illnesses are self-limiting, and almost all patients recover.

Respiratory Syncytial Virus

Respiratory syncytial virus (RSV) is an important cause of bronchiolitis and pneumonia in young children, and it is believed to be the most prevalent cause of lower respiratory tract infections in infants. RSV is so prevalent that up to 50 percent of 1-year-olds and almost all 5-year-olds possess antibodies against it. Outbreaks of RSV infections typically occur in the winter and spring. Symptoms can be mild and limited to nasal congestion, cough, sore throat, fever, conjunctivitis, and hoarseness, though serious infections lead to bronchiolitis and pneumonia, which may require hospitalization.

Nosocomial RSV infections pose a major problem for hospitals since the virus is readily spread within the nursery. Most youngsters recover from RSV infections and possess antibodies against the virus, though reinfection can occur. Very young children and children with cardiopulmonary or congenital disorders are at serious risk, and mortality is high in this group. Chemotherapy with aerosols of ribavirin is being tested and shows some promise for treating RSV infections.

MUMPS

The massive swelling of the parotid gland(s) is the unique sign of mumps. Although the symptoms of mumps were first reported in the writings of Hippocrates (fifth century B.C.), the causative agent was not demonstrated until 1934, when the symptoms were reproduced in laboratory monkeys. The virus is now cultivated in tissue cultures to make the live attenuated mumps virus vaccine, which has been available since 1967.

The Mumps Virus

The mumps virus is morphologically and chemically similar to the parainfluenza viruses, both of which belong to the Paramyxoviridae family. Its nucleocapsid is

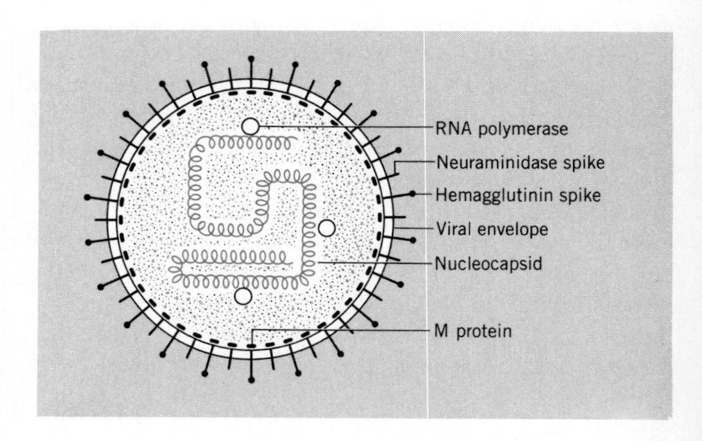

FIGURE 23.20 Diagram of the nucleocapsid and the envelope of a typical paramyxovirus.

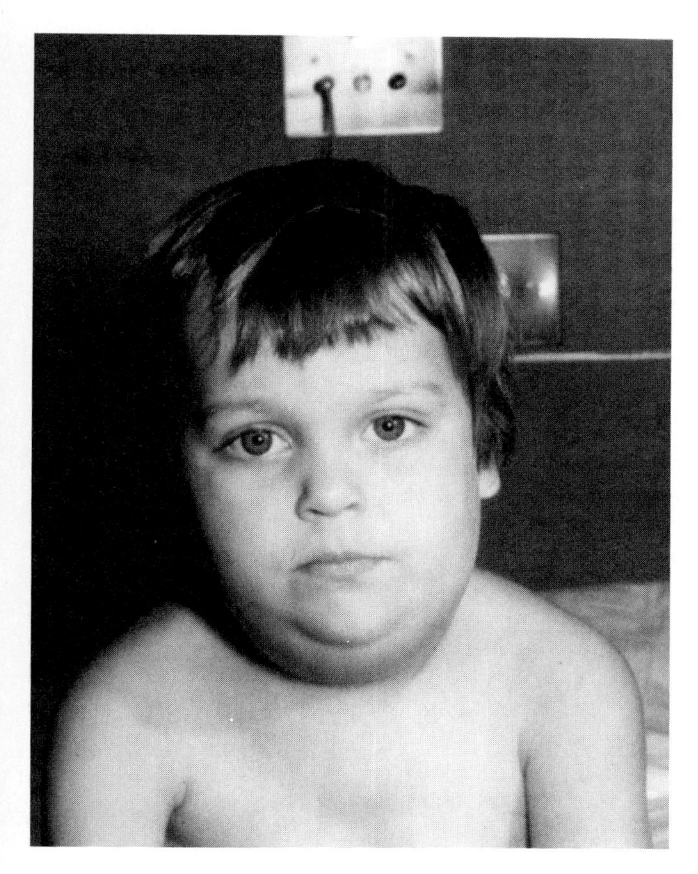

FIGURE 23.21 A patient with mumps showing extreme swelling of the parotid glands (parotitis). (From A. J. Zuckerman, J. E. Bandtvala, and J. R. Pattison (eds.), *Principles and Practices of Clinical Virology*, John Wiley & Sons, 1987.)

composed of a single-stranded RNA and nucleoprotein combined with an RNA polymerase (Figure 23.20). The ribonucleoprotein core is surrounded by an envelope that contains hemagglutinin, neuraminidase, and hemolysin. There is only one serotype of the mumps virus, and humans are its only natural host.

Symptoms

The mumps virus enters through the respiratory tract, where it infects the epithelial cells. It replicates in the upper respiratory tract and lymph nodes before giving rise to a generalized viremia. After a 16- to 18-day incubation period, the virus infects the salivary glands, causing great swelling of the parotid salivary glands (Figure 23.21). The swelling is often accompanied by an earache and a temperature of 37° to 40°C. The symptoms subside during the week following the maximum swelling. Complications of mumps include orchitis (or-ki'tis), an inflammation of the testes; oophoritis (o'of-o-ri'tis), an inflammation of the ovaries; meningitis; and encephalitis. Long-term consequences arising from these complications are rare.

Epidemiology and the Mumps Vaccine

The mumps virus infects the respiratory membranes of susceptible persons through airborne droplet nuclei or possibly by fomites. Before the wide use of the mumps vaccine, epidemics occurred from January through May, every two to five years. Children 5 to 9 were mainly involved as the virus spread readily through schools. Mumps was less likely to spread among older persons who had gained immunity through a previous clinical or subclinical case. Once the live attenuated mumps virus vaccine was licensed in the United States in 1967, the incidence fell dramatically. In recent years, however, there has been an increase in the number of mumps cases similar to that seen with measles; this may be a result of a lack of immunization among some population segments. It is recommended that the vaccine (MMR vaccine) be administered at 15 months of age.

SYSTEMIC MYCOSES

The **systemic mycoses** usually begin as pulmonary lesions when the fungal spores are inhaled in aerosols generated from soils that harbor natural populations of the fungus. The fungus establishes its primary infection in the lung before spreading to other organs and tissues. The major systemic mycoses and their causative agents are listed in Table 23.6.

The incidence of systemic mycoses is greatly influenced by the geographic distribution of the pathogenic fungi. *Histoplasma* (his'to-plaz'mah) preferentially

TABLE 23.6 Pathogenic fungi and their mycoses

DISEASE	FUNGUS	ECOLOGY/COLONIZATION (DISTRIBUTION)
	Systemic Mycoses	
Cryptococcosis[a]	*Cryptococcus neoformans*	Soil, pigeon droppings (worldwide)
Blastomycosis	*Blastomyces dermatitidis*	Soil (?) North America
Histoplasmosis	*Histoplasma capsulatum* (*H. duboissii* in Africa)	Soil, droppings of bats, birds (Mississippi and Ohio river valleys, Africa)
Coccidioidomycosis	*Coccidioides immitis*	Desert soils (northern Mexico, southwestern United States
Paracoccidioidomycosis	*Paracoccidioides brasiliensis*	Natural habitat unknown (South America, Mexico)
	Opportunistic Fungi	
Candidiasis (systemic)	*Candida albicans*	Mouth, vagina, skin, heart, gastrointestinal tract
Aspergillosis	*Aspergillus fumigatus*	Lung, ears, eyes, nervous system

[a]Known as *European blastomycosis* in European literature.

grows in soil enriched with bat or bird droppings and is naturally found in the Mississippi and Ohio River valleys; most cases of **histoplasmosis** occur in people living in one of those river valleys. The normal habitat of *Coccidioides immitis* (kok-sid′e-oi′des im′i-tis) is the desert soils of northern Mexico and southwestern

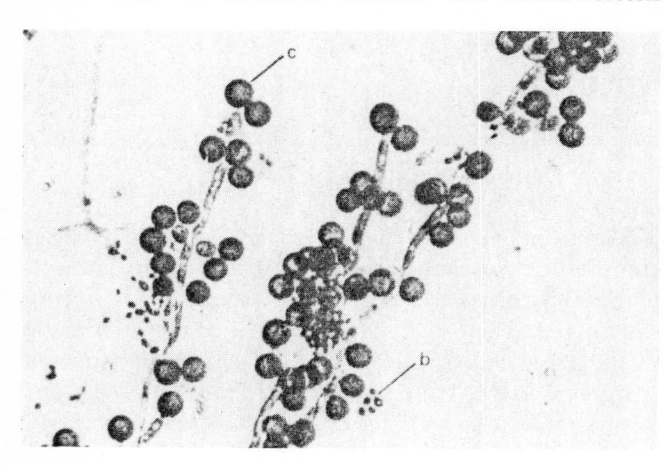

FIGURE 23.22 Pseudohyphae of *Candida albicans* with two types of asexual spores: clamydospores (c) and blastospores (b). (Courtesy of Dr. N. L. Goodman.)

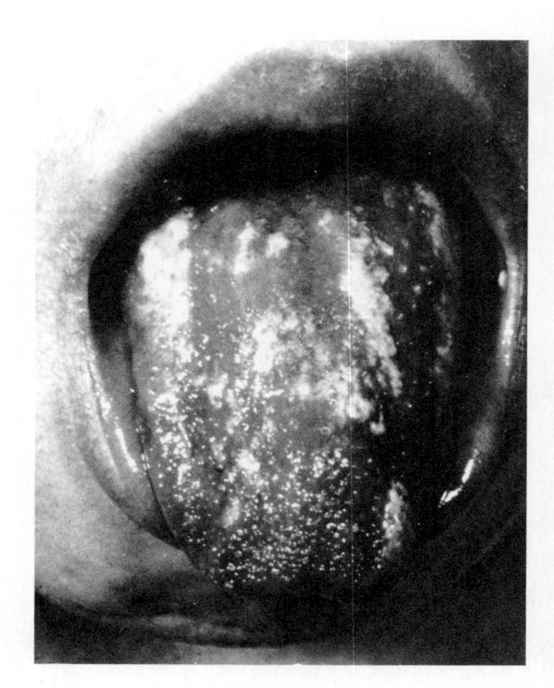

FIGURE 23.23 Typical oral candidiasis (oral thrush) with curd-like white patches over tongue. (Courtesy of Dr. Arnold Geravitch.)

United States, so most cases of **coccidioidomycosis** (kok-sid'e-oi'do-mi-ko'sis) are seen in these regions. *Cryptococcosis* (krip'to-kok-o'sis) occurs worldwide because *Cryptococcus neoformans* (krip'to-kok'us ne-o-for'manz) is found in soils and pigeon droppings worldwide (Table 23.6).

Opportunistic fungi (Table 23.6) colonize organs in both normal and compromised persons. Species of *Candida* (kan'di-dah) and *Aspergillus* (as'per-jil'us) are opportunistic fungi that can cause systemic mycoses. *Candida albicans* (Figure 23.22) is a normal inhabitant of the mouth, gastrointestinal tract, and vagina in women. It causes **oral candidiasis** (kan'di-dy'ah-sis) in newborns and persons with AIDS (Figure 23.23), and in the vagina it causes *Candida*-induced vaginitis. Disseminated *Candida* infections can affect the nervous system to give rise to meningitis, and can affect the heart. *Aspergillus* is an opportunistic fungus that grows on vegetable matter such as hay and stored grains (Figure 23.24). *Aspergillus fumigatus* (fu-mi-gat'us) can infect the lungs following inhalation of the conidia, to cause **pulmonary aspergillosis.** Aspergillosis occurs most often in compromised patients.

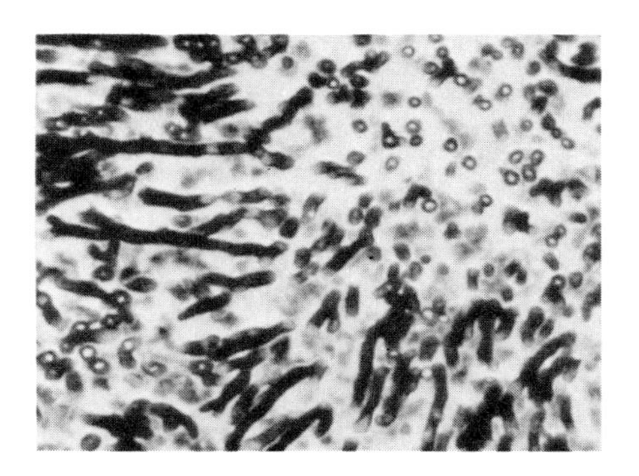

FIGURE 23.24 *Aspergillus* hyphae in bronchial washings. (Courtesy of M. C. Campbell. From M. C. Campbell and J. L. Stewart, *The Medical Mycology Handbook,* Wiley, New York, 1980. Copyright © 1980, by John Wiley & Sons, Inc.) Reprinted with permission.

Most systemic mycoses are treated with amphotericin B. Though it can have serious side effects, amphotericin B is required for what otherwise would be fatal infections. Ketoconazole, a newer broad-spectrum antifungal, has low toxicity so it can be taken orally with few side effects. Unfortunately, it is not as effective as amphotericin B for treating life-threatening mycoses.

Summary Outline

THE RESPIRATORY TRACT

1. The upper respiratory tract—which includes the mouth, tonsils, nasal cavity, nasopharynx, and throat—is protected against infectious agents by mucous secretions and saliva that contain IgAs and lysozyme.
2. The lower respiratory tract, consisting of the trachea, bronchi, and lungs, is free of foreign matter and microbes in healthy individuals.
3. The nature of the diseases of the respiratory tract ranges from self-limiting infections with lasting immunity to progressive destructive infections by intracellular parasites.

BACTERIAL RESPIRATORY TRACT INFECTIONS

1. Diphtheria is an infectious disease caused by lysogenized strains of *Corynebacterium diphtheriae* that infect the human nasopharynx and produce the diphtheria toxin.
2. The *tox* gene of the β prophage codes for diphtheria toxin, which inactivates EF-2, thereby inhibiting eucaryotic cell protein synthesis.
3. Untreated diphtheria patients die from the necrotic effects of the toxin on their heart, lungs, and kidneys. Diphtheria patients can be passively immunized with antitoxin; antibiotic chemotherapy prevents the carrier state.
4. *Bordetella pertussis* is found only in the human respiratory tract, where it grows preferentially on the ciliated epithelial cells of the bronchi and causes pertussis (whooping cough).
5. Children are protected against diphtheria and per-

tussis following active immunization with DTP shots.

6. Tuberculosis is a debilitating bacterial disease of the lungs caused by *Mycobacterium tuberculosis.* This bacterium grows intracellularly in macrophages and stimulates the formation of tubercles in the host's lungs.

7. New cases of tuberculosis originate when susceptible patients are infected by active tubercular patients or by the reactivation of dormant tubercles in previously infected patients.

8. The cell-mediated immune response to *M. tuberculosis* is the basis for the tuberculin reaction and for human resistance to tuberculosis. Tuberculosis can be cured by long-term chemotherapy.

BACTERIAL PNEUMONIA

1. Encapsulated *Streptococcus pneumoniae* are the major cause of bacterial pneumonia in humans. Streptococcal pneumonia can cause death in elderly or compromised patients, but is readily cured with penicillin in otherwise healthy patients.

2. The wall-less bacterium, *Mycoplasma pneumoniae,* causes atypical or walking pneumonia, which is treated with erythromycins or tetracyclines.

3. Capsulated strains of *Klebsiella pneumoniae* infect the lungs of compromised patients to cause bacterial pneumonia (3 percent of human cases). They also causes nosocomial infections in surgical wounds and the urinary tract.

4. *Haemophilus influenzae* causes bacterial pneumonia in children 2 to 5 years old and in virus-debilitated patients. It contributes to the increased mortality among elderly patients recovering from influenza.

5. Q fever is a lung infection caused by *Coxiella burnetii* when it is inhaled from its natural reservoir in domestic animals.

6. Parrots, turkeys, and other birds are the source of *Chlamydia psittaci,* which causes ornithosis in humans.

7. Legionnaires' disease is a pneumonic human febrile disease that occurs sporadically or as a point-source epidemic caused by *Legionella pneumophila.* This potentially fatal disease is treated with erythromycin.

8. *Legionella pneumophila* is a member of the world-wide aquatic microbial community that infects humans from aerosols generated by air-conditioning cooling towers and contaminated potable water supplies.

9. More than 22 other species of *Legionella* cause human febrile illnesses generally referred to as legionellosis.

VIRAL DISEASES OF THE RESPIRATORY TRACT

1. The common cold is a viral infection of the mucous membranes of the nasal passages characterized by sneezing and a nasal discharge of watery to thick mucus.

2. Rhinoviruses (fall–spring) and coronaviruses (winter) are RNA viruses that cause the common cold. Humans develop immunity to each infecting virus, but they remain susceptible to the others.

3. Colds are transmitted by direct contact with an infected patient and are treated with aspirin, codeine, and decongestants.

4. Adenoviruses infect the upper respiratory tract to cause common cold symptoms, pharyngitis, acute respiratory disease, pneumonia, and epidemic keratoconjunctivitis. At least half of the human adenoviral infections are asymptomatic and all are self-limiting.

5. Influenza type A virus is responsible for pandemic influenza in humans. It causes recurrent infections following antigenic changes in its hemagglutinin and neuraminidase that occur by antigenic drift or antigenic shift.

6. The severe symptoms of influenza last two to three days, but weakness may persist for one to two weeks. Some patients die from primary viral pneumonia; others are debilitated by influenza, which predisposes them to secondary bacterial infections.

7. Parainfluenza viruses cause respiratory illnesses in young children that range in severity from bronchitis with cold symptoms to croup. Croup is characterized by a barking cough and hoarseness that may be accompanied by respiratory blockage.

8. The very prevalent respiratory syncytial virus is a major cause of bronchiolitis. Infants with car-

diopulmonary or congenital disorders are at serious risk from RSV infections.

9. The mumps virus infects the human respiratory tract, causes a viremia, and after a 16- to 18-day incubation period causes a massive swelling of the parotid gland(s).
10. Complications of mumps include orchitis, oophoritis, meningitis, and encephalitis. Mumps is now controlled by the administration of a live, attenuated vaccine to children.

SYSTEMIC MYCOSES

1. Systemic mycoses begin when fungal spores in dust or aerosols are inhaled. They are treated with amphotericin B or ketoconazole.
2. Histoplasmosis is prevalent in the Ohio and Mississippi River valleys, where *Histoplasma capsulatum* is present in soil enriched with bird and bat excreta.
3. *Coccidioides immitis* is found in the arid regions of North and Central America and causes coccidioidomycosis.
4. Cryptococcosis occurs throughout the world and is caused by *Cryptococcus neoformans,* which grows in soil, especially soil enriched with pigeon droppings.
5. Species of *Candida* are opportunistic pathogens that cause thrush, *Candida*-induced vaginitis, skin infections, and infections of the gastrointestinal tract. Pulmonary aspergillosis is caused by the opportunistic pathogen *Aspergillus fumigatus.*

Questions and Topics for Study and Discussion

QUESTIONS

1. Describe how the diphtheria toxin is able to recognize target cells and inhibit protein synthesis.
2. What steps can be taken to prevent the spread of diphtheria within a population and what are the recommended treatment(s) for patients with diphtheria?
3. Describe the host–parasite relationships involved in the prolonged, insidious disease known as tuberculosis. How is tuberculosis treated and why is this different from treatment for bacterial pneumonia?
4. Describe the epidemiology of Q fever and ornithosis. In what way is *Coxiella burnetii* different from the other rickettsias?
5. Describe the discovery of the causative agent of Legionnaires' disease. Is this a new disease or one that hadn't been recognized before?
6. How can one differentiate between *Streptococcus pneumoniae* and *Mycoplasma pneumoniae* as the cause of pneumonia in a patient? How would the diagnosis affect the choice of antibiotic used to treat the infection?
7. Describe the transmission and seasonal incidence of head colds and explain why people often contract more than one viral cold during a given year.
8. Describe the antigenic structure of the influenza A virus and explain why this virus is capable of causing pandemics.
9. What significant diseases are caused by the paramyxoviruses? What groups of humans are most susceptible to these infections and why?
10. What are the clinical signs of mumps? What other parts of the body can be affected by the mumps virus?
11. Describe a systemic mycosis that is common in the continental United States. What epidemiological factors contribute to the incidence of this disease?
12. Explain why *Candida albicans* is known as an opportunistic fungal pathogen. What other fungal diseases are caused by opportunistic fungi?
13. Make a chart of the major diseases of the respiratory tract that indicates the causative agent,

mode of transmission, clinical symptoms, and treatment or prevention.

DISCUSSION TOPICS

1. Diphtheria is a bacterial disease of humans caused by an exotoxin that is coded for by a gene carried by a temperate bacteriophage. Can you think of why or how this complex situation evolved?
2. What are the mechanisms for transmitting the different respiratory tract infections? Which respiratory diseases can be prevented by interrupting the transmission of the infectious agents and which respiratory diseases cannot be controlled in this fashion? Explain.
3. Can genetic engineering be used to develop an effective vaccine against all the antigenic varieties of the infleunza virus?

Further Readings

BOOKS

Belshe, R. B., (ed.), *Textbook of Human Virology,* PSG Publishing, Littleton, Mass., 1984. The authors of separate chapters cover the history, properties, epidemiology, human diseases, prevention, and control of the viruses they discuss.

Beneke, E. S., and A. L. Rogers, *Medical Mycology Manual with Human Mycoses Mongraph,* 4th ed., Burgess, Minneapolis, 1980. A manual that can be used in the clinical laboratory and in medical mycology courses.

Volk, W. A., D. D. Benjamin, R. J. Kadner, and J. T. Parsons, *Essentials of Medical Microbiology,* 3rd ed., J. B. Lippincott, Philadelphia, 1986. A college-level textbook in medical microbiology.

ARTICLES AND REVIEWS

D'Allessio, D. J., J. A. Peterson, C. R. Dick, and E. C. Dick, "Transmission of Experimental Rhinovirus Colds in Volunteer Married Couples," *J. Infect. Dis.,* 133:28–36 (1976). A research paper on the transmission of two strains of rhinovirus.

Fraser, D. W., and J. E. McDade, "Legionellosis," *Sci. Am.,* 241:82–99 (October 1979). The story of the discovery of *Legionella pneumophila* as the causative agent of Legionnaires' disease.

Islur, J., C. S. Anglin, and P. J. Middleton, "The Whooping Cough Syndrome: A Continuing Pediatric Problem," *Clin. Pediatrics* (Phila), 14:171–176 (1975). A review of the incidence of whooping cough in Canada and the United States.

Kendrick, P. L., "Can Whooping Cough Be Eradicated?" *J. Infect. Dis.,* 132:707–712 (1975). An analysis of the results of using the DTP vaccine on the incidence of pertussis.

Middlebrook, J. L., and R. B. Dorland, "Bacterial Toxins: Cellular Mechanisms of Action," *Microbiol. Revs.,* 48:199–221 (1984). Contains an excellent discussion of the mechanisms of action of the two peptide toxins, including the diphtheria toxin.

Palese, P., and J. F. Young, "Variation of Influenza A, B, and C Viruses," *Science,* 215:1468–1474 (1982). Recombination of genes, deletions, insertions, and point mutations are considered as mechanisms for the observed changes in the structure of the influenza viruses.

Raeburn, P., "The Houdini Virus," *Science 85,* 52–57 (December 1985). This article highlights the significance of respiratory syncytial virus and the difficulties in controlling RSV infections in infants.

Razin, S., "The Mycoplasmas," *Microbiol. Revs.,* 42:414–470 (1978). An extensive review of the bacteriology (including mycoplasma viruses) of the mycoplasmas.

C H A P T E R

24

Infections of the Genitourinary System: Sexually Transmitted Diseases

OUTLINE

FOCUS OF THIS CHAPTER

Chapter 24 is concerned with infectious diseases of the genitourinary system. While some of these diseases are associated with the birth process, or with waste elimination, others are exclusively transmitted sexually. The medical, social, and economic costs of these infections mandate that we all understand their severity and contribute to their control and prevention.

Mechanical intervention in the genitourinary system, as occurs with insertion of a catheter, almost always affords pathogens access to the genital tract and often results in infection. Certain genitourinary infections are transmitted between sexual partners. The same infections in pregnant women can be transmitted to their newborns during the birthing process.

THE GENITOURINARY SYSTEM

The bean-shaped kidneys make urine by removing salts, toxins, pigments, hormones, and metabolic products from the blood (Figure 24.1). The urine collected from the kidneys is transported by the ureters to the bladder where it is stored until sufficient pressure builds up to stimulate urination through the urethra. Because the kidneys play such an integral role in overall body functioning, analysis of urine may be the single most important clue to a patient's illness.

In a healthy person, the kidneys, ureters, and urinary bladder are free of microorganisms. In both men and women, however, the anterior urethra is contaminated by microbes from the skin and/or feces. The male urethra extends approximately 8 cm from the bladder to the urethral opening at the tip of the penis; its function is to eliminate urine and deliver sperm. The female urethra is about 4 cm in length; its only function is to eliminate urine (Figure 17.3). Because the opening of the female urethra is external, it affords easy access to the flora of the vaginal and perianal (per′e-a′nal) regions, giving rise to many more cases of urethritis by these opportunistic pathogens in females as compared to males.

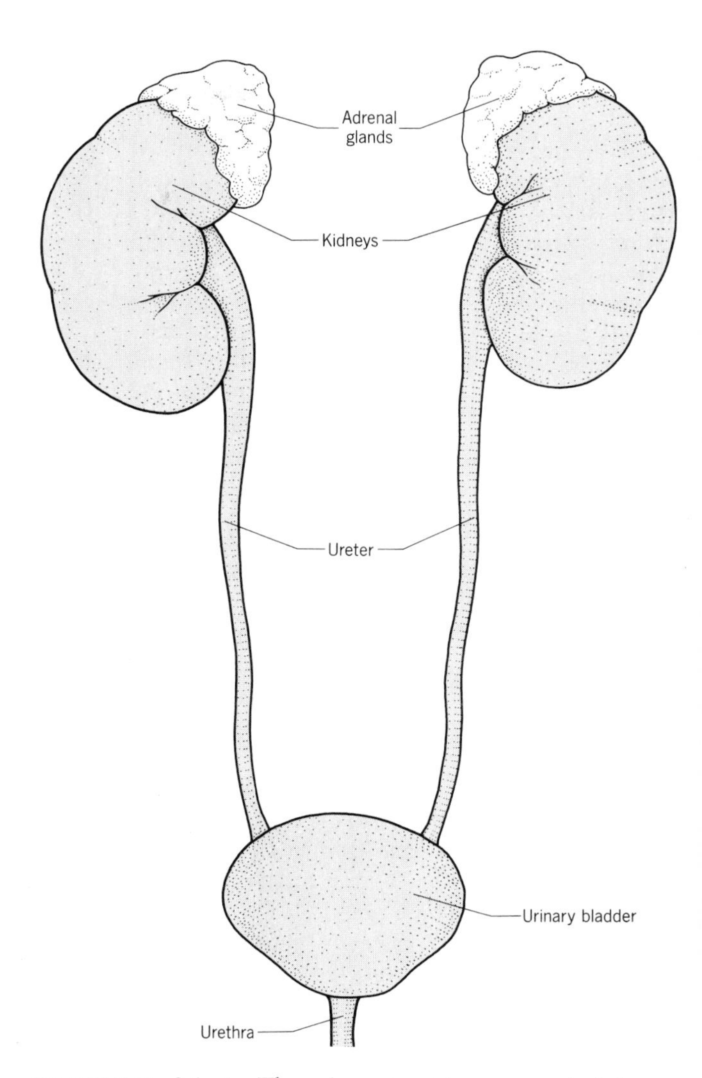

FIGURE 24.1 The urinary tract is composed of the bladder, kidneys, and ureters. These structures are sterile in the healthy human.

BACTERIAL GENITOURINARY INFECTIONS

Most urinary tract infections are caused by opportunistic pathogens that infect the anterior urethra or ascend the urethra to infect the prostate gland in men **(prostatitis),** the bladder **(cystitis),** or a kidney **(pyelonephritis).** The presence of large numbers of bacteria in the urine indicates that there is an infection of the urethra known as **urethritis** (u're-thri'tis). The population of microbes in the anterior urethra is normally controlled by urination with its flushing action. Any situation that causes a decrease in daily flow is a predisposing factor to infection. Since some bacteria are normally present in urine, the person charged with diagnosing a urinary infection must be able to differentiate normal flora from the pathogens.

NORMAL FLORA

Escherichia coli is commonly found in the perianal region and is a frequent cause of urogenital infections. Mechanical transfer of *E. coli* to the urethra occurs between partners during sexual intercourse and in the hospital setting during catheterization of a patient. As previously described, women are more susceptible to urethral contamination by *E. coli* than men (see Figure 17.3). *Escherichia coli* is able to colonize the urethra and establish an infection in part because of its ability to resist the inhibitory effects of vaginal fluids; and its fimbriae enable it to attach to urethral epithelial cells. *Corynebacterium, Fusobacterium, Gardnerella vaginalis, Mycobacterium, Mycoplasma, Neisseria, Peptostreptococcus, Staphylococcus aureus,* streptococci of the viridans group, and *Candida albicans* may also inhabit the anterior urethra or external genitalia.

The vaginal secretion from a healthy adult female contains between 10^8 and 10^9 bacteria per gram. Although the flora changes during sexual maturation and is influenced by hormones, it consists largely of aerobic and anaerobic gram-positive lactic acid bacteria. These bacteria metabolize carbohydrates to acidic products, which lower the pH of the vagina. The lactic acid bacteria are predominantly members of the genus *Lactobacillus,* with smaller numbers of *Streptococcus agalactiae* (a'gal-lac'ti-e), viridans streptococci, and *Peptostreptococcus.* Other microbes in the vagina include gram-negative rods, *Staphylococcus epidermidis,* diphtheroids, *Bacteroides* species other than *B. fragilis,* yeasts, and occasionally *Trichomonas vaginalis* (trik'o-mo'nas vag-in-ah'lis).

VAGINITIS

Both the normal flora and the opportunistic microbes can cause a vaginal inflammation called **vaginitis** (vaj'i-ni'tis). *Candida albicans* and *Trichomonas vaginalis* are common causes of vaginitis; together with nonspecific causes, they account for 90 percent of the cases.

Candida is a yeast (Chapter 16) commonly present in the vagina and on the skin. Changes in the vaginal environment enable *Candida* to outgrow the normal flora and cause *Candida* vaginitis, characterized by a cheesy discharge. Changes inducive to this infection can occur during pregnancy, use of broad-spectrum antibiotics, and steroid therapy. *Candida*-induced vaginitis can be effectively treated with clotrimazole, miconazole, or nystatin, but it often recurs. *Trichomonas vaginalis* is a flagellated, anaerobic protozoan that causes both vaginitis and trichomoniasis (see later).

Women with **nonspecific vaginitis** often complain of a malodorous discharge. Although *Gardnerella vaginalis* (gard-ne-rel'la vaj-in-ah'lis) is isolated from virtually all cases of nonspecific vaginitis, it is also present in 40 percent of women without vaginitis. Evidence suggests that nonspecific vaginitis is caused by the combination of *Gardnerella vaginalis* and bacteria of the *Bacteroides* genus.

BACTERIAL URETHRITIS

Although normal urine is presumed to be sterile, voided urine contains up to 1000 bacteria per ml, which presumably originate from the anterior urethra. When urinalysis is called for, the urine should be collected in such a manner that contamination is avoided. Bacteriurea (bak-te're-u're-ah) is defined as the presence of 100,000 or more bacteria per milliliter of a cleanly voided specimen.

Escherichia coli is the most common cause of urinary infections, which occur in women and infants (because infants wear diapers) more often than in men. Besides instrumentation, as by insertion of catheters, hospitalized patients are predisposed to infections caused by *Proteus, Providencia* (pro-vi-den'ci-ah), *Pseudomonas, Klebsiella, Enterobacter, Serratia,* staphylococci, and enterococci. These infections are often difficult to treat because antibiotics do not fully penetrate the urinary tract and the bacteria readily develop resistance to them.

Escherichia coli infections can be treated with sulfisoxazole and trimethoprim alone or in combination with ampicillin and amoxicillin. The antibiotic sensitivity of other pathogens must be determined so that appropriate treatment can be offered.

KEY POINT

It is the anatomy of the female genital tract that renders a woman far more vulnerable than a man to urethral infections caused by the normal body flora.

UROGENITAL MYCOPLASMAL INFECTIONS

Ureaplasma urealyticum (u-re′ah-plaz′ma u-re′ah-lit′i-kum) is a wall-less bacterium found in the genital tract of sexually active adults. This organism colonizes 80 percent of sexually active persons who have more than three sex partners. *Ureaplasma urealyticum* has been implicated as a cause of nongonococcal, nonchlamydial urethritis; however, since many asymptomatic persons are colonized by this bacterium, it is difficult to associate the microbe with any specific disease. *Mycoplasma hominis* (mi′ko-plaz′mah ho′mi-nis) also inhabits the urogenital tract and in females is associated with 10 percent of the cases of postabortion and postpartum fever. This bacterium may also be associated with pelvic inflammatory disease (see discussion of gonorrhea). These mycoplasmal infections can be treated with tetracyclines.

PUERPERAL SEPSIS

Puerperal sepsis—usually called childbed fever—was once the cause of a tremendous number of maternal deaths. It was not until the middle of the nineteenth century when Ignaz Semmelweis in Vienna and Oliver Wendell Holmes in Boston simultaneously discovered the underlying causes of this dreaded fever associated with childbirth. What happened was this. Obstetricians and medical students routinely performed both deliveries and autopsies without washing their hands between these tasks. The newly delivered women were thus being heavily contaminated by *Streptococcus pyogenes* (from the autopsied bodies) and died from the resulting septicemia. Antiseptic obstetrical technique and the use of antibiotics has now virtually eliminated puerperal sepsis. Nevertheless, the student of microbiology is reminded by this bit of history of the public health problems that can arise owing to unchecked proliferation of pathogenic microorganisms—some of which are yet to be understood and controlled.

STREPTOCOCCUS AGALACTIAE

Streptococcus agalactiae, a normal inhabitant of the genital and intestinal tracts, is now recognized as an important cause of meningitis, pneumonia, and bacteremia in newborns. Female carriers are the major reservoir for *S. agalactiae,* which is present in the genital tract of 15 to 30 percent of women. Newborns are presumably infected during the birth process or during their stay in the hospital nursery. Infants infected during the first three days of life may develop pneumonia, meningitis, or bacteremia, and those who have not received maternal antibodies are at greatest risk. These infections are responsive to treatment with the penicillins.

SEXUALLY TRANSMITTED BACTERIAL DISEASES

Syphilis, gonorrhea, and chlamydial urethritis are sexually transmitted bacterial diseases that can be treated with antimicrobial agents. Among the viral diseases, genital and anal warts (condyloma acuminatum) can be treated with chemicals or removed. These successes are, at this writing, heavily overshadowed by genital herpes and acquired immune deficiency syndrome (AIDS), viral diseases for which treatment and prevention are not available. Sexually transmitted diseases are a major public health problem throughout the world and cause severe and sometimes fatal illness in both children and adults.

GONORRHEA

In this country gonorrhea is the most common reportable infectious disease, with an average of close to one million cases per year from 1980 to 1984. And this figure is probably an underestimate, since it is a given that many cases treated by private physicians go unreported.

Gonorrhea is transmitted by sexual contact between both homosexual and heterosexual partners. Both men and women can be asymptomatic carriers who are unlikely to seek treatment. Others experience a painful urethritis that compels them to seek treatment. Gonorrhea is caused by a gram-negative coccus, *Neisseria gonorrhoeae* (nys-se′re-ah gon′o-re′eye), found only in humans. Many strains of *N. gonorrhoeae* possess fimbriae, which enhance their ability to colonize mucous surfaces and inhibit phagocytosis.

Symptoms

Neisseria gonorrhoeae infects the columnar epithelium of the urethra, the membranes lining the cervical neck, called the endocervix, and the anal canal. In males, the gonococci penetrate the urethra, causing pain and a purulent penile discharge (Color Plate II). Symptoms usually develop two to seven days after sexual contact with an infected partner. If the infection is not treated, the gonococci can infect the prostate and the epididymis (ep′i-did′i-mis). The host responds by depositing fibrous tissue that can block the sperm ducts and cause sterility. There is no lasting immunity to gonorrhea, so reinfection can occur.

The symptoms are less pronounced in women. Although the urethra can be infected, causing painful urination, the gonococci usually enter the vagina and migrate to the endocervix. Women often become unknowing carriers since there are few associated symptoms. The only visible sign is a purulent vaginal discharge.

Acute **pelvic inflammatory disease,** often called PID, describes a clinical syndrome attributed to the ascending spread of microorganisms from the vagina and endocervix to the endometrium and fallopian tubes. It is seen most often in sexually active, menstruating, nonpregnant women. The most common causes are *N. gonorrhoeae* and *Chlamydia trachomatis,* though other aerobic and anaerobic bacteria can be involved. Sterility can result when *N. gonorrhoeae* invades the fallopian tubes. Here they cause damage that leads to blockage as fibrous material is deposited during wound repair. As a consequence, there is a greater likelihood of an ectopic pregnancy, in which the embryo becomes implanted in a fallopian tube. This is a potentially lethal complication. In light of these possibilities, it is very important that asymptomatic gonorrheal infections be diagnosed and treated.

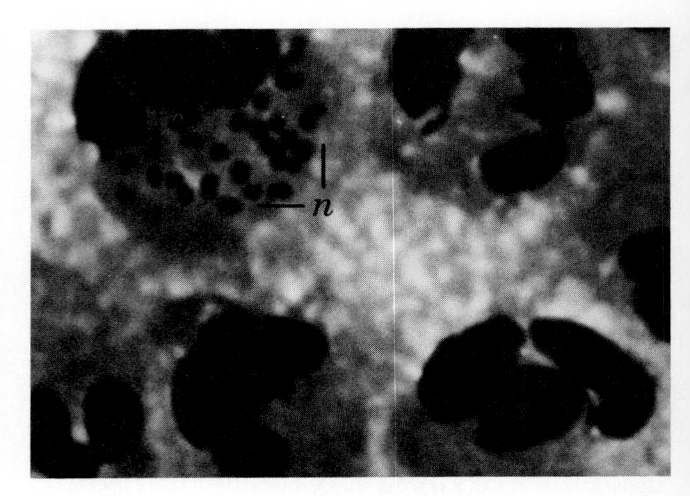

FIGURE 24.2 Typical Gram stain of a urethral smear from a male patient with gonorrhea. *Neisseria gonorrhoeae (n)* appear as a gram-negative diplococci inside the polymorphonuclear neutrophils (PMN). (Reproduced by permission from E. J. Bottone, R. Girolami, and J. M. Stamm (eds.), *Schneierson's Atlas of Diagnostic Microbiology,* 8th ed., Abbott Laboratories, North Chicago, Ill., 1982. Copyright © 1982, Abbott Laboratories.)

The gonococci can also infect the mouth and the rectum in both sexes. In females the rectum is a frequent site of infection, presumably because *N. gonorrhoeae* can be transmitted to the rectum in vaginal secretions. Rectal infections also occur among male homosexuals, and gonococci infect the pharynx when transmitted by oral–genital contact. Gonococcal ophthalmia was described in Chapter 22.

Diagnosis

If gonorrhea is suspected, a Gram stain of the vaginal or penile exudate may reveal intracellular gram-negative diplococci within the neutrophils (Figure 24.2). A patient whose Gram stain is negative may nevertheless be infected; therefore it is standard practice to culture the bacterium on Thayer-Martin medium or another suitable medium.

Epidemiology

Gonorrhea is epidemic in the United States (Figure 24.3). The epidemic gained momentum in the late 1960s and appears to have peaked in 1978. Sociologists

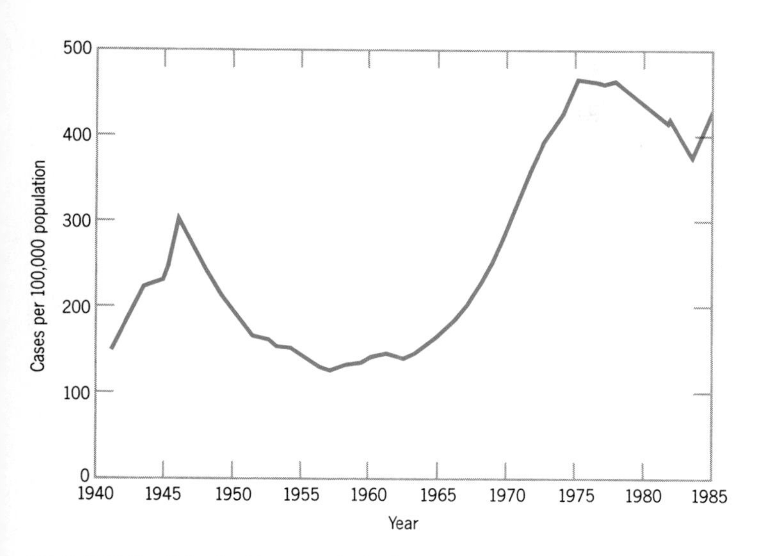

FIGURE 24.3 Reported civilian cases of gonorrhea in the United States from 1941 to 1985. (Centers for Disease Control, "Annual Summary, 1984: Reported Morbidity and Mortality in the United States," *Morbidity and Mortality Weekly Report,* Vol. 32, No. 54, 1986.)

and public health officials attribute the increase in gonorrhea to more frequent casual sex in part caused by the availability of birth control pills.

In the early 1980s, the number of gonorrhea cases dropped significantly even though the population increased—a decline that coincides with widespread publicity about genital herpes and AIDS. A similar decline in Sweden during the 1970s followed the institution of extensive public health measures.

Public Health, Prevention, and Treatment

Prevention depends not only on educating the public and instituting the necessary public health measures, but also on personal responsibility for taking protective measures. The use of condoms is an effective way to prevent transmission, as is the screening and treatment of asymptomatic carriers.

Most gonococcal infections are currently treated with a penicillin (amoxicillin) followed by a tetracycline. The tetracycline is given to control the widespread occurrence of *C. trachomata* infections in sexually active individuals. Since 1976, penicillin-resistant strains of *N. gonorrhoeae* have been isolated with increasing frequency both here and abroad. Patients infected with penicillin-resistant *N. gonorrhoeae* are treated with spectinomycin or the cephalosporin (ceftriaxone).

SYPHILIS

The origin of syphilis is unknown, though it was recognized under other names in ancient times. In a later time, during the sixteenth century, there occurred an epidemic in Europe, and it was only then that the disease was understood to be sexually transmitted.

Syphilis is caused by a tightly coiled spirochete classified as *Treponema pallidum* subspecies *pallidum* (Figure 24.4). The bacterium was first observed in 1905 in exudates of syphilitic lesions. Since that time, many microbiologists have attempted to grow *T. pallidum* in laboratory media—all without success. Researchers have been able to grow *T. pallidum* through several generations in tissue culture, but have not been able to transfer them to new tissue cultures. Infected rabbit testicles are the chief source of the bacteria for use in research and diagnosis.

Course of Syphilis

Treponema pallidum penetrates the skin or mucous membranes following direct contact with an infected person. Infections most often develop in the genitalia, but can also occur in the mouth or rectum. Since the treponemes survive for only a short time outside a host, direct contact with an infected person is required

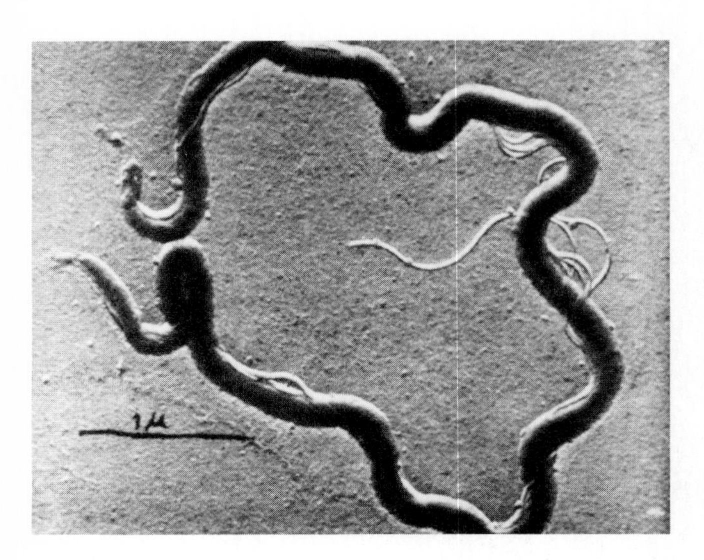

FIGURE 24.4 Electron micrograph of *Treponema pallidum,* 36,000×. (Courtesy of the Centers for Disease Control.)

for transmission. Untreated syphilis progresses through primary, secondary, and in some cases a tertiary stage (Table 24.1) that can lead to death.

During the primary stage of syphilis the treponemes reach the lymph nodes and bloodstream whence they are disseminated throughout the body. The first visible sign of syphilis is a primary lesion called a **chancre** (shang'ker) that usually appears at the site of infection about 21 days after infection (Table 24.1). The absence of a primary lesion is not uncommon; in fact, only 11 percent of syphilitic women have an identified primary lesion. The chancre contains motile spirochetes that can be observed in the darkfield microscope (Figure 24.5). The patient is infectious during the primary stage and should receive penicillin. If the patient is not treated the disease will progress to the secondary stage, even though the chancre heals spontaneously and disappears within four weeks. At this point, an uninformed patient may believe (s)he is cured completely.

Actually, however, the treponemes have multiplied and disseminated to other body parts, including the blood, eyes, joints, bones, mouth, and central nervous system. The classic signs of secondary syphilis are a mild, generalized rash, which may be accompanied by cutaneous lesions (Color Plate II) or lesions in the oral mucous membranes and the genitalia. The disease is still contagious. The symptoms may clear with or without treatment, and some patients do recover. It is risky; however, to assume that a "cure" has been effected because many second-stage cases will enter the dangerous and potentially fatal third and final stage.

The treponemes persist in the tissues and form soft lesions called gummas (gum'ah). A **gumma** is a nonspecific granulomatous lesion, which may be a large, tumorous mass or of microscopic size. Gummas of the skin or bones cause little damage, but gummas of the central nervous system, the eyes, or the heart can, by disrupting normal organ function, lead to insanity and blindness, ultimately contributing to the patient's death. Tertiary syphilis is a slowly progressive disease that can develop years after the initial infection.

Congenital Syphilis Congenital syphilis is contracted *in utero* after the fourth month of gestation from a syphilitic woman, who is usually in the secondary stage of syphilis. The fetus is either spontaneously

TABLE 24.1 Symptoms of syphilis

STAGE	INTERVAL TO NEXT STAGE	CHARACTERISTICS	OUTCOME
Primary	10 to 30 days after infection (average of 3 weeks)	Chancre at site of infection, contagious by sexual contact	Disappears within 4 weeks
Secondary	2 to 12 weeks after chancre	Generalized rash, cutaneous lesions, congenital syphilis, contagious	Many recover
Tertiary	Months or years after secondary, one-third of untreated patients develop tertiary symptoms	Slow, progressive, destructive lesions, gummas, blindness, insanity	Gummas in soft tissue, heart, and nervous tissue contributing to possible death

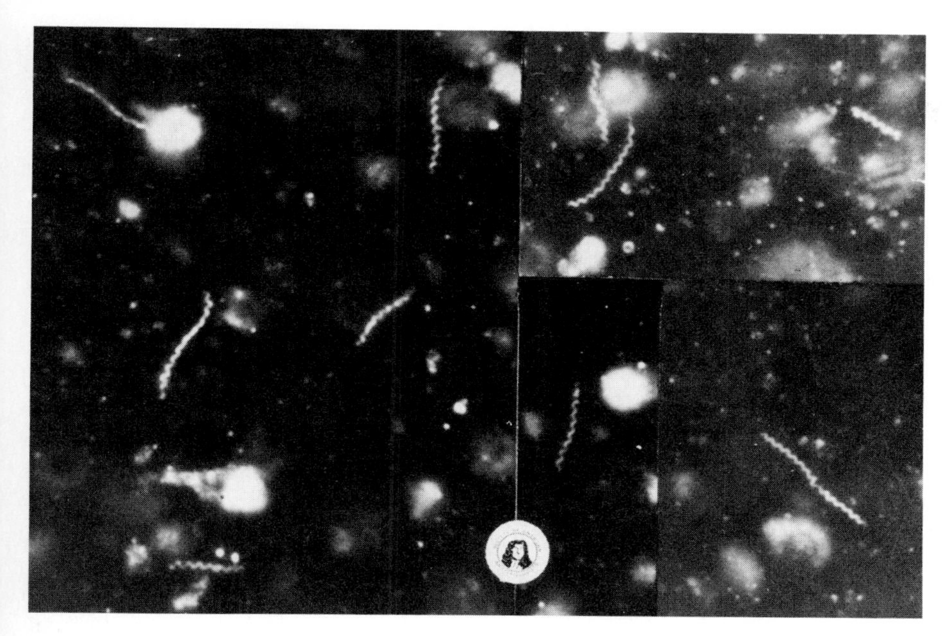

FIGURE 24.5 Dark-field micrograph of *Treponema pallidum,* showing the tightly coiled morphology of this pathogenic spirochete. (Courtesy of the American Society for Microbiology slide collection.)

aborted, or if liveborn, bears the signs of secondary syphilis.

Immunity to Syphilis

A person may develop resistance to reinfection by *T. pallidum,* but will never become fully immune. Humoral antibodies are produced that react specifically with treponemes, whereas others react with by-products of the infection process. In some patients these immune reactions are protective; however, in patients with tertiary syphilis, *T. pallidum* persists even in the presence of the immune response.

Diagnostic Tests

The most direct method for diagnosing syphilis is the microscopic observation of the spirochetes in samples taken from lesions occurring during the primary or secondary stage. The treponemes appear as thin, motile, and tightly coiled spirochetes in dark-field microscopy (Figure 24.5). The spirochetes present in smears taken from lesions or in tissue sections taken from biopsy material can be identified specifically with a fluorescent antibody procedure.

Serological tests are needed when a patient has no obvious syphilitic lesions. The **Rapid Plasma Reagin** (RPR) test and the **Venereal Disease Research Laboratory** (VDRL) test are screening procedures in which treponemes play no part. These tests were developed because the treponemes are so difficult to obtain. A disadvantage of the nontreponemal tests is that they result in a small percentage of "biologic false positive" reactions, so results should be confirmed by a direct treponemal test.

The **fluorescent treponemal antibody absorption** (FTA-ABS) test detects antibodies that react with cells of *T. pallidum* grown in rabbit testicular lesions. Other tests are based on absorption of *T. pallidum* antigens to the surfaces of red blood cells. The **microhemagglutination assay** for antibodies to *T. pallidum* (MHA-TP) and the **hemagglutination treponemal test** for syphilis (HATTS) are direct treponemal tests based on hemagglutination. Positive test detect antitreponemal antibodies in patient serum and are used to confirm the results of the RPR and VDRL tests.

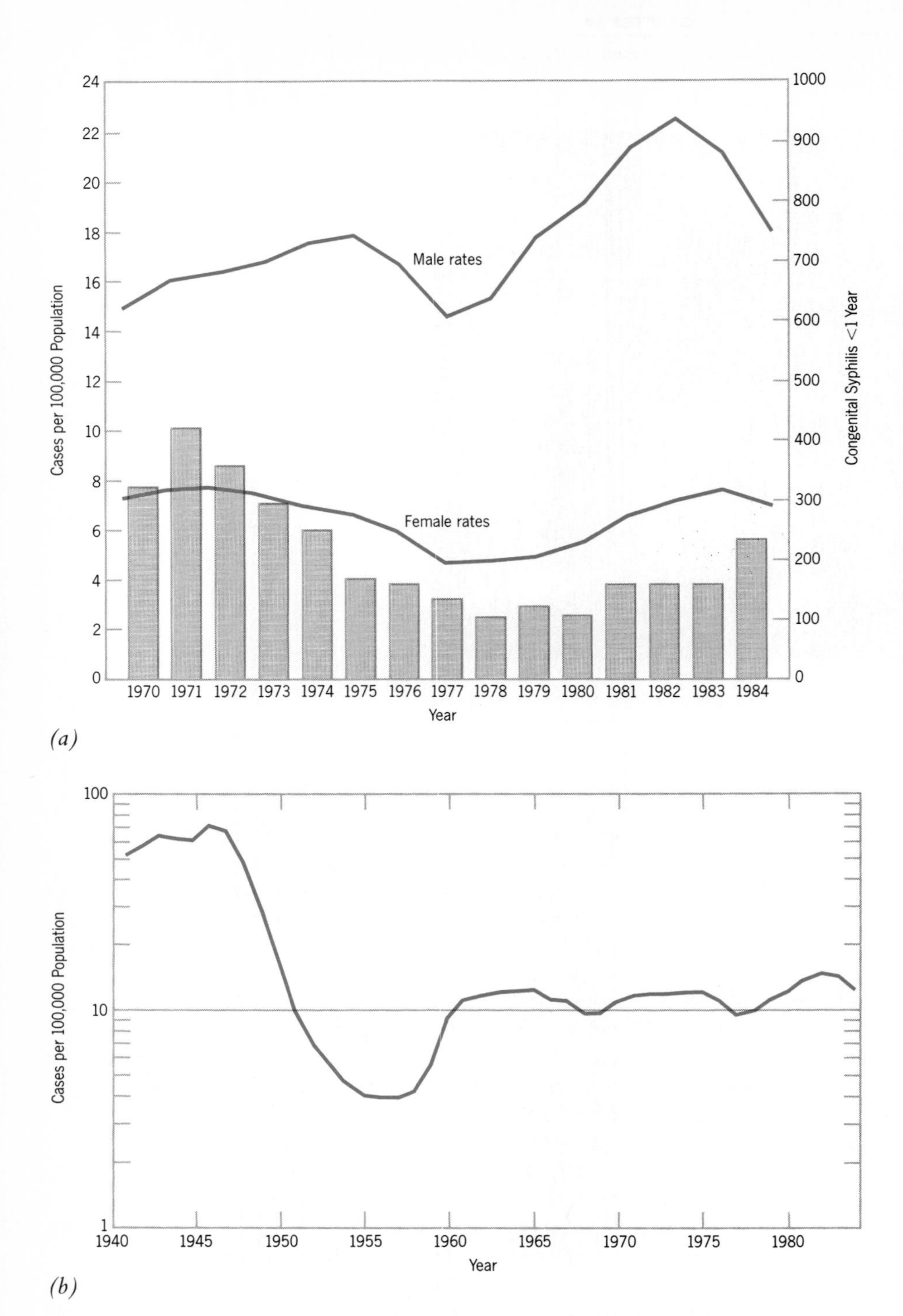

FIGURE 24.6 Epidemiology of syphilis in the United States. *(a)* The number of syphilis cases (primary and secondary) in males and females for the years 1970 through 1984. The bar graph represents the incidence of congenital syphilis during this period. *(b)* The incidence of syphilis (primary and secondary) in the civilian population from 1941 through 1984. (Centers for Disease Control, "Annual Summary 1984: Reported Morbidity and Mortality in the United States," *Morbidity and Mortality Weekly Report,* Vol. 32, No. 54, 1986.)

Epidemiology

Syphilis is transmitted between heterosexual and homosexual partners upon contact with a primary lesion, usually occurring in the genital region, but sometimes in the mouth and on the skin, and in the rectum. The incidence of primary and secondary syphilis is higher in males than females (Figure 24.6*a*) and 50 percent of the males are either homosexual or bisexual. Congenital syphilis parallels syphilis in females in occurrence.

The number of cases of syphilis peaked during World War II, then declined dramatically following the widespread use of penicillin (Figure 24.6*b*). Cases of gonorrhea have consistently and heavily outnumbered those of syphilis in recent years.

Treatment and Prevention

Primary and secondary syphilis are effectively treated with penicillin; erythromycin or tetracycline is prescribed for patients who are allergic to penicillin. The major means of controlling the spread of syphilis are avoidance of direct contact with a syphilitic person and the proper use of condoms.

Public health measures include required reporting of new cases and the screening of every patient's contacts. Congenital syphilis is prevented by testing pregnant women and by treating them if they are infected. Many states also require a premarital test as part of their public health screening program.

KEY POINT

Since *T. pallidum* cannot be grown in culture, screening tests rely on detecting antibodies against nontreponemal antigen. False positive results are possible, therefore confirmation by direct tests is required. Syphilis occurs only in humans and is curable with antimicrobials, so it is now possible theoretically to eliminate this pathogen.

CHANCROID: *HAEMOPHILUS DUCREYI*

Haemophilus ducreyi (he-mo'fi-lus du-kray'i) causes a rare, sexually transmitted disease called **chancroid** (shang'kroid) or soft chancre. It is most common among nonwhite, uncircumcised men; only about 10 percent of the cases occur in women. The infection begins as a soft lesion in the genital or perianal region, and then spreads to the adjacent lymph glands, resulting in painful swelling. Individual ulcers may grow to a diameter of 20 mm. A diagnosis of chancroid is commonly made on clinical grounds followed by the isolationg of *H. ducreyi* on special media. Chancroid is treated with trimethoprim-sulfamethoxazole or erythromycin.

CHLAMYDIA TRACHOMATIS INFECTIONS

Infections by *Chlamydia trachomatis* are now recognized in this country as the most prevalent, nonreportable sexually transmitted bacterial disease. These chlamydia are carried by sexually active men and women and are transmitted by sexual contact.

Chlamydia is an obligate intracellular bacterium that replicates by binary fission in animal cells (Figure

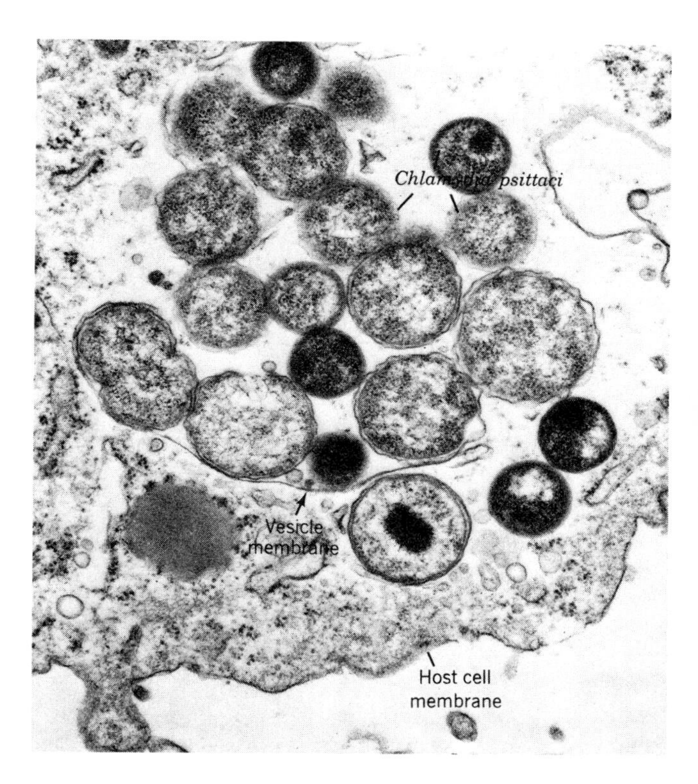

FIGURE 24.7 Electron micrograph of a ruptured vacuole containing *Chlamydia* cells in various stages of their reproductive process. The dark cells are the elementary bodies and the less dense cells are the reticulate bodies. The species shown is *C. psittaci*. (Courtesy of Dr. R. C. Cutlip. From R. C. Cutlip, *Infection and Immunity,* 1:499–502, 1970, with permission of the American Society for Microbiology.)

24.7). The reproduction cycle progresses from an infectious elementary body to a metabolically active reticulate body which is capable of division. *Chlamydia* depends on the host cells for all of its energy, so it is known as an "energy parasite."

The main reservoir for *C. trachomatis* is asymptomatic females who have chlamydial infections of the genital tract. In the past, laboratory identification of *C. trachomatis* relied on growth of the bacterium in cell cultures; it is now possible to detect the bacterium with a monoclonal fluorescent antibody assay or by a specific enzyme immunoassay.

Chlamydia trachomatis causes significant infections in both males and females. In women, *C. trachomatis* is responsible for pelvic inflammatory disease, cervicitis, and urethritis. Newborns that become infected during the birth process often develop inclusion conjunctivitis (see Chapter 22) and pneumonia. Approximately 50 percent of the reported cases of nongonococcal urethritis in males are caused by *C. trachomatis*. It is also responsible for about 50 percent of the estimated 500,000 annual cases of epididymitis in the United States.

The risk of being infected by chlamydias increases directly with the number of sex partners, and in women the risk is inversely related to age. Sexually active females under 20 years of age have two to three times more chlamydial infections than women 20 to 29 years of age, and those older than 30 experience an even lower rate. Chlamydial infections are prevalent because so many gonorrhea patients (25 to 50 percent) are also infected with *C. trachomatis,* and penicillin alone cures only the gonorrheal infection.

Chlamydial infections can be treated with tetracyclines, which are effective against both *N. gonorrhoeae* and *C. trachomatis*. Infections between sexual partners can be reduced by using a barrier method of contraception, such as condoms.

Lymphogranuloma Venereum

Three serotypes of *Chlamydia trachomatis* not associated with other chlamydial diseases cause lymphogranuloma venereum (lim′fo-gran′u-lo′mah ve-ne′re-um). This sexually transmitted disease occurs primarily in South America and Africa; fewer than 400 cases annually are reported in the United States. Lymphogranuloma venereum begins as a small papule or ulcer in the genital region. Unlike other chlamydial diseases, the bacteria spread to the regional lymph nodes and cause inguinal (groin) lymphadenopathy (ing′gwi-nal lim-fad′e-nop′ah-the). The swollen lymph nodes are painful and many develop draining fistules. Lymphogranuloma venereum is treated with tetracycline.

KIDNEY INFECTIONS: LEPTOSPIROSIS

Leptospiras (lep′to-spi′rah) are free-living or parasitic spirochetes found in ponds, streams, or infected animals. They are distinct from other spirochetes, being aerobic and having one or more hooked ends visible in dark-field microscopy (Figure 24.8). They cause leptospirosis, which is primarily a zoonosis of wild and domestic animals.

Many leptospira have been grown in artificial media containing heated rabbit serum or albumin and long-chain fatty acids. Pathogenic leptospiras incorporate the fatty acids into their cellular material, since they cannot manufacture these acids *de novo,* and use the fatty acids as a source of energy. The known serotypes of leptospira are assigned to one of two species: *Leptospira interrogans,* (in-ter′o-ganz) which includes all the pathogenic forms, and *Leptospira biflexa* (bi-fleks′ah), which includes the nonpathogenic leptospira.

LEPTOSPIROSIS IN LOWER ANIMALS

Rodents and domestic animals are the major reservoirs of the pathogenic leptospiras. The spirochetes establish

FIGURE 24.8 Scanning electron micrograph of *Leptospira interrogans* showing the helical coiling of a typical spirochete. (Courtesy of Dr. N. Charon. From O. Carleton, N. Charon, P. Allender, and S. O'Brien, *J. Bacteriol.*, 137:1413–1416, 1979, with permission of the American Society for Microbiology.)

a primary infection in the kidneys where they grow in the proximal convoluted tubules, and are excreted in urine. Infected cattle, swine, and dogs contaminate soil, pond water, lakes, and streams, which are sources of the *Leptospira* involved in human leptospirosis. Animal reservoirs experience a mild, sometimes lifelong disease, but they rarely die of leptospirosis.

LEPTOSPIROSIS IN HUMANS

The number of reported cases in the United States is usually less than 100 annually. Ninety percent are mild, but the remaining 10 percent are serious illnesses, causing a form of jaundice known as **Weil's** (vilz) **disease.** The leptospiras present in contaminated soil or water enter the host through abrasions on the skin or through mucous membranes. A high fever is followed by nausea, headache, muscular pain, and nosebleeds after a 6- to 12-day incubation period. The leptospiras are present in the blood and the cerebrospinal fluid before they infect the kidneys during the later stages. Kidney failure appears to be the immediate cause of human death. The fatality rate is from 4.6 to 32 percent. Those who recover possess a solid immunity to subsequent infection.

Treatment remains unsatisfactory even with antibiotics. In the early stages, the infection responds to penicillin or tetracyclines; unfortunately, antimicrobial treatment in the later stages of the disease, when most cases are diagnosed, is not very effective.

VIRAL DISEASES OF THE GENITOURINARY SYSTEM

Genital herpes and acquired immune deficiency syndrome (AIDS) are sexually transmitted viral diseases that are currently epidemic in the United States. AIDS, a relatively new viral disease, occurs primarily among homosexuals and intravenous drug users. About 50 percent of the patients diagnosed in the United States as having AIDS during the first seven years of the epidemic have died. Genital herpes is also a serious disease, since the virus causes recurrent symptoms and an affected person may carry the virus for life. The ability of the herpes simplex virus (Figure 24.9) to generate latent infections will probably result in an ever-increasing number of infected persons.

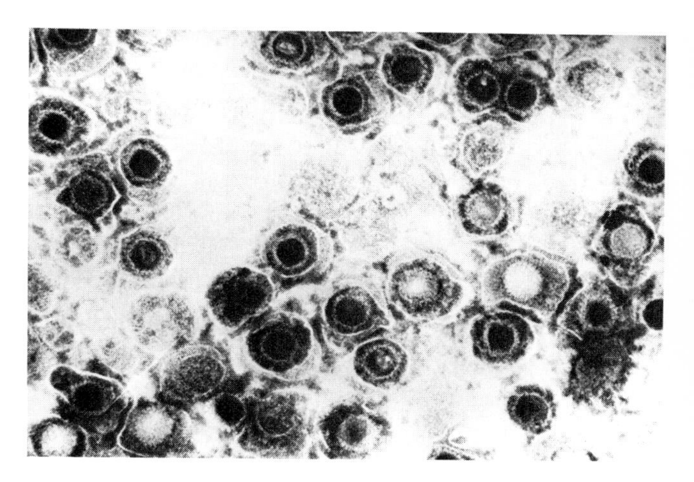

FIGURE 24.9 Electron micrograph of herpes simplex virus. (Courtesy of the Centers for Disease Control.)

SYMPTOMS OF GENITAL HERPES

Genital herpes occurs with equal frequency in males and females. Initial contact with the herpes simplex virus (HSV) results in a primary or first-episode infection. The clinical signs are the eruption of multiple (20 to 30) lesions in the genital area 2 to 14 days after contact with the infecting virus. The lesions may be present on the external genitals or on the cervix or endocervix. The primary infection is usually accompanied by a fever, headache, malaise, and muscle pain.

HSV is shed from the lesion for about 12 days as it develops from a vesicular papule to a wet open lesion. Shedding ceases when a dry crust forms over the lesion, and healing is complete approximately three weeks after the appearance of the first symptoms. HSV type 2 is responsible for more than 80 percent of first-episode genital herpes infections.

Recurrent genital herpes occurs in at least 60 percent of patients infected with HSV-2, who experience an average of four to five episodes every year. The symptoms are sporadic and milder than those associated with primary infections. Episodes of recurrent herpes usually last only 10 to 14 days before the lesions heal completely.

Although antibodies against the HSV are present following primary infection the viruses remain in the host. After the lesions heal, HSV-2 enters a latent stage in the dorsal nerve root ganglia, whence it emerges to initiate recurrent herpes. The mechanisms of latency and reactivation are not understood.

Complications of HSV-2 Infections

Neonatal herpes afflicts infants born of women who harbor HSV-2 in vaginal lesions. Skin lesions appear on the newborn, and the disease may progress to a disseminated infection that involves the central nervous system and internal organs. The fatality rate in untreated disseminated neonatal herpes is about 75 percent. Many babies who recover from neonatal herpes suffer long-term sequelae, including encephalitis, a severe inflammation of the brain.

A woman with primary genital herpes poses a 40 percent possibility that her infant will be infected. Pregnant women with genital herpes are advised to inform their physician about their history, and delivery by cesarean (se-sa're-an) section may be needed.

There is increasing evidence that women with genital herpes are at high risk for developing cervical cancer. This is based on the observation that many cervical cancer patients produce more antibodies against HSV-2 than other women. The Papanicolaou smear, popularly called the Pap test, is desirable for all women, especially those with recurrent genital herpes.

Diagnosis, Prevention, and Treatment

Genital herpes is usually diagnosed by its symptoms, although a definitive diagnosis is be made by culturing the virus in tissue cultures. The virus grows rapidly, producing cytopathic effects within 24 to 48 hours. Herpes infections can also be diagnosed by the demonstration of multinuclear giant cells at the base of the vesicle, although this test is relatively insensitive.

At this writing, no adequate control mechanism is at hand; however, accessible lesions can be treated with topical agents. Acyclovir (a-si'klo-vir) is approved for such use. Treatment of first-episode lesions with acyclovir reduces healing time, but does not prevent recurrences. Infants with disseminated herpes have been successfully treated with adenine arabinoside administered intravenously or with oral acyclovin.

Avoidance of direct contact with the herpes simplex virus is the best means of preventing infection. Medical and dental personnel should wear gloves when dealing with a herpes patient. Genital herpes is presently an incurable disease; thus, if a sexual partner has an active case or a history of recurrent genital herpes, there is a likelihood that s(he) is infectious.

A type of genital herpes called HSV-1 can be contracted by autoinoculation from fever blisters, a nonsexual means of transmitting genital herpes. HSV-1 causes 10 to 20 percent of the cases of first-episode herpes—infections that presumably originate from autoinoculation. Patients with genital herpes should avoid autoinoculation of fingers, lips, buttocks, and thighs since both HSV-1 and HSV-2 can infect these sites.

KEY POINT

Following a primary infection, HSV-2 remains in the body as a latent virus residing in the dorsal nerve root ganglia. The latent virus emerges to initiate recurrent genital herpes, and it is the latency that portends an ever-increasing number of infected individuals.

ACQUIRED IMMUNE DEFICIENCY SYNDROME (AIDS)

AIDS is a pandemic immunosuppressive viral disease that predisposes patients to life-threatening infections by opportunistic organisms and to certain cancers. Since AIDS was first reported in the United States in the spring of 1981, the incidence has continued to increase, not only here (see Figure 20.3) but throughout the world. Beginning with 200 cases in 1981, the number of cases has skyrocketed to more than 12,400 cases reported in the United States during 1986. These numbers are chilling because the fatality rate is extremely high, as previously mentioned.

The AIDS Retrovirus

AIDS is caused by a retrovirus (Figure 24.10) that infects the T4 lymphocytes known as helper T cells, which play a central role in regulating the immune system. After infection, the virus remains dormant inside its T4-lymphocyte host until the lymphocyte is immunologically stimulated by a secondary infection. This causes the AIDS virus to replicate rapidly and kill the T4 lymphocyte. The resulting depletion of the T4-cell population creates an immune deficiency, and it is this that paves the way for opportunistic pathogens to attack. The AIDS virus also infects macrophages, monocytes, and brain cells.

In 1983 Robert Gallo and his colleagues at the National Institutes of Health in Bethesda isolated a retro-

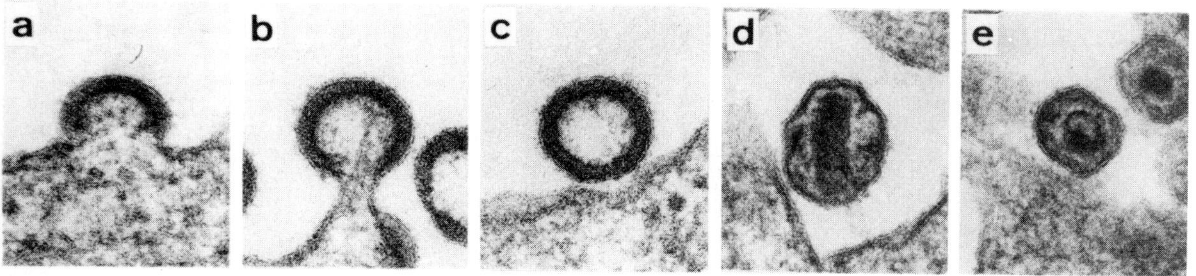

FIGURE 24.10 AIDS is caused by human immunodeficiency virus (HIV). These electron micrographs show the reproduction of HIV in T lymphocytes: *(a, b)* bud formation; *(c)* free, immature, extracellular virus particle; *(d)* virus particle with bar-shaped nucleoid; *(e)* mature HIV particle with condensed circular nucleoid core. (Courtesy of Dr. Matthew A. Gonda. From M. A. Gonda, F. Wong-Staal, R. C. Gallo, J. E. Clements, O. Narayan, and R. V. Gilden, *Science,* 227:173–177 (1985). Copyright © 1985, American Society for the Advancement of Science.)

virus from an infected AIDS patient. Gallo's group named this virus the human T-lymphotropic virus type III (HTLV-III). At the same time, Luc Montagnier and his colleagues at the Pasteur Institute in Paris isolated a similar virus to which they gave the name lymphadenopathy associated virus (LAV). These viruses turn out to be closely related strains of the AIDS virus, which is now officially called the **human immunodeficiency virus (HIV).**

Isolation of the HIV has not been followed by a cure, but it has enabled the two research groups to develop immunoassays by which antibodies against HIV may be detected. Since the virus kills T lymphocytes, Gallo's group had to grow HIV in T cells derived from a leukemia patient. These cultures produced HIV antigen that is now the basis for an enzyme-linked immunosorbent assay (ELISA) by which antibody against HIV is measured. This assay is used to screen donated blood for the presence of HIV antibodies and has virtually eliminated the risk of contracting AIDS through blood transfusions.

Symptons

The symptoms of AIDS may not appear for months or years after infection by HIV. There are many more persons who are seropositive for HIV than there are diagnosed AIDS patients; and only time will reveal whether seropositive patients can carry the virus without developing symptoms.

The initial symptoms of AIDS are extreme fatigue, fever, night sweats, and chronic diarrhea. In Africa the disease is called "slim disease" because of the rapid weight loss that occurs. The lymph glands are enlarged, especially in the neck, groin, and armpits. After the T4 lymphocytes are depleted, the patient is extremely vulnerable to opportunistic infectious diseases, and in many patients, cancers develop, most frequently one called Kaposi's (kap′o-seez) sarcoma. **Kaposi's sarcoma** is a fast-spreading skin cancer characterized by purplish blotches or bumps. In AIDS patients the frequency of this otherwise rare cancer is high. Other types of neoplasms are seen also. A rare form of pneumonia caused by the widespread, but normally harmless protozoan, *Pneumocystis carinii,* is also frequent.

Many AIDS patients experience brain damage, manifest in the early stages by forgetfulness, loss of ability to concentrate, and slowed mentation. Dementia may follow as the disease progresses. Patients finally die when the immune system ceases to protect them.

Epidemiology: Populations at Risk

The AIDS virus is transmitted not by casual contact, but by direct contact with body fluids of AIDS patients. AIDS occurs primarily in homosexual and bisexual males and in intravenous drug users (Table 24.2). It is transmitted in body fluids including semen, blood, and blood products. As of 1986, most AIDS patients in this country are males; however, it is clear

Antiviral Treatment for AIDS

Azidothymidine (AZT) was first synthesized in the early 1960s as an anticancer drug by Jerome P. Horowitz of the Michigan Cancer Foundation. Renewed interest in AZT developed after *in vitro* tests showed that AZT inhibits replication of the human immunodeficiency virus (HIV) responsible for AIDS. AZT was given to selected AIDS patients beginning in February 1986 in a placebo-controlled clinical trial. After seven months only one of 145 AIDS patients receiving the drug died compared to 16 of 137 AIDS patients receiving the placebo. These results convinced the Food and Drug Administration to call off the placebo test and make AZT available to AIDS patients who had had at least one bout of *Pneumocystis carinii* pneumonia.

AIDS patients are infected with an RNA retrovirus that uses its reverse transcriptase to make DNA from its RNA genome. AZT is an analog of deoxythymidine that interrupts the replication of HIV. AZT is phosphorylated by cellular kinases after it enters an HIV-infected cell, and this phosphorylated AZT is preferentially used by the viral reverse transcriptase to make DNA. Once AZT is incorporated, DNA synthesis stops because AZT has an azido (N_3) group at its 3' position instead of a hydroxyl. The AZT prevents attachment of additional nucleotides, and the chain ceases to elongate.

AZT offers AIDS patients a temporary reprieve from what is considered a fatal disease. This drug is not a cure for AIDS because it does not kill HIV, nor does it eliminate the virus from the host; AZT only prevents HIV replication. Patients taking AZT regain some immune function through increased levels of T4 cells, have a healthier appearance, and gain weight. A serious side effect is the suppression of blood cell production in bone marrow, an effect seen with many anticancer agents. This decreases the immune response, especially when the drug is taken for prolonged periods.

At the time of writing, AZT is in limited supply and is being made available to selected AIDS patients. This is the first drug to have a positive effect on the progressive course of AIDS. Other drugs and approaches to curing this frightening disease are aggressively being sought by scientists in many countries.

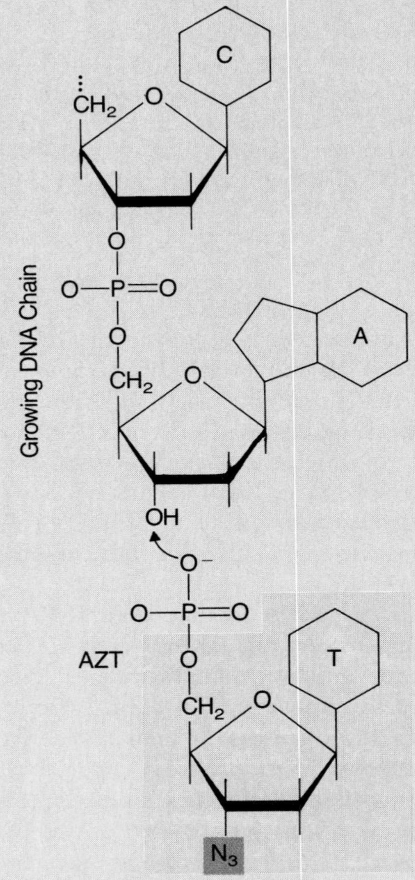

The N_3 group at the 3′ position of AZT prevents the attachment of additional nucleotides, so the DNA chain stops growing.

that AIDS can be transmitted by heterosexual contact, and in Africa the disease is more evenly distributed between the sexes.

Hemophiliacs and others who require blood transfusions were at risk at one time; this group represents about 2 percent of all AIDS patients. About 1 percent of AIDS victims are children, most of them born to parents who are either bisexual or intravenous drug

TABLE 24.2 Acquired immune deficiency syndrome patients by group and date of report[a]

PATIENT GROUP	BEFORE 1983 (Percent)	MAY 1983–APRIL 1984 (Percent)	MAY 1984–APRIL 1985 (Percent)	TOTALS (Percent)
Homosexual/Bisexual	992 (70.6)	2070 (71.5)	4199 (73.7)	7261 (72.6)
IV drug users	233 (16.6)	510 (17.6)	942 (16.5)	1685 (16.9)
Hemophilia patient	11 (0.8)	17 (0.6)	37 (0.6)	65 (0.7)
Heterosexual contact	13 (0.9)	23 (0.8)	45 (0.8)	81 (0.8)
Transfusion recipient	12 (0.9)	34 (1.1)	88 (1.5)	134 (1.3)
Children	19 (1.3)	40 (1.4)	54 (1.0)	113 (1.1)
Other unknown	126 (8.9)	202 (7.0)	333 (5.9)	661 (6.6)
Total	1406	2896	5698	10000

[a]*Morbidity and Mortality Weekly Report, Vol. 34,* No. 18 (May 10, 1985).

users. A small number of children with AIDS are hemophiliacs who received contaminated blood.

Intravenous drug users and homosexuals are at highest risk of contracting AIDS. Intravenous drug users may share needles, which became contaminated when used by an AIDS-infected persons. For homosexual males, the risk increases with the number of sexual partners and the practice of anal intercourse; the virus is believed to be transmitted by semen into the circulation through abrasions in traumatized tissue.

Transmission between heterosexual partners is of serious concern especially since studies of African prostitutes indicate that heterosexual transmission does occur. Sexual contact with prostitutes who have a large number of heterosexual contacts appears to be a risk factor in the United States also. Women with AIDS are advised not to become pregnant, since 60 percent of the infants born of women with AIDS become victims themselves. And most of them die.

Detection of antibody against HIV in stored serum samples has given epidemiologists a better picture of AIDS. Among homosexuals, the average interval between seroconversion (presence of antibody) and the diagnosis of AIDS exceeds three years. A person may have antibodies against HIV for up to 69 months without developing symptoms. The HIV is apparently able to generate a persistent infection that can last for months. In one study, the ratio of seropositive males to AIDS patients was between 50:1 and 100:1, indicating that many more persons are infected than actually have symptoms. All seropositive persons are considered to be infectious since HIV can be isolated from most of these persons.

Prevention and Treatment

"Safe sex" is the only way to avoid being infected. This means the proper use of condoms and restriction of sexual partners to persons who are not infected. A vigorous search is on for vaccines and drugs that will either prevent the spread of AIDS or treat the infection. As of this writing, azidothymidine (AZT) is the only drug available to treat AIDS. AZT appears to arrest the rapid progression of the disease, but it is not a cure. The drug acts by preventing the formation of DNA from the viral RNA genome and thereby stops the viral replication cycle.

PROTOZOAN DISEASE: TRICHOMONIASIS

Every year, over one million American women are infected by *Trichomonas vaginalis* (trik'o-mo'nas vag-in-ah'lis), a sexually transmitted flagellated protozoan that causes **trichomoniasis** (trik'o-mo-ni'ah-sis). Trichomoniasis also occurs in males, but at a very much lower rate.

Trichomonas vaginalis is a large, motile, pear-shaped flagellate (Figure 24.11) that divides by longitudinal fission. Sexual reproduction is not known to occur, and

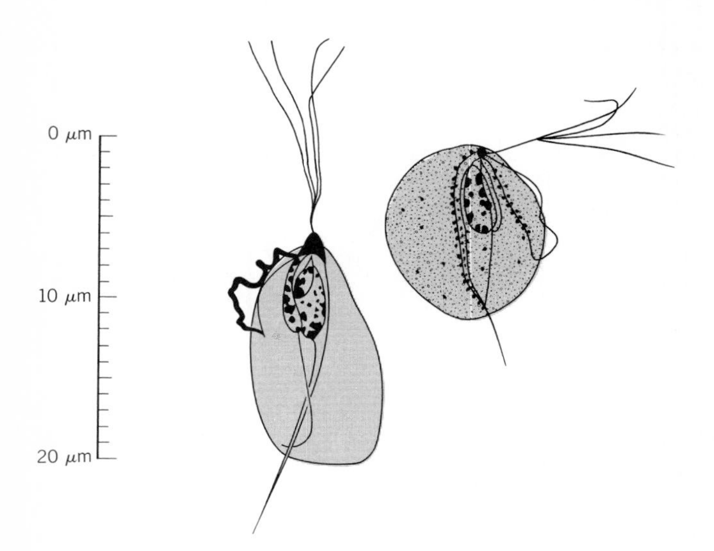

FIGURE 24.11 Trophozoites of the pear-shaped flagellate, *Trichomonas vaginalis*. (Courtesy of Dr. B. M. Honigberg. From B. M. Honigberg and V. M. King, *J. Parasitology*, 50: 345–364, 1964. Copyright © 1964. With permission of the American Society of Parasitologists.)

cysts do not form. Each flagellate possesses an undulating membrane and four anterior flagella that arise from a single site. These protozoa are identified by their characteristic twitching movement, which is visible in the light microscope. *Trichomonas vaginalis* is an anaerobic protozoan that grows on carbohydrates, bacteria, and erythrocytes; mitochondria are notably absent.

Trichomonas vaginalis multiplies in the vagina after an incubation period of between 5 and 28 days. The main symptoms are vulvovaginal soreness and vaginal discharge. The symptoms usually increase during the menstrual period, probably because the pathogen replicates rapidly at this time. Approximately 25 percent of infected females are asymptomatic, and males rarely have symptoms even when *T. vaginalis* is present in the urethra.

Infected patients and their sexual partners are treated with oral doses of metronidazole in a seven-day regimen. Approximately 95 percent of the cases are cured with treatment; however, the disease will recur in patients who are reinfected.

Summary Outline

THE GENITOURINARY SYSTEM

1. The healthy person's kidneys, ureters, and urinary bladder are free of microorganisms.
2. The anterior urethra of both males and females is naturally contaminated by microbes from the skin or feces.

BACTERIAL GENITOURINARY INFECTIONS

1. Urinary infections are common in females and in hospitalized patients who have been catheterized. Bacteria can ascend the urinary tract to infect the bladder (cystitis), prostate (prostatitis), or kidneys (pyelonephritis).
2. Vaginitis is caused by *Candida* (yeast), *Trichomonas vaginalis* (protozoan), and *Gardnerella vaginalis* together with anaerobic *Bacteroides* species (nonspecific vaginitis).
3. The presence of high concentrations of bacteria in the urine (bacteriurea) is an indication of a urinary tract infection. Urinary tract infections are treatable with antimicrobials.
4. *Ureaplasma urealyticum* has been implicated in urethritis, and *Mycoplasma hominis* may be associated with some cases of pelvic inflammatory disease.
5. *Streptococcus pyogenes* was at one time a major cause of death from puerperal sepsis. *Streptococcus agalactiae* is common in the vagina; it causes meningitis, pneumonia, and bacteremia in newborns.

SEXUALLY TRANSMITTED BACTERIAL DISEASES

1. Gonorrhea, the most common reportable infectious disease in the United States, is treated with a penicillin followed by a tetracycline.
2. *Neisseria gonorrhoeae* causes gonorrhea when it infects the genital tract, mouth, or rectum of males and females, and can cause sterility in both sexes.
3. Female carriers of *Neisseria gonorrhoeae* can contract pelvic inflammatory disease, with blockage

of fallopian tubes and ensuing ectopic pregnancy.

4. Syphilis is caused by a spirochete, *Treponema pallidum,* that has never been grown in culture.
5. Primary syphilis is characterized by a chancre that develops at the site of infection about 21 days after infection.
6. The spirochetes are found throughout the body in secondary syphilis, which is characterized by cutaneous lesions and a skin rash. Gummas develop during tertiary syphilis and can cause blindness, insanity, and finally death.
7. Congenital syphilis occurs when a fetus is infected with *T. pallidum,* usually when the mother is in the secondary stage of syphilis. The fetus is either spontaneously aborted or born with secondary syphilis.
8. Patients develop resistance to reinfection by *T. pallidum,* but never become fully immune. Antibodies produced against the spirochete are detected by diagnostic techniques.
9. Syphilis is prevented by avoiding contact with infected persons and receiving treatment with penicillin.
10. *Haemophilus ducreyi* causes a rare disease known as chancroid, most common among nonwhite, uncircumcised men.
11. *Chlamydia trachomatis* is carried in the female vagina and is transmitted by sexual contact, causing pelvic inflammatory disease, cervicitis, and urethritis. Other strains of *Chlamydia* cause lymphogranuloma venereum, which is characterized by swollen inguinal lymph glands.

KIDNEY INFECTIONS: LEPTOSPIROSIS

1. Leptospirosis is a zoonosis in domestic and wild animals caused by *Leptospira interrogans.*
2. In humans, *Leptospira* infects the kidneys and can cause death from kidney failure. Antimicrobial chemotherapy in the later stages is not very effective.

VIRAL DISEASES OF THE GENITOURINARY TRACT

1. Genital herpes is a sexually transmitted disease caused by HSV-2, with lesions at the site of infection. Recurrent genital herpes occurs when the latent HSV-2 emerges from the dorsal nerve root ganglia.
2. Recurrent herpes occurs four to five times a year in 60 percent of HSV-2 infected patients. These patients are infectious until the lesions heal, in 10 to 14 days.
3. Neonatal herpes is a serious complication that can result in disseminated herpes and herpes encephalitis.
4. There is no cure for genital herpes. Avoiding direct contact with the HVS will prevent its spread.
5. Acquired immune deficiency syndrome (AIDS) is a pandemic immunosuppressive viral disease that predisposes patients to life-threatening infections by opportunistic organisms and certain cancers.
6. AIDS is caused by a retrovirus that infects and kills T4 lymphocytes, and this in turn destroys the immune system. Persons at high risk of acquiring AIDS include homosexuals, bisexuals, intravenous drug users, hemophiliacs, and infants born to infected mothers.
7. Since AIDS was first diagnosed in 1981, it has caused an explosive epidemic in the United States as well as many other parts of the world.
8. There is no cure for AIDS, and the only recommended treatment is with azidothymidine (AZT), which helps to arrest the progression of AIDS. Safe sex and avoidance of direct contact with body fluids from high risk groups are recommended.

PROTOZOAN DISEASE: TRICHOMONIASIS

1. Trichomoniasis, a sexually transmitted protozoan disease that affects both males and females, is treated with metronidazole.

Questions and Topics for Study and Discussion

QUESTIONS

1. Make a chart depicting diseases of the female genital tract that can be cured by penicillin, tetracycline, metronidazole, and/or chlortrimazole. Name the pathogen responsible for each disease.
2. What is puerperal sepsis and why is it no longer a serious health threat?
3. Describe the symptoms of gonorrhea in males and females and explain the long-term consequences of *N. gonorrhoeae* infections.
4. Why are indirect treponemal tests for syphilis used? How do these differ from the direct treponemal tests?
5. Describe the clinical signs associated with the three stages of syphilis. During which stage(s) is the patient infectious?
6. Describe the transmission and symptoms of congenital syphilis. How does this disease compare to neonatal herpes?
7. Why is HSV-2 a serious public health problem? What is the long-term outlook for dealing with genital herpes? Explain.
8. Under what circumstances does HSV-1 cause genital herpes?
9. Describe a zoonosis that causes a kidney infection in humans. How are humans infected and why is this a serious disease?
10. Why is the mortality rate among AIDS patients so high?
11. Which groups of people are at greatest risk of contracting AIDS? Explain why they are susceptible?
12. What questions would you ask potential blood donors to determine if they are in a high-risk category for carrying HIV?

DISCUSSION TOPICS

1. In Michigan, 16 cases of syphilis were detected by the 173,000 syphilis tests performed in 1979. These tests were required by a law that prohibits the granting of a marriage license in the absence of submission of the results of a test for syphilis. What is your opinion of the law? Are other laws needed to control the spread of sexually transmitted diseases?
2. Assume that you are the pediatrician treating the 6-month-old AIDS son of a prostitute. What is the probability that the mother has AIDS and how could she be diagnosed? What steps should/could society take to limit the spread of AIDS in such situations?
3. What responsibility do people have to inform their sexual partners that they have had genital herpes, or are in a high-risk group for AIDS?

Further Readings

BOOKS

Johnson, R. C. (ed), *The Biology of the Parasitic Spirochetes,* Academic Press, New York, 1976. A well-organized collection of articles with extensive coverage of the biology of the spirochetes.

ARTICLES AND REVIEWS

Becker, Y., "The Chlamydia: Molecular Biology of Procaryotic Obligate Parasites of Eucaryocytes," *Microbiol. Revs.,* 42:274–306 (1978). A review of the biology of the chlamydias that details the life cycle of these bacteria.

Curran, J. W., W. M. Morgan, A. M. Hardy, H. W. Jaffe, W. W. Darrow, and W. R. Dowdle, "The Epidemiology of AIDS: Current Status and Future Prospects," *Science,* 229:1352–1357 (1985). An analysis of the data collected from the San Francisco CDC cohort study leads these authors to predict an explosive AIDS epidemic in the United States.

Eschenbach, D., H. M. Pollock, and J. Schachter, "Laboratory Diagnosis of Female Genital Tract Infections," in S. J. Rubin (ed.), CUMITECH 17, ASM, 1983. Detailed summary of the normal flora and pathogens of the female genital tract.

Fitzgerald, T. J., "'Pathogenesis and Immunology of *Treponema pallidum*," *Ann. Rev. Microbiol.*, 35:29–54 (1981). A review of the pathogenesis of *T. pallidum* and the immune response to infections caused by this pathogenic spirochete.

Gallo, R. C., "The AIDS Virus," *Sci. Am.*, 256:46–56 (January 1987). The second of a two-part article on human retroviruses explains the molecular biology and clinical effects of the AIDS virus.

Larsen, S., "Current Status of Laboratory Tests for Syphilis," in *Diagnostic Immunology: Technology Assessment,* College of American Pathologists, 1983. A review of the methods used to diagnose syphilis that details their advantages and disadvantages.

Laurence, J., "The Immune System in AIDS," *Sci. Am.*, 253:84–93 (December 1985). The way the AIDS virus alters the function of the T4 lymphocytes is related to a search for an AIDS vaccine

Wolontis, S., and S. Jeansson, "Correlation of Herpes Simplex Virus Type 1 and 2 with Clinical Features of Infection," *J. Infect. Dis.*, 135:28–33 (1977). A research report on the symptoms associated with more than 300 herpes simplex viruses isolated from human patients.

C H A P T E R

25

Gastrointestinal Infections and Food Poisoning

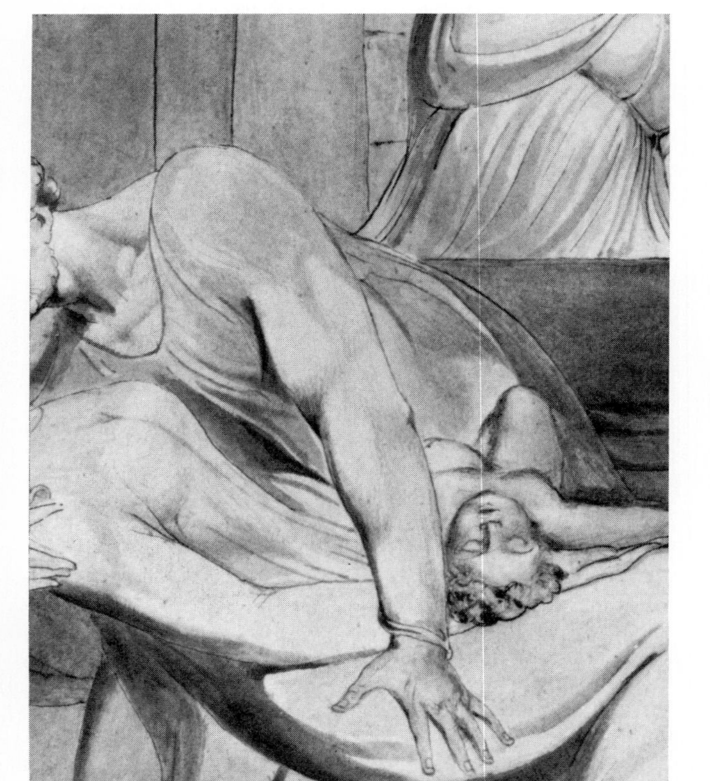

OUTLINE

FOCUS OF THIS CHAPTER

As a rule, the digestion and assimilation of food are taken for granted, until a gastrointestinal upset occurs and the system malfunctions. The cause is often an infection from pathogenic microorganisms, which can cause illness ranging in intensity from very mild to life-threatening. The pathogens, their routes of transmission, and their effects are the themes of this chapter.

The purpose of the gastrointestinal system is to digest food, absorb the nutrients for utilization by the body, and get rid of solid waste. Foods properly handled at every step of the journey from farm to table can be safely eaten by most people. Problems arise when food is improperly handled, stored, or cooked, or when the water supply is contaminated.

THE GASTROINTESTINAL SYSTEM

The digestive system (Figure 25.1) consists of the intestinal tract—the esophagus, stomach, small intestine, large intestine—and the mouth, teeth, salivary glands, gall bladder, and pancreas. Digestion begins in the mouth as food is chewed and mixed with saliva from the salivary glands. The saliva moistens the food and contains mucus, which lubricates the food particles and makes them easier to swallow. The pharynx and esophagus provide a connection between the mouth and the stomach, though they contribute nothing to digestion.

The contents of a meal are temporarily stored in the stomach while hydrochloric acid and proteolytic enzymes further the process of digestion. The partially digested food leaves the stomach and enters the duodenum. Here food is mixed with bile from the liver and pancreatic enzymes, which enter through the common bile duct. The bile and enzymes hydrolyze the partially digested food into simple carbohydrates, amino acids, fatty acids, and nitrogenous bases. These nutrients are then absorbed into the circulatory system through the villi that line the jejuneum (je-joo′num) and upper ileum (il′e-um). Water is reabsorbed as the material being digested moves through the large intestine. Waste is excreted as feces.

Some bacteria, viruses, and helminths that enter the digestive system with food and water are inactivated by the acid of the stomach and duodenum; however, others survive to colonize the ileum and large intestine. Most of these microbes are harmless, simply contributing fecal bulk and causing no illness. But a few are pathogens, which can invade the intestinal mucosa and produce enterotoxins, or in some other manner cause intestinal disease.

MICROBIAL POPULATIONS OF THE DIGESTIVE TRACT

The microbial population of the intestinal tract increases dramatically as food progresses from the stomach to the large intestine. The stomach contains very few microorganisms because its low pH (2.0 or less) prevents most of them from multiplying, and the survivors are either enmeshed in food or are extremely acid-tolerant. Most microbes do not colonize the duodenum and jejunum; food moves rapidly through these organs, washing away the microbes faster than they can reproduce. Only the anaerobic terminal ileum and the large intestine contain appreciable concentrations of microorganisms (Figure 25.2).

The small quantities of oxygen present in foodstuffs are quickly used by aerobic microbes as the food passes

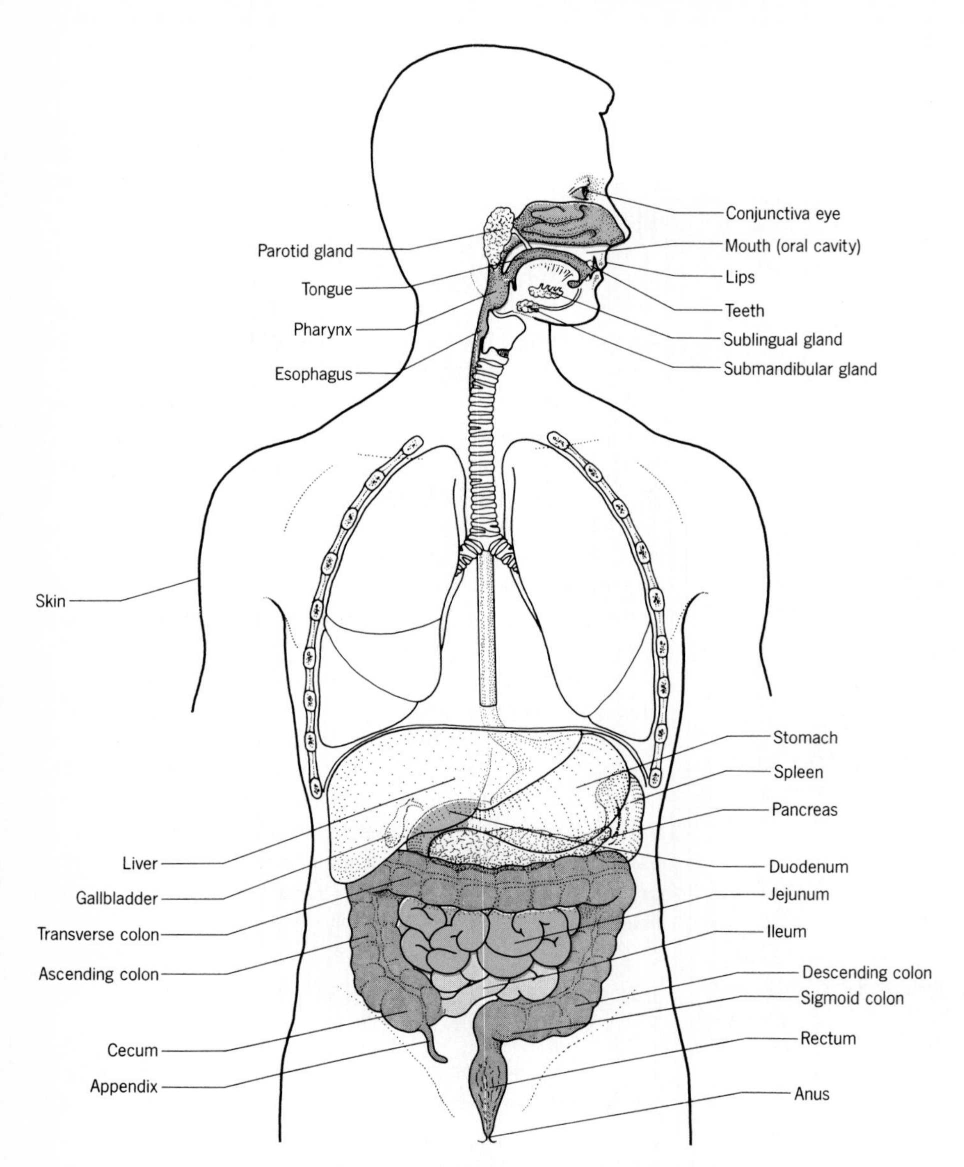

FIGURE 25.1 The digestive system. Shading indicates regions containing a normal flora.

through the upper intestinal tract. This decrease in oxygen as food moves away from the stomach is accompanied by an increase in pH. From a low of 2.0 near the stomach, the pH in the large intestine rises to between 6 and 7. The terminal ileum and the large intestine therefore are ideal environments for the growth of anaerobic bacteria, which contribute most of the biomass to the feces.

In the lower ileum (Figure 25.2), the predominant bacteria are species of *Escherichia, Lactobacillus,* yeasts, and *Bacteroides*. The obligate anaerobes in the large intestine are dominated by species of *Bacteroides;* other fermentative bacteria include *E. coli, Streptococcus* spp., and *Lactobacillus*.

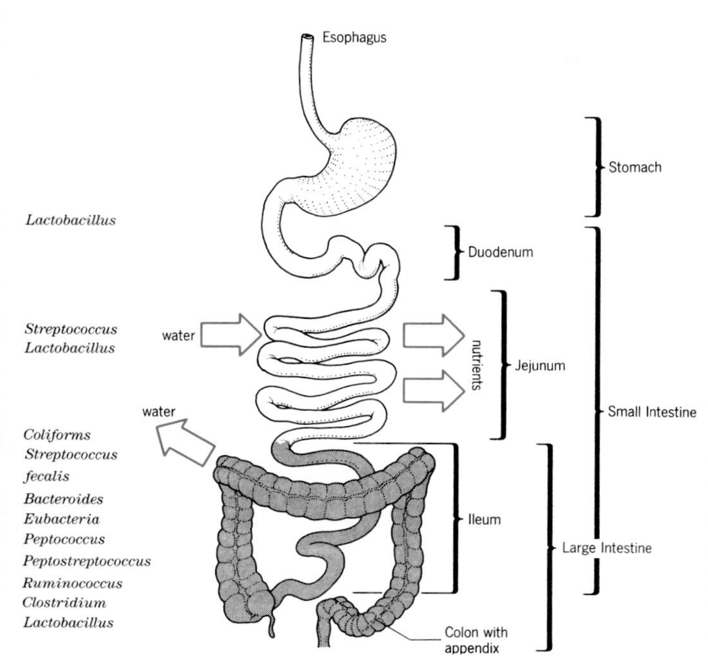

FIGURE 25.2 Location of the bacterial species comprising the normal intestinal population. Arrows indicate fluid movement and nutrient absorption. (Adapted from B. S. Drasar and P. A. Barrow, *Intestinal Microbiology,* Aspects of Microbiology 10, American Society for Microbiology, 1985.)

EFFECTS OF INTESTINAL MICROBES

A newborn's intestinal tract is colonized by bacteria soon after it begins to ingest food. The bacteria in the feces of breast-fed infants are mainly species of *Bifidobacterium* (bi′fid-o-bak-te′re-um), with lower concentrations of *Lactobacillus acidophilus, Bacillus* spp., *Clostridium* spp., coliforms, and enterococci. Children develop a normal adult intestinal flora as their diet expands to include solid foods.

The presence of microoganisms in the intestine does not lead to a disease state except in unusual circumstances. The microorganisms can cause disease following surgery, a rupture of the appendix, or malfunction of the gall bladder, or the intestine may be invaded by pathogens. Peritonitis, cholecystitis, and gastroenteritis may be the consequence.

KEY POINT

Soon after a newborn begins to feed, its intestinal tract becomes colonized by bacteria. Most of the bacteria are beneficial or harmless, merely participating in the digestive process and contributing to the bulk of the feces.

BACTERIAL INFECTIONS

A gastrointestinal illness may be a pure intoxication resulting from the ingestion of a toxin, an invasive disease in which the microbe colonizes the intestinal tract, or a toxico-infection resulting from a combination of these disease mechanisms. Cholera will serve as an example of toxico-infection. Though cholera is rare in the developed countries, it is widespread in underdeveloped countries, where sanitation and sewage treatment are often primitive at best. Since these countries constitute one-third of the world, the student can understand why it is important to learn about this disease.

CHOLERA: THE PANDEMIC DISEASE

Cholera is a life-threatening diarrheal disease that has plagued humankind for centuries. As described in Chapter 20, John Snow discovered that cholera was transmitted by contaminated water. Although cholera may be limited to local populations, as was the case in Snow's London, it is often spread from one country to another by human carriers, and a pandemic results.

The most recent cholera pandemic began in Macao and Hong Kong in 1960–61, spread to India, and reached the Near East during the late 1960s. It then spread to Africa and parts of Europe during the early 1970s. The disease is only a minor problem in the United States with fewer than 20 cases annually (usually restricted to foreign travelers).

Cholera: *Vibrio cholerae*

Vibrio cholerae is a gram-negative, polar-flagellated, facultative anaerobe that grows readily in alkaline media (pH 8.0 to 9.5) and 3 percent NaCl. A sheathed polar flagellum (see Figure 3.14) rapidly propels it through a liquid medium. Most toxin-producing strains of *V.*

cholerae are isolated from feces or environmental sources that have come in contact with human wastes. Environmental reservoirs do not appear to be a major factor in cholera transmission.

Symptoms

Cholera patients become severely dehydrated (Figure 25.3) as they lose fluid from the purging diarrhea of the disease. Infection develops following ingestion of a large number of *V. cholerae* in contaminated food or water. The bacteria colonize the entire intestinal tract from the jejunum to the colon. Here they produce the cholera toxin, which causes an intense, severe diarrhea. The stools are heavily contaminated with vibrios, sloughed-off epithelial cells, and mucus. This type of diarrhea is described as "rice-water stools" because the flecks of mucus in the diarrheal fluid look like grains of rice. The main threat to the patient is the tremendous loss of water and electrolytes, which if untreated is followed by dehydration, prostration, and eventually death.

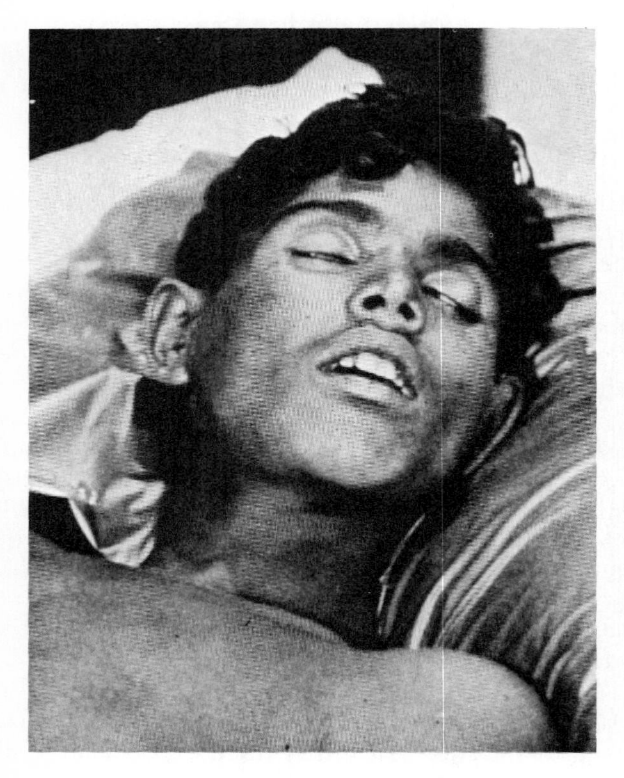

FIGURE 25.3 Severely dehydrated patient with cholera. (From G. Mandell, J. E. Bennett, and R. G. Douglas, Jr. (eds.), *Principles and Practices of Infectious Diseases,* Wiley, New York, 1985, p. 1211, Figure 3.)

The Cholera Toxin

The mechanism of cholera toxin action (Figure 25.4) serves as a model by which one can understand the effects of intestinal enterotoxins. The cholera enterotoxin is a heat-labile protein (MW 84,000) composed of three subunits, types, A_1, A_2, and B. The B subunits (MW 10,000 each) bind the toxin to a cell receptor; the A_2 subunit (MW 7000) translocates the A_1 subunit possessing the toxic activity into the cytoplasm. The A_1 subunit (MW 29,000) enzymatically activates adenylate cyclase of eucaryotic cells to cause an increase in the intracellular concentration of cyclic AMP (cAMP) as follows

$$\text{ATP} \xrightarrow{\text{adenylate cyclase}} \text{cAMP} + \text{P—P}$$

High concentrations of cAMP stimulate the intestinal mucosa to excrete chloride and bicarbonate ions (Figure 25.4). The change in the electrolyte balance results in the outpouring of Na^+, K^+, and copious amounts of water into the lumen of the intestine—in other words, diarrhea.

Treatment and Control

Cholera is treated by replacing fluids and electrolytes to prevent severe dehydration and electrolyte imbalance. Up to 25 liters of an alkaline–saline solution containing bicarbonate and lactate is administered daily, either orally or intravenously. With this simple treatment the fatality rate falls to 1 percent, whereas without it mortality may be as high as 50 percent. Tetracyclines may decrease the duration of the diarrhea and the amount of fluid needed to restore the patient to health, but they have no direct effect on the mortality rate.

The vibrios purged from infected patients reach the local water supplies and are transmitted to other persons when they drink contaminated water, bathe in it, or eat contaminated food. The only succesful method of preventing cholera is adequate sewage treatment along with purification of drinking water.

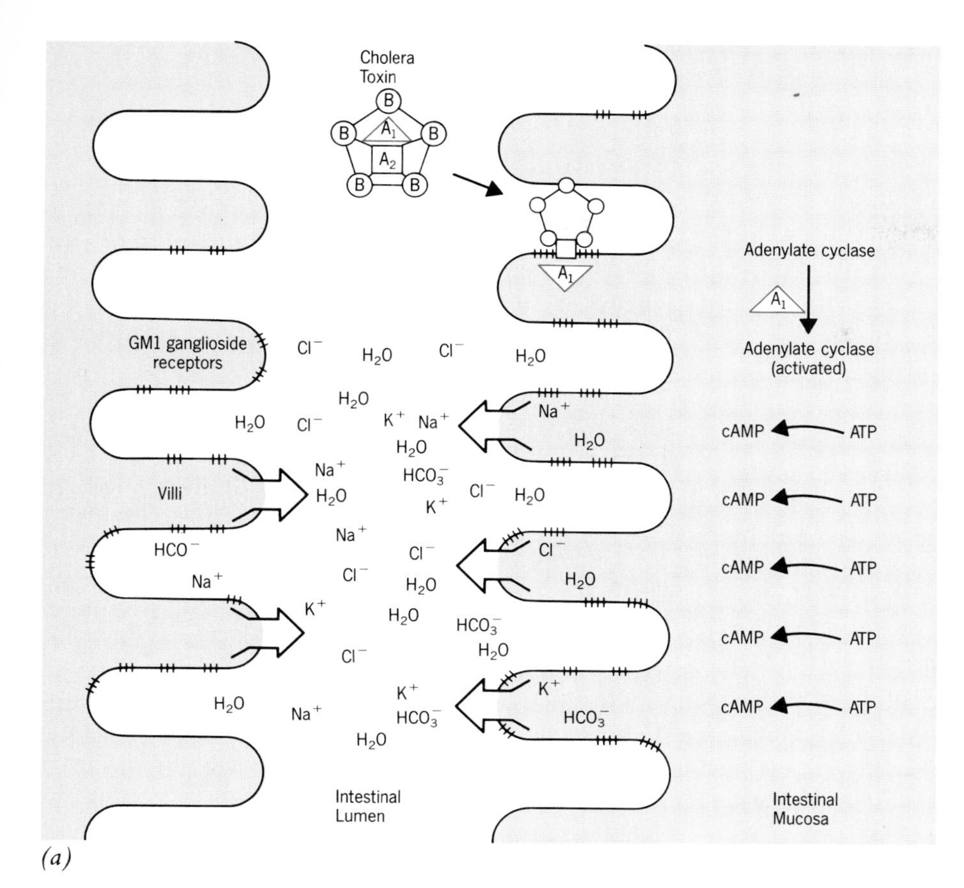

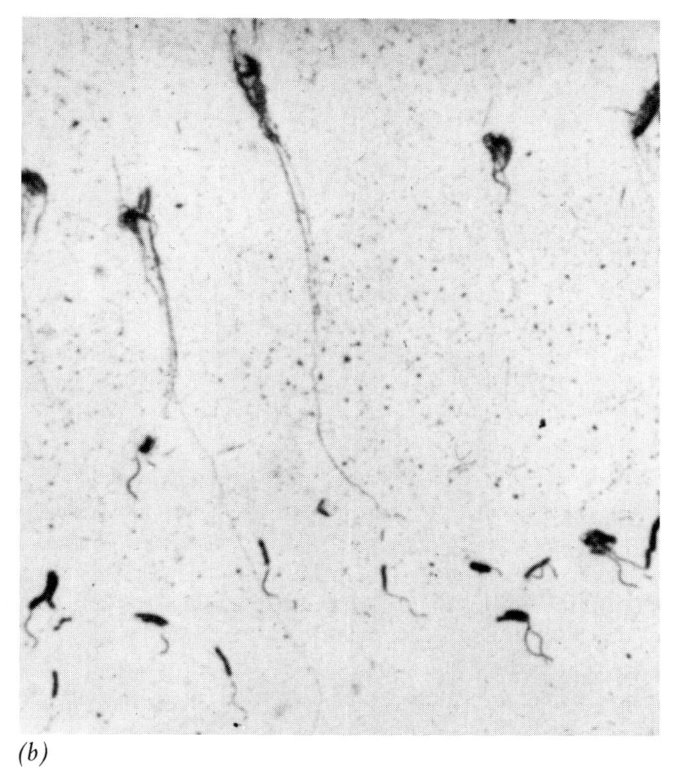

FIGURE 25.4 *(a)* Action of the cholera toxin on the intestinal mucosa. Increased concentrations of cAMP in the intestinal mucosa cause an outpouring of fluid into the intestinal lumen. *(b)* A flagella stain of *V. cholerae* demonstrating polar flagellation. (Courtesy of Centers for Disease Control.)

DISEASES CAUSED BY *ESCHERICHIA*

Coliforms, the lactose-fermenting, gram-negative bacteria that inhabit the intestinal tract, include members of the genera *Escherichia, Klebsiella, Enterobacter,* and *Citrobacter.* These bacteria are classified in the *Enterobacteriaceae* family because they are rod-shaped, gram-negative, oxidase negative, facultative anaerobes.

Escherichia coli is a normal inhabitant of the terminal ileum and large intestine. "Normal" strains of *E. coli* are opportunistic pathogens since they have the potential to cause disease in compromised patients and are a significant contributor to urinary tract infections, as discussed in Chapter 24. Other strains carry virulence plasmids that code for toxins and pili and are outright pathogens. The *ent* plasmids acquired by the pathogenic strains code for one or more enterotoxins. Other plasmids code for the pili that enable *E. coli* to colonize the intestinal mucosa.

Enterotoxic *E. coli*

Plasmid-carrying strains of *E. coli* that produce one or more enterotoxins and possess pili are designated enterotoxic *E. coli* (ETEC). ETEC colonizes the proximal region of the small intestine, where it produces the protein enterotoxins responsible for the symptoms of infantile diarrheal disease; in adults the disorder is called traveler's diarrhea.

ETEC produce a **heat-labile toxin** (LT), which has the same mechanism of action as the cholera toxin. This large molecular weight protein specifically binds to receptors on the surface of epithelial cells in the intestinal mucosa, and activates adenylate cyclase. The resulting intracellular increase in cAMP stimulates the intestinal mucosa to excrete chloride and bicarbonate ions, creating an electrolyte imbalance and an outpouring of water into the bowel—the immediate cause of diarrhea. Antibodies made against the cholera toxin inactivate both the LT of *E. coli* and cholera toxin, so these protein toxins are very similar in structure even though they are produced by different bacteria.

KEY POINT

Vibrio cholerae **and ETEC produce similar heat-labile enterotoxins, which increase the concentration of cAMP in the cells of the intestinal mucosa. A severe electrolyte imbalance results, with an outpouring of sodium, potassium, and water into the intestinal lumen. The body becomes depleted of essential fluids and electrolytes in the purging diarrhea that follows.**

ETEC can also produce a **heat-stable toxin** (ST), which is a low molecular weight (MW about 2000) polypeptide that activates guanylate cyclase. Guanylate cylase in turn increases the intracellular concentration of cyclic GMP (cGMP), which causes an outpouring of water and ions into the bowel.

Enteroinvasive *E. coli*

Enteroinvasive *E. coli* (EIEC) invades the epithelial cells of the large intestine, causing dysentery (see below). The symptoms include fever, cramps, and some diarrhea, but they are less severe than those experienced with ETEC infections. When EIEC invade the intestinal mucosa, they cause ulceration and inflammation, with release of blood and pus into the stool. Outbreaks of EIEC infections are sporadic and relatively uncommon.

Prevention and Treatment

Acute infantile diarrhea is a serious problem, because babies have little resistance to it and the pathogens can move swiftly through a hospital nursery. In developing countries, early death from acute infantile diarrhea is common. Infections are prevalent where sewage treatment facilities are inadequate or nonexistent.

Adults infected by ETEC suffer a debilitating disease known generally as travelers' diarrhea, and in Mexico as Montezuma's revenge. Travelers in developing countries are usually advised to drink beverages prepared with boiled water (tea and coffee); canned or bottled carbonated beverages, or beer or wine; to avoid using ice cubes made with contaminated water and glassware washed in contaminated water; and not to eat uncooked foods such as raw vegetables, salads, unpasteurized milk, and cheese.

Treatment of diarrheal disease is aimed at replenishing the lost fluids. Rehydration and establishment of normal electrolytic balance are especially important for affected infants. Recovery from this self-limiting disease follows the purging diarrhea, which removes the noxious substances from the bowel.

Most strains of *E. coli* are susceptible to a broad range of antibiotics including ampicillin, cephalosporins, tetracyclines, carbenicillin, and the sulfanilamide drugs. However, most *E. coli* intestinal infections are self-limiting and antibiotics are not required.

KEY POINT

An enterotoxic *E. coli* strain is created when a normal *E. coli* acquires one or more virulence plasmids that code for enterotoxins and pili. The disorder popularly called travelers' diarrhea results when ETEC colonizes the small intestine and produces enterotoxin(s).

SHIGELLA: SHIGELLOSIS

Shigella (shi-gel'ah) can infect the intestinal tract to cause a severe diarrheal and dysenteric syndrome known as *shigellosis* (shi-gel'lo'sis) or **bacillary dysentery.** These bacteria are prevalent throughout the world and are transmitted by ingestion of feces-contaminated water and foods. Although shigellosis is a serious, debilitating illness, it is self-limiting and most patients recover without treatment.

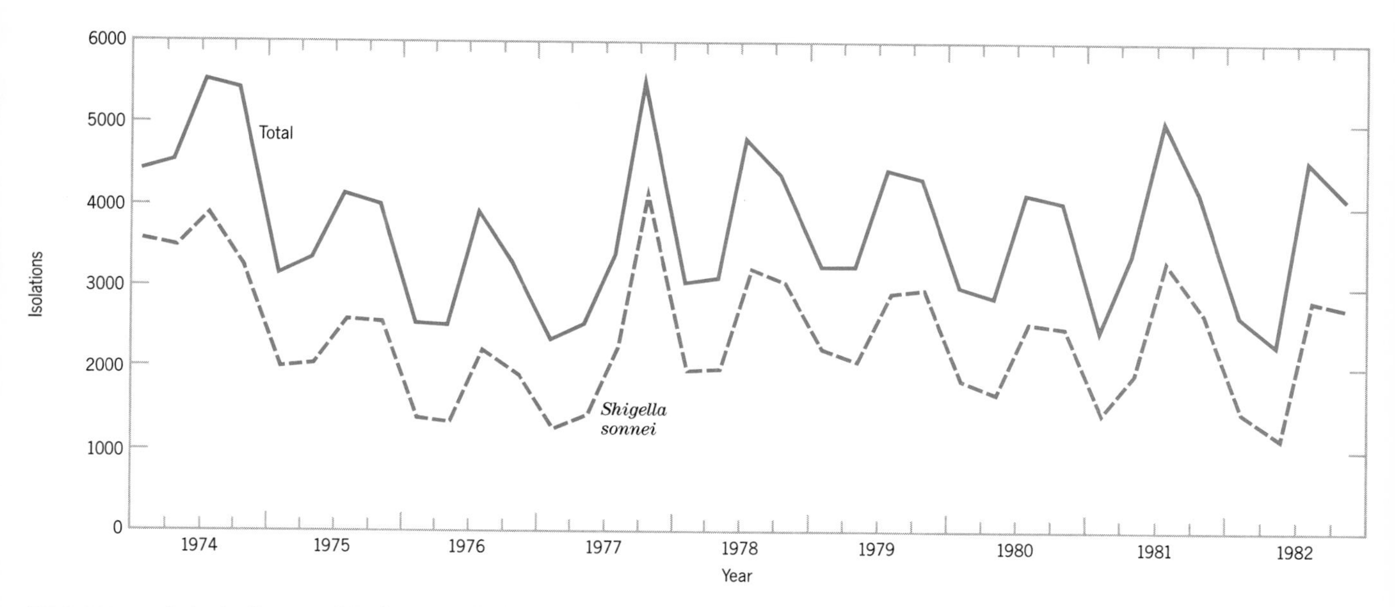

FIGURE 25.5 Reported isolations of *Shigella* dysentery in the United States between 1974 and 1982. Most cases were caused by *S. sonnei*. (From *Morbidity and Mortality Weekly Report,* Vol. 32, No. 34, September 1983.)

Characteristics of *Shigella*

Shigella is biochemically and antigenically similar to *Escherichia;* however, *Shigella* does not ferment lactose, produces no gas from glucose, and is not motile. The four species of *Shigella* are distinguished by biochemical and serological typing. *Shigella dysenteriae* (dys-en-te′ri-eye) causes a severe dysentery that is seen mostly in tropical countries. *Shigella sonnei* (son-′ne-i) causes about 70 percent of the cases of shigellosis in this country (Figure 25.5), and *S. flexneri* (flex′ner-i) is responsible for most of the remaining cases. *Shigella boydii* (boy′di-i) is the fourth recognized species. Infections are diagnosed by isolation of the causative bacterium from rectal swabs or stool specimens that have been plated on MacConkey or Hektoen agar plates.

Shigella Toxin and Virulence

Shigella produces a protein exotoxin having cytotoxic, neurotoxic, and enterotoxic activities. The **shigella toxin** inhibits protein synthesis in eucaryotic cells by inactivating the 60S ribosomal subunit. How this toxin selects its target cell, enters the cytoplasm, and inactivates the ribosome is not yet understood. *Shigella* invades the intestinal mucosa to cause local ulceration and inflammation. Invasiveness appears to be plasmid mediated, at least in *S. sonnei*. Like other gram-negative bacteria, *Shigella* possesses an endotoxin (see Chapter 17) that presumably contributes to the fever seen in about 50 percent of patients with shigellosis.

Symptoms of Shigellosis (Bacillary Dysentery)

Shigella infects the terminal ileum of the gastrointestinal tract, where it multiply rapidly. The initial symptoms of cramps and diarrhea begin as early as 12 hours after infection; the diarrhea is attributed to the bacterium's enterotoxicity. One to four days after infection, *Shigella* moves into the large intestine where it invades the cells of the intestinal wall. Here it causes acute inflammation with production of shallow ulcers in the mucosa. Blood and mucus are passed in the stool—an important sign of shigella dysentery.

Symptoms include the rapid onset of abdominal cramps, diarrhea, and fever. Each bout of diarrhea is brief, in contrast to the purging diarrhea of cholera. The symptoms advance from cramps to bloody stools over a period of 100 hours. The illness is self-limiting, and most patients recover without antibiotic therapy. In a severe case the patient may receive tetracyclines or trimethoprim-sulfamethoxazole; however, antimicrobial therapy is problematic since *Shigella* can acquire resistance plasmids from the *E. coli* coexisting in the intestinal tract.

KEY POINT

Many diarrheal diseases are self-limiting because the outpouring of fluid into the bowel flushes out the pathogen and its exotoxins. Replenishment of electrolytes and fluids is usually the only treatment that is needed.

Epidemiology

Shigellosis occurs most often in children under 10 years of age, with a peak incidence among 2-year-olds. The "four Fs" are recognized as the means of transmitting *Shigella:* "food, feces, fingers, and flies." Depending on the species of *Shigella,* infection can be established by as few as 180 viable bacteria, though it usually follows ingestion of some 5000 or more bacteria. Since human feces are the only important source of *Shigella,* prevention of shigellosis relies on proper hygiene and the adequacy of local sewage treatment.

TYPHOID FEVER, A SALMONELLA BACTEREMIA

Salmonella are pathogenic members of the *Enterobacteriaceae* family responsible for typhoid fever, salmonella septicemia, and food poisoning. Infections limited to the intestinal tract are called **salmonella gastroenteritis** or **salmonella food poisoning** (see below), whereas species able to penetrate the intestinal mucosa and enter the circulatory system cause **enteric fevers** or salmonella bacteremias. Typhoid fever is a classic example of an enteric fever.

Humans are the only source of *Salmonella typhi,* which causes typhoid fever when it is ingested in contaminated food and water. Transmission begins with carriers who contaminate food during its preparation and excrete *S. typhi* in feces and urine, which contaminates the drinking water. Huge numbers (10^5 to 10^7) of organisms must be consumed before the infection can gain a foothold, because few of the bacteria survive in the acid environment of the stomach. Those that do survive pass through the stomach to colonize the intestinal tract, penetrate the mucosa, and enter the blood.

The symptoms begin about 10 to 14 days after the infection is established. The patient appears acutely ill with a prolonged fever, slow pulse, malaise, headache, and loss of appetite. These symptoms are accompanied by abdominal pain and a "rose spots" rash on the trunk. Oddly enough, diarrhea is uncommon, even though *S. typhi* infects through the intestinal tract. The illness peaks about the third week (Figure 25.6), then subsides.

Antibodies begin to appear after the first week of infection. As the antibodies increase, it becomes diffi-

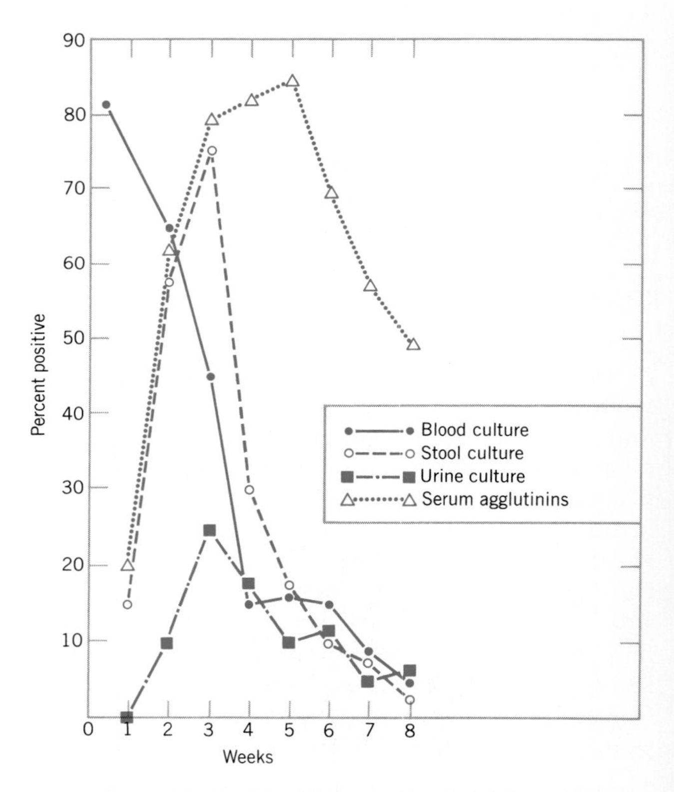

FIGURE 25.6 The kinetics of typhoid fever. The relative frequency of positive blood, urine, and stool cultures during the course of the disease. Notice that the peak concentration of serum agglutinins corresponds to the decrease in the percentage of positive cultures. (From H. R. Morgan, "The Enteric Bacteria," in R. J. Dubos and J. G. Hirsch (eds.), *Bacterial and Mycotic Infections of Man,* J. B. Lippincott, Philadelphia, 1965. Copyright © 1965 by J. B. Lippincott Company)

cult to isolate *S. typhi* from blood; it can be isolated, however, from the stool after the first week, and later from the urine. In carriers, the bacteria multiply in the biliary tract whence they are released into the intestine to contaminate the feces. Typhoid fever is an acute illness that should be treated with chloramphenicol or ampicillin.

Typhoid fever is rare in this country (about 500 cases annually), but it is frequent in underdeveloped countries where the climate is warm and the sewage treatment facilities are inadequate. Typhoid fever can be brought under control by proper sewage disposal; pasteurization of milk; maintenance of unpolluted water supplies; identification and treatment of infected persons; and monitoring of foodhandlers. This is a very large and costly undertaking in poor countries.

THE OBLIGATE ANAEROBES

The anaerobic environment of the lower intestinal tract is selective for obligate anaerobic bacteria, which comprise 95 to 99 percent of the microbial population in the large intestine.

Bacteroides and *Fusobacterium*

Species of *Bacteroides* and *Fusobacterium* (fu'so-bak-te're-um) are strictly anaerobic, gram-negative rods that are found in feces and to a lesser extent in the vagina, external genitalia, and mouth. *Bacteroides fragilis*, *B. distasonis*, and *B. vulgatus* are present in the lower intestinal tract. *Bacteroides fragilis* (Figure 25.7), the most common anaerobe associated with soft-tissue infections, has been isolated from patients with appendicitis, peritonitis, heart valve infections, rectal abscesses, and surgical wounds. This anaerobe is often resistant to penicillin, but is usually sensitive to clindamycin and selected cephalosporins.

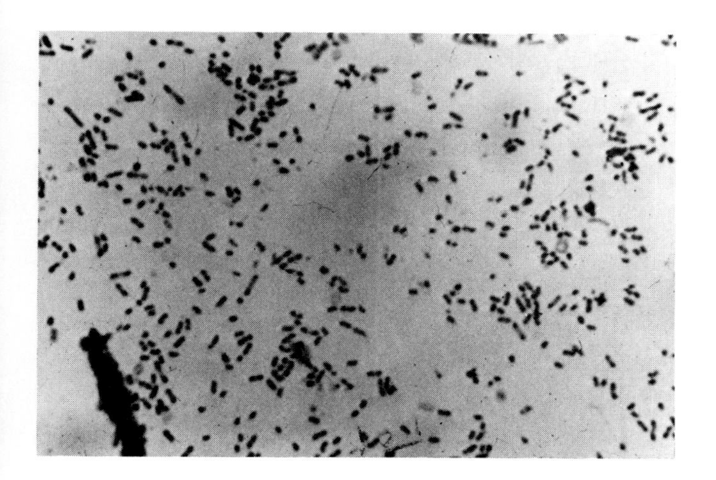

FIGURE 25.7 *Bacteroides fragilis* in a blood smear. (Courtesy of Dr. L. LeBeau.)

Fusobacterium spp. are isolated from the oral cavity and the upper respiratory tract of humans and warm-blooded animals. The pathogenic species cause soft-tissue infections and are secondary invaders of gangrenous tissue.

BACTERIAL FOOD POISONING: INTOXICATIONS

Food may serve as a vehicle for the transmission of infectious agents or toxins, as was indicated earlier. Food poisoning is the general name given to a group of acute illnesses caused by the ingestion of contaminated food. It is characterized by the occurrence of vomiting, diarrhea, gastroenteritis, and prostration. Infectious bacterial food poisoning develops when ingested bacteria grow in the intestinal tract and cause disease (Table 25.1).

The discussion that follows is limited to bacterial toxins involved in foodborne intoxications.

STAPHYLOCOCCAL FOOD POISONING

Many picnics and church suppers have come to a disastrous end when home-prepared dishes served not only as food for the guests, but also as the vehicles for transmitting staphylococcal food poisoning. The sequence of events leading to this disease often begin when food becomes contaminated by *Staphylococcus aureus* present on the preparer's hands. The bacteria have ample opportunity to multiply and produce their enterotoxin when the food is stored at too high a holding temperature; a picnic table bathed in sunlight, a banquet table at a community supper, and an uncooked, stuffed turkey left unrefrigerated overnight are excellent examples. Common sources of this food poisoning include beef, ham, chicken, turkey, and baked goods. The enterotoxin causes nausea, vomiting, and diarrhea within

TABLE 25.1 Characteristics of some types of food poisoning

ETIOLOGICAL PATHOGEN	PERCENT TOTAL OUTBREAKS[a]	INCUBATION TIME (usual)	SYMPTOMS	COMMON FOOD SOURCES
		Intoxications		
Staphylococcus aureus	12.7	2–4 hr	Gastrointestinal syndrome	Meats, poultry, baked goods
Clostridium botulinum	9.5	12–48 hr	Neuromuscular paralysis	Meat, fish, fruit, vegetables
Bacillus cereus (vomiting toxin)	3.6	1–6 hr	Vomiting, some diarrhea	Rice dishes
(diarrheal toxin)		6–24 hr	Diarrahea	Meats, sauces
		Infections		
Salmonella spp.	25.0	6–48 hr	Gastrointestinal syndrome	Milk, beef, chicken, turkey
Shigella spp.	1.8	15–20 hr	Gastrointestinal syndrome	Chicken, potato salad
Campylobacter jejuni	0.9	4–7 days	Gastrointestinal syndrome	Milk
Clostridium perfringens	10.0	9–15 hr	Diarrhea	Beef, pork, chicken, turkey
Vibrio parahaemolyticus	1.4	4–30 hr	Gastrointestinal syndrome	Shellfish, fruits, vegetables
Hepatitis A virus	8.5	10–45 days	Hepatitis	Multiple foods

Source: Centers for Disease Control, *CDC Surveillance Summaries,* 1986:35(No. 1SS).
[a]Percentages are based on 220 outbreaks recorded in 1982.

2 to 4 hours after being ingested. Recovery follows within 24 to 48 hours.

About 50 percent of *S. aureus* strains produce enterotoxins capable of causing food poisoning. The enterotoxins are a mixed group of single-chain polypeptides (MW 28,000–35,000), relatively stable to heat; thus heating contaminated food, even to the boiling point, does not necessarily inactivate the toxin.

BOTULISM FOOD POISONING

Botulism is a critical intoxication caused by ingestion of the botulism toxin in food, usually canned fruits, vegetables, fish, and meats. The toxin is one of the most poisonous substances known and quickly kills its host through its action on the nervous system. The toxin is produced by *Clostridium botulinum,* an anaerobic, spore-forming bacterium present in soil, mud, and the intestinal tract of many animals. It produces toxin as it grows in an anaerobic environment such as mud at the bottom of a pond or food improperly canned (generally home canned).

Botulism Toxin

There are at least seven types of botulism toxin produced by different strains of *C. botulinum* (Figure 25.8). All carry temperate phage, but to date only two botulism toxins have been shown conclusively to be produced by lysogenized strains.

The botulism neurotoxins possess extraordinary toxicity; they cause muscle paralysis by preventing the release of acetylcholine (as′e-til-kol′len) from the synapses of the peripheral nervous system. Less than 1 μg of toxin is estimated to be a lethal dose for a human. These exotoxins are large (MW 150,000), heat-labile proteins that are not inactivated by the acidity and degradative enzymes of the stomach. They are readily absorbed into the circulatory system from the intestine.

Symptoms

Botulism food poisoning is an intoxication that occurs sporadically in the United States with about 50 cases each year. Over 90 percent of confirmed cases originate

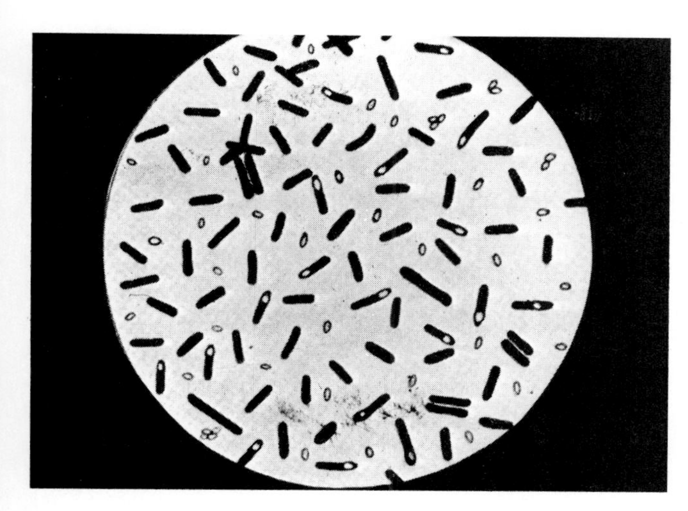

FIGURE 25.8 Photomicrograph of *Clostridium botulinum* stained with gentian violet. (Courtesy of the Centers for Disease Control.)

in home-processed fruits, vegetables, fish, and other foods. If the preparation of the food has been inadequate for its purpose, clostridial spores will survive the canning process, germinate in the anaerobic environment of the canned food, and then produce the toxin. Outward signs of contamination include the usual swelling of the can due to the gas produced during fermentation, and the foul odor of putrefaction. "When in doubt throw it out" is a safe rule to follow whenever the presence of botulism toxin is suspected.

Symptoms usually appear within 12 to 48 hours after contaminated food has been consumed; a stiff neck, nausea, vomiting, double vision, difficulty in swallowing, and muscle paralysis are characteristic. Most fatalities are caused by suffocation due to paralysis of the respiratory muscles. The prognosis for survival is good if the disease is recognized early and the victim is treated with a polyvalent antitoxin that neutralizes the toxins.

KEY POINT

Most cases of botulism food poisoning develop following ingestion of cold or warmed-over foods. The toxin is heat-labile and is destroyed by heating the food.

INFECTIOUS BACTERIAL FOOD POISONING

INFANT BOTULISM

Infant botulism affects babies between 3 and 26 weeks old. The symptoms include unexplained weakness, difficulty in swallowing, respiratory arrest, and paralysis of the ocular muscles. Until as recently as 1976, *C. botulinum* was not suspected since it wasn't known to cause infectious disease. In that year, however, researchers found *C. botulinum* in the feces of an infant experiencing the symptoms described previously, and identified botulism toxins as the cause. The disease is a toxico-infection, in which toxin-producing *C. botulinum* grows and produces its toxin in the infant's intestine. More than 60 cases of **infant botulism** were reported in the United States during each of the four years beginning in 1980.

No common source of infection has been found; however, raw honey contains spores of *C. botulinum* and has been implicated as a factor in a number of cases. On the other hand, some of the infected infants have been breast fed and have received no solid foods. Affected infants probably ingest spores of *C. botulinum,* which germinate in the intestine, multiply, and produce the toxin, which then enters the blood. Recovery without the administration of antitoxin is the rule, though some infants require assisted breathing. The infant may excrete *C. botulinum* for several weeks following recovery.

SALMONELLA FOOD POISONING

Salmonellas (sal'mo-nel'ah) are the most frequent cause of foodborne disease in the United States. Raw foods of animal origin, especially fresh pork and poultry, are the most likely carriers. Common slaughterhouse practices encourage the spread of salmonellas from one animal carcass to another. Other vehicles are shellfish, milk and milk products, and salads. Many outbreaks of salmonella food poisoning originate in foods cooked and eaten at home; others originate in foods prepared in restaurants, schools, and other public facilities.

In 1986, a severe outbreak of salmonellosis in Illinois was traced to a well-run, carefully regulated dairy. Although the implicated milk had been pasteurized, either the pasteurization process was inadequate or the milk was contaminated after pasteurization. This allowed the *Salmonella* to proliferate in the milk. Several thousand people were sickened by drinking the infected milk.

Large numbers of *Salmonella* must be ingested before the infection becomes established. The symptoms begin 6 to 48 hours after ingesting contaminated food, with the sudden onset of headache, chills, and abdominal pain, followed by nausea, vomiting, and diarrhea. The salmonellas are excreted in the feces and rarely enter the circulatory system. Patient recovery is usual and antibiotics are not recommended, except for the very young and for persons over 60.

Classification of *Salmonella*

Classification of *Salmonella* is confusing and controversial since virtually hundreds of "species" have been described. Even though experts agree that only a few true species of *Salmonella* should be recognized, no acceptable classification scheme is available for simplifying the systematics.

Serological typing using agglutination reactions (see Chapter 18) is an effective tool for identifying the over 1400 serotypes of *Salmonella*. The Kaufmann–White scheme systematically assigns numbers and letters to the different O, H, and Vi (virulence) antigens (Figure 25.9) to create distinctive serotypes. Epidemiologists who need to find the source(s) of a foodborne and waterborne salmonella outbreak need to know which serotypes are causing the epidemic. *Salmonella typhimurium* (ty-phi-mu′ri-um) and *Salmonella enteritidis* (en-te-ri′ti-dis) are among the species most often isolated in foodborne outbreaks. Once the source is found, steps can be taken to stop its spread.

Humans can also acquire *Salmonella* directly from contaminated animals. The small green turtles once sold in pet stores (now illegal in many states) are known carriers of *Salmonella*. Children would play with the turtle in the turtle bowl, contaminate their fingers, and then ingest the *Salmonella* after putting their fingers in their mouths.

CAMPYLOBACTER FOOD POISONING

Campylobacter (kam′pi-lo-bak′ter) is a slender, tightly coiled, spiral bacterium that commonly inhabits the reproductive and intestinal tracts of animals, birds, and humans. These bacteria probably caused human infections for years before the late 1970s, when microbiologists developed techniques for selectively isolating *Campylobacter*. *Campylobacter* is now recognized as an important cause of diarrheal disease in humans, especially among teenagers.

Campylobacter jejuni, the species most often isolated from infected humans, is carried by cattle, sheep, chickens, turkeys, ducks, and seagulls. Humans are usually infected when they consume contaminated meat or raw (unpasteurized) milk. The most common symptom of a *C. jejuni* infection is an acute enteritis that lasts one day to more than one week. The principal symptoms are fever, malaise, abdominal pain, and a diarrhea that can vary from loose stools to massive watery stools containing blood. Nausea and vomiting may occur. *Campylobacter* enteritis is frequently a self-limiting illness that lasts three to five days, although the symptoms may persist for one to two weeks.

CLOSTRIDIUM PERFRINGENS FOOD POISONING

Inadequately cooked meats, gravy, and poultry are the major sources of clostridial food poisoning, which is caused by enterotoxin-producing strains of *Clostridium perfringens*. The bacterium colonizes the small intestine and liberates its enterotoxin when it has been ingested in large numbers. Nine to 15 hours after infection, the person experiences abdominal cramps and diarrhea. Recovery is rapid after a bout of gastroenteritis.

VIBRIO PARAHAEMOLYTICUS FOOD POISONING

Vibrio parahaemolyticus is a halophilic bacterium that causes an acute gastroenteritis always associated with consumption of seafoods. This type of food poisoning is most prevalent in countries where the custom of eating raw fish or shellfish is followed. *Vibrio parahaemolyticus* poisoning was first described in Japan, where the practice of eating raw fish is common. A self-limiting gastroenteritis is usual, with nausea, vomiting, abdominal cramps, diarrhea, chills, and a low-grade fever.

BACILLUS CEREUS FOOD POISONING

Bacillus cereus, an aerobic sporeformer, causes two types of food poisoning, a diarrheal type and an intestinal type. The diarrheal symptoms are usually associated with consumption of contaminated meats and

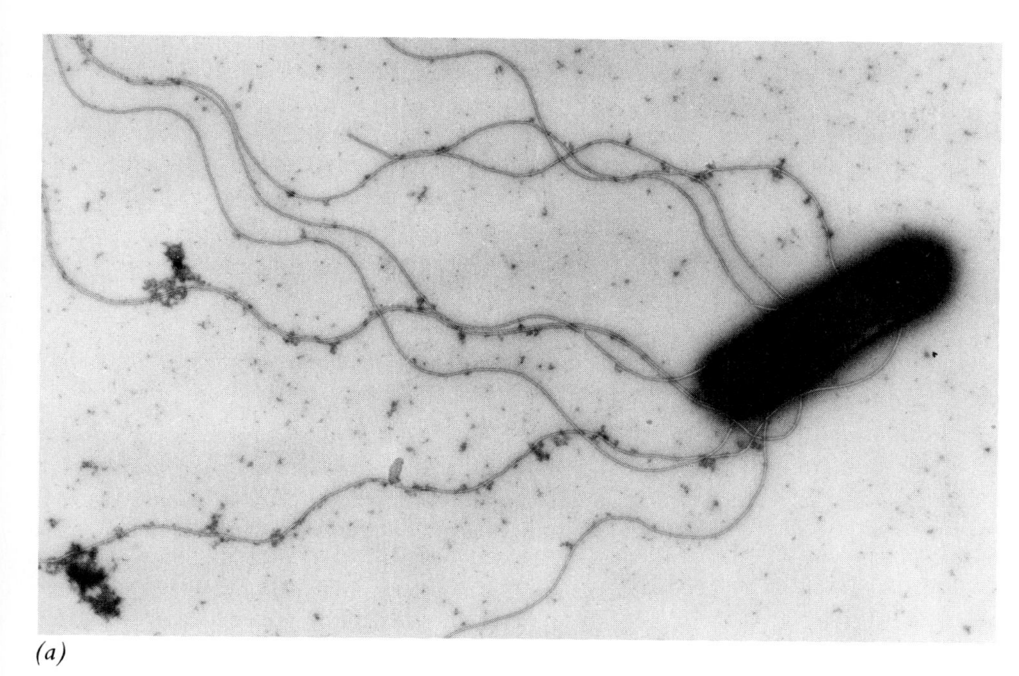

(a)

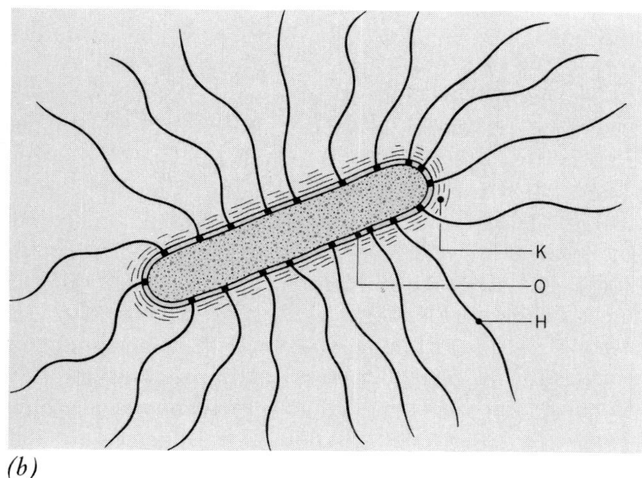

(b)

FIGURE 25.9
(a) This electron micrograph of *Salmonella* shows its long peritrichous flagella. *(b)* It's antigenic structure includes the capsular antigens K, somatic antigens O, and flagella antigens H. (Photograph courtesy of Dr. W. L. Dentler/ BPS.)

sauces, and commence 6 to 24 hours after their ingestion (Table 25.1). Vomiting is usually associated with fried rice or rice dishes, and develops within 1 to 6 hours after these foods have been eaten.

LISTERIOSIS

Listeria monocytogenes (lis-te′re-ah mo-no-si-to′je-nez) is widely dispersed in soil, animals, water, and sewage. People are infected upon direct contact with infected animals or ingestion of contaminated meat, milk, or cheese. Milk products can be contaminated by cows with *Listeria* mastitis. The symptoms of human *Listeria* infections range from mild to severe and can include septicemia, meningoencephalitis, and focal infections. Special consideration should be given to infected pregnant women because *L. monocytogenes* can cross the placenta to infect the fetus, resulting in abortion, stillbirth, or premature birth. Infections by *L. monocytogenes* can be treated with ampicillin either with or without gentamicin.

VIRAL GASTROENTERITIS

The rotavirus and the Norwalk viruses are important causes of viral gastroenteritis. The Norwalk agents are responsible for outbreaks of gastroenteritis among socially interactive groups of people. The rotaviruses cause infections of the small intestine that vary in intensity from a mild gastroenteritis to an acute diarrheal disease (in children).

ROTAVIRUSES: ACUTE INFANTILE DIARRHEAL DISEASE

Rotaviruses are a major cause of infant diarrheal disease and probably of infant mortality in many parts of the world. These viruses are carried by a variety of animals

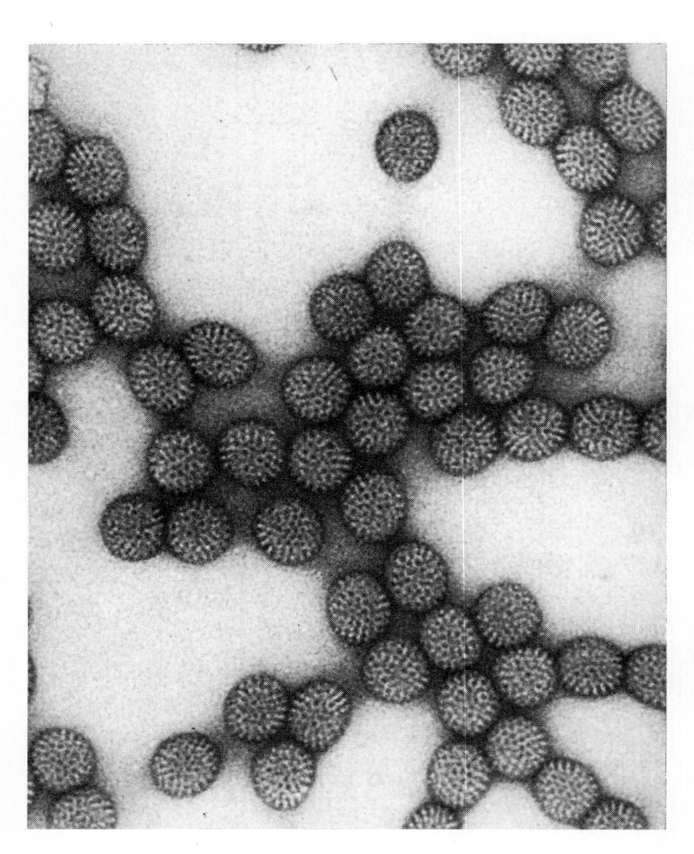

FIGURE 25.10 Electron micrograph of a rotavirus isolated from a human with acute infantile gastroenteritis. (Courtesy of the Centers for Disease Control.)

including calves, mice, piglets, foals, lambs, rabbits, and antelopes. At least four serotypes of *Rotavirus* cause human illness. Rotaviruses isolated from human feces contain 11 segments of double-stranded RNA and are surrounded by a double-shelled, wheel-like capsid measuring 70 nm in diameter (Figure 25.10). The unique organization of these viruses justifies their classification into a separate genus.

The symptoms of rotaviral infections vary from a mild diarrhea to a severe and sometimes fatal gastroenteritis. Babies between 6 and 24 months of age are most likely to experience the severe form of the disease, which is called **acute infantile diarrheal disease.** The illness begins with vomiting followed by diarrhea that lasts about five days. Death can result from the severe dehydration and electrolyte imbalance that occurs. Rehydration is the method of treatment.

Rotaviral infections are transmitted by the oral–fecal route. These infections occur primarily in the winter and early spring in temperate climes and throughout the year in the tropics. Almost all American children possess antibodies against rotaviruses before they are 3 years old. Nevertheless, the widespread infection rate in this country suggests that adequate sewage facilities alone are not sufficient to prevent the spread of the viruses.

NORWALK AND RELATED AGENTS OF GASTROENTERITIS

Epidemic diarrhea, viral gastroenteritis, winter vomiting disease, and acute nonbacterial gastroenteritis are names assigned to the diseases caused by the Norwalk and related viruses. Norwalk agents were first observed in stool samples taken from patients involved in a gastroenteritis epidemic in Norwalk, Ohio. Electron micrographs of the samples revealed a viral particle having a diameter of between 27 and 32 nm. Since that episode at least six similar agents have been identified, though the agents have yet to be grown in tissue cultures.

The Norwalk virus causes an epidemic form of gastroenteritis that begins with abdominal cramps and nausea after an incubation period of 24 to 72 hours. The symptoms, including vomiting, diarrhea, and a low-grade fever, last from 48 to 72 hours, after which the patient recovers without experiencing sequelae. These illnesses predominate during the colder months, have a short incubation period, and occur as outbreaks among groups of people who have close personal contact, as in schools and recreational camps and on cruise ships. Researchers presume that these agents are transmitted by the fecal–oral route. The illnesses are self-limiting and patients recover without treatment. A greater understanding of these forms of viral gastroenteritis and their agents is necessary before action to prevent them can be taken.

VIRAL HEPATITIS

Epidemics of **infectious hepatitis** (hep′ah-ti′tis) have been reported among military and civilian populations since the Middle Ages. Hepatitis is characterized by

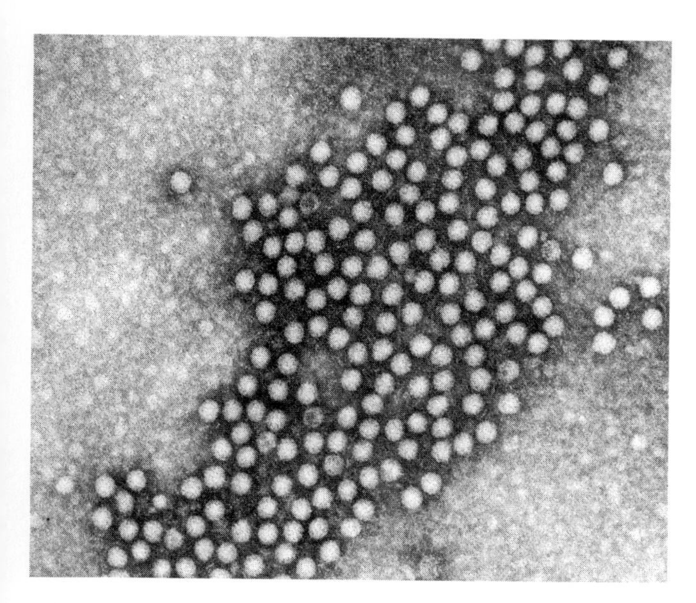

FIGURE 25.11 Heptatitis A virus particles (24 to 29 nm in diameter) isolated from feces of infectious hepatitis patients. (Courtesy of Dr. A. J. Zuckerman. From A. J. Zuckerman and C. R. Howard, *Hepatitis Viruses of Man*, Academic Press, London, 1979. Copyright © 1979, Academic Press Inc., Ltd, London.)

liver damage and the accompanying jaundice (icterus). The viral cause of hepatitis was first recognized in the mid-1940s, and we now know of at least three types of viral hepatitis, each being recognized by differences in epidemiology, symptoms, and the immunological properties of the etiological agents. The following discussion is confined to infectious hepatitis. Other forms of hepatitis, transmitted by blood and blood products, are discussed in Chapter 26.

HEPATITIS A: INFECTIOUS HEPATITIS

The infectious hepatitis seen in humans, monkeys, and chimpanzees is a viral disease transmitted by the fecal–oral route and by contaminated food. Recovery is complete in most cases and results in lasting immunity.

Hepatitis A Virus

Infectious hepatitis is caused by a spherical, naked RNA virus, **hepatitis A virus** (HAV) (Figure 25.11), which initially was identified by its hepatitis A antigen (HA). Similar particles have been localized in the liver cells of infected animals by an immunofluorescent technique capable of detecting the HA antigen. More recently the hepatitis A virus has been grown in tissue cultures, which is the first step in understanding the virus and may lead to the production of a vaccine.

Symptoms

Many persons infected by HAV have a mild (subclinical) response, while others experience the full range of symptoms following an incubation period of three to six weeks. The chronological development of hepatitis is depicted in Figure 25.12. Symptoms include malaise and weakness, intermittent nausea and vomiting, and a dull right upper quadrant pain. These are followed by the onset of jaundice and/or dark urine, which usually brings the patient to the doctor. The main signs of hepatitis are hepatomegaly (enlargement of the liver), and elevated serum aminotransferase levels resulting from liver damage. These symptoms last for 8 to 13 days in children; in adults the jaundice may last for 30 days. The virus, which reproduces in the liver, enters the feces by way of the bile duct and is transmitted by the fecal–oral route. Patients recover and have lasting immunity against future hepatitis A infections.

Diagnosis and Prevention

The presence of anti-HAV IgM in the serum at the time the clinical symptoms appear is diagnostic. Prevention focuses on interrupting the fecal–oral transmission of the virus and passively immunizing susceptible persons. Foodborne outbreaks originate in family settings or from contaminated food served in restaurants or institutions. These outbreaks can be curtailed by identifying the source of the virus—usually a food handler.

Passive immunization of exposed persons significantly reduces the incidence of infectious hepatitis, due largely to the long incubation period of the virus.

KEY POINT

The fecal–oral route is the most common means by which gastrointestinal pathogens are spread. Handwashing is often sufficient to break the chain of transmission.

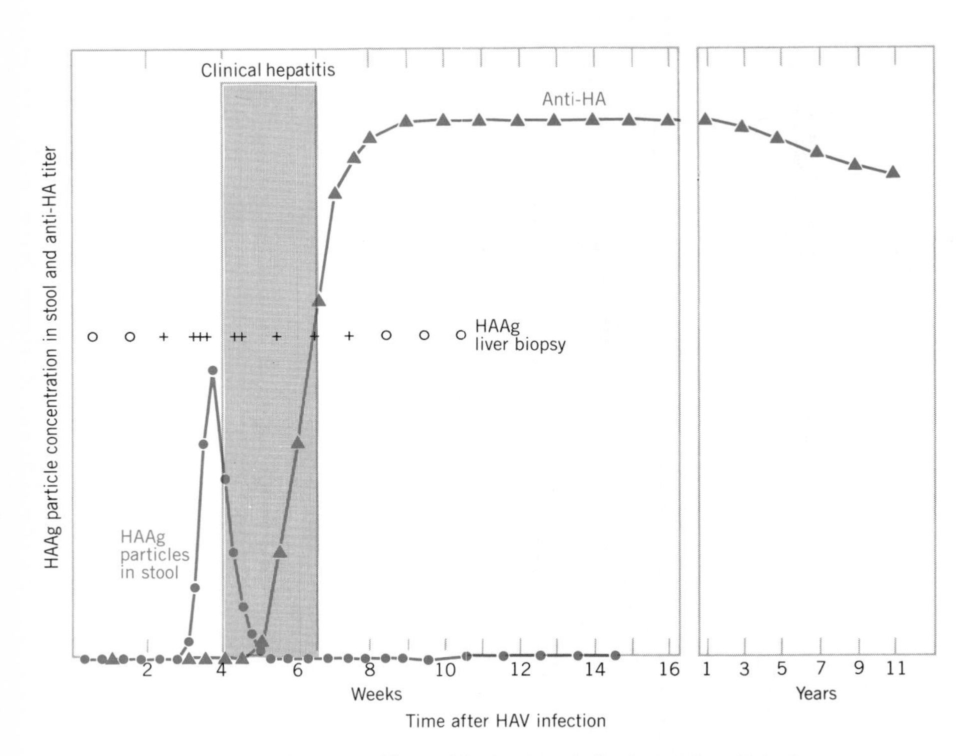

FIGURE 25.12 Course of hepatitis A virus infection. The clinical symptoms appear after a three- to six-week incubation period. Hepatitis A virus particles are excreted in feces and are present in the liver for one week before and one week after the clinical symptoms appear. (Courtesy of Dr. W. S. Robinson. From W. S. Robinson "Hepatitis A Virus," in G. L. Mandell, R. G. Douglas, Jr., and J. E. Bennett (eds.), *Principles and Practice of Infectious Diseases,* Wiley, New York, 1979. Copyright © 1979, by John Wiley & Sons, Inc.)

PROTOZOAN DISEASES

A few protozoa infect the gastrointestinal system during part of their life cycle. There they multiply until they are forced to form cysts, which are passed in the feces and survive outside the host until they are ingested by another person. Cysts survive passage through the upper digestive tract and then excyst into vegetative forms that multiply in the intestine. Amebas, flagellates, and one ciliate cause gastrointestinal infections.

AMEBIC DYSENTERY

Amebas are widely dispersed in soil, in water, and on plants, and some species are parasitic. The major ameba that is pathogenic to humans is *Entamoeba histolytica* (en'tah-me'bah his'to-lyt'i-cah) (Figure 25.13). It causes **amebic dysentery** or **amebiasis** (am'e-bi-o'sis), and is identified by its large size (up to 50 μ) and its granular cytoplasm.

Amebiasis: *Entamoeba histolytica*

Entamoeba histolytica grows as a trophozoite (trof'o-zo'it) in the large intestine where it multiplies by binary fission. The feces of an infected person contain trophozoites and cysts. The cysts are the only form of the ameba to survive outside the host, and they remain viable for up to 30 days in water and for days or months in moist soil. After being ingested, they pass unchanged through the stomach to the lower ileum, where their nuclei divide and produce eight tropho-

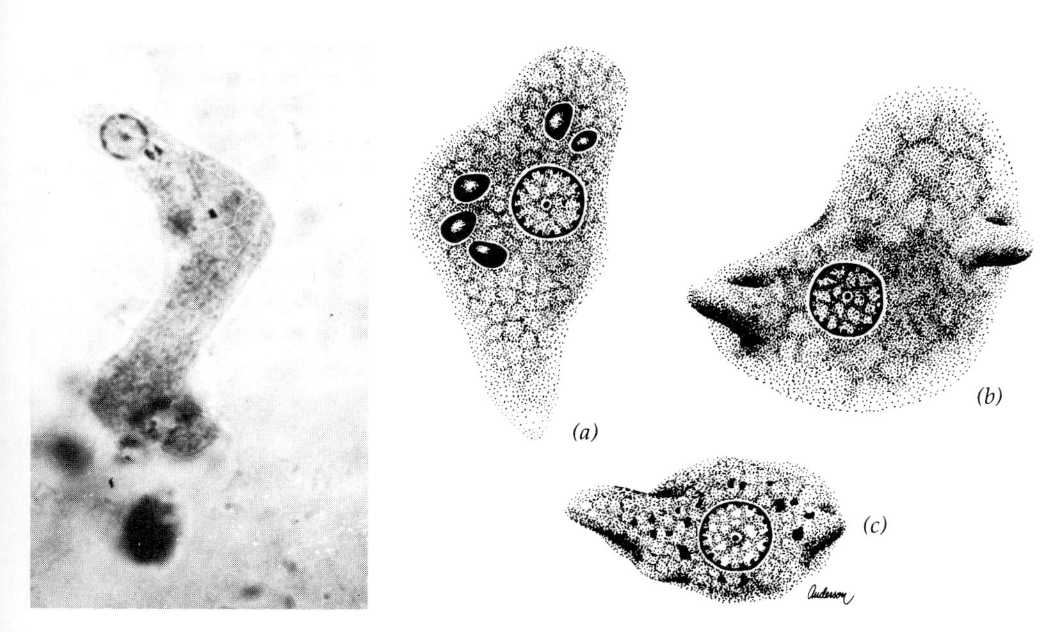

FIGURE 25.13 Micrograph and drawing of *Entamoeba histolytica* trophozoites: *(a)* ingested erythrocytes, *(b)* pseudopodia, and *(c)* ingested bacteria. (Reproduced with permission. From J. W. Beck and J. E. Davis, *Medical Parasitology,* 3rd ed., C. V. Mosby Company, St. Louis.)

zoites. These emerge through an opening in the cyst wall and invade the colon.

Although most *E. histolytica* infections are asymptomatic, some result in serious dysenteric disease, sometimes death. Those strains of *E. histolytica* that are able to invade the intestinal mucosa and produce lesions cause disease. There is abdominal pain and the production of loose stools that may contain mucus and blood. Infections confined to the intestine are not life-threatening; however, the amebas can penetrate the colon to cause **extraintestinal amebiasis,** which has life-threatening consequences.

Amebiasis is diagnosed by identifying *E. histolytica* on the basis of the size, number of nuclei in the cysts, and their motility as observed in specimens. *Entamoeba histolytica* is morphologically very similar to common nonpathogenic amebas, so experience in making the diagnosis is very important. Iodoquinol and paromomycin are drugs for treating intestinal amebiasis.

Epidemiology

The incidence of amebic dysentery is highest in the tropics where the cysts are more likely to survive. In some tropical countries upward of 40 percent of the population are infected, while in the United States only some 3 percent or so are infected. Local sanitation and personal hygiene within a community are directly related to the prevalence of amebiasis. The amebas are passed from asymptomatic carriers to others by the fecal–oral route, often in contaminated water or raw vegetables. To prevent amebiasis in a community at risk, drinking water should be boiled for 10 to 15 minutes and vegetables treated with a strong detergent, then soaked in a vinegar solution.

BALANTIDIASIS

Balantidium coli (bal'an-tid'e-um ko'li), the only ciliate known to infect humans, is a common ciliate of the intestinal tract of swine (Figure 25.14). Humans are infected by ingesting the cysts, which then pass unchanged through the upper digestive tract and excyst in the intestine. Once established in the intestine, the ciliates cause **balantidiasis** (bal-an'ti-di'ah-sis). Dysentery and, in some cases, appendicitis follow. The dysentery usually lasts one to four weeks, but may recur. The causative agent is identified in fecal specimens

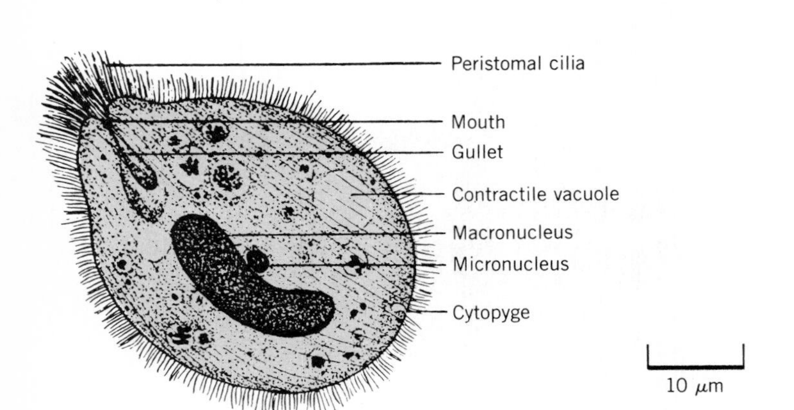

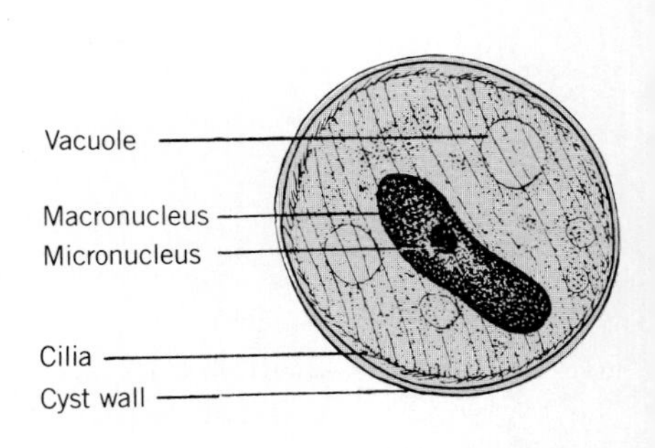

FIGURE 25.14 The trophozoite and the cyst of the ciliate *Balantidium coli*. (Reproduced with permission. From E. R. Noble and G. A. Noble, *Parasitology: The Biology of Animal Parasites,* 4th ed., Lea and Febiger, Philadelphia, 1976.)

or in scrapings from ulcers in the wall of the colon. Tetracyclines and iodoquinol are prescribed.

GIARDIASIS

Giardia lamblia (je-ar-di'ah lam'blee-ah), the most common human intestinal flagellate, is recognized as a major cause of waterborne infections in the United States (see Table 20.2). *Giardia* is a pear-shaped flagellate possessing four pairs of flagella arising from its four surfaces (Figure 25.15). It attaches to the host's intestinal mucosa by means of the flattened adhesive disk on its ventral surface. *Giardia* divides by longitudinal binary fission in the host's intestinal tract and forms cysts that survive in water to become the infective stage of the life cycle.

The trophozoites attach to the cells lining the duodenum, where the mucus secretions provide sustenance for their multiplication. After an incubation period of approximately 15 days, the patient experiences the sudden onset of watery, foul-smelling diarrhea, abdominal distension, flatulence (gas), nausea, and anorexia that lasts three to four days. The presence of trophozoites or cysts in the stools is diagnostic of giardiasis. Treatment is with quinacrine hydrochloride or metronidazole.

The extent of *G. lamblia* infections in human populations runs from a low of 2 to 5 percent to a high of 60 percent or more, depending on the geographic region. Most infections are asymptomatic; however, outbreaks of giardiasis do occur, and in recent years the incidence of giardiasis has increased significantly in the United States. Epidemics have been reported in New York, New Hampshire, and some western states. Hik-

FIGURE 25.15 *Giardia* trophozoites adhering to the microvillous surface of the mouse jejunum. (Courtesy of Dr. R. L. Owen. From R. R. Owen, P. C. Nemanic, and D. P. Stevens, *Gastroenterology,* 76:757–769, 1979.)

ers camping in remote areas of the Rocky Mountains have also contracted giardiasis, presumably because beavers and other wild animals are reservoirs for this flagellate.

KEY POINT

Protozoan diseases of the intestinal tract share certain commonalities: formation of cysts; presence in water, food, or other contaminated vehicles; ingestion of the cysts by the host; excystment into the intestine; and production of dysenteric disease.

HELMINTHIC DISEASES

Food contaminated by the eggs or larvae of parasitic worms, (helminths), causes infection after being consumed. Most of these pathogens are roundworms or flatworms (see Table 25.2; and Chapter 16). Meat inspection in itself is not sufficient to prevent these infections, because raw or undercooked beef, pork, and fish are effective vehicles for the transmission of larvae to the human host.

TAPEWORM INFECTIONS

Tapeworms attach to the intestinal mucosa by means of the scolex, or head, and grow by adding segments, called proglottids (Figure 25.16). A tapeworm can grow to a great length and may contain several thousand proglottids. The presence of proglottids in a person's perianal region or in the feces is a positive indication of tapeworm infection. Tapeworms have a distinctive life cycle that involves an intermediate host or hosts. Undercooked or raw infected meat serves as the transmission vehicle by which a person becomes infected, and thus becomes the definitive host.

Fish Tapeworm: Diphyllobothriasis

Diphyllobothrium latum (di-fil'o-both're-um lat'um), can attain a length of 3 to 10 m, with 1000 to 3000 proglottids when it grows in the definitive host, which may be humans, dogs, cats, and less frequently other mammals. Eggs produced in the worm's bisexual proglottids are passed into fresh water in feces. Once in fresh water, the eggs hatch into free-swimming larvae, which are eaten by tiny marine animals that in turn are consumed by freshwater fish. The larvae establish themselves in the muscle of the fish and are transmitted to humans who eat the infected fish. Fish

TABLE 25.2 Helminthic infections

HELMINTH (common name)	DISEASE	COMMON SOURCES	DIAGNOSIS/TREATMENT
		Platyhelminthes (flatworms)	
Diphyllobothrium latum	Fish tapeworm	Freshwater fish	Proglottids in feces or perianum/ niclosamide, paromomycin
Taenia saginata (beef tapeworm)	Taeniasis	Undercooked or raw beef	Proglottids in feces or perianum/ niclosamide, paromomycin
Taenia solium (pork tapeworm)	Taeniasis	Undercooked or raw pork	Proglottids in feces or perianum/ niclosamide, paromomycin
		Aschelminthes (Roundworms)	
Ascaris lumbricoides	Ascariasis	Fecal–oral	Eggs in feces/piperazine citrate
Enterobius vermicularis	Enterobiasis (one type of pinworm)	Fecal–oral in perianal region	Adult worms or eggs/mebendazole
Trichinella spiralis	Trichinosis	Undercooked or raw pork	Muscle biopsy/supportive treatment

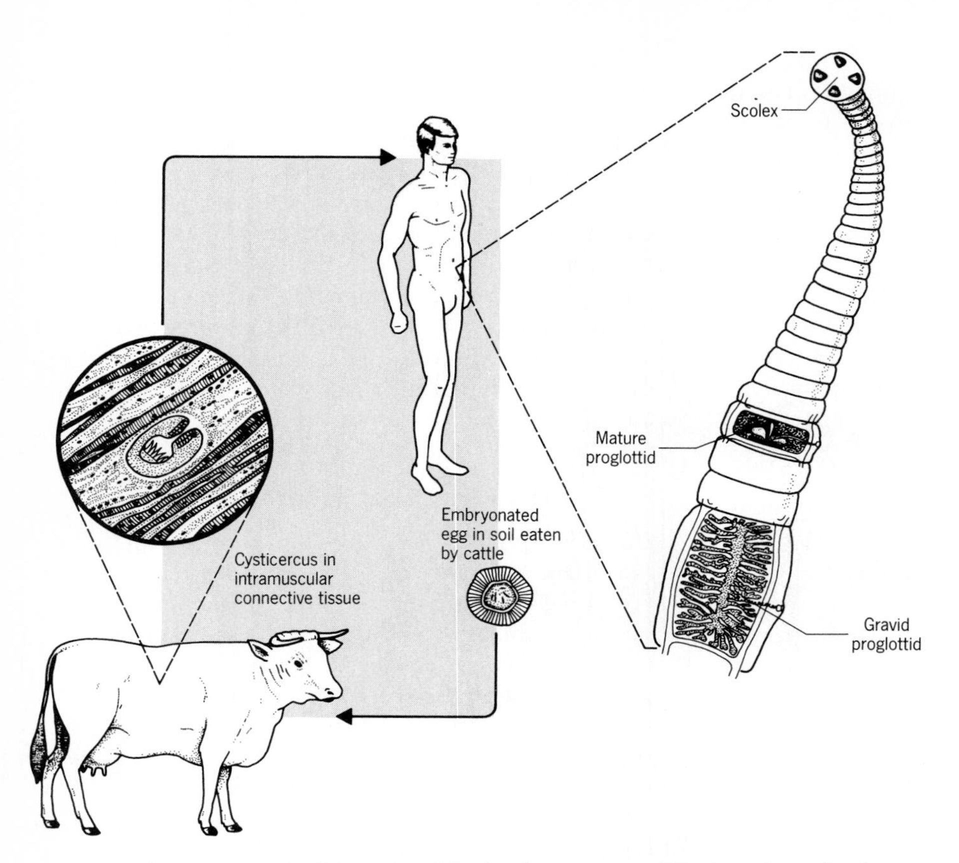

FIGURE 25.16 The life cycle of the beef tapeworm, *Taenia saginata,* begins when a human is infected with the larva present in raw or undercooked beef. The larva excyst and grows into an adult tapeworm in the small intestine of the human in about three months. The proglottids of the mature tapeworm produce eggs that are excreted in the human's feces. Cattle ingest the eggs, which hatch into larva that reach the muscle via the circulatory system. Here *T. saginata* develops into a hollow-fluid filled larva known as cysticercus bovis. The cycle repeats itself when a human consumes undercooked beef.

tapeworm is prevalent in temperate zones where freshwater fish are consumed, including North America; the Baltic countries; parts of Russia, Japan, Chile; and Argentina.

Beef Tapeworm and Pork Tapeworm

Humans, the only definitive host, expel proglottids and eggs of the beef (Figure 25.16) and the pork tapeworms in their feces. The eggs are ingested by an intermediate host, such as grazing cattle or hogs, while the animals are eating contaminated grass, food, or water. The eggs are carried in the circulation to the animal's muscles, where cysts called cysticerci form. Humans are again infected by eating this contaminated meat. The cysticerci hatch in the stomach to form a scolex, which attaches to the mucosa of the jejunum and develops in a few months into a mature worm.

The incidence of these infections in the United States is rare because, for the most part, human excrement is not used as a fertilizer, and hog farming is regulated to prevent hogs from consuming excrement-contaminated food or scraps from hog processing. Pork tapeworm infections are common, however, in

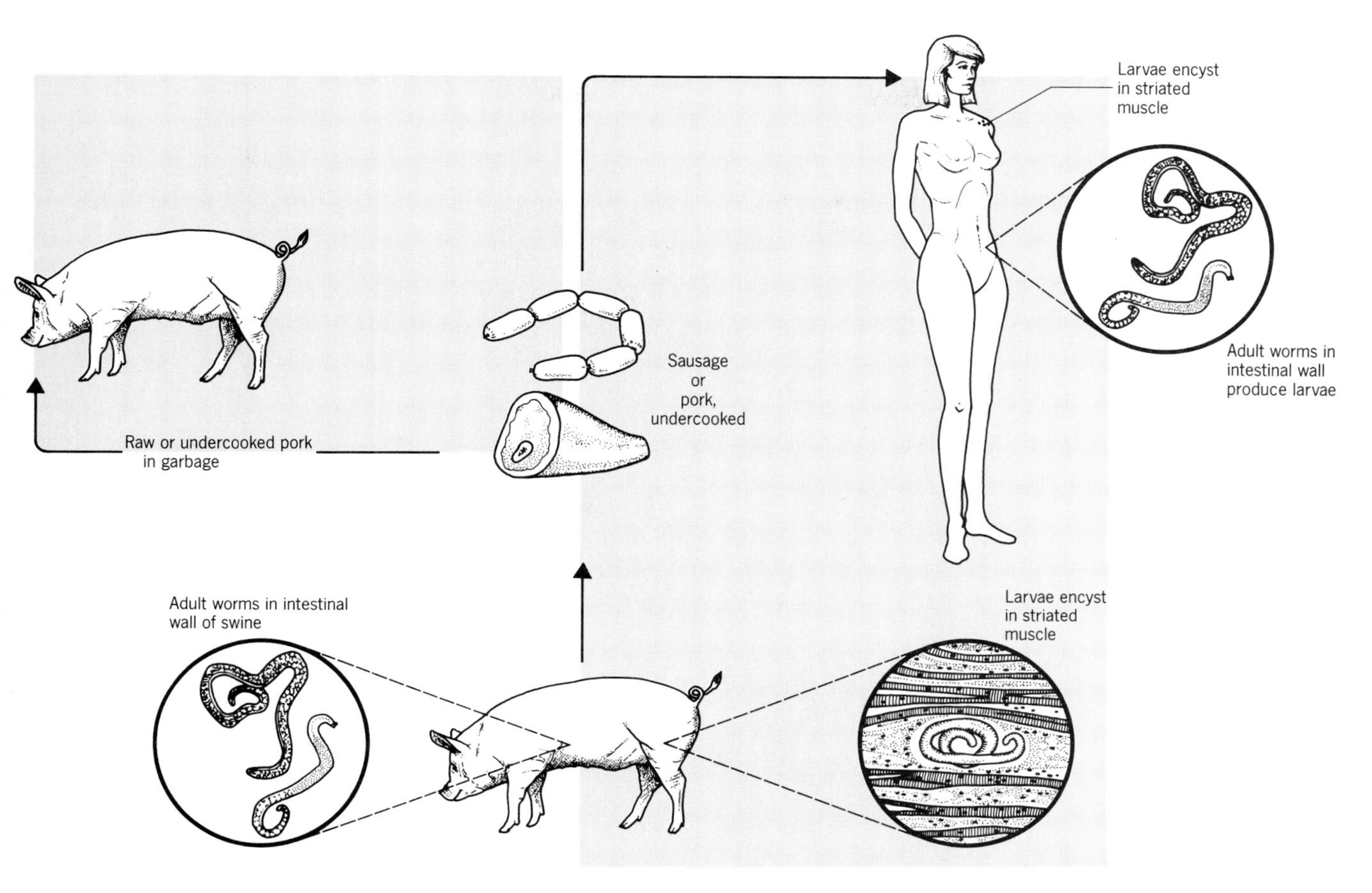

FIGURE 25.17 Trichinosis is caused by *Trichinella spiralis*. The larvae of this nematode encyst in the striated muscle of swine after the swine are infected by eating garbage containing contaminated, uncooked pork scraps. Humans are infected when they consume undercooked pork.

South and Central America, Africa, India, and China, where regulation is lax or nonexistent.

TRICHINOSIS

Trichinosis (trik-i-no′sis) is caused by the roundworm, *Trichinella spiralis* (trik′i-nel′ah spi-ra′lis), whose life cycle begins when meat with encysted larvae is ingested by a human, wild animal, or domestic animal. In humans, the larvae excyst in the small intestine and invade the intestinal mucosa, where both male and female worms are required for reproduction (Figure 25.17). After fertilization, the female invades the intestinal wall and releases her larvae, which are carried to all parts of the body and eventually burrow into muscle tissue. Here the larvae form capsules, which eventually become calcified.

Trichinosis occurs when a human ingests undercooked pork from hogs that have been fed infected garbage. The illness is uncommon in the United States since, as stated earlier, hogs are not fed garbage. At present there is no cure for trichinosis, and avoidance of undercooked or raw meat that harbors *T. spiralis* is the only way to prevent it.

ASCARIASIS

Ascaris lumbricoides (as′kah-ris lum-bri-koi′des) lives in the human small intestine where both male and female worms consume semidigested food that is passing through the digestive tract. The female produces eggs, which hatch into larvae that penetrate the intestinal wall, enter the circulation, and migrate to the liver, heart, and the bronchioles. In the bronchioles the mature worms may grow to a length of 1.5 mm. The worms are coughed up into the pharynx and swallowed, reinfecting the small intestine; and the cycle is repeated.

Female *Ascaris* can grow to a length of 22 to 35 cm. Typically, a total of 10 to 20 worms will reside in the host, yet, remarkably, the host may complain only of vague abdominal pain, or the infection may go unnoticed.

Ascariasis is a widespread disease that is estimated to affect 900 million people worldwide. In this country, the infection rate is highest among poor people living in rural, mountainous regions. The infection is spread by the fecal–oral route and is prevented by proper sanitation and personal hygiene. Treatment is with piperazine citrate, which paralyzes the worm so it can be excreted in feces.

PINWORM

Humans are the only hosts for *Enterobius vermicularis* (en′ter-o′be-us ver-mi′cu-la′ris), the cause of pinworm (enterobiasis). From the intestine the female migrates to the perianal region and lays its eggs. The perianal region itches; the person scratches; and the eggs are transferred to the hands. The hands then transmit the eggs directly, or indirectly (usually by inanimate objects and food), to the mouth. In a heavily infected household the eggs can be transmitted by aerosol.

Pinworm is a widespread helminthic disease, even in developed countries, where it is common among children. Diagnosis is made by microscopically observing worms or eggs collected on a clean sticky tape from the perianal region. Patients with enterobiasis are treated with mebendazole.

Summary Outline

GASTROINTESTINAL SYSTEM

1. The function of the gastrointestinal system is to digest food, absorb the nutrients for use by the body, and rid the body of solid waste.
2. Digestion begins when food is acted upon by the saliva. It continues in the stomach, where hydrochloric acid and proteolytic enzymes further the process before food enters the duodenum.
3. Absorption of nutrients into the circulatory system is through the villi lining the jejuneum and upper ileum. Water is reabsorbed from material in the large intestine.
4. The normal microbial population of the intestinal tract is greatest in the terminal ileum and the large intestine. This anaerobic environment is dominated by species of *Bacteroides,* with lesser numbers of *Escherichia, Lactobacillus,* and *Streptococcus.*
5. Most of these microbes contribute to the bulk of the feces and cause no illness to the host.
6. Certain pathogenic bacteria colonize or invade the intestinal mucosa or other parts of the digestive system to cause peritonitis, cholecystitis, or gastroenteritis.

BACTERIAL INFECTIONS

1. *Vibrio cholerae* colonizes the intestine and produces the cholera toxin, which causes a purging diarrhea. Cholera is a pandemic disease carried by infected persons and spread by feces-contaminated water.
2. The cholera toxin increases the intracellular concentration of cAMP in the intestinal mucosa, causing an electrolyte imbalance and resulting in an outpouring of water into the bowel.
3. Fluid and electrolyte restoration is done to prevent death from dehydration.
4. Enterotoxic *E. coli,* which contains plasmids that code for a heat-labile toxin, a heat-stable toxin, and/or pili, colonize the intestinal tract and cause a diarrheal disease.
5. The heat-labile toxin has the same mechanism of action as the cholera toxin. The heat-stable toxin

is a low molecular weight polypeptide that stimulates guanylate cyclase and causes diarrhea.

6. Enteroinvasive *E. coli* causes a dysenteric disease characterized by fever, cramps, and some diarrhea.
7. *Shigella* causes a similar invasive disease, known as bacillary dysentery or shigellosis, that is characterized by abdominal cramps, diarrhea, and the presence of blood and mucus in the stools. EIEC and shigellosis are self-limiting diseases transmitted by the fecal–oral route.
8. *Salmonella* causes enteric fevers, gastroenteritis, and bacteremia. *Salmonella typhi* penetrates the intestinal mucosa to cause a bacteremia known as typhoid fever.
9. *Bacteroides fragilis* is the most common species of anaerobic bacteria isolated from soft-tissue infections, such as peritonitis, rectal abscesses, and surgical wounds.

BACTERIAL FOOD POISONING: INTOXICATIONS

1. Staphylococcal and botulism food poisoning are examples of foodborne intoxications. Staphylococcal enterotoxins are heat-stable exotoxins responsible for foodborne gastroenteritis.
2. The botulism toxins are potent neurotoxins consumed in contaminated canned foods. They are counteracted by passive immunization with botulism antitoxins.

INFECTIOUS BACTERIAL FOOD POISONING

1. *Clostridium botulinum* colonizes the intestinal tract of infants and produces botulism toxin responsible for infant botulism.
2. *Salmonella,* the most frequent cause of foodborne disease in this country, causes a self-limiting disease called salmonella gastroenteritis. Sources of salmonella food poisoning include beef, shellfish, milk, milk products, and salads.
3. Humans are the only source of *S. typhi,* which infects the biliary tract of chronic carriers, is excreted in feces, and is transmitted by the fecal–oral route.
4. *Campylobacter* is a common inhabitant of the intestinal tract of animals, birds, and humans. *Campylobacter jejuni* is transferred to humans in contaminated meat and raw milk and causes acute enteritis.
5. *Bacillus cereus* causes diarrheal symptoms when associated with contaminated meats and sauces, and vomiting symptoms (usually) when associated with contaminated fried rice or rice dishes.
6. Other causes of foodborne gastroenteritis include *Clostridium perfringens* in inadequately cooked boiled meats, gravy, and poultry, and *V. parahaemolyticus,* which is associated only with seafoods.

VIRAL GASTROENTERITIS

1. The rotaviruses cause gastrointestinal infections that vary from a mild diarrhea to a fatal gastroenteritis.
2. Epidemic diarrhea, viral gastroenteritis, winter vomiting disease, and acute nonbacterial gastroenteritis are diseases caused by the Norwalk and related viruses.
3. Infectious hepatitis is caused by the hepatitis A virus, an RNA virus transmitted by the fecal–oral route or contaminated food. Infectious hepatitis is controlled by passive immunization.

PROTOZOAN DISEASES OF THE GASTROINTESTINAL SYSTEM

1. Protozoa that cause gastrointestinal tract infections include *Entamoeba histolytica* (amebiasis), *Giardia lamblia* (giardiasis), and *Balantidium coli* (blantidiasis).
2. Each protozoan forms an infective cyst that is excreted into the environment. Ingested cysts pass to the intestine where they grow as trophozoites and cause disease symptoms.
3. *Entamoeba histolytica* can penetrate the intestinal mucosa to cause extraintestinal amebiasis, which has life-threatening consequences.
4. Giardiasis is a major cause of waterborne infections in the United States, both in urban areas and in wilderness areas where *Giardia* is carried by beavers and other wild animals.

HELMINTH DISEASES

1. Helminths that infect the gastrointestinal system include the tapeworms of beef, pork, and fish, and

the roundworms responsible for ascariasis, pinworm, and trichinosis.

2. The tapeworms attach to the intestinal mucosa by their scolex and increase in length by generating proglottids. An intermediate host is necessary to complete their life cycle.

3. Roundworms, which cause ascariasis, enterobiasis, and trichinosis, reproduce in the human intestine. Trichinosis is transmitted by ingestion of undercooked or raw pork, while ascariasis and enterobiasis (a form of pinworm) are transmitted by the fecal–oral route.

Questions and Topics for Study and Discussion

QUESTIONS

1. How do the different regions of the intestinal tract vary with regard to oxygen content, pH, and bacterial flora?
2. Describe the diseases caused by enterotoxic *Escherichia coli*. How does this strain differ from the strains of *E. coli* present in the intestine of healthy humans?
3. Why and how do the symptoms of cholera dictate treatment? How important is this disease in your community, and in the world?
4. What is the mode of action of the cholera toxin? In what way does the cholera toxin relate to the toxins produced by enterotoxic *E. coli?*
5. What information would you seek in order to identify the source, pathogen, and seriousness of an epidemic of gastroenteritis occurring in a medium-sized university?
6. What are the clinical signs of shigella dysentery that differentiate it from cholera and salmonella gastroenteritis?
7. How does a person become a carrier of *Salmonella typhi?* How would you detect and treat such a person?
8. Describe the various foodborne clostridial diseases. Explain the seriousness of these infections and the methods used to prevent and treat them.
9. What types of food poisoning are resistant to the normal heating associated with food preparation? How can these illnesses be prevented?
10. What steps could you take to prevent further spread of infectious hepatitis in an institution for the mentally retarded?
11. Which viruses infect the intestinal tract? How serious are these infections and how are they treated?
12. Describe the life cycle of *Entamoeba histolytica* and explain the importance of cyst formation by this protozoan.
13. Discuss the epidemiology of giardiasis. Are you safe from this parasite if you drink water from a mountain stream? Explain.

DISCUSSION TOPICS

1. The acquisition of multiple-drug resistance in bacteria was first observed in *Shigella dysenteriae* isolated during an outbreak of dysentery. Under what environmental conditions would bacteria acquire multiple-drug resistance plasmids?
2. What public health steps are important in preventing the spread of gastrointestinal diseases caused by bacteria, viruses, protozoa, and helminths?

Further Readings

BOOKS

Beck, J. W., and J. E. Davies, *Medical Parasitology,* 3rd ed., C. V. Mosby Company, St. Louis, 1981. A well-illustrated paperback textbook that covers protozoa, helminths, and the medically important arthropods involved in disease transmission.

Brown, H. W., and F. A. Neva, *Basic Clinical Parasitology,* 5th ed., Appleton-Century-Crofts, Norwalk, Conn., 1983. Excellent coverage of the diseases caused by roundworms and tapeworms.

Drasar, B. S., and P. A. Barrow, *Intestinal Microbiology, Aspects of Microbiology,* American Society for Microbiology, Washington, D.C., 1985. The authors provide insight into the microbial complexity of the intestinal flora of humans and domestic animals.

DuPont, H. L., and L. K. Pickering, *Infections of the Gastrointestinal Tract,* Plenum, New York, 1980. This monograph contains chapters on the major bacterial and viral infections of the human intestine.

Stephen, J., and R. A. Pietrowski, *Bacterial Toxins, Aspects of Microbiology,* 2nd ed., American Society for Microbiology, Washington, D.C., 1986. This short paperback discusses the bacterial toxins and is an excellent supplement to the material in the text.

ARTICLES AND REVIEWS

Cukor, G., and N. R. Blacklow, "Human Viral Gastroenteritis," *Microbiol. Revs.,* 48:157–179 (1984). This review concentrates on the rotavirus and Norwalk virus as causes of gastroenteritis.

Gangarosa, E. J., et al., "Botulism in the Unted States 1899–1969," *Am. J. Epidemiology,* 93:93–101 (1971). An historic review of the incidence, morbidity, mortality, and geographical distribution of botulism in the United States.

Hirschhorn, N., and W. B. Greenough III, "Cholera," *Sci. Am., 225:* 15–21 (August 1971). This article describes the mechanism of cholera toxin, cholera epidemiology, and its treatment.

Middlebrook, J. L., and R. B. Dorland, "Bacterial Toxins: Cellular Mechanisms of Action," *Microbiol. Rev.,* 48:199–221 (1984). A current review of shigella toxin, cholera toxin, botulism toxin, and *E. coli* enterotoxin.

Nakamura, M., and J. A. Schulze, *"Clostridium perfringens* Food Poisoning," *Ann. Rev. Microbiol.,* 24:359–372 (1970). The pathogenicity of *C. perfringens* in animal and human food poisoning.

CHAPTER

26

Infectious Diseases of Blood, Liver, and Lymph

OUTLINE

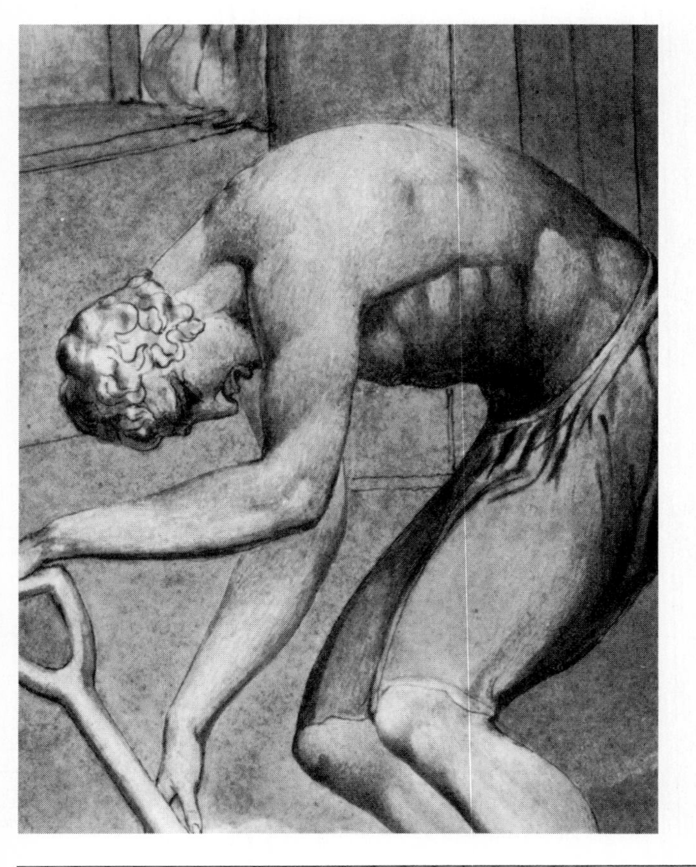

FOCUS OF THIS CHAPTER

Viruses, protozoa, or pathogenic bacteria that gain entry to the body and invade the blood or lymph can cause serious, often life-threatening, disease. In this chapter we focus on the circumstances that lead to infections of the circulatory and lymphatic systems, and the treatment and prevention of these diseases.

Blood and lymph are vehicles that transport nutrients and gases throughout the body. The circulatory system is driven by the heart, a muscular organ that pumps blood through the arteries and veins. Closely related to the circulatory system is the lymphatic system, or simply the lymphatics, which carries tissue fluid through its series of vessels and nodes.

FUNCTION OF THE CARDIOVASCULAR SYSTEM

The heart is a pump consisting basically of four chambers, which function as follows. The right atrium receives venous blood returning from the tissues. This blood is carried in the veins, the vessels leading to the heart from the tissues. The right ventricle pumps the venous blood coming from the right atrium, and sends it to the lungs. The left atrium receives blood that is high in oxygen as it returns from the lungs. The left ventricle pumps the oxygenated blood to all parts of the body via arteries. Valves are located at the entrance and the exit of each ventricle, to prevent backflow of blood into either the right or left atrium.

The flow of blood is through the arteries, into the smaller arterioles, thence to the smallest vessels, the capillaries. The capillary walls, being a single layer of cells, permit the delivery of nutrients and oxygen and the uptake of wastes from the cells. The capillaries connect with veins that transport deoxygenated blood back to the right atrium. The cycle is repeated endlessly throughout life.

The **lymphatic system** is a network of vessels and nodes that collects fluid and cells from the intercellular spaces, processes this fluid (now called lymph), then transports it to the circulation. The lymph is pumped by the muscles through the lymphatic network and returned to the blood just before the blood enters the right atrium. The tonsils, thymus, and spleen are also part of the lymphatic system.

Cells in the lymphatic system function in the immune response and in removing debris picked up by the lymph. The B lymphocytes in the lymphatic system are converted to antibody-producing plasma cells, and histiocytes attached to lymphatic tissue remove tissue debris and particulate matter from lymph by phagocytosis. Once in the histiocyte, this matter is destroyed and its component parts recycled as nutrients.

One would expect that the defense mechanisms of the blood and lymph would be equipped to protect against infectious circulatory diseases, and to a large extent this is correct. There are a few bacteria, viruses, and protozoa, however, that can cause circulatory diseases.

BACTERIAL DISEASES OF BLOOD

A **bacteremia,** the presence of bacteria in the blood, is a clear indication of an infection since blood is normally free of microbes. A bacteremia may be transient and self-limiting or may constitute a life-threatening emergency. Many bacteremias arise as a complication of an underlying disease that ultimately will prove to be fatal especially in older patients.

GRAM-NEGATIVE ROD BACTEREMIA

Opportunistic gram-negative rods can be the pathogens isolated from bacteremic patients. They enter the blood from a focus of infection somewhere in the body, for example a urinary tract infection or burn wound. Species of *Escherichia, Klebsiella, Enterobacter, Serratia, Proteus,* or *Pseudomonas aeruginosa* may be involved. The resulting symptoms include chills, fever, hyperventilation, skin lesions, and changes in mental status. The fever is caused by the stimulation of endogenous pyrogen (see Chapter 17) by the endotoxin present in the bacterial cell wall. The clinical symptoms create a state of endotoxemia, a disease that is hard if not impossible to treat. Blood culture isolation of the responsible bacteria, aggressive antimicrobial chemotherapy, and supportive care are necessary for successful treatment.

The symptoms caused by these opportunistic pathogens are different from the distinctive symptoms caused by the pathogenic *Yersinia* (plague), *Borrelia, Francisella,* and *Brucella,* which readily penetrate the body's defenses and cause infection.

BORRELIAS: RELAPSING FEVER AND LYME DISEASE

Borrelia (bo-rel′e-ah) is a parasitic spirochete having coarse, uneven, or irregular coils (Figure 26.1) that infects through an abrasion or bite of an arthropod vector. The infection is diagnosed by detecting the spirochetes in a blood smear. This spirochete is differentiated from the treponemes and leptospiras by its size and number of coils.

Borrelias are microaerophilic bacteria whose reservoirs include small rodents and other mammals, insects, and birds. Ticks and head lice are the arthropod vectors for borrelias that cause relapsing fever and Lyme disease.

Relapsing Fever

Epidemic relapsing fever is a febrile disease, occurring primarily in Africa, characterized by the cyclic appearance of spirochetes in the blood. *Borrelia recurrentis* (re-kur-ren′tis) multiplies in the blood of humans (septicemia) and is transmitted between patients by the head louse, *Pediculus humanus humanus.* Humans are the

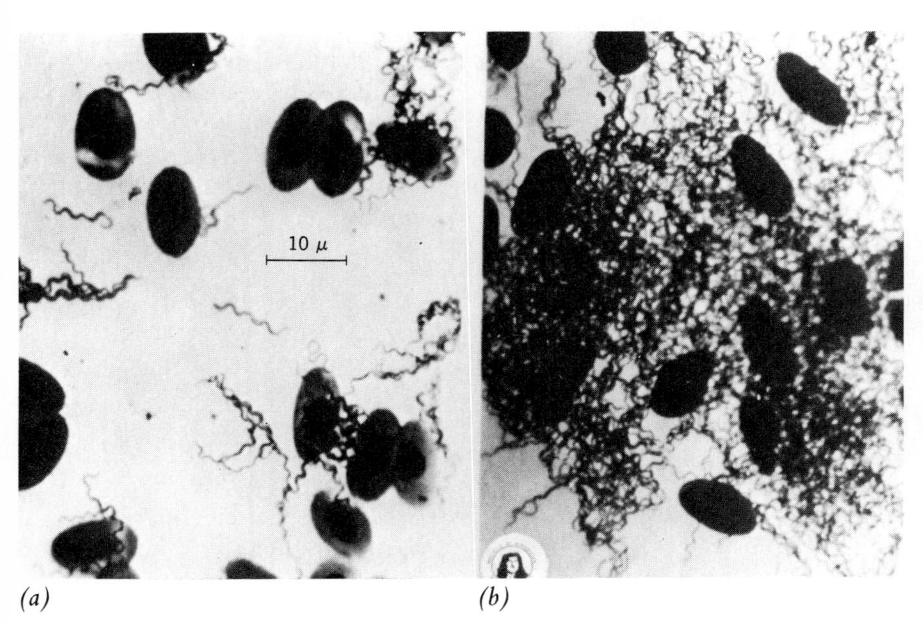

FIGURE 26.1 Photomicrograph of *Borrelia* in blood smears. (Courtesy of the American Society for Microbiology slide collection.)

only reservoir for *B. recurrentis,* thus the lice must feed on an infected human in order to transmit the disease.

After the louse is infected, the spirochetes penetrate the gut epithelium of the louse and multiply in its hemolymph. (Hemolymph is the name given to the blood-like fluid of invertebrates). A person is infected by squashing the infected louse (or lice), which permits the spirochetes in the hemolymph to penetrate the skin (whether broken or unbroken). A febrile septicemia ensues. Crowded living conditions make it easy for the louse to be transmitted between individuals.

In the United States *Borrelia* is transmitted by ticks, causing tick-borne (endemic) relapsing fever. It is rare here, only 325 cases having been reported between 1959 and 1983. Two species, *Borrelia hermsii* and *B. turicatae,* infect all tissues of the tick and are maintained in the tick population by transovarian passage.

Relapsing Fever in Humans The clinical symptoms of louse-borne and tick-borne relapsing fever are very similar, except that the latter is less severe. Illness begins with a sudden onset of fever that develops after an incubation period of 2 to 15 days. The fever continues for about 10 days, during which time the borrelias are present in the blood (Figure 26.1) and urine. As the fever subsides, the number of spirochetes in the blood decreases and they are less motile. After another 3 to 10 days the fever recurs, and once again the blood is teeming with spirochetes. In all, tick-borne relapsing fever progresses through 3 to 10 sequential fever-relapse cycles, with each febrile stage becoming less severe.

Relapses occur because the borrelias alter the structure of their surface antigens—essentially, they change coats. Their genes for surface antigens mutate at a high rate, and in their mutated state are able to overcome the immune response. New antibodies form against the new surface antigens; and again the infection subsides. Eventually the host manufactures enough antibodies against enough surface antigens to destroy the borrelias.

Lyme Disease

Lyme disease takes its name from the town of Lyme, Connecticut, where in 1975 it was first reported. Its characteristics include distinctive skin lesions, along with a stiff neck, malaise, fatigue, and swollen lymph nodes. The causative agent is *Borrelia burgdorferi.* Lyme disease is transmitted by minute ticks belonging to a species of *Ixodes* (iks-o′dez) found in the northeast, midwest, and in the western United States. About 500 cases are reported annually. Relapsing fever and Lyme disease are treated with a tetracycline or penicillin.

BACTERIAL DISEASES OF THE LYMPHATIC SYSTEM

THE PLAGUE BACILLUS

Historical documents, literary accounts, and paintings are grim reminders of the devastation wrought by the plague. Bubonic plague is aptly named; it causes severe swellings, or buboes, of the lymph nodes. During the fourteenth century the name Black Death was equally appropriate, because dark splotches (due to the then-unidentified hemorrhages) appeared under the skin before the victim died. The epidemic of 1346 to 1350 was a disaster in which one-fourth of the total population of Europe succumbed. The horror of the plague is captured in *A Journal of the Plague Years,* written by Daniel Defoe in 1722, relating the story of the London plague of 1665. More recently, *A Distant Mirror,* by Barbara Tuchman, has given us a gripping account of the epidemic of the fourteenth century. Fortunately plague is virtually a thing of the past, as will be described.

Epidemiology

The **urban cycle** of plague, which existed at the time of the Black Death, coincides with a rise in the rat population in cities. Rats are the reservoir for the plague bacillus and fleas are its vector. The bacilli grow in the foregut of fleas after the flea has fed on an infected rat. The rat succumbs to plague, and the flea moves on to another warm body from which it obtains its next blood meal. The bacilli in the flea's foregut are regurgitated as the flea bites another host, rat or human, thereby completing the transmission cycle.

If the plague gains a foothold in the human population, a few people will develop pneumonic plague. Not only are these patients very sick; they pose a threat to others since they generate infective aerosols, and people attending them are easily infected. Fortunately the urban cycle is largely a thing of the past.

In the United States, plague occurs in the **sylvan cycle.** The reservoirs are ground squirrels, prairie dogs, or other animals that live in the wild. A person may come in contact with plague bacilli when s(he) attempts to care for a plague-infected animal, or unwittingly camps or hikes in regions where plague is endemic. Infected fleas carried by these animals transmit the bacilli to humans.

Yersinia pestis Causes Plague

Yersinia pestis (yer-sin′e-ah pes′tis), the causative agent of plague, is a gram-negative, nonmotile, facultative coccobacillus. It is primarily a rodent pathogen that multiplies in its animal reservoir and is found in their blood, feces, and urine. Western ground squirrels are the most common reservoirs of *Y. pestis* in this country. *Yersinia pestis* also infects the black house rat, *Rattus rattus;* the gray sewer rat, *Rattus norvegicus* (nor-ve′ji-cus); and the less common Egyptian or roof rat, *Rattus alexandrinus* (al-ig-zan′dri-nus).

In humans, *Y. pestis* is able to survive and multiply in the circulating monocytes. The bacteria produce toxins, but exactly what role these toxins play in the disease is not clear. The few patients who do recover from plague have a high state of immunity to reinfection.

Bubonic and Pneumonic Plague

Bubonic plague is characterized by an incubation period of two to five days, followed by infection of the regional lymph nodes. Buboes appear in the inguinal (groin) area. As the bacteria multiply, there is an abrupt high fever, heralding a septicemia. The bacteria migrate to the liver, spleen, lungs, and sometimes the meninges. The patient either dies or recovers within ten days following the onset of symptoms. Between 60 and 90 percent of untreated patients die.

In the later stages of bubonic plague the bacteria attack the lungs, causing pneumonic plague. The bacteria swarm in the patient's sputum and are transmitted to others in droplet nuclei expelled when the patient coughs. During the epidemic of 1665, a decree from the mayor of London required households to be quarantined as soon as one member became ill. If the patient developed pneumonic plague, the bacteria were readily transmitted throughout the family, and everyone in the family died.

Treatment and Control

Streptomycin, chloramphenicol, and the tetracyclines are effective in treating both bubonic and pneumonic plague. However, treatment must be started within 15 hours of the onset of fever, or the patient will probably die. Eliminating the rodent reservoir and using appropriate insecticides are effective in prevention. Isolating an infected person is necessary to prevent aerosol transmission. Very few cases are reported in the United States.

FRANCISELLA: TULAREMIA

Tularemia (too′lah-re′me-ah) is an enzootic disease of small animals that is very similar to plague in its epidemiology and clinical manifestations. The bacteria are transmitted to humans indirectly by ticks and deer flies, directly through abrasions of the skin, by ingestion of large numbers of bacteria in contaminated meat, or by inhalation.

Francisella tularensis (fran-si-sel′ah) was first isolated from infected ground squirrels in Tulare County, California in 1912. It is named for Edward Francis, a microbiologist who studied it. This bacterium is a natural inhabitant of stream waters and small game animals such as rabbits and ground squirrels. Muskrats, beavers, opossums, coyotes, deer, red foxes, woodchucks, skunks, rats, mice, dogs, cats, and lambs can also be infected.

Tularemia Syndromes

Francisella tularensis causes three distinct diseases depending on its method of entry into the host. (1) Localized infections of the skin cause an ulcer to form at the infection site. This form of the disease is called **ulceroglandular tularemia.** The localized infection follows the bite of an infected deer fly or tick or bacterial contamination of a skin wound or abrasion. This form may develop on the hands of hunters and trappers after field dressing small game or fur-bearing animals. (2) A person who inhales *F. tularensis* may contract **pneu-**

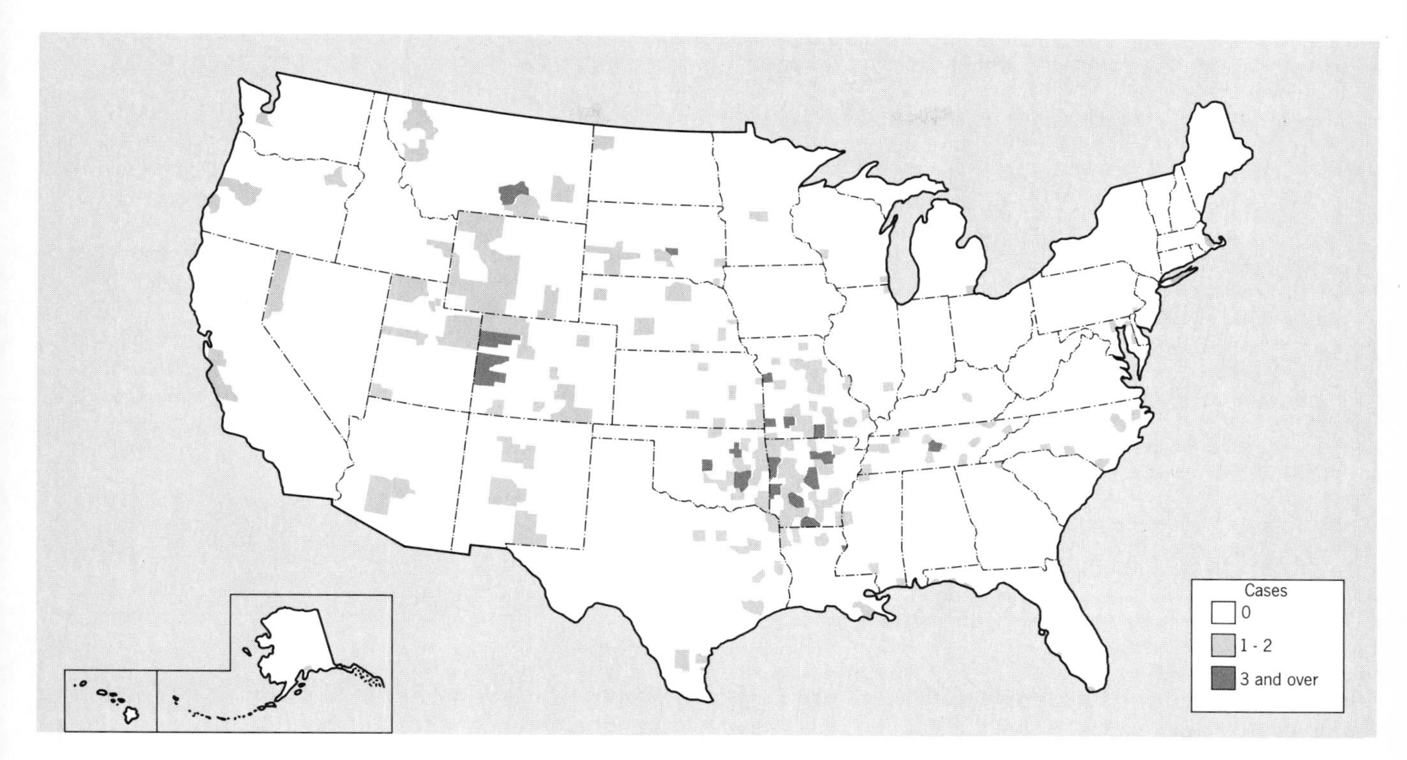

FIGURE 26.2 Geographical distribution of tularemia in the United States by county in 1984. (Centers for Disease Control, "Annual Summary 1984: Reported Morbidity and Mortality in the United States," *Morbidity and Mortality Weekly Report,* Vol. 32, No. 54, 1986.)

monic tularemia. (3) Ingestion of numerous bacteria in undercooked, contaminated meat can cause **typhoid tularemia.**

After penetrating the host, the bacteria infect the regional lymph nodes, which become enlarged, tender, and suppurative. From here the bacteria enter the blood, causing a transitory bacteremia, before infecting the lungs, liver, and spleen. This acute phase of tularemia is associated with headache, fever, and general malaise. The pneumonic and typhoid forms of untreated tularemia carry a greater fatality rate than the ulceroglandular form.

Tularemia is diagnosed by the demonstration of agglutinating antibodies in serum. Streptomycin is the antibiotic of choice for treating all forms of tularemia.

Epidemiology

Tularemia is restricted to geographical areas where infected animals live. In recent years, most cases have occurred in the south central states of Arkansas, Missouri, and Oklahoma (Figure 26.2). Tularemia persists throughout the year, with epidemics corresponding to the rabbit-hunting season in the winter and the tick and deer fly seasons in the summer. It appears to be restricted to the Northern Hemisphere, for it has not been reported south of the equator. Between 234 and 310 cases were reported annually in the United States during the early 1980s.

BRUCELLOSIS

Brucella (broo-sel'ah) is a natural parasite of domestic animals, which in turn are the main source of human infections. It is the causative agent of **brucellosis,** a febrile disease that plagued British troops stationed on the island of Malta in 1887. Sir David Bruce, an army scientist, isolated *Brucella* from the spleens of British soldiers who died of a fever called, at the time, Malta fever. He traced the source of the infection to the goat

Tularemia—New Jersey

On November 9 an 18-year-old hunter shot two rabbits behind his home in Gloucester County, New Jersey. After eviscerating the animals, he gave them to his neighbors, a 67-year-old woman and her 64-year-old husband, who skinned and froze the rabbits. During the summer the young man had noticed several dead rabbits around his house and had attributed their deaths to insecticide that had been sprayed on local fields. One of the two rabbits he shot was found to be losing its fur.

Two days after dressing out the rabbits, the young man became ill with an ulcerated hand lesion, axillary lymphadenopathy, and a fever. He was examined at the local hospital. No diagnosis was made, but he was treated with antipyretics, drugs that reduce fever. On November 23 his neighbors who had received the rabbits were admitted to the same hospital. They both had sepsis and hand lesions and were treated with gentamicin and cefazolin. The woman was also given insulin for uncontrolled, late-onset diabetes. On November 26 the young hunter was recalled to the hospital, and all three patients were treated with streptomycin. The hunter and the husband recovered; however, the woman died on December 3 from tularemia. The two rabbits were sent to CDC for analysis. Cultures from the bone marrow of both animals grew *Franciscella tularensis.*

Adapted from *Morbidity and Mortality Weekly Report,* Vol. 35, No. 48, December 5, 1986.

milk that was drunk by the troops. The disease disappeared after the troops stopped drinking milk taken from the island's goat herd.

Brucellas are aerobic, gram-negative, short, spherical to rod-shaped bacteria that multiply preferentially in the uterus or the mammary glands of domestic animals—hence their presence in milk. They cause abortion in domestic animals but not in humans. *Brucella melitensis* (me-li-ten'sis) infects sheep and goats, *B. abortus* causes abortion in cattle, *B. suis* infects domestic and wild pigs, and *B. canis* (ca'nis) infects dogs.

Brucellosis in Humans

The incubation period in human brucellosis is usually several weeks or even months after the brucellas have entered through broken skin, the digestive tract, or the conjunctiva. They invade the lymphatic system, where they multiply within the polymorphonuclear lymphocytes. At this stage of the disease, the bacteria either are destroyed or are carried by the blood to infect the liver, spleen, bone marrow, and kidneys.

Symptoms appear gradually and include fever, chills, malaise, aching, sweating, and headache. Because the fever is intermittent, another name, **undulant fever,** has often been used. Diagnosis is based on isolating *Brucella* from the blood or urine, or from a biopsy specimen of liver, bone marrow, or lymph node. The illness lasts approximately four weeks.

Epidemiology

At one time, milk and milk products were the major source of brucellosis, hence it occurred primarily among rural families who consumed unpasteurized dairy products. It is now most frequent among livestock handlers and workers in the meat-processing industry and only a small number of cases (about 9 percent) are due to contaminated dairy products.

Treatment and Control

Control of human brucellosis depends heavily on eliminating the bacterium from cattle, sheep, and goats. This goal has largely been accomplished through a state–federal eradication program that has been in effect since 1931. Inspectors work to identify and slaughter all infected lambs, cattle, and goats, and to vaccinate noninfected animals. A person diagnosed as having brucellosis is treated with a tetracycline for a period of three to four weeks, which is adequate to prevent relapses.

RICKETTSIAS: OBLIGATE INTRACELLULAR BACTERIA

Rickettsias (ri-ket'se-ah) are obligate intracellular bacteria that are maintained in nature as parasites of arthropods and warm-blooded animals (Figure 26.3). Rickettsias are usually transmitted from their animal reservoir to people by arthropod vectors.

Mice are used as the hosts for laboratory experiments with *Rickettsia*. In mice, the rickettsias enter the vascular epithelium (by an as yet unknown mechanism), multiply, and cause a dramatic increase in the vascular permeability, which often kills the mouse. Rickettsias have a similar preference for invading the vascular epithelium of people.

Although five species of *Rickettsia* cause diseases in humans (Table 28.1), flea-borne typhus fever, tick-borne typhus fever (Rocky Mountain spotted fever), and Q fever (see Chapter 23) are the only significant ones in the United States. Rickettsialpox occurs sporadically in U.S. cities, but it is a mild disease without serious consequences. Scrub typhus occurs in Asia, the Indian subcontinent, Australia, and the Pacific Islands where the mites that transmit *Rickettsia tsutsugamushi* (soo-soo-ga-moo'shee) are found. Scrub typhus is seen in the United States only in travelers returning from these regions.

Rickettsial infections are diagnosed by symptoms, the patient's history of exposure to vectors, and the seasonal and geographical incidence of the diseases. A number of serological techniques, including immuno-

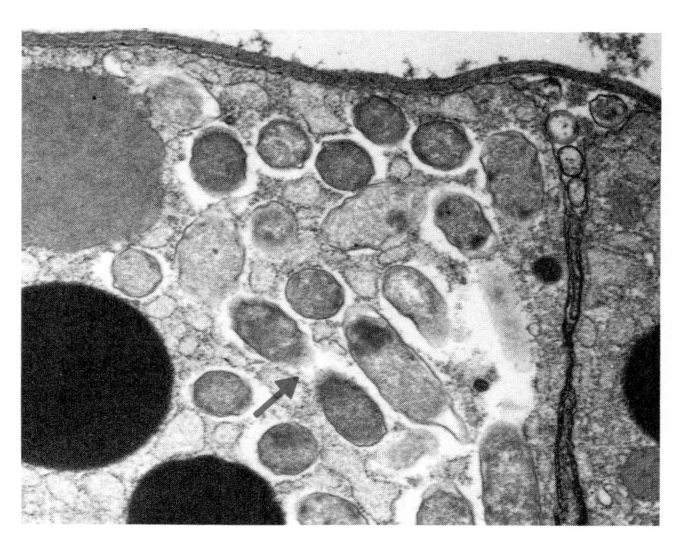

FIGURE 26.3 The spotted fever agent, *Rickettsia rickettsii,* in salivary gland tissue of the vector tick, *Dermacentor andersoni.* The arrow points to a dividing cell (24,750×). (Courtesy of Dr. W. Burgdorfer, Rocky Mountain Laboratories, Hamilton, Mont.)

fluorescence, agglutination tests, and ELISAs, are used to confirm the diagnosis. A rickettsia is rarely isolated because this is a specialized and potentially dangerous task.

FLEA-BORNE TYPHUS FEVER

Flea-borne typhus fever (once called endemic typhus) is caused by *Rickettsia typhi*, which is carried by rats

TABLE 28.1 Human rickettsial diseases

DISEASE	FLEA-BORNE (endemic) TYPHUS FEVER	SCRUB TYPHUS	TICK-BORNE TYPHUS (Rocky Mountain Spotted fever)	RICKETTSIALPOX
Causative agent	*R. typhi*	*R. tsutsugamushi*	*R. rickettsii*	*R. akari*
Natural cycle				
Reservoir	Small rodents	Rodents	Rodents	House mouse
Vector	Rat flea	Mite	Tick	Blood-sucking mite
Rash	Trunk to extremities	Trunk to extremities	Extremities to trunk	Vascular, trunk to extremities
Distribution	Worldwide	Pacific islands, Asia, Australia	Western Hemisphere	Europe, North America

and rodents and transmitted to humans by fleas. Flea-borne typhus occurs worldwide with between 50 and 100 cases annually in the United States (most occur in Texas, where 70 percent of the cases in 1984 occurred). The bacterium is spread among rats by the rat flea, which is infected when *R. typhi* enters the flea's gut during feeding. It multiplies in gut epithelial cells and is transmitted to humans in the flea's feces that contaminate skin abrasions.

Symptoms of flea-borne typhus fever include a sudden onset of chills, fever, headache, generalized pains, and exhaustion. A macular rash usually appears on the trunk after four to seven days and then spreads to the extremities. Prostration, stupor, and delirium develop in severe cases. Upon recovery, patients have lasting resistance against *R. typhi* infections. Control measures include eliminating the rat population and using insecticides to prevent the rat fleas from seeking alternative hosts. Treatment is with tetracycline or chloramphenicol.

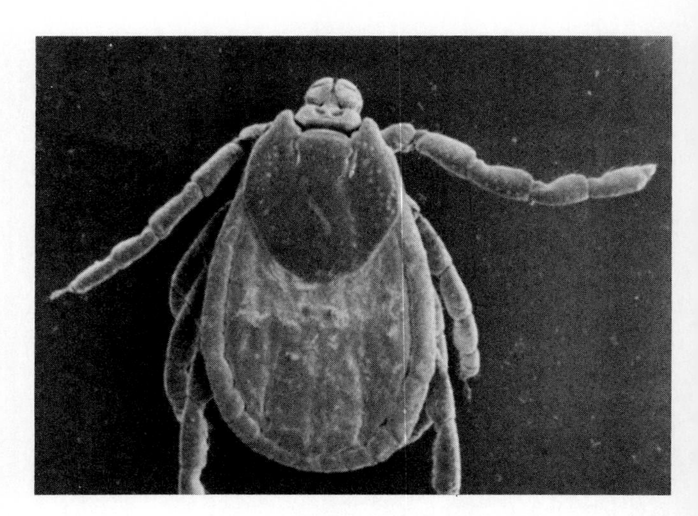

FIGURE 26.4 Rocky Mountain spotted fever is transmitted by the dog tick, *Dermacentor andersoni,* in the eastern United States. (Courtesy of the Rocky Mountain Laboratory, Hamilton, Mont.)

TICK-BORNE TYPHUS FEVER (ROCKY MOUNTAIN SPOTTED FEVER)

Tick-borne typhus, or Rocky Mountain spotted fever (RMSF), is caused by *Rickettsia rickettsii,* which occurs only in the Western Hemisphere. This rickettsia is maintained by **transovarian passage** in ticks, meaning that it is passed from one generation of ticks to the next in infected tick eggs. It is also maintained in animal reservoirs, including rabbits, opossums, dogs, and rodents. Adult ticks transfer *R. rickettsii* from animal reservoirs to humans, who are incidental hosts—they do not serve as reservoirs.

The peak incidence of RMSF coincides with the May–September tick season. Patients develop a fever, chills, prostration, and a rash one to two weeks after being bitten by an infected tick. Unlike other rickettsial rashes, this one begins on the extremities (ankles, wrists) before it spreads to the trunk. Recovered patients have immunity against reinfection. Patients are treated with tetracyclines or chloramphenicol.

Epidemiology, Prevention, and Control

RMSF is our most significant rickettsial disease with over 850 cases annually. It was first recognized in the western mountain states of Idaho and Montana, where the wood tick, *Dermacentor andersoni* (der-ma-sen′tor an-der-soh′neye), is the vector and reservoir. Now, 50 to 69 percent of the cases occur in a band of counties running from north Texas and Oklahoma eastward to the Atlantic coast from South Carolina north to Maryland (see Figure 20.7), where the dog tick, *Dermacentor variabilis* (Figure 26.4), is the major vector. People living or traveling through these regions may be exposed.

VIRAL INFECTIONS OF BLOOD AND LYMPH

Some of the most serious infectious diseases of the present decade are caused by viruses that infect blood or lymph. As described in Chapter 24, acquired immune deficiency syndrome (AIDS) is caused by a retrovirus that infects helper T lymphocytes and destroys the immune system. Hepatitis B virus is a major cause of liver infection and its presence increases the risk that the patient will later develop carcinoma of the liver. Both Epstein–Barr virus and cytomegalovirus cause latent infections of leukocytes that are manifest as infectious mononucleosis or another disease entity.

INFECTIOUS MONONUCLEOSIS: HISTORICAL PERSPECTIVE

Burkitt's lymphoma is a cancer of the lymphatics of the head and neck, seen primarily in African children. While studying Burkitt's lymphoma in 1964, three scientists, Epstein, Achong, and Barr, succeeded in culturing white cells from excised cancer tissue. This was the first time lymphocytes had been grown in culture, since normal lymphocytes are in a terminal stage of cell differentiation and do not divide. The researchers then discovered a virus in the cultures (which is now named the Epstein–Barr virus [EBV]). EBV transforms lymphocytes into a continuously dividing cell line, therefore it is classified as an oncogenic virus (as described in Chapter 12). The EBV was the first oncogenic virus to be isolated from a human tumor.

The link between EBV and infectious mononucleosis was revealed accidentally. While working with EBV, a laboratory technician became ill with symptoms of infectious mononucleosis. Analysis of blood samples drawn sequentially during her illness demonstrated an increase in antibody titer against EBV. This finding provided the link between infectious mononucleosis and its viral cause, EBV.

Infectious mononucleosis is characterized by fatigue, fever, pharyngitis, lymphadenopathy, and splenomegaly. Similar symptoms, collectively known as infectious mononucleosis syndrome, are due to cytomegalovirus and *Toxoplasma* infections.

Characteristics of the Epstein–Barr Virus

Epstein–Barr virus is a typical herpes virus: an enveloped DNA virus that causes latent infection in humans and primates. The virus infects and transforms B lymphocytes, which carry EBV as a provirus. These infected cells can be induced to enter a lytic phase upon exposure to antimetabolites, during which EBV particles are released into the culture medium.

Though EBV was the first human oncogenic virus to be isolated from a human tumor, its role in Burkitt's lymphoma is still unknown. Since Burkitt's lymphoma occurs predominantly in Africa, it is possible that malaria or some other tropical disease is a contributing factor.

Symptoms

The symptoms of infectious mononucleosis appear 30 to 50 days after the patient is infected through intimate personal contact, as in mouth-to-mouth kissing. Fever, pharyngitis, and swollen cervical lymph nodes are common symptoms, which appear abruptly and last for 10 to 14 days. There may also be malaise, headache, muscle ache, hepatomegaly (enlarged liver), and splenomegaly (enlarged spleen). In half of the cases, those with splenomegaly (sple′no-meg′ah-le), there is a danger that the enlarged spleen will rupture if the person participates in contact sports. Most patients recover without complications after a prolonged but mild illness.

Epidemiology

During their first two years of life, many babies develop antibodies against EBV in response to asymptomatic infections and symptoms of infectious mononucleosis are rarely seen in this group. Most diagnosed cases occur in the 15- to 24-year-age group. In both developed and developing countries over 90 percent of the population older than 30 are seropositive for EBV antibody.

The virus is most probably transferred in saliva. One notable study done in 1955 at the United States Military Academy at West Point established the incubation time for infectious mononucleosis. The incidence of the disease peaked approximately 45 days after the all-male corps had returned from vacation. All cadets who had contracted infectious mononucleosis had engaged in mouth-to-mouth contact with other persons during their vacation. So transmission is by direct contact and the incubation period averages 45 days.

Since EBV causes latent infections, it is not surprising that the virus is shed from one-fourth of seropositive healthy adults, in 50 to 100 percent of infectious mononucleosis patients, and by some critically ill cancer patients. The symptoms are treated with aspirin or acetaminophen (as′et-am′i-no-fen). Corticosteroids have been advocated, but their use in treating infectious mononucleosis remains controversial.

CYTOMEGALOVIRUS INFECTIONS

Prenatal cytomegalovirus (si′to-meg′ah-lo-vi′rus) infections are believed to be an important cause of virus-induced mental retardation in infants born in the western world. Most cytomegalovirus (CMV) infections that occur during childhood are asymptomatic; however, infections of young adults can give rise to the infectious mononucleosis syndrome.

Characteristics of Cytomegalovirus

Cytomegalovirus is an enveloped double-stranded DNA virus that causes latent infections like other herpesviruses. There are at least three antigenically distinct human strains of CMV in addition to the many strains that infect animals. Although the mechanism of transmission is poorly understood, CMV has been isolated from urine, saliva, and other excretions of asymptomatic, seropositive patients, and it appears to be transmitted among persons who engage in intimate contact.

Symptoms

A small percentage of newborns (0.5–2.5 percent) shed the CMV at birth, most without experiencing disease symptoms. About 5 percent of these newborns display the symptoms of **cytomegalic inclusion disease** (CID): jaundice, hepatomegaly, splenomegaly, and rash. There may also be nervous system involvement with microcephaly (abnormally small head) and motor disability. These newborns are lethargic; they experience respiratory distress; and they may have convulsive seizures. Newborns who survive CID may be mentally retarded, and have motor deficits.

The virus is transferred from an asymptomatic pregnant woman to the fetus either transplacentally or vaginally (during the birth process). The likelihood that a woman will bear a CID neonate is greater if she has a primary cytomegalovirus infection during the pregnancy. The most practical means for a pregnant woman to prevent a CMV infection is to be scrupulous in matters of personal hygiene, especially while in close contact with infants. Young adults infected with CMV can experience symptoms of infectious mononucleosis, but they recover without complications.

Epidemiology

Cytomegalovirus can be transmitted across the placenta, by oral contact, in cervical and other excretions, and during blood transfusion and organ transplantations. A remarkably high number of CMV infections follow renal and bone marrow transplantation and there is presumptive evidence that cytomegalovirus is transmitted by direct sexual contact.

HEPATITIS B VIRUS INFECTIONS

The hepatitis B virus causes a spectrum of liver diseases ranging in intensity from extremely mild to fatal. It is estimated that 100,000 Americans are infected annually, and that 6 to 8 percent of them become persistent carriers of the hepatitis B virus. These people transmit the virus to others in their blood or blood products, saliva, semen, and possibly other body fluids.

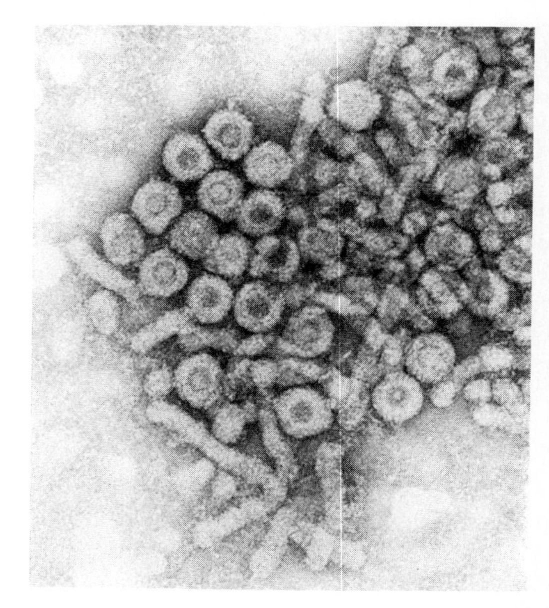

FIGURE 26.5 Electron micrograph of an aggregate of hepatitis B virus isolated from the blood of patients with serum hepatitis. Both spherical (42 nm) and tubular forms are present. (Reproduced with permission. From A. J. Zuckerman and C. R. Howard, *Hepatitis Viruses of Man,* Academic Press, London, 1979. Copyright © 1979, Academic Press Inc., Ltd, London.)

Hepatitis A and B viruses were first recognized by the presence of their antigens in serum and feces: antigen A, in the serum and feces of patients with infectious hepatitis (see Chapter 25); hepatitis B surface antigen (HBsAg), in the blood of patients infected with hepatitis B virus. These viruses are not related, though both cause hepatitis. A third type of hepatitis is caused by one or more agents possessing neither the A antigen nor the B antigen, so it is called **non-A, non-B hepatitis.**

At one time, any person whose serum contained HBsAg was characterized as having **serum hepatitis,** because transfer was primarily by blood transfusions and blood products. Now that only healthy people are permitted to donate blood and this blood is screened to detect the hepatitis B virus, blood is not a significant vehicle, and the disease goes by the name **hepatitis B.**

Hepatitis B Virus

The hepatitis B virus has never been grown in culture, so the serum of persistent carriers is the ony source of the virus. As originally isolated from serum, the virus was known as a Dane particle and identified by the HBsAg associated with it (Figure 26.5). The particles are indeed viral in nature, having a core nucleocapsid surrounded by a lipid-containing outer envelope. The nucleocapsid contains a circular DNA that is double-stranded over two-thirds of its length. The virus has the capacity to randomly integrate into the host's genome, which predisposes an infected patient to develop liver cancer.

Self-Limiting Versus Persistent Hepatitis B Infections

A hepatitis B infection may be self-limiting, or may induce a persistent viral infection that lasts for years. The incubation period of the **self-limiting infection** is about six weeks, after which the HBsAg can be detected in the blood (Figure 26.6). Dane particles appear in the blood (they are not demonstrable in feces) after the tenth week, and symptoms of hepatitis begin during the fourteenth week. After some four weeks, the symptoms clear concomitantly with the disappearance

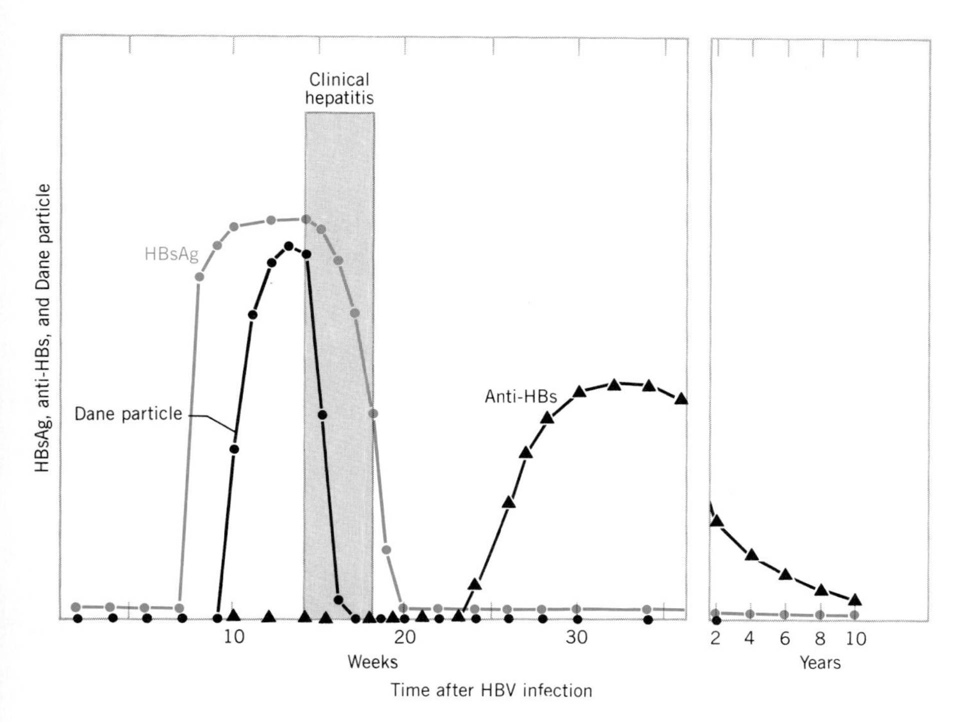

FIGURE 26.6 Course of self-limiting hepatitis B virus infection. Dane particles and HBsAg are present in the patient's blood during the later part of the 6- to 10-week incubation period. (Courtesy of Dr. W. S. Robinson. From W. S. Robinson, "Hepatitis B Virus," in G. L. Mandell, R. G. Douglas, Jr., and J. E. Bennett (eds.), *Principles and Practice of Infectious Diseases,* Wiley, New York, 1979. Copyright © 1979, John Wiley.)

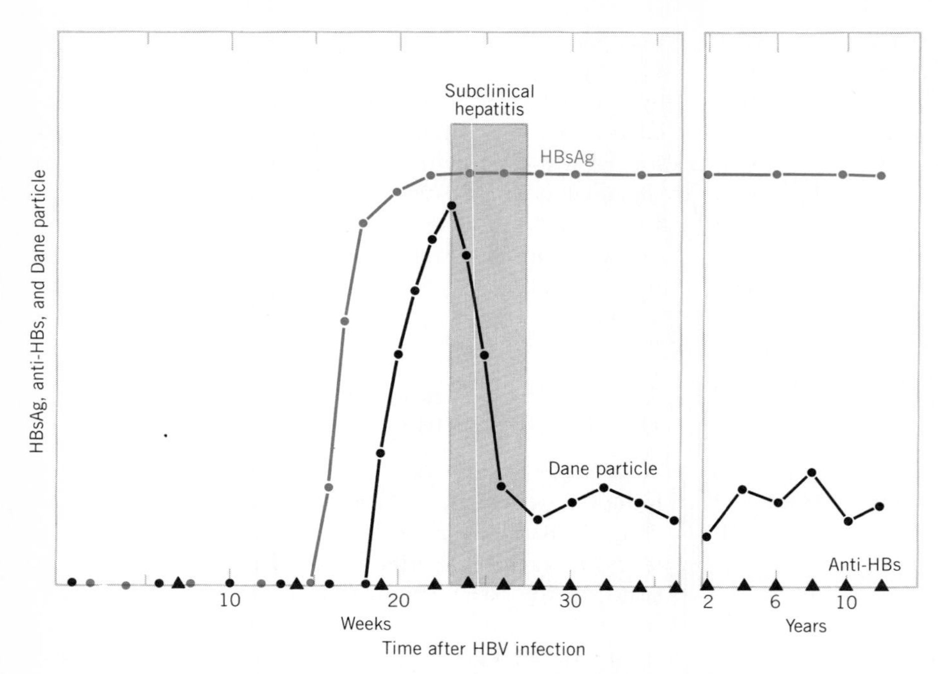

FIGURE 26.7 Persistent hepatitis B infection. The incubation period is longer than in other forms of viral hepatitis, and no anti-HBs antibodies are found in the patient's serum. (Courtesy of Dr. W. S. Robinson. From W. S. Robinson, "Hepatitis B Virus," In G. L. Mandell, R. G. Douglas, Jr. and J. E. Bennett (eds.), *Principles and Practice of Infectious Diseases,* Wiley, New York, 1979. Copyright © 1979, Wiley & Sons, Inc.

of HBsAg from the blood. Patients with this form of infection rid their bodies of virus by mounting an antibody response.

Persistent HBV infections have a longer incubation period and apparently last for years (Figure 26.7). The HBsAg appears in the serum about 15 weeks after infection and persists indefinitely. Dane particles appear in the serum 2 weeks later and reach a peak concentration coincident with the symptoms. In contrast to what happens in the self-limiting infection, serum antibodies are not made. Persistent HBV infections represent between 5 and 10 percent of acute hepatitis cases, which are favored by a low infecting dose, a young age, and a mild or asymptomatic case.

Symptoms

The symptoms of hepatitis B infections include dark urine, malaise, anorexia, nausea, fever, vomiting, headache, and jaundice. As many as 50 percent of HBV infections result in mild illness, and jaundice is absent. Liver damage caused by the hepatitis virus results in an increase in the patient's serum aminotransferase activity.

A very small number of patients suffer a prolonged bout of acute fulminant hepatitis, in which the fatality rate is high. One finding suggests a positive correlation between hepatitis B infections and primary liver cancer—an ominous note, since the latter is the most common visceral malignancy in humans.

Epidemiology

Hepatitis B is endemic in parts of Africa and Asia where more than 10 percent of some population groups have HBsAg in their serum. Contrast this figure with that in the United States, where the percentage of HBsAg persons is under 0.2 percent. As mentioned, blood transfusions and blood products are a

Hepatitis B and the Dentistry Profession

During the first six months of 1986, four clinical cases of hepatitis B were reported in a city in New Hampshire. Each patient had been seen by the same oral surgeon, to have one or more teeth extracted three to five months before becoming ill. Three had had multiple extractions during single office visits. All four patients denied other risk factors for hepatitis B virus infection.

Of the four patients, one remained seropositive for hepatitis B surface antigen (HBsAg) for more than six months and became a chronic hepatitis B carrier. He was tested and found to have the same HBsAg subtype as the oral surgeon. Ten other cases of hepatitis B were reported in the city during the first six months of 1986. Two of the patients were intravenous drug users; two were contacts of patients with unreported cases of hepatitis; and six had no identified risk factors. None of these ten patients had been treated by a dental professional or had undergone surgery.

The oral surgeon had been practicing in the city (population 75,000) for 25 years. His practice was limited to dental extractions, usually performed with a combination of intravenous sedation and local anesthesia. He had never had any symptoms suggestive of a hepatitis B infection prior to the outbreak. In July 1986 he was seropositive for HBsAg and HBeAg and negative for IgM antibody to hepatitis B core antigen, indicating that he probably was a hepatitis B carrier. He was not aware of having had any skin lesions on his hands in the past year. Although he was careful to scrub his hands between surgical procedures, he did not wear gloves.

The oral surgeon discontinued his practice when the epidemic outbreak was discovered. Letters were sent to all patients whom he had treated after January 1, 1985, informing them of their possible exposure to hepatitis B virus and offering free testing for hepatitis B serological markers.

Comment

Eight other outbreaks of hepatitis B traceable to dentists or oral surgeons have been reported since 1974. In each outbreak the implicated dentist or oral surgeon was seropositive for HBsAg and (if tested) for HBeAg, was unaware of a chronic infection, and did not use gloves during dental or surgical procedures. Traumatic procedures such as surgery and extractions have a higher infection risk than nontraumatic fillings and denture fittings. Transmission of hepatitis B by dental practitioners is preventable by immunization with one of the hepatitis B vaccines, and by wearing gloves during all procedures requiring hand contact with patients' mouths.

Adapted from *Morbidity and Mortality Weekly Report,* Vol. 36, No. 9, March 13, 1987

minor route of hepatitis B transmission in developed countries because of blood screening.

HBV infections are attributed to hemodialysis, ear piercing, tattooing, intravenous drug abuse, promiscuous heterosexual activity, male homosexual activity, and accidental puncture with a contaminated needle. The virus can also infect through open skin abrasions and direct contact with mucous membranes, but these sources of infection are rare. Contaminated instruments and all samples taken from infected patients should be sterilized to prevent the spread of the virus.

The first hepatitis B vaccine was made by purifying the virus particles (HBsAg) from the serum of chronic hepatitis B carriers. Extreme caution had to be taken to insure that all the viruses present were killed before the vaccine was used to immunize humans. Now there is a genetically engineered hepatitis B vaccine (described in Chapter 30), which is recommended for adults who are at risk of exposure, including homosexual males, users of illicit injectable drugs, household and sexual contacts of carriers, health care workers who must handle blood and blood products, hemodialysis patients, recipients of blood products, and morticians. Passive immunization with hepatitis B immune globulin, alone or in combination with hepatitis B vaccine, is recommended as prophylaxis following exposure to the virus.

NON-A, NON-B HEPATITIS

Earlier we touched on a type of hepatitis designated non-A, non-B hepatitis, because neither HAAg nor

HBsAg is involved. Non-A, non-B hepatitis is an acute, chronic form of hepatitis for which there is probably more than one etiological agent. Epidemiological evidence indicates that non-A, non-B hepatitis is the main form of hepatitis transmitted by blood and blood products, now that blood donors with hepatitis B are prevented from giving blood. The incubation period of non-A, non-B hepatitis is four to five weeks, which is intermediate to the incubation periods for hepatitis A and hepatitis B. The symptoms are the same for all types of acute hepatitis. Controlling non-A, non-B hepatitis is a serious problem since there are no accurate tests for detecting the etiological agent(s).

ARTHROPOD-BORNE VIRAL DISEASES OF BLOOD

People are the chief reservoir for the yellow fever and dengue (deng′ e) viruses, which are spread by *Aedes aegypti*. This people-biting mosquito nests in standing water in old tires, flower pots, water dishes used by pets, and discarded tin cans. Public health efforts to control mosquitoes in tropical urban centers were largely abandoned in the late 1960s when the programs became too expensive and laborious. Reestablishment of *Aedes aegypti* in tropical countries resulted in a resurgence of epidemic dengue.

DENGUE AND HEMORRHAGIC DISEASE

Dengue and hemorrhagic dengue occur in Asia, tropical Africa, the Southwest Pacific (including Hawaii), the Caribbean, Central America, and South America. They are caused by the four serotypes of the dengue virus that infect *A. aegypti* as it feeds on an infected person. The virus multiplies in the mosquito's salivary glands before it is transmitted to the next person.

Dengue is a pathological term that describes an infectious, eruptive fever of warm climates. The fever is accompanied by a frontal headache; pain in the back, limbs, joints, and behind the eyes; and lymphadenopathy. A maculopapular rash begins on the chest and spreads outward. The fever is often biphasic (peak–remission–peak) and lasts from two to seven days.

Hemorrhagic dengue has been reported in Southeast Asia since 1958 and on islands in the Southwest Pacific since 1971. Patients with this syndrome suffer visible signs of hemorrhage in addition to the just-cited clinical symptoms. Hemorrhages in the skin (purpura), blood in the urine (hematuria), gastrointestinal hemorrhage, and varying degrees of shock are its clinical symptoms.

Epidemics of dengue occur in the tropics during the rainy season. When a new virus serotype is introduced into an isolated region, the attack rates can exceed 50 percent of the population. People can protect themselves against infection by wearing protective clothing and using mosquito repellent. Live, attenuated viral vaccines are being evaluated.

YELLOW FEVER

Epidemics of yellow fever were first seen in the seventeenth century coincident with the development of trade between the colonists and South and Central America. Yellow fever was introduced to the continental United States by returning sailors, and was dreaded by the colonists because of its high mortality rate.

Major Walter Reed and the Yellow Fever Commission

Yellow fever was a chief obstacle to establishing trade with tropical countries and for many years prevented the construction of the Panama Canal. At the beginning of the twentieth century, Major Walter Reed was assigned to lead the U.S. Army Yellow Fever Commission to study the disease in Cuba.

Preliminary studies indicated that mosquitoes were the vector for yellow fever. The commission scientists began by raising adult mosquitoes from eggs in stoppered glass tubes. These adults were permitted to bite a yellow fever patient by exposing the unstoppered end of the glass tube against his skin. When mosquitoes thus infected fed on a unexposed volunteer, the person soon became ill with yellow fever. After further studies, the commission confirmed that yellow fever was transmitted by mosquitoes.

Control of the mosquito population in Cuba, and subsequently in other parts of the world, resulted in a significant reduction in the disease. Now yellow fever is limited to tropical areas of Africa, South America, and Central America (Figure 26.8).

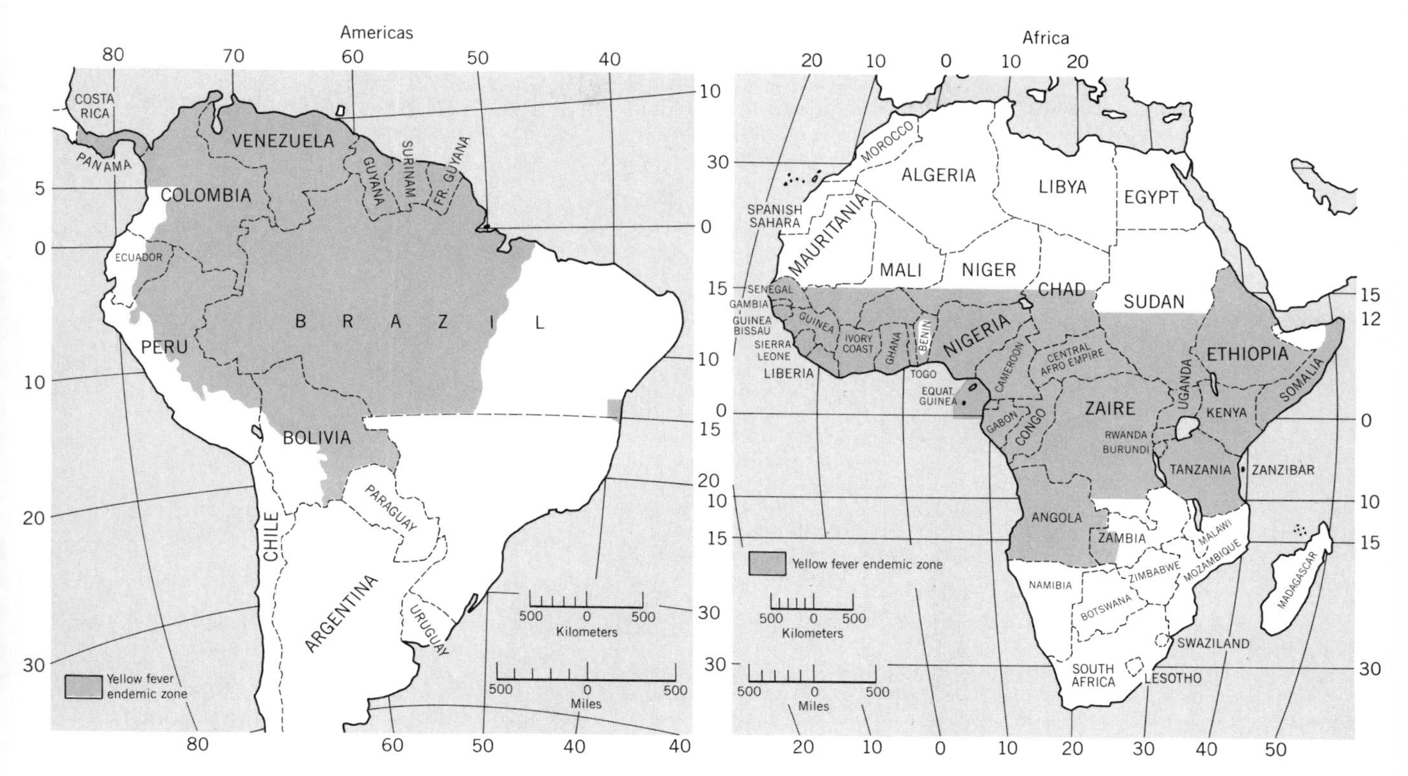

FIGURE 26.8 Yellow fever occurs in endemic zones where the *Aedes aegypti* mosquito is found. (From the World Health Organization, Geneva.)

Natural Transmission Cycles

Monkeys are the reservoir for **jungle yellow fever.** The wild virus is enzootic in tree-dwelling monkeys (howler monkeys and marmosets) and is transmitted, by infected mosquitoes, to people who enter or live near tropical forests (Figure 26.9). The jungle cycle of yellow fever results in sporadic cases of the disease, whereas epidemics of yellow fever occur in an **urban cycle** in which the yellow fever virus is transmitted between humans (reservoir) by *A. aegypti*.

Symptoms

The virus grows in the salivary glands of an infected mosquito, which is infectious throughout its six- to eight-week lifespan. Four to five days after transmission, the virus causes viremia. The symptoms of yellow fever are an influenza-like illness of one week duration, with or without jaundice. Fatality rates range from 10 to 60 percent, depending on the geographical distribution of the disease.

Prevention and Control

Mosquito control has been effective in urban areas where the jungle yellow fever cycle does not exist, but is impractical in the tropics. The yellow fever vaccine, made from the only known serotype of the virus, provides long-lasting immunity. The vaccine is given to people traveling or living in areas where yellow fever is endemic (Figure 26.8).

COLORADO TICK FEVER

Colorado tick fever is a viral disease of the mountainous regions of the western United States and Canada. The virus winters over in the nymph and adult forms of the tick, *Dermacentor andersoni*. The nymphs

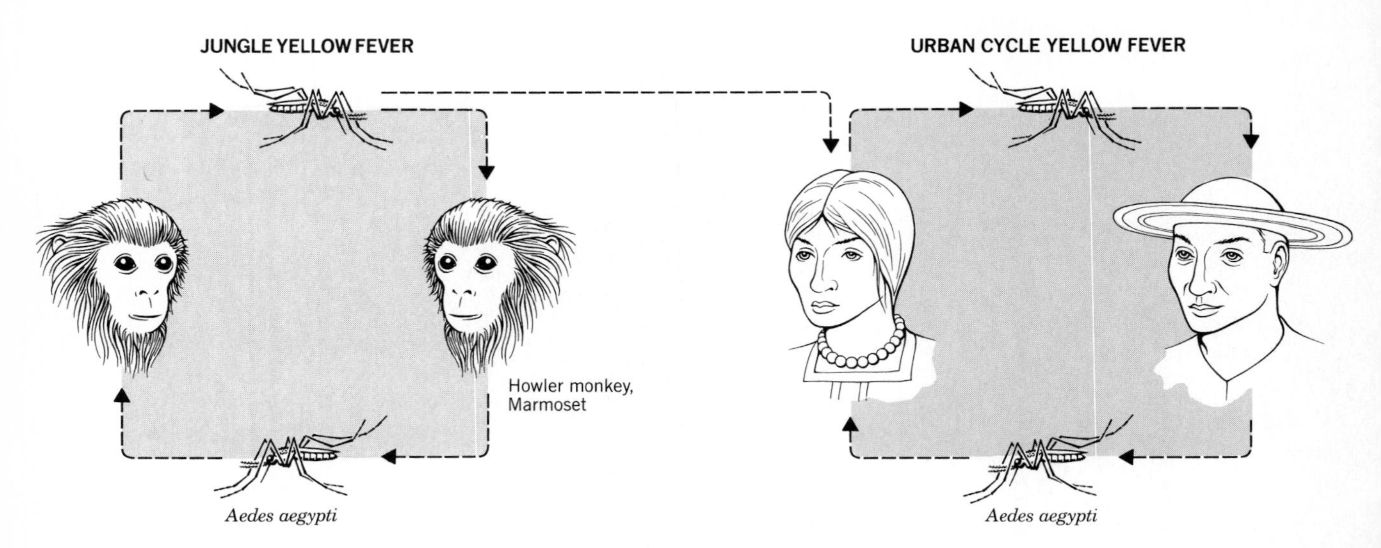

FIGURE 26.9 Yellow fever occurs in an urban cycle, in which humans are the reservoir for the virus, and a jungle cycle, in which monkeys also serve as the reservoir. In both cycles the virus is transmitted between hosts by *Aedes aegypti* mosquitoes.

infect ground squirrels and chipmunks, which become viremic and pass the virus to other ticks in a spring amplification cycle. People are infected when bitten by adult ticks. The result is a febrile disease with accompanying chills, headache, ocular pain, nausea, and vomiting. The acute phase of the illness lasts seven to ten days, with half of the patients having a "saddle-back" fever. Recovery is usually complete.

Residents of enzootic areas who work in outdoor professions or who pursue recreational activities in the Rocky Mountain states are most likely to be infected. Colorado tick fever can be prevented by the use of insect repellents, wearing protective clothing, and by the removal of ticks before they attach and feed. Campers and campsite planners should avoid areas where Colorado tick fever is a problem.

PROTOZOAN DISEASES OF BLOOD

Protozoan diseases of livestock and humans take an immense toll on human welfare. There are many protozoa that cause debilitating human illnesses and even death, especially among young children; and because of protozoan diseases of animals, the raising of domestic livestock is severely restricted. In Africa, for example, only certain breeds of cattle that are resistant to trypanosomal diseases can be raised in the areas where the tsetse fly is prevalent.

Many protozoan diseases are transmitted between animal reservoirs and humans by arthropod vectors. Their incidence depends on the distribution of the etiological agent, the resistance of the population to it, and the geographical distribution of vectors and reservoirs.

MALARIA

Malaria is estimated to afflict about 300 million people and to cause 2 to 4 million deaths annually, mostly among children less than 5 years old. Malaria was known to the Egyptians as early as 1550 B.C., and derived its name from the bad air (It. *mal* bad, *aria* air) that in ancient times was believed to be the cause of the devastating illness. The species of *Plasmodium* (plaz-mo′de-um) that cause malaria are widely distributed in the tropical regions of the world (Figure 26.10).

Plasmodias have a complex life cycle that involves both a vertebrate and an invertebrate host. *Anopheles* (ah-nof′e-lez) mosquitoes are the invertebrate hosts and

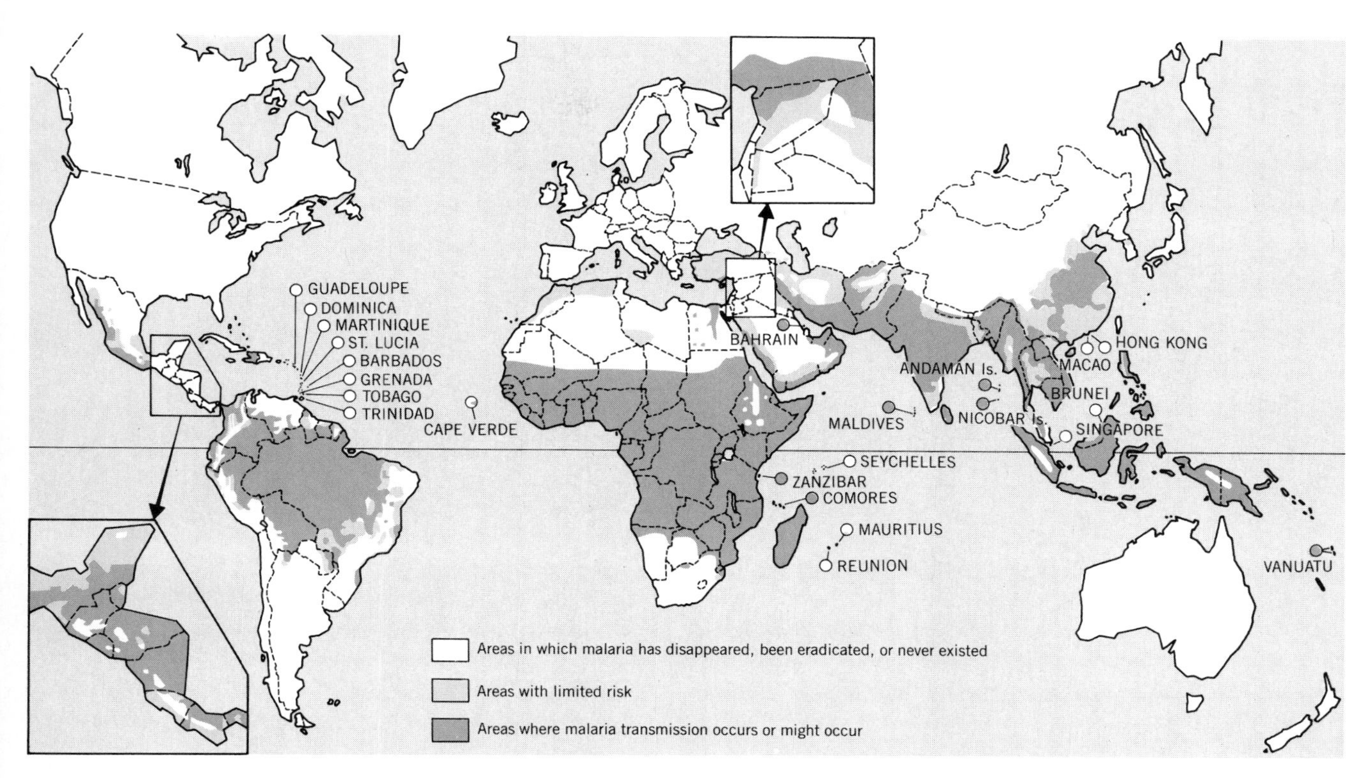

FIGURE 26.10 Areas of risk for malaria, December 1977. (WHO Weekly Epidemiological Record No 22, 1979.)

vectors of the plasmodia, while reptiles, birds, and mammals are the vertebrate hosts.

Life Cycle of *Plasmodia*

The sporozoites of *Plasmodium* spp. in the salivary glands of an infected female mosquito are transferred to a person when the mosquito bites and takes a blood meal (Figure 26.11). The sporozoites are transported in the blood to the liver, where they reproduce by an asexual process of multiple fission called schizogony. This results in the emergence of *merozoites* (mer′o-so′it), which infect either liver cells or erythrocytes.

The Erythrocyte Cycle Once in the erythrocyte, the parasites are called trophozoites (Color Plate II). In this form they again reproduce by schizogony. Now, as merozoites (Figure 26.11), they are released into the blood at synchronized intervals. Release causes the paroxysms of fever, followed by chills, that are the hallmark of malaria. Asexual replication continues and the plasmodia may persist in the peripheral blood at low levels for years, or the merozoites may enter other erythrocytes and start a new cycle of reproduction.

Sexual Reproduction in the Mosquito Sexual reproduction begins in the erythrocytes of the vertebrate host and is completed in the mosquito. *Plasmodium* forms microgamonts and motile macrogamonts in infected erythrocytes, where they remain until ingested by a female mosquito. In the mosquito's stomach, the gamonts emerge from the erythrocyte and join to form the zygote of sexual reproduction.

The zygote differentiates into a worm-like **ookinete,** which penetrates through the midgut wall, undergoes meiosis and asexual reproduction (sporogony) to produce infective sporozoites. The sporozoites then make their way to the salivary glands, ready to initiate a new reproduction cycle when they are injected into a vertebrate host.

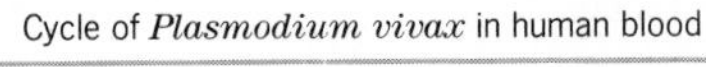

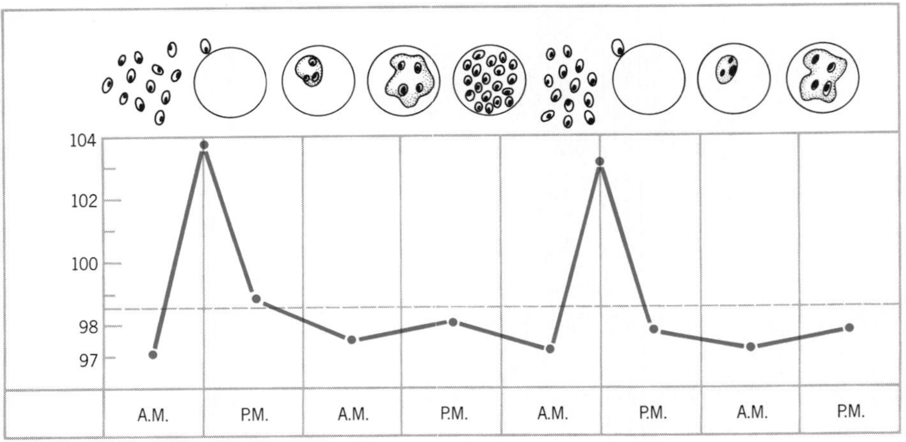

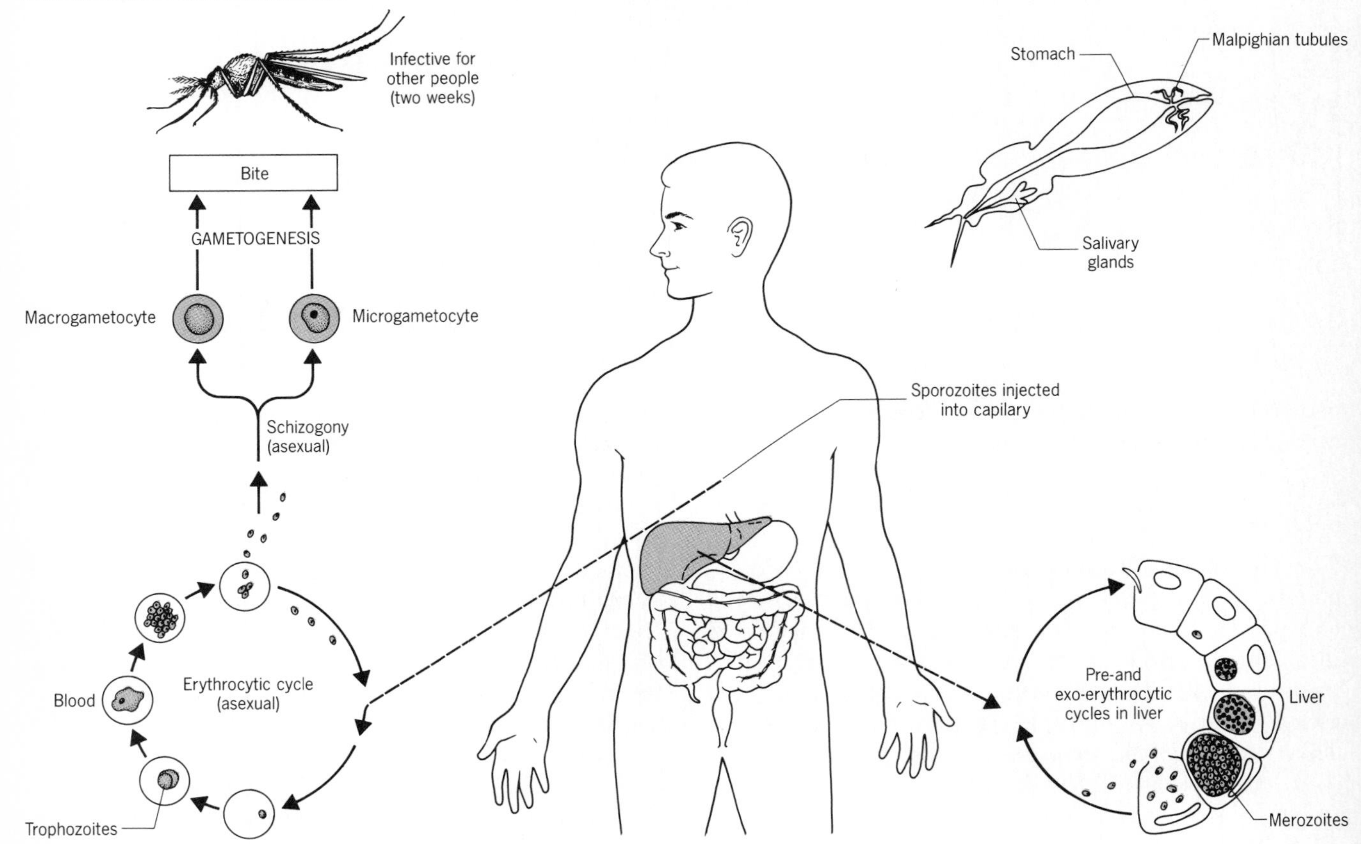

FIGURE 26.11 The life cycle of *Plasmodium vivax*. Sexual replication occurs in an invertebrate host (mosquito) and asexual reproduction (schizogony) occurs in the vertebrate host. Malaria is diagnosed by observing the reproductive phases of the parasite in a patient's red blood cells.

The Clinical Picture

The patient's fever is extreme; it may rise to 40° or 40.5°C, as the merozoites are released from the infected erythrocytes. The patient feels unbearably hot during the peak of the fever, then begins to sweat as the fever declines and is followed by chills. The cycle of chills and fever often recurs in a definite pattern of 48 or 72 hours depending on the species of *Plasmodium* responsible. Untreated cases last for three to six weeks, and patients may be asymptomatic between the chills-and-fever episodes. Relapse in untreated patients can occur after a two- to ten-month latency period.

Humans develop a partial, short-lived immunity to malaria attributed to the ability of macrophages to phagocytize infected erythrocytes; however, the parasite can remain dormant in host cells for years, causing late relapses. *Plasmodium falciparum,* the most virulent of all the species to infect humans, causes virtually all fatalities from malaria. Fortunately, persons with the sickle-cell trait have an innate resistance to *P. falciparum,* which cannot replicate in sickled erythrocytes.

Diagnosis and Treatment

Diagnosis is based on observations of the reproductive stages of *Plasmodium* in the blood. Quinine has been the anchor of treatment since earliest times, and its basic structure has served as a model for synthetic antimalarials (chloroquine and quinacrine hydrochloride). Travelers to a malarious area are advised to take chloroquine for two weeks before entering and for six weeks after leaving the region.

ARTHROPOD-BORNE FLAGELLATES

Trypanosomes (tri-pan′o-som) are flagellated, pathogenic protozoa transmitted among humans, or to humans from animal reservoirs, by blood-sucking insects. They are most prevalent in tropical climates where they cause serious illnesses.

The uniflagellated trypanosomes that infect wild vertebrates, domestic livestock, and humans are identified morphologically in blood smears from infected animals (Figure 26.12). The distinctive flagellum of a trypanosome is part of an undulating membrane that runs most of the cell's length.

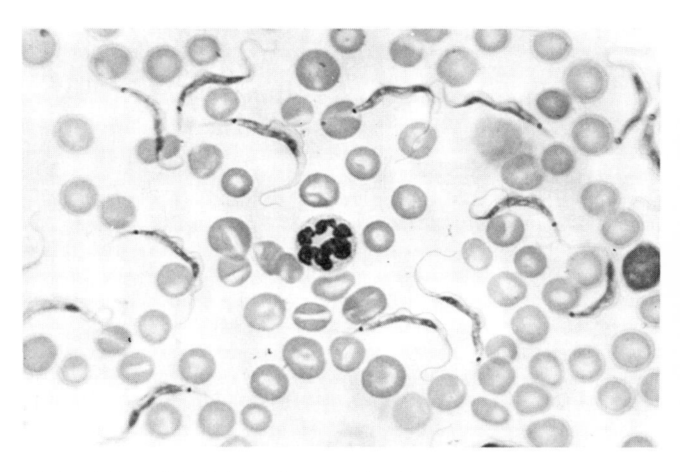

FIGURE 26.12 Micrograph of *Trypanosoma lewis* in blood. (Courtesy of Carolina Biological Supply Company.)

American Trypanosomiasis (Chagas' Disease)

American trypanosomiasis or Chagas' (chag′as) disease is caused by *Trypanosoma cruzi* (kruz′ee). It is transmitted to humans by blood-sucking insects, called "kissing bugs." The insect lives in cracks and crevices of primitive dwellings, primarily in South and Central America. It emerges at night to take a blood meal from the lips of the sleeping person. Trypanosomes replicate in the insect's gut and are excreted in its feces, which are deposited during feeding. They infect the bite wound, or the sleeping victim inadvertently rubs or scratches the bite thus transferring the typanosomes to mucous membranes or the conjunctiva.

The trypanosomes multiply in lymphoid tissue before spreading to the muscle fibers and cells of the nervous system. American trypanosomiasis occurs mainly in children with symptoms of fever, malaise, anorexia, and edema. Death from heart failure or acute meningoencephalitis occurs in 5 to 10 percent of acute cases. These infections do not respond well to chemotherapy.

African Trypanosomiasis (Sleeping Sickness)

Trypanosoma brucei causes the disease popularly known as African sleeping sickness. It is seen in the eastern and western regions of Africa, and is transmitted by tsetse flies (*Glossina* sp.). In the western regions, humans are the only blood source for the tsetse flies; in the east, humans, antelopes, and wild hogs are sources.

Trypanosomes are transmitted to humans when an infected tsetse fly takes a blood meal. In humans, the trypanosomes replicate in the blood and lymph to cause a systemic infection. Swollen lymph nodes, especially those on the back on the neck, are known as **Winterbottom's sign** and are diagnostic. Eventually, the trypanosomes infect the central nervous system, causing the clinical symptoms of apathy, somnolence, and malaise—hence the popular name. Coma is finally followed by death.

Diagnosis is made on the basis of trypanosomes present in blood or in specimens from the lymph glands. Organic arsenicals are effective when prescribed during the early stages of the infection. Eradication of the tsetse fly from areas inhabited by humans is the only effective means of control.

Summary Outline

FUNCTION OF THE CARDIOVASCULAR SYSTEM

1. The heart is a four-chambered pump. It receives blood into its left and right atria and pumps blood through arteries to the lungs and other parts of the body.
2. Blood flows through the arteries into the smaller arterioles, then into capillaries, which, being one cell thick, permit the delivery of nutrients and oxygen and uptake of wastes.
3. Blood transports oxygenated red cells, soluble nutrients, salts, phagocytes, immunoglobulins, lymphokines, hormones, and other substances to the tissues of the body.
4. The lymphatic system is a network of vessels and nodes that collects fluid, called lymph, from intercellular spaces, processes it, and transports it to the veins.
5. The lymphatic system includes the lymph nodes, the tonsils, the thymus, and the spleen.

BACTERIAL DISEASES OF BLOOD

1. Gram-negative rods, the opportunistic pathogens most often isolated from bacteremias, enter the blood from a focus of infection. Endotoxin present in their cell walls contributes to the symptoms and creates a state of endotoxemia.
2. Endemic or louse-borne relapsing fever is caused by the parasitic spirochete, *Borrelia recurrentis*. Tick-borne (endemic) relapsing fever is caused by *B. hermsii* and *B. turicatae*.
3. Relapses occur as borrelias reappear after a mutation in which the surface antigens are altered.
4. Lyme disease is a tick-borne spirochete infection caused by *B. burgdorferi*.

BACTERIAL DISEASES OF THE LYMPHATIC SYSTEM

1. *Yersinia pestis* causes bubonic plague, an enzootic disease of rodents and a fatal disease of humans if not treated. The urban cycle of plague occurs when the bacteria are transmitted to humans by fleas from the bacterial reservoir in rats.
2. Plague in the United States occurs in a sylvan cycle; the plague bacillus is transmitted from its reservoir in ground squirrels, prairie dogs, and other animals to humans by fleas.
3. *Yersinia pestis* infects the lymph nodes, producing buboes, and causing black hemorrhages to form under the skin.
4. In the later stages of bubonic plague, the bacteria attack the lungs, causing pneumonic plague, a form of the disease transmitted person to person in aerosols.
5. Tularemia is an enzootic disease of small animals caused by *Francisella tularensis*. The bacterium is transmitted directly by the bite of ticks and deer flies, and to cuts infected during field dressing of game or fur-bearing animals; and indirectly by ingestion of contaminated meat or inhalation.
6. Brucellas are transmitted in contaminated milk or

meat, causing brucellosis (undulant fever). Brucellosis is controlled by eliminating infected cattle, sheep, and goats and by treating patients with streptomycin.

RICKETTSIAS: OBLIGATE INTRACELLULAR BACTERIA

1. Rickettsias are obligate intracellular parasites maintained in nature as parasites of arthropods and warm-blooded animals.
2. Flea-borne (endemic) typhus fever occurs at a low rate in regions of the United States that harbor rats infected with *Rickettsia typhi.*
3. Tick-borne typhus fever (Rocky Mountain spotted fever), the most common rickettsial disease in the United States is caused by *R. rickettsii,* which is maintained by transovarian passage in ticks.
4. Rickettsial diseases are prevented by avoiding the vectors and are treated with a tetracycline or chloramphenicol.

VIRAL INFECTIONS OF BLOOD AND LYMPH

1. Epstein–Barr virus (EBV), originally isolated from Burkitt's lymphoma, was the first oncogenic virus to be isolated from a human tumor.
2. EBV causes infectious mononucleosis, characterized by fatigue, fever, pharyngitis, lymphadenopathy, and splenomegaly. It is transmitted between persons having close contact, as in kissing, with most cases occurring in 15- to 24-year olds.
3. A small percentage of newborns display the symptoms of cytomegalic inclusion disease, including lethargy, respiratory distress, and convulsive seizures. Newborns who survive may be mentally retarded and have motor deficits.
4. Most CMV infections of young children are asymptomatic, whereas young adults can experience symptoms of infectious mononucleosis upon primary infection. Both EBV and CMV cause latent infection.
5. The hepatitis B virus causes a spectrum of diseases in humans ranging from asymptomatic illness to fatal infections of the liver.
6. The hepatitis B virus is transmitted from carriers to others by blood, blood products, hemodialysis, tattooing, intravenous drug abuse, and sexual contact.
7. HBV may cause a self-limiting infection in which anti-HBsAg antibodies are formed, or it may cause a persistent infection, lasting years.
8. Persons at high risk of exposure should be immunized with the hepatitis B vaccine. Postexposure treatment with immune globulin is recommended.
9. Non-A, non-B hepatitis, a similar disease caused by one or more yet unidentified agents, is transmitted by blood and blood products.

ARTHROPOD-BORNE VIRAL DISEASES OF BLOOD

1. *Aedes aegypti* is the vector for transmitting the dengue viruses between their human reservoirs. Dengue is an infectious, eruptive fever of warm climates caused by four serotypes of the dengue virus.
2. Tree-dwelling monkeys are the reservoir for the jungle cycle of yellow fever, while people are the reservoir for the urban cycle of yellow fever. *Aedes aegypti* is the vector.
3. The live, attenuated yellow fever virus vaccine provides long-lasting immunity and is given to people traveling or living in areas where yellow fever is endemic—currently parts of Africa and South America.
4. Ticks and ground squirrels are the reservoirs, and ticks are the vectors for the Colorado tick fever virus. Infected people are definitive hosts for this virus, found in the mountainous regions of the western United States.

PROTOZOAN DISEASES OF BLOOD

1. *Plasmodium* spp. are widely distributed in the tropical regions of the world and cause malaria, one of the most common infectious diseases of humans.
2. *Plasmodium* is a sporozoan that replicates asexually in a vertebrate host and sexually in *Anopheles* mosquitoes, its insect vector.
3. The fever of malaria is associated with the release of merozoites from infected mammalian erythrocytes. Humans develop partial immunity to malaria. This disease is treated with antimalarials.

4. American trypanosomiasis Chagas' disease is caused by *Trypanosoma cruzi,* which is transmitted to humans by the "kissing bugs" that live in the cracks and crevices of primitive dwellings, primarily in South and Central America.

5. *Trypanosoma brucei* causes African trypanosomiasis (often called African sleeping sickness) in both west and east Africa; it is transmitted by the tsetse fly, which exists on the blood of humans, antelopes, and wild hogs.

Questions and Topics for Study and Discussion

QUESTIONS

1. Describe the febrile response seen in relapsing fever and explain its biological basis.
2. What are the similarities between tularemia and plague?
3. What groups of persons are most likely to contract brucellosis? What public health measures were instituted to control and prevent brucellosis in the United States?
4. How does obligate parasitism affect the incidence and distribution of rickettsial diseases in humans?
5. Explain the life cycles of *R. typhi* and *R. rickettsii.* Are humans necessary for the survival of these bacteria? Explain.
6. What virus causes most cases of infectious mononucleosis and how was it discovered?
7. Describe the symptoms of infectious mononucleosis and explain how the incubation period of this viral disease was determined? Could you conduct a similar experiment on your campus?
8. Compare congenital cytomegalic inclusion disease in newborns to congenital rubella syndrome. How do these diseases differ from herpes infections of newborns?
9. Describe the epidemiology of the three types of viral hepatitis. What steps can be taken to prevent their spread?
10. What precautions should be taken to protect against yellow fever and dengue during a trip to the tropics?
11. What biological process is responsible for the occurrence of the chills-and-fever cycle in malaria?
12. Describe the life cycle of the protozoan responsible for African trypanosomiasis. Why is this disease geographically localized?

DISCUSSION TOPICS

1. The rickettsias were once thought to be viruses. Do you think they are more like viruses or bacteria?
2. The first hepatitis B virus vaccine was made from virus particles isolated from pooled human plasma. Are there risks associated with this vaccine that are not present when the genetically engineered hepatitis B vaccine is used?
3. The Epstein–Barr virus was originally isolated from a human lymphoma. Do you think that oncogenic viruses are an important cause of human cancer? If so, is cancer an infectious disease?
4. What impact did banning the use of DDT have on controlling arthropod-borne infectious diseases? If you were in a position of authority to make a decision on the use of DDT, what factors would you be compelled to consider?

Further Readings

BOOKS

Bean, W., *Walter Reed,* University Press of Virginia, Charlottesville, 1982. A biography of Major Walter Reed.

Belshe, R. B. (ed.), *Textbook of Human Virology,* PSG Publishing, Littleton, Mass., 1984. Includes excellent chapters on the Epstein–Barr virus, cytomegalovirus, and the hepatitis viruses.

Brown, H. W., and F. A. Neva, *Basic Clinical Parasitology,* 5th ed., Appleton-Century-Crofts, Norwalk, Conn., 1983. A good source of material on malaria and African sleeping sickness.

Ho, M., *Cytomegalovirus, Biology and Infection,* Plenum, New York, 1982. This monograph covers both the virology and clinical impact of this important human virus.

ARTICLES AND REVIEWS

Barbour, A. G., and S. F. Hayes, "Biology of Borrelia Species," *Microbiol. Revs.,* 50:381–400 (1986). Reviews the structure, metabolism, and genetics of the borrelia that cause relapsing fever and Lyme disease.

Donelson, J. E., and M. J. Turner, "How the Trypanosome Changes Its Coat," *Sci. Am.,* 252:44–51 (February 1985). Describes the parasite's ability to remain in the bloodstream and evade the immune system.

Epstein, M. A., B. G. Achong, and Y. M. Barr, "Virus Particles in Cultured Lymphoblasts from Burkitt's Lymphoma," *Lancet,* 702–703 (March 1964). The first report of the isolation of an oncogenic virus from human tissue.

Godson, G. N., "Molecular Approaches to Malaria Vaccines," *Sci. Am.,* 252:52–59 (May 1985). Repeating antigenic sites on the outer coat of plasmodia may decoy the immune response.

Habicht, G. S, G. Beck, and J. L. Benach, "Lyme Disease," *Sci. Am.,* 257:78–83 (July 1987). The history of Lyme disease and discovery of the responsible tick-borne spirochete.

Monath, T. P., "Glad Tidings from Yellow Fever Research," *Science,* 229:734–735 (1985). A historical account of the yellow fever vaccine and a perspective on the importance of the sequencing of the yellow fever virus genome.

Norman, C., "The Unsung Hero of Yellow Fever?" *Science,* 223:1370–1372 (1984). The role played by Jesse Lazear in the early work on mosquito transmission of yellow fever.

Weiss, E., "The Biology of Rickettsiae," *Ann. Rev. Microbiol.,* 36:345–370 (1982). A review of the ecology, morphology, metabolism, pathogenesis, and immunobiology of the rickettsias.

CHAPTER

27

Infectious Diseases of the Nervous System

OUTLINE

FOCUS OF THIS CHAPTER

Humans are susceptible to nervous system diseases caused by bacterial toxins, infectious bacteria, viruses, and protozoa. This chapter focuses on diseases of the nervous system, the manner in which they are acquired, their clinical consequences, and the procedures available for their prevention and treatment.

The nervous system responds to the sensory stimuli it receives from all parts of the body and uses this information to coordinate body functions. Disturbances of the nervous system can lead to mental disorders, paralysis, coma, and even death.

THE NERVOUS SYSTEM

The nervous system is made up of billions of nerve cells organized into two parts, the **central nervous system,** which consists of the brain and the spinal cord, and the **peripheral nervous system,** consisting of the nerves branching off the brain and spinal cord. The afferent nerves transmit impulses, or signals, from the periphery to the brain. The brain processes the signals and relays them through the efferent nerves to the muscles, so they can take appropriate action.

The delicate brain and spinal cord are protected by the membranes called the **meninges** (Figure 27.1) and by the cerebrospinal fluid (CSF). The clear CSF surrounds the spinal cord and fills the four cavities (ventricles) of the brain, which literally floats in CSF.

The head is richly supplied with blood, but this

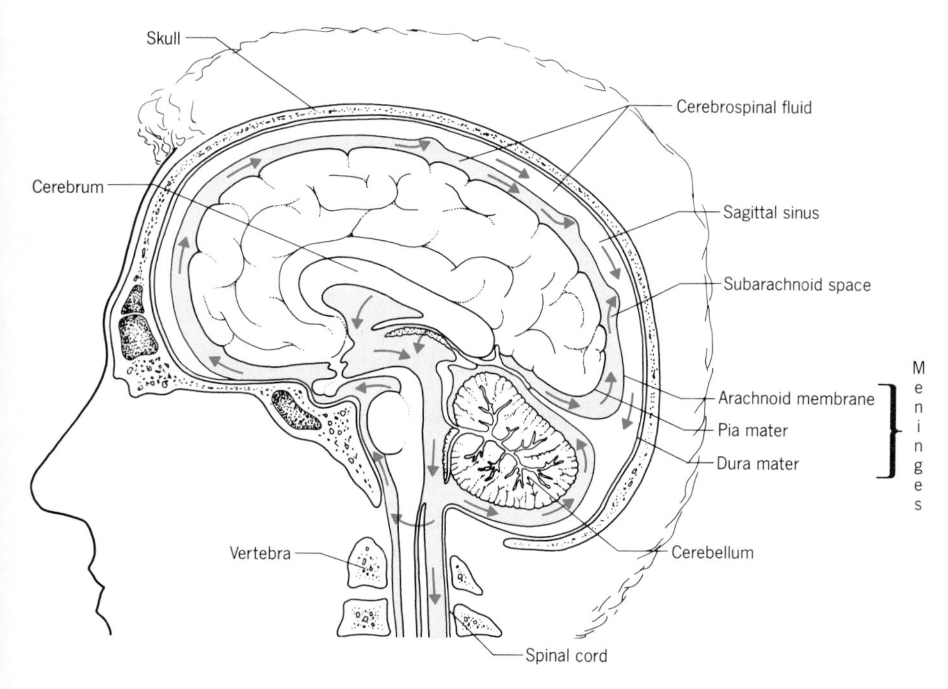

FIGURE 27.1 The central nervous system. The brain and the spinal column are surrounded by membranes collectively called the meninges.

blood never mixes with the CSF, since the tissues of the central nervous system are avascular. Essential nutrients are transferred into the CSF across a blood–brain barrier. The blood–brain interface is a complex group of anatomical barriers and physiological transport systems, which control the kinds of substances that enter the CSF and protect the central nervous system against harmful substances. Oxygen and glucose are transferred from the blood to the CSF, but microbes, blood cells, and many proteins including immunoglobulins cannot cross, so they are found in CSF only in diseased people. The CSF circulates between the nervous tissue and the meninges so that all parts of the central nervous system are supplied with nutrients.

Before a neurotropic pathogen can exert its effects, it will usually have caused damage to the blood–brain barrier. It is then that viruses, pathogens, and immunoglobulins are able to gain access to the central nervous system. Most neurotropic viruses enter through the blood; however, a few viruses move within peripheral neurons until they reach the central nervous system, a process of **neurotropic spread.** The chief manifestations of a central nervous system infection are **meningitis** (men′in-ji′tis), an inflammation of the meninges, and **encephalitis,** an inflammation of the brain.

BACTERIAL DISEASES OF THE NERVOUS SYSTEM

Bacterial meningitis is caused by organisms that first infect or are carried in the upper respiratory tract. From here the bacteria enter the bloodstream, which carries them to the blood–brain barrier of the central nervous system, where they stimulate the inflammatory response. This changes the permeability of the blood–brain barrier and permits pathogens to enter the cerebrospinal fluid and inflict damage. These changes also make the brain accessible to leukocytes, immunoglobulins, and some antibiotics.

The bacteria most likely to cause meningitis are *Neisseria meningitidis, Haemophilus influenzae,* and *Streptococcus pneumoniae.* Acute bacterial meningitis requires the utmost haste in diagnosis and treatment if the patient is to be saved.

MENINGOCOCCAL MENINGITIS

Humans are the only known reservoirs for two pathogenic species of *Neisseria: N. gonorrhoeae* causes gonorrhea (see Chapter 24) and *N. meningitidis,* commonly referred to as meningococcus, is a major cause of bacterial meningitis. Meningococcal meningitis is primarily a disease of children and young adults. It may be rapidly progressive—in fact, approximately 50 percent of patients hospitalized die within 24 hours after symptoms appear.

A preliminary diagnosis is made on the basis of clinical signs and the presence of bacteria in cerebrospinal fluid. A Gram stain of CSF showing gram-negative, nonspore-forming, bean-shaped diplococci (Figure 27.2) is positive for *Neisseria,* whereas the absence of bacteria in the Gram stain would point to aseptic meningitis (see below). A more specific diagnosis can be made using latex agglutination tests (see Chapter 18).

The presence of *Neisseria* is confirmed by plating CSF samples on chocolate blood-agar in an atmosphere of 3 to 10 percent CO_2 and 50 percent humidity. *Neis-*

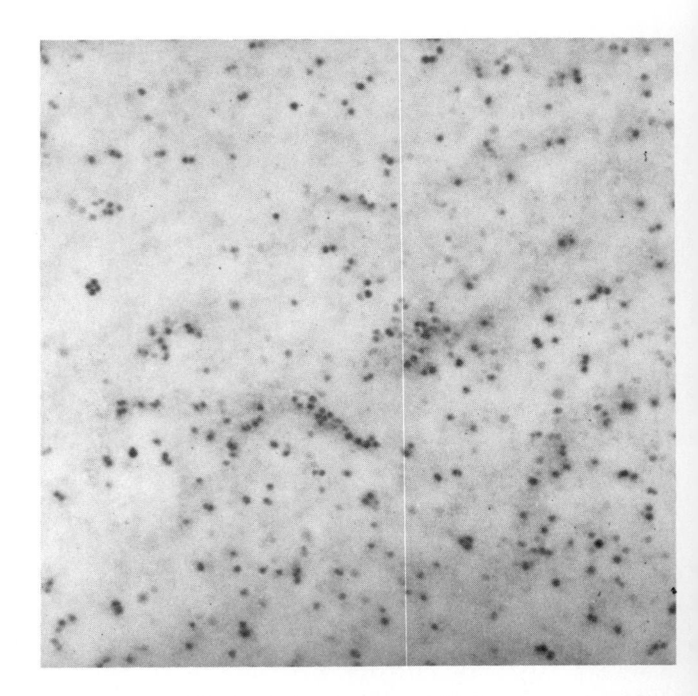

FIGURE 27.2 *Neisseria meningitidis* is a bean-shaped gram-negative coccus that causes meningitis. (Centers for Disease Control.)

seria survive poorly outside the host, so culturing has to be done with dispatch on samples taken before chemotherapy is initiated.

Epidemiology and Symptoms

In 1983, 52 percent of the reported U.S. cases of meningococcal meningitis developed in children younger than 5 years, with the highest incidence in children under the age of 1. (At one time, the disease affected military recruits who lived in confined quarters, but its incidence in this population now is low.)

About 5 percent of the population are carriers of *Neisseria meningitidis,* though the numbers rise significantly in advance of an epidemic. Physically fatigued persons living in close contact appear to have poor resistance to meningococcal infections, as do young children who have yet to produce an immune response.

As the meningococci invade the circulatory system, a bacteremia known as meningococcemia is created. From the blood, the bacteria penetrate the meninges and infect the spinal fluid. Just how they are able to gain entrance to the meninges is not known. Symptoms of vomiting, headache, slight fever, lethargy, confusion, and stiff neck follow. Diagnosis must be made quickly and therapy begun immediately. Patients between 2 months and 6 years are treated with ampicillin and chloramphenicol, and those above 6 years are given penicillin (or chloramphenicol if hypersensitive to penicillin).

The virulence of *N. meningitidis* is dependent on the presence of a polysaccharide capsule, which is antiphagocytic. Serological tests define the different polysaccharide capsules. At present most meningococcal illness is caused by antigenic groups B, C, and Y. Vaccines are available, but because only a few cases are seen today, the vaccines are not widely used.

HAEMOPHILUS MENINGITIS

Haemophilus influenzae is the most common cause of bacterial meningitis between birth and 6 years of age. The bacterium exists only as a common resident of the upper respiratory tract of humans, and no other host is

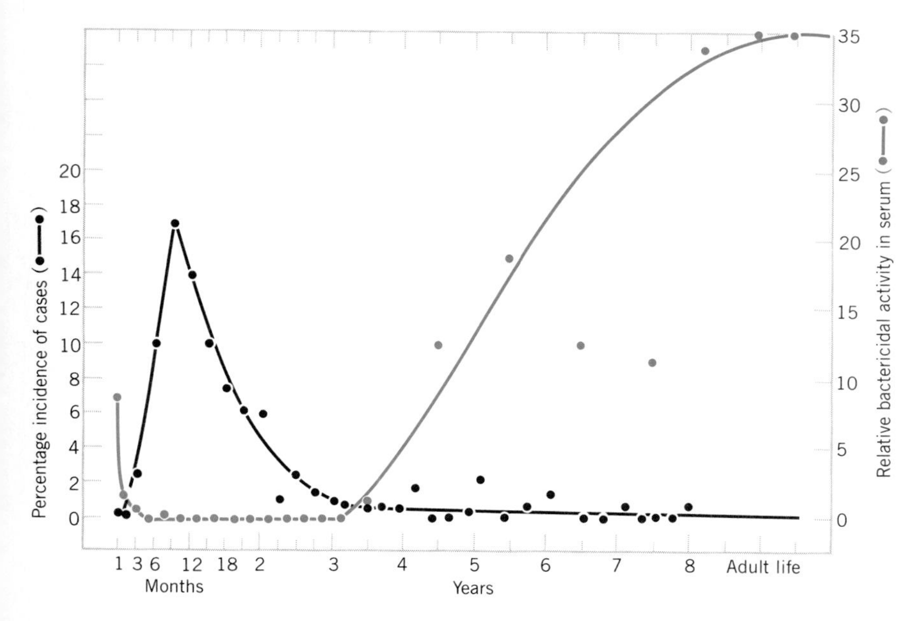

FIGURE 27.3 The influence of age on *Haemophilus influenzae* meningitis and the presence of bactericidal serum antibodies. Children under 2 years of age are the primary victims. (From L. D. Fothergill and J. Wright, *J. Immunology*, 24:273–284, 1933, copyright © 1933, The Williams and Wilkins Company.)

known. Transmission is by airborne droplets from person to person, and thus most infants become colonized during the first month of life.

Haemophilus influenzae spreads from the respiratory tract to the blood to the meninges. Babies under 6 months have a certain degree of resistance owing to the maternal antibodies that continue to circulate. These antibodies are bactericidal for *H. influenzae* (Figure 27.3); however, they disappear as the child gets older and are not replenished by the child until age 3 or so. It follows, then, that meningitis will be most frequent in youngsters 6 to 18 months of age, a period when antibodies are lacking. Untreated *H. influenzae* meningitis carries a tremendous fatality rate—between 90 and 100 percent; but with immediate care and careful treatment the fatality rate falls to 5 to 10 percent. In the United States there are about 8000 cases annually.

Children 15 to 18 months old can now be immunized with the *Haemophilus* b (Hib) vaccine, which protects them against *H. influenzae* caused meningitis. Since 1986, public health authorities have recommended that preschool children, especially those attending day-care centers, be immunized with this vaccine.

Ampicillin and chloramphenicol are effective agents for treating *Haemophilus* meningitis. Some strains of *H. influenzae* are resistant to ampicillin because they produce a β lactamase, therefore both drugs are recommended for initial therapy. Some children who recover are left with severe neurological deficits, including mental retardation and learning disabilities.

KEY POINT

The progression of *H. influenzae* meningitis is so swift that only immediate administration of antimicrobials can be effective. The need for rapid treatment is so great that serological tests are done after treatment is begun.

HANSEN'S DISEASE (LEPROSY) AFFECTS THE SKIN AND NERVES

Leprosy is one of the oldest-known human diseases, yet it remains a worldwide health problem. Historically, victims of the disease were cast out from society and forced to live in segregated colonies. People feared and hated this deforming disease; yet there was no basis for their fears, or for the isolation the victims were compelled to suffer, because leprosy is one of the least contagious diseases known. The bacterial cause of leprosy was first described in 1880 by Gerhard H. Hansen. The name **Hansen's disease** came to be preferred because of the negative connotations that have been associated with "leprosy."

Between 12 and 15 million people, living mainly in Southeast Asia and Africa, have the disease. Its special nature has led the U.S. Public Health Service to develop a central treatment facility, which is located in Carville, Lousiana. Most of the American patients, numbering some 3200, are treated there.

Mycobacterium leprae Is Slow-Growing

Mycobacterium leprae is an acid-fast rod that grows in skin and nervous tissue at sites having a relatively low temperature, such as the toes and fingers, the face, and the testicles in men. *Mycobacterium leprae* infections follow an indolent course, and symptoms may not appear for up to 6 years following infection—sometimes not until 40 years later. An early sign is the presence of light patches on the skin in areas that have lost their normal sensitivity to sensation.

As the infection progresses, the nerve endings in the hands and feet are destroyed, with consequent loss of sensation. Because the patient cannot feel pain in the extremities, physical injuries are common, and burns, swelling, and even forceful injury such as a crushed finger are not sensed.

With progressive destruction of bone and nerve tissue of the hands and feet comes loss of the use of these parts, as muscles fail to function. For example, the claw hand, as depicted in Figure 27.4, can be corrected only by surgery. With this relentless deterioration, there is shortening of fingers and toes, and finally the digits may be entirely lost. Facial disfigurement follows bacterial invasion. The eyebrows are destroyed, the nose collapses, and the skin becomes so thickened that the patient takes on the countenance of a lion.

Diagnosis

Mycobacterium leprae has not yet been grown in laboratory media, although it can be grown in the footpads of mice and in armadillos. Experimentation with *M. leprae* is very difficult because it grows selectively in tissues having a low temperature and its generation

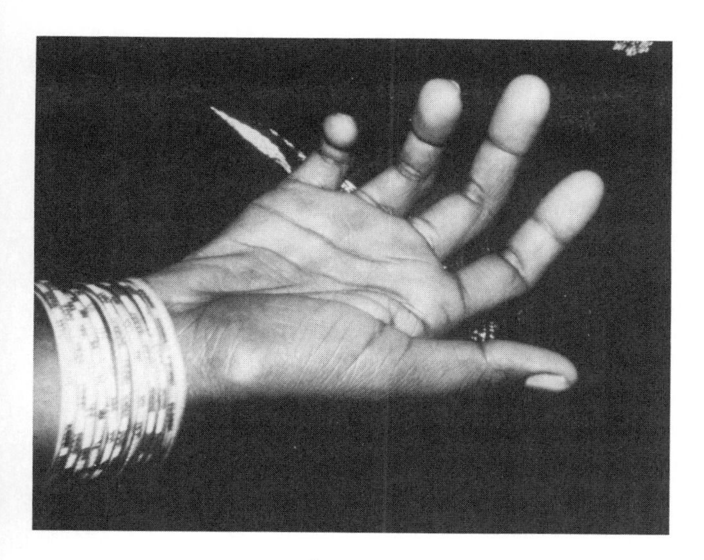

FIGURE 27.4 *Mycobacterium leprae* infects the nerves of the extremities to cause structural deformities such as the claw hand. (Centers for Disease Control.)

time is exceptionally long. The disease is diagnosed by demonstration of acid-fast bacilli in a biopsy, scrapings, or exudates from suspected lesions.

Epidemiology and Treatment

The epidemiology of leprosy is still poorly understood. Humans are the only known natural host of *M. leprae,* though natural infection in armadillos has been reported. Bacterial spread is very slow, and is presumed to be by person-to-person contact. In the past 25 years, only one staff member at the Carville Hospital has contracted Hansen's disease.

In the early stages, leprosy can be effectively treated with dapsone, with or without rifampin and clofazimine (klo-fah'zi-meen). The reason for the combination therapy is that resistant strains are known to develop during the lengthy treatment, which must continue throughout the patient's life. Only in this way can relapses be forestalled.

Unfortunately, drug treatment does not correct any damage that has already occurred, but it does prevent progression of the disease. Dapsone is the drug of choice; it is cheap, and it renders the patient noninfectious. Rifampin and clofazimine are effective against dapsone-resistant strains of *M. leprae,* but are much more expensive than dapsone.

TETANUS

Tetanus (tet'ah-nus) or "lockjaw" is a toxigenic disease caused by *Clostridium tetani,* a spore-forming bacterium commonly found in soil and the lower intestine. The infection begins when a severe or deep wound, made anaerobic by insufficient blood supply, is contaminated with the spores of *C. tetani*. Farm accidents, auto accidents, contamination of necrotic surface wounds, or deep puncture wounds from a contaminated rusty nail or wood splinter are starting points for tetanus. Vegetative cells of *C. tetani* multiply in the damaged tissue and produce proteases, responsible for tissue necrosis, and the tetanus toxin. Two to 50 days may pass before the pathological consequences of tetanus appear.

Clostridial Neurotoxins

The symptoms of tetanus and botulism are due to the neurotoxins produced by *Clostridium tetani* and *C. botulinum*. Both toxins are synthesized as single polypeptide chains with molecular weights of about 150,000. As the toxin is excreted, it is nicked into two nonidentical fragments that are joined by a disulfide bond. Structurally and functionally, these toxins are similar to diphtheria toxin, in which one polypeptide binds to the target cell receptor and the second polypeptide causes toxicity (Chapter 17).

Tetanus toxin, officially called **tetanospasmin** (tet'ah-no-spaz'min), is a protein exotoxin (MW 150,000) that binds to motor nerve endings. It then moves by neurotropic spread from the infection site to the central nervous system where it stimulates the motor neurons, presumably by inhibiting the release of a neurotransmitter. Consequently, the motor neurons are continuously excited and spastic paralysis follows. The lethal dose for the average-sized human is less than 0.2 μg, making tetanus toxin one of the most lethal substances known.

Tetanus in humans is characterized by spasms of the voluntary muscles, especially of the neck and jaw. The trivial name of "lockjaw" is derived from the patient's inability to open his or her mouth after the toxin acts on the nerves of the jaw muscles. Tetanus toxin eventually affects the diaphragm and kills by asphyxiation.

Neonatal tetanus, occurs primarily in developing countries. *Clostridium tetani* infects the umbilicus during delivery when insufficient care is taken to exclude soil and dust from the delivery area.

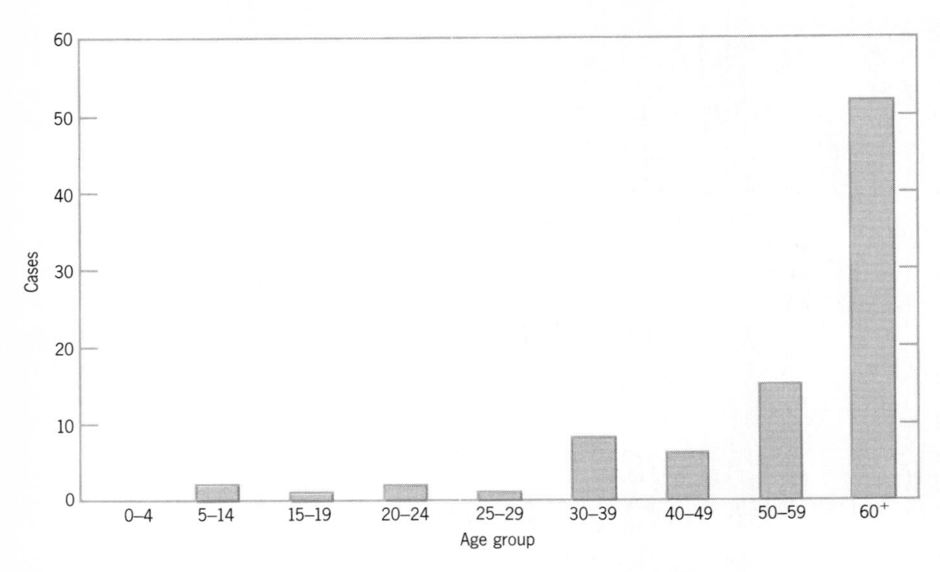

FIGURE 27.5 Tetanus cases reported in the United States, 1983. Of the 91 cases reported, 57 percent occurred in patients 60 years of age or older. Persons in this age group are likely to have inadequate levels of circulating antitoxin. (Centers for Disease Control, "Annual Summary 1983: Reported Morbidity and Mortality in the United States," *Morbidity and Mortality Weekly Report,* Vol. 32, No. 54, 1984.)

Prevention and Treatment

People immunized with tetanus toxoid produce circulating antitoxin able to neutralize tetanus toxin before it enters the nerve axon. Once the toxin is in the nerve, the antitoxin cannot neutralize it. People with tetanus are passively immunized with antitoxin, and about 50 percent of these patients are cured. Most patients who survive the first four days of tetanus recover completely.

The tetanus toxoid is given to children as part of the DTP vaccine when they are 2 to 3 months old. This is followed by a series of four "shots," with the last one given just before the child starts school (5 to 6 years of age). Booster "shots" are recommended for individuals who sustain serious wounds and who have not had a booster within the preceding five years. The incidence of all types of tetanus is very low in the United States because most people are immunized. Tetanus is seen primarily in older patients (Figure 27.5) who have inadequate levels of circulating antitoxin. Health care providers who treat the elderly should ensure that their patients are vaccinated against tetanus.

WOUND BOTULISM

Wound botulism occurs when *C. botulinum* infects severe wounds, such as those sustained in an automobile or farm accident. The severely damaged tissue lacks an adequate blood and oxygen supply, creating an environment in which *C. botulinum* multiplies and produces botulism toxin. This exotoxin binds to gangliosides (lipids of the peripheral nerves) and blocks the release of acetylcholine which causes muscle paralysis. Acetylcholine must be released from the sensory neurons into the synaptic cleft in order for normal muscle excitation to occur. Wound botulism is rare in the United States, but must be considered a problem in treatment of war-related wounds. The same toxin is responsible for foodborne botulism as described in Chapter 25.

VIRAL DISEASES OF THE NERVOUS SYSTEM

The small RNA viruses of the Picornaviridae family are classified in two genera, *Enterovirus* and *Rhinovirus*

TABLE 27.1 The Picornaviridae family

GENUS	VIRUSES	NUMBER OF SEROTYPES	HOST RANGE, GROWTH
Enterovirus	Polioviruses	3	Humans, primates, tissue culture
	Coxsackievirus group A	23	Humans: A9, A16 in tissue culture and adult mice; all others in suckling mice
	Coxsackievirus group B	6	Humans, suckling mice, tissue culture
	Echoviruses	30	Humans, tissue culture
Rhinovirus	Rhinoviruses	89	Humans

(Table 27.1). The rhinoviruses infect the respiratory tract (see Chapter 23), whereas the enteroviruses infect either by the respiratory or by the intestinal tract. The enteroviruses are discussed here because they can cause nervous system disorders.

POLIOMYELITIS

Poliomyelitis (po′le-o-mi′e-li′tis) is usually a mild asymptomatic disease, but in a small percentage of cases permanent paralysis, sometimes with deformity, follows. Poliomyelitis, or polio, as it usually is called, is an ancient affliction. We know that the early Egyptians were acquainted with polio from the reliefs and sculptures they created depicting its deformed victims.

Control of polio had its scientific beginning in 1908 when laboratory monkeys were infected with human polio. Although S. Flexner and P. A. Lewis discovered the poliovirus in 1909, nearly 40 years elapsed before John Enders and his colleagues succeeded in culturing poliovirus in tissue cultures. The tissue culture techniques developed by Enders and his group enabled virologists to isolate and grow the three strains of poliovirus now used in making the poliovirus vaccines.

Poliovirus

The polioviruses are small (20 to 30 nm), nonenveloped, icosahedral, single-stranded, plus RNA viruses. The RNA serves as the message (mRNA) for the synthesis of a single polypeptide chain that is cleaved into functional proteins (see Figure 12.9). Each of the three poliovirus serotypes elicits a distinct immunological response, so all have a place in vaccine preparations.

Poliomyelitis in Humans

The polioviruses are transmitted between humans by the fecal–oral route or by respiratory secretions from infected persons. The viruses multiply in the mucosa of the oropharynx and the intestine before spreading to the lymphatic system. An infection that does not progress beyond this stage results in a mild form of polio known as **inapparent poliomyelitis** (Figure 27.6), which represents between 90 and 95 percent of known cases. These infections are detected by isolation of the virus or by observation of a rising antibody titer.

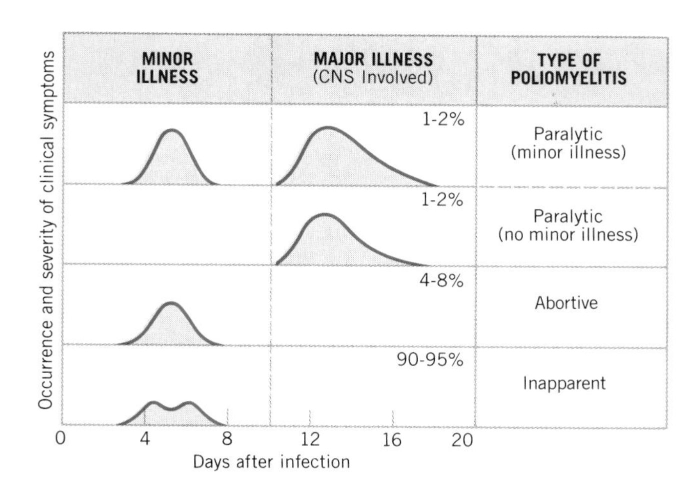

FIGURE 27.6 Types of poliovirus infections in humans. The occurrence of each type of polio within a nonimmunized population is indicated as a percentage. (Modified from D. M. Horstmann, *Yale J. Biol. Med.*, 36:5, 1963.)

In another 4 to 8 percent, a viremic stage follows virus replication in lymphatic tissue. This viremia is accompanied by symptoms of headache, fever, sore throat, anorexia, vomiting, and abdominal muscle pain. The symptoms begin two days after infection and last for two to three days (Figure 27.6). Complete recovery follows this type of infection, termed **abortive poliomyelitis** because the poliovirus does not affect the central nervous system.

The most serious form, **paralytic poliomyelitis,** develops in only 1 to 2 percent of cases. Symptoms appear about 10 days after infection either with or without the precursory symptoms of the abortive illness. The patient shows signs of central nervous system dysfunction, including meningitis, headache, fever, malaise, stiff neck, and vomiting. The presence of muscle pain, most commonly in the neck or lumbar spinal region, occurs as a precursory symptom of paralysis. If the minor illness (Figure 27.6) occurs, the patient appears to make a recovery for a period of two to five days, before the onset of paralytic symptoms.

Unlike other neurological spinal disorders, paralytic polio is characterized by asymmetric distribution of paralysis; that is, the muscle groups of only one leg or one arm are affected. The legs are affected more often than the arms, the proximal muscles of the extremities more often than the distal muscles. The onset of paralysis can be very rapid, within hours, but it usually develops over a period of several days. If the ninth and tenth cranial nerves are involved, swallowing and respiration become difficult or impossible. Adults are more likely than children to experience these problems.

Epidemiology

During the early decades of this century, infants less than 5 years of age were the primary victims of paralytic polio, and the disease was known as **infantile paralysis.** We now have evidence that poliovirus infections actually were widespread, but most people experienced only mild illness, and paralysis was rare.

The situation changed in the late 1940s, when children 5 to 9 years old became the chief victims. The incidence of paralytic polio increased, and epidemics occurred in developed countries, most commonly in the summer and fall months. From 1951 to 1955 between 10,000 and 21,000 cases of paralytic poliomyelitis were reported annually.

A dramatic fall in the number of cases followed the introduction in 1955 of the Salk vaccine (Figure 27.7). Salk vaccine (named for Jonas Salk, who had developed it) was made with killed polioviruses. It was followed in 1962 by the Sabin vaccine (named after Albert Sabin), made with attenuated polioviruses. As a result of the wide use of these vaccines, paralytic polio has been virtually eliminated in this country; in 1986 only

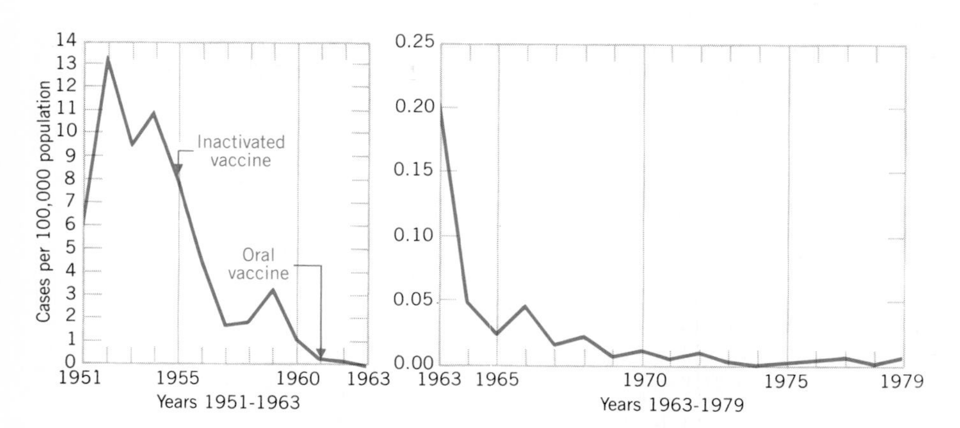

FIGURE 27.7 Number of cases of paralytic poliomyelitis yearly in the United States from 1951 to 1983. The ordinate scale is changed after 1963 (right graph) to reflect the decrease number of cases.

two cases were reported. Epidemics of polio occur sporadically within religious groups who are opposed to vaccination.

Poliovirus Vaccines

The **inactivated polio vaccine** (IPV) developed by Jonas Salk contains three strains of killed poliovirus. This vaccine is effective in eliciting circulating antibodies against all three serotypes (see Table 27.1) and cannot cause paralytic polio because the viruses are dead. It carries several disadvantages, however: it is expensive, multiple injections are required, total inactivation of the virus must be assured, and it does not stimulate secretory immunoglobulins of the intestine.

Trivalent oral polio vaccine (TOPV), developed by Albert Sabin, is composed of attenuated (living) strains of the three poliovirus serotypes. Oral administration of the vaccine initiates an active poliovirus infection that results in an asymptomatic to mild illness, and the viruses are excreted in the feces. The sequence of events is worth noting: the vaccine stimulates development of the immune response to all three serotypes; the excreted viruses reinfect immunized persons; their immune response is enhanced; and the end result is a high degree of immunity within the population. TOPV has several advantages over IPV: its cost is low; refrigeration is not needed; and it elicits both secretory and circulating antibodies. Unfortunately, however, a few cases of paralytic polio have been associated with administration of TOPV—a disadvantage that is obviously of great concern.

Under current guidelines, TOPV is recommended for immunization of infants beginning at 2 months and IPV is recommended for immunization of immunodeficient children and for immunosuppressed patients. Unimmunized adults who are at risk from exposure to wild polioviruses are given IPV because the risk of TOPV-associated paralysis is higher in adults than children. Poliovirus vaccine is not routinely administered to persons past teen age.

ENTEROVIRUSES: COXSACKIEVIRUS AND ECHOVIRUS

In 1948 researchers isolated an unidentified virus from the feces of two children who were suffering polio-like symptoms. The children lived in Coxsackie, New York, thus the name **coxsackievirus** (kok-sak′e-vi′rus), was conferred. Subsequently, many strains of coxsackievirus were isolated by infecting suckling laboratory mice with specimens from ill humans.

As tissue culture techniques became available in the 1950s, it was possible to isolate other enteric viruses from the feces of healthy children. These viruses exerted cytopathic effects in tissue cultures, but they did not grow in suckling mice. The name "orphan" was given these viruses becauses they apparently caused no clinical symptoms. The acronym **echoviruses** (ek′o-vi′rus) is now universally accepted—**e**nteric **c**ytopathic **h**uman **o**rphan viruses. Virologists have identified 29 serotypes of coxsackievirus and 30 serotypes of echovirus (Table 27.2). These members of the the Picornaviridae family are small RNA viruses that infect humans through the mouth or nose and are transmitted by the fecal–oral route.

Most coxsackievirus and echovirus infections occur in the summer and cause only asymptomatic or mildly febrile illnesses. Sometimes, however, there are symptoms, and the ensuing disease can range from a mild head cold to aseptic meningitis or myocarditis. In this text, echoviruses are included in the discussion of neurotropic viruses because some serotypes are responsible for aseptic meningitis.

Aseptic Meningitis

Aseptic meningitis is an inflammation of the meninges caused by a noncellular infectious agent. Over half the cases are caused by coxsackieviruses or echoviruses. Although the CSF in aseptic meningitis is free of bacteria, it often contains lymphocytes, whose presence is indicative of an infection. The causative virus can often be isolated immediately after the onset of symptoms. There are seven serotypes of coxsackievirus, and many echovirus serotypes, that cause aseptic meningitis (Table 27.2); poliovirus, mumps virus, flaviviruses, alphaviruses, bunyaviruses, and herpesviruses can also be responsible.

Enteroviruses, the most common cause of aseptic meningitis, predominantly infect children less than 1 year of age. Symptoms include nausea and vomiting, often accompanied by a sore throat. The symptoms last for only a few days to a week, after which the patient recovers without suffering sequelae. The disease is more severe in newborns, who become irritable,

TABLE 27.2 Clinical symptoms of coxsackievirus and echovirus infections

CLINICAL SYMPTOMS	COXSACKIEVIRUS GROUP A	COXSACKIEVIRUS GROUP B	ECHOVIRUSES
Asymptomatic infection	All	All	All
Febrile illness without coryza	All	All	All
Encephalitis	A9	B5	E9, E18
Aseptic meningitis	A7, A9	B1-5	E3, 4, 6, 7, 9, 11, 27, 29, 32
Hand-foot-and-mouth disease	A16 (also A5, A10)	B2, B5	None
Carditis	Some	B1-B5	Some
Rubelliform rash	A9	None	9 (especially)

drowsy, and reluctant to feed; in some, permanent neurological defects follow recovery.

KEY POINT

Aseptic meningitis is diagnosed by the clinical symptoms and the absence of bacteria in a Gram stain of CSF. Though usually a mild, self-limiting illness, aseptic meningitis can carry a dangerous risk in a severe case, because there are no drugs available to treat it.

Other Clinical Manifestations

The illnesses caused by coxsackieviruses and echoviruses are too numerous to detail here (see Table 27.2). A few must be noted, however, because they may be confused with other viral diseases. Echovirus 9 is often associated with a rubelliform rash that may be confused with symptoms of rubella. Coxsackieviruses type A16 and, to a lesser extent, B2 and B5, cause vesicular skin lesions that resemble herpes simplex or the lesions of varicella–zoster. These viruses cause **hand-foot-and-mouth disease,** characterized by the presence of lesions in the mouth, on the hands, and on the feet. Hand-foot-and-mouth disease can be distinguished from chickenpox (varicella–zoster virus) by the distribution of the lesions and by the failure of the lesions to scab over or scar. To complicate things further, one or more of the enteroviruses is implicated in congenital abnormalities, acute hemorrhagic conjunctivitis, diabetes, diarrhea, and carditis. Because of their numerous immunotypes and the wide spectrum of symptoms to which they give rise, diagnosis in a specific instance may be extremely difficult.

RABIES, AN ENZOOTIC VIRAL DISEASE

Rabies is a horrible, fatal disease of humans and animals. The rabies virus is transmitted between domestic and wild animals and to humans by animal bites, and causes a violent illness that almost invariably has a fearsome end.

Rabies has been known—and feared—since ancient times. In nearer times, the story of how Louis Pasteur developed the original vaccine, and in 1885 saved the life of young Joseph Meister, has been recounted any number of times (and was the subject some years ago of a popular movie).

The Rabies Virus

The cause of rabies is a bullet-shaped RNA virus classified in the Rhabdoviridae family (Figure 27.8). This virus is complex, being made up of lipids, at least four proteins, one single-stranded minus RNA molecule, and some carbohydrates. The spike-like glycoprotein structures that extrude through the viral envelope function to bind the virus to the host cell. The glycoprotein in these spikes is the major viral antigen, which, when incorporated into a vaccine, elicits the neutralizing antibody formation.

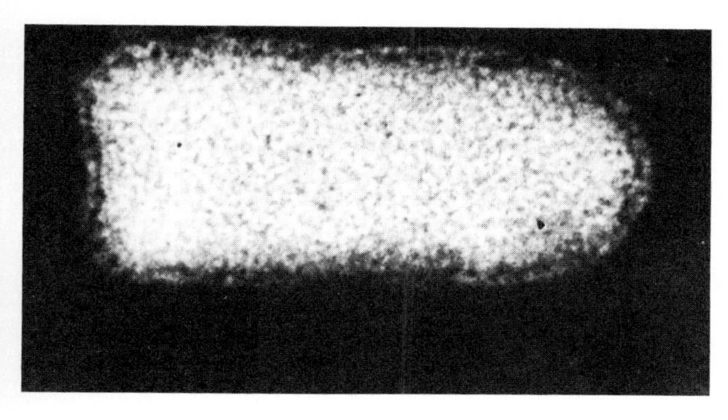

FIGURE 27.8 Electron micrograph of the bullet-shaped rabies virion. 300,000×. (Courtesy of Drs. K. Hummeler and N. Tomassini. From K. Hummeler and N. Tomassini, "Rhabdoviruses" in A. J. Dulton and F. Haguenau (eds.), *Ultrastructure of Animal Viruses and Bacteriophages: An Atlas,* Academic Press, New York. Copyright © 1973, Academic Press.)

Rabies in Animals

Rabies is an enzootic disease of both domestic and wild mammals. Wild animals, predominantly skunks, raccoons, and foxes are the usual victims, though coyotes, bobcats, weasels, and bats may also be involved. Domestic dogs, cats, cows, horses, and pigs can contract rabies as well, and any of these infected animals can transfer the virus to humans.

An infected animal secretes rabies virus in its saliva before falling ill with symptoms. One difficulty is that an apparently healthy animal can transmit the virus during the period when the animal seems well and is not suspected of being rabid. This virus secretion continues for up to 3 days in dogs, up to 2 days in cats, and up to 18 days in skunks. For this reason, a cat or dog that bites a human should, without exception, be confined for 10 days, to ensure that it is not rabid.

RABIES IN HUMANS

Most cases develop following direct contact with the saliva of a rabid animal, usually through a bite. A few cases have been traced to viral aerosols in laboratories or in bat caves, but these are rare. The virus infects muscle and connective tissue at the site of entry, multiplies and then travels along the peripheral nerves to reach the central nervous system. The virus is carried not in the circulating blood, but by neurotropic spread, thus accounting for its lengthly incubation period. After multiplying in the brain, it descends again by neurotropic spread to the animal's salivary glands.

Incubation Period

As just indicated, the virions remain localized for days before infecting nervous tissue, which causes the outward signs of rabies. The incubation period depends on the number of virions in the inoculum and the proximity of the wound to the central nervous system. Observed incubation periods have varied from 25 and 48 days for a bite on the head, to 46 and 78 days for bites on an arm or a leg. In most cases, the incubation period is long enough to allow for postexposure immunization.

Symptoms

Initial symptoms of human rabies include fever, nausea, vomiting, headache, malaise, hydrophobia (fear of water), and lethargy. There is paralysis of the swallowing reflex, and the victim becomes hydrophobic because (s)he is unable to swallow any fluid, and chokes while attempting to drink. The physical manifestations continue for a period of two to ten days, to be followed by neurological manifestations of hyperactivity, disorientation, hallucinations, seizures, or paralysis. This acute phase is followed by coma. The patient may remain in a coma for a few hours to 14 days before succumbing to seizures and respiratory paralysis. Although few cases of rabies occur in humans, those that do occur almost always end in death. Only three persons are positively known to have survived an attack of rabies, and each of them received either preexposure prophylaxis or postexposure treatment.

Prevention and Treatment

Every year some 20,000 persons need to be vaccinated against rabies, either because of their occupations or because they were exposed to a rabid or possibly rabid animal. Whether or not exposure has occurred is determined by observation for the appearance of symptoms in the involved animal while it is confined. Animal rabies is confirmed by the demonstration of *Negri bodies* in brain tissue sections made after the animal has been killed.

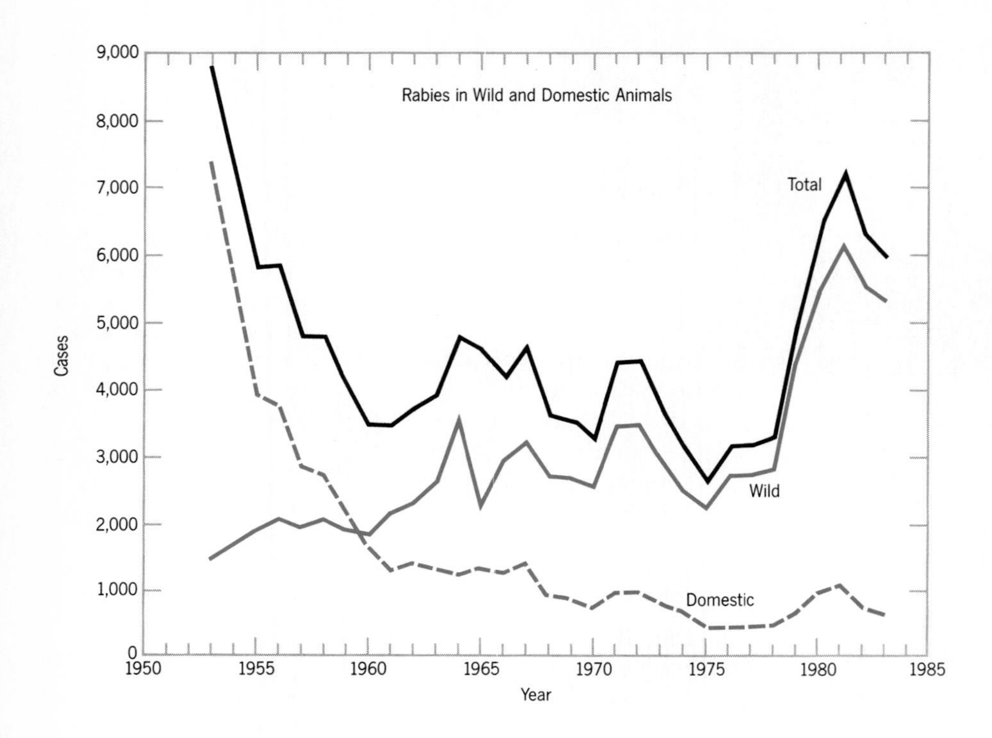

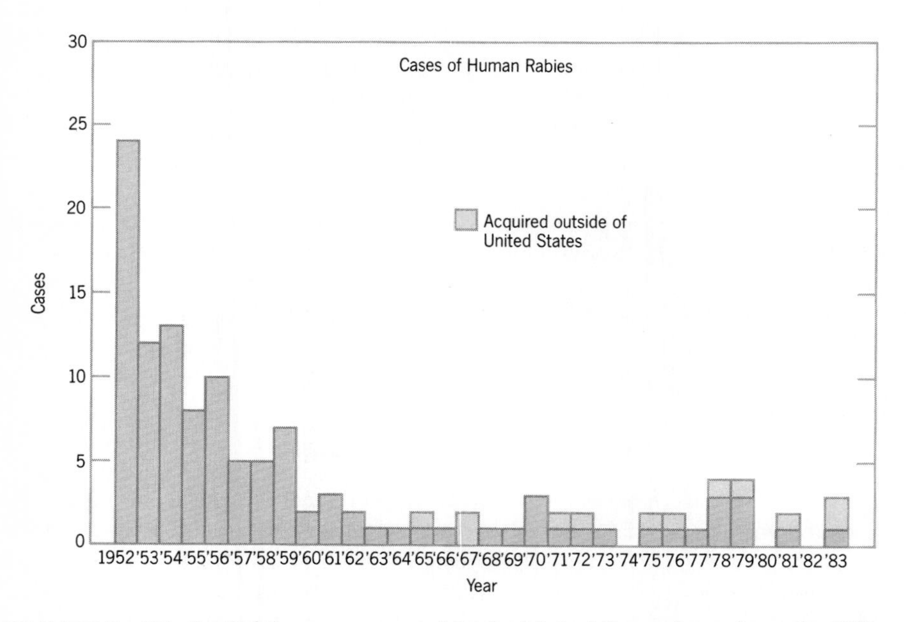

FIGURE 27.9 Rabies cases reported in the United States from the early 1950s through 1983. (Centers for Disease Control, "Annual Summary 1983: Reported Morbidity and Mortality in the United States," *Morbidity and Mortality Weekly Report,* Vol. 32, No. 54, 1984.)

Postinfection vaccination is effective because humans make antibodies in response to the vaccine before the rabies virus completes its long incubation period. The vaccine of choice is **human diploid cell rabies vaccine** (HDCV), which has been used since 1980. An injection of rabies immune globulin is administered together with one dose of the vaccine, followed by vaccine on days 3, 7, 14, and 30. HDCV is easily administered, effectively prevents rabies, and has no neurological consequences.

Local governments should initiate and maintain effective programs for vaccinating all dogs and cats. Stray and unwanted animals should be removed, especially in areas where rabies is epizootic. Preexposure vaccination is recommended for high-risk groups such as veterinarians, animal handlers, and certain laboratory workers. Professional trappers and hunters in endemic areas should receive preexposure vaccination. Sport hunters or trappers are at little risk, but they should be careful when handling carcasses of possibly rabid animals. Everyone should make an effort to avoid being bitten by a fur-bearing animal that appears sick and unafraid.

KEY POINT

The neurotropic spread of rabies virions from the site of infection to the brain takes a number of weeks, which is sufficient to actively immunize an exposed person.

Epidemiology

Since 1951 there has been a sharp drop in the number of human rabies cases nationally with fewer than five cases reported (Figure 27.9) annually. Immunization of domestic dogs and cats has greatly lessened rabies transmission by them, but rabies in animals continues to occur, with over 5000 cases reported annually.

ARTHROPOD-BORNE VIRUSES: ENCEPHALITIS

The arthropod-borne viruses are endemic in those geographical regions of the world where their reservoir and vector coexist (Table 27.3). Many of these diseases occur seasonally as the vector population peaks. Arthropod-borne viruses are classified in the Bunyaviridae, Togaviridae, and Flaviviridae family according to their size, structure, and chemical composition. Many of these arthropod-borne viruses cause **encephalitis** (Figure 27.10), an inflammation of the brain that can lead to coma and death. Recovery from encephalitis may leave the patient with permanent neurological damage.

California Encephalitis

The California encephalitis group viruses of the Bunyaviridae family are carried by ground squirrels and chipmunks and transmitted to humans by the *Aedes* mosquitoes (Table 27.3). Among this group, the **La**

TABLE 27.3 Classification, vectors, and reservoirs of arthropod-borne viruses affecting the nervous system

FAMILY (GENUS)	VIRUS	RESERVOIR	VECTOR
Bunyaviridae	California encephalitis viruses	Ground squirrels, chipmunks	*Aedes triseriatus, A. melanimon*
Togaviridae (*Alphavirus*)	Eastern equine encephalitis virus	Wild birds (quail, pigeons, pheasants)	*Culiseta melanura, Aedes* spp.
	Western equine encephalitis virus	Wild birds	*Culiseta melanura, Culex tarsalis*
	Venezuelan equine encephalitis virus	Rodents and horses	*Culex* spp., *Aedes aegypti*
Flaviviridae (*Flavivirus*)	St. Louis encephalitis virus	Perching birds	*Culex tarsalis, C. pipiens, C. nigripalpus*

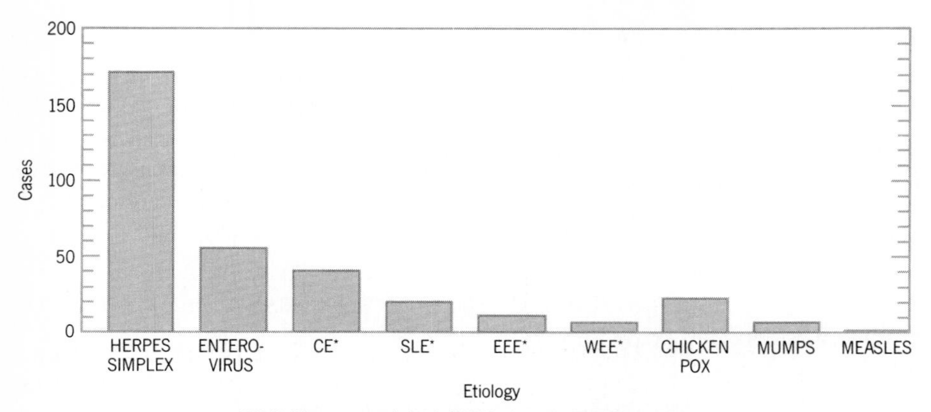

FIGURE 27.10 Encephalitis cases reported by etiology in the United States, 1983. In addition to the arthropod-borne viruses, encephalitis can be caused by herpes simplex viruses, enteroviruses, chickenpox virus, mumps virus, and measles virus. (Centers for Disease Control, "Annual Summary 1983: Reported Morbidity and Mortality in the United States," *Morbidity and Mortality Weekly Report,* Vol. 32, No. 54, 1984).

Crosse encephalitis (LE) **virus** causes the most cases of mosquito-borne encephalitis in the United States, except during years with large epidemics of St. Louis encephalitis (see below). Children living near or playing in forested areas that harbor *Aedes* mosquitoes have the highest incidence. This disease occurs predominantly in the north central and eastern United States, where the forest-dwelling mosquito, *Aedes triseriatus* (tri-ser-i-at'us) lives.

The LE virus is maintained in mosquitoes by transovarian passage and survives severe winters in mosquito eggs. In summer, the eggs hatch and form adult insects, which transmit the LE virus to chipmunks and squirrels. Here the virus replicates, causing an asymptomatic viremia, and the infection is amplified as the virus is transmitted to other squirrels and chipmunks. People living near or entering forested areas are at risk of infection. Cases of California encephalitis have occurred in the north central states of Michigan, Wisconsin, Minnesota, and Iowa.

Equine Encephalitis Viruses

Members of the *Alphavirus* genus are transmitted by arthropod bites, mainly mosquitoes and ticks (Table 27.3), among wild birds, rodents, and horses. The equine encephalitis viruses occur only in the Americas and there they have a restricted distribution.

The three major strains of alphaviruses are the western equine encephalitis (WEE) virus, the eastern equine encephalitis (EEE) virus, and the Venezuelan equine encephalitis (VEE) virus. Each virus has been isolated from the brains of sick horses (equines). Horses are a major reservoir for the VEE virus; however, they are definitive hosts (Figure 27.11) for the WEE or EEE viruses.

The **eastern equine encephalitis virus** is transmitted between many species of wild birds by the swamp mosquito, *Culiseta melanura* (ku-li-se'tah mel-an-u'ra). Because this mosquito rarely feeds on humans, species of *Aedes* probably act as the vector for human infections. Horses, people, domestic pheasants, and quail are incidental hosts. Epidemics of EEE peak between late summer and early autumn in northern climates and most often involve young children; the incidence of EEE in the United States is very low.

The **western equine encephalitis virus** is endemic in the western United States where it is maintained in a bird–mosquito cycle very similar to the EEE. Both WEE and EEE can coexist in the same geographical location in the same vectors and host birds. The WEE virus is transmitted between birds by *Culiseta melanura* and *Culex tarsalis* (ku'leks). The incidence of WEE in people peaks in June and July.

South America, Central America, and the southern United States harbor species of the *Culex* mosquito

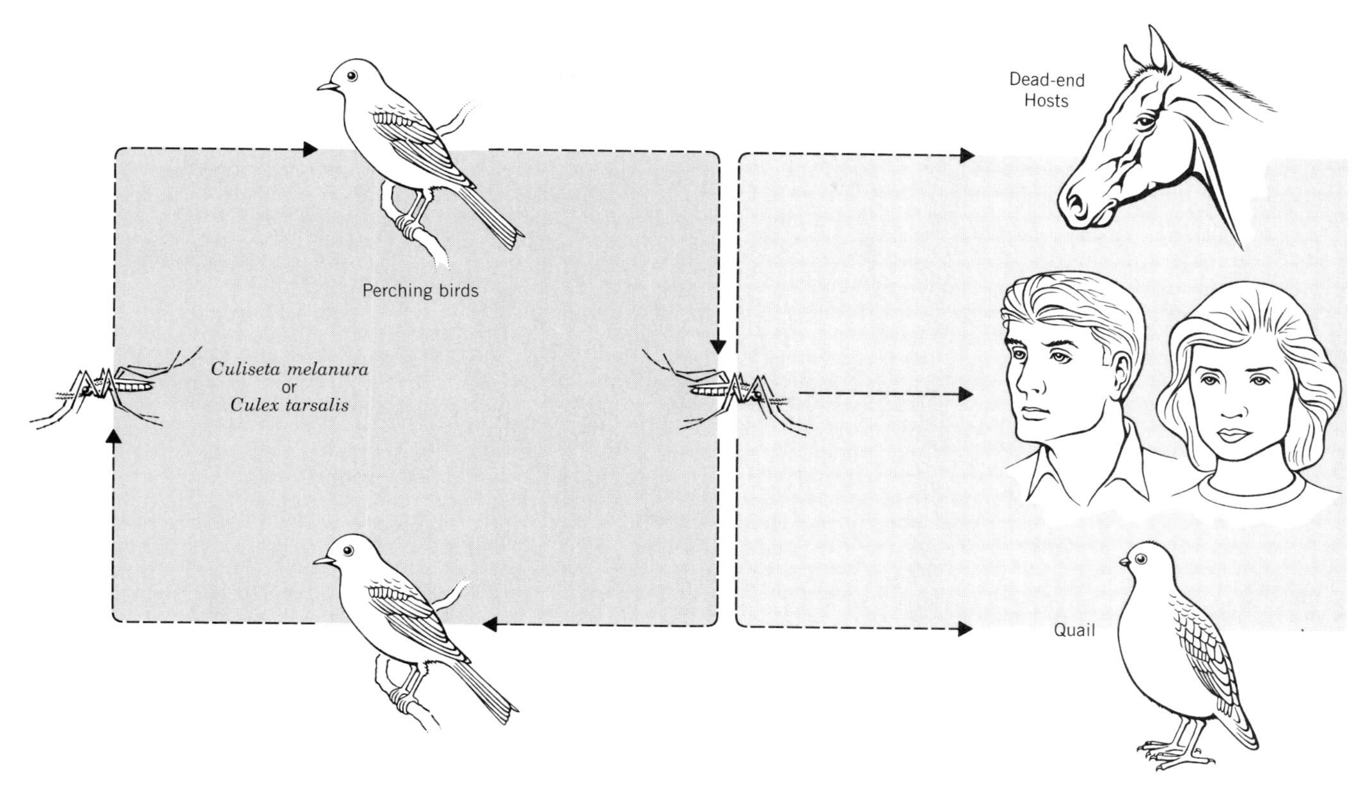

FIGURE 27.11 Epidemiology of western equine encephalitis and eastern equine encephalitis. Horses, humans, and domestic quail are definitive hosts for these viruses.

that are the vectors for the **Venezuelan equine encephalitis virus** (Figure 27.12). This virus causes a serious encephalitis in horses before it is transmitted to humans. Venezuelan equine encephalitis is rare in the United States, although a major outbreak occurred in 1971 that took the lives of more than 10,000 Texas horses. Major human epidemics of Venezuelan equine encephalitis have been reported in Venezuela and Ecuador. Mosquitoes transmit the virus between its reservoirs in horses and rodents and occasionally to humans who are incidental hosts.

Human Manifestations The equine encephalitis viruses cause a viremia accompanied by fever, chills, headache, vomiting, nausea, and body aches. The severity of the encephalitis that follows depends on the infecting virus. EEE virus causes a serious encephalitic disease with case fatality rates between 50 and 70 percent, and 30 percent of those who recover have neurological disorders that may require institutional care. WEE virus causes a milder disease in adults; however, infected children often experience a serious illness with neurological sequelae, including behavioral changes, convulsive disorders, and paralysis. The symptoms of VEE in adults are influenza-like and rarely progress to encephalitis; however, infections in children can result in a severe encephalitis that rapidly progresses to shock, coma, and death. As with other alphaviruses, children who recover are likely to have permanent neurological sequelae.

St. Louis Encephalitis Virus

The St. Louis encephalitis virus is the most important *Flavivirus* present in the United States. It is transmitted from perching birds to humans by *Culex* mosquitoes (Table 27.3). St. Louis encephalitis has been reported from most of the contiguous 48 states, with the notice-

FIGURE 27.12 Epidemiology of Venezuelan equine encephalitis. Horses and rodents are animal reservoirs for the VEE virus.

able absence of the New England states and South Carolina. Most cases occur in the Ohio–Mississippi Valley, Florida, Texas, California, and Washington. Major human epidemics are seen in the mosquito season of late summer and autumn; the last one occurred in August and September of 1975 when 1815 cases resulted in 140 deaths. Usually, fewer than 400 cases are reported.

Prevention and Treatment

The clinical consequences of viral encephalitis can be a fulminant encephalitis and death, so avoidance is of primary importance. People should avoid exposure to mosquitoes, especially in areas where encephalitis is endemic or where arthropod-borne encephalitis has been reported. Insecticides are used to control the mosquito population. There are no specific treatments available, so therapy is supportive. Vaccines are available for EEE and WEE and are used primarily for laboratory workers. Fortunately, arthropod-borne viral encephalitis is a rare disease.

SLOW-DEVELOPING NEUROLOGICAL DISEASES

Creutzfeldt–Jakob (kroits′felt-yak′ob) disease (CJD) and kuru are slow-developing diseases of the human central nervous system. Both diseases have been transmitted to laboratory chimpanzees; however, the purification and physical characterization of the infectious agents have not been reported. These diseases are proposed to be caused by unconventional infectious agents called prions. A **prion** is a small proteinaceous infectious particle that is resistant to inactivation by nucleic acid modification—so we think they are proteinaceous particles without nucleic acid. Prions are the proposed agents of scrapie (skra′pe), a degenerative neurological disease of sheep and goats.

Kuru and the Fore Tribe

Kuru is of interest because of the cannibalistic social customs of the Fore tribe needed for its transmission

Transmission of Creutzfeldt–Jakob Disease During Surgery

In mid-November 1986 a 28-year-old woman developed irregular muscle coordination 19 months after surgical removal of a cyst-like mass lying close to her brain. During surgery she received a commercially prepared graft of human brain material. By early December she required assistance in walking and had developed difficulty in speaking. Two weeks later she gave inappropriate responses to questions and developed visual hallucinations. By early January 1987 she had uncontrollable muscle contractions and was demented. Diagnosis of Creutzfeldt–Jakob disease (CJD) was confirmed by brain biopsy. She had no family history of degenerative neurologic disease nor had she received cadaveric, pituitary-derived human growth hormone. No patients with known CJD had had surgery in the same neurosurgical suite in the three months before this woman's operation.

Comment

The rare Creutzfeldt–Jakob disease is diagnosed most frequently in patients older than 50 years of age; rarely in patients less than 30. Physician-caused (iatrogenic) transmission of CJD has occurred in one patient through corneal transplant from an infected donor, in two patients exposed to intracerebral electrodes previously used in a CJD patient, in four patients operated on in neurosurgical suites where procedures on CJD patients had been performed, and in four patients receiving cadaveric, pituitary-derived growth hormone. CJD is thought to be caused by a prion. Prions are infectious agents that are extremely difficult to destroy. Treatment with ethylene oxide, ultraviolet light, ethanol, formaldelyde, and other chemicals is ineffective. Recommended sterilization procedures for prions are autoclaving for 4.5 hours at 121°C, or treatment with 1*N* NaOH followed by autoclaving for 1.5 hours. Donated body organs and tissues from patients suspected of having CJD or other slow-developing infectious diseases should not be transplanted into humans.

Adapted from *Morbidity and Mortality Weekly Report,"* Vol. 36, No. 26, February 1987.

(see Chapter 20), and the slow development of its neurological symptoms. The symptoms of kuru develop 5 to 20 years after infection, so the kuru agent is labeled a "slow virus." The symptoms are characterized by cerebellar ataxia and shivering tremors of the head, trunk, and extremities. As the symptoms progress the patient becomes unable to walk and cannot move without uncontrolled tremors, and dies three to nine months later. Since the Fore tribe has stopped the ritual of cannibalism, kuru is no longer seen in either children or adolescents; however, it still occurs in adults—a fact that attests to the long incubation period of kuru.

Creutzfeldt–Jakob Disease

Creutzfeldt–Jakob disease (CJD) is an uncommon dementing illness of humans that has a worldwide distribution. It occurs with equal frequency among men and women, and it generally strikes persons between 50 and 70 years old. The onset is rapid and is characterized by profound dementia and diffuse, rapid muscle spasms. The patient moves through a period of behavioral changes, loss of memory and reasoning, and a loss in visual acuity. Health deteriorates rapidly, with delirium, coma, and, invariably, death.

Though the etiology and epidemiology of CJD are unclear, we do know that it is infectious because it has been transmitted to chimpanzees. It is believed that both CJD and scrapie are due to similar agents. Scrapie, a transmissible degenerative nervous system disease of sheep and goats, causes a disease indistinguishable from Creutzfeldt–Jakob disease when transmitted to primates.

PROTOZOA: AMEBIC MENINGOENCEPHALITIS

About 100 cases of amebic meningoencephalitis (me-ning′go-en-sef′ah-li′tis), an inflammation of the meninges and brain, have been reported. Amebic meningoencephalitis occurs most often in children or young adults who have been swimming in fresh water. This disease begins with a severe frontal headache, lethargy, and fever; confusion, convulsions, coma, and death ensue. The disease is diagnosed by the presence of amebas and erythrocytes in cerebrospinal fluid combined with an elevated protein and a decreased sugar level. The causative agent in most cases has been identified as an ameba belonging to the genus *Naegleria* (nagle'ri-ah). Some cases of *Naegleria*-induced meningoencephalitis have been successfully treated with amphotericin B and miconazole, but most cases of this amebic disease are fatal.

Summary Outline

THE NERVOUS SYSTEM

1. The nervous system contains billions of nerve cells organized into the central nervous system (brain and spinal cord) and the peripheral nervous system (nerves connected to the brain and spinal cord).
2. The brain and spinal cord are protected by membranes, the meninges, and by the cerebrospinal fluid. The CSF supplies the avascular central nervous system with glucose and oxygen.
3. The central nervous system has access to a rich blood supply across the blood–brain barrier, which prevents harmful substances, microbes, blood cells, and many proteins from reaching neural tissue.
4. An infection of the meninges resulting in inflammation is called meningitis; encephalitis is an inflammation of the brain.

BACTERIAL DISEASES OF THE NERVOUS SYSTEM

1. Bacteria that cause meningitis enter the cerebrospinal fluid from the blood.
2. The bacteria most often responsible for meningitis are *Neisseria meningitidis, Haemophilus influenzae,* and *Streptococcus pneumoniae.* Acute bacterial meningitis demands rapid diagnosis and treatment with antibiotics.
3. *Neisseria meningitidis* from the oropharynx enters the blood (meningococcemia) before penetrating the meninges and infecting the cerebrospinal fluid. Death can result in a few hours if the patient is not treated with antibiotics.
4. *Haemophilus influenzae* is the most common cause of bacterial meningitis in children between the neonatal period and 6 years of age. Patients who survive may suffer long-term neurological sequelae, including mental retardation and learning disabilities. Children 15 to 18 months old can be immunized with the Hib vaccine.
5. Over 12 million people worldwide have Hansen's disease (leprosy), which is caused when *Mycobacterium leprae* infect and multiply in skin and nerves of the cooler body regions.
6. Hansen's disease results in a progressive destruction of the nerve endings in the feet and hands; this leads to physical injury. Patients must receive lifelong drug therapy.
7. *Clostridium tetani* multiplies in tissue wounds and produces the tetanus toxin, which binds to motor nerve endings and causes spastic paralysis. People are immunized against this disease with the DTP vaccine.
8. Wounds infected with *C. botulinum* are sources of botulism toxin, which causes muscle paralysis by inhibiting the release of acetylcholine at the nerve synapse.

VIRAL DISEASES OF THE NERVOUS SYSTEM

1. The polioviruses are transmitted by the fecal–oral route and cause inapparent, abortive, and paralytic poliomyelitis in humans. Paralytic poliomyelitis (1 to 2 percent of cases) causes paralysis of the limbs and can affect ability to breathe and swallow.

2. Poliomyelitis is prevented by the Sabin trivalent oral polio vaccine (TOPV) or the Salk inactivated polio vaccine (IPV).
3. Coxsackieviruses and echoviruses cause aseptic meningitis, with the more severe symptoms occurring in children. These viruses cause other clinical illnesses, including a rubelliform rash, hand-foot-and-mouth disease, and carditis.
4. Rabies is enzootic in wild animals, especially skunks, raccoons, and foxes, and is transmitted, rarely, to people by the saliva of a rabid animal.
5. The rabies virus infects the muscle tissue before it multiplies and travels by neurotropic spread to the brain.
6. Postexposure vaccination with human diploid cell rabies vaccine is effective in treating rabies, which has a long incubation period. Untreated human rabies is always fatal.
7. Viruses belonging to the Bunyaviridae, Togaviridae, and Flaviviridae families exist in animal reservoirs and are transmitted between animals and to humans by arthropod vectors.
8. Ground squirrels and chipmunks are reservoirs for the La Crosse encephalitis virus, which is transmitted by *Aedes triseriatus* to people who live or play near forested areas.
9. The EEE virus and the WEE virus are transferred between their reservoirs in wild birds by *Culiseta melanura;* people, horses, and domestic quail are definitive hosts for these viruses.
10. The VEE virus is present in rodents and is transmitted to horses by *Culex* mosquitoes. Epidemic encephalitis in horses normally precedes VEE in humans.
11. Perching birds are the reservoir for the St. Louis encephalitis virus, which is transmitted to humans by species of *Culex* mosquito.

SLOW-DEVELOPING NEUROLOGICAL DISEASES

1. Kuru and Creutzfeldt–Jakob disease and are slow-developing human diseases of the central nervous system thought to be caused by unconventional infectious agents known as prions.
2. Kuru, a progressive cerebellar degenerative disease, was transmitted by the cannibalistic custom of consuming dead tribe members as a rite of mourning.
3. Creutzfeldt–Jakob disease (CJD) is an uncommon, dementing illness of humans with a worldwide distribution. It has a rapid onset in late middle age and is characterized by profound dementia and diffuse, rapid muscle spasms.

PROTOZOA: AMEBIC MENINGOENCEPHALITIS

1. Amebic meningoencephalitis is a rare brain disease caused by *Naegleria* acquired from fresh water. Treatment is unsatisfactory. Most victims experience seizures, become comatose, and die.

Questions and Topics for Study and Discussion

QUESTIONS

1. Describe the natural barriers that can prevent infectious agents from entering the central nervous system.
2. What are the signs of bacterial meningitis? How would you diagnose and treat it? How might it be prevented?
3. Why are infants between the ages of 6 and 18 months most susceptible to meningitis caused by *Haemophilus influenzae?*
4. Why has it not been possible to follow Koch's postulates to prove the cause of Hansen's disease? How does this affect treatment and prevention?

5. A patient with Hansen's disease can lose fingers or toes. Why?
6. Describe the clinical symptoms of tetanus. What measures are taken to prevent tetanus and how can patients with tetanus be treated?
7. Describe three possible clinical outcomes of a poliovirus infection. What are the most serious consequences of polio? How are they treated?
8. Coxsackieviruses and echoviruses cause clinical signs that may be confused with two childhood diseases. Explain.
9. How is the population protected against rabies? What steps are taken after a person is bitten by a possibly rabid animal?
10. Which arthropod-borne viral diseases are important to humans living in North America? Describe the clinical nature of these diseases, their reservoirs and vectors, and their seasonal and geographical incidence.

DISCUSSION TOPICS

1. The Sabin vaccine is used in the United States to immunize children against polio. Would there be a reason for physicians to use the Salk vaccine instead? Explain.
2. The incidence of human rabies in the United States is low. If the practice of vaccinating dogs and cats against rabies were abolished, would you expect an increase of human rabies?

Further Readings

BOOKS

Belshe, R. B. (ed.), *Textbook of Human Virology,* PSG Publishing, Littleton, Mass., 1984. Contains excellent reviews of rabies virus and of the enteroviruses discussed in this chapter.

Leive, L., and D. Schlessinger (eds.), *Microbiology 1984,* American Society for Microbiology, Washington, D.C., 1984. Papers presented at a symposium on Hansen's disease and a symposium covering its immunology are included in this volume.

Paul, J. R., *A History of Poliomyelitis,* Yale University Press, New Haven, Conn., 1971. A comprehensive history.

ARTICLES AND REVIEWS

Gajdusek, D. C., "Unconventional Viruses and the Origin and Disappearance of Kuru," *Science,* 197:943–960 (1977). The lecture Carleton Gajdusek delivered upon acceptance of the Nobel Prize in Physiology or Medicine.

Kaplan, M. M., and H. Koprowski, "Rabies," *Sci. Am.,* 242:120–134 (January 1980). A review of the history of rabies.

Salk, J., and D. Salk, "Control of Influenza and Poliomyelitis with Killed Virus Vaccines," *Science,* 195:834–847 (1977). A review article that supports the Salk vaccine in preference to the TOPV.

Wilfert, C. M., B. A. Lauer, M. Cohen, M. L. Costenbader, and E. Myers, "An Epidemic of Echovirus 18 Meningitis," *J. Infect. Dis.,* 131:75–78 (1975). A research paper describing an outbreak of meningitis in North Carolina.

PART

5

Environmental Microbiology, Applied Microbiology, and Biotechnology

Microorganisms participate in processes that help maintain the chemical balance and biological composition of natural environments. The microbial activities involved in the natural cycling of carbon, nitrogen, and sulfur are harnessed in the designs of sewage and industrial waste treatment facilities. Microorganisms are also vital in the preparation and preservation of many foods and beverages. The products of microbial metabolism are commercially valuable; without them there would be no antibiotics, and many amino acids, vitamins, and enzymes would be unavailable or too expensive. The 1980s have witnessed an advanced biotechnology in which breakthroughs in molecular biology have enabled genetic manipulation of biologic systems. This technology has the potential to significantly alter the relationship between humankind and all the rest of the living world, on a scale so vast that its dimensions are only beginning to be understood.

C H A P T E R

28

Environmental Microbiology

OUTLINE

FOCUS OF THIS CHAPTER

Microbial populations play an essential role in maintaining an ecological balance between the biomass and the available nutrients. Chapter 28 focuses on microbes in aquatic and soil environments, their participation in the cycling of essential biological elements, and their involvement in the recycling of societie's wastes.

Our natural environment is taken for granted by most of us, until it is altered in some way, as by overuse or pollution. Preservation of a natural environment is heavily dependent on the role microbes play in maintaining the chemical balance between available nutrients and in metabolizing waste products. Pollutants—most often the chemical residues of human activities—can disrupt this balance and cause havoc to the environment.

ECOSYSTEMS

An **ecosystem** is the total community of organisms in a physically defined space. Ecosystems may be small, for example swamps, ponds, or streams; or huge, for example the Chesapeake Bay estuary or one of the Great Lakes. An ecosystem is confined by geological formations—the shores of the pond, the banks of the river, the bottom of a bay or lake (Figure 28.1). Within this confined space exists a chemical environment that greatly influences both the quantity and composition of its **biota** (sum total of the living organisms in any designated area). These physical and chemical characteristics are affected by rains that swell rivers or streams, by tides that cause cyclical changes in the oceans and estuaries, by the addition of nutrients from runoffs, industrial wastes, and sewage, and so on. The dynamic interactions of the microbes with the physical and chemical makeup of the world's many ecosystems are the subject of environmental microbiology.

FIGURE 28.1 Narragansett Bay is an estuary in Rhode Island. Providence and the Providence River are located at its north end. Heavy rains overload the Providence sewer system because the storm drains empty into the sewer. This causes an overflow and raw sewage is discharged into the bay. The upper one-third of Naraganssett Bay, once a rich shellfish resource, is now closed to all shellfishing because of fecal contamination. Fishing for finfish is prohibited in the northern half of the bay for the same reason. (Save the Bay, Inc. Providence, R. I.)

Microbial ecology is the study of the relationships between populations of microorganisms and their environments. Each ecosystem is composed of many microbial communities made distinct by local physical and chemical parameters. Temperature, pH, and availability of oxygen and nutrients dictate the species and numbers of microbes in these communities. The physical space or location where a species lives is its **habitat,** and every microorganism must have at least one natural habitat, otherwise it would not survive the rigors of its world.

ECOLOGY OF AQUATIC ENVIRONMENTS

Aquatic environments provide a variety of microbial habitats. These habitats being of different depths, there are differences in light penetration, temperature, oxygen concentration, substrate concentration, and, in extremely deep water, hydrostatic pressure. Seasonal changes also influence the habitats of all but the deeper aquatic environments.

FRESHWATER LAKES

The **biomass** (total weight of all of the living organisms) of a lake depends on the nutrients available for growth. Mountain lakes with underlying bedrock are usually described as clear and pristine because their low nutrient content supports a sparse biomass. These **oligotrophic** lakes have low concentrations of one or more of the nutrients essential for the photosynthetic growth of algae and plants that comprise the base of the food chain. Lakes surrounded by developed or agricultural land are often nutrient rich, and support extensive plant growth in the form of bottom weeds and planktonic algae. These are **eutrophic** lakes. A eutrophic lake evolves over many years, as the biomass produced in the water column or by the encroaching shoreline habitat fills in the lake with decomposing vegetation in a slow, natural process called **eutrophication.** Eutrophication is accelerated when the residue of human activities enters the lake ecosystem.

Deep lakes in temperate regions are stratified (in distinct layers) with respect to oxygen, biomass, and temperature during the warmer months of the year. The surface waters, called the **epilimnion,** are heated by the sun and are oxygenated by wave action. These waters are the habitat for photosynthetic cyanobacteria and green algae (Figure 28.2), which contribute oxygen and biomass to the ecosystem.

Lakes stratify into water layers of decreasing temperatures as the warm surface water (less dense) heated by the sun floats on the colder water below. This stratification creates a sharp transition between the upper warm and lower cold waters of a lake or pond called the **metalimnion** (Figure 28.2). During warm summer months, the surface water mixes by wave action, wind-generated currents, and thermal convection, while the cold water below the metalimnion, known as the **hypolimnion,** remains essentially static. Very little mixing occurs between the epilimnion and the hypolimnion except during severe storms. The deep water of the hypolimnion has a low oxygen tension in all but the most oligotrophic lakes, and has a constant low temperature.

Production and Decomposition in Aquatic Ecosystems

Aquatic systems are productive when they generate new biomass. The **primary productivity** of an ecosystem is measured as the amount of carbon dioxide converted to organic carbon, most of which is contributed by photosynthetic CO_2 fixation. On sunny days, surface waters are saturated with the light required in photosynthesis by the oxygenic green algae and cyanobacteria. These organisms derive minerals and carbon dioxide from the metabolic activities of bacteria, zooplankton (Figure 28.3), and large animals in the epilimnion. The biomass thus created provides the organic matter for the chemoorganotrophic growth of the ecosystem.

The anoxygenic photosynthetic bacteria (see Chapter 14) multiply in the lower realms of the photic zone. Their photosynthetic pigments utilize light not absorbed by other photosynthetic organisms and they grow in the water column where anoxic conditions prevail. In eutrophic lakes and ponds these conditions exist in shallow water beneath the green algae and cyanobacteria.

In eutrophic lakes, the water column beneath the metalimnion is anoxic, since any oxygen generated is quickly consumed by aerobic metabolism. An anaero-

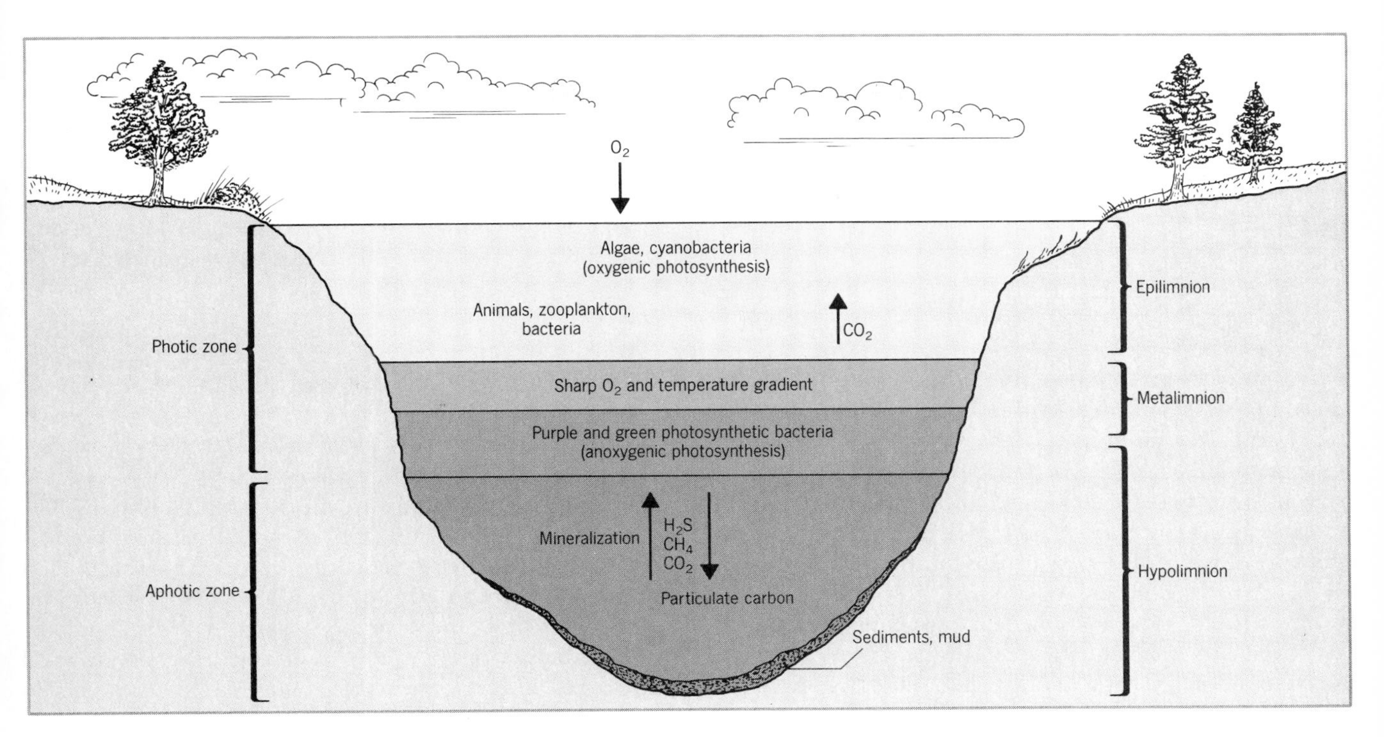

FIGURE 28.2 Deep freshwater lakes become stratified into distinct zones during the summer months. Primary productivity occurs in the waters of the epilimnion where currents and wave action mix nutrients and aerate the water. The anoxic water column beneath the metalimnion provides environments for anoxygenic photosynthetic bacteria and other anaerobic microorganisms.

bic environment is thus created for the anoxygenic photosynthetic bacteria (Figure 28.3) within the photic zone. The wavelengths of light available at these depths are affected by the color and turbidity of the water and by its depth. The anoxygenic photosynthetic bacteria seek out that environment where the wavelengths of light, absence of oxygen, and availability of nutrients released from the sediments create an ideal habitat. Here they multiply using either reduced sulfur (H_2S) or organic compounds as substrates for bacterial photosynthesis. They utilized the energy from photosynthesis to fix CO_2 into organic compounds and in so doing are responsible for **anaerobic primary productivity.**

Decomposition by anaerobic processes also occurs beneath the metalimnion. Fermentations, methanogenesis, and anaerobic respiration using sulfate, nitrate, and nitrite occur in the deep water and in the upper layers of the sediments (see Chapter 6). Bacteria metabolize organic substrates in the particulate carbon that

FIGURE 28.3 Zooplankton are the microscopic animals that drift with the currents in the upper layers of the water column. (Courtesy of W. D. Russell-Hunter.)

settles through the water column. This carbon is derived from animal fecal pellets and detritus (dead plant and animal material in an ecosystem) produced in the epilimnion. In turn, these bacteria produce CO_2, H_2, H_2S, organic acids, and reduced nitrogen, which in turn become nutrients for the growth of other microorganisms.

STREAMS AND OTHER SHALLOW AQUATIC ENVIRONMENTS

Many microorganisms colonize the surfaces of particulate matter suspended in the water column, or colonize structures adhering to the bottom. The slippery brown or green algal growth that covers the rocks in stream beds poses a challenge to fishermen who wade streams in search of trout. Microbes colonize the stream substratum in part because nutrients tend to adsorb to surfaces in aquatic environments. This colonization process can be studied by suspending clean microscope slides in aquatic environments. After a few days, enough bacteria bind to the slide to be visible under the phase microscope.

Caulobacter (cau-lo-bak'ter) and *Gallionella* (gal'li-o-nel'la) are examples of aquatic bacteria that possess stalks for adhering to sessile (attached to a base) objects. The stalk or prostheca of *Caulobacter* (Figure 28.4) is an extension of the cell's cytoplasm surrounded by a complete membrane and cell wall. Adhesive material is present on the distal end of the stalk and anchors the bacterium to the substratum. The kidney-shaped cells of *Gallionella* are attached to stalks of twisted fibrils that fasten the bacteria to the surface. They grow in aquatic environments containing iron, which is often deposited on the stalk as iron hydroxide. *Gallionella* is noted for its propensity to clog drains, water pipes, and wells with deposits of iron oxides.

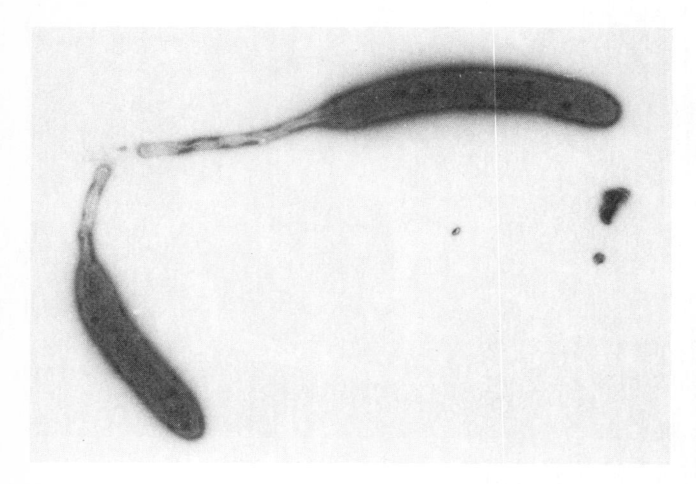

FIGURE 28.4 Electron micrograph of *Caulobacter* showing its prostheca, which terminates in a holdfast.(BPS.)

SOIL MICROBIOLOGY

Soil is a complex mixture of inorganic matter consisting of mineral particles, organic matter from decaying biomass, microorganisms, plants, and animals. The uppermost layer of soil (topsoil) supports the most life, including plants that send their roots into the topsoil from which they derive essential nutrients. In addition, topsoil is the home of many insects, worms, burrowing animals, invertebrates, and microorganisms. Soil microbes interact with the plants and the animals in soil and play a vital role in the cycling of nutrients.

THE PHYSICAL NATURE OF SOIL

Soil is composed of mineral particles of different sizes. Large soil particles are derived from weathered rocks, intermediate-sized **silt particles** are derived from primary minerals such as quartz, and the small **clay particles** (greater than 2 μm in diameter) are composed of secondary minerals such as kaolinite. Clay particles have a crystalline plate-like structure that gives them an enormous surface area. They are also negatively charged, so they bind cations, while anions such as NO_2^-, NO_3^-, SO_4^{2-}, and Cl^- leach out of the soil.

SOIL WATER AND GASES

Soils vary greatly in their ability to retain the water that is essential to biological systems. Water tends to run off clay-based soils since these soils are tightly packed and contain few air spaces for water retention. When the content of the soil is enriched with organic matter, the soil becomes more porous, and more water is retained. The spaces between the soil particles also permit penetration and release of gases, including oxygen, nitrogen, and carbon dioxide.

ORGANIC MATTER

Soil contains organic matter derived from microorganisms, animal wastes, and plants, which by far are the largest contributors of organic matter to soil. Much of this plant material is recycled by the biochemical activities of microorganisms into usable nutrients, and what remains, combines with mineral particles to form humus. **Humus** is the dark-colored material of soil that is composed largely of decay-resistant organic matter. The humus content of soil greatly increases its ability to retain water and air, and is beneficial to growing plants. Peat moss is one source of humus that is often added to flower beds and vegetable gardens to improve the soil.

The microbial biomass in the top layer of agricultural soils averages only 2.5 percent (range 0.27 to 4.8) of the total carbon present, so soil microorganisms make but a small contribution to the organic content. However, the capacity of the microbial populations to perform metabolic reactions is essential for the cycling of nutrients in soil.

SOIL ORGANISMS

Soil is home to many animals ranging in complexity from the insect pests that infect the roots of vegetable gardens to the moles that burrow beneath the topsoil. Many soil animals aerate the soil through their feeding and burrowing behaviors. Plant roots penetrate deep into the soil, establishing a plant root environment known as the **rhizosphere**—the habitat for many microorganisms. Most soils contain a rich and heterogeneous population of protozoa including flagellates, amebas, and a few ciliates. Populations of between 10,000 and 100,000 protozoa per gram of soil are typical. Photosynthetic microorganisms, including cyanobacteria, green algae, and diatoms, live in soil where light penetration and moisture levels create a habitat suitable to photosynthesis.

Bacteria are the predominant microorganisms in most soils (Table 28.1), and most are found in the upper 25 cm; few microbes live at a depth beneath 65 cm. Although species vary from one soil to another, the following genera are representative: *Arthrobacter* (5 to 60 percent of bacteria present), *Bacillus* (7 to 67 percent), *Actinomycetes* (10 to 33 percent), *Pseudomonas* (3 to 15 percent), *Agrobacterium* (1 to 20 percent), *Alcaligenes* (2 to 12 percent), and *Flavobacterium* (2 to 10 percent.

The fungi present in soil (Table 28.1) are detected by the presence of their spores. Their actual protoplasmic contribution may be greater than indicated by spore counts (see Chapter 16). Fungi prefer to grow in slightly acid soil, so more are found in forest soils with its decaying plant material than in field soils.

The long-term survival of soil fungi and certain bacteria is augmented by the spores, endospores, or cysts they form. Fungal spores can survive in soils for periods ranging from two months to more than ten years, and endospores of *Bacillus* and *Clostridium* will survive for many years in soil. In addition, many of the actinomycetes produce asexual spores; the fruiting myxobacteria produce resting stages known as myxospores; and some *Azotobacteraceae* produce cysts. These resting stages enable the organisms to withstand the summer heat of sundrenched sands and soils, and to survive long periods of drought.

TABLE 28.1 Distribution of microorganisms in various horizontal layers of soil

	ORGANISMS/GRAM OF SOIL				
DEPTH (CM)	AEROBIC BACTERIA	ANAEROBIC BACTERIA	ACTINOMYCETES	FUNGI	ALGAE
3–8	7,800,000	1,950,000	2,080,000	119,000	25,000
20–25	1,800,000	379,000	245,000	50,000	5,000
35–40	472,000	98,000	49,000	14,000	500
65–75	10,000	1,000	5,000	6,000	100
135–145	1,000	400	——	3,000	——

Source: M. Alexander, *Introduction to Soil Microbiology*, 2nd ed., Wiley, New York, 1977.

PATHOGENS IN THE SOIL

A few soil microorganisms are able to cause disease. Cryptococcosis and histoplasmosis are human diseases caused by fungal spores inhaled in soil dust (see Chapter 23). These fungi favor soils contaminated by bird droppings—chicken yard soil, soil under blackbird roosts, and soil contaminated by pigeon droppings.

Spore-forming soil bacteria cause anthrax, botulism, and tetanus. *Bacillus anthracis* multiplies in soil and contaminates the grasses of pastureland. Animals that ingest the spores of *B. anthracis* become ill with anthrax, and many die. Spores of *B. anthracis* can remain in the contaminated field for years. The spores also contaminate the hide and wool of animals in these pastures, and will infect people who come in contact with contaminated animal products such as wool or hides (see Chapter 22).

Soil-borne spores of *Clostridium botulinum* present on garden vegetables germinate and multiply in the anaerobic environment of home-canned vegetables or other foods to cause botulism food poisoning (see Chapter 25). Another soil-borne sporeformer is *C. tetani,* the causative agent of tetanus (see Chapter 27).

BIOCHEMICAL NUTRIENT CYCLES

Except for the continuing input of energy from the sun, the earth's resources are limited by what is physically present on our planet; "Spaceship Earth" was how the scientist and writer Buckminster Fuller described our world and its environs. Some resources are renewable, that is, they are recycled through the interactions of natural processes and used again and again. Others are **nonrenewable.** Oil and gas are nonrenewable resources because they cannot be replenished by natural means in a reasonable time frame (such as one's lifetime). It behooves humankind to utilize nonrenewable resources with discretion.

Cotton, trees, foods derived from animals and plants, and oxygen are examples of **renewable resources.** Living things extract nutrients from the earth to make the substances they need for growth. Plants and animals would long ago have depleted the essential biological elements if microbes had not, from time immemorial, participated in the intricate web of metabolic activities by which biological material is recycled. All the essential biological elements are recycled through these complex processes to maintain an ecological balance that enables "Spaceship Earth" to support future generations.

THE OXYGEN CYCLE

Think of the elements present in cells: Oxygen is a key element; in fact, when water is included, oxygen comprises 70 percent of a cell's weight. The hydrosphere and atmosphere (23 percent oxygen) are the chief reservoirs of oxygen available to biological systems. Reservoirs of lesser quantities of biochemically available oxygen occur in mineral deposits such as salts of carbonates, nitrates, and sulfates. The oxygen that is combined in silicates, aluminates, and mineral oxides, accounting for more than 80 percent of the oxygen on earth, is unavailable to biological systems.

Oxygen is cycled between molecular oxygen (O_2), organic compounds, and water by the metabolic activities of biological systems (Figure 28.5). As will be recalled (Chapter 14), green plants and cyanobacteria produce oxygen from water by the process of oxygenic photosynthesis. The electrons and ATP thus generated are used in the Calvin cycle for fixing CO_2, while the oxygen escapes into the atmosphere from the surfaces of green plants or is dissolved in the water surrounding aquatic plants.

Heterotrophs use oxygen to oxidize organic compounds, and chemolithotrophs use it as the oxidant for inorganic substrates. Water is a product of both of these aerobic processes, and so water is again made available for photosynthesis.

THE CARBON CYCLE

Mineral deposits of carbonate and dissolved carbonates in aquatic systems comprise 70 percent of the earth's carbon; only 20 percent of the earth's carbon is in organic compounds, and less than 1 percent is in fossil fuels and biomass. Unlike oxygen, most of the earth's carbon is available for biochemical reactions.

Carbon dioxide, whether generated from carbonates in aqueous systems or present as a gas in the atmosphere, is fixed by biological systems into organic compounds. Carbon dioxide exists in equilibrium with

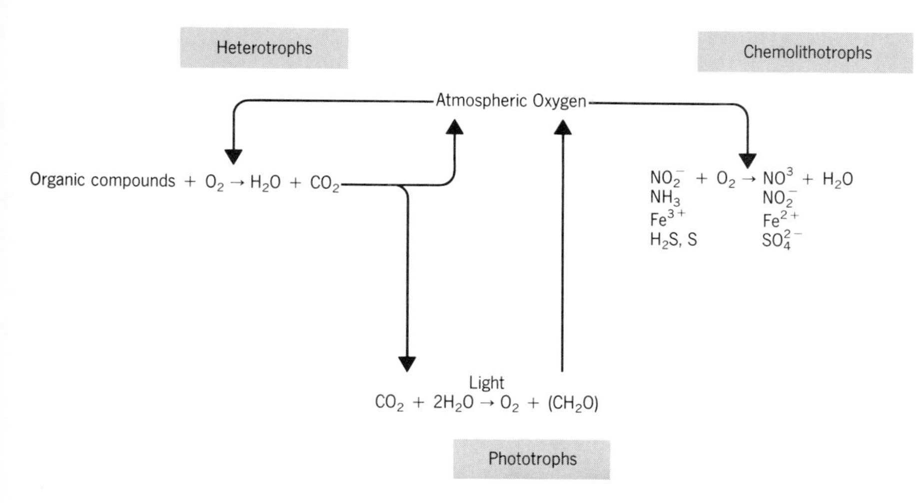

FIGURE 28.5 Oxygen cycle. Oxygen is generated by oxygenic photosynthesis and is utilized as an oxidant by heterotrophs and chemolithotrophs.

carbonates in aqueous systems according to the following reactions:

$$\begin{array}{lcr} CO_2 \text{ Atmospheric} & & Ca^{2+} \leftarrow CaCO_3 \\ \uparrow & & \downarrow \\ CO_2 + H_2O \leftrightharpoons H_2CO_3 \leftrightharpoons & H^+ + HCO_3^- \leftrightharpoons & CO_3^{2-} + 2H^+ \end{array}$$

Dissolved

Carbonic acid Bicarbonate Carbonates

The relatively insoluble carbonates (at the right) slowly form the soluble bicarbonate ion, which then produces carbon dioxide and water. These reactions are dependent on the pH of the aquatic system and the activities of microorganisms.

Carbon dioxide is also produced during the burning of fossil fuels, biological respiration, and microbial decomposition (Figure 28.6). Photosynthetic organisms and chemosynthetic bacteria fix CO_2 into organic carbon (CH_2O) in the Calvin cycle, which fixes carbon dioxide into ribulose-1,5-diphosphate at the expense of ATP and $NADPH_2$ (see Chapter 6) or with the **C_4 carboxylic acid pathway.** The C_4 pathway combines carbon dioxide with a three-carbon acid (pyruvic or phosphoenolpyruvic acid) making a four-carbon intermediate of the TCA cycle (oxaloacetic acid). The C_4 pathway is used by some bacteria, and many green plants have both the C_4 pathway and the Calvin cycle. Once fixed, the carbon becomes part of the biomass of the ecosystem.

Primary productivity, the amount of carbon fixed in these metabolic pathways, varies greatly depending on the season and the availability of nutrients within the ecosystem. A secondary source of carbon is the methane generated anaerobically from CO_2 and H_2 by **methanogenic archaebacteria.** Some of this methane is trapped in geological formations, some escapes to the atmosphere, and the remainder is metabolized by the aerobic **methanotrophic bacteria** as their source of carbon and energy.

KEY POINT

The primary productivity of an ecosystem is attributed to oxygenic photosynthesis and, to a lesser extent, to chemoautotrophic bacteria. The carbon fixed in primary production becomes a substrate for the chemoorganotrophic microorganisms and animals.

Most of the organic matter present in soil originates as plant material—leaves, rotting trees, decaying roots, and other plant tissue. Soil bacteria and fungi are largely responsible for recycling this carbon. Without

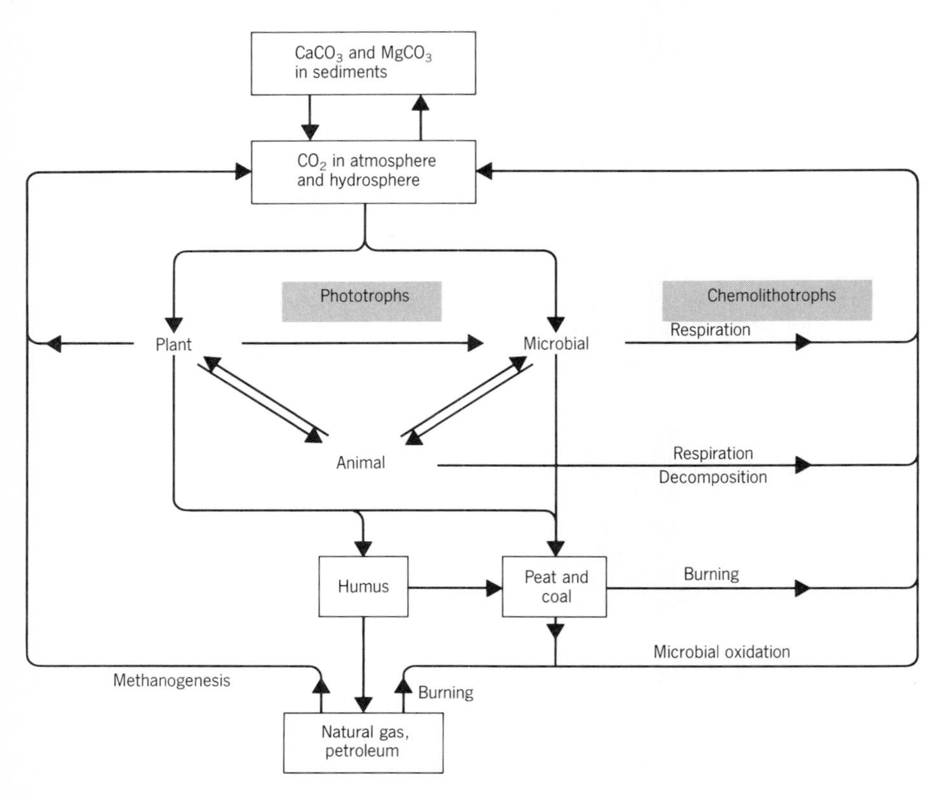

FIGURE 28.6 Carbon cycle. Carbon is converted from its inorganic forms of carbonates and carbon dioxide into organic carbon by biological processes. Organic carbon can be reconverted to carbon dioxide by respiration, decomposition, and burning. The cycling time for some forms of carbon—for example coal and oil—may be millions of years, whereas other forms of carbon are recycled much faster.

the cycling action of these microorganisms, life on this planet would suffer an irreversible decline as the nutrients essential for life became tied up in complex molecules. Insoluble dead vegetable matter is first acted on by extracellular microbial enzymes such as cellulase and chitinase to release soluble products, making them available to microbes and plants. The major contribution of the fungi is the cycling of carbon, whereas the bacteria participate in both the cycling of carbon and the mineralization (release) of nitrogen, phosphorus, and sulfur, making these elements available to root systems of plants.

Undigested organic plant and animal matter becomes part of the soil humus, a process highly visible in peat bogs where decaying vegetation retains a high organic content. Coal also originates from plant material buried in primeval swamps eons ago. Both coal and peat are converted to CO_2 when they are burned as fuel (Figure 28.6).

Complex food webs originate with the biomass generated in primary production. Single-cell algae and bacteria that use CO_2 fixation to make organic material are eaten by protozoa and zooplankton, which in turn are eaten by larger animals. All other dead plant and animal tissue as well as the excreta of animals is decomposed eventually by microbial processes. The end result is the complete cycling of organic carbon from its major reservoirs as CO_2 in the atmosphere and as carbonates in sedimentary rocks through living systems and then back to the reservoirs.

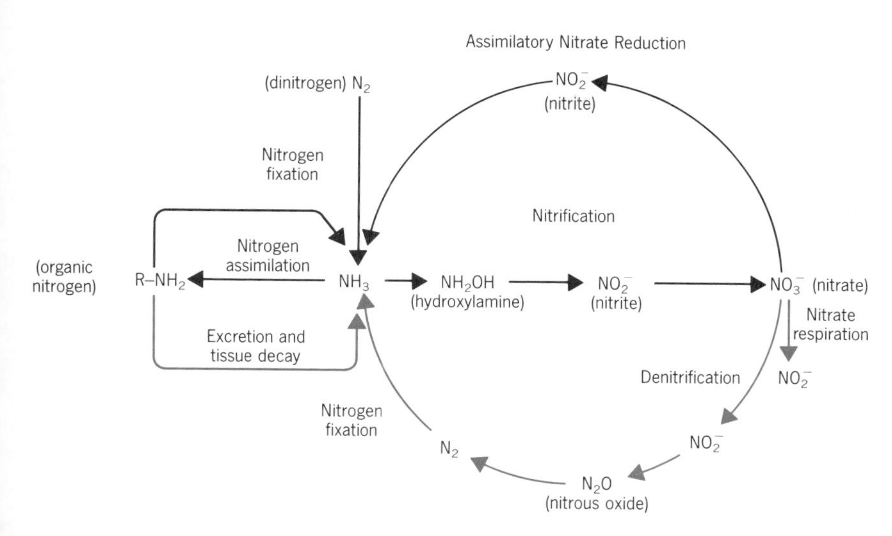

FIGURE 28.7 Nitrogen cycle. Nitrogen-fixing bacteria reduce N_2 to ammonia, which they assimilate into organic nitrogen. Nitrification is the process by which chemolithotrophs oxidize ammonia to nitrate. Nitrate is a major source of nitrogen for plants, fungi, and some bacteria. Under anaerobic conditions (colored arrows), bacteria reduce nitrate to nitrite (nitrate respiration), and some bacteria can reduce nitrite to gaseous products (denitrification).

THE NITROGEN CYCLE

Nitrogen, an essential cellular element of amino acids, purines, pyrimidines, and some coenzymes, accounts for between 9 and 15 percent of a cell's dry weight. Some cells require an organic nitrogen source such as an amino acid or a purine, while others obtain their nitrogen from inorganic compounds such as nitrogen gas (N_2), ammonia, or nitrate (NO_3^-). Nitrate and nitrogen gas are major sources of inorganic nitrogen for plants and microorganisms, and both must be reduced to ammonia (NH_3) before the nitrogen can be assimilated into biomolecules. As previously described, ammonia assimilation occurs by an enzymic pathway that forms glutamate and glutamine (see Chapter 6). The nitrogen in these amino acids can be used to form all of the cell's other nitrogenous compounds.

The availability of nitrogen for biological systems in part dictates which organisms can multiply within an environment. Atmospheric N_2 is the main reserve of nitrogen on earth, while the nitrogen in nitrate accounts for approximately 60 percent of the nitrogen in seawater; other forms of nitrogen are present in lesser concentrations. These inorganic forms of nitrogen are interconverted by the metabolic activities of microorganisms, which collectively constitute the nitrogen cycle (Figure 28.7). This cycle operates in all ecosystems to make available the different forms of nitrogen and to maintain the natural nitrogen balance.

Nitrogen Fixation

Nitrogen gas (N_2) is reduced to ammonia by free-living nitrogen-fixing bacteria and by bacteria that grow in symbiosis with leguminous plants, alder trees, and certain grasses (Figure 28.8). Only certain bacteria can use N_2 as a source of nitrogen; specifically, the bacteria that make **nitrogenase** can reduce N_2 to ammonia.

$$N_2 + 3H_2 \longrightarrow 2NH_3$$

This oxygen-sensitive enzyme reduces N_2 to ammonia at the expense of ATP and a reduced substrate, such as H_2.

Nitrogen fixation is essential to agriculture since nitrogen is a limiting nutrient for plant growth in many soils. Farmers take advantage of N_2-fixing legumes such as alfalfa, vetch, and clover to enrich their fields with fixed nitrogen (see Chapter 14). As an example,

FIGURE 28.8 The vigorous nodulated peanut plants (left) are infected with *Rhizobium,* which fixes nitrogen gas; the nonnodulated (right) peanut plants are not infected. (Courtesy of Robert Howell, ARS, USDA.)

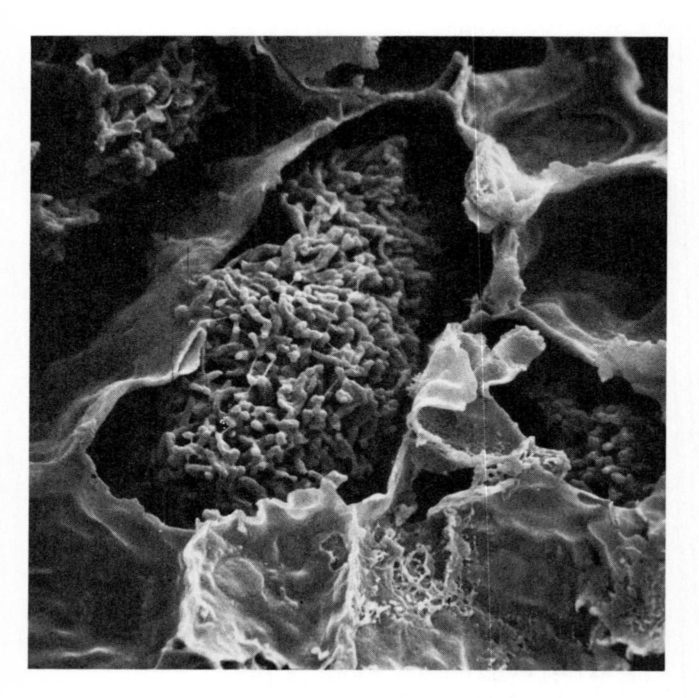

FIGURE 28.9 *Rhizobium trifolii* infects the root hairs of white clover and causes them to form nodules. This scanning electron micrograph shows the nitrogen-fixing *Rhizobium* bacteroids inside a root nodule cell of white clover. (Courtesy of J. H. Troughton.)

soybeans are cash-crop legumes that obtain 40 percent of their required nitrogen through nitrogen fixation; soil provides the remaining 60 percent. As previously described, leguminous plants form a symbiotic relationship with a specific species of *Rhizobium*. The nodules on the plant roots contain nitrogen-fixing bacteroids (Figure 28.9) that generate ammonia for plant growth. For a discussion of the microbiology of nitrogen fixation, refer to Chapter 13.

Nitrate Assimilation

The ability to use nitrate as a nitrogen source is found in plants, fungi, and some bacteria. These organisms have **assimilatory nitrate reductase,** which reduces nitrate to nitrite (NO_2^-), and **nitrite reductase,** which reduces nitrite to ammonia (Figure 28.7). The ammonia is assimilated into glutamate and glutamine. The nitrate available for this pathway comes from geological nitrate deposits or is generated by the biological process of nitrification (see below).

Ammonification

The organic nitrogen generated through nitrogen fixation and nitrate reduction enters the biomass as organic nitrogen: amino acids, purines, pyrimidines. In dead animals and plants this organic nitrogen is recycled by the process called ammonification (Figure 28.7). The proteins and nucleic acids are metabolized to amino acids, purines, and pyrimidines, and then to organic acids, volatile gases, and ammonia. The ammonia produced by ammonification is released into the environment, where it is utilized as a primary nitrogen source by plants, animals, fungi, and bacteria.

The excretory products of animals and birds pose special recycling problems. Urea, the major excretory product of terrestrial animals, is metabolized by bacteria to carbon dioxide and ammonia, products that make the soil alkaline. This in turn converts ammonium ions into ammonia gas, which is lost to the atmosphere. Ammonification is a special problem in animal feedlots (Figure 28.10) since the large concentration of animal urine makes the soil and runoff from the soil very alkaline.

Flocks of ducks, chickens, turkeys, and other birds create another problem since their chief excretory

FIGURE 28.10 Cattle feedlots are a source of nitrogen and phosphorus wastes. Runoff from this 4000 head cattle feedlot operation in Lubbock, Texas is intercepted by holding ponds at the top of the picture. (Photograph by Fred S. White, USDA.)

product is the water-insoluble purine, **uric acid.** The insoluble nature of uric acid is apparent in the large white deposits of **guano** (bird excrement) found on offshore rocky islands inhabited by seabirds. In low quantities, guano is a good nitrogen fertilizer, which is metabolized by soil organisms to urea and glyoxylic acid. However, large quantities of guano make soil very alkaline since CO_2 and ammonia are by-products of the microbial degradation of uric acid.

The ammonia produced by ammonification is either incorporated into cell biomass or becomes the substrate for another step in the nitrogen cycle—nitrification.

Nitrification

Nitrification is the aerobic oxidation of ammonia to nitrate (NO_3^-) by two distinct groups of nitrifying bacteria (see Chapter 14); fungi and plants are not involved. The ammonia oxidizers are chemosynthetic autotrophs, represented by *Nitrosomonas* and *Nitrosococcus,* which oxidize ammonia to nitrite in an aerobic energy-yielding process represented by the following equation:

$$2NH_3 + 3\ 1/2\ O_2 \longrightarrow 2NO_2^- + 3H_2O + \text{Energy}$$

The nitrite generated by these bacteria is oxidized to nitrate by a second group of chemosynthetic autotrophs, the nitrite oxidizers, represented by *Nitrobacter* and *Nitrococcus*. They oxidize nitrite to nitrate by the following equation:

$$NO_2^- + 1/2\ O_2 \longrightarrow NO_3^- + \text{Energy}$$

The nitrifying bacteria can grow in a completely inorganic medium, gaining their energy from the oxidation of ammonia or nitrite and utilizing carbon dioxide as a carbon source.

Nitrification occurs in soils, fresh water, and marine environments where species of nitrifying bacteria are common. These chemolithotrophs are like miniature factories, producing large quantities of nitrite or nitrate while gaining energy for growth and reproduction. They have long generation times, in part because they must oxidize enormous quantities of substrate to manufacture enough cell material for multiplication. The nitrate produced during nitrification is an important ni-

TABLE 28.2 Representative denitrifying bacteria

ORGANISM	REMARKS
Rhodopseudomonas spheroides	Also capable of N_2 fixation
Cytophaga johnsonae	Strain dependent
Hyphomicrobium spp.	Budding bacterium
Aquaspirillum itersonii	Produces N_2O
Spirillum lipoferum	Strain dependent
Pseudomonas spp.	Many species can denitrify
Alcaligenes faecalis	All strains denitrify nitrite
Rhizobium japonicum	Also capable of N_2 fixation
Chromobacterium lividum	Visible gas produced
Paracoccus denitrificans	Produce nitrous oxide and N_2
Thiobacillus denitrificans	Obligate denitrifier
Bacillus spp.	Many strains, produce nitrous oxide and N_2
Propionibacterium acidi-propionici	Strain dependent

trogen source for plants and is the substrate for dissimilatory nitrate reduction (see below).

Nitrate Respiration

The nitrogen cycle is complete at this point, except for the regeneration of nitrogen gas, which is the substrate for nitrogen fixation. Many bacteria can grow anaerobically using nitrate as a terminal electron acceptor in place of oxygen. The reduction of nitrate to nitrite under anaerobic conditions, **nitrate respiration,** is performed with an oxygen-sensitive, membrane-bound nitrate reductase. In true nitrate respiration, nitrite accumulates in the medium and is not reduced further, whereas other bacteria (the denitrifiers) reduce the nitrite to gaseous products. Both nitrite respiration and denitrification are sensitive to oxygen, so they occur only in anaerobic environments.

Denitrification

Denitrification, the biological production of gaseous nitrogen from the anaerobic reduction of nitrate, is performed by bacteria (Table 28.2) that use this process to grow by anaerobic respiration (see Chapter 6). The denitrifying bacteria possess enzymes that reduce nitrite to nitrous oxide (N_2O) and then to N_2 when either nitrate or nitrite is available as a terminal electron acceptor.

$$NO_3^- \longrightarrow NO_2^- \longrightarrow N_2O\ (g) \longrightarrow N_2\ (g)$$

The nitrate present in anaerobic soils is first reduced to nitrite before gaseous products, nitrous oxide (N_2O) and N_2, are formed.

Denitrification is an important geochemical and environmental process since it is the route by which "fixed nitrogen" is returned to the atmosphere (Figure 28.7). This process is also important in agriculture since the advantage of applying fertilizer nitrogen is lost when the nitrogen is reduced to gaseous forms that quickly leave the soil and enter the atmosphere. Denitrification will occur in waterlogged soils, which provide the necessary anaerobic environment.

KEY POINT

Efforts by humans to manipulate the inorganic nitrogen cycle have concentrated on nitrogenase and the process of nitrogen fixation. If this process were to be transferred to nonleguminous plants, such as corn, there would be no need to use expensive nitrogen fertilizers in growing food crops.

THE SULFUR CYCLE

Sulfur is an essential cellular element found in two amino acids and certain coenzymes. Most cells make

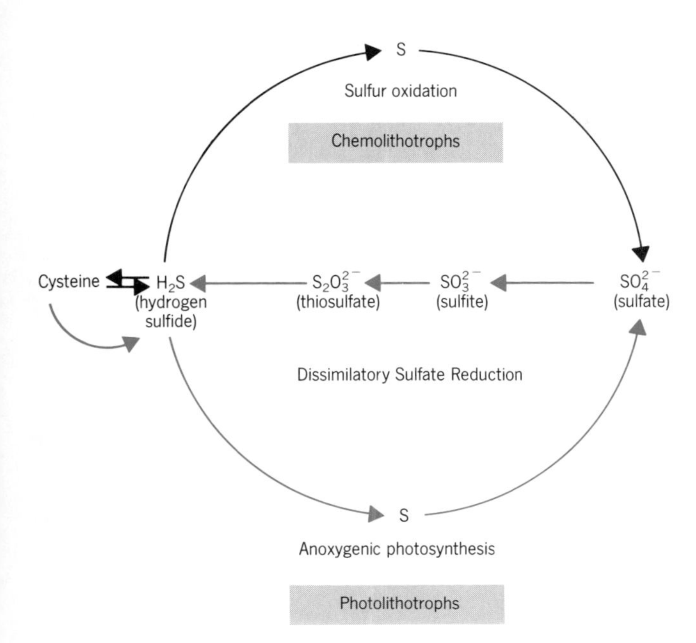

FIGURE 28.11 The sulfur cycle. Reduced-sulfur compounds (H_2S) are oxidized by the aerobic chemolithotrophs and by anoxygenic photosynthetic bacteria to sulfate. Sulfate is reduced under anaerobic conditions (colored arrows) to hydrogen sulfide by sulfate-reducing bacteria. Both aerobic and anaerobic cells assimilate sulfur into cysteine.

these sulfur-containing organic compounds from inorganic sulfur. Cells need only small quantities of sulfur since it comprises only 1 percent of a cell's dry weight. The soluble forms of biologically available sulfur are sulfate (SO_4^{2-}), thiosulfate ($S_2O_3^{2-}$), and sulfide (S^{2-}). Insoluble sulfur complexes of calcium and other divalent cations are found in soils, clay, and rocks. Sulfur is biochemically converted between its reduced forms, present in organic molecules (—SH) and oxidized forms found in the environment, by the bacteria of the sulfur cycle (Figure 28.11).

Dissimilatory Sulfate Reduction

The **dissimilatory sulfate-reducing bacteria** grow in strictly anaerobic environments by the process of anaerobic respiration. This morphologically diverse group of bacteria lives in mud, anaerobic layers of aquatic systems, and in deep wells. These bacteria gain energy by transferring electrons from organic compounds to sulfate, which in the process is reduced to hydrogen sulfide (H_2S). The dissimilatory sulfate-reducing bacteria can grow on many organic substrates, including fatty acids, lactate, succinate, and primary alcohols.

Sulfide is soluble in water at alkaline and neutral pHs,

$$H_2S \longleftrightarrow HS^- \longleftrightarrow S^{2-}$$

but it becomes a foul-smelling gas (rotten egg smell) at acidic pHs. Sulfides are toxic to animals in high concentration and cause additional problems because of the ferrous sulfide (FeS) they form by reacting with ferrous iron (Fe^{2+}). This black compound contributes to the color of anaerobic mud at the bottom of aquatic environments. Ferrous sulfide also taints domestic water drawn from deep anaerobic wells, especially when the water is transported through iron pipes.

Sulfur Oxidation

The oxidation of reduced sulfur occurs anaerobically in the light by the photosynthetic sulfur bacteria and aerobically by the colorless sulfur bacteria (Figure 28.12). Both groups play a key role in the sulfur cycle.

Phototrophs Purple sulfur and green sulfur photosynthetic bacteria grow anaerobically using either H_2 or H_2S as their photosynthetic electron donor. They oxidize H_2S to elemental sulfur (S°), which is usually deposited in intracellular granules before the S° is completely oxidized to sulfate. This oxidation occurs in anaerobic environments where light is available for photosynthesis, such as the bottom of streams and ponds, where the bacteria grow in the top layers of sand overlying black, sulfide-rich mud. Photosynthetic bacteria are also found suspended in the water column of deep lakes, and position themselves where the optimum concentration of H_2S and light converges beneath the oxygenated surface layers.

Colorless Sulfur Bacteria Colorless sulfur bacteria oxidize H_2S and other reduced-sulfur compounds to sulfate in aerobic environments. Representatives include the chemolithotrophs *Thiobacillus* and *Thiomicro-*

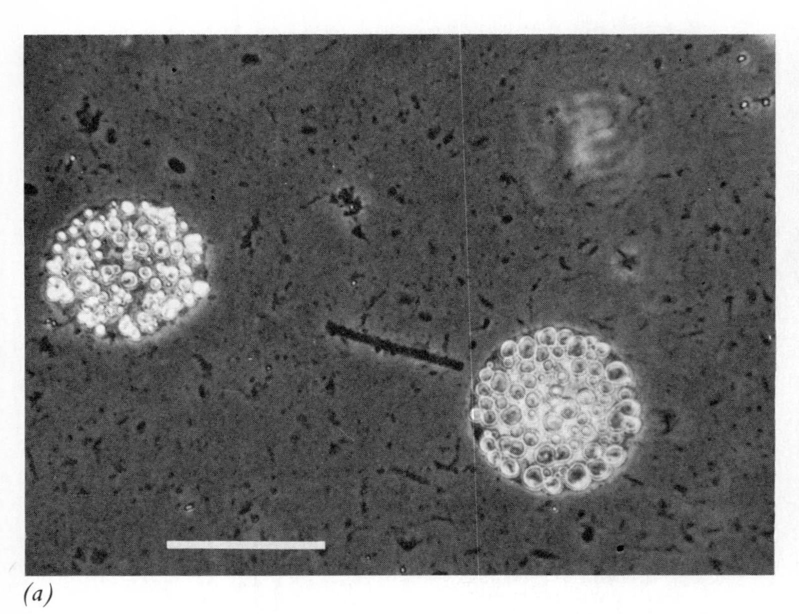

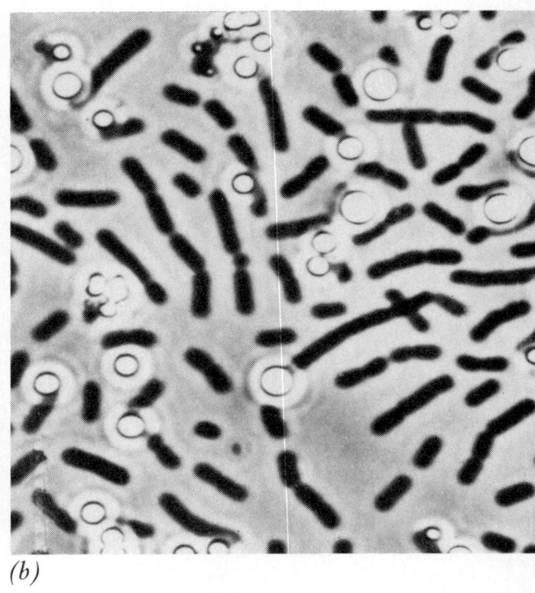

(a) (b)

FIGURE 28.12 Sulfur-oxidizing bacteria and their sulfur granules. *(a) Achromatium* deposits S intracellularly, and *(b) Chlorobium limicola* deposits S extracellularly. (Courtesy of *(a)* Dr. K. Schmidt, *(b)* Dr. N. Pfennig.)

spira (thi′oh-my-kroh-spi′rah), which grow aerobically by oxidizing reduced sulfur and fixing CO_2. Other colorless sulfur oxidizers include *Beggiatoa, Thiothrix,* and *Achromatium,* bacteria that are morphologically conspicuous and grow aerobically in the presence of hydrogen sulfide (see Chapter 13).

The colorless sulfur bacteria are found in environments where inorganic sulfur is present. Some colorless sulfur bacteria are tolerant of low acidity or high temperatures, or both; they grow in acid-mine waters where the pH can be as low as 1 or in thermal springs at temperatures of 50°C. Examples of sulfur-oxidizing thermoacidophiles are *Sulfolobus* and *Thioparus,* archaebacteria that grow in hot-acid springs. The physiological diversity of the colorless sulfur bacteria ensures that reduced-sulfur compounds will be oxidized in all environments, regardless of extremes in acidity or temperature.

Sulfate Assimilation

Sulfate, the major source of cellular sulfur for most bacteria, is assimilated after it is reduced to sulfide. The first step is the reduction of the sulfur atom to sulfite (SO_3^{2-}). Next, sulfite is reduced to sulfide, the form of sulfur enzymatically incorporated into L-serine by cysteine synthase.

$$\text{L-serine} + H_2S \longrightarrow \text{cysteine} + H_2O$$

KEY POINT

The reduced forms of sulfur (H_2S) and ammonia (NH_3) are made available to the cell's metabolic machinery as amino acids; cysteine for sulfur and glutamine and glutamate for ammonia. These amino acids donate sulfur or nitrogen for the synthesis of other cellular constituents.

SOURCES OF CELLULAR PHOSPHORUS

Microorganisms use both organic and inorganic sources of phosphate to synthesize nucleotides, nucleic acid, phospholipids, and phosphorylated proteins. Plant material is the major source of organophosphate in organic-rich environments. Soluble phosphates are released from leaf litter, plowed-under crops, detritus, and decaying aquatic organisms during biodegradation.

Deep-Sea Vents

Oceanographers began searching in the late 1970s for evidence of undersea volcanos postulated by the theory of plate tectonics. This theory views the earth's crust as a series of continental-sized plates, which slowly move in relation to each other. The rifts created when these plates spread at undersea juncture points are locations subject to undersea volcanism. The location of the rifts in the seafloor is known from oceanographic maps of the seafloor.

In 1977 the Galapagos Rift was studied with a towed camera/data system (ANGUS) that sent information back to the ship thousands of meters above. The camera took pictures of hydrothermal vents spewing black "smoke" and abundant animal life, including large tubeworms and clams. This was very exciting, since the amount and diversity of life contrasted markedly with other parts of the deep-water ocean floor, which are essentially barren of life. Most deep-water environments are inhospitable to living creatures because of the cold, great hydrostatic pressure, and the absence of significant organic matter. Most organic matter is synthesized, consumed, and recycled in the euphotic zone, and only a small fraction sinks to the deep sea. Marine biologists were truly surprised by the abundant life around the deep-sea vents.

The vents appear to lie above a large chamber of molten lava. Seawater entering this chamber reacts with minerals in the earth's crust and is heated by the molten lava. The hot, chemical-enriched seawater rises to the seafloor and emerges through hot (160° to 350°C) water vents or warm (8° to 25°C) water vents. Since the ambient pressure is over 250 atmospheres, the water temperature at 2,500 meters below the surface can be 350°C without boiling.

The original pictures from ANGUS revealed the presence of meter-long red tube worms, *Riftia pachyptila,* and giant white clams, *Clyptogena magnifica,* some 24 cm long, growing around the vents. Scientists could not exlain how these organisms lived without the benefit of a photosynthetic food supply until bacteriologists realized that this was a new kind of ecosystem, one driven by chemosynthesis.

Holger Jannasch and his colleagues discovered a preponderance of sulfur-oxidizing chemolithotrophic bacteria around the hydrothermal vents. These bacteria use the energy they gain from oxidizing reduced-sulfur compounds to fix carbon dioxide into organic matter, and this form of chemosynthesis provides the foundation of life at the vents. The milky-bluish water coming from warm vents contains masses of *Thiomicrospira,* a chemolithotroph that oxidizes the hydrogen sulfide and thiosulfate present in the vent water. *Beggiatoa,* another sulfur-oxidizing bacterium, was also found near the vents. In addition to the sulfur oxidizers, microbiologists have found bacteria that oxidize geothermal methane (methylotrophic bacteria), hydrogen, and possibly iron and manganese.

Animals living near the deep-sea vents are dependent on symbiotic bacteria for their source of carbon. The tube worm transports hydrogen sulfide to its trophosome (body), which is densely packed with bacteria that generate organic matter through chemosynthesis. A similar symbiotic relationship appears to exist between chemolithotrophic bacteria and the giant white clam.

The deep-sea vents are the first environment discovered in which chemosynthesis, as opposed to photosynthesis, is responsible for primary productivity. New undersea vents supporting similar forms of life have now been discovered in other oceans. The exciting discovery of the deep-sea vents has given microbial ecologists an entirely new ecosystem to study and understand.

(a)

(b)

(a) Giant tube worms photographed near *(b)* a deep-sea vent located in the Galapagos Rift. (Photographs by Kathleen Crane and Dudley Foster, courtesy of WHOI.)

Inorganic phosphorus exists as insoluble complexes of calcium, magnesium, and iron. Phosphorus is released from these complexes in acidic environments, created by the sulfuric acid (H_2SO_4) formed during sulfur oxidation and the acid products of nitrification (HNO_2, HNO_3). Cells assimilate inorganic phosphate as the phosphate anion (PO_4^{2-}), which is metabolically incorporated into organophosphate, using the energy-trapping mechanisms that were discussed in Chapter 6.

Phosphorus as a Limiting Nutrient

Phosphorus appears to be the limiting nutrient in freshwater lakes and in some marine ecosystems. When excess nitrogen is available from biological nitrogen fixation, the density of the phytoplankton population is a direct function of the available phosphorus. In such environments, the addition of phosphorus stimulates primary productivity, which occurs naturally when runoff from lawn fertilization, sewage overflow, excessive phosphorus in waste-water effluent, wastes from animal feedlots, and runoff from agricultural lands enrich the water. During the 1970s, phosphate detergents were major contributors of phosphate to the country's environment, and this compelled some states to pass laws prohibiting the use of phosphate detergents.

WASTE-WATER TREATMENT

Human activities generate a tremendous volume of waste water that must be treated before it is discharged into natural waterways. Treatment systems vary in size and complexity depending on the amount of waste to be treated and its composition. A rural home may have a simple septic tank and a percolating field, whereas large cities or industrial parks must make capital investments in waste-treatment facilities.

The purpose of waste-water treatment is to remove as many pollutants as is economically feasible. Waste water typically contains both inorganic pollutants in the forms of nitrogen, phosphorus, and heavy metals; and organic pollutants in the forms of sedimentable sludge, biodegradable organic compounds, and nonbiodegradable chemical wastes such as chlorophenyls and polychlorobiphenyls. Another objective of waste-water treatment is to remove or inactivate pathogenic microorganisms and viruses. Waste-water treatment facilities employ physical, biological, and chemical techniques (Figure 28.13) to produce water that will not markedly alter the environment into which it is discharged.

BIOCHEMICAL OXYGEN DEMAND

One measure of effective waste-water treatment is the BOD biological assay. The **biochemical oxygen demand** (BOD) is the requirement for molecular oxygen that accompanies the molecular oxidation of biodegradable substances in waste by microorganisms. One method of determining the BOD is to measure the oxygen concentration in the water before and after it has been incubated in an air-tight stoppered bottle with microorganisms (which may have to be added) for a set amount of time (usually five days) at a temperature between 5 and 20°C. The difference in the dissolved oxygen measurements is used to determine the BOD. The higher this difference, the greater is the BOD of the sample.

Effective waste-water treatment significantly decreases the BOD of the effluent so that it does not pollute the environment. The highest BOD values in waste waters are found in wastes from food processing, agriculture, and some industries (Table 28.3). Processing wastes with high BODs requires continued aeration to maintain high oxygen concentrations during treatment. Effluents with high BODs pollute the environment into which they are discharged by indirectly depleting the oxygen available for plant and animal life. Ecologically balanced streams, rivers, and lakes can become anaerobic when high BOD effluents are discharged into them. Such habitats select for fermenting bacteria and those that grow by anaerobic respiration. These same environments become unsuitable for aerobic animal life, which die or leave the habitat.

PRIMARY WASTE TREATMENT

The first step in waste-water treatment is to remove suspended particulate matter in settling tanks or settling basins (Figure 28.14); this step is called **primary sewage treatment.** The sedimented solids that accumulate in settling basins become part of the sand-gravel bottom. Periodically these basins are drained, and the sludge is collected and either burned or buried

Microbial Ecology

IIIA. 1 Bloom of purple sulfur bacteria on the shore of the Baltic Sea. (Courtesy of X. H. Kaltwasser.)

IIIA. 3 The sheathed bacterium *Sphaerotilus* grows in brown mats colored by the deposition of iron in the sheath material. Photograph taken in a Michigan swamp.

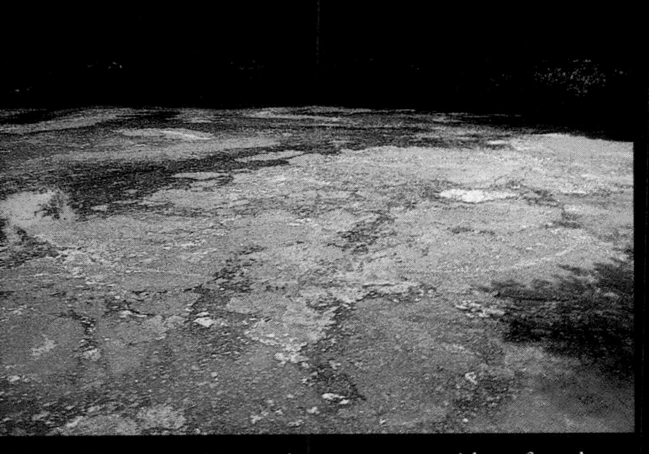

IIIA. 5 This eutrophic pond is overgrown with surface algae

IIIA. 2 Red tide is caused by dinoflagellates; they produce a water-soluble toxin that affects the nervous system of animals. (VU/Sanford Berry.)

IIIA. 4 Fucus is a brown alga that attaches itself to rocks by a hold-fast.

Nitrogen Fixation — Citrus Chancre

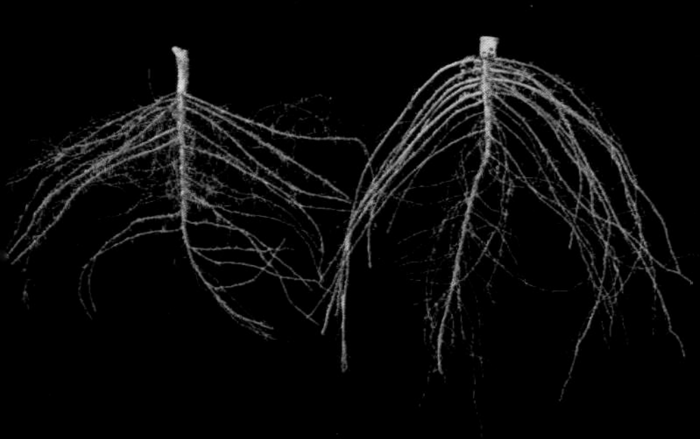

IIB. 1 Nonnodulated roots of a peanut plant. R. K. Howell, USDA.)

IIIB. 2 Nonnodulated peanut plants growing between rows of nodulated plants. (R. K. Howell, USDA.)

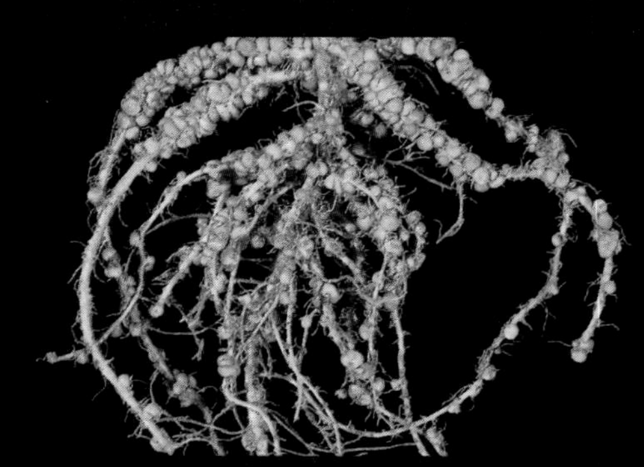

IB. 3 Nodulated roots of a peanut plant. R. K. Howell, USDA.)

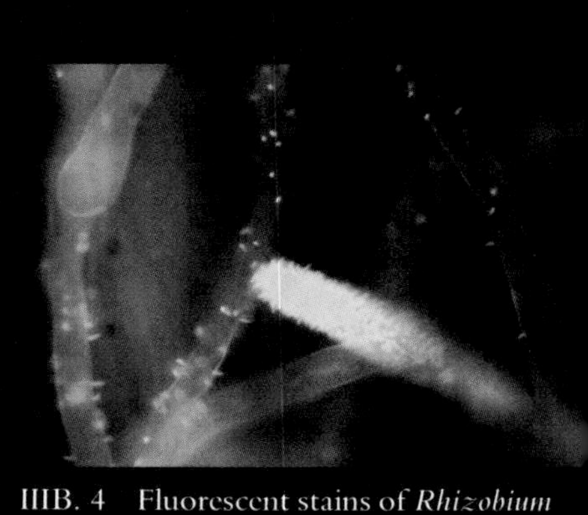

IIIB. 4 Fluorescent stains of *Rhizobium trifolii* colonizing the roothairs of clover. (B. B. Bohlool.)

IIIB. 5 Fluorescent stain of *Rhizobium trifolii* as it forms an infection thread in a clover roothair. (B. B. Bohlool.)

Production of Limulus *Amebocyte Lysate*

IIIC. 1 New England trawler used to dredge for *Limulus*.

IIIC. 2 Landing of the dredge.

IIIC. 3 *Limulus* on the deck.

IIIC. 4 *Limulus* bleeding rack at Associates of Cape Cod (ACC).

(Photographs 4 through 7 were taken at Associates of Cape Cod, Inc., Falmouth, Mass.)

IIIC. 5 The blue *Limulus* blood is collected through a needle into a depyrogenated centrifuge bottle. After bleeding, the crabs are returned to the water from which they were caught. Most of the crabs survive.

IIIC. 6 The amebocytes are centrifuged out of the blood and washed in a processing room.

IIIC. 7 Amebocyte lysate is prepared in a laminar-flow hood before it is packaged and sold as a reagent for measuring bacterial endotoxin.

Industrial Microbiology

IIID. 1 These enormous outdoor fermentation tanks are used for the microbial production of glutamic acid and lysine. Each tank is roughly 100 feet tall and holds 63,420 gallons. These tanks were built at the Kyowa Hakko Hogyo Company plant located in Hofu, Japan. (Photograph courtesy of Kyowa Hakko Hogyo Company, Ltd., Tokyo.)

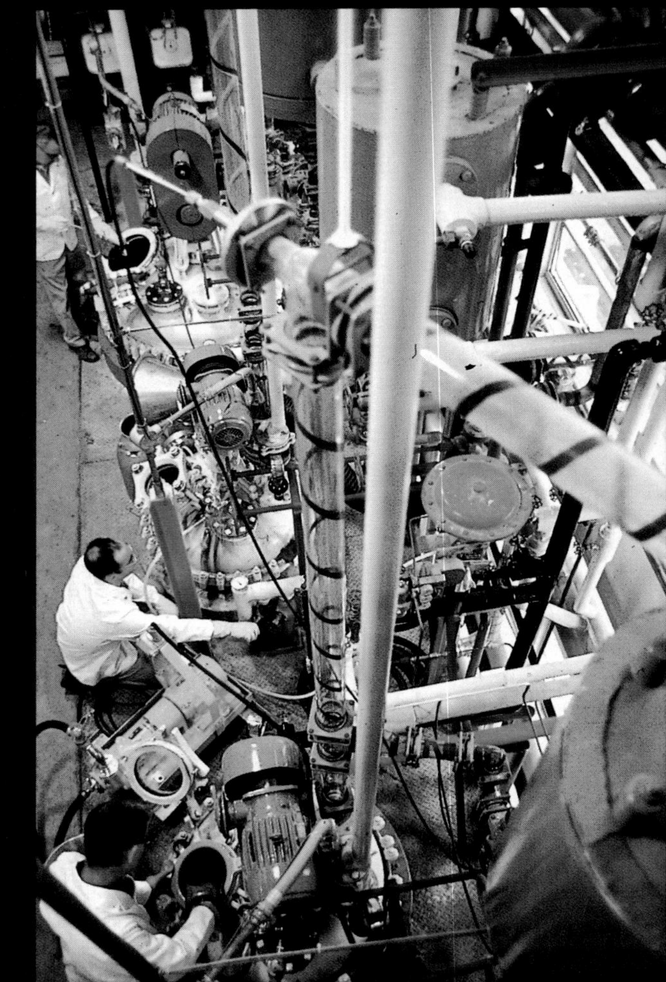

IIID. 2 Pilot fermentation plants are used to develop fermentation processes. (Courtesy of Pfizer.)

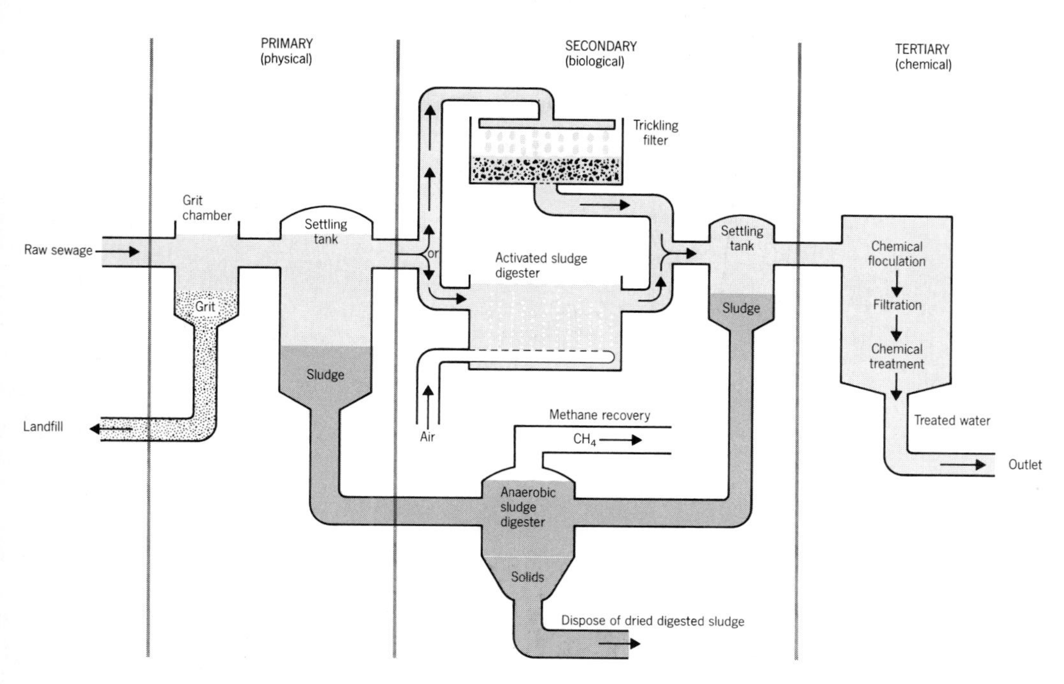

FIGURE 28.13 Waste water is treated by a sequence of purification procedures. Primary treatment physically removes particulate matter, secondary treatment is the biological process of metabolizing degradable compounds, and tertiary treatment chemically removes inorganics from treated water.

TABLE 28.3 BOD of domestic, agricultural, and industrial wastes

TYPE OF WASTE	BOD (mg/liter)
Domestic sewage	165–300
Textile, paper, and chemical industries	300–800
Petroleum, rubber, coal, and plastic industries	30–50
Piggery effluents	25,000
Cattle shed effluents	20,000
Food processing	750

in landfills. The sludge that accumulates in settling tanks is usually treated in an anaerobic sludge digester before the nondigestible matter is disposed of.

SECONDARY WASTE TREATMENT

The purpose of **secondary sewage treatment** is to reduce the BOD of both the sludge and liquid wastes from the primary treatment process. This is accomplished by anaerobic or aerobic processes, or both, in which microorganisms metabolize the waste material.

Anaerobic Sludge Digester

Anaerobic sludge digesters are large vats in which insoluble organic matter (sludge) is digested by microorganisms (Figure 28.14). Large macromolecules such as cellulose, lipids, and proteins are broken down to soluble substances by extracellular enzymes. The soluble compounds are then fermented to organic acids, alcohols, H_2, and CO_2, which in turn can serve as substrates for the methanogenic bacteria. Methane is the major usable product of the anaerobic sludge digester and is often collected and burned as a fuel for operating municipal waste-treatment plants. The remaining sludge is removed and burned, or buried in landfills, while the liquid effluent is cycled through a tertiary treatment process or back into the secondary trickling filter (see below). Waste treatment in the anaerobic

FIGURE 28.14 Aerial view of the sewage treatment facility at Bowery Bay. The circular anaerobic sludge digestors are on the left of the control building. The long rectangular troughs in the center of the photo are aeration basins where aerobic digestion occurs. (Photograph by Vincent J. Lopez.)

sludge digester is a slow process, since effective reduction in the BOD requires a period of two weeks to one month.

Aerobic Secondary Sewage Treatment

Aerobic treatment is a faster method of handling large volumes of waste water. After the solid wastes are removed by the settling process, the fluid wastes are aerated by one of two mechanisms. In a **trickling filter** liquid wastes are sprayed over a deep bed of crushed inert material such as rocks, plastics, blast furnace slag, or redwood bark. The surfaces of this bed material are colonized by aerobic bacteria, fungi, and protozoa, which oxidize the waste material in the water. The organic compounds in the liquid wastes are oxidized to CO_2 and water with the concomitant mineralization of NH_3 and PO_4^{2-}. The treated effluent is collected from the bottom of the trickling filter and transported to a settling tank, where the biomass generated in the trickling filter settles out.

A second approach is to treat liquid wastes in an **activated sludge digester.** The wastes are vigorously aerated with either pure oxygen or compressed air. Degradation occurs on flocs formed by slime-forming bacteria such as *Zoogloea ramigera* (zo-o-gloe′a ram-i′ger-a). Other bacteria, fungi, and protozoa attach to these flocs, which serve as the substratum for biodegradation. The reduction in the BOD in an activated sludge digester depends on the temperature, pH, aeration rate, and transition time. A transit time of 5 to 10 hours is sufficient to reduce the BOD by 75 to 90 percent. The effluent from the activated sludge digester then passes into a settling tank. Clear effluent is discharged from the settling tank, while the settled sludge is removed for disposal.

TERTIARY WASTE TREATMENT

Mineral pollutants and nonbiodegradable organic pollutants are removed during **tertiary sewage treatment** (Figure 28.15). The effluent from the secondary treatment contains soluble nitrogen and organophosphorus that must be removed to prevent pollution. The phosphate released from organic compounds during metabolism may precipitate as solids with the

FIGURE 28.15 The Advanced Waste Treatment Pilot Plant at Blue Plains employs activated carbon columns to remove dissolved organic contaminants from wastewater. The inflowing water is filtered, pH-stabilized by the addition of carbon dioxide in the pH control tank (foreground), and then passed through three activated carbon columns. (Courtesy of Environmental Protection Agency.)

sludge or it may remain in solution. The soluble phosphates can be removed from secondary effluents by precipitation with calcium or iron, either during secondary treatment or in a separate tertiary step. Phosphorus is a limiting nutrient in many natural environments, and effluents containing phosphate can stimulate algal blooms and lead to premature eutrophication.

Free ammonia released from organic compounds can be oxidized to nitrate during nitrification, incorporated into the biomass, and removed in sludge, or it can remain in solution. Soluble ammonia is a significant contributor to the BOD because it stimulates the aerobic process of nitrification. The ammonia can be driven off by aeration after the effluent is adjusted to an alkaline pH. Since ammonia is volatile at a high pH, this process drives it into the atmosphere.

Some nondegradable organic compounds remain, such as chlorophenols and polychlorinated biphenyls. These can be removed by adsorption to activated charcoal filters.

SURVIVORS OF WASTE-WATER TREATMENT

A major purpose of waste-water treatment is to remove or kill pathogens that can cause human disease. Waterborne human diseases can be caused by protozoa (*Giardia*), viruses (hepatitis, rotaviruses), and bacteria (*Shigella, Campylobacter*) (Table 28.4). Many of these

TABLE 28.4 Waterborne disease outbreaks by etiology and type of water system—1980–1982, United States

	PUBLIC WATER SYSTEMS				PRIVATE WATER SYSTEMS	
	COMMUNITY		NONCOMMUNITY			
	OUTBREAKS	CASES	OUTBREAKS	CASES	OUTBREAKS	CASES
AGI[a]	19	14,115	36	2,767	3	67
Giardia	22	2,451	5	127	1	4
Chemical	6	2,189	2	231	6	24
Norwalk agent	6	2,440	3	224	0	0
Shigella	1	19	3	410	0	0
Campylobacter	2	881	0	0	0	0
Hepatitis	2	93	0	0	2	58
Rotavirus	1	1,761	0	0	0	0

Source: Water-Related Disease Outbreaks Surveillance, Annual Summary 1980–1982, Centers for Disease Control.
[a]Acute gastrointestinal illness of unknown etiology.

infectious agents are removed when they become part of the biomass that settles during sedimentation.

Most of the remaining pathogens are killed by disinfecting the effluent with hypochlorite [$NaOCl$ or $Ca(OCl)_2$] or chlorine gas (Cl_2). The hypochlorite (OCl^-) produced when these compounds are mixed with water breaks down into hydrogen peroxide (H_2O_2) and HCl. Both hydrogen peroxide and hypochlorite are strong oxidants that chemically alter essential biomolecules (see Chapter 8). The concentration of dissolved organic compounds, ammonia, and sulfur compounds should be low to achieve efficient disinfection, because these compounds react with hypochlorite and reduce its effectiveness. Another concern is the formation of toxic chlorinated organic compounds during disinfection. These concerns do not outweigh the efficacy of using chlorine to disinfect large volumes of water.

DOMESTIC SEPTIC SYSTEM

Many rural dwellings do not have access to municipal sewage treatment facilities, so they rely on a less expensive, yet effective system designed around the septic tank (Figure 28.16). Raw household sewage enters the septic tank, which acts as both a settling tank and an anaerobic digester. The solids settle to the bottom and add to the sludge. The soluble wastes are digested by microorganisms or are carried in the outflow through a septic field composed of tiles buried beneath the ground. The effluent seeps out through holes in the tiles and percolates through the soil where bacteria and other microbes oxidize the nutrients not assimilated by the plants growing above the septic field. There is no assurance with the domestic septic system that all pathogens will be killed or removed; therefore, the septic tank and field should be located away from the domestic supply of well or spring water.

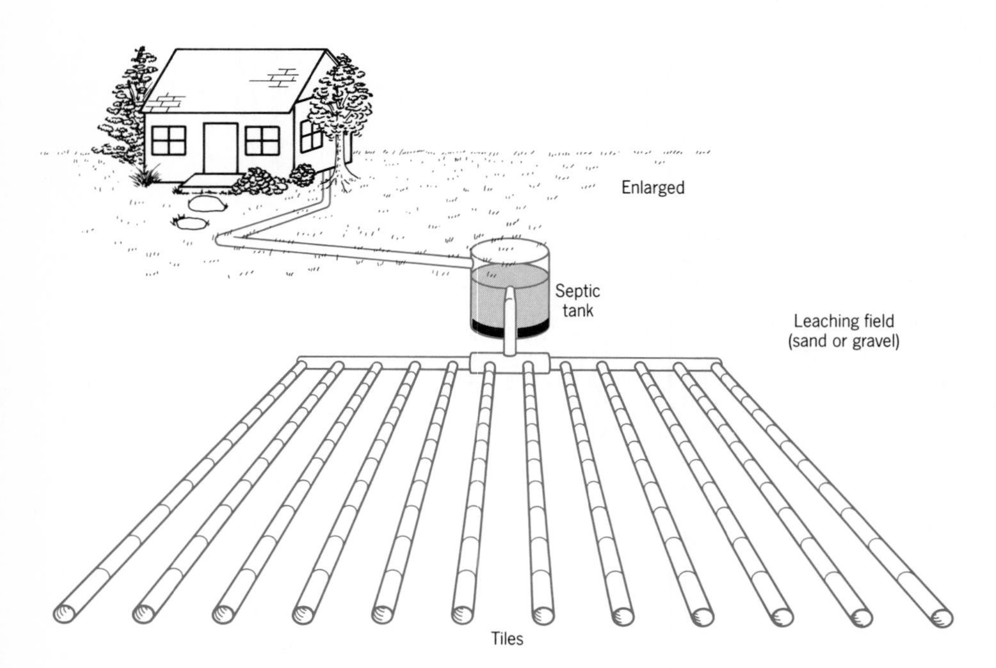

FIGURE 28.16 A septic tank of a typical domestic septic system functions as both a settling tank and an anaerobic digester. The effluent is dispersed into the ground through a network of tiles buried beneath the surface.

Summary Outline

ECOSYSTEMS

1. An ecosystem is the total community of organisms in a physically defined space. The chemical and physical characteristics of an ecosystem greatly influence both the quantity and composition of its biota.
2. Microbial ecology is the study of the relationship between populations of microorganisms and their environments.

ECOLOGY OF AQUATIC ENVIRONMENTS

1. Freshwater environments provide a variety of microbial habitats that change with the seasons. During warm months, the water column of large lakes becomes stratified to form the epilimnion and the hypolimnion.
2. Primary productivity of an aquatic ecosystem occurs primarily in the epilimnion where oxygenic photosynthetic organisms grow.
3. The anoxic hypolimnion provides an environment for the anoxic photosynthetic bacteria, and bacteria that grow by fermentation, methanogenesis, or anaerobic respiration.

SOIL MICROBIOLOGY

1. The organic content of soil is composed largely of decaying plant material with the microbial biomass making only a small contribution. Soil is a major site for the cycling of nutrients.
2. Significant numbers of bacteria, fungi, and protozoa are found in soil together with some viruses and algae. Spore-forming bacteria and fungi are able to survive for years in soil.
3. Bacteria are largely responsible for the cycling of nitrogen and sulfur, whereas both bacteria and fungi participate in cycling soil carbon.
4. Soil is a reservoir for the bacterial pathogens responsible for anthrax, tetanus, and botulism and for the fungi that cause cryptococcosis and histoplasmosis.

BIOCHEMICAL NUTRIENT CYCLES

1. Oxygen is cycled between molecular oxygen, organic matter, and water by biological reactions.
2. The molecular oxygen generated by plants, algae, and cyanobacteria is used by heterotrophs for metabolizing organic compounds and by chemolithotrophs for oxidizing their inorganic substrates.
3. Carbon dioxide from the air or generated from carbonates is converted to organic carbon in the Calvin cycle or C_4 pathway of phototrophs and chemolithotrophs (primary productivity).
4. Methane, a secondary source of carbon, is generated anaerobically from CO_2 and H_2 by methanogenic archaebacteria. Some of this methane is oxidized aerobically by methanotrophic bacteria.
5. Atmospheric N_2 is reduced to ammonia by nitrogenase in the bacterial process of nitrogen fixation.
6. Nitrate is reduced to ammonia by plants, bacteria, and fungi that assimilate nitrate. Nitrate is also the substrate for nitrate respiration in which nitrite is the product and for denitrification in which gaseous forms of nitrogen (N_2O, N_2) are products.
7. Nitrification is the aerobic oxidation of ammonia to nitrate by the nitrifying bacteria. The *Nitrosomonas* group oxidizes ammonia to nitrite and the *Nitrobacter* group oxidizes nitrite to nitrate.
8. Sulfate is reduced to sulfide by the dissimilatory sulfate-reducing bacteria, which grow by anaerobic respiration.
9. Sulfide is a substrate for the anoxygenic, sulfur photosynthetic bacteria and for the aerobic sulfur-oxidizing chemolithotrophs. Both groups of bacteria oxidize sulfide to molecular sulfur and then to sulfate.
10. Sulfate, the major source of cellular sulfur, is reduced to sulfide before being incorporated into the amino acid cysteine.

12. Microorganisms use both organic and inorganic sources of phosphorus to manufacture nucleotides, nucleic acid, phospholipids, and phosphorylated proteins.
13. Phosphorus becomes the limiting nutrient in freshwater lakes and in some marine ecosystems

when biological nitrogen fixation increases the availability of nitrogen to aquatic life.

14. Laws have been passed against the use of phosphate detergents to prevent phosphate pollution of aquatic environments.

WASTE-WATER TREATMENT

1. The purpose of waste-water treatment is to remove as many pollutants as is economically feasible and thus to decrease the BOD of the effluents.
2. Primary sewage treatment removes particulates from sewage by passing it through settling tanks or settling basins.
3. The purpose of secondary sewage treatment is to reduce the BOD of the sludge and liquid wastes by anaerobic or aerobic microbial digestion.
4. Tertiary treatment removes mineral pollutants and nonbiodegradable organic pollutants before the effluent is treated, to kill potential pathogens by disinfection with hypochlorite or chlorine gas.
5. Household waste treatment systems commonly use a septic tank for waste digestion and a tile field for secondary treatment of effluent.

Questions and Topics for Study and Discussion

QUESTIONS

1. Describe how a deep lake provides environments for both the oxygenic cyanobacteria and the anoxygenic photosynthetic sulfur bacteria.
2. Trace the biochemical pathways for converting atmospheric carbon dioxide into methane.
3. Trace the biochemical pathways for converting atmospheric N_2 into cellular material and from cellular nitrogen to its eventual release as nitrous oxide or N_2.
4. Define the following terms:

denitrification	oligotrophic
ecosystem	nitrification
epilimnion	primary productivity
eutrophic	secondary sewage treatment
habitat	

5. How can modern agriculture use what is known about the nitrogen cycle to improve crop yields?
6. Describe the environments in which you would find the sulfate-reducing bacteria. What chemical changes do these bacteria cause in their growth environment?
7. Explain how phosphorus can be a limiting nutrient for primary productivity. How does this principle affect eutrophication of lakes and other aquatic environments?
8. Explain the concept of biochemical oxygen demand. What impact do effluents with high BODs have on aquatic ecosystems?

DISCUSSION TOPICS

1. Some communities spray raw sewage on agricultural lands as their major approach to sewage treatment. What would be the advantages and disadvantages of such a process?
2. What steps has your community taken to control eutrophication of the local water resources?

Further Readings

BOOKS

Alexander, Martin, *Introduction to Soil Microbiology,* 2nd ed., Wiley, New York, 1977. A short textbook that provides extensive coverage of biochemical cycling of carbon, nitrogen, sulfur, and phosphorus.

Delwiche, C. C. (ed.), *Dentrification, Nitrification, and Atmospheric Nitrous Oxide,* Wiley, New York, 1981. Review articles cover the ecology, genetics, biochemistry, and practical problems associated with inorganic nitrogen metabolism.

Grant, W. D., and P. E. Long, *Environmental Microbiology,* Wiley, New York, 1981. A paperback written for the advanced undergraduate student interested in microbial ecology.

Mitchell, R. (ed.), *Water Pollution Microbiology,* Vol. 2, Wiley, New York, 1978. The applications of advances in microbiology to the control of water pollution are discussed by invited experts.

Postgate, J. R., *The Fundamentals of Nitrogen Fixation,* Cambridge University Press, Cambridge, 1982. A comprehensive book on the enzymology, physiology, genetics, ecology, and evolution of nitrogen fixation.

Postgate, J. R. *The Sulphate-Reducing Bacteria,* Cambridge University Press, Cambridge, 1984. A comprehensive book on the metabolism, evolution, and ecology of the sulfate-reducing bacteria.

Rheinheimer, G., *Aquatic Microbiology,* 3rd ed., Wiley, New York, 1986. An excellent introduction to aquatic microbiology.

ARTICLES AND REVIEWS

Jannasch, H. W., "Chemosynthesis: The Nutritional Basis of Life at Deep-Sea Vents," *Oceanus* 27:73–78 (1984). This first-hand account describes the evidence supporting the chemosynthetic basis for life found in the environment of the deep-sea vents.

Pfennig, N., "Microbial Behavior in Natural Environments," in D. P. Kelly and N. G. Carr (eds.), *The Microbe 1984, Part II Prokaryotes and Eukaryotes,* Thirty-Sixth Symposium of the Society for General Microbiology, pp. 23–50, Cambridge University Press, Cambridge, 1984. This stimulating essay discusses the behavior of microorganisms in their efforts to survive and flourish in their natural habitats.

Schlegel, H. G. and H. W. Jannasch, "Prokaryotes and Their Habitats," in M. P. Starr, H. Stolp, H. G. Truper, A. Balows, and H. G. Schlegel (eds.), *The Prokaryotes,* Vol. 1, pp. 43–82, Springer-Verlag, Berlin, 1981. This article is an excellent introduction to the ecology of microorganisms.

CHAPTER

29

Food and Industrial Microbiology

FOCUS OF THIS CHAPTER

The contributions of microbes to the production of commercially important products is the focus of Chapter 29. This encompasses microbial involvement in the production of foods and beverages, primary and secondary metabolic products, uses for microbial enzymes, and the production of antibiotics.

FIGURE 29.1 Bread baking and beer production were integral components of Egyptian society when this relief was incorporated into the Egyptian chapel of Hetepherakhet (about 2400 B.C.). The top panel depicts the grinding of the grain, making and kneading the dough, and baking it in a hot oven. The bottom panel shows the steps in beer making. (Rijksmusem Van Oudheden, Leiden.)

For thousands of years microorganisms have contributed to the production and preservation of foods. The Sumerians and Babylonians used yeasts in the making of beer as early as 6000 B.C.; and leavened bread was made in Egypt earlier than 4000 B.C. By 2400 B.C., bread and beer making were everyday activities, as depicted in Egyptian artwork (Figure 29.1). Ancient cultures preserved the food value of milk by making it into cheese and other milk products, as is done today. Wine manufacturing began about 3000 B.C., probably in northern Iran where wild grapes were abundant. The methods of such food preparation were passed down from generation to generation, as part of a people's folk traditions—as lore, rather than science. The food and beverage industries are now guided by scientific principles and knowledge, in which microbiology plays a key role.

PRODUCTION OF FOOD AND ALCOHOLIC BEVERAGES

FOOD SPOILAGE

Although many microbes contribute to the desired taste, color, and consistency of foods, other microbes are viewed with distate because of their association with food spoilage. Spoilage, rendering foods undesirable or unsafe for consumption, can result from physical, chemical, or microbial damage. The storage life of bruised fruits is shortened, to say nothing of their altered appearance; residual pesticides can make food unsafe; microbes may cause food deterioration. On the other hand, many foods can be stored for long periods of time, in part due to the techniques of food preservation developed by food microbiologists.

Sugar, flour, and cereals are nonperishable foods; apples and potatoes are semiperishable foods and under suitable conditions can be stored for as long as one year. Perishable foods, such as milk, meats, eggs, fish, and most vegetables will spoil rapidly in the absence of appropriate preservation; salts, drying, pasteurization, in canning, refrigeration, and storage at temperatures below freezing are all preservation methods.

TABLE 29.1 Origin and manufacture of cheeses

CHEESE (origin)	MICROORGANISMS	METHOD OF RIPENING
	Hard Cheeses (38 to 40% moisture content)	
Cheddar (Britain)	*Streptococcus lactis, S. durans, S. cremoris*	Cured at 2° to 10°C for 60 days to 1 year.
Colby (United States)	*S. lactis, S. cremoris, S. durans*	Cured for 60 days or more.
Edam (Netherlands)	*S. lactis, S. cremoris*	Shelved in layers at 10° to 15.5°C for 6 to 8 weeks.
Emmentaler (Switzerland)	*S. thermophilus, Lactobacillus helveticus, L. lactis, L. bulgaricus, Propionibacterium shermanii*	Formation of eyes at 22°C at 80 to 85% relative humidity for 3 to 4 weeks. Ripened at 4.5°C and higher for 2 to 10 months.
Gouda (Netherlands)	*S. lactis, S. cremoris, S. diacetylactis*	Cured at 10° to 15.5°C for 2 to 6 months.
Gruyère (Switzerland)	*S. lactis, S. thermophilus, Lactobacillus helveticus*	Formation of eyes at 15.5°C for 1 month, cured at 10° to 15.5°C for 80 days.
Parmesan (Italian)	*S. lactis, S. thermophilus, S. cremoris, L. bulgaricus*	Shelved for 10 months at 10°C at 85% relative humidity.
Swiss (Switzerland)	*S. lactis, S. thermophilus*	Ripened with *Propionibacterium shermanii* and *P. freudenreichii*, propionic fermentation produces CO_2 and flavor.
	Semisoft Cheeses (41 to 44% moisture content)	
Blue (France)	*S. lactis, S. cremoris*	Mold-ripened with *Penicillium roqueforti*, cured for 3 months at 9°C and 95% relative humidity.
Brick (United States)	*S. lactis, S. cremoris*	Surface-cured by *Brevibacterium linens* for 14 days. Stored for 2 to 3 months at 4.5°C.
Gorgonzola (Italy)	*S. lactis, S. cremoris*	Mold-ripened by *P. roqueforti*, cured at 4.5° to 10°C at 80% relative humidity for 30 days.
Monterey (United States)	*S. lactis, S. cremoris*	Cured for 6 weeks or more at 15.5°C and 80% relative humidity.
Muenster (United States)	*S. lactis, S. cremoris*	Cured for 7 weeks at 10° to 12.8°C and 80° relative humidity.
Roquefort (France)	*S. lactis, S. cremoris*	Mold-ripened by *P. roqueforti*, after the wheels are salted and stored in caves at Roquefort. Made from ewe milk.
	Soft Cheeses (45 to 55% moisture content)	
Brie (France)	*S. lactis, S. cremoris*	Mold-ripened by *P. candidum* for 8 to 11 days at 11°C and 95% relative humidity. Distributed within 14 days.
Camembert (France)	*S. lactis, S. cremoris*	Mold-ripened by *P. candidum* at 12.8°C and 95% relative humidity for 12 days. Distributed within 21 days.
Limburger (Belgium)	*S. lactis, S. cremoris*	Surface-ripened by *Brevibacterium linens*, cured for 3 weeks at 12.8°C and 95% relative humidity.

However, before these methods were known, perishable food was either consumed quickly or preserved by microbial processes. Milk was converted to cheese; fruit juices were made into wines and then into vinegar; and vegetables were preserved as sauerkraut, olives, and pickles.

CHEESE MANUFACTURING

Fresh milk contains acid-producing bacteria that will cause it to sour unless their metabolic activities are suppressed by refrigeration or prevented by pasteurization. The souring of milk is the first step in making the curd from which cheese is manufactured.

There are some 20 classes of cheese, each with a unique flavor and consistency, characterized as soft, semisoft, or hard (Table 29.1). The characteristics of each cheese depend on the source of the milk, the microorganisms involved, and the ripening process. The hard cheeses are ripened for periods of two months to one year and can be stored, whereas many soft cheese must be consumed or refrigerated soon after their manufacture.

The color of cheese depends on the milk source (cow, goat, sheep, camel) and on the animal's feed. Cheeses made from white milk are often colored by adding β carotene, annatto (a dye extracted from seeds of the annatto tree), or paprika. A starter culture of the bacteria used in cheese manufacturing is then added (Figure 29.2). Starter cultures of *Streptococcus lactis* and *S. cremoris* are used to make cheeses cooked between 20° and 37°C, whereas cheeses cooked at temperatures above 37°C use temperature-tolerant lactic acid bacteria, including *Streptococcus thermophilus, Lactobacillus bulgaricus* (bul-ga′ri-cus), and *L. helveticus* (hel-ve′ti-cus). These bacteria ferment milk lactose to lactic acid, which helps in the curdling process, and are essential to cheese ripening, so they must survive cooking.

Milk curd is formed by enzymatic action of rennin on the milk proteins and by the lactic acid produced by bacteria in the starter culture. **Rennin,** also called **chymosin,** is an enzyme preparation obtained from the fourth stomach of unweaned calves or microbial cultures. The enzymes in rennin hydrolyze milk casein without attacking other milk proteins such as lactalbumin and lactoglobin. The resulting polypeptides react with calcium to form a lattice structure, which is the matrix for the **curd** (Figure 29.2), composed of the solids from the coagulated milk.

In cooking, the curd contracts and expels the liquid **whey.** The moisture content of the final product is influenced by the duration and temperature of cooking. Cheddar curds are cooked at 37°C, whereas Emmentaler and Gruyère curds are cooked at 54°C for 1 to 1.5 hours. After the whey is removed, the Cheddar curd is raked, turned, and compacted to yield its characteristic texture (Figure 29.3). Chemical changes from bacterial metabolism (starter culture) continue to influence the flavor of the cheese throughout the process.

Salt and microorganisms are added to the curd before it is pressed into forms that give shape to the cheese and extract more whey. More microbes can be added; for example, the surface of the pressed curd is inoculated with bacteria or molds (Table 29.1) in the manufacture of blue cheese or Brie cheese. The young cheese is then placed in a controlled environment and allowed to ripen.

During ripening, bacterial lysis releases enzymes into the pressed curd. These enzymes together with the rennin remaining from the coagulation step break down the fats, carbohydrates, and proteins present in the curd and produce by-products that impart the characteristic flavor to the cheese.

YOGURT

Yogurt is a fermented milk product having a semiliquid consistency, like custard (Figure 29.4). It is manufactured from whole or skim milk whose solid content has been increased by the removal of water or the addition of powdered milk. The milk is fermented above 40°C by *Streptococcus thermophilus* or *Lactobacillus bulgaricus,* which produces both lactic acid and acetaldehyde. The tart flavor of yogurt is attributed to the acetaldehyde, but in most of the yogurt sold in the United States, fruit or fruit flavoring is added to mask the tartness. Yogurt made from skim milk is regarded as a "health food" because it is low in fat and high in protein. The gram-positive lactic acid bacteria in yogurt are believed to exert a beneficial effect on the intestinal flora.

BUTTERMILK, SOUR CREAM, KUMISS, AND KEFIR

Other milk products produced by the action of lactic acid bacteria are better known to peoples outside the United States. **Buttermilk** is made from skim milk by fermentation with *Streptococcus lactis* and *Leuconostoc cre-*

(a)
(b)
TEL-TRU
°F
(c)
(d)
(e)
(f)

(*g*)

(*h*)

FIGURE 29.2 The manufacture of Cheddar cheese. Lactic acid bacteria are added to prepared cow's milk before rennet is added *(a)*. The rennet curdles the milk before it is cooked *(b)* for about 1 hour and then dried by separating it from the whey *(c)*. The curd is raked, compacted *(d)*, and turned repeatedly in the cheddaring step that helps to create the characteristic texture of Cheddar cheese. The resulting slabs are milled *(e)*, before they are salted *(f)*, wrapped in cloth *(g)*, and pressed in hoops *(h)* to expel the remaining whey. The immature cheese is cured for from 2 to 12 months at temperatures in the range of 2° to 10°C. (Photographs courtesy of James Killkelly/DOT.)

FIGURE 29.3 Cheddar cheese represents about two-thirds of the cheese produced in the United States. Wisconsin Cheddar, mild, medium, aged, and uncured. (Courtesy of Wisconsin Milk Marketing Board, Inc., Madison.)

FIGURE 29.4 Yogurt is produced by fermenting milk into a semiliquid having the consistency of custard.

moris. These bacteria are also used to convert cream to **sour cream.**

The nomadic Kazakhs of western China and south central Russia make a fermented drink from mare's milk called **kumiss.** After the spring migration to summer pastures, the Kazakhs collect the mares' milk and with it make a mildly alcoholic beverage, consumed to celebrate the new season. A similar drink, **kefir,** is made by fermenting milk with the lactic acid bacterium *Lactobacillus brevis* and the yeast *Saccharomyces unisporus*.

Some adults develop an intolerance for milk lactose when they lose the ability to digest it. The lactose they consume is then available as a substrate for the intestinal flora, which metabolize it, producing gas. The person experiences flatulence and considerable discomfort. A lactose-intolerant person may be able to digest fermented milk products such as buttermilk and acidophilus milk *(L. acidophilus)* in which the lactose has been metabolized by bacteria.

LEAVENING OF BREAD

Grinding wheat into flour for bread making probably originated in Egypt, first as unleavened or flat bread. Unleavened bread is simply made by mixing flour and water into a dough and baking it. If yeasts are added a **leavened bread** is produced. The yeasts metabolize carbohydrates to produce CO_2, which increases the volume of the dough and makes the baked product light and airy.

Bread making requires special strains of yeasts (Figure 29.5). It is made from flour, usually wheat flour, that has been mixed with water, salt, sugar, and shortening. The resulting dough is inoculated with a starter culture of the yeast *Saccharomyces cerevisiae* (sak-ah-ro-mi'sez se-ri-vis'e-eye). These yeasts ferment the sugar to produce carbon dioxide and a small quantity of ethanol, which evaporates during baking. The yeasts also change the structure of the dough by modifying the gluten in the flour.

A popular specialty bread is the San Francisco **sourdough bread;** it is made with a starter culture of *Lactobacillus* spp. and the yeast *Torulopsis homii* (tor'u-lop'sis). The heterofermentative *Lactobacillus* produces carbon dioxide, ethanol, and lactic acid from sugar in the dough. Lactic acid imparts the sour flavor to the bread, while the carbon dioxide causes the dough to rise before the bread is baked. **Rye** bread also starts as a sour dough made with lactic acid bacteria.

FERMENTED VEGETABLES, RICE, AND SOYBEANS

Vegetables can be preserved by a combination of fermentative mechanisms and pickling—for example, olives, pickles, and sauerkraut. Rice and soybeans can be converted into sufu and tempeh, which add variety to the diet and are considered to be more palatable than plain rice or soybeans.

Olives

Fresh olives contain a bitter phenolic compound that must be removed with a dilute solution of sodium hydroxide before the olives are pickled in a fermentation brine. The pickling process involves a slow lactic acid fermentation by *Leuconostoc, Lactobacillus plantarum,* and *L. brevis,* which gives high concentrations of lactic acid, up to 7.1 percent. Acidity of the pickling juice and the salt in the brine combine to preserve the olives.

Pickles

Pickles are made by fermenting cucumbers in a brine (Figure 29.6) that supports a lactic acid fermentation by *Lactobacillus plantarum* (plan-ta'rum) and *Pediococcus cerevisiae*. The lactic acid thus produced imparts the charcteristic sour taste. Salt is added as the fermentation progresses until a final concentration of about 16 percent is reached. *Lactobacillus plantarum* is the dominant bacterium throughout most of the pickling process. Vinegar may be added to increase the acidity and herbs are added to flavor the pickles.

Cabbage

Sauerkraut, a fermentation product of cabbage, is made by mixing shredded cabbage with salt (2.5 percent) to extract the soluble sugar. The sugars are fermented by the microbial flora of the cabbage, including *Leuconostoc mesenteroides* (leu-co-nos'toc me-sen-ter-oi'des) and *Lactobacillus plantarum*. They produce lactic acid, which reaches a final concentration of between 1.5 and 2 percent, imparts an acid flavor to the sauerkraut, and acts as a preservative.

FIGURE 29.5 Commercial bread manufacturing is a major industry that uses equipment capable of handling large volumes of dough. *(a)* The dough is shaped into a lightly floured ball by a "rounder." *(b)* The panned loaf is conveyed to the automatic proofer for its final rising (about 1 hour). *(c)* The baked bread coming from the oven passes through a vacuum cup depanner, which places the loaf on a conveyer that goes to the cooler. *(d)* The bread is passed through a slicer before it is wrapped in moisture-proof paper. (Photographs courtesy of Pepperidge Farm, Norwalk, Conn.)

Tempeh, Sufu, and Soy Sauce

Tempeh, sufu, and soy sauce are made by fermenting plant material, primarily soybeans. **Tempeh** is made from soybeans, kedele, and cocounts. The water-soaked plant material is boiled or steamed and then drained before being mixed with a starter culture of tempeh containing the mold *Rhizopus*. The cakes of

FIGURE 29.6 Pickles are made by fermenting pickling cucumbers in a brine solution in large vats.

tempeh are then incubated until the mold penetrates the interior. Tempeh is a high-protein food used widely in Indonesia and other eastern countries.

The Chinese make a soft cheese-like product called **sufu** by fermenting soybean curd with molds, principally *Mucor.* Soy sauce is also a Chinese invention, now made in Japan and throughout the western world. **Soy sauce** is made by fermenting a salted mixture of soybeans and wheat with the mold *Aspergillus oryzae.* The resulting product is mixed with salt, then fermented at low temperatures in large tanks for a period of 8 to 12 months by *Pediococcus soyae, Saccharomyces rouxii,* and species of the yeast *Torulopsis.* The soy sauce is extruded from the vegetable matter, bottled, and used in cooking.

ALCOHOLIC BEVERAGES

Wineries, breweries, and distilleries are among the largest businesses that depend on microbes to produce the products they sell. Wineries produce wine by fermenting the juices of grapes and other fruits. Breweries produce beer by fermenting cereal grains, mainly barley, and distilleries produce concentrated alcoholic beverages by distilling of the alcohol produced during fermentation of grain (whiskeys), fruits (brandies), or molasses (rums).

Wine Manufacturing

The most popular wines are those made from grapes, which grow in temperate climates around the world.

(a)

(b)

FIGURE 29.7 Wine production. *(a)* Skilled hands gather ripe clusters of grapes when their sugar content is just right for making wine. *(b)* Modern Wilmes presses extract as much juice as possible from crushed grapes without breaking the seeds. A rubber bag inside the press inflates to squeeze the crushed grapes gently against the inside walls. *(c)* In California wineries, special wine yeasts are added to ferment the must, which occurs in wine tanks *(d)*. *(e)* The centrifuge (domed machine at left) clarifies young wines, while the barrels are used to age wine. *(f)* Wine is bottled by machine and corked. (Courtesy of Wine Institute)

Special varieties of red grapes, such as Cabernet Sauvignon, Gamy, Pinot Noir, and Zinfandel; and of white grapes such as Chardonnay, Chenin Blanc, and Riesling are grown by wineries. The quality of the wine depends on the region in which the grapes are grown and on the climatic conditions during the growing year. The grapes are harvested when the sugar content of the grape juice reaches about 20 percent.

The first step in wine making is to destem and crush the grapes to produce raw grape juice or **must** (Figure 29.7). The natural flora in the must is usually killed at this stage by pasteurization or by sulfur dioxide treatment. Sulfur dioxide is added as a gas or is generated in the must by adding metabisulfite ($Na_2S_2O_5$), sulfite (Na_2SO_3), or bisulfite ($NaHSO_3$). Sulfur dioxide controls undesirable microorganisms, inhibits the "browning" enzymes (those that turn sliced apples brown), and is an antioxidant.

The high acidity and sugar content of the must prevents the growth of indigenous bacteria, while providing an ideal environment for the *Saccharomyces cerevisiae* alcoholic fermentation. In this process, each molecule of grape sugar—glucose and fructose—is converted to two molecules of ethanol and two molecules of carbon dioxide.

To make a white wine, the juice from either white or red grapes is separated quickly from the solids or **pomace,** composed of the grape stems, skins, seeds, and part of the pulp. This prevents the pigments in the grape skins from being solubilized and coloring the product. (In making red wines, this separation occurs later.) The fermentation is begun by adding a starter culture of *Saccharomyces cerevisiae* var *ellipsoideus,* and proceeds until most of the sugar is metabolized. The fermenting wine is kept anaerobic in large vats or by insertion of a gas trap that permits carbon dioxide to escape, while preventing oxygen from entering (Figure 29.8). Once the fermentation is complete, the wine is ready for clarification and aging.

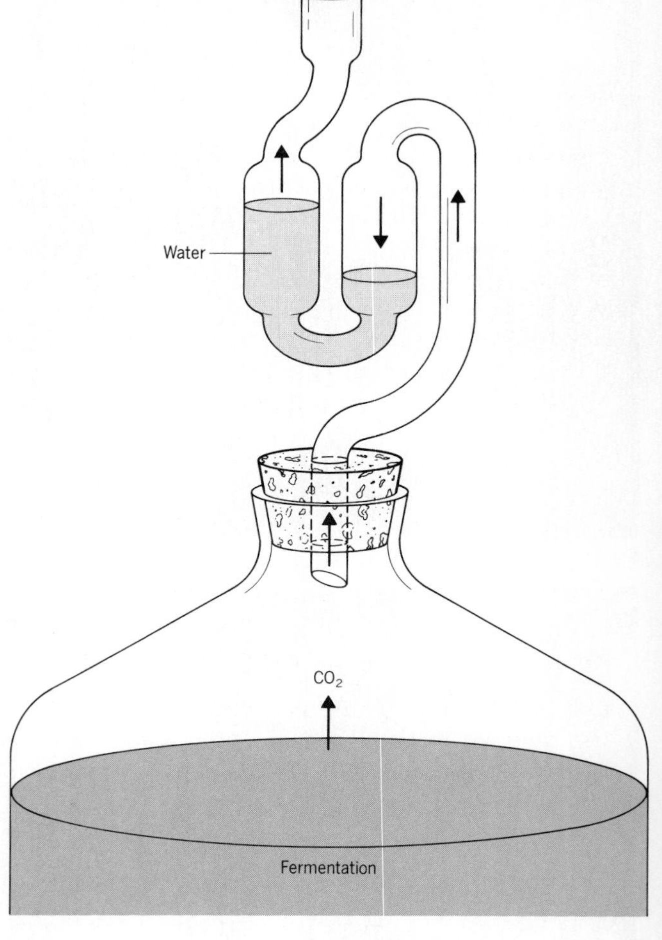

FIGURE 29.8 An air trap is used by amateur wine and beer makers to keep the fermentation vessel anaerobic. Carbon dioxide produced during the ethanol fermentation escapes by bubbling through the water in the trap, but air cannot enter the vessel.

Yeast cells suspended in the wine after fermentation settle out during **clarification.** Wine should remain undisturbed during clarification for it is essential that these cells not autolyze. After clarification, the wine is decanted or pumped out of the vessel and racked in containers, to be aged.

Aging is a complex microbiological and chemical process that greatly improves the quality of the wine, particularly red wines. Wines contain malic acid (a dicarboxylic acid), which during aging is fermented by species of *Pediococcus, Leuconostoc,* or *Lactobacillus* to lactic acid and carbon dioxide. This **malo-lactic acid fermentation** decreases the wine's acidity by converting a dicarboxylic acid (malic acid) to a monocarboxylic acid (lactic acid). Other chemical changes occur while the wine is stored at low temperatures in wine cellars and caverns. Storage in wooden casks or barrels augments this process by allowing oxygen slow access to the wine. The aging process continues even after bottling, especially with red wines, which continue to improve with aging even after ten years. Dry white wines rarely improve with age after four to seven years, and many should be consumed within two to four years after being bottled.

Sparkling wines or carbonated wines, such as champagne, can be produced by adding sugar at the time of bottling; fermentation of the sugar produces the carbon dioxide that carbonates the wine. Because this natural process results in a yeast sediment, most commercial sparkling wines are carbonated chemically.

Red wines are made from red grapes by allowing the fermentation to start before the skins are removed. The red pigments are solubilized from the skins by the heat generated by fermentation and by the chemical action of ethanol. After sufficient color is extracted, the wine is separated from its pomace and fermentation is continued. The intermediate color of **rosé** table wines is attained by fermenting the must from red grapes for a short time before the grape skins are removed.

KEY POINT

Either white or red grapes can be used to make white wines as long as the grape juice is separated from the pomace early in the process. The color of red and rosé wines comes from the pigmented skins of red grapes.

Diseases of Wines Bottled wines can be stored on their sides for long periods of time as long as the cork remains wet so that air does not enter the bottle. If oxygen enters the bottle, acetic acid bacteria will oxidize the alcohol in the wine to acetic acid, turning the wine into vinegar (see chapter 13). Lactic acid bacteria, especially *Lactobacillus trichodes* (tri-cho′des), ferment the residual sugar in the wine and turn it "mousey." These wine "diseases" can be prevented by pasteurization; however, this diminishes the quality of the wine and interferes with the aging of red wines. Manufacturers of inexpensive bulk wines filter them to remove microorganisms, or add chemicals, such as sulfur dioxide, to prevent bacterial spoilage.

Beer Brewing

Worldwide production of beer exceeds 21 billion gallons per year with the highest per capita consumption in West and East Germany, Czechoslovakia, and Belgium. Beer is an alcoholic beverage produced by a batch fermentation of the carbohydrate in barley malt and adjunct grains (Figure 29.9). Before fermentation can proceed, the starch in barley is converted to simple carbohydrates by **saccharification,** a process that begins with the formation of the malt wort.

To make malt barley (malting), the barley grains are soaked (or steeped) in water and allowed to germinate, during which amylases are induced and the internal structure of the barley kernels is modified by enzymes. The barley kernel becomes softer as the proteases and β gluconases hydrolyze the internal cell-wall structures, making the starch more accessible to the amylases. These kernels are kiln-dried to produce the stable product. Malted barley is produced in barley-growing regions of the United States, primarily Minnesota, North Dakota and South Dakota, and Washington state, then shipped to breweries.

KEY POINT

The malting process increases the levels of the α- and β-amylases present in barley. These enzymes hydrolyze the starch into simple sugars—mainly maltose with smaller amounts of glucose, sucrose, and maltotriose—during the brewing process.

The malted barley is crushed; combined with an adjunct grain such as corn grits, rice, or corn syrup; mixed with water; and heated to temperatures at which the enzymes present will break starch into simple sugars, and proteins into amino acids. The liquid portion of this digested mash, called **malt wort,** is separated from the spent grains in a round **lauter tub** (Figure 29.9). The spent grains are removed and sold as animal feed.

The malt wort is boiled in brew kettles to precipitate proteins, stop further enzymatic activity, and kill any microorganisms present. Hops, the dried ripe cones of the female flowers of the hop vine, are added toward the end of the boil. They contain the organic acids that impart the clean bitterness of beer, as well as oils that flavor beer with a hoppy aroma.

The hopped wort is cooled and transferred to large fermenters. Breweries now use closed fermenters, which may be horizontal cylindrical tanks (Figure 29.9) or the newer cylindroconical fermentation tanks with

(a)
(b)
(c)
(d)
(e)
(f)

(g) *(h)* *(i)* *(j)* *(k)*

FIGURE 29.9 Beer brewing is a batch process that begins with the malting of barley. Barley is germinated and then dried in the malting process before it is delivered to the brewery *(a)*. The adjunct grain together with some malt (10 percent) is ground in a malt mill *(b)*, mixed with water, and heated in the adjunct cooker. This is then mixed with the remainder of the malt in the mash vessel *(c)* where the malt enzymes break down the starch to simple sugars. The spent grains are removed from the slurry in the lauter tub *(d)*. The wort is transferred to the boiling kettle *(e)* where it is boiled for about 90 minutes. Hops *(f)* are added toward the end of the boil. The hot wort is transferred to the wort whirlpool tank *(g)* before being run through a cooler *(h)*. Yeast cells are transferred from the yeast tubs *(h)* to the fermenter *(i)*. After fermentation, lagering takes place in a refrigerated tank *(j)* for from 10 to 30 days. The beer is filtered *(k)* to remove the colloidal particles that form during lagering. (Photographs courtesy of Dr. Karl Siebert, Stroh Brewery Company, Detroit.)

inverted cone bottoms. The cylindroconical fermenters are often installed outdoors and have capacities as large as 5000 barrels (31 gallons per barrel). Most fermenters are designed with cooling bands or coils for temperature control.

Between 10 and 15 million yeast cells are added per milliliter of hopped wort to begin the fermentation. The yeast use up the available oxygen before they begin to produce ethanol and CO_2 by fermenting the simple amino acids and the sugars of three glucose units or smaller. Larger sugars (dextrins) remain (see below). The duration of the fermentation depends on the temperature. **Ale** fermentations occur at temperatures near 20°C and may be completed in 3 days, whereas **lager** fermentations typically occur at temperatures between 6° and 10°C and often require 7 to 12 days. At the end of the fermentation, the yeast cells settle and the beer is decanted. The cylindroconical fermentation tanks are designed in such a manner that the yeast cells settle in the narrow cone, from which they are removed before the beer is decanted. Any remaining yeast cells are often removed from the beer by centrifugation.

The beer is then pumped into cold-storage tanks and maintained at temperatures slightly above freezing for 10 to 30 days (ales are sometimes stored for only 5 days). This chill-storage or **lagering** process is necessary to clarify the beer. A haze-forming substance expands by aggregation during lagering, and either precipitates or is removed in the final filtration step (Figure 29.9). Chill-storage also alters the taste of beer by lowering the concentrations of hydrogen sulfide, acetaldehyde, and other organic compounds generated during fermentation.

Commercial beers are filtered to remove the last of the yeast cells and suspended material; bottled or canned; and then pasteurized (Figure 29.10) to kill bacteria that could cause spoilage. Keg beer is not pasteurized; it is refrigerated and consumed shortly after being packaged. The difference in taste between bottled beer and keg beer is attributed to the length of storage rather than to the pasteurization process.

Most U.S. breweries use bottom yeasts *(Saccharomyces cerevisiae)* to produce lager beers, which typically contain 3.8 percent alcohol by weight. Other beers produced in this country include low alcohol beers with about 1.9 percent alcohol, "light" beer with about 3.1 percent alcohol, super-premium beers with about 4.2 percent alcohol, and malt liquors with as much as 5 to 7 percent alcohol by weight. The heavy **English ales** are made with **top-fermenting yeasts** from dark malts containing more hops than the typical U.S. beer. The top-fermenting yeast cells are active fermenters at 20°C and appear to float on the top of the vessel. Many British ales are produced in three days and typically contain less alcohol than U.S. lager beers.

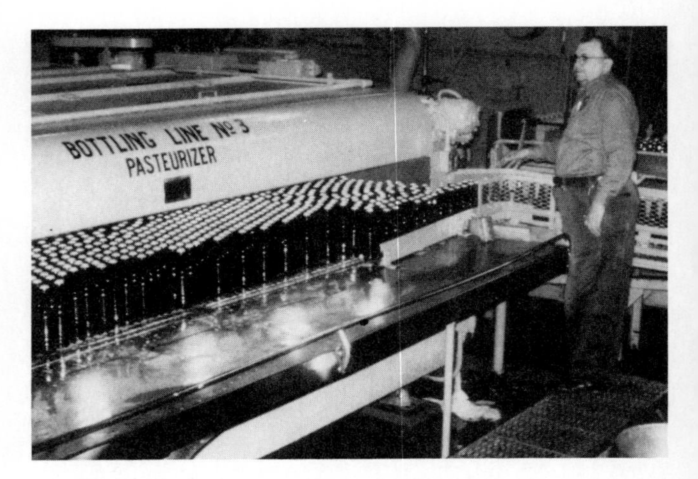

FIGURE 29.10 Beer is bottled and then pasteurized. Keg beer is not pasteurized, so it must be refrigerated until it is consumed. (Courtesy of Dr. Karl Siebert, Stroh Brewery Company, Detroit.)

"Light" Beer Malt wort contains dextrins that are not metabolized by the yeasts, and these carbohydrates add to the caloric content of the finished product. Under proper conditions, dextrins are broken down into glucose by fungal glucoamylase (see below). To produce a low carbohydrate or **"light" beer,** the brewmaster adds fungal glucoamylase to the fermenter with the yeast. The glucoamylase breaks down the dextrins into simple sugars, which are then fermented to alcohol by the yeast cells. This results in a higher alcohol content, so, to compensate, specially treated water is added to dilute the ethanol before the beer is packaged. The dilution also reduces the number of calories per serving—thus, light beers.

KEY POINT

Juices of grapes, apples, apricots, and other fruits contain sugar that can be directly fermented by yeasts to make wines. The starch in barley and adjunct grains must be bro-

ken down into simple sugars by the process of saccharification before yeasts can ferment the carbohydrate to produce beer.

Sake and Pulque

Japanese sake, a fermented rice beverage, involves the use of two microbes: a mold to break down rice starch to glucose and a yeast to ferment glucose to ethanol. The starch in steamed rice is metabolized by the amylases of *Aspergillus oryzae* during a five- to six-day incubation period, and the resulting product is called **koji.** The glucose in koji is fermented by the sake yeast strain of *S. cerevisiae* for as long as three weeks. Saccharification continues during fermentation, and the final beverage may be as much as 20 percent ethanol.

Pulque is a Mexican drink made by fermenting the juice of the agave plant. The fermentation is performed by *Saccharomyces* spp. and the bacterium *Zymomonas mobilis,* a bacterium that ferments glucose to ethanol and carbon dioxide. Tequila is a distillation product of pulque.

Vinegar Production

Vinegar is a sour liquid made by the bacterial oxidation of ethanol in fermented wine, beer, or cider and used as a condiment and preservative. Wine and fermented cider often contain *Acetobacter* and *Gluconobacter,* which oxidize ethanol to acetic acid in the presence of oxygen (see Chapter 14).

$$CH_3CH_2OH + O_2 \longrightarrow CH_3COOH + H_2O$$

These acetic acid bacteria produce vinegar that is between 4 and 7 percent acetic acid. Acid-tolerant species used commercially in vinegar production include *Acetobacter aceti, A. pasteurianus, A. peroxidans,* and *Gluconobacter oxydans.*

Vinegar is made commercially in submerged fermenters (Figure 29.11), which are vessels designed to control the temperature through heat exchangers and to provide sufficient aeration. Air is pumped into the fermenter and dispersed from the bottom of the vessel. Wine, beer, or fermented cider provides the ethanol that is converted to acetic acid. In the submerged fermenter, 98 percent of the alcohol can be converted to acetic acid in a semicontinuous, automatic process with a cycling time of 35 hours.

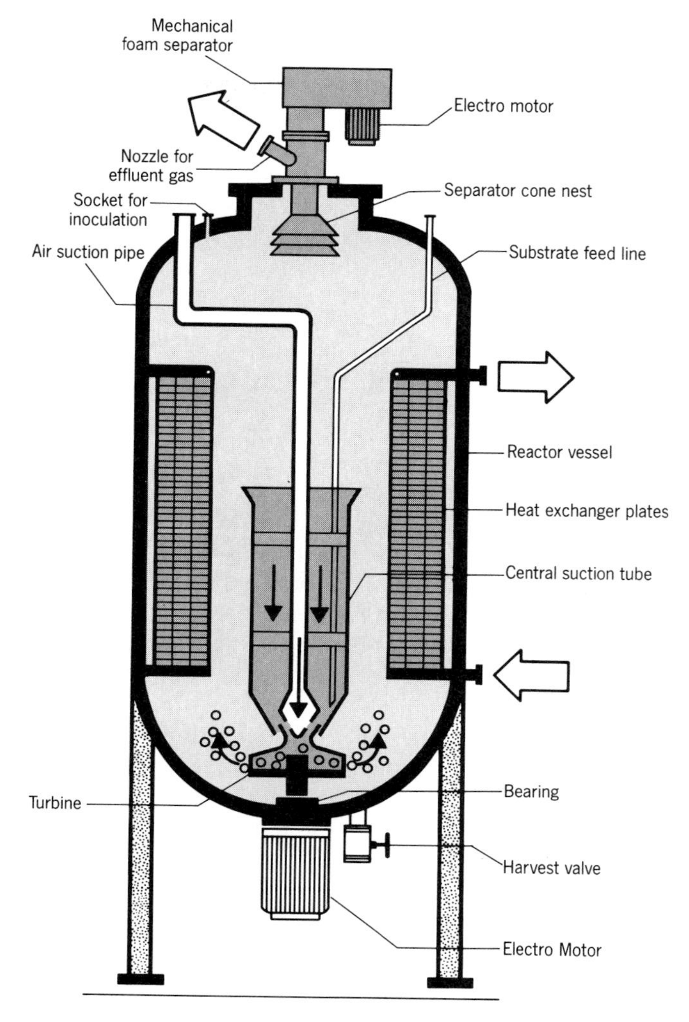

FIGURE 29.11 Diagram of a submerged fermenter used in the production of vinegar. Mixing is accomplished by a special turbine that also disperses the oxygen into the medium. The exchangers dissipate the heat produced in the reaction, and the foam produced is controlled by a mechanical foam separator. (From W. Crueger and A. Crueger, *Biotechnology, A Textbook of Industrial Microbiology,* Sinauer Associates, Sunderland, Mass., 1984.)

INDUSTRIAL MICROBIOLOGY

Enzymes, amino acids, vitamins, antibiotics, organic acids, and alcohols are commercially produced by mi-

TABLE 29.2 Commercially important primary microbial products

PRODUCT	ORGANISM	PRODUCTION METHOD	APPLICATIONS (TONS/YEAR)
Citric acid	*Aspergillus niger*	Direct fermentation	Food and beverage additive, chemical industry (250,000)
L-Glutamic acid	*Corynebacterium glutamicum*	Direct fermentation	Food additive (350,000)
L-Lysine	*C. glutamicum, Brevibacterium flavum*	Direct fermentation	Animal feed (45,000)
L-Methionine	*C. glutamicum*	Direct fermentation	Animal feed (80,000)
Vitamin B_{12}	*Propionibacterium freudenreichii, P. shermanii*	Fermentation	Animal feed, pharmaceutical industry (10,000 kg)

crobial processes. Enzymes, amino acids, and vitamins are **primary microbial products**—cellular constituents of microorganisms. By-products of metabolic activities not used by the cell for growth, such as antibiotics and the organic acids and alcohols produced as fermentation products, are known as **secondary microbial products.**

The annual commercial value of primary and secondary microbial products is measured in billions of dollars, with sales of antibiotics alone totaling more than $4 billion annually. This large industry is a primary employer of microbiologists, who are involved in many aspects of the commercial production of microbial products.

PRIMARY PRODUCTS: AMINO ACIDS, CITRIC ACID, AND VITAMINS

Commercially important amino acids produced by microbial processes include glutamic acid, lysine, and methionine (Table 29.2). Monosodium glutamate (MSG) is a flavor intensifier, while methionine and lysine are essential amino acids that humans must obtain from their diet. Microbial production of amino acids is often more expedient than chemical synthesis, since microbes produce only the L-isomers of the amino acids used to make proteins (see Chapter 2).

L-Lysine and L-Methionine

Amino acids are produced in microorganisms by metabolic pathways, as described in earlier chapters. The pathways can be altered or manipulated to yield commercial quantities of desired amino acids. The bacteria used to produce L-lysine and L-methionine are either selected for from natural sources or are created by genetic alteration. The characteristics sought are the bacterium's ability to overproduce and excrete the amino acid. Some of the better producers are mutants that contain altered enzymes and regulatory mechanisms for amino acid synthesis.

Mutants of *Corynebacterium glutamicum* (glu-tam'ic-um) and *Brevibacterium flavum* (brev-e-bak-te'ri-um flav'um) have been selected for their ability to produce L-lysine. One common metabolic pathway controlled by feedback inhibition produces L-lysine, L-threonine, and L-methionine, all from L-aspartic acid (Figure 29.12). The production of L-lysine is elevated in homoserine dehydrogenase mutants since hom^- cells cannot produce L-threonine to act as a feedback inhibitor on aspartate kinase. The aspartate kinase can also be altered by mutation such that it is not allosterically affected by L-lysine (Figure 29.12). These regulatory mutations enable these strains to produce large quantities of L-lysine.

These mutants are very effective producers of amino acids; for example, mutant strains of *B. flavum* produce up to 75 g of L-lysine per liter of medium when grown on acetate, and mutant strains of *C. glutamicum* produce up to 39 g of L-lysine per liter when grown on glucose. Other mutants of *C. glutamicum* are used to produce L-methionine, although most of the worldwide demand for methionine is satisfied with chemically produced DL-methionine.

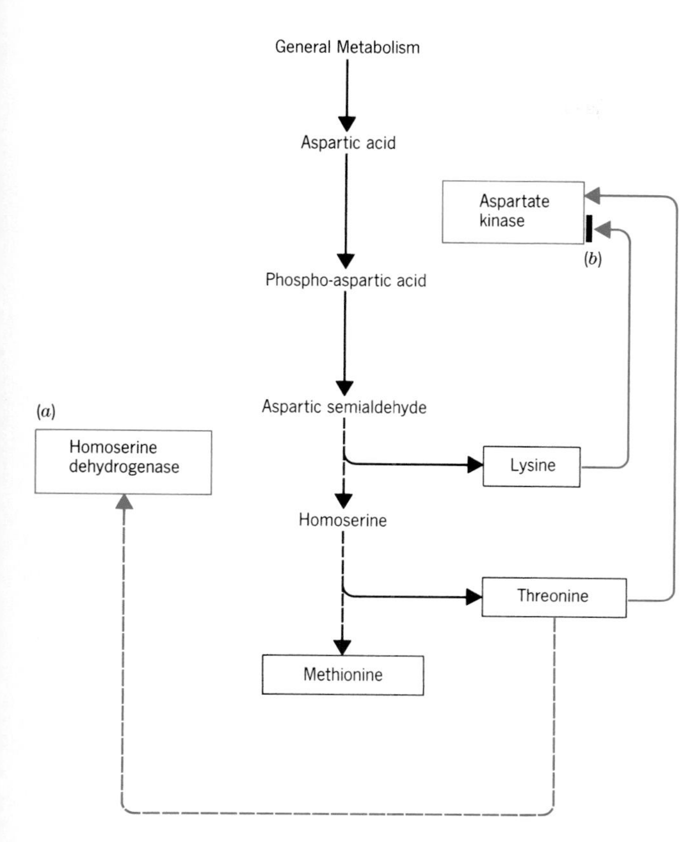

FIGURE 29.12 L-Lysine is produced commercially by *Corynebacterium glutamicum* using the synthetic pathway for the production of L-lysine, L-methionine, and L-threonine. Inactivation of homoserine dehydrogenase *(a)* and alteration of aspartate kinase *(b)* negates the feedback inhibition control over this pathway.

L-Glutamic Acid

Monosodium glutamate (MSG) is manufactured by microbial processes to avoid the DL-isomer mixture present in chemically produced amino acids. When grown on glucose, the wild type strain of *Corynebacterium glutamicum* produces more than 100 g of L-glutamic acid per liter. Glutamic acid is metabolically generated from α-ketoglutarate, an intermediate in the TCA cycle (see Figure 6.7), and ammonia. By slowing down the TCA cycle, α-ketoglutarate builds up, making more available for synthesizing monosodium glutamate. Over 350,000 tons of microbe-produced MSG are sold each year.

Citric Acid

Almost all of the world's output of citric acid, an intermediate in the tricarboxylic acid cycle, is produced microbiologically. Citric acid is a food preservative and flavoring agent of fruit juices, candy, ice cream, marmalade, and other products. It is also used in pharmaceuticals as an iron chelator (binder), as a preservative for stored blood, and in ointments and cosmetics. The chemical industry uses about 25 percent of the citric acid produced as a feedstock for making a variety of products.

Citric acid is commercially produced from sugar-beet and sugar-cane molasses by *Aspergillus niger,* which is capable of converting 42 precent of the raw sucrose to citric acid. Because of complex regulatory mechanisms, citric acid accumulates as an intermediate, and is excreted. The citric acid in the culture filtrate is precipitated by adjusting the pH to 7.2 ± 0.2 at 70° to 90°C. Over 250,000 tons of citric acid are produced annually.

Vitamin B_{12} and Riboflavin

A deficiency in **vitamin B_{12}** (cyanocobalamin) leads to pernicious anemia. Since humans cannot make vitamin B_{12}, they must meet their B_{12} requirement from the diet. Vitamin B_{12} is produced as a supplement for animal feeds, for use in the pharmaceutical industry, and as a food additive and vitamin supplement. It is a primary product, commercially produced by *Propionibacteruim freudenreichii* (freu-den-reich'i-i), which makes between 19 and 23 mg of cell-bound vitamin B_{12} per liter of fermentation medium. The vitamin is extracted from the cells by a heat treatment.

Most commercially produced **riboflavin** (vitamin B_2) is made chemically and sold as a vitamin supplement or added to vitamin-enriched breads. Deficiencies in riboflavin are manifest as dermatitis. Riboflavin can be produced microbiologically by the yeast *Ashbya gossypii* (ash'by-ah gos-syp'i-i), which produces up to 7 g of riboflavin per liter of growth medium.

SECONDARY PRODUCTS: LACTIC ACID AND ETHANOL

Lactic acid, ethanol, and acetic acid (Table 29.3) are commercially significant bacterial by-products. Microbiological processes are still used to produce lactic acid

TABLE 29.3 Commercially important secondary metabolic products

PRODUCT	ORGANISM	PRODUCTION METHOD	APPLICATIONS (TONS/YEAR)
Acetic acid	*Gluconobacter oxydans, Acetobacter aceti, A. pasteurianus, A. peroxidans*	Partial oxidation submerged fermenter, trickling generator	Vinegar
Acetic acid	None	Chemical	Industrial feedstock (1,430,000[a] in U.S.)
Ethanol	*Saccharomyces cerevisiae*	Direct fermentation	Fuel and chemical feed stock (433,000)
Lactic acid	*Lactobacillus delbrueckii, L. bulgaricus*	Direct fermentation	Food preservative; tanning and textile industries (22,000 Europe)

[a]Amount made by chemical synthesis in the United States in 1980.

and ethanol, and acetic acid is biologically produced as the major component of vinegar (see above); however, most of the acetic acid used as a chemical feedstock is chemically synthesized by the carbonylation of methanol.

Lactic Acid

Lactobacillus delbrueckii (del-bruec'ki-i) growing on glucose and *L. bulgaricus* growing on whey are homofermentative lactic acid bacteria (see Chapter 14) employed commercially to produce the lactic acid used as a food preservative, in tanning animal hides, and in the textile industry. Though virtually all of the lactic acid produced in the United States is made by chemical synthesis, European countries produce over 22,000 tons annually by fermentation.

Ethanol

Ethanol is a fuel, a fuel additive, a solvent, an antifreeze, and an organic feedstock in the chemical industry. It can be produced chemically from petroleum, or microbiologically from fermentable carbohydrates. Economic factors weigh heavily in choosing the method of ethanol production. When petroleum and natural gas prices are low, ethanol can be economically produced from petrochemical feedstocks; however, when petrochemicals are selling at a premium price, microbial production of ethanol from corn, molasses, sugar cane, or other plant material becomes more economical.

Ethanol is produced in the anaerobic fermentation of glucose by *Saccharomyces cerevisiae* (Figure 29.13). High concentrations of ethanol are inhibitory to yeasts, so only ethanol-tolerant strains are used. Cell growth usually ceases when the ethanol concentration reaches 5 percent (v/v)* even though the cells continue to produce ethanol until the concentration is between 6 and 10 percent (v/v). This inhibition can be reduced through continuous removal, and recovery, of the ethanol by vacuum distillation.

Corn and other starchy feedstocks must undergo saccharification before they can be fermented. The carbohydrate in cornstarch is enzymatically released by treating the ground corn with α amylase and glucoamylase. Molasses made from sugar beets and the juices from sugar cane are fermented directly without saccharification. Wood is a potentially inexpensive source of fermentable carbohydrate; however, the microbial process of converting hemicellulose, lignin, and cellulose into fermentable monosaccharides has not been developed for commercial applications.

Ethanol is produced in batch-fermentation systems using either hydrolyzed starch or molasses. The maximum yields of about 10 percent ethanol (v/v) are produced in a period of 5 to 12 hours, depending on the concentration of *Saccharomyces cerevisiae*. Up to 27 g of cell mass per liter are generated in the process. The cells are removed by sedimentation or centrifugation before the ethanol is purified by distillation.

*Expressed as volume of ethanol per 100 volumes of solution.

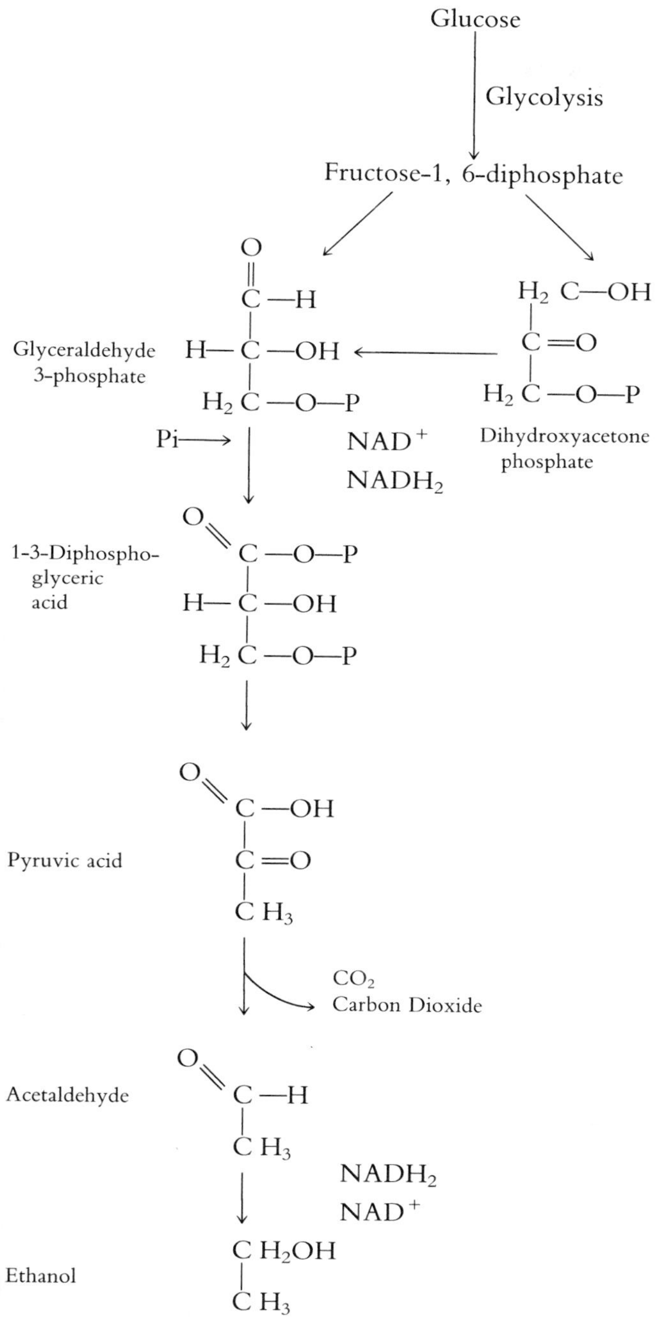

Glucose
Glycolysis
Fructose-1, 6-diphosphate

Glyceraldehyde 3-phosphate: $O=C-H$, $H-C-OH$, H_2C-O-P
Dihydroxyacetone phosphate: H_2C-OH, $C=O$, H_2C-O-P

Pi → NAD^+ → $NADH_2$

1-3-Diphosphoglyceric acid: $O=C-O-P$, $H-C-OH$, H_2C-O-P

Pyruvic acid: $O=C-OH$, $C=O$, CH_3

CO_2 Carbon Dioxide

Acetaldehyde: $O=C-H$, CH_3

$NADH_2$ → NAD^+

Ethanol: CH_2OH, CH_3

FIGURE 29.13 Yeasts produce ethanol from the pyruvic acid generated during glycolysis. Yeasts decarboxylate pyruvic acid to produce carbon dioxide and acetaldehyde, which is then reduced to ethanol at the expense of $NADH_2$.

COMMERCIAL PRODUCTION OF MICROBIAL ENZYMES

Microbial enzymes have commercial applications, in laboratory analysis, and in clinical settings to treat diseases. Those of industrial importance are usually extracellular enzymes produced in large quantities (tons) and marketed at a relatively low purity (Table 29.4). For example, proteases are incorporated into detergents; microbial chymosin (rennin) is used in cheese manufacturing; pectinases clarify fruit juices; and penicillin amidase is used in the manufacture of semisynthetic penicillins. Other industrial enzymes include glucoamylase, α amylase, and glucose isomerase, which convert complex carbohydrates into metabolizable substrates.

Small quantities of analytical enzymes and enzymes required in clinical applications are produced in highly pure, often crystalline form. L-Asparaginase is an example. This enzyme is clinically important as an antitumor agent against leukemias and lymphomas. It breaks down asparagine and thereby starves cancer cells, which have a high requirement for this amino acid.

Alkaline Proteases

Alkaline proteases hydrolyze proteins at an alkaline pH. They are effective cleaning agents since they solubilize stains caused by blood, grass, and other biological substances. Commercial preparations of alkaline proteases that contain high concentrations of phosphate were added to many detergents before 1971, when the phosphate pollution problem became a concern. Another problem was the occurrence of allergic reactions in production workers and consumers, resulting from exposure to these preparations. Proteases continue to be added to about 80 percent of laundry detergents sold in Europe, and to a few detergents marketed in the United States.

Many species of *Bacillus,* including *B. licheniformis* (li-che-ni-for'mis), *B. amyloliquefaciens, B. firmus,* and *B. megaterium* produce extracellular, alkaline proteases that are stable at alkaline pHs and at elevated temperatures. These proteases are produced commercially in aerated cultures of *Bacillus,* converted into a nonallergenic encapsulated product, and then added to the detergents. Microbial **neutral proteases** and **acid proteases** are too unstable to be used in detergents; however, neutral proteases have some applications in the leather and food industries.

TABLE 29.4 Commercial uses of microbial enzymes

ENZYME	SOURCE	USES (TONS/YEAR)
α Amylases	*Bacillus* spp.	Liquefaction of starch (300)
Glucoamylase	*Aspergillus niger, Rhizopus niveus*	Starch saccharification (300)
Glucose isomerase	*Bacillus coagulans, Actinoplanes missouriensis, Streptomyces* spp.	Fructose syrup production, glucose to fructose (50)
Pectinases	*Aspergillus niger, A. wentii, Rhizopus* spp.	Clarification of fruit juices (10)
Penicillin amidase	*Escherichia coli*	Production of 6-amino penicillanic acid from penicillin G (small quantities)
Proteases	*Bacillus licheniformis, B. amyloliquefaciens, B. firmus, B. megaterium*	Detergent additive (300)
Rennin (chymosin)	*Endothia parasitica, Mucor miehei*	Milk coagulation in cheese production (10)

Amylases

Amylases attack specific bonds in plant starches (Figure 29.14) and are used to hydrolyze starch into fermentable mono- and disaccharides. The **α amylases** have application in alcohol production, baking, paper manufacturing, the brewing and textile industries, and the commercial production of glucose from starch. The α amylases are extracellular enzymes commercially produced using *Bacillus amyloliquefaciens* and *B. licheniformis* grown on starch-containing media. Crude preparations of α amylase are made after the cells are removed from the spent medium.

Glucoamylase hydrolyzes the glucose residues from the nonreducing end of the plant starch molecule (Figure 29.14). This enzyme is produced by the molds *Aspergillus niger* and *Rhizopus niveus* grown in submerged cultures and is an important enzyme in the production of "light" beers (see above).

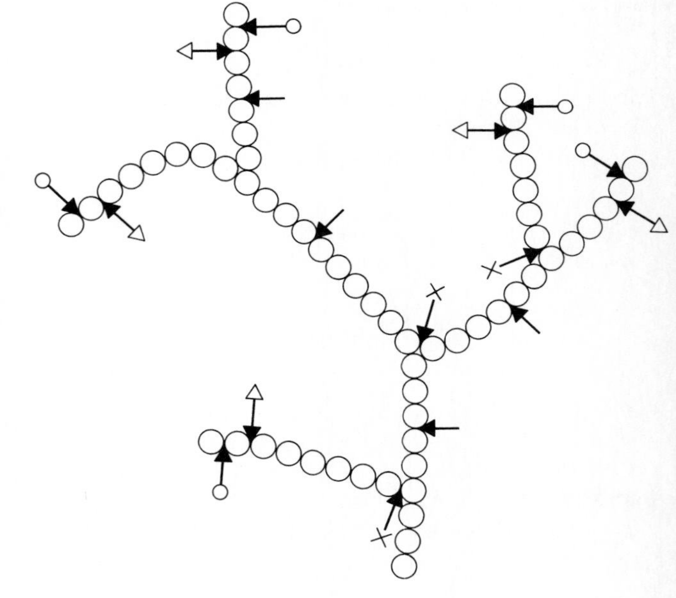

FIGURE 29.14 Starch is a complex polymer of glucose that must be hydrolyzed before it can be fermented by yeasts. Four distinct amylases are needed to break down starch.

Glucose Isomerase

Fructose, a monosaccharide, is twice as sweet as sucrose. High-fructose syrup is used extensively as a food sweetener and is gaining acceptance in the soft-drink industry as a sucrose substitute. Fructose is manufactured by enzymatically converting glucose into fructose through the action of **glucose isomerase** (Figure 29.15). Bacteria, including *Bacillus coagulans, Actino-*

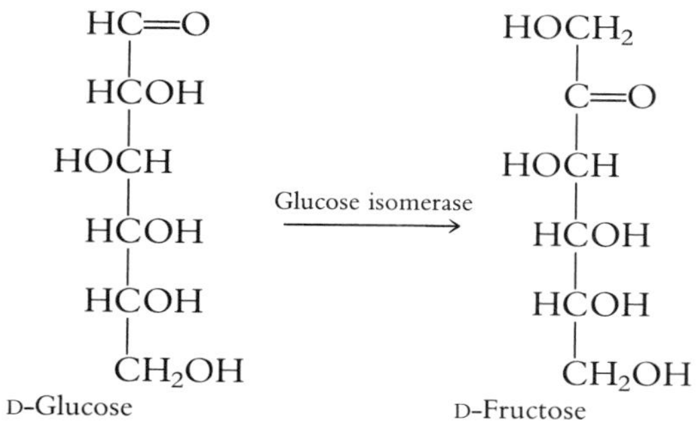

HC=O		HOCH$_2$
HCOH		C=O
HOCH	Glucose isomerase →	HOCH
HCOH		HCOH
HCOH		HCOH
CH$_2$OH		CH$_2$OH
D-Glucose		D-Fructose

FIGURE 29.15 Fructose, an important sweetener used in soft drinks, is produced from corn starch glucose by glucose isomerase.

planes missouriensis, and species of *Streptomyces,* are the source of glucose isomerase for this process. Isolation of glucose isomerase from bacteria is not economically feasible, so the entire enzyme-containing cell is used. In order to treat economically large volumes of glucose, the cells are immobilized with cross-links to produce large cellular aggregates that can be used again and again. Corn is a major source of glucose for the production of high-fructose syrup in the soft-drink industry.

Pectinases

Pectins are large, insoluble plant polymers of methyl-D-galacturonate that bind the cellulose fibers in the plant cell wall. They contribute to the turbidity of plant juices. **Pectinases** split pectins at different sites and thereby solubilize these large macromolecules. Mixtures of pectinases are prepared commercially from microbes such as *Aspergillus niger, A. wentii,* and *Rhizopus* spp. and are used to clarify fruit juices.

ANTIBIOTIC PRODUCTION

Antibiotics are secondary microbial metabolites that lack an obvious physiological role in the growth or maintenance of the organism that produces them. Of the more than 5000 antibiotics that have been discovered, only about 100 have been marketed, and of these, 69 are produced by filamentous soil bacteria of the genus *Streptomyces.* Among the most widely sold antibiotics are the penicillins and cephalosporins, produced by fungi, and the tetracyclines, produced by bacteria of the genus *Streptomyces.*

Penicillins: Production and Biosynthesis

Penicillins and cephalosporins are β-lactam, fungal antibiotics (Figure 29.16), which specifically inhibit peptidoglycan formation during bacterial cell-wall synthesis (see Chapter 21). The naturally produced penicillins are penicillin G, penicillin V, and a limited amount of penicillin O. Synthetic penicillins are made by chemical modification of 6-aminopenicillanic acid (6-APA).

Highly productive penicillin G-producing strains of *Penicillium chrysogenum* have been developed through mutation and selection. Penicillin G is produced in submerged aerobic cultures and is stimulated by feeding the cultures glucose, molasses, and phenylacetic acid, a precursor of the antibiotic. The penicillins are purified from the spent medium by a special solvent-extraction method.

6-Aminopenicillanic acid is made from this raw product by enzymatically cleaving off phenylacetic acid (Figure 29.16) with **penicillin amidase** (Table 29.4). The specificity of this enzyme varies depending on the organisms from which it is isolated; fungal type I penicillin amidases act on penicillin V, while the bacterial type II penicillin amidases act on penicillin G. The 6-APA thus produced is used in manufacture of the semisynthetic penicillins, including methicillin, propicillin, oxacillin, ampicillin, and carbenicillin (see Figure 21.4).

Cephalosporin is another β-lactam antibiotic whose basic structure is chemically modified to produce a family of semisynthetic antibiotics. Cephalosporin is produced by *Cephalosporium acremonium* (sef′ah-lo-spor′re-um ak-re-mo′ne-um), as well as by other fungi. Semisynthetic cephalosporins are widely used in the treatment of nosocomial gram-negative sepsis.

Tetracycline Production

Tetracyclines have broad applications in human and veterinary medicine and as nutritional supplements in cattle and swine production (see Chapter 21). They are broad-spectrum antibiotics produced by various species of *Streptomyces.* Chlortetracycline and oxytetracycline are naturally produced by *Streptomyces* spp. grown on sucrose and corn steep liquor. These antibiotics are purified from the culture liquor after removal of the

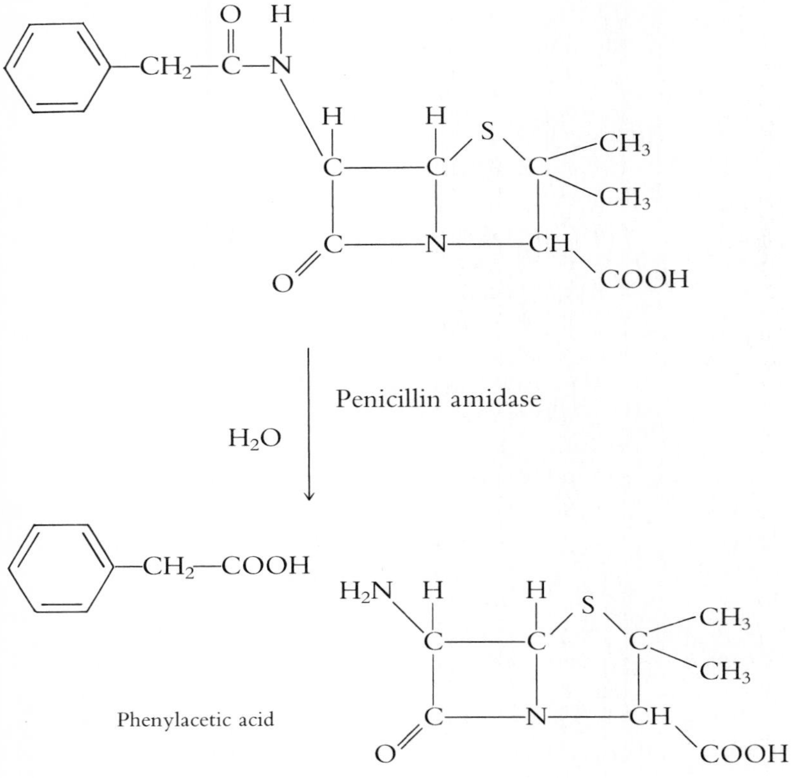

Penicillin amidase

H_2O

Phenylacetic acid

6 aminopenicillanic acid

FIGURE 29.16 Penicillanic acid is used to manufacture the penicillin family of antibiotics. Penicillanic acid is produced from penicillin G by penicillin amidase.

cells. Tetracyclines are widely used as animal feed additives since they stimulate growth.

GEOCHEMICAL LEACHING BY MICROORGANISMS

Leaching is a microbial process by which minerals are recovered from low-grade ores; it is an important process in both copper and uranium mining. The low-grade ores are first mined, then deposited in huge piles or leaching dumps (Figure 29.17). The insoluble metals are converted into water-soluble ions by bacteria-mediated oxidation reactions. The soluble ions are extracted from the dump by acidic water as it percolates through the dump pile. The mineral-rich liquid from the dump site is collected and either recycled over the ore bed or processed to remove the solubilized metal ions.

Thiobacillus ferrooxidans and *Thiobacillus thiooxidans* are chemolithotrophic bacteria involved in the leaching. These bacteria oxidize reduced sulfur to sulfate, which in aqueous environments forms sulfuric acid. The presence of this acid and the oxidation of the copper and uranium are essential to the leaching process.

Copper ores are composed of insoluble chalcocite (Cu_2S) and covellite (CuS). The copper in this ore is solubilized when it is oxidized to copper sulfate ($CuSO_4$), which in solution exists as cupric (Cu^{2+}) and

FIGURE 29.17 A leaching dump of low-grade copper ore from which copper is being recovered as the cupric ion. (Courtesy of Kennecott Copper Corporation.)

sulfate (SO_4^{2-}) ions. The soluble cupric ions are recovered from the leaching solution and used in production of copper metal.

Tetravalent uranium (UO_2) is insoluble in aqueous solutions until it is oxidized to the uranyl ion (UO_2^{2-}). This oxidation is done with the help of *T. ferrooxidans*. Organic solvents are used to extract the uranyl ion from the leaching fluid.

Summary Outline

PRODUCTION OF FOOD AND ALCOHOLIC BEVERAGES

1. Food spoilage, a change in food that makes it undesirable or unsafe for consumption, can result from physical, chemical, or microbial damage.
2. Sugar, flour, and cereals are nonperishable foods; apples and potatoes are semiperishable foods; and milk, meats, eggs, fish, and many vegetables are perishable foods.
3. Methods of preserving foods include curing meats with salts, drying, pasteurization, sterilization through canning, refrigeration, and storage at temperatures below freezing.
4. Cheeses are made from milk curds formed by the fermentations of lactic acid bacteria and the enzymatic action of rennin on milk casein.

5. The solids from coagulated milk comprise the milk curd, which is separated from the whey and then cooked. The cooked, pressed curd is ripened by the action of bacteria or fungi.
6. Yogurt is a semiliquid, fermented milk product made by inoculating milk with lactic acid bacteria.
7. Lactic acid bacteria are used to produce sour cream and buttermilk; kumiss and kefir are sour, mildly alcoholic beverages produced with lactic acid bacteria and yeasts.
8. Bread is leavened by the carbon dioxide generated by yeast fermenting sugar in dough. Both yeasts and lactic acid bacteria are used to produce sourdough bread.
9. Olives and cucumbers are preserved by fermentation in a brine. Cabbage is converted into sauerkraut when lactic acid bacteria ferment the cabbage juices extracted by salt.
10. Tempeh, sufu, and soy sauce are foods produced by the action of molds or yeasts on plant material.
11. Wines, fermented juices of grapes and other fruits, are products of the alcoholic yeast fermentation.
12. Steps in wine production include crushing the grapes to make the must, treatment with sulfur dioxide to kill the microbes present, fermentation by *Saccharomyces cerevisiae,* clarification, bottling (racking), and aging.
13. White wines are made by fermenting the must of either white or red grapes; red wines are made by fermenting the pomace and must of red grapes until enough red color is extracted from the grape skins.
14. Beer is produced by batch fermentation of the starch in barley after it is converted to simple carbohydrates by saccharification, a process that uses the amylases in the barley kernels.
15. The malt wort is boiled in brew kettles, mixed with hops, cooled, and transferred to large fermenters. Between 10 and 15 million yeast cells are added per milliliter of hopped wort to begin the fermentation.
16. At the end of the fermentation, the yeast cells settle and the beer is decanted in cold-storage tanks where the lagering process occurs. The beer is filtered to remove the remaining yeast cells and suspended material; bottled or canned; then pasteurized.
17. Lager beers are made with bottom-fermenting yeasts and ales are made with top-fermenting yeasts. Other alcoholic beverages include sake, pulque, and tequila.
18. *Acetobacter* and *Gluconobacter* oxidize the alcohol in wine or fermented cider to acetic acid, during vinegar production, done commercially in submerged fermentors.

INDUSTRIAL MICROBIOLOGY

1. Commercially important primary microbial products include amino acids, citric acid, vitamins, and enzymes.
2. Both L-lysine and L-methionine are essential amino acids that can be obtained only from the diet.
3. DL-methionine is made chemically, whereas L-lysine is commercially produced by *Corynebacterium glutamicum.* This bacterium is also used to produce L-methionine and the flavor intensifier, monosodium glutamate.
4. Citric acid, an intermediate of the TCA cycle, has many industrial and food uses and is produced by *Aspergillus niger.*
5. Vitamin B_{12} (cyanocobalamin) is a primary product produced commercially by *Propionibacterium freudenreichii* as a food supplement.
6. The microbial enzymes produced commercially include amylases, glucoamylase, glucose isomerase, pectinases, penicillin amidase, proteases, and rennin.
7. Ethanol is produced as a fuel and a chemical feedstock by microbial fermentation of corn, molasses, or other feedstock.
8. Lactic acid, used as a food preservative, in tanning animal hides, and in the textile industry, is commercially produced by growing *L. delbrueckii* on glucose and *L. bulgaricus* on whey.
9. Penicillin, cephalosporins, and tetracyclines are

antibiotics produced by microorganisms and then chemically modified or marketed directly by pharmaceutical companies.

10. Penicillin G, produced by *Penicillium chrysogenum,* is converted enzymatically to 6-APA, which is the basic molecule used to produce the synthetic penicillins.

GEOCHEMICAL LEACHING BY MICROORGANISMS

1. Copper and uranium are leached from low-grade ores through the biological activities of *Thiobacillus ferrooxidans* and *T. thiooxidans.*

2. The copper in chalcocite (Cu_2S) and covellite (CuS) ore is solubilized when oxidized to copper sulfate ($CuSO_4$). Tetravalent uranium (UO_2) is insoluble in aqueous solutions until it is oxidized to the uranyl ion (UO_2^{2-}) by the action of microbes.

Questions and Topics for Study and Discussion

QUESTIONS

1. Describe how to make yogurt from raw materials you could purchase in the grocery store.

2. Describe the role of lactic acid bacteria in the production of cheese, buttermilk, sourdough bread, and pickles.

3. How are molds used in cheese manufacturing? What other important commercial products are produced by molds?

4. What are the major differences between the microbiological processes used in making wine and beer?

5. Briefly describe how the following are made:

curd	malt wort
high-fructose syrup	must
koji	pomace
lager beer	whey

6. What characteristics make each of the following microbes commercially important?
Aspergillus niger
Corynebacterium glutamicum
Gluconobacter oxydans
Lactobacillus delbrueckii
Propionibacterium freudenreichii
Streptococcus thermophilus

7. Name five primary microbial products and describe how they are produced commercially.

8. Name five commercially important secondary microbial products and describe how each is used.

9. Devise a flow diagram of the steps needed to produce ethanol from corn.

10. Name five microbial enzymes of commercial importance and describe their common uses. Which ones are found in products used in the home?

11. Describe the control mechanisms that influence the production of L-lysine by *Corynebacterium glutamicum.*

DISCUSSION TOPICS

1. How can the principles of biotechnology be applied to producing commercially successful microbial products? What class of compounds has the greatest profit potential?

2. What strains of bacteria and fungi could you purchase in the grocery store? How would you proceed to isolate them in pure culture?

Further Readings

BOOKS

Amerine, M. A., and V. L. Singleton, *Wine, An Introduction,* University of California Press, Berkeley, 1977. This book covers all aspects of wine making from growing the grapes to producing table wines. Contains valuable information for the home wine maker.

Broad, R. G., *A Modern Introduction to Food Microbiology,* Blackwell, Oxford, England, 1983. An introduction to food microbiology for students of microbiology, food science, food technology, and related disciplines.

Crueger, W., and A. Crueger, *Biotechnology, A Textbook of Industrial Microbiology,* Sinauer Associates, Sunderland, Mass., 1984. This concise textbook is a rich source of information on the industrial uses of microorganisms.

Peppler, H. J., and D. Perlman, *Microbial Technology,* Vols. 1–2, Academic Press, New York, 1979–1980. Separate volumes cover microbial technology and fermentation technology.

Rose, A. H. (ed.), *Economic Microbiology,* Vols. 1–7, Academic Press, London, 1979–1982. Each volume is devoted to one of the following topics: alcoholic beverages, primary metabolic products, secondary metabolic products, microbial biomass, microbial enzymes, microbial biodeterioration, and fermented foods.

ARTICLES AND REVIEWS

Blonston, Gary, "The Biochemistry of Bacchus," *Science 85,* 6:68–75. Research activities at a California winery are discussed.

Brierley, C. L., "Microbiological Mining," *Sci. Am.,* 247:44–53 (1982). The role of bacteria in the leaching of copper from low-grade ore is explained.

Knorr, D., and A. J. Sinskey, "Biotechnology in Food Production and Processing," *Science,* 229:1224–1229 (1985). Reviews the uses of microorganisms in food production and processing and details the research opportunities available in this field.

Kosikowski, F. V., "Cheese," *Sci. Am.,* 252:88–99 (1985). A straightforward introduction to the production of cheese emphasizing the process for making Cheddar cheese.

Scientific American, Vol. 245 (September 1981). This entire issue is devoted to industrial microbiology with excellent articles on the history of industrial microbiology, commercial production of antibiotics, and the microbiological production of food and drink.

CHAPTER

30

Genetic Engineering and Biotechnology

OUTLINE

FOCUS OF THIS CHAPTER

Knowledge of molecular biology is being used to manipulate the genetic material of bacteria, viruses, and other organisms. In this final chapter the focus is on the techniques of gene manipulation that have spawned the era of genetic engineering. The impact of genetic engineering on biotechnology is explained, with examples of its applications.

Biotechnology, broadly defined as the use of living organisms and their components in industrial processes, entered a new era in the 1980s with the development of molecular techniques for manipulating genetic information, called **genetic engineering.** Through this technology we can better study the functional organization of genetic information, develop microorganisms that produce mammalian proteins, such as insulin and human growth factor, make vaccines from microbial and viral genes, and accelerate the generation of new strains of microbes and plants. Understanding the technology involved begins with an explanation of gene cloning.

Genetic engineering is based on a number of biological concepts, which were described earlier in this book. To augment the presentation we have included a new feature in this chapter—Concept Review—which will alert the reader to the background information needed to understand the new material.

GENE CLONING

Cellular genes are units of heredity preserved in the nucleotide sequences of double-stranded DNA. In bacteria, this molecule is huge, much too large to study as a single molecule, but we can study short pieces of DNA, as long as we have enough material. Isolating and copying fragments of DNA is accomplished by **gene cloning,** which uses microorganisms to produce millions of identical copies of a specific region of DNA.

OVERVIEW: TYPICAL CLONING EXPERIMENT

There is nothing magical about gene cloning—the techniques employed are often learned in the three- to six-week courses offered for that purpose. Here we want to highlight the important steps as outlined in the typical cloning experiment of Figure 30.1. In this experiment we are not particular about the gene(s) we clone; later we will discuss methods for cloning specific genes.

In a typical cloning experiment, foreign DNA is isolated and then cut into fragments by restriction endonucleases. These bacterial enzymes make cuts in the center of the DNA molecule (endo) breaking it into small fragments. The resulting DNA fragments are combined with a plasmid vector that has been treated with the same restriction endonuclease to create a space into which the foreign DNA can be inserted (Figure 30.1). This is now a recombinant DNA molecule since the plasmid contains a specific region of foreign DNA. The plasmid is now used as a **cloning vector,** a small plasmid or viral DNA that carries a DNA fragment from the test tube into a living cell.

Plasmid cloning vectors enter a host cell by artificial transformation. Here the foreign DNA will be copied many times as the plasmid is replicated. To clone a gene is to make multiple copies of a DNA molecule derived from a single ancestor. This general description of gene cloning describes common steps in genetic engineering experiments that use plasmid vectors, and serves as a framework for the detailed discussions that follow.

DNA Isolation

Gram-negative bacterial cells are pretreated with the chelating agent EDTA and then gently lysed in a saline solution containing the detergent sodium dodecyl sulfate. The released DNA is purified by removing proteins, RNA, and other cellular debris, and then isolated by sucrose density gradient centrifugation.

Plasmids are separated from the chromosomal DNA

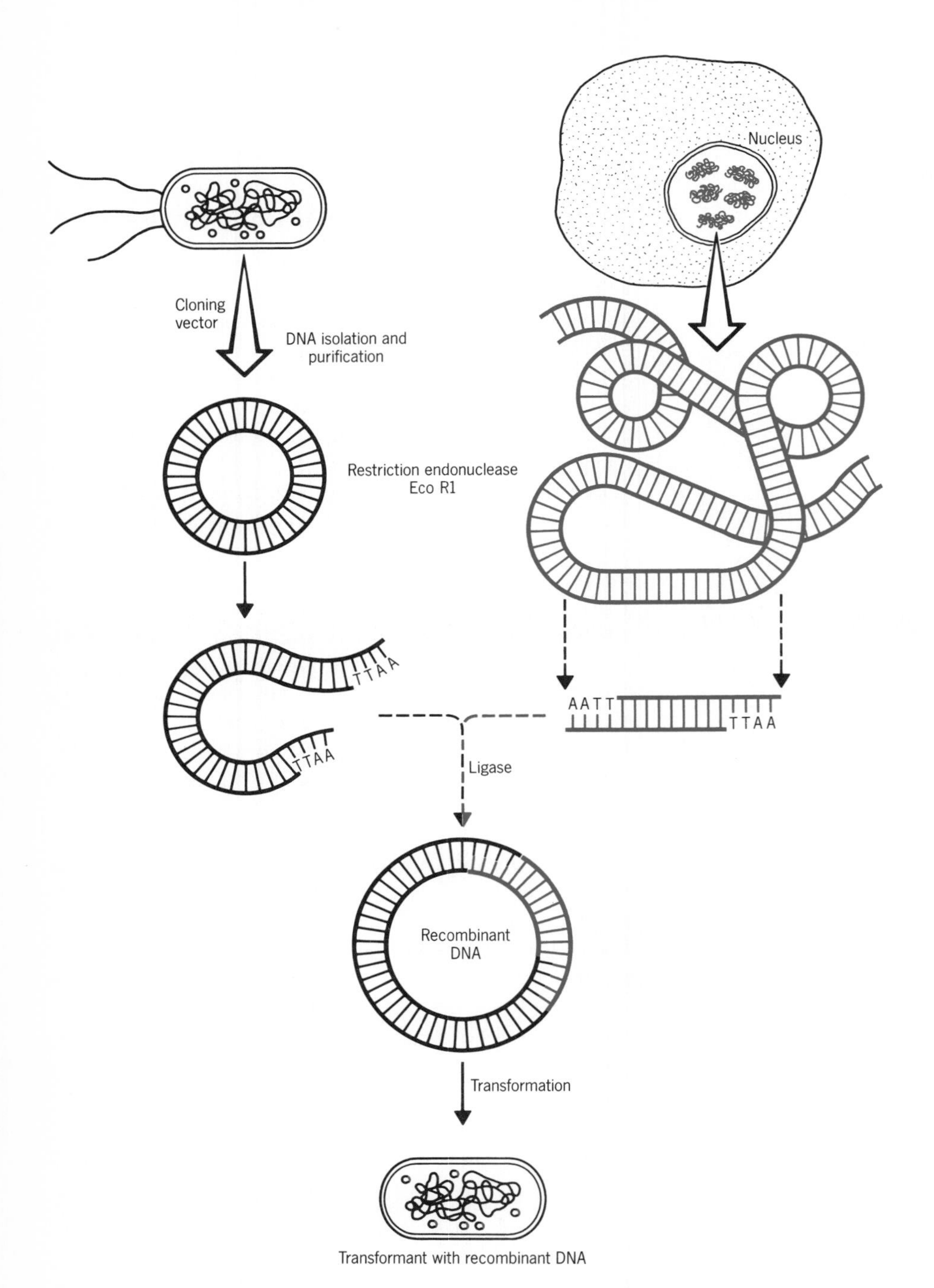

FIGURE 30.1 Typical gene-cloning experiment. Purified foreign DNA is inserted into a plasmid cloning vector after both are treated with the same restriction endonuclease. These treated DNAs have complementing sticky ends that anneal to form a recombinant DNA. After the nicks in the DNA are sealed by DNA ligase, the recombinant DNA is taken into a cell by artificial transformation. The foreign genes are cloned in this cell when multiple copies of the plasmid are made.

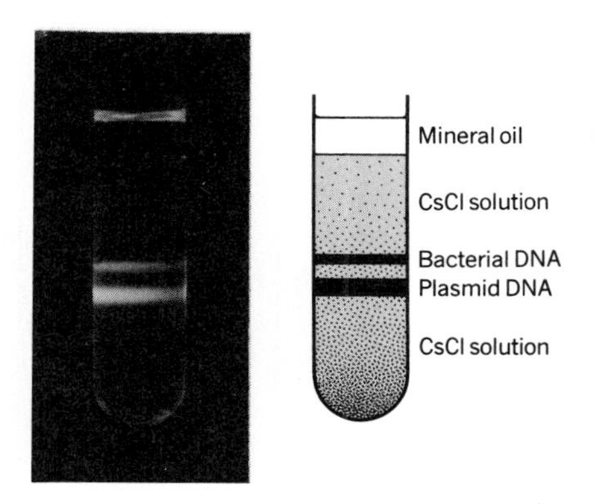

FIGURE 30.2 Bacterial DNA can be separated into plasmid and chromosomal DNA by cesium chloride (CsCl) density gradient centrifugation in the presence of ethidium bromide. Mineral oil layered on top of the gradient prevents the tube from collapsing during 48 hours of centrifugation at 35,000 rpm. Illumination with ultraviolet light reveals the presence of the linear DNA as bright orange bands. (Photograph courtesy of Karl Drlica.)

by techniques that exploit their closed-circular structure. Circular plasmids twist back on themselves (see Chapter 9) to form supercoiled molecules. Such plasmids are separated from the linear DNA fragments of the bacterial chromosome in cesium chloride ($CsCl_2$) density gradients in the presence of ethidium bromide (Figure 30.2), a fluorescent compound that inserts itself (intercalcates) between the base pairs of DNA. Linear DNA molecules bind more ethidium bromide than do closed-circular plasmid DNAs, so the heavier plasmid DNA can be separated from the lighter linear DNA (Figure 30.2). The bands of DNA are isolated by collecting fractions as they drop through a hole punctured in the bottom of the centrifuge tube.

When viral DNA is used as a cloning vector, it is isolated and purified from viral particles. Since viruses contain a single type of DNA, elaborate purification procedures are not required.

Restriction Endonucleases

Bacterial **restriction endonucleases** are the scissors that cut double-stranded DNA (see Chapter 11) into pieces called **restriction fragments.** These fragments vary in length, depending on the number of restriction endonuclease sites in the DNA. Each nuclease recognizes a specific sequence of 4, 5, or 6 nucleotides (Table 30.1) where cleavage occurs. Many bacteria have unique or different restriction endonucleases, so an entire family of these enzymes is available for cutting the DNA into unique fragments.

TABLE 30.1 Selected restriction endonucleases

			NUMBER OF SITES IN DNA FROM	
ENZYME	SOURCE	ACTIVE SITE	λ	pBR322
BamHI	*Bacillus amyloliquefaciens* H	G↓GATCC	5	1
BluI	*Brevibacterium luteum*	C↓TCGAG	1	0
BpeI	*Bordetella pertussis*	A↓AGCTT	7	1
EcoRI	*Escherichia coli* RY13	G↓AATTC	5	1
HaeI	*Haemophilus aegyptius*	$^{(A)}_{(T)}$GG ↓ CC$^{(A)}_{(T)}$		7
HindIII	*Haemophilus influenzae* R_d	A↓AGCTT	7	1
HinfI	*Haemophilus influenzae* R_f	G↓ANTC	>50	10
HpaII	*Haemophilus parainfluenzae*	C↓CGG	>50	26
HsuI	*Haemophilus suis*	A↓AGCTT	7	1
PstI	*Providencia stuartii* 164	CTGCA↓G	31	1
PvuII	*Proteus vulgaris*	CAG↓CTG	15	1
RshI	*Rhodopseudomonas sphaeroides*	CGAT↓CG	4	1
RruI	*Rhodospirillum rubrum*	AGT↓ACT	4	1
SalI	*Streptomyces albus* G	G↓TCGAC	2	1
SlaI	*Streptomyces lavendulae*	C↓TCGAG	1	0
XbaI	*Xanthomonas badrii*	T↓CTAGA	1	0
XmaIII	*Xanthomonas malvacearum*	C↓GGCCG	2	1

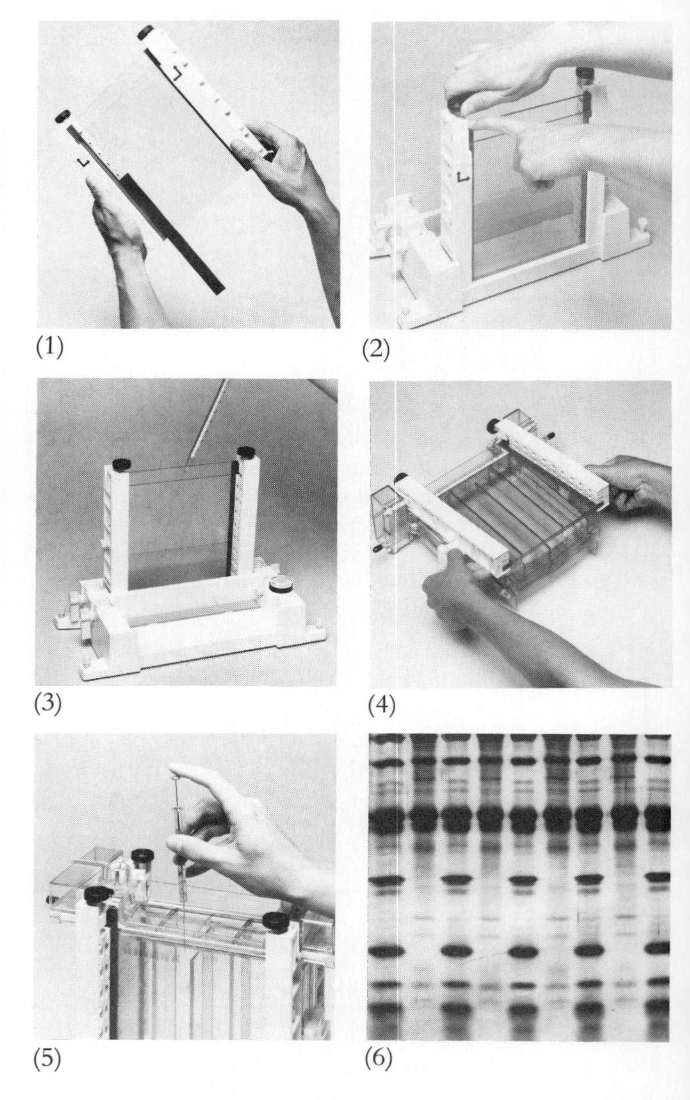

FIGURE 30.3 Slab gel electrophoretic cell used to separate different molecular weight nucleic acids. Glass sandwiches are (1) assembled and (2) aligned in preparation for making the slab gel. After (3) the gel is poured, (4) the sandwiches are attached to the cooling core, and (5) the samples are applied to the top of the gel slab. Separation of alternating high and low molecular weight standards run on a polyacrylamide gel (6) are visible after staining. (Courtesy of BIO-RAD.)

Any DNA molecule completely digested with a pure restriction endonuclease yields a reproducible population of restriction fragments. The size and number of restriction fragments are dictated by the number of cleavage sites in that DNA molecule. After complete digestion, the restriction fragments can be isolated, cloned, and characterized by sequence analysis.

Electrophoresis of DNA

Gel electrophoresis is used to separate DNA molecules of different sizes. Molecules such as plasmids and restriction fragments migrate through a gel of agarose or acrylamide toward the anode (positive pole) when placed in an electric field (Figure 30.3). They separate according to size with the larger fragments at the top and the smaller fragments moving toward the bottom of the gel (Figure 30.4).

The size of a double-stranded DNA molecule is measured as the number of nucleotide base pairs it contains. DNAs are commonly measured as **kilobase pairs,** a unit equal to 1000 base pairs (kilobase is abbreviated **kb**). (For reference, the size of the typical bacterial genome is about 5×10^3 kb.) Small DNA fragments (less than 4 kb) are separated by electrophoresis on acrylamide gels, while DNA molecules between 4 and 200 kb are separated on agarose gels (Figure 30.4). The separated DNA molecules are visualized by staining the gels with ethidium bromide and observing the fluoresence under ultraviolet light.

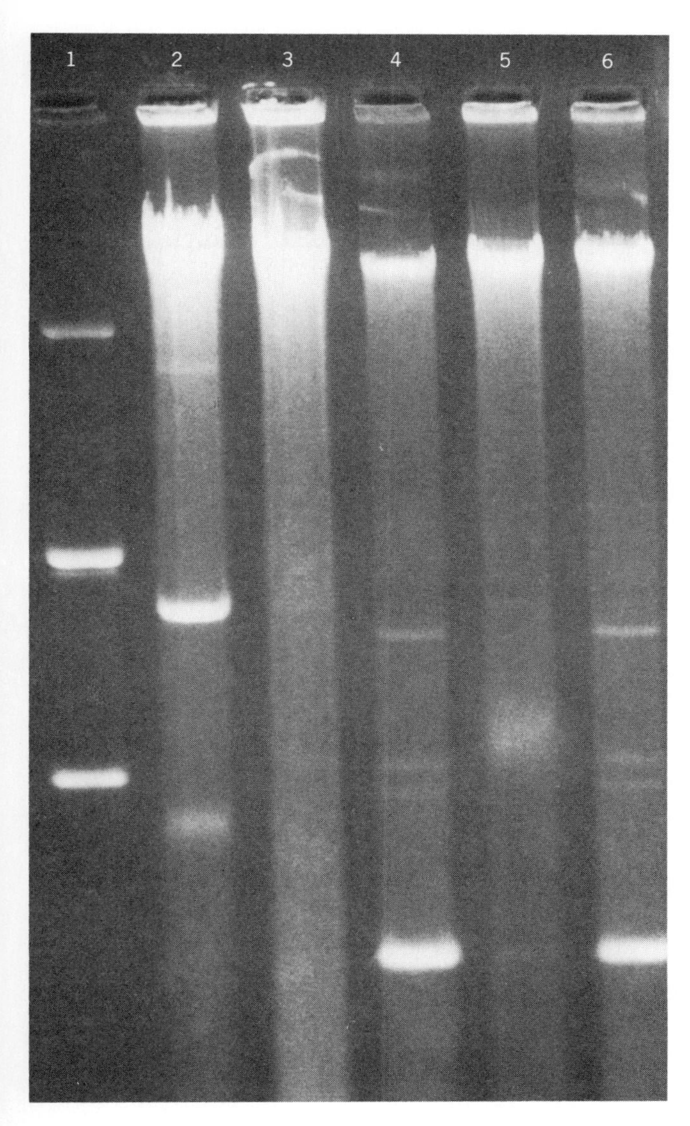

FIGURE 30.4 Photograph of plasmid separation using agarose gel electrophoresis. (Courtesy of A.P.I.)

Ligases

The final step in constructing a recombinant DNA molecule is to covalently seal the nicks or breaks in the DNA with DNA ligase (Figure 30.5). Cells use ligases in DNA synthesis (see Chapter 9) to form the phosphodiester bonds between the 3′-hydroxyl and the 5′-phosphate of two adjacent nucleotides. Three nick patterns occur in damaged DNA: *(a)* a single nick in one strand, *(b)* adjacent nicks in both strands resulting in blunt ends, and *(c)* nicks in both strands at nonaligned points (Figure 30.5). Although both bacterial and bacteriophage DNA ligases are used in genetic engineering, the DNA ligase from bacteriophage T_4 is more versatile. It repairs all three types of nicks shown in Figure 30.5.

CONCEPT REVIEW

Transformation is a natural process for transferring naked (free) DNA into competent bacterial cells of *Streptococcus, Bacillus, Acinetobacter, Azotobacter, Moraxella, Neisseria, Pseudomonas,* and *Haemophilus* (see p. 230). Although natural transformation has not been demonstrated in *Escherichia coli,* this species will take up free DNA under special laboratory conditions.

Artificial Transformation

Artificial transformation in *E. coli* occurs after the bacteria are placed in a hypotonic solution of $CaCl_2$, which alters the cell wall and converts it into a spherical cell (Figure 30.6). A side effect of this process is the loss of proteins, including nucleases, from the periplasmic space (see Chapter 4). Added "free or naked" DNA combines with Ca^{2+} to form a hydroxyl–calcium phosphate DNA complex that resists degradation by DNase. This DNA complex binds to the cell surface and then is taken into the cell during a short heat treatment. The spherical cells regain their normal cell wall structure and ability to multiply after they are transferred to a nutrient medium. Only a few of the many cells exposed to the recombinant DNA are actually transformed.

CLONING VECTORS

A cloning vector is a DNA molecule that accepts foreign genes or DNA fragments and carries them into a host cell where they are replicated. Since ideal cloning vectors do not exist in nature, they are designed and constructed from plasmid and viral DNAs. The important characteristics of a cloning vector are: it must be able to (1) accept foreign DNA, (2) carry the foreign DNA into a cell, and (3) replicate within the cell to make multiple copies. Although there are many cloning vectors, we will limit our discussion to the single plasmid cloning vector designated pBR322 and lambda bacteriophage.

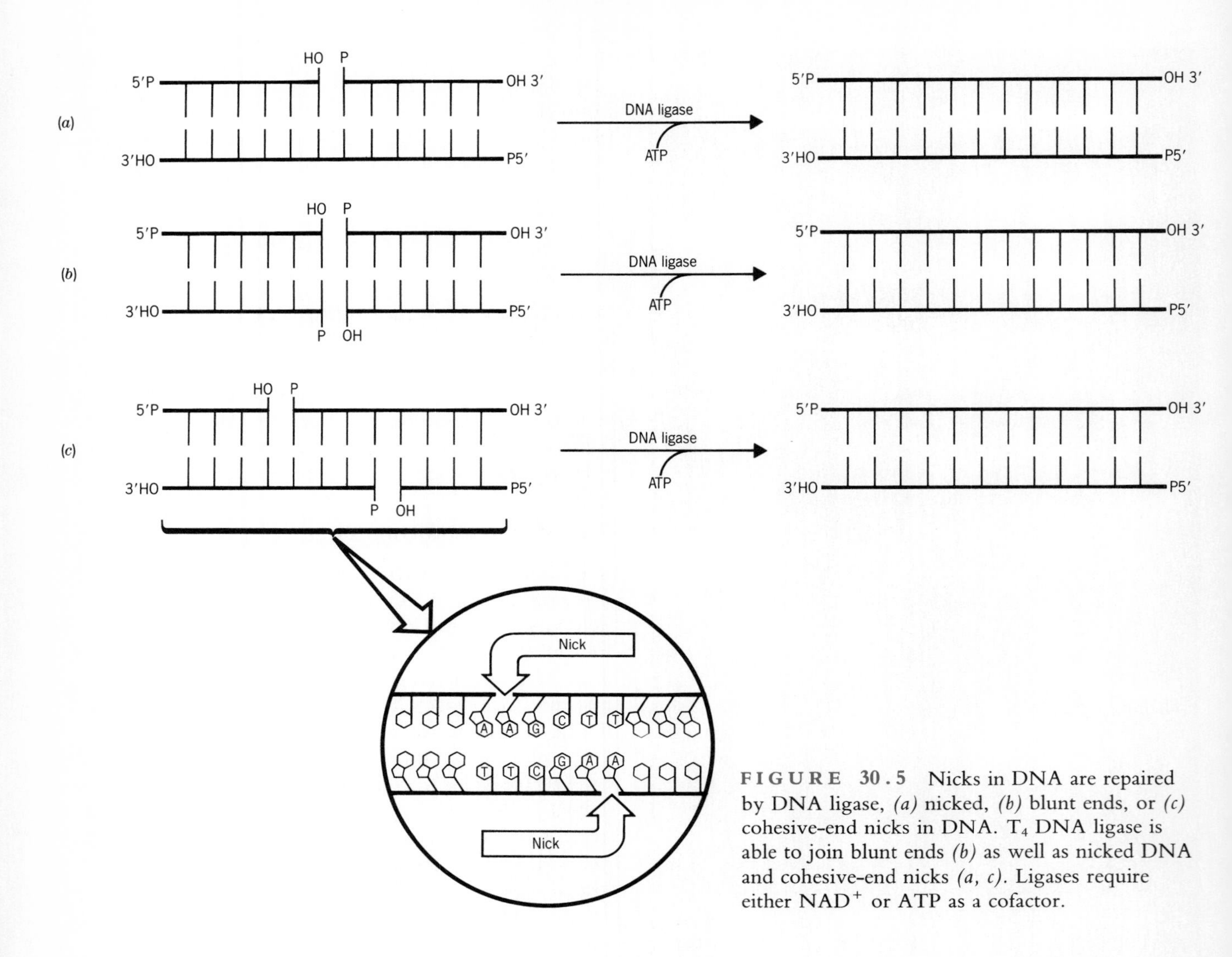

FIGURE 30.5 Nicks in DNA are repaired by DNA ligase, *(a)* nicked, *(b)* blunt ends, or *(c)* cohesive-end nicks in DNA. T_4 DNA ligase is able to join blunt ends *(b)* as well as nicked DNA and cohesive-end nicks *(a, c)*. Ligases require either NAD^+ or ATP as a cofactor.

CONCEPT REVIEW

Plasmids are small, extrachromosomal, circular DNA molecules that replicate independently of the bacterial chromosome—they are replicons (see p. 232). The plasmids that code for multiple antibiotic resistance are the R plasmids from which most plasmid cloning vectors are derived.

PLASMID CLONING VECTORS

With the correct modification, bacterial plasmids make excellent vectors for gene cloning. Selected R plasmids have been modified to be the proper size, contain a replication origin, possess known restriction endonuclease sites, and have markers by which they can be found in transformed cells.

Plasmid Size

Plasmid cloning vectors should be as small as possible since small plasmids are more easily transformed into bacterial cells. Large R plasmids have been reduced in size by cutting out extraneous segments of DNA with restriction endonucleases. The remaining genes should code for antibiotic resistance and the origin of replication.

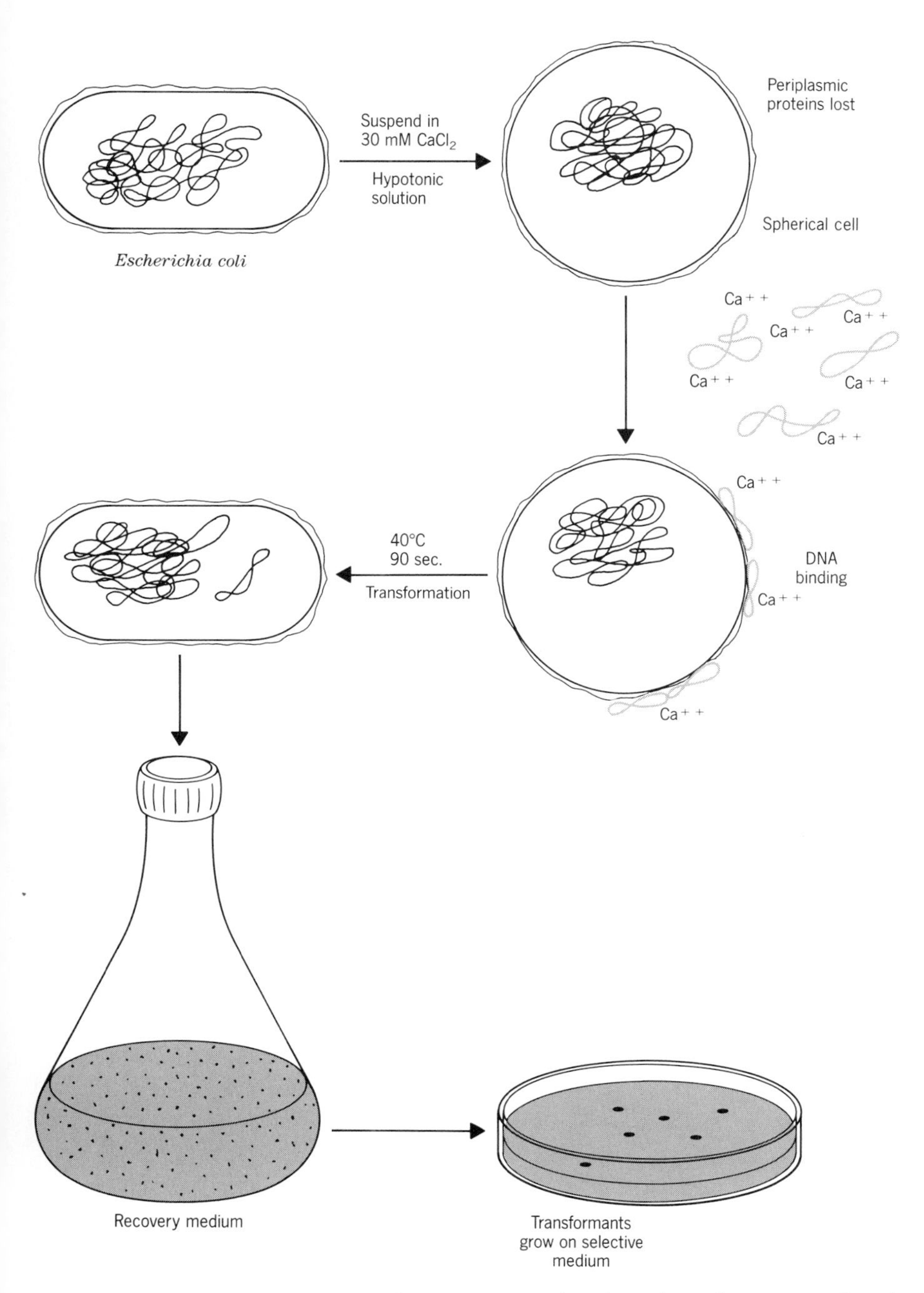

FIGURE 30.6 Artificial transformation in *Escherichia coli*. Cells are treated with hypotonic solutions of $CaCl_2$. The cells become spherical and Ca^{2+} binds free DNA to make it DNase resistant. This DNA is taken into the cells during heat treatment. After recovery in a special medium, transformants are identified by their ability to grow on a selective medium containing antibiotics.

Restriction Endonuclease Sites

An important property of a cloning vector is the number and location of its restriction endonuclease sites. Ideally there should be a single cleavage site for a restriction endonuclease in a detectable gene; a cleavage site in an antibiotic resistance gene is best, as we shall see. Foreign DNA will be inserted into this cleavage site during the formation of recombinant DNA.

Plasmid Amplification

An ideal cloning vector is capable of replication in the host cell to make many copies of the cloned gene—**plasmid amplification.** The replication of some plasmids is not controlled by the cell, and these produce over 50 copies per cell. This type of replication, called relaxed control, exists in natural plasmids, such as the Col E1. This property is so desirable that geneticists transferred the genes controlling replication from Col E1 into an R plasmid to make the pBR322, which replicates under a relaxed control mechanism.

KEY POINT

Replication of bacterial plasmids under stringent control requires protein synthesis and DNA polymerase III, and produces between one and five copies per cell. Bacterial plasmids under relaxed control are replicated by DNA polymerase I and produce over 50 copies per cell.

Selection of Transformed Clones

After the cloned vector is transformed into a host cell, the experimenter must be able to select for the transformants, and, if possible, identify transformants that possess the cloned genes. Artificial transformation results in three genetic cell types: (1) cells that were not transformed, (2) cells transformed by the cloning vector alone (no foreign DNA), and (3) cells transformed by recombinant DNA.

CONCEPT REVIEW

Antibiotic-resistant cells can be selected for by plating on a medium containing the antibiotic. Sensitivity to an antibiotic can be detected by replica plating colonies on a master plate to media containing the antibiotic (see p. 227).

The pBR322 cloning vector (Figure 30.7) carries the gene for ampicillin resistance (Amp^r). When ampicillin-sensitive cells (Amp^s) are transformed with pBR322 and

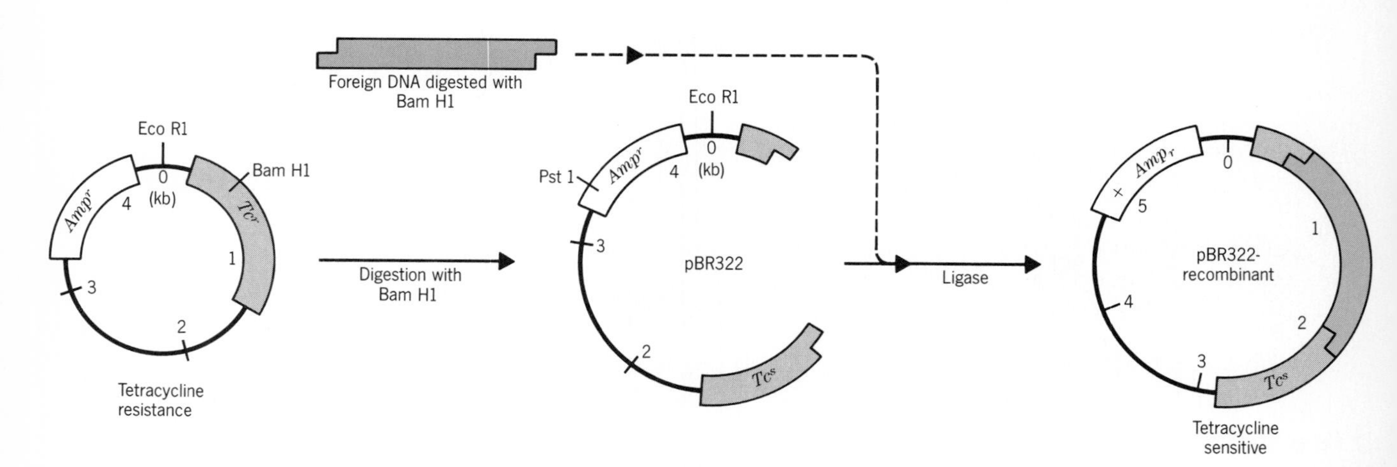

FIGURE 30.7 pBR322 is a plasmid cloning vector. Tetracycline resistance is inactivated upon the insertion of foreign DNA into pBR322 at the BamHI restriction endonuclease site located within the gene coding for tetracycline resistance. Tetracycline-sensitive cells transformed by this recombinant DNA remain sensitive to tetracycline.

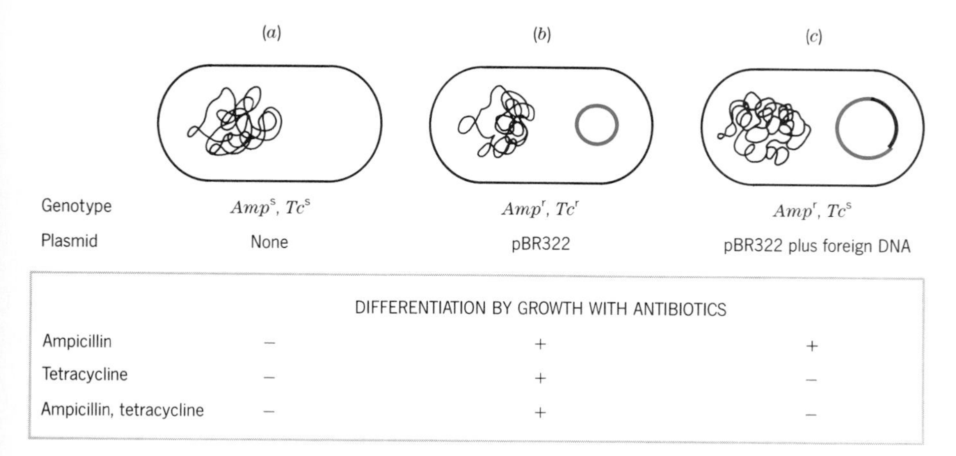

FIGURE 30.8 The possible results from antificial transformation with pBR322 are *(a)* plasmid not acquired *(Amp^s, TC^s)*, *(b)* pBR322 acquired without foreign gene(s) *(Amp^r, Tc^r)*, and *(c)* recombinant pBR322 acquired with an insertion mutation *(Amp^r, Tc^s)*.

plated on ampicillin-containing media, only the transformants will grow because they now have the *Amp^r* gene. The plasmid free host cells (Figure 30.8*a*) remain sensitive to ampicillin and will not grow in its presence.

How do we distinguish between the other two types, recipients transformed with the cloning plasmid and those transformed with recombinant DNA? By cleverly designing the cloning vector, we can distinguish between these two transformants. The single restriction site in pBR322 for the Bam HI restriction endonuclease is in the center of the gene for tetracycline resistance (Figure 30.7). When foreign DNA is inserted into this site, it disrupts this DNA sequence and causes an **insertion mutation.** Cells transformed with unaltered pBR322 plasmid will be resistant to both ampicillin and tetracycline (Figure 30.8*b*). Cells transformed with the recombinant DNA will be ampicillin-resistant but sensitive to tetracycline. So differences between types 2 and 3 are detected by replica-plating transformants (ampicillin resistant in Figure 30.8) on a medium containing tetracycline and identifying clones containing just the cloning vector (*Amp^r*, *Tc^r*) and clones that contain recombinant DNA (*Amp^r*, *Tc^s*).

LAMBDA BACTERIOPHAGE AS A CLONING VECTOR

In one important way plasmid DNA is like the DNA of a bacteriophage; both are replicated in the cytoplasm of a host cell to form multiple copies. An advantage of the bacteriophage is that it injects its DNA into its host as a natural part of its replication cycle. These properties of bacteriophages make them ideal cloning vectors. Next we describe how bacteriophages are used to clone genes, using lambda (λ) as our example.

CONCEPT REVIEW

Lambda is a temperate bacteriophage because it can exist as a replicative form in the cell's cytoplasm or integrated into the *Escherichia coli* chromosome as a prophage (see p. 257). When λ bacteriophage is used as a cloning vector, only its replicative cycle is important.

Head-Full Requirement

The λ genome is a linear double-stranded DNA molecule with sequences of 12 nucleotide bases (single-stranded) at each end. These cohesive *(cos)* ends join to make a circular replicative form (see Figure 11.14) after it enters the host's cytoplasm. To be virulent, the λ genome must have the *cos* sequences at both ends. There is also a **head-full requirement**—to be virulent the λ head must contain a DNA molecule of at least 36.8, but not longer than 51.5 kilobase pairs. The genome of wild type λ is 49 kilobase pairs long, so it satisfies this head-full rule (Figure 30.9).

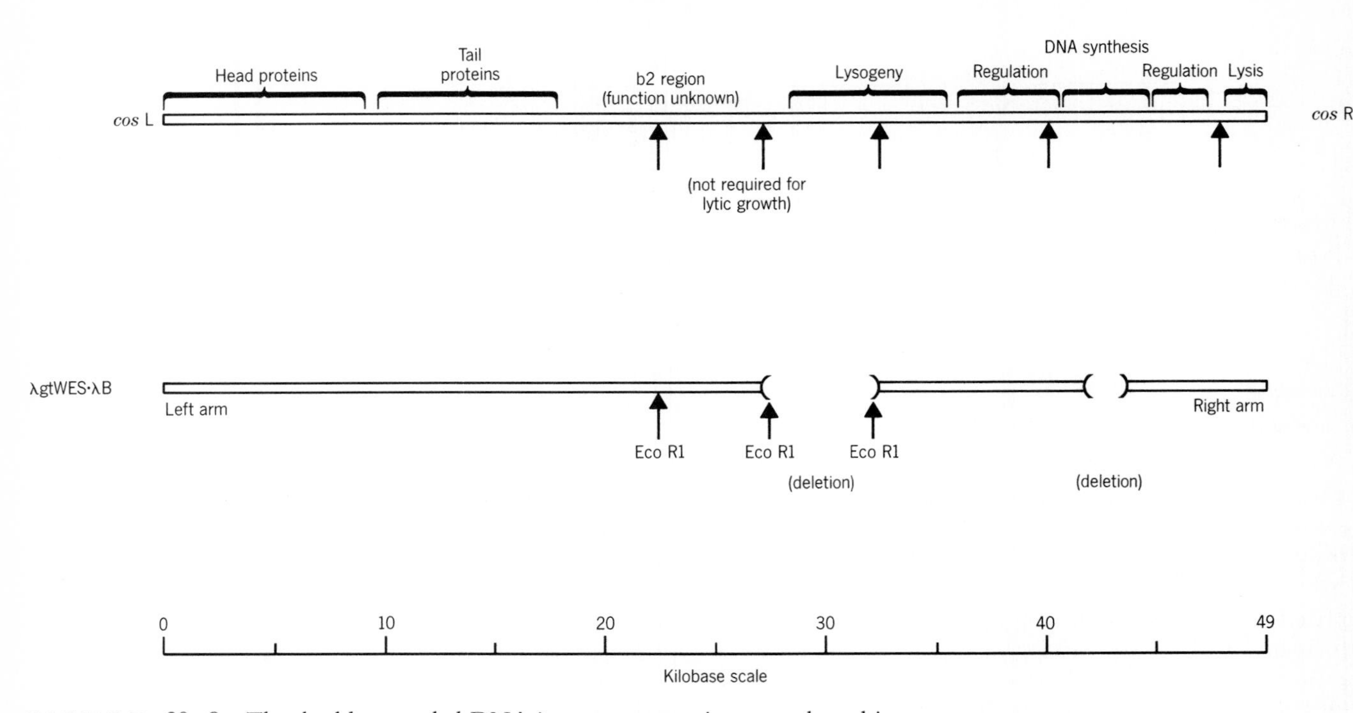

FIGURE 30.9 The double-stranded DNA λ genome contains *cos* ends and is 49 kb long. Lambda DNA can be made into a generalized transducing bacteriophage by deleting the genes coding for insertion and the nin^5 region of the right arm. This bacteriophage becomes a cloning vector when foreign DNA is inserted between the right and left arms of the λ genome.

Modification of the Lambda Genome

The organization of the λ genome lends itself to modification. The genes needed for bacteriophage replication on the left arm code for head and tail proteins, and on the right arm code for DNA synthesis, regulation, and lysis (Figure 30.9). The center of the genome contains genes for integrating λ into the host genome; these must be deleted because we need λ to undergo a lytic replication cycle. The genes for integration lie between two Eco RI sites and are easily removed with Eco RI endonuclease, reducing the λ genome to 38.4 kb. An additional deletion is required to make the vector less than 36 kb long. A 36-kb λ genome will not be packaged into empty heads unless it acquires foreign DNA to make its size fall within the head-full size requirement of 36.8 to 51.5 kb.

Cloning with Lambda

Several steps are needed to clone DNA with the modified lambda genome (Figure 30.10). (1) The λ vector DNA is digested to completion with the appropriate restriction endonuclease to remove the unnecessary "stuffer" DNA. (2) The right arm and the left arm of the λ genome are separated from "stuffer" DNA by electrophoresis or centrifugation. (3) Recombinant DNA is made by mixing the right and left λ arms with foreign DNA fragments prepared using the same restriction endonuclease. Ligase then repairs the nicks, forming a family of linear DNA molecules, each containing the essential λ genes for replication, a piece of foreign DNA, and a *cos* sequence at both ends. (4) This recombinant DNA is packaged *in vitro* into λ heads to form infective bacteriophages. (5) Cells of *E. coli* are infected with the modified λ bacteriophages, and plaques are produced that contain progeny λ.

In vitro Packaging

Special strains of bacteriophage λ are grown in *E. coli* under conditions that allow the infected cell to produce virus tails and empty virus heads. These cells are har-

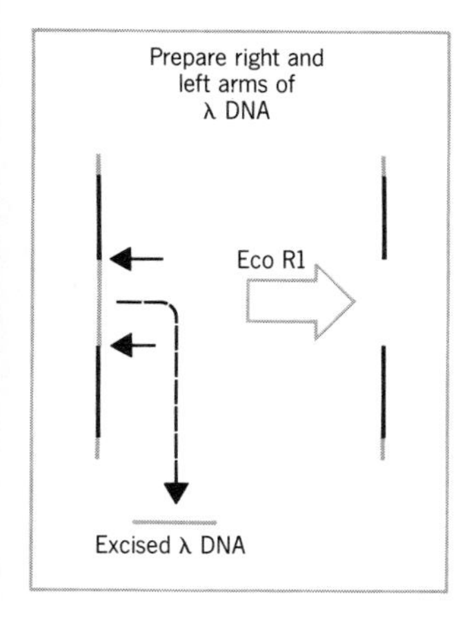

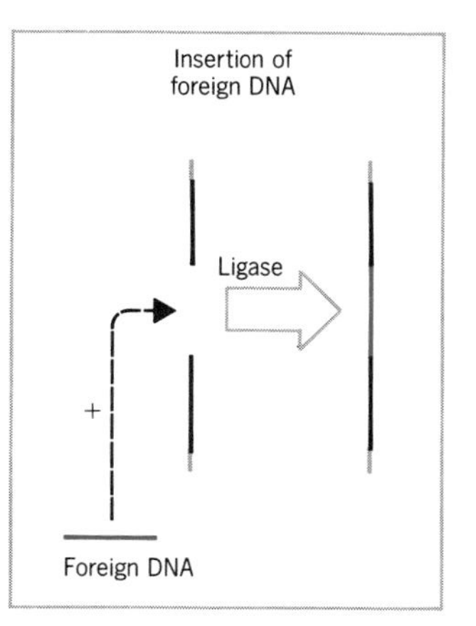

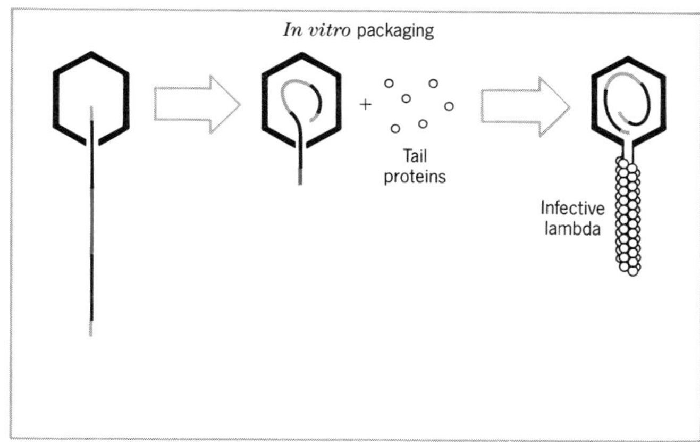

FIGURE 30.10 Steps for cloning genes with λ. The righ and left arms of the λ cloning vector are purified. Foreign DNA is inserted and ligated by ligase to make a genome between 36.8 and 51.5 kb long. This recombinant DNA is then inserted into λ heads *in vitro* to form infectious virions.

vested and stored in the freezer. To package λ DNA into these heads, the cultures are thawed and mixed with the recombinant λ DNA in test tubes, and incubated until the DNA is packaged into the λ heads. The tails then assemble to make each head an infectious particle.

Bacteriophage Clones

The λ bacteriophages are mixed with *E. coli* and then plated on a growth medium seeded with host cells. Lambda goes through a lytic replicative cycle producing many progeny viruses, which infect other cells, forming a plaque on the lawn of bacteria (see Chapter 11). Each plaque represents an individual clone, containing multiple copies of the foreign DNA. The λ bacteriophage cloning vector is about 100 times more effective in producing clones (plaques) than the plasmid-based system that depends on the artificial transformation of *E. coli*.

KEY POINT

When employed as a cloning vector, only the replicative cycle of lambda is important. Transduction and lysogeny are not involved.

CLONING STRATEGIES

Since it is impossible, for reasons having to do with available labor and economic factors, to clone and analyze all genes, we must decide which genes we want to clone and devise a strategy to accomplish the task. There are two basic strategies for cloning genes: (1) specific genes or DNA fragments can be cloned if they are available in pure form, or (2) we can use the "shotgun" approach and clone the cell's entire genome and then select the clones of interest. The shotgun approach results in about a million clones representing the **genomic library** of the cell's DNA.

GENOMIC LIBRARY

A genomic library is made by cloning millions of fragments of the cell's DNA in λ cloning vectors. The cell's DNA is randomly sheared into fragments by treating it with special restriction endonucleases. DNA fragments with an average size of 20 kb are separated from this digest by sucrose density gradient centrifugation or gel electrophoresis. This foreign DNA is inserted between the right and left arms of λ to form the recombinant λ genome that satisfied the head-full requirement. *In vitro* packaging (Figure 30.10) of the recombinant DNA into λ heads creates a population of virulent bacteriophages, each containing a piece of foreign DNA.

The λ bacteriophages produced by this *in vitro* process represent the genomic library. They can be stored for later use, or they are separated into distinct clones by the plaque plating system. Large numbers of clones can be produced on baking dish-sized agar plates that will accommodate between 10,000 and 450,000 plaques.

CONCEPT REVIEW

Information in DNA is transcribed into messenger RNA, which attaches to ribosomes and is translated into polypeptides. Retroviruses reverse this flow of information by making DNA from their RNA genome using reverse transcriptase (see p. 281). This enzyme can be used to make DNA from any mRNA.

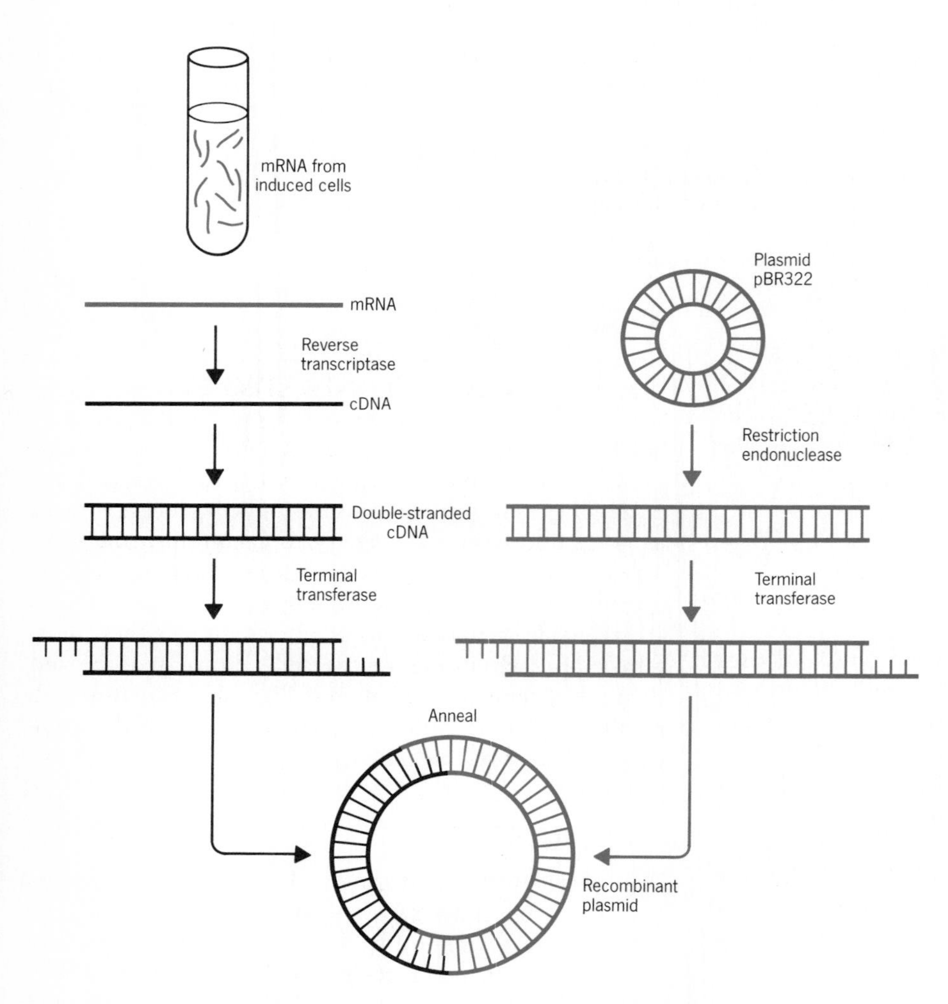

FIGURE 30.11 Manufacture of interferon genes from messenger RNA. Messenger RNA is isolated from cells induced to form interferon. Reverse transcriptase makes complementary DNA using the mRNAs as a template. Cohesive ends are added to the cDNA and to the opened pBR322 cloning vector by the action of a terminal transferase. The cDNA molecules are then inserted into the plasmid to make a recombinant plasmid.

CLONING cDNA

A second strategy for cloning genes is to isolate a specific gene or DNA fragment. For practical reasons this is done by (1) using the information in mRNA to form cDNA or (2) chemically synthesizing the gene from a known nucleotide sequence.

Sometimes it is easier to isolate the mRNA than to isolate a specific gene from DNA. Messenger mRNAs cannot be cloned directly, but they can be copied into DNA and then cloned. Purified mRNAs act as the template for making complementary DNA or **cDNA,** using the enzyme reverse transcriptase (Figure 30.11). Double-stranded cDNA contains the information necessary to make a protein and can be cloned by the procedures described above.

Genes can be synthesized chemically if the nucleotide sequence is known or if the sequence of amino acids in the gene product is known. Nucleotide sequences for a number of genes have been reported in the literature and more are being reported each month. A synthesized gene or the cDNA copy of a gene is usually small enough to be cloned in a plasmid such as pBR322. Once the gene is cloned, the investigator uses a screening technique to identify the correct clone.

SCREENING FOR DESIRED CLONES

The cloned genes now reside in a colony of bacteria containing the autonomously replicating plasmid or in a clone of bacteriophage λ generated during plaque formation. The next step is to identify clones possessing the gene(s) of interest. Genes that are not expressed or that do not measurably alter the cell's phenotype can be identified by direct screening methods, using nucleic acid probes or by detecting the physical presence of the gene product.

CONCEPT REVIEW

Under appropriate conditions two complementary single-stranded nucleic acids will spontaneously form base pairs and become double-stranded (see p. 309). These complementary pairs can be DNA/DNA or RNA/DNA hybrids.

DIRECT SCREENING METHODS

Hybridization probes and monoclonal antibodies are used in direct screening methods for detecting genes or gene products. The choice of technique depends on the availability of a hybridization probe or a specific antibody that will react with the gene product (see section on monoclonal antibodies, Chapter 18).

In Situ Colony Hybridization, Monoclonal Antibodies

Once a gene has been isolated, either in the form of mRNA or DNA, it can be used as a hybridization probe. The **probe** is a radioactively labeled, single-stranded RNA or DNA that is complementary to the gene(s) being sought. These probes are used to screen a genomic library for those clones containing specific gene sequences.

The bacterial colonies (or bacteriophage plaques) are grown on a master plate; then, part of each clone is transferred to a nitrocellulose filter (Figure 30.12). When the filter is exposed to NaOH, the cells lyse and release denatured, single-stranded DNA that binds to the filter. After the NaOH is neutralized with dilute acid, the filter is exposed to the single-stranded probe that is radioactive. The probe binds indirectly to the filter by attachment to a complementary region of single-stranded DNA. After unbound probe is washed off, the location of the bound probe is determined by measuring the radioactivity. Spots of radioactivity will correspond to colonies or plaques on the master plate (Figure 30.12).

Monoclonal antibodies that react with a specific gene product can also be used to detect cloned genes. This is a specific test because monoclonal antibodies react with a single antigen. By detecting the antibody–antigen reaction, one can determine which clones possess the gene of interest.

SUMMARY OF GENE CLONING

Almost any gene from any organism or virus can be cloned using recombinant DNA technology. The **shotgun** approach to gene cloning forms a genomic DNA library of a cell's total DNA. Selective genes are then identified and analyzed. Alternatively, genes can be chemically synthesized or enzymatically copied from mRNA using reverse transcriptase to make

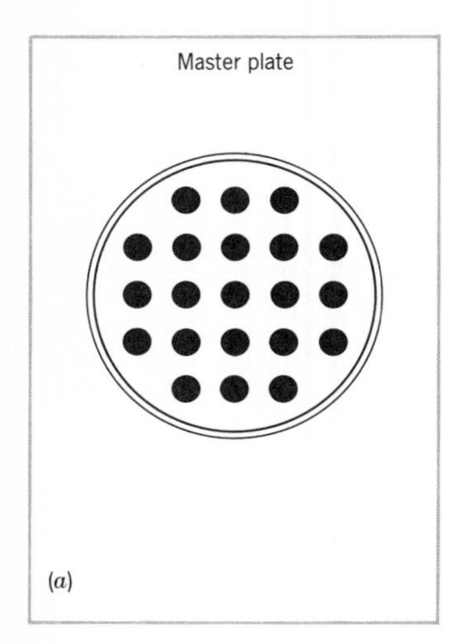

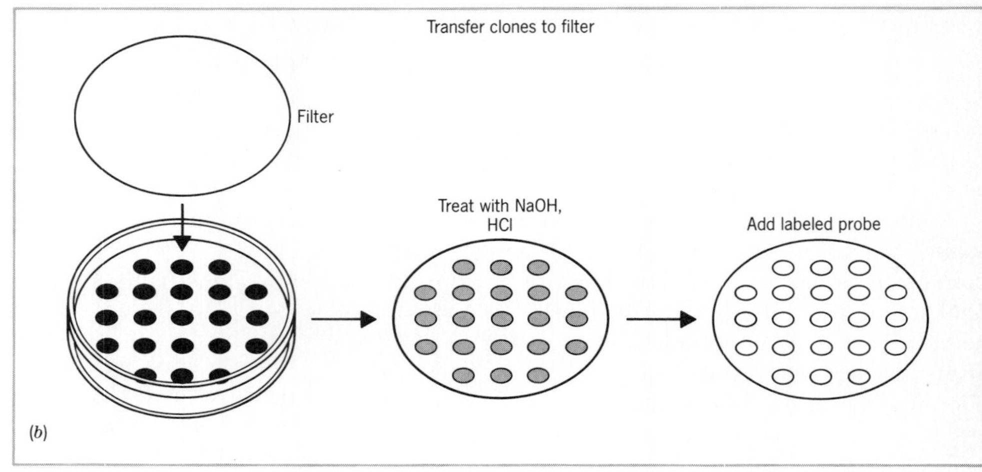

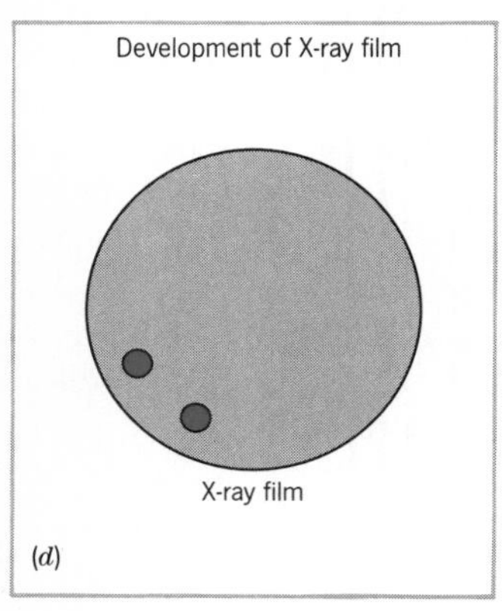

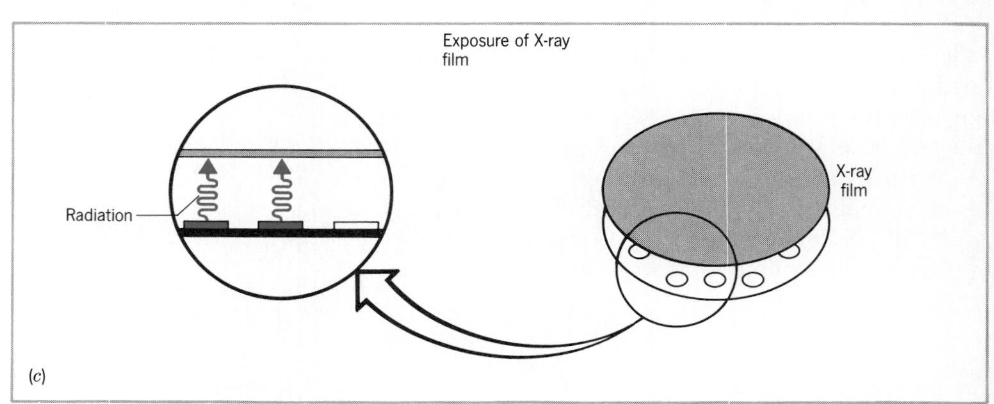

FIGURE 30.12 Hybridization probes are used to detect cloned genes. Transformants are *(a)* grown on a master plate before they are *(b)* transferred to a nitrocellulose filter. The filter is treated with NaOH, which lyses the cells and releases denatured DNA that binds to the filter. The NaOH is neutralized before the filter *(c)* is incubated with the radioactive probe. Unbound probe is washed off and the filter *(d)* is then incubated with an X-ray film. The bound probe exposes the X-ray film, indicating areas on the filter that picked-up cloned DNA from the master plate. Thus, colonies on the master plate that contain cloned DNA are identified.

cDNA. These foreign genes are inserted into plasmid or bacteriophage DNA and cloned in a host cell. Bacteriophage λ is used to produce genomic libraries and pBR322 is used to clone genes in bacterial cells. The clone of interest is identified directly with a nucleic acid probe or a monoclonal antibody that reacts with the gene product.

APPLICATIONS OF GENETIC ENGINEERING

Geneticists are using genetic engineering to analyze both genes and the DNA "sites" that control gene expression. These DNA sites are fundamental units of

TABLE 30.2 Products produced by recombinant DNA technology

PRODUCTS	USES OR POTENTIAL USES
Hormones	
Insulin	Treatment of diabetes
Human growth hormone	Treatment of dwarfism
Lymphokines	
Interferons	Treatment of viral infections, cancer
Interleukin-1	Experimental investigations
Interleukin-2	Experimental cancer therapy
Vaccines	
Foot-and-mouth disease	Immunize cloven-hoofed animals
Hepatitis B	Immunize health care professionals
Human immunodeficiency virus (HIV)	Assay for antibody in blood donors and patients
Influenza virus	Immunize high-risk patients
Varicella–zoster virus	Immunize children
Herpes simple virus	Immunize population
Malaria, schistosomiasis	Immunize susceptible population
Enzymes	
Tissue plasminogen activator (T-PA)	Dissolve blood clots
Enzyme pathway for 2-keto-L-gluconic acid	New microbial pathway used to produce L-ascorbic acid—vitamin C

the genome that are smaller than genes. A site that affects the expression of genetic material is roughly 100 base pairs long. Examples of sites include promoters, operators, terminators, attenuators, cap sites, enhancers, origins of replication, and ribosomal binding sites (see Chapter 9). The ability of molecular biologists to dissect the functional structure of the eucaryotic genome has solved some perplexing biological problems, such as understanding the mechanism of antibody diversity.

Many proteins of pharmacological importance can be produced by genetically engineered organisms. One example is insulin, a human peptide hormone now commercially produced by genetically engineered bacteria. Other important commercial applications of gene cloning include the manufacture of immunoproteins, vaccines, and enzymes (Table 30.2). Altering the genetic information of plants and animals through genetic engineering is in its infancy; however, progress is being made even in this complex area.

GENETIC ENGINEERING AND THE PHARMACEUTICAL INDUSTRY

The first industry to benefit from the application of genetic engineering is the pharmaceutical industry. Major markets exists for human peptide hormones; vaccines against viruses, bacteria, and protozoa; and for lymphokines. Other products with commerical potential include enzymes produced from cloned human genes.

Human Peptide Hormones

Insulin is a peptide hormone prescribed for some 4 million diagnosed diabetics who take daily insulin injections. The genes for the two insulin polypeptides have been cloned in *Escherichia coli,* which is used to produce synthetic insulin (see below). **Human growth hormone** could be used to treat dwarfism if sufficient quantitites of pure peptide hormone were available. The genes for the human growth hormone have been cloned in bacteria, which can produce it in sufficient quantities to make clinical trials possible. Bovine growth hormone extracted from bovine pituitary glands stimulates milk production in a well-managed dairy herd by 10 to 30 percent and increases the growth of large animals by 10 to 15 percent. Until recently this hormone was in low supply, but now through the use of genetic engineering four companies are producing bovine growth hormone in significant quantities. Once it is approved by the Food and Drug Administration, bovine growth hormone could have a major impact on the dairy industry.

Lymphokines

Lymphokines are produced in the body in small quantities by lymphocytes. **Interferons** are produced by animal cells in response to viral infection, and protect other cells against viral infection. The genes for the three types of human interferon have been cloned and enough interferon is available for testing their anti-viral and anti-cancer effects. Activated lymphocytes produce **interleukin-2,** which stimulates T lymphocytes and B lymphocytes to divide during the immune response. Interleukin-2 is now produced by genetic engineering and is undergoing clinical trials in certain forms of cancer therapy. **Interleukin-1,** a small peptide produced by macrophages that stimulate T lymphocytes to divide, is also being produced by this technology and is undergoing clinical tests. Molecular cloning of the genes coding for these human proteins is the only way commercial quantities can be produced.

Vaccines Against Infectious Diseases

Vaccines against viruses and bacteria are traditionally composed of killed or attenuated strains of the infectious agent. These vaccines have been invaluable in preventing certain infectious diseases; however, the effectiveness of some vaccines is limited because they cause allergic reactions in a small percentage of patients, and some vaccines made with attentuated strains actually cause the diseases they are designed to prevent. We cannot prevent all infectious diseases with this approach because (1) some viruses cause latent infections and we don't know their long-term consequences and (2) for technical reasons we are unable to produce sufficient quantities of the pathogen to make the vaccine.

Production of viral, bacterial, or protozoan antigens by genetic engineering is an ideal solution to these problems. Bacterial vaccines for protecting humans against dysentery, typhoid fever, cholera, and travelers' diarrhea have been proposed and viral vaccines against influenza A, varicella–zoster (chickenpox), and human immunodeficiency virus are candidates for genetic cloning. Viral vaccines composed of cloned polypeptide antigens and produced by genetic-engineering technology have been introduced to prevent foot-and-mouth disease and hepatitis B (see below). Modified viral vaccines against porcine diarrhea have also been produced. The market for animal viral vaccines is potentially very large, for these diseases cause serious economic losses.

Immunological approaches to controlling protozoan diseases have been unsuccessful in large measure because of the great difficulty in obtaining sufficient quantities of the parasite. This problem has been overcome in the case of malaria by cloning the genes for an antigen found on the surface of *Plasmodium falciparum,* the circumsporozoite protein. Antigenic preparations made from this protein protect mice against malaria infections, and perhaps this or a similar vaccine will also protect humans against malaria. The preliminary success with malaria opens the door to the genetic engineering of vaccines for controlling schistosomiasis, Chagas' disease, and African trypanosomiasis (sleeping sickness).

Enzymes

Molecular cloning lends itself to the commercial production of human enzymes, enzymes that otherwise would have to be purified from urine or autopsy material. The best example of these substances is **tissue plasminogen activator** (T-PA), a natural protein found in minute quantities in the body's circulatory system. This protein travels to blood clots where it converts plasminogen to its active form, plasmin. The plasmin breaks down clots into soluble proteins that disperse. T-PA has been produced by genetic engineer-

First Genetically Engineered Vaccine for Human Use

Hepatitis B is a worldwide public health problem, causing persistent illness with links to hepatocellular carcinoma. Health professionals estimate that 200 million people are infected with the hepatitis B virus. Since there is no effective treatment for hepatitis B infections, prevention is essential.

The hepatitis B virus has not been grown in tissue culture, so vaccines must be made by unconventional means. The first hepatitis B vaccine was made from the hepatitis B virus surface antigen (HBsAg), which was purified from the plasma of human carriers. Quantities of this vaccine are limited by the supply of human plasma containing the HBs antigen. In addition, stringent procedures must be followed to kill not only the hepatitis B virus but all other infectious agents that could originate from the plasma.

The genetically engineered hepatitis B vaccine circumvents these problems. This vaccine is made from the viral HBsAg produced in a genetically engineered yeast. The vaccine was made by first isolating DNA from hepatitis B virus. This DNA was cleaved into short segments with restriction endonucleases. The fragment coding for the HBs protein was isolated and inserted into a plasmid vector before the vector was cloned in a yeast cell.

Cultures of the transformed yeast produce HBsAg, which specifically reacts with antibodies against the viral antigen. HBsAg produced by these yeast cells aggregates into virus-sized particles resembling the Dane particles present in the plasma of hepatitis B patients. This purified protein is treated with formaldehyde to stabilize it and to kill extraneous living agents that might be present.

The hepatitis B vaccine is the first genetically engineered vaccine approved for human use in the United States. Early clinical trials indicate that it will be an effective weapon against this serious viral disease.

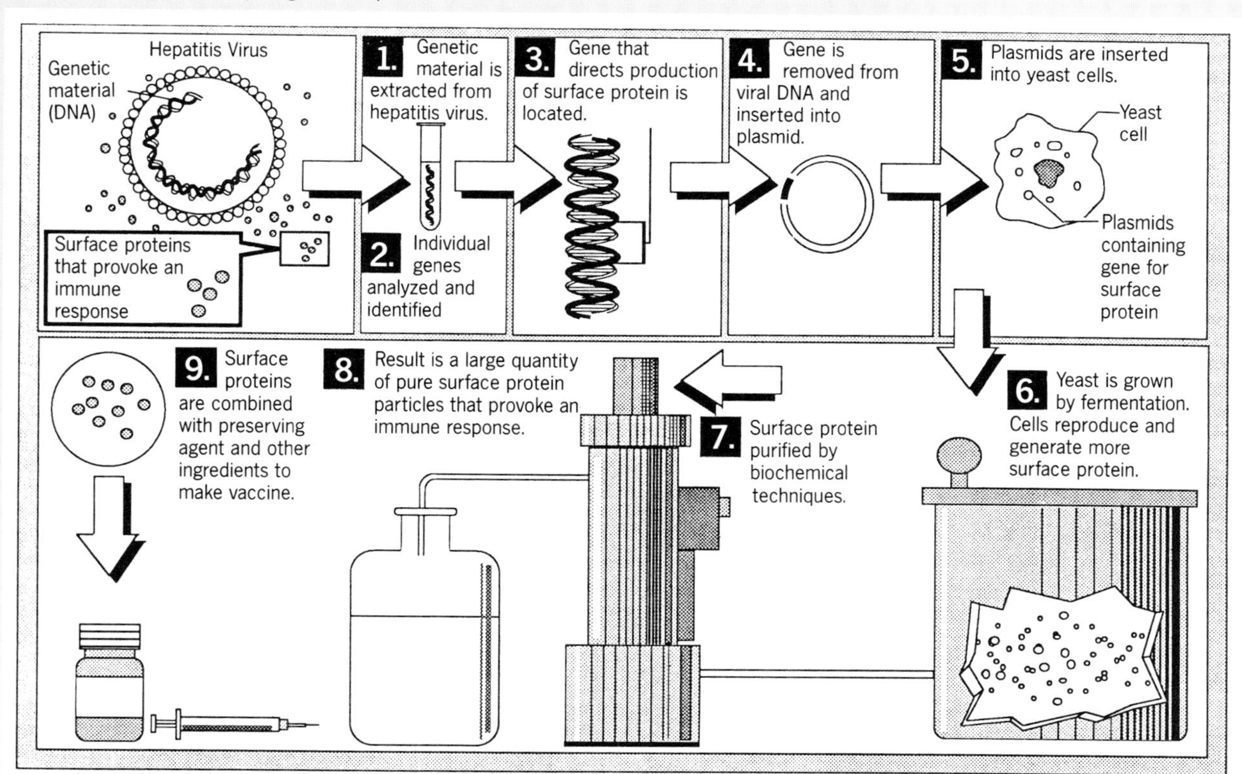

Making a genetically engineered vaccine. (Courtesy of the Philadelphia Inquirer/Kirk Montgomery.)

ing technology and is undergoing FDA approval for treating patients with myocardial infarction ("heart attacks"), pulmonary embolism, deep-vein thrombosis, and peripheral arterial occlusion.

GENETIC CLONING AND COMMERCIAL PRODUCTION OF HUMAN INSULIN

Insulin became the first health product of genetic engineering when approval was obtained, in 1982, to market human insulin produced by cultures of *E. coli* K12. Insulin is an animal peptide hormone composed of an A and a B polypeptide joined by disulfide bridges (Figure 30.13), and prescribed for treating diabetics.

The cloning of the human insulin genes was a collaborative effort between a major drug company, a biotechnology firm, and a medical center. The human insulin genes were chemically synthesized from the known amino acid sequence of both the A and the B chains. These synthetic genes were then cloned in *E. coli* K12, which produced either the A or the B chain as products.

Genetically engineered *E. coli* yields about 0.5 mg of pure insulin per liter of fermentation medium when grown in 10,000 gallon fermenters. This cloned insulin has passed clinical tests in Europe and the United States and is now available to consumers. Before this product was available, insulin was prepared solely from the pancreas of animals. Synthetic insulin is a superior drug because it is human insulin (porcine and bovine insulin are slightly different) and the supply will never be limited by the availability of animals.

GENETIC ENGINEERING OF THE FOOT-AND-MOUTH DISEASE VACCINE

Foot-and-mouth disease is a highly contagious febrile disease of cloven-hoofed animals (Figure 30.14) that occurs outside of North America. Approximately 800 million doses of foot-and-mouth disease virus (FMDV)

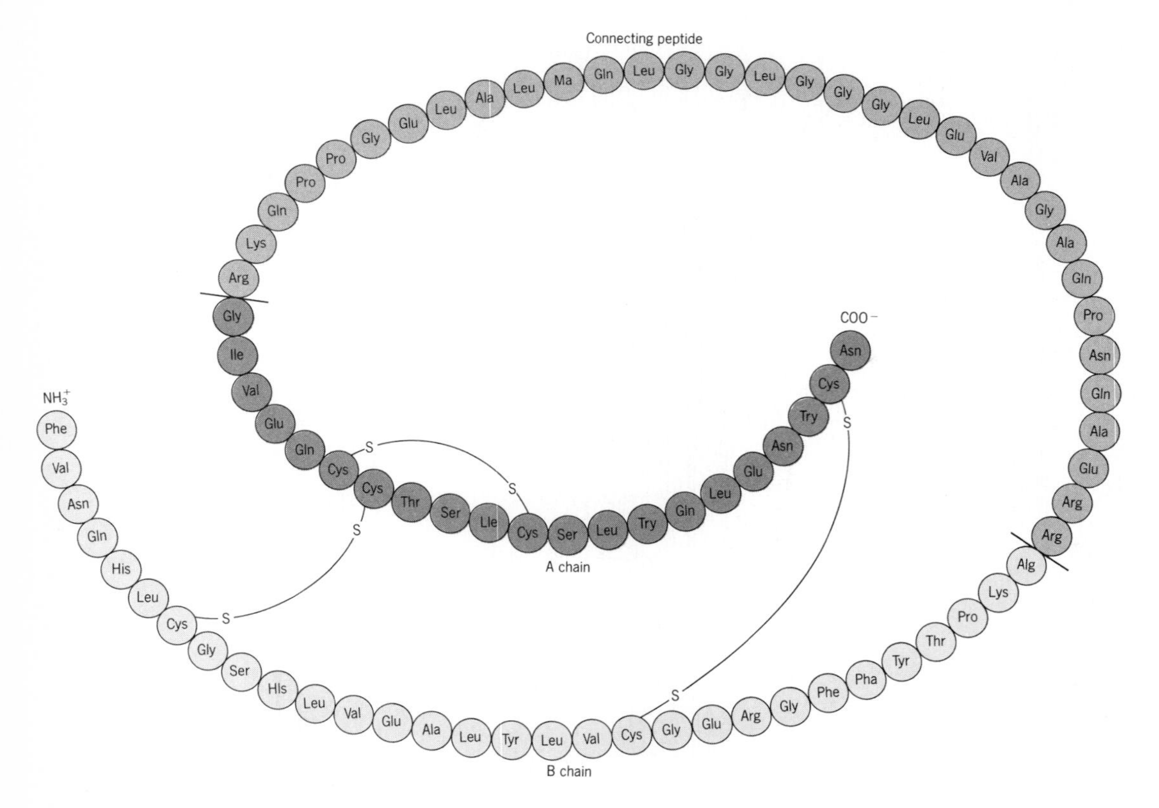

FIGURE 30.13 Proinsulin is composed of 84 amino acids. The A chain is separated from the B chain by a connecting peptide that is removed from the active enzyme. (*Source:* Office of Technology Assessment.)

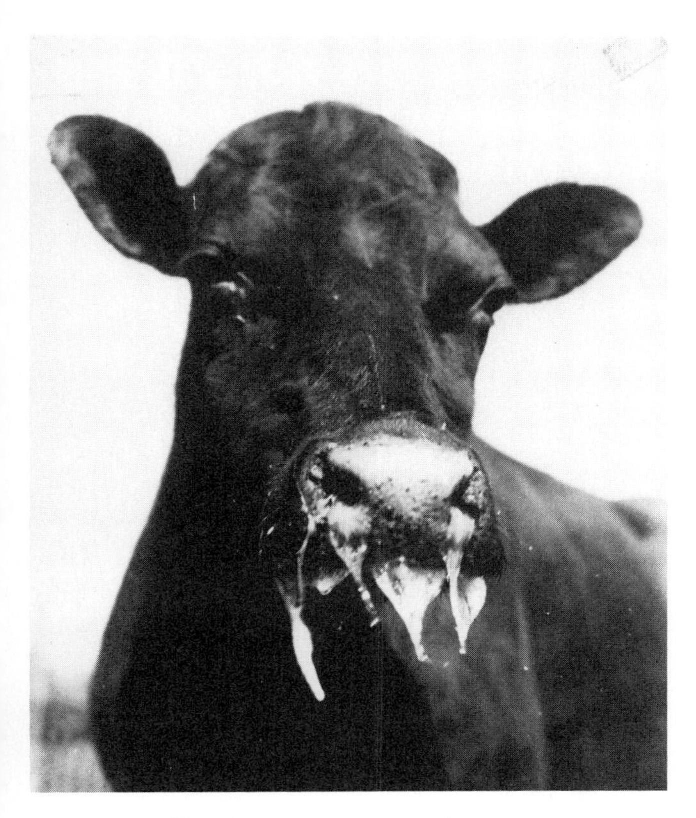

FIGURE 30.14 Foot-and-mouth disease symptoms in an infected cow include excessive flow of saliva from its mouth. (Courtesy of USDA Office of Information.)

vaccine are administered annually worldwide. Production of the vaccine from live viruses presents containment problems (see Chapter 8), and many outbreaks have been attributed to incompletely inactivated vaccines.

The gene coding for a viral protein coat (VP_3) antigen was cloned (Figure 30.15, see page 742) in *E. coli* by a joint venture between the Plum Island Animal Disease Center of the U.S. Department of Agriculture and a company doing extensive work in genetic engineering. Foot-and-mouth disease is caused by an RNA virus of the Picornavirus family. To clone the viral antigens, FMDV RNA was isolated and used as the template to make a cDNA gene. The gene coding for the viral protein (VP_3) was inserted into a modified pBR322 plasmid, which was transformed into *E. coli*. Transformants producing VP_3 protein were selected and used to make a vaccine that confers protection against foot-and-mouth disease in animals.

One advantage of the FMDV vaccine is that it can be produced without growing the infectious virus. This facilitates vaccine production and decreases the risk of accidental outbreaks. The vaccine can never cause accidental outbreaks of foot-and-mouth disease since it contains no viral RNA.

The important antigens for many viral vaccines are composed of complex glycoproteins. These are more difficult to manufacture, but molecular biologists are fervently working on them. One area that shows great promise is the use of genetic engineering to produce vaccines against the serious protozoan diseases that affect people.

CLONING PLANT AND ANIMAL GENES

Animals and plants have traditionally been bred to generate desirable traits in offspring. It is theoretically possible to accelerate the development of new plant and animal strains or to modify existing strains using recombinant DNA technology. This is not a simple task since the genetics of plant and animal cells is quite complex; however, there is progress being made and the potential benefits are great.

CLONING VECTORS FOR PLANTS

Characteristics that botanists would like to create in "new" plants include (1) increased resistance to a variety of pests, (2) ability to use nitrogen gas in place of fertilizer nitrogen, (3) ability to grow at climatic extremes of dry–wet and hot–cold, and (4) high-yielding plants that have greater food value. The first step in genetically engineered plant tissue is to develop a cloning vector capable of transferring recombinant DNA into the plant tissue. The species of *Agrobacterium* responsible for crown gall and hairy root diseases in plants provides one solution to this problem.

The Ti Plasmid

Crown galls in dicotyledons are initiated when the root crown tissue is infected by *Agrobacterium tumefaciens* carrying the Ti (tumor-inducing) plasmid. Once the plasmid enters the cells's cytoplasm, a segment of the plasmid known as the **T-DNA** is transferred into the cell's genome, resulting in unrestricted cell division and

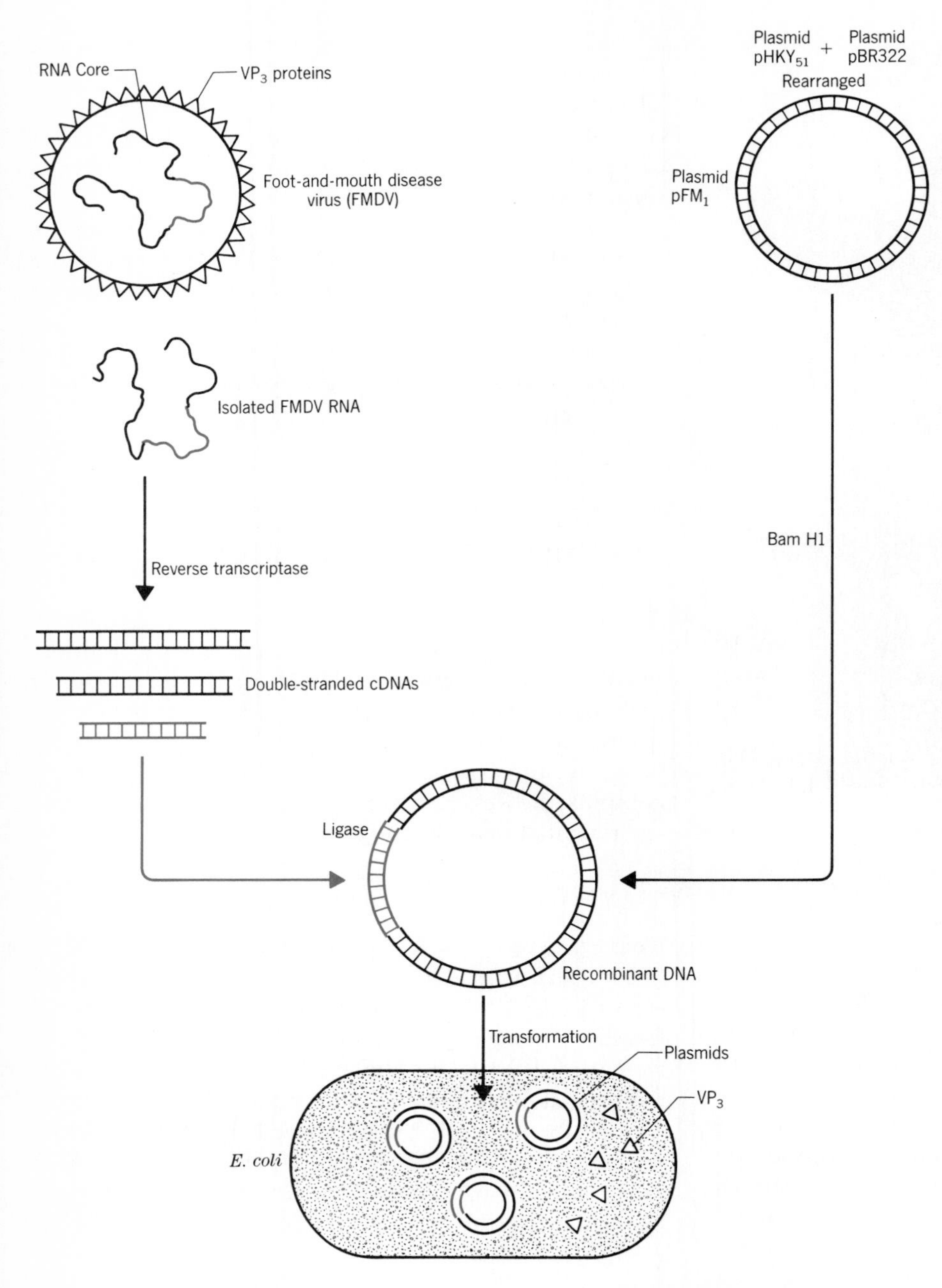

FIGURE 30.15 Genetic engineering of the foot-and-mouth disease viral vaccine. RNA isolated form the viral genome was used to make complementary DNA using reverse transcriptase. The cDNA fragments produced were cloned in a modified pBR322 cloning plasmid. Transformants producing the VP_3 PMDV protein were identified. These clones were then used to make the VP_3 protein, which elicits the production of protective antibodies against the foot-and-mouth disease virus.

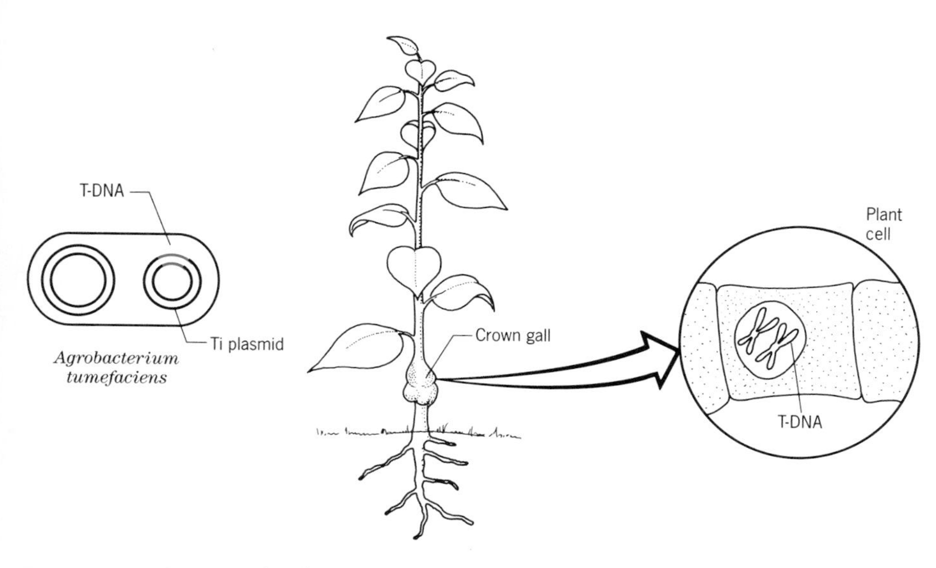

FIGURE 30.16 *Agrobacterium tumefaciens* infects dicotyledons to cause growths known as crown galls. The T-DNA in the Ti plasmid of *A. tumefaciens* is transferred to the plant genome during infection. The Ti plasmid can act as a plant cloning vector when foreign DNA is attached to the T-DNA segment.

the formation of the tumorous gall. Each transformed cell possesses the T-DNA segment (Figure 30.16) and now can be grown in culture.

It is possible to transfer genes into these plant cells by inserting foreign DNA into the T-DNA segment of the plasmid. Such a transfer has been accomplished using transposon DNA-carrying markers for antibiotic resistance. Although the Ti plasmid is a potential cloning vector for plants, it has some drawbacks. The DNA that is transferred by the T-DNA segment appears to be lost during seed production, so only plants cultivated by cuttings can be effectively modified. Another problem is that *Agrobacterium tumifaciens* infects only dicotyledons, and many of our most important food crops, such as wheat and corn, are monocotyledons.

CLONING VECTORS FOR ANIMALS

Oncogenic viruses, such as the simian virus 40 (SV_{40}), have been used to transfer foreign genes into animal cells. This double-stranded DNA virus originally isolated from monkey kidney cells is harmless to monkeys, but causes tumors in hamsters and mice. The infectious DNA of SV_{40} is capable of transforming animal cells in culture, so SV_{40} is an oncogenic virus.

SV_{40} cloning vectors have been developed that infect and transform animal cells, including human cell lines. The foreign DNA is integrated into the SV_{40} genome in such a fashion that transformants can be detected. With this vector, almost any gene can be transferred into animal tissue.

What will be the impact of genetic engineering on human health? One would hope that this powerful technology could be used to correct genetic diseases of humans, but there are few such diseases that lend themselves to manipulative genetic correction. Part of the problem is that defective genes are found in all tissues even though they are expressed only in certain tissues. The genetic engineer has the overwhelming task of inserting the corrected gene into all (or almost all) of the cells producing the defective gene product. One possible exception to this dilemma is treatment of certain types of gammaglobulinemia. The gammaglobulin diseases may respond to replacing bone marrow with "new" cells that reverse the defect in gammaglobulin production.

QUESTIONS OF RISK

Scientists began to struggle with the "question of risk" emanating from recombinant DNA work soon after the technology was discovered. What potential damage could these "new" organisms do to the human population or the environment and how could these risks be minimized? People were concerned about the spread of antibiotic resistance to pathogenic bacteria and about the cloning of oncogenic viruses in bacteria. A typical scenario went as follows:

> SV_{40} transforms animal cells in culture and makes them cancerous.
>
> SV_{40} can be cloned in *E. coli,* which is a normal inhabitant of the human intestine.
>
> Humans exposed to SV_{40} cloned in *E. coli* spread the infectious agent through their feces.
>
> An unpreventable cancer epidemic ensues.

If this scenario is even a remote possibility, should scientists be allowed to clone SV_{40} in *E. coli* or other bacteria that can colonize humans? For many years there was a moratorium on joining DNA of cancer-causing or other viruses to *E. coli* DNA. These concerns led to the establishment by the National Institutes of Health (NIH) of the Guidelines for Research Involving Recombinant DNA Molecules, which cover all recombinant DNA experiments in the United States and its territories. The director of NIH is charged with oversight of the Guidelines and is advised by the Recombinant DNA Advisory Committee (RAC) of the NIH, established in December 1979. The Guidelines prohibit certain types of experiments (Table 30.3) that could cause uncontrollable problems. Other recombinant DNA experiments, such as the field testing of genetically engineered bacteria for use in agriculture, are widely debated.

TABLE 30.3 Recombinant DNA experiments requiring federal review before initiation: 1984 Guidelines[a]

1. Deliberate formation of recombinant DNAs containing genes for the biosynthesis of toxic molecules lethal for vertebrates.
2. Deliberate transfer of a drug resistance trait to microorganisms that are not known to acquire it naturally.
3. Deliberate release of any organism containing recombinant DNA into the environment.
4. Deliberate transfer of recombinant DNA or DNA derived from recombinant DNA into human subjects.

[a]The Guidelines continue to be modified.

Although few biologists believe that recombinant DNA work is free from risk, the risks appear to be minimal as long as the Guidelines are followed. Containment facilities are required for high-risk experiments and special bacterial strains unable to grow outside the laboratory are widely used for molecular cloning.

Summary Outline

GENE CLONING

1. Gene cloning requires the isolation of one or more genes from foreign DNA, inserting them into a cloning vector, and then transferring this recombinant DNA into a cell that will make multiple copies of it.

CLONING VECTORS

1. The ideal plasmid cloning vector is small, possesses a single cleavage site for as many restriction endonucleases as possible, carries antibiotic resistance markers, and is under a relaxed control mechanism of replication.

2. The cloning plasmid is opened at a single restriction endonuclease site, preferably within an antibiotic resistance gene. Insertion of a foreign DNA fragment causes an insertion mutation of the antibiotic-resistance genes.

3. The nicks in recombinant DNA are made whole by the T_4 ligase before the plasmid is taken into a cell by artificial transformation.

4. Cells transformed with recombinant DNA are identified by their antibiotic sensitivity.
5. Lambda (λ) is a double-stranded linear DNA virus made into a cloning vector by deleting genes unimportant for its replication.
6. The right arm and the left arm of the λ genome are physically separated from "stuffer" DNA and then mixed with foreign DNA fragments to make recombinant λ genomes.
7. Recombinant λ genomes that fall within the head-full requirement of 36.8 to 51.5 kb are packaged *in vitro* to form virulent bacteriophages.
8. Clones of the λ are formed as plaques on lawns of suspectible bacteria. This system clones foreign genes 100 times more efficiently than artificial bacterial transformation.

CLONING STRATEGIES

1. The shotgun approach to gene cloning creates about one million clones, each containing a different fragment of the cell's genome. A genomic library is made by cloning millions of fragments of the cell's DNA in λ cloning vectors.
2. The alternative approach is to clone chemically synthesized genes or to clone cDNA produced from mRNA with reverse transcriptase. Synthesized genes are cloned in plasmids such as pBR322.

SCREENING FOR DESIRED CLONES

1. Hybridization probes and monoclonal antibodies are used in direct screening methods for detecting genes or gene products.
2. Isolated genes can be used as a hybridization probe. The probe is a radioactively labeled, single-stranded RNA or DNA that is complementary to the gene(s) being sought.

APPLICATIONS OF GENETIC ENGINEERING

1. The pharmaceutical industry will benefit from genetically engineered human peptide hormones, insulin, bovine growth hormone; vaccines against viruses, bacteria, and protozoa; and lymphokines.
2. The genes for the A and B polypeptide of insulin have been cloned and used to produce synthetic insulin.
3. Human growth hormone can be used to treat dwarfism, and bovine growth hormone to stimulate cows to produce more milk.
4. Interferons, interleukin-1, and interleukin-2 are lymphokines now being produced by genetically engineered cells in quantities suitable for clinical trials.
5. Genetically engineered vaccines for hepatitis B and foot-and-mouth disease are now available and a genetically modified viral vaccine against porcine diarrhea has been produced.
6. The tissue plasminogen activator that dissolves blood clots is being produced by genetic engineering technology.

CLONING ANIMAL AND PLANT GENES

1. The Ti plasmid of *A. tumefaciens* contains T-DNA that integrates into the genome of plant cells and can be used as a plant cloning vector.
2. Simian virus 40 has been modified to transfer cloned genes into the genome of the animal cells.

QUESTIONS OF RISK

1. National Institutes of Health (NIH) guidelines cover all recombinant DNA experiments in the United States and require approval for experiments that pose potential hazards.

Questions and Topics for Study and Discussion

QUESTIONS

1. Describe the enzymes used in recombinant DNA technology and explain their natural roles.
2. What are the principles behind the electrophoretic separation of DNA fragments?

3. Explain the phenomenon of insertion mutation. How is this process used in gene cloning?
4. Define or explain the following:

cDNA	*in vitro* packaging
genomic library	plasmid amplication
head-full requirement	restriction fragments
hybridization probe	T-DNA
kilobase pair	tissue plasminogen activator

5. What are the advantages of using the cloning vector pBR322 instead of a lambda bacteriophage to clone foreign DNA?
6. Describe the process of cloning the genetic information contained in a pure sample of messenger RNA.
7. Explain how DNA and RNA probes are used to detect the desired clones of bacteriophage that contain recombinant DNA.
8. What are the advantages of using molecular cloning to produce viral and protozoan vaccines?
9. Explain the dangers of two recombinant DNA experiments that are specifically prohibited by the 1982 Guidelines.

DISCUSSION TOPICS

1. Should your college or university place restrictions on recombinant DNA research? If your answer is yes, what restrictions do you advocate and how should they be enforced?
2. Cloning human genes will make available genetic probes capable of detecting genetic defects in human fetuses. What ethical problems are associated with the use of these probes?

Further Readings

BOOKS

Cherfas, J., *Manmade Life: An Overview of the Science, Technology, and Commerce of Genetic Engineering,* Pantheon Books, New York, 1983. A detailed history of genetic engineering with discussions of regulation and potential contributions to society.

Drlica, K., *Understanding DNA and Gene Cloning, A Guide for the Curious,* Wiley, New York, 1984. This well-illustrated introduction to gene cloning is written for the beginning science student.

Hanson, E. D., (ed.), *Recombinant DNA Research and the Human Perspective,* American Chemical Society, Washington, D.C., 1983. A collection of essays dealing with the human aspects of genetic engineering.

Koshland, D. E., *Biotechnology The New Frontier,* American Association for the Advancement of Science, 1986. A collection of papers on biotechnology published in *Science.*

Wade, N., *The Ultimate Experiment, Man-Made Evolution,* Walker, New York, 1977. This is an excellent account of the controversy over genetic engineering written at the time that the first regulations were being promulgated.

Watson, J. D., J. Tooze, and D. T. Kurtz, *Recombinant DNA: A Short Course,* Freeman, San Francisco, 1983. An informative, well-written, and well-illustrated introduction.

MANUALS

Perbal, B., *A Practical Guide to Molecular Cloning,* Wiley, New York, 1984. This guide introduces beginning students to the cloning techniques needed to clone any piece of DNA.

Rodriguez, R., and R. C. Tait, *Recombinant DNA Techniques, An Introduction,* Addison-Wesley, Reading, Mass., 1983. The authors provide excellent discussions of the background material necessary to understand the experimental designs of recombinant DNA technology.

ARTICLES AND REVIEWS

Abelson, P., "Biotechnology," *Science,* February 11, vol. 219, 1983. This entire issue of *Science* is devoted to biotechnology with important coverage of the impact of genetic engineering on the field.

Aharonowitz, Y., and G. Cohen, "The Microbiological Production of Pharmaceuticals," *Sci. Am.,* 245:141–152 (September 1981). The early applications of genetic engineering to the production of human hormones are presented.

Godson, G. N., "Molecular Approaches to Malaria Vaccines," *Sci. Am.,* 252:52–59 (May 1985). The story of how the circumsporozoite protein of *Plasmodium* was cloned and how it protects the parasite against the immune response.

Nester, E. W., and T. Kosuge, "Plasmids Specifying Plant Hyperplasias," *Ann. Rev. Microbiol.,* 35:531–65 (1981). The involvement of *A. tumefaciens* in crown gall disease is reviewed.

Pestka, S., "The Purification and Manufacture of Human Interferons," *Sci. Am.,* 249:36–43 (August 1983). This article describes the complex scientific steps needed to bring a genetically engineered product to the stage of clinical testing.

APPENDIX 1

Units of Measure and Mathematical Expressions

Length

1 meter = 39.3700 inches, 3.2808 feet
1 meter = 100 centimeters (cm)
1 centimeter = 10 millimeters (mm)
1 micrometer (μm) = 10^{-6} meter
1 nanometer (nm) = 10^{-9} meter
1 angstrom (Å) = 10^{-10} meter

Volume

1 liter = 1.0567 quarts
1 liter = 1000 milliliters (ml)
1 milliliter = 1 cm^3
1 microliter (μl) = 10^{-6} liter
1 gallon = 4 quarts = 3.7854 liter
1 quart = 0.9463 liter

Weight

1 ton (short) = 2000 pounds (lb)
1 pound = 16 ounces (avdp.) = 453.5923 grams (g)
1 ounce = 28.3495 grams

Temperature Conversion

°C = 5/9 (°F − 32)   °F = 9/5 (°C + 32)

LOGARITHMS

The large cell numbers encountered in microbial growth experiments can be mathematically handled with logarithms. Quantities in the millions, billions, or trillions are easily expressed as exponents of 10. For example one million is writen as 10^6, or ten to the sixth power. Said another way, ten multiplied by itself ten times equals one million. The mathematics of exponents follow set rules:

Division: 10^{12} divided by $10^6 = 10^{(12-6)}$ or 10^6
Multiplication: 10^3 times $10^5 = 10^{(3+5)}$ or 10^8

The rules of exponents are simplified by logarithms. A logarithm is the exponent to which the base (10) is raised in order to obtain a real number. The basic expression of a logarithmic function is: if $N = a^b$ then $\log_a N = b$. Other operations are

Exponents: $\log_{10} N^x = x \log_{10} N$
Subtractions: $\log_{10} 10^x - \log_{10} 10^y = x - y$, since $\log_{10} 10 = 1$
Multiplications: $\log_{10} (x \cdot y) = \log_{10} x + \log_{10} y$

The bacterial growth equation uses logarithms to express cell numbers:

$$g = \frac{0.301\ (t - t_0)}{\log_{10} B_t - \log_{10} B_0}$$

where g = Generation time
B_0 = Bacteria at time zero
B_t = Bacteria at time t
t_0 = Zero time
t = Any time after t_0
n = Number of generations

APPENDIX 2

Classification of Bacteria

The classification of bacteria changes as new data are compiled and new relationships between species are established. These changes are monitored by the Editorial Board and Trustees of Bergey's Manual Trust, who publish *Bergey's Manual of Systematic Bacteriology,* the authoritative reference work on bacterial classification. This manual is revised about every ten years to keep up with the changes in bacterial classification. The table of contents of the most recent four-volume edition is outlined below.

KINGDOM PROCARYOTAE

Divided into four divisions

DIVISION I. GRACILICUTES
Procaryotes with thin cell walls, implying a gram-negative type of cell wall.

DIVISION II. FIRMICUTES
Procaryotes with thick and strong skin, indicating a gram-positive type of cell wall.

DIVISION III. TENERICUTES
Procaryotes of a pliable and soft nature, indicating the lack of a rigid cell wall.

DIVISION IV. MENDOSICUTES
Procaryotes with faulty cell walls, suggesting the lack of conventional peptidoglycan.

Bergey's Manual of Systematic Bacteriology

VOLUME 1

Edited by Noel R. Krieg

SECTION 1
THE SPIROCHETES

Order I. *Spirochaetales*

Family I. *Spirochaetaceae*

Genus I.	*Spirochaeta*
Genus II.	*Cristispira*
Genus III.	*Treponema*★
Genus IV.	*Borrelia*★

Family II. *Leptospiraceae*

Genus I.	*Leptospira*★

SECTION 2
AEROBIC/MICROAEROPHILIC, MOTILE, HELICAL/VIBRIOID GRAM-NEGATIVE BACTERIA

Genus I.	*Aquaspirillum*
Genus II.	*Spirillum*
Genus III.	*Azospirillum*
Genus IV.	*Oceanospirillum*
Genus V.	*Campylobacter*★
Genus VI.	*Bdellovibrio*
Genus VII.	*Vampirovibrio*

SECTION 3
NONMOTILE (OR RARELY MOTILE), GRAM-NEGATIVE CURVED BACTERIA

Family I. *Spirosomaceae*
- Genus I. *Spriosoma*
- Genus II. *Runella*
- Genus III. *Flectobacillus*

Other genera:
- Genus *Microcyclus*
- Genus *Meniscus*
- Genus *Brachyarcus*
- Genus *Pelosigma*

SECTION 4
GRAM-NEGATIVE AEROBIC RODS AND COCCI

Family I. *Pseudomonadaceae*
- Genus I. *Pseudomonas*★
- Genus II. *Xanthomonas*
- Genus III. *Frateuria*
- Genus IV. *Zoogloea*

Family II. *Azotobacteraceae*
- Genus I. *Azotobacter*
- Genus II. *Azomonas*

Family III. *Rhizobiaceae*
- Genus I. *Rhizobium*
- Genus II. *Bradyrhizobium*
- Genus III. *Agrobacterium*
- Genus IV. *Phyllobacterium*

Family IV. *Methylococcaceae*
- Genus I. *Methylococcus*
- Genus II. *Methylomonas*

Family V. *Halobacteriaceae*
- Genus I. *Halobacterium*★
- Genus II. *Halococcus*★

Family VI. *Acetobacteraceae*
- Genus I. *Acetobacter*
- Genus II. *Gluconobacter*

Family VII. *Legionellaceae*
- Genus I. *Legionella*

Family VIII *Neisseriaceae*
- Genus I. *Neisseria*★
- Genus II. *Moraxella*
- Genus III. *Acinetobacter*
- Genus IV. *Kingella*

Other genera:
- Genus *Beijerinckia*
- Genus *Derxia*
- Genus *Xanthobacter*
- Genus *Thermus*
- Genus *Thermomicrobium*
- Genus *Halomonas*
- Genus *Alteromonas*
- Genus *Flavobacterium*
- Genus *Alcaligenes*
- Genus *Serpens*
- Genus *Janthinobacterium*
- Genus *Brucella*★
- Genus *Bordetella*★
- Genus *Francisella*★
- Genus *Paracoccus*
- Genus *Lampropedia*

SECTION 5
FACULTATIVELY ANAEROBIC GRAM-NEGATIVE RODS

Family I. *Enterobacteriaceae*
- Genus I. *Escherichia*★
- Genus II. *Shigella*★
- Genus III. *Salmonella*★
- Genus IV. *Citrobacter*
- Genus V. *Klebsiella*★
- Genus VI. *Enterobacter*★
- Genus VII. *Erwinia*
- Genus VIII. *Serratia*★
- Genus IX. *Hafnia*
- Genus X. *Edwardsiella*
- Genus XI. *Proteus*★
- Genus XII. *Providencia*★
- Genus XIII. *Morganella*
- Genus XIV. *Yersinia*★

Other genera of the family *Enterobacteriaceae:*
- Genus *Obesumbacterium*
- Genus *Xenorhabdus*
- Genus *Kluyvera*
- Genus *Rahnella*
- Genus *Cedecea*
- Genus *Tatumella*

Family II. *Vibrionaceae*
- Genus I. *Vibrio*★
- Genus II. *Photobacterium*
- Genus III. *Aeromonas*
- Genus IV. *Plesiomonas*

Family III. *Pasteurellaceae*
- Genus I. *Pasteurella*★
- Genus II. *Haemophilus*★
- Genus III. *Actinobacillus*

Other genera:

Genus	*Zymomonas*
Genus	*Chromobacterium*
Genus	*Cardiobacterium*
Genus	*Calymmatobacterium*
Genus	*Gardnerella*★
Genus	*Eikenella*
Genus	*Streptobacillus*

SECTION 6
ANAEROBIC GRAM-NEGATIVE, STRAIGHT, CURVED, AND HELICAL RODS

Family I. *Bacteroidaceae*

Genus I.	*Bacteroides*★
Genus II.	*Fusobacterium*★
Genus III.	*Leptotrichia*
Genus IV.	*Butyrivibrio*★★
Genus V.	*Succinimonas*
Genus VI.	*Succinivibrio*
Genus VII.	*Anaerobiospirillum*
Genus VIII.	*Wolinella*
Genus IX.	*Selenomonas*
Genus X.	*Anaerovibrio*
Genus XI.	*Pectinatus*
Genus XII.	*Acetivibrio*
Genus XIII.	*Lachnospira*★★

SECTION 7
DISSIMILATORY SULFATE- OR SULFUR-REDUCING BACTERIA

Genus	*Desulfuromonas*
Genus	*Desulfovibrio*
Genus	*Desulfomonas*
Genus	*Desulfococcus*
Genus	*Desulfobacter*
Genus	*Desulfobulbus*
Genus	*Desulfosarcina*

SECTION 8
ANAEROBIC GRAM-NEGATIVE COCCI

Family I. *Veillonellaceae*

Genus I.	*Veillonella*
Genus II.	*Acidaminococcus*
Genus III.	*Megasphaera*

SECTION 9
THE RICKETTSIAS AND CHLAMYDIAS

Order I. *Rickettsiales*

Family I. *Rickettsiaceae*

Tribe I. *Rickettsieae*

Genus I.	*Rickettsia*★
Genus II.	*Rochalimaea*
Genus III.	*Coxiella*★

Tribe II. *Ehrlichieae*

Genus IV.	*Ehrlichia*
Genus V.	*Cowdria*
Genus VI.	*Neorickettsia*

Tribe III. *Wolbachieae*

Genus VII.	*Wolbachia*
Genus VIII.	*Rickettsiella*

Family II. *Bartonellaceae*

Genus I.	*Bartonella*
Genus II.	*Grahamella*

Family III. *Anaplasmataceae*

Genus I.	*Anaplasma*
Genus II.	*Aegyptianella*
Genus III.	*Haemobartonella*
Genus IV.	*Eperythrozoon*

Order II. *Chlamydiales*

Family I. *Chlamydiaceae*

Genus I.	*Chlamydia*★

SECTION 10
THE MYCOPLASMAS

Dividion *Tenericutes*

Class I. *Mollicutes*

Order I. *Mycoplasmatales*

Family I. *Mycoplasmataceae*

Genus I.	*Mycoplasma*★
Genus II.	*Ureaplasma*★

Family II. *Acholeplasmataceae*

Genus I.	*Acholeplasma*

Family III. *Spiroplasmataceae*

Genus I.	*Spiroplasma*

Other genera:

Genus	*Anaeroplasma*
Genus	*Thermoplasma*

SECTION 11
ENDOSYMBIONTS

A. Endosymbionts of Protozoa

Genus I.	*Holospora*
Genus II.	*Caedibacter*
Genus III.	*Pseudocaedibacter*
Genus IV.	*Lyticum*
Genus V.	*Tectibacter*

B. Endosymbionts of Insects

Genus	*Blattabacterium*

C. Endosymbionts of Fungi and Invertebrates other than Arthropods

Bergey's Manual of Systematic Bacteriology

VOLUME 2
Edited by Peter H. A. Sneath

SECTION 12
GRAM-POSITIVE COCCI

Family I. *Micrococcaceae*

Genus I.	*Micrococcus*
Genus II.	*Stomatococcus*
Genus II.	*Planococcus*
Genus IV.	*Staphylococcus*★

Family II. *Deinococcaceae*

Genus I.	*Deinococcus*

Other genera:

Genus	*Streptococcus*
Genus	*Leuconostoc*
Genus	*Pediococcus*
Genus	*Aerococcus*
Genus	*Gemella*
Genus	*Peptococcus*
Genus	*Peptostreptococcus*
Genus	*Ruminococcus*
Genus	*Coprococcus*
Genus	*Sarcina*

SECTION 13
ENDOSPORE-FORMING GRAM-POSITIVE RODS AND COCCI

Genus	*Bacillus*★
Genus	*Sporolactobacillus*
Genus	*Clostridium*★
Genus	*Desulfotomaculum* (gram negative)
Genus	*Sporosarcina*
Genus	*Oscillospira* (gram negative)

SECTION 14
REGULAR, NONSPORING, GRAM-POSITIVE RODS

Genus	*Lactobacillus*
Genus	*Listeria*
Genus	*Erysipelothrix*
Genus	*Brochothrix*
Genus	*Renibacterium*
Genus	*Kurthia*
Genus	*Caryophanon*

SECTION 15
IRREGULAR, NONSPORING, GRAM-POSITIVE RODS

Genus	*Corynebacterium*
	Plant pathogenic species of *Corynebacterium*
Genus	*Gardnerella*★★
Genus	*Arcanobacterium*
Genus	*Arthrobacter*
Genus	*Brevibacterium*
Genus	*Curtobacterium*
Genus	*Caseobacter*
Genus	*Microbacterium*
Genus	*Aureobacterium*
Genus	*Cellulomonas*
Genus	*Agromyces*
Genus	*Arachnia*
Genus	*Rothia*
Genus	*Propionibacterium*
Genus	*Eubacterium*
Genus	*Acetobacterium*
Genus	*Lachnospira*★★
Genus	*Butyrivibrio*★★
Genus	*Thermoanaerobacter*
Genus	*Actinomyces*
Genus	*Bifidobacterium*

SECTION 16
THE MYCOBACTERIA

Family I. *Mycobacteriaceae*

Genus	*Mycobacterium*★

SECTION 17
NOCARDIOFORMS

Genus	*Nocardia*★
Genus	*Rhodococcus*
Genus	*Norcardioides*
Genus	*Pseudonocardia*
Genus	*Oerskovia*
Genus	*Saccharopolyspora*
Genus	*Micropolyspora*
Genus	*Promicromonospora*
Genus	*Intrasporangium*

Bergey's Manual of Systematic Bacteriology

VOLUME 3
Edited by James T. Staley

SECTION 18
GLIDING, NONFRUITING BACTERIA

Order I. *Cytophagales*

Family I. *Cytophagaceae*

Genus I.	*Cytophaga*
Genus II.	*Sporocytophaga*
Genus III.	*Capnocytophaga*
Genus IV.	*Flexithrix*
Genus V.	*Flexibacter*
Genus VI.	*Microscilla*
Genus VIII.	*Saprospira*
Genus VIII.	*Herpetosiphon*

Order II. *Lysobacterales*

Family I. *Lysobacteraceae*

Genus I.	*Lysobacter*

Order III. *Beggiatoales*

Family I. *Beggiatoaceae*

Genus I.	*Beggiatoa*
Genus II.	*Thioploca*
Genus III.	*Thiospirillopsis*
Genus IV.	*Thiothrix*
Genus V.	*Achromatium*

Family II. *Simonsiellaceae*

Genus I.	*Simonsiella*
Genus II.	*Alysiella*

Family III. *Leucotrichaceae*

Genus I.	*Leucothrix*

Families and genera *incertae sedis*

Genus	*Toxothrix*
Genus	*Vitreoscilla*
Genus	*Chitinophagen*
Genus	*Desulfonema*

Family IV. *Pelonemataceae*

Genus	*Pelonema*
Genus	*Achroonema*
Genus	*Peloploca*
Genus	*Desmanthus*

SECTION 19
ANOXYGENIC PHOTOTROPHIC BACTERIA

PURPLE BACTERIA

Family I. *Chromatiaceae*

(purple sulfur)

Genus I.	*Chromatium*
Genus II.	*Thiocystis*
Genus III.	*Thiospirillum*
Genus IV.	*Thiocapsa*
Genus V.	*Amoebobacter*
Genus VI.	*Lamprobacter*
Genus VII.	*Lamprocystis*
Genus VIII.	*Thiodictyon*
Genus IX.	*Thiopedia*

Family II. *Ectothiorhodospiraceae*

Genus I.	*Ectothiorhodospira*

(purple nonsulfur)

Genus	*Rhodospirillum*
Genus	*Rhodopseudomonas*
Genus	*Rhodobacter*
Genus	*Rhodomicrobium*
Genus	*Rhodopila*
Genus	*Rhodocyclus*

GREEN BACTERIA

(green sulfur)

Genus	*Chlorobium*
Genus	*Prosthecochloris*
Genus	*Anacalochloris*
Genus	*Pelodictyon*
Genus	*Chloroherpeton*

(multicellular filamentous green)

Genus	*Chloroflexus*
Genus	*Heliothrix*
Genus	*Oscillochloris*
Genus	*Chloronema*

SECTION 20
BUDDING AND/OR APPENDAGED BACTERIA

PROSTHECATE BACTERIA

(budding bacteria)

Genus	*Hyphomicrobium*
Genus	*Hyphomonas*
Genus	*Pedomicrobium*
Genus	*"Filomicrobium"*
Genus	*"Dicotomicrobium"*

Genus	*"Tetramicrobium"*
Genus	*Stella*
Genus	*Ancalomicrobium*
Genus	*Prosthecomicrobium*

(nonbudding bacteria)

Genus	*Caulobacter*
Genus	*Asticcacaulis*
Genus	*Prosthecobacter*
Genus	*Thiodendron*

NONPROSTHECATE BACTERIA

(budding bacteria)

Genus	*Planctomyces*
Genus	*Pasteuria*
Genus	*Blastobacter*
Genus	*Angulomicrobium*
Genus	*Gemmiger*
Genus	*Ensifer*
Genus	*Isophaera*

(nonbudding stalked bacteria)

Genus	*Gallionella*
Genus	*Nevskia*

Morphologically unusual budding bacteria involved in iron and manganese deposition

Genus	*Seliberia*
Genus	*Metallogenium*
Genus	*Caulococcus*
Genus	*Kuznezovia*

SECTION 21
ARCHAEBACTERIA

METHANOGENIC BACTERIA

Genus	*Methanobacterium*
Genus	*Methanobrevibacter*
Genus	*Methanococcus*
Genus	*Methanomicrobium*
Genus	*Methanospirillum*
Genus	*Methanosarcina*
Genus	*Methanococcoides*
Genus	*Methanothermus*
Genus	*Methanolobus*
Genus	*Methanoplanus*
Genus	*Methanogenium*
Genus	*Methanotrix*

EXTREME HALOPHILIC BACTERIA

Genus	*Halobacterium***
Genus	*Halococcus***

EXTREME THERMOPHILIC BACTERIA

Genus	*Thermoplasma***
Genus	*Sulfolobus*
Genus	*Thermoproteus*
Genus	*Thermofilum*
Genus	*Thermococcus*
Genus	*Desulfurococcus*
Genus	*Thermodiscus*
Genus	*Pyrodictium*

SECTION 22
SHEATHED BACTERIA

Genus	*Sphaerotilus*
Genus	*Leptothrix*
Genus	*Haliscominobacter*
Genus	*Lieskeella*
Genus	*Phragmidiothrix*
Genus	*Crenothrix*
Genus	*Clonothrix*

SECTION 23
GLIDING, FRUITING BACTERIA

Order I. *Myxobacterales*

Family I. *Myxococcaceae*

Genus	*Myxococcus*

Family II. *Archangiaceae*

Genus	*Archangium*

Family III. *Cystobacteraceae*

Genus I.	*Cystobacter*
Genus II.	*Melittangium*
Genus III.	*Stigmatella*

Family IV. *Polyangiaceae*

Genus I.	*Polyangium*
Genus II.	*Nannocystis*
Genus III.	*Chondromyces*

SECTION 24
CHEMOLITHOTROPHIC BACTERIA

NITRIFIERS

Family I. *Nitrobacteraceae*

Genus I.	*Nitrobacter*
Genus II.	*Nitrospina*
Genus III.	*Nitrococcus*
Genus IV.	*Nitrosomonas*
Genus V.	*Nitrosospira*
Genus VI.	*Nitrosococcus*
Genus VII.	*Nitrosolobus*

SULFUR OXIDIZERS

Genus	*Thiobacillus*
Genus	*Thiomicrospira*
Genus	*Thiobacterium*
Genus	*Thiospira*
Genus	*Macromonas*

OBLIGATE HYDROGEN OXIDIZERS

Genus	*Hydrogenbacter*

METAL OXIDIZERS AND DEPOSITORS

Family I. *Siderocapsaceae*

Genus I.	*Siderocapsa*
Genus II.	*Naumanniella*
Genus III.	*Ochrobium*
Genus IV.	*Siderococcus*

MAGNETOTACTIC BACTERIA

SECTION 25
CYANOBACTERIA AND OTHERS

Order I. *Prochlorales*

Family I. *Prochloraceae*

Genus	*Prochloron*

Bergey's Manual of Systematic Bacteriology

VOLUME 4

Edited by Stanley T. Williams

SECTION 26
ACTINOMYCETES THAT DIVIDE IN MORE THAN ONE PLANE

Genus	*Geodermatophilus*
Genus	*Dermatophilus*
Genus	*Frankia*
Genus	*Tonsilophilus*

SECTION 27
SPORANGIATE *ACTINOMYCETES*

Genus	*Actinoplanes*
Genus	*Streptosporangium*
Genus	*Ampullariella*
Genus	*Spirillospora*
Genus	*Pilimelia*
Genus	*Dactylosporangium*
Genus	*Planomonospora*
Genus	*Planobispora*

SECTION 28
STREPTOMYCETES AND THEIR ALLIES

Genus	*Streptomyces*
Genus	*Streptoverticillium*
Genus	*Actinopycnidium*
Genus	*Actinosporangium*
Genus	*Chainia*
Genus	*Elytrosporangium*
Genus	*Microellobosporia*

SECTION 29
OTHER CONIDIATE GENERA

Genus	*Actinopolyspora*
Genus	*Actinosynnema*
Genus	*Kineospora*
Genus	*Kitasatosporia*
Genus	*Microbispora*
Genus	*Micromonospora*
Genus	*Microtetrospora*
Genus	*Saccharomonospora*
Genus	*Sporichthya*
Genus	*Streptoalloteichus*
Genus	*Thermomonospora*
Genus	*Actinomadura*
Genus	*Nocardiopsis*
Genus	*Excellospora*
Genus	*Thermoactinomyces*

*Pathogenic species of the genera are discussed in Part IV of the text.

**Genera that have been placed in more than one volume: *Gardnerella, Lachnospira,* and *Butyrivibrio* have thin, gram-positive walls but stain as gram-negatives. *Halobacterium, Halococcus,* and *Thermoplasma* are Archaebacteria that are also included among the GRAM-NEGATIVE RODS AND COCCI or THE MYCOPLASMAS.

APPENDIX 3

Common Word Roots Used in Scientific Terminology

Prefixes, suffixes, Latin roots, and Greek roots are used to construct many scientific terms. Definitions of these terms make more sense and are easier to remember if one understands their origin. In addition, certain scientific terms retain the singular and plural endings of their Latin roots. Latin nouns have singular and plural endings that depend on gender. The chart below lists some examples.

Gender
Feminine
Singular -a (fimbria, hypha)
Plural -ae (fimbriae, hyphae)
Masculine
Singular -us (coccus, pilus)
Plural -i (cocci, pili)
Neuter
Singular -um (bacterium, medium)
Plural -a (bacteria, media)

a-, an- without, not. Examples: *a*biotic, in the absence of life; *an*aerobic, in the absence of air.

actin- pertaining to radial structure, ray, radius. Example: *Actin*omycetes, bacteria that form star-shaped colonies.

aer- air. Examples: *aer*obe, organism that lives in the presence of air; *aer*ation, to supply with air.

Common Word Roots Used in Scientific Terminology

albo- white. Example: *Micrococcus* **alb***us,* grow as white colonies.

amphi- both sides, doubly. Example: amphitrichous, tufts of flagella at both ends of a cell.

ant-, anti- against, counter. Examples: *anti*body, protein that inactivates its antigen; *anti*biotics, work against living microbes.

arch- ancient. Example: *Arch*aebacteria, first procaryotes derived from a unique lineage.

auto- self. Examples: *auto*troph, grows on inorganic media—self-feeder.

bio- life. Example: micro*bio*logy, the study of small living organisms.

blast- bud, child. Example: *blast*ospore, spores formed by budding.

bovi- ox or cattle. Example: *Mycobacterium* **bovi***s,* a bacterium isolated from cattle.

brevi- short. Example: *Bacillus* **brev***is,* a short rod-shaped bacterium.

caryo- nut, kernal. Example: eu*cary*ote, a cell with a true nucleus.

caul- stem or stalk. Example: **Caul***obacter,* appendaged or stalked bacteria.

cen- shared. Example: *ceno*cytic, many nuclei together in same compartment (cell).

chlor- green. Examples: *chlor*ophyll, green light-absorbing pigment found in chloroplasts; **Chlor***obium,* green sulfur, photosynthetic bacterium.

chro- color. Examples: *chro*matin, histone DNA complex in eucaryotes that stains easily; metachromatic granules, colored cytoplasmic structures.

chryso- of a golden color. Example: *Chryso*phyta, golden algae.

-cid killing. Example: bacteri*cid*al agent, a substance that kills bacteria.

co-, with, together, joint. Example: *co*enzyme, acts with an enzyme.

cocc- seed, pill. Example: gono*cocc*us, spherical bacterium.

coni- dust. Example: *con*idia, asexual spores formed at the end of an aerial hyphae.

cyan- bluish color. Example: *cyano*bacteria, blue–green oxygenic, photosynthetic bacteria.

cyst- bladder. Example: *cyst*itis, inflammation of the urinary bladder.

cyt- cell. Example: *cyt*ology, the study of cells.

de- down from, remove. Example: *de*hydrogenase, enzyme that removes hydrogen atoms.

di-, diplo- two, double. Examples: *di*morphism, two forms of a fungi; *di*ploid, two chromosome copies in a cell.

dys- bad, defective, faulty, painful. Examples: *dys*entery, disease of the large intestine.

ec-, ex- out of. Example: *ex*crete, to remove materials from the body.

en-, em- in, on. Examples: *en*demic, disease outbreak in local environment; *en*zootic, disease occurring in animals.

end- within, inside of. Examples: *endo*toxin, structural component of cell; *endo*spore, spore formed within a cell.

enter- intestine. Example: *enter*ics, bacteria found in the intestine; gastro*enter*itis, inflammation of stomach and intestine.

epi- upon, after. Example: *epi*dermis, the outer covering of the skin.

erythr- red. Example: *erythr*ocyte, red blood cell.

eu- good, normal. Example: *eu*caryote, cell with a normal nucleus.

exo- outside of, external. Example: *exo*toxin, protein poison found outside of the producing cell.

extra- outside of, beyond. Example: *extra*cellular, outside the cells of an organism.

fil- thread. Example: *fili*form, thread-shaped.

-fy to become. Example: clari*fy*, to make clear.

galact- milk, milky (G *galaktos* milk). Example: *galact*ose, monosaccharide from milk sugar.

gastr- stomach. Example: *gastro*enteritis, inflammation of the stomach and intestine.

-gen that which produces. Example: path*ogen,* microbe that produces a disease state.

-genesis development. Example: patho*genesis,* development of a disease.

-gony production of. Example: schizo*gony,* an asexual division of *Plasmodium* that produces merozoites.

haem-, hem- blood. Example: *hem*orrhage, escape of blood from vessels; *hem*olysis, escape of hemoglobin from cell.

halo- salt. Example: *hal*ophile, an organism that requires high concentrations of salt.

hepat- liver (G *hepatos* liver). Example: hepatitis, inflammation of the liver.

hetero- different, other. Example: *hetero*troph, grows on organic carbon sources—not self-sustaining.

homo- same, like. Example: *homo*zygous, possessing identical alleles at a given locus or loci.

hydr- water. Examples: *hydro*carbon, organic molecules containing carbon and water (CH_2O); *hydro*phobia, fear of water.

hyper- above, excess. Example: *hyper*sensitivity, oversensitive to an antigen or allergen.

hypo- under, below, deficient, Example: *hypo*limnion, cold water beneath the thermocline.

inter- among, between. Example: *inter*stitial, between the tissues.

intra- within, inside of. Example: *intra*cellular, inside the cell.

is- equal, same. Example: *is*otope, one of two or more varieties of an element.

-itis inflammation of. Examples: encephal*itis,* inflammation of the brain; mas*titis,* inflammation of the female breast.

kerat- horn. Example: *kerat*in, the horny substance of skin and nails.

leuk- white. Example: leukocyte, white blood cell.

lip- fat, lipid. Example: *lip*opolysaccharide, carbohydrate combined with a fat.

-logy the study of. Example: sero*logy,* the study of blood sera.

lopho- tuft. Example: *lopho*trichous, having a group of flagella on one side of the cell.

lute- yellow. Example: *Micrococcus luteus,* a bacterium that forms yellow colonies.

ly- loose, dissolve. Examples: *ly*sis, dissolution of cells; hydro*ly*sis, chemical decomposition upon the addition of water.

macr- long, large. Example: *macro*molecules, large molecules.

meso- situated in the middle. Example: *meso*phile, an organism whose optimum temperature is between 20° and 50°C.

meta- after, changed, altered. Example: *metabolism*, all the chemical changes taking place in an organism.

micr- small, minute. Examples: *micro*biology, study of small organisms; microtubules, small hollow proteinaceous filaments.

mon- single, one. Examples: *mono*trichous, having one flagellum; *mon*oxide, having a single oxygen atom.

morph- form, shape. Example: *morph*ology, the study of form and structure.

multi- much, many. Example: *multi*cellular, having many cells.

mut- to change. Example: *mut*agen, an agent that causes changes in DNA.

myc-, myces- fungus. Examples: *my*coses, disease caused by a fungus; *Streptomyces,* a bacterium that resembles a fungus.

myx- slime, mucus. Example: Ortho*myxo*viruses, viruses (influenza) that stimulate nasal mucus secretion.

necr- corpse, the dead. Example: *necr*osis, cell or tissue death.

oculo- eye. Example: bin*ocul*ar, two eye pieces.

-oid form, resembling. Example: muc*oid,* resembling mucus.

-onc bulk, mass. Example: *onc*ogenic, capable of forming a tumor.

-osis, -sis condition of. Example: symbio*sis,* the condition of living together.

pan- all, universal. Example: pandemic, a disease that affects more than one continent.

peri- around, encircling. Example: *peri*plasmic space, area around the outside of certain bacteria.

phaeo- brown. Example: *Phaeo*phyta, brown algae.

phag- eat, consume. Examples: bacterio*phag*es, viruses that eat bacteria; *phag*ocyte, a cell that engulfs particles and/or entire cells.

-phil loving, preferring. Example: baro*phil*e, an organism that prefers high pressures; eosino*phil,* white blood cell that binds eosin.

-phor bears, produces. Example: conidio*phor*e, a hyphal structure that bears conidia.

-phyll leaf. Example: chloro*phyll,* color in leaf.

pil- a hair. Example: pili, hair-like projections from a cell.

plast- organized part. Example: chloroplast, the cytoplasmic site of photosynthesis in eucaryotes.

pod- foot. Example: pseudo*pod*ium, foot-like structure of an ameba.

poly- much, many. Example: *poly*ribosome, more than one ribosome bound to a mRNA.

post- after, behind. Example: postnatal, occurring after birth.

pre-, pro- before, ahead of. Examples: *pro*caryote, cell type presumed to exist before the nucleus developed; *pre*gnant, before birth.

pseudo- false. Example: *pseudo*podium, not a true foot.

py- pus. Example: *py*ogenic, pus-forming.

pyr- fire, heat. Example: *pyr*ogen, fever-causing substance, such as endotoxin.

rhin- nose. Example: *Rhin*oviruses, RNA viruses that cause inflammation of the nasal membranes.

rhizo- root. Examples: *Rhizobium,* bacteria that grow in plant roots; rhizosphere, environment around the roots of plants.

rhodo- red. Example: *Rhodopseudomonas,* red-pigmented rod-shaped photosynthetic bacterium.

saccharo- sugar. Example: *Saccharomyces,* a sugar-fermenting yeast.

sapr- rotten, decomposing. Example: *saprophyte*, an organism that lives on decomposing living matter.

sarco- flesh. Example: *sarco*ma, a tumor of muscle or connective tissue.

scop- look at. Example: micro*scope*, an instrument used to magnify small specimens.

semi- not full, half. Example: *semi*conservative replication, the single strand of DNA (template) is retained in the double-stranded DNA copy.

sept- wall off, fence. Example: *sept*um, a cross wall in a fungal hypha.

spiro- spiral, coiled. Example: *spiro*chete, helical bacterium motile by periplasmic flagella.

spor- seed, spore. Example: *spor*ulation, formation of resting form of cell.

staphyl- bunches of grapes. Example: *Staphylococcus,* spherical bacteria that grow in cell clusters.

-stasis arrest, stoppage. Example: bacterio*stasis,* cessation of bacterial growth.

strep- twisted. Example: **Strep***tococcus,* twisted chains of cocci.

sub- below, under. Example: *sub*stratum, a lower layer of material.

super- above, beyond. Example: *super*natant, liquid layer floating above a centrifuge pellet.

sym-, syn- together, with. Example: *syn*thesis, to bring together from component parts; *sym*biosis, living together.

tax- order. Example: *tax*onomy, the orderly categorizing of organisms.

therm- heat. Example: *therm*ophile, an organism that requires a high temperature to grow.

thi- sulfur. Example: *thi*osulfate, a sulfur compound.

tox- poison. Example: *tox*igenic, characterized by poison production.

trans- across, through. Example: *trans*ovarian, transmission through infected insect ova.

tri- three. Example: *tri*angle, three-sided figure.

trich- hair. Examples: *trich*inosis, infection by a hair-like nematode; mono*trich*ous, hair-like projection from cell.

-troph food, nourishment. Example: photo*troph,* makes its own food using light as an energy source.

un- one. Example: *un*ivalent, as opposed to bivalent or multivalent.

undul- wavy. Example: *undul*ant fever, rising and falling fever characteristic of brucellosis.

vacc- cow. Example: *vacc*ination, immunization against smallpox using cowpox virus derived from cow.

vacu- empty. Example: *vacu*oles, intracellular spaces containing fluids or gases.

vitr- glass. Example: *vitr*eous humor, transparent (glassy) tissue of eyeball.

xantho- yellow. Example: *Xanthomonas,* produces yellow colonies.

zoo- animal. Example: en*zoo*tic, affecting animals, said of diseases.

zyg- yoke, joining. Example: *zyg*ote, product of sexual reproduction, resulting from the fusion of two cells.

-zym ferment. Example: en*zym*e, a protein catalyst, some of which are involved in fermentations.

GLOSSARY

Abiogenesis *See* spontaneous generation.

Acid-Fast A differential stain in which a basic dye, such as carbolfuchsin, is retained when the cells are washed with acidified alcohol.

Acids Substances that increase the concentration of hydronium ions (H_3O^+) in the liquid to which they are added.

Acne An eruptive skin disease occurring mainly on the face, chest, and back and characterized by the development of pimples.

Acquired Immunity Resistance to a specific infectious agent developed by an individual in response to a disease (natural) or following an injection of a specific antigen (artificial).

Actinomycin D An antibiotic produced by a species of *Streptomyces* that inhibits transcription in eucaryotic cells.

Active Transport The energy-dependent movement of molecules across a semipermeable membrane mediated by a carrier protein. The molecules are moved from a location of low concentration to one of high concentration against a diffusion gradient.

Aerobic Requiring free oxygen for respiration.

Aerobic Respiration The process of transforming the chemical energy in organic compounds into ATP that employs the pathways of glycolysis, the TCA cycle, and the electron-transport chain and uses free oxygen as the final electron acceptor.

Agar The polysaccharide extracted from red algal seaweeds that is used to solidify microbiological media.

AIDS (acquired immune deficiency syndrome) A pandemic immunosuppressive viral disease that predisposes patients to life-threatening infections by opportunistic organisms and to certain cancers.

Algae A heterogeneous group of unicellular and multicellular, eucaryotic, oxygenic photosynthetic organisms that lack vascular tissue.

Allografts Tissue transplants between two genetically dissimilar individuals of the same species.

Allosteric Enzyme An enzyme possessing a specific regulatory site that is physically and functionally different from its catalytic site.

Alopecia Hair loss; can be caused by fungal infections.

Amino Acid An organic compound containing a carboxyl (—COOH) and an amino (—NH_2) functional group.

Amphitrichous Flagella Attached at both ends.

Anaerobe An organism that grows in the absence of oxygen.

Anaerobic Respiration An energy-yielding process utilizing an electron-transport system in which an inorganic compound other than oxygen (nitrate, nitrite, sulfate, others) serves as the terminal electron acceptor.

Analogs Compounds that have similar chemical structures but different biochemical functions.

Anamnestic Response The accelerated formation of antibody on the second and subsequent exposures to an antigen. *Also* called the secondary or memory response.

Anaphylactic Shock The most serious type of immediate hypersensitivity caused by the direct injection of an allergen into a sensitized individual. The individual experiences difficulty in breathing, becomes cyanotic, and may die.

Anaphylatoxins Vasoactive polypeptides derived from the cleavage of the C3 and C5 components of complement.

Angstrom (Å) One ten trillionth of a meter or 10^{-10} meter.

Anion An element able to accept one or more electrons into its outer shell to become a negatively charged ion.

Annealing Joining by hydrogen bonding of two single-stranded complementary nucleic acid molecules.

Antibiotic Any microbial product which, in low concentrations, is capable of inhibiting or killing susceptible microorganisms.

Antibody A protein produced by the body in response to a foreign substance (antigen) that will specifically react with that substance in a demonstrable way.

Anticodon The sequence of three nucleotide bases in tRNA; it is complementary to a sequence of bases (codon) on the mRNA.

Antigen A foreign substance, usually a large protein or polysaccharide, that stimulates a cell to synthesize antibodies that can specifically react with that antigen.

Antigenic Drift The slow, sequential change in the antigenic nature of one or more viral surface proteins.

Antigenic Shift A major change in the antigenic nature of two or more viral surface proteins to create a serologically distinct virus.

Antiseptics Chemicals used to inhibit the growth of, or kill microorganisms on human skin, mucous membranes, or other living tissue.

Aplastic Anemia The inability to form a normal complement of white and red blood cells.

Arachnida The class of animals that includes mites and ticks.

Arthrospores Asexual fungal spores produced within a hypha separated by a cross wall (septum).

Artificial Transformation Uptake of naked DNA replicons by a cell that lacks a natural process of transformation. Used to clone plasmid genes in *E. coli*.

Ascospores Fungal spores produced by sexual reproduction in the class *Ascomycetes*.

Aseptic Absence of contaminating organisms or infectious agents.

Aseptic Meningitis An inflammation of the meninges caused by a noncellular infectious agent.

Asexual Reproduction Production of progeny without the joining of two nuclei.

Asters The array of fibers radiating from the centrioles in all directions that appear to merge with the spindle fibers.

Asthma A disease resulting from hypersensitivity to inhaled allergens. Symptoms include wheezing and difficulty in breathing.

Atomic Number The number of protons in the nucleus of an element.

Atomic Weight The number of protons plus the number of neutrons in the nucleus of an element.

Atopy The hypersensitivity reactions of hay fever, asthma, and eczema.

Autotrophy The ability of an organism to grow on a completely inorganic medium with carbon dioxide as a carbon source. *See* also Chemolithotroph and Photolithotroph.

Auxotrophs Mutants requiring nutritional supplements in their growth medium.

Axenic Culture A culture that contans only one species of organism.

Axoneme The core structure of cilia and flagella that is composed of a group of microtubules present in the 9 pairs plus 2 arrangement.

Bacteremia The presence of bacteria, that are not multiplying, in the blood.

Bactericidal Able to kill bacterial cells.

Bacteriocins Bacterial substances that kill or inhibit the growth of closely related bacteria.

Bacteriolytic Able to kill bacteria by lysis.

Bacteriophage (phage) A virus that infects a bacterium.

Bacteriostatic Able to prevent bacterial growth.

Bacteriurea An infection of the urinary tract diagnosed by the presence of 100,000 or more bacteria per milliliter of urine.

Balanced Growth Condition in which all the components of the cells in a culture are changing at a constant rate.

Barm The yeast residue from beer making; used for centuries to make leavened bread.

Barophiles Organisms that preferentially grow at high hydrostatic pressures; for example marine bacteria from the deep ocean.

Base A substance that releases hydroxyl (OH^-) ions upon dissociation. The pH of a base in a pure neutral solvent is more than 7.

Basidiospores Fungal spores produced by sexual reproduction in the class *Basidiomycetes*.

Basophils Circulating white blood cells having an irregularly shaped nucleus and numerous large cytoplasmic granules. They participate in allergic reactions.

Bayer's Junctions Zones of adhesion between the outer membrane of the cell wall and the plasma membrane in gram-negative bacteria.

BCG Vaccine The bacille Calmette Guérin vaccine made from an attenuated strain of *Mycobacterium*.

Bilharzia A name for schistosomiasis derived from the name of the German doctor (Theodorsis M. Bilharz) who discovered the responsible blood fluke.

Binary Fission Division of a cell into two equal daughter cells.

Binomial System of Nomenclature The naming of organisms by the assignment of a genus and a species name.

Biochemical Oxygen Demand The requirement for molecular oxygen that accompanies the microbial oxidation of biodegradable substances in water.

Biota The sum total of the living organisms in a designated area.

Biotechnology The use of living organisms and their products in industrial processes.

Biovar (Biotype) Defines a specific trait of a species.

Bivalence The property of antibodies that allows them to combine with two separate, but identical, antigenic determinants.

Blastospores Asexual spores that are formed by budding.

Broad-Spectrum Antibiotic An antibiotic that is effective against both gram-positive and gram-negative bacteria.

Brownian Movement Random movement of microscopic particles caused by molecular collisions at the submicroscopic level.

Buboes Inflamed and enlarged lymph nodes, often resulting from an infectious disease such as plague.

Buffer A compound that chemically resists major changes in pH by neutralizing either acids or bases.

Burst Size The increase in the number of virions (infective centers) produced during a one-step growth cycle.

Calvin Cycle The metabolic pathway for the fixation of carbon dioxide that utilizes ribulose 1, 5-diphosphate, CO_2, 3ATP, and 2NADPH.

Capsid The protein coat or shell that either surrounds the viral nucleic acid or complexes with it to form the nucleocapsid.

Capsomere Any of the proteins that comprise the capsid or the protein part of the nucleocapsid of a virion.

Capsule The uniform layer of high molecular weight polymers attached to the outside of the cell wall.

Carbohydrate An organic compound having the chemical formula $(CH_2O)n$.

Carboxysomes Polyhedral inclusion bodies observed in organisms that use CO_2 as a sole carbon source.

Carcinogen Any compound that causes cancer in animals.

Carrier An individual infected by a pathogen who does not show clinical signs of the infection.

Cation An element that gives up one or more electrons from its outer shell to become a positively charged ion.

Cell The fundamental unit of living organisms, limited by a continuous membrane surrounding its cytoplasm.

Cell Cycle The sequence of steps through which a eucaryotic cell undergoes division. The steps include mitosis, growth 1, synthesis, and growth 2.

Cell Theory The theory that living organisms are composed of membrane-limited units called cells.

Cenocytic Having many nuclei in the same cellular compartment.

Central Nervous System The brain, spinal cord, cranial nerves, and the spinal nerves.

Chancre A lesion or hard sore caused in syphilis by *Treponema pallidum* during the primary stage of infection.

Chemiosmotic Coupling The theory that the proton gradient established across a membrane during electron transport creates a membrane potential sufficient to drive ATP formation.

Chemolithotroph An organism that can gain all its energy from the oxidation of inorganic compounds.

Chemoorganotroph An organism that gains energy from the oxidation of organic compounds.

Chemotaxis Movement of cells toward chemical attractants or away from chemical repellents.

Chemotherapeutic Agent Any chemical substance used to treat infectious disease in animals or humans.

Chloramphenicol An antibiotic that binds to the 50S ribosomal subunit of bacterial ribosomes and prevents the peptidyl transferase reaction.

Chloroplast A eucaryotic membrane-bound, chlorophyll-containing organelle that is the site of photosynthesis.

Chlorosomes The cigar-shaped vesicles in the cytoplasm of certain green photosynthetic bacteria.

Cholecystitis An inflammation of the gallbladder.

Chromatic Aberration The distorted image caused by the different colors (wavelengths) of light passing through a lens.

Chromatids The duplicate structures of the eucaryotic chromosome seen during prophase.

Chromatin The complex of DNA and histones in the eucaryotic nucleus.

Chromatophores The spherical photosynthetic membranes seen in certain purple nonsulfur photosynthetic bacteria.

Chromosome The DNA macromolecule that contains essential cellular genes.

Cilia Proteinaceous protrusions from eucaryotic cells that beat with a rhythmic motion and function as organs of motility.

Ciliophora The taxonomic group of the Protozoa containing the ciliated unicellular organisms.

Clone A group of cells all arising from the same parent cell.

Cloning Vector A plasmid or bacteriophage that can be recombined with a foreign piece of DNA before it is replicated (cloned) in a host cell.

Coccus (pl. cocci) The general name given to a spherical or nearly spherical bacterium.

Codon Any sequence of three nucleotide bases in mRNA that are read by the anticodon in transfer RNA.

Coenocytic *See* cenocytic.

Coenzyme A small, reusable, organic molecule that plays an accessory role in one or more enzymatic reactions (NAD, FAD, and CoA are common coenzymes).

Coliform A general term that describes the lactose fermenting, gram-negative, rod-shaped bacteria that inhabit the human intestine.

Colony A visible accumulation of cells that usually arises from the growth of a single microorganism.

Complement A nonspecific, heat-labile group of seventeen (human) serum proteins that act in concert with one another, with antibody, and with membranes to cause cell lysis.

Condylomata Acuminata Clusters of warts that occur on the genitalia or perineum. Also called venereal warts.

Congenital Rubella Syndrome The developmental problems and the permanent defects that result from infection of a fetus by the rubella virus.

Congenital Syphilis The disease of human infants caused by *Treponema pallidum* passed from mother to fetus.

Conidium (pl. conidia) An asexual spore formed by a fungus.

Conjugation The transfer of genetic information from a donor to a recipient cell while the two cells are physically joined.

Conjugative Plasmids Extrachromosomal, double-stranded, closed-circular DNA replicons that can be transferred directly between cells.

Conjunctivitis Inflammation of the mucous membranes that line the eyelid.

Constitutive Protein A protein that is synthesized continuously by growing cells.

Contact Dermatitis Delayed hypersensitivity to small molecules that contact the skin.

Contact Inhibition The growth property of animal cells that causes them to stop dividing when they touch adjacent cells in a monolayer.

Coryza Inflammation of the mucous membranes of the nasal passages; also known as a head cold.

Covalent Bond The attractive force between two atomic nuclei that are sharing an electron.

Covalent Modification Enzymatic attachment of a small molecule(s) to an enzyme that alters its activity and controls the amount of product formed.

Coxsackieviruses A group of enteroviruses belonging to the Picornaviridae family; they are named for Coxsackie, New York.

Cutaneous Mycoses Fungal infections of the keratinous structures of human hair, nails, and epidermis.

Cyst The resting form of a microorganism derived from the entire cell. A cyst is more resistant to chemical and physical stress than the vegetative cell or trophozoite from which it originated.

Cystitis An inflammation of the urinary bladder.

Cytopathic Effect Localized or generalized degenerative changes or abnormalities of cells in culture or tissue, such as embryonated eggs, caused by the activities of a virus.

Cytoplasm The material surrounded by the plasma membrane exclusive of the nucleus.

Cytosol The nonparticulate fluid portion of the cytoplasm.

Dalton A measure of mass equal to the weight of a hydrogen atom (atomic weight = 1.007).

Dane Particles Virus-like particles found in the serum of serum hepatitis patients that possess the HBs antigen. These particles are composed of a nucleocapsid (28 nm) surrounded by an outer envelope (42 nm).

Decapsidation The process by which an animal virus's nucleic acid is released from its capsid once the virion is inside its host cell. *Also* called uncoating.

Defined Medium A water-based nutrient solution in which the concentration of each chemical constituent is known.

Definitive Host Host from which or from whom a pathogen cannot be transferred.

Denitrification The biological production of gaseous nitrogen products from the anaerobic reduction of nitrate.

Dermatophytes Fungi responsible for cutaneous mycoses.

Desensitization *See* hyposensitization.

Detergents Surface-cleansing compounds that act as wetting agents; cationic detergents are effective disinfectants.

Differential Medium A growth medium that selectively changes in response to the presence and growth of specific organisms.

Diffusion The random movement of molecules from one location to another.

Dimorphism The morphological property of a fungus that causes it to grow as yeasts in tissue (37°C) and as hyphae at room temperature (23°C).

Diphtheria An intoxication caused by the protein toxin produced by lysogenized strains of *Corynebacterium diphtheriae*.

Disease Any continuing disturbance in the structure or function of an organism that causes it damage.

Disinfectant A substance used on inanimate objects to inhibit the growth of, or kill microorganisms (not necessarily effective against spores).

DNA Hybrid A double-stranded DNA molecule formed by annealing single-stranded DNA originating from different organisms.

Echoviruses A group of enteroviruses belonging to the Picornaviridae family. The name is an acronym, *e*nteric *c*ytopathic *h*uman *o*rphan viruses.

Eclipse The time interval following bacteriophage

infection when no complete virions are present in the host cell.

Ecosystem The total community of organisms in a physically defined space.

Ectopic Pregnancy A pregnancy in which the fetus develops outside the womb–in a fallopian tube or the abdominal cavity.

Efficacy The combined qualities of a compound that contribute to its value as a chemotherapeutic agent.

Electronegativity The relative attractiveness possessed by an atomic nucleus for its electrons.

Electron-Transport Chain The membrane-bound components involved in oxidation–reduction reactions that result in the formation of energy-rich phosphate bonds.

Element One of the 93 naturally occurring atomic substances that have an invariant nucleus.

Elementary Body The infectious form of *Chlamydia*.

ELISA (Enzyme-linked-immunosorbent assay) Any of the procedures used to measure antigens or antibodies by assaying the amount of bound enzyme linked to a specific immunoglobulin.

Encephalitis An inflammation of the brain that may lead to coma and death.

Endemic Disease Any infectious disease affecting a geographically localized population at a low but persistent rate.

Endergonic Reaction An energy-utilizing reaction; it is expressed as a positive ΔG.

Endocarditis An inflammation of the membrane (endocardium) that lines the chambers of the heart.

Endocervix The mucous membranes (endometrium) lining the neck (cervix) of the womb.

Endocytosis The process by which cells form a membrane vesicle around a substance and take the material into the cell. This is the general term for phagocytosis and pinocytosis.

Endoflagella *See* periplasmic flagella.

Endoparasitic An organism that lives within a larger organism to the latter's disadvantage.

Endoplasmic Reticulum The cytoplasmic flattened membrane sacs and tubules involved in biosynthesis, transport, and packaging in eucaryotes.

Endospore The resting form of certain bacteria, formed asexually within (endo) a vegetative cell during sporulation, that is very resistant to heat, certain chemicals, and some forms of radiation.

Endotoxin An animal poison derived from the cell wall of a gram-negative bacterium, specifically the lipid A portion of the outer membrane.

Energy The capacity to do work.

Enterics Bacteria belonging to the genera of the gram-negative, facultative-anaerobic organisms of the intestinal flora.

Enterobacteriaceae The family of oxidase-negative, facultative-anaerobic, gram-negative rod-shaped bacteria.

Enterotoxin A group of bacterial exotoxins that act on the intestine.

Enzootic A disease that occurs or persists primarily in animals.

Enzyme A protein that functions as a catalyst to speed up the rate of a chemical reaction.

Enzyme Induction A control process over transcription that regulates the formation of the proteins necessary to metabolize a substrate present in the medium. The substrate often serves as the inducer.

Enzyme Repression The mechanism by which the end product(s) of a metabolic pathway prevent transcription of enzymes needed to form the product.

Eosinophils White blood cells that stain with eosin (acid stain) and are involved in resistance to helminth infections.

Epidemic Any illness caused by a specific agent that occurs in a large segment of a population.

Epidemiology The systematic study of factors involved in the occurrence of disease within a population.

Epilimnion The surface layers of a body of water in which oxygen and biomass are generated by photosynthesis.

Epiphytes Organisms (usually plants) that grow on plants without receiving nourishment from the plants.

Episome A plasmid that can exist in an autonomous, self-replicating state or can be integrated into the host cell's chromosome.

Erythroblastosis Fetalis The hemolytic disease of Rh-positive newborns born of Rh-negative mothers; the Rh problem.

Erythrogenic Toxin The toxin produced by *Streptococcus pyogenes* that is responsible for the red rash of scarlet fever.

Eucaryotes The large group of organisms (plants and animals) that are composed of cells possessing a membrane-bound nucleus.

Eutrophic Lake A nutrient-rich body of water that supports extensive plant growth.

Exergonic Reaction An energy-releasing reaction; it is expressed as a negative ΔG.

Exonuclease An enzyme that cleaves bonds at the free ends of single-stranded or double-stranded nucleic acids.

Exponential Phase The growth phase of a culture during which the maximum growth rate (minimum generation time) for a given set of growth conditions is maintained.

F′ strain A cell containing an extrachromosomal F plasmid that was formed by the excision of the F plasmid from the chromosome of an Hfr cell.

Facilitated Diffusion The carrier-mediated transport mechanism that is driven by a diffusion gradient.

Facultative Anaerobe A microbe that is able to grow in either the presence or the absence of oxygen.

Fermentation An anaerobic, energy-yielding process in which a product of the substrate acts as the terminal electron acceptor; the products are neither more oxidized nor more reduced than the substrate.

Fimbriae *See* pilus.

Firmicutes The division of the Procaryotae-containing bacteria with gram-positive cell walls.

Flagellum (pl. flagella) An external thread-like structure that occurs singly, in groups, or tufts in bacteria and is an organ of motility. Eucaryotic flagella are membrane-bound motility organelles possessing the 9 pairs plus 2 arrangement of microtubules.

Fluorescence The property of a substance that absorbs light at one wavelength (short) and emits longer wavelength light after a delay.

Fomite Any inanimate object that can participate in the transfer of an infectious agent.

Frustule The hard cell walls of diatoms arranged as two overlapping halves or valves.

Fungi The morphologically diverse group of spore-bearing, achlorophyllus, usually nonmotile organisms that possess a cell wall.

Furuncle An extended boil that penetrates into the dermis.

Gamont The gamete formed in the asexual reproduction of sporozoa.

Gastroenteritis An inflammation of the mucous membrane of the stomach and the small intestine.

Gene The sequence of nucleotides in DNA that are transcribed into RNA. Most genes are translated into a polypeptide; however, the rRNA and tRNA genes code only for RNA molecules.

Generalized Transduction The transfer of randomly selected small pieces of host-cell DNA by defective bacteriophages.

Generation Time The interval between divisions of a growing microorganism or the interval needed by a culture to double its cell number.

Genetic Code The DNA language written in triplicate base sequences that is read into mRNA by transcription and is expressed by the polypeptides made during translation.

Genetic Engineering The biochemical manipulation of genes to create an organism with genetic information it would not naturally have.

Genetics The study of how specific traits are passed from parent(s) to offspring.

Genome All the genetic information possessed by a cell or a virus.

Genomic Library A cloning strategy that results in distinct clones of overlapping segments of a cell's genetic information.

Genus (pl. genera) A taxonomic grouping of closely related species.

Germination Activation process in which a spore becomes a vegetative cell.

Gibb's Free Energy The amount of energy (measured in calories) released or utilized during a chemical reaction.

Gingivostomatitis The cold sores or fever blisters of the lips and mouth resulting from infection by herpes simplex virus type I.

Glucose Effect The preferential use of glucose by cells through the cAMP-mediated repression of inducible enzymes. *Also* called catabolic repression.

Glycocalyx General term for polysaccharide components outside of the bacterial cell wall. *See* also capsule and slime layer.

Glycolysis A sequence of enzymatic reactions by which glucose is metabolized to 2 moles of pyruvic acid, 2 moles of ATP, and 2 moles of reduced NAD.

Golgi Apparatus The membranous organelle that functions as the packaging center in the cytoplasm of eucaryotic cells.

Gracilicutes The division of the Procaryotae containing bacteria with gram-negative cell walls.

Gummas Soft lesions characteristic of the tertiary stage of syphilis and formed in response to a *Treponema pallidum* infection.

Habitat The physical space or location where a species lives.

Halophiles Organisms that require NaCl concentrations of 12 percent or higher to grow.

Haptens Low molecular weight foreign compounds that are too small to elicit antibody formation and yet can function as antigenic determinants that react with specific antibodies.

Head-Full Requirement To be infective, the lambda bacteriophage head must contain a DNA molecule between 36.8 kb pairs and 51.5 kb pairs long.

Helminths The parasitic worms, most of which are roundworms (Aschelminthes) or flatworms (Platyhelminthes).

Hemagglutination The clumping of red blood cells caused by an antibody or a virus.

Hematuria Blood in the urine.

Hemolysin One of a number of bacterial protein toxins that lyse red blood cells.

Herpes Labialis A human disease characterized by lesions (cold sores) on the lips caused by herpes simplex virus type 1.

Heterophile Antibodies Antibodies that react with erythrocyte antigens found on red blood cells in many mammals. These antibodies are present in the serum of 90 percent of the patients with infectious mononucleosis.

Hfr Strains Cells that contain a fertility plasmid integrated into their chromosome.

Humoral Antibodies Antibodies in fluids of the body, including the blood and lymph.

Hybridization The relatedness between different nucleic acids as measured by the ability of single-stranded molecules to anneal and form double-stranded complexes.

Hybridomas The fused cell lines derived from a cancerous myeloma cell and an antibody-producing plasma cell. Hybridomas produce monoclonal antibodies.

Hydrogen Bond The weak chemical bond formed between a negatively charged atom and a bound hydrogen atom.

Hydrolase An enzyme that cleaves a covalent bond through the addition of a water molecule.

Hydrolysis The process of cleaving a covalent bond with the concomitant addition of water to the reactant(s).

Hydrophilic Water-attracting.

Hydrophobia Fear of water.

Hydrophobic Lacking affinity for water.

Hypersensitivity An immune-mediated inflammatory reaction caused by a normally innocuous antigen that adversely affects the patient.

Hypha (pl. hyphae) A tubular filament containing protoplasm that is surrounded by the fungal cell membrane and cell wall.

Hypolimnion The cold water in the bottom layers of a stratified aquatic system.

Hyposensitization (desensitization) The repeated intradermal injections (allergy "shots") of increasing doses of allergens to reduce a patient's sensitivity to those allergens.

Icosahedron A symmetrical structure composed of 20 equilateral triangles that enclose a central space.

Icterus Jaundice caused by liver disease.

Immunity An individual's or a population's increased resistance to an infectious agent.

Immunoglobulins A class of globular animal proteins involved in the reactions of immunity and hypersensitivity.

Infectious Disease A disturbance in structure and/or function of a host caused by an invading organism or virus.

Inflammation Response to damaged tissue that results in redness, swelling, pain, heat, and tenderness.

Innate Immunity The repertoire of defenses, both immunological and nonimmunological, that exists prior to exposure to environmental antigens.

Inorganic Compounds The ions and gases that do not contain carbon; carbon dioxide and carbonate are considered inorganic even though they contain carbon.

Insertion Mutation The integration of foreign DNA within a gene causes a mutation; intentionally done in gene cloning, naturally done by transposons.

Interbridges Short polypeptide spacers found in the peptidoglycan of gram-positive bacteria.

Interferons Animal cell glycoproteins, produced by most vertebrate cells, that protect other cells from viral infection.

Interphase The combined times of the G_1, S, and G_2 stages of the cell cycle.

Intoxication Disease attributable to the effects of toxins: poisoning.

Invasive Disease Damage to a host caused by microorganisms and the products they produce during their multiplication within the host.

In Vitro Outside of the living system; in the test tube.

***In Vitro* Packaging** Insertion of recombinant DNA into empty bacteriophage heads. Lambda is the bacteriophage often used.

Ionic Bonds The attractive forces between atoms caused by dissimilar charges.

Ionizing Radiation Electromagnetic waves possessing sufficient energy to alter the structure of atoms.

Isografts Tissue transplants between two genetically identical individuals, such as identical twins.

Isohemagglutinins Blood group antibodies that react with red blood cell antigens.

Isomerase An enzyme that catalyzes geometric or structural changes within one molecule.

Isotope One of two or more varieties of an element, almost identical in chemical properties, but differing in atomic weight. An isotope has an unequal number of protons and neutrons in its nucleus.

Kaposi's Sarcoma A fast-spreading skin cancer characterized by purplish blotches or bumps on the skin. AIDS victims have a high frequency of this rare cancer.

Karyoplasm The material located within the nuclear membrane of eucaryotic cells.

Keratitis Infection of the cornea; can lead to blindness.

Kilobase (kb) Describes a strand of nucleic acid that is 1000 bases long.

Kilocalorie The amount of heat required to raise 1 liter of water 1°C.

Koch's Postulates The experimental approach for demonstrating the causal relationship between an infectious agent and a disease.

Koplik Spots Blue-gray specks highlighted by an inflamed background that appear on the inside of the cheek before the measles rash develops.

Krebs Cycle *See* tricarboxylic acid cycle.

Lag Phase The growth phase during which a microbe adapts to a new medium.

Lagering The chill-storage process of beer manufacture during which clarification occurs.

Leptospirosis The disease caused by *Leptospira interrogans. Also* called Weil's disease or infectious jaundice.

Lethal Dose 50 Percent (LD_{50}) The amount of toxin, or the number of bacteria, which will cause death in 50 percent of a susceptible population of exposed animals.

Leukocytes The white cells present in blood, lymph, and body tissues.

L-Forms (L-Phase Variants) Bacteria that have lost their ability (often temporarily) to form a complete cell wall.

Ligases The class of enzymes that catalyze the joining of two molecules coupled with the hydrolysis of ATP.

Light Repair The light-activated enzyme system that repairs UV damage to DNA.

Limulus Amebocyte Lysate The cytosol from the blood cell (amebocyte) of *Limulus polyphemus* (horseshoe crab). This preparation coagulates when exposed to minute concentrations of endotoxin (pyrogen).

Lipid Biomolecules that are soluble in nonpolar solvents, including waxes, steroids, neutral lipids, and phospholipids.

Lophotrichous Flagellation Tufts of flagella located at the pole of a cell.

Lyase A class of enzymes that cleave covalent bonds in the absence of an oxidation–reduction reaction and without the addition of water.

Lymph The plasma-like fluid that bathes the interstitial spaces of the body's tissues.

Lymphadenopathy Swelling of the lymph nodes.

Lymphocytes Leukocytes involved in antibody production (B lymphocytes) or in cell-mediated immunity (T lymphocytes).

Lymphocytosis Increased number of lymphocytes in the blood indicative of an infectious disease.

Lymphokines A large group of animal cell products, primarily glycoproteins, that affect other cells. They are involved in the inflammatory response and in cellular defense mechanisms.

Lyophilization The process of drying frozen biological material under vacuum.

Lysogen A bacterium that carries a prophage in its genome.

Lysosomes Membrane-bound eucaryotic organelles formed by the Golgi apparatus. They contain degradative enzymes that function in cellular digestion, destruction of foreign matter, and recycling.

Lytic Cycle The viral replication that results in the lysis of the host cell and the release of virulent virions.

Macrophage A phagocytic white blood cell, derived from a monocyte, that is present in animal tissue spaces.

Maculopapular Skin Rash Spotty blemishes with raised inflamed centers.

Malting Barley is steeped in water and allowed to germinate to induce the amylases that later participate in saccharification.

Mast Cells Granulated cells present in tissues that are directly involved in reactions of immediate hypersensitivity.

Mastigophora The taxonomic group of protozoa containing the unicellular flagellates; these protozoa are either zooflagellates or phytoflagellates.

Mastitis An inflammation of the female breast.

Meiosis The process of reductive division to form haploid (1n) eucaryotic gametes.

Mendosicutes The division of the Procaryotae, containing bacteria with widely varying cell-wall chemistries.

Meninges The membranes surrounding the brain and the spinal cord.

Meningitis An inflammation of the membranes surrounding the brain and the spinal column.

Merozoites Certain sporozoans reproduce asexually by schizogony. The schizont divides to form a number of uninucleated cells termed merozoites.

Mesophile An organism that has an optimal growth rate at temperatures between 20° and 50°C.

Mesosomes The bacterial membranes associated with the cell septum.

Metachromatic Granules (volutin) Inclusion bodies composed of polyphosphate.

Metalimnion The layer in a stratified aquatic system characterized by a sharp temperature transition.

Methanogens Bacteria that produce methane in anaerobic environments from molecular hydrogen and carbon dioxide, acetic acid, and some amines.

MIC Minimum concentration of an antibiotic that will inhibit the growth of a microorganism (*m*inimum *i*nhibiting *c*oncentration).

Microaerotolerant (microaerophilic) A microbe that grows best in environments where the concentration of oxygen is low.

Micron One millionth of a meter or 10^{-6} meter.

Microtubules Hollow proteinaceous filaments that are structural components of centrioles, eucaryotic flagella, cilia, and the architectural framework of the eucaryotic cytoplasm.

Mitochondrion (pl. mitochondria) Eucaryotic membrane-bound organelle responsible for ATP formation during aerobic respiration.

Mitosis The process of nuclear division in somatic eucaryotic cells in which chromosomes are duplicated and distributed to daughter cells.

Mixotrophs Chemolithotrophs that can use either carbon dioxide or organic compounds as carbon sources.

Mole The gram-molecular weight of a substance.

Monera An outdated term for the kingdom Procaryotae.

Morbidity The incidence of both fatal and nonfatal disease in a population.

Multiple Sclerosis A progressive degenerative disease of the central nervous system characterized by the loss of certain brain and spinal cord functions.

Must The raw juice produced from crushed fruit, which is fermented during wine making.

Mutagen A chemical or physical agent that increases the rate of mutation above the spontaneous rate.

Mutation An inheritable change in a cell's normal complement of DNA or in the RNA of an RNA virus.

Mycelium (pl. mycelia) An extensive, multibranched, interwinding mat of fungal hyphae.

Mycology The systematic study of the fungi.

Myopericarditis An inflammation of heart muscle (myocarditis) and the pericardium (pericarditis).

Negri Bodies An aggregation of rabies virus particles in the brain tissue of a rabid animal.

Neuraminidase An enzyme localized in the surface of certain enveloped virions that cleaves the bond joining sialic acid to cell receptor molecules.

Neurotropic Spread Transmission of an infectious agent with an animal body by movement within neurons.

Neutron Uncharged (neutral) particles in the nucleus of atoms.

Neutrophils *See* polymorphonuclear neutrophils.

Nitrate Respiration Reduction of nitrate to nitrite under anaerobic conditions; nitrite is not reduced further.

Nitrification The bacterial oxidation of ammonia to nitrate in a two-step process, ammonia to nitrite and nitrite to nitrate.

Nosocomial Infection An infectious illness acquired during a hospital stay.

Nuclear Region The area in the cytoplasm of a procaryotic cell that contains the cell's complement of DNA.

Nucleocapsid The protein–nucleic acid complex of a virus. Some nucleocapsids are surrounded by a lipid-containing viral envelope.

Nucleoid The isolated intact bacterial chromosome; 80 percent DNA, 10 percent RNA, and 10 percent protein.

Nucleolus Dark-staining area of the karyoplasm containing ribosomal precursors.

Nucleoside A nitrogenous base plus a sugar (nucleotide minus the phosphate).

Nucleotide A compound composed of a nitrogenous base, a sugar, and a phosphate group. Nucleotides are the basic units of RNA and DNA, and are structural components of some coenzymes.

Nucleus The membrane-bound structure of eucaryotic cells that contains cellular DNA.

Numerical Aperture The property of an objective lens defined by $n \sin \theta$ where n is the refractive index of the substance between the lens and the specimen.

Numerical Taxonomy Organization of species according to the number of similar and different traits; each trait is given equal weight.

Obligate Pathogen An infectious agent that requires a host and causes damage to the host in which it grows.

Okazaki Fragments The short pieces of single-stranded DNA that comprise the discontinuous lagging strand of semiconservative DNA synthesis.

Oligotrophic Lake A clear pristine body of water that has low concentrations of one or more nutrients essential for photosynthesis.

Oncogenes The genes found in animal cell or viral genomes responsible for cancerous growth of eucaryotic tissue.

Oncogenic Virus A virus that induces tumor formation in the host tissue it has infected.

Oophoritis Inflammation of the ovaries.

Oospore A fungal spore produced by sexual reproduction in the division Mastigomycotina.

Operon A contiguous sequence of genes, controlled by a regulator protein, that codes for proteins involved in a specific cellular function.

Opportunistic Pathogen An organism that can, but does not necessarily, cause a disease state; often a member of the normal flora of the host.

Orchitis Inflammation of the testes.

Organelle Any membrane-bound cytoplasmic structure that performs a specialized function.

Organic Compound Any compound whose formulation includes carbon and hydrogen. In addition, these compounds may contain sulfur, phosphorus, and/or nitrogen.

Osmophile An organism that is able to grow in media containing a high solute concentration.

Osmosis The movement of water molecules across a semipermeable membrane in a direction that will equalize the concentration of solute molecules on both sides of the membrane.

Osmotic Pressure The pressure necessary to prevent the movement of water across a semipermeable membrane because of the differences in solute concentrations.

Otitis An inflammation of the ear.

Oxidation The loss of an electron by an atom.

Oxidation–Reduction Reaction A two-substrate reaction in which one compound donates one or more electrons and the other accepts the electrons.

Oxidative Phosphorylation The formation of energy-rich bonds in ATP by an oxidative process involving the electron-transport chain.

Oxidoreductase An enzyme that catalyzes an oxidation–reduction reaction.

Pandemic An infectious disease that spreads throughout the world.

Parasite An organism that causes damage to its host.

Passive Immunity Resistance to a specific infectious agent developed in another individual before being transferred to the patient. Acquisition of this immune state can be by natural or artificial means.

Pasteurization Preservation of food materials by heating them at temperatures that do not alter the quality of the food.

Pathogen A microorganism capable of causing an infectious disease.

Pediococcus Spherical organisms that divide in two planes to form tetrads.

Pelvic Inflammatory Disease The clinical syndrome attributed to the ascending spread of microorganisms from the vagina and endocervix to the endometrium and fallopian tubes.

Peptidoglycan The sugar–peptide polymer of bacterial cell walls that contains *N*-acetyl glucosamine, *N*-acetyl muramic acid, and D-amino acids. This rigid polymer helps to maintain the cell's shape.

Periodontal Disease The progressive loss of supporting tissue for teeth.

Periplasmic Flagella The fibrillar structures of locomotion that lie beneath the outer membrane of a spirochete's cell wall. Also called axial filaments and endoplasmic flagella.

Periplasmic Space The area between the cytoplasmic membrane and the outer membrane of the gram-negative bacteria.

Peritonitis An inflammation of the membranes that line the abdominal cavity.

Peritrichous Flagellation Flagella dispersed around the outside of a bacterial cell.

Pertussis A severe cough; also called whooping cough, the childhood disease caused by *Bordetella pertussis*.

Phage Conversion Alterations of a host cell's phenotype resulting from the expression of one or more prophage genes.

Phagocytosis The process by which white blood cells engulf unwanted matter from tissue spaces.

Phenol Coefficient A test that compares the antibacterial activity of a disinfectant to the ability of phenol to inhibit bacterial growth.

Phenotype The totality of the observable structural and functional characteristics of an organism.

Photolithotroph An organism that grows in inorganic medium and uses light as a source of energy.

Photosynthesis The biological conversion of light energy into chemical energy (ATP) that involves one of the chlorophylls.

Phototaxis Movement of cells in response to light.

Phototrophs Organisms that use light as a source of energy.

Pili Proteinaceous protusions from bacterial cell walls that function in adherence, transfer of genetic information between cells, and in viral infections. Called fimbriae by some authors.

Pinocytosis Uptake of soluble substances (macromolecules) in vesicles formed by the plasma membrane.

Plankton A term that refers collectively to the microorganisms (chiefly microalgae and protozoans) and small metazoans that drift passively in the pelagic zone of lakes, seas, and other bodies of open water.

Plantae The kingdom in which plants are classified.

Plaque The discrete clear region in the lawn or monolayer of cells in which some or all of the cells have been lysed following the intracellular reproduction of a virus or other parasite, such as *Bdellovibrio*.

Plaque Reduction The decreased ability of a virus to form plaques when an interfering substance (interferon) is present.

Plasma The liquid part of blood that remains after the cells have been removed.

Plasmid An extrachromosomal, closed-circular, double-stranded DNA molecule that replicates autonomously in the bacterial cytoplasm.

Plasmolysis The collapse of a cell because of the high external solute concentration.

Pleomorphism The existence of different morphological forms in the same species or strain of microorganism.

Plus RNA Single-stranded RNA that functions either as mRNA or as a template for making a replicative (minus) RNA strand.

Pneumonia A disease of the lungs in which the tissue is inflamed, hardened, and watery.

Polyhydroxybutyric Acid A polymer of a lipid-like substance that exists in cytoplasmic inclusions present in certain bacteria.

Polymorphonuclear Granulocytes (PMN) Circulating phagocytic leukocytes characterized by a multilobed nucleus and numerous cytoplasmic granules.

Polyribosomes A complex of mRNA and two or more ribosomes engaged in translation.

Polysaccharides Organic molecules composed of combinations of three or more sugar molecules.

Pomace The solid components of crushed grapes present in grape juice (must). White wines are made by fermenting the juice from which the pomace has been removed.

Primary Productivity The conversion of carbon dioxide to organic carbon by biological processes, mainly photosynthesis.

Prion An unconvential infectious agent that is resistant to inactivation by nucleic-acid modifying enzymes and is extremely resistant to heating. Infections by prions are confined to the central nervous system and are associated with slow-developing diseases.

Probe A radioactively labeled, single-stranded RNA or DNA that is used to locate clones of complementary genes.

Procaryotae The kingdom of organisms containing the cyanobacteria and the bacteria.

Procaryotes The large group of organisms (bacteria) whose nucleic acid is found in a nuclear region (no nucleus is present).

Propagated Epidemic An illness in a population characterized by a gradual change in person-to-person transmission of the infectious agent.

Prophage The genome of a temperate phage when it is integrated into its host cell's DNA.

Prostatitis Inflammation of the prostate gland.

Protista The kingdom of microorganisms excluding the bacteria, fungi, and the differentiated plants and animals.

Protons The positively charged particles in an atom's nucleus.

Protoplast The spherical cell formed by treating a gram-positive bacterium with lysozyme.

Prototroph The parent cell, either the wild type or a mutant strain, from which a mutant is derived.

Protozoa The single-celled animals.

Pseudohyphae Yeast cells that grow into an extended septated tubular protoplasm and reproduce by asexual budding.

Pseudopodium A temporary cytoplasmic protrusion from an ameboid cell that functions in locomotion and feeding.

Psychrophiles Organisms that usually grow at 0°C and grow optimally below 20°C.

Puerperal Sepsis (childbed fever) A bacterial infection of the mother's endometrium and surrounding structures that follows delivery or abortion.

Pure Culture A culture that is grown from a single cell. Axenic cultures contain only one type of organism. Often the terms pure and axenic are used interchangeably.

Purpura Hemorrhages in the skin.

Pyelonephritis An inflammation of the kidney.

Pyogenic Pus producing.

R Determinants The genes encoding drug resistance on an R plasmid.

R Plasmid An autonomously replicating, double-stranded DNA molecule that codes for multiple drug resistance.

Radiation Emission and propagation of energy through space or through a substance in the form of waves.

Recombinant (Chimeric) DNA A molecule of deoxyribonucleic acid that contains two or more regions of DNA from different origins.

Reduction The gain of an electron by an atom.

Relapsing Fever A human disease caused by tick-borne or louse-borne *Borrelia* spp. and characterized by alternating febrile and recovery stages.

Rennin (chymosin) An enzyme preparation that hydrolyzes milk casein and is used to form the curd during cheese manufacture.

Reservoir A healthy animal that harbors an infectious agent capable of causing disease in other animals.

Resolving Power The distance between two objects (points) that can be visualized using a lens system.

Restriction Endonuclease A bacterial enzyme that cleaves double-stranded DNA at specific sites in the middle (as opposed to the ends) of the molecule. It functions to destroy DNA that is foreign to the cell.

Restriction Map The DNA segments resulting from restriction endonuclease digestion can be used to reconstruct a diagram of the starting DNA molecule.

Reticulate Body The intracellular form of chlamydias that is capable of growth and division.

Reticuloendothelial System The dispersed array of phagocytic cells associated with the connective tissue of the liver, spleen, and lymph nodes.

Reverse Transcriptase An enzyme that makes DNA from an RNA template.

Rhizosphere The environment around the roots of plants.

Ribosome The cytoplasmic particles containing RNA (60 percent) and protein (40 percent) on which protein synthesis occurs.

Rickettsias Bacteria that are obligate intracellular parasites of eucaryotic cells.

Rifamycins Antibiotics that inhibit transcription by binding to DNA-directed RNA polymerase and prevent mRNA formation.

Ringworm Cutaneous mycoses (tinea) caused by three genera of fungi and named for the anatomical location infected.

Rise Period The time interval between the release of the first and the last bacteriophage from infected bacteria in a one-step growth experiment.

Rod The name given to a microorganism that is longer than it is wide.

Saccharification The process of breaking down starch in grains to those simple sugars that can be metabolized by yeasts.

Saddle-Back Fever A biphasic fever with an initial peak, remission, and then another peak.

Sarcina A spherical organism that divides in three planes to form packets of eight cells.

Sarcodina The taxonomic group of Protozoa containing the single-celled animals that produce pseudopodia.

Schizogony In protozoa, the asexual reproduction of trophozoites in host cells resulting in progeny known as merozoites.

Selective Enrichment A culture established to enable one microbe or group of microbes to outgrow all other microbes present.

Semiconservative Replication The process of nucleic acid synthesis in which each template strand becomes one-half of a newly formed, double-stranded nucleic acid molecule.

Septicemia The presence of multiplying microorganisms in the blood.

Septum (pl. septa) The structural division between two adjoining cellular compartments.

Serology The branch of immunology concerned with analysis of blood sera to determine the presence and concentrations of antigens and antibodies.

Serum Blood minus all the cells and the clotting components.

Serum Sickness The formation of soluble immune complexes in the blood following passive immunization or an inappropriate transfusion.

Signal Hypothesis A hypothesis to explain how membrane proteins are manufactured, positioned within a membrane, and/or transported through a membrane.

Silent Mutation Alterations in DNA that are never expressed because of the redundancy of the genetic code.

Slime Layer Materials adhering to the cell wall in a diffused arrangement.

Specialized Transduction The bacteriophage-mediated transfer of host-cell genes that are located on the chromosome adjacent to the integration site of a prophage.

Species (bacteria) One morphological kind of organism that is able to reproduce its own kind.

Species Name The specific epitaph that modifies the genus name to describe an organism; one part of the binomial system of nomenclature.

Species Resistance Absence of susceptibility to infectious disease possessed by an entire group of animals or plants.

Spherical Aberration The inability of a lens to be focused on the entire field.

Spheroplast The spherical cell formed by the partial removal of the cell wall of a gram-negative bacterium.

Spirillum (pl. spirilla) A curve-shaped bacterium possessing polar flagella.

Spirochetes Helical organisms motile by periplasmic flagella.

Spontaneous Generation The theory (incorrect) that living organisms can arise from nonliving matter. Also called abiogenesis.

Spontaneous Mutations Inherited alterations in a cell's DNA that occur in the absence of a mutagenic agent.

Spores Resting stages of microorganisms that are resistant to desiccation, heat, and certain chemicals. See also endospores.

Sporozoa The taxonomic group of Protozoa containing the nonmotile, parasitic, single-celled animals.

Sporozoite The progeny produced from a sporozoan zygote after it undergoes meiosis.

Stationary Phase The growth phase of a culture in which the number of cells dividing is equal to the number of cells dying.

Sterile Devoid of all living things.

Sterilization The process used to free a substance or object of all living things.

Strain A designation that indicates a specific isolate of an organism.

Streptococci Spherical bacteria that grow in chains.

Streptokinase A streptococcal enzyme that breaks down plasminogen to plasmin.

Streptolysins Substances produced by streptococci that lyse blood cells.

Sty A small, inflamed swelling at the base of the eyelash, usually resulting from an infection by *Staphylococcus*.

Subcutaneous Mycoses Fungal infections of the skin, subcutaneous tissue, connective tissue, or bone.

Substrate-Level Phosphorylation An enzymatic reaction that catalyzes the transfer of inorganic phosphate (H_3PO_4) from an organic compound to ADP to form an energy-rich phosphate bond without involving the electron-transport chain.

Sulfonamides Drugs containing the group SO_2NH_2 that competitively inhibit folic acid synthesis in microorganisms and are bacteriostatic. Also called "sulfa" drugs.

Superoxide Oxygen that has accepted an additional electron (O_2^-).

Svedberg Unit A measure of the relative size of a macromolecule that is determined by ultracentrifugation.

Symbiosis The mutually beneficial relationship between two organisms living in cohabitation.

Syphilis A sexually transmitted bacterial disease caused by *Treponema pallidum*.

Systematics The study of organisms for the purpose of organizing them into groups according to their natural and evolutionary relationships.

Systemic Mycoses Fungal infections that affect the internal organs.

Teichoic Acids The negatively charged acid polymers of glycerol or ribitol joined by phosphate esters present in gram-positive bacterial cell walls.

Temperate Bacteriophage A bacterial virus that exists in both a replicative form capable of causing cell lysis and in a genetic form (prophage) that is passed on to the host's daughter cells.

Tenericutes The division of the Procaryotae containing bacteria without cell walls.

Therapeutic Index The minimum dose of a drug that is toxic to the host divided by the minimum dose required for antimicrobial activity.

Thermophiles Organisms that require temperatures of 50°C or above to grow.

Tinea Capitis Ringworm of the head.

Tinea Corporis Ringworm of the body.

Tinea Cruris Ringworm of the groin ("jock itch").

Tinea Pedis Ringworm of the foot ("athlete's foot").

Titer The reciprocal of the highest dilution of serum that will bring about a demonstrable antibody–antigen reaction.

Topoisomerase (gyrase) An enzyme that inserts supercoils into closed-circular DNA molecules.

Toxigenic Disease Damage to the host caused by the presence of a toxic substance.

Toxin A natural substance that chemically damages cells or tissue to cause disease.

Toxoid A modified bacterial exotoxin that has lost its toxicity, but retains its ability to stimulate the formation of antibodies.

TPI *Treponema pallidum* immobilization test for syphilis.

Transcription The cellular process by which the code contained in DNA is chemically rewritten into RNA.

Transduction The virus-mediated genetic transfer of genes from a host cell to a recipient cell.

Transferase An enzyme that takes a chemical group bound to one molecule and attaches it to another molecule.

Transformation The mode of genetic transfer in which an environmental fragment of cellular DNA from a donor cell is taken up and recombined into the genome of a competent recipient cell.

Transformed Animal Cell A tissue culture cell that grows in an unrestricted manner after losing the property of contact inhibition.

Translation The process of forming a polypeptide using the information contained in messenger RNA.

Transovarian Passage Transfer of an infectious agent from an adult vector (tick) to its progeny in the infected egg of the adult.

Transposon A DNA segment composed of a core sequence flanked by IS-like sequences that moves within a cell's genome.

Tricarboxylic Acid Cycle The enzymatic process for the oxidation of acetyl-CoA to two molecules of CO_2, GTP, three $NADH_2$, and two $FADH_2$. Also called the Krebs cycle.

Trichomes The wide filaments composed of individual bacterial cells.

Trophozoite The vegetative form of a protozoan.

Tubercle The histological structure formed in response to the presence of cells and growth products of *Mycobacterium* spp.

Tuberculin Test A screening test for sensitivity to *M. tuberculosis* done by giving intradermal injections of purified protein derivative isolated from *M. tuberculosis*. A positive result is indicated by a wheal and flare which develop 48 hours after the injection.

Tuberculosis (miliary) A mycobacterial disease characterized by the presence of tubercles throughout the body.

Tuberculosis (pulmonary) A progressive destructive disease of the lungs caused by *Mycobacterium tuberculosis* and, to a lesser extent, by *Mycobacterium bovis*.

Tyndallization Solutions to be treated are heated at 80° to 100°C for several minutes and then incubated, at 30° or 37°C, for 24 hours on three successive days. The solution becomes sterile when all bacterial spores have germinated into vegetative cells, which in turn are killed during the next heating.

Undulating Ridges Structures observed in certain sporozoans of unknown function.

Urethritis An infection of the urethra.

Urogenital Herpes A human disease characterized by lesions on the human genitalia caused by herpes simplex virus type 2.

Vaccination An inoculation with a specific vaccine to lessen or prevent the effects of an infectious agent.

Vaginitis Inflammation of the vagina.

Variolation Inoculation of patients with powdered smallpox crusts or fluid from a smallpox lesion to induce immunity against smallpox.

Vector An agent that is capable of transferring a pathogen from one animal to another.

Venereal Diseases Sexually transmitted diseases that are transferred between humans during sexual contact.

Vibrio A curve-shaped bacterium.

Viral Envelope The outer layer of certain animal viruses that contains lipids derived from a host-cell membrane.

Viral Induction Alterations in the cell-surface structure or cellular production of proteins in response to a viral infection.

Virion A single virus particle.

Viropexis The engulfment of an adsorbed animal virion by the formation of a vesicle derived from the host cell.

Virulence The relative ability of an infectious agent (microorganism or virus) to cause disease.

Virus A complex macromolecular form of life that uses its genetic information, encoded in either DNA or RNA, to produce replicas of itself in its host cell.

Weil's Disease A serious form of leptospirosis.

Yaw The open primary lesion in the skin disease (yaws) caused by *Treponema pertenue*.

Yeast A fungus that grows as individual cells, may reproduce by budding, and produces moist to pasty colonies.

Zoonoses Infectious diseases occurring primarily in animals and occasionally in humans.

Zoster The disease that results from the activation of a latent infection of the varicella–zoster virus in patients who have recovered from chickenpox.

Zygospores Fungal spores produced by sexual reproduction in the division Zygomycotina.

Zygote The product of sexual reproduction brought about by the fusion of two gametes.

INDEX

Throughout the index the following notations are used: **bold face** numbers indicate tables, *italicized* numbers indicate figures.